Principles of Bacterial Detection: Biosensors, Recognition Receptors and Microsystems

Principles of Bacterial Detection: Biosensors, Recognition Receptors and Microsystems

Edited by

MOHAMMED ZOUROB

Biophage Pharma Inc.
Montreal, Canada

SOUNA ELWARY

Consultant to Biophage Pharma Inc.
Montreal, Canada

ANTHONY TURNER

Cranfield University
Bedfordshire, UK

Editors

Mohammed Zourob
Biophage Pharma Inc.
Montreal
Canada
m.zourob@biophagepharma.net

Souna Elwary
Consultant to Biophage Pharma Inc.
Montreal
Canada
selwary@yahoo.com

Anthony Turner
Cranfield University
Bedfordshire
UK
a.p.turner@cranfield.ac.uk

ISBN: 978-0-387-75112-2   e-ISBN: 978-0-387-75113-9

Library of Congress Control Number: 2007941938

Printed on acid-free paper

9 8 7 6 5 4 3 2 1

springer.com

Preface

Bacterial contamination of food and water resources, as well as the increasing incidence of nosocomial infections, has us on our toes, looking for ways of recognizing these elements. In addition, the recent and growing threats to personal and territorial securities make this task even more urgent. Therefore, accurate assessment of the state of current technologies is a prerequisite for undertaking any course of action towards future improvements. In particular, development of new detection and identification technologies for the plethora of bacterial agents has become increasingly important to scientists and to regulatory agencies. In recent years, there has been much progress in the field of bacterial agents detection, resulting in the development of more accurate, fast, analyte-specific, robust, and cost effective techniques by incorporating emerging technologies from various disciplines.

Principles of Bacterial Detection: Biosensors, Recognition Receptors and Microsystems presents a significant and up-to-date review of various integrated approaches for bacterial detection by distinguished engineers and scientists. This work is a comprehensive approach to bacterial detection, presenting a thorough knowledge of the subject and an effective integration of disciplines in order to appropriately convey the state-of the-art fundamentals and applications of the most innovative approaches.

The book consists of four parts. The first part (Chapters 1–4) is an introduction to pathogenic bacteria and sampling techniques and provides an overview of the rapid microbiological methods. The second part (Chapters 5–20) describes the different transducers used for bacterial detection. It covers the theory behind each technique and delivers a detailed state-of-the-art review for all the new technologies used. The third part (Chapters 21–29) covers the different recognition receptors used in the latest methods for the detection of bacteria. It describes in detail the use of immunoassays, nucleic acids, oligonucleotide microarrays, carbohydrates, aptamers, protein microarrays, bacteriophage, phage display, and molecular imprinted polymers as recognition elements. The fourth part (Chapters 30–36) covers the different microsystems used for detection/identification and bacterial manipulations, mainly bacteria lysis in microfluidics, PCR in microfluidics, dielectrophoresis, ultrasonic manipulation techniques, and mass spectrometry.

We anticipate that the book will be helpful to academicians, practitioners, and professionals working in various fields, including biomedical sciences, physical sciences, microsystems engineering, nanotechnology, veterinary science and medicine, food QA, bioterrorism and security as well as allied health, healthcare and surveillance. Since the fundamentals are also reviewed, we believe that the book will appeal to advanced undergraduate and graduate students who study in areas related to bacterial detection.

We gratefully acknowledge all authors for their participation and contributions, which made this book a reality. We give many thanks to Olivier Laczka and Joseph Piliero for the book cover design.

Mohammed Zourob
Souna Elwary
Anthony Turner
June 2008

Contents

4. Overview of Rapid Microbiological Methods

Jeanne Moldenhauer

Part II   Biosensors

5. Surface Plasmon Resonance (SPR) Sensors for the Detection of Bacterial Pathogens

Allen D. Taylor, Jon Ladd, Jiří Homola and Shaoyi Jiang

6. Bacterial Detection Using Evanescent Wave-Based Fluorescent Biosensors

Kim E. Sapsford and Lisa C. Shriver-Lake

7. Fiber Optic Biosensors for Bacterial Detection

Ryan B. Hayman

8. Integrated Deep-Probe Optical Waveguides for Label Free Bacterial Detection

Mohammed Zourob, Nina Skivesen, Robert Horvath, Stephan Mohr, Martin B. McDonnell and Nicholas J. Goddard

9. Interferometric Biosensors

Daniel P. Campbell

10. Luminescence Techniques for the Detection of Bacterial Pathogens

Leigh Farris, Mussie Y. Habteselassie, Lynda Perry, S. Yanyun Chen, Ronald Turco, Brad Reuhs and Bruce Applegate

11. Porous and Planar Silicon Sensors

Charles R. Mace and Benjamin L. Miller

12. Acoustic Wave (TSM) Biosensors: Weighing Bacteria

Eric Olsen, Arnold Vainrub and Vitaly Vodyanoy

13. Amperometric Biosensors for Pathogenic Bacteria Detection

Ilaria Palchetti and Marco Mascini

14. Microbial Genetic Analysis Based on Field Effect Transistors

Yuji Miyahara, Toshiya Sakata and Akira Matsumoto

15. Impedance-Based Biosensors for Pathogen Detection

Xavier Muñoz-Berbel, Neus Godino, Olivier Laczka, Eva Baldrich, Francesc Xavier Muñoz and Fco. Javier Del Campo

16. Label-Free Microbial Biosensors Using Molecular Nanowire Transducers

Evangelyn Alocilja and Zarini Muhammad-Tahir

17. Magnetic Techniques for Rapid Detection of Pathogens

Yousef Haik, Reyad Sawafta, Irina Ciubotaru, Ahmad Qablan, Ee Lim Tan and Keat Ghee Ong

18. Cantilever Sensors for Pathogen Detection

Raj Mutharasan

19. Detection and Viability Assessment of Endospore-Forming Pathogens

Adrian Ponce, Stephanie A. Connon and Pun To Yung

20. Label-Free Fingerprinting of Pathogens by Raman Spectroscopy Techniques

Ann E. Grow

Part III   Recognition Receptors

21. Antibodies and Immunoassays for Detection of Bacterial Pathogens

Padmapriya P. Banada and Arun. K. Bhunia

22. Rapid Nucleic Acid-Based Diagnostics Methods for the Detection of Bacterial Pathogens

Barry Glynn

23. Oligonucleotide and DNA Microarrays: Versatile Tools for Rapid Bacterial Diagnostics

Tanja Kostic, Patrice Francois, Levente Bodrossy and Jacques Schrenzel

24. Pathogenic Bacterial Sensors Based on Carbohydrates as Sensing Elements

Haiying Liu

25. Aptamers and Their Potential as Recognition Elements for the Detection of Bacteria

Casey C. Fowler, Naveen K. Navani, Eric D. Brown and Yingfu Li

26. Protein Microarray Technologies for Detection and Identification of Bacterial and Protein Analytes

Christer Wingren and Carl AK Borrebaeck

27. Bacteriophage: Powerful Tools for the Detection of Bacterial Pathogens

Mathias Schmelcher and Martin J. Loessner

28. Phage Display Methods for Detection of Bacterial Pathogens

*Paul A. Gulig, Julio L. Martin, Harald G. Messer, Beverly L. Deffense and
Crystal J. Harpley*

29. Molecular Imprinted Polymers for Biorecognition of Bioagents

Keith Warriner, Edward P.C. Lai, Azadeh Namvar, Daniel M. Hawkins and Subrayal M. Reddy

Part IV   Microsystems

30. Microfluidics-Based Lysis of Bacteria and Spores for Detection and Analysis

Ning Bao and Chang Lu

31. Detection of Pathogens by On-Chip PCR

Pierre-Alain Auroux

32. Micro- and Nanopatterning for Bacteria- and Virus-Based Biosensing Applications

David Morrison, Kahp Y. Suh and Ali Khademhosseini

33. Microfabricated Flow Cytometers for Bacterial Detection

Sung-Yi Yang and Gwo-Bin Lee

34. Bacterial Concentration, Separation and Analysis by Dielectrophoresis

Michael Pycraft Hughes and Kai Friedrich Hoettges

35. Ultrasonic Microsystems for Bacterial Cell Manipulation

Martyn Hill and Nicholas R. Harris

36. Recent Advances in Real-Time Mass Spectrometry Detection of Bacteria

Arjan L. van Wuijckhuijse and Ben L.M. van Baar

Contributors

Evangelyn Alocilja
Biosystems and Agricultural Engineering
Michigan State University
East Lansing, Michigan
USA

Bruce Applegate
Department of Food Science
Purdue University
West Lafayette, Indiana
USA

Pierre-Alain Auroux
National Institute for Standards and Technology
EEEL, Semiconductor Electronics Division
Gaithersburg, Maryland
USA

Ben L. M. van Baar
TNO Defence
Security and Safety
Rijswijk, The Netherlands

Eva Baldrich
Centro Nacional de Microelectronica
IMB-CNM-CSIC
Esfera UAB
Campus Universidad Autónoma de Barcelona
Barcelona, Spain

Padmapriya P. Banada
Molecular Food Microbiology Laboratory
Department of Food Science
Purdue University
West Lafayette, Indiana
USA

Ning Bao
Department of Agricultural and Biological Engineering
School of Chemical Engineering
Birck Nanotechnology Center
Bindley Bioscience Center
Purdue University
West Lafayette, Indiana
USA

Arun K. Bhunia
Molecular Food Microbiology Laboratory
Department of Food Science
Purdue University
West Lafayette, Indiana
USA

Manfred Biebl
Profos AG
Regensburg
Germany

Levente Bodrossy
Department of Bioresources
Austrian Research Centres
Seibersdorf, Austria

Carl AK Borrebaeck
Department of Immunotechnology
and
CREATE Health
Lund University
Lund, Sweden

Eric D. Brown
Department of Biochemistry and Biomedical
   Sciences and
Department of Chemistry
McMaster University
Hamilton, Canada

Daniel P. Campbell
Georgia Tech Research Institute
Atlanta, Georgia
USA

Fco. Javier Del Campo
Instituto de Biotecnología y Biomedicina
Departamento de Microbiología y Genética
Universidad Autónoma de Barcelona
Barcelona, Spain

S. Yanyun Chen
Department of Food Service
Purdue University
West Lafayette, Indiana
USA

Irina Ciubotaru
QuarTek Corporation
Greensboro, North Carolina
USA

Stephanie A. Connon
California Institute of Technology
and
Jet Propulsion Laboratory
Pasadena, California
USA

Beverly L. Deffense
Department of Molecular Genetics and Microbiology
University of Florida College of Medicine
Gainesville, Florida
USA

Souna Elwary
Consultant to Biophage Pharma Inc.
Montreal, Canada

Leigh Farris
Department of Food Service
Purdue University
West Lafayette, Indiana
USA

Casey C. Fowler
Department of Biochemistry and Biomedical
    Sciences and
Department of Chemistry
McMaster University
Hamilton, Canada

Patrice Francois
Genomic Research Laboratory
Division of Infectious Diseases
University of Geneva Hospitals
Geneva, Switzerland

Barry Glynn
The National Diagnostics Centre
National University of Ireland
Galway, Ireland

Nicholas J. Goddard
School of Chemical Engineering and Analytical
    Science (CEAS)
The University of Manchester
Manchester, UK

Neus Godino
Centro Nacional de Microelectronica
IMB-CNM-CSIC
Esfera UAB
Campus Universidad Autónoma de Barcelona
Barcelona, Spain

Ann E. Grow
Biopraxis, Inc.
San Diego, California
USA

Paul A. Gulig
Department of Molecular Genetics and Microbiology
University of Florida College of Medicine
Gainesville, Florida
USA

Mussie Y. Habteselassie
Department of Food Service
Purdue University
West Lafayette, Indiana
USA

Yousef Haik
Department of Mechanical Engineering
United Arab Emirates University
Al Ain, United Arab Emirates
and
Center of Research Excellence
    in Nanobioscience
University of North Carolina
Greensboro, North Carolina
USA

Crystal J. Harpley
Department of Molecular Genetics
    and Microbiology
University of Florida College of Medicine
Gainesville, Florida
USA

Nicholas R. Harris
School of Electronics and Computer Science
The University of Southampton
Southampton
UK

Daniel M. Hawkins
University of Surrey
School of Biomedical and Molecular Sciences
Guildford, Surrey
UK

Ryan B. Hayman
Department of Chemistry
Tufts University
Medford, MA
USA

Martyn Hill
School of Engineering Sciences
The University of Southampton
Southampton
UK

Kai Friedrich Hoettges
Centre for Biomedical Engineering
University of Surrey
Guildford
Surrey, UK

Jiři Homola
Institute of Photonics and Electronics
Academy of Sciences
Prague
Czech Republic
and
Department of Chemical Engineering
University of Washington
Seattle, Washington
USA

Robert Horvath
Nanotechnology Centre
Cranfield University
Bedfordshire, UK

Michael Pycraft Hughes
Centre for Biomedical Engineering
University of Surrey
Guildford
Surrey, UK

Shaoyi Jiang
Department of Chemical Engineering
University of Washington
Seattle, Washington
USA

Barbara Jones
National Institute of Standards and Technology
Chemical Sciences and Technology Laboratory
Biochemical Science Division
Gaithersburg, Maryland
USA

Ali Khademhosseini
Center for Biomedical Engineering
Department of Medicine
Brigham and Women's Hospital
and
Harvard Medical School
Harvard-MIT Division of Health Sciences
   and Technology
Massachusetts Institute of Technology
Cambridge, Massachusetts
USA

Tanja Kostic
Department of Bioresources
Austrian Research Centres GmbH - ARC
Seibersdorf, Austria

Jan W. Kretzer
Profos AG
Regensburg
Germany

Olivier Laczka
Centro Nacional de Microelectronica
IMB-CNM-CSIC
Esfera UAB
Campus Universidad Autónoma de Barcelona
Barcelona, Spain

Jon Ladd
Department of Chemical Engineering
University of Washington
Seattle, Washington
USA

Edward P.C. Lai
Ottawa-Carleton Chemistry Institute
Department of Chemistry
Carleton University
Ottawa, Ontario
Canada

Gwo-Bin Lee
Department of Engineering Science
National Cheng Kung University
Tainan, Taiwan

Yingfu Li
Department of Biochemistry and Biomedical
   Sciences and
Department of Chemistry
McMaster University
Hamilton, Canada

Haiying Liu
Department of Chemistry
Michigan Technological University
Houghton, Michigan
USA

Martin J. Loessner
Institute for Food Science and Nutrition
Zurich, Switzerland

Tracey Elizabeth Love
Defence Science and Technology
   Laboratory
Porton Down, Wiltshire
UK

Chang Lu
Department of Agricultural
   and Biological Engineering
School of Chemical Engineering
Birck Nanotechnology Center
Bindley Bioscience Center
Purdue University
West Lafayette, Indiana
USA

Charles R. Mace
University of Rochester
Rochester, New York
USA

Julio L. Martin
Department of Molecular Genetics
   and Microbiology
University of Florida College of Medicine
Gainesville, Florida
USA

Marco Mascini
Dipartimento di Chimica
Università di Firenze
Sesto Fiorentino, Italy

Akira Matsumoto
Department of Bioengineering
Graduate School of Engineering
The University of Tokyo
Tokyo, Japan

Martin B. McDonnell
Defence Science and Technology Laboratory
Porton Down, Wiltshire
UK

Harald G. Messer
Department of Molecular Genetics
    and Microbiology
University of Florida College of Medicine
Gainesville, Florida
USA

Benjamin L. Miller
University of Rochester
Rochester, New York
USA

Stefan Miller
Profos AG
Regensburg
Germany

Yuji Miyahara
Biomaterials Center
National Institute for Materials Science
Tsukuba
Japan
and
Department of Materials Engineering
Graduate School of Engineering
The University of Tokyo
Tokyo
Japan

Stephan Mohr
School of Chemical Engineering and Analytical
    Science (CEAS)
The University of Manchester
Manchester, UK

Jeanne Moldenhauer
Excellent Pharma Consulting
Mundelein, Illinois
USA

David Morrison
Harvard-MIT Division of Health Sciences
    and Technology
Massachusetts Institute of Technology
Cambridge, Massachusetts
USA

Zarini Muhammad-Tahir
Biosystems and Agricultural Engineering
Michigan State University
East Lansing, Michigan
USA

Xavier Muñoz-Berbel
Centro Nacional de Microelectronica
IMB-CNM-CSIC
Esfera UAB
Campus Universidad Autónoma de Barcelona
Barcelona, Spain

Francesc Xavier Muñoz
Centro Nacional de Microelectronica
IMB-CNM-CSIC
Esfera UAB
Campus Universidad Autónoma de Barcelona
Barcelona, Spain

Raj Mutharasan
Department of Chemical and Biological Engineering
Drexel University
Philadelphia, Pennsylvania
USA

Azadeh Namvar
Department of Food Science
University of Guelph
Guelph, Ontario
Canada

Naveen K. Navani
Department of Biochemistry and Biomedical
    Sciences and
Department of Chemistry
McMaster University
Hamilton, Canada

Eric Olsen
Clinical Investigation Facility
David Grant USAF Medical Center
Travis Air Force Base, CA
USA

Keat Ghee Ong
Department of Biomedical Engineering
Michigan Technological University
Houghton, Michigan
USA

Ilaria Palchetti
Dipartimento di Chimica
Università di Firenze
Sesto Fiorentino, Italy

Lynda Perry
Department of Food Service
Purdue University
West Lafayette, Indiana
USA

Adrian Ponce
California Institute of Technology
Jet Propulsion Laboratory
Pasadena, California
USA

Ahmad Qablan
The Hashemite University
Zarqa, Jordan

Subrayal M. Reddy
School of Biomedical and Molecular Sciences
University of Surrey
Guildford, Surrey
UK

Brad Reuhs
Department of Food Service
Purdue University
West Lafayette, Indiana
USA

Toshiya Sakata
Department of Materials Engineering
Graduate School of Engineering
The University of Tokyo
Tokyo, Japan

Kim E. Sapsford
George Mason University
Manassas, Virginia
USA
and
Center for Bio/Molecular Science
and Engineering
U.S. Naval Research Laboratory
Washington, D.C.
USA

Reyad Sawafta
QuarTek Corporation
Greensboro, North Carolina
USA

Mathias Schmelcher
Institute for Food Science and Nutrition
Zurich, Switzerland

Jacques Schrenzel
Genomic Research Laboratory and Clinical
    Microbiology Laboratory
Service of Infectious Diseases
University Hospital of Geneva
Department of Internal Medicine
Geneva, Switzerland

Lisa C. Shriver-Lake
Center for Bio/Molecular Science
    and Engineering
U.S. Naval Research Laboratory
Washington, D.C.
USA

Nina Skivesen
Inano
Interdisciplinary Nanoscience Center
University of Aarhus
Aarhus, Denmark

Kahp Y. Suh
School of Mechanical and Aerospace Engineering
Seoul National University
Seoul, Korea

Ee Lim Tan
Department of Biomedical Engineering
Michigan Technological University
Houghton, Michigan
USA

Allen D. Taylor
Department of Chemical Engineering
University of Washington
Seattle, Washington
USA

Ronald Turco
Department of Food Service
Purdue University
West Lafayette, Indiana
USA

Anthony Turner
Cranfield University
Bedfordshire
UK

Arnold Vainrub
Department of Anatomy, Physiology,
    and Pharmacology
Auburn University
Auburn, Alabama
USA

Vitaly Vodyanoy
Department of Anatomy, Physiology,
    and Pharmacology
Auburn University
Auburn, Alabama
USA

Keith Warriner
Department of Food Science
University of Guelph
Guelph, Ontario
Canada

Christer Wingren
Department of Immunotechnology
and
CREATE Health
Lund University
Lund, Sweden

Arjan L. van Wuijckhuijse
TNO Defence
Security and Safety
Rijswijk, The Netherlands

Sung-Yi Yang
Department of Engineering Science
National Cheng Kung University
Tainan, Taiwan

Ahmed E. Yousef
Professor of Food Microbiology
Department of Food Science and Technology
and
Department of Microbiology
Parker Food Science Building
Ohio State University
Columbus, Ohio
USA

Pun To Yung
California Institute of Technology
Pasadena, California
USA

Mohammed Zourob
Biosensors Division
Biophage Pharma Inc
Montreal, Canada

Introduction

I

Introduction to Pathogenic Bacteria

Tracey Elizabeth Love and Barbara Jones

Abstract

This chapter is a brief introduction to pathogenic microorganisms and also discusses virulence factors. An understanding of virulence factors is important, as they represent potential targets for the detection of microbial pathogens. Sources and routes of infection are also briefly discussed with reference to specific examples. There are a number of ways in which infection could be acquired, including via contaminated food and water; hospital acquired infection; "naturally acquired" infection; and intentional infection, for example, through the use of biological warfare agents. The focus of the review is predominantly on human pathogens. However, there are a range of other microbial pathogens of particular importance in other areas; for example, animal and plant pathogens, which will not be discussed. Finally, a brief overview of the detection of pathogenic bacteria is presented.

1. Pathogenic Microorganisms

Over many years there has been considerable debate as to the exact definitions of pathogenicity and virulence. These two words are often used interchangeably, but pathogenicity has been defined as the ability of an organism to cause disease and virulence as the relative severity of the disease caused by the organism (Watson and Brandly 1949). It has become increasingly apparent that virulence is highly complex and is dependent on the interaction between the host and the microorganism (Casadevall and Pirofski 2001). Taking into account the problems associated with defining virulence, virulence factors have also been difficult to characterise. Two definitions that have been put forward are that a virulence factor is (1) a "component of a pathogen that when deleted specifically impairs virulence but not viability" (Wood and Davis 1980); or (2) a "microbial product that permits a pathogen to cause disease" (Smith 1977). However, these often do not apply to infections caused by commensal or opportunistic pathogens, where often classic virulence determinants do not exist. Furthermore, the definitions may not account for host tissue damage that has been caused by the induction of a particular part of the host's immune response, such as cytokine synthesis (Henderson et al. 1996). Therefore an understanding of virulence factors is important, as these can often be used to specifically detect pathogenic microorganisms. Classical virulence factors include factors that aid in a number of stages of infection:

1) host cell attachment;
2) entry to the host cell;

Tracey Elizabeth Love • Defence Science and Technology Laboratory, Porton Down, Wiltshire, UK.
Barbara Jones • National Institute of Standards and Technology, Chemical Sciences and Technology Laboratory, Biochemical Science Division, Gaithersburg, Maryland, USA.

M. Zourob et al. (eds.), *Principles of Bacterial Detection: Biosensors, Recognition Receptors and Microsystems*,
© Springer Science+Business Media, LLC 2008

3) evasion of detection by the hosts immune system;
4) intracellular or extracellular replication and inhibition of phagocytosis.

Virulence factors can be either requisite, where the gene products discriminate between pathogenic and nonpathogenic species; or contributory factors, that alter the severity of the disease. Again, the ability to cause disease and its severity will also be dependent on the immune status of the host. Contributory virulence factors do not fulfil the definition of virulence factors; nor do they separate pathogenic from nonpathogenic species, as they may be found in a wide range of microorganisms but still have a role in damage to host cells. A general consensus of opinion has been that regardless of the function of a gene product, if its expression leads to damage of the host cell then it is a virulence factor. Therefore Casadevall and Pirofski (1999) suggest that virulence factors should be defined as "attributes that mediate host damage." Bacterial pathogens usually possess a number of virulence factors that are essential in establishing infection and causing disease. Classical virulence factors include toxins, as well as molecules that are involved in adherence, invasion of the host, evasion of the host's immune response, and iron acquisition.

1.1. Toxins

Some microorganisms (e.g., *Bacillus anthracis*) produce toxins that are the major cause of clinical symptoms observed in patients. Toxins can be integral parts of the bacterium, such as lipopolysaccharide (endotoxins), or secreted molecules (exotoxins). Toxins often perform other functions, such as the making of adhesins (Tuomanen and Weiss 1985). Toxin secretion may also be regulated as part of an orchestrated response by the bacterium. The lipopolysaccharide (LPS) content of pathogenic Gram-negative cell walls is contained within a microorganism and usually released when the cell dies or is broken down (by autolysis or by the host's immune response).

Unlike exotoxins, endotoxins are believed not to have any direct enzymatic action; and it is the lipid A portion, usually embedded within the bacterial membrane, that is believed to be the toxic component. As LPS is released from the bacterial cell, a number of host molecules involved in the inflammatory response are released (e.g., cytokines). One of the most important cytokines released is tumour necrosis factor-α (TNF-α). This molecule usually prevents the spread of a localised infection. However, the rapid stimulation of high levels of TNF-α within the bloodstream results in fever, damage to host tissue, an alteration of metabolism, and the production of further cytokines (IL-6, IL-8, IL-1, and PAF, platelet activating factor). These cytokines produce further damage to host cells and tissue resulting in a dramatic decrease in blood pressure and reduced blood flow to major organs leading to multiple organ failure (Tracey and Cerami 1993, Rink and Kirchner 1996). Exotoxins can be divided into a number of broad categories summarised below.

1.2. Adherence

Another important factor in establishing an infection is the ability of a microorganism to attach to a host cell or to an extracellular matrix. The macromolecules and structures involved in specific attachments to host cell receptors are often referred to as adhesins. Proteinaceous adhesins can be classified into two groups: afimbrial adhesins (sometimes termed nonpilus adhesins) and fimbriae (or pili). However, not all adhesins are essential for microbial virulence (Krogfelt 1991).

Fimbrial adhesins or pili can be observed by electron microscopy as hair-like structures that are present predominantly on the surface of Gram-negative bacteria (nearly all Gram-positive organisms do not possess pili). A number of Gram-negative pathogens utilise pili for adherence such as *Vibrio cholerae* and *Neisseria gonorrhoeae*. Originally it was suggested that pili were homopolymeric, composed of a number of repeating pilin (or fimbrin) subunits. However, it has become apparent that for many pili (e.g., Type I and Pap pili) these structures are heteropolymeric. Furthermore, the protein subunits located either at the tip or the base of these structures are important for a particular function. For example, the minor protein subunits located at the distal end of the pilus (the tip fibrillum) are frequently involved in attachment to the host cell receptor (Hultgren et al. 1993). Pili have the capacity to attach to a number of different receptors (Table 1.1) and genetic variation in the tip adhesin confers differences in binding affinity to host cell receptors, which allows for differences in the tissue tropism. Bacterial DNA may encode multiple operons for fimbrial expression; for example, *Salmonella. typhimurium* encodes four fimbrial operons (*fim, lpf, pef,* and *agf*). Deletion of individual fimbrial operons resulted in a minor reduction in virulence, but a quadruple mutant demonstrated significant attenuation compared to the wild type strain (Van der Velde et al. 1998). This suggests that deletion of a single component of virulence can be compensated by the presence of related virulence factors in the case of some microorganisms.

Curli are another form of pilus type adhesins found in some strains of *Escherichia coli* and *Salmonella entereditis* spp. They are highly stable, thin, irregular surface structures that facilitate binding to host proteins such as plasminogen and fibronectin. All pili are assembled in a highly ordered manner. and although the assembly mechanism may vary, common characteristics are often observed (Soto and Hultgren 1999). Nonpilus adhesins can be found in Gram-negative, Gram-positive, and mycobacterial pathogens. Examples of nonfimbrial adhesins are summarised in Table 1.2.

Table 1.1.   Examples of bacterial toxins and modes of action (compiled from Merrit and Hol 1995, Schiavo et al. 1992, Welch 1991, Savarino et al. 1993, Falzano et al. 1993, and Schmitt et al. 1999).

Type of Toxin	Mode of Action	Example Organisms
A-B Toxins (Type III toxins)	The A subunit has enzymatic activity that mediates toxicity. The B-subunit binds to the host cell receptor and allows for delivery of the A-subunit.	*Vibrio cholerae* (cholera toxin), *Bordetella pertussis* (pertussis toxin), *Corynebacterium diptheriae* (diptheria toxin).
Proteolytic Toxins (Type II Toxins)	Proteolytic cleavage of host cell molecules.	*Clostridum tetani* (tetanus toxin), *Clostridium botulinum* (botulinum toxin)
Pore-forming Toxins (Type II Toxins)	Creation of a pore within the plasma membrane of the host cell that leads to cell lysis.	*E. coli* (hemolysin), *Bordetella pertussis* (adenylate cyclase)
Interference with Host Cell Function	One group affects the host cell cytoskeleton by modification of Rho family (small GTP binding proteins) resulting in various detrimental effects on actin polymerisation (toxin dependent).	*E. coli* (CNF, cytotoxic necrotising factor), *Clostridium difficile* (toxins A and B)
IgA Protease Type Proteins	Bacteria that secret these toxins have their own secretion systems encoded within the bacterial DNA. The mode of action of many of these toxins has yet to be elucidated.	*Neisseria meningitidis, Haemophilus influenzae*
Heat Stable Toxins	Heat stable toxins – binding of the toxin to its receptor stimulates activation of guanylate cyclase increasing intracellular GMP in turn causing dramatic ion flux changes.	Enterotoxinogenic *E. coli, Yersinia enterocolitica, Vibrio cholerae.*
Superantigens (Type I Toxins)	Immunostimulatory toxins that bind to MHC class II molecules stimulating the production of T-cells and triggering the release of cytokines involved in the inflammatory response. This causes fever, shock, and erythematous rash.	*Staphlococcus aureus, Streptococcus pyogenes, Staphylococcal enterotoxin.*

Table 1.2.   Examples of Gram negative pilin adhesins (modified
from Wizemann et al. 1999).

Adhesin	Strain	Host Receptor
PapG	*Klebsiella pneumoniae*	Gal α(1–4)Gal
FimH	*Escherichia*	Mannose oligosaccharides
MrkD	*Escherichia coli*	Type V collagen
HifE	*Haemophilus influenzae*	Sialylyganglioside-GM1

Bacteria can also produce other molecules that are involved in host cell attachment, such as intimins. These allow for close associations between the pathogens and result in rearrangement of the cytoskeleton, interference with host cell signalling, and possibly in bacterial internalisation. A more novel mechanism of host cell attachment is a pathogen secreted receptor, which is endocytosed by the host cell and subsequently presented on the host cell surface in a phosphorylated form. This then functions as a receptor for the bacterial cell, for example in enteropathogenic *E. coli* (Kenny et al. 1997).

Other surface structures may also be involved in specific and nonspecific adherence to a host cell, such as a slime layer, capsule, LPS, techoic acid and lipotechoic acid. For example, the capsular glucose and mannan polysaccharides of *Mycobacteria* species adhere to complement receptor 3 and the mannose receptors of host cells (Daffe and Etienne 1999). The techoic acids of *Staphylococcus* and *Streptococcus* spp can also function as adhesins (Walker 1998). Slime layers and capsules are usually composed of polysaccharides, but can be made of polypeptides. If the surrounding material is unorganised and loosely attached to the cell wall it is referred to as a slime layer, whereas an organised layer that is firmly attached to the bacterial cell is termed a capsule. Both may mediate specific or nonspecific attachment, but do not necessarily have a role in pathogenicity or virulence. Many adhesins have also demonstrated important roles in evasion of the host immune system (e.g., capsules, LPS, and techoic and liptotechoic acids).

Both Gram-negative and Gram-positive pathogens often express and utilise a large repertoire of adhesins (Tables 1.1 and 1.2). Adhesion can occur as a protein-protein or protein-carbohydrate interaction, and a vast array of host molecules are used as adhesin targets (Table 1.3). For example, surface immunoglobulin, glyocproteins, glycolipids, and extracellular matrix proteins such as fibronectin, collagen, or laminin have been shown to interact with adhesins (Finlay and Falkow 1997).

Table 1.3.   Examples of afrimbrial adhesins in Gram-negative
and Gram-positive bacteria (Brubaker 1995).

Adhesin	Strain	Host Cell or Receptor
Pertactin	*Bordetella pertussis*	Integrin
HMW1/HMW2	*Haemophilus influenzae*	Human epithelial cells
Envelope antigen F1	*Yersinia pestis*	Not known
Le[b] binding adhesion	*Helicobacter pylori*	Fucosylated Le[b] histocompatibility blood group antigens
CpbA/SpsA/PbcA/PspC	*Streptococcus pneumoniae*	Cytokine activated epithelial and endothelial cells.
P1, Pac	*Streptococcus mutans*	Salivary glycoprotein
FnbA, FnbB	*Staphyloccus aureus*	Fibronetin

1.3. Invasion

As discussed previously, intimins often mediate a close association with the host cell, which can lead to internalisation of the pathogenic microorganism. Classes of adhesin that facilitate entry into the host cell are termed invasins and are the most common mode of entry of a bacterial pathogen into both phagocytic and nonphagocytic cells (Finlay and Falkow 1997). In phagocytic cells, pseudopod formation occurs as cell signals mediate cytoskeletal rearrangements, resulting in polymerisation and depolymerisation of actin. Internalisation of a bacterial pathogen by cells other than professional phagocytes is mediated by invasin-activated actin rearrangements. This has the same effect as in a phagocytic cell, resulting in forced phagocytosis of the pathogen (Bliska et al. 1993, Rosenshine and Finlay 1993). Some pathogenic species also utilise host microtubules (polymerised tubulin) to enter nonphagoctyic cells e.g., *N. gonorrrhoae* or *K. pneumoniae*. The exact mechanism is undefined, but they do not appear to utilise this mechanism as an essential virulence factor. Some pathogens target phagocytes and may use phagocytic pathways for internalisation. Once the pathogen is internalised, it resides initially within a membrane vesicle. The pathogen can either remain within or escape from the vesicle. Many pathogenic species remain within the vesicle and have evolved mechanisms to evade the cellular response of the host. For example, capsules and LPS can serve as a protective barrier for internalised pathogens. Other factors, such as the secretion of enzymes that neutralise oxygen radicals and proteolytic enzymes that can degrade host cell lysosyme, are also important for intracellular survival. Exploitation of host cell signals may also occur (e.g., acidic pH) to activate replication and the initiation of the expression of other virulence factors or cascades required for intracellular survival. Pathogenic bacteria may also have the ability to replicate within the host cell and spread to other host cells.

1.4. Evasion of the Host Immune Response

Many surface elements of bacterial pathogens serve to aid in the evasion of the host's immune response. Capsules consisting of a mix of polysaccharide, protein, and glycoprotein prevent complement activation by inhibition of the assembly of C3 convertase on the bacterial cell surface using a variety of mechanisms. Prevention of C3 convertase assembly on the bacterial surface inhibits the efficiency of phagocytosis, as opsonisation of the pathogen is less likely to occur. C3 convertase may assemble beneath the capsule, and the C3b molecule may also be able to diffuse through; however, the capsular network blocks subsequent contact with phagocytic receptors (Taylor and Roberts 2005). Lack of C3 convertase on the surface also reduces the probability of the formation of a membrane attack complex (MAC) on the underlying bacterial surface. The capsule itself may provoke an immune response. One way by which a number of pathogenic species have overcome this is to produce a nonimmunogenic capsule composed of polysaccharides similar to host polysaccharides—for example, sialic or hyaluronic acid. LPS also aids in the evasion of complement activation and phagocytosis. The LPS O-antigen blocks C3 convertase assembly through the binding of sialic acid. Variation in the length of the LPS O-antigen side chain prevents the assembly of an effective MAC conferring serum resistance, which is important for establishing systemic infections. Another example of evasion of complement activation and phagocytosis is the production of enzymes and toxins to prevent the migration of phagocytes to the infected site (for example, through the enzymatic degradation of a the chemoattractant C5a) (Taylor and Roberts 2005).

Toxins may also protect pathogens against phagocytosis by killing phagocytes and reducing the production of toxic reactive oxygen intermediates (oxidative burst). Many highly virulent pathogenic bacteria target phagocyte receptors and also have the ability to survive phagocytosis by polymorphonuclear leukocytes (PMNs), macrophages, and monocytes. Many strategies exist for surviving within these cells, such as:

1) prevention of phagosome-lysosome fusion;
2) release from the phagosome prior to lysosome fusion;
3) secretion of molecules that reduce the toxic effects of the components of the lysosome into the phagolysosome (e.g., the expression of enzymes such as superoxide dismutase that detoxify oxygen radicals).

Other mechanisms include the production of cell walls that are resistant to degradation by lysosyme, and interference with signalling pathways through the production of enzymes that are homologous to host cell enzymes.

Pathogens can also evade the host's immune response by inhibiting the production of antibodies by utilising a range of strategies including phase variation (constant switching of the expression of surface antigens). Use of this strategy allows the microorganism to avoid antibodies, as those produced will only be effective against the forms previously expressed (Meyer 1991). Strain variation of immunodominant and highly expressed proteins may confer a selective advantage to pathogenic species. Furthermore, surface components such as LPS, capsules, S-layers, flagella, and outer membrane proteins may demonstrate antigenic variation (Brunham et al. 1993). As mentioned previously, nonimmunogenic structures or layers such as capsular polysaccharides or carbohydrates may be produced that resemble those of the host and even mimic the function of host cell proteins (Stebbins and Galán 2001). Bacteria may also be coated in host cell proteins, such as fibronectin and collagen. The coating of bacteria with host antibodies can prevent opsonisation, possibly through prevention of recognition by specific antibodies to surface located antigens or the inhibition of complement assembly. One example of this is protein A of *S. aureus* and protein G of *S. pyogenes*, which bind to the Fc portion of immunoglobulins, thus covering the pathogen in antibody. Some pathogenic species also express specific iron binding receptors. Although the primary function of these receptors is iron acquisition, it has also been suggested that they may have a role in masking surface antigens.

1.5. Iron Acquisition

Iron is required for bacterial replication; however, low iron concentrations are often found within a host, particularly within humans. Therefore for survival and growth, some form of iron acquisition mechanism is required. Siderophores are high affinity iron binding low molecular weight compounds, secreted by the bacteria to chelate iron (Neilands 1995). The iron-siderophore complex is then bound by siderophore receptors located on the bacterial surface. It has been demonstrated that this type of iron acquisition mechanism may contribute to bacterial virulence, but this has not always been found to be the case. In humans most iron is bound to proteins such as ferratin, lactoferratin, hemin, or transferrin. It has been shown that some pathogens are able to utilise these as an iron source through binding to receptors on the bacterial surface, although the exact mechanism of the removal of the iron has not been clearly defined. Another potential method of iron acquisition is the production of other virulence factors such as exotoxins, invasins, adhesins, and outer membrane proteins where expression is activated by low iron concentrations (Litwin and Calderwood 1993). It has been suggested that exotoxins kill host cells, thus releasing iron stores which can then be utilised by the pathogen. Many bacteria have more than one method of iron acquisition, therefore deletion may be compensated by another system.

1.6. Regulation of Virulence Factors

Infection of a host with a pathogen presents the microorganism with adaptive changes required for survival and replication within different environments. These changes may be in response to a number of biochemical and physical parameters such as temperature, pH, ion concentration, growth phase, osmolarity, oxygen, calcium, and iron levels (Gross 1993). Many virulence factors will be required only when an organism is within a host, others may also be essential for survival outside the host. Once a pathogen is within a host, environmental signals will induce the switching on and off of various virulence genes, dependent on the stage of infection. It is common for a single regulatory element to control the expression of numerous virulence factors that may not be related; these are sometimes termed global regulators. It has also become apparent that virulence factor expression is a complex interdependent process relying on various cues from the host and the pathogen. The expression of a number of virulence factors may be coordinated simultaneously by several different regulatory elements; alternatively, a single virulence factor can come under the control of several regulatory elements.

2. Sources and Routes of Infection

2.1. Natural Infection

Pathogenic bacteria are widely distributed and can be found in the soil, other animals or humans, food, or water, depending on bacterial species. All these represent potential sources of infection. Infection can be via a number of routes, such as inhalation, ingestion, abrasion to the skin, contaminated blood, or the bite of an insect vector. The way in which infection is established will again depend on the microorganism of concern and may also have an effect on the predicted outcome of the disease. For example, an organism such as *B. anthracis*, the aetiological agent of anthrax, has a number of different forms. In the case of humans, there are three main forms of the anthrax disease: cutaneous, inhalational, and gastrointestinal. Any of these types of infection can result in systemic anthrax, which is nearly always fatal (Mock and Fouet 2001). The severity of disease is usually greatest with inhalational anthrax, and if untreated this form of the disease has a mortality rate approaching 100% (Turnbull 1991, Webb 2003). Cutaneous anthrax is often self-limiting with or without the appropriate treatment (Hambleton and Turnbull 1990). There are a number of diseases that are of importance throughout the world; however, there are three areas for which the rapid sensitive detection of pathogenic microorganisms has been addressed in this review: food- and water-borne pathogens, hospital acquired infections, and the intentional use of pathogenic bacteria for biological warfare or bioterrorism. These are discussed mainly in the context of human infections, but obviously there is a range of pathogenic microorganisms that can infect animals and plants and are of importance, but that will not be discussed in detail within this review.

2.2. Food and Water

There are many different types of food-borne pathogens such as bacteria, viruses, parasites, prions, and bacterial toxins. The symptoms of illness usually comprise mild or severe gastrointestinal discomfort such as nausea, vomiting, and/or diarrhoea, but can also extend to life-threatening renal (kidney), hepatic (liver), and neurological complications. Infection from food-borne pathogens are often unreported due to generic diagnoses such as "stomach flu" and the fact that many individuals do not seek medical treatment unless symptoms become severe. The duration of illness varies and is typically short for bacterial infection, but can be chronic for viral or parasitic infections. Well known factors contributing to bacterial contamination of foods include the improper handling and storage of foods, inadequate cooking or reheating, cross-contamination between raw and cooked foods or fresh produce, and poor hygiene of food service workers. Overall, food-borne diseases are estimated to cause more than 76 million illnesses each year in the United States (Mead et al. 1999). Common foodborne pathogens identified by the FDA are *Escherichia coli, Salmonella, Shigella, Campylobacter, Yersinia enterocolitica* and *Yersinia pseudotuberculosis, Vibrio, Listeria monocytogenes, Staphylococcus aureus, Bacillus cereus, Clostridium perfringens, Clostridium botulinum,* as well as yeasts, moulds, and mycotoxins. Food-borne pathogens differ widely in their incidence, infectiveness, and symptoms of disease. Some food-borne infections are merely uncomfortable, while some can become life-threatening.

The issue of the presence of food-borne pathogens is exacerbated by the low threshold of infectious organisms required to cause severe illness. For example, as few as 15 Salmonellosis organisms can lead to a severe condition (US Food and Drug Administration 2007). Therefore, the choice of method for detection and analysis of the pathogens is paramount. Two distinct groups are generally reported: conventional methods and rapid detection methods. Though many rapid detection methods are used in the initial screening of suspected foods, confirmatory testing for positive results using conventional methods is usually performed (Arora et al. 2006). Conventional methods for detecting food-borne pathogens require selective enrichment of the pathogen, plating and characterization from colony and organism morphology, and traditional procedures such as sugar fermentation and immunoprecipitation. These conventional methods are often approved as a method of identification by regulatory bodies but are often labour intensive and can require up to five days in the case of some culture methods. However, they remain the standard for food-borne pathogen detection and identification. The term "rapid method" is used to describe an array of tests including: polymerase chain reaction (PCR) and DNA hybridization, real-time polymerase chain reaction (RQ-PCR), nucleic acid-based sequence amplification (NASBA), enzyme-linked immunosorbent assay (ELISA), and restriction enzyme analysis (REA), among others.

2.3. Hospital Acquired Infections

Hospital acquired infections occur throughout the world in both developed and nondeveloped countries, are one of the most common causes of morbidity in patients (Ponce de Leon 1991), and have a significant economic impact (Ponce de Leon 1991, Plowman et al. 1999, Wakefield et al. 1988, Coella et al. 1993, Wenzel 1995). Infections are usually in the lower respiratory tract, in surgical wounds or the urinary tract, and a WHO study indicated that the highest frequency of infection was in intensive care and acute surgical and orthopaedic wards (WHO 2002). The patient population is often immunocompromised due to age, the presence of another disease, or immunosuppressive treatments such as chemotherapy, making individuals much more suspectible to infection, particularly with opportunistic pathogens. (Geddes and Ellis 1985). As in the case of food-borne pathogens, there are a number of organisms which can cause infection—including bacteria, viruses, fungi, and parasites. An individual could be infected by person-to-person transmission, by bacteria in their own flora, or by contaminated objects they have come into contact with. The majority of hospital acquired infections caused by bacteria are *Staphylococcus aureus,* coagulase-negative staphylococci, enterococci, or enterobacteriacae (WHO 2002). One other factor that has been highlighted in recent years is the emergence of bacteria that are resistant to many antimicrobial therapies, sometimes resulting in multidrug-resistant strains or "super bugs." One of the overriding reasons for this is thought to be the widespread indiscriminate use of antibiotics to treat infections. Bacteria that have developed multidrug resistance include strains of staphylococci, pneumococci, enterococci, and tuberculosis (Longworth 2001).

2.4. Intentional Infection—Biological Warfare

Biological agents with potential use in biological warfare (BW) can broadly be divided into three categories: toxins (that have a range of sources such as animals, bacteria, and plants), viruses, and bacteria. The number of agents that could be utilised is extensive; however, of these a significantly smaller proportion could be effectively disseminated through the aerosol route, considered the most likely route for a large-scale attack (Christopher et al. 1997, Christopher et al. 1999, Eitzen 1997, Eitzen et al. 1998, Franz 1997, Kortepeter and Parker 1999, Peters and Dalrymple 1990). Microbial pathogens that could be used include: *Bacillus anthracis, Yersinia pestis,* Venezuelan equine encephalitis (VEE), *Francisella tularensis, Variola* virus, and the haemorrhagic fever viruses (arenaviruses, filoviruses, flaviviruses and bunyaviruses). Toxins also represent a threat from potential use as BW agents; examples include those that could be isolated from *Clostridium botulinum, Ricinus communis,* trichothecene mycotoxins, or staphylococcal enterotoxins (Hawley and Eitzen 2001). Some of these, such as VEE, are considered incapacitating agents, and others, such as *Bacillus anthracis,* as lethal agents (Hawley and Eitzen 2001). Of all the lethal agents the most serious to the human host are haemorrhagic fever viruses such as the Ebola virus, Lassa fever, or the Marburg virus, as no prophylaxis or vaccines are currently available (Koch et al. 2000). However, these viruses are much more difficult to produce than toxins or bacteria. Virus propagation requires more sophisticated equipment and a higher degree of expertise such as tissue culture. Bacteria by comparison are much easier to grow once a source of the pathogen has been found. In the case of BW agents, often the LD_{50} or ID_{50} in humans is predicted to be low, and any potential use needs to be identified quickly and at a relatively low level, making the requirement for rapid, specific, and sensitive detection of these pathogenic microorganisms essential.

3. Detection of Pathogenic Microorganisms

The rapid, sensitive, and specific detection of pathogenic microorganisms is essential if effective treatment is to be provided to a susceptible population. In the case of bacteriological testing, traditional microbiology has proved a time consuming procedure. Organisms have to be isolated and grown, and usually a series of biochemical tests must be completed for identification (Helrich 1990, Kaspar and Tartera 1990). Techniques such as the polymerase chain reaction (PCR) used for the amplification of pathogen-specific DNA sequences have proved to be sensitive. However, when using environmental samples, a degree of sample preparation is required since impurities contained within the sample may inhibit the PCR. Furthermore, the use of small sample volumes (sometimes 1 µl) means that the sample often has to be concentrated to obtain the desired sensitivity (Radstrom et al. 2004).

Biosensors are particularly attractive as a means to detect and identify potential pathogenic microorganisms due to their potential specificity and sensitivity (although this is also governed by the choice of recognition element), together with the provision of information in near real time. Biosensors also allow the analysis of complex sample matrices (Hobson et al. 1996, Ivnitski et al. 1999). To provide protection, i.e., timely warning of the presence of a pathogen, environmental samples are often analysed using biosensors. This presents an additional problem, in that other microorganisms will also be present within the sample. The detector needs to be able to discriminate the pathogen of interest from the background, and this can be achieved in a number of ways. These include (a) detection of an increase in the number of particles, (b) detection of an increase in biological particles, (c) detection of pathogenic biological agents, or (d) the specific identification of a biological agent. Specific detection is dependent on the interaction of the target analyte (e.g., a protein) with a recognition element (e.g., an antibody). The use of biosensors for sensitive specific detection of a pathogenic microorganism still remains a significant challenge, and success is often dictated by the nature of the detection element (the specific ligand) and the choice of target analyte (Labadie and Desnier 1992).

4. Conclusions

There are a wide range of pathogenic microorganisms and a number of environments in which they may need to be detected. In the case of bacteriological testing, traditional microbiology has proved a time consuming procedure. Organisms have to be isolated and grown, and usually a series of biochemical tests must be completed for identification (Helrich 1990, Kaspar and Tartera 1990). Therefore, a range of new methods, particularly those using biosensors, is being developed in order to provide rapid, sensitive, and specific detection of pathogenic microorganisms for effective treatment to be provided to a susceptible population.

References

Arora K, Chand S, Malhotra BD (2006) Anal Chim Acta 568:259–74

Bliska JB, Galan JE, Falkow S (1993) Signal transduction in the mammalian cell during bacterial attachment and entry. Cell 73:903–920

Brubaker RR (1985) Mechanisms of bacterial virulence. Ann Rev Microbiol 39:21–50

Brunham RC, Pummer FA, Stephens RS (1993) Bacterial antigenic variation host immune response and pathogen-host coevolution. Infect Immun 61:2273–2276

Cassadevall A, Pirofski L (1999) Host-pathogen interactions: redefining the basic concepts of virulence and pathogenicity. Infect Immun 67(8):3703–3713

Cassadevall A, Pirofski L (2001) Host-pathogen interaction: the attributes of virulence. J Infect Dis 184:337–344

Christopher GW, Cieslak TJ, Pavlin JA, Eitzen EM (1997) Biological warfare: a historical perspective. JAMA 278(5):412–417

Coella R et al (1993) The cost of infection in surgical patients: a case study. J Hosp Infect 25:239–250

Daffe M, Etienne G (1999) The capsule of Mycobacterium tuberculosis and its implications for pathogenicity. Tubercul Lung Dis 79:153–169

Dietsch KW, Moxon ER, Wellems TE (1997) Shared themes of antigenic variation and virulence in bacteria, protozoal and fungal infections. Microbiol Mol Biol Rev 61:281–293

Eitzen EM (1997) Use of biological weapons. In: Sidell SR, Takafugi ET, Franz DR (Eds) Medical aspects of chemical and biological warfare. TMM Publications, Washington, DC, pp 437–450

Falk IS (1928) A theory of microbiologic virulence. In: Jordan EO, Falk IS (Eds) The newer knowledge of bacteriology and immunology. University of Chicago Press, Chicago IL, pp 565–575

Falkow S (1997) What is a pathogen? ASM News 63:136–169

Falzano L, Fiorenti C, Donelli G, Michel E, Kocks C, Cossart P, Cabanie L, Oswald E, Boquet P (1993) Induction of phagocytic behaviour in human epithelial cells by Escherichia coli cytotoxic necrotising factor type I. Mol Microbiol 9:1247–1254

Finlay BB, Falkow S (1997) Common themes in microbial pathogenicity revisited. Microbiol Mol Biol Rev 61(2):136–169

Geddes AM, Ellis CJ (1985) Infection in immunocompromised patients. Q J Med 55(216):5–14

Gross R (1993) Signal transduction and virulence regulation in human and animal pathogens. FEMS Microbiol Rev 10:301–326

Hambleton P, Turnbull PC (1990) Anthrax vaccine development: a continuing story. Adv Biotechnol Processes 13:105–122

Hava DL, Camilli A (2002) Large scale identification of serotype 4 Streptococcus pneumoniae virulence factors. Mol Microbiol 45(5):1389–1405

Hawley RJ, Eitzen EM (2001) Biological weapons—a primer for microbiologists. Ann Rev Microbiol 55:235–253

Helrich K (Ed) (1990) Official methods of analysis of the association of official analytical chemists, 15th Edition. Association of Analytical Chemists Inc, Gaithersburg, MD, pp 449–450

Henchal EA et al. (2001) Current laboratory methods for biological threat agent identification. Clin Lab Med 21(3):661–+

Henderson B, Poole S, Wilson M (1996) Bacterial modulins: a novel class of virulence factors which cause host tissue pathology by inducing cytokine synthesis. Microbiol Rev 60:316–341

Hobson NS, Tothill I, Turner AP (1996) Microbial detection. Biosens Bioelectron 11(5):455–477

Hultgren SJ, Abraham SN, Capron M, Falk P, St Geme JW, Normark S (1993) Pilus and nonpilus bacterial adhesins assembly, function, and recognition. Cell 73:887–901

Ivnitski D, Abdel-Hamid I, Atanasov P, Wilkins E (1999) Biosensors for the detection of pathogenic bacteria. Biosens Bioelectron 14:599–624

Koch S, Wolf H, Danapel C, Feller KA (2000) Optical flow-cell multichannel immunosensor for the detection of biological warfare agents. Biosens Bioelectron 14:779–784

Kenny B, Devinney R, Stein M (1997) Enteropathogenic *E. coli* (EPEC) transfers its receptor for intimate adherence into mammalian cells. Cell 91:511–520

Krogfelt KA (1991) Bacterial adhesion genetics, biogenesis, and its role in pathogenesis of afrimbrial adhesins of *Escherichia coli*. Rev Infect Dis 13:721–735

Kortepeter MG, Parker GW (1999) Potential biological weapons threats. Emerg Infect Dis 5(4):523–527

Labadie J, Desnier I (1992) Selection of cell wall antigens for the rapid detection of bacteria by immunological methods. J Appl Bacteriol 72(3):220–226

Litwin CM, Calderwood SB (1993) Role of iron in regulation of virulence genes. Clin Microbiol Rev 6:137–149

Longworth DL (2001) Microbial drug resistance and the roles of the new antibiotics. Clev Clin J Med 68(6):496–504

Mead PS, Slutsker L, Dietz V, Mccaig LF, Bresee JS, Shapiro C, Griffin PM, Tauxe RV (1999) Emerg Infect Dis 5:607–625

Merrit EA, Hol WG (1995) AB5 toxins. Curr Opin Struc Biol 5:165–171

Meyer KF et al. (1974) Plague immunization. VI. Vaccination with fraction 1 antigen of *Yersinia pestis*. J Infect Dis 129(Suppl):S41–S45

Meyer TF (1991) Evasion mechanisms of pathogenic Neisseriae. Behring Inst Mitt 88:194–199

Mock M, Fouet A (2001) Anthrax. Ann Rev Microbiol 55:647–671

Neilands JB (1995) Siderophores: structure and function of microbial iron transport compounds. J Biol Chem 270:26723–26726

Ponce-de-Leon S (1991) The needs of developing countries and the resources required. J Hosp Infect 18 (Suppl):376–381

Plowman R et al. (1999) The socioeconomic burden of hospital-acquired infection. London public health laboratory service and London school of hygiene and tropical medicine

Rink L. Kirchner H (1996) Recent progress in the tumour necrosis-alpha field. Int Arch Allergy Immunol 111:199–209

Radstrom P, Knutsson R, Wollffs P, Lovenklev M, Lofstrom C (2004) Pre-PCR processing strategies to generate PCR-compatibile samples. Mol Biotechnol 26(2):133–146

Rosenshine I, Finlay BB (1993) Exploitation of host signal transduction pathways and cytoskeletal functions by invasive bacteria. Bioessays 15:17–24

Savarino SJ, Fasano A, Watson J, Martin BM, Levine MM, Guandalini S, Guerry P (1993) Enteroaggregative *Escherichia coli* heat stable enterotoxin 1 represents another subfamily of E coli heat stable toxin. P Natl Acad Sci USA 90:3093–3097

Schiavo G, Benfenati F, Poulain B, Rosetto O, Polverino De Laureto P, Dasgupta BR, Montecucco C (1992) Tetanus and Botulinum-B neurotoxin block neurotransmitter release by proteolytic cleavage of synaptobrevin. Nature 359:832–835

Schmitt CK, Meysick KC, O'Brien AD (1999) Bacterial toxins: friends or foes.Emerg Infect Dis 5(2):224–234

Smith H (1977) Microbial surfaces in relation to pathogenicity. Bacteriol Rev 41:475–500

Soto GE, Hultgren SJ (1999) Bacterial adhesins: common themes and variation in architecture and assembly. J Bacteriol 181(4):1059–1071

Stebbins CE, Galan JE (2001) Structural mimicry in bacterial virulence. Nature 412:701–705

Taylor CM, Roberts IS (2005) Capsular Polysaccharides and their Role in Virulence. In Russell W, Herwald H (Eds). Concepts in Bacterial virulence. Contrib Microbiol Basel, Karger, pp 55–66

Tracey K, Cerami A (1994) Tumor necrosis factor: a pleiotrophic cytokine and therapeautic agent. Ann Rev Med 45:491–503

Tuomanen E, Weiss A (1985) Characterisation of two adhesins of *Bordatella pertussis* for human ciliated respiratory-epithelial cells. J Infect Dis 152:1028–125

Turnbull PCB (1991) Anthrax vaccines: past, present, and future. Vaccine 9:533–539

US Food and Drug Administration Center for Food Safety and Applied Nutrition. Foodborne pathogenic microorganisms and natural toxins handbook. http://www.cfsan.fda.gov, June 14, 2007

Van der Velde AWM, Bäumer AJ, Tsolis RM, Heffron F (1998) Multiple fimbrial adhesins are required for full virulence of *Salmonella typhimurium* in mice. Infect Immun 66(6):2803–2808

Walker T (1998) Microbiology. WB Saunders, Philadelphia

Watson DW, Brandly CA (1949) Virulence and pathogenicity. In: Clifton CE, Raffel S, Baker HA (Eds) Annual review of microbiology. Annual Reviews Inc, Stanford CA, pp 195–220

Webb GF (2003) A silent bomb: the risk of anthrax as a weapon of mass destruction. P Natl Acad Sci USA 100(7):4346–4351

Welch RA (1991) Pore-forming cytolysins of Gram-negative bacteria. Mol Microbiol 5:521–528

Wizemann TM, Adamou JE, Langermann S (1999) Adhesins as targets for vaccine development. Emerg Infect Dis 5(3):395–403

Wood WB, Davis BD (1980) Host-parasite relations in bacterial infections. In: Davies BD, Dulbecco R, Eisen HN, Ginsberg HS (Eds) Microbiology. Cambridge. pp 551–571

Wenzel RP (1995) The economics of nosocomial infections. J Hosp Infect 31:79–87

Wakefield DS et al (1988) Cost of nosocomial infection: relative contributions of laboratory antibiotic, and per diem cost in serious *Staphylococcus aureus* infections. Am J Infect Control 16:185–19211

World Health Organisation (2002) Prevention of hospital-acquired infections, 2nd edition. WHO/CDS/EPH/2002/12. Available from http:///www.who.int/csr/resources/publications/drugresist/WHO_CDS_CSR_EPH_2002_12/en

2

Sample Preparation: An Essential Prerequisite for High-Quality Bacteria Detection

Jan W. Kretzer, Manfred Biebl and Stefan Miller

Abstract

Rapid microbial testing is more and more preferred worldwide. Conventional time-consuming methods with detection times taking up to several days are being replaced by rapid tests that take only a few hours. With the development of new, rapid, and accurate methods for the detection of bacterial contaminants, the requirements for sample preparation techniques are more and more challenging. In fact, sample preparation is the critical step with respect to the applicability of novel methods. Sample preparation comprises sampling/sample drawing, sample handling, and sample preparation. To fulfil the demands of modern microbiology the ideal procedure should permit rapidly providing the processed sample in a small volume which contains the analyte in the highest concentration possible. The analyte has to be free of substances interfering with the detection method to be finally applied. Additionally, sample processing procedures used should not result in any loss of the bacterial analyte, thereby enabling quantitative measurements.

Techniques for the preparation of samples subjected to microbiological examination are described, especially focusing on the methods applied to investigate the occurrence of pathogenic organisms in foods as well as in the food processing environment.

Sample drawing methods for the monitoring of air and surfaces are outlined. Moreover, different sample preparation methods intended to be carried out prior to the detection of intact bacterial cells or bacterial nucleic acids are discussed in detail. Special attention is paid to magnetic particle-based separation methods, as these tools have gained increasing importance due to their outstanding advantages.

1. Introduction

The rapid detection of pathogenic bacteria is getting more and more into focus and new, accurate, and rapid methods for detection of pathogenic bacteria are constantly being developed. Conventional detection techniques usually take a few days, whereas rapid methods aim to provide results within hours. Globalisation, together with the awareness of high hygienic standards, being an essential requirement for food safety and general public health; and this causes a permanent demand for the development of new and rapid methods. Biodefense programs strengthen these efforts towards the development of fast and efficient detection methods for pathogenic bacteria. This increasing awareness is also driven by the knowledge of dose-response relationships of specific organisms that may turn out to be infective at very low levels.

Jan W. Kretzer, Manfred Biebl and Stefan Miller • Profos AG, Josef Engert Str. 11, 93053 Regensburg, Germany.

M. Zourob et al. (eds.), *Principles of Bacterial Detection: Biosensors, Recognition Receptors and Microsystems*, © Springer Science+Business Media, LLC 2008

However, improved detection methods depend more and more on the quality of the sample preparation, since:

- Samples are *heterogeneous*.
- The target bacteria may be of *low concentration*.
- The food/feed matrices may be *incompatible* with the analytical methods.
- *Pooling* of samples, facilitating a high throughput analysis, could be enabled.
- Detection methods normally use only *small measurement volumes*.

Therefore, pretest sample preparation must be considered as an important step to achieve a result within a reasonable amount of time thereby avoiding or reducing the need for time-consuming culture enrichment steps. Thus, the increasing need for speed and precision in new detection methods illustrates the importance of sophisticated methods for sampling and sample preparation within the overall process. The proper development and adaptation of sample preparation towards the end-point detection method applied is essential for exploiting the whole potential of the complete workflow of any diagnostic method.

The overall goals of sample preparation are:

- to *concentrate the target cells,* so they fit the practical operating range of the detection method;
- to *remove or reduce the effects of inhibitory substances*; and
- to *reduce the heterogeneity of samples* in order to ensure negligible variations between repeated sampling.

Sampling comprises three main aspects that are of particular importance when applying any kind of method to detect microbial contaminants. These are:

- sample drawing;
- appropriate sample handling; and
- adequate sample preparation, addressing the specific requirements of the detection method applied.

The most critical points of these consecutive steps in sampling will be addressed in the following chapter, with the focus on food testing and monitoring of food processing environments.

2. The Sample

In most if not all cases, microbiological examination of a specific item or matrix cannot be accomplished by examining the object in total. Therefore it is necessary to define a reasonable amount of that matrix to be subjected to the diagnostic microbiological method that will be applied. This clearly defined part is referred to as the sample.

Different diagnostic methods used to analyse these samples will address different questions (e.g., qualitative or quantitative measurement of pathogenic bacteria within the analyte), therefore the specific properties of a sample provided for measurement have to be oriented to these problems.

A representative sample addressing the needs for microbiological examination should:

- contain the organism of interest at the same concentration and physiological state as the whole matrix;
- be of a reasonable size, permitting carrying out the sample preparation and the subsequent diagnostic method; and
- enable conclusion of the original level of contamination in the case of quantitative measurements.

The development or the implementation of new diagnostic methods always demands an accurate investigation with respect to the optimum properties of the sample. Therefore sample preparation is a tool of rapidly increasing importance to really exploit the full potential provided by new, fast, and efficient detection methods for pathogenic bacteria.

3. Sampling

3.1. Sample drawing

Sampling points have to comply with the requirements of in-house quality assurance concepts, such as the Hazard Analysis of Critical Control Point (HACCP) concepts of the food and feed processing factories. It is also known that a more frequent sampling can increase reliability. Pooling of samples is widely used and can result in higher throughput, but has to be achieved without loss of sensitivity.

On a practical basis one has to be aware, drawing a sample, that the appropriate sample size is a key issue for the successful detection of a specific contaminant.

The appropriate sample size depends on:

- the expected number of organisms,
- the contamination level of interest, and
- the detection limit of the diagnostic method intended to be applied.

The probable level of contamination directly depends on the environmental conditions of the sample matrix, while the contamination level of interest is usually related to the intended purpose of the whole diagnostic procedure. Diagnostic methods can either be used to investigate the general presence or absence of a specific organism (e.g., testing infant formula for the presence/absence of *Enterobacter sakazakii*), or they are applied to determine the overall degree of bacterial contamination (e.g., determining the number of coliform bacteria in water bodies as an indicator for water quality).

The specific method applied to draw a sample is directly linked to the environmental niche that is supposed to be examined.

4. Microbiological Examination of Foods

Most protocols applied to microbiological testing of foods specify in detail the amount of material to be tested as well as the procedure applicable to draw a sample. The most common procedures recommend amounts of 10–25 g of solid foods and volumes of 10–25 ml of liquid foods/beverages as a representative sample size. All other detailed parameters directly depend on the specific food, and of course the organism sought after.

Results obtained when performing a microbiological examination of surface ripened cheese will, for example, turn out to be completely different depending on the site of the sampling. Samples taken from the surface will typically contain yeasts and moulds, while samples from the inside will mainly contain lactic acid bacteria.

Because of these facts there is no general procedure applicable for food sampling. The overall procedure depends on the specific issue to be addressed.

5. Microbiological Examination of Surfaces

By far the most popular method for testing the hygienic properties of surfaces is the Replicate Organism Detection and Counting (RODAC) method. Applying RODAC, an agar plate is placed upon the area to be tested and pressed on with moderate pressure for 3 s. Plates are subsequently incubated for 48 h to detect all culturable organisms present. The method is limited regarding its applicability with respect to the level of contamination, the size of the area that can be sampled, and the conditions of the surface to be tested. It was reported to be not applicable for uneven, rough, or heavily contaminated surfaces. (Jay et al. 2005).

In order to expand the range of applications the RODAC method can be modified by using selective media.

To enable testing of larger areas, surfaces can be swabbed using different kinds of swabs or sponges. Sampling devices can subsequently be incubated in nutritional media of varying compositions to estimate the level of contamination.

Swab surface sampling methods represent the current state of the art, primarily applied. An introduction of new materials examined by Österblad and coworkers, intended to expand the applicability of swabs with respect to an improvement of the total volume absorbable or the survival rates of specific organisms, did not result in a significantly better performance (Österblad et al. 2003). Nevertheless the basic principle of surface swabbing was further developed, driven by political incidents. In order to allow surface testing of larger areas for substances implicated in bioterrorism, novel sampling kits were developed. A remarkable expansion of the limits of standard surface sampling is represented by a study undertaken by Buttner and coworkers (Buttner et al. 2004). Under the experimental conditions described, an area of $1 m^2$ was successfully sampled, reaching a detection threshold of $40–100 CFU/m^2$ of bacterial spores.

Testing surfaces by using adhesive tape as a sample collector was also successfully applied (Mossel et al. 1966), but has not resulted in a broad application so far.

6. Microbiological Examination of Air

An essential prerequisite of safe food production is the thorough monitoring of the processing environment. The most challenging topic in this field is the reliable estimation of the biological burden of air. A common method to allow at least a statement of the quality of air with respect to the presence/absence of culturable microorganisms is the gravity settling plate (GSP) method. Performing this method, an agar plate is exposed horizontally to the air for a duration of 15 min up to several hours (Al-Dagal and Fung 1990). By subsequent incubation and cell count a qualitative result is obtained. As this method does not allow quantitatively measuringing the number of organisms present, it cannot provide the adequate information required for an assessment of a contamination risk. Moreover, the GSP method utilizes cultivation on a nutritional medium, thereby making the detection of nonculturable organisms (VBNC = viable but not culturable) impossible.

Processing a defined portion of air is required to enable quantification of the organisms present in a specific volume. Numerous different techniques to estimate the number of organisms present have been developed. These methods include:

- Filtration using filters with various pore sizes to classify the airborne particles by diameter. Filters of a specific size can subsequently be rinsed and the water used can be subjected to culture enrichment or any other kind of detection method (Edmonds 1979). A remarkable improvement of this filtration method is the liquid trap method, by which the air to be examined is drawn directly through a liquid (buffer solution or nutritional medium) to retain and thereby sample delicate microorganisms (Gregory 1971, Kingsley 1967).
- The all glass impinger (Figure 2.1), which has been chosen as a reference sampler for comparison in the development of new air samplers (Tyler and Shipe 1959, Tyler 1959, Shipe et al. 1959, Brachman et al. 1964, Lembke et al. 1981). Using this device, the air is drawn through a curved tube and through a jet which is 30 mm from the impinger base. The

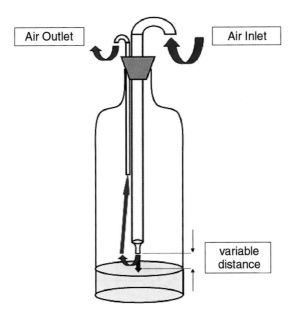

Figure 2.1. AGI-30 impinger.

end of the tube has a smaller diameter, which helps impinge the airborne particles at high velocity. The particles are collected in a liquid (buffer solution or nutritional medium) at the bottom. The distance between the tube end and the impinger can be manipulated in favor of collecting different sizes of airborne particles (Al-Dagal and Fung 1990).

- The Anderson six-stage microbial impactor (Figure 2.2), demonstrating another effort made regarding the enumeration and simultaneous classification of airborne particles. Applying the Anderson sampler, the air to be tested is sucked through sieves at six succeeding stages. From the top to the bottom the velocity of the air increases due to the decreasing pore size of the sieves. At each stage, particles of a specific size and density are retained and collected. Particles can be collected on solid agar as well as in fluids (buffer solution or nutritional medium; Lembke et al. 1981, Al-Dagal and Fung 1990).
- Electrostatic precipitation was shown to be suitable for large volumes of air. The principle of the system is that the air is sucked through the device, passing an electrostatic sampler. Here, near the entrance, airborne particles are charged and attracted to an electrode of the opposite

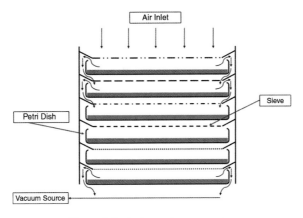

Figure 2.2. Anderson 6-stage sampler.

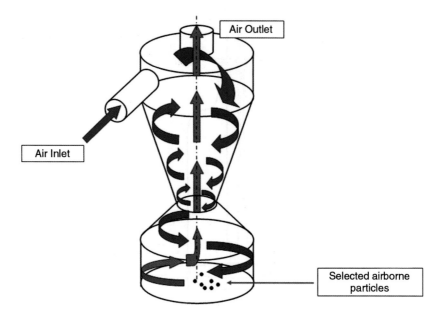

Figure 2.3. Centrifugal sampler.

charge inside the apparatus (Gregory 1973, Kingsley 1967). Particles can be collected in fluids or on agar plates close to the electrode.

- The centrifugal sampler (Figure 2.3), which today is probably the most common device used for sampling airborne microorganisms. In this device the air is sucked into a cyclone in which the particles present are accelerated and finally separated at the wall of the cyclone by centrifugal force. This principle can be applied with agar plates or buffer-containing Petri dishes as sample collecting vessels at the bottom of the cone (Errington and Powell 1969).

Since most new rapid tests try to circumvent cultural enrichment steps and additionally are carried out on a small scale, only sampling methods applicable with small volumes of sampling buffer, like the centrifugal sampler, seem to be compatible with these new methods.

7. Sample Handling

Sample handling is critical mainly due to potential cross-contamination. Cross-contamination can cause massive problems in pathogen detection—especially in food microbiology, as most foods provide excellent nutritional conditions for numerous bacteria. Cross-contaminating organisms like nonpathogenic staphylococci originating from human skin can overgrow and thereby mask the contaminating target organism during enrichment, eventually inhibiting its detection. Usually cross-contamination is prevented by handling the samples with sterile equipment. Samples that are not directly subjected to analysis should always be stored at temperatures from 0–4°C to avoid alteration of the composition of the sample flora as far as possible. Although storage at this temperature range prevents or at least limits the growth of most microorganisms, extended storage duration should be avoided, particularly when applying culture enrichment-based methods. Depending on the specific organism, cells might be sublethally injured due to low temperatures. Additionally, it has to be mentioned that prolonged exposure to sunlight might also influence test results due to the impact of UV radiation on the organisms present, as well as on their respective genetic material.

8. Sample Preparation

Appropriate sample preparation represents the most challenging step in any sampling procedure. Most protocols are very dedicated to the detection method intended to be applied, whereas methods with a broad spectrum of applications are quite rare. This section, on the one hand, deals with sample preparation for detection of intact bacterial cells, enabling a broad variety of specific treatments to be carried out subsequently. On the other hand, sample preparation protocols applied for subsequent detection of nucleic acids are addressed. This is due to the fact that tremendous scientific work has been dedicated during the last two decades to the development of organism-specific detection methods based on the polymerase chain reaction (PCR), real-time polymerase chain reaction, nucleic acid sequence-based amplification (NASBA), and nucleic acid-based biochips.

9. Sample Preparation for Detection of Intact Bacterial Cells

Most sample preparation protocols focusing on the detection of intact, viable, or nonviable bacterial cells are applied to address three main aspects:

1. increasing the number of organisms of interest present in the sample,
2. concentration and subsequent separation of the organism from the sample matrix, and
3. removing residual sample components and contaminating background flora that may interfere with the detection method applied afterwards.

In order to elevate the number of viable target organisms present, a sample is usually subjected to culture enrichment. Enrichment of food as well as of environmental samples is performed by homogenizing the sample physically (e.g., using paddle blenders, etc.), dilution in an organism-specific medium, and subsequent incubation at appropriate temperatures. Culture enrichment of samples requires the suppression of the concomitant flora by the addition of selective agents. The choice of selective agents as well as of adequate sources for carbon, proteins/amino acids, and essential supplements addressing specific nutritional requirements of the target organism represents an adequate basis for selective enrichment. The medium composition and the growth conditions (temperature, aerobic/anaerobic, pH, etc.) can be regarded as different hurdles, excluding organisms from growth, other than the target.

The range of temperatures enabling multiplication of most bacteria associated with intoxications or infections in humans is not very broad. Nevertheless, even temperature can be utilized as a selective condition.

Historically, a cold enrichment method to enrich Listeriae in samples has been used for decades. Performing cold enrichment, a sample suspected to contain *Listeria* spp. is incubated at 4°C for weeks to months. The ratio of Listeriae to background flora is increased due to the ability of *Listeria* to multiply at this low temperature (Gray et al. 1948). However, these time-consuming steps of course cannot be used for rapid detection of Listeriae.

Although culture enrichment is probably the most popular method applied, its application is limited to viable and culturable organisms. In the case of nonviable or nonculturable targets, increasing the concentration of organisms must be performed by alternative methods. This is necessary if the detection method applied is limited with respect to the overall time-to-result.

To allow fast and efficient concentration and separation of bacterial cells from a sample matrix, different procedures were developed.

Concentrating and selectively recovering cells from the sample suspension can be achieved by differential centrifugation (Meyer et al. 1991, Niederhauser et al. 1992, Rodrigues-Szulc et al. 1996), size-exclusion filtration (Besse et al. 2004), or selective immobilization of cells.

Immunomagnetic separation (IMS) (Figure 2.4) was shown to be an ideal tool for the selective immobilization of bacteria—a tool that was developed originally for the isolation of blood cells (Lea et al. 1985). Performing immunomagnetic separation, paramagnetic particles coated with organism specific ligands are added to the liquid sample that has to be analyzed. After binding of the immunomagnetic particles to the target organisms, these can be easily separated by applying a magnetic field.

Within two decades immunomagnetic separation had successfully been applied to the isolation of numerous different pathogenic organisms from food and environmental samples (Skjerve et al. 1990, Fratamico et al. 1992, Fluit et al. 1993, Uyttendaele et al. 2000). It is noteworthy that the use of IMS allowed the detection of as few as 2×10^2 CFU/ml of *Listeria monocytogenes* in enriched foods, according to the study by Skjerve and coworkers (Skjerve et al. 1990). Although the term "immunomagnetic" was directly linked to the application of either monoclonal or polyclonal antibodies in the original application of the method, other affinity ligands for immobilization of cells on paramagnetic particles were also developed. An alternative class of ligand-binding molecules for immobilization of bacteria are different lectins (Patchett et al. 1991, Payne et al. 1992). However, these sugar-binding molecules lack the required specificity for a given type of bacterial cell surface and are therefore not suitable for separating target cells from concomitant microflora. Moreover, because of their polymeric nature and binding properties, lectins generally promote strong agglutination of the paramagnetic particles.

Bennett was the first to use immobilized bacteriophages as a specific ligand for separation of bacterial cells (Bennett et al. 1997). As most bacteriophages are highly specific to their host organism, the use of bacteriophages or bacteriophage-derived agents for specific detection of bacteria presents a promising alternative to antibodies. Expanding this application, immobilized bacteriophages were employed in a magnetic separation procedure by Sun (Sun et al. 2001). Despite the ingenious concept of this assay, it contains some important limitations:

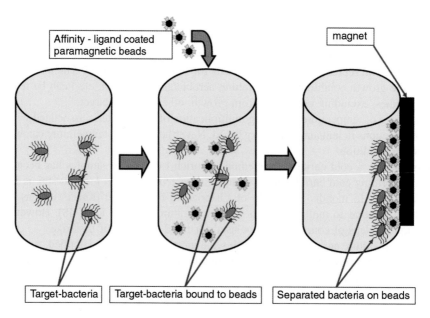

Figure 2.4. Affinity magnetic separation.

The specificity of the bacteriophages is not only a result of the specificity binding to the receptors of the host cell, but is also the sum of other interactions like the restriction modification system of the host. Using intact bacteriophages in the separation of bacteria may impair further morphological, biochemical, or genetic analysis of the captured cells. Due to the infection cycle of the bacteriophage subsequent to binding, the bacterial cells may be lysed within a short period of time. Phages like bacteriophage T4 were also shown to degrade the bacterial DNA at the beginning of the infection cycle (Warner et al. 1970).

A further step towards the development of a new generation of bacteriophage-derived ligands for use in *affinity magnetic separation* was done applying bacteriophage lytic enzymes for this purpose by Kretzer and coworkers (Kretzer et al. 2007). The application of recombinant variants of cell wall binding domains of bacteriophage endolysins for the separation of *Listeria monocytogenes*, *Bacillus cereus,* and *Clostridium perfringens* was successfully demonstrated in this study. In combination with 24-hour selective culture enrichment, detection of very low levels of *Listeria monocytogenes* down to 0.1–1.0 CFU/g in varying foods was enabled. The beneficial properties of these bacteriophage encoded enzymes were previously studied in detail by Loessner and coworkers (Loessner et al. 2002).

Although these proteins show high affinity binding, their applicability is limited to Gram-positive bacteria. Phage-derived tail-fibre proteins, originally involved in phage-host cell recognition, represent an alternative technology, applicable for both Gram-positive and Gram-negative bacteria (Schuch et al. 2006, Miller et al. unpublished). Regarding the specific separation and concentration of bacterial cells, the utilization of bacteriophage-derived ligands can be regarded as a promising future perspective representing an alternative to antibody-based techniques.

Table 2.1, based on a recent review by Stevens and Jaykus (Stevens and Jaykus 2004), is intended to provide an overview of the different attempts carried out to concentrate bacterial cells from various matrices.

10. Sample Preparation for Detection of Bacterial Nucleic Acids

Detection methods targeting nucleic acids have gained more and more importance; therefore, the main aspects of sample preparation with respect to these methods will be discussed here. This will include the preparation of samples that were already subjected to methods for enrichment, concentration, and separation of bacterial cells as described above and will additionally include alternative methods for direct extraction or purification of the genetic material.

Sample preparation methods for nucleic acid-based assays are carried out mainly to release the bacterial DNA/RNA and to further remove substances that are potentially inhibiting the target-amplifying enzymatic reactions. A broad summary of substances responsible for (1) the inactivation of polymerases, (2) the degradation or capture of nucleic acids, and (3) the interference with cell lysis was described by Wilson (1997).

According to a review by Rådström and coworkers (Rådström et al. 2004), sample preparation methods for assays detecting nucleic acids can roughly be divided into (1) biochemical methods, (2) immunological methods, (3) physical methods, and (4) physiological methods.

A basic and very simple procedure widely used to disrupt the cell wall of the organism examined is performed by heating the cells to about 100 °C in buffer solutions that may contain detergents like Triton X100, sodium dodecylsulfate, Tween 20, etc. Subsequently, lysed cells are subjected to further processing. This simple method is rather ineffective when applied to Gram-positive bacteria. Here the addition of specific phage lytic proteins was shown to drastically increase the lysis efficiency (Loessner et al. 1995).

The application of biochemical methods is usually put into practice by applying ready-to-use DNA/RNA extraction kits. Numerous commercially available kits are preferentially used

Table 2.1. Overview: methods to concentrate bacterial cells from various matrices.

Method	Principle	Application	Advantage/Efficiacy	Comment	Reference
Adsorption – ion exchange resin	Binding bacteria on cationic exchange resin – release by pH alteration	*Burkholderia cepacia* from soil	Rapid; relatively inexpensive; nonspecific Recovery : 35%	Sample pretreatment recommended; pH manipulation destroys cell viability	Jacobsen and Rasmussen 1992
Adsorption – metal hydroxides	Hydroxides of zirconium, titaneous or hydroxyapatite; Used in conjunction with centrifugation	*L. monocytogenes* from nonfat dry milk (I) *E.coli* from beef (II)	Rapid, inexpensive,simple non-specific, amenable to large sample sizes Recovery $65 - 96\%$ (I) Recovery 9–99% (II)	Sample pretreatment required. Appears to work best on simple sample matrices	Lucore et al. 2000 (I) Berry and Siragusa 1997 (II)
Aeoqueous two-phase partitioning	Cells partition in one of two immiscible liquid phases	*L. monocytogenes* from sausage	Rapid, inexpensive, simple nonspecific, Recovery 56–90%	Phase partitioning manipulated using pH, polymer concentration and addition of salt	Pedersen et al. 1998
Centrifugation – simple centrifugation	Low speed centrifugation ($< 1000 \times g$) to sediment debris; High speed centrifugation ($> 8000 \times g$) to sediment bacteria. Used with or without coagulation or flocculation	*E.coli*; *Listeria* spp.; *Bacillus* spp.; *Shigella* spp.; *Yersinia* spp. and *Salmonella* spp. from seafood and soft cheese	Rapid, inexpensive, simple nonspecific, amenable to large sample sizes	Bacteria adhere to and sediment with matrix components. Best if preceded by an elution step.	Wang et al. 1997
Centrifugation – differential centrifugation	Low speed centrifugation ($< 1000 \times g$) followed by high speed centrifugation ($> 8000 \times g$). Used with or without coagulation or flocculation	*L. monocytogenes* from meat homogenate	Rapid, inexpensive, simple nonspecific, amenable to large sample sizes 1000-fold improvement in detection limit of subsequent PCR	Bacteria adhere to and sediment with matrix components; few products available to promote desorption without destroying cell viability	Niederhauser et al. 1992

Table 2.1. (Continued)

Method	Principle	Application	Advantage/Efficiacy	Comment	Reference
Centrifugation – density gradient centrifugation	Cell separation by centrifugation within a density gradient; chemical additives to establish a gradient required	*E. coli* O157:H7 from beef homogenate	Can be designed to separate different species from one another Recovery 20–45%	Expensive; difficult to perform; osmotic strength of gradient destroys cell viability; fat entraps bacteria at interface	Lindquist 1997
Filtration – crude filtration	Filtration using cheesecloth; filter paper; homogenizing bags with filter	*E. coli* from milk	Rapid, inexpensive, simple nonspecific, amenable to large sample sizes Recovery 10–95%	High particulate foods clog filters; bacterial cells can adsorb to the filter or retentate	Fernandez-Astorga et al. 1996
Filtration – electro +/− filtration	Bacteria tend to have a net negative charge. Filtration by using electropositive filters: sample prefiltration to remove debris frequently required	Yeast and Lactic acid bacteria from wine and beverages	Rapid, inexpensive, simple nonspecific, Recovery 74–100%	Rapid filter clogging even if samples are pre-filtered; applicable to small volumes only; desorption of bacteria from filters frequently inefficient	Thomas 1988
Affinity separation (AS) lectin based AS	Immobilization of bacterial cells on lectin-coated solid support	*L. monocytogenes* from ground-beef and milk	Rapid, simple, nonspecific Recovery 13–50%	Expensive, sample pre-treatment recommended; best applied to small sample volumes	Payne et al. 1992

Table 2.1. (Continued)

Method	Principle	Application	Advantage/Efficiacy	Comment	Reference
Affinity separation (AS) antibody (polyclonal or monoclonal)-based AS	Immobilization of bacterial cells on antibody-coated solid support	*L. monocytogenes* from cheese	Rapid, simple, highly-specific; standard method for some foods. Detection threshold at 2×10^2 CFU/ml of food-enrichment	Expensive, sample treatment to remove debris recommended for many foods; Applicable to small volumes	Skjerve et al. 1990
Affinity separation (AS) bacteriophage-based AS	Immobilization of bacterial cells on bacteriophage-coated solid support	*S. enteritidis* from pure culture	Rapid, simple, highly-specific	Expensive, sample treatment to remove debris recommended for many foods Applicable to small volumes Confirmation of identification of separated organism not possible	Sun et al. 2001
Affinity separation (AS) bacteriophage-derived ligand-based AS	Immobilization of bacterial cells on bacteriophage-derived, ligand-coated solid support	*L. monocytogenes* from different foods	Rapid, simple, highly-specific Detection threshold at 0.1×1.0 CFU/g of food	Inexpensive recombinant affinity ligand; sample treatment to remove debris recommended for many foods. Applicable to small volumes	Kretzer et al. 2007

in scientific as well as routine diagnostic laboratories. Most of these ready-to-use kits need enrichment or concentration of the bacterial target prior to use. Immunological methods for concentration and separation of cells as well as culture enrichment-based methods, applicable to ensure sufficient cell densities, are described and discussed in detail under 9.

Physical methods unless not already described in 9 include buoyant density gradient centrifugation, aqueous two-phase extraction systems, and dilution.

Discontinuous buoyant density gradient centrifugation was successfully applied as a pre-PCR preparation step by Wolffs and coworkers (Wolffs et al. 2005). According to this study, centrifugation permitted the detection of as little as 8.6×10^2 CFU/ml of *Campylobacter* spp. in chicken rinse samples. Furthermore, it was demonstrated that centrifugation allowed the selective recovery and quantification of viable cells.

Separation of target cells from inhibitory substances by aqueous two-phase systems can, when optimized for the specific application, enable efficient and fast elimination of unwanted material from the sample. The performance of this method, when applied to specific, challenging problems was impressively demonstrated by Lantz and coworkers. The extraction of the sample matrix prior to PCR, applying an aqueous two-phase system, permitted the lowering of the detection threshold of *L. monocytogenes* in soft cheese by four orders of magnitude (Lantz et al. 1996).

Reducing the concentration of inhibitory substances can also successfully be achieved by introducing dilution steps of the sample present. A prerequisite for this strategy is a sufficient sensitivity of the detection method applied. Dilution of samples was successfully used in clinical microbiology when performing PCR with different clinical specimens (Verkooyen et al. 1996, Biel et al. 2000, Chui et al. 2004). In a study performed by Waage and coworkers, enrichment cultures of *Campylobacter jejuni* and *Campylobacter coli* derived from environmental water, sewage, and food samples were diluted with sterile broth to enable detection of the target organisms by nested PCR (Waage et al. 1999). The detection threshold achieved in this study turned out to be as low as 3–15 CFU/100 ml of sampled water. Another method to remove the PCR inhibitors, the use of ultrasound standing waves (USW) for cell washing, is described in chapter 35 by Martyn Hill and Nicolas R. Harris within this book.

Physiological methods comprise the culture enrichment as well as the biosynthesis of genetic material serving as the amplification template. In the case of the PCR, biosynthesis of DNA is directly linked to the multiplication of the specific organism due to the constant ratio of copies of a specific genetic element per cell. In the case of RNA-based detection methods like NASBA, biosynthesis of RNA is dependent on the general growth conditions of a cell as well as the state of expression of a specific gene.

According to Rådström and coworkers (Rådström et al. 2004), the methods listed in Table 2.2 provide a comparison of the performance of different pre-PCR processing methods.

Nucleic acid-based methods for the detection of bacteria are a relatively fast and powerful tool. However, the outcome of these methods is very much dependent on the quality of the sample preparation applied. Affinity magnetic separation using proteins derived from bacteriophages would allow efficient and specific concentration on the bacteria of interest. Due to a high affinity binding of the bacteria to the magnetic beads, inhibitory substances can be removed by a number of washing steps, and after a lysis step the nucleic acid is provided with a quality sufficient for subsequent detection.

11. Conclusions and Future Perspectives

As the development of new and fast detection methods is a steadily ongoing process, as is the improvement of existing procedures, the requirements for the sampling techniques applied are challenging. For these new, rapid, and precise methods a high homogeneity, purity,

Table 2.2. Comparison of different pre-PCR processing methods.

Sample preparation method	Product of sample preparation	Homogeneity	Concentration	Removal of inhibitors	Time required	Cost
Biochemical: extraction	DNA/RNA	+	−/+	+	−/+	−
Immunological: immunomagnetic capture	Cell / DNA/RNA	−/+	−/+	−/+	−/+	−
Physical: buoyant density centrifugation	Cell	−/+	+	−/+	+	−/+
Physiological: enrichment	Cell	−	+	−	−	+

(+ = advantageous; −/+ = average; − = disadvantageous)

and concentration of the sample is an essential prerequisite to sufficiently detect microbial contaminations.

Drawing a representative sample per se remains a challenging step in diagnostic procedures, particularly with respect to quantitative measurements. Sample preparation addressing the specific requirements of any diagnostic method is a key issue of the complete procedure, and therefore must be customized towards the end point detection assay. Nevertheless, there are some basic requirements that have to be taken into account. Sample preparation procedures intended to be applied in microbiological examinations should elevate the number of detectable targets while lowering the amount of inhibitory or interfering matters. Furthermore, sample preparation also should be a fast process in order not to lose the advantage of rapid diagnostic methods.

To enable processing of large sample quantities by high throughput analysis, the sample size to be processed must be reduced. Most modern methods like PCR, ELISA, biochips, or biosensor-based methods are performed in a microscale format. Thus the need for sample concentration is evident. Besides physical techniques like filtration or centrifugation, magnetic particle-based separation has been broadly applied, addressing several challenging problems throughout the past twenty years. With the development of new, highly specific binding ligands, magnetic particle-based separation of cells represents a promising method to efficiently concentrate target cells and simultaneously separate and remove interfering materials of the sample matrix.

To fully exploit the potential provided by new, fast, and efficient detection methods, these have to be considered and implemented as part of a complete workflow which starts with, and is strongly influenced by, sample preparation.

References

Al-Dagal M and Fung DYC (1990) Aeromicrobiology – a Review. Crit. Rev. Food Sci. Nutr. 29:333–340

Bennett AR, Davids FG, Vlahodimou S, Banks JG, and Betts RP (1997) The use of bacteriophage-based systems for the separation and concentration of Salmonella. J. Appl. Microbiol. 83:259–65

Berry ED, and Siragusa GR (1997) Hydroxyapatite adherence as a means to concentrate bacteria. Appl. Environ. Microbiol. 63:4069–4074

Besse NG, Audinet N, Beaufort A, Colin P, Cornu M and Lombard B (2004) A contribution to the improvement of Listeria monocytogenes enumeration in cold-smoked salmon. Int. J. Food Microbiol. 91:119–27

Biel SS, Held TK, Landt O, Niedrig M, Gelderblom HR, Siegert W, and Nitsche A (2000) Rapid quantification and differentiation of human polyomavirus DNA in undiluted urine from patients after bone marrow transplantation. J. Clin. Microbiol. 38:3689–3695

Buttner MP, Cruz P, Stetzenbach LD, Klima-Comba AK, Stevens VL, and Emanuel PA. Evaluation of the Biological Sampling Kit (BiSKit) for large-area surface sampling. Appl Environ. Microbiol. 70:7040–5

Brachman PS, Ehrlich R, Eichenweald HF, Gabelli VJ, Kethley TW, Madin SH, Maltman JR, Middlebrook G, Morton JD, Silver IH, and Wolfe EK (1964) Standard sampler for assay of airborne microorganisms. Science 144:1295

Chui LW, King R, Lu P, Manninen K, and Sim J (2004) Evaluation of four DNA extraction methods for the detection of Mycobacterium avium subsp. paratuberculosis by polymerase chain reaction. Diagn. Microbiol. Infect. Dis. 48:39–45

Errington FP and Powell EO (1969) A cyclone separator for aerosol sampling in the field. J Hyg. (Lond.) 67:387–99

Edmonds RL (1979) Aerobiology – The Ecological System Approach. Dowden, Hutchinson & Ross, Stroudsburg, PA

Fernandez-Astorga A, Hijarrubia MJ, Lazaro B and Barcina I (1996) Effect of the pre-treatment for milk samples filtration on direct viable cell counts. J. Appl. Bacteriol. 80:511–516

Fluit AC, Torensma R, Visser MJC, Aarsman CJM, Poppelier MJJG, Keller BHI, Klapwijk P and Verhoef J (1993) Detection of *Listeria monocytogenes* in cheese with the magnetic immuno-polymerase chain reaction assay. Appl. Environ. Microbiol. 59:1289–1293

Fratamico PM, Schulta FJ and Buchanan RL (1992) Rapid isolation of *Escherichia coli* O157:H7 from enrichment culture of foods using an immunomagnetic separation method. Food Microbiol. 9:105

Gray ML, Stafseth HJ, Thorp F Jr, Sholl LB, Riley WF Jr (1948) A new technique for isolating *Listerellae* from the bovine brain. J. Bacteriol. 55:471–476

Gregory PH (1971) Airborne microbes: their significance and distribution. Proc. R. Soc. Lond. B. Biol. Sci. 177:469–83

Jacobsen C and Rasmussen O (1992) Development and application of a new method to extract bacterial DNA from soil based on a separation of bacteria from soil with cation-exchange resin. Appl. Environ. Microbiol. 58(8):2458–2462

Jay JM, Loessner MJ and Golden DA (2005) Modern Food Microbiology. Springer Science & Business Media Inc., New York, NY

Kingsley VV (1967) Bacteriology Primer in Air Contamination Control. University of Toronto Press, Toronto, Canada

Kretzer JW, Lehmann R, Schmelcher M, Banz M, Kim KP, Korn C, and Loessner MJ (2007) Use of high-affinity cell wall-binding domains of bacteriophage endolysins for immobilization and separation of bacterial cells. Appl. Environ. Microbiol. 73:1992–2000

Lantz PG, Tjerneld F, Hahn-Hagerdal B and Rådström P (1996) Use of aqueous two-phase systems in sample preparation for polymerase chain reaction based detection of microorganisms. J. Chromatogr. B. Biomed. Appl. 680:165–170

Lea T, Vartdal F, Davies C and Ugelstad J (1985) Magnetic monosized polymer particles for fast and specific fractionation of human mononuclear cells. Scand. J. Immunol. 22:207–16

Lembke LL, Kniseley RN, van Nostrand RC and Hale MD (1981) Precision of the all-glass impinger and the Andersen microbial impactor for air sampling in solid-waste handling facilities. Appl. Environ. Microbiol. 42:222–225

Lindqvist R (1997) Preparation of PCR samples from food by a rapid and simple centrifugation technique evaluated by detection of *Escherichia coli* O157:H7. Int. J. Food Microbiol. 37:73–82

Loessner MJ, Schneider A, and Scherer S (1995) A new procedure for efficient recovery of DNA, RNA, and proteins from Listeria cells by rapid lysis with a recombinant bacteriophage endolysin. Appl. Environ. Microbiol. 61:1150–2

Loessner MJ, Kramer K, Ebel F, and Scherer S (2002) C-terminal domains of *Listeria monocytogenes* bacteriophage murein hydrolases determine specific recognition and high-affinity binding to bacterial cell wall carbohydrates. Mol. Microbiol. 44:335–49

Lucore LA, Cullison MA and Jaykus LA (2000). Immobilization with metal hydroxides as a means to concentrate foodborne bacteria for detection by cultural and molecular methods. Appl. Environ. Microbiol. 66(5):1769–1776

Meyer R, Luthy J and Candrian U (1991) Direct detection by polymerase chain reaction (PCR) of *Escherichia coli* in water and soft cheese and identification of enterotoxigenic strains. Lett. Appl. Microbiol. 13:268

Mossel DAA, Kampelmacher EH, and Van Noorle Jansen LM (1966) Verification of adequate sanitation of wooden surfaces used in meat and poultry processing. Zentralbl. Bakteriol. 201:91–104

Niederhauser C, Candrian U, Hofelein C, Jermini M, Buhler HP and Luthy J (1992) Use of polymerase chain reaction for detection of *Listeria monocytogenes* in food. Appl. Environ. Microbiol. 58:1564–1568

Österblad M, Järvinen H, Lönnqvist K, Huikko S, Laippala P, Vilijanto J, Arvilommi H and Huovinen P (2003) Evaluation of a New Cellulose Sponge-Tipped Swab for the Microbiological Sampling: a Laboratory and Clinical Investigation. J. Clin. Microbiol. 41: 1894–1900.

Patchett, R.A., A.F. Kelly, R.G. Kroll. 1991. The adsorption of bacteria to immobilized lectins. J. Appl. Bact. 71:277–284

Payne MJ, Campbell S, Patchett RA and Kroll RG (1992) The use of immobilized lectins in the separation of *Staphylococcus aureus, Escherichia coli, Listeria* and *Salmonella* spp. from pure cultures and food. J. Appl. Bacteriol. 73:41–52

Pedersen LH, Skouboe P, Rossen L and Rasmussen OF (1998) Separation of *Listeria monocytogenes* and *Salmonella berta* from a complex food matrix by aqueous polymer two-phase partitioning. Lett. Appl. Microbiol. 26:47–50

Rådström P, Knutsson R, Wolffs P, Lövenklev M and Löfström C (2004) Pre-PCR Processing. Mol. Biotechnol. 26:133–146

Rodrigues-Szulc UM, Ventoura G, Mackey BM and Payne MJ (1996) Rapid physicochemical detachment, separation and concentration of bacteria from beef surfaces. J Appl Bacteriol. 80:673–81

Schuch R and Fischetti VA (2006) Detailed genomic analysis of the Wß and γ phages infecting *Bacillus anthracis*: Implication for evolution of environmental fitness and antibiotic resistance. J. Bacteriol. 188:3037–3051

Shipe EL, Tyler ME, and Chapman DN (1959) Bacterial aerosol samplers. II. Development and evaluation of the Shipe sampler. Appl. Microbiol. 7:349

Skjerve E, Rorvik LM, Olsvik O (1990) Detection of *Listeria monocytogenes* in foods by immunomagnetic separation. Appl. Environ. Microbiol. 56:3478–3481

Sun W, Brovko L and Griffiths M (2001) Use of bioluminescent *Salmonella* for assessing the efficiency of constructed phage-based biosorbent. J. Ind. Microbiol. Biotechnol. 2:126–128

Thomas DS (1988) Electropositively charged filters for the recovery of yeasts and bacteria from beverages. J. Appl. Bacteriol. 65:35–41

Tyler ME, Shipe EL and Painter RB (1959) Bacterial aerosol samplers. III. Comparison of biological and physical effects in liquid impinger samplers. Appl. Microbiol. 7:355–62

Tyler ME and Shipe (1959) Bacterial aerosol samplers. I. Development and evaluation of the all-glass impinger. Appl. Microbiol. 7:337–49

Verkooyen RP, Luijendijk A, Huisman WM, Goessens WH, Kluytmans JA, van Rijsoort-Vos JH, and Verbrugh HA (1996) Detection of PCR inhibitors in cervical specimens by using the AMPLICOR Chlamydia trachomatis assay. J. Clin. Microbiol. 12:3072–3074

Waage, AS, Vardund T, Lund V and Kapperud G (1999) Detection of small numbers of *Campylobacter jejuni* and *Campylobacter coli* cells in environmental water, sewage, and food samples by a seminested PCR assay. Appl. Environ. Microbiol. 65:1636–1643

Wang RF, Cao WW and Cerniglia CE (1997) A universal protocol for PCR detection of 13 species of foodborne pathogens in foods. J. Appl. Microbiol. 83:727–736

Warner HR, Snustad P, Jorgensen SE, and Koerner JF (1970) Isolation of bacteriophage T4 mutants defective in the ability to degrade host deoxyribonucleic acid. J. Virol. 5:700–708

Wilson IG (1997) Inhibition and Facilitation of Nucleic Acid Amplification. Appl. Environ. Microbiol. 63:3741–3751

Wolffs P, Norling B, Hoorfar J, Griffiths M and Rådström P (2005) Quantification of Campylobacter spp. In chicken rinse samples by using flotation prior to real-time PCR. Appl. Environ. Microbiol. 71:5759–5764

3

Detection of Bacterial Pathogens in Different Matrices: Current Practices and Challenges

Ahmed E. Yousef

Abstract

Successful pathogen detection depends on analyst's understanding of the nature of the matrix and the properties of the targeted microorganism. The matrix could be simple (e.g., drinking water) and easy to analyze for pathogens, or complex (e.g., fermented meat products or fecal samples) and requires an elaborate method to isolate the targeted microorganism. Some pathogens are recovered easily on common laboratory media but others may need time-consuming resuscitation on specialized media with incubation under strictly controlled conditions. Currently used methods for detecting pathogens rely on culture, immunological, genetic, and other techniques. These methods often include a preliminary step to amplify the pathogen's population or a signal representing this microorganism. Enrichment is the most commonly used, but highly unpopular, technique to accomplish the amplification just described. In culture-based detection methods, the targeted pathogen is isolated from the enrichment using selective and differential media, then identified on the basis of multiple biochemical properties. Alternatively, the identification is accomplished by immunological or genetic techniques. Identification as commonly done does not prove the pathogenicity of the targeted organism, a deficiency that needs to be rectified in future detection methods. Rapid detection of pathogens in real time by means that are not destructive to the matrix is an idealistic goal that may materialize in near future.

1. Introduction

Infectious and toxigenic bacteria cause a great deal of human suffering and death. Protecting the public against disease-causing bacteria depends on the efficiency and reliability of the methods designed to detect these pathogens. In general terms, identifying the disease-causing bacteria is the first and most critical step toward the cure of the patient and the protection of healthy individuals. In case of disease outbreaks, identifying the causative agent is a time-sensitive matter. The ability of analysts to link a pathogenic isolate from a patient with that from a source of infection, whether it is food, water, animals, or environment, helps decrease the risk of similar outbreaks. Reliable detection methods are essential not only for tracking pathogens associated with outbreaks, but also for assessing the microbiological quality of food, water, and medications, resulting in the increased safety of consumers.

Ahmed E. Yousef • Professor of Food Microbiology, Department of Food Science and Technology and Department of Microbiology, The Ohio State University.

M. Zourob et al. (eds.), *Principles of Bacterial Detection: Biosensors, Recognition Receptors and Microsystems*,
© Springer Science+Business Media, LLC 2008

2. Analytical Tools and Methods: A Historical Perspective

The discoveries that made microbiology the advanced science we know today were the outcome of unique methods and innovative tools developed during the past few centuries (Wistreich and Lechtman 1984). The microscope was first used in the mid-1600s to view small forms of life. A few decades later, Antonie van Leeuwenhoek published his observations that revealed the world of microbiology. The microscope remains one of the most valuable tools for modern microbiologists. Subsequent to Leeuwenhoek's era, progress in microbiology was slowed by the advance of the spontaneous generation theory, stating that microorganisms arise from non-living media. In the 1800s, Louis Pasteur dismissed this theory and proved that bacteria cause food spoilage. His pioneering work is the basis for today's aseptic technique. Pasteur even proposed the "germ theory of disease," but couldn't prove it experimentally. In 1876, Robert Koch used an innovative technique to establish the link between germs and diseases. He isolated *Bacillus anthracis* from animals suffering from anthrax and used the pure culture to infect healthy animals and replicate the disease symptoms. His procedure encompasses research rules that are famously known as Koch's postulates. His work was inspirational to scientists who developed selective and differential media for isolating microorganisms from infected hosts or complex environments. Basic tools and methodologies developed by Leeuwenhoek, Pasteur, Koch, and many other biologists, led to great discoveries in microbiology.

3. Defining the Terms

Some of the most commonly used microbiological terms have interpretations that vary with context. Following are definitions of some of these terms, as used in this chapter. The word "**sample**" will cover both laboratory and analytical samples, unless indicated otherwise. For a definition of sample, the reader may consult one of many available resources (e.g., AOAC International 2005a). An "**isolate**" refers to a bacterium that has been cultured and purified using microbiological media, but has not been identified. Once identified, the designated scientific name (i.e., genus and species) will be used.

Detecting a pathogen in a matrix requires the analyst to follow an acceptable analytical method. A "**method**" is a collection of techniques and steps that are required to analyze a sample. A "**technique**" is a basic unit of analysis and it usually has multiple uses. With proper adaptation, a technique is included in a particular method to meet specific goals. The technique is made of a number of well-defined "**analytical steps.**" The title of a detection method commonly highlights one of its critical techniques. For example, the AOAC official method 996.08 (AOAC International 2005c) is titled "*Salmonella* in foods: Enzyme-linked immunofluorescent assay screening method." This method encompasses two main techniques: culture (for enrichment) and immunoassay (for identification); however, the latter was included in the method's title. The basic immunoassay technique was originally designed to detect antigens whether these originate from microbial or non-microbial sources. However, the immunoassay in the method just mentioned includes the use of monoclonal antibodies that are specific for *Salmonella* spp. This adaptation makes the immunoassay useful in accomplishing the goals intended for this particular method.

4. Matrix Complexity and Pathogen Detection

Complexity of a detection method depends greatly on the matrix that contains the targeted pathogen. Matrices vary greatly in physical properties, chemical composition, and microbial diversity. Drinking water is probably the simplest matrix to analyze since it is predominantly water with a very small amount of minerals, and it ideally contains only few live microbial cells per liter. On the contrary, fecal matter is a complex semisolid mixture of undigested or partially digested food components as well as live and dead microbial cells. Microbial counts in fecal matter can be as high as 10^{10} cfu/g (van Houte and Gibbons 1966). Although water and feces represent extreme examples, analysts are often challenged by the complexity of samples presented for analysis. Detection methods are typically developed with the assumption that pathogens are present in a complex matrix. The following is a grouping of matrices that are commonly analyzed for pathogenic bacteria, and examples representing each category:

- water (drinking, and natural or artificial bodies of waters)
- food (solid or liquid)
- clinical (tissues, blood, feces, urine)
- environment (air, surfaces, drains, waste solids or fluids)

Which of these matrices is analyzed more often than others for pathogenic contaminants? The frequency of publishing information about given pathogen-matrix combinations may serve as an indicator of the relative importance of pathogen detection to the field analyzing this matrix. Using this concept, Lazcka et al. (2007) concluded that most of the pathogen detection publications are applicable to the food industry (38%), clinical diagnosis (18%), and water and environment quality control (16%). Additionally, these authors concluded that *Salmonella*, *Escherichia coli*, and *Listeria* are the most commonly targeted pathogens in published detection methods.

5. Techniques Currently Used in Pathogen Detection Methods

When a new and promising analytical technique is developed, biologists quickly incorporate it into pathogen detection methods. For example, the advent of polymerase chain reaction (PCR) as a DNA amplification technique led to the emergence of methods that rely on PCR for the detection of various pathogens. Although these methods are PCR-based, they normally include other familiar (i.e., conventional) techniques such as culturing and microscopic examination. Novel techniques are always needed to help microbiologists resolve current and emerging scientific problems and fill knowledge gaps in this dynamic field. These emerging techniques are the basis for the methods of the future.

Microscopic, culture, biochemical, immunological, and genetic techniques are used in various pathogen detection methods (Table 3.1). Most pathogen detection methods include enrichment (a culture technique) and enzyme-linked immunoassay (an immunological technique) or polymerase chain reaction (a genetic technique). Therefore, these three types of techniques will be presented in some detail. Applications of these techniques in methods for detecting *Salmonella* spp. will be used as examples throughout this section.

5.1. Culture Techniques

Culturing in microbiology refers to the transfer of an organism from its ecological niche (e.g., water stream), transient vehicle (e.g., food), or storage medium (as in case of stock culture), into a growth-permitting laboratory medium. The inoculated medium is then incubated, at an optimum growth conditions and for a suitable length of time, to allow cell multiplication, resulting in a culture of the organism. The laboratory medium could be non-selective, selective, or differential, depending on the goal of the culture technique.

Table 3.1. Selected techniques commonly used in pathogen identification methods.

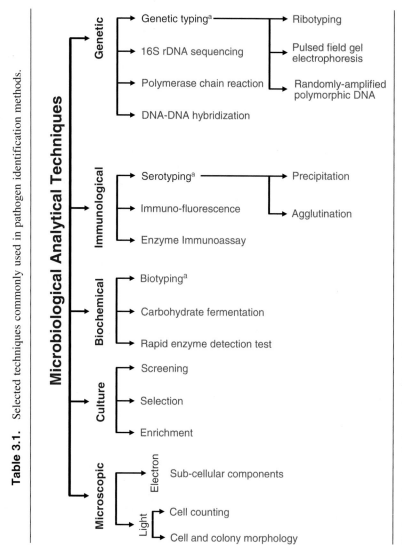

[a]Typing techniques mostly for analysis at the strain level.

Following is a description of these forms of culturing.

- **Non-selective culturing** relies on using growth-permitting non-selective media. "Enrichment," which is used extensively in pathogen detection methods, is a non-selective culture technique. Buffered peptone water, for example, is used to enrich environmental samples in *Salmonella* spp. (Thomason et al. 1977).
- **Selection**, which refers to the use of laboratory media that contain selective agents; these permit the growth of the target bacterium and inhibit or kill other members of the microbial population. Commonly used selective agents include antibiotics, salts, and acids. Selective enrichment broth or selective agar media are often used in culture-based conventional detection methods. For example, the bacteriological analytical manual (BAM) method for detection of *Salmonella* spp. in food includes a selection step using tetrathionate broth (Andrews and Hammack 2006).
- **Screening** is a culture technique to distinguish target from non-target microorganisms. Laboratory media supplemented with differential agents are used in screening. These agents allow analysts to detect visually the target microorganisms in a microbial population. Acid-producing bacteria, for example, are distinguished from nonacid-producers when suitable pH indicators are included in carbohydrate-containing agar media.
- **Selection and screening** are executed simultaneously using selective-differential media. Xylose lysine desoxycholate agar is a selective-differential medium often used in culture-based methods for detection of *Salmonella* spp. This medium contains sodium desoxy-cholate that selects for *Enterobacteriaceae*, and phenol red, a pH indicator that reveals acid production by non-*Salmonella* isolates (Andrews and Hammack 2006). When executed properly, selection and screening, done sequentially or simultaneously, enable analysts to isolate target microorganisms as pure cultures.

5.2. Enzyme-Linked Immunoassay

Immunoassay technique is based on the interaction between an antigen and antibody, and the technology required for detecting or quantifying this interaction. There is a large variety of immunoassays, and these techniques are used broadly in many fields including toxicology and pathogen detection. Enzyme-linked immunosorbent assay (ELISA) and enzyme-linked fluorescent immunoassay are immunological techniques that can be adapted to detect antigens that originate specifically from the targeted bacterial pathogen. Hence, these techniques are useful screening or identification tools in pathogen detection methods. For basics of immunoassay, chapters by Carpenter (2007) or Lam and Mutharia (1994) may be reviewed.

Salmonella spp. are detected in food by an official method that relies on an enzyme-linked immunofluorescent assay (AOAC International 2005c). Multiple enrichment steps should precede the immunoassay; these amplify the antigenic signal of the targeted pathogen. A portion of the enrichment is boiled to release *Salmonella* antigens. Subsequently, the method is run in an automated fashion. *Salmonella* monoclonal antibodies are coated on a pipette tip-like device. This device serves as an antibody holder and a pipette. The boiled enrichment is mixed with the appropriate reagents and cycled in and out of the pipette tip for a specified time. *Salmonella* antigens, if present, bind to monoclonal antibodies coating the interior of the tip, and unbound compounds are washed away. Antibody-alkaline phosphatase conjugates are cycled in and out of the tip; these bind to any *Salmonella* antigen bound to the interior of the pipette tip. Unbound conjugate is washed away, and a substrate (4-methyl umbelliferyl phosphate) for the bound enzyme is introduced. The substrate is converted by the enzyme on the tip wall to a fluorescent product (4-methyl umbelliferone) and fluorescence intensity is measured. Presence or absence of *Salmonella* in the food sample is determined after the results are analyzed by an equipment-integrated computer. According to this reference (AOAC International 2005c), positive results by the immunoassay just described must be confirmed by culture-based procedures.

5.3. Polymerase Chain Reaction (PCR)

The PCR is a molecular technique for in vitro amplification of a DNA fragment via enzymatic replication. Products of PCR amplification (amplicons) are separated on agarose gel, stained and the resulting fluorescent DNA bands are detected. Alternatively, real-time PCR technique allows for the detection of PCR amplification products while they are formed. The PCR technique, in its multiple formats, is widely used in various fields of biology, including detection of human pathogens. For background information about the PCR technique, an introductory article by Winter (2005) and book chapters by Nolte and Caliendo (2007) and Atlas and Bej (1994) can be reviewed.

An official method for detection of *Salmonella* (AOAC International 2005d) makes use of an automated PCR technique (BAX, DuPont Qualicon, PA, USA). A food sample is enriched in *Salmonella* using non-selective enrichment broth, as described earlier. Bacteria, in an aliquot of the enrichment, are lysed to release their DNA. The cells' lysate is mixed with DNA polymerase, nucleotides, *Salmonella*-specific primers, and a fluorescent dye. The specific primers should anneal to a unique DNA sequence in the *Salmonella* genome. The reaction mixture just described is transferred to the thermocycler of the automated system, where the mixture is subjected to repeated heating and cooling. Heating denatures the DNA, separating it into single strands. During cooling, the specific primers anneal to complementary sequences in the DNA of *Salmonella*, if the bacterium was originally present in the sample. At the right temperature, annealed primer is extended through the addition of nucleotides by the catalytic action of the polymerase. This creates a copy of the targeted DNA sequence. When heating and cooling is repeated, this initiates another cycle of DNA denaturing, primer annealing, and primer extension, and thus additional copies of the target DNA sequence are made. The fluorescent dye binds with the double-stranded DNA, as amplicons are formed. Amplification is completed when sufficient heating and cooling cycles are applied, and then the detection step begins. Detection involves denaturing double-stranded DNA (mainly, the PCR amplification product) by raising the temperature of the reaction mixture. This also releases the dye and thus decreases the fluorescent signal. The change in fluorescence is interpreted by the system's computer software, and the results are reported as positive or negative for presence of the pathogen in the sample. According to this AOAC protocol, presumptive positive samples must be confirmed by culture-based procedures.

6. Basics of Pathogen Detection

Bacterial pathogens are occasionally detected in food, water, feces, blood, and other body fluids or tissues. When present in these matrices, only a small population of a given pathogen may be expected. Natural pathogenic contaminants in most environments (e.g., *Listeria monocytogenes* in food) rarely exceed 10^3 cfu/ml or g matrix (Gombas et al. 2003). Simple microbiological culturing techniques, such as direct plating on selective agar media, do not reveal such small numbers because of the interference of the matrix or its microbiota, or the inherent detection limitation of the technique. It is necessary to increase the pathogen population to levels detectable by the analytical method, whether it is conventional or rapid. Therefore, the matrix is commonly enriched in the pathogen of interest before any further analysis. Enrichment techniques usually increase not only the pathogen of interest but also the sample's interfering microbiota. After this multiplication step, a selection step may be needed to increase the prevalence of the pathogen in the microbial population of the analyzed sample. This may be followed by screening the selected population for some of the morphological, biochemical, or genetic characteristics of the targeted pathogen. In culture-based methods, screening is carried out using differential agar media. If these steps are performed successfully, it should be relatively easy to isolate the microorganism in question as a pure culture and to speculate its generic (from genus) part of the name. However, when analyzing for pathogenic microorganisms, it is crucial to identify the putative isolate at the species, or even the subspecies and infrasubspecies levels. Conventionally, identification is accomplished by analyzing the isolate for a large number of biochemical properties such catalase, oxidase, and nitrate reductase reactions. The biochemical tests, in combination with key morphological properties (e.g., Gram-straining), may be sufficient to identify the pathogen. However, analysts often resort to immunological or genetic techniques for detecting or confirming the identity of the isolate. Since members of the same species may vary considerably in pathogenicity (e.g., *Escherichia coli*), it may be necessary to demonstrate the virulence or toxicity of the isolate. For example, confirming the association of a food with botulism requires testing the isolated anaerobic spore-former for production of *Clostridium botulinum* neurotoxin, using mouse bioassay (Solomon and Lilly 2001).

The detection exercise succeeds if the suspect bacterium in the sample survives the selective pressure of the methods, produces a typical reaction or response during screening, is positively identified as the organism targeted by the analysis, and exhibits signs of pathogenicity. The following is a discussion of pathogen detection basics. Samples of techniques used at various phases of these analyses are presented in Table 3.2.

6.1. Sampling

The nature of the matrix dictates the type of sampling technique. The following are considerations for sampling different matrices, in preparation for pathogen detection.

6.1.1. Air Sampling

Although viruses and fungi are responsible for a great number of air-transmitted diseases, many bacteria cause diseases through this route. For example, *Mycobacterium tuberculosis* (Mastorides et al. 1999), *Legionella pneumophila* (Dondero et al. 1980), and *Bacillus anthracis* (Meehan et al. 2004) are responsible for serious air-transmitted diseases. The microbiological quality of air in hospitals, nursing homes, food processing facilities, and many other locations may be monitored for presence of bacterial pathogens. A simple method to measure air quality is to monitor sedimentation of microorganisms on agar media plates that are left exposed to air for a predefined period of time (settle plates). These plates are incubated and the colonies may be screened for the presence of a given pathogen. Air, in a particular environment, may be mechanically pumped onto the surface of agar media before the plates are incubated. Alternatively, sampling may be efficiently accomplished by filtering an air stream through a microfilter. Microorganisms are released from the filter using a suitable diluent and the microbial analysis is carried out. Many useful reports addressing various aspects of environmental contamination have been published (e.g., Sehulster and Chinn 2003).

6.1.2. Surfaces Sampling

Surface samples from clinical (e.g., throat or other body sites), food (e.g., meat or chicken skin), and environmental (e.g., floor or drains) sources are often taken to determine the infection or contamination of these surfaces with pathogens. Analysts should plan to collect samples that reasonably represent the tested surface. Swabbing is the most common technique to collect surface samples (Evancho et al. 2001). This is accomplished using sterile cotton swabs, usually in combination with wetting media. For larger surfaces, particularly when only a low level of the target bacterium is expected, the sponge technique is used. Replicate organism direct agar contact (RODAC) plates may be used to swab surface samples from easily accessible flat

Table 3.2. Pathogen detection steps and examples of techniques used in culture-based and alternative methods.

Step	Culture-based	Alternative Immunological	Genetic	Miscellaneous
Sampling/sample preparation	Equally needed in all methods			
Pathogen amplification	Enrichment	_[a]	_[a]	For simple liquid sample: – Centrifugation – Filtration
Selection and screening	Selection: – Selective enrichment – Selective plating Screening: – Differential media – Selective differential media	Immunomagnetic separation. Capture of pathogen's antigen on solid phase (e.g., immunoassay methods)	Selective amplification of target's unique sequence (PCR) Use of capture DNA probe (hybridization)	Chromogenic media (e.g., Rainbow agar for *Escherichia coli*)
Identification	Morphological examination Biochemical testing Serotyping	Immunofluorescence assay Enzyme immunoassay: – Enzyme-linked Fluorescence Assay – Enzyme-linked immunosorbent assay	PCR. Sequencing 16S rRNA gene. Typing: – Ribotyping – Pulsed-filed gel electrophoresis Microarray	Cellular fatty acid profile. Multilocus enzyme electrophoresis Fourier Transformed Infrared (FTIR) Spectroscopy
Pathogenicity	– Koch's postulate – Mammalian cell culture	Immunoassay for toxins or other virulence factors	Detection of virulence genes or their transcription product	
Specific trait testing: – Antibiotic resistance	Antibiotic selective media		Detection of antibiotic-resistance genes	

[a]Enrichment is commonly performed because there are no feasible alternatives.

surfaces. RODAC plates contain an agar medium suitable for culturing the targeted pathogen. The RODAC plates should contain enough agar medium so that the convex surface of the medium rises above the rim of the plate. Rinsing is another means of collecting surface samples. This is particularly suitable for sampling a whole chicken or turkey.

6.1.3. Bulk Sampling

A sample is a small portion that ideally represents the bulk to be tested. Sampling a large lot or batch of a heterogeneous product could be challenging. Heterogeneity of the source causes sampling errors, and thus the reliability of the results become questionable. When the sample is removed from a flowing river, circulating blood, conveyor belt, or similar dynamic system, it is described as a "specimen" (AOAC International 2005a). Sampling errors in these case may not be avoidable. Sampling for pathogen detection is complicated by the fact that many pathogens are present in very small numbers and therefore sample size and sample uniformity are major considerations. For in-depth coverage of sampling and sample handling, other publications should be consulted (e.g., Andrews and Hammack 2003, Evancho et al. 2001, Miller et al. 2007).

6.2. Sample Preparation

Samples should be quickly and efficiently prepared for analysis to minimize changes in microbial contents or profile. Preparations may include thawing, if the sample was delivered frozen; partitioning, shredding, or grinding of solid samples; or just thorough mixing. These preparations also ease the transfer and measuring (volume or weight) of analytical samples. Measured samples may be mixed with diluents or media before analysis. For solid samples, this mixing may be combined with homogenization in a blender or stomacher. Homogenization in a suitable diluent may also help break bacterial clumps and release pathogens from the solid to the aqueous phase (Yousef et al. 1988).

6.3. Pathogen Amplification

A method for pathogen detection may vary in its minimum detection limit, depending on the matrix analyzed. A PCR-based method, for example, may favorably detect a few bacterial cells when these are suspended in a simple matrix such as water or buffer. However, using the same method to analyze a complex food (e.g., cheese or sausage) may require a pathogen population of 10^6 cfu/g or greater (e.g., Lantz et al. 1994). Similar scenarios have been reported when novel techniques are incorporated into methods for detecting pathogens in complex matrices (Jaykus 2003).

The presence of a small population of a pathogen in a physically- or chemically-complex matrix is a great challenge to analysts. Although only a few pathogen cells may be present, fermented foods, for example, often contain large non-pathogenic microbial populations originating from a starter culture added during product manufacture. Secondary or commensal microbes are also expected in this type of food. Finding a pathogen of concern in such a matrix is often likened to a search for a "needle in a haystack." Simple microbiological techniques, such as direct plating of the sample on selective agar media, commonly fail to detect such a small population. Hence, it is crucial that detection methods include steps that increase pathogen population, which is often accomplished through cells' multiplication.

Enrichment is currently the most reliable step to amplify microbial populations and improve the detectability of pathogens in samples. It is assumed that enrichment of a 25-g sample enables the analyst to detect a minimum of one pathogen cell/25-g sample. Unfortunately, enrichment is the most time-consuming (typically taking 24–48 hr to complete) and the least desirable technique in detection protocols. Enrichment jars are bulky and they occupy a large space in laboratory incubators or refrigerators. During enrichment, strong off odors develop due to the growth of most matrix biota and the metabolism of the nutrient-rich mix of sample and broth. Therefore, most analysts seek alternatives to the enrichment process. When molecular techniques were integrated into detection methods, many analysts hoped that amplification of a characteristic DNA sequence in the pathogen's genome would make the PCR a viable alternative to the culture-based enrichment technique. Unfortunately, this goal remains elusive. Inhibition of PCR by media components and interference of non-target DNA (from sample microbiota or matrix cells) hindered the analyst's ability to rely on this technique for detection of a small population of a pathogenic contaminant. Enzyme immunoassay techniques such as ELISA also embody a signal amplification process. An enzyme molecule, conjugated to an antigen on the targeted cell, catalyzes the conversion of a large number of chromogenic substrate molecules into a colored product. Despite this amplification process, immunoassay-based detection methods often display a minimum detection limit that is higher than that for culture-based methods. Many researchers feel that including one or more enrichment techniques in pathogen detection methods is inevitable (Feng 2001, Yousef and Carlstrom 2003).

6.4. Selection and Screening

After amplification (i.e., enrichment), a selection step is needed to increase the concentration of the pathogen, relative to the total microbial population of the analyzed sample. To achieve this goal, selective enrichments and selective plating are used in culture-based detection methods. Selection and screening, as culture techniques, have been discussed earlier in this chapter.

Selection suppresses most unwanted bacteria, but it does not preclude them totally. It is not uncommon that analysts find numerous colonies on selective media that are subsequently proven to be non-targets. In this case, screening on differential media gives the analyst a chance to discriminate between the pathogen and non-target contaminants. Analysts may carefully design the screening process to benefit from unique morphological, biochemical, or even genetic characteristics of the targeted pathogen. In culture-based methods, selection and screening may be combined by using selective-differential media. If selection and screening are successful, it should be relatively easy to isolate the microorganism in question as a pure culture and to speculate about its genus affiliation. Screening samples for foodborne pathogens has been addressed in several references (e.g., Yousef and Carlstrom 2003).

Immunomagnetic separation (IMS), which is patented for Bisconte De Saint Julien (2000), may be considered an alternative selection technique. The IMS is based on the use of magnetic beads coated with antibodies that specifically react with antigens of the targeted pathogen. Therefore, the beads selectively capture the targeted pathogen, leaving behind other bacteria. When the beads are removed from the mixture, this separates the pathogen from interfering media components and other non-target microorganisms. Many researchers implemented the immunomagnetic technique, along with PCR or enzyme immunoassay, to develop rapid detection methods (Benoit and Donahue 2003).

6.5. Identification

Identification is determining the identity of an isolate by matching its characteristics with those of a previously identified species. Occasionally, the properties of an isolate cannot be matched; this could lead to a discovery of a new species. If pursued further, the researcher should define the taxonomic group of this unmatched isolate and assign a scientific name according to international rules (Brenner et al. 2005). Identification of bacteria requires the knowledge of their morphological, biochemical, physiological, and genetic characteristics. Collectively, these characteristics can be grouped as phenotype and genotype. It is advisable that identification schemes start with broad categorization (e.g., Gram-staining) and progress to more specific tests. Identification leads to determining the isolate's genus and species.

There is a growing need to identify isolates at the subspecies and infrasubspecies (i.e., strain) levels; the latter is commonly referred to as "typing." Stains of a pathogenic species may vary considerably in virulence (Myers et al. 2006) or susceptibility to drugs (Struelens 2006). Consequently, typing becomes essential in epidemiological studies and disease treatment. Similar to identification at the species level, typing can be accomplished by analyzing the morphology, biochemical characteristics, or antigenic makeup of an isolate, to determine its morphotype, biotype, or serotype, respectively (Brenner et al. 2005). Additionally, susceptibility to phage is used to determine the isolate's phage-type. Many genetic techniques are used in typing isolates at the strain level (Table 3.1). These techniques include restriction fragment length polymorphism (RFLP), low frequency restriction fragment analysis, which is determined by pulse-field gel electrophoresis (PFGE), and ribotyping (Gillis et al. 2005). Following are properties commonly used in identification of bacterial isolates.

6.5.1. Morphological Characteristics

Bacterial colony morphology includes colony shape, dimension, pigmentation, and others. Cell morphology, as observed under the microscope, includes Gram reaction (positive or negative), shape (e.g., coccus or rod), organization (e.g., single or chain), presence and properties of endospore (e.g., central or terminal), flagellation (e.g., polar or peritrichous), and others. Unlike eucaryotes, bacteria have simple morphological properties that cannot be relied upon as a means for classifying these organisms (Brenner et al. 2005). Although bacterial isolates cannot be identified solely on the basis of their morphological characteristics, revealing these traits may guide the analyst in developing a sound identification scheme.

6.5.2. Biochemical and Physiological Traits

A large number of biochemical and physiological characteristics have been used in bacterial identification. Results of these tests may be used in bacterial identification on the basis of numerical taxonomy (Sneath 2005). Identification of a bacterial isolate may require 100–200 assays and the proper algorithm to analyze these data and predict the isolate's identity (Brenner et al. 2005).

Biochemical and physiological traits may be grouped as follows:

- **Rapid biochemical reactions**. These indicate the presence of a single enzyme or enzyme complex. Reactions catalyzed by surface or extracellular enzymes are easily and rapidly analyzed. Other reactions may require incubating the isolate with enzyme substrate for several hours. Representing this category are reactions catalyzed by catalase, oxidase, nitrate reductase, amylase, β-galactosidase, thermonuclease, and urease.
- **Carbohydrate fermentation**. Analysts may test the isolate's ability to utilize a certain carbohydrate as the only carbon source. The carbohydrate utilization profile could be useful in identifying a bacterial isolate. Fermentation of glucose, sorbitol, and mannitol are examples of tests under this category.
- **Miscellaneous physiological features**. Growth at different temperatures, pH values, salt concentrations, and gaseous environments are some of the physiological properties used in identifying bacteria. Isolates may also be tested for growth in the presence of various antimicrobial substances (e.g., antibiotics).

6.5.3. Serological Properties

Bacterial isolates may be tested for their characteristic antigens, which interact with specific antibodies. A bacterial cell may carry one or more of these antigens:

- Somatic (O) antigens. In Gram-negative bacteria, O-antigens are made of the O-polysaccharide, a component of the lipopolysaccharide in the cell's outer membrane. Gram-positive bacteria may also carry somatic antigens, but these are generally less defined, in comparison with those of the Gram-negatives.
- Flagellar (H) antigens which are heat-labile proteins.
- Capsular (K) antigens which are made of carbohydrates of cell capsule.

Isolates that belong to a given genus may be classified into species on the basis of their antigenic makeup. Similarly, isolates of a given species may be typed into serovars (also described as serotypes) that differ antigenically. Species with well defined antigens are relatively easy to identify on the basis of serology. In *E. coli*, 173 O-antigens, 56 H-antigens, and 103 K-antigens are currently known; these are valuable in serotyping members of this species (http://ecoli.bham.ac.uk/path/sero.html). Salmonellae have been divided into >2000 serovars based on their O and H antigens (Krieg 2005).

Serological analyses, at various levels of complexity, are common bacterial identification tests. Antigen-antibody reactions are detected as agglutination, precipitation, color change (e.g., ELISA), or immuofluorescence. Automated immuoassays, as tools to identify bacterial isolates, are commercially available. One of these automated assays was discussed previously in this chapter. According to Krieg (2005), these tests offer a means for achieving quick, presumptive identification of bacteria.

6.5.4. Genetic Characteristics

Phenotypic traits of bacteria, which include their morphological, biochemical, and physiological characteristics, have been used for many decades for identifying bacterial isolates. When a sufficient number of these characteristics are defined, it becomes feasible to apply numerical taxonomy principles to predict isolate identity (Sneath 2005). This approach, however, was not always adequate to set taxonomic boundaries between different species. Analysts often observe phenotypic continua that make it difficult to separate species on the basis of phenotypic characteristics (Brenner et al. 2005). Genotyping is a very useful approach to overcome this ambiguity.

In an early implementation of genetic techniques in bacterial identification, analysts matched isolates to be identified with known species using the mole% G+C value of genomic DNA. This approach is a crude measure of genetic relatedness and was not very useful as an identification tool. Subsequently, DNA hybridization was introduced to determine relatedness among bacteria (Brenner et al. 2005). DNA hybridization is used as an identification technique by measuring the similarity in sequences between DNA of an isolate and that of bacteria with known identity. In this technique, double-stranded DNA of an isolate and a known bacterium are heat-denatured into single strands. When these DNA strands are mixed and cooled 25–30°C below their denaturing point, complementary sequences from these two organisms will hybridize and DNA heteroduplexes are formed. Sequence complementation between the two bacteria can be measured by separating single and double strands on hydroxyapatite, or digesting the unpaired regions using single-strand specific nuclease (Crosa et al. 1973). Additionally, the proportion of unpaired bases may be determined, as % divergence, by comparing the thermal stability of homologous (i.e., complementary DNA strands of the same bacterium) and heterologous duplexes. It has been agreed that a species is defined as strains with $\geq 70\%$ DNA-DNA relatedness (as determined by DNA-DNA hybridization techniques) and $\leq 5\%$

divergence (% unpaired bases). Unfortunately, accurate measurement of whole-genome DNA-DNA hybridization is questionable; this somewhat marginalizes the value of this technique as an identification tool.

Nucleotide sequence-based identification of bacteria is currently the most reliable approach. Sequence analysis of the gene encoding the 16S rRNA is becoming the backbone of phylogenetic classification of bacteria (Brenner et al. 2005). Universal primers for this conserved gene can be used to amplify its sequence (Weisburg et al. 1991). To identify an isolate, its 16S rRNA gene nucleotide sequence is matched with known sequences in the widely-available genomic databases (e.g., http://www.ncbi.nlm.nih.gov). Stackebrandt and Goebel (1994) concluded that strains having < 97% similarity in their 16S rRNA gene sequence represent different species. Furthermore, strains with ≥ 97% similarity in this gene's sequence may or may not belong to the same species; in this case, DNA relatedness must be assessed before making a final determination.

6.6. Pathogenicity Testing

After identifying an isolate, it may be necessary to prove its link to a disease case or outbreak. Pathogenesis is a complex process that cannot be explained on the basis of the phenotypic and genotypic properties of a bacterium. Therefore, demonstrating an isolate's pathogenicity is the ultimate proof of its link to an infection. Although most analytical laboratories do not complete pathogenicity testing, addressing this topic concisely in this chapter is appropriate. The following are select techniques to determine an isolate's pathogenic properties.

6.6.1. Koch's Postulates

Koch's postulates, or rules of proof, define the criteria for establishing a causal relationship between a suspect pathogen and a disease (Wistreich and Lechtman 1984). Considering bacterial pathogens, these postulates can be stated as follows:

i. The suspect bacterium must be found in all organisms suffering from the disease, but not in healthy ones.
ii. The suspect bacterium must be isolated from a diseased organism, and grown in pure culture.
iii. The isolated bacterium should cause disease when introduced into a healthy organism.
iv. The bacterium must be reisolated from the diseased organism and identified as closely identical to the original causative agent.

Although Koch successfully used these postulates to establish the cause of anthrax, they were not applicable to all diseases. Currently, a number of infectious agents are accepted as the cause of disease despite the failure to fulfill all of Koch's postulates. If the detected bacterium matches known pathogens in phenotypic and genotypic characteristics, demonstrating its pathogenicity may not require the stringency imposed by Koch's postulates.

6.6.2. Mammalian Cell Culture (Tissue Culture)

The human intestinal Caco-2 cell line, human cervical HeLa cells, and others are often used in demonstrating bacterial virulence and invasiveness (AOAC International 2005b, Langendonck et al. 1998, Sambuy et al. 2005). The mammalian cells are grown as a monolayer and inoculated with the isolate in question. Virulent pathogens may penetrate and grow inside the cell monolayer or cause cell damage. If these events are detected, virulence and invasiveness of the pathogen are demonstrated. Caco-2 (Langendonck et al. 1998) and HeLa (AOAC International 2005b) cells were used to demonstrate the virulence of *L. monocytogenes* and *E. coli*, respectively.

6.6.3. Virulence Genes and Gene Expression Products

Inference of pathogenicity of an isolate is possible when the bacterium is analyzed for the presence of virulence genes or gene expression products (e.g., Reckseidler et al. 2001). If a bacterium causes intoxication, detection of its toxin in the culture or contaminated sample may prove the bacterium-disease link (Bennett 2001).

6.7. Testing for Specific Traits

During the analysis of clinical samples, antibiotic resistance is one of the most targeted traits in pathogenic isolates. Methicillin-resistant *Staphylococcus aureus*, vancomycin-resistant enterococci, and multidrug-resistant *Mycobacterium tuberculosis* are some of many problematic strains, and the rapid demonstration of their antibiotic resistance can save lives (Tenover 2007). A culture technique is often used to determine the antibiotic resistance of an isolate. This involves using antibiotic-supplemented media (e.g. Landman et al. 1996). The culture technique requires 24 hr, at least, and the sensitivity of these tests has been questioned. Alternative rapid molecular techniques for detecting antibiotic-resistance seem promising (Tenover 2007).

7. Challenges to Current Detection Methods

7.1. Pathogen Quantification Problems

Results of detection methods are not very amenable to quantitative interpretations. Even the most practiced culture methods are not as quantitative as some analysts may have anticipated (Sutton 2006). The enumeration technique, for example, produces a skewed estimate of cell population; only cells able to form colonies under the conditions of the test are counted. Furthermore, a colony that appears on the medium may not represent a single cell; it may have arisen from a clump of cells. Microbiologists wisely consider these counts as estimates and report them as colony-forming units (cfu), not cells. There is also confusion between the "limit of detection," which is 1 cfu when only one colony appears on the plate, and the "limit of quantification" which is 25 cfu for a plate with the least countable colonies without declaring the count an estimate. Therefore, the accuracy required for microbiological criteria or specifications are not attainable by current detection methods.

7.2. Can a Small Bacterial Population be Detected Rapidly and Reliably?

Bacteria are small organisms with a cell volume in the range of $0.01–7 \, \mu m^3$ (Norland et al. 1987). To illustrate a method's detection limits in a quantitative sense, consider some pathogenic bacteria having $0.1–1.0 \, \mu m^3/cell$. To simplify the quantitation, one may assume that bacterial cell density is not much different from that of the aqueous surrounding. Based on this scenario, the presence of one bacterial cell/g matrix would be equivalent to $1.0^{-13}–1.0^{-12} \, g$ bacterial fresh weight/g matrix. This means a method with a detection limit of one bacterium/g medium is detecting a cell's heterogeneous organic matter in the heterogamous matrix at 0.1–1.0 part per trillion (ppt) levels. The actual detection limit is likely smaller than these estimates for many reasons. The analysts may target a cell component (e.g., a DNA sequence or a chemical marker) instead of the whole cell, or they may be interested in bacterial cell dry weight, instead of the fresh weight; in these cases, the minimum detection limit would be even smaller than 0.1–1.0 ppt. Considering that some researchers report methods with detection of 1–5 cells per 25 g sample (e.g., Chen et al. 1998, Jeníková 2000), the wt/wt detection limit of such a method is much smaller than that just estimated. This may illustrate the difficulty in detecting or quantifying small bacterial populations without cell propagation steps.

Amplification of the bacterial cell population, or any cellular components targeted by the analysis, is a prerequisite for successful detection. As described earlier, enrichment has been used reliably to augment the pathogen's population, but this technique is becoming the most derided and time-consuming step in the analysis. Efforts are underway to replace the enrichment process with an analyst-friendly technique, but no satisfactory alternative has been developed yet.

7.3. Which Traits to Analyze, and How Many Tests are Needed for Identifying a Bacterial Pathogen?

Being prokaryotic single-celled organisms, bacteria have a simple morphology which cannot be used as a basis for their classification or identification. However, analysts should carefully consider these morphological characteristics before they develop a battery of identification tests. Biochemical and physiological characteristics have been very useful in identifying an isolate, and analysts may employ up to 200 tests to identify unknown isolates using elaborate numeric techniques (Brenner et al. 2005). Luckily, commercial kits, equipment, and computer software are available to automate this tedious task (Feng 2001). However, it is sometimes impossible to identify a bacterium reliably on the basis of biochemical traits, since these would survey only a small portion of the bacterial genome. A battery of 300 tests would assess only 5–20% of the genetic potential of bacteria (Brenner et al. 2005).

Serological tests are used extensively in pathogen identification, and some immunoassay techniques have been automated. It is however difficult to correlate serological properties of an isolate with its genotypic or other phenotypic traits. Researchers, for example, used *E. coli* somatic and flagellar antigens to identify the O157:H7 serotype. Unfortunately, diseases that were originally associated with *E. coli* O157:H7 (i.e., hemorrhagic colitis and hemolytic uremic syndrome) were found to be caused by many other serotypes of *E. coli* (Nataro et al. 2007). It is therefore unreliable to use these serological techniques to identify this causative agent or track these diseases.

Genetic techniques were introduced in pathogen detection methods by targeting characteristic sequences in bacterial genomes. For example, identification of *L. monocytogenes* was made easier by introducing a PCR technique that targets the pathogen's *iap* gene (Bubert et al. 1992). Mutiplex PCR technique allows simultaneous detection of multiple genes within the same genome and thus improves pathogen identification (Fagan et al. 1999). There is no doubt that molecular techniques are valuable in modern detection methods, but several problems remain. Separating PCR amplification products by electrophoresis and matching sizes of amplicons on agarose gel is not a precise exercise. Luckily, real-time PCR and similar automated techniques resolved this issue. Inhibition of PCR by matrix components has been the greatest drawback in using this technique. Application of 16S rRNA gene sequencing has evolved as the most reliable pathogen identification technique (Janda and Abbott 2002). Primers for highly-conserved regions of this gene are in use, and universal primers for amplifying the complete gene, in preparation for full gene sequencing, are available. It is generally accepted that an isolate whose 16S rRNA gene sequence is <97% similar to those of the isolate's closest phylogenetic neighbors constitutes a new species. In spite of the many advantages of the 16S rRNA gene sequencing as an identification tool, it is not easily implemented as a routine test in analytical laboratories. The technique is time-consuming and may not be automated easily. Sequencing 16S rRNA gene surveys only a small portion of the bacterial genome; therefore, isolates with > 97% sequence similarity may belong to different species as judged by DNA-DNA hybridization analysis.

It is generally agreed that the most acceptable approach for identifying an isolate is to integrate all available data on its phenotypic and genotypic traits. Assembling and assimilating all data on an isolate's diverse traits should aid the analyst in making a sound judgment about its identity. This approach is the basis of polyphasic taxonomy (Brenner et al. 2005, Gillis et al. 2005, Janda and Abbott 2002). With better automation of detection techniques, large data sets may be gathered and polyphasic analysis may become feasible.

7.4. Real-Time Detection

There is an urgent need for improving current detection methods, particularly in issues related to speed. Rapid detection of pathogens is critical for ensuring the safety of the public. In more tangible terms, rapid detection saves time, effort, and cost. The ultimate goal is the ability to detect pathogens in real time and by means that are not destructive to the matrix analyzed. This may seem an idealistic goal. However, considering the rapid advances in science and technology, non-destructive and non-disruptive detection methods that are now out of reach may become the "conventional" methods of the future.

References

Andrews WH and Hammack TS (2003) Food sampling and preparation of sample homogenate. Bacteriological Analytical Manual Online (http://www.cfsan.fda.gov), accessed July 10, 2007

Andrews WH and Hammack TS (2006) *Salmonella*. Bacteriological Analytical Manual Online (http://www.cfsan.fda.gov), accessed July 10, 2007

AOAC International (2005a) Guide to method format. The official methods of analysis, 18th ed., AOAC International, Gaithersburg, MD

AOAC International (2005b) Official Method 982.36: Invasiveness of mammalian cells by *Escherichia coli*. The official methods of analysis, 18th ed., AOAC International, Gaithersburg, MD

AOAC International (2005c) Official method 996.08: *Salmonella* in foods, enzyme-linked immunofluorescent assay screening method. The official methods of analysis, 18th ed., AOAC International, Gaithersburg, MD

AOAC International (2005d) Official Method 2003.09: *Salmonella* in selected foods, BAX® automated system. The official methods of analysis, 18th ed., AOAC International, Gaithersburg, MD

Atlas RM and Bej AK (1994) Polymerase chain reaction. p. 418–435. In: Gerhardt P (ed.) Methods for General and Molecular Bacteriology. American Society for Microbiology, Washington, D.C.

Bennett RW (2001) Staphylococcal enterotoxins: micro-slide double diffusion and ELISA-based methods. Bacteriological Analytical Manual Online (http://www.cfsan.fda.gov), accessed July 10, 2007

Benoit PW and Donahue DW (2003) Methods for rapid separation and concentration of bacteria in food that bypass time-consuming cultural enrichment. J. Food Prot., 66:1935–1948

Bisconte De Saint Julien J-C (2000) Process and installations for separation of magnetic particles in a fluid for biological analysis, and application of said process. US Patent 6143577

Brenner DJ, Staley JT and Krieg NR (2005) Classification of prokaryotic organisms and the concept of bacterial speciation, p. 27–32. In: Brenner DJ, Krieg NR and Staley JT (eds), Bergey's Manual of Systematic Bacteriology, 2nd ed. vol. 2, Part A. Springer, New York, NY

Bubert A, Kohler S and Goebel W (1992) The homologous and heterologous regions within the *iap* gene allow genus- and species-specific identification of *Listeria* spp. by polymerase chain reaction. Appl. Environ. Microbiol. 58:2625–2632

Carpenter AB (2007) Immunoassays for the diagnosis of infectious diseases. In: Murray PR, Baron EJ, Landry ML, Jorgensen JH and Pfaller MA (eds) Manual of Clinical Microbiology, 9th ed., Am. Soc. Microbiol., Washington, D.C., p 257–270

Chen S, Xu R, Yee A, Wu KY, Wang C-N, Read S, and De Grandis SA (1998) An automated fluorescent PCR method for detection of shiga toxin-producing *Escherichia coli* in foods. Appl. Environ. Microbiol. 64:4210–4216

Crosa JH Brenner DJ, and Falkow S (1973) Use of a single-strand specific nuclease for analysis of bacterial and plasmid Deoxyribonucleic acid homo- and heteroduplexes. J. Bacteriol. 115:904–911

Dondero TJ, Rendtorff RC, Mallison GF, Weeks RM, Levy JS, Wong EW and Schaffner W (1980) An outbreak of Legionnaires' disease associated with a contaminated air-conditioning cooling tower. New Engl. J. Med. 302(7):365–370

Evancho GM, Sveum WH, Moberg LJ and Frank JF (2001) Microbiological monitoring of the food processing environment, p. 25 35, In: Downes FP and Ito K (ed), Compendium of Methods for the Microbiological Examination of Foods, 4th ed. Am. Public Health Assoc., Washington, D.C.

Fagan PK, Hornitzky MA, Bettelheim KA and Djordjevic SP (1999) Detection of shiga-like toxin (*stx1* and *stx2*), intimin (*eaeA*), and enterohemorrhagic *Escherichia coli* (EHEC) hemolysin (EHEC *hlyA*) genes in animal feces by multiplex PCR. Appl. Environ. Microbiol. 65:868–872

Feng P (2001) Rapid methods for detecting foodborne pathogens. Bacteriological Analytical Manual Online (http://www.cfsan.fda.gov), accessed July 10, 2007

Gillis M, Vandamme P, Vos PD, Swings J and Kersters K (2005) Polyphasic taxomony, p. 43–48. In: Brenner DJ, Krieg NR and Staley JT (eds), Bergey's Manual of Systematic Bacteriology, 2nd ed. vol. 2, Part A. Springer, New York, NY

Gombas DE, Chen Y, Clavero RS and Scott VN (2003) Survey of *Listeria monocytogenes* in ready-to-eat foods. J. Food Prot. 66:559–569

Janda JM and Abbott SL (2002) Bacterial identification for publication: when is enough enough? J. Clin. Microbiol. 40:1887–1891

Jaykus L (2003) Challenges to developing real-time methods to detect pathogens in foods. ASM News 69:341–347

Jeníková G, Pazlarová J and Demnerová J (2000) Detection of *Salmonella* in food samples by the combination of immunomagnetic separation and PCR assay. Int. Microbiol. 3:225–229

Krieg NR (2005) Identification of prokaryotes, p. 33–38. In: Brenner DJ, Krieg NR and Staley JT (eds) Bergey's Manual of Systematic Bacteriology, 2nd ed., vol. 2, Part A. Springer, New York, NY

Lam JS and Mutharia LM (1994) Antigen-antibody reactions. p. 104–132. In: Murray RGE (ed), Methods for General and Molecular Bacteriology. Am. Soc. Microbiol., Washington, D.C.

Landman D, Quale JM, Oydna E, Willey B, Ditore V, Zaman M, Patel K, Saurina G and Huang W (1996) Comparison of five selective media for identifying fecal carriage of vancomycin-resistant enterococci. J. Clin. Microbiol. 34:751–752

Langendonck NV, Bottreau S, Bailly L, Tabouret M, Marly J, Pardon P, Velge P (1998) Tissue culture assays using Caco-2 cell line differentiate virulent from non-virulent *Listeria monocytogenes* strains. J. Appl. Microbiol. 85:337–346

Lantz PG, Tjerneld F, Borch E, Hahn-Hagerdal B and Radstrom P (1994) Enhanced sensitivity in PCR detection of *Listeria monocytogenes* in soft cheese through use of an aqueous two-phase system as a sample preparation method. Appl. Environ. Microbiol. 60:3416–3418

Lazcka A, Campo FJD and Muñoz FX (2007) Pathogen detection: a perspective of traditional methods and biosensors. Biosens. Bioelectron. 22:1205–1217

Mastorides SM, Oehler RL, Greene JN, Sinnott JT, Kranik M and Sandin RL (1999) The detection of airborne *Mycobacterium tuberculosis* using micropore membrane air sampling and polymerase chain reaction. Chest 115:19–25

Meehan PJ, Rosenstein NE, Gillen M, Meyer RF, Kiefer MJ, Deitchman S, Besser RE, Ehrenberg RL, Edwards KM and Martinez KF (2004) Responding to detection of aerosolized *Bacillus anthracis* by autonomous detection systems in the workplace. Morbid. Mortal. Week. Rep. 53(RR07):1–12. (http://www.cdc.gov/mmwr/preview/mmwrhtml/rr5307a1.htm), accessed July 10, 2007

Miller JM, Krisher K and Holmes HT (2007) General principles of specimen collection and handling. In: Murray PR, Baron EJ, Landry ML, Jorgensen JH and Pfaller MA (eds) Manual of Clinical Microbiology, 9th ed., American Society of Microbiology, Washington, D.C., p 43–54

Myers GSA, Rasko DA, Cheung JK, et al. (2006) Skewed genomic variability in strains of the toxigenic bacterial pathogen, *Clostridium perfringens*. Genome Res. 16:1031–1040

Nataro JP, Bopp CA, Fields PI, Kaper JB and Strockbine NA (2007) *Escherichia*, *Shigella*, and *Salmonella*, p. 670–687. In: Murray PR, Baron EJ, Landry ML, Jorgensen JH and Pfaller MA (eds) Manual of Clinical Microbiology, 9th Am. Soc. Microbiol., Washington, D.C.

Nolte FS and Caliendo AM (2007) Molecular detection and identification of microorganisms, p. 218–244. In: Murray PR, Baron EJ, Landry ML, Jorgensen JH and Pfaller MA (eds) Manual of Clinical Microbiology, 9th ed., Am. Soc. Microbiol., Washington, D.C.

Norland S, Heldal M and Tumyr O (1987) On the relation between dry matter and volume of bacteria. Microb. Ecol. 13:95–101

Reckseidler SL, DeShazer D, Sokol PA and Woods DE (2001) Detection of bacterial virulence genes by subtractive hybridization: identification of capsular polysaccharide of *Burkholderia pseudomallei* as a major virulence determinant. Infect. Immun. 69: 34–44

Sambuy Y, Angelis ID, Ranaldi G, Scarino ML, Stammati A and Zucco F (2005) The Caco-2 cell line as a model of the intestinal barrier: influence of cell and culture-related factors on Caco-2 cell functional characteristics. Cell Biol. Toxicol. 21:1–26

Sehulster L and Chinn RYW (2003) Guidelines for environmental infection control in health-care facilities. Morbid. Mortal. Week. Rep. 52(RR10);142 (http://www.cdc.gov/mmwr/preview/mmwrhtml/rr5210a1.htm), accessed July 10, 2007

Sneath PHA (2005) Numerical taxonomy. In: Brenner DJ, Krieg NR and Staley JT (eds) Bergey's manual of systematic bacteriology, 2nd ed., vol. 2, Part A. Springer, New York, NY, p 39–48

Solomon HM and Lilly T (2001) *Clostridium botulinum*. Bacteriological Analytical Manual Online (http://www.cfsan.fda.gov), accessed July 10, 2007

Stackebrandt E and Goebel BM (1994) Taxonomic note: a place for DNA-DNA reassociation and 16S rRNA sequence analysis in the present species definition in bacteriology. Int. J. Sys. Bacteriol. 44:846–849

Struelens MJ (2006) Rapid identification of methicillin-resistant *Staphylococcus aureus* (MRSA) and patient management. Clin. Microbiol. Infect. 12:23–26

Sutton S (2006) Counting colonies. Pharm. Microbiol. Forum Newsl. 12 (9):1–5

Tenover FC (2007) Rapid detection and identification of bacterial pathogens using novel molecular technologies: infection control and beyond. Med. Microbiol. 44:419–423

Thomason BM, Dodd DJ and Cherry WB (1977) Increased recovery of salmonellae from environmental samples enriched with buffered peptone water. Appl. Environ. Microbiol. 34:270–273

van Houte J and Gibbons RJ (1966) Studies of the cultivable flora of normal human feces. Antonie van Leeuwenhoek 32:212–222

Weisburg WG, Barns SM, Pelletier DA and Lane DJ (1991) 16S ribosomal DNA amplification for phylogenetic study. J. Bacteriol. 173:697–703

Winter PC (2005) Polymerase chain reaction (PCR). Encyclopedia of Life Sciences, John Wiley and Sons, Ltd. (http://www3.interscience.wiley.com/cgi-bin/home), accessed July 10, 2007

Wistreich GA and Lechtman MD (1984) Microbiology. MacMillam Publishing Company, New York

Yousef AE and Carlstrom C (2003) Food microbiology: A Laboratory Manual. John Wiley and Sons, Inc., Hoboken, NJ

Yousef AE, Ryser ET and Marth EH (1988) Methods for improved recovery of *Listeria monocytogenes* from cheese. Appl. Environ. Microbiol. 54:2643–26

4

Overview of Rapid Microbiological Methods

Jeanne Moldenhauer

Abstract

In recent years there have been significant advances in microbiology, achieved through the sister sciences of chemistry, molecular biology, and computer aided imaging. These have resulted in a significant increase in the methods available for the detection, enumeration, and identification of microorganisms in the laboratory. This chapter provides a brief overview of the types of technologies available and the premise of how they work.

1. Introduction

The typical microbiology laboratory is a complex operation and involves the performance of a variety of laboratory testing methods. Many of the methods typically used in a microbiology laboratory have their basis in the works of Drs. Lister, Pasteur, and Koch. Many of these methods are over a hundred years old. In many cases, these methods are called conventional or traditional methods. In recent years, there has been an emphasis on updating or revising these methods to provide more sensitive, more accurate, or faster methods. This has resulted in a number of alternative or rapid microbiological methods (RMMs). Some countries also refer to these types of methods as "modern methods." While these methods are called rapid microbiological methods, it may be more correct to refer to them as alternative methods, i.e., a different way of performing the methods.

It should be noted that many of these methods may be more expensive than their conventional counterparts, either in the direct cost of the test or in the time it takes to prepare the sample for testing. It is believed by many vendors of this equipment and some users that obtaining results more quickly is worth the cost of these systems.

The tragic events of 9/11 resulted in a large amount of money being invested by the United States Department of Defense in the development of methods to detect bioterrorism agents. Significant advances have been made as a result of this financial investment. Many of the developers are actively looking for other opportunities to use these systems to generate income. Since these types of systems were designed to detect very low levels of contamination, many of them can be applied for use in a pharmaceutical manufacturing environment.

This chapter provides information on the history of rapid microbiological methods, some of the alternative (rapid) microbiological methods that may potentially be used to replace the conventional methods (overview of the technologies), and some information on identification systems. The terms "alternative microbiological methods" and "rapid microbiological methods

Jeanne Moldenhauer • Excellent Pharma Consulting, Mundelein, IL

M. Zourob et al. (eds.), *Principles of Bacterial Detection: Biosensors, Recognition Receptors and Microsystems*,
© Springer Science+Business Media, LLC 2008

are used synonymously" in this chapter. It should be noted that thousands of systems are in some stage of development; therefore, only a limited number of technologies could be described in this chapter.

2. A History of Rapid Microbiological Methods: Industry Reluctance to Accept These Methods

While science moved forward in development of rapid/alternative microbiological methods, the industry has been slow to accept and implement these methods. One of the greatest fears of industry came from a concern that regulators would not recognize or accept these methods in place of traditional methods. Another concern was that companies would not be allowed to change test limits based upon the test method, i.e., they would use a superior method that was likely to detect more organisms and not be allowed to adjust the limits to accommodate the sensitivity of the new method.

3. Types of Microbial Testing Performed

Current testing methods are divided into three basic categories:

– Is something there?—Qualitative Testing
– If there, how much is there?—Quantitative Testing
– If there, what is it?—Identification Testing

While there are newer methods in all of these categories, there is an ongoing debate on whether identification testing should be considered as an alternative method or in a separate category, due to the differences in validation methods. No one questions whether it is newer, potentially faster, or better.

4. Types of Rapid Microbiological Methods

The classification systems frequently used for alternative methods are based upon how the technology works, e.g., growth of microorganisms, viability of microorganisms, presence/absence of cellular components or artifacts, nucleic acid methods, traditional methods combined with computer-aided imaging (which might also be considered automation of an existing method), and combination methods.

4.1. Growth-Based Technologies

These methods are based upon the measurement of biochemical or physiological parameters that reflect the growth of the microorganisms. Examples of these types of methods include: ATP bioluminescence, colorimetric detection of carbon dioxide production, measurement of change in head space pressure, impedance, and biochemical assays.

4.2. Viability-Based Technologies

Viability-based technologies do not require growth of microorganisms for detection. Differing methods are used to determine if the cell is viable, and if viable cells are detected, they can be enumerated. Examples of this type of technology include solid phase cytometry, flow fluorescence cytometry, and optical imaging with NADH detection (Moldenhauer 2005).

4.3. Cellular Component or Artifact-Based Technologies

These technologies are based on the detection of a specific cellular component or artifact within the cell for detection and/or identification. Examples of these systems include: fatty acid profiles, mass spectrometry (e.g., MALTI-TOF), ELISA, fluorescent probe detection, and bacterial endotoxin-Limulus Amebocyte Lysate Test (Moldenhauer 2005).

4.4. Nucleic Acid-Based Technologies

Nucleic acid methods are used as the basis for operation of these types of technologies. Examples of this type of technology include: DNA probes, ribotyping/molecular typing, and polymerase chain reaction (PCR). Chapter 22 describes the nucleic acid-based technologies for bacterial detection (Moldenhauer 2005).

4.5. Automated Methods

This type of method involves using a classical method for most of the processing of a sample, and then using imaging software or perhaps an LED sensor and a colorimetric change to detect the growth earlier than methods requiring visual detection of growth. In most cases, detection of growth using human vision typically requires growth to 10^5 or 10^6 cells. These types of technologies typically replace the human detection methods with machine detection. Computer-aided imaging can detect growth at much lower levels of cellular growth, e.g., <100 cells. Examples of this computer-aided imaging type of system include: COVASIAM and Growth Direct. An example of another automated system is BacT/ALERT (Moldenhauer 2005).

4.6. Combination Methods

Some systems utilize more than one type of methodology or test to achieve a final result, e.g., a system that tells whether an organism is present and then is also capable of identifying the microorganism. Examples of combination systems include: Lab-on-a-Chip (LOC) technologies, mass arrays, and so forth (Moldenhauer 2005).

5. Overview of Rapid Technologies and How They Work

This section includes methods used for the detection of microorganisms. As discussed in the previous section, alternative methods are classified based upon a variety of different types, e.g., growth, viability, artifact-based, nucleic acid, automated, or combination technologies.

5.1. Adenosine Tri-Phosphate (ATP) Bioluminescence

Adenosine triphosphate (ATP) is a growth-based type of technology. ATP is present in all living cells. In the presence of the substrate D-luciferin, oxygen, and magnesium ions, the enzyme Luciferase utilizes the energy from ATP to oxidise D-luciferin and produce light. (This is the same substrate that causes a firefly to light up.) See Chapter 10 for more information about the luminescence techniques.

The quantity of light or bioluminescence produced can be measured by sensitive lumenometers, and is proportional to the amount of ATP present in the sample. The emitted light is usually expressed as relative light units (RLU) rather than direct estimates of microbial numbers. This is because the levels of ATP within cells will vary from organism to organism and also with the metabolic state of the microbe (Moldenhauer 2005). ATP bioluminescence reduces the test time by approximately one third of that taken by the traditional method. ATP bioluminescence can be used to screen both filterable and non-filterable samples.

The PallCheck (The Pall Company), Milliflex (Millipore Corporation), Rapiscreen (Celsis), novaLUM (Charm), are all luminometers. Hygiena also has a small hand-held luminometer. The PallCheck can detect as few as 10–100 cells/100 mL in a filtered sample (Will 2003).

The qualitative test systems are used to detect the presence or absence of microorganisms after preenrichment of the sample in media. The majority of ATP-based systems currently available are qualitative. Quantitative systems are also available, which enumerate microorganisms present in a sample preparation. This is done by detection of microcolonies using a combination of the ATP-bioluminescence and the Most Probable Number (MPN) techniques. MPN is a very old technique, and is explained in detail in the FDA's Bacteriological Analytical Manual (BAM), which is available online at the FDA's website (FDA January 2001).

5.2. Adenylate Kinase

Adenylate kinase is a growth-based technology. It is a cellular component that allows for microbial detection. The adenylate kinase released from cells reacts with ADP to form ATP. The ATP is detected using an ATP bioluminescence method. Using these technologies significantly lowers the level of detection for the system. The assay has a detection level of about 34 cells. The detection level can be reduced further (to about 4 cells) by extending the incubation for ATP generation time (Kricka 2003) . The limit of detection stated by the vendor is 1 organism per sample (Celsis Fact Sheet 003 2005). The AKuScreen by Celsis uses this technology.

5.3. Autofluorescence

This technology is growth-based and utilizes the fact that cells fluoresce when illuminated with blue light (Billinton and Knight 2000). This autofluoresence may be due to the response of flavins to the illumination (Aubin 1979, Benson et al. 1979). By coupling a high-resolution imaging system to a blue-light laser, the technology allows enumeration of microcolonies long before they would become visible to the human eye. The autofluorescence can be coupled with a sensitive camera and imaging system (Rapid Micro Biosystems) (Moldenhauer 2005). Another methodology, the BioVigilant Instantaneous Microbial Detector (IMD) uses optical assemblies to measure particle size with another assembly to detect laser-induced fluorescence (coming from specific cellular metabolites) and specific algorithms to distinguish viable and non-viable particles (Bolotin 2005).

COVASIAM, developed at the University of Mexico (Corjidi et al. 2003) and Growth Direct (Rapid Micro Biosystems) counts microcolonies on a membrane surface (Moldenhauer 2005).The Growth Direct system can detect microcolonies of approximately 50 cfu's, with a range of approximately 10–120 cells (Straus 2007). BioVigilant's IMD (Air) counts cells in the air using autofluorescence combined with optical detection and sizing. IMD (Liquid) counts cells in liquids using a combination of optical detection and sizing coupled with autofluorescence.

5.4. Biochemical Assays and Physiological Reactions

These types of assays are based upon the types of cellular components present. These are typically referred to as phenotypic methods. Pure culture suspensions are tested with a series of biochemical substrates or subjected to analysis to generate a spectrum. The microorganisms have specified reactions to these test conditions. The results are compared to a database of expected results. These comparisons allow the user to identify the microorganism. Systems vary widely from manual to highly automated. A common difficulty with these systems is that many of them require performance of a Gram stain prior to further evaluation. The accuracy of the Gram stain performed can significantly influence the results obtained (Moldenhauer 2005). Commonly used systems include: API Systems (bioMerieux), BIOLOG Systems (Biolog), VITEK and VITEK[2] (bioMerieux).

5.5. Biosensors and Immunosensors

Various types of technologies may be used for this type of testing, depending upon the type of sensor used. Part II of the book describes the different transducers used up to the present for bacterial detection. Typically, recognition receptors such as the immunological reagents are combined with various sensor detection systems to produce an immunosensor for pathogens (including bioterrorism organisms; [Kricka 2003] (Moldenhauer 2005)). There are a variety of test kits, and sensors are specific for the type of application being monitored.

5.6. Carbon Dioxide Detection

This technology is growth-based and allows for continuous monitoring for contamination using a fluorescent carbon dioxide system. A pH sensitive, fluorescent CO_2 sensor is poured into the bottom of each container. A series of computer algorithms are necessary to assess an increased rate of change and a sustained increase in CO_2 production (Meszanos 2003). Some types of commercial media formulations may not be available using these technologies (Moldenhauer 2005). The Bactec (Becton Dickinson and Company) system uses this technology.

5.7. Changes in Headspace Pressure

This is a growth-based technology in which electronic transducers are used to measure positive or negative pressure changes in the head space of each culture bottle. These changes are caused by microbial growth. If the growth produces significant production and/or consumption of gas, the samples are flagged as positive. Large quantities of samples can be placed into these instruments for testing with frequent monitoring of the head space pressure. These systems are based on non-invasive, continuous, automated monitoring of microbial cultures (Moldenhauer 2005).

The BacT/Alert (bioMerieux) and the ESP Microbial Detection System (AccuMed) use this type of technology. The BacT/Alert can detect low levels of microorganisms. Presentations given by Genzyme, as part of their approval of an automated sterility test, indicated that challenges of <50 cfu's were detected successfully by the system (Prinzi 2007).

5.8. Colorimetric Detection of Carbon Dioxide Production

This is a growth-based technology and works on the principle that as microorganisms grow, they produce carbon dioxide. In this technology, the test samples are placed in culture bottles for monitoring. The samples are incubated, agitated, and monitored for the presence of microorganisms. These systems use colorimetric detection of CO_2 production from the growth of organisms. Some of the systems commercially available will detect color change, flag a positive test sample, and notify the user. These systems are often considered to be non-invasive microbial detection systems and can accommodate a large number of samples. Although commonly used clinically for blood cultures, Genzyme received approval from FDA (CBER) in 2004 for the use of this method (with the BacT/Alert System) for sterility testing (as a 14-day automated sterility test methodology). There are many who believe that this type of technology is useful for slow-growing microorganisms like *Mycobacteria* (Pharmeuropa 2004, Moldenhauer 2005). The BacT/Alert (bioMerieux) and the ESP Microbial Detection System (AccuMed InternationaI) use this type of technology.

5.9. Concentric Arcs of Photovoltaic Detectors with Laser Scanning

This type of technology combines optical particle size technology with laser spectral analysis. The system is comprised of five concentric arcs of photovoltaic detectors, i.e., an orb-like platform. The sample being evaluated is suspended in a liquid or gas inside a vial or sample collection device, placed near the center of the orb. A laser beam of red, solid-state composition is passed through the sample. The scattered light intensities generate a spectrum that is compared to a library of known scatter patterns using a statistical classification algorithm. Contamination can be identified in seconds. This light scattering pattern becomes a fingerprint type of identification for the microorganism. It includes the size of the particle, the shape of the particle, and the optical characteristics. Light patterns are evaluated at multiple angles to detect and differentiate the size of the microorganism almost instantaneously. Identification occurs within a few milliseconds after the particle passes through the beam (DeSorbo, Moldenhauer 2005).

A US patent has been issued for the MIT System (MicroImaging Technologies, Inc.). The product claims include enumeration of microorganisms present down to the single cell level, determination of the size of microorganisms present, identification of the microorganisms, and the ability to handle mixed cultures (DeSorbo, Moldenhauer 2005).

5.10. Direct Epifluorescent Filter Technique (DEFT)

DEFT is a viability-based technology. The samples are filtered and stained using a fluorescent viability indicator. Acridine orange was the original stain utilized, but more recent applications utilize 4', 6-diamidino-2-phenylindole (DAPI). Epifluorescence microscopy is used to detect fluorescing microorganisms. The sensitivity of the technique depends on the volume filtered and the number of fields viewed under the microscope. Microcolony formation can further enhance the accuracy of the technique and the detecting of cell viability. Automated and semiautomated systems have been developed which can speed up the process and increase the accuracy of the technique. A concern using this type of technology is the ability to differentiate between fluorescing organisms and autofluorescing particles. Robustness may be affected by how the microorganisms are distributed on the membrane. Low viscosity fluids work best with this technology, although the possibility of using prefiltration provides for the testing of other solutions. (Pharmeuropa 2004, Moldenhauer 2005). There are a variety of microscopes available with epifluorescence capabilities.

5.11. DNA Sequencing

Clearly the method with the least risk of variability for genotypic identification would be to sequence the entire chromosome of the unknown organism and compare the sequence to others in an identification database. It is possible to restrict the sequencing to a limited area of the chromosome that provides a reliable means of comparison among different bacteria. Sequencing of the initial 500 base-pairs of the 16S codon has been established as a reasonable method for this comparison. MicroSeq markets a software package and database allowing the identification of unknown microorganisms based on this information. The sequences are generated by a modification of the polymerase chain reaction, and the relatedness of the derived sequence to others in the proprietary database is determined as the basis for the identification.

This technology has several advantages. It can be used to identify filamentous fungi, bacteria, and yeast (Woo et al. 2003, Hall et al. 2003, Patel et al. 2003, Cloud et al. 2004). It can also be used to identify slow-growing organisms, or even those which cannot be cultured. The major disadvantage of the system is the high costs associated with it in terms of dedicated facilities, personnel training, and consumables. The available library is extensive, containing over 1,400 bacterial entries (with extensive coverage of Gram-negative non-fermenters, *Bacillus, Coryneforms, Mycobacteria*, and *Staphylococcus*), as well as over 900 mold entries (Moldenhauer 2005).

5.12. Endospore Detection

Chapter 19 discusses the different detection and viability assessments of endospore forming pathogens. This method is a cellular component-based technology. It is based upon the knowledge that a major component of the spore case is calcium dipicolinate [Ca (dpa)]. Dipicolinate anions are only present in bacterial endospores. Ca (dpa) and dipicolinate anions (dpa^2), when dissolved, do not photoluminesce. Data have shown that Terbium (Tb^{3+}) is able to complex with dpa^{2-} forming a photoluminesecent complex (Rosen et al. 2003, Moldenhauer 2005). Spore detection using this method is part of US patent #5,876,960 (issued to the US Army Research Laboratory). The methodology is described in Rosen et al. (2003).

5.13. Enzyme Linked Immunosorbent Assay (ELISA)

ELISA is a cellular component-based technology. This technology uses an antigen-antibody reaction to detect unique microorganisms or cellular components. It has been used in clinical applications for many years (Moldenhauer 2005). VIDAS and Mini-VIDAS (bioMerieux), Tecra Salmonella ELISA (International Bioproducts), and Salmonella Tek ELISA (Organon Teknika) utilize this technology. Assay times vary from 24–52 hours for bacteria and 4–24 hours for microbial toxins. Enrichment may be necessary, as most immunological tests require 10^4 cells/mL to ensure detection (Newby and Johnson 2003).

5.14. Flow Cytometry

Flow cytometry is viability-based. Microorganisms are labeled in solution with a non-fluorescent marker. The marker is taken up into the cell and cleaved by intracellular enzymatic activity to produce a fluorescing substrate. The labeled sample is automatically injected into a quartz flow cell, which passes each microorganism one by one past a laser excitation beam for detection. The staining and detection mechanisms are similar to the solid phase cytometry systems except for detection of organisms in solution and not in the solid phase, allowing for non-filterable solutions to be tested. Results are typically obtained within 1.5–2 hours, although the limit of detection is approximately 100 cfu per mL (Moldenhauer 2005). Several systems have been developed for the pharmaceutical market, which range from simple manual systems with single test capability to highly automated units with high test throughput potential. D-Count (AES Chemunex) uses this technology. Chapter 33 covers all aspects of flow cytometry, including the miniaturized systems, focusing, and the detection techniques.

5.15. Fluorescent Probe Detection

This technology uses a combination of cellular component and nucleic acid technologies to detect the presence of microorganisms. The nucleic acid probes are designed to bind to specific target sites on or in cells. The probes contain a molecule that is capable of fluorescing when stimulated by an energy source like a laser. See the nucleic acid probes section for additional information on how the probes work. One limitation of this type of system is the size of the sample that may be tested. The RBD3000 (AATI) uses this technology. The next generation RBD, the Micro PRO by AATI, utilizes a laser-based optical system for rapid detection of microorganisms. The microorganisms are labeled with a fluorescent dye, allowing detection and enumeration. This system has a dynamic range of 10^1–10^6 cfu/mL (AATI Micro PRO™ 2007).

5.16. Fatty Acid Profiles (Fatty Acid Methyl Esters, FAMEs)

This is a cellular component-based technology and is considered a phenotypic method. The fatty acids present in microorganisms can be used to determine the identification of the microorganisms. The fatty acid composition has been shown to be homogeneous within different taxonomic groups. Isolates are grown on standard media and selected for testing. The testing procedure includes saponification of fatty acids, methylation, and extraction, resulting in FAMEs. The FAMEs are measured using gas chromatography and are compared to a library of known organisms for identification (Pharmeuropa 2004). This technology requires that the methods used for organism growth be standardized (Moldenhauer 2005). Additionally, gas chromatographs should be calibrated and standardized regularly. The Sherlock Microbial Identification System, more frequently referred to as the MIDI system, uses this technology.

5.17. Fourier Transformed Infrared Spectroscopy (FTIR)

This is a cellular component-based technology and is considered a phenotypic method. A FTIR can be used to generate an infrared spectrum of microorganisms. The patterns generated have been shown to be stable across taxonomic groups. The patterns are compared to a database of spectra of known microorganisms to identify the microorganism. Like some other technologies, standardization of the growth conditions is critical (Moldenhauer 2005).

FTIR can also be used for the measurement of biomass, using a fiber optic probe in which the two ends of the fibers are inserted into the bioreactor. The fibers must be inserted in a way that allows for a small gap between them. While FTIR can be used for this measurement, in most cases, better results can be obtained using the Mid Infrared Range (MIR) spectral analysis. Another way to resolve the issue is to use an attenuated total reflection (ATR) cell. In this application, the beam reflects inside the crystal and penetrates about 1 μm into the sample on the sample/crystal side of the interface. This technology (FTIR-ATR) can be used for in situ monitoring of fermentation processes (Känsäkoski et al. 2006). A variety of systems are commercially available using this technology.

5.18. Gram Stains (Rapid Method)

This technology is cellular component-based, considered a phenotypic method, and uses a single solution, without fixatives and washes. Results are obtained in a few minutes. Syto-9 stain and red-fluorescent hexidium iodide nucleic acid stain are used. The method can be used with mixed cultures. Gram-positive organisms stain a reddish-orange and Gram-negative organisms stain green. The fluorescent stains can be viewed/assessed using a fluorescent microscope (with a standard fluorescein long pass optical filter set) or using flow cytometry. The reagents have been designed to show low background stain (intrinsic). Dead cells do not show a predicted staining pattern. There are also procedures specified for use with direct epifluorescence filter techniques (DEFT). This technology significantly reduces the variability seen with traditional test methods (Moldenhauer 2005).

A second staining kit, "ViaGram Red[+] Bacterial Gram Stain and Viability Kit," is similar to the first kit described, but it uses two stains and three colors so that viable and non-viable cells can be readily detected in addition to the Gram reaction. Plasma membrane integrity is used as the distinguishing factor of live bacterial cells. Intact membranes are detected with a blue stain, while damaged membranes stain green. The red stain is evidence for Gram-positive bacteria. The LIVE Bac Light Bacterial Gram Stain Kit and the "ViaGram Red[+] Bacterial Gram Stain" use this technology.

5.19. Impedance

Impedance is a growth-based technique (also referred to as an electrochemical method). The premise is that growing microorganisms metabolize large complex constituents such as proteins and carbohydrates and convert them to smaller charged by-products such as amino acids, carbon dioxide, and acids. These smaller by-products of metabolism build up and eventually change the electrical conducting properties of the supporting growth medium. When an alternating current is applied across electrodes to this growth medium, a change in impedance can be observed. The resistance to the flow of an alternating current through a conducting material is defined as impedance (Moldenhauer 2005). Chapter 15 describes the impedance-related techniques.

Two types of microbial detection systems are used with this technology, direct and indirect impedance. Direct impedance systems work by detecting changes in electrical conductivity of growth media when an AC current is passed across two electrodes. Indirect impedance systems detect carbon dioxide produced by metabolizing organisms using chemical sinks such as potassium hydroxide. As the carbon dioxide is ionized, changes in impedance result. There is no direct contact between the electrodes and the microorganisms under investigation. When microorganisms multiply, a *detection threshold* is reached, above which an electrical signal is detected by both types of test system. Generally this detection limit is around the 10^6 cfu/mL level for many microbial species. The lower the initial population is, the longer the time taken to reach the detection threshold. The following systems use this technology: Bactometer (bioMerieux), BacTrac (Sy-Lab), RABIT (Don Whitley Scientific Ltd), PDS® (Biophage Pharma), and the Malthus Microbial Detection System (Malthus Diagnostics).

5.20. Immunological Methods

One can use an antigen-antibody reaction, i.e., a cellular component-based technology, to detect unique microorganisms or cellular components. These types of systems are useful for pathogen detection and may be used for identification as well. In some cases, the systems may not be able to distinguish whether the cells detected are viable (Pharmeuropa 2004). A variety of pathogen detection kits, such as Enzyme Linked Immunosorbent Assays (ELISA), are available based upon the type of pathogen. Chapter 21 describes the different immunological assays in more details.

5.21. Lab-on-a-Chip (LOC), Arrays, Microarrays and Microchips

These applications use a combination of different types of methodologies for detection and/or identification. An array is defined as an orderly arrangement of data. For example, a microtiter well plate can be considered an array of rows and columns of data. Using an array of data to perform a single test or series of tests has been used in many applications in microbiology (Moldenhauer 2005). Microarrays are fabricated with different technologies—for example, printing with fine-point pins onto glass slides, photolithography, inkjet printing, electrochemistry, in situ synthesis, and the like. One such device is the Combi Marix Custom Array (Miller 2006). Each microchip is like a miniature laboratory, and some scientists refer to them as a "lab-on-a–chip" device. The chips have become miniature analyzers, i.e., small versions of analytical instrumentation.

Typical microbiological reagents include oligonucleotides, proteins, DNA, and so forth (Kricka 2003). One application of this technology is the antibody dot or microspot assay. A small amount of antibody, typically 10–100μm, is placed on the bottom surface of a plastic well. This antibody dot is used as the capture antibody in a microimmunoassay (Kricka 2003). This technology can be used for a variety of applications, e.g., microbial virulence factors

(Chizhikov et al. 2001), antimicrobial resistance and identification, and bacterial discrimination (Busit et al. 2002).

Currently, many of these technologies are very expensive. As more commercial systems become available, the costs may be reduced. Various sources are available for these technologies. Others choose to "build" a chip to meet specific laboratory requirements. These technologies are described in detail in parts II (Recognition Receptors) and III (Microsystems) in this book.

5.22. Limulus Amebocyte Lysate (LAL) Endotoxin Testing

LAL is a cellular component-based technology. Amebocyte lysate recovered from horseshoe crabs (*Limulus)* have similar blood coagulation properties to those of humans. This similarity has allowed for the use of this reagent with samples to detect the presence of bacterial endotoxins. The quantitation of endotoxin present requires the use of standards that are appropriately certified or licensed. Three different methods are available: gel clot, kinetic turbidometric, and chromagenic. The gel clot method is an endpoint determination of the amount of endotoxin present. Several different dilutions are evaluated as part of the test to determine the lowest concentration of endotoxin at which a clot forms. This type of test can require a larger test sample to perform the various dilutions. The kinetic turbidometric test method allows for faster handling and smaller sample sizes. Another variation of this test method is the chromogenic method. One should note that the results can be affected by various parameters, including pH, composition, ions present, and so forth (Moldenhauer 2005).

The Pyrogent Gel Clot (BioWhittaker), Pyrotell (Associates of Cape Cod), BioTek, and the handheld unit from Charles Rivers Endosafe are examples that use this technology. These systems have widespread acceptance by regulators as a replacement for the rabbit pyrogen test. The gel clot method is sensitive to about 0.03 EU/mL. The chromogenic method is sensitive at approximately 0.005 EU/mL, and the turbidometric test is sensitive to 0.001 EU/mL (Novitsky and Hochstein 2003).

5.23. Mass Spectrometry (Matrix-Assisted Laser Desorption-Time of Flight (MALTI-TOF))

This is a cellular component-based technology. When microbial isolates are heated in a vacuum, the gaseous breakdown products can be analyzed using mass spectrometry. A spectrum is generated. The spectrum is compared to a database of known organisms for identification. Intact cells subjected to intense ionization (matrix-assisted laser desorption time of flight) release charged particles in distinct patterns. These patterns are compared to a database of known microorganisms to complete the identification. A current limitation of this technology for identification is the size of the database of known microorganisms. MALTI-TOF (Waters MicrobeLynx, Kratos Analytical Systems, and Perspective Biosystem Voyager) is used to perform this analysis.

A variation of this testing method is SELDI-TOF-MS, an acronym for surface enhanced laser desorption ionization time of flight mass spectroscopy. It utilizes Protein Chip Arrays and mass spectroscopy. In this technology a combination of test methods and miniaturization are used in a single platform to capture the proteins of interest. Thus, protein profiling is incorporated into the characterization scheme, minimizing other sources of variability (Dare 2005, Shah et al. 2005). Real-time mass spectrometry techniques for bacterial election will be discussed in details in Chapter 36.

5.24. Microcalorimetry

This technology is viability-based but also requires catabolic activity. The process of microbial catabolism results in heat that can be measured by microcalorimetry. The sample is placed into a sealed ampoule with media inside a calorimeter. The instrumentation can be used to establish growth curves. When high levels of contamination are present, one may need to use flow calorimetry (Pharmeuropa 2004). This technology is not appropriate for distinguishing if a single contaminant is present or a mixed contaminant (Pharmeuropa 2004).

5.25. Micro-Electro-Mechanical Systems (MEMS)

This type of system integrates mechanical, electrical, fluidic, and optical elements with sensors and actuators. They are manufactured on a single piece of silicon using microfabrication techniques. The various lab-on-a-chip technologies described in detail above are part of this type of application (Miller 2006). See Section III (Microsystems) for more information.

5.26. Nanotechnology

This type of technology operates of the atomic, molecular or macromolecular range of 1–100 nanometers. It utilizes high voltage electron beam (E-beam) lithography. The E-beam is used to Scan across a surface that has a thin film or resist. The electrons chemically change the resist allowing a pattern to be formed (Miller 2006).

5.27. Near Infrared Spectroscopy (NIRS)

NIRS is an optical spectroscopic method for quantitative measurements. It has been very useful on the chemistry side, due to the strength of the absorption bands in the analysis range. In the NIR range of 2.0–2.5 μm absorbance occurs, allowing detection of many important biological bonds, for example, aliphatic C-H, aromatic or alkene C-H, amines N-H, and O-H bonds. This method has been applied to determine biomass in a variety of processes (Känsäkoski et al. 2006).

Vaidyanathan et al. (2001) studied this process in terms of its applicability to biomass determination. They developed methods for calibration using stepwise multiple linear regression (SMLR) and partial least squares (PLS) regression. They used three different wavelengths. These types of calculation models were determined for *Strepomyces fradiae, Penicillium chrysogenum*, total sugars, and ammonium. These models were updated by Crowley et al. (1991) for key parameters including process analytes, biomass, glycerol, and methanol, allowing them to be used with *P. pastoris*. Subsequent refinements by a variety of individuals have allowed for use of this technology with *E. coli, Vibrio cholerae, CHO-K1* animal cell cultures, and so forth.

5.28. Nucleic Acid Probes

As the name of this technology implies, it is a nucleic acid-based, genotypic technology. The data available from nucleic acid sequencing is used to select a desired nucleic acid. The acids are extracted, immobilized to a solid phase, and hybridized to a labeled probe. Alternatively, the extracted nucleic acids can be labeled and hybridized to an immobilized probe (Rudi 2003). This technology is frequently used for characterization or identification of microorganisms, pathogen detection, and the like (Moldenhauer 2005). Chapter 22 covers the different nucleic acid assays. The systems available, Gene-Trak Systems (Gene-Trak) and Gene-Probe Systems (Gene-Probe), are dependent upon the type of outcome desired.

5.29. Optical Particle Detection

An application of MIE scattering is used to detect the total number of particles present, including both viable and non-viable particulates. Light scattering is the phenomenon in which the light is disturbed by the presence of particles. If the particle size is less than the wavelength of light, the scattering is not very sensitive to the particle size. In those cases where the particles are comparable to the wavelength of incident light, a complex pattern of scattering occurs. This pattern is very dependent upon the size and morphology of the particle, and is called MIE scattering. There are three key properties associated with MIE scattering: (1) The scattered light is concentrated in a forward direction. (2) The distribution (angular) of the scattered light intensity is very sensitive to the size of the particle. (3) the cross-section (i.e., the scattered portion of the light intensity) is proportional to the particle size (using a monotonic and complex manner). Visible light provides the ideal situation for MIE scattering of particles of a size typically seen in airborne particulates, i.e., 0.5–50 microns (Jiang 2005).

This technology also uses the fact that microorganisms contain certain metabolites that are able to autofluoresce under ultraviolet illumination. The wavelengths of light chosen for determining the presence of viable microorganisms is based upon the presence of NADH, which is only present in viable cells. This autofluorescence can be used as a marker for the presence of microorganisms. Combining the optical sizing and autofluorescence capabilities, it is possible to differentiate other biological particulates from microorganisms. This type of system detects individual particles whether viable or non-viable, sizes the particles with a resolution of ± 0.25 micron, and simultaneously determines if the particles are non-viable or biological (i.e., NADH present). Particles within the size range from 0.5 to in excess of $20 \, \mu m$ can be detected using this technology. The system can detect bacteria, yeast, molds, and spores with single cell detection. It cannot detect viruses. It is not able to identify the microorganisms detected (Jiang 2005). The Instantaneous Microbe Detector (IMD) by BioVigilant uses this technology. The system utilizes a probe that can be placed in the area for detection. Both air and liquid probe configurations are available.

5.30. Polymerase Chain Reaction (PCR)

PCR is a nucleic acid-based, genotypic technology and operates like a photocopier machine, making copies of nucleic acid fragments. Nucleic acid fragments are amplified using polymerization techniques as used in the cell, i.e., a mirror-image copy is made that is then used to make subsequent copies of the original target. A variety of PCR methods may be used— reverse transcripterase PCR (PCR-RT), nucleic acid sequence based amplification (NASBA), or transcription mediated amplification (TMA; Moldenhauer 2005).

The BAX® Microbial Identification System (Qualicon) and the Probelia System (BioControl Systems) are examples of systems using this technology. The Dupont BAX system evaluates samples at the genetic level, and polymerase chain reaction (PCR) is used to detect bacterial DNA sequences by homology to a supplied primer. PCR utilizes the requirement of DNA synthase for a specific primer in the initiation of DNA synthesis. By supplying the primer, the technician controls where in the bacterial chromosome the initiation occurs. The BAX system uses 96-well microwell plates, allowing for a high throughput. Results are available overnight, and are analyzed by an associated software program which provides a simple positive or negative report. The PCR reagents are supplied as tablets, minimizing human manipulation and enhancing shelf life. There are a variety of primers available, allowing the user to search for a particular organism of interest. Chapter 31 describes the PCR on-chip.

5.31. Rep-PCR

Rep-PCR is a type of polymerase chain reaction. PCR Bacterial Barcodes is a system marketed by Diversilabs. It utilizes the polymerase chain reaction (PCR) and primers homologous to repetitive sequences in the bacterial chromosome (Versalovic et al. 1991, Lupski and Weinstock 1992, Waterhouse and Glover 1993). This technique then results in a series of amplified fragments, separable by electrophoresis, which is unique to each organism. The technology has been applied to the discrimination of a variety of microorganisms (Healy et al. 2004, Shutt et al. 2005, Cangelosi 2004, Pounder et al. 2005, Chau et al. 2004).

5.32. Raman Spectroscopy

A Raman spectrophotometer can be used to generate a spectrum of microorganisms. It is considered a cellular component-based phenotypic system. The patterns generated are stable across taxonomic groups. The patterns are compared to a database of spectra of known microorganisms. Studies have been performed in clinical settings indicating that microbial identifications could be made with cultures with about 5 hour's incubation. The test is non-destructive and the cultures could be further incubated for genotypic identification (Moldenhauer 2005).

Raman spectroscopy has also been used to monitor fermentation processing. This can be used online, or semi-online (Känsäkoski et al. 2006). Raman spectrophotometers are commercially available. Chapter 20 describes the surface-enhanced Raman spectroscopy (SERS) and its applications for biological agent detection.

5.33. Ribotyping/Molecular Typing

This technology is a nucleic acid-based, genotypic method. It utilizes restriction fragments of nucleic acids from bacterial genomes. The size-separated fragments are hybridized to a ribosomal RNA probe. A chemiluminescent substrate is applied. Camera technology is used to convert the luminescing DNA fragments to digital information. The digital information is captured and the data extracted. A pattern is generated and compared to a database of known patterns for identification. The ribotype is a stable epidemiological marker and provides definitive taxonomic information. Molecular typing is considered the "gold standard" in the identification of microorganisms. As a result, *Bergey's Manual* has undergone numerous changes in the last few years, as the cultures listed have been characterized using genotypic methodologies (Moldenhauer 2005). The MicroSeq 16S rDNA Bacterial Identification System (Applied Biosystems), and the Riboprinter (DuPont Qualicon) utilize this technology.

5.34. Solid Phase Laser Scanning Cytometry

This technology detects the presence of viable cells. Solid phase cytometry uses membrane filtration to capture potential microbial contaminants from filterable samples prior to labeling the captured cells with a universal viability substrate. Once within the cytoplasm of metabolically active microorganisms, the non-fluorescent substrate is enzymatically cleaved to release free fluorochrome by the hydrolytic enzyme esterase. Only viable microorganisms with membrane integrity have the ability to retain the marker used in the assay. A laser-based detector then automatically scans the membrane, and the number of fluorescently-labeled cells is immediately reported (Moldenhauer 2005).

Solid phase cytometry eliminates the need for cell multiplication. Sensitivity to the single-cell level is possible independent of the volume of the sample filtered. In addition to vegetative cells, the technique is also able to detect spores (bacterial and fungal), stressed organisms, and fastidious organisms. Near real-time results are obtained, typically within 2–5 hours of sample preparation. The detection limit for this type of system is a single cell. (Note: It is not a cfu, as growth is not required for detection; Moldenhauer 2005).

Solid Phase Cytometry was accepted for pharmaceutical grade water testing by the FDA in February 2004 and was also accepted by the United Kingdom in 2000. It was accepted for sterility testing in the USA by the FDA in 2006. Additional applications for these types of uses have also been accepted.

Regulatory agencies have indicated concerns with those organisms that are viable but not culturable (VBNC), i.e., they are metabolically active but cannot reproduce by the conventional methods used for the growth of microorganisms. Since this type of technology can detect microorganisms that are not currently detected using conventional methods, it may be necessary to change the values of the existing limits to accommodate the newer type of detection. Changing the limits for sterility testing would not be considered appropriate, as any confirmed microbial contamination would be considered inappropriate for a sterile pharmaceutical product. Single cell detection is possible with this instrument. The Scan*RDI* (AES-Chemunex) instrument uses this technology.

5.35. Southern Blotting/Restriction Fragment Length Polymorphism

The RiboPrinter, marketed by Qualicon, is the primary product in this category. This equipment is an automated Southern Blot apparatus, where a loopful of the unknown colony is place in a lysis buffer then inserted into the machine. Liberated cellular DNA is cleaved by a restriction endonuclease (usually EcoRI) and then the fragments are electrophoretically separated on an agarose gel. The DNA fragments are converted from double-stranded to single-stranded, and then transferred to a hybridization filter where they are exposed to the 16S rDNA probe for hybridization. The end result of the process is a banding pattern that arises from the migration of those fragments that contain sequences homologous to the 16S rDNA codon. The strengths of this method include the high degree of automation, and the discriminatory powers it provides the technician. It can be used for identification, and the standard database contains over 200 bacterial genera and more than 1400 species and serotypes that are critical to the food and pharmaceutical industries. The user can also create custom identification libraries.

5.36. Spiral Plating

Spiral plating is a growth-based technology. Using this method, a continuous dilution of bacteria is spread onto an agar plate in a spiral pattern. The system uses a proprietary algorithm to determine the original concentration of CFU in the suspension. For example, the Spiral Plater by Spiral Biotech reduces the number of plates used in bacterial enumeration by performing the entire 6 log_{10} unit dilution series on a single plate utilizing its proprietary algorithms (Gilcrist et al. 1973, Trine et al. 1983, Schalkowsky 1996). The range of detection for this type of system is from 30 to 4×10^5 cfu/mL (Silley and Sharpe 2003).

5.37. Turbidimetry

Turbidity is a growth-based technology. As microorganisms grow, one can detect changes in the opacity of the growth medium. Optical density measurements can detect the differences in opacity at specified wavelengths, using a spectrophotometer (usually in the range of 420–615 nm). Another version of this methodology uses microtitre plate readers with continuous detectors, to detect organism growth earlier (Pharmeuropa 2004). Difficulty in determining the level of opacity is possible, due to the presence of viable and non-viable cells. Common usage for this type of test is to determine microbiological suspension or inocula sizes. There are a variety of spectrophotomers available.

6. Potential Areas of Application of Rapid Microbiological Methods

Table 4.1 provides a list of some of the ways rapid microbiological methods can be applied in a pharmaceutical environment. The information in this table summarizes the rapid microbiological applications for which the literature has indicated methods of use. Endnotes

Table 4.1. Rapid Microbiological Methods (RMMs) overview of potential applications.

Microbiological Test	Potential Technologies for Use	Systems Available	Published Literature	Comments[*]
Viable Aerobic Counts				
Viable Aerobic Count	Electrochemical Methods		PHARMEUROPA 2004	Yeasts and molds may require use of an indirect technique to determine the counts.
Viable Aerobic Count (Bioluminescence may be determined in a tube or microtiter plate)	Bioluminescence	Pall Check (Pall) RapiScreen (Celsis) AkuScreen (Celsis) Milliflex (Millipore)	PHARMEUROPA 2004	Product interference is rare, but should be assessed during validation of the system.
Viable Aerobic Count (Bioluminescence determined on a membrane)	Bioluminescence	Pall Check (Pall) RapiScreen (Celsis) AkuScreen (Celsis) Milliflex (Millipore)	PHARMEUROPA 2004	Product interference is rare, but should be assessed during validation of the system.
Quantification of Microorganisms				
Quantification	Real-time Quantitative PCR (RT-Q-PCR)	Microcompass™ (Lonza)	PHARMEUROPA 2004 Lonza Product information (formerly known as TVO or Micro Alert™ from Cambrex Bioscience)	May be appropriate for quantification of specific organisms or groups of microorganisms. Uses PreCellys sample preparation unit to isolate bacteria, yeasts and spores. Uses both 16s and 18s RNA. Note non-viable RNA has a half life of <30 minutes and can be differentiated from viable RNA. The theoretical limit of 1 cfu is feasible in samples. Time to results is <4hrs.

Table 4.1. (Continued)

Microbiological Test	Potential Technologies for Use	Systems Available	Published Literature	Comments*
	RT-PCR and hybridization probes to quantitate and detect mutation	LightCycler™ (Roche)	Caplin et al. 1999	Sensitivity from 1 organism per vial to 10^8 organisms per vial. Less than 14-hour test duration.
	Carbon Source Utilization (Optical Systems)	Solaris 32 or 128 (Neogen)	Neogen	
	Fluorescent Cell Labeling and Laser Excitation	Bacti Flow (AES Chemunex) Micro PRO™ (AATI)	AES Product Literature AATI Micro PRO Product Literature	
Determination of Cell Mass (Biotech applications—e.g., biomass determinations during bioprocesses)	Light absorbance using optical density or turbidity	Spectrophotometers may be used	Känsäkoski et al. 2006	Note: A sample is taken from the bioreactor and optical density is measured. The problem is that with this method alone, one cannot directly correlate the counts to the cells or cell mass.
	Measurement of wet or dry weight	Balances	Känsäkoski et al. 2006	
	Flow Cytometry	Coulter Counter	Känsäkoski et al. 2006	
	Electronic Counting		Känsäkoski et al. 2006	
	Indirect methods	Oxygen uptake, carbon dioxide evolution rate, ATP-production rate, NAD sensor, heat production, stirrer power input, redox level, pH value/base addition, and green fluorescent protein (GFP)		Typical methods could be fluorescence or calorimetry.
	Direct Methods	Electrical Optical Methods		Capacitance, impedance Fluorescence, light absorbance, light scattering, and real time imaging.
		Acoustic Resonance densitometry Nuclear magnetic resonance spectroscopy.	Känsäkoski et al. 2006	

Table 4.1. (Continued)

Microbiological Test	Potential Technologies for Use	Systems Available	Published Literature	Comments*
	In-line sensors	Flow Injection Analysis (FIA)-like technology. (Spectrophotometric) can be used with flow cytometry for additional testing.	Känsäkoski et al. 2006	May be sequential injection analysis (SIA), which is a computer controlled and programmable, bi-directional flow methodology. It uses very small volumes.
	In situ sensors	Laser cell density monitor, e.g., LA-300, ASR (Tokyo)	Känsäkoski et al. 2006	
	In situ sensors, impedance, capacitance, or permittivity	Impedance, e.g., Biomass Monitor 220®, Aber Instruments	Känsäkoski et al. 2006	Can be used for high density cultivations, but they are affected by changes in the culture broth, and can show low sensitivity at high cell densities.
	Near Infrared Spectroscopy (NIRS)	Used with various calculation models and wavelengths to determine cell mass.	Känsäkoski et al. 2006	Note: Changes in the type of microorganism may necessitate changes in the calculation models used.
Finished Product Sterility Testing	Microscopy Cell Counts Laser Scanning Solid Phase Cytometry	Various Microscopes ScanRDI	Känsäkoski et al. 2006 Gressett Presentation	
Purity Analysis Assess purity of single stranded RNA and DNA oligonucleotides and double stranded RNA interference (RNAi) products.	96-well system using parallel capillary gel electrophoresis with on-line fixed wavelength UV absorbance detection	Oligo PRO™ (AATI)	AATI Oligo PRO Product Literature	Low cost, size based and minimal preparation required.
Routine Bioburden Evaluations Bioburden Present (Raw Materials, In-process, and/or Finished Products)	Direct Epifluorescent Filtration Technique (DEFT)			Use is limited to low viscosity fluids. This method has been used successfully for the food industry, toiletries and aqueous pharmaceuticals.

Table 4.1. (Continued)

Microbiological Test	Potential Technologies for Use	Systems Available	Published Literature	Comments*
	Solid Phase Cytometry	Scan*RDI* (AES Chemunex)	McDaniel Presentation PHARMEUROPA 2004	It may be difficult to distinguish false positives from stressed organisms using this methodology. This technology cannot be used with non-filterable samples. Samples must be filterable, or soluble and then filtered.
	Flow Cytometry	RBD-3000 (AATI)		Can be used in the presence of significant particulate matter. Test can only be used for presence/absence tests when pre-incubation is required. Note: The allowable sample size may be small.
	Optical Scanning with Fluorescent dyes to label microorganisms	Micro PRO™ (AATI)	AATI Micro PRO™ Product Literature	
	Bioluminescence	PallCheck (Pall)	Pall Literature and Product Brochure	This technology is used for Presence/Absence Test for Microbial Limits Test per USP <61> and EP <2.6.12>. Method approved by FDA and EMEA for GSK submissions. This technology is used for product bioburden for terminally sterilized products.
		RapiScreen and AKuScreen (Celsis) Milliflex (Millipore)	Celsis Product Literature	
	Conventional Methods with Computer-aided Imaging	Growth Direct™ (Rapid Microbiolgoy BioSystems)		
	Real-time Quantitative PCR (RT-Q-PCR)	Microcompass™ (Lonza)	Lonza (formerly known as Micro Alert™ or TVO from Cambrex) product information	Presentations indicate this system can be used with powders.
Injured Cell Detection				
Viable Cells Present Following Injury	Microcalorimetry		PHARMEUROPA 2004	This method requires catabolic activity and minimum number of organisms.

Table 4.1. (Continued)

Microbiological Test	Potential Technologies for Use	Systems Available	Published Literature	Comments*
Viable Biological Indicators Post Sterilization	Solid Phase Laser Scanning Cytometry	ScanRDI (AES Chemunex)	Moldenhauer Presentation	Note: Requires that incubation temperature be changed to optimal for BI growth in order to show equivalent or better growth. Uses 3.5 hour incubation.
	Bioluminescence Conventional Methods with Computer-aided Imaging Carbon Source Utilization	PallCheck (Pall) Growth Direct™ (Rapid Micro Biosystems) Solaris Vials (Neogen)	Pall Literature and Brochure Straus Presentation Neogen product literature	Note: One must use oxidative biological indicator spore strips.
Determination of Inoculation Concentration				
Determination of Size of Inoculum	Turbidimetry		PHARMEUROPA 2004	Limited to healthy organisms. Cannot distinguish between viable and non-viable particulates.
Determination of Biological Indicator Control Counts (not heat stressed)	Solid Phase Laser Scanning Cytometry Flow Cytometry Conventional Methods with Computer-aided Imaging Bioluminescence	ScanRDI (AES Chemunex) RBD-3000 (AATI) Growth Direct™ (Rapid Micro Biosystems) Pall Check (Pall)	Moldenhauer Presentation Hawkins Presentation Straus Presentation Pall Literature and Brochure	Note: Requires that incubation temperature be changed to optimal for BI growth to show equivalence. Uses 3.5 hour incubation.
Presence/Absence of Specific Organism (Pathogenic and Non-Pathogenic Methods)				
Presence/Absence of Specific Microorganisms (Note: May or may not be pathogenic)	Phage-based Methods RT-PCR	Warnex™ Real Time PCR System (Warnex)	PHARMEUROPA 2004 Numerous kits available. Warnex product literature and brochures.	Sample composition can have a significant impact on detection.

Table 4.1. (Continued)

Microbiological Test	Potential Technologies for Use	Systems Available	Published Literature	Comments*
	Media Development to Improve Detection		PHARMEUROPA 2004	Require minimal or no confirmation tests for identification.
	Media Development to Improve Detection and VIDAS System Kits	Several different VIDAS kits (dependent upon pathogen) bioMerieux	bioMerieux Literature and VIDAS Brochures	AOAC Approved Methods.
	Immunological Methods, e.g., ELISA		PHARMEUROPA 2004	Method does not ensure that positive results are due to the presence of viable microorganisms.
	Enrichment / Amplification Methods	GeneGen (Sy-Lab)	Sy-Lab Product Literature	Tests 1-4 genes.
	Fluorescent Cell Labeling and Laser Excitation	BactiFlow (AES Chemunex)	AES Product Literature	Linear to 10^5 cells.
Pathogen Detection	Fourier Transformed Infrared (FTIR) Spectroscopy		PHARMEUROPA 2004	Isolates tested must be grown on standard media with standard incubation conditions. Conditions for operation of the FTIR should be appropriately standardized.
	LightCycler™ and RT-PCR	Bordella pertussus Detection (Roche)	Noppen et al. 2001	
	Mass Spectrometry Malti-TOF	Malti-TOF (Waters)	Dare et al. Presentation PHARMEUROPA 2004	Isolates must be cultured prior to analysis. Usefulness is dependent upon the size of the database available.
	Electrospray ionization spectrometry (ESI) and base composition analysis of PCR amplification and specific primers	PCR/ESI-MS	Ecker, et. al. May 31, 2005	Naval Health Research Center Protocol NHRC.2001.008
	Biochemical Assays based upon Physiological Reactions		PHARMEUROPA 2004	Requires use of a pure culture, less than 3 days old.
	Self-contained Polymer Partition	Desktop Testing Unit (DTU) (Pathogen Detection Systems)	Pathogen Detection Systems Product Literature	E. coli and Total Coliform detection. Detects contamination < 1 cfu in 16 hours, higher contamination levels, e.g., 10^6 cfu in 4-6 hours.

Table 4.1. (Continued)

Microbiological Test	Potential Technologies for Use	Systems Available	Published Literature	Comments*
	Carbon Source Utilization and Optical Detection	Soleris 32 and 128 (Neogen)	Neogen Product Literature	Applicable to a large number of food pathogens.
Detection of Methicillin Resistant Staphylococcus aureus	Adenylate Kinase (Bioluminescence)	Rapid AK™ (AKuScreen) Celsis	Acolyte Biomedica Product Specifications	
		BacLite®FLEX (Accolyte Biomedica)	Acolyte Biomedica Product Specifications	
Presence / Absence of Microorganisms (Sterility Test)	Gas Production or Consumption	BacT/Alert 3D (bioMerieux)	bioMerieux product literature and brochure	This system has been approved by FDA (submitted by Genzyme to CBER).
Mycoplasma Detection	Nucleic Acid Amplification Techniques		PHARMEUROPA 2004	Several different versions of the techniques are available. Each technique has its own limitations.
	Nucleic Acid Amplification Techniques with Gen Probe Assay Hybridization Techniques	Mycoplasma Tissue Culture Non-Isotopic (MTC-NI) Rapid Detection System (Pall)	Pall Product Literature and Brochure	Detection occurs in <75 minutes
		MycoAlert™ (Lonza)	Lonza Presentation	
Microbiological Assays				
Microbiological Assay of Antibiotics	Electrochemical Methods		PHARMEUROPA 2004	Yeasts and molds may require use of an indirect technique.
Preservatives Testing (e.g., Antimicrobial Efficacy, Efficacy, Sensitivity)				
Antimicrobial Preservative Efficacy Testing	Microcalorimetry		PHARMEUROPA 2004	This technology requires catabolic activity and a minimum number of organisms present.

Table 4.1. (Continued)

Microbiological Test	Potential Technologies for Use	Systems Available	Published Literature	Comments*
	Solid Phase Laser Scanning Cytometry Flow Cytometry	ScanRDI (AES Chemunex)	Michael Miller Presentation AES Product AES Literature	
Preservative Efficacy Testing	Electrochemical Methods	D-count (AES Chemunex)	PHARMEUROPA 2004	Yeasts and molds may require use of an indirect technique.
	Bioluminescence	Pall Check (Pall) Note: May also be used with Biotrace.	Pall Literature and Brochure See also Biotrace Product Literature.	Product interference is rare, but must be assessed during validation of the system.
Disinfectant Efficacy				
Disinfectant Efficacy	Bioluminescence	Pall Check (Pall)	Pall Literature and Brochure PHARMEUROPA 2004	Limited to healthy organisms. Cannot distinguish between viable and non-viable particulates.
Preservative Sensitivity of Pharmacopeial Challenge Organisms from Formulated Products	Turbidimetry (automated mode)			
Detection of Slow-Growing Microorganisms				
Detection of Slow-Growing Microorganisms	Consumption or Production of Gas	BacT/Alert 3D (bioMerieux)	bioMerieux Product Literature	A direct relationship cannot be made between the original microbial bioburden and the detectable end-point.
Testing Filterable Products				
Filterable Products	Bioluminescence		PHARMEUROPA 2004	Product interference is rare, but should be assessed during validation.
	Solid Phase Laser Scanning Cytometry	ScanRDI (AES Chemunex)	AES Chemunex Product Literature	
Non-filterable Products				
Non-filterable Products	Bioluminescence	Pall Check (Pall)	Pall Product Literature and Brochure PHARMEUROPA 2004	Product interference is rare, but should be addressed during validation.

Table 4.1. (Continued)

Microbiological Test	Potential Technologies for Use	Systems Available	Published Literature	Comments*
Environmental Monitoring and Identification of Organism Applications				
Environmental Monitoring – Surface Sampling	Bioluminescence	Pall Check (Pall)	Pall Product Literature and Brochure PHARMEUROPA 2004 Straus 2007	Product interference is rare, but should be assessed during validation.
	Conventional Methods with Computer-Aided Imaging	Growth Direct™ (RAPID MICRO BIOSYSTEMS)		
	Solid Phase Laser Scanning Cytometry using Polym'air	ScanRDI (AES Chemunex)	Moldenhauer and Yvon 2005	
Environmental Monitoring – Air Sampling	Conventional Methods with Computer-Aided Imaging	Growth Direct™ (Rapid Micro Biosystems)	Straus 2007	
	Solid Phase Laser Scanning Cytometry using Polym'air	ScanRDI (AES Chemunex)	Moldenhauer and Yvon 2005	
	Optical Sensors with Riboflavin Metabolism	Instantaneous Microbial Detection (IMD) (BioVigilant)	Jiang 2005	
Environmental Monitoring-Water Monitoring (Pharmaceutical Grade)	Solid Phase Cytometry		PHARMEUROPA 2004	May require adjustment of the process numerical limits. Spore testing/monitoring may require additional procedures.
Water Hygiene	Bioluminescence	PallCheck (Pall)	PHARMEUROPA 2004	Product interference is rare, but should be assessed during validation.
		RapiScreen or AkuScreen (Celsis)	PHARMEUROPA 2004	
		Milliflex (Millipore)	PHARMEUROPA 2004	

Table 4.1. (Continued)

Microbiological Test	Potential Technologies for Use	Systems Available	Published Literature	Comments*
		SystemSURE^II And UltraSnap^TM (Hygiena)	Hygiena Product Literature and Brochure	Product interference is rare, but should be assessed during validation. Results in 15–20 seconds. Handheld unit that stores 500 results.
		Pi102 Luminometer (Hygiena)	Hygiena Product Literature PHARMEUROPA 2004	Product interference is rare, but should be assessed during validation. Desktop System

Identification Methods

There are too many systems to include all of them in this listing. Some of the most common systems are included. The articles referenced include more of these systems.

Microbiological Test	Potential Technologies for Use	Systems Available	Published Literature	Comments*
Identification of Environmental Isolates and Product Flora	Fatty Acid Analysis (FAMES)	Sherlock (MIDI)	Clinical Microbiological Reviews 2005 PHARMEUROPA 2004	Isolates tested must be grown on standard media with standard incubation conditions. Conditions for operation of the gas chromatograph should be appropriately standardized.
	Fatty Acid Analysis and DNA Sequencing	Sherlock 6.0 (MIDI) Note: Sequencer is not included	Clinical Microbiological Reviews 2005 PHARMEUROPA 2004	Isolates tested must be grown on standard media with standard incubation conditions. Conditions for operation of the gas chromatograph should be appropriately standardized. Database has about 2500 bacteria and 1140 fungi.
	Fourier Transformed Infrared (FTIR) Spectroscopy		PHARMEUROPA 2004	Isolates tested must be grown on standard media with standard incubation conditions. Conditions for operation of the FTIR should be appropriately standardized.
	Mass Spectrometry Malti-TOF	MicrobeLynx^TM (Waters)	Waters Literature and article in press PHARMEUROPA 2004	Isolates must be cultured prior to analysis. Usefulness is dependent upon the size of the database.
	Biochemical Assays based upon Physiological Reactions	Biolog and Omnilog (Biolog) Uses Carbon Source Utilization	PHARMEUROPA 2004 Biolog Product Literature	Requires use of a pure culture, less than 3 days old.

Table 4.1. (Continued)

Microbiological Test	Potential Technologies for Use	Systems Available	Published Literature	Comments*
		Vitek 2 Compact (bioMerieux) Enzymatic Method Phoenix (BD) Enzymatic Method		Each different technique available has its own limitations.
	Nucleic Acid Based Amplification Techniques Genetic Fingerprinting (Ribotyping)	MicroSeq (Applied Biosystems) Riboprinter (Dupont Qualicon)	PHARMEUROPA 2004 PHARMEUROPA 2004	This method is currently used for identification of bacteria.
	Triangulation Identification of Genetic Evolutaton of Risks (TIGER)	Tiger (ISIS Pharmaceuticals)	Minkel August 1, 2005	Uses 2-tier approach to determine the bacterial identification and subsequent strain determination. Takes about 5 hours.
	Real Time PCR and Light Cycler	LightCycler™ (Roche)	Emrich 2000 a, b	
Cleaning and Sanitization				
Control of Cleaning and Sanitization	Bioluminescence	Pall Check (Pall)	Pall Product Literature and Brochure	
Cleaning Validation	Solid Phase Laser Scanning Cytometry	ScanRDI (AES Chemunex)	McDaniel Presentation	
Endotoxin Testing				
In-line Testing	Chromagenic	Pyrosense with Automated Sampler (Lonza)	Lonza Product Literature	
Handheld Unites		Several Available (Charles Rivers Endosafe)	Charles Rivers Product Literature	

Table 4.1. (Continued)

Microbiological Test	Potential Technologies for Use	Systems Available	Published Literature	Comments*
Media Fills				
Bulk Container Media Fills (e.g., fermentation, bulk containers)	Solid Phase Laser Scanning Cytometry	ScanRDI (AES Chemunex)	McDaniel Presentation	
Finished Product Containers	Solid Phase Laser Scanning Cytometry	ScanRDI (AES Chemunex)	Dalmasso Presentation	
Surface Swab Analysis				
Evaluation of surfaces post decontamination	Optical system with fluorescent dyes to label microorganisms	Micro PRO™ (AATI)	AATI Micro PRO™ Product Literature	System is in beta testing for this application

provide the source of information used to determine applicability. Identification systems have been included. In many cases the comments provided come from information in compendial chapters regarding the technology. One should also note that some of the material comes from the vendor's product information.

Singer and Cundell (2003) present a strong argument on the advantages of rapid methods. They describe 18 steps or operations in an idealized manufacturing and control process for an aseptically produced parenteral product. Of these 18 steps, five involve microbiological analysis; but these five steps typically take longer than all the other steps combined, due to the long incubation periods required for the reproduction of the bacteria.

7. Disclaimer

There are reports of thousands of rapid microbiology systems that are in some stage of development for use in place of traditional microbiological methods. This article introduces some of the technologies available. Inclusion or exclusion of available methods is not meant to confer credibility, endorsement, or acceptance of some methods over other methods.

8. Conclusions

The advent of rapid microbiological methods has revolutionized the pharmaceutical, medical device and food industries. Things thought only to exist in science fiction have become reality. While significant work may be required to prove them effective for regulated industry applications, those who have preserved have achieved significant financial and technical benefits. The systems presented in this chapter are current at the time of this publication. However the science is continuing to advance and new systems are available every day. With the significant advancements that have occurred in the last ten years, it is hard to conceive what may happen in the next decade.

Acknowledgments

Much of the information in this chapter was gained from knowledge developed as part of the task force working on PDA Technical Report Number 33, chaired by Dr. Tony Cundell; the task force working on Rapid Methods for USP (USP WG18) chaired by Dr. Michael Miller; the vendors of the equipment; and presentations given at various RMUG conferences. Additional thanks are extended to Mark Pfitzenmaier, Jeff Little, and Casey Costello, who aided in the collection of product literature on rapid microbiological methods. Kudos to all of you!

References

Acolyte Biomedica Product Literature. Accessed at www.acolytebiomedica.com (downloaded 27 Sep 2007)

AATI Micro PRO™ Product Literature. Accessed at http://www.aati-us.com/systems/rbd3000.html (downloaded 10 July 2007)

AATI Oligo PRO™ Product Literature. http://www.aati-us.com/systems/oligo.html (downloaded 10 July 2007)

AES Product Literature. http://www.aeschemunex.com/Pages/rapidintro.htm (27 Sep 2007)

Aldridge C et al. (1977) Automated microbiological detection/identification system. J Clin Microbiol. 6 (4): 406–413

Anon. (January 2005) Clinical Microbiological Reviews p 147–162

Anon. Tools for microbiology. Bioprobes 43:11–13 www.probes.com (downloaded 15 Dec 2006)

Aubin JE (1979) Autofluorescence of viable cultured mammalian cells. J. Histochem Cytochem. 27:36–43

Benson RC, Meyer RA, Zaruba ME and McKhann GM (1979) Cellular autofluorescence—is it due to flavins? J Histochem Cytochem 27:44–48

Billinton N and Knight AW (2000) Seeing the wood through the trees: A review of techniques for distinguishing green fluorescent protein from endogenous autofluorescence. Anal Biochem. 291:175–197

bioMerieux and VIDAS Product Literature. www.biomerieux-industry.com/servlet/srt/bio/industry-microbiology/dynPage?node=biopharma_applications (downloaded on 15 Sep 2007)

Bolotin C (December 2005) Instantaneous Microbial Detection. Controlled Environments Magazine® (Electronic) http://www.cemag.us/articles.asp?pid=564 (downloaded on 27 Sep 2007)

Biotrace (Division of 3M) Product Literature and Brochure. www.biotrace.co.uk/content.php?hID=5 (downloaded on 15 Sep 2007)

Brown B and Leff L (1996) Comparison of fatty acid methyl ester analysis with the use of API 20E and NFT strips for identification of aquatic bacteria. Appl Environ Microbiol. 62(6):2183–2185

Bruch CW (1972) Objectionable microorganisms in nonsterile drugs and cosmetics. Drug Cosmet Ind. 11:51

Busit E, Bordoni R, Castiglioni B, Monciardini P, Sosio M, Donadio S, Consolandi C, Bernardi LR, Battaglia C and De Bellis G (2002) Bacterial discrimination by means of a universal array approach mediated by LDR (ligase detection reaction). BMC Microbiology. 2:27(www.biomedcentral.com/1471-2180/2/27)

Bussey DM and Tsuji K (1986) Bioluminescence for USP sterility testing of pharmaceutical suspension products. Appl Environ Microbiol. 51:349–355

Cangelosi GA (2004) Evaluation of a high-throughput tepetitive-sequence-based PCR System for DNA fingerprinting of *Mycobacterium tuberculosis* and *Mycobacterium avium* complex strains. J Clin Microbiol. 42(6):2685–2693

Caplin BC, Rasmussen RP, Bernard PS, Wittwer CT (1999) Light Cycler Hybridization Probes. Biochemica No.1, p 5–8

Celsis (2005) Fact Sheet 003 Limit of Detection of Celsis AKuScreen, FS047–2. www.celsis.com, (downloaded 9 July 2007)

Charles Rivers Laboratories Product Literature. http://www.criver.com/endotoxin_and_rapid_microbiological_products/literature.html. (downloaded on 15 Sep 2006)

Chau AS, et al. (2004) Application of real-time quantitative PCR to molecular analysis of *Candida albicans* strains exhibiting reduced susceptibility to azoles. Antimicrob Agents Chemother. 48(6):2124–2131

Chizhikov V, Rasooly A, Chumakov K and Levy DD (July 2001) Microarray analysis of microbial virulence factors. 67(7):3258–3263

Cloud JL, et al. (2004) Evaluation of partial 16S ribosomal DNA sequencing for identification of nocardia species by using the MicroSeq 500 system with an expanded database. J Clin Microbiol. 42(2):578–584

Corkidi G, Trejo M and Nieto-Sotelo J (2003) Automated colony counting using image-processing techniques, In: Olson WP and Godalming (eds) Rapid Analytical Microbiology: The Chemistry and Physics of Microbial Identification Parenteral Drug Association and Davis Horwood International Publishing, Bethesda, MD and Surrey, UK.

Craythorn J, et al. (1980) Membrane filter contact technique for bacteriological sampling of moist surfaces. J Clin Microbiol. 12(2):250–255

Dalmasso G (April 2007) Presentation Implementation and validation of the laser scanning cytometry for the control of sterile products." A.I.M.I.F. Conference, San Malo, France

Dare D (2005) Microbial identification using Maldi-TOF MS. In: Miller MJ (ed) Encyclopedia of Rapid Microbiological Methods, vol. 3. PDA/DHI Publishers, Bethesda, MD, pp. 19–56

Dare D, et al. (2005) ASMS Poster Presentation. Fast Reliable Identification of Staphylococcus haemolytica by Matrix Assisted laser Desorption / Ionisation Time of Flight Mass Spectrometry

Davidson CA (1999) Evaluation of two methods for monitoring surface cleanliness-ATP bioluminescence and traditional hygiene swabbing. Luminescence 14:33–38

DeSorbo MA (1 Aug 2002) Rapid Contamination Detection Technology Patent Granted. http://cr.pennet.com/display-article/150543/15/ARTCL/none/none/1 (downloaded on 13 Jan 2008)

Duchaine C, et al. (2001) Comparison of endotoxin exposure assessment by Bioaerosol Impinger and filter sampling methods. Appl Environ Microbiol. 67(6):2775–2780

Ecker DJ, et al. (May 31, 2005) Rapid identification and strain typing of respiratory pathogens for epidemiological surveillance. PNAS 102(2):8012–8017

EMEA (1997a) Note for guidance on validation of analytical procedures: definitions and terminology. CPMP/ICH/381/95

EMEA (1997b) Note for guidance on validation of analytical procedures: methodology. CPMP/ICH/281/95

Emrich T (2000a) Detection of telomerase components by quantitative real time PCR online PCR analysis with the Light Cycler Biochemica. 4:16–19

Emrich T (2000b) The Light Cycler Instrument and MagNA Pure LC: An automated system for the evaluation of telomerase expression by quantitative RT-PCR. Biochem. 4:10–13

FDA (January 2001) Bacteriological analytical manual online. www.cfsan.fda.gov/~ebam/bam-toc.html (downloaded on 22 Jul 2006)

FDA (February 2003) Guidance for industry: comparability protocols—chemistry, manufacturing, and controls information. http://www.fda.gov/cber/gdlns/cmprprot.pdf

FDA (August 2003) Part 11: Electronic records, electronic signatures—scope and application. http://www.fda.gov/cder/guidance/5667fnl.htm

FDA (September 2004a) Guidance for industry: sterile drug products produced by aseptic processing—current good manufacturing practice. Department of Health and Human Services. U.S. Food and Drug Administration, Washington, D.C.

FDA (2004b) Guidance for industry—PAT a framework for innovative pharmaceutical development, manufacture and quality assurance. Department of Health and Human Services. U.S. Food and Drug Administration, Washington, D.C.

FDA (September 2004c) Pharmaceutical cGMPs for the 21st Century—A Risk-Based Approach. Department of Health and Human Services. U.S. Food and Drug Administration, Washington, D.C.

Funke G and Funke-Kissling P (2004) Evaluation of the new VITEK 2 card for identification of clinically relevant Gram-negative rods. J Clin Microbiol. 42(9):4067–4071

Funke G and Funke-Kissling P (2005) Performance of the new VITEK 2 GP card for identification of medically relevant Gram-positive cocci in a routine clinical laboratory. J Clin Microbiol. 43(1):84–88

Gilchrist JE, et al. (1973) Spiral plate method for bacterial determination. Appl Microbiol. 25(2):244–252

Gordon and Watson (1994) A note on sample size determination for comparison of small probabilities. Controlled Clinical Trials. 15:77–79

Gressett G (April 2007) Presentation. FDA Approval of an Alternative Sterility Test Method for a Sterile Ophthalmic Product. A.I.M.I.F.Conference, San Malo, France

Hall L, et al. (2003). Experience with the MicroSeq D2 large-subunit ribosomal DNA sequencing kit for identification of commonly encountered, clinically important yeast species. J Clin Microbiol. 41(11):5099–5102

Hall L, et al. (2003b) Evaluation of the MicroSeq System for Identification of Mycobacteria by 16S ribosomal DNA sequencing and its integration Into a routine clinical Mycobacteriology laboratory. J Clin Microbiol. 41(4):1447–1453

Hall L, et al. (2004) Experience with the MicroSeq D2 large-subunit ribosomal DNA sequencing kit for identification of filamentous fungi encountered in the clinical laboratory. J Clin Microbiol. (2):622–626

Hawkins K, (2005) The RBD 3000. Presentation at IVT Microbiology Event of the Year. Arlington, VA.

Healy M, et al. (2004) Identification to the species level and differentiation between strains of Aspergillus clinical isolates by automated repetitive-sequence-based PCR. J Clin Microbiol. 42(9):4016–4024

Heikens E, et al. (2005) Comparison of genotypic and phenotypic methods for species-level identification of clinical isolates of coagulase-negative Staphylococci. J Clin Microbiol. 43(5):2286–2290

Hygiena Product Literature and Brochure, including SystemSURE[II] and UltraSnap[™]. www.hygiena.net (downloaded on 8 Feb 2007)

Jiang JP (2005) Instantaneous Microbial Detection Using Optical Spectroscopy. pp. 121—142. In: Miller MJ (ed) Encyclopedia of Rapid Microbiological Methods, vol. 3. PDA/DHI Publishers, Bethesda, MD

Jimenez L (2001) Rapid methods for the microbiological surveillance of pharmaceuticals. PDA J Pharm Sci Tech. 55(5):278–285

Jimenez L, et al. (2000) Use of PCR analysis for sterility testing in pharmaceutical environments. J Rapid Method & Automation In Microbiol. 8:11–20

Juozaitis A, et al. (1994) Impaction onto a glass slide or agar vs impingement into a liquid for the collection and recovery of airborne microorganisms. Appl Environ Microbiol. 60(3):861–870

Känsäkoski M, Kurkinen M, von Weymarn N, Niemelä P, Neubauer P, Juuso E, Erikäinen T, Turunen S, Aho S and Suhonen P ESPOO (2006) Process analytical technology (PAT) needs and applications in the bioprocess industry: a review. (http://www.vtt.f.i/publications/index.jsp)

Kricka LJ (2003) New technologies for microbiological assays. In: Easter MC (ed) Rapid Microbiological Methods in the Pharmaceutical Industry. Interpharm/CRC, Washington, D.C., pp. 233–248

Lonza PowerPoint Presentation. http://www.lonzabioscience.com/Content/LAL.asp (Note: during the writing of this chapter, the Cambrex systems were purchased by Lonza)

Lupski JR and GM Weinstock (1992) Short, interspersed repetitive DNA sequences in prokaryotic genomes. J Bacteriol. 174(14):4525–4529

Manu-Tawiah W, et al. (2001) Setting Threshold Limits for the Significance of Objectionable Microorganisms In Oral Pharmaceutical Products. PDA J Pharm Sci Tech. 55(3):171–175

Mason DJ, et al. (1998) A fluorescent Gram stain for flow cytometry and epifluorescence Microscopy. Appl Environ Microbiol. 64(7):2681–2685

McDaniel A (April 2007) Solid phase cytometry for detection of bacterial contamination in mammalian cells culture systems. A.I.M.I.F Conference Presentation, San Malo, France

Meszaros A (2003) Alternative technologies for sterility testing. In: Easter MC (ed) Rapid Microbiological Methods in the Pharmaceutical Industry..Interpharm/CRC, Washington, D.C., pp. 179–185

Miller MJ (2006) Rapid microbiological methods for a new generation. www.pharmamanufacturing.com/articles/2006/019.htmo?page=print (downloaded on 8 Sep 2007)

Minkel JR (August 1, 2005) Tiger catches pathogens by the toe. The Scientist 19(15):31.

Moldenhauer J (2002) Feasibility of ScanRDI for Biological Indicators. Barnette International Conference Presentation, Caribe Hilton, Puerto Rico

Moldenhauer J (September 2005) Rapid microbiological methods and PAT initiative. Guide to Biopharmaceutical Advances: The Biopharm International Guide Supplement. Advanstar 11–20

Moldenhauer J and Sutton SVW (2004) Towards an improved sterility test. PDA J Pharm Sci Tech. 58(6):284–286

Moldenhauer J and Yvon P (2005) Environmental monitoring using Scan RDI Polym'air. In: Moldenhauer, J (ed) Environmental Monitoring: A Comprehensive Handbook. PDA/DHI, River Grove, IL and Bethesda, MD, pp. 249–260

Neogen Product Literature and Solaris Brochure. http://www.neogen.com (Downloaded on 10 Sep 2007)

Newby PJ and Johnson B (2003) Overview of alternative rapid microbiological technologies. In: Easter MC (ed) Interpharm/CRC, Washington, D.C., pp. 41–60.

Newby P, et al. (2004) The introduction of qualitative rapid microbiological methods for drug-product testing. Pharm Technol. (PAT Supplemental Issue) p 6–12

Niskanen A and Pohja MS (1977) Comparative studies on the sampling and investigation of microbial contamination of surfaces by the contact plate and swab methods. J Appl Bacteriol. 42:53–63

Noppen C, Martinato I, Reischl U and Schaefer C (2001) High speed purification and detection of Bordetela pertussus: A straight forward application for MagNA pure LC and the Light Cycler System in microbiological research. Biochemica. 1:17–19

Novitsky TJ and Hochstein HD (2003) Limulus Endotoxin Test. In: Easter MC (ed) Rapid Microbiological Methods in the Pharmaceutical Industry. Interpharm/CRC, Washington, D.C., pp. 187–210

Odlaug TE, et al. (1982) Evaluation of an automated system for rapid identification of Bacillus biological indicators and other Bacillus Species. PDA J Parenteral Sci Tech. 36(2):47–54

Olive DM and Bean P (1999) Principles and applications for DNA-Based typing of microbial organisms. J Clin Microbiol. 37(6):1661–1669

PallCheck Product Literature and Brochure. www.pall.com/datasheet_biopharm_38304.asp (Downloaded on 5 June 2007)

Parenteral Drug Association (May/June 2000) Technical Report 33: Evaluation, validation and implementation of new microbiological testing methods. J. Pharm. Sci. Technol. 54(3), Suppl. TR33

Patel JB, et al. (2000) Sequence-based identification of Mycobacterium species using the MicroSeq 500 16S rDNA Bacterial Identification System. J Clin Microbiol. 38(1):246–251

Pathogen Detection Systems Product Literature www.pathogendetect.com/ (Downloaded on 1 Sep 2007)

PHARMEUROPA (October 2004) 5.1.6.: Alternative methods for control of microbiological quality. PHARMEUROPA 16(4) pp. 555–565 (Note: Subsequently finalized July 2006)

Poletti L (1999) Comparative efficiency of nitrocellulose membranes vs. RODAC plates. In Microbial Sampling on Surfaces. J Hosp Infect. 41:195–201

Pounder JI, et al. (2005) Repetitive-sequence-PCR-based DNA fingerprinting using the DiversiLab System for identification of commonly encountered Dermatophytes. J Clin Microbiol. 43(5):2141–2147

Powers EM (1995) Efficacy of the Ryu nonstaining KOH technique for rapidly determining Gram reactions of food-borne and waterborne bacteria and yeasts. Appl Environ Microbiol. 61(10):3756–3758

Prigione V, et al. (2004) Development and use of flow cytometry for detection of airborne fungi. Appl Environ Microbiol. 70(3):1360–1365

Prinzi S (January 21–23, 2007) Cutting edge technology for the future of quality control using BacT/ALERT automated microbial detection system. Presentation at RMUG™, Arlington VA

Rosen DL, Fell Jr NF and Pellegrino PM (2003) Spectroscopic detection of bacterial endospores using terbium cation reagent. In: Olson WP (ed) Rapid Analytical Microbiology: The Chemistry and Physics of Microbial Identification. edParenteral Drug Association and Davis Horwood International Publishing, Ltd., Bethesda, MD and Godalming Surrey, UK, pp. 230–235

Rudi K (2003) Application of nucleic acid probes for analyses of microbial communities. In: Olson WP (ed) Rapid Analytical Microbiology: The Chemistry and Physics of Microbial Identification Parenteral Drug Association and Davis Horwood International Publishing, Ltd., Bethesda, MD and Godalming Surrey, UK, pp. 13–40

Schalkowsky S (1996) Predictive antimicrobial preservative effectiveness testing. Pharm Forum. 22(4):2690–2695

Shah HN et al. (2005). Surface enhanced laser desorption/ionisation time of flight mass spectrometry (Seldi-TOF-MS): A potentially powerful tool for rapid characterisation of microorganisms. In: Miller MJ (ed) Encyclopedia of Rapid Microbiological Methods, vol 3. PDA/DHI Publishers, Bethesda MD, pp. 57–95

Shutt CK, et al. (2005) Clinical evaluation of the DiversiLab microbial typing system using repetitive-sequence-based PCR for characterization of Staphylococcus aureus strains. J Clin Microbiol. 43(3):1187–1192

Silley P and Sharpe AN (2003) Labor saving devices and automation of traditional methods,. In: Easter MC (ed) Rapid Microbiological Methods in the Pharmaceutical Industry. Interpharm/CRC, Washington D.C., pp. 61–72

Singer DC and Cundell AM (2003) The role of rapid microbiological methods within the process analytical technology initiative. Pharmacopeial Forum. 29(6):2109–2113

Straus D (January 21–23, 2007) The Growth Direct™ System—A rapid non-destructive method for microbial enumeration. Presentation at RMUG™, Crystal City, Virginia

Straus D (Sep 2007) Chief Science Office. Rapid Micro Biosystems. Personnel Communication.

Sutton SVW and AM Cundell (2004) Microbial identification in the pharmaceutical industry. Pharm Forum. 30(5):1884–1894

Sutton SVW (2005) Validation of alternative microbiology methods for product testing: quantitative and qualitative assays. Pharm Technol. 29(4):118–122

Sy-Lab Product Literature. www.sylab.com/# (Downloaded on 1 Oct 2006)

Tang S (1998) Microbial limits reviewed: the basis for unique Australian regulatory requirements for microbial quality of non-sterile pharmaceuticals. PDA J Pharm Sci Tech. 52(3):100–109

Trinel PA, et al. (1983) Automatic diluter for bacteriological samples. Appl Environ Microbiol. 45(2):451–455

Vaidyanathan S, et al. (2001) Assessment of near-infrared spectral information for rapid monitoring of bioprocess quality. Biotechnology and Bioengineering 74(5):376–388

Versalovic J, et al. (1991) Distribution of repetitive DNA sequences in eubacteria and application to fingerprinting of bacterial genomes. Nuc Acids Res. 19(24):6823–6831

Warnex Product Literature and Brochures. www.warnex.ca/en/index.php (Downloaded on 7 July 2007)

Waterhouse RN and Glover LA (1993) Identification of procaryotic repetitive DNA suitable for use as fingerprinting probes. Appl Environ Microbiol. 59(5):1391–1397

Waters Literature and Article in Galley Proof form. www.waters.com/watersdivision/Contentd.asp?ref=CEAN-5KUSS8 (Downloaded on 1 July 2007)

Westin L, Miller C, Vollmer D, Canter D, Radtkey R, Nerenberg M and O'Connell JP (March 2001) Antimicrobial resistance and bacterial identification utilizing a microelectronic chip array. Journal of Clinical Microbiology 39 (3):1097–1104

Will K (2003) ATP bioluminescence and its use in pharmaceutical microbiology. In: Easter MC (ed) Rapid Microbiological Methods in the Pharmaceutical Industry. Interpharm/CRC, Washington, D.C., pp. 88–98

Woo PCY, et al. (2003) Usefulness of the MicroSeq 500 16S ribosomal DNA-based bacterial identification system for identification of clinically significant bacterial isolates with ambiguous biochemical profiles. J Clin Microbiol. 41(5):1996–2001

II

Biosensors

5

Surface Plasmon Resonance (SPR) Sensors for the Detection of Bacterial Pathogens

Allen D. Taylor, Jon Ladd, Jiří Homola and Shaoyi Jiang

Abstract

Modern biosensor technologies can provide rapid quantification of bacterial pathogens. Surface plasmon resonance (SPR) sensors are an optical platform capable of highly sensitive and specific measuring of biomolecular interactions in real-time. This label-free technology can quantify the kinetics, affinity and concentration of surface interactions. SPR sensors have been used to detect bacterial pathogens in clinical and food-related samples. This chapter discusses the fundamental theory behind SPR sensors and state-of-the-art SPR instrumentation, surface chemistries, molecular recognition elements and detection strategies, as well as specific challenges associated with bacterial detection using SPR sensors. SPR-based detections of bacterial cells, genetic markers and antibody biomarkers are reviewed and discussed.

1. Introduction

Modern sensor technologies can provide rapid and sensitive detection of bacterial pathogens. One such cutting-edge sensor technology is based on surface plasmon resonance (SPR). SPR sensors provide sensitive, label-free, and real-time monitoring of reactions and can quantify the characteristics of biomolecular (e.g., oligonucleotides, proteins, bacteria) interactions on a surface, including their kinetics, affinity, and concentration. SPR sensors have been successfully applied to environmental monitoring, biotechnology, medical diagnostics, drug screening, food safety, and homeland security (Homola 2003, Rich and Myszka 2005b).

This chapter discusses SPR theory, state-of-the-art SPR sensor instrumentation, surface chemistries, molecular recognitions elements, and detection strategies. Also, a review is presented on the application of SPR sensors to the detection of bacteria, genetic markers, and antibody biomarkers related to food safety.

2. Fundamentals of Surface Plasmon Resonance Biosensing

A surface plasmon resonance (SPR) biosensor consists of an optical system in which surface plasmons are excited and interrogated, a coating incorporating biorecognition elements which interact with an analyte in a liquid sample, and a fluidic system consisting of a flow-cell

Allen D. Taylor, Jon Ladd, Shaoyi Jiang • Department of Chemical Engineering, University of Washington.
Jiří Homola • Institute of Photonics and Electronics, Academy of Sciences, Prague, Czech Republic and Department of Chemical Engineering, University of Washington, Seattle, Washington, USA

M. Zourob et al. (eds.), *Principles of Bacterial Detection: Biosensors, Recognition Receptors and Microsystems*,
© Springer Science+Business Media, LLC 2008

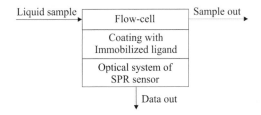

Figure 5.1. Surface plasmon resonance (SPR) biosensor system.

or cuvette for sample confinement at the SPR sensor surface and a fluid-handling system (Figure 5.1). Surface plasmons are special modes of the electromagnetic field propagating along a metal-dielectric interface (Raether 1988). Surface plasmons are TM-polarized waves (the vector of the magnetic field lies in the plane of the interface and is perpendicular to the direction of propagation of the surface plasmon). The electromagnetic field of a surface plasmon is localized at the interface and decays evanescently into both the metal and dielectric. The field of a surface plasmon decays exponentially in the direction perpendicular to the direction of propagation and is characterized by means of the penetration depth (the plane parallel to the metal-dielectric interface at which the amplitude of the field falls to 1/e of its value at the surface). The penetration depth into a dielectric increases with the wavelength, and for a surface plasmon at the interface of gold and an aqueous environment and wavelengths within 600–1000 nm, it ranges from 150 to 600 nm. The propagation constant of a surface plasmon (Raether 1988) can be expressed as:

$$\beta = \frac{\omega}{c}\sqrt{\frac{\varepsilon_d \varepsilon_m}{\varepsilon_d + \varepsilon_m}} = k\sqrt{\frac{\varepsilon_d \varepsilon_m}{\varepsilon_d + \varepsilon_m}}, \tag{5.1}$$

where $\varepsilon_m = \varepsilon'_m + i\varepsilon''_m$ and $\varepsilon_d = \varepsilon'_d + i\varepsilon''_d$ are permittivities of the metal and dielectric, respectively; ε'_j and ε''_j are real and imaginary parts of ε_j (j is m or d) and $i = \sqrt{-1}$; ω is the angular frequency; c is the speed of light in vacuum; and $k = 2\pi/\lambda$ is the free-space wavenumber, where λ is the free-space wavelength (Boardman 1982, Raether 1988). For lossless metal and dielectric ($\varepsilon''_m = \varepsilon''_d = 0$), equation (5.1) represents a guided mode providing that the permittivities ε'_m and ε'_d are of opposite signs, and $\varepsilon'_m < -\varepsilon'_d$. For metals obeying the free-electron model (Born and Wolf 1999), this condition can be fulfilled for frequencies lower than the plasma frequency of the metal (Boardman 1982). Metals such as gold and silver exhibit a negative real part of permittivity in the visible and near-infrared region of the spectrum and therefore can support surface plasmons at these frequencies. In SPR biosensors, gold is the most frequently used metal due to its good chemical stability.

A light wave can excite a surface plasmon at a metal-dielectric interface if the component of the light wavevector parallel to the interface matches that of the surface plasmon. As follows from equation (5.1), the real part of the propagation constant is larger than that which can be provided by the component of the wavevector of light in the dielectric. Therefore, the light wavevector needs to be enhanced to match that of the surface plasmon. The most commonly used method for the excitation of surface plasmons is attenuated total reflection in prism couplers (Raether 1988, Homola 2003). The process of excitation of surface plasmons by the attenuated total reflection method is illustrated in Figure 5.2.

A light wave passes through an optical prism (prism permittivity $\varepsilon_p > \varepsilon_d$) and is made incident on the metal film (typically about 50 nm thick) at an angle of incidence larger than the critical angle for the prism-dielectric system. The light wave is totally reflected, giving rise to an evanescent wave propagating along the metal film. If the propagation constant of the evanescent wave matches that of the surface plasmon at the outer boundary of the metal film,

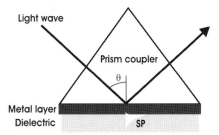

Figure 5.2. Excitation of a surface plasmon on a thin metal film by means of the attenuated total reflection (ATR) method.

the coupling between the light wave and the surface plasmon can occur (Raether 1988). This condition can be written as:

$$k_x = \frac{2\pi}{\lambda} \sqrt{\varepsilon_0} \sin(\theta) = \mathrm{Re}\left\{ \frac{2\pi}{\lambda} \sqrt{\frac{\varepsilon_1 \varepsilon_2}{\varepsilon_1 + \varepsilon_2}} \right\}, \tag{5.2}$$

where θ denotes the angle of incidence, k_x denotes the component of the wavevector of the light which is parallel to the interface, and λ is the wavelength of the light wave in vacuum. A portion of light energy is transferred into the energy of a surface plasmon and dissipated in the metal film, resulting in a drop in the intensity of reflected light. The coupling condition can be fulfilled for multiple combinations of the angle of incidence and the wavelength. Therefore, a characteristic dip associated with the excitation of the surface plasmon can be observed both in angular (θ variable, λ fixed) and spectral (λ variable, θ fixed) domains (Figure 5.3). As the propagation constant of the surface plasmon depends on the refractive index of the dielectric (5.1), a change in the refractive index of the dielectric results in a change in the propagation constant and consequently in a change in the SPR dip in the spectrum of reflected light (Figure 5.3). SPR biosensors exploit the sensitivity of the propagation constant of a surface plasmon to refractive index changes to measure binding-induced changes in the refractive index occurring within the evanescent field of the surface plasmon. In SPR biosensors, a surface plasmon is excited on a thin metal film to the surface of which a biorecognition element is attached. The binding of analyte in solution to the biorecognition element on the SPR sensor surface produces a local increase in the refractive index (Figure 5.4). The refractive index change gives rise to a change in the propagation constant of the surface plasmon, which is subsequently measured as a change in the coupling angle of incidence (SPR sensors with angular modulation) or the coupling wavelength (SPR sensors with wavelength modulation).

3. SPR Sensor Instrumentation

In the optical system of a SPR sensor, surface plasmons are optically excited and the SPR signal is encoded into a light wave interacting with the surface plasmons. Based on the method of excitation of the surface plasmons, SPR sensors can be classified as employing (1) prism couplers (Matsubara, Kawata, and Minami 1988; Liedberg, Lundstrom, and Stenberg 1993; Homola, Pfeifer, and Brynda 1997), (2) grating couplers (Jory, Vukusic, and Sambles 1994; Jory et al. 1995; Dostalek, Homola, and Miler 2005), (3) optical fibers (Jorgenson and Yee 1993; Homola 1995; Piliarik et al. 2003), or (4) integrated optical waveguides (Van Gent et al. 1990; Mouvet et al. 1997; Harris et al. 1999).

Currently, most SPR sensor platforms are based on the attenuated total reflection method and prism coupling and use either wavelength (Homola, Pfeifer, and Brynda 1997) or angular

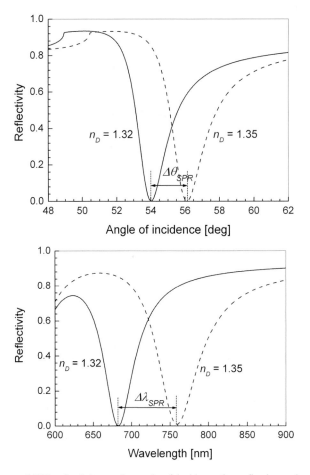

Figure 5.3. Dependence of TM reflectivity on the angle of incidence for a fixed wavelength and two different refractive indices of the dielectric, $\lambda = 682$ nm (top). Dependence of TM reflectivity on the wavelength for a fixed angle of incidence and two different refractive indices of the dielectric, $\theta = 54$ deg (bottom). Geometry: SF14 glass prism, 50 nm thick gold layer, and dielectric with a refractive index of 1.32.

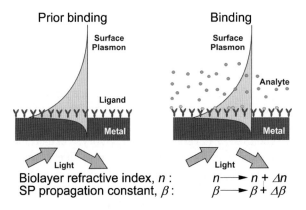

Figure 5.4. Principle of SPR biosensing.

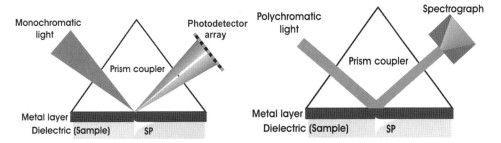

Figure 5.5. SPR sensor based on the attenuated total reflection and angular modulation (left). SPR sensor based on the attenuated total reflection and wavelength modulation (right).

(Matsubara, Kawata, and Minami 1988; Liedberg, Lundstrom, and Stenberg 1993) modulation. An operating principle of a typical SPR sensor with angular modulation is illustrated in Figure 5.5. A convergent monochromatic beam passes through a prism coupler. For a certain angle of incidence, the coupling between the light wave and the surface plasmon occurs. A change in the refractive index at the sensor surface gives rise to a change in the angular position of the SPR dip, which can be measured by a position-sensitive photodetector (e.g., CCD or photodiode array). In a SPR sensor based on wavelength modulation, surface plasmons are excited using a collimated polychromatic beam (Figure 5.5, right), and a change in the refractive index at the sensor surface is determined from the spectrum of reflected light by means of an optical spectrograph.

SPR sensors for the analysis of complex matrices in realistic environmental conditions need to discriminate between the refractive index changes due to the specific interaction between the biorecognition element and the analyte, and changes due to background refractive index variations. Fluctuations in the background refractive index are typically caused by changes in composition of the sample (e.g., residual matrix components) and by temperature variations. Discrimination between these changes can be achieved by using SPR sensor architectures with measuring and reference channels (Nenninger et al. 1998). A typical configuration of a multi-channel SPR sensor is illustrated in Figure 5.6, which shows a SPR sensor with four parallel sensing channels and angular modulation. In this sensor, a divergent beam produced by a LED is collimated and focused by means of a cylindrical lens to produce a wedge-shaped beam of light which is used to illuminate a thin gold film on the back of a glass prism containing several sensing areas (channels). The imaging optics consists of one imaging and one cylindrical lens ordered in such a way that the angular spectrum of each sensor channel is projected on a

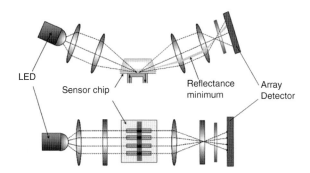

Figure 5.6. SPR sensor with angular modulation and four parallel sensing channels (provided by, S. Löfås, Biacore AB.).

Figure 5.7. Portable SPR sensor system with wavelength modulation and eight sensing channels developed at the Institute of Photonics and Electronics, Prague.

separate row (or rows) of the array detector. This optical design has been adopted by BIACORE (Pharmacia Biosensors AB; since 1996, BiacoreAB) and resulted in a family of commercial SPR sensors (http://www.biacore.com; Karlsson and Stahlberg 1995; Nice and Catimel 1999) offering high performance (resolution down to 1×10^{-7} RIU; RIU—refractive index unit) and multiple sensing channels (up to 4) for simultaneous measurements. An alternative approach to discriminating a specific sensor response from the interfering effects consists of the deconvolution of contributions to the sensor response originating at the sensor surface and in the bulk material using multi-surface-plasmon-spectroscopy (Homola, Lu, and Yee 1999; Homola et al. 2005; Adam, Dostalek, and Homola 2006). The SPR sensors employing multiple surface plasmons excited at different sensing channels by different wavelengths are commonly referred to as SPR sensors with wavelength division multiplexing (Homola et al. 2005).

In recent years, we have witnessed an extensive effort in research laboratories worldwide aiming to bring the SPR method out of the laboratories into the field and create a rapid, sensitive detection technology that meets the growing bioanalytical needs in medical diagnostics, environmental monitoring, and food safety and security. Various miniaturized SPR optical platforms have been developed (Kukanskis et al. 1999; Thirstrup et al. 2004). Figure 5.7 shows an 8-channel portable SPR sensor with prism coupling and wavelength modulation combining parallel channel architecture with a wavelength division multiplexing of sensing channels. The sensor system incorporates the optical platform, fluidic unit, temperature stabilization and supporting electronic hardware. The sensor is capable of resolving refractive index changes smaller than 5×10^{-7} RIU.

4. Surface Chemistries and Molecular Recognition Elements

While instrumentation determines the inherent sensor sensitivity, surface chemistries determine the overall performance of a sensor. A typical surface consists of molecular recognition elements (MRE) immobilized on a nonfouling background. The properties of a good MRE include high specificity (the ability to discriminate an individual target analyte from non-target analytes), affinity (the ability to bind strongly to an analyte), and stability (the ability for long-term storage and for detection under various conditions). SPR sensors have been functionalized with a wide range of biorecognition elements, including oligonucleotides (Nelson et al. 2002), aptamers (Balamurugan et al. 2006), antibodies (Minunni and Mascini 1993; Mouvet et al. 1997; Shimomura et al. 2001; Oh et al. 2003a), proteins (Usami, Mitsunaga, and Ohno 2002; Asano et al. 2004), and whole cells (Choi et al. 2005). Antibodies are the most prevalent MRE due to their high affinity, specificity, and commercial availability (Mullett, Lai, and Yeung 2000). Other MREs used with SPR sensors include molecularly imprinted polymers (Lotierzo et al. 2004) and organic synthetic receptors (Wright et al. 1998), which have the potential for high stability under varying environmental conditions. However, their affinity and specificity have been significantly lower than that achieved by biorecognition elements.

The successful implementation of a specific, label-free binding event requires a surface chemistry that retains the highest activity of the immobilized MRE and minimizes non-specific binding to the surface. The highest surface activity is achieved by maximizing the amount of the MRE in a favorable orientation and conformation on the surface. The simplest method used to immobilize MREs is direct adsorption onto a metal surface. However, this leads to a loss in activity and uncontrolled exchange of MREs on the surface. A more commonly used surface chemistry which helps maintain protein conformation involves self-assembled monolayers (SAMs). A SAM is a densely packed monolayer formed via the spontaneous adsorption of either thiols or disulfides on the metal surface. The monolayers result in a robust surface with properties determined by its exposed functional end groups, such as positive charge ($-NH_3^+$), negative charge ($-COO^-$), hydrophobicity ($-CH_3$), and hydrophilicity ($-OH$). MREs immobilized on SAMs are generally chemically linked to the surface. The most commonly used chemistry is 1-ethyl-3-[3-dimethylaminopropyl]carbodiimide hydrochloride (EDC) and n-hydroxysuccinimide (NHS), which crosslinks an amine group of a biomolecule to a carboxylic acid functional group of a SAM. Studies have shown that protein orientation also greatly affects the activity of an immobilized MRE. Orienting the binding epitopes of an antibody away from the surface increases the bioactivity of the antibody over a randomly immobilized antibody (Chen et al. 2003). Several methods have been used to orient proteins including charged surfaces (Chen et al. 2003), protein G or A (Fratamico et al. 1998), and conjugation chemistries that target the carbohydrate region of monoclonal antibodies (Fleminger et al. 1990).

Besides MRE bioactivity, another important concern in sensor performance is minimizing non-specific adsorption to obtain maximum specificity for a given assay. Non-specific binding can cause background signals, thus convoluting responses for specific interactions. Extensive research has been performed to develop surfaces that resist non-specific adsorption for sensor and biomedical applications. Integration of non-fouling materials such as oligo(ethylene glycol) (OEG) has become a standard practice for sensing surfaces (Nelson et al. 2001; Boozer et al. 2003; Boozer et al. 2004; Ladd et al. 2004). These surfaces have proven to provide good non-fouling characteristics in simple protein solutions. However, these surfaces often suffer from considerable non-specific adsorption from complex matrices, such as blood or bacteria. Recently, zwitterionic-based materials have been reported to be superlow fouling for proteins, blood plasma, and bacteria using a SPR sensor (Chen et al. 2005; Zhang, Chen, and Jiang 2006).

The stability of MREs is a concern for chip storage and applications under harsh environmental conditions. The storage characteristics of 11 protein arrays were studied by Angenendt (Angenendt et al. 2002), showing that protein arrays consisting of five different antibodies could be stored for eight weeks without significant loss of bioactivity. A novel method of creating protein arrays by converting a DNA array into a protein array via DNA-directed immobilization has recently been demonstrated (Boozer et al. 2004; Ladd et al. 2004; Boozer et al. 2006). Single-stranded DNA (ssDNA) sequences were immobilized on the sensor surface, either via SAM formation or through a streptavidin linkage. An antibody was conjugated to a ssDNA whose sequence was complementary to one of the surface-immobilized ssDNA strands. The conjugate was then immobilized on the surface through DNA hybridization. Because of the high specificity of DNA hybridization, numerous protein-DNA conjugates can be immobilized simultaneously in specified locations, thus quickly transforming a DNA array into a protein array. This method of protein immobilization allows the proteins to remain in solution until the time of assay, maintaining their stability and conformation. Sensor sensitivity using this immobilization method has shown an increase of up to 50-fold over conventional protein immobilization techniques (Boozer et al. 2004; Ladd et al. 2004). Dehybridization of the DNA duplex regenerates the ssDNA surface, allowing the re-immobilization of proteins through site-directed immobilization. Thus, a single chip can be used for the detection of multiple interactions, simply by immobilizing a different conjugate following regeneration of the ssDNA surface.

Sensor surfaces that are recyclable are convenient and desirable to reduce the cost and time of detection assays. Detection of chemical and biological analytes is usually performed with high affinity MREs, where the binding interaction is, under normal conditions, irreversible. Several regeneration methods have been demonstrated by which the analyte-antibody binding is reversed, leaving the sensor surface available for subsequent measurements. Demonstrated regeneration methods include changing pH (Yu et al. 2005), using detergents (Lotierzo et al. 2004), or enzymes (Mouvet et al. 1997; Gobi and Miura 2004). Typically, tens of regeneration cycles have been possible without significant loss of activity of the sensor surface (Gobi et al. 2005; Yu et al. 2005; Shankaran et al. 2006).

5. Detection Formats

Because chemical and biological analytes vary widely in mass, SPR sensors use various detection formats based on the mass of the target analyte. Direct detection of medium and large analytes ($>10,000\,Da$) has been demonstrated with reasonable detection limits (Homola et al. 2002; Oh et al. 2003a). However, the direct binding of low molecular weight analytes ($<1,000\,Da$) at the sensor surface does not produce a sufficient change in refractive index. Typically, a sandwich (Minunni and Mascini 1993), competitive (Shimomura et al. 2001), or inhibition (Mouvet et al. 1997) assay is used to improve detection limits for these analytes.

In a sandwich assay, as shown in Figure 5.8a, a primary MRE is first immobilized on the sensor surface. Sample is then flowed over the sensor surface allowing the MRE to specifically capture analyte. The sensor response from this direct detection will vary depending on the mass and amount of the analyte being bound. Subsequently flowing a secondary MRE will cause the sensor response to be amplified, thus improving detection limits as well as verifying the captured analyte.

The principle of a competitive assay, as shown in Figure 5.8b, is that two analytes compete for the same binding sites of an immobilized MRE. A low molecular weight analyte is conjugated to a larger protein (e.g., bovine serum albumin) capable of producing a significant sensor response when bound to the surface. This conjugate is used in two different approaches: (1) the conjugate is captured on the surface at a fixed concentration, and subsequently the analyte to be detected can compete the conjugate off the surface; or (2) the analyte to be detected is mixed with the conjugate and the mixture is flowed over the sensor surface, the two components compete for the same binding sites. In the first format, the sensor response is negative, corresponding to a reduction in refractive index at the sensor surface as the conjugate is replaced on the surface by the low molecular weight analyte. In the second format, the sensor response is inversely proportional to the concentration of the low molecular weight analyte in the mixture.

An inhibition assay, as shown in Figure 5.8c, is an indirect method where no portion of the analyte is detected directly by the sensor. A derivative of the analyte is immobilized on

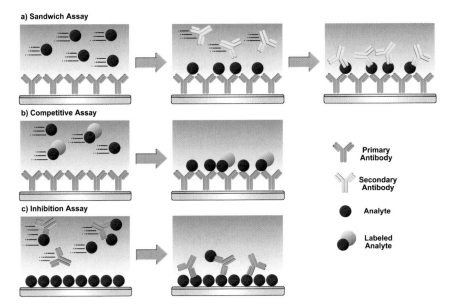

Figure 5.8. Illustration of three typical detection assay formats used with a SPR sensor: a) in a sandwich assay, the analyte is first detected directly by capture on a sensing surface immobilized with a molecular recognition element (MRE), and subsequently amplified using a secondary antibody; b) in a competitive assay, the target analyte is mixed with an analyte conjugated to a larger protein and are allowed to compete for the same binding sites of the MRE on the surface; c) in an inhibition assay, the target analyte is incubated with a fixed concentration of a MRE, then the analyte-free MRE is detected on a surface immobilized with a derivative of the analyte.

the sensor surface. A known quantity of MRE is incubated with the analyte and allowed to bind in solution. The incubated sample is then flowed over the sensor surface and MREs with free binding epitopes are captured on the sensor surface. As the concentration of analyte in the incubated sample increases, the concentration of free MRE binding epitopes decreases. Thus, the sensor response is inversely proportional to the analyte concentration.

6. Quantification of Bacteria Cells

Bacterial detection is typically used to diagnose infections or assess the risk of pathogenic organisms in food and the environment. SPR sensors can detect analytes in complex samples (e.g., blood, urine, stool extract, fruit juices, and food extracts) with limited or no sample preparation. This section discusses the challenges for SPR sensors to quantify bacterial cells and the effects of sample treatment on detection limits. A review of up-to-date research is presented.

6.1. Challenges for the Detection of Whole Bacteria by SPR

The size and morphology of live bacteria create a number of challenges for SPR sensors. These include the limited penetration depth of the electromagnetic field of a surface plasmon compared to the typical size of a bacterium; the low refractive index contrast between the bacterium cytoplasm and the aqueous environments in which detection is usually performed; the availability and accessibility of corpuscular antigens on the bacterium surface binding to biorecognition elements; and the limited diffusion of bacteria towards the sensor surface in the typical hydrodynamic conditions of the microfluidic flow-channels of SPR sensors.

The electromagnetic field of a surface plasmon decays exponentially away from the metal surface. The penetration depth of a surface plasmon depends on the refractive index of the sample and the wavelength, and is typically ~150–600 nm, while the dimensions of bacterial cells are on the order of 1–3 μm. Typical surface chemistries used in SPR sensors immobilize biorecognition elements with a distance of 10–100 nm from the sensing surface, thus limiting the whole bacteria's exposure to the evanescent wave. Recently, it was demonstrated that the probing depth of SPR sensors can be increased well beyond 1 μm by employing a special type of surface plasmon—the long-range surface plasmon (Slavik and Homola 2007).

SPR sensors detect changes in the refractive index within the evanescent field as compared to the background running medium. As a result, a SPR sensor will measure the refractive index change caused by the "dry mass" of a bacterium and will exclude the water enclosed in the bacterium. Other label-free methods based on vibrational phenomena, such as quartz crystal microbalance (QCM), microcantilevers, and surface acoustic wave (SAW), detect the "wet mass" of the bound bacterium.

SPR sensors rely predominantly on microfluidics to bring solutions to the chip surface, which creates several challenges for the detection of whole bacteria. Because the typical dimensions of the fluid delivery system are on the order of microns, clogging of the fluidics by whole bacteria cells becomes possible. The large dimensions of a bacterium also make diffusion to the surface slow. Since detections with a SPR sensor are dependent on the ability of the analyte to reach the surface and bind to an immobilized MRE, this slow diffusion could limit the sensor response. Flow conditions complicate the detection further because particles tend to align in the center of the flow channel by a phenomenon called hydrodynamic focusing. Flow also produces shear forces that may prevent the analyte from binding to the MRE. The flow rates typically used in SPR sensor experiments result in low Reynolds number flow, and limited mixing occurs to affect the aforementioned issues. However, in recent years researchers have shown that peristaltic action introduced by some delivery systems can cause mixing in low Reynolds number flow (Jackson et al. 2002; Truesdell et al. 2003).

6.2. Effect of Bacteria Sample Treatment

Numerous sample treatment methods have been presented to overcome some of the challenges discussed above. Taylor et al. compared five sample treatment methods of *Escherichia coli* O157:H7 for detection with a SPR sensor (Taylor et al. 2005; Taylor et al. 2006). The sample treatment methods that were tested and the corresponding direct limit of detections (LOD) were: live: 10^7 cfu/ml, heat-killed: 10^6 cfu/ml, heat-killed and soaked in 70% ethanol cells: 10^6 cfu/ml, heat-killed and detergent lysed: 10^5 cfu/ml, and heat-killed and ultrasonicated: 10^5 cfu/ml. The detection curves for direct detection using three of the sample treatment methods are shown in Figure 5.9. The LOD for each of the sample treatment methods was improved by one order of magnitude by using a secondary antibody in a sandwich assay. The difference in sensor response for varying sample treatment methods is attributed to the effect of each method on the size and morphology of the analyte. Lysing ideally breaks the bacterium cell into smaller pieces, thus improving mass-transport, flow limitations, and possibly bringing more of the bacterium closer to the evanescent wave region of the sensor surface. This study shows that the sample treatment method and assay format significantly affect SPR sensor response.

6.3. Examples of Bacteria Detection

The following section is a review of SPR detection of bacteria cells by either direct detection, sandwich assay, or inhibition assay, using various sample preparations. A review of bacteria detections by SPR sensor is shown in Table 5.1.

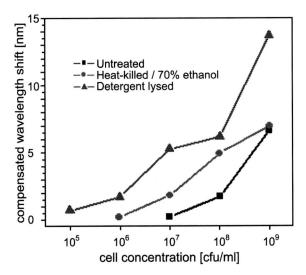

Figure 5.9. Resonant wavelength shift versus concentration of bacteria for the detection of *E. Coli* O157:H7 comparing untreated, heat-killed, then ethanol soaked, and detergent-lysed samples by direct detection (reprinted from Taylor et al. (2005), with permission of Elsevier).

6.3.1. *Escherichia coli*

E. coli is a bacterium that lives in the intestines of all animals. While most strains are harmless, several strains, such as *E. coli* O157:H7, can cause serious illness. *E. coli* O157:H7 produces verotoxins that cause severe damage to the lining of the intestines, leading to bloody diarrhea and sometimes kidney failure. Most illness is caused by ingesting undercooked ground beef, but other sources include raw milk, bean sprouts, lettuce, spinach, and sewage-contaminated water (Center for Food Safety and Nutrition & United States Food and Drug Administraton 2007).

The first SPR sensor detection of bacterial cells was published by Fratamico et al. in 1998. A sandwich assay was developed using immobilized monoclonal antibody (MAb) and a secondary polyclonal antibody (PAb). Figure 5.10 shows the detection of 5×10^9 cfu/ml viable *E. coli* O157:H7 by sandwich assay and the specificity of the sensing surface exhibited by no significant response to either *S. typhimurium* or *Y. enterocolitica*. The detection limit was $5–7 \times 10^7$ cfu/ml and the surface was regenerated and used for at least 50 measurements. Antibodies were immobilized by three methods. Capture of the MAb Fc region by amine-coupled protein A or protein G did not enhance detection compared to amine-coupling of the MAb directly to the surface. In 1999, Fratamico et al. tried to improve detection limits using an inhibition assay to detect *E. coli* O157:H7 (Fratamico et al. 1999). A known concentration of PAb to *E. coli* was incubated with viable *E. coli*. The bacteria were then centrifuged out and the supernatant containing unbound antibody was flowed across the sensing surface functionalized with an anti-Fab antibody. Using an inhibition assay to quantify whole cells should improve the mass transport and flow-related limitations of detecting of whole cells. The detection limit for this assay was between 10^6 and 10^7 cfu/ml.

In 2002, Oh et al. used a Multiskop SPR sensor for direct detection of *E. coli* O157:H7 with a detection limit of 10^4 cfu/ml (Oh et al. 2002). The sensing surface was functionalized by a MAb captured by immobilized protein G on a SAM. In 2003, they improved their detection limit to 10^2 cfu/ml using the same sensor system, by optimizing the SAM composition (Oh et al. 2003b). Neither study reported a treatment method or amplification protocol.

In 2005, Meeusen et al. used a Spreeta sensor for direct detection of viable *E. coli* O157:H7 obtaining a LOD of 8.7×10^6 cfu/ml without amplification (Meeusen, Alocilja, and

Table 5.1. Overview of SPR sensor detections of bacteria.

Analyte	Sensor	Sample Treatment	Detection Matrix	Limit of Detection [cfu/mL]	Assay Format	References
Bacteria cells:						
Escherichia coli O157:H7						
	Biacore	Live	HBS pH 7.4	$5 - 7 \times 10^7$	Sandwich	(Fratamico et al. 1998)
	Biacore	Live	HBS pH 5.0	$10^6 - 10^7$	Inhibition	(Fratamico et al. 1999)
	Multiskop	Live		10^4	Direct	(Oh et al. 2002)
	Multiskop	Live		10^2	Direct	(Oh et al. 2003b)
	Spreeta	Heat killed	PBS pH 7.4	10^6	Direct	(Su and Li 2005)
	Spreeta	Live	BPW	8.7×10^6	Direct	(Meeusen, Alocilja, and Osburn 2005)
	Custom-built	Live	PBS pH 7.4	10^7	Direct	(Taylor et al. 2005)
				10^6	Sandwich	
		Heat killed		10^6	Direct	
				10^5	Sandwich	
		Heat killed & ethanol soaked		10^6	Direct	
				10^5	Sandwich	
		Detergent lysed		10^5	Direct	
				10^4	Sandwich	
	Custom-built	Heat killed & ultrasonicated	PBS pH 7.4	10^4	Sandwich	(Taylor et al. 2006)
			Apple juice pH 7.4	10^4	Sandwich	
			Apple juice pH 3.7	10^4	Sandwich	
	Reichert SR7000	Live	PBS pH 7.4	10^6	Direct	(Subramanian, Irudayaraj, and Ryan 2006a)
				10^3	Sandwich	
	Reichert SR7000	Live	Apple juice	10^8	Direct	(Subramanian and Irudayaraj 2006)
				10^6	Sandwich	
***Salmonella* spp.**						
S. enteritidis	Custom-built	Heat killed & ethanol soaked	PBS pH 7.4	10^6	Direct	(Koubova et al. 2001)
½ S. enteritidis & ½ S. typhimurium	Biacore	Heat killed	HBS pH 7.4	1.7×10^5	Sandwich	(Bokken et al. 2003)
S. paratyphi	Multiskop	Direct		10^2	Direct	(Oh et al. 2004b)

Table 5.1. (Continued)

Analyte	Sensor	Sample Treatment	Detection Matrix	Limit of Detection [cfu/mL]	Assay Format	References
S. typhimurium	Multiskop	Direct	PBS pH 7.4	10^2	Direct	(Oh et al. 2004a)
S. choleraesuis	Custom-built	Heat killed & ultrasonicated	Apple juice pH 7.4	5×10^4	Sandwich	(Taylor et al. 2006)
			Apple juice pH 3.7	5×10^4	Sandwich	
				5×10^4	Sandwich	
Listeria monocytogenes						
	Custom-built	Heat killed	PBS pH 7.4	10^6	Direct	(Koubova et al. 2001)
	Biacore 3000	Heat killed	PBS pH 7.3	$1 - 2 \times 10^5$	Inhibition	(Leonard et al. 2004; Leonard et al. 2005)
	Biacore 3000	Live	PBS pH 7.4	10^7	Direct	(Hearty et al. 2006)
	Custom-built	Heat killed & ultrasonicated	PBS pH 7.4	3.5×10^3	Sandwich	(Taylor et al. 2006)
			Apple juice pH 7.4	3.5×10^3	Sandwich	
			Apple juice pH 3.7	3.5×10^3	Sandwich	
Campylobacter jejuni						
	Custom-built	Heat killed & ultrasonicated	PBS pH 7.4	10^5	Sandwich	(Taylor et al. 2006)
			Apple juice pH 7.4	10^5	Sandwich	
			Apple juice pH 3.7	10^5	Sandwich	
Bacillus subtilus						
	Biacore 2000	Spores	HBS pH 7.4	10^7 spores/mL	Direct	(Perkins and Squirrell 2000)
	Custom-built w/ light scattering	Spores	HBS pH 7.4	10^7 spores/mL	Direct	
Helicobacter pylori						
		Live		10^9	Direct	(Nishimura et al. 2000)
		Ultrasonicated		2×10^7	Direct	
Legionella pneumophila						
	Multiskop	Live		10^2	Direct	(Oh et al. 2003a)
Yersinia enterocolitica						
	Multiskop	Live		10^2	Direct	(Oh et al. 2005a)
Staphylococcus aureus						
	Reichert SR7000	Live	PBS pH 7.4	10^7	Direct	(Subramanian, Irudayaraj, and Ryan 2006b)
		Live		10^5	Sandwich	
Vibrio cholerae O1						
	Multiskop	Live		3.7×10^5	Direct	(Jyoung et al. 2006)

Table 5.1. (Continued)

Analyte	Sensor	Sample Treatment	Detection Matrix	Limit of Detection [cfu/mL]	Assay Format	Reference
Genetic markers: *Escherichia coli*						
Verotoxin-2 subunit A gene	Biacore 2000	PCR	*E. coli O157:H7* TES pH 7.5	1.5×10 M DNA	Direct	(Kai et al. 1999)
Shigatoxin-2 subunit A gene	Biacore 2000	PCR	*E. coli O157:H7* Stool sample	10^2 cfu/0.1g	Direct	(Kai et al. 2000)
Verotoxin-2 subunit gene	Biacore 2000	PCR	*E. coli O157:H7*		Direct	(Miyachi et al. 2000)
16s ribosomal RNA	Custom-Built Imaging	Total cellular RNA or PCR	*E. coli*		Direct	(Nelson et al. 2002)
Salmonella spp.						
invA gene	Biacore 2000	PCR	*S. enteritidis & typhimurium*		Direct	(Miyachi et al. 2000)
Bacillus subtilus						
16s ribosomal RNA	Custom-built Imaging	Total cellular RNA or PCR	*B. subtilis*		Direct	(Nelson et al. 2002)
Antibody markers: **Salmonella spp.**						
Serum antibodies to *S. enteritidis* H:g,m & *S. typhimurium* H:I & H:1,2	Biacore 3000	Lyophilized	Chicken sera : HBS pH 7.4 [1/40 v/v]		Direct	(Jongerius-Gortemaker et al. 2002)
Egg yolk antibodies to *S. enterica* LPS	Biacore 3000		Egg yolk : HBS pH 7.4 [1/5 v/v]			(Thomas et al. 2006)

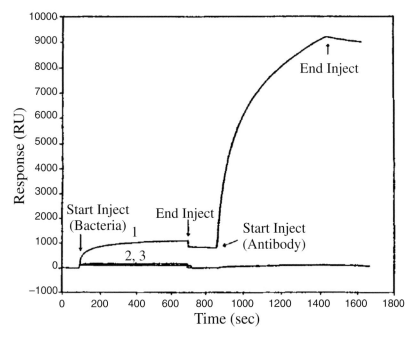

Figure 5.10. Overlay plots of sensorgrams showing the interaction of monoclonal antibody 8-9H (ligand) with *E. coli* O157:H7 (1), *S. typhimurium* (2), and *Y. enterocolitica* (3) followed by injection of polyclonal antibody at $50\,\mu g/ml$. The bacteria were injected at about 5×10^9 cfu/ml (reprinted from Fratamico et al. (1998), with kind permission of Springer Science and Business Media).

Osburn 2005). Su et al. compared the detection of heat-killed *E. coli* O157:H7 from a Spreeta SPR sensor to a quartz crystal microbalance (QCM) sensor (Su and Li 2005). In both sensors PAb was captured on a protein A surface. The detection ranges for SPR and QCM were 10^6–10^8 cfu/ml and 10^5–10^8 cfu/ml, respectively. The report concludes that QCM had a lower LOD and better signal to noise ratio than the Spreeta SPR sensor, which is attributed to the differences in effective detection thickness.

In 2005, Taylor et al. reported a comparison of SPR sensor detection of *E. coli* O157:H7 prepared by several sample treatment methods and was previously discussed (Taylor et al. 2005). Subramanian et al. used a Reichert SR7000 SPR sensor for a sandwich assay detection of viable *E. coli* O157:H7 in buffer (Subramanian, Irudayaraj, and Ryan 2006a) and apple juice (Subramanian and Irudayaraj 2006). The sensing surface was functionalized with an amine-coupled antibody on an OEG-SAM. In buffer the LOD was 10^6 cfu/ml for direct detection and 10^3 using a sandwich assay. The LOD in apple juice was 10^8 cfu/ml for direct detection and 10^6 using a sandwich assay.

6.3.2. *Salmonella* spp.

Salmonella spp. is a food pathogen usually found in raw meat, poultry, and seafood, as well as eggs and dairy products. (Center for Food Safety and Nutrition & United States Food and Drug Administraton 2007). In 2001, Koubova et al. demonstrated the detection of heat-killed then ethanol soaked *S. enteritidis* using a custom-built SPR sensor. The sensing surface was created by physisorbing a double layer of antibodies on bare gold, then crosslinking the proteins with gluteraldehyde. The LOD without amplification was 10^6 cfu/ml. Bokken et al. used a Biacore system in 2003 to demonstrate the detection of heat-killed *Salmonella* strains from groups A, B, D, and E, according to Kauffmann-White typing. Antibodies were immobilized on a carboxymethylated dextran chip by EDC/NHS coupling. 53 different *Salmonella* serovars and 30 non-*Salmonella* species at 10^7 cfu/ml were analyzed for specificity. The LOD for a solution consisting of half *S. enteritidis* and half *S. typhimurium* using a sandwich assay was 1.7×10^5 cfu/ml. In 2003 and 2004, Oh et al. detected *S. typhimurium* (Oh et al. 2004a) and *S. paratyphi* (Oh et al. 2004b), respectively. The method for detecting *S. typhimurium* was similar to their study of *E. coli* O157:H7. The sensing surface for the detection of *S. paratyphi* was formed by capturing the antibody with self-assembled thiolated protein G. Both studies had the same detection range of 10^2–10^7 cfu/ml. As with previous studies, no amplification or treatment method was reported.

6.3.3. *Listeria monocytogenes*

Listeria monocytogenes is a food pathogen that has been found in raw fish, cooked crab, raw and cooked shrimp, raw lobster, surimi, and smoked fish (Center for Food Safety and Nutrition & United States Food and Drug Administraton 2007). In 2001, Koubova et al. demonstrated the detection of heat-killed *L. monocytogenes* down to 10^6 cfu/ml. This was published in conjunction with the study of *S. enteritidis* and the same sensing surface chemistry, a physisorbed double layer of antibody crosslinked by gluteraldehyde, was used.

In 2004, Leonard et al. used an inhibition assay to detect heat-killed *L. monocytogenes* with a Biacore 3000 sensor (Leonard et al. 2004). The detection was achieved by incubating heat-killed *L. monocytogenes* with an inhibition assay similar to their study of *E. coli* O157:H7 in 2003. The surface was dextran-coated with anti-rabbit PAb immobilized by EDC/NHS chemistry. The LOD was 10^5 cfu/ml. In 2005, Leonard et al. improved the inhibition assay method by producing a PAb to Internalin B (InlB) from *L. monocytogenes* and expressing the InlB protein in *E. coli* to be purified and immobilized on the sensing surface (Leonard et al. 2005). The same method was used as in the previous study, but the InlB protein immobilized surface was specific to the antibody incubated with *L. monocytogenes*, as opposed to the anti-rabbit PAb which would react with all antibodies produced from rabbit.

In 2006, Hearty et al. produced and characterized a novel MAb for virulent *L. monocytogenes* and demonstrated its capability to specifically detect the bacteria with a Biacore 3000 SPR sensor (Hearty et al. 2006). The antibody was immobilized on a carboxymethylated dextran surface by EDC/NHS chemistry. Figure 5.11 shows the SPR sensor response to various concentrations of viable *L. monocytogenes*. The LOD for the direct detection assay was 10^7 cfu/ml.

6.3.4. Other Bacteria

SPR sensors have also been used to detect other species of bacteria, including *Campylobacter jejuni* (Taylor et al. 2006), *Bacillus subtilus* (Perkins and Squirrell 2000), *Helicobacter pylori* (Nishimura et al. 2000), *Legionella pneumophila* (Oh et al. 2003a), *Yersinia enterocolitica* (Oh et al. 2005a), *Staphylococcus aureus* (Subramanian, Irudayaraj, and Ryan 2006b), and *Vibrio cholerae* O1 (Jyoung et al. 2006). Perkins et al. in 2000 coupled a custom SPR sensor to a light scattering sensor to improve the detection of *Bacillus subtilus* spores compared to conventional SPR (Perkins and Squirrell 2000). They then compared the result to that obtained using a Biacore 2000 sensor. The study concluded that spore suspensions that were readily detectable by light scattering at 10^7 spores/ml did not produce significant SPR sensor responses. In 2003 and 2005, Oh et al. detected *L. pneumophila* (Oh et al. 2003a) and *Y. enterocolitica* (Oh et al. 2005a), respectively. The method was similar to their study of *E. coli* O157:H7. Both studies had the same detection range of 10^2–10^7 cfu/ml, which is also the same for previous studies from their group. As with previous studies, no amplification or treatment method is reported. In 2006, Subramanian et al. compared SAM surfaces from OEG terminated mono-

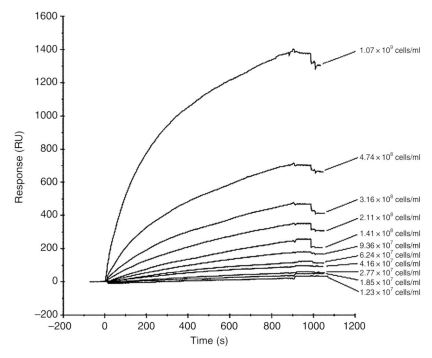

Figure 5.11. Overlayed binding profile for increasing concentrations of *L. monocytogenes* cells. *L. monocytogenes* cells were passed over the mAb2B3-CM5 sensor chip surface at a flow rate of 1 μl/min for 15 min. The 'bound' response levels were taken 60 s after the end of the sample injection (reprinted from Hearty et al. (2006), with permission of Elsevier).

and dithiols for the detection of viable *S. aureus* (Subramanian, Irudayaraj, and Ryan 2006b). The dithiol surface immobilized slightly more antibody; however, the monothiol surface had a higher SPR response to the bacteria. Both mono- and dithiol surfaces had the same LOD of 10^7 cfu/ml for direct detection and 10^5 cfu/ml with a sandwich assay. In 2006, Jyoung et al. demonstrated an LOD of 3.7×10^5 cfu/ml for *V. cholerae* O1 using a similar method to that of the study of *E. coli* by Oh et al. in 2002 (Jyoung et al. 2006).

6.3.5. Detection of Multiple Bacteria

A majority of studies relating to the SPR detection of bacteria have focused on improving detection limits and proving specificity for a single species of bacteria. Expanding SPR sensors to multi-channel or array formats would enable the high-throughput detection of multiple species of bacteria simultaneously. Two studies have been published to demonstrate the detection of four different species of bacteria on a single SPR sensor chip.

In 2005, Oh et al. reported the detection of *E. Coli* O157:H7, *S. typhimurium*, *L. pneumophila*, and *Y. enterocolitica* on a multi-channel SPR sensor (Oh et al. 2005b). Different monoclonal antibodies were bound to a protein G immobilized surface in each of four flow channels. The four individual bacteria were then flowed in succession across the sensing channels at 10^5 cfu/ml. Good specificity was demonstrated to the individual bacteria.

In 2006, Taylor et al. demonstrated the quantitative and simultaneous detection of *E. coli* O157:H7, *S. choleraesuis*, *L. monocytogenes*, and *C. jejuni* in buffer and apple juice using a custom-built 8-channel SPR sensor (Taylor et al. 2006). The bacterial samples were treated by heat-killing and ultrasonicating prior to detection. Figure 5.12 shows the sandwich assay detection curves of amplification antibody response versus analyte concentration for the individual species of bacteria in buffer at pH 7.4, apple juice at native pH 3.7, and apple juice at adjusted pH 7.4, as well as in a mixture of all four species of bacteria in buffer. The detection of individual species of bacteria in a mixture of all four species correlated well with the detection of the individual species of bacteria in buffer. Adjusting the pH of apple juice from the native pH of 3.7 to a physiological pH of 7.4 increased the sensor response in all cases. The increase in sensor response was attributed to the pH-dependence of the antibody-antigen binding. The difference of the sensor response in apple juice from that in buffer was attributed to the matrix effects (e.g., sugars in apple juice), but the limits of detection were similar for all cases. In buffer the LOD for *E. coli* was 1.4×10^4 cfu/ml, *S. choleraesuis* was 4.4×10^4 cfu/ml, *C. jejuni* was 1.1×10^5 cfu/ml, and *L. monocytogenes* was 3.5×10^3 cfu/ml.

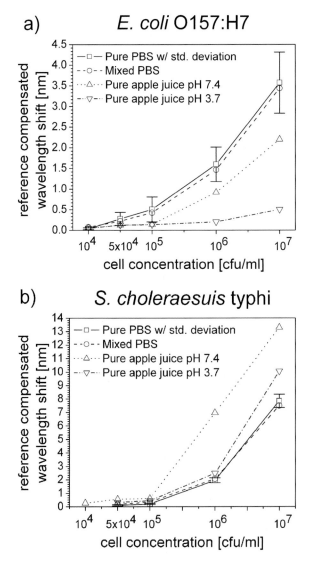

Figure 5.12. SPR resonant wavelength shift vs. concentration of bacteria for the detection of (a) *E. coli* O157:H7, (b) *S. choleraesuis* serotype typhimurium,

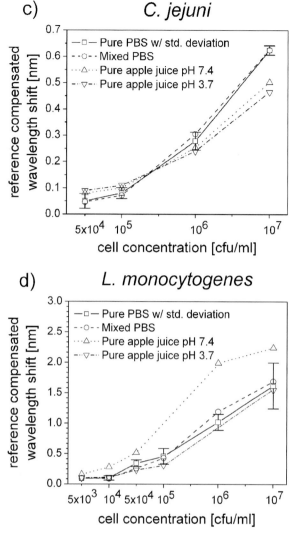

Figure 5.12. *continued* (c) *C. jejuni*, and (d) *L. monocytogenes* in various matrices. Samples contain either one bacterial species in buffer (pure PBS), all four species of bacteria at equal concentrations in buffer (mixed PBS), one bacterial species in apple juice at adjusted pH 7.4 (pure apple juice 7.4), or one bacterial species in apple juice at native pH 3.7 (pure apple juice 3.7). The error bars shown for pure PBS data represent standard deviations for at least five detections at each concentration (reprinted from Taylor et al. (2006), with permission of Elsevier).

7. Genetic Markers

Genetic markers are the known oligonucleotide sequences that can be correlated to identifying characteristics of bacteria. Detection of genetic markers has been successfully used to identify bacteria in clinical (Gouvea et al. 1990; Ramotar et al. 1995) and food samples (Vanpoucke 1990). Most detection assays for oligonucleotides use polymerase chain reaction (PCR) to amplify the specific target sequences. PCR enzymatically replicates oligonucleotides at an exponential rate, thus creating thousands of identical copies from a small amount of starting material. Incorporating PCR amplification into detection assays has enabled detection of a single oligonucleotide from the starting material. SPR sensors have been demonstrated as a real-time, label-free, and high-throughput method for the detection of oligonucleotides and have been used to identify bacteria by the detection of oligonucleotides from unamplified total cellular RNA and PCR-amplified RNA and DNA.

In 1999, Kia et al. demonstrated SPR (Biacore 2000) detection of oligonucleotides from asymmetric PCR amplification of an *E. coli* O157:H7 genetic marker for the verotoxin-2 subunit A gene (Kai et al. 1999). A biotinylated DNA probe sequence was immobilized on a streptavidin coated surface. The asymmetric PCR product, a double stranded target DNA with a single stranded target probe at the 3' terminus, was detected with a LOD of 1.5×10^{-7} M DNA (LOD for bacteria not reported). In 2000, Kia et al. demonstrated the detection of *E. coli* O157:H7 in stool samples using the same method, reporting a LOD of 10^2 cfu/0.1 g stool sample (Kai et al. 2000). In 2000, Miyachi et al. used the same SPR sensor method to detect both the verotoxin-2 subunit A gene from *E. coli* O157:H7 and the *invA* gene, a virulence determinant for *Salmonella*, from both *S. enteritidis* and *S. typhimurium* (Miyachi et al. 2000). The PCR amplification method was modified to use a chimeric RNA-DNA primer rather than a RNAase to degrade the RNA part of the double stranded PCR product to produce the double stranded DNA with a single stranded target probe at the 3' terminus.

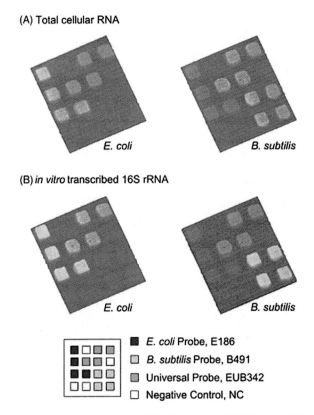

Figure 5.13. SPR images showing the representative hybridization of total RNA isolated from a cell culture (A) or in vitro transcribed rRNA (B) onto DNA arrays. Hybridization onto the array is indicated by a change in the percent reflectivity of incident light. The pattern used for immobilization of single stranded DNA probes is shown in the legend. A) A 35 mg sample of total RNA from *E. coli* was first exposed to the array for 1 h (top). After measurements, the array was denatured using 8 M urea. The experiment was repeated with the same amount of total RNA from *B. subtilis* (bottom). B) A 50 mg sample of in vitro transcribed 16S ribosomal RNA from *E. coli* was first exposed to the array for 1 h (top). After denaturation, the experiment was repeated with the same amount of in vitro transcribed 16S rRNA from *B. subtilis*. The array used was hybridized and denatured more than 20 times before the data shown were logged. A 4 × 4 section of the same 10 × 12 array is shown for all images in the figure (reprinted from Nelson et al. (2002), with permission from Blackwell Publishing).

In 2002, Nelson et al. demonstrated the label-free detection of 16s ribosomal RNA (rRNA) using SPR imaging of a DNA array (Nelson et al. 2002). The array contained DNA probes designed to be complementary to 16s rRNA of *E. coli*, *Bacillus subtilis*, and a highly conserved sequence found in most bacteria. The DNA array was produced by a multistep procedure, using UV-photopatterning techniques to produce spots on a SAM with a reactive functional end group for linking 5' thiol-modified DNA probes. The area surrounding the DNA spots was subsequently modified with an OEG group to provide non-fouling characteristics to the background. Figure 5.13 shows the results from the detection of total cellular rRNA or PCR-enriched rRNA from both *E. coli* and *B. subtilis* on a 4×4 spot section of a 10×12 DNA array. The surface exhibited good specificity and could be regenerated more than 20 times. A detection limit of $0.2\,\mu g/ml$ rRNA was determined for fragmented *E. coli* total cellular RNA.

8. Antibody Biomarkers

Antibodies are soluble proteins, produced by the humoral immune response of an animal, that identify and neutralize foreign agents like bacteria by specifically binding to antigens that are unique to the target. Testing animals for antibodies that bind to bacterial antigens can establish a history of infection. Antibodies to infectious agents can often persist for a long period of time after infection, and serological methods can be used to diagnose infections in humans as well as identify risks in food sources. SPR sensors can be used to test sera and blood for antibodies to bacteria by immobilizing antigens from infectious agents on a non-fouling surface and then detecting antibodies in the samples.

The direct detection of animal sera antibodies against *S. enteritidis* (H:g,m flegellin) and *S. typhimurium* (H:i and H:1,2 flagellins) was carried out using a Biacore 3000 sensor (Jongerius-Gortemaker et al. 2002). The antigens, expressed as fusion proteins in *E. coli*, were purified and immobilized using EDC/NHS chemistry on a dextran surface. The chicken sera were diluted to 2.5% in buffer to reduce non-specific binding. The method positively differentiated the sera of chickens infected by *S. enteritidis* and *S. typhimurium* from the sera of chickens with no history of infection or those infected with *S. infantis*, *S. pullorum*, or *S. gallinarum*.

A SPR sensor assay was developed to detect antibodies in egg yolk that would indicate infection by *S. enterica* serovar *enteritidis* or screen vaccination efficacy in layer hens (Thomas et al. 2006). The sensor surface was coated with lipopolysaccharide (LPS) O antigen isolated from *S. enteritidis*. The egg yolk was diluted to 20% in buffer to reduce non-specific binding. The method was used to analyze 163 egg yolk and combined egg white and yolk samples from chickens exposed to 10^8 cfu *S. enteritidis* and 90 egg yolk and combined egg white and yolk samples from uninfected chickens. The method gave a positive detection for chickens infected with *S. enteritidis*, *S. gallinarum*, or *S. typhimurium*, while producing negative responses for uninfected control samples and chickens infected with *S. infantis*. The method was unable to differentiate infections from serovars *enteritidis*, *gallinarum*, and *typhimurium*, which share the O 9 and O 12 somatic antigens.

9. Conclusions and Future Perspectives

SPR sensors have been used to detect bacteria in both clinical and food-related samples. In this review, we discussed the application of SPR sensors to the detection of bacterial cells, genetic markers, and antibody markers. In addition, SPR sensors have been used to detect numerous metabolic products of bacteria, including *Staphylococcal enterotoxin* A (Evenson et al. 1988; Rasooly and Rasooly 1999; Medina 2006) and B (Nedelkov, Rasooly, and Nelson 2000; Homola et al. 2002; Naimushin et al. 2002; Slavik, Homola, and Brynda 2002;

Naimushin et al. 2003; Nedelkov and Nelson 2003; Medina 2005), Botulinum neurotoxins, Cholera toxin (Phillips et al. 2006), *Clostridium perfringens* β-toxin (Hsieh et al. 1998), *E. coli* shiga-like toxin (Kanda et al. 2005), and *E. coli* enterotoxin (Spangler et al. 2001). They have also been applied to the fundamental studies of bacterial interactions with surfaces (Rich and Myszka 2005a; 2005b) and the monitoring of enzyme expression as a continuous online sensor for bioreactor cultivations (Vostiar, Tkac, and Mandenius 2005). However, these applications were considered outside the scope of this review and not discussed at length.

SPR sensors are a cutting-edge technology capable of label-free and real-time monitoring of reactions, which enables the quantification of macromolecular interaction properties such as kinetics, affinity, specificity, and concentration. Trends in SPR instrumentation are pushing towards automation, miniaturization, and high-throughput capabilities of sensor systems. Automation and miniaturization are necessary to improve the feasibility of SPR sensors for food-related and environmental monitoring in the field. Development of SPR sensors capable of parallelized quantitative analysis of large numbers of molecular interactions will have a major impact on medical diagnostics and drug development. These capabilities will ultimately decrease the labor and cost of the SPR biosensor systems and enable the even deeper penetration of this technology.

Acknowledgments

The authors acknowledge the financial support of the National Science Foundation (CBET-0528605), Grant Agency of the Academy of Sciences of the Czech Republic (IAA400500507), Academy of Sciences of the Czech Republic (KAN200670701), and the United States Food and Drug Administration.

References

Adam P, Dostalek J and Homola J (2006) Multiple surface plasmon spectroscopy for study of biomolecular systems. Sensors and Actuators B-Chemical 113:774–781

Angenendt P, Glokler J, Murphy D, Lehrach H and Cahill DJ (2002) Toward optimized antibody microarrays: a comparison of current microarray support materials. Analytical Biochemistry 309:253–260

Asano K, Ono A, Hashimoto S, Inoue T and Kanno J (2004) Screening of endocrine disrupting chemicals using a surface plasmon resonance sensor. Analytical Sciences 20:611–616

Balamurugan S, Obubuafo A, Soper SA, McCarley RL and Spivak DA (2006) Designing highly specific biosensing surfaces using aptamer monolayers on gold. Langmuir 22:6446–6453

Boardman AD (1982) Electromagnetic surface modes. Wiley, Chichester, New York

Bokken G, Corbee RJ, van Knapen F and Bergwerff AA (2003) Immunochemical detection of Salmonella group B, D and E using an optical surface plasmon resonance biosensor. Fems Microbiology Letters 222:75–82

Boozer C, Yu QM, Chen SF, Lee CY, Homola J, Yee SS and Jiang SY (2003) Surface functionalization for self-referencing surface plasmon resonance (SPR) biosensors by multi-step self-assembly. Sensors and Actuators B-Chemical 90:22–30

Boozer C, Ladd J, Chen SF, Yu Q, Homola J and Jiang SY (2004) DNA directed protein immobilization on mixed ssDNA/oligo(ethylene glycol) self-assembled monolayers for sensitive biosensors. Analytical Chemistry 76:6967–6972

Boozer C, Ladd J, Chen SF and Jiang ST (2006) DNA-directed protein immobilization for simultaneous detection of multiple analytes by surface plasmon resonance biosensor. Analytical Chemistry 78:1515–1519

Born M and Wolf E (1999) Principles of optics : electromagnetic theory of propagation, interference and diffraction of light. Cambridge University Press, Cambridge

Center for Food Safety and Nutrition and United States Food and Drug Administraton (2007) "Bad Bug Book" Foodborne pathogenic microorganisms and natural toxins handbook. http://www.cfsan.fda.gov/~mov/badbug.zip

Chen SF, Liu LY, Zhou J and Jiang SY (2003) Controlling antibody orientation on charged self-assembled monolayers. Langmuir 19:2859–2864

Chen SF, Zheng J, Li LY and Jiang SY (2005) Strong resistance of phosphorylcholine self-assembled monolayers to protein adsorption: Insights into nonfouling properties of zwitterionic materials. Journal of the American Chemical Society 127:14473–14478

Choi JW, Park KW, Lee DB, Lee W and Lee WH (2005) Cell immobilization using self-assembled synthetic oligopeptide and its application to biological toxicity detection using surface plasmon resonance. Biosensors & Bioelectronics 20:2300–2305

Dostalek J, Homola J and Miler M (2005) Rich information format surface plasmon resonance biosensor based on array of diffraction gratings. Sensors and Actuators B-Chemical 107:154–161

Evenson ML, Hinds MW, Bernstein RS and Bergdoll MS (1988) Estimation of Human Dose of Staphylococcal Enterotoxin-a from a Large Outbreak of Staphylococcal Food Poisoning Involving Chocolate Milk. International Journal of Food Microbiology 7:311–316

Fleminger G, Hadas E, Wolf T and Solomon B (1990) Oriented Immobilization of Periodate-Oxidized Monoclonal-Antibodies on Amino and Hydrazide Derivatives of Eupergit-C. Applied Biochemistry and Biotechnology 23:123–137

Fratamico PM, Strobaugh TP, Medina MB and Gehring AG (1998) Detection of Escherichia coli O157 : H7 using a surface plasmon resonance biosensor. Biotechnology Techniques 12:571–576

Fratamico PM, Strobaugh TP, Medina MB and Gehring AG (1999) A Surface Plasmon Resonance Biosensor for Real-Time Immunologic Detection of *Escherichia Coli* O157:H7. In: Tunick M, Fratamico PM, Palumbo SA (eds) New Techniques in the Analysis of Foods. Kluwer Academic, New York, pp 103–111

Gobi KV and Miura N (2004) Highly sensitive and interference-free simultaneous detection of two polycyclic aromatic hydrocarbons at parts-per-trillion levels using a surface plasmon resonance immunosensor. Sensors and Actuators B-Chemical 103:265–271

Gobi KV, Tanaka H, Shoyama Y and Miura N (2005) Highly sensitive regenerable immunosensor for label-free detection of 2,4-dichlorophenoxyacetic acid at ppb levels by using surface plasmon resonance imaging. Sensors and Actuators B-Chemical 111:562–571

Gouvea V, Glass RI, Woods P, Tanguchi K, Clark HF, Forrester B and Fang ZY (1990) Polymerase Chain-Reaction Amplification and Typing of Rotavirus Nucleic-Acid from Stool Specimens. Journal of Clinical Microbiology 28:276–282

Harris RD, Luff BJ, Wilkinson JS, Piehler J, Brecht A, Gauglitz G and Abuknesha RA (1999) Integrated optical surface plasmon resonance immunoprobe for simazine detection. Biosensors & Bioelectronics 14:377–386

Hearty S, Leonard P, Quinn J and O'Kennedy R (2006) Production, characterisation and potential application of a novel monoclonal antibody for rapid identification of virulent Listeria monocytogenes. Journal of Microbiological Methods 66:294–312

Homola J (1995) Optical-Fiber Sensor-Based on Surface-Plasmon Excitation. Sensors and Actuators B-Chemical 29:401–405

Homola J, Pfeifer P and Brynda E (1997) *Optical biosensing using surface plasmon resonance spectroscopy.* Proc. SPIE 3105

Homola J, Lu HB and Yee SS (1999) Dual-channel surface plasmon resonance sensor with spectral discrimination of sensing channels using dielectric overlayer. Electronics Letters 35:1105–1106

Homola J, Dostalek J, Chen SF, Rasooly A, Jiang SY and Yee SS (2002) Spectral surface plasmon resonance biosensor for detection of staphylococcal enterotoxin B in milk. International Journal of Food Microbiology 75:61–69

Homola J (2003) Present and future of surface plasmon resonance biosensors. Analytical and Bioanalytical Chemistry 377:528–539

Homola J, Vaisocherová H, Dostálek J and Piliarik M (2005) Multi-analyte surface plasmon resonance biosensing. Methods 37:26–36

Hsieh HV, Stewart B, Hauer P, Haaland P and Campbell R (1998) Measurement of Clostridium perfringens beta-toxin production by surface plasmon resonance immunoassay. Vaccine 16:997–1003

Jackson WC, Kuckuck F, Edwards BS, Mammoli A, Gallegos CM, Lopez GP, Buranda T and Sklar LA (2002) Mixing small volumes for continuous high-throughput flow cytometry: Performance of a mixing Y and peristaltic sample delivery. Cytometry 47:183–191

Jongerius-Gortemaker BGM, Goverde RLJ, van Knapen F, and Bergwerff AA (2002) Surface plasmon resonance (BIACORE) detection of serum antibodies against Salmonella enteritidis and Salmonella typhimurium. Journal of Immunological Methods 266:33–44

Jorgenson RC and Yee SS (1993) A Fiber-Optic Chemical Sensor Based on Surface Plasmon Resonance. Sensors and Actuators B 12:213–220

Jory MJ, Vukusic PS and Sambles JR (1994) Development of a Prototype Gas Sensor Using Surface-Plasmon Resonance on Gratings. Sensors and Actuators B-Chemical 17:203–209

Jory MJ, Bradberry GW, Cann PS and Sambles JR (1995) A Surface-Plasmon-Based Optical Sensor Using Acous-tooptics. Measurement Science & Technology 6:1193–1200

Jyoung JY, Hong SH, Lee W and Choi JW (2006) Immunosensor for the detection of Vibrio cholerae O1 using surface plasmon resonance. Biosensors & Bioelectronics 21:2315–2319

Kai E, Sawata S, Ikebukuro K, Iida T, Honda T and Karube I (1999) Detection of PCR products in solution using surface plasmon resonance. Analytical Chemistry 71:796–800

Kai E, Ikebukuro K, Hoshina S, Watanabe H and Karube I (2000) Detection of PCR products of Escherichia coli O157 : H7 in human stool samples using surface plasmon resonance (SPR). Fems Immunology and Medical Microbiology 29:283–288

Kanda V, Kitov P, Bundle DR and McDermott MT (2005) Surface plasmon resonance imaging measurements of the inhibition of Shiga-like toxin by synthetic multivalent inhibitors. Analytical Chemistry 77:7497–7504

Karlsson R and Stahlberg R (1995) Surface-Plasmon Resonance Detection and Multispot Sensing for Direct Monitoring of Interactions Involving Low-Molecular-Weight Analytes and for Determination of Low Affinities. Analytical Biochemistry 228:274–280

Koubova V, Brynda E, Karasova L, Skvor J, Homola J, Dostalek J, Tobiska P and Rosicky J (2001) Detection of foodborne pathogens using surface plasmon resonance biosensors. Sensors and Actuators B-Chemical 74:100–105

Kukanskis K, Elkind J, Melendez J, Murphy T, Miller G and Garner H (1999) Detection of DNA hybridization using the TISPR-1 surface plasmon resonance biosensor. Analytical Biochemistry 274:7–17

Ladd J, Boozer C, Yu QM, Chen SF, Homola J and Jiang S (2004) DNA-directed protein immobilization on mixed self-assembled monolayers via a Streptavidin bridge. Langmuir 20:8090–8095

Leonard P, Hearty S, Quinn J and O'Kennedy R (2004) A generic approach for the detection of whole Listeria monocytogenes cells in contaminated samples using surface plasmon resonance. Biosensors & Bioelectronics 19:1331–1335

Leonard P, Hearty S, Wyatt G, Quinn J and O'Kennedy R (2005) Development of a surface plasmon resonance-Based immunoassay for Listeria monocytogenes. Journal of Food Protection 68:728–735

Liedberg B, Lundstrom I and Stenberg E (1993) Principles of Biosensing with an Extended Coupling Matrix and Surface-Plasmon Resonance. Sensors and Actuators B-Chemical 11:63–72

Lotierzo M, Henry OYF, Piletsky S, Tothill I, Cullen D, Kania M, Hock B and Turner APF (2004) Surface plasmon resonance sensor for domoic acid based on grafted imprinted polymer. Biosensors & Bioelectronics 20:145–152

Matsubara K, Kawata S and Minami S (1988) Optical Chemical Sensor Based on Surface-Plasmon Measurement. Applied Optics 27:1160–1163

Medina MB (2005) A biosensor method for a competitive immunoassay detection of staphylococcal enterotoxin B (SEB) in milk. Journal of Rapid Methods and Automation in Microbiology 13:37–55

Medina MB (2006) A biosensor method for detection of Staphylococcal enterotoxin A in raw whole egg. Journal of Rapid Methods and Automation in Microbiology 14:119–132

Meeusen CA, Alocilja EC and Osburn WN (2005) Detection of E-coli O157: H7 using a miniaturized surface plasmon resonance biosensor. Transactions of the Asae 48:2409–2416

Minunni M and Mascini M (1993) Detection of Pesticide in Drinking-Water Using Real-Time Biospecific Interaction Analysis (Bia). Analytical Letters 26:1441–1460

Miyachi H, Yano K, Ikebukuro K, Kono M, Hoshina S and Karube I (2000) Application of chimeric RNA-DNA oligonucleotides to the detection of pathogenic microorganisms using surface plasmon resonance. Analytica Chimica Acta 407:1–10

Mouvet C, Harris R, Maciag C, Luff B, Wilkinson J, Piehler J, Brecht A, Gauglitz G, Abuknesha R and Ismail G (1997) Determination of simazine in water samples by waveguide surface plasmon resonance. Analytica Chimia Acta 338:109–117

Mullett WM, Lai EPC and Yeung JM (2000) Surface plasmon resonance-based immunoassays. Methods 22:77–91

Naimushin A, Soelberg S, Bartholomew D, Elkind J and Furlong C (2003) A portable surface plasmon resonance (SPR) sensor system with temperature regulation. Sensors and Actuators B 96:253–260

Naimushin AN, Soelberg SD, Nguyen DK, Dunlap L, Bartholomew D, Elkind J, Melendez J and Furlong CE (2002) Detection of Staphylococcus aureus enterotoxin B at femtomolar levels with a miniature integrated two-channel surface plasmon resonance (SPR) sensor. Biosensors & Bioelectronics 17:573–584

Nedelkov D, Rasooly A and Nelson RW (2000) Multitoxin biosensor-mass spectrometry analysis: a new approach for rapid, real-time, sensitive analysis of staphylococcal toxins in food. International Journal of Food Microbiology 60:1–13

Nedelkov D and Nelson RW (2003). Detection of staphylococcal enterotoxin B via Biomolecular interaction analysis mass spectrometry. Applied and Environmental Microbiology 69:5212–5215

Nelson BP, Liles MR, Frederick KB, Corn RM and Goodman RM (2002) Label-free detection of 16S ribosomal RNA hybridization on reusable DNA arrays using surface plasmon resonance imaging. Environmental Microbiology 4:735–743

Nelson KE, Gamble L, Jung LS, Boeckl MS, Naeemi E, Golledge SL, Sasaki T, Castner DG, Campbell CT and Stayton PS (2001) Surface characterization of mixed self-assembled monolayers designed for streptavidin immobilization. Langmuir 17:2807–2816

Nenninger GC, Clendenning JB, Furlong CE and Yee SS (1998) Sensors and Actuators B 51:38

Nice EC and Catimel B (1999) Instrumental biosensors: new perspectives for the analysis of biomolecular interactions. Bioessays 21:339–352

Nishimura T, Hifumi E, Fujii T, Niimi Y, Egashira N, Shimizu K and Uda T (2000) Measurement of Helicobacter pylori using anti its urease monoclonal antibody by surface plasmon resonance. Electrochemistry 68:916–919

Oh BK, Kim YK, Bae YM, Lee WH and Choi JW (2002) Detection of Escherichia coli O157 : H7 using immunosensor based on surface plasmon resonance. Journal of Microbiology and Biotechnology 12:780–786

Oh BK, Kim YK, Lee W, Bae YM, Lee WH and Choi JW (2003a) Immunosensor for detection of Legionella pneumophila using surface plasmon resonance. Biosensors & Bioelectronics 18:605–611

Oh BK, Lee W, Lee WH and Choi JW (2003b) Nano-scale probe fabrication using self-assembly technique and application to detection of Escherichia coli O157: H7. Biotechnology and Bioprocess Engineering 8:227–232

Oh BK, Kim YK, Park KW, Lee WH and Choi JW (2004a) Surface plasmon resonance immunosensor for the detection of Salmonella typhimurium. Biosensors & Bioelectronics 19:1497–1504

Oh BK, Lee W, Kim YK, Lee WH and Choi JW (2004b) Surface plasmon resonance immunosensor using self-assembled protein G for the detection of Salmonella paratyphi. Journal of Biotechnology 111:1–8

Oh BK, Lee W, Chun BS, Bae YM, Lee WH and Choi JW (2005a) Surface plasmon resonance immunosensor for the detection of Yersinia enterocolitica. Colloids and Surfaces a-Physicochemical and Engineering Aspects 257–58:369–374

Oh BK, Lee W, Chun BS, Bae YM, Lee WH and Choi JW (2005b) The fabrication of protein chip based on surface plasmon resonance for detection of pathogens. Biosensors & Bioelectronics 20:1847–1850

Perkins EA and Squirrell DJ (2000) Development of instrumentation to allow the detection of microorganisms using light scattering in combination with surface plasmon resonance. Biosensors & Bioelectronics 14:853–859

Phillips KS, Han JH, Martinez M, Wang ZZ, Carter D and Cheng Q (2006) Nanoscale glassification of gold substrates for surface plasmon resonance analysis of protein toxins with supported lipid membranes. Analytical Chemistry 78:596–603

Piliarik M, Homola J, Manikova Z and Ctyroky J (2003) Surface plasmon resonance sensor based on a single-mode polarization-maintaining optical fiber. Sensors and Actuators B-Chemical 90:236–242

Raether H (1988) Surface-Plasmons on Smooth and Rough Surfaces and on Gratings. Springer Tracts in Modern Physics 111:1–133

Ramotar K, Waldhart B, Church D, Szumski R and Louie TJ (1995) Direct-Detection of Verotoxin-Producing Escherichia-Coli in Stool Samples by Pcr. Journal of Clinical Microbiology 33:519–524

Rasooly L and Rasooly A (1999) Real time biosensor analysis of Staphylococcal enterotoxin A in food. International Journal of Food Microbiology 49:119–127

Rich RL and Myszka DG (2005a) Survey of the year 2003 commercial optical biosensor literature. Journal of Molecular Recognition 18:1–39

Rich RL and Myszka DG (2005b) Survey of the year 2004 commercial optical biosensor literature. Journal of Molecular Recognition 18:431–478

Shankaran DR, Matsumoto K, Toko K and Miura N (2006). Development and comparison of two immunoassays for the detection of 2,4,6-trinitrotoluene (TNT) based on surface plasmon resonance. Sensors and Actuators B-Chemical 114:71–79

Shimomura M, Nomura Y, Zhang W, Sakino M, Lee KH, Ikebukuro K and Karube I (2001) Simple and rapid detection method using surface plasmon resonance for dioxins, polychlorinated biphenylx and atrazine. Analytica Chimica Acta 434:223–230

Slavik R, Homola J and Brynda E (2002) A miniature fiber optic surface plasmon resonance sensor for fast detection of staphylococcal enterotoxin B. Biosensors & Bioelectronics 17:591–595

Slavik R and Homola J (2007) Ultra-high resolution long range surface plasmon-based sensor. Sensors and Actuators B: Chemical 123:10–12

Spangler BD, Wilkinson EA, Murphy JT and Tyler BJ (2001) Comparison of the Spreeta (R) surface plasmon resonance sensor and a quartz crystal microbalance for detection of Escherichia coli heat-labile enterotoxin. Analytica Chimica Acta 444:149–161

Su XL and Li Y (2005) Surface plasmon resonance and quartz crystal microbalance immunosensors for detection of Escherichia coli O157 : H7. Transactions of the Asae 48:405–413

Subramanian A, Irudayaraj J and Ryan T (2006a) A mixed self-assembled monolayer-based surface plasmon immunosensor for detection of E-coli O157: H7. Biosensors & Bioelectronics 21:998–1006

Subramanian A, Irudayaraj J and Ryan T (2006b) Mono and dithiol surfaces on surface plasmon resonance biosensors for detection of Staphylococcus aureus. Sensors and Actuators B-Chemical 114:192–198

Subramanian AS and Irudayaraj JM (2006) Surface plasmon resonance based immunosensing of E. coli O157: H7 in apple juice. Transactions of the Asabe 49:1257–1262

Taylor AD, Yu QM, Chen SF, Homola J and Jiang SY (2005) Comparison of E-coli O157: H7 preparation methods used for detection with surface plasmon resonance sensor. Sensors and Actuators B-Chemical 107:202–208

Taylor AD, Ladd J, Yu QM, Chen SF, Homola J and Jiang SY (2006) Quantitative and simultaneous detection of four foodborne bacterial pathogens with a multi-channel SPR sensor. Biosensors & Bioelectronics 22:752–758

Thirstrup C, Zong W, Borre M, Neff H, Pedersen HC and Holzhueter G (2004) Diffractive optical coupling element for surface plasmon resonance sensors. Sensors and Actuators B-Chemical 100:298–308

Thomas E, Bouma A, van Eerden E, Landman WJM, van Knapen F, Stegeman A and Bergwerff AA (2006) Detection of egg yolk antibodies reflecting Salmonella enteritidis infections using a surface plasmon resonance biosensor. Journal of Immunological Methods 315:68–74

Truesdell RA, Vorobieff PV, Sklar LA and Mammoli AA (2003) Mixing of a continuous flow of two fluids due to unsteady flow. Physical Review E 67

Usami M, Mitsunaga K and Ohno Y (2002) Estrogen receptor binding assay of chemicals with a surface plasmon resonance biosensor. Journal of Steroid Biochemistry and Molecular Biology 81:47–55

Van Gent J, Lambeck PV, Kreuwel HJM, Gerritsma GJ, Sudholter EJR, Reinhoudt DN and Popma TJA (1990) Optimization of a Chemooptical Surface-Plasmon Resonance Based Sensor. Applied Optics 29:2843–2849

Vanpoucke LSG (1990) Salmonella-Tek, a Rapid Screening Method for Salmonella Species in Food. Applied and Environmental Microbiology 56:924–927

Vostiar I, Tkac J and Mandenius CF (2005) Intracellular monitoring of superoxide dismutase expression in an Escherichia coli fed-batch cultivation using on-line disruption with at-line surface plasmon resonance detection. Analytical Biochemistry 342:152–159

Wright JD, Oliver JV, Nolte RJM, Holder SJ, Sommerdijk N and Nikitin PI (1998) The detection of phenols in water using a surface plasmon resonance system with specific receptors. Sensors and Actuators B-Chemical 51:305–310

Yu QM, Chen SF, Taylor AD, Homola J, Hock B and Jiang SY (2005) Detection of low-molecular-weight domoic acid using surface plasmon resonance sensor. Sensors and Actuators B-Chemical 107:193–201

Zhang Z, Chen SF and Jiang SY (2006) Dual-functional biomimetic materials: Nonfouling poly(carboxybetaine) with active functional groups for protein immobilization. Biomacromolecules 7:3311–3315

6

Bacterial Detection Using Evanescent Wave-Based Fluorescent Biosensors

Kim E. Sapsford and Lisa C. Shriver-Lake

Abstract

Detection and identification of bacteria is an important aspect of our world today. Outbreaks of pathogenic bacteria, either occurring naturally in food or possibly being used as weapons by bioterrorists for contamination of food, air, and water, are constantly in the news. Identifying the specific bacteria responsible for these outbreaks and their potential source is of great importance. Biosensors, specifically evanescent wave-based fluorescence biosensors, are evolving to meet these challenges. In evanescent wave sensors, light is launched into an optical waveguide at such an angle that the light is internally reflected completely at the interface of the waveguide and the surrounding medium. An electromagnetic wave is generated at the surface that penetrates 100–200 nm into the surrounding medium (air, buffer). Fluorophores that are bound within this region by the target bacteria and a recognition molecule are excited. Many different configurations of this method have been developed and will be discussed in this chapter.

1. Introduction

Bacteria are found throughout our world: in our environment (air, soil, and water), our food (plant and animal), and even in our bodies. Most of the time the bacteria are useful, but several strains and their associated toxins are known to be hazardous to our health and the quality of our food supply. Bacteria cannot be observed by the unaided eye, smelt, or tasted, and their toxic effects are not normally immediately identifiable. Bacteria can enter our bodies through ingestion of contaminated food and water, through inhalation, and through cuts in the skin. While most healthy individuals have varying degrees of illness following exposure, the elderly, young children and the immune compromised are susceptible to extreme illness and possibly death. Today, concerns also exist over the intentional contamination of food, water, and air by terrorist groups. Typical methods for detection of bacteria include culturing, followed by visual examination of stained cells or immunoassays in solutions. The associated bacterial toxins are usually identified by ELISA-based immunoassays or mouse toxicity assays. In most cases, it takes a day or more to identify the organism responsible for a health hazard. Most of the detection methods employed today are for a single analyte, i.e., either bacteria or toxins, but

Kim E. Sapsford • George Mason University, 10910 University Blvd, MS 4E3, Manassas, VA 20110 USA, Center for Bio/Molecular Science and Engineering, Code 6900, U.S. Naval Research Laboratory, Washington, DC 20375-5320 USA. **Lisa C. Shriver-Lake** • Center for Bio/Molecular Science and Engineering, Code 6900, U.S. Naval Research Laboratory, Washington, DC 20375-5320 USA.

M. Zourob et al. (eds.), *Principles of Bacterial Detection: Biosensors, Recognition Receptors and Microsystems,*
© Springer Science+Business Media, LLC 2008

not both. Identification of harmful bacteria and their associated toxins to prevent illness is critical to the food industry (plant and animal), homeland security, and in clinical diagnosis and therapeutics. The ability to perform rapid identification of these pathogens from a variety of clinical, food, and environmental samples for multiple bacteria and/or their toxins is strongly desired in many arenas.

Taking cues from nature, with its exquisite specificity and sensitivity, biosensors are detection systems that use a biological recognition molecule coupled with a transducer to record the interaction of the recognition molecule with its target analyte (Rogers 2006; Scheller and Schmid 1992; Turner et al. 1987; Wise and Wingard Jr 1991). The biorecognition molecule can provide both very narrow specificity, as in DNA-based systems, and very broad specificity, such as cell-based sensors. The majority of today's biosensors use antibodies or DNA probes for bacterial detection, although systems using carbohydrates and protein arrays are currently being developed. In Part III of the book, extensive descriptions of different types of biorecognition molecules will be presented.

As broad as the choices for biorecognition molecules are, so too are the choices for signal transduction. Electrochemical techniques (amperometric, potentiometric, impedimetric) previously investigated for chemical sensors, coupled to enzymes and antibodies, were employed by the first biosensors (Wingard Jr and Ferrance 1991). These types of biosensors are still being investigated for the detection of bacteria (Boyaci et al. 2005; Muhammad-Tahir and Alocilja 2003; Radke and Alocilja 2005a; Radke and Alocilja 2004; Radke and Alocilja 2005b; Ruan et al. 2002; Susmel et al. 2003). Areas of concern with electrochemical transduction include understanding the response of the electrode interfaces and avoiding electrical interference. Electrochemical biosensors will be discussed in greater detail in Chapters 13–16. In piezoelectric-based systems, a potential is generated when the piezoelectric material is stressed, thereby producing resonant waves. Interaction of the analyte with the biorecognition molecule causes a change in the frequency of these resonant waves, which can be measured. Basic types of piezoelectric sensors are bulk-wave surface acoustic wave (SAW), and piezoelectric cantilevers (Araya-Kleinsteuber et al. 2006; Berkenpas et al. 2006; Campbell and Mutharasan 2005; Deobagkar et al. 2005; Lin and Tsai 2003; Zhao et al. 2005). However, changes in mass due to nonspecific adsorption can be an issue with these types of sensors.

In addition to changes in electrical properties, other transducers are based on changes in light properties. Optical-based transduction systems encompass a wide variety of sensors based on the change in a property of light upon interaction of the recognition molecule and the target analyte. These properties include amplitude (adsorption), wavelength (fluorescence, surface plasmon), polarization, and time dependence (time-resolved fluorescence). Gauglitz (2005) published a detailed review of direct optical sensor principles (Gauglitz 2005).

In many optical sensor arrangements, light is launched into an optical waveguide at greater than a critical angle, such that the light is reflected at the interface of the waveguide and a surrounding matrix of lower refractive index. The critical angle is determined by the refractive indices of the waveguide and the surrounding media (usually water, air, or buffer). When the critical angle is exceeded, the light is considered to be totally internally reflected. While the light appears to reflect at the interface, an electromagnetic wave in the lower refractive index material is generated which is referred to as the evanescent wave. The power in this wave decreases exponentially away from the interface and its depth is on the order of 100–200 nm. A major advantage of this narrow sensing range is the lower response due to bulk effects from the sample solution.

There are several different types of sensors that use the change in refractive index within the evanescent field for transduction. Figure 6.1 illustrates three different measurement techniques: attenuated reflectance (ATR), total internal reflectance (TIRF), and surface plasmon resonance (SPR), used with evanescent wave sensors. SPR has been a popular transduction method for biosensors measuring bacteria and their associated toxins (Bergwerff and van

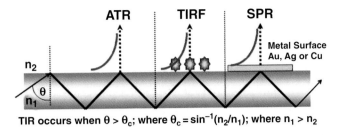

Figure 6.1. Biosensing formats that use evanescent wave technology (adapted from Gauglitz 2005).

Knapen 2006; Leonard et al. 2005; Medina 2006; Pattnaik and Srivastav 2006; Subramanian et al. 2006). A detailed description of SPR biosensing, including commercial instruments, is presented in Chapter 5. Bacterial assays have also been developed using attenuated reflectance (ATR; Nivens et al. 1993), resonant mirrors (Lathrop et al. 2003), grating couplers (Horvath et al. 2003; O'Brien et al. 2000), and interferometry. Several of these evanescent sensors be discussed in detail in Chapters 6–9 of this book. This chapter will focus on evanescent wave-based fluorescence detection.

Herron and coworkers list three primary advantages of employing fluorophores and evanescent wave excitation for detection (Herron et al. 2006). First, enhanced sensitivity is achieved due to the light intensity in the evanescent field versus that of transmitted light. Second, the selectivity of the evanescent field surface reduces matrix effects due to the limited range of the evanescent wave (100–200 nm). Finally, because of the limited penetration depth of the evanescent field, precipitation, filtration, or wash steps, normally performed prior to sample analysis to reduce matrix effects, are not required, thereby reducing assay time and complexity. The fluorophores can be intrinsic or extrinsic in nature, although introduction of an extrinsic fluorophore may be considered a disadvantage due to the increased steps required for the reagent preparation/assay.

Both bacterial cells and the bacterial protein toxins they produce contain a number of intrinsic fluorophores such as the amino acid residues tryptophan, tyrosine, and phenylalanine; and the fluorescent cofactors flavin adenine dinucleotide (FAD) and β-nicotinamide adenine dinucleotide phosphate (NADPH; Lakowicz 1999). However, these intrinsic fluorophores can have fairly low quantum yields and require excitation in the UV-to-low-visible region leading to high background fluorescence signals and large excitation sources. The majority of the studies highlighted in this chapter make use of extrinsic fluorophores ($\lambda_{EXCITATION}$ 450–700 nm), of which there are numerous possibilities available and numerous methods for biomolecular attachment. Commonly used fluorophores include fluoroscein, rhodamine, cyanine dyes (Amersham Biosciences 2006), AlexaFluor (Invitrogen Corporation 2006a), HiLyte (AnaSpec Inc 2006), and Dylight (Pierce Biotechnology Inc 2006). These dyes come with reactive groups that permit simple attachment to a tracer recognition probe. The Invitrogen handbook, *A Guide to Fluorescent Probes and Labeling Technologies*, available free on the Molecular Probes website is a great resource for choosing appropriate extrinsic fluorescent probes and suitable bioconjugation techniques (Invitrogen Corporation 2006b). Commercial availability of fluorescently-labeled recognition molecules has grown significantly over the last ten years. Excitation of these fluorophores using small light sources (i.e., diode lasers) while collecting the emission in a wavelength region with low intrinsic fluorescence from the sample matrix has greatly advanced the development of biosensors. Careful choice of the appropriate fluorophore and sensing arrangements, such as fluorescence resonance energy transfer (FRET), can provide larger signal-to-noise ratios, improving sensitivity and leading to the expansion of biosensing capabilities (Sapsford et al. 2004a).

2. Current State of Bacterial Fluorescent TIRF Biosensors

There are a number of evanescent wave-based fluorescent biosensors which detect bacteria both directly by measuring cells, and indirectly through the recognition of bacterial protein toxins. Surfaces which use the principle of TIRF can be non-planar, such as fiber optics and capillaries, or planar in origin (Sapsford et al. 2004d). Each type of biosensor will be addressed either in the following sections or in later chapters.

2.1. Non-Planar Substrates

2.1.1. Fiber Optics

There are various fiber optic-based methods that employ fluorescence for the detection of bacterial cells and bacterial toxins that have been reviewed in a number of papers (Lim 2003; Taitt et al. 2005; Taitt et al. 2002). However, this subject is covered in detail in Chapter 7 within this book and so will not be discussed here.

2.1.2. Capillaries

There are two capillary configurations that utilize the evanescent wave format (Dhadwal et al. 2004; Wolfbeis 1996). In one configuration the inner capillary surface provides evanescent field excitation with detection of the radiating fluorescence emission. Dhadwal et al. (2004) explored this configuration for a nucleic acid-based biosensor. To date, detection of bacteria or their toxins using this format has not been demonstrated. In the other configuration, direct illumination of the fluorophore causes excitation with the detection of fluorescence in the guided modes of the capillary walls. Ligler et al. (2002) used this configuration for the detection of bacterial toxins as did Zhu and coworkers for the detection of *E. coli* O157 (Zhu et al. 2005).

2.2. Planar Substrates

Planar waveguide TIRF-based biosensors as a field of research largely grew out of the area of fiber optics. The waveguides themselves come in a variety of forms, from internal reflection elements (IRE) to integrated optical waveguides (IOWs). When the thickness of the waveguide is much larger than the wavelength of the reflected light, as is the case for the NRL Array Biosensor which uses standard bulk glass microscope slides, the waveguide is referred to as IRE. The excitation light can be coupled into IRE-based waveguides using simple end firing, diffraction gratings, or optical prisms. If the thickness of the waveguide is decreased such that it approaches the wavelength of the incident light, the path length between points of reflection becomes increasingly shorter. At the thickness where the standing waves overlap and interfere, a continuous streak of light appears across the waveguide and the IRE becomes an IOW. IOWs are commonly prepared by depositing an optically transparent thin film of high refractive index material, such as $SiON_x$ and Ta_2O_5, onto the surface of a glass substrate, and light is typically coupled into the waveguide using a diffraction grating or prism.

Planar surfaces offer the possibility of creating patterns of immobilized capture molecules on the surface of the waveguide allowing a single surface to be used in the screening of multiple samples for multiple target analytes. It is this capability for simultaneous multiple-analyte detection that has been the driving force behind the development of a number of technologies in this field, with applications ranging from food safety to homeland security.

2.2.1. NRL Array Biosensor

One of the leaders in the field of planar waveguide-based TIRF-based biosensors for the detection of bacteria and bacterial toxins has been the group led by Dr. Frances Ligler at the Naval Research Laboratory (NRL). The developed system, the NRL Array Biosensor, has demonstrated its utility in a variety of applications and is currently being transitioned into the commercial sector for manufacture. The technical aspects of the NRL Array Biosensor, including optical design and assay development, have been reviewed in a number of articles (Golden et al. 2005a; Golden et al. 2005b; Ligler et al. 2007; Sapsford et al. 2004c; Sapsford et al. 2004d; Taitt et al. 2004a). Briefly, PDMS flow cells are used both to create patterns of immobilized capture agents (typically antibodies) on the surface of glass microscope slides and to run assays for target detection (using antibodies labeled with Cy5 or Alexa Fluor647). Initial assay development and optimization was carried out offline and the final slide imaged using the NRL Array Biosensor. In a number of these studies the simultaneous detection of both bacterial cells and bacterial toxins in multiple samples has been demonstrated, including a study that investigated patterning of capture antibodies using a noncontact microarray printer (Delehanty and Ligler 2002; Rowe et al. 1999b; Taitt et al. 2002). As the NRL Array Biosensor has evolved into its present day fully automated, portable version, assays are frequently developed and run online, and a number of blind field trials have been carried out using this system (Shriver-Lake et al. 2007a).

2.2.1.1. Bacterial Protein Toxin Detection

Antibodies, because of their inherent selectivity and sensitivity, have been the first choice of capture/tracer agents for the development of rapid bacterial identification methods using the NRL Array Biosensor. Some of the first studies with the prototype NRL Array Biosensor demonstrated the system's ability to measure physiological bacterial markers, such as the F1 antigen of *Yersinia pestis*; and bacterial protein toxins, including SEB (secreted by *Staphylococcus aureus*), in a multi-analyte format using sandwich-based immunoassays (Rowe et al. 1999a; Rowe et al. 1999b; Wadkins et al. 1998). It is the protein toxins produced by many of the food-infecting bacteria that are responsible for causing severe illness. However, diagnosis of the specific disease based on toxin detection can be complicated by the fact that many of the symptoms of infection are similar for different bacterial toxins. Therefore, the ability to test for multiple bacterial protein toxins simultaneously aids in rapid diagnosis, so that appropriate, timely treatment can be administered—which is extremely important to prevent severe illness or potentially death. SEB and F1 antigen assays were demonstrated not only in buffer but also in clinical samples, including saliva, nasal, urine, blood, and serum fluids with little effect on the immunoassay performance. The simultaneous detection of SEB and cholera toxin (produced by the bacteria *Vibrio cholerae*) was demonstrated, along with four other biohazards both in buffer and in various environmental samples (Rowe-Taitt et al. 2000c). The first automated version of the NRL Array Biosensor was demonstrated using optimized SEB immunoassays, while the same study measured the bacterial toxins SEB, cholera toxin, and the botulinum toxoids A and B (potent neurotoxins produced by *Clostridium botulinum*) using the non-automated format (Rowe-Taitt et al. 2000b).

Due to public health concerns, much of the recent work in assay development, using the NRL Array Biosensor, has concentrated on the rapid detection of foodborne contaminants. SEB, for example, has been measured in a number of spiked food samples, including homogenates of fruit and meat, beverages, and carcass washings (Shriver-Lake et al. 2003). Immunoassays taking < 20 min were demonstrated with little if any sample pretreatment and no preconcentration, resulting in 0.5 ng/ml LODs for all spiked sample matrices tested. Immunoassays were later developed for the simultaneous detection of SEB and botulinum toxoid A in a variety of spiked food samples (Sapsford et al. 2005). Dose-response curves for botulinum toxoid A were found to be more sensitive to matrix effects than the corresponding SEB assays, with LODs ranging from 20 ng/ml–500 ng/ml, depending on the sample matrix.

While much of the focus of the NRL array biosensor has been in the development of immunoassays using antibodies as capture agents; alternative capture agents, including simple carbohydrates, a cellular receptor, and antimicrobial peptides, have also been investigated. The cellular receptor ganglioside G_{M1} was successfully used to capture cholera toxin, with a 40 ng/ml LOD achieved using antibody tracers (Rowe-Taitt et al. 2000a). The monosaccharides N-acetylneuraminic acid (Neu5Ac) and N-acetylgalactosamine (GalNAc) have been used to capture both cholera and tetanus toxins with different affinities (Ngundi et al. 2006b). The binding of cholera toxin to immobilized Neu5Ac was also investigated in real time, giving an association constant of $1.3 \times 10^8 \, M^{-1}$, comparable to the antibodies studied (Ngundi et al. 2006a). Cowpea mosaic virus (CPMV) was investigated as a tracer for immunoassays using the NRL Array Biosensor. The ability to label the virus capsid with antibody and up to 60 fluorescent dyes (without quenching) resulted in an improved LOD in SEB sandwich immunoassays, when the virus nanoscaffold was used as a tracer, relative to a mole equivalent of dye-labeled antibody (Sapsford et al. 2006c). While indirect, a method of measuring exposure to bacteria toxins has been demonstrated by screening sera, from donors, for human antibodies against tetanus toxin (from *Clostridium tetani*), diphtheria toxin (*Corynebacterium dihtheriae*) and SEB (Moreno-Bondi et al. 2006). Such a biosensor has clinical applications both in monitoring exposure to pathogens and measuring the efficacy of vaccination.

2.2.1.2. Bacterial Detection

In addition to detecting bacterial toxins individually or in a multi-analyte system, the NRL Array Biosensor has been employed for the detection of bacteria themselves. As with bacterial toxins, assays developed for bacteria primarily use antibodies as the recognition molecule and are developed individually. Concerns for pathogenic bacterial contamination involve homeland security (biothreat agents), clinical diagnostics, and food safety. Early work with the NRL Array Biosensor focused on the detection of biothreat bacteria such as *Bacillus anthracis* and *Franciscella tularensis* in clinical and environmental samples (Rowe-Taitt et al. 2000c; Taitt et al. 2002). Using this evanescent wave sensor, the matrix effects were minimal and detection limits of $10^3 - 10^5$ cells/ml were achieved. Recent work has focused on the detection of pathogenic bacteria in food samples whether intentionally contaminated (bioterrorism) or not. In today's world, food is transported all over the globe, exposing people to a variety of bacteria. Rapid detection of pathogenic bacteria at ports of entry, storage sites, and arrival at processing facilities is necessary to reduce chances of an epidemic. Assays for *Salmonella typhimurium*, *Escherichia coli* O157:H7, *Camplylobacter jejuni*, *Listeria monocytogenes*, and *Shigella* species have been developed and demonstrated in a variety of food matrices (meat, fruit, vegetables, alcoholic and non-alcoholic beverages) (Sapsford et al. 2006b; Sapsford et al. 2004c; Taitt et al. 2004a; Taitt et al. 2004b; Shriver-Lake et al. 2007b). These bacteria were detected with the portable automated NRL Array Biosensor in less than 30 minutes at limits of detection of $10^3 - 10^6$ cells/ml, depending on the antibody-analyte pair, with very limited pretreatment and no enrichment. One advantage of the NRL Array Biosensor is the ability to detect both toxins and bacteria. Figure 6.2 is an image of simultaneous bacterial and bacterial toxin detection.

Antimicrobial peptides (AMP) are the first line of defense against microbial invasion in a host immune system. These peptides have also been utilized as biorecognition molecules with the NRL Array Biosensor for the detection of specific bacteria (Kulagina et al. 2006; Kulagina et al. 2005). Unlike antibodies, these peptides recognize several strains of bacteria. Kulagina et al. used several AMPs (magainin, polymyxin B, polymyxin E, cecropin A,

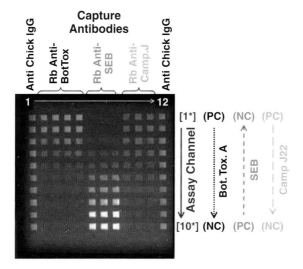

Figure 6.2. The NRL Array Biosensor—the final CCD image of a sandwich immunoassay used for the detection of bacterial cells and bacterial protein toxins (adapted from Golden et al. 2005b).

parsin) for the detection of *Escherichia coli, Salmonella typhimurium, Coxiella burnetti, and Brucella melitensis* (Kulagina et al. 2006; Kulagina et al. 2007; Kulagina et al. 2005). The limits of detection (LODs) were similar with these peptides to those observed with an analogous antibody-based assay. Each AMP had differing affinities to the bacteria. Using a mixed array of AMPs provides identification of the bacteria present based on the pattern of binding.

2.2.2. Other Optical Waveguides

Other systems, based on planar optical waveguides coupled with TIRF detection, have been described in the literature (Taitt et al. 2005). However, only a few have currently been applied to bacterial and bacterial protein toxin detection.

2.2.2.1. Bacterial Protein Toxin Detection

Integrated optical waveguides coupled with fluorescence-based detection have been developed for a number of bacterial protein toxins. Membrane-based assays were developed for measuring the direct binding of cholera toxin (Kelly et al. 1999). IOW surfaces were functionalized with a bilayer membrane containing the ganglioside G_{M1} labeled with either a donor or acceptor fluorescent dye. Binding of cholera toxin to the fluid membrane surface brought the donor/acceptor dyes into close proximity, resulting in fluorescence resonance energy transfer (FRET) and a distinct change in the fluorescence signal from both dyes. In later studies, sandwich immunoassays were incorporated into the bioactive membrane to measure the protective antigen (PA). When PA is paired with both the lethal and edema factors in *Bacillus anthracis* the lethal and edema toxins are formed (Martinez et al. 2005). The performance of IOWs formed from sol-gel and $SiON_x$ materials was compared, with the sol-gel waveguide resulting in lower background luminescence and improved sensitivities.

2.2.2.2. Bacterial Detection

The use of fluorescence excited by the evanescent wave for the screening of pathogenic bacteria such as *Escherichia coli* O157:H7 was developed by Rahman et al. (2006). After the sample was mixed with Cy5-labeled anti-*E. coli* 0157:H7, the mixture was passed through a membrane filter where the bacteria were captured. The membrane was dried and analyzed with a diffuse reflection spectrophotometer. Detection limits down to 5×10^5 cells/ml were obtainable in less than 45 minutes. Zourob and coworkers have developed a system using a metal-clad leaky waveguide (MCLW) to extend the evanescent field (Zourob et al. 2003), with ultrasound waves used to move the bacteria within a fluid (Zourob et al. 2005a) on the surface of the waveguide. *Bacillus subtilis* var. *niger* (BG) were detected at $10^3 - 10^4$ spores/ml with MCLWs (Zourob et al. 2005b; Zourob et al. 2005c). Instead of antibodies, Hollars et al. have explored the use of DNA probes binding DNA targets from bacteria in microfluidic channels (Hollars et al. 2006). There are other types of planar waveguides that have been employed for bacteria detection, but not using fluorescence transduction, and they will therefore not be discussed here.

2.2.3. TIRF-Microscopy

Due to recent advances in technology, TIRF-based fluorescent microscopy (TIRFM) has proven an increasingly popular tool for studying single molecule processes that occur at surfaces (Perkel 2006; Semiatin 2006) and within cells (Mashanov et al. 2003). In TIRFM, fluorescent molecules on a planar surface are excited by the evanescent wave and visualized through a microscope. In an effort to understand the formation of bacterial biofilms, which can lead to infection, TIRFM has successfully been used for studying the attachment of motile *E. coli* cells to surfaces (Smith et al. 2002; Vigeant et al. 2002; Vigeant et al. 2001). However, reports concerning the application of TIRFM as a method of detecting, rather than studying, bacterial cells and bacterial protein toxins have only recently appeared in the literature, but hold great promise for sensitive detection. Gardin et al. described a slightly different type of microscopy referred to as waveguide excitation fluorescence microscopy (WExFM) (Grandin et al. 2006). The proposed setup could potentially improve target sensitivity and surface specificity, as well as provide large analysis areas and multicolor imaging capabilities. Although they have not yet used the system for bacterial or bacterial toxin detection, they suggest that this would be a possible application for the technology.

2.2.3.1. Bacterial Protein Toxin Detection

Using hydrogel-based protein microchips employing standard fluorescence microscopy for detection, Rubina and coworkers were able to screen for a number of biotoxins simultaneously (Rubina et al. 2005). However, to date there are limited studies using TIRFM for the detection of bacterial protein toxins. Supported lipid bilayers, containing either the ganglioside G_{T1b} or G_{M1}, were patterned onto substrates, and the binding of fluorescently labeled cholera toxin B subunit or tetanus toxin C fragment was investigated (Moran-Mirabal et al. 2005). Multicolor TIRFM was used to measure two bacterial toxins, diphtheria and tetanus toxin, simultaneously plated onto poly-L-lysine coated surfaces (Hoshino et al. 2005). The two toxins were distinguished from one another using antibodies labeled with different color quantum dots (QDs).

2.2.3.2. Bacterial Detection

Imaging fluorescently-labeled bacteria is routinely performed with TIRFM but not the specific detection of unlabeled whole cells. Detection of DNA targets from bacterial cells using DNA probes is a method that could be indirectly employed for bacteria detection with microscopy. Hollars et al. (2006) developed a bio-assay for detection of co-localized fluorescent DNA probes onto DNA targets. The fluorescence was excited using TIRF within a planar microfluidic channel. They were able to achieve analysis of DNA targets in the low pM range in under a minute.

3. Future Aspects of Bacterial Fluorescent TIRF Biosensors

TIRF-based technology could be a benefit to, and likewise benefit from, a number of the methods and ideas discussed in this book. Current issues concerning certain applications, such as sensitivity, sample preparation, and multiplexing, may be addressed through use of integrated Lab-on-a-Chip/microfluidics technologies and/or alternative identification agents for detection. TIRF-based excitation of a surface, for example, has proven to be more sensitive than epi-illumination, typically used by confocal scanning microscopes, as it is intrinsically less prone to background interference (Lehr et al. 2003).

Future developments with TIRF biosensing technologies will be driven by the application, the required skill level of the operator, and the LODs required. For example, field deployable instruments, used for environmental and military applications, will have different operational requirements compared to an instrument that remains in the lab and is used in a more clinical setting. While TIRF-based technology employing IOWs are more sensitive than the corresponding IRE arrangements (Brecht et al. 1998; Lehr et al. 2003), the optics associated with coupling laser light into IOWs are much more sensitive to alignment issues, therefore making IOW-based technologies potentially less suitable for field deployable devices, where instrument transportation may occur in less than ideal conditions.

The NRL Array Biosensor is probably the most developed TIRF-based system for bacterial cell and bacterial toxin detection described in this chapter (Ligler et al. 2007). Single and simultaneous multi-analyte detection of a number of bacterial species has been demonstrated using sandwich immunoassays. Antibodies, known for their high specificity, and other specific species such as DNA are excellent molecular recognition elements for the detection of predetermined targets. Multi-analyte detection using these highly specific species has clearly been demonstrated; however, biosensor technology would benefit from the ability to measure potentially unknown targets. Arrays of selective and semi-selective agents, such as carbohydrates and peptides (discussed in Part III of the book), coupled with pattern recognition software may represent the future of bacterial biosensors, allowing the user to screen for both predetermined and potentially unknown targets. The NRL Array Biosensor has demonstrated initial studies using semiselective capture agents such as monosaccharides and antimicrobial peptides; however, the assays currently require either direct fluorescent labeling of the target or the use of tracer antibodies (Kulagina et al. 2006; Kulagina et al. 2007; Kulagina et al. 2005; Ngundi et al. 2006a; Ngundi et al. 2006b).

Sensitivity, while dependent on the application, is an important issue for bacterial biosensors. Often, the desired LODs are as low as 1 cfu/ml for bacterial cells and sub- ng/ml levels for bacterial protein toxins. Current buffer LODs, using sandwich immunoassays with the NRL Array Biosensor, range from 10^3–10^6 cfu/ml for the bacterial targets and 0.1–200 ng/ml for the bacterial protein toxins (Ligler et al. 2007). LODs could be improved by up to a factor of 10 by increasing the assay time (Sapsford et al. 2004c; Taitt et al. 2004b), and by optimizing the antibody capture and tracer combinations (monoclonal versus polyclonal; also different species). Improving the sensitivity may also be achieved using one or a combination of the following: the introduction of preconcentration or amplification steps prior to assay detection, alternative capture agents, and alternative waveguides.

Fluorescence coupled with DNA detection for bacterial species is common, such as the high density gene-chip arrays which are typically read using confocal scanners with epi-illumination of the surface (Epstein et al. 2002); although as mentioned TIRF excitation may improve sensitivity (Lehr et al. 2003). One area that may greatly improve the LODs using TIRF-based detection for bacterial cells is the use of DNA coupled with PCR, either prior to or after initial capture on the sensing surface. DNA extraction from bacterial cells followed by PCR amplification and detection on a DNA modified waveguide could offer improved sensitivities. Performing PCR confirmation, after antibody capture and detection using a fiber optic biosensor, has also been demonstrated for *E. coli* O157:H7 (Simpson and Lim 2005). The use of PCR amplification after capture reduced the impact of PCR inhibitors commonly found in complex sample matrices.

The use of magnetic beads, normally referred to as immunomagnetic separation (IMS), to extract, clean up, and concentrate samples prior to analysis is common (Gijs 2004; Magnani et al. 2006) and may improve sensitivity. Due to the large size of the magnetic beads, bacterial detection is typically determined using sandwich-based immunoassays coupled with either simple solution fluorescence (Yang and Li 2006), fluorescence microscopy (Morozov and Morozava 2006; Steingroewer et al. 2005), or flow cytometry (Hibi et al. 2006) measurements. The use of magnetic beads coupled with TIRF detection has recently been described but is still in the proof-of-concept stage (Wellman and Sepaniak 2006). Here, magnetic beads functionalized with anti-rabbit-IgG were exposed in solution to rabbit IgG and then dye labeled anti-rabbit-IgG. An external magnet was then used to drive the magnetic beads to the prism surface, where those containing the resulting fluorescent sandwich complex were excited using TIRF; the resulting LODs were in the low ng/ml range.

Another area which may be considered when addressing sensitivity is the specifics of the waveguide. As mentioned, TIRF-based technology employing IOWs is more sensitive than the corresponding IRE (Brecht et al. 1998; Lehr et al. 2003), although the corresponding optics for IOWs are often more complicated and less robust. The use of metal-enhanced fluorescence to improve sensitivity is a fairly recent development for TIRF-based detection systems. The technology takes advantage of the observed fluorescence enhancement of dyes that occurs in the proximity of silver films (Matveeva et al. 2004; Matveeva et al. 2005) and can result in up to a 40-fold enhancement in signal. The origin of this observed enhancement is the result of two simultaneous phenomena: a local field enhancement that increases the efficiency of the dye excitation, and an increase in the fluorescent dye's radiative decay rate (Matveeva et al. 2004). Metal-clad waveguides, such as the MCLW mentioned previously, use thin films rather than island formations to produce enhanced fluorescence signals, when compared to conventional fluorescence detection configurations (Minardo et al. 2006; Zourob et al. 2003; Zourob et al. 2005c).

The use of quantum dots (QDs) as fluorescent probes may also improve sensitivity and multiplexing capabilities. QDs offer a number of advantages over conventional fluorescent dyes, including high photostability, size-tunable color, narrow emission profiles, and the ability to excite all sizes (colors) of QDs using UV wavelengths (Medintz et al. 2005b; Sapsford et al. 2006c). Simultaneous, multiplexed immunoassays using different color QDs as fluorophores have been demonstrated for both bacterial cell detection (Yang and Li 2006) and bacterial protein toxins (Goldman et al. 2004; Hoshino et al. 2005). Only one study however used TIRF-based technology to image the QDs (Hoshino et al. 2005). The possibility of patterning QDs onto waveguide surfaces and imaging the immobilized QDs with a slightly modified version of the NRL Array Biosensor has been demonstrated (Medintz et al. 2006; Medintz et al. 2005a; Sapsford et al. 2004b), leading to the possibility of using QDs as fluorophores in bacterial detection with this TIRF-based system.

QDs, as well as conventional fluorescent dyes, can play a more active role in analyte detection through the process of fluorescence resonance energy transfer (FRET). There are numerous potential materials that could be used in FRET-based assays and the technique has been used for a variety of applications (Sapsford et al. 2004a; Sapsford et al. 2006a). One of the benefits of FRET-based detection, when designed correctly, is that it allows direct measurement of the target analyte without the need for additional tracer steps (Figure 6.3). Although FRET combined with TIRF is still in the early stages of development (Kelly et al. 1999; Ko and

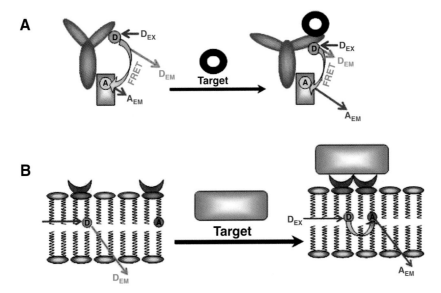

Figure 6.3. Various biosensing designs for FRET based detection of target analytes. (A) Conformational change-based FRET—the antibody is labeled with the donor in the Fab region, and a carrier molecule that targets the Fc portion of the antibody, is labeled with the acceptor species. The donor-acceptor pair undergo FRET. Binding of the antigen induces a conformational change in the antibody Fab region that changes the efficiency of the FRET. (B) FRET sensor of cholera toxin—G_{M1} is labeled with either a fluorescent donor or acceptor dye and anchored in a fluid-like biomimic membrane surface. Binding of cholera toxin causes lateral diffusion within the membrane surface, bringing the donor and acceptor into close proximity resulting in FRET. The resulting sensor platforms are effectively reagentless because introduction of the target causes a direct change in the amplitude of acceptor emission without additional steps (adapted from Sapsford et al. 2004a).

Grant 2006), it promises to be an excellent combination for rapid direct detection biosensors, capable of multiplexed assays for target detection.

4. Conclusions

Numerous methods have been developed for the detection of bacteria and their associated toxins employing fluorescence-based evanescent wave biosensors. Without pre-enrichment or pre-concentration, the desired sensitivity of 1 bacterium/ml has yet to be accomplished, but LODs of $10^3 - 10^4$ cells/ml have been achieved. In addition to detecting bacteria, several of the fluorescence-based sensors have detected bacterial toxins to sub-ng/ml levels. The NRL Array Biosensor has demonstrated detection of bacteria and their associated toxins individually or together in a variety of matrices (environmental, clinical, and food). As the fluorescent-based planar evanescent wave biosensors continue to evolve, improving sensitivity, assay performance, and miniaturization, they will appear in a variety of applications, including point-of-care testing of clinical samples, monitors for homeland security, and in the food processing industry for food safety.

Acknowledgements

The views expressed here are those of the authors and do not represent those of the U.S. Navy, the U.S. Department of Defense or the U.S. Government.

References

Amersham Biosciences (2006) Fluorescence screening reagents guide. www.amershambiosciences.com/applic/upp00738. nsf/vLookupDoc/242718157-B653/$file/FSR_brochure.pdf (Accessed 12/1/2006)

AnaSpec Inc. (2006) Superior fluorescent labeling dyes - hilytefluor. www.anaspec.com/content/pdfs/c_literature57.pdf (Accessed 12/1/2006)

Araya-Kleinsteuber B, Roque ACA, Kioupritzi E, Stevenson AC and Lowe C (2006) Magnetic acoustic resonance immunoassay (MARIA): A multifrequency acoustic approach for the non-labeled detection of biomolecular interactions. J. Mol. Recognit. 19:379–385

Bergwerff AA and van Knapen F (2006) Surface plasmon resonance biosensors for detection of pathogenic microorganisms: Strategies to secure food and environmental safety. J. AOAC Int 89:826–831

Berkenpas E, Millard P and Pereira da Cunha M (2006) Detection of Escherichia coli O157:H7 with langasite pure shear horizontal surface acoustic wave sensors. Biosens. Bioelectron. 21:2255–2262

Boyaci IH, Aguilar ZP, Hossain M, Halsall HB, Seliskar CJ and Heineman WR (2005) Amperometric determination of live Escherichia coli using antibody-coated paramagnetic beads. Anal. Bioanal. Chem. 382:1234–1241

Brecht A, Klotz A, Barzen C, Gauglitz G, Harris RD, Quigley GR, Wilkinson JS, Sztajnbok P, Abuknesha RA, Gascon J, Oubina A and Barcelo D (1998) Optical immunoprobe development for multiresidue monitoring in water. Anal. Chim. Acta 362:69–79

Campbell GA and Mutharasan R (2005) Detection of pathogen Escherichia coli O157:H7 using self-excited PZT-glass microcantilevers. Biosens. Bioelectron. 21:462–473

Delehanty JB and Ligler FS (2002) A microarray immunoassay for simultaneous detection of proteins and bacteria. Anal. Chem. 74:5681–5687

Deobagkar DD, Limaye V, Sinha S and Yadava RDS (2005) Acoustic wave immunosensing of Escherichia coli in water. Sens. Actuator B-Chem. 104:85–90

Dhadwal HS, Kemp P, Alle, J and Dantzler MM (2004) Capillary waveguide nucleic acid based biosensor. Anal. Chim. Acta 501:205–217

Epstein JR, Biran I and Walt DR (2002) Fluorescence-based nucleic acid detection and microarrays. Anal. Chim. Acta 469:3–36

Gauglitz G (2005) Direct optical sensors: Principles and selected applications. Anal. Bioanal. Chem. 381:141–155

Gijs MAM (2004) Magnetic bead handling on-chip: New opportunities for analytical applications. Microfluid. Nanofluid. 1:22–40

Golden JE, Shriver-Lake LC, Sapsford KE and Ligler FS. (2005a) A "Do-it-yourself" Array biosensor. Methods 37:65–72

Golden JP, Taitt CR, Shriver-Lake LC, Shubin YS and Ligler FS (2005b) A portable automated multianalyte biosensor. Talanta 65:1078–1085

Goldman ER, Clapp AR, Anderson GP, Uyeda HT, Mauro JM, Medintz IL and Mattoussi H (2004) Multiplexed toxin analysis using four colors of quantum dot fluororeagents. Anal. Chem. 76:684–688

Grandin HM, Stadler B, Textor M and Voros J (2006) Waveguide excitation fluorescence microscopy: A new tool for sensing and imaging the biointerface. Biosens. Bioelectron. 21:1476–1482

Herron JN, Wang H-K, Tan L, Brown SZ, Terry AH, Durtschi JD, Simon EM, Astill ME, Smith RS and Christensen DA (2006) Planar waveguide biosensors for point-of-care clinical and molecular diagnostics, in Fluorescence Sensors and Biosensors, R. B. Thompson, Ed., 283–332 Boca Raton, FL: Taylor and Francis Group, LLC

Hibi K, Abe A, Ohashi E, Mitsubayashi K, Ushio H, Hayashi T, Ren H and Endo H (2006) Combination of immunomagnetic separation with flow cytometry for detection of Listeria monocytogenes. Anal. Chim. Acta 573-574:158–163

Hollars CW, Puls J, Bakajin O, Olsan B, Talley CE, Lane SM and Huser T (2006) Bio-assay based on single molecule fluorescence detection in microfluidic channels. Anal. Bioanal. Chem. 385:1384–1388

Horvath R, Pedersen HC, Skivesen N, Selmeczi D and Larsen NB (2003) Optical waveguide sensor for on-line monitoring of bacteria. Opt. Lett. 28:1233–1235

Hoshino A, Fujioka K, Manabe N, Yamaya S-L, Goto Y, Yasuhara M and Yamamoto K (2005) Simultaneous multicolor detection system of the single-molecular microbial antigen with total internal reflection fluorescence microscopy. Microbiol Immunol 49:461–470

Invitrogen Corporation. (2006) The Alexa fluor dye series. www.probes.invitrogen.com/handbook/boxes/0442.html (Accessed 12/1/2006)

Invitrogen Corporation (2006) The handbook: A guide to fluorescent probes and labeling technologies. http://probes.invitrogen.com/handbook/ (Accessed 12/1/2006)

Kelly D, Grac, KM, Song X, Swanson BI, Frayer D, Mendes SB and Peyghambarian N (1999) Intergrated optical biosensor for detection of multivalent proteins. Opt. Lett. 24:1723–1725

Ko S and Grant SA (2006) A novel fret-based optical fiber biosensor for rapid detection of Salmonella typhimurium. Biosens. Bioelectron. 212:1283–1290

Kulagina N, Shaffer KM, Anderson GP, Ligler FS and Taitt CR (2006) Antimicrobial peptide-based biosensor assays for *E. coli* and *Salmonella* cells. Anal. Chim. Acta 575:9

Kulagina N, Shaffer KM, Ligler FS and Taitt CR (2007) Antimicrobial peptides, new recognition molecules for challenging targets. Sens. Actuator B-Chem. 121:150–157

Kulagina NV, Lassman ME, Ligler FS and Taitt CR (2005) Antimicrobial peptides for detection of bacteria in biosensor assays. Anal. Chem. 77:6504–6508

Lakowicz JR (1999) Principles of fluorescence spectroscopy, New York: Kluwer Academic/Plenum Publishers

Lathrop AA, Jaradat ZW, Haley T and Bhunia AK (2003) Characterization and application of a *Listeria monocytogenes* reactive monoclonal antibody c11e9 in a resonant mirror biosensor. J. Immunol. Methods 281:119–128

Lehr H-P, Brandenburg A and Sulz G (2003) Modeling and experimental verification of the performance of TIRF-sensing systems for oligonucleotide microarrays based on bulk and integrated optical planar waveguides. Sens. Actuator B-Chem. 92:303–314

Leonard P, Hearty S, Wyatt G, Quinn J and O'Kennedy R (2005) Development of a surface plasmon resonance-based immunoassay for *Listeria monocytogenes*. J. Food Prot. 68:728–735

Ligler FS, Breimer M, Golden JP, Nivens DA, Dodson JP, Green TM, Haders DP and Sadik OA (2002) Integrating waveguide biosensor. Anal. Chem. 74:713–719

Ligler FS, Sapsford KE, Golden JP, Shriver-Lake LC, Taitt CR, Dyer MA, Barone S and Myatt CJ (2007) The Array Biosensor: Portable, automated systems. Anal. Sci. 23:5–10

Lim DV (2003) Detection of microorganisms and toxins with evanescent wave fiber-optic biosensors. IEEE Proceedings 91:902–907

Lin H-C and Tsai W-C (2003) Piezoelectric crystal immunosensor for the detection of staphylococcal enterotoxin b. Biosens. Bioelectron. 18:1479–1483

Magnani M, Galluzzi L and Bruce IJ (2006) The use of magnetic nanoparticles in the development of new molecular detection systems. J. Nanosci. Nanotechnol. 6:2302–2311

Martinez JS, Grace WK, Grace KM, Hartman N and Swanson BI (2005) Pathogen detection using single mode planar optical waveguides. J. Mater. Chem. 15:4639–4647

Mashanov GI, Tacon D, Knight AE, Peckham M and Molloy JE (2003) Visualizing single molecules inside living cells using total internal reflection fluorescence microscopy. Methods 29:142–152

Matveeva E, Gryczynski Z, Malicka J, Gryczynski I and Lakowicz JR (2004) Metal-enhanced fluorescence immunoassays using total internal reflection and silver island-coated surfaces. Anal. Biochem. 334:303–311

Matveeva E, Gryczynski Z, Malicka J, Lukomska J, Makowiec S, Berndt KW, Lakowicz JR and Gryczynski I (2005) Directional surface plasmon-coupled emission: Application for an immunoassay in whole blood. Anal. Biochem. 344:161–167

Medina MB (2006) A biosensor method for detection of staphylococcal enterotoxin A in raw whole egg. J Rapid Meth Aut Mic 14:119–132

Medintz IL, Sapsford KE, Clapp AR, Pons T, Higashiya S, Welch JT and Mattoussi H (2006) Designer variable repeat length polypeptides as scaffolds for surface immobilization of quantum dots. J. Phys. Chem. B 110:10683–10690

Medintz IL, Sapsford KE, Konnert JH, Chatterji A, Lin T, Johnson JE and Mattoussi H (2005a) Decoration of discretely immobilized cowpea mosaic virus with luminescent quantum dots. Langmuir 21:5501–5510

Medintz IL, Uyeda HT, Goldman ER and Mattoussi H (2005b) Quantum dot bioconjugates for imaging, labeling and sensing. Nat. Mater. 4 435–446

Minardo A, Bernini R, Mottola F and Zeni L (2006) Optimization of metal-clad waveguides for sensitive fluorescence detection. Optics Express 14:3512–3527

Moran-Mirabal JM, Edel JB, Meyer GD, Throckmorton D, Singh AK and Craighead HG (2005) Micrometer-sized supported lipid bilayer arrays for bacterial toxin binding studies through total internal reflection fluorescence microscopy. Biophys. J. 89:296–305

Moreno-Bondi MC, Taitt CR, Shriver-Lake LC and Ligler FS (2006) Multiplexed measurement of serum antibodies using an array biosensor. Biosens. Bioelectron. 21:1880–1886

Morozov VN and Morozava TY (2006) Active bead-linked immunoassay on protein microarrays. Anal. Chim. Acta 564:40–52

Muhammad-Tahir A and Alocilja EC (2003) A conductometric biosensor for biosecurity. Biosens. Bioelectron. 18:813–819

Ngundi MM, Taitt CR, McMurray SA, Kahne D and Ligler FS (2006a) Simultaneous determination of kinetic parameters for the binding of cholera toxin to immobilized sialic acid and monoclonal antibody using an array biosensor. Biosens. Bioelectron. 22:124–130

Ngundi MM, Taitt CR, McMurry SA, Kahne D and Ligler FS (2006b) Detection of bacterial toxins with monosaccharide arrays. Biosens. Bioelectron. 21:1195–1201

Nivens DE, Chambers JQ, Anderson TR, Tunlid A, Smit J and White DC (1993) Monitoring microbial adhesion and biofilm formation by attenuated total reflection/fourier transform infrared spectroscopy. J. Microbial. Methods 17:199–213

O'Brien T, Johnson LH, Aldrich JL, Allen SG, Liang LT, Plummer AL, Krak SJ and Boiarski AA (2000) The development of immunoassays to four biological threat agents in a bidiffractive grating biosensor. Biosens. Bioelectron. 14:815–828

Pattnaik P and Srivastav A (2006) Surface plasmon resonance - applications in food science research: A review. J Food Sci Technol 43:329–336

Pierce Biotechnology Inc. (2006) Dylight 549, 649, 680, and 800 reactive dyes and conjugates. www.piercenet.com/Objects/View.cfm?type=ProductFamily&ID=8A6A6F77-93B0-4460-983F-6D15AA329CF8 (Accessed 12/1/2006)

Radke SA and Alocilja EC (2005a) A high density microelectrode array biosensor for detection of E. coli o157:H7. Biosens. Bioelectron. 20:1662–1667

Radke SM and Alocilja EC (2004) Design and fabrication of a microimpedance biosensor for bacterial detection. IEEE Sensors 4:434–440

Radke SM and Alocilja EC (2005b) A microfabricated biosensor for detecting foodborne bioterrorism agents. IEEE Sensors 5:744–750

Rahman S, Lipert RJ and Porter MD (2006) Rapid screening of pathogenic bateria using solid phase concentration and diffuse reflectance spectroscopy. Anal. Chim. Acta 569:83–90

Rogers KR (2006) Recent advances in biosensor techniques for environmental monitoring. Anal. Chim. Acta 568:222–231

Rowe-Taitt CA, Cras JJ, Patterson CH, Golden JP and Ligler FS (2000a) A ganglioside-based assay for cholera toxin using an array biosensor. Anal. Biochem. 281:123–133

Rowe-Taitt CA, Golden JP, Feldstein MJ, Cras JJ, Hoffman KE and Ligler FS (2000b) Array biosensor for detection of biohazards. Biosens. Bioelectron. 14:785–794

Rowe-Taitt CA, Hazzard JW, Hoffman KE, Cras JJ, Golden JP and Ligler FS (2000c) Simultaneous detection of six biohazardous agents using a planar waveguide array biosensor. Biosens. Bioelectron. 15:579–589

Rowe CA, Scruggs SB, Feldstein MJ, Golden JP and Ligler FS (1999a) An array immunosensor for simultaneous detection of clinical analytes. Anal. Chem. 71:433–439

Rowe CA, Tender LM, Feldstein MJ, Golden JP, Scruggs SB, MacCraith BD, Cras JJ and Ligler FS (1999b) Array biosensor for simultaneous identification of bacterial, viral, and protein analytes. Anal. Chem. 71:3846–3852

Ruan C, Yang L. and Li Y (2002) Immunobiosensor chips for detection of Escherichia coli O157:H7 using electro-chemical impedance spectroscopy. Anal. Chem. 74:4814–4820

Rubina AY, Dyukova VI, Dementieva EI, Stomakhin AA, Nesmeyanov VA, Grishin EV and Zasedatelev AS (2005) Quantitative immunoassay of biotoxins on hydrogel-based protein microchips. Anal. Biochem. 340:317–329

Sapsford KE, Berti L and Medintz IL (2004a) Fluorescence resonance energy transfer. Concepts, applications and advances. Minerva Biotec. 16:247–273

Sapsford KE, Berti L and Medintz IL (2006a) Materials for fluorescence resonance energy transfer analysis: Beyond traditional donor-acceptor combinations. Angewandte Chemie-Int Ed 45:4562–4588

Sapsford KE, Medintz IL, Golden JP, Deschamps JR, Uyeda HT and Mattoussi H (2004b) Surface-immobilized self-assembled protein-based quantum dot nanoassemblies. Langmuir 20:7720–7728

Sapsford KE, Ngundi MM, Moore MH, Lassman ME, Shriver-Lake LC, Taitt CA and Ligler FS (2006b) Rapid detection of foodborne contaminants using an array biosensor. Sens. Actuator B-Chem. 113:599–607

Sapsford KE, Pons T, Medintz IL and Mattoussi H (2006c) Biosensing with luminescent semiconductor quantum dots. Sensors 6:925–953

Sapsford KE, Rasooly A, Taitt CR and Ligler FS (2004c) Rapid detection of Campylobacter and Shigella species in food samples using an array biosensor. Anal. Chem. 76:433–440

Sapsford KE, Shubin YS, Delehanty JB, Golden JP, Taitt CR, Shriver-Lake LC and Ligler FS (2004d) Fluorescence-based array biosensors for detection of biohazards. J. Appl. Microbiol. 96:47–58

Sapsford KE, Taitt CR, Loo N and Ligler FS (2005) Biosensor detection of botulinum toxoid A and staphylococcal enterotoxin B in food. Appl. Environ. Microbiol. 71:5590–5592

Scheller F and Schmid RD (1992) Biosensors: Fundamentals, technologies, and applications. VCH Publishers, Inc., New York, NY

Shriver-Lake LC, Shubin YS and Ligler FS (2003) Detection of staphylococcal enterotoxin B in spiked food samples. J. Food Prot. 66:1851–1856

Shriver-Lake LC, Turner S and Taitt CR (2007a) Rapid detection of Escherichia coli O157:H7 spiked into food matrices. Anal. Chem. Acta 584:66–71

Shriver-Lake LC, Erickson JS, Sapsford KE, Ngundi MM, Shaffer KM, Kulagina NV, Hu JE, Gray SA, Golden JP, Ligler FS, Taitt CR (2007b) Blind laboratory trials for multiple pathogens in spiked food matrices. Anal. Letters 40:3219–3231.

Simpson JM and Lim DV (2005) Rapid PCR confirmation of E. Coli O157:H7 after evanescent wave fiber optic biosensor detection. Biosens. Bioelectron. 21:881–887

Smith LV, Tamm LK and Ford RM (2002) Explaining non-zero separation distances between attached bacteria and surfaces measured by total internal reflection aqueous fluorescence microscopt. Langmuir 18:5247–5255

Steingroewer J, Knaus H, Bley T and Boschke E (2005) A rapid method for the pre-enrichment and detection of *Salmonella typhimurium* by immunomagnetic separation and subsequent fluorescence microscopial techniques. Eng. Life Sci. 5:267–272

Subramanian A, Irudayaraj J and Ryan T (2006) A mixed self-assembled monolayer-based surface plasmon immunosensor for detection of *E. Coli* O157:H7. Biosens. Bioelectron. 21:998–1006

Susmel S, Guilbault GG and O'Sullivan CK (2003) Demonstration of labelless detection of food pathogens using electrochemical redox probe and screen printed gold electrodes. Biosens. Bioelectron. 18:881–889

Taitt CR, Anderson GP and Ligler FS (2005) Evanescent wave fluorescence biosensors. Biosens. Bioelectron. 20:2470–2487

Taitt CR, Anderson GP, Lingerfelt BM, Feldstein MJ and Ligler FS (2002) Nine-analyte detection using an array-based biosensor. Anal. Chem. 74:6114–6120

Taitt CR, Golden JP, Shubin YS, Shriver-Lake LC, Sapsford KE, Rasooly A and Ligler FS (2004a) A portable array biosensor for detecting multiple analytes in complex samples. Microb. Ecol. 47:175–185

Taitt CR Shubin YS, Angel R and Ligler FS (2004b) Detection of *Salmonella enterica serovar typhimurium* by using a rapid, array-based immunosensor. Appl. Environ. Microbiol. 70:152–158

Turner APF, Karub, I and Wilson GS (1987) Biosensors: Fundamentals and applications. Oxford University Press, Oxford

Vigeant MA-S, Ford R, Wagner M and Tamm LK (2002) Reversible and irreversible adhesion of motile *Escherichia coli* cells analyzed by total internal reflection aqueous fluorescence microscopy. Appl. Environ. Microbiol. 68:2794–2801

Vigeant MA-S, Wagner M, Tamm LK and Ford R (2001) Nanometer distances between swimming bacteria and surfaces measured by total internal reflection aqueous fluorescence microscopy. Langmuir 17:2235–2242

Wadkins RM, Golden JP, Pritsiolas LM and Ligler FS (1998) Detection of multiple toxic agents using a planar array immunosensor. Biosens. Bioelectron. 13:407–415

Wellman AD and Sepaniak MJ (2006) Magnetically-assisted transport evanescent field fluoroimmunoassay. Anal. Chem. 78

Wingard Jr LB and Ferrance JP (1991) Concepts, biological components, and scope of biosensors, in Biosensors with Fiberoptics, D. L. Wise and J. L. B. Wingard, Eds., 1–27 Clifton, New Jersey: Humana Press

Wise DL and Wingard Jr LB (1991) Biosensors with fiberoptics, Clifton, NJ: Humana Press

Wolfbeis OS (1996) Capillary waveguide sensors. Trac-Trends Anal. Chem. 15:225–232

Yang L and Li Y (2006) Simultaneous detection of *Escherichia coli* O157:H7 and *Salmonella typhimurium* using quantum dots as fluorescence labels. The Analyst 131:394–401

Zhao J, Zhu W and He F (2005) Rapidly determining *E. coli* and *P. aeruginosa* by an eight channels bulk acoustic wave impedance physical biosensor. Sens. Actuator B-Chem. 107:271–276

Zhu P, Shelton DR, Karns JS, Sundaram A, Li S, Amstutz P and Tang C-M (2005) Detection of water-borne *E. coli* O157:H7 using the integrating waveguide biosensor. Biosens. Bioelectron. 21:678–683

Zourob M, Hawkes JJ, Coakley WT, Treves Brown BJ, Fielden PR, McDonnell M and Goddard NJ (2005a) Optical leaky waveguide sensor for detection of bacteria with ultrasound attractor force. Anal. Chem. 77:6163–6168

Zourob M, Mohr S, Treves Brown BJ, Fielden PR, McDonnell M and Goddard NJ (2003) The development of a metal clad leaky waveguide sensor for the detection of particles. Sens. Actuator B-Chem. 90:296–307

Zourob M, Mohr S, Treves Brown BJ, Fielden PR, McDonnell M and Goddard NJ (2005b) Bacteria detection using disposable optical leaky waveguide sensors. Biosens. Bioelectron. 21:293–302

Zourob M, Mohr S, Treves Brown BJ, Fielden PR, McDonnell M and Goddard NJ (2005c). An integrated metal clad leaky waveguide sensor for detection of bacteria. Anal. Chem. 77:232–242

7

Fiber Optic Biosensors for Bacterial Detection

Ryan B. Hayman

Abstract

Rapid and specific identification of bacteria is critical for clinical and biosafety applications. Fiber optic biosensors (FOBs) are increasingly being applied to the detection of bacteria in food and water supplies, food processing facilities, and homeland security operations. These biosensors can be used for multiplexed pathogen detection or to confirm the results of other techniques, often in less than one hour. FOBs offer several advantages over conventional culture-based techniques, or polymerase chain reaction (PCR)-based assays, in terms of speed, specificity, and depth of information content. In addition, some sensor platforms have been developed into portable systems capable of emergency field deployment. In this chapter, we will discuss the detection of bacteria using fiber optic immunosensors, nucleic acid-based FOBs in various assay formats, and several applications of these technologies.

1. Fiber Optic Biosensors

The biosensor systems discussed here generally consist of recognition elements immobilized on the core of optical fibers, a source of excitation light, and a detector (Figure 7.1). Most of the fibers are constructed of pulled silica glass, with optical properties modified through the use of dopants such as GeO_2, Al_2O_3, B_2O_3, and F (Taitt et al. 2005). When the excitation light from a laser or white light travels through the waveguide via total internal reflection, the evanescent wave generated at the surface of the fiber core excites the fluorophores (Bhunia et al. 2007). A photo-detector can then capture the emission light, either from the surface of planar waveguides, or at the end of the fiber in the case of most FOBs. Imaging is typically done with a CCD camera, CMOS camera, or photodiode array (Taitt et al. 2005).

There are a variety of recognition elements that can be attached to the core of the optical fiber. Many groups have developed immunoassays for whole-cell bacteria detection (DeMarco et al. 1999; Ferreira et al. 1999; Anderson et al. 2000; Rowe-Taitt et al. 2000; DeMarco and Lim 2001; Ferreira et al. 2001; Tims et al. 2001; DeMarco and Lim 2002a; Tims and Lim 2003; Geng et al. 2004; Kramer and Lim 2004; Taitt et al. 2004; Tims and Lim 2004; Simpson and Lim 2005; Geng et al. 2006; Ligler et al. 2007). Several antibody immobilization strategies have been studied, including physical adsorption, covalent binding, and linking via intermediates (Taitt et al. 2005). Physical adsorption, while simple, is not widely used for antibody conjugation, due to the reversibility of immobilization and irregularity in

Ryan B. Hayman • Walt Lab, Department of Chemistry, Tufts University.

M. Zourob et al. (eds.), *Principles of Bacterial Detection: Biosensors, Recognition Receptors and Microsystems*,
© Springer Science+Business Media, LLC 2008

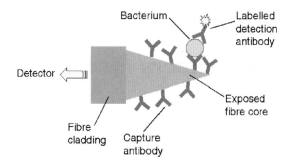

Figure 7.1. Schematic of a fiber optic sandwich immunoassay. Capture antibodies have been immobilized on the fiber core that has been exposed by acid etching. Emission from the excited fluorophore travels back through the proximal end to the detector.

coverage density. Covalent binding is typically done through the use of silanes. Amine- or thiol-terminated silanes are activated with cross-linkers, such as glutaraldehyde. Activated silanes with aldehyde or epoxy groups have also been used (Taitt et al. 2005). Capture antibodies can bind the target, to which fluorescently-labelled signal antibodies then bind in a sandwich immunoassay format. Oligonucleotide probes can be used to detect nucleic acids lysed from the bacteria either in a sandwich format (Ahn and Walt 2005; Ahn et al. 2006) or by hybridization to targets fluorescently labelled during PCR amplification (Song et al. 2006). Amine-modified oligos are typically covalently attached to glutaraldehyde-activated silica microbeads with cyanuric chloride (Epstein et al. 2002; Epstein et al. 2003; Ahn and Walt 2005; Shepard et al. 2005; Ahn et al. 2006; Song et al. 2006). Optimization of the recognition elements and attachment chemistries are critical to the sensitivity and specificity of FOBs for the detection of bacteria.

1.1. Whole-Cell Detection

1.1.1. Evanescent-Field Sensing

Fiber optic biosensors have the potential to detect bacteria in aerosolized form with greater speed and specificity than current microbiological methods. In 1999, Ferreira et al. demonstrated one of the first applications of FOBs to the detection of aerosolized bacteria, a common cause of nosocomial infections (Ferreira et al. 1999). This application is particularly important in the critical-care units of hospitals, where it has been suggested that microorganism levels should not exceed 10 CFU per cubic meter. A section of silica fiber was partially etched with HF, creating a sensing region of approximately $0.33\,cm^2$ with reduced cladding for evanescent coupling with the surroundings. The sensing area of this fiber was then wound into a single loop and placed in a Petri dish with selective culture media. A stream of air was sampled and introduced to the dish for 30 s at a rate of 100 L/min. The resulting bacterial growth caused a change in medium colour and a release of metabolic products. Ferreira et al. observed a decrease in the output power of the light due to changes in the degree of coupling between the evanescent field and the medium. Colonies of *Streptococcus pneumoniae* and methicillin-resistant *Staphylococcus aureus* (MRSA) were detected in 13 and 6 hr, respectively. In 2001, Ferreira and co-workers used FOB to detect *E. coli* O157:H7 (Ferreira et al. 2001). Scanning electron microscopy revealed that the bacteria had grown around the sensing region of the fiber. While these results were obtained more quickly than culture-based techniques, the FOBs relied on the specificity of the media and not on an intrinsic recognition element. The reliance on media also limits the ability to detect multiple pathogens per dish. Detection speed was dependent on the bacterial growth rate. The detection of airborne bacteria is highly time-sensitive and would likely benefit from further FOB development.

1.1.2. Sandwich Immunoassays

The Lim group have used the four-channel Analyte 2000 system (Research International, Monroe, WA) developed in conjunction with the Naval Research Laboratory to use the sandwich assays on polystyrene or silica fiber optics (DeMarco et al. 1999; DeMarco and Lim 2001, 2002b; Tims and Lim 2003; Kramer and Lim 2004; Tims and Lim 2004; Simpson and Lim 2005). Briefly, 7.5 cm sections of cladding from 600 μm diameter silica optical fibers were removed and then tapered by acid etching. These silica fibers and tapered injection-moulded polystyrene waveguides were then functionalized with streptavidin and conjugated to biotinylated antibodies. To create a light dump, the ends of the polystyrene probes were capped with flat black paint. Using Cy5-labelled signal antibodies, DeMarco and Lim have tested apple juice and ground beef spiked with *Escherichia coli*. The researchers observed a limit of detection of 3.2×10^3 CFU/mL ± 1 log for polystyrene and 6.0×10^3 CFU/mL ± 1 log for silica fibers in unpasteurized apple juice in 15 min (DeMarco and Lim 2001). At the upper values of these ranges, no false positive signals were observed. DeMarco and Lim were able to detect 5.2×10^2 CFU/g of *Escherichia coli* in 10 g samples of ground beef using polystyrene waveguides and 9.0×10^3 CFU/g in 25 g samples using the silica-based sensors without false positives (DeMarco and Lim 2002b). The FOB in this iteration was not sensitive enough for the detection of a sufficiently low concentration of *E. coli* cells for real-world food contamination testing but would likely need only a short enrichment. In 2004, Geng et al. used the Analyte 2000 to detect *Listeria monocytogenes* cells in hot dog and bologna (Geng et al. 2004). Careful selection of capture and signal antibodies resulted in a limit of detection of 4.3×10^3 CFU/mL. Recently, Geng et al. used signal antibodies labelled with Alexa Fluor 647 dye and a four-hour enrichment (Geng et al. 2006). These changes decreased the limit of detection of their Analyte 2000 to 1 CFU/mL, a result which, when combined with the ability to complete analyses within one working day, demonstrates the viability of FOBs for rapid and inexpensive food testing.

Tims and Lim have utilized the Analyte 2000 system to detect *Bacillus anthracis* spores in various powder samples with minimal preparation, an application of intense interest since the 2001 anthrax spore mailings (Tims and Lim 2004). Biotinylated capture antibodies were immobilized on tapered polystyrene fibers and used with Cy5-labelled anti-*B. anthracis* signal antibodies in a sandwich assay format. Suspensions of 1mg/mL talc, cornstarch, baking soda, and *Bacillus thuringiensis*-containing pesticides were made in phosphate-buffered saline and tested with and without spiked *B. anthracis* spores. Talc was observed to cause the highest non-specific signal above background and was used as a universal baseline. Spores were detected in all of the powders at a level of 3.2×10^5 spores/mL with no false positives in less than 1 hr. These results are notable due to the difficulty of detecting spores in powders with molecular techniques.

A portable, fully automated FOB dubbed the RAPTOR (originally MANTIS) has been developed by the Naval Research Laboratory and commercialized (Research International, Monroe, WA) (Anderson et al. 2000). Similar to the Analyte 2000, the RAPTOR utilizes four sandwich assay FOBs contained in a coupon that contains all necessary fluidic channels. Powered by a 12-volt battery, the instrument incorporates a laser diode for excitation and photodiode for emission light quantification from Cy5-labelled signal antibodies. Anderson et al. reported the ability to simultaneously detect *B. anthracis* Sterne and *Francisella tularensis* at concentrations of 50 CFU/mL and 5×10^4 CFU/mL, respectively, in 10 min. King and coworkers used the RAPTOR to simultaneously detect down to 5×10^4 spores/mL of *Bacillus globigii* (a surrogate species for *B. anthracis*) and 10^7 CFU/mL of *Erwinia hericola* (King et al. 2000).

Kramer and Lim used the RAPTOR to detect *Salmonella* Typhimurium in spent alfalfa sprout irrigation water, a typically labour-intensive process required by the U.S. Food and Drug Administration (Kramer and Lim 2004). The disposable cartridges of four polystyrene fibers could be used for up to 40 assays, provided the target was not detected. The limit of detection for *Salmonella* Typhimurium in buffer and spent irrigation water, an inherently complicated matrix with high bacterial load, was found to be $\sim 5 \times 10^5$ CFU/mL, equivalent to levels detected with seeds spiked with bacteria at 50 CFU/g of seeds. The assays required 20 min, far less than typical microbiological methods. It may be feasible to integrate this automated system into irrigation water supply lines. For commercial viability, the assays would have to be designed for a greater number of bacteria, among which would be the other thousands of serovars of *Salmonella*.

The rapid detection of bacteria with FOBs can be confirmed with conventional microbiological methods. The Lim group has developed rapid PCR confirmation techniques to be used in fiber optic biosensor detection. Their PCR confirmation has been demonstrated with (Tims and Lim 2003) and without (Simpson and Lim 2005) an enrichment step. Enrichment involved submerging the polystyrene waveguide in Luria-Bertani broth for approximately 6 h, after which centrifugation was used to remove the bacteria. Early attempts to recover the bacteria directly by boiling the waveguide in water were unsuccessful, possibly due to the release of PCR inhibitors when the polystyrene was exposed to high temperatures. Direct PCR of the waveguide was realized only by removing the blackened waveguide tips. The waveguide biosensor offers an immunospecific concentration of bacteria in tandem with rapid detection, should conventional plating be desired (Simpson and Lim 2005). The main advantage to using the approach taken by Lim et al. was that it combined complementary techniques with the speed of an immunosensor, confirmed by sensitive nucleic acid amplification. Due to the high likelihood of cross-reactivity of the antibodies with antigenically related organisms, such as less virulent forms of *E. coli*, direct PCR from a small section of the waveguides offered the ability to specifically confirm the presence of bacteria (Simpson and Lim 2005). Also, by using only a small section of the waveguide, at least 50% of the fiber optic biosensor was left unaltered for more conventional enrichment and plating.

1.2. Bead-Based Arrays

Fiber optic microarrays offer high feature density and discrete, parallel analysis of thousands of sensors. The solid support typically consists of a 1 mm diameter bundle of approximately 50,000 hexagonally close-packed fibers (Epstein et al. 2003). When exposed to acid, the core of each fiber is etched at a faster rate than the cladding, producing a well approximately 3 μm in diameter. Oligonucleotide-coupled microspheres applied to this etched face self-assemble into a randomly ordered nucleic acid sensing array. By encoding several oligonucleotide-labelled beads with different fluorescent dyes, multiple analytes can be simultaneously detected. This procedure results in multiple copies of each sensor type, reducing the chance of false positive or false negative results. A modified epi-fluorescence microscope has been used to image the fiber optic microarrays (Epstein et al. 2003). A xenon arc lamp provides excitation light. The optical channel is then selected by excitation and emission filters and a dichroic mirror. Fibers inserted into a chuck are positioned with an X-Y stage and imaged with a charge-coupled device (CCD) camera. With megapixel resolution (1280×1024), the CCD camera uses multiple pixels per fiber and is able to resolve 3-μm diameter sensors.

Fiber optic bundles composed of thousands of individual waveguide cores sharing a common cladding have been utilized to create microsphere-based FOBs. Each microsphere, or bead, is optically addressable and able to detect probe-target hybridization, usually through labelling with fluorescent reporter groups (Ferguson et al. 2000). These fiber optic microarrays have been shown to be sensitive down to zeptomolar levels of target DNA (Epstein et al. 2002), and offer highly multiplexed, redundant sensors. Advances in bead-based FOBs have made them highly suited for the detection of nucleic acids extracted from bacterial pathogens (Ahn and Walt 2005; Ahn et al. 2006; Song et al. 2006). Nucleic acid FOBs have become increasingly focused on multiplexed arrays amenable to high throughput assays. Multiplexed FOBs have been used for a variety of nucleic acid sensing experiments, including probing variations in entire genomes (Brogan and Walt 2005), detecting antibiotic resistance, and binary barcoding for strain differentiation (Shepard et al. 2005). These arrays can contain tens of thousands of sensing elements, which can detect a vast number of DNA sequences in parallel (Epstein et al. 2003; Ahn and Walt 2005; Song et al. 2006).

1.3. Nucleic Acid Sandwich Assays

A DNA sandwich assay format offers several advantages over conventional microbiological techniques, namely speed and specificity. In a sandwich assay, the target hybridizes to a capture probe, followed by labelling with a signal probe, which anneals to a separate site on the target. Culture-based screening often requires 3–6 days to detect bacteria and is labour intensive (Lammerding and Fazil 2000). While PCR-based assays are inherently sensitive, visualization of stained products on agarose gels lacks specificity. Real-time PCR utilizes a labelled internal probe to assist in quantification and can be highly strain-specific, but often takes 1–2 hrs. and requires expensive specialized thermocyclers (Mothershed and Whitney 2006). Cross-reactivity in immunological methods often results in lowered sensitivity and decreased quantification (Ahn et al. 2006). Ahn et al. have used oligonucleotide microarrays in a sandwich assay to rapidly detect both food pathogens (Ahn and Walt 2005) and multiple harmful algae bloom (HAB) species (Ahn et al. 2006).

Ahn and coworkers developed biosensor arrays by first chemically etching the distal end of 500-μm diameter fiber optic bundles containing 6,000 of 3-μm diameter fibers to form an array of microwells (Ahn and Walt 2005). Oligonucleotide capture and signal probes specific to four *Salmonella* spp. virulence genes (*InvA*, *invE*, *spvB*, and *agf*) were designed using commercially-available software. These capture probes were coupled to 3.1 μm diameter amine-modified polystyrene microspheres with cyanuric chloride-glutaraldehyde chemistry. Prior to coupling, the microspheres were encoded with various concentrations of Eu dye, each corresponding to a different probe sequence, or "bead type." The resulting discrete levels of image brightness were then used to decode the bead types.

Using this setup, Ahn et al. first verified the specificity of single bead types using Cy3-labelled synthetic targets with sequences complementary to the capture probe (Ahn and Walt 2005). Solutions of synthetic targets were detected down to 10 fM. Optimization of the hybridization times for the capture and signal probes was accomplished by varying each incubation from 1 to 60 min using solutions of autoclaved *Salmonella* cultures. Based on these experiments, the researchers chose capture and signal hybridization times of 40 and 20 min, respectively. Detection limits for the four bead types varied from 10^3 to 10^8 CFU/mL of purified DNA from *S. enteriditis* cultures. Interestingly, Agf probes performed well in experiments involving synthetic target solutions, but failed to produce adequate signal intensity when testing DNA from culture. Ahn et al. hypothesized that this disparity was due to the lower than intended T_m of the Agf signal probe. The heated stringency washes employed may have removed the target due to the reduced hybridization stability of lower T_m.

The sandwich assay microarrays were evaluated with common food pathogens and "mock" samples (Ahn and Walt 2005). Five *Salmonella* strains, *S. enteriditis*, *S. gaminarai*, *S. hartford*, *S. rubislaw*, and *S. typhimurium*, were detected at concentrations of 10^4 and 10^7 CFU/mL. Several other common foodborne pathogens were also tested, but at a higher, 10^8 CFU/mL,

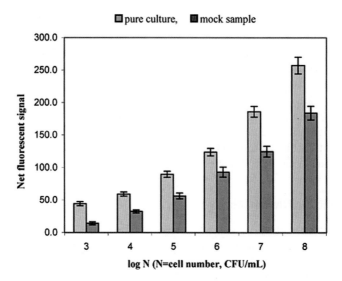

Figure 7.2. Arrays of invA2 probes were tested with pure cultures (light gray) and mock samples (dark gray) for concentrations ranging from 10^3 to 10^8 cfu/mL in triplicate. Decreased signal intensities were observed for mock samples with *E. coli* and nontarget organisms cultured from open air (Reprinted with permission from Ahn et al. (2005), Copyright 2005 American Chemical Society).

concentration. No false positive signals were observed. "Mock" samples were generated by exposing growth media to open air for 12 hr and spiking the broth with *E. coli* prior to incubation. These cultures were then spiked with *S. enteriditis* DNA and tested with the fiber optic microarrays. Net fluorescent signals for pure cultures were compared to those for the mock samples (Figure 7.2).

Even at high concentrations of closely related organisms, probe specificity to *Salmonella* spp. was maintained (Ahn and Walt 2005). Probe specificity is due to several factors. First, sandwich assays employ two probes, both of which are needed to generate a signal, reducing the effect of non-specific hybridization. The chance of false positive and false negative signals was reduced by the use of ~100 × redundancy for each bead type. The use of multiple bead types designed to detect several virulence genes per pathogen added additional redundancy and were a significant advantage to using bead-based fiber optic microarrays. Additional bead types can readily be added to expand the capability of the arrays while maintaining the specificity of existing probes.

Ahn et al. have also applied their fiber optic microarrays to multiplexed HAB detection (Ahn et al. 2006). Using identical instrumentation, the group designed oligonucleotide probes specific to the rRNA of three phytoplankton species, *Alexandrium fundyense*, *Alexandrium ostenfeldii*, and *Pseudo-nitzschia australis*. Using optimized capture and signal probe hybridization times of 30 and 15 min, respectively, the rRNA lysed from as few as 5 cells was detected. This limit of detection is equivalent to approximately 4.3×10^7 molecules, assuming 8.6×10^6 molecules/cell. The presence of non-target plankton in seawater concentrate added to the sample matrix was not observed to change the detection limit.

The three capture probes, NA1S, AO2, and auD1S, were assembled into a multiplexed array and tested with single- and multiple-target samples (Ahn et al. 2006). Probe specificity was first tested with synthetic targets. The detection limits were determined to be 100fM for all three probes. The multiplexed array was then tested individually with four target organism strains: *A. fundyense* GTCA28; *A. ostenfeldii* HT-240D2 and HT-120D6; and *P. australis* 1BA. Responses to these organisms are presented in Figure 7.3 and represent the net fluorescence signal from samples containing 5,000 cells. Similar signal patterns were achieved with 5-,

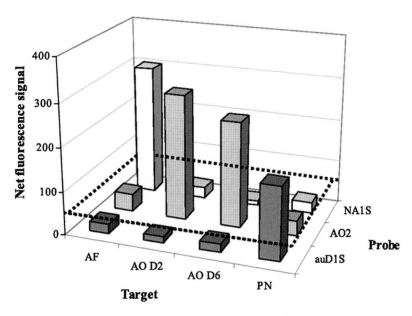

Figure 7.3. Multiplexed array response of three probe types—NA1S, AO2, and auD1S—to 5000 cell lysates of individual algae *A. fundyense* GTCA28 (AF); *A. ostenfeldii* HT-240D2 (D2) and HT-120D6 (D6); and *P. australis* 1BA (PN). The limit of detection is represented by the dashed line (Reprinted with permission from Ahn et al. (2006), Copyright © 2006, the American Society for Microbiology)

50-, and 500-cell samples. Combinations of two and three HAB species were tested with the microarray in triplicate, the results of which are shown in Figure 7.4 (Ahn et al. 2006). No cross-reactivity was observed with these closely related organisms. Using fiber optic microarrays in a sandwich assay format Ahn et al. have demonstrated the ability to rapidly (~45 min) detect low numbers (~5 cells) of multiple HAB species in culture and spiked seawater samples. The simplicity, sensitivity, and specificity of the analyses with this platform make it attractive to more automated approaches and field deployment.

1.4. Nucleic Acid Direct Hybridization

In a direct hybridization detection format, targets are labelled prior to hybridization to oligonucleotide probes, typically through PCR amplification in which one primer is modified with a fluorescent dye. Song et al. used this strategy to generate a highly multiplexed array capable of simultaneously detecting six biological warfare agents (BWAs) and 1 BWA stimulant organism in cultured and spiked wastewater samples (Song et al. 2006). These BWAs included *Bacillus anthracis* (BA), *Yersinia pestis* (YP), *Francisella tularensis* (FT), *Brucella melitensis* (BM), *Clostridium botulinum* (CB), and *Vaccinia* virus (VA); the BWA stimulant was *Bacillus thuringiensis kurstaki* (BTK). Two primer pools were developed, each containing 1 primer pair per microorganism. Reverse primers were labelled with Cy3 dye.

A multiplexed array of 18 50-mer oligonucleotide probes was created with the method used by Ahn et al. (Song et al. 2006). This high number of probes increased the level of encoding complexity in Song's array and required two dyes, Eu-dye and fluorescein, which were used in six concentrations (0.01, 0.025, 0.05, 0.1, 0.25, 0.5, and 1M). Examples of decoding images are presented in Figure 7.5, where the fluorescein concentration is constant and Eu-dye concentration is increased. Song noted that while the fluorescence intensity of encoding dye slowly decreased with time, likely due to photobleaching, the normalized intensities remained comparable.

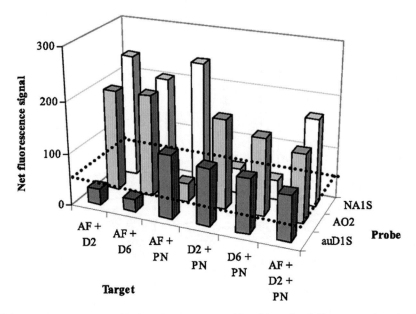

Figure 7.4. A multiplexed array with three bead types—NA1S, AO2, and auD1S—was used to simultaneously detect two or three of the following HAB species: *A. fundyense* GTCA28 (AF); *A. ostenfeldii* HT-240D2 (D2) and HT-120D6 (D6); and *P. australis* 1BA (PN). In all cases, only specific signals above the limit of detection (dashed line) were observed (Reprinted with permission from Ahn et al. (2006), Copyright © 2006, the American Society for Microbiology).

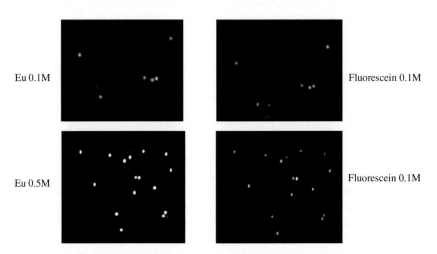

Figure 7.5. Images of two combinations of microsphere encoded with Eu-dye and fluorescein. Using separate optical channels, these two dyes were used in six concentrations to encode 18 different bead types (Reprinted with permission from Song et al. (2006), Copyright 2006 American Chemical Society).

To test the multiplexed array, Song et al. used the two multiplexed primer pools to amplify targets from culture or spiked wastewater samples, and pooled the PCR products prior to hybridization (Song et al. 2006). In this way, each array was able to simultaneously register signals for up to 18 probes specific to BWAs. Mixed bacteria samples were created by combining BA and BTK or YP and FT in several ratios. The results from this experiment are presented in Figure 7.6. In all cases, highly specific signals were observed for multiple probes per organism. Simulated environmental samples were then created by spiking wastewater with

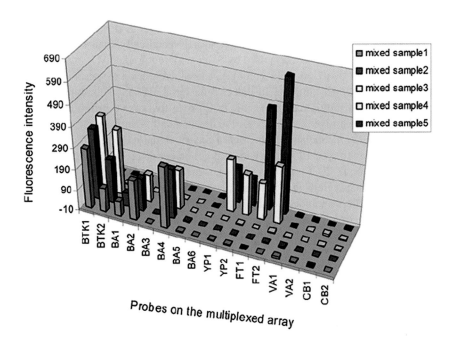

Figure 7.6. Direct hybridization detection of BWAs in samples created by combining two BWAs in different ratios. Samples consisted of (1) BA/BTK 1:1, (2) BA/BTK 1:5, (3) BA/BTK 1:9, (4) YP/FT 1:1, and (5) YP/FT 1:9. Primer pools were combined before hybridization (Reprinted with permission from Song et al. (2006), Copyright 2006 American Chemical Society).

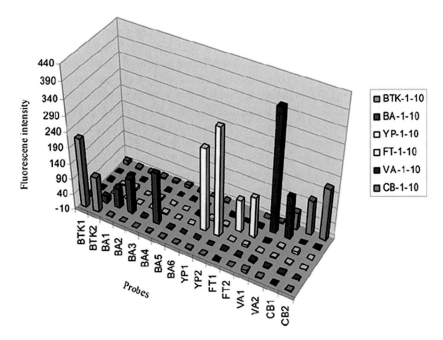

Figure 7.7. Parallel detection of multiple BWAs spiked 1:10 in wastewater samples. Each experiment consisted of two multiplexed PCR pools that were combined before hybridization to the array (Reprinted with permission from Song et al. (2006), Copyright 2006 American Chemical Society).

autoclaved BWA cultures. As shown in Figure 7.7, the PCR products of the two primer pools were detected in parallel by the multiplexed array. This research demonstrates the powerful ability to identify PCR products more sensitively and specifically than visualization by gel electrophoresis. The microsphere arrays used by Song et al. required only 30 min to detect 10 fM target concentrations in 50 μL sample volumes. Using dye-encoded sensors, a high number of probes was simultaneously utilized to detect 6 bacterial BWAs. This approach is limited, however, by the level of multiplexing possible in PCR, in this case 6 primer pairs.

Optical fiber microarrays have been used to detect and discriminate closely related strains of bacteria such as *E. coli.* Shepard et al. used a binary signal/no signal multilocus sequence typing (MLST) strategy to type 12 strains with 6 oligonucleotide probes in a direct hybridization experiment (Shepard et al. 2005). Probes were designed to be cross-reactive to PCR products amplified from virulence regions, generating a patterned response. To maximize differentiation, polymorphic regions were included in the interior of the probe sequences. Those sequences that were the perfect complements to the probes produced a signal as expected. With high stringency conditions, PCR products with sequences that were not perfect complements to the probes did not hybridize, resulting in no signal. Sequences with high variability in the primer regions did not amplify and produced no signal. In principle, this mismatch-intolerant binary response was able to discriminate 2^n strains, where n is equal to the number of probes. One advantage to using this microarray strategy is that emerging pathogenic strains with highly mutated or missing target genes will produce a signature similar to known bacteria. This fiber optic microarray MLST strategy takes advantage of the high variability of virulence regions and could be applied to the detection and typing of disease outbreaks such as avian flu and severe acute respiratory syndrome.

1.5. Extension Reactions

Several assays that rely on extension/ligation or single base extension reactions have been developed. Intended for high throughput human whole-genome genotyping, these fiber optic assays are capable of upwards of 1520 genotyping calls per fiber bundle, depending on assay format (Fan et al. 2003; Gunderson et al. 2005; Gunderson et al. 2006; Peiffer et al. 2006; Steemers et al. 2006). An allele-specific oligonucleotide (ASO) undergoes extension and ligation to a locus-specific oligo (LSO), creating a template that is then amplified with fluorescently-labelled "universal" primers via PCR. Hybridization of the PCR products to microarrays results in a two-colour genotyping call (Fan et al. 2003). A single base extension assay was developed that circumvents the need for PCR. Instead of using allele-specific primers, capture sequences on two different bead types were designed to differ by only one base on their 3' ends. After extension with biotinylated nucleotides and staining with streptavidin-modified fluorophores, the targets were hybridized to the beads and only the beads with perfect complements produced signals. Both of these extension assays can theoretically be used for bacterial detection and genotyping and offer the advantage of simplifying the target amplification needed. Universal PCR primers eliminate the time-consuming development of multiplexed PCR protocols.

2. Conclusions and Future Perspectives

Fiber optic biosensors based on sandwich immunoassays and nucleic acid hybridization have developed into highly specific and sensitive tools for the rapid detection of bacteria. In general, fiber optic microarray sensors are more specific, but slower than evanescent wave sensors (Iqbal et al. 2000). Using a sandwich assay format, waveguide biosensors have been used to analyze dirty samples such as *B. anthracis* spores mixed with powders (Tims and

Lim 2004) with little sample pre-treatment and are generally simpler systems than oligonu-cleotide microarrays. Such platforms are amenable to integration in microfluidic devices and can be adapted for flow-through sampling of environmental water samples (Kramer and Lim 2004). Recent improvements to waveguide sensitivity (Zourob et al. 2005b) and speed (Zourob et al. 2005c) will reinforce the advantages of these inexpensive, disposable (DeMarco and Lim 2002b; Zourob et al. 2005a) substrates, though the assays are still limited by the efficacy of the capture and signal antibodies. Results from rapid waveguide identification of bacteria can be confirmed by culture enrichment and/or PCR (Tims and Lim 2003; Kramer and Lim 2004; Simpson and Lim 2005). While typically only one type of antigen is detected per waveguide, bead-based fiber optic microarrays are capable of simultaneously detecting tens to thousands of nucleic acid targets (Gunderson et al. 2005; Song et al. 2006). Nucleic acid sandwich assays, particularly RNA-based approaches, offer increasingly multiplexed, rapid bacteria detection without amplification (Ahn and Walt 2005; Ahn et al. 2006). By employing PCR amplification, direct hybridization sensors have decreased the limit of detection to less than 10 fM (Song et al. 2006). Extension/ligation reaction strategies that have been applied to high throughput human genotyping are promising for bacterial pathogen detection. These assays use universal primers to circumvent typical PCR multiplexing issues and can theoretically be applied to strain differentiation and virulence gene identification.

The advantages of fiber optic biosensors for the detection of bacteria are compelling, as the assays that have been developed over the past ten years are sensitive, specific, and rapid. Portable, automated systems such as the RAPTOR have been rigorously field tested. There are also some disadvantages for FOBs. Microarray cost-effectiveness and data processing are increasingly a concern and will require further innovations and standards to exploit the advantages of such highly parallel assays (Garaizar et al. 2006). Enrichment is often required for immunoassays to reach a 1 CFU/mL limit of detection, though the time needed for this procedure can be brief. Microarray complexity, sample processing, and speed of analysis will undoubtedly improve in the coming years. Currently, no one platform fully captures the potential for the field. The challenge in the future will be how to balance the cost and complexity of the platform with the demands of the analytical problems.

Acknowledgements

The authors wish to thank Dr. Christopher LaFratta for his input and assistance in the preparation of this manuscript.

References

Ahn S, Kulis DM, Erdner DL, Anderson DM and Walt DR (2006) Fiber-optic microarray for simultaneous detection of multiple harmful algal bloom species. Appl. Environ. Microbiol. 72:5742–5749

Ahn S and Walt DR (2005) Detection of *Salmonella* spp. using microsphere-based, fiber-optic DNA microarrays. Anal. Chem. 77:5041–5047

Anderson GP, Rowe-Taitt CA and Ligler FS (2000) Raptor: A portable, automated biosensor. Proceedings of the First Conference on Point Detection for Chemical and Biological Defense

Bhunia AK, Banada P, Banerjee P, Valadez A and Hirleman ED (2007) Light scattering, fiber optic- and cell-based sensors for sensitive detection of foodborne pathogens. Journal of Rapid Methods and Automation in Microbiology. 15:121–145

Brogan KL and Walt DR (2005) Optical fiber-based sensors: Application to chemical biology. Current Opinion in Chemical Biology. 9:494–500

DeMarco DR and Lim DV (2001) Direct detection of *Escherichia coli* O157:H7 in unpasteurized apple juice with an evanescent wave biosensor. Journal of Rapid Methods and Automation in Microbiology. 9:241–257

DeMarco DR and Lim DV (2002a) Detection of *Escherichia coli* O157:H7 in 10- and 25-gram ground beef samples with an evanescent-wave biosensor with silica and polystyrene waveguides. J. Food Prot. 65:596–602

DeMarco DR and Lim DV (2002b) Detection of *Escherichia coli* O157:H7 in 10- and 25-gram ground beef samples with an evanescent-wave biosensor with slica and polystyrene waveguides. J. Food Prot. 65:596–602

DeMarco DR, Saaski EW, McCrae DA and Lim DV (1999) Rapid detection of *Escherichia coli* O157:H7 in ground beef using a fiber-optic biosensor. J. Food Prot. 62:711–716

Epstein JR, Lee M and Walt DR (2002) High-density fiber-optic genosensor microsphere array capable of zeptomole detection limits. Anal. Chem. 74:1836–1840

Epstein JR, Leung APK, Lee K-H and Walt DR (2003) High-density, microsphere-based fiber optic DNA microarrays. Biosensors and Bioelectronics. 18:541–546

Fan J-B, Oliphant A, Shen R, Kermani BG, Garcia F, Gunderson KL, Hansen M, Steemers F, Butler SL, Deloukas P, Galver L, Hunt S, McBride C, Bibikova M, Rubano T, Chen J, Wickham E, Doucet D, Chang W, Campbell D, Zhang B, Kruglyak S, Bentley D, Haas J, Rigault P, Zhou L, Stuelpnagel J and Chee MS (2003) Highly parallel SNP genotyping. Cold Spring Harbor Symp. Quant. Biol. LXVIII:69–78

Ferguson JA, Steemers FJ and Walt DR (2000) High-density fibre-optic DNA random microsphere array. Anal. Chem. 72:5618–5624

Ferreira AP, Werneck MM and Ribeiro RM (1999) Aerobiological pathogen detection by evanescent wave fiber optic sensor. Biotechnol. Tech. 13:447–452

Ferreira AP, Werneck MM and Ribeiro RM (2001) Development of an evanescent-field fiber optic sensor for *Escherichia coli* O157:H7. Biosensors and Bioelectronics. 16:399–408

Garaizar J, Rementeria A and Porwollik S (2006) DNA microarray technology: A new tool for the epidemiological typing of bacterial pathogens? FEMS Immunol. Med. Microbiol. 47:178–189

Geng T, Morgan MT and Bhunia AK (2004) Detection of low levels of *Listeria monocytogenes* cells by using a fiber-optic immunosensor. Appl. Environ. Microbiol. 70:6138–6146

Geng T, Uknalis J, Tu S-I and Bhunia AK (2006) Fibre-optic biosensor employing alexa-fluor conjugated antibody for detection of *Escherichia coli* O157:H7 from ground beef in four hours. Sensors. 6:796–807

Gunderson KL, Steemers FJ, Lee G, Mendoza LG and Chee MS (2005) A genome-wide scalable SNP genotyping assay using microarray technology. Nature Genetics. 37:549–554

Gunderson KL, Steemers FJ, Ren H, Ng P, Zhou L, Tsan C, Chang W, Bullis D, Musmacker J, King C, Lebruska LL, Barker D, Oliphant A, Kuhn KM and Shen R (2006) Whole-genome genotyping. Methods Enzymol. 410:359–376

Iqbal SS, Mayo MW, Bruno JG, Bronk BV, Batt CA and Chambers JP (2000) A review of molecular recognition technologies for detection of biological threat agents. Biosensors and Bioelectronics. 15:549–578

King KD, Vanniere JM, Leblanc JL, Bullock KE and Anderson GP (2000) Automated fiber optic biosensor for multiplexed immunoassays. Environ. Sci. Technol. 34:2845–2850

Kramer MF and Lim DV (2004) A rapid and automated fiber optic-based biosensor assay for the detection of *Salmonella* in spent irrigation water used in the sprouting of sprout seeds. J. Food Prot. 67:46–52

Lammerding AM and Fazil A (2000) Hazard identification and exposure assessment for microbial food safety risk assessment. International Journal of Food Micriobiology. 58:147–157

Ligler FS, Sapsford KE, Golden JP, Shriver-Lake LC, Taitt CR, Dyer MA, Barone S and Myatt CJ (2007) The array biosensor: Portable, automated systems. Anal. Sci. 23:5–10

Mothershed EA and Whitney AM (2006) Nucleic acid-based methods for the detection of bacterial pathogens: Present and future considerations for the clinical laboratory. Clin. Chim. Acta. 63:206–220

Peiffer DA, Le JM, Steemers FJ, Chang W, Jenniges T, Garcia F, Haden K, Li J, Shaw CA, Belmont J, Cheung SW, Shen RM, Barker DL and Gunderson KL (2006) High-resolution genomic profiling of chromosomal aberrations using infinium whole-genome genotyping. Genome Research. 16:1136–1148

Rowe-Taitt CA, Hazzard JW, Hoffman KE, Cras JJ, Golden JP and Ligler FS (2000) Simultaneous detection of six biohazardous agents using a planar waveguide array biosensor. Biosensors and Bioelectronics. 15:579–589

Shepard JRE, Danin-Poleg Y, Kashi Y and Walt DR (2005) Array-based binary analysis for bacterial typing. Anal. Chem. 77:319–326

Simpson JM and Lim DV (2005) Rapid pcr confirmation of *E. coli* O157:H7 after evanescent wave fibre optic biosensor detection. Biosensors and Bioelectronics. 21:881–887

Song L, Ahn S and Walt DR (2006) Fibre-optic microsphere-based arrays for multiplexed biological warfare agent detection. Anal. Chem. 78:1023–1033

Steemers FJ, Chang W, Lee G, Barker DL, Shen R and Gunderson KL (2006) Whole-genome genotyping with the single-base extension assay. Nature Methods. 3:31–33

Taitt CR, Anderson GP and Ligler FS (2005) Evanescent wave fluorescence biosensors. Biosensors and Bioelectronics. 20:2470–2487

Taitt CR, Shubin YS, Angel R and Ligler FS (2004) Detection of *Salmonella enterica* serovar Typhimurium by using a rapid, array-based immunosensor. Appl. Environ. Microbiol. 70:152–158

Tims TB, Dickey SS, DeMarco DR and Lim DV (2001) Detection of low levels of *Listeria monocytogenes* within 20 hours using an evanescent wave biosensor. American Clinical Laboratory. 28–29

Tims TB and Lim DV (2003) Confirmation of viable *E. coli* O157:H7 by enrichment and PCR after rapid biosensor detection. Journal of Microbiological Methods. 55:141–147

Tims TB and Lim DV (2004) Rapid detection of *Bacillus anthracis* spores direcly from powders with an evanescent wave fibre-optic biosensor. Journal of Microbiological Methods. 59:127–130

Zourob M, Mohr S, Brown BJT, Fielden PR, McDonnell MB and Goddard NJ (2005a) Bacteria detection using disposable optical leaky waveguide sensors. Biosensors and Bioelectronics. 21:293–302

Zourob M, Mohr S, Brown BJT, Fielden PR, McDonnell MB and Goddard NJ (2005b) An integrated metal clad leaky waveguide sensor for detection of bacteria. Anal. Chem. 77:232–242

Zourob M, Mohr S, Brown BJT, Fielden PR, McDonnell MB and Goddard NJ (2005c) An integrated optical leaky waveguide sensor with electrically induced concentration system for the detection of bacteria. Lab on a Chip. 5:1360–1365

8

Integrated Deep-Probe Optical Waveguides for Label Free Bacterial Detection

Mohammed Zourob, Nina Skivesen, Robert Horvath, Stephan Mohr, Martin B. McDonnell and Nicholas J. Goddard

Abstract

Rapid, specific, and sensitive detection of pathogenic bacteria is very important in areas like food safety, medical diagnostics, hospital infection, and biological warfare. Optical evanescent wave sensors are evolving to meet these challenges. Evanescent wave biosensors generate an electromagnetic wave at the sensor surface that penetrates 100–200 nm into the surrounding medium, and have proven to be a highly sensitive tool to monitor interactions in the close vicinity of the sensor surface. However, the use of such waveguides for bacterial detection is problematic for several reasons. These include the short penetration depth of the evanescent field of these waveguides (100–200 nm) compared to the typical size of a bacterium (1–5 μm), which places the majority of the bound cell outside the evanescent field. In addition, the low refractive index contrast between the bacterium cytoplasm and the aqueous environments in which detection is usually performed, as well as the availability and accessibility of antigens on the bacterium surface binding to the biorecognition elements. Finally, the sensor performance can be limited due to (1) the mass transport of large analytes like bacteria, which limits the binding to the immobilized recognition receptors; (2) non-specific binding; and (3) long analysis time.

This chapter will focus on the development of different configurations of deep-probe optical evanescent wave sensors such as metal-clad leaky waveguides (MCLW) and waveguide sensors with low-index substrates for bacterial detection. In addition, two complete detection systems integrated with physical force fields to overcome these problems will be presented. These sensor systems are based on MCLW sensors and integrated with, respectively, an electric field and ultrasound standing waves as a physical force to concentrate and enhance the capture of bacteria spores into immobilized antibodies on the sensor surface. The integration improves the detection limit by a few orders of magnitude and shortens the analysis time significantly.

1. Introduction

Bacterial contamination of food, water, and air is widespread and poses a continuous and expanding problem. U.S. estimates have placed the economic impact of food-borne illnesses associated with bacterial contamination of meat, poultry, and eggs as high as $14 billion per year (Buzby et al. 1996; Mead et al. 1999). In the U.K., the public laboratory service has

Mohammed Zourob • Biosensors Division, Biophage Pharma, Montreal (QC). **Nina Skivesen** • Inano, Interdisciplinary Nanoscience Center, University of Aarhus, Denmark. **Robert Horvath** • Nanotechnology Centre, Cranfield University, Bedfordshire, UK. **Stephan Mohr and Nicholas J. Goddard** • School of Chemical Engineering and Analytical Science (CEAS), The University of Manchester, UK. **Martin B. McDonnell** • Defence Science and Technology Laboratory, Porton Down, Wiltshire, UK.

M. Zourob et al. (eds.), *Principles of Bacterial Detection: Biosensors, Recognition Receptors and Microsystems*,
© Springer Science+Business Media, LLC 2008

indicated that, in 2001, food poisoning notifications had increased by 600% since 1982. Today, infectious diseases caused by bacteria account for as many as 40% of the 50 million annual deaths worldwide; and microbial diseases constitute the major causes of death in many developing countries (Abdel-Hamid et al. 1999). Another serious problem is hospital infections by opportunistic enterobacteria, as a result of their high resistance to antibiotics (Whyte et al. 1983). The annual worldwide cost of combating biofouling and microbial induced corrosion, which poses serious problems in industrial water-handling systems, i.e., corrosion in fluid conduits, mechanical parts, and other construction materials, has been estimated to be billions of U.S. dollars (Zeikus et al. 1991).

Conventional microbiological methods for determining the cell counts of bacteria employ selective culture, biochemical, and serological characterization. Although these achieve sensitive and selective bacterial detection, they typically require days to weeks to yield a result. The capability of instantaneous detection of pathogenic bacterial agents has increased tremendously based on the advent of sensor technologies capable of detection in real-time with high specificity, selectivity, and portability, using ultra-small volumes and low-cost assays.

Hitherto, a large number of different technologies had been developed for detection of bacteria utilizing optical, electrochemical, and mechanical detection methods. Optical biosensors have been the subject of intense interest over the past two decades. This is due to the numerous advantages provided by optical methods; e.g., they can be miniaturized, have multiplexing capabilities, and can combine rapid response times with high sensitivity for analyte evaluation. Hence, the use of such sensors in the real-time detection of bacteria appears promising.

A wide range of optical sensor systems have been employed for bacterial detection including; high index waveguide sensors (Kim et al. 2007; Adanyi et al. 2006), surface plasmon resonance sensors (Medina 1997; Fratamico et al. 1998; Perkins and Squirrell 2000; Leonard et al. 2004; Leonard et al. 2005; Taylor et al. 2006; Balasubramanian et al. 2007), resonant mirror sensors (Watts 1994; Hirmo et al. 1998; Hirmo et al. 1999), fiber optical techniques (Ferreira et al. 1999; Ferreira et al. 2001; DeMarco et al. 1999; DeMarco et al. 2001; Shepard et al. 2005), and interferometric sensors (Seo et al. 1999; Campbell et al. 2003). Most optical sensor systems are based on evanescent wave sensing, where perturbations in the refractive index (RI) close to the sensor surface are probed by the exponentially decaying wave.

The basic application of optical evanescent wave sensors is to apply the sample to be analyzed on the sensor surface, where the biorecognition molecules of interest are immobilized, forming a layer which has specific binding sites for the object in question. For biosensing the samples are typically aqueous solutions, where the analytes in question have a different RI from the buffer and hence the surface attachment of the analytes results in a change in RI at the sensor surface. The "extension" of the evanescent field away from the sensor surface is referred to as the penetration depth, d_P, and only RI changes within this depth can be detected by the sensor. More precisely, d_P is the decay length of the evanescent wave intensity.

Evanescent wave biosensors have proven to be a highly sensitive tool to monitor interactions in the close vicinity of the sensor surface. However, they generally have a limited d_P in the order of 100–150 nm, and thus they are suitable mostly for detection of interactions with small targets such as viruses (10–100 nm), and proteins (1–20 nm). Detection of large targets such as bacterial cells (0.5–5 μm) is more problematic and requires a sensor supporting a larger extension of the evanescent field. Moreover, it would be desirable to be able to tune the penetration depth and therefore fit the probing depth to a given analyte size.

In Figure 8.1 the principle of evanescent wave sensing is illustrated for evanescent fields with penetration depths of (a) 100 nm and (b) 1 μm in comparison to the biologically significant targets to be sensed.

This chapter will focus on the deep-probe optical evanescent wave sensors based on planar optical waveguides. Several designs for optical waveguide sensors with tuneable,

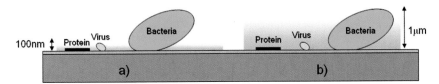

Figure 8.1. Schematics of the evanescent field of (a) conventional evanescent wave sensors of 100 nm penetration depth and (b) a deep-probe evanescent wave sensor, in comparison to biologically significant targets to be sensed.

and in principle unlimited, penetration depth have been suggested, which include low-index substrate waveguides (so-called reverse-symmetry waveguides) and metal-clad leaky waveguides (MCLWs).

1.1. Planar Optical Waveguides

Planar optical waveguides (POW) are structures that confine and guide electromagnetic radiation in a thin film deposited on a solid substrate (Figure 8.2). A POW at its simplest consists of a three-layer structure: a substrate (S); a thin, transparent dielectric film (F) deposited on the substrate; and a cover medium (C) on top of the thin film with refractive indices (RIs) n_S, n_F, and n_C, respectively. For sensor applications the cover medium will constitute the sample to be analyzed.

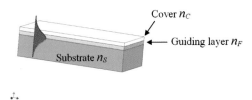

Figure 8.2. Schematic showing a standard POW, along with the transverse intensity profile of a guided mode.

The interesting feature of a POW is that light can be guided in the film, also called the waveguide film, by total internal reflection (TIR) if the RI of the film is higher than the RIs of the surrounding media, $n_F > \max\{n_S, n_C\}$ and if the light propagates in the film above a certain angle called the critical angle, θ_C.

In the following we summarize the basics of TIR, the main physical features of evanescent waves, and the most important configurations when these waves are used for sensing with POWs.

1.2. Total Internal Reflection and Evanescent Waves

Figure 8.3 illustrates different situations in which light is reflected from and transmitted (refracted) across a smooth and planar interface between two different non-absorbing and uniform materials having refractive indices n_1 and $n_2 (n_1 > n_2)$, depending on the incident angle θ_I, in the material with RI n_1. The angle of the refracted light, θ_R, is given by Snell's law: $n_1 \sin(\theta_I) = n_2 \sin(\theta_R)$. At a specific angle of incidence, the angle of refracted light reaches $\theta_R = 90°$. This is called the critical angle: $\theta_C = \sin^{-1}(n_2/n_1)$. The critical angle only exists when the light is incident on the boundary from the high RI medium. If the incident angle is small and below the critical angle, a fraction of light is reflected back into medium n_1 at an angle equal to the incident angle, and the rest is transmitted into medium n_2 at the angle

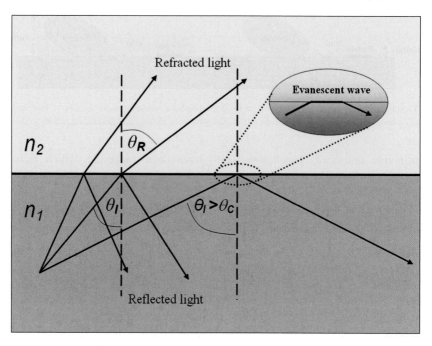

Figure 8.3. Illustration of refracted and reflected light at a specularly smooth and planar interface of two different nonabsorbing materials $(n_1 > n_2)$. The special case of total internal reflection (TIR) is also illustrated.

θ_R. Increasing θ_I also results in an increase of θ_R, and at θ_C the transmitted wave propagates parallel to the interface. When the angle of incidence exceeds the critical angle, no refraction occurs; instead, the light is reflected back into the medium of incidence and TIR is obtained.

However, deeper analysis shows that during TIR the light penetrates into the low index (cover) media before "turning back" into the media with higher RI. When optical rays are considered, the reflected optical ray is shifted parallel to the interface, as shown in Figure 8.3. This lateral shift is called the Goos-Hänchen shift (Born and Wolf 1999; Tien 1977). The "short journey" in the low index media also means that the light experiences a phase shift during TIR. But it is important to emphasize that under the TIR condition, no net flow of energy occurs across the interface; all of the light intensity reflects back and the interface in many senses acts like a perfect mirror.

When considering light as optical waves instead of as rays, the light that penetrates and propagates in the low index media is called the evanescent wave. More detailed analysis shows that this wave travels parallel with the interface (just like the optical ray shown in Figure 8.3) and has an exponentially decaying amplitude away from the interface. The field amplitude of the evanescent wave is described by $A = A_0 \exp(-z/d_P)$, where $z > 0$ is the distance from the interface (see Figure 8.4). The word "evanescent" refers to this decaying light amplitude perpendicular to the surface. This amplitude results in an intensity decrease away from the interface, and thus the most sensitive area is that closest to the sensor surface. RI variations inside the bulk are therefore not detected. It is important to note that the penetration depth can be controlled by an appropriate choice of the ratio of the refractive indices of the two materials, the angle of incidence, and the wavelength of the incident light.

In many practical situations it is advantageous to apply the optically denser medium as a thin film instead of the semi-infinite medium, as shown in Figure 8.2. This film is usually applied on a solid substrate with a refractive index less than the RI of the film. In this case the

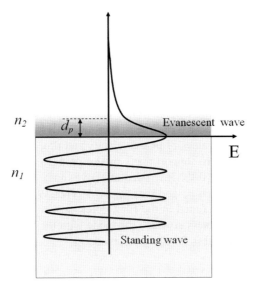

Figure 8.4. Schematic of the field structure when TIR occurs at a smooth planar interface.

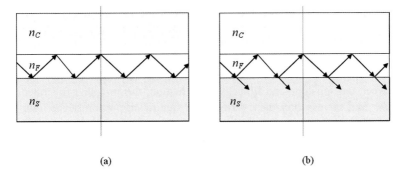

(a)   (b)

Figure 8.5. Illustration of (a) guided waveguide modes and (b) leaky waveguide modes in a three-layered waveguide structure.

light can travel by total internal reflection inside the film, in a zig-zag manner (see Figure 8.5a), creating the evanescent wave both in the aqueous solution and in the substrate (Tien 1977).

1.3. Waveguide Modes

When the traveling light experiences TIR both at the film/cover and film/substrate boundaries the "light propagation form" is called a *waveguide mode* (see Figure 8.5a). Waveguide modes only arise when the reflected light from the film/cover and the film/substrate interfaces achieve constructive interference. To understand this, it has to be emphasized that when light propagates in the waveguide film it experiences a phase shift due to both the distance travelled in the film and the TIR at the interfaces. To achieve guided modes, incident and twice-reflected light at any position in the film must be in phase, which requires the magnitude of the total phase change after a complete cycle in the film to be an integer multiple of 2π (Tien 1977). Thus not all arbitrary beams at incident angles beyond the critical angle can achieve a guided mode, but only waves with specific discrete incident angles. When considering the phase change at every reflection, these angles are called the resonant angles. For a more comprehensive review on the above mentioned interference phenomena and waveguide resonances, the reader can consult a review article by Tien (1977).

Typically, the waveguide modes are used as the sensing feature in POW sensors. When the RI of the cover is changed, the incident angle for which the waveguide modes are excited shifts, thus the resonant angle corresponds to a given RI of the cover medium for the specific waveguide design. The resonant angle of the waveguide mode is also influenced by the RI of the film and the surrounding medium, and the modes can be characterized by a so-called effective refractive index, N, given by $n_F \cos(\theta_I)$, representing the refractive index of an equivalent bulk medium having identical values of phase velocity and propagation constant.

A POW can be designed to support many modes propagating along the guide axis, each with a different mode order m (0,1,2,3,....) and effective index N_m. The effective index of all the guided modes varies between n_F and $\max\{n_S, n_C\}$ (Tien 1977). Generally, the number of waveguide modes that can be excited in the film increases with increasing film thickness, d_F and n_F. The film's lowest thickness at which the mode with the mth mode order can no longer exist is called the cut-off thickness for the given mode.

1.4. Frustrated Total Internal Reflection, Leaky Modes

In some cases the light experiences TIR at the film/cover boundary only and the light energy leaks out into the substrate (see Figure 8.5b). These types of modes are called *leaky modes*. This happens if the RI of the substrate is larger than the RI of the film (TIR is not possible at the film/substrate boundary); however in these situations usually the light energy leakage into the substrate is large, and the propagating mode damps significantly during propagation.

One way to get a relatively small leakage is to introduce a thin film of metal or absorbing layer such as polymer or dye with a complex dielectric function or a low refractive index material between the substrate and film. In this situation, the so-called frustrated TIR (FTIR) happens at the film's lower boundary, reflecting a significant portion of the light energy back into the film. These types of leaky modes are very popular in practical applications and, as with the waveguide modes, they are broadly used in the evanescent wave sensor systems.

Leaky modes can also be characterized by a mode order, m, and an effective index, N. When these modes are used for sensing, the mode's effective refractive index, N_m, is shifted due to the analytes, and this shift is followed, as with the waveguide modes. However, it is noted that for leaky modes the effective refractive index is usually complex, and the imaginary part reflects the damping of the mode due to the energy leakage into the substrate.

1.5. Literature on Waveguides for Bacterial Detection

A number of optical waveguide biosensors have been employed for bacterial detection, including high index waveguide sensors (Kim et al. 2007; Adanyi et al. 2006), surface plasmon resonance sensors (Medina 1997; Fratamico et al. 1998; Perkins and Squirrell 2000; Leonard et al. 2004; Leonard et al. 2005; Taylor et al. 2006; Balasubramanian et al. 2007), resonant mirror sensors (Watts 1994; Hirmo et al. 1998; Hirmo et al. 1999), fiber optical techniques (Ferreira et al. 1999; Ferreira et al. 2001; DeMarco et al. 1999; DeMarco et al. 2001; Shepard et al. 2005), and interferometric sensors (Seo et al. 1999; Campbell et al. 2003). Various configurations and assays using surface plasmon resonance, optical fibre sensors, and planar interferometric platforms have been employed for the detection of bacterial cells and bacterial toxin. However, these subjects are covered in detail in Chapters 5, 7, and 9 of this book and so will not be discussed here.

A resonant mirror biosensor, an IAsys instrument (Affinity Sensors, Cambridge, UK), has been used to distinguish between bacterial strains on the basis of the difference in cell surface proteins. *Staphylococcus aureus* (Cowan-1), which produces protein A at the cell surface, was detected by binding to human IgG. The LOD for *S. aureus* (Cowan-1) was quoted as 8×10^6–8×10^7 cells/ml. Conversely, the *S. aureus* strain (Wood-46), which does not express protein A, was not detected. The sensitivity of detection was increased 100-fold when using a human IgG-colloidal gold complex (30 nm in diameter) in a sandwich assay format (Watts et al. 1994). The resonant mirror was also used for the study of the interaction of *Helicobacter pylori* with immobilized silylglycoconjugates (Hirmo et al. 1998) and human gastric mucin (Hirmo et al. 1999). The grating waveguide sensors have been used for bacterial detection, e.g., Microvacuum chips (OWLS, Microvacuum, Hungary) using refractive index detection. The advantage of this configuration is that the grating coupler is integrated into the structure and forms an integral part of the sensor. The MicroVaccum grating platform has been used for label-free detection of *Salmonella* using antibodies. The reported limit of detection was 1.3×10^3 cfu/ml (Kim et al. 2007; Adanyi et al. 2006).

The NRL fluorescence planar format has been developed by Rowe et al. (1999). The system utilizes a standard sandwich immunoassay format with antigen-specific "capture" antibodies immobilized in a patterned array using a PDMS flow-cell on the surface of a planar waveguide. The assay has been demonstrated for the detection of different bacteria such as *E. coli* O157:H7, *Listeria monocytogenes, Anthracis, B. globigii, Salmonella typhimurium, Shigella* species, and *Campylobacter jejuni* bacteria in buffer, and a variety of food and beverage samples including milk, yogurt, sausages, carcass, and apple juice. The reported detection limit was 10^3–10^6 cells/ml in less than 30 min (Sapsford et al. 2004; Sapsford et al. 2006; Taitt et al. 2004a; Taitt et al. 2004b; Shriver-Lake et al. 2007). Antimicrobial peptides (AMP) have also been utilized as recognition molecules with the NRL Array Biosensor for the detection of *E. coli* O157:H7 and salmonella bacteria. The limits of detection (LODs) using the AMP were similar with their analogous antibody-based assay. LODs for *E. coli* O157:H7 and *S. typhimurium* obtained with AMPs during sandwich assays were in the ranges of 5×10^4–5×10^5 and 1×10^5–5×10^6 cells mL^{-1}, respectively (Kulagina et al. 2006). Monosaccharide arrays such as *N*-acetyl galactosamine (GalNAc) and *N*-acetylneuraminic acid (Neu5Ac) have been investigated using the NRL system for the detection of bacterial toxins. The reported LODs for cholera toxin and tetanus toxins were at 100 ng/ml (Ngundi et al. 2006). An evanescent waveguide Mark 1.5 instrument (IVD systems, Santa Barbara, CA) was used to detect BG spores and *Erwinia herbicola* (EH) (live bacterium) as mimics for biological warfare agents. The method used involves mixing the analytes with a fluorescent-labeled antibody that is brought in contact with the capture antibody array. The LOD for BG and EH was 5×10^5 cfu/mL (Sipe et al. 2000).

Ligler et al. developed another format of integrated waveguide biosensors for the detection of *E. coli* O157 using fluorescent sandwich immunoassays performed inside a glass capillary. The waveguide glass capillary was illuminated at 90 degrees and subsequent collection of fluorescence from the end of the waveguide. The capillary tube serves as a waveguide and as an incubation vessel for the growth of bacterial pathogens after capture, allowing for confirmation of viability, as well as amplification and retrieval for further characterization (Zhu et al. 2005).

2. Deep-Probe Optical Waveguide Sensors with Tunable Evanescent Field

The basic conventional optical waveguide sensor comprises, as already mentioned, a three-layer structure; and this sensor type is most suited for measuring interaction on a small scale, due to the limited penetration depth of the evanescent field into the sample. A larger penetration depth and the ability to tune the penetration depth of the evanescent field to a given distance will result in sensors for specific purposes with exactly the probing depth needed for a given sample.

In order to achieve a deep-probe waveguide sensor element, a different design of the waveguide has to be used, which can be achieved either by choosing materials with other properties or by adding additional layers to the conventional waveguide structure. In this section, two waveguide designs will be introduced, based on the low-index-substrates (in the literature also called the reverse-symmetry waveguide) and the metal-clad leaky waveguide (MCLW) (in the literature also called the metal-clad waveguide (MCWG) or coupled plasmon-waveguide resonator (CPWR)).

2.1. Waveguide Modes, Light Coupling and Sensing Depths of Evanescent Waves

In order to optimise the design of the waveguide structure to achieve a penetration depth of the evanescent wave in the μm-range it is necessary to analyse the influence of different waveguide parameters on the electromagnetic field in the waveguiding structure.

As a plane wave propagates as a zig-zag ray along the dielectric waveguide film by TIR, it undergoes a phase shift of $\phi_{F,S}$ and $\phi_{F,C}$ upon reflection at the substrate and cover boundaries, respectively. Furthermore, as the wave travels once across the thickness (d_F) of the guide, it undergoes a phase change, $\phi_{\Delta S} = 2d_F k \sqrt{n_F^2 - N^2}$, where $k = 2\pi / \lambda$ is the wave number in vacuum. For all the multiple reflected waves to interfere constructively, the total phase accumulated in one cycle between the two guide boundaries must be an integral multiple of 2π, and thus the basic criteria for waveguide modes can be written as:

$$2\pi m = \phi_{\Delta S} + \phi_{F,C} + \phi_{F,S}, \tag{8.1}$$

where $m = 0, 1, 2 \ldots$ is the mode order.

The mode order of a waveguide mode determines the electromagnetic field in the waveguide, and to get a complete picture of the electromagnetic field in the waveguide structure the zig-zag beam description is not comprehensive enough; therefore descriptions based on the electromagnetic fields in the different waveguide layers have been used. The profile of the electromagnetic field in the waveguide can be calculated from Maxwell's equations and the boundary conditions.

In Figure 8.6 the electromagnetic fields in the waveguide layers are represented by plane waves, where A_I and B_I, ($I = S, F, C$) are the amplitudes of the up- and down-going waves, respectively, in the individual medium.

The eigenmodes of the three-layer (substrate-film-cover) structure can be obtained by solving Maxwell's equations for the structure, initially assuming that the waveguide structure is

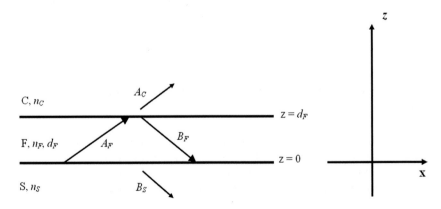

Figure 8.6. The electromagnetic field in a three-layer structure with a waveguide film of thickness d_F.

left alone without any field incident from the outside—which is the usual procedure to identify the eigenmodes of a system. The solutions are obtained by using the solution ansatz ψ for the electromagnetic field, based on the fields illustrated in Figure 8.6.

$$\psi = \begin{cases} A_C \exp\left[i\left(k_x x + k_{z,C} z\right)\right] & \text{(in the cover)} \\ A_F \exp\left[i\left(k_x x + k_{z,F} z\right)\right] + B_F \exp\left[i\left(k_x x - k_{z,F} z\right)\right] & \text{(in the film)} \\ B_S \exp\left[i\left(k_x x - k_{z,S} z\right)\right] & \text{(in the substrate),} \end{cases} \tag{8.2}$$

where ψ represents the E-field for TE-polarized light and the H-field for TM-polarized light, k_x is the x-component of the wave vector, and $k_{z,I} = k\sqrt{n_I^2 - N^2}$ is the wave vector component along z in medium I. From the boundary conditions, (I)ψ and $\partial_z^2 \psi$ are continuous across a boundary for TE-polarized light and (II)ψ and $n^{-2}\partial_z^2 \psi$ are continuous across a boundary for TM-polarized light; it is possible for each polarization to obtain four equations relating the four amplitudes in Eq. 8.2. By introducing a polarization index $\rho = 0$, 1 for the TE and TM polarized case, respectively, the two systems of equations can be written in the form:

$$\overline{\overline{A_\rho}}\, \vec{\psi} = \vec{0}, \tag{8.3}$$

where $\vec{\psi} = \{A_C, A_F, B_F, B_S\}$ and $\overline{\overline{A_\rho}}$ is given by:

$$\overline{\overline{A_\rho}} = \begin{bmatrix} \exp\left[ik_{z,C}d_F\right] & -\exp\left[ik_{z,F}d_F\right] & -\exp\left[-ik_{z,F}d_F\right] & 0 \\ \dfrac{k_{z,C}}{n_C^{2\rho}}\exp\left[ik_{z,C}d_F\right] & -\dfrac{k_{z,F}}{n_F^{2\rho}}\exp\left[ik_{z,F}d_F\right] & \dfrac{k_{z,F}}{n_F^{2\rho}}\exp\left[-ik_{z,F}d_F\right] & 0 \\ 0 & 1 & 1 & -1 \\ 0 & \dfrac{k_{z,F}}{n_F^{2\rho}} & -\dfrac{k_{z,F}}{n_F^{2\rho}} & \dfrac{k_{z,S}}{n_S^{2\rho}} \end{bmatrix}.$$

In order to get non-trivial solutions, the determinant of $\overline{\overline{A_\rho}}$ needs to be zero: $Det\left[\overline{\overline{A_\rho}}\right] = 0$, which leads to the mode equation:

$$m\pi = d_F k \left(n_F^2 - N_m^2\right)^{0.5} - \arctan\left[\left(\frac{n_F}{n_S}\right)^{2\rho}\left(\frac{N_m^2 - n_S^2}{n_F^2 - N_m^2}\right)^{0.5}\right] - \arctan\left[\left(\frac{n_C}{n_S}\right)^{2\rho}\left(\frac{N_m^2 - n_C^2}{n_C^2 - N_m^2}\right)^{0.5}\right]. \tag{8.4}$$

Equation 8.4 is equivalent to the solution found by the ray-tracing approach in Eq. 8.1, when the phase shifts at the boundaries are written as:

$$\phi_{F,J} = -2\arctan\left[\left(\frac{n_F}{n_J}\right)^{2\rho}\left(\frac{N_m^2 - n_J^2}{n_F^2 - N_m^2}\right)^{0.5}\right]. \tag{8.5}$$

The mode equation includes the value N_m, which is referred to as the effective refractive index of the mode and corresponds to the resonance angle of the propagating beam in the film, $N_m = n_F \sin(\theta_m)$. If $N_m > n_F$, all three values of $k_{z,I}$, $(I = S, F, C)$ are imaginary and the solution is non-physical.

For $n_F > N_m > n_{\max}$ exponential decaying fields are formed in the cover and substrate, and a discrete set of sinusoidal fields inside the film are generated. If $n_{\max} > N_m > n_{\min}$, the field will be exponentially decreasing in the cover layer, and sinusoidal in the film and substrate layers, if $n_S > n_C$. These modes, known as substrate radiation modes, propagate into the substrate and damp out over a short distance in the z-direction. The case in which $n_{\min} > N_m$ results in an oscillatory field in all three regions. These non-guided modes lose energy through both substrate and cover regions and are called air radiation modes (Tien 1977).

For waveguide modes in a pure dielectric structure, N_m varies between the values $n_{max} = Max\{n_C, n_S\}$ at $d_F = d_{CUTOFF}$ and n_F at $d_F \to \infty$. d_{CUTOFF} is defined as the film thickness at which the field's propagation angle in the film gets below the critical angle at either the film-cover or film-substrate interface, in which case the light escapes from the waveguide. The fact that $n_{max} < N_m < n_F$ implies that $k_{z,C} = k \left(n_C^2 - N_m^2\right)^{0.5}$ and that $k_{z,S} = k \left(n_S^2 - N_m^2\right)^{0.5}$ are always purely imaginary and the electromagnetic fields in the substrate and cover medium are therefore evanescent. However, in the film, the real value of $k_{z,F} = k \left(n_F^2 - N_m^2\right)^{0.5}$ results, in that the electromagnetic field consists of two propagating waves. Hence, the solutions assume characteristic mode power profiles as shown in Figure 8.7.

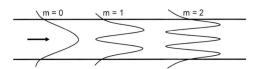

Figure 8.7. The three lowest-order mode-power profiles in a three-layer waveguide.

From Eq. 8.3 the exact electromagnetic field can be calculated for a given mode order, a given waveguide structure, and a given angle (for example, the resonance) by assuming one of the field amplitudes known, e.g., $A_F = 1$.

From the three-layer mode equation, Eq. 8.4, the values d_{CUTOFF} and the penetration depth of the evanescent wave in the cover medium, $d_{P,C}$, can be found for the three-layer structure as:

$$d_{CUTOFF} = \frac{1}{k\sqrt{n_F^2 - n_{max}^2}} \left[\arctan \left(\frac{n_F^{2\rho}}{n_{min}^{2\rho}} \frac{\sqrt{n_{max}^2 - n_{min}^2}}{n_F^2 - n_{max}^2} \right) + 2\pi m \right], \tag{8.6}$$

$$d_{P,C} = k^{-1} \sqrt{N_m^2 - n_C^2} \left[\left(N_m/n_F\right)^2 + \left(N_m/n_C\right)^2 - 1 \right]^{\rho}. \tag{8.7}$$

From the cut-off thickness it is also possible to calculate the number of modes $(m + 1)$ that can be excited in a waveguide film of a given thickness.

2.1.1. Light Coupling Techniques

Four techniques can be employed to couple an optical beam which is propagating in free space into a thin-film optical waveguide. These are: direct focusing, end-fibre, grating, and prism coupling (see Figure 8.8a-d).

For the direct focusing and end-fibre coupling the light is coupled into the waveguide film at the end of the device. Direct focusing describes the situation where the beam is focused at the end of the waveguide film through a lens, while for end-fibre coupling the light is coupled from a fibre into the end of the waveguide film using a fibre with a core dimension comparable to the thickness of the waveguide film. Both techniques are quite simple; however, the coupling losses can be large, due to scattering and reflections from the interface between the air and the waveguide film arising from different angles of light incident at the interface and roughnesses and variations across that interface.

Briefly, prism and grating coupling elements in many practical sensor devices make possible the "communication" of the outside world with the waveguide and are used to monitor the mentioned effective refractive indices.

When a grating is used to excite the modes, usually the waveguide film is modulated at the film/cover or film/substrate interface (or both) to form a surface relief grating coupler.

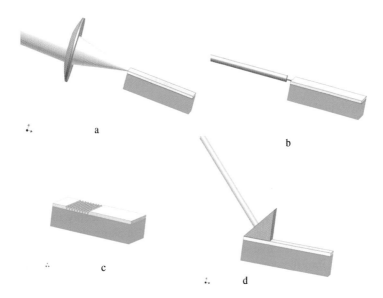

Figure 8.8. Schematic of (a) direct focusing, (b) end-fire coupling, (c) grating coupling, and (d) prism coupling techniques.

The incident beam illuminates this grating, and one of the diffraction orders excites the modes. The mode's effective refractive index can be calculated from the so-called grating equation (Tein 1977; Tiefenthaler and Lukosz, 1989):

$$N_m = n_{air} \sin{(\theta_m)} + l\frac{\lambda}{\Lambda}, \tag{8.8}$$

where n_{air} is the RI of air, λ is the vacuum wavelength of the incident light, Λ is the grating period, and l is the diffraction order.

The condition for prism coupling is similar to the grating condition. For prism coupling the prism often constitutes the substrate of the waveguide structure, or the RI of the prism and that of the substrate are matched, and thus the correspondence between the effective refractive index of the waveguide and the parameters for the prism coupling is given by the following:

$$N_m = n_{prism} \sin{(\theta_{prism})}, \tag{8.9}$$

where n_{prism} is the RI of the substrate and θ_{prism} is the incident angle inside the prism at the interface of the prism and the remaining waveguide structure.

In both cases the incident angle is monitored using a goniometer or a CCD and converted to effective refractive index values using the above equations.

In contrast to the grating coupling, the prism coupling results in direct coupling of the light into the waveguide structure. Thus the prism coupling is often used for reflectance measurements from the waveguide structure rather than measurements of the guided light in the waveguide structure. For prism coupling, Fresnell's reflection laws can therefore be applied for calculating the intensity ratio between the reflected light and the incoupled light. In general, the total reflectance from a layered structure can be calculated using Fresnel's reflection laws, here for a three-layered structure:

$$R_{123} = |r_{123}|^2 = \left| \frac{r_{12} + r_{23} \exp\left[i2k_{z,2}d_2\right]}{1 + r_{12}r_{23} \exp\left[i2k_{z,2}d_2\right]} \right|^2, \tag{8.10}$$

$$r_{IJ} = \frac{n_J^{2\rho}k_{z,I} - n_I^{2\rho}k_{z,J}}{n_J^{2\rho}k_{z,I} + n_I^{2\rho}k_{z,I}}, \, k_{z,I} = k\sqrt{n_I^2 - N^2}, \, k = \frac{2\pi}{\lambda}.$$

2.2. Waveguide Designs Based on Low-Index Substrates

A three-layer, planar optical waveguide typically consists of a high-refractive-index film with refractive index (RI) n_F and thickness d_F sandwiched between two lower-refractive-index materials referred to as the substrate and cover medium with RIs n_S and n_C, respectively. The cover medium is usually an aqueous solution with refractive index around 1.33 containing, for example, the bacteria to be detected.

In conventional waveguides the substrate is usually glass or polymer, which have refractive indices larger than that of the (aqueous) cover sample being investigated, typically 1.53 for glass. This fact causes the evanescent field to extend deeply into the substrate and shortly into the cover (Horvath et al. 2002a). However, making the refractive index of the substrate smaller than the refractive index of the aqueous cover solution, the waveguide mode will penetrate deeper into the cover medium than into the substrate. The mode's penetration depth into the aqueous solution can be tuned, in principle, up to infinity. This type of waveguide geometry is often termed a reverse symmetry waveguide and was introduced by Horvath and coworkers (Horvath et al. 2002a; Horvath et al. 2002b) and analyzed in several studies (Horvath et al. 2003a; Horvath et al. 2005a; Horvath et al. 2005b; Horvath et al. 2006).

To understand the extended mode penetration depth using reverse symmetry waveguides, Eq. 8.7 needs to be revisited for the mode's penetration depth into the cover medium. As can be seen, the penetration depth depends on the wavelength, the cover medium index, and the effective refractive index of the mode (N_{eff}). For waveguide modes the effective refractive index is by definition less than the refractive index of the waveguiding film and larger than the maximum value of the refractive index of the substrate and cover medium: $n_F > N_m > n_{max}$. The effective refractive index can be tuned in this range by changing, for example, the thickness of the waveguiding film (see Eq. 8.4).

Figure 8.9 shows the cover penetration depth as a function of the effective refractive index. It is immediately seen that for $n_S > n_C$, the penetration depth is limited and reaches its maximum value at $N = n_S$. For $n_S = 1.53$ (glass substrate), the maximum penetration depth is

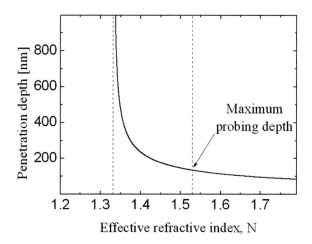

Figure 8.9. Cover penetration depth as a function of the effective refractive index. Using a substrate with RI less than the RI of the aqueous cover solution, the penetration depth into the aqueous cover can be tuned up to infinity. While using glass as a substrate with RI of 1.53, the probing depth has a maximum value around 180 nm.

around 180 nm. Using such a waveguide, only a very small part of the whole surface-attached bacteria volume is detected. However, by choosing a substrate with a refractive index *less* than the refractive index of the cover medium $n_C > n_S$, the allowed range of effective refractive indices is between n_C and n_F. Since at $N = n_C$ the cover penetration depth goes to infinity (referred to as the cover cut-off), now the penetration depth can be tuned in principle up to infinity (see Figure 8.9). Penetration depths fitted to the size of the detected bacteria are possible using this type of waveguide geometry.

Reverse symmetry waveguides can be made by using freestanding waveguiding films (air as a substrate) (Horvath et al. 2002a; Horvath et al. 2003b; Skivesen et al. 2003) or by using nanoporous silica substrates (Horvath et al. 2005b). This material contains nanometer-sized air pores embedded in glass, lowering the average (or effective) refractive index of the material without introducing significant loss effects. Using nanoporous silica, a waveguide substrate refractive index around 1.2 has been demonstrated, and such a waveguide was tested for bacteria detection (Horvath et al. 2003a).

2.2.1. Bacteria Detection Using Reverse Symmetry Waveguides

The reverse symmetry waveguide was demonstrated for bacterial detection using non-specific electrostatic interaction and a positively charged layer of poly-L-lysine coating on the sensor surface. 3×10^7 *E. coli* K12 cells per ml were used in phosphate buffer, and flow above the waveguide using a fluid system (Horvath et al. 2003a).

The waveguide applied for sensing consisted of a glass support, a micron-thick nanoporous silica layer (acting as the waveguide substrate), and a polystyrene (PS) waveguiding film. For light coupling a surface relief grating was heat-embossed into the PS film. The waveguiding film was so thin (approx. 150 nm) that only the lowest order TE and TM modes were supported. The detailed fabrication steps can be found in Horvath et al. (2005b).

After exposing the waveguide surface to the bacterial solution, a shift of the incoupled resonant peak positions to higher values was noticed due to the attachment of the bacterial cells to the surface of the grating coupler (Figure 8.10a). This shift to higher coupling angles indicates that the RI of a bacterial cell is slightly larger compared to the RI of buffer. After flushing the cuvette with pure buffer again the peak positions remained, indicating that the

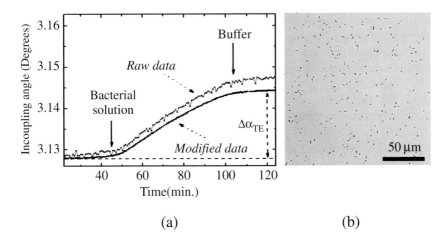

(a)   (b)

Figure 8.10. Measured peak shift versus time due to bacterial adhesion to the waveguide surface (a), and the microscope image of the waveguide surface after the exposure to *E. coli* cells (b). (Reproduced with permission from Optics Letters 2003, 28, 1233–1235. Copyright 2003. Optical Society of America).

bacterial cells were already attached to the surface (Figure 8.10b). Using the nanoporous silica-based reverse symmetry waveguide, noise-corrected detection limits of 78 and $60 \, \text{cells/mm}^2$ for TE and TM mode were demonstrated (Horvath et al. 2003a). In this system neither specific bacterial detection using a specific recognition element, e.g., antibodies, nor the detection limit in terms of cells per ml has been documented.

Even so, the presented reverse symmetry waveguide sensor shows great potential in terms of sensitivity, due to its enhanced penetration depth into the sample. The detection of larger biological objects such as bacterial cells ($\sim 1 \, \mu\text{m}$) and eukaryotic cells ($\sim 10 \, \mu\text{m}$) will benefit dramatically from the extended evanescent field. By choosing the correct film thickness, a tailor-made reverse symmetry waveguide for a sample of a certain size can, in principle, be constructed. Detection of eukaryotic cell attachment and spreading has been also demonstrated in Horvath et al. (2005a). These results indicate that the reverse symmetry waveguide can be particularly useful for the monitoring of cell behaviour, such as in toxicological screening and drug-testing, one of the fastest growing biosensing activities at the moment.

2.3. Waveguide Designs Based on Metal- and Dye-Clad Substrates—Leaky Modes

Deep-probe waveguide sensors can also be designed using more prevalent materials than presented in the previous section for the reverse symmetry waveguide, namely by introducing a fourth layer to the waveguide structure. The extra layer should comprise a material with a complex RI value. This can be, for example, a metal layer, a polymer film, or a dye layer. In this section, bacterial cell detection will be presented using waveguide designs including a metal layer and a dye layer. In the literature these structures are referred to in various ways: for example, the metal-clad leaky waveguide (MCLW), metal-clad waveguide (MCWG), coupled plasmon resonance waveguides (CPRW), and for the dye layer, dye-clad leaky waveguide (DCLW). MCLW will be used *as a general term*; and for a specific waveguide design, the material used for the cladding will be stated as part of the abbreviation by replacing the "M" with the specific material used—for example, AG-CLW for a MCWG where silver is used for the cladding and DCLW for a dye cladding.

The deep-probe four-layered waveguides comprise a planar structure of a substrate S, a thin cladding of metal or dye or other complex-valued materials M, a dielectric waveguide film F and a cover medium C, with refractive indices and layers thickness n_I and d_I, respectively. The structure of these waveguides is thus similar to that of a conventional dielectric waveguide sensor, but with an extra layer introduced between the substrate and the waveguide film. The MCLW can also be compared to the structure of an SPR-sensor. The difference between the two structures is that a dielectric waveguide film is added on top of the metal-coated layer for the MCLW structure.

The MCLW sensor can be operated in a number of setups. Here, two different setups will be discussed: reflection mode and scattering mode. In both situations light is incident on the structure through the substrate, but for the reflection mode the reflected intensity is measured, while for the scattering mode the intensity of the light scattered from captured biological particles on the sensor surface through the cover medium is measured. The light is coupled into the waveguide structure through a prism or a grating coupler. As with the dielectric waveguide, a waveguide mode can be excited in the waveguide film; however, for the MCLW it will be a leaky waveguide mode.

In reflection mode the angle of incidence is scanned and thus the propagation angle of light in the waveguide film is varied, while monitoring the intensity of the reflected light by a detector. The measured sensorgram shows the reflectance vs. the angle of incidence, and a dip in reflectance is identified at the resonance angle, θ_m, of the waveguide mode. The operation of the MCLW-sensor is similar to the operation of the dielectric waveguide sensor; the measured change in position of the resonance angle corresponds to a given change in n_C (cover medium). As with the dielectric waveguide, waveguide modes are excited in the MCLW and are guided in the waveguide film. However, in contrast to the dielectric waveguides, where the guided light experiences total internal reflection at both the film/cover and the film/substrate boundaries, the guided light in the MCLW is totally internally reflected only at the film/cover boundary. At the film/metal boundary, a normal reflectance of the light occurs, in which the light is partly transmitted into the metal cladding and the substrate and partly reflected back into the waveguide film at the film/metal interface, which gives rise to an attenuation of the guided light in the waveguide film.

In scattering mode, the incident light is fixed at the resonance angle, θ_m, and scattered light from the sensor surface due to the presence of biological objects within the evanescent field is imaged in real time using a CCD camera. At the resonance angle, the light is reflected by TIR from the film/cover interface, and an evanescent wave exists in the cover medium. The presence of bacterial cells on the surface will cause a scattering of light. Evanescent wave light illumination is a favourable method of generating scattering and fluorescence excitation of objects near the sensor surface. This technique gives a dark background and provides a significant improvement in spatial resolution for objects near surface structures, in comparison with other microscopy techniques (Rohrbach 2000). Scattering and fluorescence are very sensitive techniques, which are capable of detecting low concentrations of bacteria where refractive index sensors give a negligible response (Perkins and Squirrell 2000). This combination affords a higher sensitivity than that which might be achieved using detection by RI changes.

For a deeper analysis of the MCLW structure the field distribution in the layers should be considered, similar to the analysis of the three-layer structure in the previous section. The field description of the three-layered structure in Eq. 8.2 can be expanded and applied for the four-layer structure to derive a mode equation in the same manner as the three-layer mode equation (Eqs. 8.2 and 8.3), e.g., by adding the electromagnetic fields in layer M to the solution ansatz ψ in Eq. 8.2:

$$A_M \exp\left[i\left(k_x x + k_{z,M} z\right)\right] + B_M \exp\left[i\left(k_x x - k_{z,M} z\right)\right].$$

This results in a mode equation quite similar to the three-layer mode equation, where only the term for the phase shift at the film/substrate boundary is interchanged with that for a new phase shift at the film/metal/substrate boundary:

$$\phi_{F,M,S} = 2\arctan\left[i\frac{\left(1 - r_{F,M}\right)\left(1 - r_{M,S}\exp\left[i2k_{z,M}d_M\right]\right)}{\left(1 + r_{F,M}\right)\left(1 + r_{M,S}\exp\left[i2k_{z,M}d_M\right]\right)}\right],$$

where $r_{I,J} = \dfrac{n_J^{2\rho}k_{z,I} - n_I^{2\rho}k_{z,J}}{n_J^{2\rho}k_{z,I} + n_I^{2\rho}k_{z,J}}$ and $k_{z,I} = \pm k\sqrt{n_I^2 - N^2}$.

The sign of $k_{z,I}$ should be shifted to "-" if N is complex and $\mathrm{Re}\left[N\right] > \mathrm{Re}\left[n_I\right]$. From the mode equation, the allowed value of $N = N_{RE} + iN_{IM}$ is derived for which the *ansatz* ψ for the electromagnetic field is a solution to Maxwell's equations. Generally, N is complex due to the complex RI of the cladding.

The phase shift at the film/cover boundary for the four-layer mode-equation is given in exactly the same way as for the three-layer mode equation; and thus again a change in n_C results in a change in the effective mode RI, N_m of the MCLW and in the resonance angle of the waveguide mode. However, here the waveguide modes are leaky in the cladding and substrate due to the complex RI of the cladding layer added to this structure; and hence, the intensity of the waveguide mode is attenuated along the waveguide film.

The dip in the reflectance spectrum recorded in the reflection mode depends strongly on the material used for the cladding, where the complex value of the dielectric constant influences the width of the resonance dip. In general, the total reflectance from a layered structure can be calculated using Fresnell's reflection laws. For a four-layered structure of layers *1*, *2*, *3* and *4* for which the lower (*1*) and upper (*4*) media are considered semi-infinite, the reflectance is given by (Born and Wolf et al. 1999):

$$R_{1234} = |r_{1234}|^2 = \left| \frac{r_{12} + r_{234} \exp\left[i2k_{z,2}d_2\right]}{1 + r_{12}r_{234} \exp\left[i2k_{z,2}d_2\right]} \right|^2, \tag{8.11}$$

where $R_{123} = |r_{123}|^2 = \left| \dfrac{r_{12} + r_{23} \exp\left[ik_{z,2}d_2\right]}{1 + r_{12}r_{23} \exp\left[ik_{z,2}d_2\right]} \right|^2$, $\quad r_{ij} = \dfrac{n_J^{2\rho}k_{z,I} - n_I^{2\rho}k_{z,J}}{n_J^{2\rho}k_{z,I} + n_I^{2\rho}k_{z,I}}$, $\quad k = \dfrac{2\pi}{\lambda}$.

Here, r_{ij} is the amplitude reflection coefficient between layers *I* and *J*, $k_{z,I}$ the normal wave vector component along z in medium *I*, n_I the RI of medium *I*, and λ is the light wavelength in vacuum. The reflectance from a layered structure with any given number of layers can be calculated by continuing the recursive procedure already used in expanding Eq. 8.10 for the four-layer structure.

Actually, the MCLW sensors can be divided into two types of sensors, dip-type and peak-type, where the name refers directly to the measured sensorgrams (Skivesen et al. 2005b). The two types of sensors are differentiated by the choice of cladding material and the thickness of the cladding used in the sensor design. Applying, for example, a metal with permittivity $\varepsilon_M = \varepsilon_M' + i\varepsilon_M''$, for which ε_M' has a high value and a thickness of the metal layer in the range of 1–10 nm, results in a peak-type MCWG (Skivesen et al. 2005a; Skivesen et al. 2005b; Zourob et al. 2005a; Zourob et al. 2005b); while applying a metal with a low value of ε_M'' and a metal layer with a thickness in the range of 40–60 nm results in a dip-type MCLW.

From the literature (Tein 1977) it is known that the imaginary part of the metal permittivity ε_M'' is responsible for the losses in MCLWs; and thus it can be expected that sharp resonances (sharp dips) arise for MCLWs with small values of ε_M'' and that the dips become less sharp when ε_M'' is increased, which is exactly the property exploited in the peak-type MCLW. For the peak-type MCLW the sensing operation is slightly different, as the sensorgram for this sensor shows a reflectance maximum at the critical angle due to a very broad resonance dip for the waveguide mode. This maximum is used as a detection parameter.

Two sensorgrams for a peak- and a dip-type MCLW, respectively, are shown in Figure 8.11, along with two sensorgrams for a silver-coated SPR sensor. All three configurations are optimized for a sharp and distinct appearance of the peak/dip in the sensorgram. The sensorgrams are for $n_C = 1.330$ and $n_C = 1.333$, respectively, and by analyzing the change in angle position of the dips/peaks in response to the change in n_C, the sensors' cover sensitivity $\partial N_m / \partial n_C$ can be found:

Ag-SPR: $\partial N_m / \partial n_C = 1.24$
Dip-type Ag-CLW: $\partial N_m / \partial n_C = 0.74$
Peak-type Ti-CLW: $\partial N_m / \partial n_C = 1.$

Here, the Ag-SPR sensor shows a higher sensitivity than the two MCLW sensors. However, from Figure 8.11 it can be seen that the sensorgrams for the MCLWs have a very sharp resonance compared to the SPR resonance. Comparing the full-width-half-max (FWHM) of the Ag-CLW resonance and the Ag-SPR resonance shows that the former is a factor of 10 smaller than the latter. Also, the FWHM of the Ti-CLW peak is smaller than the SPR

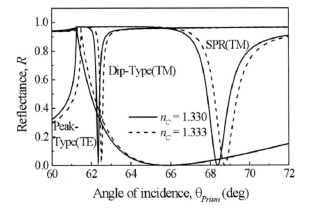

Figure 8.11. Sensorgrams for SPR, Peak- and dip-type MCLW sensors for cover RI, $n_C = 1.330$ (solid line) and $n_C = 1.333$ (dotted line). The parameters used for the three sensors are: SPR: Silver, $n_S = 1.517$, $d_M = 56$ nm, $\rho = 1$. Dip-type: Silver, $n_S = 1.517$, $d_M = 60$ nm, $n_F = 1.59$, $d_F = 330$ nm, $\rho = 1$. Peak-type: Titanium, $n_S = 1.517$, $d_M = 5$ nm, $n_F = 1.47$, $d_F = 240$ nm, $\rho = 0$. (Reprinted from Skivesen et al. (2007), with permission of Elsevier)

resonance FWHM. A more narrow resonance dip or peak improves the possibility of separating the individual resonances for small changes in n_C, and thus the normalized cover sensitivity of both MCLW sensors exceeds that of the SPR sensor.

In Figure 8.12 the penetration depth of the evanescent field into the cover $d_{P,C}$ is calculated versus the film thickness for the lowest waveguide modes and also for the SPR using gold and silver for comparison. It is seen that the penetration depths of the waveguide modes are infinite at the cut-off points. The penetration depths of the waveguide modes are quite different from those of the SPR modes, where the penetrations are limited to a depth of 200 nm. For the peak-type MCLW the theoretical penetration depth is infinity, because the peak position equals the critical angle of reflectance in the waveguide film.

For the peak-type MCWG it should be noted that the resonance peak changes characteristics according to which materials are being used and the exact position of the resonance compared to the critical angle. If the peak is excited at the critical angle, as shown in Figure 8.12, the peak is a sharp spike and the penetration depth of the evanescent field is infinite; however, if the resonance is excited at an angle slightly above the critical angle, the peak becomes rounded and the penetration depth becomes finite (Zourob et al. 2005a).

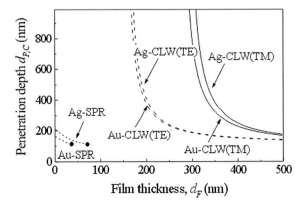

Figure 8.12. Calculated cover penetration depth for silver and gold substrate MCWG. The RIs used are: $n_S = 1.517$, $n_F = 1.47$, $n_C = 1.33$, $n_M = 0.065 + i4$ (silver) and $n_M = 0.15 + i3.2$ (gold). (adapted from Skivesen et al. 2005a)

The high normalized sensitivity for the MCLW sensors and the large probing depth of the evanescent field into the cover medium makes the sensors suitable for detection of micron-scale biological objects, such as bacterial cells, since an infinite cover penetration depth can be achieved for both the dip- and peak-type sensor. However, for the dip-type MCLW the cover penetration depth can be tuned by adjusting the film thickness; and thus the MCLW can also be applied for measurements of thinner adlayer thicknesses, with high normalized adlayer sensitivity [Skivesen et al. 2007].

The large penetration depth of the MCLWs makes possible the detection of larger particles, as the light will interact more with the cell body than is the case with conventional waveguides. It was found that the scattering and fluorescence intensity from MCLW sensors is up to three times stronger than that observed with the SPR sensor (Zourob et al. 2005a).

The MCLW sensor was simplified by developing dye-cladded sensors which have some advantages compared with metal-coated sensors. The cladding can be spin coated onto the waveguide structure rather than using vacuum deposition for the metals, which results in cheaper and faster fabrication processes. The main problem with the metal clad waveguide is that the interface between the metal layer and the dielectric layer for many metals is unstable (Podgorsek et al. 1998) and applying an electric field to the structure can oxidize the metal layer (Zourob et al. 2005a).

2.3.1. Results

Here we present a few examples of experimental results using the MCLW sensor. Cell detection was demonstrated using an Ag-CLW (dip-type) sensor and a Ti-CLW (peak-type) sensor operating in reflection mode, and a Silica waveguiding-CLW sensor was applied for bacterial detection, where the sensor was operating in scattering mode. Previously the MCLW sensor was applied for measurements on lipid bilayers by Salamon (Salamon et al. 1997; Salamon et al. 1999). The measurements included characterization of thickness, RI, extinction coefficients, and optical anisotropy of lipid bilayers (Salamon et al. 1994; Salamon and Tollin 2001; Salamon and Tollin 2004); they also performed measurements of integral and surface proteins in lipid bilayers (Salamon et al. 2003). They applied the MCLW using the name CPWR.

2.3.1.1. Reflection Mode: Detection of Bacterial Cells

The measurement of RI is a favorable technique to study biomolecular interactions, allowing real-time analysis of biospecific reactions without the need for biomolecular labeling. Zourob et al. (2005a) employed two different approaches for bacterial detection using RI detection with the MCLW (glass slide n = 1.51, 8.5 nm Ti, and 260 nm silica) chip. The first approach used the anti-BG antibody immobilized via an EDC/NHS-modified dextran surface. Unreacted groups were then blocked, and the sensor was exposed to different concentrations of BG spores (10^3–10^8 spores/mL). Figure 8.13 shows the MCLW sensor response for different concentrations (1×10^5, 3×10^5, and 7×10^5 spores/mL). The limit of detection was estimated to be 8×10^4 cells/mL when 5 mL of BG spores suspension was pumped through for 20 min. This value has been taken as the concentration corresponding to a response of the intercept plus three times the standard deviation of this value from the calibration curve. The second approach used the immobilization of 1×10^5 spores/mL onto the EDC/NHS-activated dextran prior to the binding of a specific antibody. It was noticed that the response resulting from the bacteria immobilization was similar to the response observed for binding of the antibody. Suspensions of different bacterial spores (1×10^5, 3×10^5, and 7×10^5 spores/mL) were immobilized onto the EDC/NHS-modified dextran surface followed by 100 μg/mL anti-BG antibody. The increase in RI from the antibody addition was related to the concentration of the bacterial spores

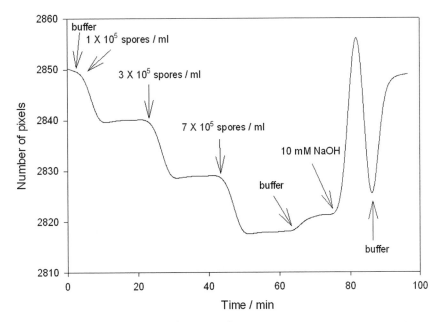

Figure 8.13. RI sensorgram for 1×10^5, 3×10^5, and 7×10^5 spores/mL bacteria using MCLW sensor (arrows indicate the times at which various solutions were added) (Smoothed data for clarity). (Reprinted with permission from Zourob et al. (2005), Copyright 2005 American Chemical Society).

immobilized onto the sensor surface. This has been confirmed by measuring the fluorescence emission from the binding of Cy5-labeled anti-BG to the immobilized BG. This indicates that there is proportionality between the immobilized bacteria and the captured antibodies. It can be concluded from the result that extending the evanescent field to the whole volume of the cell improves the detection limit for bacterial monitoring.

Skivesen et al. (2007) compared the peak- and dip-type MCLW using two different types of skin cells, NHDF (normal human dermal fibroblast) and HaCaT (human ceratinocyte cell line). The waveguides structures were in the case of the dip-type MCLW, a glass substrate (n = 1.517), silver cladding (60 nm), followed with a spin-coated polystyrene waveguiding layer (n = 1.59, d = 330 nm); and in the peak-type MCLW, a glass substrate (n = 1.517), coated with titanimum cladding (8.5 nm), followed by a silica waveguiding layer (n = 1.47, d = 260 nm). The sensors were operated in reflection mode and the two types of MCLW sensors both showed a very sharp and distinct resonance. Both sensors had a large probing depth of the evanescent field in the cover medium. The dip-type MCLW can be tuned to obtain either a low or a high penetration depth by adjusting the film parameters, and thus the sensor can be used for measurements on both thin adlayer and micron-scale objects. The dip-type sensor can be operated in multimode operation or supporting both a TE- and TM-mode, resulting in modes with different penetration depths in the cover, giving a sensor that can measure at different depths in the cover medium. Thus the dip-type MCWG sensor has the same advantages as the reverse symmetry dielectric waveguide. In contrast, for the peak-type MCWG sensor it is not possible to tune the penetration depth. Detection limits of $1 \, cell/mm^2$ are achievable with both MCWG sensor types, but the dip-type shows a slightly improved detection limit compared to the peak-type sensor.

2.3.1.2. Scattering Mode: Detection of Bacteria

Zourob et al. (2005a, 2005b, 2005c) have presented bacteria detection with the peak-type using a silica guiding MCLW chip operated in scattering mode where the sensor structure is

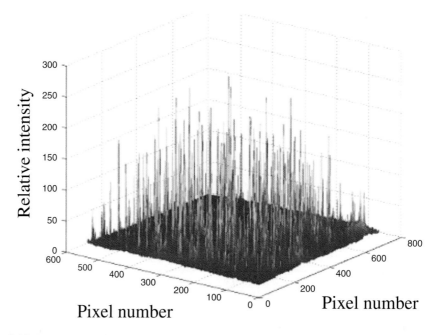

Figure 8.14. Intensity plot of scattering of BG spores captured on anti-BG spores on the silica guiding MCLW surface. (Reprinted with permission from Zourob et al. (2005), Copyright 2005 American Chemical Society).

illuminated at resonance angle for the peak. Antibodies for BG spores were immobilized via a monomer glutraldehyde-activated surface. The sensor was exposed to different concentrations of BG spores (10^2–10^8 spores/ml) to find the detection limit of the device. Figure 8.14 shows the MCLW sensor response after applying a concentration of 1×10^6 spores/ml, and it shows typical images of scattered light from captured BG bacteria spores. The sharp spikes represent individual particles on the surface. The plots were analysed to count the number of particles present on the sensor surface. The MCLW was found to generate a lower background signal than the dye-CLW, due to the smooth surface of the vacuum deposited silica on the MCLW; whereas absorbing materials-CLW produced a rougher surface, due to spin coating. The limit of detection was estimated to be approximately 1×10^4 cells/ml when 5 ml of BG spores suspension was pumped through for 20 min; this was based on the minimum response shift that could be distinguished on the instrument.

As the BG spores were flowed across the surface of the sensor chip, they appeared either as a diffuse area of moving light or as brighter, smaller points moving more slowly, and could occasionally be seen to come to an instantaneous stop, presumably having been captured by the immobilized antibody. No such attachment was observed with BG spores at control surfaces coated with Bovine serum Albumin (BSA) or Fetal Calf serum (FCS), or with *E. coli* strain on anti-BG antibody-coated surfaces. Upon stopping the flow, the diffusely emitting particles appeared to settle onto the MCLW chip, where they became brighter and more sharply defined. Capture of the spores at the surface was indicated by cessation of movement and the bacteria remaining in place when the flow was restarted (Zourob et al. 2005a).

Figure 8.15 shows the normalized area occupied on a functionalized silica surface by captured bacterial spores, as determined from surface imaging, plotted against bacterial spore concentrations ($1 \times 10^3 - 1 \times 10^{10}$ cells/mL). Three different tests were performed: BG spores binding to immobilized anti-BG spores, BG spores binding to immobilized nonspecific protein BSA (control surface), and *E. coli* binding to immobilized anti-BG spores (control surface). Only the BG/anti-BG assay showed significant BG capturing, while there was no significant

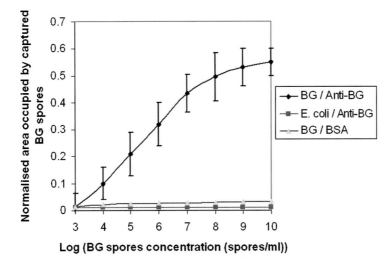

Figure 8.15. Normalized area occupation of BG and anti-BG bacterial spores vs. solution concentration of three different assays: (▲) BG/BSA functionalized control surface, (◆) BG/anti-BG immobilized surface, and (■) *E. coli*/anti-BG immobilized surface. The surfaces were exposed for 60 min to the respective solutions. Error bars represent ±1 standard deviation, n = 4. (Reprinted with permission from Zourob et al. (2005), Copyright 2005 American Chemical Society).

BG capturing on the BSA-coated surface and in the case of incubating *E. coli* on the anti-BG–coated surface. It is clear that there is a dependent relationship between the BG bacterial spores captured by the immobilized anti-BG spores and the concentration of the tested solution. The dose response was found to be linear over five decades of bacterial concentration (regression coefficient $r > 0.988$, the probability of the linear fit $p < 0.001$). The standard curve for the BG/anti-BG assay reaches a plateau at cell concentrations above 8×10^7 cells/ml, corresponding to a surface coverage of bacterial spores of 50–60%. This plateau indicates a certain degree of saturation of possible binding events in the close vicinity of the surface. This may be due to the cells already at the surface sterically hindering other spores interacting with the immobilized antibody. In addition, negative charges on the waveguide silica surface and cell-to-cell repulsive forces may prevent the close packing of cells. Indeed, as cell surfaces consist of many different charged groups, both short- and long-range interactive forces between cells would be expected (Watts et al. 1994).

2.3.1.3. Immobilization Matrices

The type of matrices used for immobilization of the recognition element for bacterial cell detection is crucial to achieve high sensitivity. Two important conditions have to be considered specifically for bacterial detection: (1) the accessibility of the recognition elements in the immobilization matrix for bacteria binding on the sensor surface, and (2) the binding of the analytes within the most sensitive region of the evanescent field, immediately adjacent to the sensor surface.

Zourob et al. (2005a) performed a comparative study of a number of surface immobilization matrices for preparation of bacteria immunosensors to find the best immobilization matrix. The matrices were evaluated by operating the MCLW sensor in scattering mode by measuring the bacteria coverage on the sensor surface along with the scattering intensity after bacteria binding for each of the matrices. The results are shown in Figure 8.16.

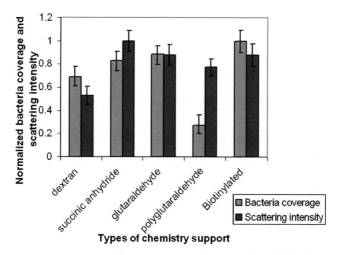

Figure 8.16. Normalised spore coverage after 1 h incubation and scattering intensity using different supports. (Reprinted with permission from Zourob et al. (2005), Copyright 2005 American Chemical Society).

Five immobilization matrices were tested and compared. These included a dextran layer on the sensor surface, a succinic anhydride-modified surface, a monomer glutaraldehyde-activated surface, a polyglutaraldehyde-activated surface, and a streptavidine/biotinylated antibody-coated surface.

The amount of bacteria captured by the immobilized antibody (bacteria coverage) on the different matrices and the resultant scattering intensity varied greatly depending upon the surface modification and immobilization chemistry used. In general, the bacterial coverage using dextran and polyglutaraldehyde gave a low bacteria coverage, while the succinic acid-modified surface and the monomer glutaraldehyde- and streptavidine/biotinylated-coated surfaces resulted in the highest bacteria coverage.

The scattering intensity for captured bacteria to the immobilized antibodies was found comparable to most of the chemistries, except that for the dextran layer. The glutaraldehyde (monomer and polymerized)-coated surface, succinic acid, and the streptavidine/biotinylated antibody-coated surface showed the highest scattering intensity due to the placement of captured cells very close to the sensor surface. However, dextran showed the lowest scattering intensity. Dextran coating is a 3D-hydrogel, commonly used method for small protein detection. However, for bacterial cell detection the applicability for the dextran layer is limited. The pore sizes in the dextran hinder the diffusion of bacteria into the dextran layer; also, dextran is very compact at the sensor surface due to its binding mechanism, and thus this hydrogel will waste the most sensitive area of the evanescent field.

3. Integrated Deep-Probe Optical Waveguides Systems

Processes that enhance molecular or bacterial interactions with a defined immobilised matrix on solid surfaces are of particular relevance in optimising the sensor performance. In general, when working with bacterial cell detection, one of the main "features" to consider is the diffusion of these large analytes to the immobilized recognition element on the sensor surface. Preferably, in order to enhance the sensor performance, the analytes of interest in the sample should be moved quickly to be in contact with the immobilized recognition elements to be captured on the sensor surface. The sensor performance can, for example, be limited due to (1) the mass transport of large analytes such as bacteria, which limits the binding of targets to the immobilized recognition receptors; (2) non-specific binding; and (3) long analysis time.

One approach to enhance the diffusion of the analyte to the immobilized recognition element is to use a physical force field that directs the cells onto an activated sensor surface in a flow-through system, where they can be captured by the recognition element. In this section, the sensor systems based on deep probe MCLW sensors will be integrated with an electric field and ultrasound standing waves, respectively, as physical force fields to concentrate and enhance the capture of bacteria spores into immobilised antibodies on the sensor surface (Zourob et al. 2005a, 2005c). The integration improves the detection limit by a few orders of magnitude and shortens the analysis time significantly.

3.1. Integration with Electric Field

The isoelectric point of bacteria can be exploited to enhance the sensor performance by integrating an electric field across the sample volume into the sensor system resulting in a physical force to push the cells to the sensor surface. The isoelectric point for most bacterial cells is ≤ 4 (DeFlaun and Condee 1997). Hence, bacteria have a net negative charged surface in deionized water (pH $= 7.0$), and a net positive charge at low pH levels (≤ 3). Thus the bacteria can be directed towards the sensor surface by applying a negative voltage to the top cover of the flow channel and a positive voltage to the metal layer of the waveguide, or vice versa depending on the pH of the buffer solution. Hence, cells are repelled from the top cover and collected on the charged metal layer on the sensor surface.

The evanescent field sensor illumination is one of the best techniques to study the cell concentration on the sensor surface movement, as it observes the cells in real time when it pushes within the evanescent field at the sensor surface as it eliminates interference from the bulk solution. Hence this technique is very powerful for observing the enrichment of individual particles onto sensor surfaces (Hughes and Morgan 1999). An important application of coupling the waveguide sensor chip with the electric field is the capability to concentrate and enhance the capture of bacteria from a diluted sample to the immobilized antibodies so that relatively low concentrations may be analysed in a shorter time.

Zourob et al. (2005b) constructed a flow cell incorporating the MCLW and ITO electrode, as shown in Figure 8.17. Initial experiments were conducted after treating the surface with BSA overnight, and the bacteria were introduced to the sensor system at a constant flow rate of $200 \, \mu l/$ min in a 50 mM histidine buffer. The sensor system was operated in real-time scattering mode using a CCD camera.

Different concentrations of BG bacterial spores (10^3–10^8 spores/ml) were introduced into the electric field sensor chip-based flow cell and an appropriate DC potential was applied so

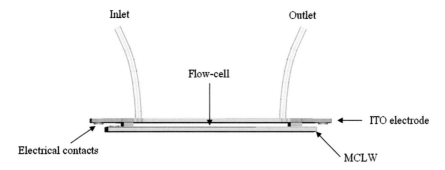

Figure 8.17. A schematic diagram for MCLW sensor chip integrated with electric field. (Reprinted from Zourob et al. (2005), with permission of The Royal society of Chemistry).

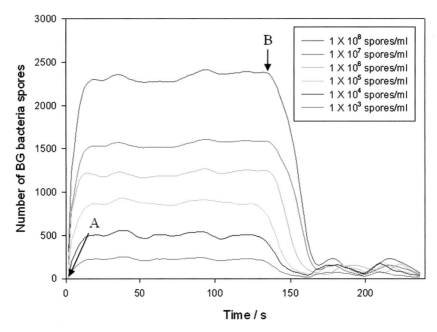

Figure 8.18. The effect of applying positive (A) and negative (B) potential to the metal layer of the MCLW sensor on deposition and repulsion of different concentrations of BG bacteria spores. (Reprinted from Zourob et al. (2005), with permission of The Royal society of Chemistry).

that BG bacterial spores were either collected onto or repelled from the MCLW sensor surface. Figure 8.18 shows the detected number of bacterial spores on the sensor surface versus time. At time **A,** a negative potential was applied to the ITO electrodes and a positive potential to the metal layer of the MCLW sensor, resulting in an increase in the number of BG bacterial spores driven to the sensor surface. At time **B**, the potential is reversed and a decrease in the number of BG bacterial spores on the sensor surface is observed, hence the bacterial spores are repelled away. The diagram shows results for six different concentrations of BG bacterial spores. The results indicate that for each of the bacteria suspensions, approximately the same number of BG bacterial spores was deposited onto the sensor surface and repelled away from the sensor surface when the potential was reversed.

The same experiments were repeated with immobilized antibodies on the MCLW sensor surface before depositing the BG bacterial spores. Figure 8.19 shows the number of BG bacterial spores concentrated on the MCLW sensor surface, with a positive potential applied to the metal layer with respect to the ITO layer at time **A**, and subsequently when reversing the potential at time **B**. The remaining bacterial spores on the MCLW sensor surface after reversing the field are captured by the immobilized antibodies as shown in Figure 8.19. The number of bacterial spores captured was estimated by calculating the number of scattering points on the surface, and this was confirmed by direct counting immediately after terminating the experiments.

The number of cells captured using the MCLW with the electric field applied for 2 min was found to be equal to the number of bacteria captured in 1 h using the same concentration of bacteria but with no applied electric field. Thus the electric field gave a 30-fold enrichment. The limit of detection was found to be 1×10^3 cells/ml when the electric field was applied for 2 min. This value has been taken as the concentration that gave a signal three times the background noise. This significant enrichment of captured bacteria spores indicates a practical way to substantially improve the detection limit and shorten the response time of future waveguide sensors for bacterial agents analysis.

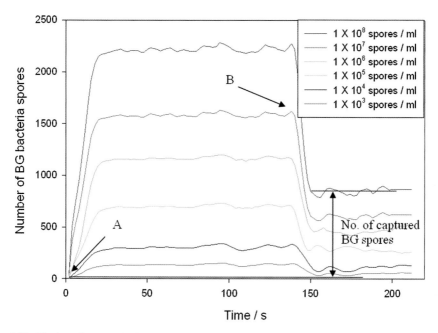

Figure 8.19. The use of an electric field potential to capture/repel BG spores and demonstration of the number of bacteria spores captured on the sensor surface. (Reprinted from Zourob et al. (2005), with permission of The Royal society of Chemistry).

3.2. Integration with Ultrasound Standing Waves (USW)

A detection system similar to the one previously described has been proposed, but in this case using the ultrasound standing waves across the flow cell instead of using an electric field to concentrate the bacterial cells on the MCLW sensor surface. USW will be discussed in more details in chapter 35 ultrasonic microsystems for cell manipulation. Basically, when particles suspended in a fluid are exposed to an ultrasonic standing wave, they experience an acoustic radiation pressure that drives them towards the (usual pressure) nodes or antinodes of the acoustic field, where they aggregate to form bands of particles in localized clumps, bounded by a clarified medium. This is because the mass density and the speed of sound in the particles are greater than the respective values of the medium. This phenomenon has been used in developing a system with a region of standing wave radiation, where bacteria are moved to a central nodal position located on the sensor surface. This enhances the capture rate by driving the bacteria onto the sensor surface, which is coated with immobilized antibodies. Ultrasound standing waves are appropriate for working with biological cells in an aqueous medium, since it is not deleterious for the particles. USW can be applied to different particle sizes and to environmental samples without problems—unlike the electric field, which suffers from the presence of the salts. The USW sensor system is amenable to miniaturization. (See chapter 35 on ultrasonic microsystems for bacterial cell manipulation.)

The sensor system with USW has been tested for the detection of BG bacterial spores using a MCLW sensor coated with immobilized antibodies operated in scattering mode. A schematic diagram of the chamber construction used in integrating the MCLW and USW is shown in Figure 8.20. The basic construction of the flow chamber was described previously by Hawkes et al. (2004) and comprised a 30 mm wide, 0.66 mm thick (i.e., 3 MHz fundamental resonance thickness) PZ26 polished ceramic ultrasonic transducer (Ferroperm, Krisgard, Denmark); a 1.5 mm thick (i.e., equivalent to a half wavelength in the metal at 3 MHz) stainless steel coupling plate glued to the transducer and separating it from the water layer; a spacer to define

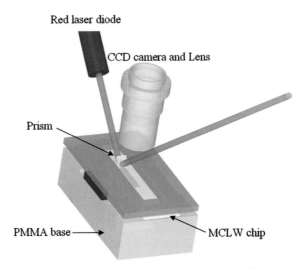

Figure 8.20. Schematic of the ultrasound-MCLW instrumental set up. (Reprinted with permission from Zourob et al. (2005), copyright 2005 American Chemical Society)

the channel dimensions; and a 1 mm thick ($\lambda/4$ wavelengths at 3 MHz) MCLW chip acoustic reflector. Generation and control of the frequency and voltage was applied to the transducer as described before (Spengler et al. 2000; Zourob et al. 2005c). The ceramic transducer was driven by an amplifier, with a sinusoidal signal from a function generator (Hewlett Packard 3326A), and a RF voltmeter was used to detect the voltage across the ceramic.

The integrated system was tested for enhancing bacterial capture by introducing a solution of bacteria spores, with a concentration of 10^6 cells/ml, at $200\,\mu l\ \text{min}^{-1}$ flow rate into the flow cell. Here, a small number of spore "clumps" was observed forming in the central region of the chamber when the ultrasound (0.5 V at 1.92 MHz frequency) was turned on. The "clumps" were held stationary against the flow and increased in size with exposure time as more spores flowed into the system. It was found that the ultrasound deposited over 96% of BG bacteria spores from the bulk solution onto the MCLW sensor surface. Thus the integration of USW enhances the capturing of BG bacteria spores to the immobilized antibodies and shortens the analysis time. The number of bacteria spores captured during ultrasound application over 3 min exceeds that of spores captured in 1 hour using the same concentration and flow rate, but without ultrasound applied to the system (Zourob et al. 2005c).

An additional experiment was performed with the sensor system (0.4 V, 2.92 MHz) in which the ultrasound was applied over a limited area of the sensor surface. The sensor area was imaged using a CCD camera by ensuring propagation of the waveguide mode along the line of the flow of the solution at the sensor surface. Two images were taken, as shown in Figure 8.21, for two different positions of incident light into the waveguide structure. In Figure 8.21a the point of incident light was chosen to be on the area of the sensor surface exposed to the bacteria detection before the ultrasound was applied. Figure 8.21b shows the scattered intensity from the same point beyond the ultrasound exposed area. It was possible to monitor the effect of the ultrasound on the surface of the MCLW sensor as seen from Figure 8.21, where the difference in depositing in relation to specific regions of the MCLW is clearly observed. Little or no depositing occurred before the spores reached the ultrasound field (Figure 8.21a, area a), whereas maximum deposition occurred within the area exposed to ultrasound (Figure 8.21, area b). A small number of BG bacteria spores were deposited in the region beyond the field, after the flow had carried the spores past the ultrasound-exposed area (Figure 8.21b, area c). Spore binding approaches saturation at the region of maximum attachment (Figure 8.21b). The

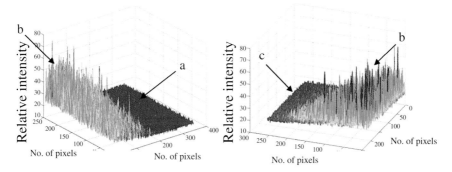

Figure 8.21. Comparison of bacteria deposited (a) before, (b) within, and(c) after the ultrasound electrode area. (Reprinted from Zourob et al. (2005), copyright 2005 American Chemical Society).

reduced amount of attachment further downstream (Figure 8.21c) may be accounted for by the reduced concentration of spores near the surface due to earlier spore removal from the suspension (Zourob et al. 2005c).

The general position of the spore clumps in the chamber did not change during the frequency scan from 2.910 to 2.920 MHz. This enabled the effect of changing the spore concentration to be examined using the MCLW sensor, while ensuring that the same areas of the chamber could be compared to each other. The regions of maximum spore adherence, at the centre of the clumping patterns, were compared using different concentrations of bacteria spores. Figure 8.22 shows the relationship between the numbers of BG bacteria spores captured onto the immobilized anti-BG over a concentration range of 10^2–10^9 spores ml^{-1}. This was achieved by counting the number of BG bacteria spores captured from the images taken for 3 min in real time over the area of mode propagation.

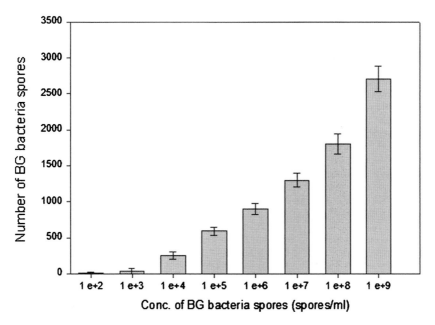

Figure 8.22. The relationship between the number of BG captured and the concentration of BG suspensions. (Reprinted from Zourob et al. (2005), copyright 2005 American Chemical Society).

4. Conclusions and Future Perspectives

In this chapter we described the development of different configurations of deep-probe optical evanescent wave sensors, such as metal-clad leaky waveguides and low-index substrate sensors, for bacterial detection. It was shown that increasing the penetration depth of the evanescent field increases the sensitivity of such sensors for bacterial detection, as it accommodates the majority of the bacterial body within the evanescent field. In addition, two physical force fields, an electric field and ultrasound standing waves, have been integrated with MCLW sensors to concentrate and enhance capturing of bacterial spores into immobilized antibodies on the sensor surface. The integration improves the detection limit by a few orders of magnitude, shortens the analysis time significantly, and reduces non-specific binding which causes false positive results.

Trends in deep-probe optical waveguide sensor instrumentation are pushing ahead with regard to automation, miniaturization, and high-throughput capabilities in order to meet the requirements for real-time sample analysis. In addition, increasing the penetration of the evanescent field in the deep-probe optical waveguide will open the door for such principles to be applied to other devices, such as interferometry, cell imaging microscopy, and Raman microscopy.

References

Abdel-Hamid I, Ivnitski D, Atanasov P, Wilkins E (1999) Highly sensitive flow-injection immunoassay system for rapid detection of bacteria. Anal. Chem. Acta 399:99–108

Adányi N, Váradi M, Kim N and Szendrö I (2006) Development of new immunosensors for determination of contaminants in food. Current Appl. Phys. 6:279–286

Balasubramanian S, Sorokulova IB, Vodyanoy VJ and Simonian AL (2007) Lytic phage as aspecific and selective probe for detection of Staphylococcus aureus-Asurface plasmon resonance spectroscopic study. Biosens. Bioelectron. 22:948–955

Born M, Wolf E (1999) Principles of Optics, 7th ed. Cambridge University Press, Cambridge, UK

Buzby JC, Roberts T, Lin J, McDonald JM (1996) Economic Report 741: Bacterial foodborne disease: medical costs and productivity losses. USDA, Washington, D.C.

Campbell DP, Gottfried DS, Scheffter SM, Beck MC and Halpern MD (2003) Interferometric Optical Waveguide Sensor for Anthrax Spore Detection. ACS National Meeting Proceedings, New York

DeFlaun MF, Condee CW (1997) Electrokinetic transport of bacteria. J. Hazard. Mater. 55:263–277

DeMarco DR, Saaski EW, McCrae DA, Lim DV (1999) Rapid detection of Escherichia coli O157:H7 in ground beef using fibreoptic biosensor. J. Food Protect. 62:711–716

DeMarco DR, Lim DV (2001) Direct detection of Escherichia coliO157:H7 in unpasteurized apple juice with an evanescent wave sensor. J. Rapid Meth. Automation Micro 9:241–257

Ferreira AP, Werneck MM, Ribeiro RM (1999) Aerobiological pathogen detection by evanescent wave fibre optic sensor. Biotechnol. Tech. 13:447–452

Ferreria AP, Werneck MM, Ribeiro RM (2001) Development of an evanescent-field fibre optic sensor for Escherichia coli O157:H7. Biosens. Bioeletron. 16:399–408

Fratamico P, Strobaugh T, Medina M, Gehring A (1998) Detection of Escherichia coli 0157:H7 using a surface plasmon resonance sensor. Biotechnol. Tech. 127:571–576

Hawkes JJ, Long MJ, Coakley WT, McDonnell MB (2004) Ultrasonic deposition of cells on a surface. Biosens. Bioelectron. 19:1021–1028

Hirmo S, Artursson E, Puu G, Wadstorm T and Nilsson B (1998) Characterization of Helicobacter Pylori interactions with sialglycoconjugates using a resonant mirror biosensor. Anal. Biochem. 257:63–66

Hirmo S, Artursson E, Puu G, Wadstorm T and Nilsson B (1999) Helicobacter Pylori interactions with human gastric mucin studied with a resonant mirror biosensor. J. Microbiol. Meth. 37:177–182

Horvath R, Lindvold LR and Larsen NB (2002a) Reverse-symmetry waveguides: Theory and Fabrication. Applied Physics B 74:383–393

Horvath R, Pedersen HC and Larsen NB (2002b) Demonstration of reverse symmetry waveguide sensing in aqueous solutions. Appl. Phys. Lett. 81:2166–2168

Horvath R, Pedersen HC, Skivesen N, Selmeczi D and Larsen NB (2003a) Optical waveguide sensor for on-line monitoring of bacteria. Optics Letters 28:1233–1235

Horvath R, Lindvold LR and Larsen NB (2003b) Fabrication of all-polymer freestanding waveguides. J. Micromech. Microeng. 13:419–424

Horvath R, Pedersen HC, Skivesen N, Selmeczi D and Larsen NB (2005a) Monitoring of living cell attachment and spreading using reverse symmetry waveguide sensing. Appl. Phys. Lett. 86, Art. No. 071101

Horvath R, Pedersen HC, Skivesen N, Svanberg C, Larsen NB (2005b) Fabrication of reverse symmetry polymer waveguide sensor chips on nanoporous substrates using dip-floating. Journal of Micromechanics and Microengineering 15:1260–1264

Horvath R, Pedersen HC, Cuisinier FJG (2006) Guided wave sensing of polyelectrolyte multilayers. Applied Physics Letters 88:111102–111104

Hughes MP, Morgan H (1999) An evanescent-field technique for dielectrophoresis studies of colloidal particles. Meas. Sci. Technol.10:759–762

Kim N, Park I-S, Kim W-Y (2007) Salmonella detection with a direct-binding optical grating coupler immunosensor, Sens. Actuators B 121:606–615

Kulagina NV, Shaffer KM, Anderson GP, Ligler FS, Taitt CR (2006) Antimicrobial peptide-based array for Escherichia coli and Salmonella screening. Anal. Chimica Acta 575:9–15

Leonard P, Heatry S, Quinn J and O'Kennedy R (2004) A generic approach for the detection of whole listeria monocytogenes cells in contaminated samples using surface plasmon resonance. Biosens. Bioelectron. 19:1331–1335

Leonard P, Heatry S, Wyatt G, Quinn J and O'Kennedy R (2005). Development of surface plasmon resonance-based immunoassay for listeria monocytogenes. J. Food Protec. 68:728–735

Mead PS, Slutsker L, Dietz V, McCaige LF, Bresse JS, Shapiro C, Griffin PM, Tauxe RV (1999) Food related illness and death in the United States. Emerg. Infec. Dis. 5:607–725

Medina M, Houten L, Cooke P, Tu S (1997) Real-time analysis of antibody binding interactions with immobilized E. coli O157:H7 cells using the BIAcore. Biotechnol. Tech. 11:173–176

Ngundi MM, Taitt CR, McMurry SA, Kahne D, Ligler FS (2006) Detection of bacterial toxins with monosaccharide arrays. Biosensors and Bioelectronics 21:1195–1201

Perkins E, Squirrell D (2000) Development of instrumentation to allow the detection of microorganisms using light scattering in combination with surface plasmon resonance. Biosens. Bioelectron. 14:853–859

Podgorsek P, Franke H (1998) Optical determination of molecule diffusion coefficients in polymer films. Appl. Phys. Lett. 73:2887–2889

Rohrbach A (2000) Observing secretory granules with a multi-angle evanescent wave microscope. Biophys. J. 78:2641–2654

Rowe C, Tender L, Feldstein M, Golden J, Scruggs S, MacCraith B, Cras J, Ligler F (1999) Array Biosensor for Simultaneous Identification of Bacterial, Viral, and Protein Analytes. Anal. Chem. 71:3846–3852

Salamon Z, Wang Y, Tollin G and Macleod HA (1994) Assembly and molecular organization of self-assembled lipid bilayers on solid substrates monitored by surface plasmon resonance spectroscopy. Biochim. Biophys. Acta 1195:267–275

Salamon Z, Macleod HA and Tollin G (1997) Coupled plasmon-waveguide resonators: A new spectroscopic tool for probing proteolipid film structure and properties, Biophys. J. 73:2791–2797

Salamon Z, Brown MI and Tollin G (1999) Plasmon resonance spectroscopy: probing molecular interactions within membranes. TIBS 24:213–219

Salamon Z and Tollin G (2001) Optical anisotropy in lipid bilayer membranes: Coupled plasmon-waveguide resonance measurements of molecular orientation, polarizability, and shape. Biophys. J. 80:1557–1567

Salamon Z, Lindblom G and Tollin G (2003) Plasmon-waveguide resonance and impedance spectroscopy studies of the interaction between penetratin and supported lipid bilayer membranes. Biophys. J. 84:1796–1807

Salamon Z and Tollin G (2004) Graphical analysis of mass and anisotropy changes observed by plasmon-waveguide resonance spectroscopy can provide useful insights into membrane protein function. Biophys. J. 86:2508–2516

Sapsford KE, Rasooly A, Taitt CR, Ligler FS (2004) Rapid Detection of Campylobacter and Shigella species in food samples using an array biosensor. Analytical Chemistry 76:433–440

Sapsford KE, Ngundi MM, Moore MH, Lassman ME, Shriver-Lake LC, Taitt CA, Ligler FS (2006) Rapid detection of foodborne contaminants using an Array Biosensor. Sens. Actuators B 113:599–607

Seo KH, Brackett RE, Hartman NF and Campbell DP (1999) Development of a Rapid Response Biosensor for Detection of Salmonella Typhimurium. J. Food Protection 62:431–437

Shepard JRE, Danin-Poleg Y, Kashi Y and Walt DR (2005) Array-Based Binary Analysis for Bacterial Typing. Anal. Chem. 77:319–326

Shriver-Lake LC, Turner S, Taitt CR (2007) Rapid detection of Escherichia Coli O157:H7 spiked into food matrices. Anal. Chimica Acta 584:66–71

Sipe DM, Schoonmaker KP, Herron JN, Mostert MJ (2000) Evanescent planar waveguide detection of biological warfare simulants. Proc. SPIE 3913:215–222

Skivesen N, Horvath R, Pedersen HC (2003) Multimode reverse-symmetry waveguide sensor for broad-range refractometry. Optics Letters 28:2473–2475

Skivesen N, Horvath R, Pedersen HC (2005a) Optimization of metal-clad waveguide sensors. Sens. Actuators B 106:668–676

Skivesen N, Horvath R, Pedersen HC (2005b) Peak-type and dip-type metal-clad waveguide sensing. Opt. Lett. 30:1659–1661

Skivesen N, Horvath R, Pedersen HC, Thinggaard S, Larsen NB (2007) Deep-probe metal-clad waveguide biosensors. Biosens. Bioelectronics 22:1282–1288

Spengler JF, Jekel M, Christensen KT, Adrian RJ, Hawkes JJ, Coakley WT (2000) Observation of yeast cell movement and aggregation in a small-scale MHz-ultrasonic standing wave field. Bioseparation 9:329–341

Taitt CR, Golden JP, Shubin YS, Shriver-Lake LC, Sapsford KE, Rasooly A, Ligler FS (2004a) A portable array biosensor for detecting multiple analytes in complex samples. Microbial Ecology 47:175–185

Taitt CR, Shubin YS, Angel R, Ligler FS (2004b) Detection of Salmonella enterica serovar typhimurium by using a rapid, array-based immunosensor. Applied and Environmental Microbiology 70:152–158

Taylor AD, Ladd J, Yu QM, Chen SF, Homola J and Jiang SY (2006) Quantitative and simultaneous detection of four foodborne bacterial pathogens with a multi-channel SPR sensor. Biosens. Bioelectron. 22:752–758

Tien PK (1977) Integrated optics and new wave phenomena. Rev. Mod.Phys. 49:361–420

Tiefenthaler K, Lukosz W (1985) Grating couplers as integrated optical humidity and gas sensors. Thin Solid Films 126:205–211

Tiefenthaler K, Lukosz W (1989) Sensitivity of grating couplers as integrated-optical chemical sensors. J. Opt. Soc. Am. 6:209–220

Watts H, Lowe C, Pollard-Knight D (1994) Optical biosensor for monitoring microbial cells. Anal. Chem. 66:2465–2470

Whyte W, Lidwell OM, Lowbury EJ, Blowers R, (1983) Suggested bacteriological standards for air in ultraclean operating rooms. J. Hosp. Infect. 4:133–139

Zeikus G, Johnson EA (eds) (1991) Mixed Cultures in Biotechnology. McGraw-Hill, New York, pp 341–372

Zhu P, Shelton DR, Karns JS, Sundaram A, Li S, Amstutz P, Tang C-M (2005) Detection of water-borne E. coli O157 using the integrating waveguide biosensor. Biosens Bioelectronics 21:678–683

Zourob M, Mohr S, Treves-Brown BJ, Fielden PR, McDonnell MB, Goddard NJ (2005a) An Integrated Metal Clad Leaky Waveguide Sensor for Detection of Bacteria. Anal. Chem.77:232–242

Zourob M, Mohr S, Treves-Brown BJ, Fielden PR, McDonnell MB, Goddard NJ (2005b) An integrated optical leaky waveguide sensor with electrically induced concentration system for the detection of bacteria. Lab Chip 5:1360–1365

Zourob M, Hawkes JJ, Coakley WT, Mohr S, Treves-Brown BJ, Fielden PR, McDonnell MB, Goddard NJ (2005c) Optical Leaky Waveguide Sensor for Detection of Bacteria with Ultrasound Attractor Force. Anal. Chem. 77:6163–6168

9

Interferometric Biosensors

Daniel P. Campbell

Abstract

Every chemical reaction or interaction causes a change in refractive index, including such bioconjugate interactions as antibody/antigen, DNA hybridization and enzyme/substrate interactions. Interferometry is an optical method for measuring refractive index changes. With the proper choice of sensing film, an interferometer can identify and quantify the presence of a biological moiety. An interferometer compares optically two almost equivalent light paths – one that interrogates the refractive index change caused by a bioconjugate interaction, and the other that serves as a reference that cancels out any nonspecific interactions. Interferometers have the capability of detecting refractive index changes of 10^{-7}, which corresponds to ppb concentrations of small molecules, pg/mL concentrations of toxins and proteins, and 100s–1000s of whole cells, viruses and spores. Several optical interferometric designs are described. Most configurations combine a bioconjugate reaction isolated on a rigid support with a long interaction length of mm to cm to achieve high sensitivity. The most common interferometric configuration utilizes a planar optical waveguide. The evanescent field associated with a wave-guided beam extends above the waveguide surface where the bioreceptor is immobilized. The bioconjugate interaction perturbs the propagating beam and the extent of this perturbation is measured by comparing the phase of the light traveling along the sensing channel with that traveling along a reference channel that is not functionalized with the bioreceptor. The phase change is measured by optically combining the two beams at the output of the interferometer to create an interference pattern, a series of dark and light fringes that is caused by constructive and destructive interference. By proper choice of receptor molecule and calibration, both the identity and the quantity of a specific bioentity can be measured with the interferometric biosensor.

1. Principles of Optical Interferometry

Optical interferometry is a means for measuring small changes that occur in an optical beam along its path of propagation. These changes can result from changes in the length of the beam's path, a change in the wavelength of the light, a change in the refractive index of the media the beam is traveling through, or any combination of these. The phase change, φ, that may occur is directly proportional to the pathlength, L, and refractive index, n, and inversely proportional to the wavelength change, as shown in equation 9.1.

$$\varphi = 2\pi L\,n/\lambda. \tag{9.1}$$

Daniel P. Campbell • Georgia Tech Research Institute, Atlanta, GA 30332 USA

M. Zourob et al. (eds.), *Principles of Bacterial Detection: Biosensors, Recognition Receptors and Microsystems*,
© Springer Science+Business Media, LLC 2008

Of the three variables, L, λ, and n, the change in refractive index measured in a planar waveguide arrangement is commonly used in biosensor applications. Displacements are also used, but typically changes in length are not very large, leading to lower sensitivity. When a bioconjugation event takes place, such as binding of one protein with another, and the propagating light beam passes through the volume where the binding event has taken place, a change in the refractive index can be observed. To measure this change, a reference propagating beam is used. This reference beam is typically placed adjacent to the sensing beam, but it does not encounter the binding event. The reference is combined with the sensing beam to create an interference pattern of alternating dark and light fringes. Whenever a chemical or physical change occurs in the sensing arm, the interference pattern will shift, producing a sinusoidal output.

Optical interference and optical interferometry have their origins in experiments conducted by Thomas Young in 1803 (Young 1804). The interference phenomena he explored were used to establish the wave nature of light. Young passed light through a pinhole which illuminated two adjacent pinholes or slits which in turn produced a series of dark and light fringes on a screen placed beyond the slits. The first pinhole provided the necessary coherence for the light passing through the subsequent pair of slits. This coherence is now easily provided by the use of a laser. Since Thomas Young's initial experiment, four basic interferometers have been developed that find application in chemical or biochemical measurements: Michelson, Fabry-Perot, Mach-Zehnder, and Young's. The Michelson interferometer lies at the heart of FTIR spectrometry, wherein the frequency components of the interferogram are converted through a Fourier transform to provide the infrared spectra used in almost all chemistry labs for spectral identification of a compound's structure. The Michelson interferometer also finds use in the measurement of the refractive and adsorption properties of gases, where a measurement is made with and without the gas or compound in one of the beam's paths. The Fabry-Perot interferometer is designed to provide very long pathlengths and therefore longer interaction lengths through the use of a resonant cavity. These cavities are formed by the placement of two mirrors facing each other and a small aperture which allows the light into the cavity, where it bounces back and forth until some of the energy is picked off through another aperture. Pathlengths of meters to kilometers are possible with mirrors only inches apart. Long optical path designs allow very sensitive measurements of a compound's adsorption and refractive properties. In the Michelson and traditional Fabry-Perot interferometers, prior reference measurements need to be taken with an empty cavity (although fiber optic and planar versions of a Fabry-Perot have been used). These traditional schemes have found limited sensor applications due to the complexity of having moving mirrors and the lack of real time referencing. However, a static cavity would find use when placed at the end of a fiber doubly-reflected porous silicon surface, or integrated into a photonic crystal cavity resonator.

A practical sensor design would have no moving parts, making it simple to implement, and would have built in self-referencing, so that real-time monitoring is possible. The Mach-Zehnder and Young's interferometers offer these two features. Both use a light source that is divided into two beams. The two beams follow similar paths, with one beam sensing the medium it is propagating through and the other beam, shielded from this same medium, serving as a reference. The Mach-Zehnder and Young's interferometers are identical except in the way the two beams are combined to form the interference pattern. The Mach-Zehnder uses optical components to direct the beams to converge, whereas the Young's relies on natural beam divergence or slit diffraction to cause the two beams to interfere with each other.

Interferometry as described above deals mainly with through-space/volume measurements. These interferometers are not the portable device one envisions when one thinks of the word sensor. The interferometric configurations which may work well for a gaseous substance would find little application for detection of a biological species or dissolved organic compounds. The reason for this is that the change in refractive index in the sensing medium occurring in a bioconjugate event is very small. For example, if a bioconjugate reaction takes place in bulk solution, be it in air or solution, the product will be indistinguishable from a mixture of the unreacted starting materials. It wasn't until the bioconjugate reaction became fixed to a surface that the idea of an interferometric biosensor could be realized. Not only does this idea find use in biosensing applications, but the idea of having a chemically selective film that concentrates a given analyte into the film provides the basis for a chemical sensor.

In 1983, Lukosz and Tiefenthaler (Tiefenthaler and Lukosz 1984), while working on grating couplers for planar waveguides, discovered that relative humidity changes affected their grating coupling experiments. The humidity changes affected the coupling angles and coupling efficiencies. These responses were due to the interaction of water vapor with the evanescent fields of these thin waveguide devices. Further experiments exploited these effects as a way to monitor gas and humidity changes (Tiefenthaler and Lukosz 1984, 1985), and the investigators proceeded to study bioconjugate interactions in these grating coupling devices. The next step was to exploit the length that the planar waveguide provides as an improvement to the short interaction of the grating couplers to produce a highly sensitive planar waveguide interferometric biosensor (Nellen et al. 1988, Lukosz and Tiefenthaler 1988, 1989, Lukosz et al. 1990) .

1.1. Optical Waveguides

The development of planar optical waveguides was a product of the telecommunication industry's switch from electrical transfer of information through copper wire to optical transmission through optical fibers. Planar waveguides that were developed for optical switching, multiplexing-demultiplexing, and a myriad of related devices opened the way for integrating the components that comprise an optical bench and putting them into a two-dimensional configuration on a piece of flat material. Not only were waveguides developed, but also methods to couple the light into and out of these waveguides, defined channels for the light to follow, the directing of light through reflective and diffractive elements, and beamsplitters to divide and combine the light. All the components one would need to fabricate interferometers were now positioned on a planar surface. In addition, fabrication methods developed for the semiconductor/computer chip industry, including photolithography, etching, and deposition, find use in making these optical structures. Diode lasers and detectors, products of the telecommunication/computer/home electronics industry supply many of the components needed to design and construct an entire optical sensing system.

Interferometry could not easily be utilized for the sensing of biological reactions until the reaction was positioned on a surface and a method was found for interrogating the surface with sufficient interaction energy to produce a detectable signal. Optical waveguides provided a way to fulfill these two requirements.

Optical waveguides allowed for the development of interferometers by bringing the light beam in close proximity to the biological species, as well as by providing a platform to concentrate the biological or chemical species in order to maximize the interaction between the light and the species to be sensed. Other methods for measuring biological events on the surface through interferometry, but without the use of waveguides, have been developed and will also be described.

The type of interferometer that is preferred for biosensing is the Mach-Zehnder and/or Young's interferometer, not the Michelson interferometer familiar from FTIR. The Mach-Zehnder/Young's works best for biosensor applications because it is preferable for a sensor to use a monochromatic light source, have no moving mirrors as in FTIR, and measure only the real part of the index of refraction (not the imaginary part such as absorption).

The comparative simplicity of the Mach-Zehnder/Young's mechanics and electronics allows for a very inexpensive and portable device. Lack of complexity in the hardware design may sacrifice some ability to discriminate, but this is made up for by the choice of chemistry in both the sensing and reference arms. The simplest form for the interferometer is one which places the chemistry on a waveguide.

Waveguides come in two basic shapes: cylindrical (optical fibers) and planar. Both designs allow for the monitoring of bioconjugate reactions either at the fiber's end, or along the surface of the planar waveguide. In the fiber, the reaction typically occurs at the fiber's distal end and the information is typically reflected back along the fiber to be analyzed. In the planar configuration, the optical paths, for sensing and reference, travel through the regions where the bioconjugate reaction occurs. The sensing path and the reference path travel along similar lines and the reference path is used to cancel out nonspecific effects. Either the interference signal produced by combining the two beams can occur on the chip, or the beams can be coupled out of the waveguide, combined, and analyzed. The waveguide provides both the conduit for the light and the platform for the bioconjugation event to occur.

Fiber optic waveguides have been used as interferometers especially in white-light interferometry and Fabry-Perot configurations. The designs for the fiber-optic interferometers are limited to reactions that can be placed either at the distal end of the fiber, or, by shaving, at the core of the fiber, exposing the evanescent field and forming the self-described D-fiber. The method of light propagation for optical fibers is analogous to that for the planar waveguides.

The planar optical waveguide allows for the design of an interferometer to monitor a bioconjugate reaction over a longer distance than the fiber interferometers. The planar surface also allows the placement of several interferometric devices on a single substrate, a feat not easily attained with the fiber. The sensitivity afforded by the planar waveguide design is a function of its length, the structure of the waveguide, and the affinity of the bioconjugate reaction. The different interferometer designs developed that use a planar configuration all have the ability to achieve the same relative sensitivity for a given bioconjugate event when employing equal interaction lengths; however, without the proper choice of waveguide materials and design, this equivalency is usually not attained. The waveguide and the design dictate the ultimate sensitivity of the planar interferometer. With the interferometer in hand, a laser source and a detector complete the transducer. Software and display turn the signals into a readable output. Sensing chemistry and flow cells connect the interferometer to the environment the interferometer is attempting to sense.

Planar waveguides provide an optical platform for interferometry. They are similar in structure to fiber optic waveguides. The fiber optic waveguides have high refractive index cores, which direct the light along the fiber by total internal reflection and a lower refractive index cladding to contain the evanescent field associated with the guided beam. The planar waveguide is basically an unfurled fiber optic where the high refractive index material is positioned atop a low index substrate. The planar waveguide has an evanescent field which extends above and below the waveguide. The evanescent field that extends above the waveguide layer is used to interrogate changes in the refractive index of the environment.

1.2. Planar Waveguide Operation

Planar waveguides are composed of thin films of a transparent dielectric material with a higher index of refraction than the substrate on which they are deposited. The light is confined in the waveguide through total internal reflection at the interfaces between the waveguide and the substrate and between the waveguide and the cover media. Optical guiding and the number of optical modes that are allowed to propagate must satisfy the transverse resonance condition in order to exist. The transverse resonance condition states that a guided beam must experience

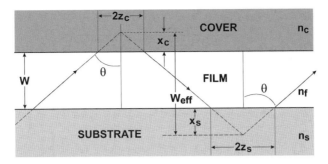

Figure 9.1. Light ray propagation in a waveguide.

a 2π phase shift between equivalent points in the cycle of propagation and reflections in the ray trace translating along the waveguide. Fig. 9.1 shows one of these cycles representing the 2π phase shift between equivalent points (first and last ray arrows). Equations 9.2 and 9.3 define the requirements for the transverse electric field TE and the transverse magnetic field TM modes of light propagation, respectively.

$$2kn_f W \cos\theta - 2\tan^{-1}\left(\frac{\left(n_f^2 \sin^2\theta - n_s^2\right)^{1/2}}{n_f \cos\theta}\right)$$
$$-2\tan^{-1}\left(\frac{\left(n_f^2 \sin^2\theta - n_c^2\right)^{1/2}}{n_f \cos\theta}\right) = 2\pi m, \tag{9.2}$$

$$2kn_f W \cos\theta - 2\tan^{-1}\left(\frac{n_f^2}{n_s^2}\frac{\left(n_f^2 \sin^2\theta - n_s^2\right)^{1/2}}{n_f \cos\theta}\right)$$
$$-2\tan^{-1}\left(\frac{n_f^2}{n_c^2}\frac{\left(n_f^2 \sin^2\theta - n_c^2\right)^{1/2}}{n_f \cos\theta}\right) = 2\pi m, \tag{9.3}$$

where m = mode number (0,1 …)
n_f = waveguide film refractive index
n_c = cover refractive index
n_s = substrate refractive index
W = waveguide thickness in nm
and k = $2\pi/\lambda$

Each equation accounts for the phase shift due to transmission through the waveguide media, ($2kn_f$ W cos θ), and the two reflections, one off the waveguide-substrate interface (the second term in the equation that contains n_s), and the other off the waveguide-cover interface (the third term in the equation that contains n_c). Note that there are lateral shifts, $2z_c$ and $2z_s$, in the phase due to the total internal reflection of the surfaces. These are known as Gooth-Hanchen shifts and they represent the optical beam penetrations into the cover and substrate media. The penetration into the cover layers is what gives rise to the sensing evanescent field, x_c. Combined with the beam penetration into the substrate, x_s, the two add to the thickness of the waveguide film, W, to give the effective thickness, W_{eff}, of the waveguide. Note in equation (9.3), an added multiplying factor appears for the TM mode. This results in the requirement for a thicker waveguide in order to support the TM mode. In turn, the TE mode becomes more buried in the

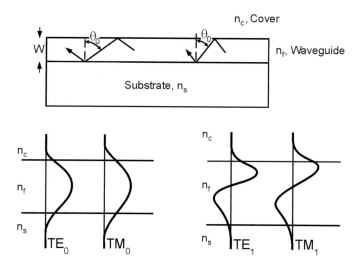

Figure 9.2. Ray and field models for different modes of planar waveguides.

waveguide as the thickness required for TM guiding is achieved. Either or both modes can be used in interferometry.

As either the waveguide film's thickness or the refractive index increases, additional modes will satisfy the transverse resonance condition and will be permitted to guide. The lowest order mode, the zeroth order mode, $m = 0$, is shown in the ray model as the ray with the shallower propagation angle, even though the angle that is of importance is the angle, θ, off the normal for propagation in Fig. 9.2. The next higher order mode, the first order mode, $m = 1$, is shown propagating at a smaller angle, θ, off the normal for propagation. The ray trace and electric field distribution for the first two modes is shown in Fig. 9.2. The TE has a larger angle off the normal than the TM mode, but this angular difference is much smaller than the angular difference between the zeroth and first order modes.

Waveguiding can occur only if a combination of refractive index differences between the waveguiding layer and the substrate layer, as well as a given waveguide thickness, are achieved. Waveguide behavior is presented in more elaborate detail in (Nishihara et al. 1985); for biosensor applications, it is only necessary to know that for light to be guided, the transverse resonance condition must be met.

Examining Fig. 9.2's depiction of the field model, it is seen for the higher order modes that the evanescent field extends further above the surface. In an analogous fashion, the TM mode for a given order extends further into the cover layer than the TE mode. Two interferometer designs take advantage of these relative differences in evanescent field extensions to produce the interferometric signal: the polarimetric interferometer which uses TE and TM modes of the same order to produce the interference signal; and the two-mode interferometer which uses the difference between a lower order and a higher order mode to create the interference. In both cases, the reference arms of the interferometer provided by the TE and the zeroth order for the two types of interferometers described are collinear with the sensing arm. These built-in references are more confined to the waveguide, which greatly helps minimize thermal and mechanical noise. They also provide some cancellation of the background refractive index changes, since there is a small fraction of the evanescent field of the reference mode that is being affected by changes in the solution.

Looking at the transverse resonance condition equations, Equations 9.2 and 9.3, it can be seen that only one term is affected in the event of a chemical or biological reaction on the waveguide surface. That term, the one reflecting the change in the cover index, n_c, appears in the final term of the transverse resonance equation. In contrast to the free space interferometry equation, Equation 9.1, where the phase change is directly proportional to the index of refraction encountered by the optical sensing beam, in the waveguide only a small fraction of the optical electromagnetic field, the evanescent field, interacts with refractive index changes occurring on the waveguide's cover layer. As a result, Equation 9.1 becomes Equation 9.4.

$$\varphi = 2\pi \, \text{L} \, \Delta n_{eff}/\lambda. \tag{9.4}$$

The change in the effective mode index, Δn_{eff}, is the refractive index that the waveguide's composite structure changes as a result of the change in the cover's refractive index. The value of n_{eff} is a function of the indexes of the cover, waveguide, and substrate, as well as of the thickness of the waveguide. n_{eff} can be calculated by the use of the transverse resonance equation, determining the angle of propagation, θ, and knowing the refractive index of the waveguide film, n_f, as shown in equation 9.5.

$$n_{eff} = n_f \sin \theta. \tag{9.5}$$

The effective mode index differs from the index of the waveguide film in that the optical field is being influenced by the index of the three or more layers that make up the waveguide structure. N_{eff} has a value that lies between the index of the substrate and cover and the index of the waveguide material. For example, a 1500Å layer of silicon nitride, n = 1.85, deposited on a fused silica substrate, n = 1.457, and having an aqueous solution serving as the cover index, n = 1.333, has an effective mode index of approximately 1.56. When a biological organism, n = 1.5, is adsorbed to the waveguide surface, it is displacing the water in that volume, increasing the cover index, and in turn increasing the effective mode index, slowing down the speed of light propagation in the waveguide, which is equivalent to retarding the phase of the light.

The greater the change in Δn_{eff} that can be produced by a biological event occurring at the waveguide surface, the more sensitive the waveguide interferometer will be. The change of the waveguide's n_{eff} upon a biological binding event can be increased by increasing the amount of evanescent field available above the waveguide surface. The sensitivity can be designed into the waveguide structure by trying to maximize the amount of evanescent field extending into the cover layer. The amount of evanescent field is dictated by the combination of the refractive index of the waveguide, the thickness of the waveguide, and the refractive index of the substrate material. Sensitivity increases as the difference between the refractive index of the waveguide and that of the substrate increases, and also as the thickness of the waveguide itself decreases.

The planar waveguide configuration also allows for easy addition of other optical components onto or into the waveguide or substrate. Gratings can be fashioned to couple light in and out; and mirrors, beamsplitters, and modulators can be used to manipulate the optical beam.

1.3. Types of Waveguides

There are two major types of planar waveguides—those with a graded index and those that are step-indexed (Fig. 9.3). In addition, a new waveguide structure has been added to these others, the photonic crystal waveguide, which looks quite different from the more traditional structures. That structure will be described later. Waveguides can be composed of any optically transparent material, such as glassy materials and polymers, depending on their structure.

The graded index waveguide has an outside surface refractive index that is high initially, and gradually tapers down to the index of the substrate. This type of waveguide is produced using a substrate (usually glass) that has monovalent ions present in its structure, such as sodium, that are exchangeable with silver, cesium, potassium, lithium, thallium, and other monovalent

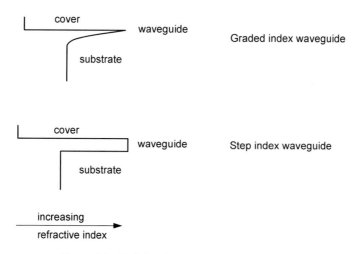

Figure 9.3. Graded and step index waveguide profiles.

ions. The exchange is achieved by placing the substrate in a molten salt bath that contains the ion to be exchanged. The concentration of the ion in the bath, the choice of that ion, the chemical makeup of the glass substrate, and the temperature of the salt bath will determine the properties of the resulting waveguide. The depth and amount of ion-exchange in the glass will determine the refractive index gradient and the number of optical modes that the waveguide will support.

Further modification of the waveguide can be achieved by applying either an electric field while the waveguide material is in the molten salt bath or an annealing process after the bath. Indices of refraction changes near the surface of 0.003 to 0.1 have been reported for graded-index waveguides (Millar and Hutchins 1978, Walker and Wilkinson 1983, Gato and Srivastava 1996) .

Step index waveguides, as the name implies, show a step change in index of refraction between the waveguide and the substrate. These waveguides are produced by depositing a high index material onto a lower index substrate. For polymer and sol-gel waveguides, the waveguide material can be spin-coated or dip-coated onto the substrate. For other glassy materials, processes borrowed from the semiconductor industry are used: chemical vapor deposition, ion-assisted deposition, plasma deposition, evaporation, and sputtering. These processes give fine control of the deposition process and allow one to tailor the properties desired in the waveguide.

The best sensitivity, where the maximum amount of evanescent field available for interaction is achieved close to cutoff, occurs at the thickness at which waveguiding becomes possible. However, being too close to cutoff can incur heightened losses due to inhomogeneities at the surface, so a little added thickness is desirable.

Photonic crystal waveguides are relatively new to planar waveguide sensing. Photonic crystals come in different designs, such as inverse-opal materials, which basically look to be comprised of a material formed from back-filling a close-packed arrangement of spheres, filling in the gaps between the spheres and then dissolving the spheres away, leaving the reverse opal photonic crystal structure. Another more simplified version consists of a material that has holes drilled in it in a very close-spaced and systematic way. These materials have unique properties for guiding light. The crystals can be self-collimating. Optical components such as mirrors, beam-splitters, and resonant cavities can be incorporated into the structure with proper hole placement.

Evaluating the sensitivity of various waveguide/substrate systems can be accomplished by simply running a salt solution or something similar over the waveguide surface and measuring the phase change interferometrically. As the concentration of the salt solution increases, so does the refractive index of the solution and, in turn, the effect on the guided mode in the waveguide. The graded-index waveguides are inherently less sensitive than step-index waveguides because only a small perturbation of the index is possible using simple ion exchange. Step-index waveguides are best for maximizing the difference between the refractive indices of the waveguide and substrate. For example, an ion exchange waveguide that was produced using a BK-7 glass substrate immersed in a molten salt bath containing 0.25 mole % $AgNO_3$ in $NaNO_3$ for 20 minutes at 325 °C, generated $0.22\,\pi$ radians of phase shift for a 0.001 change in refractive index in the cover solution over a 1 cm pathlength. A step-index waveguide that was made by depositing 1100Å of Si_3N_4 through a chemical vapor deposition technique onto a fused silica substrate produced a $7.6\,\pi$ radian phase shift for the same 0.001 change in refractive index above the waveguide over the same 1 cm pathlength. This 34-fold increase in interferometric sensitivity between the two waveguides points out the much greater extent of evanescent field associated with the step-index waveguide compared to the graded-index waveguide. The small evanescent field associated with the graded-index waveguide arises from the relatively small difference between the waveguide and the substrate and the increased thickness the waveguide assumes in order to achieve waveguiding. The ion exchange waveguides have a difference of a few thousandths of a refractive index, whereas the step-index Si_3N_4 waveguides have a 0.4 difference in refractive index between the waveguide and the fused silica substrate. Optimizing the refractive index difference between the waveguide and substrate, and decreasing the thickness of the waveguide so that the waveguide is just thick enough to guide, results in extending the evanescent field further into the cover media, increasing sensitivity. Waveguides made from Ta_2O_5, n = 2.2, have shown even greater sensitivity, more than a 40-fold increase over the ion exchange waveguides.

With the two types of waveguides, graded-index and step-index, different designs have been used to construct planar waveguide interferometers. Designs include slab waveguides and channel waveguides. Slab waveguides have no two-dimensional structure but simply a high index film deposited over a wide area. In the slab waveguide, lateral modes become effectively infinite, eliminating scattering between transverse and lateral modes. Another design, the channelized waveguide, has embedded channels or ridged channels. As the name implies, the embedded waveguides are produced by altering the refractive index of the substrate, usually by ion exchange. To fabricate, a channel is defined photolithographically using a metal mask on an ion exchangeable glass. The glass substrate is immersed in a molten salt bath where ions are exchanged in the open regions of the mask, producing a graded index profile that can be simply diffusion driven and looks somewhat semicircular in profile, or it can be electric field-assisted to produce more of a step profile.

Alternatively, the channel can be photolithographically rendered on deposited high index material and the material etched partly or completely away to give two versions of a rectangular profile step index waveguide. In the channel waveguides, the width of the channels has to be

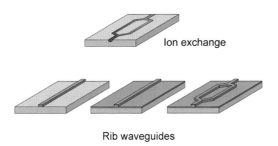

Ion exchange

Rib waveguides

Figure 9.4. Different types of channel waveguides and Mach-Zehnder interferometers.

designed to avoid the fact that at given widths, additional lateral modes can propagate. Efforts are made to limit these effects by proper control of the channel's thickness and the smoothness of the sides of the channel where points can scatter light into other possible modes than the ones desired.

2. Light Coupling Methods

To use waveguides as interferometers, light must first be coupled into and out of the waveguide. There are three commonly used methods for coupling the light into the waveguide: end-firing, prism coupling, and grating coupling (Fig. 9.5). Each method has advantages and disadvantages that become apparent when constructing an interferometric waveguide sensor.

End-firing is the simplest way to excite a guided mode in a waveguide. Light is fed into the waveguide from a smooth edge face. The beam is either focused to the waveguide's edge with a lens or fed to this edge by close positioning of a fiber optic. Maximum efficiency occurs when the beam profile closely matches the guided beam profile (Nishihara et al. 1985). For the zeroth order mode this profile is close to a simple Gaussian. The edge of the waveguide is either cleaved, as in silicon substrates, or polished to a defect-free surface in order to maximize coupling and minimize scattering. High coupling efficiencies are possible but not absolutely necessary, since the interferometric measurements only require minimum intensities because phase measurements are power independent. End-firing works well when the waveguide is thick, such as in an ion-exchange waveguide. Alignment of the input beam and waveguide becomes problematic when dealing with a submicron thick, high refractive index step-waveguide. With waveguide dimensions on the order of 0.1 micron, the position of the light to be coupled requires precise manipulation of the light source or optical beam and waveguide. This is further complicated with channelized waveguides, which require not only vertical but also horizontal positioning. The alignment problem can greatly hinder their integration into a practical sensor, especially one in which the waveguide chip is to be changed frequently.

Prism coupling also offers high light coupling efficiency. In this case, the prism is positioned in intimate contact with the waveguide surface. High refractive index prisms are used to excite the guided wave by phase-matching the incident wave with the guided mode. Prisms such as heavy flint glass (n = 2.009) and rutile ($n_o = 2.584$; $n_e = 2.872$) are typically

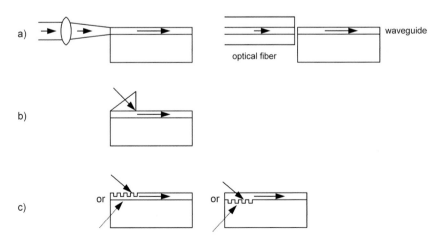

Figure 9.5. Optical waveguide coupling methods: a) end-firing focused beam and end-firing optical fiber; b) prism coupling; c) grating coupling.

used. A bit of pressure and a scrupulously clean surface are necessary for the contact. The visual appearance of a mirrored interface is an indication that good contact is achieved. Beveling the front edge of the prism provides a thin gap for the light to jump into the waveguide, stay confined, and not be coupled back out on a subsequent bounce off the waveguide substrate interface. Guiding will occur when the angle θ from the incident light in the prism equals the propagation constant β, where $β = k\,n_f\sin θ$, $k = 2π/λ$, and n_f is the index of the waveguide. Changing the angle allows one to address different modes in the waveguide, if the waveguide structure will support them.

Prisms have not proven to be useful in sensor applications. To use prisms to couple light into the waveguide, pressure has to be exerted on the prism. In the lab, this pressure is achieved using a screw mechanism to push the prism against the waveguide. This screw mechanism is required for both the input and output light coupling. The mechanism is on the top side of the waveguide, where sensing reactions are to occur, and the screws and prisms get in the way. In addition, for a practical device, prisms would need to go down on the waveguide in the same location each time a waveguide chip was changed. It has been observed that each time the setup is reassembled the angle and position of the input beam has to be adjusted in order to couple the light in. The output is not affected as much, but it may change the position of the output beam enough to need adjustment to direct the beams to the detector. However, the biggest problem is that the prisms need to be placed directly on the waveguide. This makes the flow cell needed for aqueous phase biosensor measurements difficult to incorporate into the design. The gasket around the flow cell will also be in contact with the waveguide, decoupling the light from the waveguide where it makes contact. Adding an isolation layer under the cell and gasket would not affect the guided light, but this involves additional fabrication steps while adding to the overall length of the waveguide chip.

Ideally, if the flow cell can be placed on one side of the waveguide while coupling in the light from the other side, the optics will not interfere with the cell. This brings us to the third method of light coupling, the grating coupler.

The grating coupler is a periodic structure with alternating refractive indices fabricated into the substrate or waveguide by either embossing, etching, or ion exchange. A commercially available grating can be embossed into a deformable waveguide material such as a sol-gel or polymer to produce a grating. For example, the sol-gel SiO_2-TiO_2 is typically dip-coated onto a substrate and allowed to partially dry. A surface relief grating is pressed against the coating with a pressure of 50–100 lbs for a few minutes (Gato and Srivastava 1996, Heuberger and Lukosz 1986, Ramos et al. 1986). Upon release of the pressure, the grating die is removed and the sol-gel is cured thermally. Gratings from 0.25 to 0.85 micron, with efficiencies between 10 % and 25 %, have been formed using this method. Polymers can be used in a similar manner but they are less rigid, less inert, and have larger dn/dt's, changes in refractive index with temperature which can alter the grating coupling ability.

Any resulting gratings can be sealed by adding a protective film over them, such as SiO_2, to make them immune to changes in the sensing environment, while providing the index contrast for the grating.

Alternatively, the grating can be rendered photolithographically. The grating mask can be generated in a number of ways. One method uses two overlapping laser beams. A UV laser beam is divided into two beams with a beamsplitter. The two beams are initially spread out and then angled, spatially filtered, and made to converge at a position producing an interference pattern with the desired period. The patterning of the grating region can be rendered through a shadow mask into the photoresist directly on the waveguide substrate or on a master photomask. The grating is etched with reactive ion etching or wet chemical etching (Hartman et al. 1998). The grating can be fabricated into the substrate or the waveguide but in either case, a SiO_2 overlayer is used to seal the area of the grating from the environment. Either the substrate/waveguide or the waveguide/overlayer combination provides the index contrast. These gratings have efficiencies similar to those of embossed gratings.

Gratings provide the option and advantage of having the light come in from the bottom of the waveguide, thus allowing for placement of the test cell and related fluidics on the sensing side. Gratings have less stringent alignment conditions than does the end-fire approach. Input optics can be designed to provide a cone of light with a reasonable fraction of the light at both the correct angle and position for coupling into the waveguide. This allows waveguide chips to be routinely replaced, producing a "plug and play" sensor. Gratings work well for slab waveguides; however, they have not been used with channel waveguides. The reasons may be either that the small channel dimensions make the alignment difficult or that power input may be limited, but this author has not seen any cases to date. Gratings with large widths have been used to excite several interferometers on a single chip at once, with uniform coupling efficiency (Hartman et al. 1998).

2.1. Interferometers

Waveguides have evanescent fields that is sensitive to changes occurring in the volume directly above the waveguide surface. A chemically selective film in this area can interact with the guided light beam in the waveguide. Chemical or physical interactions change the index of refraction, causing the propagating light speed, or phase, to change. To measure this change, a reference propagating beam is optically combined with the sensing beam to create an interference pattern of alternating dark and light fringes. This describes a planar waveguide non-channelized Mach-Zehnder interferometer; a schematic of this optical arrangement appears in Fig. 9.6.

When a chemical or physical change occurs in the sensing arm, the interference pattern will shift, producing a sinusoidal output. The phase shift is defined in terms of n_{eff}, the effective index change of the waveguide/sensing film/substrate combination. Over a length, L, in Equation 9.6, this phase change accumulates.

$$\varphi = 2\pi L \Delta n_{eff}/\lambda. \tag{9.6}$$

The detection level is limited by noise, both thermal and mechanical, from the waveguide system, the laser stability, and the detector characteristics. Detection limits of 10^{-6} change in refractive index, or better, are typical.

Sensitivity is a somewhat subjective issue. If one simply examines the equation for change of phase and places an arbitrary but common number for wavelength, say, 632 nm, and an

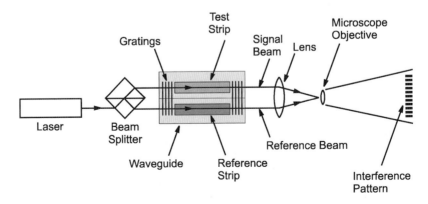

Figure 9.6. Diagram of typical Mach-Zehnder interferometer.

interaction length of, say, 10 mm, the answer to the level of detection will be based on the author's claim of how sensitive he can detect phase change. If the claim is 0.01 radians, the maximum limit with all the light being affected by the change in refractive index occurring in the bioconjugate reaction, is 1×10^{-7}. However, with the waveguide devices described, the amount of light interacting is usually on the order of 10 %, so that the minimum change of refractive index detectable with 0.01 radians resolution is 1×10^{-6} refractive index. But if the authors claim they can detect 0.001, radians, then the limit is again 1×10^{-7} for a 10 mm pathlength. Additional gain in sensitivity can be accomplished by increasing the amount of optical field interacting with the reaction to a level greater than 10 %, possibly 2–3 fold increase or more. A thousandth of a radian phase change is claimed by many, but usually that refers to their optical system, not within a given experiment. The experiment can have temperature fluctuations as well as mechanical vibrations. These can be kept to a minimum by proper design of the system and controlling the temperature accurately, but 0.001 is about the limit. The level 0.01 radians is more realistic. Some claim software that can get their signal-to-noise ratio better, but again this is usually with the system at optimal performance, not during an actual experiment. The length can be increased, but again this only improves the sensor increase linearly in sensitivity. Double the length, and the sensitivity doubles. And wavelength changes provide even less improvement in sensitivity, since they are relatively small in the visible and near infrared range (possibly a factor of two).

The real limitation is the "chemical" noise of the system that arises from changes in the solution, and in the waveguide when in the solution. These changes may occur from material in the solution, material that is either being adsorbed to the surface or being lost from the surface. The signals can arise from the porosity of the waveguide. They can arise from the movement of bioreceptors on the surface, changes in conformation, denaturation of proteins, hydrogen bonding changes, or any other chemical or physical phenomena that occur. All these can cause changes in the refractive index sensed by the interferometric sensor system. One tries to limit the differences by designing the reference arm of the interferometer of the sensor to account for all these things, but I have yet to see one that actually does. Even with nothing on the surface of the waveguide except a solution, water or air, there are changes in the signal, though much less with air than water. Add a flowing solution, and there is more fluctuation. With a bioreceptor film on the sensing arm and an inert biomolecular film on the other, there is more, even in the single channel devices. If the flow solution to be measured is a "true" biological sample such as blood, urine, or washings from a chicken carcass, there will be greater fluctuations in the signal produced. Sometimes the signal is random and sometimes it produces a response that goes in one direction. If a buffered stock solution with some proteinaceous backfilling molecule is run, it will settle down and a reasonably good stable baseline will be achieved, but it may take some time. So sensitivity measurements or claimed sensitivities may not be all that meaningful when one wants to measure a sample in a reasonable time frame and have it provide a believable number.

Size is another issue with regard to sensor systems. All sensors are desired to be small in size and measure a single cell. How realistic is this? Suppose one had a miniature Mach-Zehnder interferometer of 15 micron square size with a flow cell of a similar vertical dimension. The flow cell would have a volume of 3 picoliters. The solution being tested has just been evaluated by ELISA and found to have a concentration near its sensitivity limit, approximately 10^6 cells/ml. Using this concentration, the chance that there will be a cell within the 3 picoliter flow cell at any given time is 1 in 300. Note that the cell has to land on the sensing arm of the interferometer. The sensing arm is functionalized with an antibody made to bind with the cell. With the solution flowing through the flow cell, will the organism that passes through every 300 change-outs of volume in that cell have time to diffuse to the waveguide surface and bind to it? One might be waiting a long time. So keep in mind, small sensors are advantageous to large bulky systems, but there is a limit to what is practical if you want to detect something.

These limitations also apply to the devices that measure pathlength changes versus refractive index changes. A limit of 1 angstrom thickness change is roughly equivalent to a 0.001 phase change.

Interferometric sensors used for biological detection differ from most sensing schemes in that the sensing is accomplished by directly monitoring a bioconjugate reaction occurring within the evanescent field extending out from the interferometer's sensing channel. There are no subsequent steps (typical in many other sensing schemes) in which a second bioconjugate is reacted with the first to produce a sandwich complex. In those designs, the first bioconjugate reaction is used to capture the target antigen. The second bioconjugate contains some form of reporter molecule. The reporter can be a fluorescent tag, a radioactive tag, or an enzymatic label. For either of the first two, the reading can take place after the second bioconjugate reaction is complete. With the enzymatic tag, a chemical substrate is added that reacts with the enzyme to produce a colored or fluorescent product, which in turn is detected. All these secondary bioconjugate-involving sensing schemes have added complexity to the sensing system, with the sequential addition of test sample, washing steps, secondary bioconjugate addition step, more washings, and then readings or, in the enzymatic case, addition of the substrate and then readings.

The interferometer eliminates all these steps and allows one to monitor in real time the binding of the antigen, be it a protein, strand of DNA or RNA, virus, spore, or whole cell, to the capture compound, which could be an antibody, aptamer (synthetic DNA or RNA antibody), single strand of DNA for hybridization measurement, or other bioconjugate reaction such as biotin/avidin binding.

The use of real-time referencing afforded by the interferometric design also holds the advantage of the nulling of thermal and mechanical noise, optical cancellation of nonspecific binding, and bulk index changes. Optical integration of various components of the system provides for added stability, compact size, and low cost. Several interferometric schemes have been investigated. In each configuration, the sensing and reference arms are combined in different ways, such as with different beam-combining schemes. Each has advantages and disadvantages, with no configuration clearly eclipsing the others. However, an earlier fiber optic scheme that requires bulk optic/fiber optic hybridization appears to be somewhat cumbersome for practical sensor use. Sensors have evolved from interferometers with discrete and separate optical paths, and those with side-by-side channels on a single surface, to those with stacked and parallel arrangements of channels.

The two most common configurations are the Young's and Mach-Zehnder interferometers. The difference between the two involves only whether the two beams converge through natural divergence of the beams (Young's) or through a physically forced combination (Mach-Zehnder). Even within these different interferometers there are different methods for accomplishing the same objective.

Fiber optics was developed for the conduction of light, and planar waveguides for integration of the optical circuitry needed to manipulate optical signals. Fiber optic interferometry initially used a combination of bulk and fiber-optic components. When all-fiber interferometers were first developed their use was restricted mainly to temperature and pressure measurements. Various configurations that use fiber optic interferometry are outlined in a review article by Kersey et al. 1990. The first fiber optic interferometer to be used in a biosensing application consisted of a temperature sensor used to measure the heat produced in an enzymatic reaction (Choquette and Locascio-Brown 1984). Researchers loaded an antibody conjugate onto a fiber and placed a reference fiber adjacent to the sensing fiber. When the catalase-labeled antibody reacted with H_2O_2, heat was produced, and that in turn produced an interferometric signal.

Employing only the distal end of the fiber limits the interaction length of the interferometer. To increase this interaction, the fiber's cladding can be shaved off, leaving a D-shaped structure where the evanescent field once confined to the cladding is now exposed. An example of the D-fiber that appeared recently used the exposed area as a means for exciting a surface plasmon in a metal film deposited on the exposed area of the fiber (Chiu et al. 2005). The researchers combined the D-fiber surface plasmon resonance (SPR) with heterodyne interferometry to yield a device capable of measuring an index of refraction change of 10^{-6}.

Also in recent work, researchers have presented a dual channel low optical coherence tomography system that measures the relative path differences between the orthogonal polarization modes using polarization- maintaining fibers (Akkin et al. 2002). The system is capable of measuring pathlength changes of approximately 1 angstrom or a phase change of 10^{-3}. The bioconjugate reaction does not occur on the fiber; the fiber merely serves as a means to gather the optical information and interfere it. The researchers envisioned developing an optical imaging system to monitor cell or tissue dynamics.

Fiber-optic interferometers continue to be developed, but are used primarily for acoustic, gyroscopic, pressure, and temperature measurements. For chemical sensors, however, most fiber optics find use in absorbance and fluorescence measurements, where the interrogating or excitation wavelength is transmitted down the fiber and the signal reflected is a measure of the absorption or fluorescence taking place at the fiber's end.

Interferometers find a more versatile structure in a planar configuration. As opposed to the measurements made with the short distal end of an optical fiber, the planar configuration offers the advantage of an evanescent field interacting over a long interaction length. The planar configuration enhances the amount of evanescent field due to its thinner waveguide, and has the added ability of allowing the integration of various optical components onto or into the planar surface.

A critical consideration in waveguide interferometry lies in the design of the reference channel. Since interferometry measures the difference between the sensing and reference arms, the reference must be designed to null out as much of the background signal as possible. The background signal can be caused by thermal differences, mechanical changes, and bulk index (the solution's refractive index) effects. Local variation in the bulk index of the sampling media requires good mixing and presentation of the solution to the sensor. The chemistry should be designed to mimic all the chemical effects that could happen on the sensing arm except for the specific bioconjugate reaction being employed for sensing. This means that the sensing arm chemistry would have the same dn/dt, the same physical size, population density, and hydrophilicity/hydrophobicity, so that the sensing volume contains an equivalent amount of buffer or bulk solution within the evanescent field. The reference arm should be as close as possible to the sensing arm in structure, response, and length, the only difference being the affinity of the receptor molecule to the targeted antigen, so that the reference serves as a true nulling control.

The fiber-optic approach cannot compete with an integrated design, since the reference arm is physically discrete from the sensing arm. Planar waveguide interferometry places both arms onto the same physical structure, with spacing as little as a millimeter or less apart. The close spacing essentially negates any thermal or mechanical perturbations. However, bulk refractive index changes can still be bothersome with the close arrangement and must be taken into account in the fluid dynamics of the cell. An ideal system would have background changes happening in the bulk solution that are simultaneously sensed in both the reference and sensing arms.

2.2. Collinear or Single Channel Interferometers

There are two fundamental designs for arranging the sensing and reference arms of the interferometer: one in which the light beams travel within the same volume in the waveguide, and the other in which the two arms are arranged side-by-side. Two of these common path interferometer designs, where two modes are traveling with the same volume, are the polarimetric or difference interferometer and the two-mode interferometer. These designs attempt to minimize the unwanted effects that may be differentially affecting the sensing and reference arms when they are separate. The sensing and reference light paths in the difference or polarimetric interferometer are collinear, so that the light is passing through the same waveguide volume, thus nulling any slight variation in thermal and mechanical effects. However, note that these cancellations chiefly apply to effects originating from either the waveguide or the substrate and not from any effects originating in the cover solution, apart from the bioconjugate reaction of interest.

The collinear polarimetric or difference interferometer has its origins in the Lukosz group in Switzerland (Akkin et al. 2002, Stamm and Lukosz 1993, 1994). In this configuration (Fig. 9.7), the sensing and reference beams travel through the same volume with the waveguide. Sensing occurs due to the difference in extensions of the evanescent fields of the two polarizations, the transverse electric, TE, and the transverse magnetic, TM. The TE will propagate through a thinner waveguide than will the TM mode. When the waveguide becomes thicker, the TE mode becomes more confined to the waveguide and the waveguide starts to support a TM mode also. When both of these modes are propagating, the TM mode will have more evanescent field above the waveguide than does the TE. The difference accounts for different sensitivities to chemical or physical changes above the waveguide for the two modes. The two modes experience different changes in phase for the same reaction, which is the source of the sensor's response, but the ability to isolate the reaction of interest is limited.

To construct this polarimetric interferometer, both polarizations of light are coupled into the waveguide by rotating the laser approximately $45°$ from the normal polarization. The approximation is noted because different coupling and propagation efficacies of the two polarizations require a slight biasing for equivalent signals. Both polarizations propagate down the waveguide, with the TM mode experiencing a change in phase preferentially over the TE mode. Both modes are coupled out, then separated into different polarizations by using a beamsplitter, two Wollaston prisms, a $\lambda/2$ and $\lambda/4$ plate, and four photodetectors to measure the phase difference (Lukosz et al. 1997), and thus, the response of the sensor.

This array of optical components can be further simplified to a Wollaston prism and polarizer, or a grating, lens, and polarizer, with the resulting interference pattern imaged on a CCD array (Lukosz et al. 1997). The pattern appears orthogonal or off-axis, respectively, for

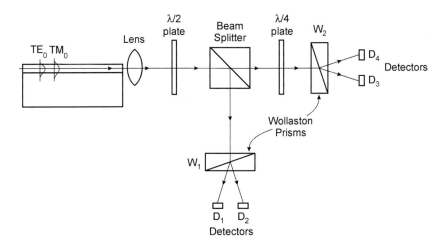

Figure 9.7. Diagram showing the guided wave output of a polarimetric (or difference) interferometer.

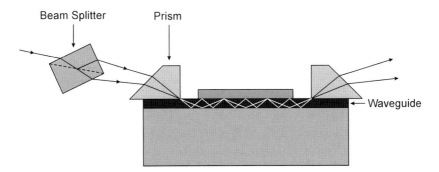

Figure 9.8. Two-mode interferometer.

the two schemes, allowing for the possibility of multiple sensors on one substrate and analysis by a single two-dimensional array.

The two-mode interferometer has two different order modes, with the same polarization traveling with the same volume of the waveguide (Fig. 9.8 (Hartman et al. 1988, Hartman 1990)). The two modes have evanescent fields that extend to different heights above the waveguide surface. The lower order mode will be more confined to the waveguide and serves as the reference, whereas the higher order mode has a field that extends farther into the cover media and serves as the sensing arm of the interferometer. As with the polarimetric interferometer, it is not possible to bury one arm completely while retaining a good deal of sensing capacity in the other. Because of this, the signal measured does not truly reflect the Δn_{eff} of the waveguide for the sensing mode. It is also not possible to differentiate a binding event from a bulk index change or temperature change, nor is it possible to segregate nonspecific binding from the targeted binding event. Even though the single channel two-mode interferometer looks rather elegant in design, implementing it requires the ability to bring in two beams at different angles to excite the different modes of the waveguide. Whether this is accomplished with a prism or grating, this represents a design hurdle unless the grating could be designed to accept a single angle input and launch the two angles of propagation necessary to excite two modes. In addition, the output also has similar considerations, since the two beams are exiting the waveguide at different angles. Some optics would be required to combine the beams to form the interference pattern.

In order to take full advantage of the polarimetric or two-mode interferometric scheme, the waveguide needs to be designed to maximize the difference in sensitivity between the two modes. One mode is buried in the substrate, while the extent into its surroundings of the other mode is maximized. To achieve the greatest differential, thin single mode step-index waveguides are used that are just slightly thicker than what is needed for the propagation of TM light, or in the case of the two-mode device, just thick enough for the second mode to be supported in the waveguide. The pioneer researchers in the polarimetric interferometer employed sol gel-based TiO_2-SiO_2 waveguides with refractive indices in the 1.8–2.0 range and an optimal thickness in the 2000 Å range.

It is not possible to completely bury the TE mode while having the TM mode's evanescent field maximized. Therefore, the reference beam of the interferometer, the TE polarized beam, is not completely blind to the bioconjugate binding events taking place on the waveguide's surface. This diminishes the phase change measured by the interferometer. Depending on the waveguide system, this decrease in response can be as great as 50 % less than that seen in the side-by-side Mach-Zehnder interferometer model or any design which completely separates the sensing from the reference response. In addition, the TE and TM modes not only respond to the binding event but also respond (and to different amounts) to nonspecific binding of

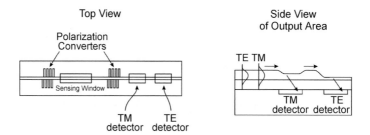

Figure 9.9. Diagram of an integrated polarimetric (or difference) interferometer.

errant biomolecules at the surface as well as index changes in the bulk media. Lack of a well-designed reference hurts the sensitivity of the difference interferometer to single out the specific binding events, as well as its ability to null variations in bulk index changes. The gain in mechanical and thermal stability in the collinear design may not be enough to compensate for these shortcomings unless bulk index changes can be minimized, and the change due to specific binding can be enhanced over any nonspecific binding.

Neither the polarimetric (or difference) interferometer nor the two-mode interferometer can distinguish between a binding event and a change in bulk refractive index and/or temperature. However, one attempt to deal with this problem uses two wavelength inputs at the same time in order to provide enough additional information to differentiate two effects, either binding and bulk index change, or binding and temperature (but not bulk index change and temperature) (Stamm et al. 1998). This design requires an additional laser; in this particular device, an Argon-ion 488 nm laser was combined with a He-Ne laser operating at 633 nm.

The need for multiple optical components required to analyze the polarimetric or difference interferometer's output is a disadvantage to compact sensor design. Subsequent work (Koster et al. 2000) has integrated the difference interferometer's polarization optics onto a waveguide chip (Fig. 9.9). The device employs a polarization converter to convert some of the singly polarized TE input beam to TM mode. The two polarizations travel under the sensing area to another polarization converter that allows mixing, followed by decoupling of the two modes in a sequential manner. The device features detectors that are integrated in the supporting silicon substrate. Only the laser remains external.

The major shortcoming of the single channel devices, the polarimetric and two-mode interferometers, is that any bulk index of refraction changes that occur in the sensing area will be sensed differently by the two modes, or polarizations. In the side-by-side interferometric design, these bulk index problems can be effectively cancelled; however, effective fluid flow is important in preventing any differential index changes resulting from nonhomogeneous mixing and flow over the two sensing areas that appear as different refractive indices of refraction to the two arms of the interferometer.

2.3. Two-Channel Interferometers

The majority of work in the area of interferometric biosensors is in the area of two-channel or two-arm interferometers. The two most common designs are the Young's and Mach-Zehnder interferometers. The Young's interferometer uses the natural divergence of light to produce the interference pattern, whereas the Mach-Zehnder design physically combines the two beams by some optical scheme. Both interferometers have true reference arms, which are highly important for biological measurements, since one wants to maximize the possible signal, and the sensing arm permits taking advantage of the whole Δn_{eff} produced in the sensing arm by the biological binding event. The separate reference channel provides a means to eliminate not only most thermal and mechanical noise by being closely positioned to the sensing arm, but also accounts both for the bulk index changes in the surrounding media and for the amount of nonspecific binding of nontargeted biological material.

The two-channeled or two-arm interferometers come in either defined channel devices or open/slab waveguide designs. The channels can be either embedded in the substrate, built on the substrate as in ribbed waveguides, or simply be areas defined by where the light propagates and the chemistry is deposited on either open or slab waveguides. The Mach-Zehnder interferometric design did not start out as a chemical sensor but as a modulator for the telecommunications industry. Ranganath and Wang (1977) designed a waveguide with Mach-Zehnder interferometric design. A responsive material was placed on one arm of the device. An electronic interaction, rather than a chemical presence, changed the phase in one arm of the interferometer. The optical modulator, in this case, had buried the waveguides in order to shield them from environmental effects in the surroundings. Sensors, on the other hand, seek to maximize the effects from the surroundings as the means to alter the phase of the guided light. Opening up areas on these channels provides regions in which to apply chemically selective films to detect specific compounds or biological entities. Opening up a single arm produces what is referred to as a buried reference interferometer. The buried reference interferometer finds use in chemical sensing. Coating the sensing area with a polymer and comparing it to the unchanging buried reference provides the interference pattern shift associated with the concentration of the analyte. Typically, the polymer sensing films are thicker than the evanescent field, approximately 5000 Å. Making the films this thick shields the sensor from any environmental effects and allows one to measure the change due to adsorbing the analyte alone. Calibration with known concentrations of analyte provides a calibration curve for using the sensor for real-life applications. Since the polymer can act as a concentrator of the analyte, owing to the effect of the partition coefficient between the analyte affinity for its media versus the polymer sensing film, sensitivities for a given analyte can be very high. For example, a polymer film of poly-(2,6-dimethylphenylene-p-oxide) has a partition coefficient for trichloroethylene (TCE) in water of approximately 5, and since partition coefficients are measured on a log scale, this means that for every molecule of TCE in a water solution, the sensing polymer has 100,000 molecules concentrated in the film at equilibrium. Sensitivities for TCE in turn are projected in the low parts per billion to high parts per trillion ranges as a result (Campbell et al. 2004).

For biosensing applications, the sensing films are typically on the monomolecular layer level and one is trying to detect the binding of a large protein or organism to this sensing film. The sensing area is going to be responsive not only to the binding of the specific target analyte but also to nonspecific interactions or a change in the refractive index of the medium. Most commonly, the change is due to some combination of all three. In order to selectively distinguish an analyte, the reference arm is opened up and is exposed to the solution being measured. By the judicious choice of surface chemistry, one can account for nonspecific binding and bulk refractive index effects. The desired effect is that the reference serves to cancel out optically any of these extraneous effects from nonspecific binding, and bulk index changes from the measured signal, directly. The design of the reference arm is crucial for maximum detection of the desired analyte.

The design of channels, and the y-junctions used to direct light down each arm of the interferometer and then bring them back together again, are also products of telecommunications research. Entire channelized Mach-Zehnder interferometers have channels fabricated in or on a planar substrate; examples are shown in Figs. 9.4 and 9.11. There are two types of channels: ion-exchange (Helmers et al. 1996, Drapp et al. 1997, Luff et al. 1998) or deposited ribs (Stamm et al. 1998, Fischer and Muller 1992, Brosinger et al. 1997, Schipper et al. 1997, Weisser et al. 1999, Lillie et al. 2004, Lechuga et al. 2003, Prieto et al. 2003a, b) (Fig. 9.4).

The ion-exchange method begins with the deposition of a patterned metal mask that has the channel areas open. Submersion in an ion-containing melt diffuses the ions into the channels and creates the waveguide structure. The sensing and reference areas are then defined by a deposited overcoat.

For ribbed channels, fabrication requires the deposition of a high-index film onto a lower-index substrate. The channels are rendered photolithographically and can be either etched or patterned using a lift-off process. The typical ribbed structure consists of a raised high-index material on a base of the same material, all deposited onto a lower-index substrate. Devices have been fabricated that work as either single-mode (Lechuga et al. 2003, Prieto et al. 2003a, b) or multimode devices (Lillie et al. 2004).

The channelized branched Mach-Zehnder interferometers require the device to be rather lengthy, because the divergence angle in a Y-junction has to be very small. When the angle is small, greater length is required to separate the channels sufficiently for chemical or biochemical functionalization. There is also a taper into the Y-junction; this angle must be designed so that the mode will propagate without converting to other lateral modes. Channel widths must be set so as not to introduce other lateral modes. However, multimode waveguides of this type have been shown to work for chemical sensing applications (Lillie et al. 2004) and should easily transition over to biosensing devices. The edges of the channels need to be smooth so as to minimize scattering, a problem seen mainly with deposited waveguides. Ion-exchanged waveguides suffer from inherent lack of sensitivity.

For both types of channel devices, the sensing and reference windows are deposited as an overcoat on the waveguide structure. Light can be end-fired into the waveguide using either a lens to focus the light, or a fiber aligned with the edge (Nishihara et al. 1985). V-groove technology for fiber alignment has been used if the waveguide is constructed on a silicon substrate (Fischer and Muller 1992). The goal for most devices of this type is not to use fibers for inputting the light but rather to incorporate the light source and detector fully integrated in the device (Lillie et al. 2004).

The channelized Mach-Zehnder interferometer provides a means for sending a point light source down two or more channels and to combine two signals into one interferometric signal. However, the output phase will start out at some random position in the phase according to the way the beams combine. When the two beams are brought together by the use of a Y-junction or by using an external beamsplitter (Heideman 1993) (Fig. 9.10), they interfere with a resultant phase which arbitrarily lies anywhere along the sinusoidal intensity of the fringe pattern, since no fringe pattern is output. It is difficult to design the phase position of the output signal to be near quadrature, the point of the interferometer's greatest sensitivity to phase change. A deliberate index change in one of the device arms is needed to allow determination of both the

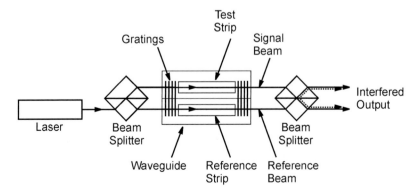

Figure 9.10. Beamspiltter combination in a slab Mach-Zehnder interferometer.

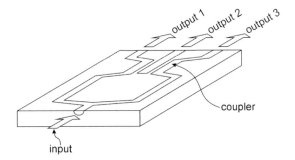

Figure 9.11. Mach-Zehnder interferometer with a three waveguide output coupler.

phase location at the start of the experiment and the min-max possible when going through a total of 2π phase shift, one entire interference fringe.

This quadrature problem with the Mach-Zehnder interferometer, in which the beam combining on the chip provides only a single point on the interference pattern, has been mitigated (Luff et al. 1998) by using a three-waveguide coupler, again borrowing from the telecom industry. The coupler (Fig. 9.11) is comprised of three adjacent waveguides: the two arms of the interferometer and a third, positioned between them. The output signals are offset in phase by $2\pi/3$, providing at least one output near quadrature. The sum of all three outputs is a constant, thus the coupler allows for the monitoring of laser light stability or absorption changes in the interferometer.

Using a slab waveguide to construct the arms of the interferometer eliminates the difficulties encountered in coupling the light into a narrow channel of the channelized Mach-Zehnder, and permits the use of either a prism or a grating coupler to inject the light into the waveguide and back out for subsequent combining and analysis. In this design, a high refractive index material is either deposited on a lower index substrate or ions are exchanged in the right type of glass to form a waveguide over the entire substrate. No channels per se are necessary; the arms of the interferometer are defined by where the light goes, and thus the sensing and reference chemistry are deposited, or a cover layer can be deposited on the slab waveguide, leaving openings where the sensing chemistry is deposited. Access to the waveguide may also be necessary for coupling in the light if prisms are to be used. If gratings are the choice for coupling the light in and out, no further openings other than for chemistry are needed. So-called channels can be relatively wide, millimeters wide, so that the alignment becomes simplified. Interaction lengths can be dictated by the size of the substrate the waveguide is deposited on, and by the opening in the isolation layer or the deposited chemistry layer. The interaction length can be centimeters in size, and with the sensitivity of the sensor being directly proportionally to its length, this is then the simplest and most straightforward way to gain a great deal of sensitivity for the interferometric sensor. The interferometer diagrammed in Figs. 9.6 and 9.10 shows two ways to assemble this type of Mach-Zehnder interferometer. Note the wide sensing channels and narrow light paths in comparison.

In the interferometer shown in Fig. 9.6, a laser light source is split into two beams by the use of a beamsplitter. One of the beams is used for sensing the biochemical reaction on the waveguide, and the other beam, the reference. The beams are coupled into the waveguide, shown here using grating couplers, but prisms could serve the same purpose. The two beams travel along the waveguide. An area on the waveguide is functionalized to interact with the target analyte, and one of the beams is directed to travel under this area. The other beam travels under a chemistry used to help cancel out any nonspecific interactions. After the beams travel the length of the waveguide they are coupled out by use of the grating coupler. The beams are focused to interact with each other by means of a simple convex lens. The interference pattern is formed where these two beams overlap. The spatial separation of the fringe, the fringe pattern, is determined by the angle at which the two beams converge. With two beams separated by 4 mm on the waveguide, and using a 30 mm convex lens to combine the beams, the fringe spacing, y, is $y = \lambda f / d$, where f is the focal length of the lens, and d is the spacing between the two parallel beams. At 632 nm the fringes are approximately 5 microns wide, too small to be discerned by most detectors. With a microscope objective placed at the focus of the interference pattern, the pattern is expanded to whatever size is convenient for the detector arrangement. At this point one can either use a slit that is narrower than the fringe or place it in front of the detector, or, to take advantage of signal averaging, image a several-fringe image on a 2D array.

A flow cell is mounted on the waveguide to allow the introduction of the aqueous solutions typically used in biosensor applications, and the test sample is flowed through the cell. When there is binding between the surface functionalized waveguide and the biological target there will be a change in the phase of the sensing arm. This will be reflected as an apparent shift in the interference pattern. The amount of shift is related to the amount of reaction taking place at the waveguide surface, and with prior calibration of the system will allow one to equate the fringe shift to a given concentration of analyte in the solution. Note that the fringe pattern is not actually moving; it is merely winking in and out of phase in a manner analogous to a theater marquee where the sequential lights provide the appearance of movement, but the lights are simply turning on and off, essentially in and out of phase.

With the integration of the optical components, it is possible to shorten the distance between the two arms of the interferometer, which in turn will increase the width of the fringes to a point where the use of a microscope objective is not needed and the lens focal length can be shortened to allow for placement of the detector within a short distance from the waveguide, condensing the entire sensor package.

Another Mach-Zehnder scheme, shown in Fig. 9.10, uses the same planar slab waveguide, containing arms of the interferometer as defined by where the beams travel and where the chemistry is deposited; and also uses a similar arrangement as described above to input the light, but a different scheme for generating the interferometric signal. After the beams are coupled out of the waveguide they are directed into a beamsplitter, which combines the beams and produces two identical signals that represent some arbitrary position on the sinusoidal interference fringe pattern (Heideman 1993). In this case, quadrature can only occur by chance and the min-max signals for the fringe pattern would not be known. One practical approach for determining min-max and setting quadrature incorporates a phase shifter on one of the interferometric arms (Heideman 1993, Heideman et al. 1994). The phase shifter is a rotatable glass slide located externally to the waveguide. In another example, Heideman et al. (1996) placed an electrically addressable electrooptic material, ZnO_2, on one of the interferometric arms on the waveguide. Application of a field alters the index of refraction and therefore the phase of the beam. This arrangement allows one to not have to measure the shift of the fringe, but only measure the amount of voltage applied to the phase shifter to return the signal to quadrature. The calibration of the amount of voltage applied to maintain quadrature for a given concentration of analyte is used for sensor response versus the measurement of phase change.

An interesting variation on the channelized Mach-Zehnder model, the Young interferometer (Brandenburg 1997, Brandenburg et al. 2000, Ymeti et al. 2002), Fig. 9.12, cuts off the device after the sensing and reference arms without recombining on the chip. The light exits out of the end of the waveguide. The exiting beams overlap at some distance from the waveguide, forming an interference pattern. The Young's interferometer requires no additional optics and allows for complete monitoring of the fringe pattern. Using a single-point detector, the min-max of the fringe pattern can be scanned by translation of another slit smaller than the fringe period and positioned in front of the detector. Alternately, an array detector can analyze the entire fringe pattern without the need of any translation.

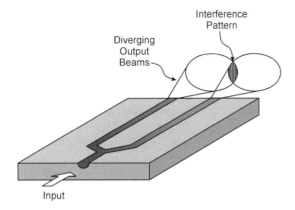

Figure 9.12. Diagram of Young's interferometer on a planar waveguide.

In one version of the Young's interferometer, the light exiting from each arm is passed through a set of slits. The light diffracts through the slits (Hradetzky and Brandenburg 2002, Schmitt et al. 2004) to form a very clean fringe pattern in a manner similar to Young's experiment in 1803.

A possible disadvantage of Young's configuration can be the distance from the waveguide output to the detector required for maximum fringe resolution. The period of the fringe is equal to $\lambda L/d$, where L is the distance from the waveguide and d is the separation of the two channels. Since the fringe pattern exists in all space beyond the beam's exit and overlap, placement of the detector can be at any distance as long as there is overlap. The fringe spacing that can be resolved is a function of either the detector's pixel pitch or the width of the slit placed in the interference pattern for a single point detector. However, in the configuration shown in Fig. 9.12, the spacing between the arms of the interferometer is very small due to the constraints of the Y-junction angle; this allows close placing of the detector to the waveguide since the fringe period is quite large.

The spacing of the channels as well as the distance from the light's exit from the waveguide affects the spacing of the fringe pattern. The fringe pattern increases with closer proximity of the channels or beams. The space between the sensing and reference beams is determined by how close one can fabricate the channels for the channelized version, how close two beams can propagate in the slab version without interfering with each other, how good the columniation of the beams is, and how fine an area of functionalization of the sensing chemistries is possible. Using a combination of photolithography and surface energy manipulation, the distance between different chemistries can be brought to less than $100\,\mu m$ range.

A stacked Young's interferometer has also been investigated (Cross et al. 1999, 2003). Here, the waveguides are layered together, with a buffer medium separating the waveguides vertically. The diffracted light creates the interference pattern as it exits the waveguide. This elegant approach could be very useful in chemical sensing, where a buried reference is required. However, for biosensors, the buried reference prevents easy accounting of bulk index changes. In the second reference, the researchers utilize an approach analogous to that of Lukosz's (1995) polarimetric interferometer approach, in which the proper choice of refractive index and thickness of the waveguide allows one to account for bulk index changes and focus on the changes occurring from the bioconjugate reaction alone. Lukosz used two wavelengths to perform this operation; whereas Cross employed a dual waveguide design providing adlayer thickness and refractive index from the differences in the two waveguides and their two modes, TE and TM. In surface plasmon resonance (SPR), an approach which detects the interference between the surface plasmon wave and the bulk wave propagating at the grazing angle at the surface has been shown to allow for the cancellation of bulk refractive index effects (Alieva and Konopsky 2004). However, although these elegant approaches allow one to measure the thickness and density of the biofilm formed by the bioconjugate reaction, and also may allow one to account for the changes in bulk index, they may not account for nonspecific binding reactions that may occur. Again, how does one account for any nonspecific binding without having a channel that experiences all the effects encountered by the sensing channel minus the specific bioconjugate reaction being measured? This remains one of the challenges of interferometry; and an even bigger challenge is encountered when one is trying to differentiate between live or viable cells and dead cells. Live cells could be fatal, whereas the dead cells, though they may interact with the sensing film, would be harmless.

In addition to the interferometers so far encountered, which take advantage of either fiber or planar optical schemes, there are devices that have been developed that have taken large traditional interferometers and condensed them down to form a series of compact and rugged interferometers. Manning Applied Technologies (Manning et al. 2004) has described three different interferometers: a nutating prism interferometer, a multiplex Fabry-Perot interferometer, and a bilithic interferometer. These devices are composed of optical components with moving mirrors in a size smaller than a shoebox.

3. Interferometric Array Sensors

The desirability of sampling for several analytes at once, or being able to analyze several samples sequentially without the need for regeneration or changing chips, can be realized by having multiple interferometers fabricated on the same substrate. Two approaches to this multi-interferometer scheme are being developed; one using a branched network of channelized interferometers (Ymeti et al. 2002), and the other placing several interferometers side-by-side on an open slab waveguide structure (Hartman 1997, Campbell 2005).

A group from the Netherlands has taken a single channel waveguide and subdivided it over and over again to produce a branched network of interferometers on the same waveguide. Each interferometer has a pair of sensing and reference arms. The device has a single light source that consists of a fiber optic that couples light into the channel where, through a route of channels and y-junctions, the light is sent down each channel of the various interferometers (Ymeti et al. 2003, 2005). The output from each pair of channels for each interferometer emerges out the end of the waveguide and naturally diverges and overlaps to form the interference pattern. The unique feature of the multiple Young's interferometer is how the output is handled so that each interferometer can be analyzed separately. If each interferometric pair had equal spacing, then the series of fringes formed would all have the same fringe period, causing difficulty in distinguishing the output from each interferometer separately without interference patterns from the others. To separate each interferometer, the relative spacing of the arms, 60, 80, and 100 um, is different for each interferometer, producing a series of fringe patterns all having different frequencies (Ymeti et al. 2003). By picking the proper spacing between the arms, each fringe frequency can be analyzed without ambiguity, using the proper software. The fringe patterns can even overlap, but the analysis can examine the separate frequency components individually and determine the phase shift of each interferometer. A schematic of the output is shown in Fig. 9.13.

The non-channelized approach to a multiple interferometric array has used two designs; one in which the beam combining is accomplished on the chip before the light is coupled out, and the other in which the beam combining is done by the use of a lens after the light beams are coupled out of the waveguide.

The design shown in Fig. 9.14 has thirteen interferometers on a single chip that measures 1×2 cm. A single laser is fanned out onto a broad input grating fabricated in the substrate.

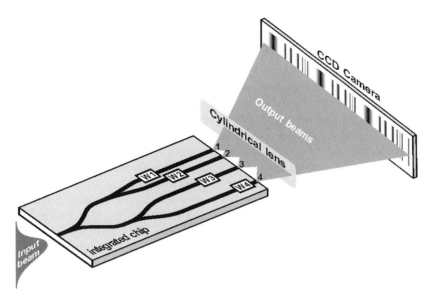

Figure 9.13. Multi-channel Young's interferometer.

The grating launches light into all the interferometers at once. The waveguide consists of a deposited slab of Si_3N_4 on glass. The channel lengths are defined by using patterned thick SiO_2. The sensing chemistry is placed within the channels. The channels are defined by the chemistry immobilized on the waveguide surface and the collection area of the total internal reflection (TIR) element located at the output end of the channels, rather than by the beams themselves. The light, after traveling under the sensing and reference chemistries, is gathered up by reflective elements and directed to a beamsplitter, where the beams from both arms of the interferometer are made to interfere. After interfering, the resulting beams are reflected again off the back side of the TIR and sent to the output grating to be coupled out and directed to the detector. A series of segmented gratings couple out the interference signals and discard the light associated with the edges of the channels. As in the previous on-chip beam

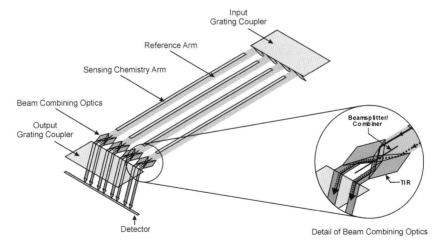

Figure 9.14. Diagram of 4 interferometers from a 13-interferometer chip; detail shows integrated optical components on the chip.

combining schemes, the output represents one point on an infinitely broad fringe. Movement of one of the arms of the interferometer is necessary in order to determine the min-max of the fringe.

The number of interferometers in this design was determined by the pixel pitch of the detector. Recent improvements in detectors with tighter pixel pitches have opened the door to increasing the number of interferometers possible on a single chip. The arrays in USB (Universal Serial Bus) cameras provide pixel pitches that are around 5–10 microns, which is equivalent to approximately fifty interferometers in the same area where thirteen were fabricated.

The on-chip beam-combining scheme shown above requires many steps to fabricate, and tight etching tolerances are required for elements such as the beamsplitter. To simplify the multiple interferometer arrays and provide an inexpensive design, an off-chip beam-combining scheme has been developed (Campbell 2005). In this design, a broad grating again is used to illuminate all the interferometer channels with the same laser beam. The channels are defined by thick SiO_2 deposited on the Si_3N_4 slab waveguide. Holes in the SiO_2 layer provide access to the waveguide surface and the area where the sensing and reference chemistry are deposited. At the end of the waveguide area are fabricated a series of segmented gratings that are staggered along the propagation direction of the waveguide. A portion of this interferometer design is shown in Fig. 9.15. The gratings are narrower than the waveguide channels so as to pick off only light that travels down the center of each channel and to minimize any effects caused by the edges of the SiO_2 buffer layer.

The grating pairs are staggered so that they are spatially separated when combined and imaged on the detector. An eight-interferometer channel device has been fabricated with a stagger of 0.5 mm between each pair of gratings. The entire output of the grating array is sent through a cylindrical planoconvex lens to combine the beams into a series of interference patterns imaged onto a CCD array. The resulting image is shown in Fig. 9.16. As seen, enough separation is present in the array of fringe patterns to fabricate additional interferometers onto the same chip.

The staggered arrangement of gratings to produce a series of separate interference patterns can also be accomplished by using a series of gratings each with a slightly different grating period to slightly alter the output angle of the beams being coupled out of the waveguide. The overall effect is the same, but the length taken up by the output gratings on the chip is reduced (Campbell 2005).

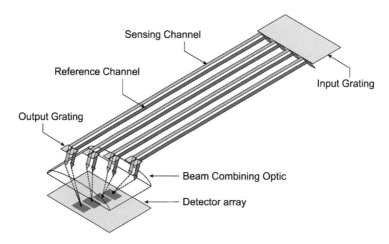

Figure 9.15. Portion of 8-interferometer chip with staggered output gratings.

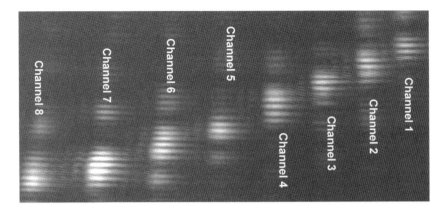

Figure 9.16. Image from 8-channel interferometer with staggered gratings.

Of the designs for waveguide interferometers described, all have advantages and disadvantages in their design, fabrication, and implementation. All can achieve roughly the same sensitivity for an equivalent pathlength. To achieve equal sensitivity for the same length, the change in Δn_{eff} has to be the same for a given bioconjugate reaction. Graded-index waveguides will never achieve the sensitivity of the step-index waveguides. Interaction length and the amount of evanescent field available for probing the surface are the key factors responsible for the sensitivity of the optical system to a given change in the refractive index.

4. Surface Plasmon Interferometry

Surface plasmon interferometry was first described by Nikitin in 1997 (Kabashin and Nikitin 1997). Nikitin used a unique approach that investigated the phase change information not usually measured in surface plasmon resonance experiments (Kabashin and Nikitin 1997, Grigorenko et al. 1999, Nikitin et al. 1999). This open-path interferometer uses the metallized surface of an SPR prism in a Mach-Zehnder arrangement with an expanded beam. This technique does not measure the angular change typically measured by SPR, but rather uses the reflected beam exiting the prism, combines it with a reference beam, and produces an interference signal. A sensitivity of 4×10^{-8} in refractive index change was estimated based on the measurement of different gases. Following this work, Nikitin proposed a system to measure spectral-phase interference. This method was based on the detection of a spectrum of reflected or transmitted light that would be modulated by interference in a sensitized receptor layer on a glass slide (Nikitin et al. 2000a). The researchers proceeded to use surface plasmon resonance interferometry for the analysis of a microarray of prefunctionalized spots (Nikitin et al. 2000b). Analogous systems were built by Ho et al. (2002), which claimed a 3×10^{-6} change in refractive index sensitivity, and by Hsiu et al. (2002), which was used to measure hybridization occurring over a DNA microarray. Nikitin improved the arrangement by changing from a Mach-Zehnder/SPR hybrid to a Fabry-Perot device, developing a spectral correlation scheme for this spectral resonance interferometer device (Nikitin et al. 2003). The system, called the "Picoscope" (Nikitin et al. 2005), shown in Fig. 9.17, is based on measurements of the correlation signal of two interferometers: a scanned Fabry-Perot interferometer, and a second interferometer produced by the interactions on a simple microscope cover slip derivatized with a sensitive layer laid out into a 12×8 spot array. The design can make multiple analytical measurements at once for a wide variety of receptor sites.

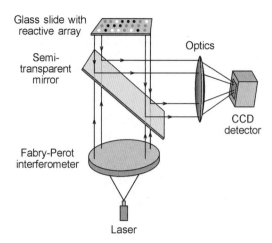

Figure 9.17. Picoscope.

In an alternate technique, common-path phase-shift interferometry surface plasmon resonance provides another optical arrangement to get more phase information from SPR (Su et al. 2005). Though somewhat complicated in arrangement, the laser light that produces the surface plasmons over a wide area is reflected off the surface, and phase information is gleaned from the reflected beam by the use of an electrooptic modulator. The sample information is imaged on a CCD array as a function of voltage applied to the electrooptic modulator, which sequentially directs s and p polarization to be imaged on the detector. The system has about the same sensitivity as most interferometric schemes, with the added bonus of permitting the imaging of several derivatized areas on the entire sensing chip.

5. Other Interferometric Methods and Designs

A few other interferometric schemes investigated during the last decade or more include the use of Fabry-Perot fringe generation with white light, and Zeeman interferometry and a spinning disc interferometer. Phase information has also been obtained from surface plasmon resonance setups.

A Fabry-Perot approach has been explored by Gauglitz et al. (1993) at the end of a fiber and on the surface of a porous silicon chip. Here, changes in thickness and index of refraction caused by binding of proteins, or by the swelling of a polymer at the end of a fiber, alter the pathlength of white light being sent down the fiber. The interference of the reflected light produces Fabry-Perot interference lines. A change in the reflection due to binding of a biomolecule, organism, or swelling of the surface film shifts these fringes. The amount of shift is indicative of the change. However, this reflectometric interference spectroscopy lacks sensitivity, because the pathlength change is so small compared to the lengths being discussed for the planar waveguides.

Reflection interference off a porous silicon substrate caused great interest when it was first presented (Lin et al. 1997), since it had unexpectedly high sensitivity. Porous silicon contains etched pores with a very high surface area. When white light is shined on the porous area, the surface will reflect off both the top of the substrate and the bottom of the pores. The two reflections interfere to produce Fabry-Perot fringes. When the pores are functionalized with antibodies, a shift is observed upon exposure to the conjugate antigen. Even though the pathlength difference between the two reflections is small and the pores are only 1–5 microns in depth, the change appears to be too large to result from mass addition alone. It is thought to arise from modification of carrier concentrations in the semiconductor due to the binding. Further analysis showed that the results were somewhat misleading; that the change seen was due to oxidation of the silicon pores (Dancil et al. 1999). However, the effect from reactions involving the pore material itself have been exploited to detect fluorophosphate nerve agents from the etching of the silicon dioxide by the HF product of nerve agent hydrolysis (Sohn et al. 2000). More recently, DNA hybridization has been performed on these porous silicon surfaces. When hybridization occurs, a shift in the reflectance spectra is observed (DeStefano et al. 2004). The true nature of this effect is unknown, but it is thought to be either related to the carrier concentration, or due to mechanical stresses induced by intermolecular interactions causing either the index of refraction or the thickness to change.

Zeeman interferometry uses a Zeeman laser and a planar waveguide. This interferometer takes advantage of the two frequencies that are generated by the Zeeman laser, with a 250 kHz difference in wavelength (Grace et al. 1997). As the two modes propagate through the waveguide, a phase difference accumulates between the modes as a result of a change in the sensing layer. The 250 kHz beat frequency sine wave, generated when the two modes are combined, is measured at the output and compared to the reference sine wave from the laser. The phase difference is indicative of a change in the coating on the waveguide. Although this example is a chemical sensor, this technique could be used for measuring a bioconjugate reaction. As is the case with the polarimetric interferometer, this collinear scheme is also prone to sensing bulk index differences.

Another device design built to handle a large number of tests on a single chip is the spinning disc interferometer, called the BioCD (Varma et al. 2005). The BioCD consists of multiple interferometers fabricated on the surface of a 2-inch mirrored disk, with gold derivatized spokes radiating out from the center well area. Antibodies are immobilized on the gold surfaces. The sample is placed in the well and spun, spreading the material to be tested out over the spokes. The disc spins at 6000 rpm, resulting in high data acquisition using lock-in amplification. The thickness controlled spokes act as wavefront-splitting interferometers, and at the 1/8 λ condition give rise to a phase difference of $\pi/2$ between the surface of the spoke and the surface of the disc, so that the difference between the reflected light is at quadrature, assuring that the sensor works in its linear response region. Averaging the response from a number of spokes gives an average signal corresponding to the amount of binding that took place. There is space on the disc to perform multiple assays for a given sample. A laser reads the spinning disc in a manner similar to a bar scan and records information from a barcode. This approach could easily find application in combinatorial genomics or proteomics without the need for a fluorescent or colorimetric label, especially when attempting to examine planar arrays of reactions/interactions such as DNA hybridizations.

A Fabry-Perot Mach-Zehnder device has also appeared. In this device, rather than just a simple sensing channel, the sensing area is replaced with a periodically segmented waveguide, which provides a resonant cavity-type device to provide for a longer pathlength (Kinrot 2006). The device provides the sensitivity of a common Mach-Zehnder but is less than 1/10 the length. How it deals with background index is not obvious, unless both arms are going to ultimately be comprised of segmented structures with only one arm functionalized with the bioreceptor chemistry and the other not.

6. Surface Functionalization

No matter what interferometric scheme is used, the sensing and reference arms of the interferometer have to be functionalized with a chemistry that will interact with the targeted biomolecule or organism. These receptors can be electrostatically bound to the surface, covalently attached, or conjugated with another biomolecule. To covalently attach these receptors, one typically utilizes the hydoxy groups that populate most of the glassy surfaces of many waveguide materials.

Glassy waveguide surfaces include materials such as Si_3N_4, TiO_2, or SiO_2. If the surface doesn't contain such groups it is possible to deposit a thin film of SiO_2 on the waveguide to provide such sites. In the case of Si_3N_4, simply cleaning the surface replaces amine groups with hydroxyls.

Protein, aptamer, and DNA attachment chemistry employ methods developed for affinity chromatography. Books have been written on the subject (Hermanson 1996). Typically organosilane chemistry is used to attach reactive functionalities to the glass surface. These silanes usually terminate in amino, carboxaldehye, or sulfhydryl groups. A protein receptor will bond to an amino group after prior oxidation, or to a carboxaldehyde by reductive amination. Sulfhydryls will form disulfide bonds with cysteine. Any number of bifunctional linkers can also be used to form a long tether to the protein, thus increasing its mobility.

Other schemes employ attachment of protein A to the surface, followed by binding to the Fc portion of an antibody. Avidin is also employed as a functionalized surface for biotinylated proteins and DNA, taking advantage of the strong binding constant of the avidin-biotin system.

Subsequent backfilling with a small protein filler is used to minimize nonspecific binding. The reference arm of the interferometer, if the transducer uses one, is handled in a similar fashion. A dummy protein is attached, and the remaining surface is backfilled with a small protein. The object is to make the reference equivalent to the sensing arm in all ways except the specific binding event that one is attempting to sense. Unfortunately, it is not that simple. Nonspecific binding of errant protein is typically quite prevalent, especially in clinical samples; therefore, the nulling out of the nonspecific signal is not easily accomplished. Much work has been done to minimize the nonspecific signal by using surfaces that resist the adsorption of proteins. Polyethylene glycols and their derivatives have garnered the most attention and appear to work relatively well. Various other surfaces with amide, phosphoamide, urea, and variations of polyethylene glycols have also been compared (Chapman et al. 2000). The ability to limit nonspecific binding sets the detection limit for the collinear polarimetric and two-mode schemes, since both specific and nonspecific interactions look identical to the transducer.

Proteinaceous antibodies for detection of the target biological species have been the workhorse of both sensor sensing surfaces as well as the laboratory-based assays such as ELISAs (enzyme linked immunosorbent assay). However, new receptors are finding applications, especially in sensors where the method may be a direct-read method, such as interferometry. These include DNA and RNA aptamers, aptamer-protein hybrids, and imprinted surfaces.

7. Sample Collection Systems

Typically the sensor itself is part of a complete sensor system, such as what is diagrammed in Fig. 9.18.

Prior to sensing, the material has to be collected. For many biosensing applications the sample is already present in a solution ready to be tested. If it is in a urine or blood specimen, a drinking water sample, or a chicken carcass washing or other foodstuff analysis, or is a mysterious white powder that arrives in a questionable envelope, the interferometer can be put to immediate use and can test the sample for some targeted moiety with a waveguide functionalized to react with the targeted antigen. Sometimes the target has to be collected from the air before it can be put in solution and be tested. For all biological testing, the antibodies, aptamers, and other bioreceptor molecules derivitized on a waveguide surface require an aqueous, or more likely a buffered aqueous, solution to be effective in capturing the antigen, since water/buffer provides stability to the receptor and essentially mediates the binding event. In some cases,

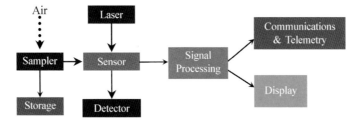

Figure 9.18. Schematic of a complete sensor system.

the sample can be collected by wiping a surface and then putting the wipe, such as a swab, in solution and introducing the solution to the sensor. However, in cases typical of a terrorist attack, the sample may be airborne and may be in very low concentrations. A collector may be necessary to concentrate the target organism from the air and then mix it into solution before testing. Different sampling methods and devices are commercially available. Some samplers, such as the collector developed by Meso Systems, work by drawing air through a very fine filter and trapping the organisms on the filter. Periodically the filter is removed and captured material is washed off and then tested. These devices have collection volumes in the 100s–1000s of liters per minute. However, the filter has to be physically removed and washed before a screening of what is captured is possible. This is an interrupted method, in which one has to decide how long to collect before stopping collection and taking a measurement.

Cyclonic collectors such as SpinCon, developed by Midwest Research, draw in 10,000–30,000 liters/minute through cyclonic action, spinning the heavier materials to the outside of a cylinder which has a thin film of water on the surface to capture the particles. The SpinCon cycles through vast quantities of air and concentrates the material in the air into between 1 and 10 ml of solution. The solution can in turn be examined with the sensor to determine if any biological threat exists. The SpinCon can repeat this cycle over and over continuously as long as it has a reservoir of solution to use. In most cases, the device is much larger and more expensive than the interferometric sensor.

An electrostatic collector called ALPES (Aerosol-to-Liquid Particle Electrostatic Sampler), developed by the Savannah River Technology Center, uses an inexpensive fan to pass air through a statically electrified water fountain. The water continuously runs over one of the electrodes and the material is drawn into the flowing water by the electric field. The collector can basically run constantly, integrating whatever may be present in the environment. In this case 100s–1000s of liters per minute can pass through the collector, and the collector can be integrated with the sensor so the system is constantly monitoring the air. Eventually, enough dirt and dust will be collected so that the system will need to be refreshed.

The real-time sensing capability of the interferometer makes real the possibility of a constant monitoring system for whole organisms in either water or with collected air samples.

Whole-cell detection is feasible, but if the detection of cellular pieces such as DNA or some proteinaceous material is required, the same whole cells that were collected can be lysed to release the DNA or proteins inside them. Dehybridization of the double-strand DNA can be accomplished thermally or chemically in a flow-through system setup.

8. Interferometric Applications for Whole-Cell Detection

Interferometric biosensors can be used for the entire range of biological entities from small components such as proteins, genetic material including DNA and RNA; small cofactors and organic compounds such as biotin, sugar, and metabolites; large species, including viruses and

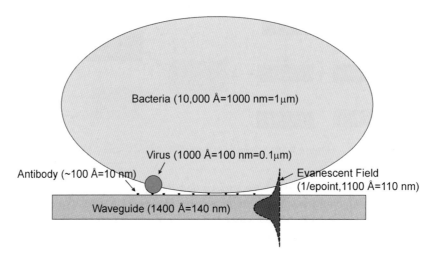

Figure 9.19. Size comparison of various biological analytes and the evanescent field.

spores; to cellular material, whole cells, and organisms. There is a problem as the species to be detected becomes larger, as in the case of the whole organism. This involves a limitation in the extent to which the evanescent field projects above the waveguide. In Fig. 9.19, the evanescent field is shown in comparison to various biological entities, and it can be seen that this field envelops the smaller protein DNA and viruses but doesn't extend enough to completely interact with whole cells. For most of the waveguide systems described in this paper, the evanescent field, which exponentially decays from the surface, has the electric field concentrated near the waveguide surface. For a typical step-index waveguide such as Si_3N_4 on fused silica, the $1/e$ point, that is, the distance within which approximately 65 % of the field is concentrated, is in the first 1100 Å from the surface of the evanescent field device.

If one calculates the extent of interaction of the exponentially decaying wave front with a spherical cell placed on the planar waveguide surface, one finds that the evanescent field is interacting with less than 10 % of the sphere. If the cell is further out from the surface by a linker and antibody, or the cell is being immobilized by some cellular extension such as a flagella, the amount of interaction becomes even less. The use of reverse symmetry waveguides, or working in the infrared where the evanescent field scales with wavelength, would allow the whole evanescent field to interact with the cell and increase the transducer's sensitivity by an order of magnitude or more.

In comparison to a typical 1-micron organism, the evanescent field is interacting primarily with only a part of the cell outer walls. And when one thinks of the three-dimensional structure of the cell, the waveguide's evanescent field is interacting with only a very small fraction of the entire organism present. The interferometer does respond to whole cell binding and achieves sensitivities comparable to if not better than ELISA methods, so much better sensitivity could be realized if the field could be extended further or if the antigen is small.

Either of these possibilities can be put into use. Though the organism cannot be made smaller, the cells could be ruptured and the sensor could examine fragments of the cell, fragments small enough to increase their interaction with the evanescent field. To accomplish this approach, the use of several different capture ligands that are designed to interact with different fragments, protein, or genetic material released from the ruptured cell, would have to be incorporated into the sensing film's receptor makeup.

The other approach may be more straightforward and requires a change in the design of the waveguide to enhance the evanescent field's extension into the volume above the waveguide.

As mentioned in the discussion of the physics of waveguides, the evanescent field is a product of the waveguide's index of refraction, the thickness of the waveguide, and the refractive indexes of both the substrate and the cover. The index of the cover is limited by the fact that almost all biological sensor measurements are taken in water, n = 1.333, thus the value of the cover index cannot be manipulated. To push the evanescent field up further above the waveguide, the index of the substrate has to be lowered to an index less than water. This is not easy to accomplish. Most glassy materials have indexes greater than 1.45. However, new materials and deposition techniques have been developed that provide refractive indexes less than that of water. These waveguide structures are collectively called reverse symmetry waveguides, since they are reversed from the standard composition of $n_c < n_s < n_f$ to become $n_s < n_c < n_f$.

The electric field distribution of the reverse symmetry waveguides reflects the inversion of the refractive indexes of the layer that makes up the waveguide. The field distribution of the reverse symmetry waveguide can push the electric field up from the waveguide so as to provide better if not complete interaction with a whole organism. The lower the refractive index of the substrate, the further the evanescent field can be extended up above the waveguide. Material such as MgF_2, which has a low index to begin with, can be lowered even further by inclusion of free volume in the MgF_2 matrix. This can be accomplished by taking MgF_2 sol-gels and casing them at different conditions and curing them at different temperatures. Materials with indexes of refraction between 1.15 and 1.4 have been produced and could in turn be used as substrate material for reverse symmetry waveguides (Murata et al. 2006). Nanoporous silica also can achieve low indexes of refraction by incorporating a large amount of free volume within the SiO_2 matrix (Horvath et al. 2002a, b). These materials provide excellent candidate substrates for these types of waveguides.

One final means to achieving an increase in the interaction of the evanescent field is obvious from the fact that the evanescent field is a function of wavelength. As the wavelength increases, the extension of the evanescent field increases proportionally. If the sensor's system is constructed at 10 times the wavelength, the evanescent field extends 10 times further out. At 6-micron wavelength, the 1/e point of the evanescent field is 1.1 microns for a comparable step-index waveguide. An extension of the field this much provides total interaction of the evanescent field with the bound organism. However, moving to infrared wavelengths introduces a whole new set of problems. New waveguide materials have been developed, but laser sources and detectors are more difficult—not to mention the problems of dealing with light one cannot align by eye. However, tolerances are reduced as the wavelength increases, and waveguide inhomogeneities are less of a problem as grating coupler dimensions become large enough to provide easy fabrication.

The literature presents several different interferometric sensing schemes. With most schemes, the objective is to show the sensitivity, compactness, and versatility of the interferometers. All interferometric devices stress the benefits of direct label-free detection systems, taking advantage of the underlying strengths of the interferometer.

With interferometry, there is no need to label the bioconjugate reaction with fluorescent, radiological, or enzymatic tags, since the interferometer can observe the initial conjugate reaction between the bioreceptor on the transducer surface and the antigen directly in the test solution, producing a measurable signal. Research in interferometry has not focused on many specific biosensor applications but merely explored the transducer's design and sensitivity. Note that the sensor can be no better than the sensing chemistry used on the transducer. Most researchers have evaluated a specific interferometer design, usually with a generic bioconjugate system. Many simply use salt or sugar to deliver an index change to the transducer. The researchers that have tried a biological system for evaluation of the transducer use a simple protein antibody/antigen system such as an IgG/anti-IgG. DNA hybridization has been shown to be observed in real time with a 21-base match (Schneider et al. 1997). Whole cells have appeared only a few times, as we shall see.

The use of a second reporter such as a microparticle with whole-cell detection would achieve little of the necessary amplification, since the binding of the microparticle to the cell's surface is occurring in a weak portion or outside the evanescent field. Also, the use of a sandwich-type assay defeats the primary advantage of the interferometer, that of being a real-time, direct assay.

Only a few references to whole-cell detection appear in the literature or have been presented at conferences over the time interferometers have been investigated. The reason may be only that the groups working in interferometry do not have a specific biological detection application in mind, but are merely presenting different optical configurations available for these applications. It is a much simpler experiment to prove that an interferometer works by testing it with a salt or sugar solution. To simply prove and then state that a new biosensor is available requires only a simple straightforward assay. The reaction of IgG with anti-IgG is well known, well studied, and inexpensive to conduct. Issues such as the size of the bioconjugate and the interrogating evanescent field are not an issue. The stability of the receptor is also not a problem. The actual handling of live cells gives rise to health concerns, and if one is going after the organisms of interest (those in the area of biological warfare), the need for BSL-2, BSL-3, other containment facilities, and proper handling limits many researchers. Some benign organisms can fill the bill for proving one's device is capable of the detection of whole organisms, but they don't carry the interest of many sponsors that fund this kind of work. To say that you can detect anthrax but have only done *Bacillus globigii* does not carry the same selling power as the real thing, nor does it carry the risk.

Other than IgG/anti-IgG, other protein-based assays, and DNA, few have reported success in detecting viruses, spores, and whole living cells with an interferometric sensor. Viruses appear to be the best candidates for whole-cell detection with evanescent field interferometry, owing to their size (100 nm) being comparable to the 1/e position on many evanescent field devices. Workers in the Netherlands were first with the recent reporting of the direct detection of HSV-1 virus using an antibody receptor (Ymeti et al. 2007). In this work, antibody-coated channels of a channelized Mach-Zehnder were used to measure increasing amounts of the HSV-1 virus. The sensor was challenged with concentrations of virus from 10^3 to 10^7 particles/mL in buffer. A response was obtained for all concentration levels. Whole serum detection for the same virus was obtained for 10^5 particles/mL.

Recently, researchers at Georgia Tech have used an open channel Mach-Zehnder with off-chip beam-combining to measure the concentration of avian influenza virus. The assay employed a hemagglutinin-specific antibody bound to the waveguide (Xu et al. 2007). Detection limits as low as 0.0005 HA units/mL were obtained. An HA unit is defined as the minimum amount needed to cause agglutination of the chicken's whole blood cells. An HA unit is typically equivalent to 10^4 virus particles (Wagner and Hewlett 2004). A sandwich assay further improved these detection results. Experiments were performed with two strains of H7 influenza subtypes with the H8 subtype to show no cross reactivity with the different antigenic subtypes.

Prior to this whole-virus work, work was done on human influenza, but rather than detect the whole-virus particle, the cells were ruptured either by heat or detergent to release a nucleocapsid protein that was detected in a simple protein-protein immunoassay (Schneider et al. 1997).

Spores and whole bacteria have sizes in the 1-micron diameter range. They do not lend themselves well as candidates for evanescent wave interferometry, where the evanescent field penetrates the cell structure to only a small degree. Other interferometric schemes where that thickness can be exploited would be a better choice, but the planar evanescent devices were the first to try whole-cell detection of bacteria. The other forms of interferometry are just beginning to explore the area.

Two papers have investigated the detection of *Salmonella* (Schneider et al. 1997, Seo et al. 1999). In both cases an antibody specific for an epitope on the cell surface is covalently attached to the waveguide surface using standard techniques. Both interferometers were of the two-channel slab waveguide Mach-Zehnder variety with off-chip beam-combining. A simple immunoprotein was bound to the reference channel. After a stable baseline was obtained, solutions containing the whole *Salmonella* were introduced to the waveguide by use of a flow cell. Phase change due to adsorption occurred almost immediately and continued unabated until the solution was switched back to buffer. Binding of whole cells was verified by inspecting the waveguide under a microscope, but coverage of whole cells was poor.

It could not be determined whether the phase response was due solely to the whole cells or to conjugate proteins or other material having leeched out of the cell. A response was detected from very concentrated solutions of 10^7 cells/ml to 5×10^4 cells/ml (Campbell et al. 2003). The 10^5 cells or cfu (colony forming units) per ml is comparable to the best ELISAs; but the interferometers required no secondary steps, just introduction of the *Salmonella* and monitoring a response. The detection of *Salmonella* was done in buffer and a more realistic applicable media, chicken carcass washings (Seo et al. 1999).

As is true for whole-cell detection as well as all biosensor applications, the affinity of the bioreceptor on the waveguide surface for the antigen being sensed dictates how well the assay will perform given equivalent sensing platforms. Another dominant factor in whole-cell detection is the limitation imposed by diffusion. In order to trap the whole cell with the antibody or any other receptor, it is necessary for the cell to diffuse to the surface of the transducer. Diffusion is not a major factor when the analyte is molecular in size. The response of the device slows down as the analyte becomes larger. And it is very much slower with a whole cell. This is a limitation encountered by all sensors where the antigen has to be immobilized on the surface.

In an attempt to optimize a waveguide interferometric assay, a study was done comparing the capture efficiency of a series of antibodies to the spore for Bacillus anthracis (Campbell et al. 2003). Four antibodies were investigated in two capture/detector configurations. The best capture antibody was covalently attached to the sensing arm of an open channel waveguide interferometer. Anthrax spores were suspended in water and total particle counts were made. Solutions containing 10^4 and 10^6 cfu/mL were introduced to the sensor through a flow cell mounted on the waveguide. Signals were detected for both samples with a total phase change of 0.2 radians for 10^4 cfu/mL and almost 2 radians for 10^6 cfu/mL within an hour's data collection time. To show the selectivity of the anthrax receptor to its intended target, a challenge of 10^6 cfu/mL of *Bacillus globigii* provided no signal within an hour.

As with the lysed viruses, other workers have used an interferometer to detect the presence of bacteria through the use of an assay that responds to the presence of a secreted enteropathogenic protein, but then applied the technique to whole-cell detection. Researchers at the University of Rochester (Horner et al. 2006) measured for the presence of *Escherichia coli* by the use of a bioconjugate reaction between arrays of the extracellular domain translocated intimin receptor with the protein intimin, Tir's natural binding partner. The transducer is a reflective interferometer which measures the reflection of polarized light off a silicon dioxide layer with the receptor bound to the surface, and the reflection from the interface between the thin silicon dioxide and silicon interface. The change in destructive interference between the two reflected beams is a function of the functionalized thickness before and after the binding event, the interrogating wavelength, and the angle and refractive indexes of the layers. The device works in detecting either the protein alone or the whole cell, and does so selectively. The advantage of such a sensor configuration is that the whole cell can interact with the light, rather than solely a portion of the evanescent field.

Reflective schemes have allowed for the interrogation of a wide array of functionalized areas by mimicking the 96-well-type ELISA format, making it planar and allowing the interrogation of all regions on the chip at once through an interferometric mode of transduction.

Whole cells fit this reflective format well, because of their size and their effect on changing the phase of a reflected light beam. Nikitin et al. (2005) has taken this format and made a 96-spot plate that is interrogated by using two interferometers. One of the interferometers is the biochip, a piece of glass with a functionalized surface and a back surface for the light to reflect off of and interfere. The second interferometer is a scanned Fabry-Perot interferometer that is piezoelectrically driven. The "Picoscope" uses the interference between two beams reflected from the glass substrate; one beam reflects off the bioconjugate surface, and the second reflects off the bottom of the glass. Note that the bottom of the glass faces the mirror, which combines the two beams from the different interferometers, and the top of the glass slide is where the bioconjugate reactions are occurring. The interference depends on the phase thickness, and the thickness and refractive index of the sensitive film. Coupled with this is a light phase modulated by means of a Fabry-Perot interferometer, which provides controlled phase information to optically couple with the phase information being generated from the functionalized array of the biochip. The schematic is shown in Fig. 9.17. The device permits information to be obtained at the designated areas and to be blind to the index of refraction of the bulk solution. *Listeria monocytogenes* was used to evaluate the scheme and was detectable down to 10^4 cells/mL. Since one is measuring the phase difference between the areas functionized for binding, not only can one observe the direct binding which causes a phase difference; but one can use a second antibody for additional verification of the organism, since the second antibody will increase the phase thickness of the cell. This could also be a means for serotyping the species present, if species-specific antibodies are used.

In a similar fashion, a complete imaging system, practically a biological spectrometer, was assembled using holographic techniques (Yeom et al. 2006). The three-dimensional sensing and image reconstruction was performed by single-exposure digital holography. A Mach-Zehnder interferometer was constructed with a microscope imaging array in one of the interferometer arms. The complex amplitude holographic images are computationally reconstructed at different depths by an inverse Fresnel reconstruction. Morphology and shape recognition allow for identification of the organism present. Experimental results have been obtained with *Sphacelaria alga*, *Tribonema aequale alga*, and *Polysiphonia alga* targets. Sequential images can be used to show movement of the cells and therefore provide a means for distinguishing between live and dead cells.

A reverse symmetry waveguide configuration to enhance the projection of the evanescent field, thereby increasing the interaction of the evanescent field with the bioorganism, was investigated (Horvath et al. 2003). Although the setup was not interferometric, it was comprised of components found in many of the devices seen in interferometry. Instead of the standard high refractive index waveguide deposited on a lower refractive index substrate, the reverse symmetry waveguide has the waveguide deposited on an even lower refractive index than the water solution being investigated above the waveguide; this configuration reverses the relative refractive indexes of the substrate and superstrate. To increase the extension of the evanescent field, the substrate was comprised of nanoporous silica with a low refractive index of 1.22. Upon this substrate, a waveguide composed of polystyrene, n = 1.57 was deposited. The aqueous solution flowing above the waveguide has an index of refraction of 1.33; thus the evanescent field's strength was inverted, with the bulk of the evanescent field now residing in the aqueous solution. Light was coupled into the waveguide from below through a grating coupler, which also provided the area of bioconjugative transduction. As the organism electrostatically bound to a naturally positively charged layer of poly-L-lysine, the angle for coupling the light into the waveguide changed. Measuring the change in coupling angle for either the TE or TM light into the waveguide, through the optimization of the light exiting out the end of the waveguide, provides a measure of organism adsorption to the waveguide. The organism measured with this arrangement was *E. coli* K12 and was measured at a concentration of 10^7 cells/mL. Detection levels are limited with this arrangement, because the interaction area is very small, probably only a couple of grating periods. The reverse symmetry configuration, which is similar to the resonant mirror devices (Watts et al. 1994), produced an almost 800 times improvement in detection limit. The researchers project a limit of detection corresponding to 78 and 60 cells/mm^2 for TE and TM modes, respectively.

Another factor that can have a bearing on the ability to detect whole cells is the size of the area in which transduction is taking place. In the case of the microcavity in the photonic crystal, the width and length of the cavity are smaller than the cell itself. The probability of the whole cell binding in this region becomes exceedingly small. In surface plasmon resonance devices, the interaction area can be only a few wavelengths long, requiring the whole cell to be captured in a very small area to produce the maximum amount of signal. SPR also relies on evanescent fields generated at a metal surface to interact with the binding event. Since the extent of the evanescent field is determined by the refractive indices of the metal and the solution, the extent of interaction with SPR would be even less than that obtained by dielectric waveguides. Diffusion and capture within a small area may offset some of the lack of sensitivity of the surface plasmon sensor for whole cell detection which has been reported to be 5×10^7 cfu/ml for *E. coli* (Fratamico et al. 1998). Direct detection with SPR has been demonstrated down to 10^6 cfu/ml for *Salmonella enteritidis* and *Listeria monocytogenes* (Koubova et al. 2001). However, the relative sensitivities of different transducers would require the same bioconjugate pair for comparative analysis, for it is well known that antibody affinities for a given target vary greatly with the source of the antibody, the method of attachment, and any other conditions that the antibody had to experience before the bioassay, such as even its growing conditions.

Other organism detection using whole cells has been done at Georgia Tech (Gottfried 2006), but has only been presented at conferences. Whole cell detection was accomplished with *Campylobacter*, 1000 cfu/mL detection limit; *E. coli*, 10^4 cfu/mL detection limit; and *Listeria* and *Yesinia Pestis*, 10^5 cfu/mL detection limits. *Campylobacter* was measured using both monoclonal and polyclonal antibodies as shown in Fig. 9.20.

For *Campylobacter* the antibody was regenerated by the use of 0.2 M glycine/HCl at pH 3 without loss of activity. The *E. coli* antibody showed cross-reactivity with *Salmonella*.

In an attempt to gain additional sensitivity, a sandwich assay was used. In this scheme, a second antibody is added after the first has reacted with the antigen. The second antibody might be linked to a microparticle such as gold or polystyrene. These microspheres provide some amplification by adding large index of refraction changes (Schneider et al. 2000).

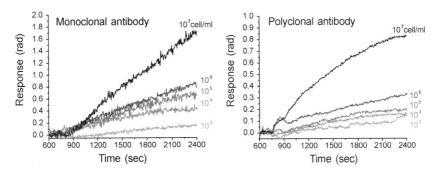

Figure 9.20. Detection of *Campylobacter* with mono- and polyclonal antibodies.

9. Advantages and Limitations

Many different interferometric and reflectometric sensing designs have been presented. Each was evaluated by exploring a bioconjugate binding reaction. Most were tested with a protein-protein assay. A few showed the ability to monitor DNA hybridization. Only a small number were tested with whole organisms. The reason for this could stem from the lack of sensitivity of most schemes, due to the lack of evanescent field available to interact with the dimensions of the cell. The probability of the cell being bound in the interaction area is limited as well. The different methods utilize different bioconjugate systems with differing affinities making any comparison unclear. The one parameter than is not easy to control is the binding affinity of the antibody or other bioreceptor with the target organism. The use of a specific aptamer with a consistent structure and binding constant would permit the use of a bioconjugate reaction that would be close to equivalent no matter what transducer it is performed on. However, aptamers for whole organisms are few, and they do not have optimized affinities. For the waveguide-based interferometers, the inherent sensitivities in the design prove them to be essentially equivalent. The reflectometric and fiber optic-based interferometers are far less sensitive because of their limitations: the small pathlength change in the reflectometric model, and the minimal evanescent field in the fiber-optic model.

Interferometers can be prepared using different materials for both the waveguide and the substrate. The choice is ultimately limited by the investigator's available facilities. If all researchers could employ the same materials and thicknesses, there would be no distinction in detection limits, based on waveguide sensitivity and the change in n_{eff}. What finally limits detection is physical: the detection scheme, the ability to suppress mechanical and thermal noise, bioconjugate binding constants, and fluidics. The researcher can have some control of other aspects as well, such as the mechanics of sample introduction, diffusion, and bulk index variation.

The major advantage of an interferometer is that it can detect any chemical reaction, since all reactions and interactions produce a change in refractive index. Therefore there is no need for labels, color change, or resonance effects to produce a response. The interferometer provides a universal platform for measuring any chemical event. But this advantage is also its major limitation. Being sensitive to any chemical event, the interferometer will respond to any change, be it the specific reaction, any nonspecific adsorption, temperature change, or background bulk index changes. Thus the design of the sensing and reference chemistry must be specific and also must diminish any interference. The reference arm should not respond to the specific biological event, but should be able to quantify any competing and background events that cause refractive index changes and to produce an equivalent phase change as the sensing arm, and optically cancel these effects.

Any limitation due to the expense of lasers and detectors has evaporated over the years, and should provide little barrier to the development of interferometric sensors, as well as other optical sensor schemes. For the design of the waveguide itself, an array of thin film deposition systems is available at most institutions. If not, the sol-gel waveguide is an inexpensive and adaptable alternative. The limiting factors for most optical devices are the electronics and software development needed for signal processing and data interpretation. It may be quite a leap for many to change from having a computer tethered to a sensor to having a completely self-contained sensor unit.

10. Potential for Improving Current Performance

Except for the polarimetric scheme, which requires four detectors, sensor configurations have employed one detector. Single point detection has held down sensor costs. But the inability to know one's position on the sine curve of the interference pattern with a single point limits an exact quantification of the phase change. Scanning a fringe requires a mechanical device to shift the fringe to quadrature. Electrooptic phase shifters incorporated in the device allow phase shifting while suppressing any mechanical noise. However, having more points sampled along the fringe curve would further improve detection limits. Three points define a sine wave, but a detector array would allow even more points to be sampled. Multiple fringes can be examined with the small pixel pitch detector arrays available. With the drop in the price of detector arrays from around a thousand dollars to fewer than fifty for a USB camera, there is no reason to use anything else. Processing of a multiple fringed pattern shift has yielded phase resolution to less than 0.01 radian. The multiple fringe data analysis can also differentiate between a change in intensity versus a change in phase, while eliminating any laser fluctuations or sample absorbance changes.

As detector resolution is increased, other sources of error will in turn need to be suppressed in order for the interferometer to reach its full potential. Controlling the temperature of the sample and of the transducer can minimize thermal noise. Mechanical noise can be suppressed by integrating as much of the optical path as possible onto the waveguide. Gratings, reflectors, beam splitters, and detectors have all been integrated onto one chip; only the laser remains external. It is only a matter of time before the laser is also incorporated onto the chip, at least in some hybrid form.

Fluidics for sample introduction has yet to be engineered to reduce chemical noise and deliver the organisms in a uniform fashion. As the sample enters the cell above the waveguide, there occurs a change in refractive index. It may be a very slight change, but both arms of the interferometer must see it simultaneously or there will be a phase difference between them. Some researchers propose ignoring the initial data, and focusing only on the subsequent phase changes from binding events. This could be a valid approach, unless one is automating the screening of numerous samples, in which the solution and flow may be changing often. Close proximity of the sensing arms will minimize the differential. The collinear polarimetric interferometer is uniquely qualified for lateral index variation, but not for vertical variation.

The rate-determining step in detection is the rate of diffusion of the analyte to the sensor surface. Diffusion of large proteins is slow and may require up to an hour for the sensor to reach equilibrium. Whole organisms are even slower. Slopes obtained from the rate of response provide a better means to obtain an analysis in a reasonable period of time. Fluidic design could increase the rate of delivery of the analyte to receptors on the surface. If uniform delivery of an isotropic mixture can be achieved (with a minimization of the noise that always seems to appear at the initial injection of the test sample, from a variety of possible mechanisms), interferometric sensing methods will provide the fast real-time results that these techniques promise.

At present, the two-dimensional sensor area of the interferometer does not fully exploit the evanescent field. The development of a three-dimensional sensing matrix would take advantage of a larger share of the evanescent field and shorten detection times. The sensor would benefit from more receptor volume, even though the major restriction still would be the slow diffusion to that sensing region. The response after the organism binds will benefit greatly from an extended evanescent field. The promise of reverse symmetry waveguides will push the evanescent field to interact totally with the organism. This is especially so if the sensing layer becomes larger and three-dimensional.

The development of interferometric sensors appears to be proceeding along two divergent paths, each driven by applications. One path leads toward the integration of all optical components, source, waveguide, and detector, on a single chip. This approach adds cost to the waveguide in that, economically, it requires long-term use. This type of sensor would find applications as a long-term stand-alone used for in-line food monitoring and bioterrorism detection.

The other path leads to a simple waveguide design with inexpensive minimal waveguides that can be discarded or recycled after a measurement is completed. Waveguides that have only a means to couple light in and out, along with the sensing chemistry, would find use in clinical applications, where discarding is preferable to the possibility of any sequential contamination. The sensing box would contain the source, the beam-combining optics, the detector, and the processor. The waveguide would act as a "plug and play" component to an optical system.

With clever chemists to functionalize a surface to detect the required compound and to exclude interferents, the current developments in fluidic design and the advancement of signal processing and interpretation, contained in ever cheaper and smaller packages, will create a future for the sensing of biochemical reactions using "light speed."

Acknowledgements

The author would like to thank Sheree Colestock of the Georgia Tech Research Institute for generating the figures used in this chapter. Special thanks go to Professor Candice J. McCloskey of Georgia Perimeter College for selflessly editing my writing. Thanks also go to Dr. David S. Gottfried for valuable discussions.

References

Akkin T, Dave DP, Milner TE and Rylander III HG (2002) Interferometric Fiber-Based Optical Biosensor to Measure Ultra-Small Changes in Refractive Index. SPIE Proc. 4616:9

Alieva EV and Konopsky VN (2004) Biosensor Based on Surface Plasmon Interferometry Independent on Variations of Liquid's Refraction Index. Sens. Actuators B 99:90–97

Brandenburg A (1997) Differential Refractometry by an Integrated-Optical Young Interferometer. Sens. Actuators B 38:266–271

Brandenburg A, Krauter R, Kunzel M and Schulte H (2000) Interferometric Sensors for Detection of Surface-Bound Bioreactions. Appl. Opt. 39:6396–6405

Brosinger F, Freimuth H, Lacher M, Ehrfeld W, Gedig E, Katerkamp A, Spencer F, Cammann K (1997) A Label-Free Affinity Sensor with Compensation of Unspecific Protein Interaction by a Highly Sensitive Integrated Optical Mach–Zehnder Interferometer on Silicon. Sens. Actuators B 44:350–355

Campbell DP (2005) Interferometric Sensors for Monitoring Our Environment. LAT Conf. Proc., St. Petersburg, Russia

Campbell DP, Gottfried DS and Cobb-Sullivan JM (2004) Groundwater Monitoring of VOCs with an Interferometric Optical Waveguide Sensor, SPIE Proc. 5586:136

Campbell DP, Gottfried DS, Scheffter SM, Beck MC and Halpern MD (2003) Interferometric Optical Waveguide Sensor for Anthrax Spore Detection. ACS National Meeting Proc., New York

Chapman RG, Ostuni E, Takayama S, Holmlin RE, Yan L and Whitesides GM (2000) Surveying for Surfaces that Resist the Adsorption of Proteins. J. Am. Chem. Soc. 122:8303–8304

Chiu M-H, Wang S-F and Chang R-S (2005) D-Type Fiber Biosensor Based on Surface Plasmon Resonance Technology and Heterodyne Interferometry. Opt. Lett. 30:233–235

Choquette SJ and Locascio-Brown L (1984) Thermal Detection of Enzyme-Labelled Antigen-Antibody Complexes Using Fiber-Optic Interferometry. Sens Actuators B 22:89–96

Cross G, Ren Y and Freeman NJ (1999) Young's Fringes from Vertically Integrated Slab Waveguides: Applications to Humidity Sensing. J. Appl. Phys. 86:6483–6499

Cross GH, Reeves AA, Brand S, Popplewell JF, Peel LL, Swann MJ and Freeman NJ (2003) A New Quantitative Optical Biosensor for Protein Characterization. Biosensors and Bioelectronics 19:383–390

Dancil K-PS, Greiner DP and Sailor MJ (1999) A Porous Silicon Optical Biosensor: Detection of Reversible Binding of IgG to a Protein A-Modified Surface. J. Am. Chem. Soc. 121:7925–7930

DeStefano L, Moretti L, Lamberti A, Longo O, Rocchia M, Rossi AM, Arcari P and Rendina I (2004) Optical Sensors for Vapors, Liquids, and Biological Molecules Based on Porous Silicon Technology. IEEE Trans. Nanotechnology, 3:49–54

Drapp B, Piehler J, Brecht A, Granglitz G, Luff BJ, Wilkinson JS and Ingenhoff J (1997) Integrated Optical Mach-Zehnder Interferometers as Simazine Immunoprobes. Sens. Actuators B 38:277–282

Fischer K and Muller J (1992) Sensor Application of SiON Integrated Optical Waveguides on Silicon. Sens. Actuators B 9:209–213

Fratamico P, Strobaugh T, Medina M and Gehring A (1998) Detection of *Escherichia coli* 0157:H7 Using a Surface Plasmon Resonance Sensor. Biotechnol. Tech. 12:571–576

Gato L and Srivastava R (1996) Time-Dependent Surface-Index Change in Ion-Exchanged Waveguides. Opt. Commun. 123:483–486

Gauglitz GA, Brecht A, Kraus G and Nahm W (1993) Chemical and Biochemical Sensors Based on Interferometry at Thin (Multi-) Layers. Sens. Actuators B. 11:21–27

Gottfried DS (2006) Private communication

Grace KM, Shrouf K, Honkanen S, Agras P, Katila P, Leppihalme M, Johnson RG, Yang X, Swanson B and Peyghambarian N (1997) A Phase-Locked Fibre Interferometer with Intensity Noise Compensation. Electronic Lett. 33:1650–1651

Grigorenko AN, Nikitin PI and Kabashin AV (1999) Phase Jumps and Interferometric Surface Plasmon Resonance Imaging. Appl. Phys. Lett. 75:3917–3919

Hartman NF (1990) Optical Sensing Apparatus and Method. U.S. Patent No. 4 940 328

Hartman NF (1997) Integrated Optic Interferometric Sensor. U.S. Patent No. 5623561

Hartman NF, Campbell DP and Gross M (1988) Multimode Waveguide Chemical Sensor. Proc. IEEE-LEOS '88, 298

Hartman NF, Cobb JM and Edwards JG (1998) Optical System-on-a-Chip for Chemical and Biochemical Sensing: the Platform. SPIE Proc. 3537:302–309

Heideman RG (1993) PhD thesis, University of Twente, Netherlands

Heideman RG, Kooyman RPH and Greve J (1994) Immunoreactivity of Adsorbed Anti Human Chorionic Gonadotropin Studied with an Optical Waveguide Interferometric Sensor. Biosensors and Bioelectronics 9:33–43

Heideman RG, Veldhuis GJ, Jager EWH and Lambeck PV (1996) Fabrication and Packaging of Integrated Chemo-Optical Sensors. Sens. Actuators B 35:234–240

Helmers HP, Greco R, Rustad R, Kherrat R, Bouvier G and Benech P (1996) Performance of a Compact, Hybrid Optical Evanescent-Wave Sensor for Chemical and Biological Applications, Appl. Opt. 35:676–680

Hermanson GT (1996) Bioconjugate Techniques. San Diego, CA: Academic Press, USA

Heuberger K and Lukosz W (1986) Embossing Technique for Fabricating Surface Relief Gratings on Hard Oxide Waveguides. Appl. Optics 25:1499–1504

Ho HP, Lam WW and Wu SY (2002) Surface Plasmon Resonance Sensor Based on the Measurement of Differential Dhase. Rev. Sci. Instr. 73:3534–3539

Horner SC, Mace CR, Rothberg LJ and Miller BL (2006) A Proteomic Biosensor for Enteropathogenic E. coli. Biosensors and Bioelectronics 21:1659–1663

Horvath R, Linvold LR and Larsen NB (2002a) Reverse-Symmetry Waveguides: Theory and Fabrication. Appl. Phys. B. 74:383–393

Horvath R, Pedersen HC and Larsen NB (2002b) Demonstration of Reverse Symmetry Waveguide Sensing in Aqueous Solutions. Appl. Phys. Lett. 81:2166–2168

Horvath R, Pedersen HC, Skivesen N, Selmeczi D and Larsen NB (2003) Optical Waveguide Sensor for On-Line Monitoring of Bacteria. Opt. Lett. 28:1233–1235

Hradetzky D and Brandenburg A (2000) Planar Interferometric Sensor for Refractometric and Immunosensing Applications. Europtrode V 179

Hsiu F-M, Chen S-J, Tsai C-H, Tsou C-Y, Su Y-D, Lin G-Y, Huang K-T, Chyou J-J, Ku W-C, Chiu S-K and Tzeng C-M (2002) Surface Plasmon Resonance Imaging System with Mach-Zehnder Phase-Shift Interferometry for DNA Micro-Array Hybridization. SPIE Proc. 4819:167

Kabashin AV and Nikitin PI (1997) Interferometer Based on a Surface-Plasmon Resonance for Sensor Applications. Quantum Electronics 27:653–654

Kersey AD, Marrone MJ and Davis MA (1990) Polarization Insensitive Fiber Optic Michelson Interferometer. SPIE Proc. 1367:2

Kinrot N (2006) Investigation of a Periodically Segmented Waveguide Fabry-Perot Interferometer for Use as a Chemical/Biosensor. J. Lightwave Tech. 24:2139–2145

Koster TM, Posthuma NE and Lambeck PV (2000) Fully Integrated Optical Polarimeter. Europtrode V, 179

Koubova V, Brynda E, Karasova L, Skvor J, Homola J, Dostalek J, Tobiska P and Rosicky J (2001) Detection of Foodborne Pathogens Using Surface Plasmon Resonance Biosensors. Sens. Actuators B. 74:100–105

Lechuga LM, Sepulveda B, Llobera A, Calle A and Dominguez C (2003) Integrated Optical Silicon IC Compatible Nanodevices for Biosensing Applications, SPIE Proc. 5119:140

Lillie JJ, Thomas MA, Denis KA, Jokerst NM, Henderson C and Ralph SE (2004) Modal Pattern Analysis and Experimental Investigation of Multimode Interferometric Sensing: a Path to a fully Integrated Silicon-CMOS-Based Chem/Bio Sensors. 2004 IEEE LEOS Annual Meeting Conference Proceedings, LEOS 2004:352

Lin VS-Y, Motesharei K, Dancil K-PS, Sailor MJ and Ghadiri MR (1997) A Porous Silicon-Based Optical Interferometric Biosensor. Science 278:840–843

Luff BJ, Wilkinson JS, Piehler J, Hollenback U, Ingenhoff J and Fabricius N (1998) Integrated Optical Mach-Zehnder Biosensor. J. Lightwave Tech. 16:583–592

Lukosz W (1995) Integrated Optical Chemical and Direct Biochemical Sensors. Sens. Actuators B 29:37–50

Lukosz W and Tiefenthaler K (1983) Integrated Optical Input Couplers as Biochemical Sensors. IEEE Conf. Proc., 2nd Eur. Conf. Integrated Optics, Florence 227:152

Lukosz W and Tiefenthaler K (1989) Sensitivity of Grating Couplers as Integrated-Optical Chemical Sensors. J. Opt. Soc. Am. B 6:209–220

Lukosz W and Tiefenthaler K (1988) Sensitivity of integrated optical grating and prism couplers as Biochemical sensors. Sens. Actuators B 15:273–284

Lukosz W, NellenPM, Stamm C and Weiss P (1990) Output Grating Couplers on Planar Waveguides as Integrated Optical Chemical Sensors. Sens. Actuators B 1:585–588

Lukosz W, Stamm C, Moser HR, Ryf R and Dubendorfer J (1997) Difference Interferometer with New Phase-Measurement Method as Integrated-Optical Refractometer, Humidity Sensor and Biosensor. Sens. Actuators B 38:316–323

Manning C, Gross MJ, Hanashaw T, Kirlin RL and Samuels A (2004) Compact Interferometers for Chemical and Biological Agent Detection. SPIE Proc. 5268:125

Millar CA and Hutchins RH (1978) Manufacturing Tolerances for Silver-Sodium Ion-Exchange Planar Optical Waveguides. J. Phys. D; Appl. Phys. 11: 1567–1576

Murata T, Ishizawa H, Motoyama I and Tanaka A (2006) Preparation of High-Performance Optical Coatings with Fluoride Nanoparticle Films Made from Autoclaved Sols. Appl. Optics 45:1465–1468

Nellen PM, Tiefenthaler K and Lukosz W (1988) Integrated optical Input Grating Couplers as Biochemical Sensors. Sens. Actuators B 15:285–295

Nikitin PI, Beloglazov AA, Kochergin VE, Valeiko MV and Ksrenevich TI (1999) Surface Plasmon Resonance Interferometry for Biological and Chemical Sensing. Sens. Actuators B 54:43–50

Nikitin PI, Gorshkov BG, Nikitin EP and Ksenevich TI (2005) Picoscope, a new label-free biosensor. Sens. Actuators B. 111–112:500–504

Nikitin PI, Gorshkov BG, Valeiko MV and Rogov SI (2000a) Spectral-Phase Interference Method for Detecting Biochemical Reactions on a Surface. Quantum Electronics 30:1099–1104

Nikitin PI, Gorshkov BG, Valeiko MV, Savchuk AI, Savchuk OA, Steiner G, Kuhne C, Huebner A and Salzer R (2000b) Surface Plasmon Resonance Interferometry for Micro-Array Biosensing. Sens. Actuators B 85:189–193

Nikitin PI, Valeiko MV and Gorshkov BG (2003) New direct Optical Biosensors for Multi-Analyte Detection. Sens. Actuators B. 90: 46–51

Nishihara H, Haruna M and Suhara T (1985) Optical Integrated Circuits, Chapter 2. New York: McGraw-Hill p 226.

Prieto F, Sepulveda B, Calle A, Llobera A, Dominguez C, Abad A, Montoya A and Lechuga LM (2003a) Integrated Optical Interferometric Nanodevice Based on Silicon Technology for Biosensor Applications. Nanotechnology 14:907–912

Prieto F, Sepulveda B, Calle A, Llobera A, Dominguez C and Lechuga LM (2003b) Integrated Mach–Zehnder Interferometer Based on ARROW Structures for Biosensor Applications. Sens. Actuators B 92:151–158

Ramos BL, Choquette SJ and Nell Jr. NF (1986) Embossable Grating Couplers for Planar Waveguide Optical Sensors. Anal. Chem. 68:1245–1249

Ranganath TR and Wang S (1977) Ti-Diffused LiNbO$_3$ Branched Waveguide Modulators: Performance and Design, IEEE J. Quantum Electron. QE-13: 290

Schipper EF, Brugman AM, Dominguez C, Lechuga LM, Kooyman RPH and Greve J (1997) The Realization of an Integrated Mach-Zehnder Waveguide Immunosensor in Silicon Technology. Sens. Actuators B 40: 147–153

Schmitt K, Schirmer B and Brandenburg A (2004) Development of a Highly Sensitive Interferometric Biosensor. SPIE Proc. 5461:22

Schneider BH, Dickinson EL, Vach MD, Hoijer JV and Howard LV (2000) Optical Chip Immunoassay for hCG in Human Whole Blood. Biosensors Bioelectronics 15:597–604

Schneider BH, Edwards JG and Hartman NF (1997) Hartman Interferometer: Versatile Integrated Optic Sensor for Label-Free, Real-Time Quantification of Nucleic Acids, Proteins, and Pathogens. Clinical Chem. 43: 1757–1808

Seo KH, Brackett RE, Hartman NF and Campbell DP (1999) Development of a Rapid Response Biosensor for Detection of *Salmonella Typhimurium*. J. Food Protection 62:431–437

Sohn H, LetantS, Sailor MJ and Trogler WC (2000) Detection of Fluorophosphonate Chemical Warfare Agents by Catalytic Hydrolysis with a Porous Silicon Interferometer. J. Am. Chem. Soc.,122:5399–5400

Stamm C and Lukosz W (1993) Integrated Optical Difference Interferometer as Refractometer and Chemical Sensor. Sens. Actuators B 11:177–181

Stamm C and Lukosz W (1994) Integrated Optical Difference Interferometer as Biochemical Sensor. Sens. Actuators B 18:183–188

Stamm C, Dangel R and Lukosz W (1998) Biosensing with the Integrated-Optical Difference Interferometer: Dual-wavelength operation. Opt. Commun. 153: 347–359.

Su Y-D, S-J Chen and Yeh T-L (2005) Common-Path Phase-Shift Interferometry Surface Plasmon. Resonance Imaging System. Opt. Lett. 30:1488–1490

Tiefenthaler K and Lukosz W (1984) Integrated Optical Switches and Gas Sensors. Optics Lett. 9:137–139

Tiefenthaler K and Lukosz W (1985) Grating Couplers as Integrated Optical Humidity and Gas Sensors. Thin Solid Films 126:205–211

Varma MM, Peng L, Regnier FE and Notle DD (2005) Label-Free Multi-Analyte Detection Using a Bio-CD. SPIE Proc. 5699:503

Wagner EK and Hewlett MJ (2004) Basic Virology. Blackwell, Malden Massachussetts, p 125

Walker RG and Wilkinson CDW (1983) Integrated Optical Ring Resonators Made by Silver Ion-exchange in Glass. Appl. Optics 22:1029–1035

Watts H, Lowe C and Pollard-Knight D (1994) Optical Biosensor for Monitoring Microbial Cells. Anal. Chem. 66:2465–2470

Weisser M, Tovar G, Mittler-Neher S, Knoll W, Brosinger F, Greimuth H, Lacher M and Ehrfeld W (1999) Specific Bio-Recognition Reactions Observed with an Integrated Mach–Zehnder Interferometer. Biosensors and Bioelectronics 14:405–411

Xu J, Suarez D and Gottfried DS (2007) Detection of Avian Influenza Virus Using an Interferometric Biosensor. Anal. Bioanal. Chem. 389:1193–1199

Yeom S, Moon I and Javidi B (2006) Real-Time 3D Sensing, Visualization and Recognition of Dynamic Biological Micro-Organisms. IEEE Proc. 94:550–566

Ymeti A, Greve J, Laqmbeck PV, Wink T, van Hovell SWFM, Beumer TAM, Wijn RR, Heideman RG, Subramaniam V and Kanger JS (2007) Fast, Ultrasensitive Virus Detection Using a Young Interferometer Sensor. Nano Lett. 7:394–397

Ymeti A, Kanger JS, Greve J, Besselink GAJ, Lambeck PV, Wijn R and Heideman RG (2005) Integration of Microfluidics with a Four-Channel Integrated Optical Young Interferometer Immunosensor. Biosensors and Bioelectronics 20:1417–1421

Ymeti A, Kanger JS, Greve J, Lambeck PV, Wijn R and Heideman RG (2003) Realization of a Multichannel Integrated Young Interferometer Chemical Sensor. Appl. Opt. 42:5649–5660

Ymeti A, Kanger JS, Wijn R, Lembeck PV and Greve J (2002) Development of a Multichannel Integrated Interferometer Immmunosensor. Sens. Actuators B 83:1–7

Young T (1804) The Bakerian Lecture: Experiments and Calculations Relative to Physical Optics, Phil. Trans. R. Soc. 94:1

10

Luminescence Techniques for the Detection of Bacterial Pathogens

Leigh Farris, Mussie Y. Habteselassie, Lynda Perry, S. Yanyun Chen, Ronald Turco, Brad Reuhs and Bruce Applegate

Abstract

Luminescence-based techniques for the detection of microbial pathogens are extensively employed in industrial setting where the continuous monitoring of bacterial contamination is of great importance. The primary advantage of all luminescence-based assays is their rapidity and sensitivity. Here we describe two different types of luminescence systems that have been adapted for commercial use, bioluminescence (BL) and chemiluminescence (CL). BL is a naturally occurring process by which living organisms convert chemical energy into light. Light-emitting pathways have been identified in bacteria, insects, and other eukaryotic organisms. Bacterial (*lux*) systems have been extensively studied and have been engineered for a variety of purposes. In the most common adaptation of the *lux* genes for the microbial detection, luciferase reporter phages are constructed for the direct and specific identification of many bacterial species including *Salmonella* spp., *Listeria*, and *E. coli* O157:H7. Central to the *lux* reaction is that bioluminescence is dependent on higher-level energy intermediates, allowing levels of light to be correlated to changes in bacterial metabolism. The firefly (*LUC*) luciferase is also widely used in biotechnology. Since all living things possess intracellular pools of ATP, many applications of the *LUC* system capitalize on the ATP-dependency of this luminescence reaction for the detection of microbial populations in situ. The *LUC* system is also useful in determining the efficacy of sanitizing agents, as decreases in BL are proportional to the number of active bacteria within a defined matrix. Other eukaryotic luciferases, such as those from marine copepod *Gaussia princeps* and Jamican click beetle, are currently been explored as alternative means for bacterial detection in extreme environmental conditions, and in situations where the simultaneous detection of multiple bacterial species is desired. CL is generally defined as the production of light by chemicals during an exothermic reaction, and CL differs from BL in that light production is not catalyzed by biological reactions. Although not as widely used in industrial applications, CL is sometimes preferred to BL-based detection systems due to the relative simplicity of the reaction and the elimination of certain steps sometimes required for the optimization of BL. CL has been used mainly for the detection of foodborne pathogens in combination with immunoassays. Using CL-linked antibodies specific for certain bacterial antigens, allows the simultaneous detection of *E. coli* O157:H7, *Yersinia enterocolitica*, *Salmonella typhimurium*, and *Listeria monocytogenes*. Luminescence-based techniques are proven effective agents in the detection of contaminating microbial populations, and with increases in the sensitivity and simplicity of such techniques, their application in numerous industrial and commercial settings will only grow.

Leigh Farris, Mussie Y. Habteselassie, Lynda Perry, S. Yanyun Chen, Ronald Turco, Brad Reuhs and Bruce Applegate • Department of Food Science, Purdue University.

M. Zourob et al. (eds.), *Principles of Bacterial Detection: Biosensors, Recognition Receptors and Microsystems*,
© Springer Science+Business Media, LLC 2008

1. Beyond Robert Boyle's Chicken

The use of luminescence in biotechnology began as a curious observation. In 1667 Robert Boyle, English chemist and founding member of the Royal Academy of Science, discovered bioluminescence in a very unlikely place: his kitchen pantry. A startled house staffer roused Boyle late one evening after observing twenty spots of bluish-green light on chicken carcasses purchased the week prior. Boyle studied the luminescence and reported several observations: 1) the glowing chicken gave no heat; 2) he determined air was needed for the luminescence by covering part of the chicken; 3) he poured wine on the luminescent parts of the chicken and the luminescence decreased; and 4) after examining the chicken, he cooked it and reported that it was delicious, reinforcing the importance of cooking food to ensure its safety. His observation with the wine is the basic premise behind the microtox assay extensively used to analyze the toxicity of aqueous samples. This use of bioluminescence has also been exploited for examining the efficacy of antimicrobial compounds. Fig. 10.1 shows an example of the use of bioluminescence to examine in situ the effect of an antimicrobial sprayed on chicken.

Boyle's fascination with bioluminescence became one of his great scientific legacies, leading him to discover the requirement of oxygen for the emission of light by certain luminescent fungi (Boyle 1667). Nearly two centuries later, Dubois confirmed in vitro oxygen dependency and demonstrated the necessity of an enzyme (luciferase) and a luciferase substrate for light emission (Dubois 1887).

Bioluminescence is a naturally occurring process by which living organisms convert chemical energy into light. Light-emitting systems have been identified in diverse sets of organisms, including bacteria, dinoflagellates, fish, fungi, and insects (Hastings 1968, 1978; Nealson and Hastings 1979; Hastings 1983; Campbell 1989; Haygood 1990; Meighen 1991; Wilson and Hastings 1998; Haygood et al. 1999). In this energy-dependent process, light is generated during the catalytic oxidation of a substrate by a luciferase in the presence of molecular oxygen (Goodkind and Harvey 1952; Harvey 1953; Wilson and Hastings 1998). Although the processes of light emission and oxygen-dependency are shared features of all bioluminescent organisms, each luciferase system has evolved independently (Wilson and Hastings 1998). There exists no sequence homology from among the luciferases of bacteria, dinoflagellates, and coelenterates (Hastings 1968, 1978; Nealson and Hastings 1979; Hastings 1983); and differences in the regulation and the cell biology of the bioluminescent systems are also apparent, as luciferase enzymes are found in both bacteria (*lux*) and eukaryotes (*luc*). Indeed, even the emission spectra of the *luc* and *lux* systems differ (at 560 nm and 490 nm, respectively; Billard and DuBow 1998).

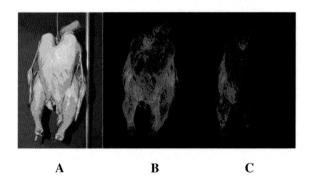

A   **B**   **C**

Figure 10.1. Chicken carcass contaminated with a bioluminescent bacteria: A) contaminated chicken in room light; B) contaminated chicken with lights off; C) contaminated chicken with the right half sprayed with an antimicrobial product.

Over the past decades, molecular biological techniques have allowed the characterization and manipulation of the luciferase genes, those genes responsible for the curious glow of Robert Boyle's chicken. Bioluminescence-based assays have also been employed as sensors, detecting the presence of harmful environmental contaminants in soils and water samples (Heitzer et al. 1992; Heitzer and Sayler 1993; Heitzer et al. 1994; Roda et al. 2004; Sakaguchi et al. 2007; Stocker et al. 2003; Kim et al. 2005) and in food (Immonen and Karp 2007; Baumann and van der Meer 2007). Luminescence genes have been used to study bacterial gene expression in animals (Contag et al. 1997; Contag et al. 1998; Zhang et al. 1999; Wu et al. 2001; Greer and Szalay 2002; Yu et al. 2003), plants (Olsson et al. 1989; Koncz et al. 1990; Schneider et al. 1990; Greer and Szalay 2002), and bacteria (Pellinen et al. 2004). Bioluminescence reporter systems also have been engineered for the in situ detection of a wide range of microorganisms (de Weger et al. 1991; Stewart and Williams 1992; Burlage and Kuo 1994; Prosser 1994; Jansson 1995; Prosser et al. 1996; Hwang and Farrand 1997).

The primary advantage of all luminescent-based assays is their rapidity and sensitivity (Roda et al. 2004). The rapid detection of pathogens is necessary in industrial settings, and bioluminescence-based methods are the most promising for the in situ detection of bacteria, especially in food processing applications where contamination risks must be monitored continuously. Minimal time is required for the synthesis of the luciferase and the emission of light due to the use of photomultiplier tubes and charge-coupled devices for the detection of very few photons (Ulitzur 1989; Stewart and Williams 1992), while bacterial detection by the standard plating procedure can take several days. In addition, luminescent detection systems are preferred over standard bacterial cultivation methods due to their ability to detect those bacteria that have entered a "viable but non-culturable state" (Billard and DuBow 1998).

Bioluminescent techniques also enjoy advantages over fluorometric techniques because no excitation wavelength is necessary for the visualization of light; and because, unlike fluorescent labeling of bacterial strains, there is an absolute energy dependency for bioluminescence emission, allowing differentiation between live and dead cells. This makes bioluminescence an especially attractive technique for in situ monitoring of microbial contamination and hygiene efficacy.

The recognition by multidisciplinary scholars of the distinct advantages of luminescence has led to its application in many biological fields (Roda et al. 2004)., It is the intent of the authors to provide a brief, though not exhaustive, introduction to the usefulness of luminescence for the in situ detection of pathogens. This chapter will focus on two major types of luminescent systems, bacterial and firefly luciferases, and how each of these systems has been applied to the in situ detection of bacteria. Special mention will made of several alternative luciferases in use today.

2. The Bacterial (*lux*) Luminescent System for Direct Pathogen Detection

Several bacterial *lux* systems, such as those from the marine bacteria *Vibrio fischeri* and *Vibrio harveyi* and from *Photorhabdus luminescens*, have been identified (Hastings 1978; 1983; Meighen 1991, 1993; Wilson and Hastings 1998). In contrast to the single polypeptide luciferase of the firefly *luc* system (described later), the luciferase of all bacterial luminescent systems is a heterodimeric protein whose subunits are encoded by *lux A* and *lux B*. In addition to these luciferase genes, light generation also requires the expression of three other genes, *lux C, D,* and *E*, which form a fatty acid reductase complex. All of these *lux* genes are organized in an operon, and are transcribed as a polycistronic message (Evans et al. 1983; Meighen 1991, 1993).

Central to the light emission of bacteria is the oxidation of a reduced riboflavin mononucleotide ($FMNH_2$) and a long-chain fatty aldehyde, such as dodecanal, in a reaction that proceeds as follows:

$$FMNH_2 + RCHO + O_2 \rightarrow FMN + RCOOH + H_2O + light$$

The biochemical aspects of the luciferase reaction are seen in Fig. 10.2. In this reaction, bioluminescence is dependent on a higher energy intermediate ($FMNH_2$; Meighen 1993), allowing changes in levels in bioluminescence to be correlated to changes in bacterial metabolism (Billard and DuBow 1998).

The induction of luminescence in certain light-emitting bacteria, such as *V. fischeri* and *V. harveyi*, is an interesting phenomenon. In these bacteria, the expression of the luciferase is autoinduced in a process called quorum sensing (Wergrzyn and Czyz 2002; Shiner et al. 2005; Waters and Bassler 2005). During the early exponential phase, cell densities are low and cultures remain "dark." As bacterial cultures grow within a confined environment, a freely diffusible small molecule is synthesized by the bacteria and accumulates in the medium. Once a threshold cell density is reached, the synthesis of the luciferase is initiated and cultures begin to luminesce, resulting in a more than 1000-fold amplification in light production (Meighen 1991). The quorum-sensing mechanisms of both bioluminescent *Vibrio* species have been studied extensively and reviewed by others (Boettcher and Ruby 1995; Miller and Bassler 2001; Fuqua and Greenberg 2002; Wergrzyn and Czyz 2002; Federle and Bassler 2003; Shiner et al. 2005; Waters and Bassler 2005).

The potential of the *lux* genes in biotechnological applications was first investigated in the early 1980s when Ulitzur and Kuhn developed a luciferase reporter phage (Ulitzur and Kuhn 1987). Phages are ideal agents for the detection and identification of pathogens due to the specificities that they display for their target bacteria. In this system, the luciferase-encoding genes *lux A* and *B* were incorporated behind phage-specific promoters into the genome

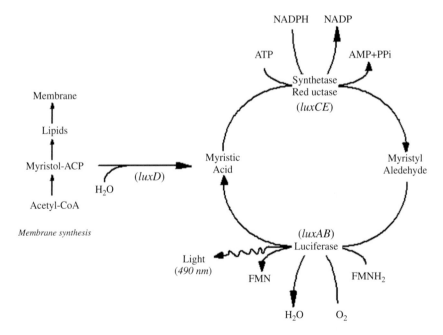

Figure 10.2. Biochemical reactions for the production of visible light from bacteria. Proteins are labeled with the appropriate genes from which they are expressed.

of bacteriophage lambda. Within the lambda phage particle, the luciferase genes remained unexpressed due to an absence of transcriptional and translational machinery. Upon infection of non-luminescent target bacteria, the luciferase genes were expressed, allowing the target bacteria to be detected by light emission.

Since that seminal work, luciferase-encoding reporter phages have been widely and successfully used in the direct detection of many bacterial species including *Salmonella* spp, *Listeria*, and *E. coli* O157:H7. All of these strains are etiological agents of food-borne illnesses and their associated contamination of food products during processing and packaging is of great concern (Fierer 1983; Khurana and Kumar 1993; Huisintveld et al. 1994; Beuchat and Ryu 1997; Ekperigin and Nagaraja 1998; de Boer and Beumer 1999; Guard-Petter 2001; Sivapalasingam et al. 2004; Lammerding 2006). Luciferase reporter phages are potentially useful in the routine screening for these pathogens, especially in foods and environmental samples. Bioluminescent reporter phages containing bacterial luciferase genes and specific for *Salmonella* spp. have been constructed, such as P22::*luxAB* (Turpin et al. 1993) (Chen and Griffiths 1996). In these assays, a plasmid-based *luxAB* cassette was introduced into *Salmonella* host strains and a generalized transducing phage P22 was used to encapsidate the luciferase encoding genes. Upon P22::*luxAB* infection of the target *Salmonella*, the *luxAB* luciferase genes were expressed and the target cells began to luminesce. A bioluminescent maxima of 10^8 photons per ml of pure culture was reached within 1 to 3 hours of phage addition. The luciferase reporter phage system also allowed *Salmonella* to be visualized within whole eggs (Chen and Griffiths 1996). Preincubation of target *Salmonella* with the luciferase reporter phages appears to increase the sensitivity of the assay; previous reports stated that as few 100 *Salmonella* cfu could be detected using this technology (Stewart 1990). Additionally, a 6-hour preincubation lowers the detection limit to as few as 10 cfu of *Salmonella* (Chen and Griffiths 1996).

A luciferase-transducing phage also has been created in order to detect *E. coli* O157:H7 (Waddell and Poppe 2000). Using a mini-Tn10 transposon, the *luxAB* genes of *V. harveyi* were introduced into the phage ΦV10 prophage genome. Phage ΦV10 is a temperate phage that infects common phage types of *E. coli* O157:H7. Initial characterization of this recombinant phage showed that ΦV10::*luxAB* transduced *n*-decanal-dependent bioluminescence to *E. coli* O157:H7, which was measurable approximately 1 h post infection (Waddell and Poppe 2000).

The use of a temperate phage in the development of luminescent pathogen detection does have disadvantages, including the generation of false-negatives. Lysogens of a parent phage used to create a luciferase-transducing phage might be present, preventing superinfection by a recombinant phage of the same origin and consequently the luminescence of target cells.

The above disadvantage could be avoided by using lytic phage, as shown in the development of a broad-spectrum phage-based luminescent detection system for *Listeria* (Loessner 1991; Zink and Loessner 1992). The identification and characterization of phage A511, a virulent myovirus that infects 95% of *L. monocytogenes* serovar 1/2 and 4, allowed the construction of a recombinant luciferase reporter phage in which a *V. harveyi* luciferase gene fusion was introduced into a defined region in the phage late genes, and expression of the *luxAB*-encoded luciferase was driven by the strong major capsid protein promoter (Loessner et al. 1996). The luciferase reporter phage A511::*luxAB* has polyvalent capabilities, detecting a wide range of *Listeria*. Exquisite sensitivity was achieved by the addition of an enrichment step; as little as one viable *L. monocytogenes* per gram of sample was identified (Loessner et al. 1996). The practical application of phage A511::*luxAB* in the detection of *Listeria* in contaminated foods was tested (Loessner et al. 1997), and detection using the luciferase reporter phage was shown to be as sensitive as the standard plate method used to screen for the presence of pathogens. The detection of *Listeria* using A511::*luxAB* was much more rapid, taking only 24 hours, compared to the 4 days required by standard plating on *Listeria* detection media (Loessner et al. 1997). Several disadvantages of A511::*luxAB* reporter phage have been noted.

Due to the lytic properties of this particular phage, the luminescence could be detected only within several hours following phage infection. Additionally, phage A511 is able to infect *Listeria* spp. in addition to *L. monocytogenes*; this cross-reactivity could lead to false-positive results. The creation of an A511:*luxAB* phage that demonstrates a more refined host-specificity ("host-range mutants") would increase its industrial usefulness (Billard and DuBow 1998).

In the luminescent reporter phage assays described above, the addition of a fatty-aldehyde (such as n-decanal) substrate is necessary for the generation of a bioluminescent response by the phage-encoded *luxAB* luciferase. Therefore, the timing of the aldehyde addition will affect light production and consequently is a potential source of error. An alternative, two-component system has been developed for the in situ detection of pathogenic bacteria (Bright et al. 2004) (Fig. 10.3). In this assay, the *luxI* gene, which encodes the quorum sensing molecule from *V. fischeri,* N-(3-oxohexanoyl)-homoserine lactone (3-oxo-C6-HSL), is introduced into a host-specific transducing phage. The second component of the assay is a whole-cell bioreporter, ROLux, that is sensitive to the quorum-sensing molecule. Following phage infection of target cells, the 3-oxo-C6-HSL molecules diffuse out of the target cell after infection and induce bioluminescence from the ROLux bioreporters. Assays incorporating the bacteriophage M13::*luxI* with the ROLux reporter and a known population of target bacteria were developed and have shown consistent detection limits of 10^5 cfu target organisms. Additionally, there appears to be an inverse relationship between the number of target bacteria and the time required for detection. At higher concentrations of target bacteria, measurable bioluminescence occurs almost immediately (Bright et al. 2004).

The two-component bioluminescent assay was applied to the detection of *S.* Typhimurium LT2 in lettuce suspensions (Kim, unpublished data). In a manner similar to that used to construct the recombinant M13::*luxI*, phage P22 was engineered to express the 3-oxo-C6-HSL molecules upon infection of target *S.* Typhimurium bacteria. As few as 10^3 *S.* Typhimurium cfu ml^{-1} of lettuce suspension were detected at 6 hours post-infection. The greatest bioluminescent response was observed at 11 hours post-incubation, and corresponded to 10^5 *S.* Typhimurium LT2 cfu ml^{-1}. One major criticism of the two-component assay is that its execution requires the introduction of a bacterial biosensor into a food or environmental sample. This could be problematic in a lab setting, as it adds a level of complexity due to the growth state of the reporter. Optimization of the two-component system also could be difficult. Because both the reporter and target bacteria must grow in the same media, initial population densities are

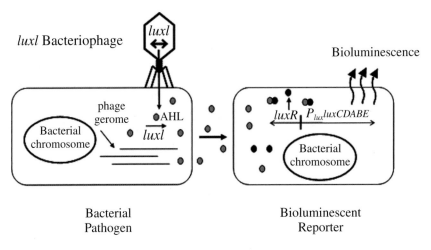

Figure 10.3. Two-component bacteriophage bioluminescent reporter system.

critical. If the numbers of target bacteria are large, nutrient limitation and oxygen limitation can limit light production, resulting in a critical limitation of the assay. High numbers of non-target bacteria can also result in the same limitation. To remove this limitation initial bacterial populations would need to be known, or the assay would need to be run with numerous dilutions of the sample. These limitations make the two-component detection assay less practical as a rapid method than other methods discussed in this chapter.

Bioluminescence-based techniques also have been used to monitor bacteria in their natural environment (Rattray et al. 1990; de Weger et al. 1991; Prosser et al. 1996; Hwang and Farrand 1997; Molina et al. 1998). Rattray et al. developed a nonextractive technique for the in situ detection of a bioluminescent *E. coli* in soils (1990). Later bioluminescence was used in the construction of *Pseudomonas* spp. that could be monitored in soils and in the rhizosphere (de Weger et al. 1991; Hwang and Farrand 1997; Molina et al. 1998).

3. The Firefly (*luc*) Luminescent System for Direct Pathogen Detection

In the firefly *luc* luminescence system, its luciferin, a heterocyclic carboxylic acid, is oxidized in an ATP-dependent manner. The luciferase itself is a 62 kDa protein that is active in its monomeric form (Hastings 1983). The general reaction proceeds as follows:

$$\text{luciferin} + \text{ATP} + \text{Mg}^{2+} + \text{O}_2 \rightarrow \text{oxyluciferin} + \text{AMP} + \text{PP}_i + \text{CO}_2 + \text{light}$$

In this system, maximum bioluminescence is reached within 0.3 seconds when the substrate, ATP, Mg^{2+}, and O_2 are in excess, and the total light output is directly proportional to the amount of ATP present intracellularly (McElroy and DeLuca 1983; Gould and Subramani 1988). Due to its ATP-dependence, the firefly luciferase could also be correlated to the physiological state of the bacteria (Billard and DuBow 1998).

The firefly luciferase is widely used in biotechnology. Based on the premise that all living things possess intracellular pools of ATP, many applications of the *luc* system capitalize on the ATP-dependency of the luminescence to detect the presence of microbial populations in situ (Thore et al. 1983; Stanley 1989; Satoh et al. 2004; Kamidate et al. 2006). 10^4 bacterial colony-forming units (cfu) generate enough ATP (approximately 10^{-14} mol) to catalyze a bioluminescent reaction that can be detected using a luminometer (Sakakibara et al. 2003; Satoh et al. 2004).

The food industry extensively employs ATP-based bioluminescence assays to monitor total microbial populations on poultry, meat, and fresh produce (Bautista et al. 1994; de Boer and Beumer 1999; Ukuku et al. 2001; Ukuku et al. 2005; Cho and Yoon 2007). The rapidity of the bioluminescent response to microbial ATP makes this system especially attractive to those who wish to comply with hazard analysis critical control point (HACCP) recommendations; however, early ATP assays were not sensitive enough for many food industry applications. A 24-hour enrichment phase to allow for microbial growth lowers the detection limit to 1 cfu g^{-1} of foodstuff. In addition, to screening for low-level contamination using an enrichment step, ATP-based bioluminescence assays also can be used without enrichment as a direct test for microbial blooms, and can be implemented in the detection of acts of agro-terrorism, when high levels of contamination of food supplies are suspected (Lim et al. 2005).

Several modifications have been made to enhance the sensitivity of the ATP-based bioluminescence assay, including the reduction of background ATP, the regeneration of ATP, and improved imaging technology. Early efforts to remove extracellular ATP, including isolation of bacteria by filtration and enzymatic treatment, were ineffective (Thore et al. 1983; Schram and Weyens-van Witzenburg 1989). Sakakibara et al. developed a dual enzymatic method to successfully remove extracellular ATP (Sakakibara et al. 1997). This treatment uses the combined effects of adenosine phosphate deaminase and apyrase and converts extracellular ATP and other adenosine-derivatives to inosine monophosphate (IMP). IMP is unable to participate in the *luc* mediated bioluminescent pathway. The resulting low levels of background bioluminescence allow a semi-quantitative estimation of microbially-derived ATP.

The strength of a microbially-derived ATP signal is dependent on assay timing as well as the number of contaminating bacteria. Peak bioluminescence generated by the *luc* luciferase is achieved within 0.3 sec and declines to approximately 10% of that original value shortly thereafter (Gould and Subramani 1988). The addition of an enzyme, such as adenylate kinase (AK) or pyruvate orthophosphate dikinase (PPDK), that is capable of regenerating ATP from low energy precursors, can result in a sustained strong bioluminescent signal and a more sensitive assay. AK amplification causes the recycling of ADP to ATP in a magnesium-dependent manner. When employed in an on-site industrial setting, AK amplification can result in bioluminescence values that are 100-fold greater than standard ATP assays (Corbitt et al. 2000). In PPDK amplification, AMP and pyrophosphate produced by the firefly luciferase are converted back into ATP (Sakakibara et al. 1999). Using this system, a single bacterial cell is detectable (Sakakibara et al. 2003). Indeed, it has been demonstrated that ATP amplification using PPDK can result in a 10,000-fold increase in sensitivity over conventional ATP assays and potentially could be extended to in situ use (Satoh et al. 2004).

Certain ATP extractants, such as cationic surfactants, can strongly inhibit the activity of luciferases (Ludin 1986; Velazquez 1997). Traditionally, dilution of the ATP extract was required prior to the bioluminescence assay to eliminate inhibition of the luciferase. This method can lower the sensitivity for ATP in the firefly luciferase assay. In order to relieve the inhibitory effects and maintain sensitivity, Kamidate et al. have developed a novel assay in which ATP extractants are delivered using liposomes (Kamidate et al. 2006). This technique has improved the detection limit of ATP in *E. coli* extracts 10-fold.

Several other factors, including pH and temperature, will interfere with the detection of bacteria on surfaces. The firefly luciferase has an optimum activity at pH 7–8 (Gould and Subramani 1988), and there have been reports that the luciferase will lose 50% of its original activity after 30 minutes at 40°C (Kajiyama 1992). To address these potential problems and optimize the ATP-assay for industrial hygiene purposes, several commercially available kits were developed (Colquhoun et al. 1998; de Boer and Beumer 1999) (Table 10.1). Kits usually require multiple reagents and the use of a portable luminometer for bioluminescence detection, making the on-site hygiene assay potentially cumbersome. Several companies have developed "assay pens" that combine the luciferase and other reagents into a single step and allow for increased ease of use in industrial settings (BioTrace International). Others have incorporated the AK enzyme in the development of more sensitive commercially available kits (Celsius).

While a powerful detection tool, the ATP-based bioluminescence assay is unable to determine the identity of the contaminating bacteria. To do so, a target-specific assay must be employed. As with the *lux* system, the *luc* genes have been used in the construction of luciferase reporter phages, and this technology has been employed as a sensitive tool for the detection of live mycobacteria (Sarkis et al. 1995). Recombinant luciferase reporter phages were constructed from the lytic mycobacteriophage TM4 for the detection of two mycobacterial species, *M. avium* and *M. paratuberculosis*. The sensitivity of the phage-based luciferase assay is poor; in culture, more than 10^4 mycobacteria are required to generate a detectable response. Bioluminescent responses were not sustainable, presumably because phage-mediated cell lysis resulted in attenuated production of the luciferase.

Temperate luciferase reporter phages have also been optimized for the detection of mycobacteria (Sarkis et al. 1995). Recombinant L5 transducing phages were constructed by the incorporation of the *luc* luciferase gene into a portion of the phage genome that is highly

Table 10.1. Commercially available kits for the detection of bacterial ATP.

Company	Available Kits
Calbiochem-Novabiochem Corporation	ATP Luminescence Assay Kit
Molecular Probes (Invitrogen)	ATP Determination Luminescence Kit
PerkinElmer	ATPlite® Cell Viability Assay
Promega	ENLIGHTEN® Total ATP Rapid Biocontamination Detection Kit
	ENLIGHTEN® ATP Luminescence Assay System
	BacTiter-Glo® Luminescent Cell Viability Assay Kit
Biothema	Microbial ATP Kit HS
Roche Applied Science	ATP Bioluminescence Assay Kit, CLS II
	ATP Bioluminescence Assay Kit, HS II
Sigma-Aldrich	ATP Bioluminescent Assay Kit
Celsius	AKuScreen®
Thermo Scientific	ATP Luminescence Assay Kit
Oxford Biomedical Research, Inc.	ATP Luminescence Assay Kit
Cambrex Bio	ViaLight® MDA Plus Cytotoxicity and Cell Proliferation Bioassay
New Horizons Diagnostics	Profile® 1 Rapid Bacterial Detection Kit
Kikkoman	CheckLite™ 250 Plus

expressed during lytic development. The use of a temperate phage resulted in a greater than 5-fold increase in bioluminescence relative to a lytic phage, and the bioluminescent response was detectable over longer time intervals. Moreover, longer incubation times led to a further increase in assay sensitivity; the L5 luciferase reporter phage was able to detect as few as 10 cfu after a 48-hour incubation (Sarkis et al. 1995).

While not as widely used as its bacterial counterpart, the firefly luciferase does provide some advantages in microbial detection (Cebolla et al. 1995). As with the bacterial *lux* system, there is an absolute energy dependency for light emission, allowing differentiation between live and dead cells. The relative simplicity of the *luc* system eases genetic manipulations. In contrast to the heterodimeric bacterial luciferase, only a single gene for the *luc* luciferase must be cloned into an expression vector (Gould and Subramani 1988). In addition, the *luc* system has the highest quantum efficiency (ratio of photons emitted to reacting materials) of all known bioluminescent reactions, possibly enhancing the detection capabilities of firefly luciferase system (Gould and Subramani 1988; Koncz et al. 1990; Lampinen et al. 1992; Hakkila et al. 2002).

Several disadvantages to the ATP assay should be noted. The ATP assays described above require the extraction of the intracellular ATP via cell lysis. The destruction of the sample precludes any further assays to discern the identity of the contaminating bacteria (Craig et al. 1991). Furthermore, the in situ usefulness of the ATP assay is limited. Luciferin must be added exogenously and therefore might hinder its usefulness as a bioreporter within matrices. *Luc*-mediated light reactions can occur in cell-free extracts and bacteria upon the addition of required compounds (Pellinen et al. 2004), but there is a question whether the luciferin substrate is able to freely diffuse across cell membranes (Billard and DuBow 1998).

4. The Use of Alternative Luciferases in Pathogen Detection

Other eukaryotic luciferases, such as those derived from members of the coelenterate family, are beginning to be used in pathogen detection and reporter technology (Wiles et al. 2005). The luciferase of *Renilla reniformans*, a sea pansy that emits light upon mechanical stimulation, was first characterized by Matthews et al. (1977). Subsequent isolation and expression in *E. coli* and other eukaryotic systems has led to its commercial availability in an assay system (Promega Corporation) (Lorenz et al. 1991) (Lorenz et al. 1996; Srikantha et al. 1996). Recently the luciferase of the marine copepod *Gaussia princeps* was shown to have increased stability at low pH and higher temperatures (Wiles et al. 2005), potentially making this luciferase preferred in industrial processing applications.

One useful characteristic of these other eukaryotic luciferases is that each emits light of a different wavelength. There are four different click beetle luciferases, which all utilize the same substrate as the firefly luciferase, and naturally emit light ranging from green (530 nm) to red (635 nm) (Hastings 1968; Wood et al. 1989; Wood et al. 1989). This unique characteristic of the eukaryotic luciferases has led to the development of bioluminescent phenotypes that allow the simultaneous detection of multiple bacterial species in a single sample (Branchini et al. 2007). The Promega Dual-Luciferase Reporter Assay system uses both the firefly and *Renilla* luciferases in one sample. Recently, two dual-color reporter systems using the red- and green-emitting luciferases from the Jamaican click beetle and the *Phrixothrix* railroad worm have been reported (Almond et al. 2003; Kitayama et al. 2004).

5. Luminescent-Based Immunoassays

The use of *luc* genes also has been described for the development of bioluminescent enzyme immunoassays (BEIA) for the rapid detection of pathogens (Fukuda et al. 2000; Valdivieso-Garcia et al. 2003; Fukuda et al. 2005). Critical to the development of these BEIAs was the creation of fusion proteins between the firefly luciferase and a binding protein (e.g., protein A, protein G, or stretavidin) that could serve as universal reagents (Tatsumi et al. 1996; Beigi et al. 1999; Karp and Oker-Blom 1999; Nakamura et al. 2004; Roda et al. 2004). In a BEIA described by Valdiviseso et al. it was shown that with enrichment, as little as 1 cfu of *Salmonella* Enteritidis and Typhimurium could be detected in 25 ml of chicken rinses (2003). Comparison of this BEIA to the standard plating technique, however, revealed no significant differences in sensitivity or specificity. Similarly, in an assay described by Fukuda et al. (2005), monoclonal antibodies against the core region of the *Salmonella* lipopolysacchride were generated and biotinylated. Upon the addition of a streptavidin-linked firefly luciferase complex, the authors were able to detect 7.3×10^2 CFU ml^{-1} *Salmonella* in pure culture, a detection limit that is similar to that of some PCR methods (Fukuda et al. 2005). Recently the luciferase of the ostracod *Cypridina noctiluna* was biotinylated, and proven to be a sensitive alternative luciferase in BEIA (Wu et al. 2007).

6. Chemiluminescence Detection Methods

Chemiluminescence (CL) can generally be defined as the production of light by certain chemicals through a highly exothermic oxidation reaction (Garcia-Campana et al. 2001). The emission of light (photons) is from unstable intermediate compounds that are created after an excitation step during the release of energy. For a production of visible light, which is important in pathogen detection systems or detection of other analytes, the reaction should be capable of releasing energy in the range of 40–70 kcal/mol, which limits the pool of chemiluminescent compounds (Rakicioglu et al. 2001).

CL, as the name implies, is different from bioluminescence (BL), in which the production of light occurs in biological systems catalyzed by enzymes (Coulet and Blum 1992; Girotti et al. 2001; Stanley 2005). Its usage has not yet been as common as BL in the food industry; but it offers some advantages over BL in that it is simpler and does not require certain steps such as the removal of extracellular ATP, inactivation of ATP hydrolyzing enzymes, or extraction of intracellular ATP, as is the case for the ATP-luciferase-luciferin BL system (Siro et al. 1982; Eschram and Witzenburg 1989; Sugiyama and Lurie 1994). Overall, however, bioluminescent detection systems are more sensitive than chemiluminescent systems (Kricka and Thorpe 1983; Yamashoji et al. 2004). The difference between CL and BL and their use as analytical techniques has been reviewed elsewhere in detail (Kricka and Thorpe 1983; Kricka 1991; Coulet and Blum 1992).

A chemiluminescent reaction can generally be described by the equation shown below:

$$\text{Chemiluminescent molecule} + \text{Oxidant} \xrightarrow{Catalyst} \text{Unstable/excited intermediate} \rightarrow \text{Product} + \text{Light}$$

Several chemiluminescent compounds are used for the reaction, including luminol [5-amino-2, 3-dihydro-1, 4-phthalazine dione], lophine [2,4,5-triphenylimidazole], and lucigenin [bis-N-methylacridinium nitrate], whereas the oxidant is commonly hydrogen peroxide (H_2O_2) (Kricka and Thorpe 1983; Tu et al. 2005). Metals (e.g., Mo, Ni^{2+}, Cr^{3+}) or enzymes (e.g., horseradish peroxidase, alkaline phosphatase, xanthine oxidase) are used as catalysts (Kricka and Thorpe 1983; Kricka 1991; Magliulo et al. 2007). Luminol is the most commonly used chemiluminescent compound for detection of food pathogens (Yamashoji and Takeda 2001; Yamashoji et al. 2004; Tu et al. 2005; Magliulo et al. 2007). This has to do with its high quantum efficiency in light production as compared to the other chemiluminescent compounds (Kricka 1991).

A common luminol-based chemiluminescent assay involves the oxidation of luminol with hydrogen peroxide in the presence of a catalyst. The resulting dinegative ion reacts with oxygen to give an excited amino-phthalate ion, which emits blue light in aqueous solution and yellow-green in DMSO (Rakicioglu et al. 2001). The emitted light is subsequently captured with a luminometer or a charge coupled device (CCD) camera. Recent developments in the instrumentation for CL have been extensively reviewed by Lerner (2001). For a pathogen detection system, the light emission can be correlated with a known number of pathogens to produce a standard curve for a quantitative analysis (Yamashoji and Takeda 2001; Magliulo et al. 2007).

CL has been used mainly for the detection of foodborne pathogens in combination with immunoassays. Immunoassays employ specific antibodies that bind with the pathogen of interest (antigen) (Siddons et al. 1992; Curiale et al. 1994; Bennette et al. 1996). The most commonly used immunoassay platform for a chemiluminescent detection system is the enzyme immunoassay sandwich (EIS) system (Fig. 10.4). The basic components of an EIS system are the capture and reporter antibodies (Swaminathan and Feng 1994; Notermans et al. 1997). The capture antibody is responsible for retrieving and concentrating the pathogen from the food matrix and is immobilized on a range of solid surfaces including magnetic beads, membranes, polystyrene tubes, or microtiter plates (de Boer and Beumer 1999; Park et al. 1999; Gehring et al. 2006). The reporter antibody also specifically binds to the pathogen, which subsequently gets sandwiched between the two antibodies. An enzyme conjugated to the reporter antibody catalyzes the oxidation of a chemiluminescent molecule that is added to the assay mixture. To avoid false positive results, the unbound reporter antibody is removed from the assay through a wash process in addition to the blockage of non-specific binding sites on the solid support system for the capture antibody.

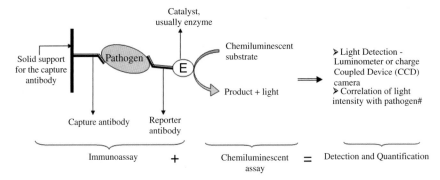

Figure 10.4. A commonly used enzyme immunoassay chemiluminescence-based foodborne pathogen detection system.

One example of the use of a CL-EIS system for the detection of *Escherichia coli* O157:O7 in ground beef was demonstrated by Tue et al. (2005). The detection limit of the system was $10^3 \, \mathrm{cfu \, mL^{-1}}$. However, with a 5 h enrichment step at 37°C, the method was able to detect $1 \, \mathrm{cfu \, g^{-1}}$ of *E. coli* O157:H7 inoculated into ground beef.

The CL-EIS system was also used by Magliulo et al. (2007) to simultaneously detect four different foodborne pathogens (*Escherichi coli* O157:H7, *Yersinia enterocolitica*, *Salmonella typhimurium*, and *Listeria monocytogenes*). They designed a special polystyrene 96-well microtiter plate for this purpose. The bottom of each well was made to have four subwells, each of which had a specific monoclonal antibody immobilized on it against an individual pathogen. Upon the addition of samples to the main well, each bacterium was bound in one of the subwells that had the antibody specific to it. A mixture of peroxidase-labeled polyclonal antibodies for all the bacteria was added to the well and incubated. Quantitative analysis of the bacteria in the samples was done with a luminol-based chemiluminescent assay based on the peroxidase activity of the bound polyclonal antibodies in each well. The validity of the method was evaluated with spiked meat and fecal samples and did not show any false positive or negative results. It was also consistent with results obtained by standard reference microbiological methods. The main advantage of the method is its ability to simultaneously detect four different pathogens in a relatively short period of time, resulting in less sample and reagent use. One disadvantage of the system is, however, its high detection limit (10^4–$10^5 \, \mathrm{cfu \, mL^{-1}}$), which requires a pre-enrichment step before the chemiluminescent assay to detect a low number of pathogens in samples. It also requires the development of very specific antibodies and the use of a highly sensitive detection system (e.g., an ultra-sensitive CCD camera), because of the low sample volume ($100 \, \mu\mathrm{L}$).

The major drawback of the standard CL-ESI system is its inability to differentiate between viable and nonviable pathogens. To address this issue, Tu et al. (2005) modified the CL-ESI method by coupling the luminol-based chemiluminescent assay with a viability assay based on detection of intracellular NAD(P)H. Immunomagnetic beads are first used to capture and concentrate *E. coli* O157:H7 from ground beef samples, followed by the addition of the membrane-permeable compound menadione, which is reduced to menadiol by intracellular NAD(P)H. Menadiol in turn generates superoxide anions (O_2^-) from molecular oxygen, which serves as the oxidant in a luminol/horseradish peroxidase chemiluminescent reaction. The chemiluminescent reaction is dependent on the existance of intracellular NAD(P)H, and hence on the viability of the bacteria. The assay can, therefore, be used in assessing the effectiveness of antimicrobial agents and heat treatment on pathogens for sanitation purposes, as a decrease in CL would indicate a decease in the number of viable of pathogens (Nishimoto and Yamashoji 1994; Yamashi and Takeda 2001; Yamashoji et al. 2001). In comparison to the CL-ESI assay, the viable assay method is less sensitive, with a detection limit of 10^4 cfu mL^{-1}, as opposed to 10^3 cfu mL^{-1} for CL-ESI (Tu et al. 2005). The sensitivity of the viable assay method can, however, be improved if Mo-EDTA complex is used as the catalyst instead of horseradish peroxidase (Yamashoji et al. 2004).

A relatively simple chemiluminescent method of detection was developed for *Listeria* spp by Vidon et al. (2001). This method exploits the natural CL of *Listeria*, which can be enhanced by luminol. Milk samples are sprayed on polyvinylidene fluoride membranes, which are incubated on agar plates for 14–16 h at 37°C to form microcolonies. The membrane with the microcolonies is then spayed with luminol-based chemiluminescent solution. Light emission from the microcolonies is captured with an ultrasensitive photon counter CCD camera. The addition of cellobiose into the agar media and UV light treatment of the microcolonies before spraying the bioluminescent solution were shown to enhance the CL from the microcolonies. This method is simpler than the previously described CL methods and does not need a microscope to count the microcolonies, but it is applicable only to *Listeria*. It also lacks the ability to discriminate between pathogenic and nonpathogenic *Listeria* spp.

The use of chemiluminescent assays for detection of reporter gene expression in biomedical and pharmaceutical industries is very common (Bronstein et al. 1996). The premise behind these assays is the provision of a chemiluminescent substrate that is specifically acted upon by an enzyme that is expressed by the gene of interest. The production of light will, therefore, indicate gene expression. One example of a reporter gene is *lacZ*, which encodes β-galactosidase (Bronstein 1996). This gene is present in the genome of *E. coli* and other coliforms (Campbell et al. 1973; Lawrence and Ochman 1998). Mathew and Alocilja (2002) developed a chemiluminescent assay to detect coliforms in food based on the presence of β-galactosidase, the action of which towards a phenylgalactosidase-substituted dioxetane substrate results in chemiluminescence. The same method was also used by Mathew et al. (2004) to detect *E. coli* O157:H7 in fresh produce.

In conclusion, CL is growing in popularity as a tool for pathogen detection and monitoring. Currently, it is generally less sensitive than BL, but offers the advantage of being simpler and faster. Its detection limit is set to improve in the future with improvements in photon (light) detection technologies and the different components of the chemiluminescent assay.

7. Conclusions and Future Perspectives

Basic research into the mechanisms of bioluminescence has yielded numerous practical applications, including methods for detection of microbial contamination. The most commonly used bioluminescence pathways are the bacterial *lux* and firefly *luc* pathways. In particular, *luc* pathway-based ATP assays have been successfully commercialized for non-specific real-time detection of microbial contamination on surfaces. These hygiene kits are used routinely on food production lines for checking the effectiveness of sanitation procedures. The *lux* pathway has been incorporated into *lux*-transducing reporter phages for detection of specific pathogens (*Salmonella* spp., *E. coli* O157:H7, and *Listeria*). Phage-based bioluminescent detection systems are promising due to their sensitivity, rapidity, and lack of response to dead target cells. Dependence on target viability, due to the requirement for high energy compounds not present in dead cells, is a major advantage of all bioluminescent detection systems.

Indirect methods of pathogen detection include immunoassays used in conjunction with either bioluminescent or chemiluminescent reactions. BEIA has been successfully used to detect *Salmonella* spp., while CL-EIS has been employed in the indirect detection of a number of foodborne pathogens. While chemiluminescence-based immunoassays enjoy several advantages over BEIA (relative simplicity and faster detection times), CL-EIS are still not as widely used due to their inability to differentiate between viable and nonviable bacteria.

In the years following Robert Boyle's discovery, the mechanisms of both bioluminescence and chemiluminescence have been meticulously studied, leading to their eventual employment in both direct and indirect methods of microbial detection. This chapter has highlighted some of those glowing successes and has also touched upon those areas in the development of luminescent technologies for wide-spread use that are lacking. The future of the application of these technologies depends on the improvement in light detection technologies, both sensitivity and reduced cost. Currently the photon multiplier tube is the primary methodology in measuring luminescence. However recent improvements in CCD technology holds promise for cost reduction and increased sensitivity.

In conclusion, luminescence-based techniques have demonstrated their effectiveness in the detection of contaminating microbial populations. Continued investigation into the optimization of such systems, as well as the development of new, more sensitive bioluminescent and chemiluminescent technologies, will cement the agent of Boyle's "curious glow" in both academic research and industry practice.

References

Almond B, Hawkins E, Stecha P, Garvin D, Paguio A, Butler B, Beck M, Wood M and Wood K (2003) Introducing Chroma-*LUC* technology. In Promega Notes, pp. 11–14

Baumann B and van der Meer JR (2007) Analysis of bioavailable arsenic in rice with whole cell living bioreporter bacteria. J Agric Food Chem 55: 2115–2120

Bautista DA, Vaillancourt JP, Clarke RA, Renwick S and Griffiths MW (1994) Adenosine triphosphate bioluminescence as a method to determine microbial levels in scald and chill tanks at a poultry abattoir. Poult Sci 73:1673–1678

Beigi R, Kobatake E, Aizawa M and Dubyak GR (1999) Detection of local ATP release from activated platelets using cell surface-attached firefly luciferase. Am J Physiol 276:C267–278

Bennette AR, MacPhee S and Betts RP (1996) The isolation and detection of *Escherichia coli* O157 by use of Immuno-magnetic separation and immunoassay procedures. Lett Appl Microbiol 22:237–243

Beuchat LR and Ryu JH (1997) Produce handling and processing practices. Emerg Infect Dis 3:459–465

Billard P and DuBow MS (1998) Bioluminescence-based assays for detection and characterization of bacteria and chemicals in clinical laboratories. Clin Biochem 31:1–14

Boettcher KJ and Ruby EG (1995) Detection and quantification of Vibrio fischeri autoinducer from symbiotic squid light organs. J Bacteriol 177: 1053–1058

Boyle R (1667) Experiments concerning the realtion between light and air in shining wood and fish. Philosophical Transactions of the Royal Society London 2:581–600

Branchini BR, Ablamsky DM, Murtiashaw MH, Uzasci L, Fraga H and Southworth TL 2007. Thermostable red and green light-producing firefly luciferase mutants for bioluminescent reporter applications. Anal Biochem 361: 253–262

Bright NG, Carroll Jr RJ and Applegate BM (2004) A model system for pathogen detection using a two-component bacteriophage/bioluminescent signal amplification assay. pp 13–19

Burlage RS and Kuo CT (1994) Living biosensors for the management and manipulation of microbial consortia. Annu Rev Microbiol 48:291–309

Campbell AK (1989) Living light: biochemistry, applications. Essays Biochem 24:41–81

Campbell JH, Lengyel JA and Langridge J (1973) Evolution of a Second Gene for ß -Galactosidase in Escherichia coli. Proc Natl Acad Sci 70:1841–1845

Cebolla A, Vazquez ME and Palomares AJ (1995) Expression vectors for the use of eukaryotic luciferases as bacterial markers with different colors of luminescence. Appl Environ Microbiol 61:660–668

Chen J and Griffiths MW (1996) Salmonella detection in eggs using Lux(+) bacteriophages. Journal of Food Protection 59:908–914

Cho M and Yoon J (2007) The application of bioluminescence assay with culturing for evaluating quantitative disinfection performance. Water Res 41:741–746

Colquhoun KO, Timms S and Fricker CR (1998) A simple method for the comparison of commercially available ATP hygiene-monitoring systems. J Food Prot 61: 499–501

Contag CH, Spilman SD, Contag PR, Oshiro M, Eames B, Dennery P, Stevenson DK, and Benaron DA (1997) Visualizing gene expression in living mammals using a bioluminescent reporter. Photochem Photobiol 66:523–531

Contag PR, Olomu IN, Stevenson DK and Contag CH (1998) Bioluminescent indicators in living mammals. Nat Med 4: 245–247

Corbitt AJ, Bennion N and Forsythe SJ (2000) Adenylate kinase amplification of ATP bioluminescence for hygiene monitoring in the food and beverage industry. Lett Appl Microbiol 30:443–447

Coulet PR and Blum LJ (1992) bioluminescence/chemiluminescence based sensors. Trends Anal Chem 11:57–61

Craig FF, Simmonds AC, Watmore D, McCapra F and White MR (1991) Membrane-permeable luciferin esters for assay of firefly luciferase in live intact cells. Biochem J 276 (Pt 3):637–641

Curiale MS, Lepper W and Robison BJ (1994) Enzyme-linked immunoassays for detection of Listeria monocytogens in dairy products, seafoods and meats: collaborative study. J Assoc Off Anal Chem 77:1472–1489

de Boer E and Beumer RR (1999) Methodology for detection and typing of foodborne microorganisms. Int J Food Microbiol 50:119–130

de Weger LA, Dunbar P, Mahafee WF, Lugtenberg BJ and Sayler GS (1991) Use of Bioluminescence Markers To Detect Pseudomonas spp. in the Rhizosphere. Appl Environ Microbiol 57:3641–3644

Dubois R (1887) Note sur la fonction photogenique chez les Pholades. COMPTES RENDUS DES SEANCES DE LA SOCIETE DE BIOLOGIE ET DE SES FILIALES 3:564–568

Ekperigin HE and Nagaraja KV (1998) Microbial food borne pathogens. Salmonella. Vet Clin North Am Food Anim Pract 14:17–29

Eschram E, Witzenburg AW (1989) Improved ATP methodology for biomass assays. J Biolumin Chemilumin 4: 390–398

Evans JF, McCracken S, Miyamoto CM, Meighen EA and Graham AF (1983) In vitro synthesis of subunits of bacterial luciferase in an Escherichia coli system. J Bacteriol 153: 543–545

Federle MJ and Bassler BL (2003) Interspecies communication in bacteria. J Clin Invest 112: 1291–1299

Fierer J (1983) Invasive Salmonella-Dublin Infections Associated with Drinking Raw-Milk. Western Journal of Medicine 138:665–669

Fukuda S, Tatsumi H, Igarashi H and Igimi S (2000) Rapid detection of Staphylococcus aureus using bioluminescent enzyme immunoassay. Lett Appl Microbiol 31: 134–138

Fukuda S, Tatsumi H, Igimi S and Yamamoto S (2005) Improved bioluminescent enzyme immunoassay for the rapid detection of Salmonella in chicken meat samples. Lett Appl Microbiol 41:379–384

Fuqua C and Greenberg EP (2002) Listening in on bacteria: acyl-homoserine lactone signalling. Nat Rev Mol Cell Biol 3: 685–695

Garcia-Campana AM, Baeyens WR and Zhang Z (2001) Chemiluminescence-based analysis: An introduction to principles, instrumentation, and applications. In: Garcia-Campana AM and Baeyens RG (eds) Chemiluminescence in Analytical Chemistry. Marcel Dekker, New York, pp 41–65

Girotti S, Ferri EN, Bolelli L, Sermasi G and Fini F (2001) Applications of bioluminescence in analytical chemistry. In: Garcia-Campana AM and Baeyens, RG (eds) Chemiluminescence in Analytical Chemistry. Marcel Dekker, New York, pp 247–284

Goodkind MJ and Harvey EN (1952) Preliminary studies on oxygen consumption of luminous bacteria made with the oxygen electrode. J Cell Physiol 39: 45–56

Gould SJ and Subramani S (1988) Firefly luciferase as a tool in molecular and cell biology. Anal Biochem 175:5–13

Greer 3rd LF and Szalay AA (2002) Imaging of light emission from the expression of luciferases in living cells and organisms: a review. Luminescence 17: 43–74

Guard-Petter J (2001) The chicken, the egg and Salmonella enteritidis. Environ Microbiol 3: 421–430

Hakkila K, Maksimow M, Karp M and Virta M (2002) Reporter genes lucFF, luxCDABE, gfp, and dsred have different characteristics in whole-cell bacterial sensors. Anal Biochem 301: 235–242

Harvey EN (1953) Bioluminescence: evolution and comparative biochemistry. Fed Proc 12:597–606

Hastings JW (1968) Bioluminescence. Annual Review of Biochemistry 37: 597–630

Hastings JW (1978) Bacterial and dinoflagellate luminescent systems. In: Herring, PJ (ed) Bioluminescence in Action. Academic Press, London, pp 129–170

Hastings JW (1983) Biological diversity, chemical mechanisms, and the evolutionary origins of bioluminescent systems. Journal of Molecular Evolution 19:309–321

Haygood MG (1990) Relationship of the luminous bacterial symbiont of the Caribbean flashlight fish, Kryptophanaron alfredi (family Anomalopidae) to other luminous bacteria based on bacterial luciferase (luxA) genes. Arch Microbiol 154: 496–503

Haygood MG, Schmidt EW, Davidson SK and Faulkner DJ (1999) Microbial symbionts of marine invertebrates: opportunities for microbial biotechnology. J Mol Microbiol Biotechnol 1:33–43

Heitzer A and Sayler GS (1993) Monitoring the efficacy of bioremediation. Trends Biotechnol 11:334–343

Heitzer A, Malachowsky K, Thonnard JE, Bienkowski PR, White DC and Sayler GS (1994) Optical biosensor for environmental on-line monitoring of naphthalene and salicylate bioavailability with an immobilized bioluminescent catabolic reporter bacterium. Appl Environ Microbiol 60: 1487–1494

Heitzer A, Webb OF, Thonnard JE and Sayler GS (1992) Specific and Quantitative Assessment of Naphthalene and Salicylate Bioavailability by Using a Bioluminescent Catabolic Reporter Bacterium. Appl Environ Microbiol 58: 1839–1846

Huisintveld JHJ, Mulder RWAW and Snijders JMA (1994) Impact of Animal Husbandry and Slaughter Technologies on Microbial-Contamination of Meat—Monitoring and Control. Meat Science 36: 123–154

Hwang I and Farrand SK (1997) Detection and Enumeration of a Tagged Pseudomonas fluorescens Strain from Soil by Using Markers Associated with an Engineered Catabolic Pathway. Appl Environ Microbiol 63: 1641

Immonen N and Karp M (2007) Bioluminescence-based bioassays for rapid detection of nisin in food. Biosens Bioelectron 22: 1982–1987

Ito Y, Sasaki T, Kitamoto K, Kumagai C, Takahashi K, Gomi K and Tamura G (2002) Cloning, nucleotide sequencing, and expression of the β-Galactosidase-encoding gene (lacA) from Aspergillus oryzae. J Gen Appl Microbiol 48:135–142

Jansson JK (1995) Tracking genetically engineered microorganisms in nature. Curr Opin Biotechnol 6: 275–283

Kamidate T, Yanashita K, Tani H, Ishida A and Notani M (2006) Firefly bioluminescent assay of ATP in the presence of ATP extractant by using liposomes. Anal Chem 78: 337–342

Karp M and Oker-Blom C (1999) A streptavidin-luciferase fusion protein: comparisons and applications. Biomol Eng 16: 101–104

Khurana R and Kumar A (1993) Occurrence of Salmonella in Eggs and Meat. Journal of Food Science and Technology-Mysore 30:447–448

Kim MN, Park HH, Lim WK and Shin HJ (2005) Construction and comparison of Escherichia coli whole-cell biosensors capable of detecting aromatic compounds. J Microbiol Methods 60:235–245

Kitayama Y, Kondo T, Nakahira Y, Nishimura H, Ohmiya Y and Oyama T (2004) An in vivo dual-reporter system of cyanobacteria using two railroad-worm luciferases with different color emissions. Plant Cell Physiol 45:109–113

Koncz C, Langridge W, Olsson O, Schell J and Szalay AA (1990) Bacterial and firefly luciferase genes in trasngenic plants: advantages and disadvantages of a reporter gene. Developmental Genetics 11: 224–232

Kricka LJ (1991) Chemiluminescent and bioluminescent techniques. Clin Chem 37:1472–1481

Kricka LJ, Thorpe GHG (1983) Chemiluminescent and bioluminescent methods in analytical chemistry. Analyst 108:1274–1296

Lammerding AM (2006) Modeling and risk assessment for Salmonella in meat and poultry. J AOAC Int 89: 543–552

Lampinen J, Koivisto L, Wahlsten M, Mantsala P and Karp M (1992) Expression of luciferase genes from different origins in Bacillus subtilis. Mol Gen Genet 232: 498–504

Lawrence JG and Ochman H (1998) Molecular archaeology of the Escherichia coli genome. Proc Natl Acad Sci 95:9413–9417

Lerner DA (2001) Recent evolution in instrumentation for Chemiluminescence. In: Garcia-Campana AM and Baeyens RG (eds) Chemiluminescence in Analytical Chemistry. Marcel Dekker, New York, pp 83–122

Lim DV, Simpson JM, Kearns EA and Kramer MF (2005) Current and developing technologies for monitoring agents of bioterrorism and biowarfare. Clin Microbiol Rev 18:583–607

Loessner MJ (1991) Improved procedure for bacteriophage typing of Listeria strains and evaluation of new phages. Appl Environ Microbiol 57:882–884

Loessner MJ, Rees CE, Stewart GS and Scherer S (1996) Construction of luciferase reporter bacteriophage A511::luxAB for rapid and sensitive detection of viable Listeria cells. Appl Environ Microbiol 62:1133–1140

Loessner MJ, Rudolf M and Scherer S (1997) Evaluation of luciferase reporter bacteriophage A511:luxAB for detection of Listeria monocytogenes in contaminated foods. Appl Environ Microbiol 63:2961–2965

Lorenz WW, Cormier MJ, O'Kane DJ, Hua D, Escher AA and Szalay AA (1996) Expression of the Renilla reniformis luciferase gene in mammalian cells. J Biolumin Chemilumin 11:31–37

Lorenz WW, McCann RO, Longiaru M and Cormier MJ (1991) Isolation and expression of a cDNA encoding Renilla reniformis luciferase. Proc Natl Acad Sci U S A 88: 4438–4442

Magliulo M, Simoni P, Guardigli M, Michelini E, Luciani M, Lelli R and Roda A (2007) A rapid multiplexed chemiluminescent immunoassay for detection of Escherichia coli O157:H7, Yersinia enterocolitica, Salmonella typhimurium, and Listeria monocytogenes pathogen bacteria. J Agric Food Chem 55:4933–4939

Mathew FP, Alagesan D and Alocilja EC (2004) Chemiluminescence detection of Escherichia coli in fresh produce obtained from different sources. Bioluminescence 19:193–198

Mathew FP and Alocilja EC (2002) Photon Based Sensing of Pathogens in Food. Proc. of 2002 IEEE Sensors Conference, Orlando, FL. June 11–15, 2002

Matthews JC, Hori K and Cormier MJ (1977) Purification and properties of Renilla reniformis luciferase. Biochemistry 16: 85–91

McElroy WD and DeLuca MA (1983) Firefly and bacterial luminescence: basic science and applications. J Appl Biochem 5:197–209

Meighen EA (1991) Molecular biology of bacterial bioluminescence. Microbiol Rev 55: 123–142

Meighen EA (1993) Bacterial bioluminescence: organization, regulation, and application of the lux genes. Faseb J 7:1016–1022

Miller MB and Bassler BL (2001) Quorum sensing in bacteria. Annu Rev Microbiol 55: 165–199

Molina L, Ramos C, Ronchel MC, Molin S and Ramos JL (1998) Construction of an efficient biologically contained pseudomonas putida strain and its survival in outdoor assays. Appl Environ Microbiol 64: 2072–2078

Nakamura M, Mie M, Funabashi H and Kobatake E (2004) Construction of streptavidin-luciferase fusion protein for ATP sensing with fixed form. Biotechnol Lett 26: 1061–1066

Nealson KH and Hastings JW (1979) Bacterial bioluminescence: its control and ecological significance. Microbiol Rev 43: 496–518

Nishimoto F, Yamashoji S (1994) Rapid assay of cell activity of yeast cells. J Ferment Bioeng 77:107–108

Notermans S, Beumer R and Rombouts F (1997) Detecting foodborne pathogens and their toxins: conventional versus rapid and automated methods. In: Doyle MP, Beuchat LR and Montville TJ (eds) Food Microbiology: Fundamentals and frontiers. ASM Press, Washington, D.C., pp 697–709

Olsson O, Escher A, Sandberg G, Schell J, Koncz C and Szalay AA (1989) Engineering of monomeric bacterial luciferases by fusion of luxA and luxB genes in Vibrio harveyi. Gene 81:335–347

Park S, Worob RW and Durst RA (1999) Escherichia coli O157:H7 as an emerging foodborne pathogen: A literature review. Crit Rev Food Sci Nutri 39481–502

Pellinen T, Huovinen T and Karp M (2004) A cell-free biosensor for the detection of transcriptional inducers using firefly luciferase as a reporter. Anal Biochem 330: 52–57

Prosser JI (1994) Molecular marker systems for detection of genetically engineered micro-organisms in the environment. Microbiology 140 (Pt 1):5–17

Prosser JI, Killham K, Glover LA and Rattray EA (1996) Luminescence-based systems for detection of bacteria in the environment. Crit Rev Biotechnol 16:157–183

Rakicioglu Y, Schulman JM and Schulman SG (2001) The application of chemiluminescence in organic analysis. In: Garcia-Campana AM and Baeyens RG (eds) Chemiluminescence in Analytical Chemistry. Marcel Dekker, New York, pp 105–122

Rattray EA, Prosser JI, Killham K and Glover LA (1990) Luminescence-based nonextractive technique for in situ detection of Escherichia coli in soil. Appl Environ Microbiol 56: 3368–3374

Roda A, Pasini P, Mirasoli M, Michelini E and Guardigli M (2004) Biotechnological applications of bioluminescence and chemiluminescence. Trends Biotechnol 22:295–303

Sakaguchi T, Morioka Y, Yamasaki M, Iwanaga J, Beppu K, Maeda H, Morita Y and Tamiya E (2007) Rapid and onsite BOD sensing system using luminous bacterial cells-immobilized chip. Biosens Bioelectron 22:1345–1350

Sakakibara T, Murakami S, Eisaki N, Nakajima M and Imai K (1999) An enzymatic cycling method using pyruvate orthophosphate dikinase and firefly luciferase for the simultaneous determination of ATP and AMP (RNA). Anal Biochem 268:94–101

Sakakibara T, Murakami S, Hattori N, Nakajima M and Imai K (1997) Enzymatic treatment to eliminate the extracellular ATP for improving the detectability of bacterial intracellular ATP. Anal Biochem 250: 157–161

Sakakibara T, Murakami S and Imai K (2003) Enumeration of bacterial cell numbers by amplified firefly bioluminescence without cultivation. Anal Biochem 312: 48–56

Sarkis GJ, Jacobs Jr WR and Hatfull GF (1995) L5 luciferase reporter mycobacteriophages: a sensitive tool for the detection and assay of live mycobacteria. Mol Microbiol 15:1055–1067

Satoh T, Kato J, Takiguchi N, Ohtake H and Kuroda A (2004) ATP amplification for ultrasensitive bioluminescence assay: detection of a single bacterial cell. Biosci Biotechnol Biochem 68:1216–1220

Schneider M, Ow DW and Howell SH (1990) The in vivo pattern of firefly luciferase expression in transgenic plants. Plant Mol Biol 14:935–947

Schram E and Weyens-van Witzenburg A (1989) Improved ATP methodology for biomass assays. J Biolumin Chemilumin 4:390–398

Shiner EK, Rumbaugh KP and Williams SC (2005) Inter-kingdom signaling: deciphering the language of acyl homoserine lactones. FEMS Microbiol Rev 29: 935–947

Siddons CA, Chapman PA and Rush BA (1992) Evaluation of an enzyme immunoassay kit for detecting cryptosporidium in faeces and environmental samples. J Clin Pathol 45:479–482

Siro M, Romar H, Lovgren T (1982) Continuous flow method for extraction and bioluminescence assay of ATP in baher's yeast. Eur J Appl Microbiol Biotechnol 15:258–264

Sivapalasingam S, Friedman CR, Cohen L and Tauxe RV (2004) Fresh produce: a growing cause of outbreaks of foodborne illness in the United States, 1973 through 1997. J Food Prot 67:2342–2353

Srikantha T, Klapach A, Lorenz WW, Tsai LK, Laughlin LA, Gorman JA and Soll DR 1996. The sea pansy Renilla reniformis luciferase serves as a sensitive bioluminescent reporter for differential gene expression in Candida albicans. J Bacteriol 178:121–129

Stanley PE (1989) A review of bioluminescent ATP techniques in rapid microbiology. J Biolumin Chemilumin 4:375–380

Stewart GS (1990) In vivo bioluminescence: new potentials for microbiology. Lett Appl Microbiol 10:1–8

Stewart GS and Williams P (1992) lux genes and the applications of bacterial bioluminescence. J Gen Microbiol 138: 1289–1300

Stocker J, Balluch D, Gsell M, Harms H, Feliciano J, Daunert S, Malik KA and van der Meer JR (2003) Development of a set of simple bacterial biosensors for quantitative and rapid measurements of arsenite and arsenate in potable water. Environ Sci Technol 37:4743–4750

Sugiyama A, Lurie KG (1994) An enzymatic fluorometric assay for adenosine 3':5'-monophosphate. Anal Biochem 218:20–25

Swaminathan B, Feng P (1994) Rapid detection of food-born pathogenic bacteria. Annu Rev Microbiol 48:401–426

Tatsumi H, Fukuda S, Kikuchi M and Koyama Y (1996). Construction of biotinylated firefly luciferases using biotin acceptor peptides. Anal Biochem 243:176–180

Thore A, Lundin A and Ansehn S (1983) Firefly luciferase ATP assay as a screening method for bacteriuria. J Clin Microbiol 17:218–224

Tue S, Uknalis J, Yamashoji S, Gehring A and Irwin P (2005) Luminescent methods to detect viable and total Escherichia coli O157:H7 in ground beef. J Rapid Meth Autom Microbiol 13:57–70

Turpin PE, Maycroft KA, Bedford J, Rowlands CL and Wellington EMH (1993) A rapid luminescent phage-based MPN method for the enumeration of Salmonella typhimurium in environmental samples. Letters in Applied Microbiology 16:24–27

Ukuku DO, Pilizota V and Sapers GM (2001) Bioluminescence ATP assay for estimating total plate counts of surface microflora of whole cantaloupe and determining efficacy of washing treatments. J Food Prot 64:813–819

Ukuku DO, Sapers GM and Fett WF (2005) ATP bioluminescence assay for estimation of microbial populations of fresh-cut melon. J Food Prot 68: 2427–2432

Ulitzur S (1989) The regulatory control of the bacterial luminescence system—a new view. J Biolumin Chemilumin 4:317–325

Ulitzur S and Kuhn J (1987) Introduction of lux genes into bacteria, a new approach for specific determination of bacteria and their antibiotic susceptibility. In: Schlomerich J, Andreesen R, Kapp A, Ernst M and Woods WG (eds) Bioluminescence and Chemiluminescence: New Perspectives. Wiley, Bristol, pp 463–472

Valdivieso-Garcia A, Desruisseau A, Riche E, Fukuda S and Tatsumi H (2003) Evaluation of a 24-hour bioluminescent enzyme immunoassay for the rapid detection of Salmonella in chicken carcass rinses. J Food Prot 66:1996–2004

Vidon DJM, Donze S, Muller C, Entzann A and Andre P (2001) A simple chemiluminescence-based method for rapid enumeration of Listeria spp. microcolonies. J Appl Microbiol 90:988–993

Waddell TE and Poppe C (2000) Construction of mini-Tn10luxABcam/Ptac-ATS and its use for developing a bacteriophage that transduces bioluminescence to Escherichia coli O157:H7. FEMS Microbiol Lett 182:285–289

Waters CM and Bassler BL (2005) Quorum Sensing: Cell-to-Cell Communication in Bacteria. Annual Reviewof Cell and Developmental Biology 21:319–346

Wergrzyn G and Czyz A (2002) How do marine bacteria produce light, why are they bioluminescent, and can we employ bacterial bioluminescence in aquatic biotechnology? Oceanologia 44:291–305

Wiles S, Ferguson K, Stefanidou M, Young DB and Robertson BD (2005) Alternative luciferase for monitoring bacterial cells under adverse conditions. Appl Environ Microbiol 71:3427–3432

Wilson T and Hastings JW (1998) Bioluminescence. Annu Rev Cell Dev Biol 14: 197–230

Wood KV, Lam YA and McElroy WD (1989a) Introduction to beetle luciferases and their applications. J Biolumin Chemilumin 4:289–301

Wood KV, Lam YA, Seliger HH and McElroy WD (1989b) Complementary DNA coding click beetle luciferases can elicit bioluminescence of different colors. Science 244:700–702

Wu C, Kawasaki K, Ogawa Y, Yoshida Y, Ohgiya S and Ohmiya Y (2007) Preparation of biotinylated cypridina luciferase and its use in bioluminescent enzyme immunoassay. Anal Chem 79:1634–1638

Wu JC, Sundaresan G, Iyer M and Gambhir SS (2001) Noninvasive optical imaging of firefly luciferase reporter gene expression in skeletal muscles of living mice. Mol Ther 4:297–306

Yamashoji S, Asakawa A, Kuwasaki S and Kuwamoto S (2004) Chemiluminescent assay for detection of viable microorganisms. Anal Biochem 333:303–308

Yamashoji S, Manome I, Ikedo M (2001) Menadione catalyzed O_2 production by Escherichia coli cells: application for rapid chemiluminescent assay to antimicrobial susceptibility testing. Microbiol Immunol 45:333–340

Yamashoji S, Takeda M (2001) Menadione-catalyzed luminol chemiluminescent assay for the viability of Escherichia coli ATCC 25922. Microbiol Immunol 45:737–741

Yu YA, Timiryasova T, Zhang Q, Beltz R and Szalay AA (2003) Optical imaging: bacteria, viruses, and mammalian cells encoding light-emitting proteins reveal the locations of primary tumors and metastases in animals. Anal Bioanal Chem 377: 964–972

Zhang W, Conta, PR, Madan A, Stevenson DK and Contag CH (1999) Bioluminescence for biological sensing in living mammals. Adv Exp Med Biol 471: 775–784

Zink R and Loessner MJ (1992) Classification of virulent and temperate bacteriophages of Listeria spp. on the basis of morphology and protein analysis. Appl Environ Microbiol 58:296–302

11

Porous and Planar Silicon Sensors

Charles R. Mace and Benjamin L. Miller

Abstract

The development of novel sensors able to produce a response directly upon binding of a target biomolecular analyte remains a major area of research in materials science and analytical chemistry. As the primary "raw material" for the microelectronics industry, and because of its biocompatibility and optical properties, silicon has drawn considerable attention in this field. This chapter discusses the current state of efforts in our group and others to develop porous and planar sensors based on silicon. Porous silicon, so-called because of its complex three-dimensional network structure, provides a high internal binding surface and allows for observation of binding by changes in the reflectivity or luminescence spectra of single- or multi-layer devices. Planar silicon sensors, exemplified here by Arrayed Imaging Reflectometry (AIR), do not have the high surface area of porous silicon yet still respond with a high degree of sensitivity to the binding of analytes to the sensing surface. Examples are presented for the use of both types of sensors for the detection of DNA, proteins, and pathogenic bacteria.

1. Introduction

Silicon has many characteristics that render it an ideal material from which to build a biosensing platform: as a commodity material of the microelectronics industry, it is inexpensive, and an immense industrial infrastructure has been built up for its transformation into complex electronic devices. It is biocompatible (obviously critical if one is interested in the eventual production of implantable sensors), and readily derivitized with a broad range of capture agents for small molecules, peptides, proteins, and nucleic acids. Perhaps most importantly, it can be fabricated in ways that allow for ready detection of subnanometer changes in its physical structure. It is this latter property that makes label-free optical (and in some cases electrical) detection of biomolecules using silicon-based sensors possible.

This chapter will describe the collaborative efforts of four research groups at the University of Rochester in the development of two platform technologies for biosensing based on silicon: one based on the optical properties of porous silicon (and in particular on one-dimensional photonic bandgap structures), and one (termed arrayed imaging reflectometry) based on the manipulation of antireflective coatings on planar silicon surfaces. We note at the outset that many other groups have made vital contributions to the development of silicon-based sensor structures, particularly in the porous silicon area; however, because of space constraints it will be necessary for us to focus primarily on our own work.

Charles R. Mace and Benjamin L. Miller • University of Rochester

M. Zourob et al. (eds.), *Principles of Bacterial Detection: Biosensors, Recognition Receptors and Microsystems*,

1.1. Porous Silicon: A Three-Dimensional Matrix for Biosensing

It has been recognized for many years that silicon, when subjected to an electrochemical etching process, becomes porous. Furthermore, the size and density of these pores may be controlled, depending on the etch conditions and type of silicon etched, to provide nanoporous (<10 nm), mesoporous (10–50 nm), or macroporous (100–500 nm) material (Fig. 11.1). Interest in porous silicon was further heightened by Canham's discovery of the room-temperature photoluminescence of some forms of mesoporous silicon (Canham 1990). As a biosensing material, the first demonstration of porous silicon's utility was provided by Sailor and Ghadiri in 1997 (Lin et al. 1997). In that report, the authors observed shifts in interference fringes in the reflection spectra of single-layer porous silicon as a function of the captured target.

Our own studies on porous silicon biosensing grew out of work by the Fauchet group in the 1990s directed towards the production of optical devices. Fauchet and coworkers found that careful alteration of the current density as a function of time during the etching process could produce Bragg mirrors, or microcavity resonators (essentially a 1-D photonic bandgap structure). As shown in Fig. 11.2, such a device consists of two multilayer mirror films on either side of an "active layer," where each period of the multilayer mirrors is made up of one high-porosity and one low-porosity layer. These efforts culminated in the Fauchet group's description of a porous silicon microcavity resonator acting as a photodiode in 1999 (Chan and Fauchet 1999).

As an initial test of the microcavity resonator as a biosensor, a mesoporous device with a surface layer of thermal oxide was derivitized with a DNA probe for lambda bacteriophage by treatment of the oxidized porous silicon surface with aminopropyl(triethoxy)silane, followed by glutaraldehyde, followed by amino-terminated DNA. As expected, we were able to observe shifts in the photoluminescence spectrum following treatment of this DNA-functionalized chip

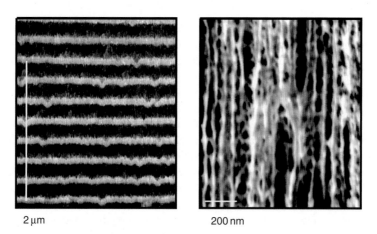

2 μm   200 nm

Figure 11.1. Mesoporous silicon. At left, one can see alternating layers of high and low porosity, obtained via time-dependent variation of the current density during etching.

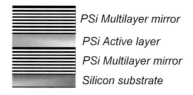

PSi Multilayer mirror

PSi Active layer

PSi Multilayer mirror

Silicon substrate

Figure 11.2. Microcavity resonator (Bragg reflector) schematic.

with the complementary DNA sequence. Importantly, exposure to a mismatched DNA sequence did not produce a shift, verifying that spectral changes depended on the specific binding of the target molecule. Extension to a more complex target was provided by incubation of the chip with a full-length *lambda*-bacteriophage, providing the first demonstration of virus detection with a porous silicon sensor (Chan et al. 2000).

While these results were encouraging, a key concern that remained was whether similar selectivity could be observed following exposure to a more complex analytical sample, for example a bacterial lysate. We tested this question by derivitizing a mesoporous microcavity with TWTCP (1), a synthetic receptor for bacterial lipid A (Hubbard, Horner and Miller 2001). Initial results with this device were disappointing: no lipid A-dependent red-shifts in the photoluminescence spectrum were observed. However, when a mixed solution of TWTCP and glycine methyl ester (a "spacer molecule") was used to prepare the sensor, we observed that this device functioned as intended. Presumably, the first device failed either because covalent attachment of all four of the reactive amino groups of TWTCP resulted in its immobilization in an inactive conformation, or simple steric crowding prevented lipid A binding. The latter has been observed by several groups in the context of planar DNA sensors (Du et al. 2005); of course, the complex three-dimensional matrix of porous silicon means that avoidance of steric crowding is an even more critical design concern.

TWTCP

While the lipid A sensor worked well in detecting purified material, would it work (and work selectively) in the complex milieu provided by a raw bacterial lysate? This would then provide a method for distinguishing Gram-(–) from Gram-(+) bacteria, since a defining characteristic of Gram-(–) bacteria is their outer cell membrane made up largely of lipopolysaccharides (of which diphosphoryl lipid A is the conserved head group). We were gratified to observe that this indeed worked well: exposure of the sensor to lysates of *Escherichia coli* or *Salmonella minnesota*, both Gram-(–) bacteria, produced red shifts in the photoluminescence spectrum, while exposure to *Bacillus subtilis* or *Lactobacillus acidophilus*, Gram-(+) bacteria, gave no response (Fig. 11.3; Chan et al. 2001). This provided the first "silicon analogue" of the Gram stain, a core procedure of bacteriology essentially unchanged since its introduction in 1884 (Young et al. 1977).

1.2. Effect of PSi Immobilization on Probe Viability: Experiments with GST

These initial experiments provided the impetus for a much more detailed examination of the physical properties and performance of porous silicon biosensors. Of particular concern was the need to produce structures that would allow immobilization and detection of higher molecular weight targets. Antibodies and proteins are obviously significantly larger than

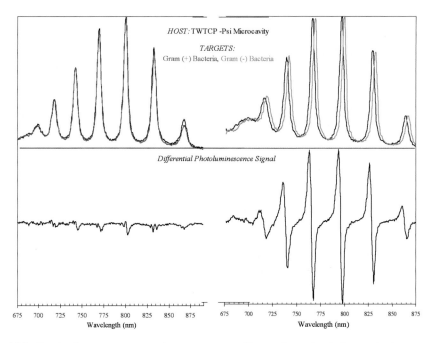

Figure 11.3. Photoluminescence spectra of a mesoporous silicon microcavity before and after exposure to bacterial lysate.

molecules like TWTCP and lipid A; unlike large molecules such as *lambda*-bacteriophage genomic DNA, preservation of their activity requires that they retain a globular structure. Would such molecules be able to penetrate into a mesoporous silicon sensor? To provide the best chance of this occurring, we enlarged the pore size of as-etched chips using an NaOH post-etch treatment (DeLouise and Miller 2004b). While this destroyed the silicon nanostructures that are the source of photoluminescence, the microcavity devices produced in this manner nevertheless possess well-defined reflectance spectra that can serve as reporters of pore infiltration (either via probe molecule immobilization or target binding).

In order to gain a greater understanding of the effect immobilization within the porous silicon matrix has on the activity of a protein, as well as to determine the theoretical mass sensitivity such a device would have, we carried out a series of experiments on porous silicon-immobilized glutathione S-transferase (GST). GST is a ubiquitous detoxifying enzyme, the activity of which may be monitored spectrophotometrically using a colorimetric substrate. As such, it was possible to use the spectrophotometric measurements on immobilized enzyme reactivity as an independent check on the measured changes in reflectance spectra (DeLouise and Miller 2004a). Immobilization of GST in the porous silicon matrix caused a two- to four-fold reduction in enzyme activity relative to that in the solution phase; this was attributed to a combination of immobilization of the enzyme in an inactive or inaccessible conformation, and limited diffusion of the substrate through the pores (DeLouise and Miller 2005). Further studies correlating enzyme immobilization and optical response for a $\lambda/2$ p+ mesoporous silicon microcavity established that the sensitivity of the device was in the range of 50–250 pg/mm^2 (DeLouise, Kou, and Miller 2005). Critically, these studies provided independent experimental confirmation that the optical response of a porous silicon sensor indeed relate predictably and quantitatively to the amount of material captured within the pores.

1.3. Toward Larger Targets: The First Macroporous Microcavity Structures

As discussed above, the use of mesoporous silicon sensors for the detection of large molecules (for example, antibodies), viruses, and whole bacteria is hampered by size limitations determined by the pore size of the material. While macroporous silicon (pore size >200 nm) has been known for some time (Janshoff et al. 1998) it is also suboptimal as a sensing material for the opposite reason: in this case, the pores are so large that capture of a target molecule does not dramatically change the refractive index, and therefore the device sensitivity is relatively low (Ouyang, Striemer, and Fauchet 2006). A porous silicon sensor with pore diameters in the 100–200 nm range would potentially provide a useful compromise in material infiltration capability and sensitivity; recently, the preparation of such a material has been accomplished, and initial tests of its properties in the context of sensors have been promising.

Typical procedures for the preparation of macroporous silicon rely on mixed etchant solutions consisting of organic solvent (often ethanol) and an oxidant such as chromic acid, in addition to the "standard" HF. In 2005, Fauchet and coworkers reported that macroporous silicon with pore diameters in the 100–150 nm range could be prepared from n type Si wafers. Importantly, initial tests verified that microcavity structures could be prepared using a similar time-dependent variation of the current density (Fig. 11.4), and these microcavities were able to detect immobilized biomolecules via observed changes in the reflectance spectra (Ouyang et al. 2005). In subsequent experiments, we have been able to use these "small macroporous" devices to detect *E. coli* expressing the extracellular domain of the enteropathogenic *E. coli*-specific protein intimin, using a portion of the intimin-binding protein tir as the capture molecule (further details of the protein-protein recognition system are provided in later sections of this chapter) (Ouyang et al. 2007). While the observed sensitivity was not particularly high (4 micromolar intimin, or a mass sensitivity of 130 fmol), this nevertheless demonstrated feasibility for the device.

1.4. Porous Silicon Bandgap Sensors in Novel Formats: "Smart Bandages" and "Smart Dust"

While sensor chips—either as single-analyte devices or as arrays—are useful in an immense range of applications, an area of research that is just beginning to emerge is the incorporation of label-free sensors into other, "nontraditional" formats. Two examples of this in the context of porous silicon sensors are Delouise et al.'s incorporation of a porous silicon sensor into a hydrogel matrix to provide a working example of a "smart bandage," and the Sailor group's description of porous silicon "smart dust."

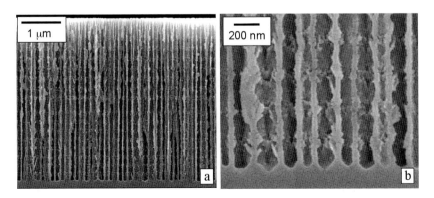

Figure 11.4. Macroporous silicon microcavity (SEM images) at low (a) and high (b) magnification.

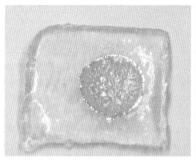

Figure 11.5. Hydrogel-immobilized porous silicon sensor, fully hydrated (left) and dry (right).

A porous silicon layer may be released from the underlying wafer by application of a short current burst at the end of the etching procedure. DeLouise and coworkers used this capability to prepare mesoporous Si microcavities that were then contact-laminated onto a commercial hydrogel matrix (Johnson and Johnson NuGel) commonly used for wound care (DeLouise et al. 2005). This hydrogel-immobilized sensor was able to detect concentration-dependent changes in the refractive index of glucose solutions. Furthermore, the hydrogel could be completely dried and rehydrated repeatedly over the course of a year without significant loss in optical performance (Fig. 11.5).

Another alternative implementation of a sensor is to produce finely divided material that retains the optical sensing properties of the "bulk" sensor. Such "smart dust" could be uniformly distributed in a solution, potentially providing significantly faster response times. Sailor and coworkers described one strategy for producing such materials in 2002 (Schmedake et al. 2002).

2. Arrayed Imaging Reflectometry—A Planar Silicon Biosensor

The previous section contained a brief survey of porous silicon one-dimensional bandgap structures. In this section, we will discuss the development and implementation of arrayed imaging reflectometry (AIR), a label-free biosensor constructed from an antireflective coating on planar silicon. Because this sensor platform has not been previously reviewed, we will discuss it in somewhat more detail.

2.1. Theory

2.1.1. Physical Rationale

How light interferes with itself is a function of many variables: the polarization, wavelength, and angle of the incident light; the optical properties of the media in which the light interacts; and the precision of the values of the aforementioned elements. At its most reduced level, we can look at the physical basis of arrayed imaging reflectometry (AIR) as a special case of interference. In order to utilize the AIR technique, a suitable substrate system and set of initial optical conditions must be fixed. The result is an antireflective (AR) coating on the base material culminating in the total destructive interference of reflected light. This null reflectance condition is characterized by being extremely sensitive to the parameters by which it was created; slight deviations will destroy the AR coating on the substrate. Subsequently, one can specifically monitor the changes in reflectance—from dark to light—as a function of

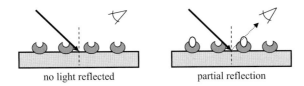

Figure 11.6. Detected reflections off a generic AIR substrate. No reflectance is observed when probes remain free, but partial reflectances occur as targets are bound at the surface (i.e., layer thickness increases).

a chosen property. For AIR, we employ this phenomenon as a biosensor, where an increase in reflectance is concomitant with a film thickness increase due to a surface biomolecular recognition event; in other words, the signal is transduced *directly* from the probe-target interaction without the aid of a secondary label (Fig. 11.6).

It is important to note that AIR is not dependent upon lateral penetration of electromagnetic radiation into the interfering media. This results in a high degree of spatial resolution, thus permitting the technique to be implemented in an arrayed format. An array can consist of separate sensing area "spots" which are then used to enhance the redundancy or quantity of individual probes. The entirety of the sensor surface, incorporating each discrete probe spot, can be monitored simultaneously by a single image of the resultant reflectance without any loss in overall sensitivity. The raw number of probe spots able to be imaged is controlled by detector (usually a CCD camera) quality and size; the probe spot density is restricted by the technique used to manufacture the array and the diffraction limit.

Other groups have also developed biomolecular sensing techniques based on reflectance (Jin et al. 1995; Liedberg, Nylander, and Lundström 1983) or other optical phenomena (Brynda et al. 2002; Martinez et al. 2005); however, the two most commonly compared to AIR are surface plasmon resonance imaging (Smith and Corn 2003) (SPRi) and reflectometric interference spectroscopy (RIfS) (Schmitt et al. 1997), even though these methods are vastly different from our own in terms of signal transduction. SPRi is a special configuration of traditional surface plasmon resonance where spot arrays, instead of individually addressed flow channels, can be monitored in real time by fixing the appropriate initial incident conditions. In this way, the output is reflectance intensity for a specific pixel set—as in AIR—rather than reflectance as a function of a shift in angle or wavelength of a spectral feature. In RIfS, a float glass slide is modified with tantalum oxide and chemical vapor-deposited silicon dioxide in order to provide an interference layer stack (Piehler, Brecht, and Gauglitz 1996). White light is then incident onto the substrate at a fixed angle, and an interference spectrum, as a function of wavelength, is acquired. The interference spectrum is monitored over the course of the experiment for shifts indicative of a surface binding event. A key difference between these two techniques and AIR is that both SPRi and RIfS rely on shifts relative to a non-zero minimum, while AIR relies on a starting condition of zero reflectivity. In principle, this provides a substantial performance advantage in terms of sensitivity and dynamic range. However, as we shall see, it also places stringent design criteria on the substrate material used in the construction of sensor chips.

2.1.2. Substrate Design

Before exploring the use of AIR as a biosensor, the rationale behind the design of our substrate must be discussed. We have chosen planar, crystalline silicon as our base substrate material for many of the same reasons that we discussed in the context of PSi biosensors. Silicon, as a major constituent of the semiconductor industry, is well characterized, readily available, optically flat, and has an associated oxidized form (silicon dioxide, SiO_2). Once a silicon wafer has been introduced to the atmosphere, a thin layer of native oxide rapidly develops (Morita et al. 1990). For most of our applications, this is not a sufficient enough thickness to work with; however, this oxide can be stripped off with a stringent cleaning process involving hydrofluoric acid (Ljungberg, Söderbärg, and Bäcklund 1993). Upon returning to the pure silicon wafer, a thicker layer of silicon dioxide can either be grown by thermal processes (Deal and Grove 1965) or deposited by sputtering (Tabata et al. 1996) or chemical vapor deposition (Batey and Tierney 1986).

Silicon dioxide, serving as the foundation for the interference layer, offers a great deal of flexibility as a material. Coupled with ellipsometry as a metrology tool, dilute HF etching of surface SiO_2 provides the ability to finely tune the thickness of the layer. The sensitivity of AIR is ultimately tied to our capability of controlling the thickness of the SiO_2 layer. Another component of the system is the attached linker and probe molecules on the oxide. For the purpose of modeling, these are treated as a single layer, while the detected target molecules are added as an individual component as well. Probe choice and immobilization techniques will be discussed later. The combination of these materials, coupled with air as the ambient incident medium, will serve as the generic substrate for our research, both theoretical and experimental.

2.1.3. Mathematical Model

Modeled as a multilayer film composed of silicon dioxide and linker and probe molecules, the propagation of light through the substrate system can be followed mathematically (Hecht 2002). Assuming transparent media, there are incident, reflected, and transmitted rays for a single interface (Fig. 11.7). The primary reflected and transmitted rays can be described by simple physical principles: the law of reflection, Snell's law, and continuity of the electric and magnetic fields across a boundary. The entire system can then be described as a function of the thickness (d) and refractive indices of the media (n), and the polarization, angle (θ), and wavelength (λ) of the incident wave.

These parameters control the phase shift of the transmitted ray in any subsequent medium. This phase shift plays an integral role in producing the total destructive interference condition. In order to produce zero reflectance from the substrate, the amplitude and phase of the primary reflectance need to be orthogonally matched by those of the resultant rays emitted due to the interactions within the interference layers.

The basis of our data simulations is directly derived from these relationships at each interface. Each correlation describing the total electric and magnetic field components produces a characteristic matrix for the interface (Eqn 11.1–11.4):

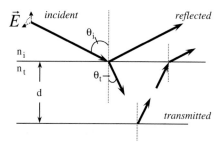

Figure 11.7. General diagram of incident light at a single interface.

$$\begin{bmatrix} E_i \\ H_i \end{bmatrix} = M \begin{bmatrix} E_t \\ H_t \end{bmatrix}. \tag{11.1}$$

$$M = \begin{bmatrix} \cos(d\beta) & i\sin(d\beta)/Y \\ iY\sin(d\beta) & \cos(d\beta) \end{bmatrix}. \tag{11.2}$$

$$\beta = \left(\frac{2\pi}{\lambda_0}\right) n_i \cos\theta_t. \tag{11.3}$$

$$Y = \sqrt{\varepsilon_0/\mu_0}\, n_i \cos\theta_t. \tag{11.4}$$

A resultant characteristic matrix for the full system, a product of those for each interface, can be constructed to simulate the propagation of light through our substrate. From the characteristic matrix, the total reflectance can be computed as a function of all relevant parameters.* It is through this simulation that we are able to predict the initial conditions that provide total destructive interference.

As an initial experiment, the interference spectrum of a nominal oxide layer (1,000 nm) on silicon was simulated in order to identify spectral characteristics. The simulation was compiled as a function of angle space for a fixed wavelength and random polarization state (Fig. 11.8a); the same simulation was run for defined polarization states (Fig. 11.8b). While any arbitrary wavelength could have been used, the 632.8 nm line of a helium-neon (HeNe) laser provided the narrow bandwidth requirement (Mielenz et al. 1968) and, as a commercially available commodity, could be easily integrated into the measurement apparatus if the suitable properties are shown.

As can be seen in Fig. 11.8b, there is a defined feature for pure s-polarized light around 73° that corresponds to an approximate reflectance of 0.01. Thus far, we have assigned lead conditions for wavelength and polarization state. Finally, we must minimize the desired feature for both incident angle and SiO$_2$ thickness. Fig. 11.9 shows this simulated minimization resulting

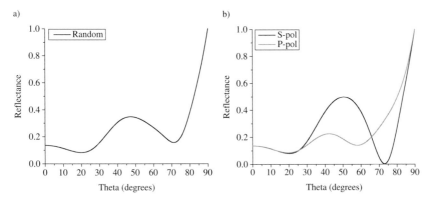

Figure 11.8. Simulated reflectance spectra over angle space at 632.8 nm for randomly polarized incident light (a) and pure s- and p-polarized incident light (b).

---

*A nice treatment, in more detail and with a full explanation of boundary conditions, can be found in Eugene Hecht's *Optics*.

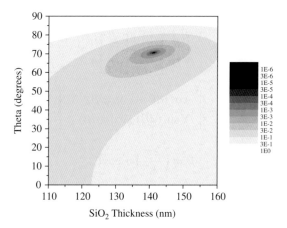

Figure 11.9. Two-dimensional simulated reflectance spectrum over angle and thickness space. The reflectance dynamic range is depicted to vary from 1E-6 to 1E+0; however, this is a misleading limitation of the simulation, as reflectance does go to zero at the minimum.

in the final parameters for total destructive interference: incident wavelength of 632.8 nm, incident angle of 70.5°, SiO_2 thickness of 141.1 nm, and s-polarized light. The reason for the angle shifting slightly lies in the altered contribution of the thinner oxide layer. Although the simulation just described employs the use of monochromatic light, an example considering polychromatic light could have easily illustrated the technique. In fact, experiments employing a polychromatic source were among the first implementations of AIR.

2.1.4. Monitoring the Null Reflectance Condition

2.1.4.1. General Apparatus Design

The general reflectance detection apparatus begins with a component acting as the source: either a tunable, multiwavelength device, or a laser. The selected beam is then passed through polarizing optics to enforce the required s-polarization. Subsequently, the beam can be enlarged through an aperture (microscope objective or other optic) and collimated at a diameter suitable to illuminate the entire sensor surface. Sufficient collimation is implemented with a shear plate. This ensures that the light is incident at a single angle. Depending on the apparatus configuration and the beam path, the sample may either be placed horizontally on a stage or held vertically by a vacuum chuck. Light reflected off the sample is then captured by a CCD camera. Data is then downloaded to a computer for analysis.

2.1.4.2. Polychromatic Platform

White light from a xenon source is made quasimonochromatic, to a bandwidth of 1 nm, using a spectrometer. This beam is kept incident at an angle of 70.6° for the duration of the experiment. The fixed angle imposes a distinct wavelength-dependent null condition. While this could create a serious limitation for monitoring raw reflectance deviations, this implementation of the technique is comparable to surface plasmon resonance (SPR) in that a spectrum is obtained and monitored for shifts (Johnston, Yee, and Booksh 1997; Peterlinz and Georgiadis 1996), albeit still in an arrayed format. A cache of images are recorded over a wavelength span and analyzed for intensity. The average intensity over a spot area is plotted as a function of the wavelength. This is performed for all aspects of the experiment, and spectra are overlayed and analyzed for shifts. The wavelength corresponding to the minimum reflectance value is

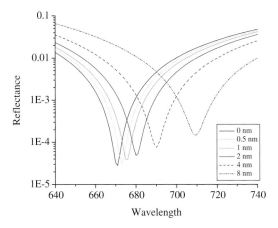

Figure 11.10. Simulated reflectance spectra of a multiwavelength AIR experiment. The initial background spectrum corresponds to 0 nm bound.

determined, and the shift is calculated from a background reference minimum. From this shift, the thickness of bound material is directly computed. Fig. 11.10 shows such an overlay of spectra derived from simulated data to mimic an actual experiment.

The background spectrum is near the global minimum, as noted by the lower total reflectance and defined lineshape. As target material is bound and the layer thickness increases, the spectrum undergoes a red shift, while the lineshape broadens. The broadening of the lineshape, and the increase in absolute intensity, stem from diverging from the null condition. A side effect of this broadening is an uncertainty in the absolute value of the minimum wavelength. If the shift in the minimum is plotted as a function of the bound target thickness, a calibration curve is obtained (Fig. 11.11). A fit to the plot shows that the wavelength minimum shift is linear with respect to the amount of target bound: for every nanometer of bound target, there is a corresponding shift of 4.8 nm of the minimum. From this, we can predict that for a system with 1 nm spectral resolution, the sensitivity is about 0.2 nm of bound material. When this instrumentation obstacle is circumvented, the sensitivity will increase.

By acquiring reflectance data over the desired range of wavelengths, a map is created. In this map, the reflected intensity for a given wavelength is stored for specific surface areas.

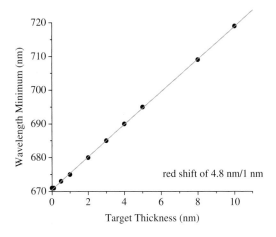

Figure 11.11. Calibration curve for multiwavelength AIR.

The reflectance spectrum is compiled for each area and the minimum is determined. From this minimum, a three-dimensional surface can be extracted where the x- and y-axes are the sensor surface (in pixels) and the z-axis is the deviation of the surface thickness (in nm) from a standard background. From these thickness maps, the thickness of bound material for an entire array can be graphically represented.

2.1.4.3. Monochromatic Platform

As inferred from the initial simulated data, the single-wavelength apparatus begins with a linearly polarized HeNe laser. While the laser operating power has not yet been discussed as an important factor in sensitivity, a nominal power of 5–10 mW is suggested. The working sensitivity of the technique depends on sampling small reflectance changes near the background reflectance of an individual probe spot. Reflectance intensity (I_r, Eqn 11.5) is a function of the input power (I_o) and absolute reflectance (R); therefore, in order to enhance the contrast of data acquired very near the minimum, a higher laser power may be required:

$$R = I_r/I_o. \tag{11.5}$$

A linearly polarized laser offers a head start on enforcing a pure s-polarized incident beam. Another polarizing element, a Glan-Thompson or linear glass polarizer, is required to further enhance the polarization ratio.

The next optical component is a spatial filter consisting of a microscope objective lens and a pinhole. Imperfections in upfield optics—polarizer, mirror, or the source itself—make this a requirement. While a typical HeNe laser is single-moded (TEM_{00}), the spatial filter reestablishes the original Gaussian distribution of the beam (Wein 1999). Due to the action of focusing the beam to its waist in order to pass through the pinhole, the spatial filter also acts as a beam enlarging element. Following proper collimation, the cleaned and expanded beam is then incident on the sample stage. Reflected light is then collected by a CCD camera for analysis.

In single-wavelength AIR, only one image is captured per condition because all variables, except for target film thickness, are fixed. Therefore, the raw reflectance intensity increase (ΔR) is a direct function of the amount of material bound. The background and experimental image intensities for a spot area are the only information needed to produce data from a tested condition. Fig. 11.12 shows a set of simulated reflectance spectra for the single wavelength implementation of AIR. The contrast value ($\Delta R/R$) for 0.1 nm of bound material is approximately 3700%. In fact, we predict a $\Delta R/R$ of 11% for a thickness increase of as small as 0.01 nm. These reflectance changes are directly represented by the intensity for each CCD pixel; therefore, calculating the intensity value for a spot area before and after target introduction provides the desired data.

2.2. Applications of AIR Biosensing

In this section, we will center on translating the theoretical foundation into biosensor applications. The first step towards the development of a relatively young technique is to identify how well experimental results match with what is theoretically expected based on the model. As a result, a few early limitations will be discussed. The remainder of the section will focus on applying AIR towards the label-free detection of pathogenic bacteria.

2.2.1. Limitations

Obviously, the simulations that we have discussed thus far represent the ideal implementation of the technique. It is not yet feasible to obtain total destructive interference by this

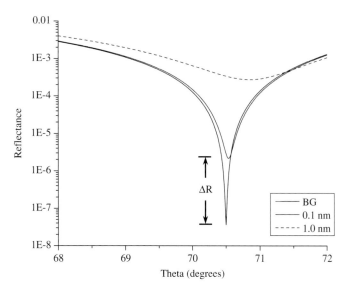

Figure 11.12. Simulated reflectance spectra of a multiwavelength AIR experiment. Change in reflectance (ΔR) is shown for binding of 0.1 nm of target.

technique in the lab. However, it is sensible to expect that the working conditions are able to provide a close enough approximation. While exact parameters are required for the null condition, the technique is sensitive enough, and robust enough, that suitable measures can be taken to approach the minimum at absolute reflectances of 1E-6. As we will show, the technique is an extremely sensitive biosensor, even operating above the reflectance zero.

The polychromatic and monochromatic platforms share common limitations while being dissimilar enough to have their own. For example, both suffer from polarization impurities. The largest s:p-polarization ratio achieved to date is approximately 30,000:1 (Mace, Striemer, and Miller 2006) as determined by intensity measurements of the incident beam.

A test to evaluate simulated performance to experimental performance at this polarization ratio utilized a suite of chips with known silicon dioxide thicknesses as determined by spectroscopic ellipsometry. Each chip was then mounted on the reflectometer and had its absolute reflectance measured. Fig. 11.13 graphically represents this experimental data, as well as simulated spectra for pure s-polarized light and 30,000:1 s:p-polarized light for comparison. The experimental data and the simulated data do not perfectly coincide, but this difference can be attributed to slight uncertainties in the position of the ellipsometer beam (thickness measurement) and the reflectometer beam (reflectance measurement). This obtainable polarization state approaches the null reflectance condition sufficiently to retain sensitivity to small surface thickness changes. Another shared limitation is in the angular divergence of the incident light. Light emitted from the chosen source, laser or otherwise, will have some finite angular divergence. Using a shear plate to enforce collimation and a rotatable stage to mount the sample confers increased accuracy in the incident angle. However, a slight error in the angle value still exists, as well as a drifting out of the light beam due to divergence. Since the null condition is based on a single angle, a broad incident angle band will cause the detected reflectance to contain the summation of information from all constituent angles. This will lead to a broadening of the reflectance minimum and a decrease in sensitivity. The sampled angle space is limited enough that the sensitivity decrease is very small. Also, the properties of the CCD camera used as a detector—namely, bit depth, dynamic range, dark current, and quantum efficiency—play a large role in limiting the reflectance values that are able to be monitored.

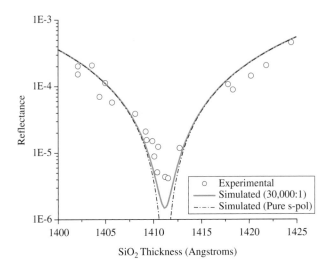

Figure 11.13. Measured reflectances for known silicon dioxide thicknesses around the minimum condition.

The main impediment for the polychromatic platform lies in the tunability of the incident wavelength. The spectrometer bandwidth used for most of our wavelength scanning experiments is 1 nm. Higher resolution spectrometers are available, and, if substituted into the reflectometer, they would both deepen the reflectance well and allow for more defined wavelength scans, thus greatly increasing the sensitivity. Monochromatic AIR, on the other hand, implicitly does not have this problem. While the wavelength-scanning variant of AIR can tolerate slight thickness irregularities in the SiO_2 layer, the fabrication of precise oxide thicknesses by dilute hydrofluoric acid etches is a constant, yet surmountable, hurdle in day-to-day experimentation for single-wavelength AIR. For the most part, chips can be batch etched to the target starting SiO_2 thickness. This is the rate limiting process for the platform. An unfortunate consequence of over-etching the SiO_2 layer beyond the target thickness is the rejection of that specific chip. Only recently have methods been developed to rebuild the lost thickness through layer-by-layer polyelectrolyte self-assembly (Gao, Lu and Rothberg 2006).

2.2.2. Probe Immobilization

A trait inherent to all biosensors, regardless of labeling status or means of signal trans-duction, is probe immobilization. The method implemented may be, in some cases, user-dependent, but for the most part is dominated by the physical properties of the sensor material. Many materials function as promoters of nonspecific adsorption (Eteshola and Leckband 2001; Li et al. 2001) or act as an anchor for reactive self-assembled monolayers (Bain and White-sides 1988; Mrksich and Whitesides 1996). The role of the terminal hydroxyl of a silicon dioxide surface is highly flexible, as it may act as a nucleophile (Bikiaris et al. 2005; Tripp and Hair 1993) or support adsorption. For this reason, silicon dioxide is readily derivitized through a variety of chemical methods. These chemical reactions result in the effective transformation of the hydroxyl group to any of a number of chemical functionalities including, but certainly not limited to, amines (Huang et al. 2001) or halides (Hergenrother, Depew, and Schreiber 2000) (Fig. 11.14). From each initial reaction, a secondary chemical can be added to further alter the surface reactivity, or the probes may be directly coupled. This is by no means a compre-hensive discussion of surface immobilization protocols, but probe layer formation strategy is

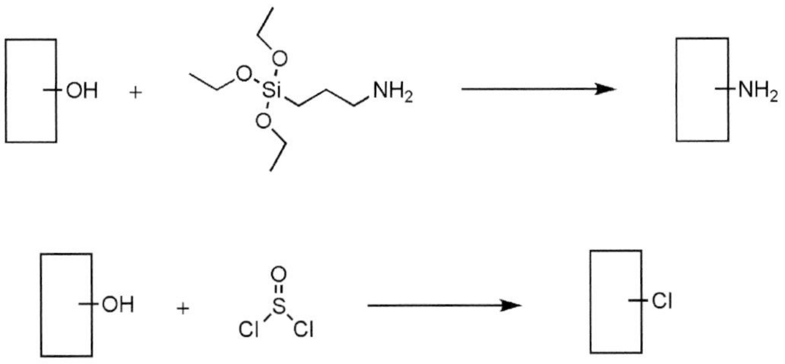

Figure 11.14. Typical methods for derivitization of silicon dioxide surfaces to terminal amines (with APTES) or halides (with thionyl chloride).

an important topic for any biosensor. For the purposes of this chapter, the coupling of amine- or biotin-containing probes will be discussed.

Amine-terminated probes—certain small molecules, synthesized oligonucleotides, peptides, antibodies—can be immobilized to the silicon dioxide surface in a variety of ways. Typically, our work utilizes an amine-terminated alkoxysilane (γ-aminopropyl triethoxysilane, APTES) to tether a homobifunctional crosslinker (glutaraldehyde) to the surface. This leaves a terminal aldehyde to react with a free amine on our probe, forming an imine. Imines are reversible in solution (Huc and Lehn 1997), but upon the assembly of the probe layer, they become effectively solvent inaccessible, and therefore stable. Another homobifunctional crosslinker that utilizes an amino-silanized surface is disuccinimidyl carbonate (DSC) (Macbeath and Schreiber 2000). DSC, through two amide bonds, creates a stable urea linker between the surface and the probe. As an alternate route, the resulting surface amine after the addition of APTES may be reacted with succinic anhydride in order to yield a terminal carboxylic acid (Diehl et al. 2001). From here any carbodiimide (e.g., EDC or DCC) or combination of carbodiimide and N-hydroxysuccinimide (NHS) will activate the carboxylic acid (Jeong and Park 2001; Nedelkov and Nelson 2001). The activated ester is now primed for attack by a free amine, which results in the formation of a stable amide bond. A more direct coupling to the SiO_2 surface is completed through the nucleophile-mediated displacement of a halide formed by a surface reaction with thionyl chloride (Hergenrother, Depew, and Schreiber 2000) (Fig. 11.14). While this approach was used for the attachment of alcohols, it is our experience that it works well with amines as well. One caveat of this approach is the high, localized surface concentration of hydrochloric acid that is released upon probe immobilization.

Biotinylated molecules are also prevalent as biomolecular probes. In order to immobilize molecules of this class, a layer of avidin or streptavidin must first be attached to the surface. This can be accomplished in one of two ways: direct coupling of avidin to the surface, or a primary layer of biotin as a tether. Routes to the former have been covered above; the latter method is accomplished through manipulations of biotin's terminal carboxylic acid. The briefly described carboxylic acid activation schemes hold here, but an alternative, synthetic biotin called sulfo-NHS biotin may be used to directly couple to the APTES-generated surface amine (Ouyang et al. 2005). From this biotin monolayer, avidin will specifically assemble on the surface. As a homotetramer, binding four molecules of biotin per molecule of avidin (Chilkoti, Tan, and Stayton 1995), biotinylated probes will readily immobilize on this layer. The biotin-avidin complex is reversible under thermodynamic control, but, due to the extremely high affinity, the interaction is considered approximately covalent under most experimental conditions.

2.2.3. Pathogen Detection

Organisms within a genus, or even subspecies variants, are distinguished and categorized by their genetic makeup. Any dissimilarity in the sequence of a gene (nucleotide polymorphisms, lack of total homology) or the specific absence/presence of a gene can allow the organism to be selectively detected by a biosensor utilizing oligonucleotide hybridization. Along this same line of thought, unique genes will encode unique proteins. Typically, these proteins or peptide sequences are completely exclusive to an organism. Specific nucleotide aptamers (Gronewold et al. 2005), antibodies (Wang et al. 2001), or recombinantly expressed binding partners (Phizicky et al. 2003) can then be employed as surface immobilized probes. Also, some bacterial subspecies can be identified by the polysaccharide chains that comprise outer membrane antigens (Campbell and Mutharasan 2005). All-in-all, nature provides an assortment of means for bacterial recognition. In this final section, we will describe probe selection routes employed by AIR for the detection of pathogenic species.

2.2.3.1. Detection by Oligonucleotide Hybridization

Oligonucleotide hybridization, single strands mating to form double stranded DNA or RNA, is a well-known process. The free energy of hybridization is dictated by the nucleotide bases involved, the strand length, and the degree of complementarity. Base mismatches influence and limit strand interactions, and this predisposition is exploited by numerous biosensing platforms (Zhong et al. 2003; Du et al. 2005), label-free or otherwise. Extending this further, if a biosensor is sensitive enough to discriminate between these single base mismatches, then oligonucleotide strands comprised of vastly different sequences should easily discern their targets. Hindering the observation of oligonucleotide hybridization is the manner in which AIR data is collected, namely on a dried substrate. Ordinarily, a high salt concentration is required to sustain double-stranded oligonucleotides in order to quench the repulsive forces due to the phosphate backbone. Drying the substrate in the presence of the salt requirement increases the percent of target hybrids that remain, but will leave streaks of salts. These salt streaks will scatter light, and may interfere with reflectance measurements. Washing with distilled water negates the salt streaks and disrupts a large percentage of oligonucleotide hybrids, thus severely decreasing the amount of the expected signal. While this would impair many techniques, AIR is sensitive enough to detect the remaining probe-target hybrids.

The first published example of AIR was the differentiation of Monkeypox virus and Smallpox virus, two separate species of the genus *Orthopoxvirus* (Lu et al. 2004). The detection strategy was the hybridization of single-strand probe DNA (surface) to single-strand complementary, synthetic DNA (solution). Each probe DNA oligonucleotide strand was biotinylated and immobilized on a biotin-anchored layer of streptavidin. Three distinct probe strands were arrayed in wells created through hydrophobic patterning on the bare oxide surface:

B1: biotin-5'-TTT TTT TTT GTT CTT CTC ATC ATC (Positive control probe)
B2: biotin-5'-TTT TTT TTT GAT GAT GAG AAG AAC (Negative control probe)
B3: biotin-5'-AAG ATG CAA TAG TAA T-3' (Smallpox probe)

In order to distance the probe strand from the sensor surface, and theoretically facilitate strand hybridization, a nine-thymidine spacer was introduced 5' to the control sequences. Experimental DNA solutions were then pipetted directly into the experimental well of choice. The positive control consisted of a full complement strand to B1; the negative control was the same complement to B1, albeit with a different probe sequence; the positive experimental was a selected sequence from a Smallpox gene; and the negative experimental was a selected sequence from a Monkeypox gene. All steps, from patterning to target exposure, were monitored

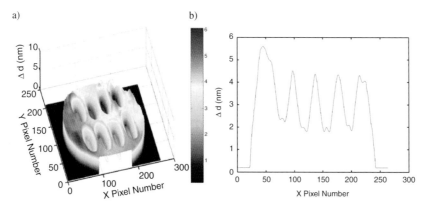

Figure 11.15. Processed data for the background of the Smallpox DNA detection experiments—(a) processed data map showing the measured thickness for all parts of the array; (b) x-axis profile of the bottom four wells (note thickness depressions for the wells).

by acquiring multiwavelength scans of the array (Fig. 11.15). Upon exposure to target strand solutions, the reflectance minimum for each well containing complementary probes exhibited the expected red shifts (B1 and B3). Likewise, negative control strands of noncomplementary DNA exhibited no shifts (B2 and B3). Quantifying the shifts using the calibration profile derived from simulations provided the increase in thickness for the layer. Based on the thickness increase, the length of the detected strand, the size of the well, and the density of the probe layer, the molar amount of bound material can be calculated. For this exercise, 40 femtomoles of complement strand DNA were detected per well. Although this was a relatively simple experiment (synthetic target molecules in purified solutions), it represented how AIR could sensitively discriminate between two different species of the same genus through selective hybridization of DNA probes. While this pathogen detection dealt with viruses, it served as an important proof-of-principle springboard for subsequent experiments. The next extension of the technique utilizing the oligonucleotide hybridization model was the detection of the bacterium *Pseudomonas aeruginosa* directly from culture (unpublished data) by monitoring for target ribosomal RNA sequences. Since the ribosomal RNA sequence is highly conserved throughout the evolution of a species, this made an attractive target for the differentiation of bacterial specimens. Indeed, we found that we were easily able to selectively distinguish *Pseudomonas aeruginosa* and *Escherichia coli* (*E. coli*) based on arrays of ribosomal RNA.

2.2.3.2. Detection by Protein Array

Others have previously reported the use of antibody arrays to detect pathogenic strains of the Gram-negative bacterium, *E. coli* (Muhammed-Tahir and Alocilja 2003; Ruan, Yang, and Li 2002). This is certainly an attractive route due to the widespread commercial availability of bacterium-specific antibodies able to discern *E. coli* serotypes. Serotype designations categorically describe strains due to the lipopolysaccharide content of the cell wall antigens (O-designation) and the motility of the cell (H-designation) (Beutin et al. 2005). Our lab, however, devised a detection assay based on a bacterial protein array. Specifically, we have been able to exploit enteropathogenic *E. coli*'s (EPEC) own means of host infection (Horner et al. 2006) as a biosensing strategy. EPEC cells, and other pathogenic strains, contain a series of genes that bestow to them their pathogenicity, called the locus of enterocyte effacement (LEE) (McDaniel et al. 1995). Two proteins encoded by the LEE are intimin and the translocated intimin receptor (tir). These proteins are required for the formation of pedestals between the host cell and EPEC cell, which ultimately lead to docking between the two cells (DeVinney et al. 2001). Intimin is constitutively expressed in the outer membrane of EPEC and is upregulated in the presence of a host cell (Knutton et al. 1997); tir is secreted from the infectious EPEC cell by the type III secretion needle complex (Thomas et al. 2005); tir is then threaded through the host cell membrane and anchors itself through the recruitment of host actin filaments (Touzé et al. 2004). The extracellular intimin binding domain of tir, which is now properly folded and presented, is available to provide a "natural" interaction site with intimin (Fig. 11.16).

In order to apply this interaction as an EPEC biosensor, an initial model system was tested. Expression vectors for the intimin binding domain of tir (tir-IBD) and the extracellular domain of intimin (intimin-ECD) were generously provided by a collaborator (Luo et al. 2000). Tir-IBD would then serve as our probe protein for initial tests utilizing purified solutions of intimin-ECD as well as the probe for detecting full length intimin naturally expressed on EPEC cell membranes. All resulting experiments were analyzed on a first-generation monochromatic reflectometer. Tir-IBD was immobilized into an arrayed pattern via the previously described APTES-glutaraldehyde method and reflectance changes were compared to a ubiquitin negative control. We found that we could detect nanomolar-range concentrations of pure intimin-ECD selectively.

Since the intimin-ECD domain was capable of binding to the tir-IBD probe, we then theorized that this probe could bind full-length, native intimin as well. While there are numerous intimin and tir isoforms, structural homology is adequately conserved to allow for the interaction to occur, albeit with slightly altered affinities (Sinclair and O'Brien 2004; Ramachandran et al. 2003). We were confident that any EPEC serotype could be detected in this fashion. For this reason, and also due to a lower likelihood of human pathogenicity, we chose rabbit-specific REPEC (serotype O15:H-, ATCC #49106) as our experimental strain. REPEC was cultured overnight in LB medium, and an aliquot was pelleted and transferred to DMEM medium to facilitate intimin upregulation (Knutton et al. 1997). Allowing sufficient time to reproduce and express intimin, the culture was pelleted and resuspended in buffer for analysis. No other preparative work (lysis, filtration, etc.) was required. Two negative control strains were studied as comparisons: *Pseudomonas aeruginosa*, a Gram-negative pathogenic bacterium used in a previous study that does not express intimin; and JM109, a benign *E. coli* strain that does not express intimin. Each cell solution, 200 μL total, was pipetted over the entire array and chip. A successful experiment would show reflectance changes for tir-IBD spots on chips exposed to REPEC solutions only (Fig. 11.17). In fact, this was the result of the experiment.

In contrast to serotype differentiation mediated by antibodies, this method of detection does not provide that level of strain selectivity. However, as a rapid assay capable of verifying the presence of pathogenic *E. coli* strains, all of which express intimin, this process works well. This approach, the *in situ* observation of natural cell surface contacts, can be expanded to include the detection of bacteria that interact with host cells either by similar means, namely translocated

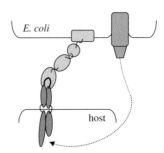

Figure 11.16. Schematic of the intimin-tir interaction. Tir is secreted by EPEC's type III secretion needle complex into the host cell. Tir then anchors into the host and presents itself to intimin for binding.

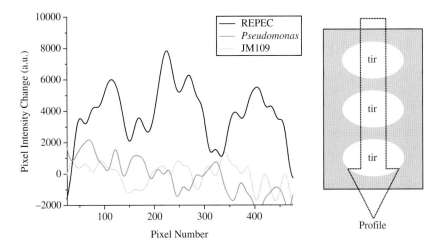

Figure 11.17. Reflectance change profile for three contiguous tir spots upon addition of solutions containing REPEC, *Pseudomonas aeruginosa*, and JM109 cells.

effector proteins, or by inherent extracellular interactions (Hooper and Gordon 2001; Schuert et al. 2002). One could also incorporate serotype-specific antibodies in the array should that level of identification be desirable.

As a diagnostic aid, there is a need to detect the presence of substances, proteins or otherwise, that are alien to our own makeup. Cell membrane-anchored proteins are not the only markers signaling the presence of a bacterium. Bacteria contain proteins in their cytosol that are unique to their species, particularly divergent when compared to their host, and many species secrete proteins into their surrounding environment using secretion systems (Hueck 1998). Therefore, another use of protein arrays is for the detection of bacteria-specific proteins. As an initial test of this, and to investigate the sensitivity limits of AIR, we again utilized the intimin-tir interaction of enteropathogenic *E. coli* (Mace, Striemer, and Miller 2006). Serial dilutions of purified intimin-ECD were prepared and screened against a tir-IBD array, again using ubiquitin controls. This battery of tests was performed on a second-generation monochromatic reflectometer, which gave crisper images and higher sensitivity. The sensitivity limit was found to be 10 pM.

For many applications, sensitivity as a concentration value is all that is required. However, in order to compare AIR to other techniques over a broad range of signal transduction methods, the concentration value must be converted into a more recognizable notation, namely as a mass/area value (DeLouise, Kou, and Miller 2005). Conversion is accomplished through the use of thermodynamic models describing the probe-target interaction. An equilibrium analysis was performed as a first approximation. Our intimin-ECD and tir-IBD protein system has been well studied, so the K_d of binding is known (Luo et al. 2000); we also know the molecular weights of both proteins. From this information, we can begin to build a model.

$$K_d = \frac{[probe]_{free}\,[analyte]_{free}}{[bound]}. \tag{11.6}$$

The formula for the K_d of a one-to-one binding interaction is displayed in equation 11.6, where tir-IBD is the probe molecule and intimin-ECD is the analyte molecule. It is then a simple exercise to rewrite the unknown "probe" and "analyte" concentrations in terms of known quantities, immobilized concentration of tir-IBD ($probe_I$) and solution concentration of intimin-ECD ($analyte_S$), and the unknown variable "bound" (Eqn 11.7). The concentration

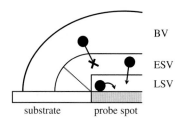

Figure 11.18. Schematic of the volumes used in the diffusion-limited model. Note that the volumes and distances are not to scale.

of immobilized tir-IBD, probe$_I$, is estimated based upon a model created from the crystal structure (Protein Data Bank ID# 1F02T), and sequence analysis and ellipsometric measurements performed in our laboratory.

$$K_d = \frac{[probe_I - bound]\,[analyte_S - bound]}{[bound]}. \tag{11.7}$$

Equation 11.7 is then rewritten as a quadratic in order to solve for the concentration of bound analyte as a function of the aforementioned known values. Using this equilibrium analysis to solve for the bound analyte concentration, and using conversions based on the intimin-ECD molecular weight and experimental parameters, yields a detected mass/area of $0.04\,pg/mm^2$.

However, since target solutions are static on an AIR chip prior to measurement, an equilibrium analysis is not the most objective route. As a result, a diffusion-limited analysis was performed. In our diffusion-limited scenario, three volumes now come into play: the bulk volume (BV), the effective sensing volume (ESV), and the local sensor volume (LSV). The BV represents the $100\,\mu L$ target solution; the ESV represents the volume surrounding that lies within one diffusion length of the probe spot, i.e., target molecules are able to reach the sensor surface in the time frame of an experiment; the LSV represents an "event horizon" volume where the effective concentration of free tir-IBD is so high that all intimin-ECD molecules bound within it will immediately rebind a neighboring probe upon reversible dissociation (Fig. 11.18). Using this analysis, 10 pM of intimin-ECD corresponds to $0.33\,pg/mm^2$ of bound material. Although this is approximately an order of magnitude higher than the equilibrium model, it is more physically accurate in describing the experiment. As a comparison, the sensitivity is on the order of other highly regarded methods (Homola, Yee, and Gauglitz 1999; Thibault et al. 2006). Assessing this result at a more basic level, this value corresponds to one bound intimin-ECD molecule per 28,000 tir-IBD molecules.

3. Conclusions and Future Perspectives

Although the work performed with protein arrays has proven to be an interesting and fruitful path, due to the manner in which they are produced—namely molecular biology, expression, and purification—the use of cloned protein domains as probes is somewhat limited. In order to fully take advantage of the technique's capability to spatially resolve information from individual sensor spots, a part of the future for AIR lies in high density and high numeracy arrays of oligonucleotides and antibodies. These probe molecules are commercially available for a plethora of individual targets, and their applications are nearly limitless if used in concert. While we have already discussed the immobilization and use of oligonucleotide arrays, we have not yet published work using antibody arrays. However, this is a straightforward process,

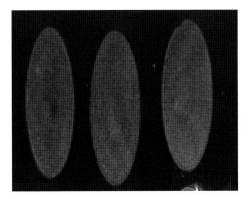

Figure 11.19. An anti-human IgG array (three identical spots) immobilized using APTES/glutaraldehyde chemistry.

and we have been able to immobilize antibodies using our standard methods (Fig. 11.19). We are currently implementing them in experimental arrays. The strengths of AIR can be seen in its simplicity, high sensitivity, and adaptability to a variety of sensing systems. With sufficient growth, and by constructing a chip with all of the aforementioned probe molecules, one could envision an AIR sensor capable of detecting molecular markers from all aspects of a bacterium: flagellar, cell surface, cytosolular, and nuclear. Thus, in a single experiment, not only could one confirm the presence of said bacterium, but could do so selectively in the presence of other organisms and with a greater amount of information acquired.

References

Bain CD and Whitesides GM (1988) Molecular-level control over surface order in self-assembled monolayer films of thiols on gold. Science 240:62–63

Batey J and Tierney E (1986) Low-temperature deposition of high-quality silicon dioxide by plasma-enhanced chemical vapor deposition. J. Appl. Phys. 60:3136–3145

Beutin L, Strauch E, Zimmermann S, Kaulfuss S, Schaudinn C, Männel A and Gelderblom HR (2005) Genetical and functional investigation of fliC genes encoding flagellar serotype H4 in wildtype strains of Escherichia coli and in a laboratory E. coli K-12 strain expressing flagellar antigen type H48BMC Microbiol. 5:4–14

Bikiaris DN, Vassilou A, Pavlidou E and Karayannidis GP (2005) Compatibilisation effect of PP-g-MA copolymer on iPP/SiO2 nanocomposites prepared by melt mixing. Eur. Polym. J. 41:1965–1978

Brynda E, Houska M, Brandenburg A and Wikerstål A (2002) Optical biosensors for real-time measurement of analytes in blood plasma. Biosens. Bioelectron. 17:665–675

Campbell GA and Mutharasan R (2005) Detection of pathogen Escherichia coli O157:H7 using self-excited PZT-glass microcantilevers. Biosens. Bioelectron. 21:462–473

Canham LT (1990) Silicon quantum wire array fabrication by electrochemical and chemical dissolution of wafers. Appl. Phys. Lett. 57:1046–1048

Chan S and Fauchet PM (1999) Tunable, narrow, and directional luminescence from porous silicon light emitting devices. Appl. Phys. Lett., 75:274–276

Chan S, Fauchet PM, Li Y, Rothberg LJ and Miller BL (2000) Porous silicon microcavities for biosensing applications. Phys. Stat. Sol. A. 182:541–546

Chan S, Horner SR, Miller BL and Fauchet PM (2001) Identification of gram negative bacteria using nanoscale silicon microcavities. J. Am. Chem. Soc. 123:11797–11798

Chilkoti A, Tan PH and Stayton PS (1995) Site-directed mutagenesis studies of the high-affinity streptavidin-biotin complex: Contributions of tryptophan residues 79, 108, and 120. Proc. Natl. Acad. Soc. USA. 92:1754–1758

Deal BE and Grove AS (1965) General relationship for the thermal oxidation of silicon. J. Appl. Phys. 36:3770–3778

DeLouise LA and Miller BL (2004a) Quantitative assessment of enzyme immobilization capacity in porous silicon. Anal. Chem. 76:6915–6920

DeLouise LA and Miller BL (2004b) Trends in porous silicon biomedical devices: tuning microstructure and performance trade-offs in optical biosensors. Proc. SPIE. 5357:111–125

DeLouise LA and Miller BL (2005). Enzyme immobilization in porous silicon: quantitative analysis of the kinetic parameters for glutathione-S-transferases. Anal. Chem. 77:1950–1956

DeLouise LA, Fauchet PM, Miller BL and Pentland AP 2005. Hydrogel-supported optical-microcavity sensors. Adv. Mater. 17:2199–2203

DeLouise LA, Kou PM and Miller BL (2005) Cross-correlation of optical microcavity biosensor response with immobilized enzyme activity. Insights into biosensor sensitivity. Anal. Chem. 77:3222–3230

DeVinney R, Puente JL, Gauthier A, Goosney D and Finlay BB (2001) Enterohaemorrhagic and enteropathogenic Escherichia coli use a different Tir-based mechanism for pedestal formation. Mol. Microbiol. 41:1445–1458

Diehl F, Grahlmann S, Beier M and Hoheisel JD (2001) Manufacturing DNA microarrays of high spot homogeneity and reduced background signal. Nucleic Acids Res. 29:e38

Du H, Strohsahl CM, Camera J, Krauss TD and Miller BL (2005) Sensitivity and specificity of metal surface-immobilized "molecular beacon" biosensors. J. Am. Chem. Soc. 127:7932–7940

Eteshola E and Leckband D (2001) Development and characterization of an ELISA assay in PDMS microfluidic channels. Sens. Actuators B. 72:129–133

Gao T, Lu J and Rothberg LJ (2006) Biomolecular sensing using near-null single wavelength arrayed imaging reflectometry. Anal. Chem. 78:6622–6627

Gronewold TMA, Glass S, Quandt E and Famulok M (2005) Monitoring complex formation in the blood-coagulation cascade using aptamer-coated SAW sensors. Biosens. Bioelectron. 20:2044–2052

Hecht E (2002) Optics, 4th ed. Addison Wesley, San Francisco

Hergenrother PJ, Depew KM and Schreiber SL (2000). Small-molecule microarrays: covalent attachment and screening of alcohol-containing small molecules on glass slides. J. Am. Chem. Soc. 122: 7849–7850

Homola J, Yee SS and Gauglitz G (1999) Surface plasmon resonance sensors: review. Sens. Actuators B. 54:3–15

Hooper LV and Gordon JI (2001) Glycans as legislators of host–microbial interactions: spanning the spectrum from symbiosis to pathogenicity. Glycobiology. 11:1R-10R

Horner SR, Mace CR, Rothberg LJ and Miller BL (2006) A proteomic biosensor for enteropathogenic E. coli. Biosens. Bioelectron. 21:1659–1663

Huang Y, Duan X, Wei Q and Lieber CM (2001) Directed assembly of one-dimensional nanostructures into functional networks. Science. 291: 630–633.

Hubbard RD, Horner SR and Miller BL (2001) Highly substituted ter-cyclopentanes as receptors for lipid A. J. Am. Chem. Soc. 123:5810–5811

Huc I and Lehn J-M (1997) Virtual combinatorial libraries: Dynamic generation of molecular and supramolecular diversity by self-assembly. Proc. Natl. Acad. Sci. USA. 94:2106–2110

Hueck CJ (1998) Type III protein secretion systems in bacterial pathogens of animals and plants. Microbiol. Mol. Biol. Rev. 62:379–433

Janshoff A, Dancil K-PS, Steinem C, Greiner DP, Lin VS-Y, Gurtner C, Motesharei K, Sailor MJ and Ghadiri MR (1998) Macroporous p-type silicon Fabry-Perot layers. Fabrication, characterization, and applications in biosensing. J. Am. Chem. Soc. 120:12108–12116

Jeong JH and Park TG (2001) Novel polymer-DNA hybrid polymeric micelles composed of hydrophobic poly(D,L-lactic-co-glycolic acid) and hydrophilic oligonucleotides. Bioconjugate Chem. 12:917–923

Jin G, Tengvall P, Lundström I and Arwin H (1995) A biosensor concept based on imaging ellipsometry for visualization of biomolecular interactions. Anal. Biochem. 232:69–72

Johnston KS, Yee SS and Booksh KS (1997) Calibration of surface plasmon resonance refractometers using locally weighted parametric regression. Anal. Chem. 69:1844–1851

Knutton S, Adu-Bobie J, Bain C, Phillips AD, Dougan G and Frankel G (1997) Down regulation of intimin expression during attaching and effacing enteropathogenic Escherichia coli adhesion. Infect. Immun. 65:1644–1652

Li J, Tan W, Wang K, Xiao D, Yang X, He X and Tang Z (2001) Ultrasensitive optical DNA biosensor based on surface immobilization of molecular beacon by a bridge structure. Anal. Sci. 17:1149–1153

Liedberg B, Nylander C and Lundström I (1983) Surface plasmon resonance for gas detection and biosensing. Sens. Actuators. 4:299–304.

Lin VS-Y, Motesharei K, Dancil K-PS, Sailor MJ and Ghadiri MR (1997) A Porous silicon-based optical interferometric biosensor. Science. 278:840–843

Ljungberg K, Söderbärg A and Bäcklund Y (1993) Spontaneous bonding of hydrophobic silicon surfaces. Appl. Phys. Lett. 62:1362–1364

Lu J, Strohsahl CM, Miller BL and Rothberg LJ (2004) Reflective interferometric detection of label-free oligonucleotides. Anal. Chem. 76:4416–4420

Luo Y, Frey EA, Pfuetzner RA, Creagh AL, Knoechel DG, Haynes CA, Finlay BB and Strynadka NCJ (2000) Crystal structure of enteropathogenic Escherichia coli intimin–receptor complex. Nature. 405:1073–1077

Macbeath G and Schreiber SL (2000) Printing proteins as microarrays for high-throughput function determination. Science. 289:1760–1763

Mace CR, Striemer CC and Miller BL (2006) Theoretical and experimental analysis of arrayed imaging reflectometry as a sensitive proteomics technique. Anal. Chem. 78:5578–5583

Martinez JS, Grace WK, Grace KM, Hartman N and Swanson BI (2005) Pathogen detection using single mode planar optical waveguides. J. Mater. Chem. 15:4639–4647

McDaniel TK, Jarvis KG, Donnenberg MS and Kaper JB (1995) A genetic locus of enterocyte effacement conserved among diverse enterobacterial pathogens. Proc. Natl. Acad. Sci. USA. 92:1664–1668

Mielenz KD, Nefflen KF, Rowley WRC, Wilson DC and Engelhard E Reproducibility of helium-neon laser wavelengths at 633 nm. Appl. Opt. 7:289–294

Morita M, Ohmi T, Hasegawa E, Kawakami M and Ohwada M (1990) Growth of native oxide on a silicon surface. J. Appl. Phys. 68:1272–1281

Mrksich M and Whitesides GM (1996) Using self-assembled monolayers to understand the interactions of man-made surfaces with proteins and cells. Annu. Rev. Biophys. Biomol. Struct. 25:55–78

Muhammed-Tahir Z and Alocilja EC (2003) Fabrication of a disposable biosensor for Escherichia coli O157:H7 detection. IEEE Sensors Journal. 3:345–351

Nedelkov D and Nelson RW (2001) Analysis of native proteins from biological fluids by biomolecular interaction analysis mass spectrometry (BIA/MS): exploring the limit of detection, identification of non-specific binding and detection of multi-protein complexes. Biosens. Bioelectron. 16:1071–1078

Ouyang H, Christophersen M, Viard R, Miller BL and Fauchet PM (2005) Macroporous silicon microcavities for macromolecule detection. Adv. Funct. Mater. 15:1851–1859

Ouyang H, Striemer CC and Fauchet PM (2006) Quantitative analysis of the sensitivity of porous silicon optical biosensors. Appl. Phys. Lett. 88:163108–163110

Ouyang H, DeLouise LA, Miller BL and Fauchet PM (2007) Label-free quantitative detection of protein using macroporous silicon photonic bandgap biosensors. Anal. Chem. 79:1502–1506

Peterlinz KA and Georgiadis R (1996) In situ kinetics of self-assembly by surface plasmon resonance spectroscopy. Langmuir. 12:4731–4740

Phizicky E, Bastiaens PIH, Zhu H, Snyder M and Fields S (2003) Protein analysis on a proteomic scale. Nature. 422:208–215

Piehler J, Brecht A and Gauglitz G (1996) Affinity detection of low molecular weight analytes. Anal. Chem. 68:139–143

Ramachandran V, Brett K, Hornitzky MA, Dowton M, Bettelheim KA, Walker MJ and Djordjevic SP (2003) Distribution of intimin subtypes among Escherichia coli isolates from ruminant and human sources. J. Clin. Microbiol. 41:5022–5032

Ruan C, Yang L and Li Y (2002) Immunobiosensor chips for detection of Escherichia coli O157:H7 using electrochemical impedance spectroscopy. Anal. Chem. 74:4814–4820

Schmedake TA, Cunin F, Link JR and Sailor MJ (2002) Standoff detection of chemicals using porous silicon "smart dust" particles. Adv. Mater. 14:1270–1272

Schmitt H-M, Brecht A, Piehler J and Gauglitz G (1997) An integrated system for optical biomolecular interaction analysis. Biosens. Bioelectron. 12:809–816

Schubert W-D, Urbanke C, Ziehm T, Beier V, Machner MP, Domann E, Wehland J, Chakraborty T and Heinz DW (2002) Structure of internalin, a major invasion protein of Listeria monocytogenes, in complex with its human receptor E-cadherin. Cell. 111:825–836

Sinclair JF and O'Brien AD (2004) Intimin types a, b, and g bind to nucleolin with equivalent affinity but lower avidity than to the translocated intimin receptor. J. Biol. Chem. 279:33751–33758

Smith EA and Corn RM (2003) Surface plasmon resonance imaging as a tool to monitor biomolecular interactions in an array based format. Appl. Spectrosc. 57:320A-332A

Tabata A, Matsuno N, Suzuoki Y and Mizutani T (1996) Optical properties and structrue of SiO$_2$ films prepared by ion-beam sputtering. Thin Solid Films. 289:84–89

Thibault G, Yudin J, Wong P, Tsitrin V, Sprangers R, Zhao R and Houry WA (2006) Specificity in substrate and cofactor recognition by the N-terminal domain of the chaperone ClpX. Proc. Natl. Acad. Sci. USA. 103:17724–17729

Thomas NA, Deng W, Puente JL, Frey EA, Yip CK, Strynadka NCJ and Finlay BB (2005) CesT is a multi-effector chaperone and recruitment factor required for the efficient type III secretion of both LEE- and non-LEE-encoded effectors of enteropathogenic Escherichia coli. Mol. Microbiol. 57:1762–1779

Touzé T, Hayward RD, Eswaran J, Leong JM and Koronakis V (2004) Self-association of EPEC intimin mediated by the β-barrel-containing anchor domain: a role in clustering of the Tir receptor. Mol. Microbiol. 51:73–87

Tripp CP and Hair ML (1993) Chemical attachment of chlorosilanes to silica: a two-step amine-promoted reaction. J. Phys. Chem. 97:5693–5698

Wang CC, Huang R-P, Sommer M, Lisoukov H, Huang R, Lin Y, Miller T and Burke J 2001. Array-based multiplexed screening and quantitation of human cytokines and chemokines. J. Proteome Res. 1:337–343

Wein GR (1999) A video technique for the quantitative analysis of the Poisson spot and other diffraction patterns. Am. J. Phys. 67:236–240

Young LS, Martin WJ, Meyer RD, Weinstein RJ and Anderson ET (1977) Gram-negative rod bacteremia: microbiologic, immunologic, and therapeutic considerations. Ann. Intern. Med. 86:456–471

Zhong X, Reynolds R, Kidd JR, Kidd KK, Jenison R, Marlar RA and Ward DC (2003) Single-nucleotide polymorphism genotyping on optical thin-film biosensor chips. Proc. Natl. Acad. Sci. USA. 100:11559–11564

Acoustic Wave (TSM) Biosensors: Weighing Bacteria

Eric Olsen, Arnold Vainrub and Vitaly Vodyanoy

Abstract

This chapter is focused on the development and use of acoustic wave biosensor platforms for the detection of bacteria, specifically those based on the thickness shear mode (TSM) resonator. We demonstrated the mechanical and electrical implications of bacterial positioning at the solid-liquid interface of a TSM biosensor and presented a model of the TSM with bacteria attached operating as coupled oscillators. The experiments and model provide an understanding of the nature of the signals produced by acoustic wave devices when they are used for testing bacteria. The paradox of "negative mass" could be a real threat to the interpretation of experimental results related to the detection of bacteria. The knowledge of the true nature of "negative mass" linked to the strength of bacteria attachment will contribute significantly to our understanding of the results of "weighing bacteria." The results of this work can be used for bacterial detection and control of processes of bacterial settlement, bacterial colonization, biofilm formation, and bacterial infection in which bacterial attachment plays a role.

1. Introduction

Rapid, specific, sensitive detection and enumeration methods for microbial pathogens have long been a subject of research. This is especially true in the area of food-related illness prevention, where it's estimated that over three million deaths occur annually worldwide at a cost of $6.5–34.9 billion (Buzby and Roberts 1997) due to the consumption of food products contaminated with bacteria, bacterial toxins, or viruses (Foodborne Diseases 1997). The perceived need for instantaneous detection of pathogenic biological agents in both simple and complex matrices has increased tremendously based on the advent of sensor technologies capable of detecting macromolecules in near instantaneous or real time. To this end, specific, selective miniaturized biosensor assays that combine reliability, speed, and portability while reducing sample size and assay costs are needed to replace conventional identification techniques.

Thousands of papers have been published describing a myriad of engineering approaches for microbial biodetection since 1962, when Clark and Lyons (1962) first published their essay on a reusable enzymatic electrode. These approaches are sometimes broadly categorized into optical, calorimetric, biological-biochemical, electrochemical, and acoustic wave-mass change methods. Of these, optical methods (e.g., SPR) and acoustic wave-mass change methods appear

Eric Olsen • Clinical Investigation Facility, David Grant USAF Medical Center, Travis Air Base, CA.
Arnold Vainrub • Department of Anatomy, Physiology, and Pharmacology, Auburn University, AL.
Vitaly Vodyanoy • Department of Anatomy, Physiology, and Pharmacology, Auburn University, AL.

M. Zourob et al. (eds.), *Principles of Bacterial Detection: Biosensors, Recognition Receptors and Microsystems*,
© Springer Science+Business Media, LLC 2008

to suitably combine speed, sensitivity, and portability for future development of rapid biosensors for microbial analyses (O'Sullivan and Guilbault 1999; Janshoff, Galla, and Steinem 2000; Skládal 2003).

The term "biosensor" is used rather broadly these days. As defined here, acoustic wave biosensors consist of two main components: a biological receptor that possesses affinity for an analyte of interest, and a piezoelectric transducer to convert the chemical signal of sample-receptor coupling to an amplified signal output that provides qualitative or quantitative assessment of their interaction. Acoustic waves in piezoelectric substrates (e.g., quartz) used as sensor platforms are based on mechanical waves created by an applied electric field. These waves propagate through the substrate and are then transformed back to an electric field for measurement. These discoveries have helped lead to the development of a wide range of acoustic wave devices (Morgan 2000) for applications including sensing of bacterial cells in solution.

In this chapter we will focus on the development and use of acoustic wave biosensor platforms for the detection of bacteria, specifically those based on the thickness shear mode (TSM) resonator. We also discuss the mechanical and electrical implications of bacterial positioning at the solid-liquid interface of a TSM biosensor, and present a model of the TSM with bacteria attached operating as coupled oscillators.

2. Historical Perspective, Theory and Background

2.1. Piezoelectricity and Acoustic Waves

Biosensors based on acoustic waves are rooted in numerous fundamental concepts, including the discovery of piezoelectricity by Pierre and Paul-Jacques Curie in 1880 (Curie and Curie 1880), the theory of acoustic waves as predicted by Lord Rayleigh in his 1885 analysis of surface waves in solids (Rayleigh 1885), and Augustus Love's work on acoustic waves published in 1911 (Love 1911), which included a description of shear surface waves having motion perpendicular to the sagittal plane. Subsequently, surface elastic waves were first measured in piezoelectric transducers in the 1940s and 1950s (Victorov 1967), and in 1965 White and Voltmer (1965) experimentally demonstrated direct piezoelectric coupling to surface elastic waves using an interdigital electrode transducer (IDT) on a piezoelectric plate.

Piezoelectricity refers to the generation of electrical charges in response to an applied mechanical stress. The converse is also true; application of a suitable electric field to a piezo-electric material (substrate) creates a mechanical stress, or as the name implies "converse piezoelectricity." While there are many different types of acoustic wave devices, all use the converse piezoelectric effect to produce acoustic waves. These waves propagate through a substrate, and are then transformed back to an electric field for measurement. This interconnection between piezoelectricity and acoustic waves has led to the development of a wide range of acoustic wave device applications.

2.2. Acoustic Wave Devices

The use of acoustic wave devices in electronics can be traced back more than 80 years (Morgan 2000; Gizeli 2002) and today includes timing and frequency control for applications that require extreme precision and stability such as mobile phones, satellite communications, and radio transmitters. Several of the emerging applications for these devices in the medical sciences (biological and chemical sensors) and industrial and commercial applications (vapor, humidity, temperature, and mass sensors) may eventually equal the demand of the telecommunications market.

Acoustic devices are generally described by way of their wave distribution, either through or on the surface of the piezoelectric substrate. Basically, acoustic waves differ in velocities and directions of particle movement within the substrate. Depending on the material and boundary conditions there can be different variants. Fig. 12.1 shows the configuration of typical acoustic

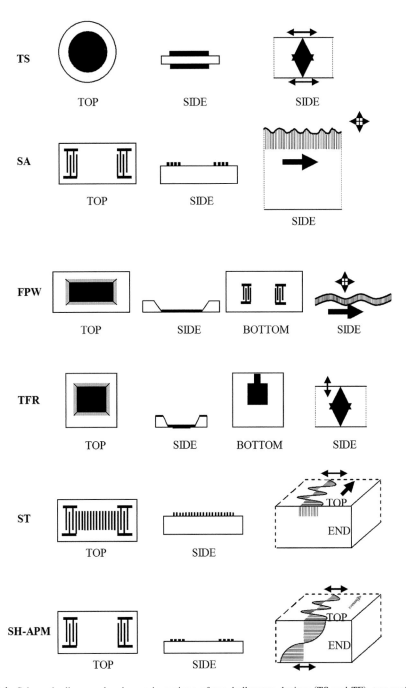

Figure 12.1. Schematic diagram showing various views of two bulk wave devices (TS and TF), two surface wave devices (SA and ST), and two plate wave devices (FPW and SH-APM). Wave motions are indicated by light arrows showing particle displacement directions and larger bold arrows showing wave propagation direction. (Grate and Frye 1996; © John Wiley & Sons Limited; reproduced with permission).

wave devices. Transverse or shear waves have particle displacements that are normal to the direction of wave propagation and which can be polarized so that the particle displacements are either parallel or normal to the sensing surface. Shear horizontal wave motion signifies transverse displacements polarized parallel to the sensing surface; shear vertical motion indicates transverse displacements normal to the surface. Some properties of selected acoustic wave devices, including TSM, transverse shear mode (i.e., QCM, quartz crystal microbalance); SAW, surface acoustic wave; STW, surface transverse wave; SH-APM, shear horizontal acoustic plate mode; FPW, flexural plate wave; and TRAW, thin rod acoustic wave, are shown in Table 12.1 (Rickert et al. 1999). A wave propagating through the substrate is called a bulk wave. The most frequently used bulk acoustic wave devices are the TSM resonator and the SH-APM. A wave propagated on the surface of a substrate is known as a surface wave. The most broadly

Table 12.1. Comparison of selected acoustic wave sensors (Rickert et al. 1999. © John Wiley & Sons Limited. Reproduced with permission)

Type	Wave type	Parameter deter- mining the resonance frequency	Typical frequency[a] (MHz)	Typical example with: material resonance frequency (MHz) thickness d of substrate (μm) wavelength $\lambda(\mu$m)	Medium of preferential use
TSM (QCM)	Volume, horizontal	Thickness d	5–30	Quartz 6 270 540	Gas, liquid
SAW	Surface, vertical	Spacing of interdigital electrodes	30–500	Quartz 158 760 20	Gas
STW	Surface, horizontal	Spacing of interdigital electrodes	30–500	Quartz 250 500 20	Liquid gas[b]
Love- mode	Surface, horizontal	Spacing of interdigital electrodes and thickness d of wave guiding layer	80–300	Quartz 110 500 40	Liquid gas
SH- APM	Plate, Horizontal	Thickness d and spacing of interdigital electrodes	25–200	Quartz 101 203 50	Liquid gas[b]
FPW	Plate, vertical	Thickness d and spacing of interdigital electrodes	2–7	Zinc oxide 5.5 3.5 100	Gas liquid[c]
TRAW	Volume, longitudinal	Frequency of coupling piezoelectric transducer	0.5–8	Au[d] 1.95 50	Liquid gas[b]

[a]Material and wave velocity influence the resonance frequency in all cases.
[b]Preliminarily designed for application in liquids, but applications in gas are possible.
[c]Possible as wave velocity is less than compressional velocity of sound in liquid.
[d]The transducer itself is not piezoelectric, but acoustic waves travel through it.

used surface wave devices are the SAW and shear-horizontal surface acoustic wave (SH-SAW) sensors, also recognized as the surface transverse wave (STW) sensor. The waves are guided by reflection from multiple surfaces. Typical representatives of plate wave devices are FPW and APM.

All acoustic wave devices are sensitive to perturbations of many different physical parameters. The change in the properties of the pathway over which the acoustic wave propagates will result in a change in output signal. While all acoustic wave devices will function in gases or vacuum, only a few operate efficiently in liquids. Whether an acoustic wave device can operate in liquid is determined by the direction of the particle displacement at the surface of the device. TSM, SH-APM, and SH-SAW devices all generate waves that propagate primarily in a shear horizontal motion. A shear horizontal wave does not radiate appreciable energy into liquid, allowing functionality without excessive attenuation. Conversely, SAW devices utilizing Rayleigh waves have a substantial surface-normal displacement that radiates compression waves into the liquid and thus cannot be employed in the liquid phase (Grate and Frye 1996). An exception to this rule occurs for devices using waves that propagate at a velocity lower than that of sound in liquid. Therefore, such modes do not couple to compressional waves in liquid and are thus relatively unattenuated (Ballantine et al. 1997).

3. TSM Biosensors

Acoustic wave devices such as the TSM are essentially highly sensitive analytical balances, capable of discriminating extremely small mass deposition events. This makes them excellent analytical tools for the study of specific molecular interactions at the solid-liquid interface in air, and under vacuum or aqueous conditions (Bunde, Jarvi, and Rosentreter 1998; Cavicacute, Hayward, and Thompson 1999; Ivnitski et al. 1999; O'Sullivan and Guilbault 1999; Kaspar et al. 2000; Stadler, Mondon, and Ziegler 2003; Yakhno et al. 2007).

The TSM resonator may be better known as the quartz crystal microbalance (QCM), because its natural resonant properties are based on the piezoelectric properties of resonators prepared normally from quartz. The QCM usually consists of a thin, round AT- or BT-cut (angular orientation in relation to internal crystallography) quartz crystal wafer with two metallic electrodes (e.g., gold, silver, or palladium) deposited uniformly onto both sides of the quartz (Grate and Frye 1996). The quartz substrate can have varying dimensions and resonant frequencies, the most common being 100 kHz and 1, 2, 4, 5, 8, and 10 MHz (Scherz 2000). In itself it comprises an oscillatory circuit that can be modeled as an extended Butterworth-van Dyke equivalent circuit depending upon load conditions (Fig. 12.2) (Janshoff, Steinem, and Wegener 2004). The piezoelectric properties of the quartz result in deformation of the crystal when an electrical potential is created across the electrodes, which in turn induces a transverse, standing wave of resonance oscillation in the quartz at a fundamental frequency (Babacan et al. 2000). AT-cut crystals displace the oscillation parallel to the resonator surface and are utilized predominantly in liquids, due to their temperature stability. Any changes in the resonance frequency of the crystal are usually attributed to the effect of added mass due to binding at the active (overlapping) area of the electrodes. Theoretical modeling of the TSM response to mass accumulation has been demonstrated under various loading conditions, including ideal mass layers (thin layers of Au and SiO_2), a semi-infinite fluid (glycerol in water), and a viscoelastic layer represented by thin layers of oil (Martin, Granstaff, and Frye 1991; Bandey et al. 1999).

According to theory (Sauerbrey 1959), when a mass, m, binds at the surface of the sensor, a corresponding proportional decrease of the resonator's oscillation frequency occurs, the total quantity of which can be solved for using Sauerbrey's (1959) equation as follows, provided

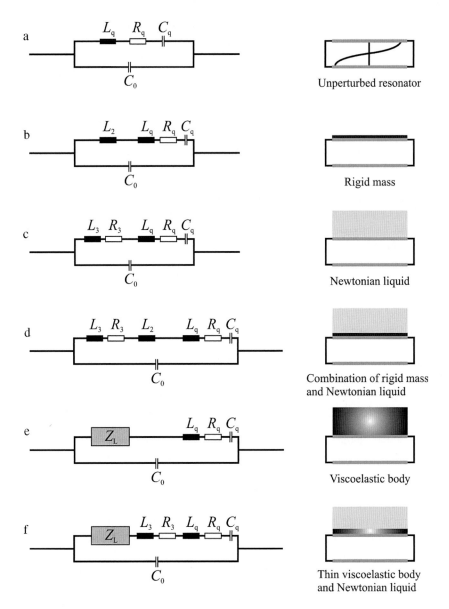

Figure 12.2. Extended equivalent circuits derived from Butterworth-van Dyke circuit for different load conditions: (a) unperturbed quartz plate; (b) rigid mass; (c) Newtonian liquid; (d) combination of rigid mass and Newtonian liquid; (e) thick viscoelastic layer; and (f) thin viscoelastic body and Newtonian liquid (adapted from Janshoff, Steinem, and Wegener 2004; with kind permission of Springer Science and Business Media).

that the mass creates a rigid, uniform film that does not slip and has the same acousto-elastic properties as quartz:

$$\Delta f = -Cf(\Delta m), \tag{12.1}$$

where Δf is the observed change in frequency (Hz) of the resonator under oscillation at its fundamental frequency due to mass loading, Cf = sensitivity factor of the resonator in Hz/ng/cm^2, and Δm = change in mass per unit area in g/cm^2.

Traditionally, the TSM has served as a mass-sensitive monitor for commercial applications such as thin-film deposition under vacuum, and electroless and electroplating processes. Sauerbrey's (1959) calculations were originally described for depositions under vacuum conditions but his theory has been extended to liquid application, as proof in concept development of sensors for biological analysis have increased dramatically in the past decade.

Acoustic wave biosensors in general have been the subject of intense research since the first analytical application reported by King (1964). As a solitary mass-sensitive transducer the device is non-specific. However, when the electrode is coated with a high affinity receptor or biorecognition component through a reliable deposition process, sample coupling between the receptor and its complementary analyte at the sensor surface can be attributed to a mass change (Rickert et al. 1999) that can be converted to a signal output, amplified, and processed to provide specific, sensitive qualitative or quantitative measurement of their interaction. Thus, a biosensor is the spatial unity of a physical transducer and a complementary biological recognition component such as an antibody, bacteriophage, DNA, or enzyme.

For more in-depth information regarding theory, the reader can consult numerous excellent references including Janshoff and Steinem (2001), and Ballantine et al. (1997).

3.1. Detection of Microorganisms

Improved characterization and modeling (Martin, Granstaff, and Frye 1991; Bandey et al. 1999) of TSM responses and functionality under liquid loading conditions have hastened development of rapid bacterial biosensors, because for the most part bacteria are naturally found under liquid conditions. There are numerous proposed applications, including use in the food industry (Leonard et al. 2003), water and environmental monitoring (Kurosawa et al. 2006), pharmaceutical sciences (Pavey 2002), bio-threat defense (Ivnitski et al. 1999; Petrenko and Vodyanoy 2003), and clinical diagnostics (Lazcka, Campo, and Munoz 2007).

The specificity of any TSM sensor is wholly dependent upon a complementary, immobilized bioreceptor. Bioreceptors for whole cell analysis generally correspond to some outside portion of the cell wall such as proteins, or possibly lipopolysaccharides or some other cell wall-associated structure (Sorokulova et al. 2005). Direct application for the detection of whole bacterial cells includes food pathogens such as *Salmonella* spp., *Escherichia coli*, and *Listeria monocytogenes;* as well as other human pathogens such as *Chlamydia trachomatis, Vibrio cholerae, Staphylococcus aureus, Pseudomonas aeruginosa, Mycobacterium tuberculosis, Fransicella tularensis, Legionella,* and *Bacillus anthracis* spores. A comprehensive list of selectively identified or quantitated bacterial organisms (e.g., biofilm formation in selective culturing media) using acoustic wave devices is given in Table 12.2. Also, acoustic wave biosensors have been used for the direct detection of other microorganisms including human, plant, and bacterial viruses such as coronavirus (Zuo et al. 2004), tobacco mosaic virus (Dickert et al. 2004), dengue virus (Su et al. 2003), hepatitis A and B (Konig and Gratzel 1995), rotavirus and adenovirus (Konig and Gratzel 1993), cymbidium mosaic potexvirus and odontoglossum ringspot tobamovirus (Eun et al. 2002), and M-13 phage (Uttenthaler et al. 2001); yeast (Muramatsu et al. 1986; Hayden and Dickert 2001; Hayden, Bindeus, and Dickert 2003); and even algae (Nakanishi et al. 1996).

Acoustic wave biosensors have also been used for indirect detection of microorganisms through the detection of corresponding: DNA from *E. coli* O157:H7 (Deisingh and Thompson 2001; Mo et al. 2002; Mao et al. 2006), hepatitis A virus (Zhou et al. 2002), and human papilloma virus (Wang et al. 2002); specific bacterial protein products for *E. coli* (Nanduri et al. 2007); antigenic proteins from dengue virus (Wu et al. 2005; Tai et al. 2006);

Table 12.2. Acoustic wave (AWD) biosensors developed for bacterial detection

Bacterium	AWD	Receptor	LLOD	Reference
Bacillus subtilis	PM	NS	–	(Ishimori, Karube, and Suzuki 1981)
Bacillus thringiensis spores	SHSAW	Ab	1764 spores	(Branch and Brozik 2004)
Chlamydia trachomatis	TSM	Ab	260 ng/mL	(Ben-Dov, Willner, and Zisman 1997)
Escherichia coli	TSM	NS	–	(Zhang et al. 2002)
Escherichia coli	TSM	NS	10 cells/ml	(He et al. 1994)
Escherichia coli	TSM	NS	–	(Otto, Elwing, and Hermansson 1999)
Escherichia coli	TSM	NS	–	(Zhao, Zhu and He 2005)
Escherichia coli	TSM	SIP	unknown	(Dickert et al. 2003)
Escherichia coli	SSBW	Ab	400 cells/ml	(Deobagkar et al. 2005)
Escherichia coli	TSM	NS	–	(He and Zhou 2007)
Escherichia coli	SHSAW	Ab	10^6 cells/ml	(Moll et al. 2007)
Escherichia coli	SHSAW	Ab	$\sim 10^9$ cells/ml	(Berkenpas, Millard and Pereira da Cunha 2006)
Escherichia coli	SHSAW	Ab	$10^5 - 10^6$ cells/ml	(Howe and Harding 2000)
Escherichia coli	TSM	Ab	10^3 cells/ml	(Su and Li 2004)
Escherichia coli	TSM	Ab	1.7×10^5 cells/ml	(Kim, Rand, and Letcher 2003)
Escherichia coli	FPW	Ab	3.0×10^5 cells/ml	(Pyun et al. 1998)
Fransicella tularensis (Ft)	TSM	Ft antigen	5×10^6 cells/ml	(Pohanka and Skládal 2005)
Klebsiella sp.	PM	NS	–	(Ishimori, Karube, and Suzuki 1981)
Legionella	SHSAW	Ab	10^6 cells/ml	(Howe and Harding 2000)
Listeria monocytogenes	TSM	Ab	1.0×10^7 cells/ml	(Vaughan, O'Sullivan, and Guilbault 2001)
Milk bacteria*	TSM	NS	–	(Chang et al. 2006)
Mixed bacteria**	TSM	NS	–	(He et al. 2006)
Mycobacterium tuberculosis	TSM	Ab	10^5 cells/ml	(He and Zhang 2002)
Mycobacterium tuberculosis	TSM	NS	2×10^3 cells/ml	(He et al. 2003)
Proteus sp.	TSM	NS	–	(Yao et al. 1998)
Proteus vulgaris	TSM	NS	120 cells/ml	(Tan et al. 1997)
Proteus vulgaris	TSM	NS	340 cells/ml	(Deng et al. 1997)
Proteus vulgaris	TSM	NS	–	(Bao et al. 1996b)
Pseudomonas aeruginosa	TSM	NS	3.3×10^5 cells cm^{-2}	(Niven et al. 1993)
Pseudomonas aeruginosa	TSM	NS	60-100 cells/ml	(Zhao, Zhu and He 2005)
Pseudomonas aeruginosa	TSM	Ab	1.3×10^7 cells/ml	(Kim, Park, and Kim 2004)
Pseudomonas aeruginosa	TSM	NS	–	(Reipa, Almeida, and Cole 2006)
Salmonella sp.	TSM	Ab	3.2×10^6 cells/ml	(Park, Kim and Kim 2000)
Salmonella serotypes A,B,D	TSM	Ab	10^5 cells/ml	(Wong et al. 2002)
Salmonella typhimurium	TSM	Ab	10^3 cells/ml	(Bailey et al. 2002)
Salmonella typhimurium	TSM	Ab	100 cells/ml	(Olsen et al. 2003)
Salmonella typhimurium	TSM	Phage	0 cells/ml	(Olsen et al. 2006)
Salmonella typhimurium	TSM	Ab	100 cells/ml	(Pathirana et al. 2000)
Salmonella typhimurium	TSM	Phage	100 cells/ml	(Olsen et al. 2007)
Salmonella typhimurium	TSM	Ab	1.5×10^9 cells/ml	(Babacan et al. 2000)
Salmonella typhimurium	TSM	Ab	10^7 cells/ml	(Babacan et al. 2000)
Salmonella typhimurium	TSM	Ab	5.3×10^5 cells/ml	(Ye, Letcher, and Rand 1997)
Salmonella typhimurium	TSM	Ab	10^5 cells/ml	(Prusak-Sochaczewski and Luong 1990)
Salmonella typhimurium	TSM	Ab	$10^5 - 10^6$ cells/ml	(Su and Li 2005)
Salmonella typhimurium	TSM	Ab	100 cells/ml	(Kim, Rand, and Letcher 2003)
Salmonella typhimurium	TSM	Ab	9.9×10^5 cells/ml	(Park and Kim 1998)
Salmonella paratyphi A	TSM	Ab	170 cells/ml	(Fung and Wong 2001)
Salmonella paratyphi A	TSM	Ab	10^5 cells/ml	(Si et al. 1997)
Salmonella enteriditis	TSM	Ab	1.0×10^5 cells/ml	(Si et al. 2001)
Salmonella enteriditis	TSM	Ab	1.0×10^5 cells/ml	(Ying-Sing, Shi-Hui, and De-Rong 2000)

Table 12.2.   (Continued)

Bacterium	AWD	Receptor	LLOD	Reference
Staphylococcus aureus	TSM	Ab	5×10^5 cells/ml	(Le et al. 1995)
Staphylococcus epidermidis	TSM	Fibronectin	100 cells/ml	(Pavey et al. 2001)
Staphylococcus epidermidis	TSM	NS	100 cells/ml	(Bao et al. 1996a)
Streptococcus mutans	TSM	NS	–	(Kreth et al. 2004)
Vibrio cholerae	TSM	Ab	4×10^4 cells/ml	(Carter et al. 1995)

Ab = corresponding antibody; NS = not selective (e.g. for biofilm monitoring) or some other means of selectivity (e.g. specific culture media) was used other than attached bioreceptor; PM = piezoelectric membranes; SIP = surface imprinted polymer layer; SSBW = surface skimming bulk wave; * = non-specific detection of bacterial growth in milk; ** = non-specific detection of bacterial growth in blood culture bottles.

antibodies from bacteria including *Helicobacter pylori* (Su and Li 2001), *Treponema palladium* (Aizawa et al. 2001), *Salmonella enteritidis* (Su et al. 2001), *Francisella tularensis* (Pohanka and Skládal 2005), and *Staphylococcus epidermidis* (Pavey et al. 1999), the helminth *Schistosoma japonicum* (Wu et al. 1999, 2006), and African swine fever virus (Uttenthaler, Kolinger, and Drost 1998); and bacterial toxins from *E. coli* including LT (Spangler et al. 2001), Stx (Uzawa et al. 2002), and an unidentified endotoxin (Qu et al. 1998), and *Staphylococcus* including SEB and C2 toxins (Harteveld, Nieuwenhuizen, and Wils 1997; Gao, Tao, and Li 1998; Lin and Tsai 2003). Additionally, new innovations such as gas chromatography coupled to SAW technology have been used for indirect detection of *Klebsiella pneumoniae, Pseudomonas aeruginosa, Escherichia coli,* and two *Candida albicans* yeast strains (Casalinuovo et al. 2005). Ion chromatography combined with TSM has been used to monitor *Lactobacillus* fermentation through lactic acid production (Zhang et al. 2001).

As shown in Table 12.2, the overwhelming majority of acoustic wave biosensors described in the literature for direct detection of whole bacterial cells is based on the TSM platform, with the most frequently targeted organism being *Salmonella,* specifically *S. typhimurium. Salmonella* is a leading etiology of foodborne illness and death in the U.S. (Mead et al. 1999).

Prominent acoustic wave sensors for *Salmonella* include those of Prusak-Sochaczewski and Luong (1990), who reported the first QCM assay for *Salmonella* with an assay time of 50–60 s, a lower detection limit of 10^5 cells/ml, and 0.5–5 hour incubation period, depending on the concentration of the microbial suspension; Park and Kim (1998), whose thiolated immunosensor possessed an assay time of 30–90 minutes, a lower detection limit of 9.9×10^5 cells/ml, and a detection range up to 1.8×10^8 cells/ml; Ye, Letcher, and Rand (1997), whose linear ($R = 0.942$) biosensor assay for *S. typhimurium* had a 25 min response time, a lower detection limit of 5.3×10^5 CFU/ml, and a range up to 1.2×10^9 CFU/ml; Pathirana et al. (2000), who developed an antibody-based TSM sensor to detect *Salmonella typhimurium* in poultry that possessed rapid analytical response times of 79 ± 20 seconds, linear ($R > 0.98$, $p < 0.01$) dose-response over 5 decades (10^2 to 10^7 cells/ml) of bacterial concentration, sensitivity of 18 ± 5 mV/decade of *S. typhimurium* concentration, and a detection range of 350 ± 150 to 10^{10} cells/ml; and the sensors of Babacan et al. (2000, 2002), Park, Kim and Kim (2000), Su and Li (2005), and Kim, Rand, and Letcher (2003).

3.2. Measurement in Liquid

TSM functionality in liquids is complex. Influences from numerous non-gravimetric contributions include liquid viscosity and density (Bandey et al. 1999); surface free energy (Thompson et al. 1991); roughness, surface charge density, and water content of biomolecules

(Janshoff and Steinem 2001); pressure and temperature (Niven et al. 1993); and the viscoelasticity and interfacial effects (Lucklum 2005) of thin films deposited in the form of bioreceptors. Therefore, the use of Sauerbrey's (1959) equation to strictly quantitate mass deposited to the solid-liquid interface under liquid conditions is controversial. Sauerbrey's equation was developed based on the oscillation of TSM in vacuum and only applies to thin, uniform, rigid masses attached tightly to the crystal. Thus, frequency response under liquid conditions cannot solely be attributed to mass deposition (Gizeli 2002; Lucklum 2005). For example, TSM sensors exposed to relatively large protein and polysaccharide molecules in solution have also been shown to give responses that did not correlate with mass changes at the solid-liquid interface (Ghafouri and Thompson 1999). The authors ascribed this phenomenon to viscoelastic and acoustic coupling at the interface. One could expect especially complicated interfacial properties when the TSM sensor is exposed to larger biological entities such as bacterial cells. Electromechanical forces created by live and moving organisms may contribute to the apparent mass of binding bacteria. Additionally, factors such as nutrition, growth, differentiation, chemical signaling, and mutagenic exposure may also factor in controlling the physiological state of binding bacteria. A bacterial cell (e.g., *E. coli*) can possess a mass of approximately 665 fg, making it one million times heavier than a typical (150 kD) antibody molecule (Neidhardt 1987) used as a bioreceptor. Bacteria carry out or are involved with various movements including flagellation, Brownian motion, chemotaxis, swimming behavior, adaptation, and other cell phenomena (Alberts et al. 1994). Bacterial binding on sensor surfaces may also depend on the presence of fimbriae (Otto, Elwing, and Hermansson 1999), flagella (Sorokulova et al. 2005), or other surface-associated adhesion factors, as well as the ability of single cells to associate and form colonies. Bacterial interaction with a biosensor may also be highly dependant upon environmental conditions (Olsen et al. 2003).

Notwithstanding, the ability to function in liquid environments conducive to bacterial growth and the fact that mass can be sensitively and specifically differentiated as a molecular recognition/binding event at the solid-liquid interface are two good reasons the TSM is being developed as a rapid detection tool. Normally, the TSM sensor is enclosed within a cell into which fluids are injected ("flow injection analysis") or flowed via a peristaltic pump. Numerous examples are available by reviewing the references in Table 12.2. While "closed systems" are prevalent and rather simple in operation, solutions can also be simply applied by pipette directly to the surface of the TSM, or what can be contrasted as an "open system," where fluids are directly applied to the sensor surface (Olsen et al. 2003, 2006). Systems have also been devised for air-borne sample-to-liquid transfer (Frisk et al. 2006) to facilitate acquisition of airborne threat agents such as *Bacillus anthracis* spores.

3.3. TSM Biosensor Characteristics

Bacterial binding as the signal output of the transducer has been measured and analyzed using many different formats to give a detailed analysis of surface/interface changes, including fundamental resonance and/or overtone frequency shift, frequency shift with dissipation, voltage, resistance and capacitance, and acoustic impedance. Absolute or differential (Δf) frequency changes alone due to binding are given by most authors and appear to be acceptable and sensitive as a measurement of sensor functionality. For example, Fig. 12.3a shows typical frequency response curves of a prepared (phage) biosensor tested with logarithmic concentrations of *S. typhimurium* ranging from 0 (PBS)–10^7 cells/ml. For each concentration, the sensor quickly comes to steady-state equilibrium within several hundred seconds following specific phage-bacteria binding at the solid-liquid interface. Plotting the mean values of steady-state frequency readings as a function of bacterial concentration (Fig. 12.3b) gives a high dose-response relationship ($R = -0.98$, p < 0.001), small signal to noise ratio (-10.9 Hz) measured as the slope

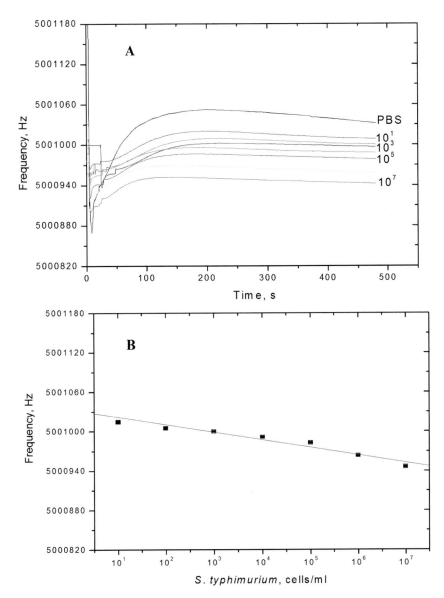

Figure 12.3. (A) Frequency responses of phage biosensor to increasing concentrations of *S. typhimurium* as a function of time. (B) Dose-response relation of mean values (n = 2800 ± 2) of steady-state output sensor frequencies as a function of *S. typhimurium* concentration. Bars are SD = 2.9 − 10.0 Hz. Curve is linear least squares fit to experimental data ($R = -0.98$, slope $= -10.9$ Hz, p < 0.001) (reprinted from Olsen et al. (2006), with permission of Elsevier).

of the linear portion of the dose-response, linearity over six decades of bacterial concentration, and a lower limit of detection at 100 cells/ml, well below the infectious dosage of *Salmonella*.

    Other authors (Otto, Elwing, and Hermansson 1999) attempt to determine dissipation in the system (ΔD) as a quantitative measure of system damping, usually due to lossy or viscoelastic films or near surface interaction of the bacteria. Resistance (R), capacitance (C), and/or impendence (L) measurements are sometimes determined (He et al. 2003; Kim, Rand, and Letcher 2003; Su and Li 2005) through a high frequency impedance analyzer based on the TSM as an RLC series equivalent circuit.

There are many other characteristics of TSM platforms that require consideration when developing and testing biosensors, including:

- **Specificity:** the strength of the interaction between a molecular probe (e.g., antibody) and an antigen (target analyte) as estimated by the dissociation constant K_d. The smaller the K_d the higher the specificity of binding. The free energy of dissociation (ΔG_d) of a ligand-receptor complex is related to its equilibrium dissociation constant K_d by the equation:

$$\Delta G_d = -kT \ln(K_d/K_0), \tag{12.2}$$

  where k is a Boltzmann constant and T is a temperature in °K. The equation refers to a standard reference state where all chemical species are 1 M (i.e., $K_0 \sim 0.6$ molecules/nm^3) and attributes a free energy of zero to a complex with a dissociation constant of 1 M (Chothia and Janin 1975).

- **Binding selectivity:** is defined by a selectivity coefficient (K). Binding selectivity can be estimated from dose responses of a biosensor to different analytes (e.g., bacteria). The signal response V as a function of the primary analyte (e.g., bacteria) concentration (C) can be represented by the following empirical equation:

$$V = A + S \log C, \tag{12.3}$$

  where C is the primary analyte concentration, A is the constant, and S is the slope of the dose response dependence, defined as the sensitivity of the sensor (Pathirana et al. 1996).

  The selectivity coefficient for any other analyte to the primary analyte (e.g., bacteria) (K) can be determined from the signal responses at different concentrations using a method similar to the matched potential method (Pathirana et al. 1996; Umezawa 1996). The selectivity coefficient is defined as the concentration ratio (R) of primary to interfering species $[\Delta(C_p)/\Delta(C_i)]$, which gives the same response change at the same condition. Using the definition of the selectivity coefficient and Eq. 12.2, the following is derived:

$$R = C_p/C_i, \tag{12.4}$$

$$K = R = S_i/S_p, \text{ when } \Delta C_p \text{ approaches zero,} \tag{12.5}$$

  where S_p and S_i are slopes of signal responses to primary and interfering species (other bacteria), respectively.

- **Sensitivity:** The change of the biosensor's output signal when the analyte content (total quantity or concentration) changes by one unit. For non-linear sensors, the sensitivity depends on the analyte level and is given by the slope of the sensor's output curve versus the analyte content.
- **Detection threshold.** The ability of the biosensor to discriminate an analyte (e.g., bacteria) from background at the lowest quantity of analyte in the testing solution.
- **Dynamic or Working Range:** The range of the analyte content over which the sensor can perform qualitative or quantitative detection.
- **Linear range:** That part of the dynamic range where the sensor's output is a linear function of the analyte content.

- **Saturation:** The level at which the sensor no longer functions correctly. For biosensors, this is usually the point where the bioreceptor has been saturated with analyte and reaches a peak signal.
- **Response Time:** The amount of time required to detect the analyte as given by the signal output.
- **Accuracy:** Closeness of the sensor measurement result to the actual quantity of cells in solution. Actual quantity of cells (usually stated in reports as cells/ml) is found from traditional plate culture of the organism. Optical counting methods are also possible (Olsen 2000).
- **Stability:** The ability of the sensor signal to give a constant, steady output signal when measuring a steady input, such as a load of cells.

3.4. Commercial TSM Microbalances

Traditionally, the TSM has served as a mass-sensitive monitor in commercial applications such as thickness monitoring and deposition rate control for thin films under vacuum, and for electroless and electroplating processes (Krause 1993). The functionality of the TSM under liquid conditions has increasingly driven adaptation to the development of extremely sensitive biosensors in the past decade. Total QCM systems are relatively inexpensive and simple in operation, requiring for the most part only the resonator crystal, external oscillatory circuit, and frequency counter. Many of the systems described in the literature for sensor developments are pieced together or custom built and may additionally include impedance analyzers, thermostatic jacketing for temperature control, and pump or flow injection equipment. With the advent of the Internet, numerous commercial QCM products including crystal resonators and holders, frequency monitors, flow cells, and even entire systems are now easily available throughout the world, making entry into this field reasonable in terms of cost and availability. A recent review of the Internet yielded numerous larger manufacturers and suppliers of complete QCM systems (Table 12.3).

One such commercially available microbalance produced by Maxtek Inc. can be used for both biosensor preparation and testing and consists of a 50 cm sensor probe connected by a tri-axial cable to a precalibrated plating monitor (Fig. 12.4). This system is often used in electroplating processes within vats, necessitating the long probe and open face exposed to solution. The plating monitor has a frequency resolution of 0.03 Hz and mass resolution of 0.375 ng/cm^2 at 5 MHz. TSM transducers are precleaned AT-cut plano-plano quartz liquid-plating resonators possessing a 5 MHz nominal resonant frequency. Resonators (2.54 cm diameter, 333 μm thickness) have gold plated electrodes evaporated onto titanium adhesion layers on both the top and bottom (Fig. 12.5). The electrodes are polished to an average surface roughness of approximately 50 Å. This minimizes liquid entrapment within the pores at the crystal surface, reducing the creation of apparent mass loadings under liquid measurement conditions. Also, resonators are pretested to assure conformance to critical accuracy specifications required for reproducibility, and rate and thickness measurements (PM-740 series operation and service manual 1996). Both the bioreceptor, during sensor preparation (Fig. 12.6a), and the analyte, during sensor testing, can be directly applied to the surface of the sensor by pipette (Fig. 12.7). Absolute frequency readings from the sensor are transferred to a PC directly from the plating monitor or via a multimeter, in which case voltage readings can be captured (Pathirana et al. 2000). The sensor probe, attached to the stand, and all necessary components of the experiment can be contained at room temperature within an Atmosbag™ gloved isolation chamber (Sigma-Aldrich, Milwaukee, WI) inflated with inert nitrogen gas during bioreceptor deposition studies to prevent possible contamination of the resonator by particulate matter.

Table 12.3. Selected commercially available QCM systems

Company	Internet URL (http://)	QCM products
Maxtek, Inc.	www.maxtekinc.com	RQCM, crystals and holders oscillators, flow/liquid cells, thin-film monitors/controllers
Q-Sense	www.q-sense.com	E4 QCM-D, D300; EQCM, crystals
Universal Sensors, Inc.	intel.ucc.ie/sensors/universal	PZ-105, crystals, flow cells
Seiko EG&G	speed.sii.co.jp/pub/segg/hp	QCM934, QCA922
Princeton Applied Research	www.princetonappliedresearch.com	QCM922, EQCM
ICM, Inc.	www.qcmsystems.com/index.html	crystals, oscillators, flow cells
QCM Research	www.qcmresearch.com	CQCM, TQCM, Mark 21 QCM Thin-film controllers
Tectra	www.tectra.de/qmb.htm	MTM-10 thin film monitors/controllers
KSV Instruments, LTD	www.ksvltd.com	QCM-Z500, crystals, EQCM flow cells, pumps, temp control unit, spin-coater/holder
SRS	www.thinksrs.com	QCM-100, QCM-200, EQCM, crystals and holders, oscillators thin-film controllers, flow cells
Masscal	www.masscal.com	G1 QCM
Faraday Labs	www.faradaylabs.com	QCM
Initium, Inc.	www.initium2000.com	Affinix Q
Sigma Instruments	www.sig-inst.com	Q-pod, SQM-160, crystals, thickness/rate monitors
Tangidyne	www.tangidyne.com	Optical crystals and holders
Technochip	www.technobiochip.com	μLibra QCM, EQCM, "Electronic Nose"

In addition to some of the previously mentioned TSM characteristics, some additional factors to consider before purchasing commercially available equipment include cost, resolution, reproducibility, reliability, ruggedness, analytical range, speed, noise, cost, power requirements, space limitations, availability, technical servicing/maintenance, life expectancy, data capture capabilities, ease of use, and other analytical capabilities and adaptabilities such as use under differing temperatures, pressures, or other environmental conditions, and adaptability to peripheral devices such as voltmeters, PC, peristaltic pumps, and thermostatic jacketing.

Like all other sensory devices, the TSM as a sensor platform has its advantages and drawbacks. In addition to addressing necessary characteristics of TSM biosensors such as speed, accuracy, precision, sensitivity, and specificity, several other factors should be considered, including incubation time of analyte, numerous steps including application of analyte and washing and drying, regeneration of the sensor surface if reusability is a factor, and total cost of assay to include resonators, reagents, bioreceptors, etc.

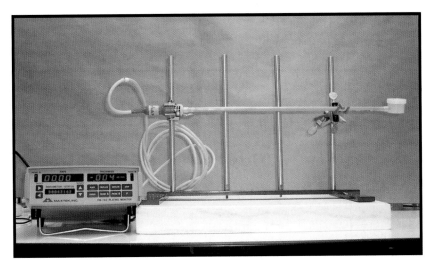

Figure 12.4. QCM platform for deposition and biosensor measurements. A plating monitor was connected to a sensor probe that was horizontally clamped to a lattice stand then tilted 10° transverse to the stand. The sensor probe and stand were positioned atop a marble slab to reduce extraneous environmental vibrations.

3.5. Immobilization of Probes onto Sensor Surface

A major drawback to the TSM as a sensor is its non-specificity. Anything that can and will attach to it under liquid loading conditions can be recognized as a molecular binding event. Therefore, application of bioreceptors is necessary in order to affect specificity towards the analyte of choice (bacteria, bacteria components, toxins, or complementary DNA, etc.). The sensing properties of a sensor depend on the physical-chemical environment of antibody and antigen-antibody complex, which are in turn determined by antibody immobilization techniques (Ahluwalia et al. 1992; Storri, Santoni, and Mascini 1998). While the TSM can be very quick in its measurement, building the sensor with bioreceptors can be a tedious, multi-step process that can take numerous hours or even days. Additionally, and possibly the greatest consideration,

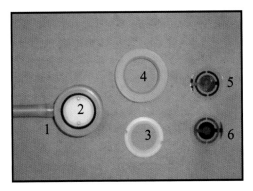

Figure 12.5. Maxtek sensor probe and associated components: (1) housing for external oscillatory circuit electrical contacts; (2) crystal holder cavity with o-ring (black) installed. The gold index pins that contact the reverse electrode of the resonator are clearly visible; (3) teflon resonator retainer ring; (4) threaded retainer ring cover; (5) sensing electrode of polished, 5 MHz AT-cut thickness shear mode quartz resonator; (6) contact electrode of quartz resonator. The "active area" of the resonator is that central portion of the sensing electrode that overlaps the contact electrode ($\approx 34.19\,\text{mm}^2$).

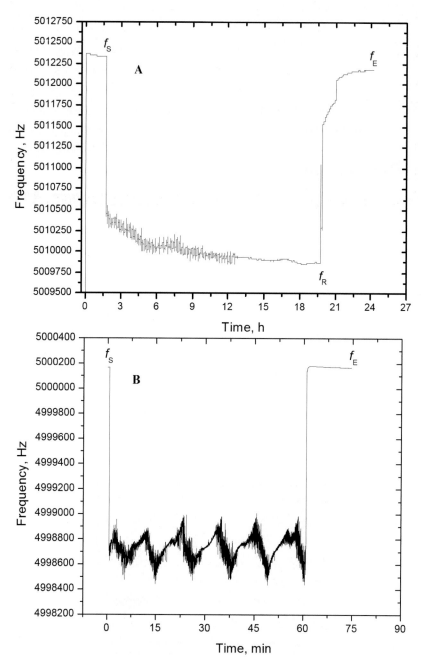

Figure 12.6. (A) Representative line graph depicting frequency change as a function of phage binding to the resonator over time. Eighteen-hour incubation period is shown. f_S: Application of phage solution to clean, dry resonator at steady state: 5,012,338 Hz. f_R: Removal of phage solution, washing, and drying of resonator. f_E: Dried resonator at steady state: 5,012,177 Hz. $\Delta f = (f_S) - (f_E) = -161$ Hz. (B) Representative line graph of a clean resonator with degassed water only (control) depicting frequency change as a function of time. One-hour incubation period is shown. f_S: Application of water to clean, dry resonator at steady state: 5,000,167 Hz. f_E: Dried resonator at steady state: 5,000,167 Hz. $\Delta f(f_S) - (f_E) = 0$ Hz (reprinted from Olsen et al. (2006), with permission of Elsevier).

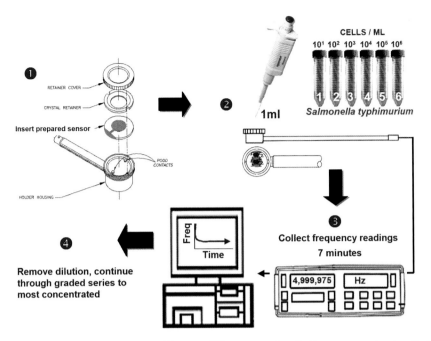

Figure 12.7. Testing scheme for biosensors: (1) Prepared biosensor was installed into sensor probe; then (2, 4) tested with a graded series of *S. typhimurium* test solutions; and (3) frequency (or voltage) output of sensor was recorded for data analysis.

is the reproducibility of the bioreceptor immobilization process. An in-depth analysis of the techniques of probe immobilization onto sensor surfaces is presented in the following sections.

3.5.1. Physical Adsorption

The most common techniques involve direct bonding of an antibody receptor to a reactive group coupled to the surface. The coupling agent and reactive group are generally selected to match the chemistry of the specific antibody. However, the adsorption process is difficult to control and the amount of protein adsorbed to most solid surfaces is usually below that which would correspond to a close-packed monolayer. Further, during the adsorption, the exposure of internal hydrophilic groups of proteins to hydrophobic surfaces causes a decrease in the activity and specificity of the protein/target interactions. In spite of these shortcomings of this method, direct physical adsorption is the simplest way of antibody immobilization on the sensor surface. This method has been successfully employed for immobilization of a wide range of biological elements directly onto piezoelectric electrodes, including anti-human serum albumin (Muratsugu et al. 1993), IgG (Minunni, Skladal, and Mascini 1994), goat anti-ricin antibody (Carter et al. 1995a), anti-*Vibrio Cholera* (Carter et al. 1995a), African swine fever virus protein (Uttenthaler, Kolinger, and Drost 1998), recombinant protein fragments of HIV specific antibodies (Aberl and Wolf 1993), filamentous phage (Sykora 2003; Olsen et al. 2006; Nanduri et al. 2007; Olsen et al. 2007), lytic phage (Balasubramanian et al. 2007), and designer peptides (Selz et al. 2006). Protein molecules adsorb strongly and irreversibly on gold surfaces due to hydrophobic actions (Horisberger 1984, 1992).

Quantitative "dip and dry" deposition experiments can be used to monitor physical adsorption of bioreceptors in the preparation of biosensors (Olsen et al. 2006). Dip and dry, as described by Prusak-Sochaczewski and Luong (1990), is the change in the resonant frequency,

Table 12.4. Quantity of filamentous phage physically adsorbed to resonators as a function of time

Incubation (min)	$-\Delta f$ (Hz)	Δm (ng)[d]	Phage adsorbed (virions)[e]
20	45[a]	795	1.80×10^{10}
40	60[b]	1065	2.41×10^{10}
60	92[c]	1625	3.68×10^{10}
1080	136	2402	5.45×10^{10}
1440	163	2880	6.50×10^{10}

[a] Mean average of 5 experiments, SD = 31.1 Hz.
[b] Mean average of 3 experiments, SD = 46.5 Hz.
[c] Mean average of 4 experiments, SD = 59.2 Hz.
[d] Adsorbed phage mass as determined by Sauerbrey equation, $\Delta f = (0.0566)(\Delta m)$.
[e] Quantity of phage deposited to the active area ($34.19 \, \text{mm}^2$) of the upper sensing electrode as calculated from $\Delta m/m_{\nu}$, where the mass of a single virion (m_{ν}) is 2.66×10^7 dal $/ \, 6.023 \times 10^{23}$ dal $= 44.1 \times 10^{-9}$ ng.

Δf, of a dry TSM resonator prior to and after mass deposition. Using Sauerbrey's (1959) equation, the physical adsorption of phage as a function of time can be determined. For example, Fig. 12.6a shows steady-state oscillation of a dry, clean resonator prior to the application of 1 ml of diluted stock phage E2 in suspension (6.7×10^{10} virions/ml) (f_S), followed by an 18 h incubation period at room temperature, removal (f_R) of the phage suspension and washing with degassed water, and finally drying, with a subsequent return to steady-state resonance (f_E). The resulting frequency change, Δf, measured as a decrease, $f_S - f_E$, was -161 Hz, indicating that phage adsorbed to the resonator. This can be contrasted to a control (Fig. 12.6b) consisting of a clean TSM resonator tested with degassed water only, which indicated no frequency change ($f_S - f_E = 0$ Hz). Resonance frequency changes due to phage adsorption were determined for periods up to 24 h (1440 min) (Table 12.4). When the quantity of adsorbed phage is graphed as a function of time the majority of phage appeared to adsorb within the first few hours after deposition was started (Olsen et al. 2006). The quantity of phage in virions can be calculated from the total adsorbed biomass, Δm, by estimating the mass of a single recombinant fd-tet phage at 2.66×10^7 daltons, based on 4000 pVIII outer coat proteins, each containing 55 amino acids with a total molecular weight of 2.35×10^7 (Kouzmitcheva 2005), and DNA with a molecular weight of 3.04×10^6 (Petrenko 2007). As shown in Table 12.4, the total number of phage particles deposited to the TSM resonator ranged from $1.8 \times 10^{10} - 6.5 \times 10^{10}$ virions as a function of exposure time (20 min –24 h, respectively) to phage in solution. Phage deposition to the TSM was confirmed in real time by fluorescence microscopy for a period of two consecutive hours and characterized by strong, non-reversible binding under aqueous conditions (Olsen et al. 2006).

3.5.2. Other Coupling Methods

To overcome disadvantages of a direct physical adsorption method, a range of immobilization methods have been suggested, including lipid bilayer entrapment (Ramsden 1997a, 1997b, 1998, 1999, 2001), thiol/disulfide exchange, aldehyde and biotin-avidin coupling (Mittler-Neher et al. 1995), photo-immobilization to photolinker-polymer-precoated surfaces (Gao et al. 1994), molecular imprinted polymer layers (Dickert et al. 2003; Dickert, Lieberzeit, and Hayden 2003; Dickert et al. 2004), and site-specific immobilization of streptavidin (Tiefenauer et al. 1997).

3.5.3. Combined Langmuir-Blodgett/Molecular Assembling Method

A more advanced approach for the immobilization of antibodies for the immunosensor coatings is through the combined Langmuir-Blodgett (LB)/molecular assembling method

(Samoylov et al. 2002a, 2002b). The method includes LB deposition (Petty 1991; Pathirana et al. 1992; Barraud et al. 1993; Pathirana 1993; Vodyanoy 1994; Bykov 1996; Pathirana et al. 1996; Sukhorukov et al. 1996; Pathirana, Neely, and Vodyanoy 1998; Olsen 2000; Pathirana et al. 2000; Olsen et al. 2003; Petrenko, Vodyanoy, and Sykora 2007; Olsen 2005; Olsen et al. 2007) of a biotinylated monolayer onto a sensor surface and non-LB, molecular self-assembling of a probe layer using biotin/streptavidin coupling (Furch et al. 1996; Volker and Siegmund 1997).

The combined LB/molecular assembling method has been demonstrated with biosensors based on phage display-derived peptides as biorecognition molecules (Samoylov et al. 2002a, 2002b). Schematic design of the peptide sensor is shown in Fig. 12.8a. Monolayers

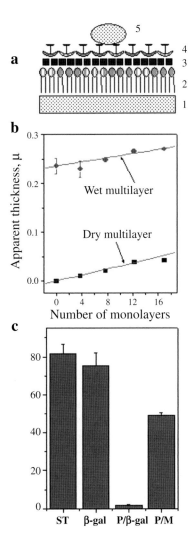

Figure 12.8. Design and functional validation of a peptide biosensor. (a) The schematic design of the peptide biosensor. The biosensor consists of four components: 1 - quartz crystal; 2 - biotinylated phospholipid; 3 - streptavidin; 4 - biotinylated peptide; 5 is a tissue vesicle. (b) Calibration of acoustic wave device with stearic acid monolayers. (c) Validation of peptide sensor preparation. **ST** – sensor was covered with biotinylated phospholipid and exposed to streptavidin. The bar represents the change of mass due to binding of streptavidin to biotinylated phospholipid. β-**gal** – sensor covered with streptavidin was exposed to β-gal solution. The bar represents the change of mass due to binding of β-gal to streptavidin. **P/β-gal** – completed sensor, covered with peptide, was exposed to β-gal. The bar represents the change of mass due to β-gal binding to the sensor. **P/M-** completed sensor, covered with peptide, was exposed to murine muscle homogenate. The bar represents the change of mass due to binding of the component of the tissue homogenate (3.8 mg/ml protein) to the peptide. (Samoylov et al. 2002b. © John Wiley & Sons Limited. Reproduced with permission).

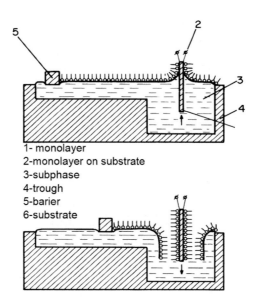

Figure 12.9. Langmuir-Blodgett (LB) monolayers. First, a monolayer is formed on a subphase surface and compressed to a desired surface pressure. A solid substrate is then moved through the monolayer vertically, so that it is dipped into and withdrawn completely out of subphase. The first monolayer is transferred to the substrate like a carpet with the tail groups toward the substrate surface during the downward movement of the substrate through the condensed monolayer. A monolayer is transferred to the substrate both when it is raised through and when it is lowered into the subphase through a compressed monolayer (reprinted from Yilma et al. (2007), with permission of Elsevier).

containing phospholipid, N-(biotinoyl)-1,2-dihexadecanoyl-sn-glycero-3-phosphoethanolamine (2), were transferred onto the gold surface of an acoustic wave sensor (1) using the Langmuir–Blodgett technique. Multilayers were obtained by successive dipping of the sensors through the monomolecular film deposited at a water-air interface (Fig. 12.9). Biotinylated peptide (4) was coupled with the phospholipid via streptavidin intermediates (3) by molecular self-assembly. Measurements of binding of target vesicles were carried out using a PM-700 Maxtek plating monitor with a frequency resolution of 0.5 Hz at 5 MHz. The device was calibrated with stearic acid monolayers. The deposition of increasing numbers of stearic acid monolayers on the surface of acoustic wave crystal resulted in linear increase of the mass (Fig 12.8b). The deposition of a single monolayer of stearic acid on the crystal resulted in additional mass of $2.6 \times 10^{-7} \mathrm{g\,cm^{-2}}$. This agrees well with the theoretical estimate based on the molecular area of stearic monolayer in the condensed state (Davies and Rideal 1963). Binding of streptavidin to biotinylated phospholipid is an important step in immobilization because concentration and orientation of streptavidin molecules determine the properties of a bound molecular probe. The change of mass due to streptavidin binding normally reached $80\,\mathrm{ng\,cm^{-2}}$, or $8 \times 10^{11}\,\mathrm{molecules\,cm^{-2}}$ (Fig. 12.8c—ST). When the samples were exposed to 500 nM biotinylated β-gal for 2 h the apparent mass change was at the level of $80\,\mathrm{ng\,cm^{-2}}$, or $3.4 \times 10^{11}\,\mathrm{molecules\,cm^{-2}}$ (Fig. 12.8c—β-gal). The completed biosensor, covered with the biotinylated peptide, no longer bound biotinylated β-gal (Fig. 12.8c—P/ β-gal), but strongly bound target vesicles (Fig. 12.8c—P/M).

The combined LB/molecular assembling method was also exercised in the immobilization of filamentous phage onto the surface of thickness shear mode (TSM) quartz sensors (Petrenko et al. 2005; Olsen et al. 2006, 2007). Monolayers containing biotinylated phospholipid were transferred onto the gold surface of the sensor using the Langmuir-Blodgett technique (Fig. 12.9). Biotinylated phage was coupled with the phospholipid via streptavidin intermediates by molecular self-assembling. The dissociation constant of 0.6 nM found by this method compares well with one found for antibodies isolated from a phage display library (Vaughan et al. 1996).

3.5.4. Solvent-Free Purified Monolayers

An important aspect of sensor preparation is defining the conditions under which monolayers prepared with bioreceptors can be successfully formed on a liquid/gas interface and then optimized in terms of sensitivity, reliability, and useful lifetime. Although some effects of pH, ionic strength, and oriented coupling on the immunosensor performance have been examined (Barraud et al. 1993; Ahmad and Ahmad 1996), detailed information about the influence of physical, chemical, and molecular environments on the antigen-antibody system remains largely unknown.

Traditional methods for forming LB films (Gaines 1966) require dissolution of monolayer forming compounds into a volatile organic solvent. As a separate phase, the organic solvent functions to prevent dissolution of the monolayer components in the aqueous phase. When the mixture is spread onto an aqueous subphase solution at the air-liquid interface, the solvent evaporates, leaving a monolayer at the interface. Unfortunately, the organic solvent often damages the monolayer components and leaves an undesirable residue (Sykora, Neely, and Vodyanoy 2004). LB films formed from such monolayers may also possess unacceptable levels of nonspecific binding (Ahluwalia et al. 1992), which is non-saturable and hampers quantitative measurement of specific binding. These problems can be solved using methods of monolayer formation that don't require use of an organic solvent (Trurnit 1960; Sobotka and Trurnit 1961; Pattus, Desnuelle, and Verger 1978; Pattus and Rothen 1981; Pattus et al. 1981) and have been demonstrated by immobilizing polyvalent somatic O antibodies specific for most *Salmonella* serovars onto gold electrodes of TSM resonators using the LB method (Pathirana et al. 2000).

Many features of antibody immobilization originate from the very nature of the antibody itself. Typical antibodies are Y-shaped molecules (2 Fab plus Fc immunoglobulin structure) with two antigen binding sites located on the variable region of the Fab fragments. All classes of antibody produced by B lymphocytes can be made in a membrane-bound form and in a soluble secreted form (Alberts et al. 1994). The two forms differ only in their carboxyl terminals; the membrane-bound form has a hydrophobic tail (Fc) that anchors it in the lipid bilayer of the B cell membrane, whereas the secreted form has a hydrophilic tail, which allows it to escape from the cell. Of these, only the form with a hydrophobic tail is capable of being held by the monolayers. Thus, it is uniquely qualified for use in the Langmuir-Blodgett technique. This form also renders it suitable for proper alignment and orientation in sensor membranes.

Antibodies derived from immunized animals in the form of antisera or purified protein preparations can be present in both membrane-bound and soluble form and may contain impurities. Organic solvents used as a spreading carrier in LB monolayer preparation may drag these impurities and both forms of antibodies into the monolayer. Furthermore, these methods may produce monolayers with high densities of antibodies but also with residuals of organic solvent, impurities, and entrapped hydrophilic antibodies that destabilize the monolayer and modulate antigen-antibody interactions. A monolayer with no solvent can be formed on the air-liquid interface by allowing the spreading solution to run down an inclined wetted planar surface that is partially submersed into subphase (Fig. 12.10) (Pathirana et al. 2000). Membrane vesicles (natural components of serum, or the artificial lipid vesicles) are positioned on a wet slide at the edge of a positive meniscus of liquid, at the liquid-air interface. The hydrophobic antibodies are bound to the vesicular membrane; hydrophilic antibodies and some impurities are suspended inside the vesicle. When surface forces rupture the vesicle, it splits into a monolayer and purification occurs. Membrane-bound antibodies are left bound to the newly created monolayer, but soluble antibodies and impurities dissipate into the subphase beneath the monolayer. Only membrane-bound antibodies surrounded by compatible lipids are left when the monolayer is compressed and transferred onto a sensor surface. Alternatively, probes can be conjugated with vesicles by covalent binding (Betageri et al. 1993). Lipid vesicles containing whole antibodies or Fab fragments can also be constructed. Large, unilamellar liposomes

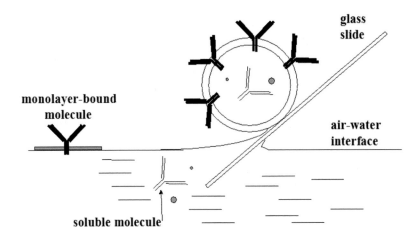

Figure 12.10. Monolayer formation from lipid vesicles. Surface forces rupture the vesicle, splitting it into a monolayer. The monolayer, with the membrane-bound molecules, is then compressed and transferred onto the sensor surface (reprinted from Pathirana et al. (2000), with permission of Elsevier).

are prepared from synthetic L-α(1,2-dipalmitoyl-sn-glycero-3-phosphocholine) (DPPC) and maleimido phenyl butyrate phosphatidylethanolamine (MPB-PE). Monoclonal antibodies (or Fabs) with specificity for *S. typhimurium* or *E. coli* O157:H7 are modified by the heterobifunctional reagent N-Succinimydyl-3(2-pyridyldithio)propionate (SPDP) in the presence of dicetyl phosphate and dithiothreitol (DTT) and conjugated to liposomes. The liposomes, containing specific antibodies, are then converted into monolayers and deposited on the sensor surface by the LB method (unpublished results).

3.5.5. Immobilization of Monolayers of Phage Coat Proteins

3.5.5.1. Phages As a Recognition Probe

A large number of bio-assays and biosensors depend on highly specialized, sensitive, and selective antibodies as recognition reagents (Goodchild et al. 2006). While antibodies frequently have the desired sensitivity and selectivity, their use is limited by many factors. For example, the binding properties of antibodies may be lost due to unfavorable environmental conditions (Olsen et al. 2003). This factor can be especially important in dealing with environmental applications, where organic solvents must be used for extraction of compounds (Ahmad and Ahmad 1996). Also, production of polyclonal antibodies requires a process that is very time- and labor-intensive, and can produce a variable product. Production of monoclonal antibodies is often even more difficult and expensive. These limitations can be addressed in part by using bacteriophage or their coat proteins as recognition elements for biosensors (Goldman et al. 2000; Petrenko et al. 2005; Nanduri et al. 2007). Both lytic and filamentous phages present reach libraries to identify proteins interacting with molecular targets. In phage display, the phage filament serves as the framework for random peptides that are fused to the N-terminus of every copy of the major phage coat protein. These random peptides form the "active site" of the landscape phage and comprise up to 25% by weight of the particle and up to 50% of its surface area (an extraordinarily high fraction compared with natural proteins, including antibodies) (Nanduri et al. 2007). A large mixture of such phages, displaying up to a billion different guest peptides, is called a "landscape library." From this library, phages can be affinity selected for specificity to a certain antigen, thus functionally mimicking antibodies. These phages can be efficiently and conveniently produced and are secreted from the cell nearly free of intracellular components in a yield of about 20 mg/ml (Nanduri et al. 2007). The purification procedure is simple and does not differ dramatically from one clone to another. The surface density of the phage binding peptides is 300–400 m^2/g, comparable to the best known absorbents and catalysts (Nanduri et al. 2007), and with thousands of potential binding sites per particle, creates a multivalency. Other advantages of phages over antibodies include the extraordinary robustness of the phage particle. It is resistant to heat (up to 70 °C), many organic solvents (such as acetonitrile), urea (up to 6M), acid, alkali, and many other stresses (Nanduri et al. 2007). Purified phages can be stored indefinitely at moderate temperatures without losing infectivity (Nanduri et al. 2007). Thus, phages may be viable as substitute antibodies in many applications such as biosensors, affinity sorbents, hemostatics, etc. Numerous examples of uses of both lytic and filamentous phages as probes for biological detection in biosensors have been reported in the literature (Chin et al. 1996; Ramirez et al. 1999; Goldman et al. 2000, 2002; Auner et al. 2003; Olsen et al. 2003; Sayler, Ripp, and Applegate 2003; Ozen et al. 2004; Petrenko, Vodyanoy, and Sykora 2007; Tabacco, Qian, and Russo 2004; Chen et al. 2005; Nanduri 2005; Petrenko et al. 2005; Sorokulova et al. 2005; Wu et al. 2005; Lakshmanan et al. 2006; Olsen et al. 2006; Balasubramanian et al. 2007; Nanduri et al. 2007).

3.5.5.2. Phage Coat Technology

A critical step in the use of whole phages or phage proteins as a bioreceptor is their immobilization to the platform area on the sensor where the analytes (bacteria, toxins, etc.) will bind. Immobilization of whole phage particles to a sensor surface presents certain difficulties with phage positioning. While multivalent, phage particles are typically assembled in bundles that may present obscurity of binding sites (Fig. 12.11). Immobilization of proteins extracted by organic solvents may present difficulties in depositing a correctly oriented layer of proteins on the sensor surface. A better way of phage protein immobilization is to use the intact protein coat of the phage particle. For example, when T2 lytic bacteriophage was subjected to osmotic shock, it lost most of its DNA, but the protein coat or "ghost" of the phage was left intact (Herriott and Barlow 1957). The coat retained the phage shape and some of the biological functions of the phage. Kleinschmidt and coworkers (Kleinschmidt et al. 1962) were able to

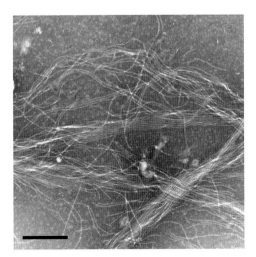

Figure 12.11. Transmission electron micrograph of bacteriophage 1G40 on a formvar, carbon coated grid of 300 mesh size using a wetting agent (0.1% BSA). The phage particles have aggregated as bundles on the grid. Bar = 200 nm (courtesy of Dr. V. Nandury).

convert T2 coat particles into protein monolayers by allowing the water suspension of coat particles to run down a wet glass rod glass surface that was partially submersed into subphase (similar to that shown in Fig. 12.10) (Trurnit 1960; Sobotka and Trurnit 1961; Pattus, Desnuelle, and Verger 1978; Pattus and Rothen 1981; Pathirana et al. 2000).

Monolayers made of the phage coats transferred to solid substrates were first described by Kleinschmidt et al. (1962). A similar approach was applied to obtain monolayers of filamentous phages. Griffith and coworkers (Griffith, Manning, and Dunn 1981; Petrenko, Vodyanoy, and Sykora 2007; Olsen 2005) demonstrated that filamentous bacteriophages transformed into hollow spherical particles upon exposure to a chloroform-water interface. These particles could then be converted into monolayers and deposited onto solid substrate by the LB method (Sykora 2003; Olsen 2005). Thus, the technology of phage coat immobilization consists of three major steps: phages are first converted into spheroids, monolayers are formed from the spheroids, and finally the monolayers are deposited onto the sensor surface by the LB method.

Phage coat monolayers made of coats of lytic and filamentous phages have been immobilized onto biosensor surfaces. When a suspension of filamentous phage protein streptavidin binder, 7b1, (Petrenko and Smith 2000) was vortexed with an equal volume of chloroform and the aqueous phase was examined by electron microscopy, spherical particles termed "spheroids" were observed along with other semicircular particles that may be intermediates in the filament to spheroid conversion (Griffith, Manning, and Dunn 1981).

Chloroform transforms the infectious phage filaments into non-infective hollow spheres. This drastically alters the surface architecture of the phage. As well, the α-helix content of pVIII decreases from 90% to 50–60% (Griffith, Manning, and Dunn 1981; Roberts and Dunker 1993). Spheroids are formed when the coat proteins contract into vesicle-like structures and two-thirds of the phage DNA is extruded (Griffith, Manning, and Dunn 1981). An electron micrograph of this is shown in Fig. 12.12 (Petrenko, Vodyanoy, and Sykora 2007). Similar hollow spheroids can be obtained by the same method from phage f8–1 that bind *Salmonella typhimurium* (Olsen 2005). Olsen formed LB monolayers from the spheroid suspension using a wetted glass rod that was partially submersed into the subphase (Trurnit 1960; Sobotka and Trurnit 1961; Kleinschmidt et al. 1962; Pattus and Rothen 1981; Pathirana et al. 2000; Petrenko, Vodyanoy, and Sykora 2007; Olsen 2005). After the vesicle slid down the glass rod and reached the air-water interface, surface forces ruptured the vesicle and split it into a monolayer.

Compression of an LB monolayer prepared from a spheroid suspension yields a pressure (Π)-area (A) isotherm (Fig. 12.13) (Sykora 2003). The curve is biphasic, having a small "kink"

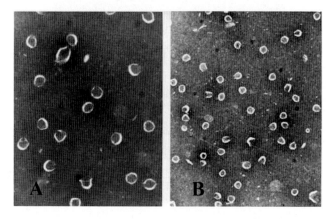

Figure 12.12. Electron micrographs of 7b1 filamentous bacteriophage following chloroform treatment. Sample was stained with 2% phosphotungistic acid. Spherical particles are called "spheroids." Mag., 302,500x (A) and 195,300x (B) (Olsen et al. 2007; reproduced by permission of The Electrochemical Society).

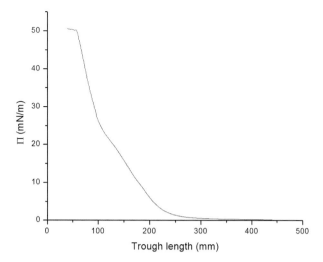

Figure 12.13. Surface pressure-area isotherm of monolayer formed when 7b1 spheroid suspension was spread at the air/water interface at 21 °C and compressed at a rate of 30 mm/min.

around 20 mN/m, followed by a steep condensed region. A pressure of ∼ 50 mN/m was noted before the barrier reached the end of its stroke. This pressure is very high for protein monolayers, indicating a very stable system (Davies and Rideal 1963; Gaines 1966). Fig. 12.14 shows elasticity versus surface pressure for the monolayer. There are two maxima in elasticity separated by a minimum around 20 mN/m (from the "kink" in the isotherm). The largest maximum reach was ∼ 50 mN/m, very high for protein monolayers (Davies and Rideal 1963; Gaines 1966), at a pressure of 30 mN/m. This pressure is optimal for transferring monolayers onto solid sensor substrates by the LB method (Petrenko, Vodyanoy, and Sykora 2007). A similar technology for immobilization of a *Salmonella* binder phage E2 onto a surface of a QCM sensor was developed by Olsen et al. (2007).

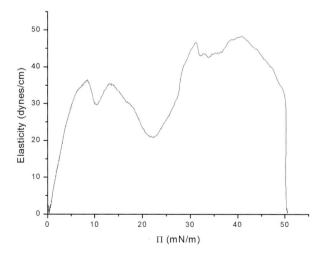

Figure 12.14. Graph of elasticity versus surface pressure (Π) for monolayer formed from spreading of spheroid suspension. Elasticity was calculated from the surface pressure-area isotherm (Fig. 8.13).

3.5.5.3. Phage Coat Protein Structure

The amino acid sequence of the pVIII coat protein from phage 7b1 with the foreign octapeptide insert is shown in Fig. 12.15. Conformation of the pVIII binding peptide at both the air/water interface and on the sensor surface can be elucidated based on the amino acid sequence. The octapeptide insert consisting of residues number 2 through 9 is located at the N-terminal region. Based on the amino acid sequence, the peptide is expected to have three α-helical regions according to both the Garnier-Robson (Garnier, Osguthorpe, and Robson 1978) and Chou-Fasman (Chou and Fasman 1974) calculation methods, as shown in Fig. 12.16. The central region of the peptide is hydrophobic, while the N-terminal and C-terminal regions are somewhat hydrophilic. This hydrophobic region is the part of the peptide that spans the bacterial cell membrane during assembly (Bashtovyy et al. 2001; Houbiers et al. 2001; Branch and Brozik 2004; Houbiers and Hemminga 2004; Aisenbrey et al. 2006). Most of the amphipathic and flexible regions of the peptides correspond with the hydrophilic regions of the peptide. According to calculations of the antigenic index, the most probable antigen-binding region lies on the N-terminus, which is the region where the octapeptide insert is located.

A hypothetical arrangement of the pVIII coat proteins at the air/water interface is shown in Fig. 12.17. Here, the hydrophilic N-terminal and C-terminal α-helices interact with the water phase while the central hydrophobic region remains at the interface. A hypothetical arrangement of these peptides on the sensor surface is subsequently shown in Fig. 12.18 (Bashtovyy et al. 2001; Houbiers et al. 2001; Houbiers and Hemminga 2004; Im and Brooks 2004; Aisenbrey et al. 2006). Here the peptides are suggested to be arranged in a conformation similar to that

N-A<u>VPEGAFSS</u>DPAKAAFDSLQASATE<u>YIGYAWAMVVVIVGATIGI</u>KLFKKFTSKAS-C
     1   2

Figure 12.15. Amino acid sequence of 7b1 filamentous bacteriophage pVIII coat protein. The foreign octapeptide insert, VPEGAFSS (underlined region 1), is located between residues 1 and 10 at the N-terminal portion (N) of the protein. The hydrophobic region of the protein is underlined. C designates the C-terminus of the peptide.

Amino Acid # (beginning at N-terminus)

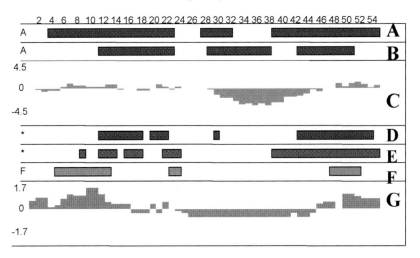

Figure 12.16. DNASTAR analysis of pVIII phage coat protein: (A) α-helical regions (Garnier-Robson method); (B) α-helical regions (Chou-Fasman method); (C) hydrophilicity plot; (D) α-helical amphipathic regions; (E) β-sheet amphipathic regions; (F) flexible regions; (G) antigenic index. (Olsen et al. 2007; reproduced by permission of The Electrochemical Society).

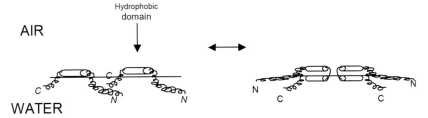

Figure 12.17. Hypothetical schematic of the arrangement of pVIII coat proteins at the air/water interface. N and C designate the N-terminus and C-terminus of the peptide, respectively, while the cylinder represents the hydrophobic domain. (Olsen et al. 2007; reproduced by permission of The Electrochemical Society).

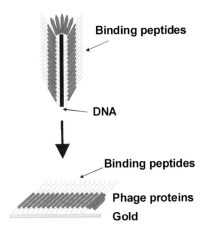

Figure 12.18. Hypothetical arrangement of skinned phage deposited to hydrophilic QCM substrates by Langmuir-Blodgett method. (Olsen et al. 2007; reproduced by permission of The Electrochemical Society).

in the phage particle, where the positively charged lysine residues of the C-terminal region interact with the negatively charged gold surface, thus allowing the N-terminal region and the octapeptide insert to be exposed to solvent.

3.5.6. Immobilization of Molecular Probes onto Porous Substrates

Immobilization of large molecular probes (antibodies, proteins, DNA, etc.) requires a complex environment in order to maintain viability and functional activity of the probes. These conditions are difficult to meet on a continuous solid sensor surface. Under natural conditions biological receptors are supported by biological membranes that are interfaced with water solutions on both sides. The Australian group of Cornell (Cornell et al. 1997, 2001) devised a multi-step assembly procedure to tether a lipid bilayer containing molecular probes linked to a gold surface. Such a tethered configuration is of interest in general for sensor technology because it creates a water reservoir between the sensor surface and membrane and serves to maintain the bilayer fluidity and facilitate the incorporation of molecular probes. Although this example clearly demonstrates the feasibility of an electrode-supported ion channel-based sensor, it suffers from several serious problems. First is the shear complexity of the synthetic approach; using thiol- and lipid-based self-assembly techniques, six different reagents are sequentially organized onto the gold surface. Second, because the tethers are randomly positioned on the electrode, weaker unsupported regions of the membrane could collapse. Finally, there appears to be no obvious patterning procedure. To overcome the above problems and in addition develop qualitatively new functionalities, thin porous substrates of silica have been developed (Fan et al. 2000; Jiang et al. 2006; Nishiyama et al. 2006; Xomeritakis et al. 2007). These films can be used as a new type of support for molecular probes in biosensors (Thust et al. 1999; Bessueille et al. 2005; Gawrisch et al. 2005; Yun et al. 2005; Dai, Baker, and Bruening 2006; Song et al. 2006). In our laboratory we have immobilized the antimycotic agent amphotericin B onto the porous silicon surface and observed ion currents associated with ion conductance of amphotericin B ion channels connected to internal and external reservoirs of liquids separated by a membrane (Yilma et al. 2007a, 2007b).

4. Problem of "Negative Mass"

The thickness shear mode quartz crystal resonator (QCM) is often considered a mass-sensitive sensor. During the sensing process, it's expected that the response (frequency change, Δf) of the sensor is directly related to any additional mass that adheres to the resonator, usually resulting in a resonance frequency decrease ($\Delta f < 0$). Sauerbrey (1959) demonstrated this for thin, rigid layers (like metal film). For mass m the frequency decrease is

$$\Delta f = -f_0 \frac{\Delta m}{M_q},$$

where, f_0 is the resonance frequency and M_q is the mass of a quartz oscillator. However, use of QCM for in situ bacteria and cell detection in fluids has revealed more complex sensor responses. For example, the observed value of Δf differs from that predicted by Sauerbrey's (1959) relation, and the signal is often small (Thompson et al. 1991; Voinova, Jonson, and Kasemo 2002).

Observed deviations from Sauerbrey's (1959) predicted mass theory have been noted during sensor testing under liquid conditions using both antibodies (Olsen et al. 2003) and phages (Olsen et al. 2006) as bioreceptors. The most peculiar results show that under certain conditions there is an appearance of a negative apparent mass; i.e., with increasing bacterial concentration there can be a dose-dependent decrease of the apparent mass. It is possible that the bacterial microenvironment and location of the antigen on the surface of a bacterium can determine the value and sign of the analytical signal generated by the acoustic wave device (Olsen et al. 2003). Bacterial positioning and binding may be very important at the solid/liquid interface of the sensor and factors such as viscoelasticity, shear forces and damping. For example, in our studies using *Salmonella* and *E. coli* antibodies as receptors (Olsen et al. 2003), when attachment between bacteria and bacteria-specific somatic O antibodies at the solid-liquid interface of a TSM resonator was rigid and strong, the sensor's output was directly proportional to the logarithmic concentration of free bacteria in suspension, and the sensor's behavior could be described as that predicted by mass theory ($\Delta f < 0$) (Fig. 12.19a). Conversely, flexible binding observed for bacteria attached by flagella to immobilized flagellar H antibodies resulted in inversely proportional sensor signals ($\Delta f > 0$) (Fig. 12.19b). This premise was affirmed by studying the responses of environmentally aged sensors. Sensor responses and binding efficiency, confirmed by dark-field microscopy, decreased as the duration of sensor environmental aging under differing conditions of temperature increased (Fig. 12.20).

Viscoelastic properties of the bacterial layer attached to the surface are anticipated to be different depending on the mechanism of binding—somatic or flagellar. Also, the viscous shear and viscous drag forces of the attached bacteria are very different. Clearly, bacteria rigidly or flexibly attached (Fig. 12.21) take different roles in the oscillation of the whole system. When binding is rigid, bacteria oscillate in unison with the sensor and therefore contribute to the effective oscillating mass of the system. This is shown by the increase of the apparent

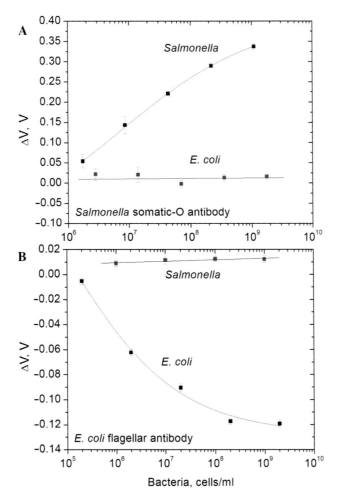

Figure 12.19. Dose responses for rigid and flexible positioning of bacteria on biosensor: (A) *Salmonella* and *E. coli* dose responses to sensor prepared with somatic O *Salmonella* antibodies; (B) *Salmonella* and *E. coli* dose responses to sensor fabricated with flagellar H-type *E. coli* antibodies. Curves are sigmoid fit to experimental data. Straight lines are the linear least squares fit. Bars are SD (reprinted from Olsen et al. (2003), with permission of Elsevier).

mass when bacterial binding concentration is increased. Conversely, in the case of flexible attachment, the oscillation of the bacteria may be not in phase with the oscillation of the sensor, resulting in a decrease in the apparent mass even when concentration of binding bacteria is increased. Additionally, we propose that the electrically charged bacterium on the surface of an acoustic wave sensor is not only engaged in the mechanical oscillations of the crystal but also directly interacts with the electric field driving the sensor crystal. This field drives the piezoelectric quartz crystal and at the same time creates an electrophoretic force applied to the electrically charged bacterium. The piezoelectric and electrophoretic forces can be of different values and directions, depending on the positioning of bacteria by the O antigen (Fig. 12.21a—firm positioning) or the H antigen (Fig. 12.21a—flexible attachment), and their combination may contribute to the change of the apparent mass of the bacteria as measured by the acoustic wave device. Obstruction of antibodies by the buildup of a biofilm during aging may cause decreased accessibility to bacterial targets (Fig. 12.21a—weak binding).

   In contrast with Sauerbrey's (1959) observations for a thin, firmly attached film, this seeming contradiction of normal mass loading theory is consistent with the observations of

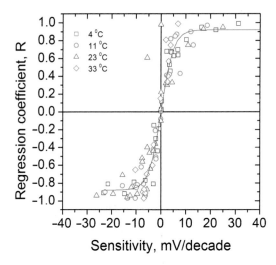

Figure 12.20. Experimental regression coefficient of individual sensors (e.g., see Fig. 12.3b) at differing temperatures as a function of sensitivity for environmentally aged *Salmonella* sensors prepared with somatic O antibodies. The linear portions of dose response signals were fitted by linear regression. Curve is the sigmoidal fit to experimental data points at indicated temperatures (reprinted from Olsen et al. (2003), with permission of Elsevier).

Dybwad (1985), who first described mass-dependent frequency increases in conjunction with particulates, such as small (10–50 µm diameter) Au spheres, under normal atmospheric conditions while loosely attached to a horizontally positioned QCM resonator. Dybwad's (1985) proposed equivalent mechanical model (Fig 12.21b) of a loosely bound particle as a coupled mass-spring system corresponds exactly with the flexible attachment of our simplified model shown in Fig. 12.21a (Olsen et al. 2003) depicting bacterial positioning at the sensor surface as the determinant factor of the sensor's analytical response. Dybwad's (1985) results were affirmed by Vig and Ballato (1998), who stated:

> "Significant deviation from the Sauerbrey equation will also occur when the mass is not rigidly coupled to the QCM surfaces. The effects of liquids have been discussed in the sensor literature [references given]; however, the effects of nonrigid coupling of solids do not seem to be well-

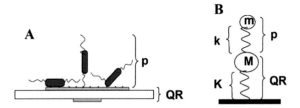

Figure 12.21. Mechanical models of analyte-resonator interaction as composite and coupled oscillators. (A) Corresponding model of Olsen et al. (2003) that shows bacterial binding positions (p) at the solid/liquid interface of the quartz resonator (QR). When binding is firm between bacteria and receptor (left), the natural frequency of the cell as an independent mass-spring system equals the frequency of the resonator, forming a composite unit that produces expected mass loading effect with corresponding frequency decrease. When binding is flexible or weak (center and right, respectively) between analyte and receptor a coupled oscillator is formed, the frequency of which is dictated by the difference in the spring constants between the oscillator and bacteria. (B) Coupled oscillator model of Dybwad (1985) depicting quartz resonator (QR) as one mass (M) spring (K) system, and a loosely attached particle (p) as a second mass (m) spring (k) system. Attachment of the loose particle causes QR to oscillate at a new, higher frequency when $k < K$. When $k = K$, a composite system is formed that produces expected mass loading effect with corresponding frequency decrease.

known. For example, when a particle is placed on an electrode of a QCM, the Sauerbrey equation predicts a decrease in the frequency of the QCM, but the frequency actually increases. When the particle on the resonator is modeled as a coupled oscillator, the model correctly predicts a frequency increase [as verified by Dybwad (1985)]."

This type of behavior has been documented in the literature on several occasions. Berg, Johannsmann, and Ruths (2002) used a single asperity contact to show that frequency shifts associated with a quartz resonator operating in shear mode increased linearly with increasing contact radius. Borovsky et al. (2001) used a nanoindenter probe in conjunction with a QCM to elicit positive frequency shifts characteristic of the contact stiffness. Sorial and Lec (2004) experienced size-dependent frequency increases with polystyrene spheres using a QCM under aqueous conditions. Otto et al. (1999) observed diminished, $\Delta f < 0$, response for weaker bacteria attachment using an *E. Coli* QCM sensor. Other documented reports of negative apparent mass using QCM-based platforms include Hayden et al. (2003) and Dickert et al. (2003), in response to loose binding of bacterial cells in yeast imprinted layers; Dickert et al. (2004), due to loose binding of non-specific compounds to tobacco mosaic virus imprinted polymers; Marxer et al. (2003), who attributed frequency increases to alterations in cytosolic viscosity of adsorbed epithelial cells; Thompson, Arthur, and Dhaliwal (1986), due to immunochemical interactions resulting in decreased acoustic transmission at the liquid/solid interface; and Pereira de Jesus, Naves, and Lucia do Lago (2002), as a result of polymeric film stiffness in the determination of boron. More recently, Lucklum (2005) described non-gravimetric contributions of viscoelastic films at the solid/liquid interface, resulting in positive frequency shifts. He clearly showed that typical elastic and energy dissipation properties are as important to frequency response, both positive and negative, as the layer's mass and therefore in many cases the added mass cannot be determined simply from the QCM response alone. The author correctly notes that the traditional name "quartz crystal microbalance (QCM)" as a technique is misleading.

Collectively, these reports support our hypothesis that positive frequency shifts can be observed under certain conditions as a result of viscoelastic changes at the solid/liquid interface from surface films, bioreceptor layers, and bacterial attachment. We contend that the continuous model of the viscoelastic layer is not directly applicable to the bacterial sensing. This is a consideration for researchers in terms of the importance of bacterial attachment schemes (bioreceptors) that result in high-affinity and multiple binding valences. In the next section, we present a coupled oscillators model that explicitly accounts for discrete events of bacteria attachment in terms of the elastic constant and the dissipation of the bond between bacteria and the sensor surface. This model is in agreement with our previous model (Olsen et al. 2003), since the parameters of an LRC series equivalent circuit model of the QCM are perfectly analogous to a damped harmonic oscillator system (Table 12.5).

Table 12.5. Analogous parameters between a mechanical spring-mass system and LRC series circuit

		System	
Harmonic Oscillator	Unit	LRC Circuit	Unit
Displacement	x	Charge	q
Velocity	v	Current	1
Force	F	Voltage	V
Mass	m	Inductance	L
Damping constant	b	Resistance	R
Spring constant	k	Capacitance^{-1}	1/C
Natural frequency	$\omega_o = (k/m)^{1/2}$	Natural frequency	$\omega_o = (1/C)^{1/2}$

5. Coupled Oscillators Model

The continuous layer model does not elucidate total understanding of the bacterial sensing process when discrete bacteria are bonded to the sensor surface. We present here a simple coupled oscillators model, depicted in Fig. 12.22.

The unloaded quartz oscillator is described by the oscillator of mass M connected to the spring with the force constant K and moving in the fluid with the viscous friction force $-\Gamma v$, where v is velocity. This oscillator models realistic assay conditions where the bacterial bioreceptors (e.g., antibodies or phages) are deposited on the surface of an immersed TSM transducer, but no bacteria are present. The bonded bacterium of mass m is connected to the oscillator by an elastic bond with the force constant k and experiences a viscous friction with coefficient γ. The equations of the motion for the quartz oscillator and each of the oscillating bacteria numbered by the index i $(i= 1, 2...n)$ are:

$$M\frac{d^2X}{dt^2} = -KX - \Gamma\frac{dX}{dt} + \sum_{i=1}^{n}k(x_i - X - a) + F_0 e^{i\omega t}$$

$$m\frac{d^2x_i}{dt^2} = -\gamma\frac{dx_i}{dt} - k(x_i - X - a).$$

(12.6)

Here $F_0 \sin \omega t = \text{Im} F_0 e^{iwt}$ is the periodic external force driving the oscillator. We want to find a stationary solution of Eqs. (12.6) when all the oscillators move with the frequency of the external force:

$$X = X_0 e^{i\omega t}, x_i - a = x_{i0}e^{i\omega t}.$$

(12.7)

Substitution of Eq. (12.7) into Eqs. (12.6) gives a system of linear algebraic equations:

$$(-\omega^2 M + i\Gamma\omega + K)X_0 - \sum_{i=1}^{n}k(x_{i0} - X_0) = F_0$$

$$(-m\omega^2 + i\gamma\omega)x_{i0} + k(x_{i0} - X_0) = 0.$$

(12.8)

To solve Eqs. (12.8) with respect to X_0, the variables x_{i0} are expressed as:

$$x_{i0} = \frac{kX_0}{-m\omega^2 + i\gamma\omega + k}.$$

(12.9)

Substitution of Eq. (12.9) into the first part of Eq. (12.8) gives:

$$X_0 = \frac{F_0[m(\omega_0^2 - \omega^2) + i\gamma\omega]}{[M(\Omega_0^2 - \omega^2) + i\omega\Gamma][m(\omega_0^2 - \omega^2) + i\gamma\omega] - nk(m\omega^2 - i\gamma\omega)}.$$

(12.10)

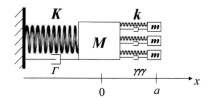

Figure 12.22. Coupled oscillator's model. Quartz resonator is presented by oscillator with mass M, spring constant K, and viscous damping Γ. Each attached bacterium is modeled as an individual coupled oscillator with mass m, spring constant k, and viscous damping γ.

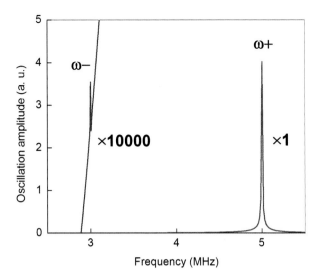

Figure 12.23. Resonance curve for the quartz resonator frequency 5 MHz and the bacteria elastic binding strength 3 MHz. Strong resonance mode ω+ (red curve) near 5 MHz is dominant where as the mode ω- (blue curve) near 3 MHz is weak (shown at 10,000X magnification).

Here,

$$\Omega_0{}^2 = K/M, \; \omega_0{}^2 = k/m, \tag{12.11}$$

denotes the resonance frequencies for uncoupled oscillators. The resonance frequencies of coupled oscillators correspond to maxima of the amplitude $|X_0|$ as a function of ω. Fig. 12.23 presents an example of the resonance curve for $\Omega_0/2\pi = 5 \cdot 10^6$ Hz, $\Gamma/2\pi M = 1$ Hz, $\omega_0/2\pi = 3 \cdot 10^6$ Hz, $\gamma/2\pi m = 1$ Hz, $n = 100$, and $m/M = 2 \cdot 10^{-7}$. As expected, the main resonance is near the quartz oscillation frequency Ω_0, because the effect of bacteria binding is small. The second weak resonance occurs near the bacterium oscillation frequency ω_0, but it is too weak to be registered experimentally.

When damping does not occur ($\Gamma = 0$, $\gamma = 0$), complete consideration of the resonances is feasible in closed analytical form. At resonance the oscillation amplitude given by Eq. (12.10) grows infinitely because the denominator is zero:

$$M(\Omega_0^2 - \omega^2)(\omega_0^2 - \omega^2) - nk\omega^2 = 0. \tag{12.12}$$

This is a square equation for ω^2 and it has two positive solutions, ω_- and ω_+. Fig. 12.24 shows ω_- and ω_+ as functions of the bacteria elastic binding strength ω_0 and $\Omega_0/2\pi = 5 \cdot 10^6$ Hz, $n = 100$, and $m/M = 2 \cdot 10^{-7}$. As discussed above and shown in Fig. 12.23, only the resonance near Ω_0 is strong and observable in a realistic case when damping occurs. Thus, in Fig. 12.24, the observable resonance corresponds to ω_+ branch when the bacteria binding is weak ($\omega_0 < \Omega_0$), but is given by ω_- branch when the binding is strong ($\omega_0 > \Omega_0$). Hence, there are two systems corresponding to the two marked rectangular areas in Fig. 12.24. First, for weak bacteria binding ($\omega_0 < \Omega_0$) the resonance frequency increases in contradiction to the intuitive expectation that the addition of the bacterial mass to the oscillator will decrease the frequency. The positive frequency shift grows when the bacteria binding ω_0 increases and approaches the frequency Ω_0. The second system corresponds to strong bacteria binding ($\omega_0 < \Omega_0$) when the resonance frequency decreases as expected for mass added to the oscillator. In this case

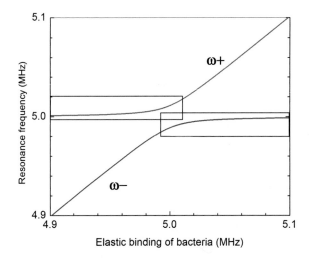

Figure 12.24. Frequency of two resonance modes ω+ and ω- as a function of the elastic binding strength of the bacteria. Each mode is strong only in the binding range denoted by the corresponding rectangle box. The mode becomes weak and finally unobservable when its frequency departs from the resonance frequency of 5 MHz.

the frequency shift becomes smaller when bacteria binding ω_0 increases. These results are also valid in the realistic case of non-zero damping ($\Gamma/2\pi M = \gamma/2\pi M = 1$ Hz), as shown in Fig. 12.25. Again the resonance frequency shift is positive for weak and negative for strong bacterial elastic binding.

The different behavior for weak and strong bacterial attachment can be understood from the bacterial oscillator equations of motion. We evaluate Eq. (12.9) at the frequency Ω_0

$$x_{i0} = \frac{X_0}{\left(1 - \dfrac{\Omega_0^2}{\omega_0^2}\right)^2 + \dfrac{\gamma^2 \Omega_0^2}{k^2}} \left(1 - \frac{\Omega_0^2}{\omega_0^2} - i\frac{\gamma\Omega_0}{k}\right). \tag{12.13}$$

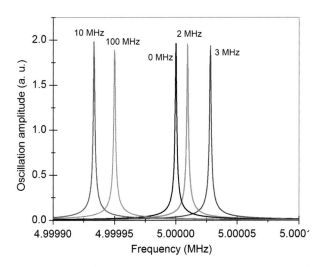

Figure 12.25. Resonance curves for different bacteria elastic binding strengths at 0 (no binding), 2, 3, 10, and 100 MHz, as noted. Notice that the shift of resonance frequency changes from negative to positive as the binding decreases below the resonance frequency of 5 MHz.

The relative phase of x_{i0} with respect to X_0 in Eq. (12.13) is the phase of the multiply

$$\left(1 - \frac{\Omega_0^2}{\omega_0^2} - i\frac{\gamma\Omega_0}{k}\right). \tag{12.14}$$

For weak attachment ($\omega_0 < \Omega_0$), the real part of this complex number is negative. This means that for small γ the bacteria and quartz surface oscillate with the phase shift close to 180°, i.e., anti-phase. Hence, the bacterial oscillator pushes the quartz oscillator towards the equilibrium position (Fig. 12.25). This increases the restoring force and thus also increases the resonance frequency of the quartz oscillator. In the case of strong bacteria attachment ($\omega_0 > \Omega_0$), the bacterial and quartz oscillators move approximately in the same phase and the frequency decreases. Therefore, the model of coupled oscillators clearly reveals the mechanism of the apparent "negative mass" effect.

To discuss the performance and rational design of the bacterial sensor we calculated the frequency response as a function of the number of attached bacteria. Fig. 12.26 shows the results for different strengths of the elastic attachment bond. The output signal (frequency shift) is considerably stronger for strong attachment. This suggests using a strong attachment method to achieve a low bacteria detection threshold. Importantly, a substantial linear range of detection occurs that is especially broad for weak attachment. This allows finding the number of attached bacteria by calibration measurement of only the initial slope of the response curve. Additionally, we modeled the effect of inhomogeneous bacterial binding strength. Fig. 12.27 shows that for two types of binding sites with 2 and 10 MHz attachment strength, the response of a 5 MHz sensor crucially depends on the distribution of bacteria between the sites. In particular, at about 0.8/0.2 distribution, the signal is very small and thus a false negative result will be measured. This underlines the importance of using the bioreceptors with a narrow distribution of strong binding strengths.

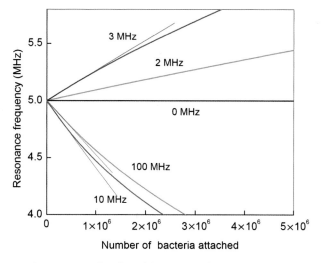

Figure 12.26. Resonance frequency as a function of the number of bacteria attached for different elastic binding strengths at 0, 2, 3, 10, and 100 MHz, as indicated. Notable is substantial linear dynamic range increasing for weaker attachment.

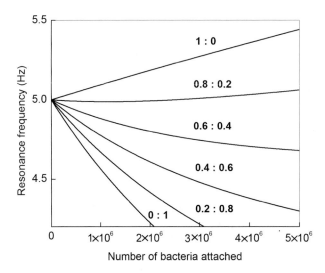

Figure 12.27. The effect of inhomogeneous bacterial binding strength on sensor. The response of a 5 MHz sensor crucially depends on distribution of bacteria between the two types of binding sites with 2 and 10 MHz attachment strength. In particular, at about 0.8/0.2 distribution, the signal is very small and thus a false negative result will be measured.

6. Conclusions

Even though acoustic wave technologies such as the TSM are not new, their adaptation to biological analysis has flourished mainly in the past decade. Most likely the next ten years will see an even larger contingent of researchers developing biosensors based on these platforms as affordability, access, sensitivity, and technical understanding increase. The most remarkable property of acoustic wave devices is their relative simplicity. The number of different devices adapted to work in the biological/medical environment is likely to expand in the future. It is therefore vital to establish a good understanding of the nature of the signals produced by acoustic wave devices when they are used for testing bacteria. The paradox of "negative mass" is a real threat to the interpretation of experimental results related to the detection of bacteria. Knowledge of the true nature of "negative mass" linked to the strength of bacteria attachment will contribute significantly to our understanding of the results of "weighing bacteria." We hope it may stimulate increased interest in the technology and motivate new experiments with a variety of microorganisms. The impact of these studies may extend beyond an appreciation of bacterial detection. One may now begin to conceive of strategies for the study and control of processes of bacterial settlement, bacterial colonization, biofilm formation, and bacterial infection in which bacterial attachment plays a role.

Acknowledgments

Our work was supported by grants from Sigma Xi Grants-in-Aid of Research (10040088), DARPA (MDA972–00–1-0011), ARO/DARPA (DAAD 19–01–10454), NIH (R21 AI055645), USDA (99–34394–7546), and Aetos Technologies Inc. We are grateful to Dr. Bryan Chin for support and dialogue, and Oleg Pustovyy for technical assistance. The views expressed in this article are those of the authors, and do not reflect the official policy or position of the United States Air Force, Department of Defense, or the U.S. Government.

References

Aberl F and Wolf H (1993) Present trends in immunosensors. LaborPraxis 70–74, 76–77

Ahluwalia A, Derossi D, Ristori C, Schirone A and Serra G (1992) A comparative study of protein immobilization techniques for optical immunosensors. Biosens. Bioelectron. 7:207–214

Ahmad A and Ahmad S (1996) Solvent effect on antibody antigen interaction. Environ. Res. 5:29–36

Aisenbrey C, Harzer U, Bauer-Manz G, Bar G, Chotimah INH, Bertani P, Sizun C, Kuhn A and Bechinger B (2006) Proton-decoupled 15N and 31P solid-state NMR investigations of the Pf3 coat protein in oriented phospholipid bilayers. FEBS Journal 273:817–828

Aizawa H, Kurosawa S, Tanaka M, Yoshimoto M, Miyake J and Tanaka H (2001) Rapid diagnosis of Treponema pallidum in serum using latex piezoelectric immunoassay. Anal. Chim. Acta 437:167–169

Alberts B, Bray D, Lewis J, Raff M, Roberts K and Watson JD (1994) Molecular biology of the cell, 3rd ed. New York: Garland Publishing

Auner GW, Shreve G, Ying H, Newaz G, Hughes C and Xu J (2003) Dual-mode acoustic wave biosensors microarrays. Proc. SPIE Int. Soc. Opt. Eng. 5119:129–139

Babacan S, Pivarnik P, Letcher S., and Rand, A.G. 2000. Evaluation of antibody immobilization methods for piezo-electric biosensor application. Biosens. Bioelectron. 15:615–621

Babacan, S., Pivarnik, P, Letcher S and Rand AG (2002) Piezoelectric flow injector analysis biosensor for the detection of Salmonella typhimurium. J. Food Sci. 61: 314–320

Bailey CA, Fiebor B, Yen W, Vodyanoy V, Cernosek RW and Chin BA (2002) Thickness shear mode (TSM) resonators used for biosensing. Proc. SPIE Int. Soc. Opt. Eng. 4575:138–149

Balasubramanian S, Sorokulova IB, Vodyanoy VJ and Simonian AL (2007) Lytic phage as a specific and selective probe for detection of Staphylococcus aureus—A surface plasmon resonance spectroscopic study. Biosens. Bioelectron. 22:948–955

Ballantine DS, White RM, Martin SJ, Ricco AJ, Zellers ET, Frye GC, Wohltjen H, Levy M and Stern R (1997) Acoustic Wave Sensors: Theory, Design, & Physico-Chemical Applications. Academic Press, San Diego

Bandey HL, Martin SJ, Cernosek RW and Hillman AR (1999) Modeling the responses of thickness-shear mode resonators under various loading conditions. Anal. Chem. 71:2205–2214

Bao L, Deng L, Nie L, Yao S and Wei W (1996a) Determination of microorganisms with a quartz crystal microbalance sensor. Anal. Chim. Acta 319:97–101

Bao L, Deng L, Nie L, Yao S and Wei W (1996b) A rapid method for determination of Proteus vulgaris with a piezoelectric quartz crystal sensor coated with a thin liquid film. Biosens. Bioelectron. 11:1193–1198

Barraud A, Perrot H, Billard V, Martelet C and Therasse J (1993) Study of immunoglobulin G thin layers obtained by the Langmuir-Blodgett method: application to immunosensors. Biosens. Bioelectron. 8:39–48

Bashtovyy D, Marsh D, Hemminga MA and Pali T (2001) Constrained modeling of spin-labeled major coat protein mutants from M13 bacteriophage in a phospholipid bilayer. Protein Sci. 10:979–987

Ben-Dov I, Willner I and Zisman E (1997) Piezoelectric immunosensors for urine specimens of Chlamydia trachomatis employing quartz crystal microbalance microgravimetric analyses. Anal. Chem. 69:3506–3512

Berg S, Johannsmann D and Ruths M (2002) Frequency response of quartz crystal shear-resonator during an adhesive, elastic contact in a surface forces apparatus. J. Appl. Phys. 92:6905–6910

Berkenpas E, Millard P and Pereira da Cunha M (2006) Detection of Escherichia coli O157:H7 with langasite pure shear horizontal surface acoustic wave sensors. Biosens. Bioelectron. 21:2255–2262

Bessueille F, Dugas V, Vikulov V, Cloarec JP, Souteyrand E and Martin JR (2005) Assessment of porous silicon substrate for well-characterized sensitive DNA chip implement. Biosens. Bioelectron. 21: 908–916

Betageri GV, Black CD, Szebeni J, Wahl LM and Weinstein JN (1993) Fc-receptor-mediated targeting of antibody-bearing liposomes containing dideoxycytidine triphosphate to human monocyte/macrophages. J. Pharm. Pharamcol. 45:48–53

Borovskya B, Krim J, Syed Asif S and Wahl K (2001) Measuring nanomechanical properties of a dynamic contact using an indenter probe and quartz crystal microbalance. J. Appl. Phys. 90:6391–6396

Branch DW and Brozik SM (2004) Low-level detection of a Bacillus anthracis simulant using love-wave biosensors on 36° YX LiTaO3. Biosens. Bioelectron. 19:849–859

Bunde RL, Jarvi EJ and Rosentreter JJ (1998) Piezoelectric quartz crystal biosensors. Talanta 46:1223–1236

Buzby JC and Roberts T (1997) Economic costs and trade impacts of microbial food-borne illness. World Health Stat. Q. 50:57–66

Bykov VA (1996) Langmuir-Blodgett films and nanotechnology. Biosens. Bioelectron. 11:923–932

Carter RM, Jacobs MB, Lubrano GJ and Guilbault GG (1995a) Piezoelectric detection of ricin and affinity-purified goat anti-ricin antibody. Anal. Lett. 28:1379–1386

Carter RM, Mekalanos JJ, Jacobs MB, Lubrano GJ and Guilbault GG (1995b) Quartz crystal microbalance detection of Vibrio cholerae O139 serotype. J. Immunol. Methods 187:121–125

Casalinuovo I, Di Pierro D, Bruno E, Di Francesco P and Coletta M (2005) Experimental use of a new surface acoustic wave sensor for the rapid identification of bacteria and yeasts. Lett. Appl. Microbiol. 42:24–29

Cavicacute BA, Hayward GL and Thompson M (1999) Acoustic waves and the study of biochemical macromolecules and cells at the sensor-liquid interface. Analyst 124:1405–1420

Chang K-S, Jang H-D, Lee C-F, Lee Y-G, Yuan C-J and Lee S-H (2006) Series quartz crystal sensor for remote bacteria population monitoring in raw milk via the Internet. Biosens. Bioelectron. 21:1581–1590

Chen M, Liu M, Yu L, Cai G, Chen Q, Wu R, Wang F, Zhang B, Jiang T and Fu W (2005) Construction of a novel peptide nucleic acid piezoelectric gene sensor microarray detection system. J. Nanosci. Nanotechnol. 5:1266–1272

Chin RC, Salazar N, Mayo MW, Villavicencio V, Taylor RB, Chambers JP and Valdes JJ (1996) Development of a bacteriophage displayed peptide library and biosensor. Proc. SPIE Int. Soc. Opt. Eng. 2680:16–26

Chothia C and Janin J (1975) Principles of protein-protein recognition. Nature 256:705–708

Chou PY and Fasman GD (1974) Prediction of protein conformation. Biochemistry 13:222–245

Clark LCJ and Lyons C (1962) Electrode systems for continuous monitoring in cardiovascular surgery. Ann. N. Y. Acad. Sci. 102:29–45

Cornell BA, Braach-Maksvytis VL, King LG, Osman PD, Raguse B, Wieczorek L and Pace RJ (1997) A biosensor that uses ion-channel switches. Nature 387:580–583

Cornell BA, Krishna G, Osman PD, Pace RD and Wieczorek L (2001) Tethered-bilayer lipid membranes as a support for membrane-active peptides. Biochem. Soc. Trans. 29:613–617

Curie J and Curie P (1880) Ann. de Chim. et Phys. 91:294

Dai J, Baker GL and Bruening ML (2006) Use of porous membranes modified with polyelectrolyte multilayers as substrates for protein arrays with low nonspecific adsorption. Anal. Chem. 78:135–140

Davies LT and Rideal EK (1963) Interfacial Phenomena. Academic Press, New York

Deisingh A and Thompson M (2001) Sequences of E. coli O157:H7 detected by a PCR-acoustic wave sensor combination. Analyst 126:2153–2158

Deng L, Tan, H, Xu Y, Nie L and Yao S (1997) On-line rapid detection of urease-producing bacteria with a novel bulk acoustic wave ammonia sensor. Enzyme Microb. Technol. 21:258–264

Deobagkar DD, Limaye V, Sinha S and Yadava RDS (2005) Acoustic wave immunosensing of Escherichia coli in water. Sens. Actuators B 104:85–89

Dickert F, Hayden O, Lieberzeit P, Palfinger C, Pickert D, Wolff U and Scholl G (2003) Borderline applications of QCM-devices: synthetic antibodies for analytes in both nm- and um-dimensions. Sens. Actuators B 95:20–24

Dickert FL, Hayden O, Bindeus R, Mann K, Blaas D and Waigmann E (2004) Bioimprinted QCM sensors for virus detection–screening of plant sap. Anal. Bioanal. Chem. 378:1929–1934

Dickert FL, Lieberzeit P and Hayden O (2003) Sensor strategies for microorganism detection–from physical principles to imprinting procedures. Anal. Bioanal. Chem. 377:540–549

Dybwad G (1985) A sensitive new method for the determination of adhesive bonding between a particle and a substrate. J. Appl. Phys. 58:2789–2790

Eun A, Huang L, Chew F, Li S and Wong S (2002) Detection of two orchid viruses using quartz crystal microbalance (QCM) immunosensors. J. Virol. Methods 99:71–79

Fan H, Lu Y, Stump A, Reed ST, Baer T, Schunk R, Perez-Luna V, Lopez GP and Brinker CJ (2000) Rapid prototyping of patterned functional nanostructures. Nature 405:56–60

World Health Organization (1997) Foodborne diseases—possibly 350 times more frequent than reported.

Frisk T, Ronnholm D, van der Wijngaart W and Stemme G (2006) A micromachined interface for airborne sample-to-liquid transfer and its application in a biosensor system. Lab. Chip 6:1504–1509

Fung YS and Wong YY (2001) Self-assembled monolayers as the coating in a quartz piezoelectric crystal immunosensor to detect Salmonella in aqueous solution. Anal. Chem. 73: 5302–5309

Furch M, Ueberfeld J, Hartmann A, Bock D and Seeger S (1996) Ultrathin oligonucleotide layers for fluorescence based DNA-sensors. Proc. SPIE Int. Soc. Opt. Eng. 2928:220–226

Gaines GLJ (1966) Insoluble Monolayers at Liquid-gas Interfaces. Interscience, New York

Gao H, Kislig E, Oranth N and Sigrist H (1994) Photolinker-polymer-mediated immobilization of monoclonal antibodies, F(ab')2 and F(ab') fragments. Biotechnol. Appl. Biochem. 20:251–263

Gao Z, Tao G and Li G (1998) Research on detection of type C2 staphylococcus enterotoxin in food with piezoelectric immunosensors. Wei Sheng Yan Jiu 27:122–124

Garnier J, Osguthorpe DJ and Robson B (1978) Analysis of the accuracy and implications of simple methods for predicting the secondary structure of globular proteins. J. Mol. Biol. 120:97–120

Gawrisch K, Gaede HC, Luckett KM, Polozov IV and Yeliseev A (2005) Solid-supported membranes inside porous substrates and their use in biosensors, WIPO Patent WO/2005/069004 http://www.wipo.int/pctdb/en/ia.jsp?IA=US2005/ 000069&LANGUAGE=EN (accessed April 11, 2007)

Ghafouri S and Thompson M (1999) Interfacial properties of biotin conjugate avidin complexes studied by acoustic wave sensor. Langmuir 15:564–572

Gizeli E (2002) Biomolecular Sensors. Taylor & Francis, Inc., New York, pp 176–206

Goldman ER, Pazirandeh MP, Charles PT, Balighian E.D and Anderson GP (2002) Selection of phage displayed peptides for the detection of 2,4,6-trinitrotoluene in seawater. Anal. Chim. Acta 457:13–19

Goldman ER, Pazirandeh MP, Mauro JM, King KD, Frey JC and Anderson GP (2000) Phage-displayed peptides as biosensor reagents. J. Mol. Recognit. 13:382–387

Goodchild S, Love T, Hopkins N and Mayers C (2006) Engineering antibodies for biosensor technologies. Adv. Appl. Microbiol. 58:185–226

Grate JW and Frye GC (1996) Acoustic wave sensors. Sens. Update 2:37–83

Griffith J, Manning M and Dunn K (1981) Filamentous bacteriophage contract into hollow spherical particles upon exposure to a chloroform-water interface. Cell 23:747–753

Harteveld JLN, Nieuwenhuizen MS and Wils ERJ (1997) Detection of staphylococcal enterotoxin B employing a piezoelectric crystal immunosensor. Biosens. Bioelectron. 12:661–667

Hayden O, Bindeus R and Dickert FL (2003) Combining atomic force microscope and quartz crystal microbalance studies for cell detection. Meas. Sci. Technol. 14: 1876–1881

Hayden O and Dickert FL (2001) Selective microorganism detection with cell surface imprinted polymers. Adv. Mater. 13:1480–1483

He F, Geng Q, Zhu W, Nie L, Yao S and Meifeng C (1994) Rapid detection of Escherichia coli using a separated electrode piezoelectric crystal sensor. Anal. Chim. Acta 289:313–319

He F and Zhang L (2002) Rapid diagnosis of M. tuberculosis using a piezoelectric immunosensor. Anal. Sci. 18:397–401

He F, Zhang X, Zhou J and Liu Z (2006) A new MSPQC system for rapid detection of pathogens in clinical samples. J. Microbiol. Methods 66:56–62

He F, Zhao J, Zhang L and Su X (2003) A rapid method for determining Mycobacterium tuberculosis based on a bulk acoustic wave impedance biosensor. Talanta 59:935–941

He F and Zhou J (2007) A new antimicrobial susceptibility testing method of Escherichia coli against ampicillin by MSPQC. J. Microbiol. Methods 68:563–567

Herriott RM and Barlow JL (1957) The protein coats or "ghosts" of Escherichia coli phage T2. Preparation, assay, and some chemical properties. J. Gen. Physiol. 40:809–825

Horisberger M (1984) Electron-opaque markers: a review. In: Polak JM and Varndell IM (eds) Immunolabelling for Electron Microscopy. Elsevier, Amsterdam

Horisberger M (1992) Colloidal gold and its application in cell biology. Int. Rev. Cytol. 136:227–87

Houbiers MC and Hemminga MA (2004) Protein-lipid interactions of bacteriophage M13 gene 9 minor coat protein. Mol. Membr. Biol. 21:351–359

Houbiers MC, Wolfs CJAM, Spruijt RB, Bollen YJM, Hemminga MA and Goormaghtigh E (2001) Conformation and orientation of the gene 9 minor coat protein of bacteriophage M13 in phospholipid bilayers. Biochim. Biophys. Acta Biomembr. 1511:224–235

Howe E and Harding G (2000) A comparison of protocols for the optimisation of detection of bacteria using a surface acoustic wave (SAW) biosensor. Biosens. Bioelectron. 15:641–649

Im W and Brooks CL (2004) De novo folding of membrane proteins: an exploration of the structure and NMR properties of the fd coat protein. J. Mol. Biol. 337:513–519

Ishimori Y, Karube I and Suzuki IS (1981) Determination of microbial populations with piezoelectric membranes. Appl. Environ. Microbiol. 42:632–637

Ivnitski D, Abdel-Hamid I, Atanasov P and Wilkins E (1999) Biosensors for detection of pathogenic bacteria. Biosens. Bioelectron. 14:599–624

Janshoff A, Galla HJ and Steinem C (2000) Piezoelectric mass-sensing devices as biosensors-an alternative to optical biosensors? Angew. Chem. Int. Ed. Engl. 39:4004–4032

Janshoff A and Steinem C (2001) Quartz crystal microbalance for bioanalytical applications. Sens. Update 9:313–354

Janshoff A, Steinem C and Wegener J (2004) Noninvasive Electrical Sensor Devices to Monitor Living Cells Online. Ultrathin Electrochemical Chemo- and Biosensors. Springer, New York

Jiang Y-B, Liu N, Gerung H, Cecchi JL and Brinker CJ (2006) Nanometer-thick conformal pore sealing of self-assembled mesoporous silica by plasma-assisted atomic layer deposition. J. Am. Chem. Soc. 128:11018–11019

Kaspar M, Stadler H, Weiss T and Ziegler C (2000) Thickness shear mode resonators ("mass sensitive devices") in bioanalysis. Fresenius J. Anal. Chem. 366:602–610

Kim G-H, Rand AG and Letcher SV (2003) Impedance characterization of a piezoelectric immunosensor part II: Salmonella typhimurium detection using magnetic enhancement. Biosens. Bioelectron. 18:91–99

Kim N and Park I-S (2003) Application of a flow-type antibody sensor to the detection of Escherichia coli in various foods. Biosens. Bioelectron. 18:1101–1107

Kim N, Park I-S and Kim D-K (2004) Characteristics of a label-free piezoelectric immunosensor detecting Pseudomonas aeruginosa. Sens. Actuators B 100:432–438

King WH (1964) Piezoelectric sorption detector. Anal. Chem. 36:1735–1739

Kleinschmidt AK, Lang D, Jacherts D and Zahn RK (1962) Preparation and length measurements of the total deoxyribonucleic acid (DNA) content of T2 bacteriophages. Biochim. Biophys. Acta 61:875–864

Konig B and Gratzel M (1993) Detection of viruses and bacteria with piezoelectric immunosensors. Anal. Lett. 26:1567–1585

Konig B and Gratzel M (1995) A piezoelectric immunosensor for hepatitis viruses. Anal. Chim. Acta 309:19–25

Kouzmitcheva G (2005) Personal communication.

Krause R (1993) Process control for Ni/Au plating with QCM technology. Circuitree 6:10–12

Kreth J, Hagerman E, Tam K, Merritt J, Wong D, Wu B, Myung N, Shi W and Qi F (2004) Quantitative analyses of Streptococcus mutans biofilms with quartz crystal microbalance, microjet impingement and confocal microscopy. Biofilms 1:277–284

Kurosawa S, Park J, Aizawa H, Wakida S, Tao H and Ishihara K (2006) Quartz crystal microbalance immunosensors for environmental monitoring. Biosens. Bioelectron. 22:473–481

Lakshmanan RS, Hu J, Guntupalli R, Wan J, Huang S, Yang H, Petrenko VA, Barbaree JM and Chin BA (2006) Detection of Salmonella typhimurium using phage based magnetostrictive sensor. Proc. SPIE Int. Soc. Opt. Eng. 6218:62180Z

Lazcka O, Campo FJD and Munoz FX (2007) Pathogen detection: A perspective of traditional methods and biosensors. Biosens. Bioelectron. 22:1205–1217

Le D, He F-J, Jiang TJ, Nie L and Yao S (1995) A goat-anti-human IgG modified piezoimmunosensor for Staphylococcus aureus detection. J. Microbiol. Methods 23:229–234

Leonard P, Hearty S, Brennan J, Dunne L, Quinn J, Chakraborty T and O'Kennedy R (2003) Advances in biosensors for detection of pathogens in food and water. Enzyme Microb. Technol. 32:3–13

Lin H-C and Tsai W-C (2003) Piezoelectric crystal immunosensor for the detection of staphylococcal enterotoxin B. Biosens. Bioelectron. 18:1479–1483

Love AEH (1911) Some problems of geodynamics. Cambridge: Cambridge

Lucklum R (2005) Non-gravimetric contributions to QCR sensor response. Analyst 130:1465–1473

Mao X, Yang L, Su X-L and Li Y (2006) A nanoparticle amplification based quartz crystal microbalance DNA sensor for detection of Escherichia coli O157:H7. Biosens. Bioelectron. 21:1178–1185

Martin SJ, Granstaff VE and Frye GC (1991) Characterization of quartz crystal microbalance with simultaneous mass and liquid loading. Anal. Chem. 63:2272–2281

Marxer C, Coen M, Greber T, Greber U and Schlapbach L (2003) Cell spreading on quartz crystal microbalance elicits positive frequency shifts indicative of viscosity changes. Anal. Bioanal. Chem. 377:578–586

Mead P, Slutsker L, Dietz V, McCaig L, Bresee J, Shapiro C, Griffin P and Tauxe RV (1999) Food-related illness and death in the United States. Emerg. Infect. Dis. 5:607–625

Minunni M, Skladal P and Mascini M (1994) A piezoelectric quartz crystal biosensor as a direct affinity sensor. Anal. Lett. 27:1475–1487

Mittler-Neher S, Spinke J, Liley M, Nelles G, Weisser M, Back R, Wenz G and Knoll W (1995) Spectroscopic and surface-analytical characterization of self-assembled layers on Au. Biosens. Bioelectron. 10:903–916

Mo X-T, Zhou Y-P, Lei H and Deng L (2002) Microbalance-DNA probe method for the detection of specific bacteria in water. Enzyme Microb. Technol. 30:583–589

Moll N, Pascal E, Dinh DH, Pillot J-P, Bennetau B, Rebiere D, Moynet D, Mas Y, Mossalayi D, Pistre J and Dejous C (2007) A Love wave immunosensor for whole E. coli bacteria detection using an innovative two-step immobilisation approach. Biosens. Bioelectron. 22:2145–2150

Morgan DP (2000) A history of surface acoustic wave devices. Int. J. High Speed Electron. Syst. 10:553–602

Muramatsu H, Kajiwara K, Tamiya E, Karube I (1986) Piezoelectric immunosensor for the detection of Candida albicans microbes. Anal. Chim. Acta 188:257–261

Muratsugu M, Ohta F, Miya Y, Hosokawa T, Kurosawa S, Kamo N and Ikeda H (1993) Quartz crystal microbalance for the detection of microgram quantities of human serum albumin: relationship between the frequency change and the mass of protein adsorbed. Anal. Chem. 65:2933–2937

Nakanishi K, Karube I, Hiroshi S, Uchida A and Ishida Y (1996) Detection of the red tide-causing plankton Chattonella marina using a piezoelectric immunosensor. Anal. Chim. Acta 325:73–80

Nanduri V (2005) Phage at the air-liquid interface for the fabrication of biosensors. Doctoral Dissertation. graduate.auburn.edu/auetd Auburn University, Alabama

Nanduri V, Sorokulova IB, Samoylov AM, Simonian AL, Petrenko VA, Vodyanoy V (2007) Phage as a molecular recognition element in biosensors immobilized by physical adsorption. Biosens. Bioelectron. 22:986–992

Neidhardt FC (1987) Escherichia coli and Salmonella. American Society For Microbiology, Washington, D.C.

Nishiyama Y, Tanaka S, Hillhouse HW, Nishiyama N, Egashira Y and Ueyama K (2006) Synthesis of ordered mesoporous zirconium phosphate films by spin coating and vapor treatments. Langmuir 22:9469–9472

Niven D, Chambers J, Anderson T and White D (1993) Long-term, on-line monitoring of microbial biofilms using a quartz crystal microbalance. Anal. Chem. 65:65–69

O'Sullivan CK and Guilbault GG (1999) Commercial quartz crystal microbalances—theory and applications. Biosens. Bioelectron. 14:663–670

Olsen EV (2000) Functional durability of a quartz crystal microbalance sensor for the rapid detection of Salmonella in liquids from poultry packaging. Masters Thesis. http://graduate.auburn.edu/auetd Auburn University, Alabama (accessed April 10, 2007)

Olsen EV (2005) Phage-coupled piezoelectric biodetector for Salmonella typhimurium. Doctoral Dissertation. http://graduate.auburn.edu/auetd Auburn University, Alabama (accessed April 10, 2007)

Olsen EV, Pathirana ST, Samoylov AM, Barbaree JM, Chin BA, Neely WC and Vodyanoy V (2003) Specific and selective biosensor for Salmonella and its detection in the environment. J. Microbiol. Methods 53:273–285

Olsen EV, Sorokulova IB, Petrenko VA, Chen IH, Barbaree JM and Vodyanoy VJ (2006) Affinity-selected filamentous bacteriophage as a probe for acoustic wave biodetectors of Salmonella typhimurium. Biosens. Bioelectron. 21:1434–1442

Olsen EV, Sykora JC, Sorokulova IB, Chen I-H, Neely WC, Barbaree JM, Petrenko VA and Vodyanoy VJ (2007) Phage fusion proteins as bioselective receptors for piezoelectric sensors. Electrochem. Soc. Trans. 2:9–25

Otto K, Elwing H and Hermansson M (1999) Effect of ionic strength on initial interactions of Escherichia coli with surfaces, studied on-line by a novel quartz crystal microbalance technique. J. Bacteriol. 181:5210–5218

Ozen A, Montgomery K, Jegier P, Patterson S, Daumer KA, Ripp SA, Garland JL and Sayler GS (2004) Development of bacteriophage-based bioluminescent bioreporters for monitoring of microbial pathogens. Proc. SPIE Int. Soc. Opt. Eng. 5270:58–68

Park I-S and Kim N (1998) Thiolated Salmonella antibody immobilization onto the gold surface of piezoelectric quartz crystal. Biosens. Bioelectron. 13:1091–1097

Park I-S, Kim W-Y and Kim N (2000) Operational characteristics of an antibody-immobilized QCM system detecting Salmonella spp. Biosens. Bioelectron. 15:167–172

Pathirana S, Myers LJ, Vodyanoy V and Neely WC (1996) Assembly of cadmium stearate and valinomycin molecules assists complexing of K+ in mixed Langmuir-Blodgett films. Supramol. Sci. 2:149–154

Pathirana S, Neely WC, Myers LJ and Vodyanoy V (1992) Interaction of valinomycin and stearic acid in monolayers. Langmuir 8:1984–1987

Pathirana S, Neely WC and Vodyanoy V (1998) Condensing and expanding the effects of the odorants (+)- and (−)-carvone on phospholipid monolayers. Langmuir 14:679–682

Pathirana ST (1993) Interaction of valinomycin and stearic acid in monolayers at the air/water interface. Doctoral Dissertation. Auburn University, Alabama

Pathirana ST, Barbaree J, Chin BA, Hartell MG, Neely WC and Vodyanoy V (2000) Rapid and sensitive biosensor for Salmonella. Biosens. Bioelectron. 15:135–141

Pattus F, Desnuelle P and Verger R (1978) Spreading of liposomes at the air/water interface. Biochim. Biophys. Acta 507:62–70

Pattus F and Rothen C (1981) Lipid-protein interactions in monolayers at the air-water interface. In: Azzi A, Brodbeck U, Zahler P (eds) Membrane Proteins. Springer-Verlag, Berlin pp 229–240

Pattus F, Rothen C, Streit M and Zahler P (1981) Further studies on the spreading of biomembranes at the air/water interface. Structure, composition, enzymatic activities of human erythrocyte and sarcoplasmic reticulum membrane films. Biochim. Biophys. Acta Biomembr. 647:29–39

Pavey KD (2002) Quartz crystal analytical sensors: the future of label-free, real-time diagnostics? Expert Rev. Mol. Diag. 2:173–186

Pavey KD, Ali Z, Olliff CJ and Paul F 1999. Application of the quartz crystal microbalance to the monitoring of Staphylococcus epidermidis antigen-antibody agglutination. J. Pharm. Biomed. Anal. 20:241–245

Pavey KD, Barnes L, Hanlon G, Olliff C, Ali Z and Paul F (2001) A rapid, non-destructive method for the determination of Staphylococcus epidermidis adhesion to surfaces using quartz crystal resonant sensor technology. Lett. Appl. Microbiol. 33:344–348

Pereira de Jesus D, Naves C and Lucia do Lago C (2002) Determination of boron by using a quartz crystal resonator coated with N-Methyl-D-glucamine-modified poly(epichlorohydrin). Anal. Chem. 74:3274–3280

Petrenko VA (2004) Personal communication.

Petrenko VA and Smith GP (2000) Phages from landscape libraries as substitute antibodies. Protein Eng. 13:589–592

Petrenko VA, Sorokulova IB, Chin BA, Barbaree JM, Vodyanoy VJ, Chen IH and Samoylov AM (2005) Biospecific peptide probes against Salmonella isolated from phage display libraries, and their diagnostic and drug delivery uses. US Patent Application 20050137136, filed Apr 29, 2004

Petrenko VA and Vodyanoy VJ (2003) Phage display for detection of biological threat agents. J. Microbiol. Methods 53:253–262

Petrenko VA, Vodyanoy VJ and Sykora JC (2007) Methods of forming monolayers of phage-derived products and uses thereof. US Patent #7, 267, 993

Petty MC (1991) Application of multilayer films to molecular sensors: some examples of bioengineering at the molecular level. J. Biomed. Eng. 13:209–214

PM-740 Series Operation and Service Manual. (1996) Maxtek, Inc., Sante Fe Springs, California

Pohanka M and Skládal P (2005) Piezoelectric immunosensor for Francisella tularensis detection using immunoglobulin M in a limiting dilution. Anal. Lett. 38:411–422

Prusak-Sochaczewski E and Luong JHT (1990) Development of a piezoelectric immunosensor for the detection of Salmonella typhimurium. Enzyme and Microb. Technol. 12:173–177

Pyun JC, Beutel H, Meyer JU and Ruf HH (1998) Development of a biosensor for E. coli based on a flexural plate wave (FPW) transducer. Biosens. Bioelectron 13:839–845

Qu X, Bao L, Su X and Wei W (1998) Rapid detection of Escherichia coliform with a bulk acoustic wave sensor based on the gelation of Tachypleus amebocyte lysate. Talanta 47:285–290

Ramirez E, Mas JM, Carbonell X, Aviles FX and Villaverde A (1999) Detection of molecular interactions by using a new peptide-displaying bacteriophage. Biosensor. Biochem. Biophys. Res. Commun. 262:801–805

Ramsden JJ (1997a) Dynamics of protein adsorption at the solid/liquid interface. Recent Res. Dev. Phys. Chem. 1:133–142

Ramsden JJ (1997b) Protein adsorption at the solid/liquid interface. Conference on Colloid Chemistry: In Memoriam Aladar Buzagh, Proceedings, 7th, Eger, Hung., Sept. 23–26, 1996, 148–151

Ramsden JJ (1998) Biomimetic protein immobilization using lipid bilayers. Biosens. Bioelectronics 13:593–598

Ramsden JJ (1999) On protein-lipid membrane interactions. Colloids Surfaces B Biointerfaces 14:77–81

Ramsden JJ (2001) Multiple interactions in protein-membrane binding. NATO Science Series, 335:244–269. IOS Press, Amsterdam

Rayleigh L (1885) On waves propagating along the plane surface of an elastic solid. Proc. London Math. Soc. 17:4–11

Reipa V, Almeida J and Cole KD (2006) Long-term monitoring of biofilm growth and disinfection using a quartz crystal microbalance and reflectance measurements. J. Microbiol. Methods 66:449–459

Rickert J, Gopel W, Hayward GL, Cavic BA and Thompson M (1999) Biosensors based on acoustic wave devices. Sens. Update 5:105–139

Roberts LM and Dunker AK (1993) Structural changes accompanying chloroform-induced contraction of the filamentous phage fd. Biochemistry 32:10479–10488

Samoylov AM, Samoylova TI, Hartell MG, Pathirana ST, Smith BF and Vodyanoy V (2002a) Recognition of cell-specific binding of phage display derived peptides using an acoustic wave sensor. Biomol. Eng. 18:269–272

Samoylov AM, Samoylova TI, Pathirana ST, Globa LP and Vodyanoy VJ (2002b) Peptide biosensor for recognition of cross-species cell surface markers. J. Mol. Recognit. 15:197–203

Sauerbrey GZZ (1959) Use of quartz vibrator for weighing thin films on a microbalance. Z. Phys. 155:206–212

Sayler GS, Ripp SA, Applegate BM (2003) Bioluminescent biosensor device. US Patent Application 20030027241, filed July 20, 2001

Scherz P (2000) Oscillators and Timers. McGraw-Hill, Inc., New York

Selz KA, Samoylova TI, Samoylov AM, Vodyanoy VJ and Mandell AJ (2006) Designing allosteric peptide ligands targeting a globular protein. Biopolymers 85:38–59

Si S, Lia X, Fungb Y and Zhub D (2001) Rapid detection of Salmonella enteritidis by piezoelectric immunosensor. Microchem. J. 68:21–27

Si S, Ren F, Cheng W and Yao S (1997) Preparation of a piezoelectric immunosensor for the detection of Salmonella paratyphi A by immobilization of antibodies on electropolymerized films. Fresenius J. Anal. Chem. 357:1101–1105

Skládal P (2003) Piezoelectric quartz crystal sensors applied for bioanalytical assays and characterization of affinity interactions. J. Braz. Chem. Soc. 14:491–502

Sobotka H and Trurnit HJ (1961) Need title of their article here. In: Alexander P, Block RJ (eds) Analytical Methods of Protein Chemistry. Pergamon Press, Oxford, UK, pp 212–243

Song M-J, Yun D-H, Jin J-H, Min N-K and Hong S-I (2006) Comparison of effective working electrode areas on planar and porous silicon substrates for cholesterol biosensor. Jpn. J. Appl. Phys. 45:7197–7202

Sorial J and Lec R (2004) A piezoelectric interfacial phenomena biosensor. Masters Thesis. http://dspace.library.drexel.edu/handle/1860/82 Drexel University, Pennsylvania (accessed April 10, 2007)

Sorokulova IB, Olsen EV, Chen IH, Fiebor B, Barbaree JM, Vodyanoy VJ, Chin BA and Petrenko VA (2005) Landscape phage probes for Salmonella typhimurium. J. Microbiol. Methods 63:55–72

Spangler BD, Wilkinson EA, Murphy JT and Tyler BJ (2001) Comparison of the Spreeta®surface plasmon resonance sensor and a quartz crystal microbalance for detection of Escherichia coli heat-labile enterotoxin. Anal. Chim. Acta 444:149–161

Stadler H, Mondon M and Ziegler C (2003) Protein adsorption on surfaces: dynamic contact angle (DCA) and quartz-crystal microbalance (QCM) measurements. Anal. Bioanal. Chem 375:53–61

Storri S, Santoni T and Mascini M (1998) A piezoelectric biosensor for DNA hybridisation detection. Anal. Lett. 31:1795–1808

Su C-C, Wu T-Z, Chen L-K, Yang H-H and Tai D-F (2003) Development of immunochips for the detection of dengue viral antigens. Anal. Chim. Acta 479:117–123

Su X-L and Li SFY (2001) Serological determination of Helicobacter pylori infection using sandwiched and enzymatically amplified piezoelectric biosensor. Anal. Chim. Acta 429:27–36

Su X-L and Li Y (2004) A self-assembled monolayer-based piezoelectric immunosensor for rapid detection of Escherichia coli O157:H7. Biosens. Bioelectron. 19:563–574

Su X-L and Li Y (2005) A QCM immunosensor for Salmonella detection with simultaneous measurements of resonant frequency and motional resistance. Biosens. Bioelectron. 21:840–848

Su X, Low S, Kwang J, Chew VHT and Li SFY (2001) Piezoelectric quartz crystal based veterinary diagnosis for Salmonella enteritidis infection in chicken and egg. Sens. Actuators B 75:29–35

Sukhorukov GB, Montrel MM, Petrov AI, Shabarchina LI and Sukhorukov BI (1996) Multilayer films containing immobilized nucleic acids. Their structure and possibilities in biosensor applications. Biosens. Bioelectron. 11:913–922

Sykora JC (2003) Monolayers of biomolecules for recognition and transduction in biosensors. Doctoral Dissertation. Auburn University, Al. http://graduate.auburn.edu/auetd (accessed April 10, 2007)

Sykora JC, Neely WC and Vodyanoy V (2004) Solvent effects on amphotericin B monolayers. J. Colloid Interface Sci. 269:499–502

Tabacco MB, Qian X and Russo J (2004) Fluorescent virus probes for identification of bacteria. US Patent Application 20040191859, filed Mar 24, 2003

Tai D, Lin C, Wu T, Huang J and Shu P (2006) Artificial receptors in serologic tests for the early diagnosis of dengue virus infection. Clin. Chem. 58:1486–1491

Tan H, Deng L, Nie L and Yao S (1997) Detection and analysis of the growth characteristics of Proteus vulgaris with a bulk acoustic wave ammonia sensor. Analyst 122:179–184

Thompson M, Arthur C and Dhaliwal G (1986) Liquid-phase piezoelectric and acoustic transmission studies of interfacial immunochemistry. Anal. Chem. 58:1206–1209

Thompson M, Kiplingt A, Duncan-Hewitt W, Rajakovic L and Cavic-Vlasak B (1991) Thickness-shear-mode acoustic wave sensors in the liquid phase: a review. Analyst 116:881–890

Thust M, Schoning MJ, Schroth P, Malkoc U, Dicker CI, Steffen A, Kordos P and Luth H (1999) Enzyme immobilization on planar and porous silicon substrates for biosensor applications. J. Mol. Catal. B: Enzym. 7:77–83

Tiefenauer LX, Kossek S, Padeste C and Thiebaud P (1997) Towards amperometric immunosensor devices. Biosens. Bioelectron. 12:213–223

Trurnit HJ (1960) The spreading of protein monolayers. J. Colloid Sci. 15:1–13

Umezawa Y (1996) CRC handbook of ion-selective electrodes: selectivity coefficients. CRC Press, Boca Raton, Florida

Uttenthaler E, Kolinger C and Drost S (1998) Quartz crystal biosensor for detection of the African Swine Fever disease. Anal. Chim. Acta 362:91–100

Uttenthaler E, Schraml M, Mandel J and Drost S (2001) Ultrasensitive quartz crystal microbalance sensors for detection of M13-Phages in liquids. Biosens. Bioelectronics 16:735–743

Uzawa H, Kamiya S, Minoura N, Dohi H, Nishida Y, Taguchi K, Yokoyama S, Mori H, Shimizu T and Kobayashi K (2002) A quartz crystal microbalance method for rapid detection and differentiation of shiga toxins by applying a monoalkyl globobioside as the toxin ligand. Biomacromolecules 3:411–414

Vaughan RD, O'Sullivan CK and Guilbault GG (2001) Development of a quartz crystal microbalance (QCM) immunosensor for the detection of Listeria monocytogenes. Enzyme Microb. Technol. 29:635–638

Vaughan TJ, Williams AJ, Pritchard K, Osbourn JK, Pope AR, Earnshaw JC, McCafferty J, Hodits RA, Wilton J and Johnson KS (1996) Human antibodies with sub-nanomolar affinities isolated from a large non-immunized phage display library. Nature Biotechnol. 14:309–314

Victorov IA (1967) Rayleigh and Lamb Waves—physical theory and applications. Plenum Press, New York

Vig J and Ballato A (1998) Comments about the effects of non-uniform mass loading on a quartz crystal microbalance. IEEE Trans. Ultrason. Ferroelectrics Freq. Contr. 45:1123–1124

Vodyanoy V (1994) Functional reconstitution of mammalian olfactory receptor. Report. ARO-27364.10-LS

Voinova MV, Jonson M and Kasemo B (2002) "Missing mass" effect in biosensor's QCM applications. Biosens. Bioelectron. 17:835–841

Volker M and Siegmund HU (1997) Forster energy transfer in ultrathin polymer layers as a basis for biosensors. EXS 80:175–191

Wang J, Fu W, Liu M, Wang Y, Xue Q, Huang J and Zhu Q (2002) Multichannel piezoelectric genesensor for the detection of human papilloma virus. Chinese Med. J. (Engl) 115:439–442

White RM and Voltmer FW (1965) Direct piezoelectric coupling to surface elastic waves. Appl. Phys. Lett. 7:314–316

Wong YY, Ng SP, Ng MH, Si SH, Yao SZ and Fung YS (2002) Immunosensor for the differentiation and detection of Salmonella species based on a quartz crystal microbalance. Biosens. Bioelectron. 17:676–684

Wu T-Z, Su C-C, Chen L-K, Yang H-H, Tai D-F and Peng K-C (2005) Piezoelectric immunochip for the detection of dengue fever in viremia phase. Biosens. Bioelectron. 21:689–695

Wu Z-Y, Shen G-L, Li Z-Q, Wang S-P and Yu R-Q (1999) A direct immunoassay for Schistosoma japonium antibody (SjAb) in serum by piezoelectric body acoustic wave sensor. Anal. Chim. Acta 398:57–63

Wu Z, Wu J, Wang S, Shen G and Yu R (2006) An amplified mass piezoelectric immunosensor for Schistosoma japonicum. Biosens. Bioelectron. 22:207–212

Xomeritakis G, Liu NG, Chen Z, Jiang YB, Koehn R, Johnson PE, Tsai CY, Shah PB, Khalil S, Singh S and Brinker CJ (2007) Anodic alumina supported dual-layer microporous silica membranes. J. Membr. Sci. 287:157–161

Yakhno T, Sanin A, Pelyushenko A, Kazakov V, Shaposhnikova O, Chernov A, Yakhno V, Vacca C, Falcione F and Johnson B (2007) Uncoated quartz resonator as a universal biosensor. Biosens. Bioelectron. 22:2127–2131

Yao S, Tan H, Zhang H, Su X and Wei W (1998) Bulk acoustic wave bacterial growth sensor applied to analysis of antimicrobial properties of tea. Biotechnol. Prog. 14:639–644

Ye J, Letcher S and Rand AG (1997) Piezoelectric biosensor for the detection of Salmonella typhimurium. J. Food Sci. 62:1067–1071, 1086

Yilma S, Cannon-Sykora J, Samoylov A, Lo T, Liu N, Brinker CJ, Neely WC and Vodyanoy V (2007a) Large-conductance cholesterol-amphotericin B channels in reconstituted lipid bilayers. Biosens. Bioelectron. 22:1359–1367

Yilma S, Liu N, Samoylov A, Lo T, Brinker CJ and Vodyanoy V (2007b) Amphotericin B channels in phospholipid membrane-coated nanoporous silicon surfaces: implications for photovoltaic driving of ions across membranes. Biosens. Bioelectron. 22:1605–1611

Ying-Sing F, Shi-Hui S and De-Rong Z (2000) Piezoelectric crystal for sensing bacteria by immobilizing antibodies on divinylsulphone activated poly-m-aminophenol film. Talanta 51:151–158

Yun D-H, Song M-J, Hong S-I, Kang M-S and Min N-K (2005) Highly sensitive and renewable amperometric urea sensor based on self-assembled monolayer using porous silicon substrate. J. Kor. Phys. Soc. 47:S445–S449

Zhang J, Xie Y, Dai X, Wei W (2001) Monitoring of Lactobacillus fermentation process by using ion chromatography with a series piezoelectric quartz crystal detector. J. Microbiol. Methods 44:105–111

Zhang S, Wei W, Zhang J, Mao Y and Liu S (2002) Effect of static magnetic field on growth of Escherichia coli and relative response model of series piezoelectric quartz crystal. Analyst 127:373–377

Zhao J, Zhu W and He F (2005) Rapidly determining E. coli and P. aeruginosa by an eight channels bulk acoustic wave impedance physical biosensor. Sens. Actuators B 107: 271–276

Zhou X, Liu L, Hu M, Wang L and Hu J (2002) Detection of hepatitis B virus by piezoelectric biosensor. J. Pharmaceut. Biomed. Anal. 27:341–345

Zuo B, Li S, Guo Z, Zhang J and Chen C (2004) Piezoelectric immunosensor for SARS-associated coronavirus in sputum. Anal. Chem. 76:3536–3540

13

Amperometric Biosensors for Pathogenic Bacteria Detection

Ilaria Palchetti and Marco Mascini

Abstract

Biosensor technology has the potential to speed the detection of food pathogen and to increase specificity and sensitivity of the analysis. Electrochemical biosensors have some advantages over other analytical transducing systems, such as the possibility to operate in turbid media, comparable instrumental sensitivity, and possibility of miniaturisation. Basically electrochemical biosensor can be based on potentiometric, amperometric or impedimetric/conductimetric transducers. In this chapter, amperometric transducers will be described in detail. In particular amperometric biosensors for food pathogen will be reviewed as microbial metabolism-based, antibody-based (immunosensor), and DNA-based biosensor.

1. Introduction

Reliable detection methods of pathogenic, toxin-producing bacteria such as *Salmonella spp, Lysteria monocytogenes, E. coli 0157:H7*, or *Staphylococcus aureus,* that are responsible for some of the major world-wide food-borne outbreaks are more and more requested (De Boer and Beumer 1999; Ivnitski et al. 1999; Ivnitski et al. 2000; Banati 2003; Alocilja and Radke 2003; Lim 2003; Leonard et al. 2003; Olsen 2000; Siragusa et al. 1995). To meet the expectations of users, analytical instruments for bacteria detection must have the specificity to distinguish between different bacteria, the adaptability to detect different analytes, and the sensitivity to detect bacteria online and directly in real samples without pre-enrichment (Ivnitski et al. 2000). The device must also be simple and inexpensive to design and manufacture. Biosensor technology is claimed to satisfy these requirements (Ivnitski et al. 2000).

Despite the wide variation in biosensors and biosensor-related techniques that have been introduced, the consensus definition for these devices has remained fairly constant—an analytical device composed of a biological recognition element directly interfaced to a signal transducer, which together relate the concentration of an analyte (or group of related analytes) to a measurable response.

Biosensors for bacterial detection involve biological recognition components such as a receptor, nucleic acid, or antibodies attached to an appropriate transducer. Ivnitiski et al. (2000) listed the requirements that a biosensor should possess to be used in bacterial detection. Firstly, sensitivity and selectivity: the ideal biosensor should be able to detect a single bacterium in a reasonably small sample volume; it should be able to distinguish not only individual bacterial

Ilaria Palchetti and Marco Mascini • Dipartimento di Chimica, Università di Firenze, Italy.

M. Zourob et al. (eds.), *Principles of Bacterial Detection: Biosensors, Recognition Receptors and Microsystems,*
© Springer Science+Business Media, LLC 2008

species in the presence of other microorganisms or cells, but it should be able to distinguish individual bacterial strains of the same species. The precision should be less than 10%. The analytical system should discriminate between live and dead cells. The assay time should be in the range 5–10 min. The measurement should be direct, without pre-enrichment. Not all of these analytical requirements are satisfied at the same time by bacterial biosensors described in the literature. However, all the different biosensor schemes meet at least one of these technical requirements: i.e., they possess a highly automated format (single-button device); no skilled personnel are necessary to use the assay; In addition, compact, portable, hand held instrumentation can be used.

Depending on the method of signal transduction, biosensors can be classified into five basic groups: optical, mass, electrochemical, thermal, and magnetic. Electrochemical biosensors have some advantages over other analytical transducing systems, such as the ability to operate in turbid media, comparable instrumental sensitivity, and the capability of miniaturisation. As a consequence of miniaturisation, small sample volumes can be accommodated. Modern electro-analytical techniques (square wave voltammetry, chronopotentiometry, chronoamperometry, differential pulse voltammetry) have a very low detection limit (10^{-7}–10^{-9}M). In situ or online measurements are both available. Furthermore, the equipment required for electrochemical analysis is simple and cheap compared to most other analytical techniques (Ivnitski et al. 2000). Electrochemical biosensors can be based on potentiometric, amperometric, or conductrimetric transducers. In this chapter, amperometric transducers will be described in detail.

2. Amperometric Biosensors

L. C. Clark, Jr., is generally considered to be the pioneer of amperometric biosensors. In 1962, Clark reported trapping an enzyme that reacted with oxygen against the surface of a platinum electrode, using a piece of dialysis membrane. He then followed the activity of the enzyme, glucose oxidase, by changes in oxygen concentration. Because glucose oxidase is highly specific, it reacts only with glucose, producing hydrogen peroxide and gluconic acid. This general analytical scheme laid the foundation for the first commercial glucose biosensor.

Basically amperometric biosensors rely on an electrochemically active analyte that can be oxidized or reduced at a working electrode. This electrode is poised at a specific potential with respect to a reference electrode. The current produced is linearly proportional to the concentration of the electroactive product.

Typical equipment for amperometric analyses includes a three-electrode cell, based on a working electrode, a reference electrode, and an auxiliary electrode; as well as the voltage source and the devices for measuring current and voltage. An example of an amperometric apparatus is presented in Fig. 13.1. If measurements of the current are carried out by scanning the potential, then we obtain a voltammetric system. In this case, measurements are performed of the changes in time (τ) in the current (I) flowing through the system of electrodes in relation to the potential (E) applied to the working electrode. The registered changes in the current allow the $I(\tau) = f[E(\tau)]$ relationship to be drawn, which is called the voltammogram.

In both amperometric and voltammetric biosensors the transducer is an electrode. Typical electrode materials are platinum (Pt), gold (Au), carbon, and for some applications, mercury. Nowadays some innovative techniques for electrode preparation have been proposed: thick- and thin-film technology, silicon technology, etc.; they are characterised by the possibility of mass production and high reproducibility. Among these, the equipment needed for thick-film technology is less complex and costly, and thus this method is one of the most often used for sensor production. Thick-film technology consists of depositing inks on a substrate in a film of controlled pattern and thickness, mainly by screen-printing. The inks may be

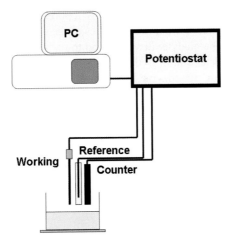

Figure 13.1. An example of a voltametric system: the working, auxiliary and reference electrodes inside the electrochemical cell; the potentiostat (data acquisition system); computer equipped with the program for recording, visualization, and archiving voltametric.

printed on several kinds of supports like glass, ceramic, or plastic. Many different types of ink are now commercially available, differing in composition and electrical behaviour. Recently, polymeric thick-film technology has been developed. In this procedure, thermoplastic-based resins are printed using a screen-printing technique. These kind of inks can be printed on plastic sheets, because they polymerise at low temperature (around 100°C), or using UV light. Screen-printing is a flexible and versatile technique, and one of its main advantages is the possibility of choosing the shape and dimensions of the sensor. The interest in these devices as electrochemical transducers in biosensor production is due to the ability to make them disposable; this characteristic arises from the low cost and the mass production of these systems. In electrochemistry a disposable sensor offers the advantage of not suffering from the electrode fouling that can result in loss of sensitivity and reproducibility. The single-use sensors have other important advantages, especially in the field of clinical analysis, such as avoidance of contamination among samples. Moreover, the micro- and nano-dimensions of these screen-printed devices are important to satisfy the needs of decentralized genetic testing. The high degree of reproducibility that is possible for these one-time use electrodes eliminates the cumbersome requirement for repeated calibration. The type of instrument used for these measurements is also very easy to obtain and can be inexpensive and compact; this allows for the possibility of in situ measurements.

Amperometric or voltammetric biosensors typically rely on an enzyme system that catalytically converts electrochemically non-active analytes into products that can be oxidized or reduced at a working electrode. Enzymes typically used in amperometric biosensors are oxidases that catalyze the following class of reactions:

$$\text{Substrate} + O_2 \rightarrow \text{Product} + H_2O_2$$

As a result of the enzyme-catalyzed reaction, the substrate concentration can be determined by amperometric detection of O_2 or H_2O_2. An example of this configuration would be an oxygen-consuming enzyme coupled to a Clark electrode. The ambient oxygen concentration is then continuously monitored as it diffuses through a semipermeable membrane and is reduced at a platinum electrode. Other common configurations include the use of oxidases specific to various substrates to produce H_2O_2, which is then oxidized at the electrode surface.

Although these devices are the most commonly reported class of biosensors, they tend to have a small dynamic range due to saturation kinetics of the enzyme, and a large over-potential is required for oxidation of the analyte; this may lead to oxidation of interfering compounds as well (e.g., ascorbate in the detection of hydrogen peroxide).

In addition to their use in enzyme-based biosensors, amperometric transducers have also been used to measure enzyme-labelled tracers for affinity-based biosensors (mainly immunosensors and genosensors). Enzymes which are commonly used include: horse radish peroxidase (HRP) and alkaline phosphatase (AP). Compounds of interest, measured using disposable amperometric electrodes, include PCBs, triazines, and various toxins.

Limitations for amperometric and voltammetric transducers include potential interferences to the response if several electroactive compounds generate false current values. These effects have been eliminated through the use of selective membranes, which carefully control the molecular weight or the charge of compounds which have access to the electrode.

2.1. Microbial Metabolism-Based Biosensors

Various combinations of biosensors based on the monitoring of microbial metabolism have been reported in the literature and summarised in some recent reviews (Invitski et al. 2000). They can be based on direct measurements of physical phenomena occurring during the biochemical reactions on a transducer surface. Parameters such as oxygen consumption can be measured by electrochemical transducers, e.g., the amperometric Clark-type oxygen electrode (Suzuki et al. 1991).

An interesting strategy to detect bacteria through their metabolic processes was based on the electrochemical detection of specific marker enzymes (mainly oxido-reductase), after incubation in an appropriate medium.

As a case study, we report the determination of the enzyme β-D-glucuronide glucuronosohydrolase (GUS) and of the enzyme β-d-galactosidase (B-GAL) for coliform detection. Current methods of *E. coli* detection using GUS or B-GAL involve incorporation of chromogenic substrates into culturing media and spectrophotometric monitoring of the liberated chromophore. Conversion of the substrate p-nitrophenyl-b-D-glucuronide (PNPG) to p-nitrophenol (PNP) and D-glucuronic acid indicates the presence of GUS, hence showing the presence of *E. coli*. Mulchandani et al. (2005) reported an electro-oxidative method for PNP detection using a Moraxella species-modified carbon paste electrode. Moraxella sp. degrades PNP and produces a more electroactive hydroquinone as an early intermediate. Hydroquinone is oxidised at a lower potential (0.3 V) than PNP. Togo et al. (2007) investigated an electrochemical method of GUS detection based on the production of PNP from PNPG and PNP degradation by a Moraxella sp. *Pseudomonas putida* JS444 was also used in initial studies because of its ability to degrade PNP faster than the Moraxella sp.

B-GAL has also been used for enumerating coliform in water media. Specifically, B-GAL catalyses the breakdown of lactose into galactose and glucose. Perez et al. (2001) described a rapid method for detection of viable *E. coli* in water samples, using amperometric detection of 4-aminophenol (4-AP) after hydrolysis of the substrate 4-aminophenyl-b-d-galactopyranoside (4-APGal) by the bacterial enzyme β-d-galactosidase (Fig. 13.2). The bacteria were recovered

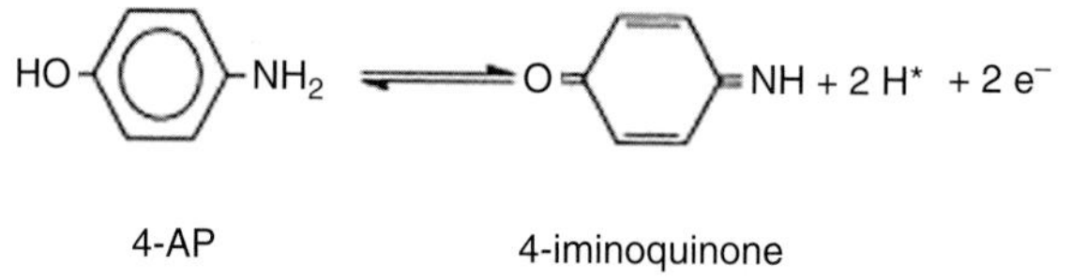

$$\text{HO}-\bigcirc-\text{NH}_2 \rightleftharpoons \text{O}=\bigcirc=\text{NH} + 2\,\text{H}^+ + 2\,\text{e}^-$$

4-AP   4-iminoquinone

Figure 13.2. Oxidation scheme of 4AP at the electrode surface.

Table 13.1. Numbers of different β-D-galactosidase positive bacteria in environmental water samples; each value is the average of triplicate measurements.

Sample	Bacterial Number ± S.D. R2A +X-gal (cfu/ml)	Bacterial Number ±S.D. Cromocult (cfu/ml)	
	X-gal+	Coliforms	E. coli
Coastal water (S1)	$3.5 \times 10^2 \pm 1 \times 10^2$	1 ± 1	0.35 ± 0.6
Coastal water (S2)	$6.0 \times 10^2 \pm 3 \times 10^1$	8 ± 2	1.0 ± 0.6
Sewage effluent (S3)	$1.6 \times 10^5 \pm 1 \times 10^4$	$8.0 \times 10^3 \pm 9 \times 10^2$	$2.0 \times 10^3 \pm 4 \times 10^2$
Harbour 1 (S4)	$2.5 \times 10^2 \pm 4 \times 10^1$	34 ± 10	10 ± 2
Harbour 2 (S5)	$4.0 \times 10^2 \pm 6 \times 10^1$	10 ± 6	1 ± 1
River (S6)	$1.0 \times 10^3 \pm 5 \times 10^1$	32 ± 6	8 ± 3

by filtration and incubated in a selective medium, lauryl sulphate broth (LSB) supplemented with the substrate 4-APGal at 44.5°C. The electrochemically active molecule 4-AP was produced after hydrolysis of 4-APGal by the enzyme b-galactosidase. 4-AP was measured by amperometry and was detected at a due concentration of *E. coli*. The time necessary for reaching that concentration was inversely related to the initial *E. coli* concentration of the sample. Environmental samples and suspensions of *E. coli* IT1 were assayed. 4-AP was detected after 7.3 and 2.0 h in samples containing initial concentrations of *E. coli* IT1 of 4.5 and 4.5×10^6 cfu ml^{-1}, respectively. For environmental samples with initial *E. coli* concentrations of 1.0 and 2.0×10^3 cfu ml^{-1}, 4-AP were detected after 10 and 6.6 h, respectively. Table 13.1 summarizes some experimental results of the proposed method.

Serra et al. (2005) reported a rapid method for the detection of faecal contamination in water, based on the use of a tyrosinase composite biosensor for improved amperometric detection of β-galactosidase activity. The method relies on the detection of phenol released after the hydrolysis of phenyl -D-galactopyranoside (PG) by Beta-galactosidase. Under the optimized PG concentration and pH values, a detection limit of 1.2×10^{-3} unit of B-galactosidase/ml was obtained. The capability of the sensor for the detection of *Escherichia coli* was evaluated using polymyxin sulfate to allow permeabilization of the bacteria membrane. A detection limit of 1×10^6 cfu of *E. coli*/ml was obtained with no preconcentration or pre-enrichment.

2.2. Immunosensors

Immunological detection with antibodies is another technology that has been successfully employed for the detection of specific microorganisms and microbial toxins. The suitability of these antibodies depends mainly on their specificity.

Immobilisation of antibodies directly on the surface of an electrode for rapid bacterial detection has been demonstrated (Brewster et al. 1996). Fewer than 100 bacteria labelled with an antibody conjugate were detectable within 5 min after capture on the electrode. Brewster and Mazenko (1998) developed an assay for rapid determination of *E. coli* O157:H7 which combined filtration capture with amperometric detection; the method had a detection limit of 5×10^3 cfu/ml with a 25 min analysis time. The major obstacle to decreasing the lower detection limit was the nonspecific adsorption of the conjugate on the membrane.

Abdel-Hamid et al. (1999) eliminated the need for pre-enrichment by the use of a flow-through immunofiltration assay combined with an amperometric sensor. They reported a low detection limit of 50 cells/ml and an overall analysis time of 30 min, but with a limited analytical range of 100–600 cells/ml. The immunosensor consists of a disposable antibody-modified filter membrane resting on top of a hollow carbon rod which acts as the working electrode. Another hollow carbon rod acts as a counter electrode, and a hollow Ag/AgCl disk as a reference electrode. The authors used a special device in which the liquid flows from the inlet of the immunosensor, through the filter membrane, and then through the hollow channel formed in the working, reference, and counter electrodes, respectively. A sandwich scheme of immunoassay was employed. The activity of the peroxidase label (captured on the membrane surface) was measured using an amperometric technique with iodide ions in a phosphate buffer solution acting as the mediator. A polarization potential of 0.0 V versus Ag/AgCl was applied between the working and reference electrodes.

Various strategies have been proposed in the literature to selectively capture the microorganism; here we want to focus on the coupling of the selectivity of antibody-coated superparamagnetic beads with the rapidity and sensitivity of electrochemical methods.

Magnetic beads are uniform microparticles comprising superparamagnetic material wrapped in a polymer shell. They have an even dispersion of magnetic material (Fe_2O_3 and Fe_3O_4) throughout the bead and are coated with a polymer that allows the adsorption or coupling of various molecules. The possibility of maintaining the beads in suspension by shaking or rotating ensures a rapid and efficient binding of the target analytes. Shape and size uniformity prevent "clumping" and nonspecific binding due to irregularly shaped particles. Their superparamagnetic properties allow the quantitative magnetic separation of the beads and ensure that they retain no residual magnetism when removed from the magnetic field. They have been used for the selective separation of bacteria and their quantisation using different methods. The inoculation of the beads on agar plates (with bacteria immobilized on their surfaces), the staining of the immobilized bacteria with a fluorescent dye, the development of ELISA tests, and the use of electrochemiluminescence were all accomplished using immunomagnetic separation.

The approach proposed by Perez et al. (1998) and developed in the author's lab is based on an amperometric flow-injection system for the measurement of viable *E. Coli* O157, measuring their respiratory activity (Fig. 13.3). The selective immunological separation of

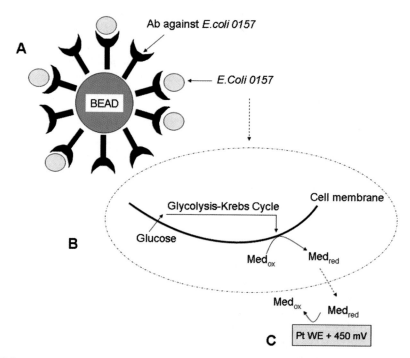

Figure 13.3. Schematic model of the whole method, performed in three separate steps: A) the selective capture of E. coli O157 using antibody derivatized magnetic particles, B) the reaction of bacteria with a mediator, and C) the electrochemical measurement of the reduced mediator using an amperometric method. (Reproduced with permission from Anal Chem. 1998, *70*, 2380–2386. Copyright 1998 American Chemical Society).

E. Coli was performed by antibody-coated magnetic particles. The kinetics and the capacity parameters regarding the attachment of the bacteria to the immunobeads were studied. The immunomagnetic separation was then used in conjunction with electrochemical detection to measure the concentration of viable bacteria. Electrochemical detection was carried out using redox mediators, potassium hexacyanoferrate(III) and 2,6-dichlorophrnolindophenol. The measurement was performed using an FIA system. A calibration curve of colony-forming units (cfu) against electrochemical response was obtained (Fig. 13.4). The detection limit was 10^5 cfu/ml, and the complete assay was performed in 2 h. This technique could easily be automated, and the analysis can be performed quickly and continuously. The renewal of the immunosensor sensing surface was accomplished by removing the magnet and washing down the magnetic particles. The immunosensor was then ready for the injection of new antibody-modified magnetic particles for another cycle. Some advantages over ELISA methods are the direct detection of viable cells (and not the total bacterial load) and the need for only one antibody (not enzyme-labelled), thus making the assay faster (only one washing step is necessary) and less expensive.

Sippy et al. (2003) described a rapid and sensitive technique to detect low numbers of the model organism *E. coli* O55, combining lateral flow immunoassay (LFI) for capture and amperometry for sensitive detection. Nitrocellulose membranes were used as the solid phase for selective capture of the bacteria, using antibodies to *E. coli* O55. Different concentrations of *E. coli* O55 in Ringers solution were applied to LFI strips and allowed to flow through the membrane to an absorbent pad. The capture region of the LFI strip was placed in close contact with the electrodes of a Clarke cell poised at $+0.7$ V for the detection of hydrogen peroxide. Earlier research showed that the consumption of hydrogen peroxide by a bacterial catalase provided a sensitive indicator of aerobic and facultative anaerobic microorganism numbers. Modification and application of this technique to the LFI strips demonstrated that the consumption of 8 mM hydrogen peroxide was correlated with the number of microorganisms presented to the LFI strips in the range of 2×10^1–2×10^7 colony forming units (cfu). Capture

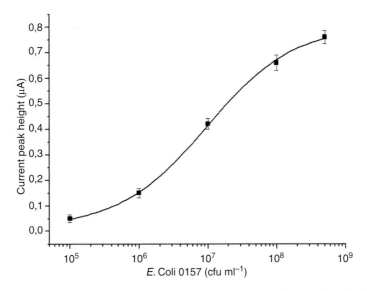

Figure 13.4. Amperometric FIA calibration curve of E. coli O157 performed by coupling the immunomagnetic separation and the further reaction with the mediator mixture (30 mM each mediator in PBS, KCl). Glucose concentration, 4 mM; temperature, 37 °C; incubation time, 1 h. Each value is the average of three measurements, and the error bars represent the standard deviation of the measurements. (Reproduced with permission from Anal Chem. 1998, *70*, 2380–2386. Copyright 1998 American Chemical Society).

efficiency was dependent on the number of organisms applied and varied from 71% at 2×10^2 cfu to 25% at 2×10^7 cfu. The procedure was completed in less than 10 min and could detect less than 10 cfu captured from a 200 ml sample applied to the LFI strip. The described method provides proof of principle for the basis of a new technological approach to the rapid, quantitative, and sensitive detection of bacteria that express catalase activity.

Another approach is that described by Mittelmann et al. (2002); these authors developed an amperometric biosensor based on the activity of beta D-galactosidase for quantifying coliforms, represented by *E. coli* and *Klebsiella pneumoniae*. The experimental system allows the simultaneous analysis of eight samples using disposable screen-printed electrodes. The specific detection of *E. coli* was achieved by using an antibody-coated electrode that specifically binds the target bacteria. The detection is based on the electrochemical measurement of B-galactosidase activity, using p-amino-phenyl-B-D-galactopyranoside as the substrate. The same electrochemical method developed for the identification of low concentrations of *E. coli* was applied by Yemini et al. to *Bacillus cereus* as a model for *B. anthracis*, and to *Mycobacterium smegmatis* as a model for *M. tuberculosis*. Enzymatic activity was determined electrochemically using as substrate para-Amino-phenyl-α-D-glucopyranoside (p-AP-α-GLU) for *B. cereus* and para-amino-phenyl-β-Dglucopyranoside (p-AP-β-GLU) for *M. smegmatis*. The product of the reaction, p-aminophenol (p-AP), is oxidized at the carbon anode at 220 mV vs. (Ag/AgCl) reference electrode. The detection procedure takes less than 8 h and the results are shown online throughout the process.

2.3. DNA-Based Biosensors

In recent years various kinds of electrochemical biosensors based on identification of the bacterial nucleic acid have been developed.

A DNA biosensor is defined as an analytical device incorporating an oligonucleotide, with a known sequence of bases, or a complex structure of DNA (like DNA from calf thymus) either integrated within or intimately associated with the electrode.

DNA biosensors can be use to detect DNA fragments or either biological or chemical species. In the first application, DNA is the analyte and it is detected through the hybridisation reaction (this kind of biosensor is also called a genosensor). In the second application, DNA plays the role of receptor of specific biologic and/or chemical species, such as target proteins, pollutants, or drugs.

The development of DNA-based biosensors for the detection of specific nucleic acid sequences consists in the immobilisation onto the surface of a chosen transducer of an oligonucleotide with a specific base sequence called a "probe." The complementary sequence ("target") present in the sample solution is recognised and captured by the probe through the hybridisation reaction. The evaluation of the extent of the hybridisation determines if the complementary sequence of the probe is present in the sample solution or not.

Mainly, DNA biosensors have been coupled to PCR, as the specific detection method of the amplified base sequence.

Electrochemical transducers have received considerable recent attention with regard to the detection of DNA hybridization. Some excellent reviews summarised the recent progress in this field (Lucarelli et al. 2004,; Kerman et al. 2004; Wang 2002; Palecek 2001; Palecek 2002; Pividori et al. 2000), and a few examples of pathogen detection using genosensors are cited.

In our recent paper we describe the simultaneous detection of different food pathogenic bacteria by means of a disposable electrochemical low density genosensor array. The analytical method relied on the use of screen-printed arrays of gold electrodes, modified using thiol-tethered oligonucleotide probes (Fig. 13.5). The samples identifying the bacteria of interest were obtained from the corresponding genomic DNAs through PCR amplification. These unmodified

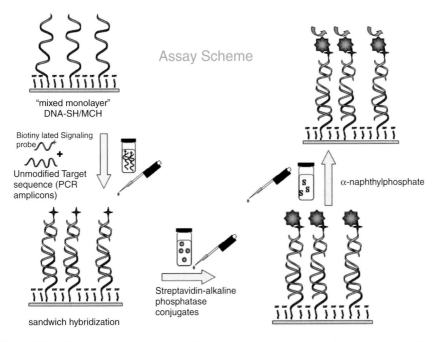

Figure 13.5. Scheme of the detection method based on a gold screen-printed electrochemical transducer.

PCR products were captured at the electrode interface via sandwich hybridisation with surface-tethered probes and biotinylated signalling probes. The resulting biotinylated hybrids were coupled with a streptavidin-alkaline phosphatase conjugate and then exposed to an α-naphthyl phosphate solution. Finally, differential pulse voltammetry was used to detect the α-naphthol signal (Fig. 13.6).

The sequences of synthetic oligonucleotide probes and targets, as well as PCR primers and PCR amplicons, are listed in Table 13.2.

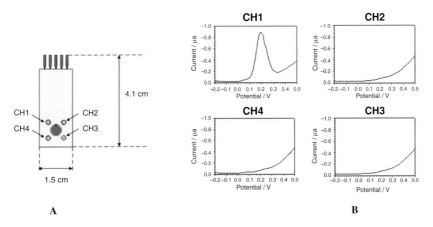

Figure 13.6. A) Scheme of the screen-printed array of electrodes comprising four gold working electrodes and a silver pseudo-reference electrode (in the center); and B) example of simultaneously acquired DPV signals, obtained after having hybridised the synthetic target sequence (20 nM) on CH1 only—CH2, CH3 and CH4 were interacted with a non-complementary oligo. (Reprinted from Farabullini et al. (2007), with permission of Elsevier).

Table 13.2. PCR Amplicons and synthetic oligonucleotide probes and targets. Within each amplicon the primers used for PCR amplification are underlined; the capture probes are indicated in bold-italics.

Salmonella spp.: PCR amplified fragment of the InvA gene (388 bp)

5'-GCC*GCGCGCGAACGG*CGAAGCGTACTGGAAA*GGGAAAGCCAGC*TTTACGGTTCCTTTGACGGTGCGAT
GAAGTTTATCAAAGGTGAC*GCTATTGCCGGC*ATCATTATTATTCTTTGTGAACTTTATTGGCGGTATTTCG
GTGGGGATGACTCGCCATGGTATGGATTTGTCCTCCGCCCTGTCTACTTATACCATGCTGACCATTGGTG
ATGGTCTTGTCGCCCAGATCCCCGCATTGTTGATTGCGATTAGTGCCGGTTTTATCGTGACCCGCGTAAA
TGGPCGATAGCGATAATATGGGGCGGAATATCATGACGCAGCTGTTGAACAACCCATTTGTATTGGTTG
TTACGGCTATTTTGACCATTTCAATGGGAACTCTGCCGGGAT-3'
Capture probes (12-mer):
CP0: 5' - HS-(CH$_2$)$_6$- GCG CGC GAA CGG - 3'
CP1: 5' - HS-(CH$_2$)$_6$- GGG AAA GCC AGC - 3'
CP2: 5' - HS-(CH$_2$)$_6$- GCT ATT GCC GGC - 3'
Signaling probe: 5' - TTT GTG AAC TTT ATT GGC GG -TEG -biotin - 3'
Synthetic target: 5' - CCG CCA ATA AAG TTC ACA AAACG CCG TTC GCG CGC - 3'
Lysteria monocytogenes: PCR amplified fragment of the InlA gene (255 bp)
5'-*CCTAGCAGGTCT*AACCGCACTCACTAACTTAGAGCTAAATGAAAATCAGTTAGAAGATATATTAGCCCAA
TTTCTAACCTGAAAAATCTCACATATTTAACGTTGTACTTTAATAATATAAGTGATATAAGCCCAGTTT
CTAGTTTAACAAAGCTTCAAAGATTATTTTTCTATAATAACAAGGTAAGTGACGTAAGCTCGCTTGCGA
ATTTAACCAATATTAATTGGCTTTCGGCTGGGCATAACCAAATTAGCGA -3'
Capture probe (12 mer): 5' - HS-(CH$_2$)$_6$- CCT AGC AGG TCT - 3'
Signaling probe: 5' - CGC TTG CGA ATT TAA CCA AT -TEG -biotin - 3'
Synthetic target: 5' - ATT GGT TAA ATT CGC AAG CGG TTA GAC CTG CTA GG - 3'
Staphylococcus aureus
Capture probe (12 mer): 5' - HS-(CH$_2$)$_6$- CAA TGT GCG GGT - 3'
Signaling probe: 5' - CGA TCA ATT TAT GGC TAG AC -TEG -biotin - 3'
Synthetic target: 5' - GTC TAG CCA TAA ATT GAT CGT TGA CCC GCA CAT TG - 3'
E. Coli O157:H7
Capture probe (12 mer): 5' - HS-(CH$_2$)$_6$-CGC GAA CAG TTC - 3'
Signaling probe: 5' - ATC AAG CAT GCC TGA TA -TEG -biotin - 3'
Synthetic target: 5' - TAT CAG GCA TGG CTC TTG ATA ACG AAC TGT TCC GG - 3'

The analytical strategy was based on the identification of toxins produced by the bacteria of interest. Thus, for each bacteria strain, the capture and signalling probes were selected within the sequence of a gene encoding a strain-specific toxin. This step was the most important in order to have a strain-specific assay. The genes were chosen by considering the frequency of the presence of the expressed toxins in contaminated foods and the virulence of the different bacterial strains. As *Salmonella enterica* is the most important cause of salmonellosis infections (Malorny et al. 2003), the gene codifying for the invasion protein A (invA) was considered in the case of this bacteria. The gene of the interlysine A (inlA) was used to design a specific probe for *Lysteria monocytogenes* (Jaradat et al. 2002), while the gene of the enterotoxin A was used for *Staphylococcus aureus* detection (Balaban and Rasooly 2000). For *E. coli* 0157:H7, the probe was designed from the gene codifying for Hemolysin A (hlyA) (Call et al. 2001).

Since the same base sequence can be present in different genes (Malorny et al. 2003), a detailed study on the specificity of the chosen base sequences was done, in order to minimise the number of false positive results. In the case of *Salmonella* and *Lysteria*, the specificity of PCR primers was controlled by using the BLAST (Basic Local Alignment Search Tool) (http://www.ncbi.nlm.nih.gov/blast) program. The sequences were aligned by means of MultAlin software (http://prodes.toulouse.inra.fr/multalin/multalin.html). Sequences of highly specific regions were selected to design toxin-specific primers. Toxin-specific oligonucleotide probes were designed within the region flanked by the primers. *E. coli*

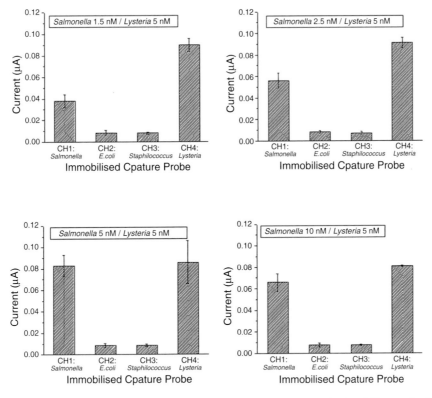

Figure 13.7. Simultaneous analysis of diluted *Salmonella and Lysteria* amplicons mixed in different ratios (1:5, 1:2, 1:1, 2:1). Each sample was analysed in triplicate. The error bars represent the standard deviation (n = 3). (Reprinted from Farabullini et al. (2007), with permission of Elsevier).

0157:H7 and *Staphylococcus aureus* synthetic probes and targets were chosen according to the literature (Call et al. 2001; Sergeev et al. 2004).

Simultaneous detection of different samples was achieved, analysing solutions of both *Salmonella* and *Lysteria* amplicons mixed in different ratios (Fig. 13.7).

As expected, the analytical signals were observed only at the positions modified with the corresponding capture probe, demonstrating the feasibility of the simultaneous detection of different analytes. The nonspecific signal observed at the WEs modified with *Staphylococcus* and *E. Coli* probes was comparably negligible, thus highlighting the selectivity of the procedure. Control experiments demonstrated that this undesired signal was essentially due to the direct nonspecific adsorption of streptavidin-alkaline phosphatase conjugate onto the electrode surface, thus excluding cross-hybridisation phenomena. The dependence of the analytical signal on the concentration of the amplicon from *Salmonella* was also shown in Fig. 13.7. The electrochemical response rose with the sample concentration up to 5 nM and then levelled off. Interestingly, the signal of 5 nM *Lysteria* amplicon appeared to be unaffected by the presence of different amounts of the other samples, also demonstrating the absence of any cross-interference.

Elsholz et al. (2006) described a low-density electrical 16S rRNA specific oligonucleotide microarray and an automated analysis system for the identification and quantitation of pathogens. The pathogens are *Escherichia coli*, *Pseudomonas aeruginosa*, *Enterococcus faecalis*, *Staphylococcus aureus*, and *Staphylococcus epidermidis*, which are typically involved in urinary tract infections. Interdigitated gold array electrodes (IDA-electrodes), which have structures in the nanometer range, have been used for very sensitive analysis. Thiol-modified oligonucleotides are immobilized on the gold IDA as capture probes. They mediate the specific recognition of the target 16S rRNA by hybridization. Additionally, three unlabeled oligonucleotides are hybridized in close proximity to the capturing site. They are supporting molecules, because they improve the RNA hybridization at the capturing site. A biotin-labelled detector oligonucleotide is also allowed to hybridize to the captured RNA sequence. The biotin labels enable the binding of avidin alkaline phophatase conjugates. The phosphatase liberates the electrochemical mediator p-aminophenol from its electrically inactive phosphate derivative. The electrical signals were generated by amperometric redox cycling and detected by a unique multipotentiostat. The readout signals of the microarray are position specific current, and change over time in proportion to the analyte concentration. If two additional biotins are introduced into the affinity binding complex via the supporting oligonucleotides, the sensitivity of the assays increases more than 60%. The limit of detection of *Escherichia coli* total RNA has been determined to be 0.5 ng/L. The control of fluidics for variable assay formats as well as the multichannel electrical readout and data handling have all been fully automated. The fast and easy procedure does not require any amplification of the targeted nucleic acids by PCR.

Wang (2002) described a genosensor for *Cryptosporidium* as well as *E. coli*, *Giardia*, and *Microbacterium tubercolosis* detection, based on the immobilisation of specific oligonucleotides onto a carbon-paste electrode and chronopotentiometry for monitoring the hybridisation events.

3. Conclusion and Future Perspectives

In this chapter we described in detail the electrochemical format developed in the author's lab. However, research in the field of detection of pathogens by means of biosensors is increasing, and excellent reviews are appearing in the literature describing various strategies. However, there are a number of practical and technical issues which must be overcome in the development of bacterial biosensors prior to their successful commercialisation. The first of these is the sensitivity of the assays in different samples.

For instance, according to the U.S. Environmental Protection Agency (EPA), the highest level of coliform that is allowed in drinking water is 5% (http://www.epa.gov/safewater/contaminants/ecoli.html), but the maximum contaminant level goal (MCLG), the level of contamination in drinking water below which there is no known or expected risk to health, is zero (no coliforms in 100 ml). Hence a biosensor must be able to provide a detection limit as low as a single coliform organism in 100 ml of drinking water, with a rapid analysis time and low cost.

Nowadays there is a great effort expended to improve technological aspects in the biosensor field, especially in the development of fully automated analytical systems based on combining multisensor technology with discriminative mathematical methods, or with remote sensing technology, in order to have a system able to detect different parameters, in a fully automated way and in remote conditions. These improvements will be also useful in the field of bacterial detection.

References

Abdel-Hamid I, Ivnitski D, Atanasov P, Wilkins E (1999) Flowthrough immunofiltration assay system for rapid detection of E. coli O157: H7. Biosensors and Bioelectronics 14:309–316

Abdel-Hamid I, Ivnitski D, Atanasov P, Wilkins E (1999) Highly sensitive flow-injection immunoassay system for rapid detection of bacteria. Analytica Chimica Acta 399:99–108

Alocilja EC, Radke SM (2003) Market analysis of biosensors for food safety. Biosensor and Bioelectronics 18:841–846

Banati D (2003) The EU and candidate countries: How to cop with food safety policies? Food Control 14:89–93

Balaban N, Rasooly A (2002) Staphylococcal enterotoxins. Int. J. Food Microbiol. 61:1–10

Brewster JD, Gehring AG, Mazenko RS, Van Houten LJ, Crawford CJ (1996) Immunoelectrochemical assays for bacteria: use of epifluorescence microscopy in development of an assay for salmonella. Analytical Chemistry 68:4153–4159

Brewster JD, Mazenko RS (1998) Filtration capture and immunoelectrochemical detection for rapid assay of Escherichia coli O157:H7. Journal of Immunological methods 211:1–18

Call DR, Brockman FJ, Chandler DP (2001) Detecting and genotyping E.Coli 0157:H7 using multiplexed PCR and nuclei acid microarrays. J. Food Microbiol. 67:71–80

De Boer E, Beumer RR (1999) Methodology for detection and typing of foodborne microorganisms. Int. J. Food Microbiol. 1-2:119–130

Del Giallo ML, Lucarelli F, Cosulich E, Pistarino E, Santamaria B, Marrazza G, Mascini M (2005) Steric factors controlling the surface hybridization of PCR amplified sequences. Anal. Chem. 77 (19), 6324–6330

Elsholz B, Wo R, Blohm L, Albers J, Feucht H, Grunwald T, Jurgen B, Schweder T, Hintsche Rainer (2006) Automated Detection and Quantitation of Bacterial RNA by Using Electrical Microarrays Anal. Chem. 78:4794–4802

Farabullini F, Lucarelli F, Palchetti I, Marrazza G, Mascini M (2007) Disposable Electrochemical Genosensor for the Simulataneous Analysis of Different Bacterial Food Contaminants. Biosensors and Bioelectronics 22:1544–1549

Invitski D, Abdel-Hamid I, Atanasov P, Wilkins E (1999) Biosensors for detection of pathogenic bacteria. Biosens. Bioelectron. 14:599–624

Invitski D, Abdel-Hamid I, Atanasov P, Wilkins E, Striker S (2000) Application of Electrochemical Biosensors for Detection of Food Pathogenic Bacteria. Electroanalysis 12:5 317–325

Jaradat ZW, Schutze GE, Bhunia AK (2002) Genetic homogeneity among Listeria monocytogens strains from infected patients and meat products from two geographic locations determined by phenotyping, ribotyping and PCR analysis of virulane genes. Int. J. Food Microbiol. 76:1–10

Kerman K, Kobayashi M, Tamiya E (2004) Recent trends in electrochemical DNA biosensor technology. Meas. Sci. Technol. 15:R1–R11

Leonard P, Hearty S, Brennan J, Dunne L, Quinn J, Chakraborty T, O'Kennedy R (2003) Advances in biosensors for detection of pathogens in food and water. Enzyme Microb. Tech. 32 (1-2):3–13

Lim DV (2003) Detection of Microorganisms and Toxins with Evanescent Wave Fiber-Optic Biosensors. Proceedings of the IEEE 91 (6):902–907

Lucarelli F, Marrazza G, Turner APF, Mascini M (2004) Carbon and gold electrodes as electrochemical transducers for DNA hybridisation sensors. Biosens. Bioelectron. 19:515–530

Malorny B, Hoorfar J, Bunge C, Helmuth R (2003) Multicenter Validation of the analytical accuracy of salmonella PCR: towards an International Standard. Appl. Environ. Microbiol. 69 (1):290–296

Manzano M, Cocolin L, Astori G, Pipan C, Botta GA, Cantoni C, Comi G (1998) Development of a PCR microplate-capture hybridization method for simple, fast and sensitive detection of Salmonella serovars in food. Mol. Cell. Probes 12:227–234

Mittelmann AS, Ron EZ, Rishpon J (2002) Amperometric quantification of total coliforms and specific detection of Escherichia coli. Analytical Chemistry 74:903–907

Mulchandani P, Hangarter CM, Lei Y, Chen W, Mulchandani A (2005) Amperometric microbial biosensor for pnitrophenol using Moraxella sp.-modified carbon paste electrode. Biosens Bioelectron 21:523–527

Olsen JE (2000) DNA-based methods for detection of food-borne bacterial pathogens. Food Res. Int. 33:257–266

Palecek E (2002) Past, present and future of nucleic acids electrochemistry. Talanta 56:809–819

Palecek E, Fojta M (2001) Detecting DNA hybridisation and damage. Anal. Chem. 73:74A–83A

Perez FG, Tryland I, Mascini M, Fiksdal L (2001) Rapid detection of Escherichia coli in water by a culture-based amperometric method. Analytica Chimica Acta 427:149–154

Perez FG, Mascini M, Tothill IE, Turner APF (1998) Immunomagnetic Separation with Mediated Flow Injection Analysis Amperometric Detection of Viable Escherichia coli O157. Anal Chem 70:2380

Perez FG, Mascini M, Tothill IE, Tuner APF (1998) Anal. Chem. 70:2380–2386

Pividori MI, Merkoci A, Alegret S (2000) Electrochemical genosensor design: immobilisation of oligonucleotides onto transducer surfaces and detection methods. Biosens. Bioelectronics 15:291–303

Sergeev N, Volokhov D, Chizhikov V, Rasooly A (2004) Simultaneous analysis of multiple stapylococcal enterotoxin genes by an oligonucleotide microrraray assay. J. Clin. Microbiol. 42:2134–2143

Serra B, Morales MD, Zhang J, Reviejo AJ, Hall EH, Pingarron JM (2005) In-a-Day Electrochemical Detection of Coliforms in Drinking Water Using a Tyrosinase Composite Biosensor. Anal. Chem. 77:8115–8121

Siragusa GR, Cutter CN, Dorsa WJ, Koohmaraie M (1995) J. Food Protect. 58:770–775

Sippy N, Luxton R, Lewis RJ, Cowell DC (2003) Rapid electrochemical detection and identification of catalase positive micro-organisms. Biosensors and Bioelectronics 18:741–749

Suzuki H, Tamiya E, Karube I (1991) Electroanalysis, disposable amperometric CO_2 sensor, employing bacteria, and a miniaturized oxygen electrode 3:53–57

Togo AC, Collins WV, Leigh LJ, Pletschke BI (2007) Novel detection of Escherichia coli b-D-glucuronidase activity using a microbially-modified glassy carbon electrode and its potential for faecal pollution monitoring. Biotechnol Lett. 29:531–537

Wang J (2002) Electrochemical nucleic acid biosensors. Anal. Chim. Acta. 469:63–71

Yang M, McGovern ME, Thompson M (1997) Anal. Chim. Acta 346:259–275

Yemini M, Yaron L, Yagil E, Rishpon J (2007) Specific electrochemical phage sensing for Bacillus cereus and Mycobacterium smegmatis. Bioelectrochemistry 70:180–184

14

Microbial Genetic Analysis Based on Field Effect Transistors

Yuji Miyahara, Toshiya Sakata and Akira Matsumoto

Abstract

In this chapter, potentiometric detection methods for microbial DNA involved recognition events by use of genetic field effect devices will be described. Fundamental principles of field effect devices and the technical background with their ongoing applications in the field of bio-sensor technologies, termed bio-FET, will be first introduced. Then concept of genetic field effect transistor will be described with emphasis on their fabrication, characteristics, and recent applications to microbial Single Nucleotide Polymorphysms (SNPs) Analysis as well as DNA sequencing. By comparing to other conventional methods, technical significance and future perspective of the genetic field effect transistor will also be discussed in detail.

1. Introduction

Rapid and tremendous advances have been achieved during the last fifty years in the field of microelectronics. Highly reliable and functional chips can be easily fabricated using a precisely controlled production process. Inexpensive personal computers with high performances are now available and various kinds of information are at our disposal. On the other hand, with the rapid increase of knowledge in the fields of medicine and biology, micro- and nano-technologies in electronics have been applied to these fields for parallel processing of information, miniaturization of analysis systems, and exploration of the molecular mechanisms of life. The biosensor is a typical example of the fusion between biotechnology and microelectronics. The biosensor consists of transducers and membranes on which biologically active substances are immobilized. Physical and chemical changes at the membrane as a result of biochemical reactions are transduced to electrical signals in the transducer. Electrochemical electrodes such as ion-selective electrodes and oxygen electrodes are commonly used as the transducer in conventional biosensors.

We have explored several approaches to using advanced semiconductor technology to fabricate intelligent microbiochemical sensing devices (Miyahara et al. 1985, 1988, 1991a, 1991b, 1994; Tsukada et al. 1990; Manz et al. 1990; Kajiyama et al. 2003). The biochemical field effect transistor (biochemical FET) is a biosensor which uses the ion-sensitive field effect transistor (ISFET) as the transducer. The FET is one of the most important and fundamental devices in

Yuji Miyahara • Biomaterials Center, National Institute for Materials Science, Tsukuba, Japan; Department of Materials Engineering, Graduate School of Engineering, The University of Tokyo. **Toshiya Sakata** • Department of Materials Engineering, Graduate School of Engineering, The University of Tokyo. **Akira Matsumoto** • Department of Bioengineering, Graduate School of Engineering, The University of Tokyo.

M. Zourob et al. (eds.), *Principles of Bacterial Detection: Biosensors, Recognition Receptors and Microsystems*,
© Springer Science+Business Media, LLC 2008

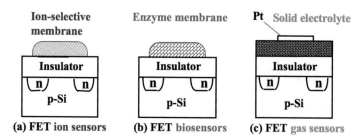

Figure 14.1. Biochemical Field Effect Transistors (a) FET ion sensors: Ion-selective membrane is formed on the surface of the gate insulator (b) FET biosensor: Immobilized enzyme membrane is formed on the surface of the gate insulator (c) FET gas sensor: Solid electrolyte thin film is deposited on the surface of the gate insulator.

the integrated circuit. In the case of FET-type biochemical sensors, the species to be detected and the sensor's selectivity to those species can be determined by the materials coated on the surface of the gate insulator. Ion sensors, biosensors, and oxygen sensors have been developed using polymer membranes, immobilized enzyme membranes, and a solid electrolyte thin film, respectively, as shown in Fig. 14.1. Recently, we have been investigating electrostatic detection of biomolecular recognition using a biologically coupled field effect transistor (bio-FET). The principle of bio-FET is based on potentiometric detection of charge-density change which is induced at a gate insulator/solution interface by specific biomolecular recognition. In this scheme, the charge-density change is directly transduced into electrical signals by the field effect. Based on this principle, various types of bio-FETs have been developed for the detection of DNA molecules (Souteyrand et al. 1997; Fritz et al. 2002; Uslu et al. 2004; Kim et al. 2004; Pouthas et al. 2004; Sakata and Miyahara et al. 2005, 2006a, 2006b).

In this chapter, the operational principles of bio-FETs and their application to genetic analyses are described. Limitations for their practical use and future perspectives for the bio-FETs are also discussed.

2. Fundamental Principles of Field Effect Devices

2.1. Metal-Insulator-Semiconductor (MIS) Capacitor

The metal-insulator-semiconductor (MIS) structure is a useful tool to understand the operation of MISFET. The MIS structure is simply a capacitor in which the insulator is placed between the metal and the semiconductor electrodes, as shown in Fig. 14.2. The MIS structure treated in this section is assumed to be ideal.

When a voltage is applied between the metal and the semiconductor electrodes, the surface of the semiconductor is modified depending on the sign and magnitude of the applied voltage. When a p-type silicon is used as a substrate, the majority carriers are positive charges, which are called holes. When no voltage is applied on the metal electrode, charge distribution in the silicon is uniform. Holes and immobile ions called the acceptors are electrically balanced. When a negative voltage is applied on the metal electrode, additional holes are attracted at the insulator-silicon interface by the Coulomb force, as shown in Fig. 14.2(a). The electric field is induced in the insulator. This condition is called accumulation, since the majority carriers, holes in this case, are accumulated at the surface of the silicon. The number of holes attracted at the insulator-silicon interface decreases as the applied voltage increases. When the applied voltage turns to positive, holes are expelled from the silicon surface, leaving behind the negatively-charged immobile acceptors, as shown in Fig. 14.2(b). This condition is called depletion, since mobile charged carriers, holes in this case, are depleted in the surface region. The electric field

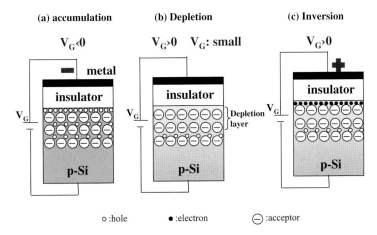

Figure 14.2. Control of carrier density at the surface of Si in the MIS structure:(a) accumulation: The majority carriers, holes, are accumulated at the surface of silicon substrate (b) depletion: Carriers are depleted at the surface of silicon substrate (c) inversion: The minority carriers, electrons, are induced at the surface of silicon substrate.

is induced both in the insulator and the depletion layer. When the applied voltage increases further in the positive direction, negatively charged mobile carriers, electrons, are induced at the insulator-silicon interface, as shown in Fig. 14.2(c). This condition is called inversion, since the number of electrons which are the minority carrier in the p-type silicon exceeds that of the holes and the surface is thus inverted to n-type. The thin layer of electrons at the interface, which is called the inversion layer, plays an important role in the operation of the MISFET.

When this principle is applied to biosensors, the metal gate electrode is replaced by an aqueous solution. Instead of the gate metal electrode, a reference electrode is placed in the aqueous solution to control the potential at the surface of the gate insulator. A charge-density change induced at the surface of the gate insulator can be detected as a result of electrostatic interaction between the charges at the gate surface and the carriers in the silicon.

2.2. Principles of Biologically Coupled Field Effect Transistors for Genetic Analysis (Genetic FETS)

The conceptual structure of the biologically coupled field effect transistor for genetic analysis (genetic FET) is shown in Fig. 14.3. Oligonucleotide probes are immobilized on the surface of the gate insulator. The principles of the genetic FET are based on the detection of charge-density change, which is induced at the gate surface by specific binding of DNA molecules. The genetic FET is immersed in a measurement solution together with an Ag/AgCl reference electrode with saturated KCl solution, as shown in Fig. 14.3. The potential of a measurement solution is controlled and fixed by the gate voltage (V_G) through the reference electrode. The thickness of the gate insulator of this kind of FET is usually on the order of 100 nm. A passivation layer of pH-sensitive material such as Si_3N_4, Al_2O_3, and Ta_2O_5 is formed on top of the SiO_2 layer. When charged species in aqueous solution are adsorbed on the surface of the gate insulator as shown in Fig. 14.4, electrons in a silicon substrate electrostatically interact with adsorbed charged species. The electrical characteristics of the bio-FETs such as the current-voltage characteristic are influenced as a result of this electrostatic interaction. In the case of the genetic FETs, when complementary DNA molecules are contained in a sample solution, hybridization occurs at the surface of the gate area. Since DNA molecules are negatively charged in an aqueous solution, the hybridization event can be detected by measuring

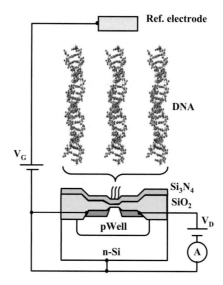

Figure 14.3. Conceptual Structure of Genetic Field Effect Transistor Oligonucleotide probes are immobilized on the surface of the gate insulator. An Ag/AgCl reference electrode with saturated KCl solution is used to control the gate voltage.

a shift of electrical characteristics such as the relationship between the gate voltage V_G and the drain current I_D, as shown in Fig. 14.4. In the V_G-I_D characteristic, the voltage at which the drain current starts to flow is called the threshold voltage, V_T. If we measure the direction and the amount of the V_T shift after the specific reactions of biomolecules, we can obtain information on the polarity and density of charges carried with the adsorbed biomolecules. When an n-channel FET is used, the V_T shifts in the positive direction in response to negatively charged DNA molecules.

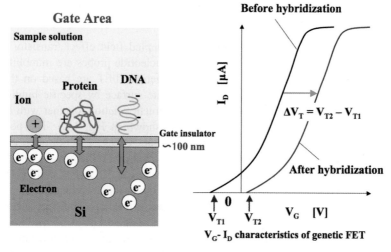

Figure 14.4. Electrostatic interaction at the gate and it's effect on the electrical characteristic of the genetic FETs. Electrostatic interaction between adsorbed charged species and electrons at the silicon surface induces a shift of the V_G-I_D characteristic of the bio-FET.

3. Fundamentals of Genetic Analysis

3.1. DNA

Deoxyribonucleic acid (DNA) is a nucleic acid that encodes genetic information for the development and function of all living organisms, except some groups of viruses which utilize RNA (ribonucleic acid) instead as a genetic substance. It is a long polymer strand composed of repeating units called nucleotides (Fig. 14.5). Four types of bases, namely, adenine (A), cytosine (C), guanine (G), and thymine (T), along with intervening sugar and phosphate that are attached together, are the components that form nucleotides. Importantly, these bases exhibit complementary base-pairings, A-T and G-C, resulting in the formation of the double helical structure of DNA. Because these base-pairings (physical bindings) involve different numbers of hydrogen bonds, two for A-T and three for G-C, the binding strengths are different between the two pairings; namely, G-C is stronger than A-T. As a result, with the increased number of G-C pairs in double-stranded DNA, its melting temperature increases, reflecting the increased thermodynamic stability. Therefore, when undergoing a reaction of DNA hybridization (base-pairings) on a DNA chip that will be minutely described in later sections, the temperature is one of the most important parameters for achieving high accuracy and good reproducibility in the DNA chip analysis. The helix of the DNA structure cycles every 3.3 nm with its width of 2.2–2.6 nm (Fig. 14.5). Since the distance between two adjacent bases is about 0.34 nm, one cycle of the helix contains 10 bases. DNA in aqueous solution is negatively charged due to a large population of phosphate groups on the molecule, an important feature enabling quantitative analysis of the molecule via electrophoresis.

3.2. Genetic Analysis

The general purpose of base sequence analysis is to determine the order of the four bases (A, G, C, T) aligned in each DNA strand. The prevailing method for reading base sequences has been a combined technique based on multiple fluorochromes and electrophoresis. In the

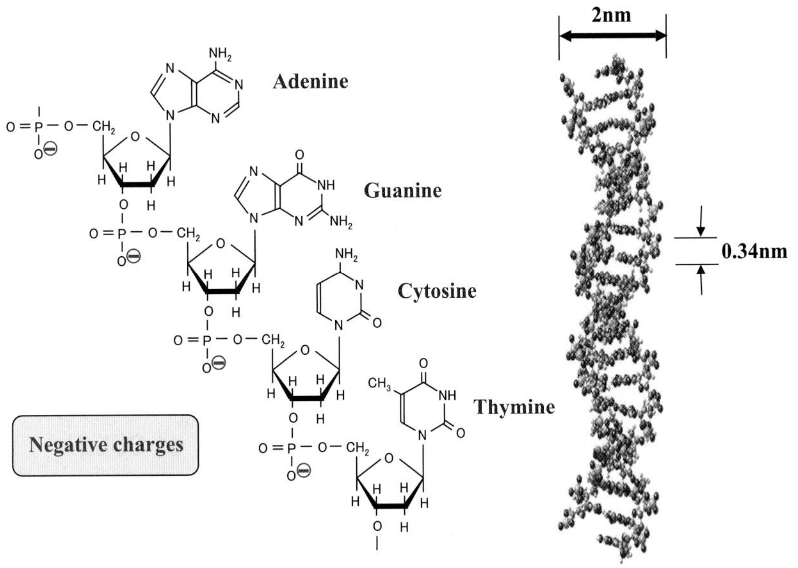

Figure 14.5. Structure of deoxyribonucleic acid (DNA).

International Human Genome Project, DNA sequencer (a combined technique of electrophoresis and fluorescence) has proved itself to be a powerful tool allowing for high-speed reading of the DNA base sequences. Such a technique, however, does not directly clarify the meanings of the genetic codes. Indeed, since the completion of the project in 2003, there has been an increasing effort directed toward a functional analysis of the decoded genes. Cells constituting our bodies contain a series of genomes, where patterns of active/inactive parts of genes differ depending upon the type of cells and tissues. Consequently, the types of proteins synthesized in each tissue differ, leading to diversities in the forms and functions of different tissues.

A technique to clarify patterns of expressed genes for certain types of cells and tissues is called expressed gene analysis. For such an analysis, DNA chips or DNA microarrays are utilized in order to comprehensively examine the expressed messenger RNAs (mRNAs) in various cells and tissues. Based on this technique, for example, identification of cancer and other disease-related genes is possible by comparing the patterns of expressed DNA with those obtained in normal cells.

3.3. DNA Chip / DNA Microarray

A DNA chip or DNA microarray is defined as a monolithic device bearing different DNA fragments immobilized on its surface (probe DNA), whose hybridization events, upon addition of a complementary "target" DNA, can be visualized through various detection/imaging techniques such as fluorescence, radio isotope, and oxidation-reduction labelings. Based on the choice of target DNA and the designed reaction protocol, the technique allows multiple gene expression analysis, sequencing, medical diagnostics, and drug development (via differential polymorphism) in a massive and parallel manner.

For such a device, the immobilization of "probe" molecules or "probe DNA" onto the surface with controlled density, chemistry (such as polarity and hydrophobicity), and morphology is a key issue for obtaining desirable system capability (Pirrung 2002; Tiefenauer and Ros 2002). While a physical method presents a convenience of procedure, a covalent immobilization method may be more favored, assuring long-term stability and thus maintenance of the device's capability. The probe density, the number of probing sites per unit area or that within an individual probe site, reflects its information density. In order to minimize array size, the probe sites and their spacing (pitch) must be as small as possible. The probe density is also important for controlling intermolecular interactions and multiple binding behaviors. The selection of linker molecules, which provide a distance between the device surface and the probing sites (probe molecules), controls the flexibility of the formed linkage governing steric effects, as well as the kinetics of the probing reactions (DNA hybridization). The reactivity of the linker molecules against the surface and the method of treatment are also crucial to the morphological control of the probe decoration onto the surface.

In the following sections, immobilization methods generally conducted for the probe biological molecules, mainly focusing on DNA, will be described. Some aspects in each method regarding current understandings and the ongoing efforts for technical improvement will also be presented.

4. Immobilization of DNA Molecules on the Surfaces of Solid Substrates

4.1. Silanization

For a source material on which a DNA array is integrated, a standard microscope glass slide, silicon wafer in forms of fused or oxidized, and gold are usually utilized. Construction of a high-density DNA array was originally attained based on a photochemical technique, which involves the in situ, spatially addressable, parallel synthesis of a probe DNA. Today, most immobilization strategies of presynthesized DNA or PCR products onto the chip surface involve chemical modification of silicone oxide-based surfaces due to oxidation, very close vicinity of the pure silicone surface is also regarded as silicone oxide (Eggers et al. 1994; Beattie et al. 1995; Guo et al. 1995; Kallury et al. 1998).

First, it is necessary to coat the chip surface with an organosilane that can covalently bind to the silicon oxide surface, a step called silanization. The other end of the organosilane bearing molecule is properly functionalized for the following conjugation with a probe DNA. In order to remove any organic contaminants from the surface that could potentially interfere with the following chemical decoration processes onto the surface, a preceding cleaning is usually applied. This cleaning procedure involves the use of detergents, strong oxidizers such as H_2SO_4/H_2O_2 ("piranha" solution), or NH_3/H_2O_2, oxygen plasma, and sonication. There are a number of organosilane coupling reagents commercially available that have been utilized for the immobilization of DNA and other biological molecules. For example, (3-glycidyloxypropyl)trimethoxysilane (3-GPS), (3-aminopropyl)triethoxysilane (3-APTES), (3-mercaptopropyl)triethoxysilane (3-MPTS), and haloacetamidosilanes have been extensively used. The first step of the silanization is a physisorption of these reagents onto the surface, followed by rapid hydrolysis of the alkoxy groups (methoxy or ethoxy groups), yielding the formation of hydroxyl groups that can covalently interact with the silanol surface (Silberzan et al. 1991). It is generally believed that cleavage of the silane group on the surface to generate a silanol group can be accelerated by the action of a strong base on the generated hydroxyl linkages between the silane and the surface. Once the silanols form, upon annealing, they can condense with the surface silanol residues to form stable polymeric siloxane linkages (Fig. 14.6).

Here the form of reaction, the obtained molecular conformation, and consequently the morphology (isotropy) could differ, depending on the type of coupling reagents utilized. Usually,

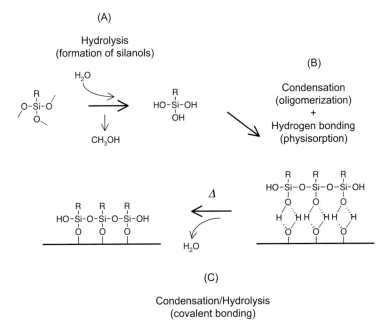

Figure 14.6. Schematic illustration of silanization reaction using trimethoxy organosiline; (A), the reactive alkoxyl groups are hydrolyzed by the surface water on a hydrated silanol surface, followed by condensation (B), where formation of silane oligoers in situ and the simultaneous physisorption onto the surface through hydrogen bonding occur, and then (C)—thermal curing of the film—promotes further condensation between free silanol groups and the surface silanols, causing covalent attachment of the film across the surface.

trialkoxyl organosilane (all reagents listed above belong to this type) is used due to its ability to form a multiple linkage with the surface, leading to a higher chemical stability compared to those obtained using mono- or dialkoxyl organosilanes. Mono- or dialkoxyl organosilanes tend to suffer from higher susceptibility to hydrolysis, resulting in a low chemical stability. However, a disadvantage of using the trialkoxyl type of organosilanes is that they prefer to react with a glass or silicone oxide using only two sianol groups and leave one group free. This remaining free silanol group causes a significant degree of polymerization on the surface as well as in solution prior to condensation with the solid substrate. This polymerization could lead to a highly heterogeneous morphology, which is a potential disadvantage when one intends to obtain a uniformly decorated surface with good reproducibility. Some attempts have been reported to optimize this issue by limiting the concentration of the alkoxysilane used, and by using an anhydrous condition and the post-silanization curing procedure (Wikstrom et al. 1988; Shlyakhtenko et al. 1999; Vandenberg et al. 1991).

These treatment conditions also affect the layer thickness, another significant factor affecting the surface morphology. For example, a monolayer of (3-aminopropyl)triethoxysilane (3-APTES), one of the most common DNA derivatizing organosilane coupling reagents, has been reported under controlled, relatively mild condition at room temperature using toluene as the reaction solvent (Vandenberg et al. 1991), while more forcing conditions end up with the production of multilayers.

The complementarily functionalized oligo-DNAs can be readily introduced either at the 3' or 5' ends onto the decorated surfaces via silanization as described above.

4.2. Thiol-Gold Bonding

A more sophisticated methodology to immobilize biomolecules onto the chip surface makes use of the interaction between the thiol group and gold. The technique is based on the facts that the thiol group is able to form a quasi-covalent bond with gold (Crumbliss et al. 1992; Parker et al. 1996), and that an alkane molecule with a chain length >8 can form an ordered as well as densely packed monolayer (self-assembling monolayers, SAM) in a proper solvent (Mrksich et al. 1996). Thus, using thioalkane molecules, the other end group of which has been functionalized for the conjugation with the probe molecules such as DNA, the SAM-mediated immobilization of the probe molecules onto the chip surface made of gold can be accomplished (Fig. 14.7). A strong advantage of using a gold substrate is that it allows for a direct observation of each production step (via measurements of the formed layer thickness), as well as binding (recognition) events, by means of surface plasmon resonance (SPR) (Thiel et al. 1997). Another advantageous feature in the SAM method is that probe molecules can be introduced onto the surface in a molecularly oriented (self-assembled) fashion, which greatly improves the reaction mode and efficiency that take place on the prepared surface.

Based on this technique, various types of molecules including antibodies (Lu et al. 1995; Duschl et al. 1996; Disley et al. 1998), proteins, DNAs (Thiel et al. 1997), and other synthetic recognition molecules (Rickert, Weiss, and Gopel 1996) have been successfully immobilized onto the gold surface, through which various types of arrays and biosensors have been developed.

4.3. Avidin, Streptavidin and Biotin

Avidin is a 66-kDa glycoprotein obtained from egg white. Streptavidin is a 60-kDa protein isolated from the culture broth of *Stretomyces avidinii*. Both have a strong avidity for biotin, a 244-kDa water-soluble vitamin H, with the dissociation constant of 10^{-15} M. This extremely strong, practically irreversible binding property provides an important option for the

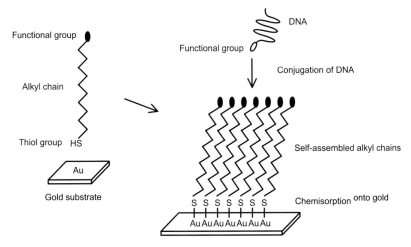

Figure 14.7. A SAM (self-assembling monolayers)-based strategy for introducing a DNA molecule onto a gold substrate utilizing a thiol-gold bonding. One end of the SAM molecule is modified with a thiol group (alkanethiol), yielding covalent attachment of the molecule with the gold substrate. The other end of the alkanethiol is functionalized as to react with a properly functionalized probe DNA.

biomolecular immobilization methodology. The conjugation with biotin molecule is accomplished by use of a series of biotinylation reagents such as sulfosuccinimidyl 6-(biotinamido) hexanoate (NHS-LC-biotin). The biotinylated molecules, accordingly, can be indirectly introduced to the substrate through the binding with avidin- or streptavidin decorated anchor molecules that are pre-immobilized onto the substrate via other methods including silanization and thiol-gold interactions (Sakahara and Saga 1999).

4.4. Others

A unique method enabling direct attachment of DNA to silicone surfaces without intervening silicone oxide layer has also been developed (Strother et al. 2000a, Strother et al. 2000b). In this method, silicone with terminal hydrogen atoms is first prepared by treating silicone oxide surface with hydrofluoric acid (HF). Here, the Si-H bond undergoes photochemical addition to alkenes to form Si-C bond. Taking advantage of the phenomenon, a Boc-protected (t-butyl carbamate) 10-amino-1-decene was introduced to the surface to obtain a Boc amine surface. Deprotection of the Boc group converts the surface to that of high chemical reactivity with an exposed amine group available for further coupling with DNA molecules (with appropriately functionalized end groups) (Silberzan et al. 1991, Kallury et al. 1998).

Another widely conducted methodology to immobilize biological molecules onto the chip or electrode surfaces involves preparation of thin polymer layers or gels. Various conducting polymer (physisorbed) layers including polypyrrole (Yon-Hin et al. 1993), polythiophen (Hiller et al. 1996), PEG-vinylferrocene (Sirkar and Pishko 1998) and silk fibroin (Qian et al. 1997) have been prepared in the form of tiny fibers, which are shown to be suitable as matrices for redox enzymes. These polymers with such morphologies (fibers) provide increased surface density. In addition, when the polymer is endowed with properly functionalized groups available for further couplings with other biofunctional molecules such as DNA, significantly increased probing (detecting) density can be achieved. A representative polymer utilized for such a purpose is polylysine (Thiel et al. 1997). This largely (positively) charged polyamine also serves as a physical, electrochemical absorber. It should be noted that these physical-attaching techniques in most cases are used in combination with other covalent-binding methodologies that are described in earlier sections, often in the form of intervening "gluing" layers.

5. Genetic Analysis Based on Field Effect Devices

5.1. Fundamental Characteristics of Genetic Field Effect Devices

5.1.1. Detection of DNA Molecular Recognition Events

A specific binding of charged biomolecules at the gate surface can be detected as a shift of the threshold voltage V_T, which can be determined in the gate voltage V_G and the drain current I_D (V_G-I_D) characteristics of the genetic field effect devices (Fig. 14.8). The V_G-I_D characteristics of the genetic field effect devices shifted along the gate voltage V_G axis in the positive direction after immobilization of oligonucleotide probes (Fig. 14.8(a)). Oligonucleotide probes immobilized were a normal type for R353Q locus of factor VII gene (Table 14.1 (Kajiyama et al. 2003)). In order to evaluate the V_T shift in more detail, the local area shown in Fig. 14.8(a) (surrounded area) was magnified (Fig. 14.8(b)). The V_T shifts after hybridization and specific binding of DNA binder are also shown in Fig. 14.8(b). When oligonucleotide probes were immobilized on the gate surface, the V_T shifted along the V_G axis by the amount of 32 mV. The positive shift is due to negative charges induced at the gate surface after immobilization process including cleaning, silanization, glutaraldehyde-treatment, immobilization of oligonucleotide probes and blocking with glycine. Immobilization of oligonucleotide probes and glycine blocking is considered to contribute to the V_T shift to a large extent. When the complementary target DNA was introduced to the gate surface and hybridized with oligonucleotide probes, the V_T shifted in the positive direction by the amount of 12 mV. This is due to increase of negative charges of the target DNA by hybridization. After hybridization, a DNA binder, Hoechst 33258 was introduced to the gate surface. The V_T shifted in the negative direction by the amount of 14 mV. The negative shift of the V_T indicates increase of positive charges at the gate surface and is due to specific binding of Hoechst 33258 to the double-stranded DNA. This is in contrast to the positive change of the

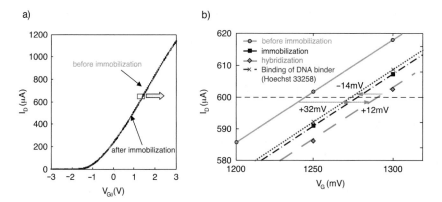

Figure 14.8. Electrical signals of molecular recognition events on the genetic field effect transistor (FET). (a) Gate voltage (V_G)-drain current (I_D) characteristics of genetic FET before and after immobilization of oligonucleotide probes. (b) Threshold voltage V_T shifts after immobilization of oligonucleotide probes, hybridization of target DNA and specific binding of Hoechst 33258. The amount of the V_T shift was determined at a constant drain current of 600 μA.

Table 14.1. Base sequences for oligonucleotide probes at the R353Q locus of factor VII gene (Kajiyama et al. 2003).

Locus	Function	Sequence	$T_m(°C)$
R353Q	R353Q-normal(N)		
	probe	5′-amino group-CCACTACC<u>G</u>GGGCACGT-3′ (17mer)	$60\,(T_{m1})$
	target	5′-ACGTGCCCCGGTAGTGG-3′ (17mer)	
	R353Q-mutant(M)		
	probe	5′-amino group-CCACTACC<u>A</u>GGGCACGT-3′ (17mer)	$57\,(T_{m2})$
	target	5′-ACGTGCCCTGGTAGTGG-3′ (17mer)	

N and M indicate normal (wild-type) and mutant allele-specific oligonucleotides, respectively. T_m shows the melting temperature.

V_T due to negatively charged DNA molecules. Thus, the charge density change at the gate surface after each molecular recognition event can be successfully detected using genetic field effect devices.

The V_T shifts for various DNA binders were investigated using genetic field effect devices (Fig. 14.9). All the DNA binders tested are ionized and carry positive charges in an aqueous solution (Fig. 14.9(a)). When each DNA binder was introduced to the gate surface of the genetic field effect devices after hybridization, the V_T shifted in the negative direction due to the positive charges of DNA binders (Fig. 14.9(b)). A single molecule of Hoechst 33258 has three positive charges in an aqueous solution, as shown in Fig. 14.9(a). Therefore, we have expected that the bigger V_T shift could be obtained based on larger charge density changes by the use of Hoechst 33258 than the use of other DNA binders except for Hoechst 33342. However, the V_T shifts due to DNA binders are more complicated. The binding between double-stranded DNA and DNA binder is dependent on charges, chemical structure of the DNA binder, base sequence of the DNA and so on. DNA binders such as Hoechst 33258, Hoechst 33342 and DAPI were reported to be bound selectively to AT-rich sites of DNA molecule in the minor groove (Sriram et al. 1992, Boger et al. 2001), and to show weak binding to GC sites (Wilson et al. 1989). On the other hand, EB and PI are known as intercalator and show high affinity toward double-stranded DNA and generally exhibit modest base pair selectivity (Rye et al. 1993, Dengler et al. 1995). The V_T shift caused by PI was approximately twice as big as that of EB (Fig. 14.9(b)). This can be explained by the difference of the number of charges between PI and EB (Fig. 14.9(a)). The use of some of the DNA binders after hybridization is effective to discriminate between the signal of double-stranded DNA and that of non-specifically adsorbed single-stranded DNA at the gate surface, because some of the DNA binders used in this study specifically react with double-stranded DNA.

DNA recognition events such as primer extension reaction can be also directly detected as electrical signal by use of genetic field effect devices. The 11-base oligonucleotide probes on the genetic field effect devices were hybridized with the 21-base target DNA at first. In order to extract small changes of the output voltages of the genetic field effect devices after extension reaction, the output voltages before the introduction of DNA polymerase for both genetic and reference field effect devices were initialized and adjusted to zero as shown in Fig. 14.10 by adding or subtracting offset voltages, although the absolute output voltages are not zero and different between active and reference field effect devices. After washing, the genetic field effect devices were immersed in a reaction mixture and thermostable DNA polymerase was introduced into the gate surface. The V_T of the field effect devices changed during primer extension reaction as shown in Fig. 14.10. Differential measurement was performed using a pair of field effect devices; one is the genetic field effect device with immobilized oligonucleotides probes, and the other is the reference field effect device without olinonucleotide probes. The

a)

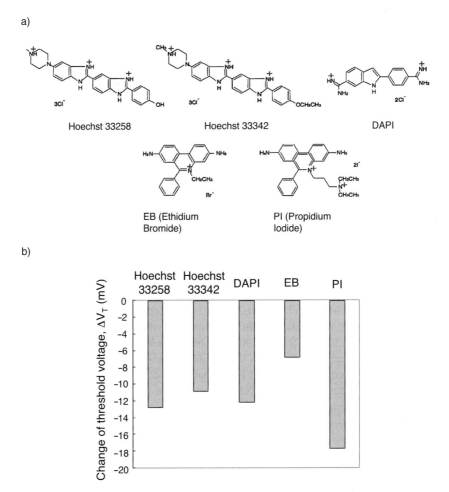

b)

Figure 14.9. Potentiometric detection of DNA binders. (a) Chemical structures of various DNA binders. All the DNA binders shown here have positive charges in an aqueous solution and react with double-stranded DNA. (b) Shifts of the threshold voltage (V_T) for various DNA binders. Each DNA binder was dissolved in deionized water at a concentration of $100\,\mu M$. The number of charges in an aqueous solution is different among the DNA binders shown in (a).

common changes of the V_T due to ambient salt and pH changes and temperature difference in the buffer solution can be cancelled out using the differential measurement. The difference of the ΔV_T between the genetic field effect device and the reference field effect device during extension reaction, ΔV_T^{diff}, increased drastically up to about 10 mV. This positive change in the ΔV_T^{diff} is rightly due to negative charges of polynucleotide extended by primer extension reaction. The ΔV_T of the reference field effect device is considered to be mainly due to the temperature change. From Fig. 14.10, we could demonstrate that the primer extension event on the gate surface was transduced directly into the electrical signal by the use of the genetic field effect device.

The effect of base length of the target DNA on the ΔV_T^{diff} was investigated (Fig. 14.11 and Table 14.2 (Kajiyama et al. 2003)). The linear relationship between the base length and the ΔV_T^{diff} was obtained up to 41 bases. The ΔV_T^{diff} after extension reaction increased to 24 mV, when target DNA with 41 bases was used (Fig. 14.11). This is because the number of charges on the gate surface increased after primer extension with increasing the template base length. However, the ΔV_T^{diff} did not follow the linear relationship, when the target DNA samples with

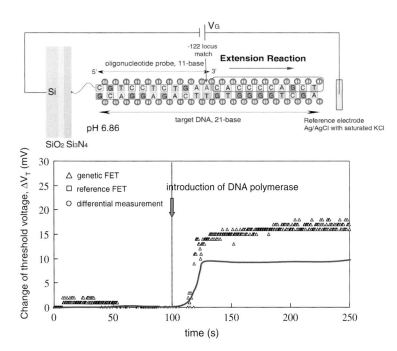

Figure 14.10. Time course of the threshold voltage shift DV_T^{diff} during primer extension reaction. The target DNA was hybridized with complementary oligonucleotide probe on the gate insulator prior to primer extension reaction. The base lengths of the target DNA and oligonucleotide probe were 21 bases and 11 bases, respectively. After hybridization, *Taq* DNA polymerase and dNTPs were introduced into the genetic and the reference FETs at 72 °C as indicated by an arrow.

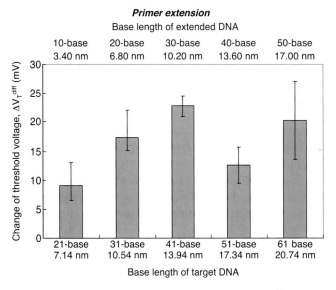

Figure 14.11. Effect of target DNA base length on the threshold voltage shift DV_T^{diff} after primer extension reaction. The base length of the complementary target DNAs were 21 bases, 31 bases, 41 bases, 51 bases, and 61 bases, respectively. The averages of the two experiments are shown.

Table 14.2. Base sequences for oligonucleotide probes and target DNA at the -122 locus of factor VII gene (Kajiyama et al. 2003). N and M indicate normal (wild-type) and mutant allele-specific oligonucleotides, respectively.

Locus	Function	Sequence
−122	−122-normal(N)	
	probe	5′-amino group-CGTCCTCTGA**A**-3′ (11mer)
	target	5′-AGCTGGGGTG**T**TCAGAGGACG-3′ (21mer)
		5′-TGCAGCTCTCAGCTGGGGTG**T**TCAGAGGACG-3′ (31mer)
		5′-GGCGGCCAGGTGCAGCTCTCAGCTGGGGTG**T**TCAGAG GACG-3′ (41mer)
		5′-CATGGCCACTGGCGGCCAGGTGCAGCTCTCAGCTGGG GTG**T**TCAGAGGACG-3′ (51mer)
		5′-GCAGGGGATGCATGGCCACTGGCGGCCAGGTGCAGC TCTCAGCTGGGGTG**T**TCAGAGGACG-3′ (61mer)
	−122-mutant(M)	
	probe	5′-amino group-CGTCCTCTGA**G**-3′ (11mer)
	target	5′-AGCTGGGGTG**C**TCAGAGGACG-3′ (21mer)

51 bases and 61 bases were used. The reason for this non-linearity is considered to be related to the width of the electrical double layer at the interface between the gate insulator and an aqueous solution. The width of the electrical double layer, the Debye length, which is expressed in equation (14.1), is about 10 nm in the diluted salt solution (approximately 1 mM) and about 1 nm in the physiological solution (approximately 100 mM).

$$\delta = (\varepsilon\varepsilon_0 kT/2z^2q^2I)^{1/2}, \tag{14.1}$$

Where δ is the Debye length, ε is the permittivity of the electrolyte solution, ε_0 is vacuum permittivity, k is the Boltzman constant, T is the absolute temperature in Kelvin, z is the valency of the ions in the electrolyte, q is elementary charge, and I is the ionic strength of the electrolyte.

The charge density change induced within the Debye length can be detected with the genetic field effect devices, while the charge density change induced outside the Debye length is shielded by counter ions and cannot be detected with the genetic field effect devices. In the present study, a 25 mM phosphate buffer solution was used for the threshold voltage measurement. The Debye length at the gate insulator surface is therefore considered to be a few nanometers. The length of the target DNA with 41 bases is 13.94 nm, when it is straight. But oligonucleotide probes and the target DNA are flexible in the aqueous solution and oligonucleotide probes are not always perpendicular to the surface of the gate insulator. It is therefore reasonable to ascribe saturation of the linear relationship between the threshold voltage shift and the base length over 51 bases to the Debye length. Since the Debye length is dependent on the ionic strength of the aqueous solution as shown in equation (14.1), it is important to optimize the buffer concentration used for the measurement of the threshold voltage shift.

5.1.2. Immobilization Density of Oligonucleotide Probes

The molecular recognition events such as hybridization and specific binding of DNA binder could be directly transduced into electrical signal using the genetic field effect devices (Fig. 14.8). The change in the surface charge density could be detected as a shift of the V_T of the genetic field effect devices. The V_T shift after hybridization, ΔV_T, can be expressed in equation (14.2), where Q_{ds-DNA} is the charge per unit area of the double-stranded DNA after hybridization, Q_{ss-DNA} is the charge per unit area of the single-stranded oligonucleotide probes, ΔQ_{DNA} is the charge difference per unit area after hybridization, and C_i is the gate capacitance per unit area.

$$\Delta V_T = (Q_{ds-DNA} - Q_{ss-DNA})/C_i = \Delta Q_{DNA}/C_i, \tag{14.2}$$

Since $\Delta V_T = 12\,mV$ and $C_i = 4.3 \times 10^{-4}\,F/m^2$ for the genetic field effect device, the amount of charges increased after hybridization is calculated to be $5.1 \times 10^{-6}\,C/m^2$. The base lengths of the oligonucleotide probe and the target DNA used in this study are both 17 bases, which corresponds to 5.78 nm in length. Negative charges derived from phosphate groups are distributed along the double-stranded DNA from the gate surface to the bulk of the sample solution. We assume that these negative charges along the DNA molecules contributed to the V_T shift equally and that all the oligonucleotide probes were hybridized with the target DNA. Under these assumptions, the number of oligonucleotide probes on the channel region can be calculated to be 2.3×10^4, which corresponds to $1.9 \times 10^8/cm^2$. The surface density of oligonucleotide probes immobilized on glass, silicon dioxide and gold has been reported to be in the order of 10^9 to $10^{13}/cm^2$, which was determined by different methods (Chrisey et al. 1996, Steel et al. 1998, Kumar et al. 2000, Guo et al. 2001, Huang et al. 2001, Peterson et al. 2001). Since the density of the oligonucleotide probes is strongly dependent on the method and materials used for a substrate and immobilization, the number of oligonucleotide probes immobilized on silicon nitride could be increased by optimizing the immobilization method. It is noted that hybridization with 2.3×10^4 target DNA molecules resulted in the V_T shift of 12 mV. Therefore, detection of DNA molecules by the use of the genetic field effect device can be very sensitive, if hybridization is carried out sufficiently.

5.2. Single Nucleotide Polymorphisms (SNPs) Analysis

For genetic analysis in clinical diagnostics, single nucleotide polymorphisms (SNPs) are the most common form of DNA variation in humans and are important markers in "tailored medicine". An extensive collection of SNPs would serve as a valuable resource for the discovery of genetic factors affecting disease susceptibility and resistance. The development of SNP genotyping based on this information has been proceeded using diverse methods in recent years (Chee et al. 1996, Landegren et al. 1998, Howell et al. 1999, Sauer et al. 2000, Tyagi et al. 2000, Syvanen et al. 2001, Jobs et al. 2003) and will enable clinicians to determine which pharmacological agent is most effective for treating a given patient's condition (Schene et al. 1995).

A number of methods for SNP detection have been developed, including restriction fragment length polymorphism (RFLP) analysis (Parsons and Heflich 1997), single-strand conformation polymorphism (SSCP) analysis (Orita et al. 1989), allele-specific oligonucleotide hybridization (ASOH) (Saiki et al. 1989), oligonucleotide ligation assay (Landegren et al. 1988), primer extension assay (Pastinen et al. 1997), Taqman assay (Livak, Marmaro and Todd 1995), molecular beacons (Tyagi, Bratu and Kramer 1998), pyrosequencing (Ronagi, Uhlen and Nyren 1998) and so on. In order to analyze reaction products, various platforms have been developed including gel electrophoresis (Pastinen, Partanen and Syvanen 1996), oligonucleotide arrays (Pastinen et al. 2000), mass spectrometry (Haff and Smirnov 1997), semiconductor-based DNA chip (Gilles et al. 1999), etc. Among these SNP genotyping methods, fluorescent detection is widely used, while chemiluminescence and mass of the reaction product are detected for pyrosequencing and mass spectrometry, respectively (Lockhart et al. 1996, Cooper et al. 2001). Here, we introduce the results of SNP genotyping using the genetic field effect devices in combination with allele specific oligonucleotide hybridization or primer extension reaction.

5.2.1. Controlling Hybridization Temperature for SNPs Analysis

We have prepared two types of the genetic field effect devices to detect single base change in the target DNA. We used the R353Q locus of factor VII gene as a model sample (Table 14.1 (Kajiyama et al. 2003)). Normal (wild type) oligonucleotide probes were immobilized on the gate surface of one of the genetic field effect devices (N-type genetic field effect device), while mutant oligonucleotide probes were immobilized on the gate surface of the other genetic field effect devices (M-type genetic field effect device). The N-type and M-type were set into a buffer solution and the temperature of the buffer solution was controlled at 60 °C for hybridization with target normal DNA in a normal sample and at 57 °C for hybridization with target mutant DNA in a mutant sample (Fig. 14.12). When normal DNA was hybridized at 60 °C with N-type and M-type, the V_T shift of the full matched N-type was bigger than that of the one-base mismatched M-type (Fig. 14.12(b-1)). On the other hand, when mutant DNA was hybridized at 57 °C with both field effect devices, the bigger V_T shift was obtained for the full matched M-type than for the one-base mismatched N-type (Fig. 14.12(b-2)). This means that it is possible to detect one-base change of target DNA using the genetic field effect devices. However, the difference of the signals between the N-type and the M-type in response to the

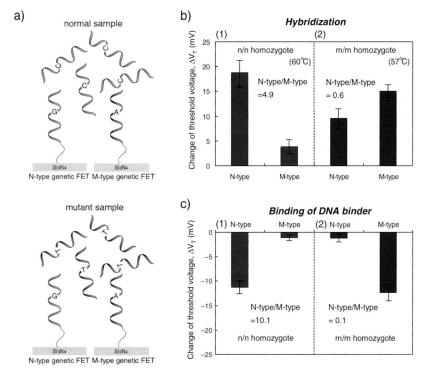

Figure 14.12. Discrimination of single nucleotide polymorphism (SNP) using genetic field effect transistor (FET). The V_T shifts obtained with three different genetic FETs are averaged for the same condition and the standard deviation is expressed as a bar. (a) Schematic representation showing the method for SNP analysis using genetic FET. The N-type and M-type genetic FETs were prepared by immobilizing oligonucleotide probes for normal (wild type) and mutant type genes at the locus of R353Q, respectively. A pair of N-type and M-type genetic FETs were hybridized with normal target or mutant target at optimum temperatures. (b) SNP detection by allele specific oligonucleotide hybridization at controlled temperatures. The change of the threshold voltage V_T after hybridization was measured for N-type and M-type genetic FETs, hybridized with normal target DNA (1) and mutant target DNA (2). Hybridization was carried out at 60 °C for a normal sample, and at 57 °C for a mutant sample, respectively. (c) SNP detection using genetic FETs in combination with DNA binder. Signal to noise ratio to distinguish one-base change drastically improved when Hoechst 33258 was used after hybridization.

mutant sample (Fig. 14.12(b-2)) was relatively smaller than that in response to the normal sample (Fig. 14.12(b-1)). The ratio of the signals between the N-type and the M-type was 4.9 in the case of the normal sample, while the signal ratio between the N-type and the M-type was 0.6 in the case of the mutant sample. Precise control and optimization of the temperatures during the hybridization and washing processes are important for the robust detection system. After hybridization, Hoechst 33258 was introduced to the gate surfaces of the N-type and the M-type (Fig. 14.12(c)). The ratio of the signals between the N-type and the M-type was 10.1 in the case of the normal sample, while the signal ratio between the N-type and the M-type was 0.1 in the case of the mutant sample. The ability to discriminate single base change could be increased by the use of the DNA binder in combination with the genetic field effect devices. It is noted that the directions of the V_T shifts were positive for hybridization with target DNA and negative for the DNA binder respectively because of intrinsic charges of DNA and DNA binder.

For the SNP typing analysis, it is preferable to use the genetic field effect devices at a constant temperature. The genetic field effect devices were evaluated for SNP typing at 57°C using the artificial samples of three different genotypes (Fig. 14.13). When the N-type and the M-type were hybridized with a normal/normal (n/n) homozygote sample, a normal/mutant (n/m) heterozygote sample and a mutant/mutant (m/m) homozygote sample respectively (Fig. 14.13(a)), the ratios of the V_T shifts of the N-type and the M-type were 2.4, 1.2, 0.6 for a n/n homozygote sample, a n/m heterozygote sample and a m/m homozygote sample, respectively (Fig. 14.13(b)). It was therefore possible to distinguish three genotypes of the target DNA by allele specific oligonucleotide hybridization using the genetic field effect devices. When a DNA binder, Hoechst 33258 was introduced and reacted with double-stranded DNA at the gate surface after hybridization, the ratios of the V_T shifts of the N-type and the M-type were 9.6, 1.2, 0.1 for a n/n homozygote sample, a n/m heterozygote sample and a m/m homozygote sample, respectively (Fig. 14.13(c)). We found that the use of DNA binder after hybridization was effective to distinguish genotypes of the target DNA more clearly than hybridization only. The results of target hybridization and specific binding of DNA binder demonstrate ability of the genetic field effect devices to distinguish the three different genotypes.

For the simultaneous analysis of many different SNPs with high precision, it is important to optimize the temperature of each oligonucleotide probe during hybridization and washing process. For this purpose, the thermal gradient DNA chip has already proposed (Kajiyama et al. 2003), and evaluated for genotyping, in which temperatures of oligonucleotide probes could be controlled independently. Unlike other types of DNA microarray, the thermal gradient DNA chip will allow multiplex loci typing using different optimal temperatures for each locus on the same chip. Since the fabrication process of the genetic field effect devices is compatible with that of the thermal gradient DNA chip, the genetic field effect devices can be integrated in a Si island of the thermal gradient DNA chip. By combining the genetic field effect devices with the thermal gradient DNA chip, more precise and reliable SNPs analysis will be realized.

5.2.2. SNPs Analysis Based on Primer Extension

Furthermore, we propose to use the genetic field effect devices in combination with primer extension for SNP genotyping. This method is based on the sequence-specific primer extension of two allele-specific oligonucleotide probes that differ at their 3'-end nucleotide defining the alleles. Normal (wild type) oligonucleotide probes were immobilized on the gate surface of N-type genetic field effect device, while mutant oligonucleotide probes were immobilized on the gate surface of M-type genetic field effect device. The target DNA is hybridized with the normal and the mutant oligonucleotide probes on the gate surface of each field effect device. The hybridization events are followed by the introduction of DNA polymerase and all four deoxynucleotides. The sequence-specific extension can be controlled by a match or mismatch

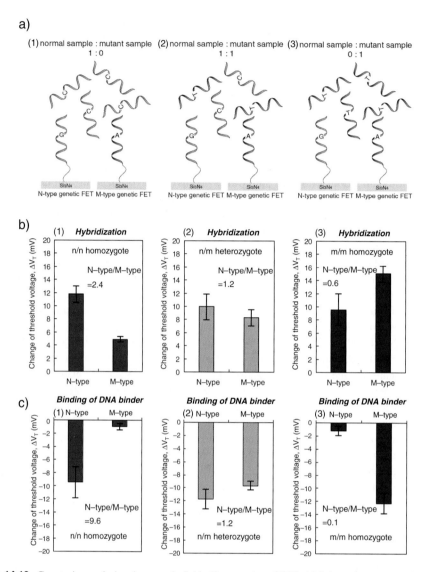

Figure 14.13. Genotyping analysis using genetic field effect transistor (FET). (a) Schematic representation showing the method for genotyping analysis using genetic FET. (b) Genotyping of R353Q locus by hybridization at 57 °C. The concentration ratios of normal sample to mutant sample were 1:0, 1:1, 0:1 for the three different genotypes, respectively. The ratios of the VT shifts of the N-type genetic FET and M-type genetic FET, the N-type/M-type ratios were 2.4, 1.2, 0.6 for three different genotypes, respectively. (c) Genotyping using genetic FETs in combination with Hoechst 33258 as DNA binder. The N-type/M-type ratios were 9.6, 1.2, 0.1 for three different genotypes, respectively.

at the 3'-end of each oligonucleotide probe. When the 3'-end of the oligonucleotide probe is complementary to the polymorphic site of the target DNA, DNA polymerase extends the immobilized oligonucleotide probes with dNTPs in a template-dependent manner. As a result of extension reaction, negative charges increase at the gate surface of the genetic field effect device, because of intrinsic negative charges of polynucleotide. This charge density change can be detected as a shift of the threshold voltage V_T of the genetic field effect device (Fig. 14.14(a)). On the other hand, when the 3'-end of the oligonucleotide probe is not complementary to the target DNA, extension reaction does not occur and the charges at the gate surface dose not

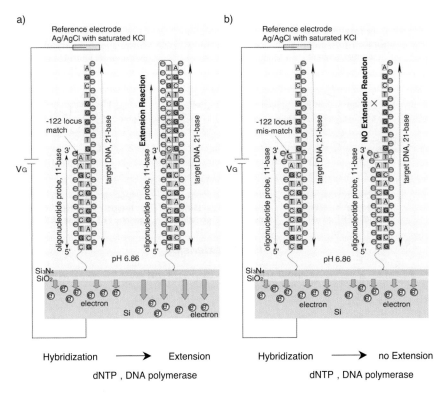

Figure 14.14. Scheme for potentiometric detection of allele-specific primer extension using genetic field effect transistor (FET). The base sequence of the oligonucleotide probe was designed so that target DNA was matched or mismatched at the 3'-end of the probe. (a) When the 3'-end of the oligonucleotide probe is complementary to the target DNA, extension reaction occurs and negative charges increase at the gate surface of the genetic FET. (b) When the 3'-end of the oligonucleotide probe is not complementary to the target DNA, extension reaction does not occur and the charges at the gate surface dose not change.

change (Fig. 14.14(b)). Thus, information on match or mismatch at the polymorphic site of the target DNA can be detected by the use of the genetic field effect devices, which allows simple and precise discrimination of SNP genotyping.

We prepared three different genotypes using synthetic oligonucleotides for the -122 locus of factor VII gene (Table 14.2 (Kajiyama et al. 2003)). The N-type and M-type genetic field effect devices were set into a buffer solution, and hybridized with a normal/normal (n/n) homozygote sample, a normal/mutant (n/m) heterozygote sample and a mutant/mutant (m/m) homozygote sample, respectively (Fig. 14.15). After introducing the genetic field effect devices into the incubated reaction mixture including thermostable DNA polymerase and dNTPs, the differential threshold voltage shifts ΔV_T^{diff} were evaluated. The ratios of the ΔV_T^{diff} between the N-type and the M-type were 9.96, 1.11, 0.08 for a n/n homozygote sample, a n/m heterozygote sample and a m/m homozygote sample, respectively (Fig. 14.15(b)). The results of allele specific extension on the gate surface demonstrate the high signal to noise ratio for SNP genotyping and the good ability of the genetic field effect devices to distinguish the three different genotypes.

5.3. DNA Sequencing

Although a number of methods for SNP analysis have been developed as described in the above section, DNA sequencing techniques are still required to be improved in terms of cost, simplicity, high throughput in order to analyze not only SNPs but also genomic variations

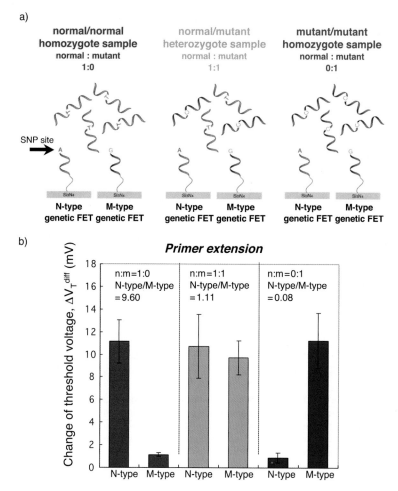

Figure 14.15. Genotyping analysis using genetic field effect transistor (FET). (a) Schematic representation showing the method for genotyping analysis using the genetic FET. A pair of N-type and M-type genetic FETs were hybridized with a normal/normal (n/n) homozygote sample, a normal/mutant (n/m) heterozygote sample and a mutant/mutant (m/m) homozygote sample, respectively. (b) Genotyping using the genetic FET in combination with primer extension. The ratios of the threshold voltage shifts DV_T^{diff} of the N-type genetic FET and M-type genetic FET were 9.96, 1.11, and 0.08 for three different genotypes, respectively. The averages of the two experiments are shown.

such as insertion/deletion, short tandem repeat, etc. A new method for DNA sequencing is introduced in this section, which is based on detection of intrinsic charges of DNA molecules using the field effect.

Oligonucleotide probes are immobilized on the Si_3N_4 gate surface. The complementary target DNA is hybridized with the oligonucleotide probes on the gate surface. The hybridization events are followed by the introduction of DNA polymerase and one of each deoxynucleotide (dCTP, dATP, dGTP or dTTP). DNA polymerase extends the immobilized oligonucleotide probes in a template-dependent manner (Fig. 14.16). As a result of extension reaction, negative charges increase at the gate surface of the field effect devices, because of intrinsic negative charges of incorporated molecules. This charge density change can be detected as a shift of the threshold voltage V_T of the field effect devices. Thus, iterative addition of each deoxynucleotide and measurement of threshold voltage allow a direct, simple and non-labeled DNA sequencing.

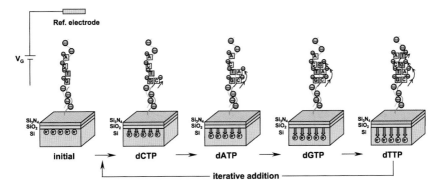

Figure 14.16. Scheme for DNA sequencing based on the FETs in combination with extension reaction. Each deoxynucleotide is incorporated into the probe-target duplex on the FET in the following order: dCTP, dATP, dGTP and dTTP.

The base sequences of factor VII gene including two SNP sites and that of hereditary hemochromatosis (Table 14.3 (Kajiyama et al. 2003)) were used to demonstrate the principle of DNA sequencing based on the field effect devices. We have paid special attention to the buffer concentration to be used for measuring charge density change at the gate surface. The potential change induced by adsorption of proteins at the gate surface was reported to be dependent on the electrolyte concentration (Nakajima, Esashi and Matsuo 1980). It is therefore important to optimize the Debye length at the gate insulator/solution interface. In the present study, a 0.025 m phosphate buffer solution was used for measuring charge density change at the gate surface, while the conventional reaction mixture was used for single-base extension reaction.

The 11-base oligonucleotide probes were immobilized on the gate surface and hybridized with the 21-base target DNA for the base sequence of −122 (Table 14.3 (Kajiyama et al. 2003)). The V_T shifts were measured after incorporation of deoxynucleotides (Fig. 14.17). When the field effect device was soaked into the DNA polymerase buffer solution containing dCTP, the V_T shifted in the positive direction by the amount of 3.8 mV after single-base extension. Next, the field effect device was soaked into the DNA polymerase buffer solution containing dATP, the V_T shifted in the positive direction further by the amount of 3.7 mV, because dATP was incorporated into the probe-target duplex on the field effect device. The positive shifts of the V_T are due to increase of negative charges of incorporated deoxynucleotides. When the measurements of the V_T shifts after single-base extension reaction were performed fifteen

Table 14.3. Base sequences for oligonucleotide probes and targets based on the R353Q and −122 of factor VII gene and the C282Y of hereditary hemochromatosis (Kajiyama et al. 2003).

Locus	Function	Sequence
R353Q	R353Q-wild type	
	probe	5′-amino group-CCACTACCG-3′ (9mer)
	target	5′-ACGTGCCCCGGTAGTGG-3′ (17mer)
−122	−122-wild type	
	probe	5′-amino group-CGTCCTCTGAA-3′ (11mer)
	target	5′-AGCTGGGGTGTTCAGAGGACG-3′ (21mer)
C282Y	C282Y-wild type	
	probe	5′-amino group-AGATATACGTG-3′ (11mer)
	target	5′-CTCCACCTGGCACGTATATCT-3′ (21mer)

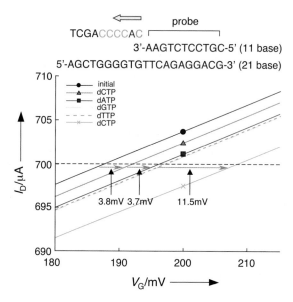

Figure 14.17. Threshold voltage V_T shifts after single-base extension reaction at the gate surface. The threshold voltages were shifted in the positive direction because of intrinsic negative charges of deoxynucleotides. The amount of the V_T shift was determined at a constant drain current of $700\,\mu A$.

times, the average V_T shift was 3.2 mV with the standard deviation of 1.1 mV. When the field effect device was introduced into the buffer solutions containing dGTP and dTTP, respectively, the V_T remained nearly constant, because the deoxynucleotides were not incorporated into the non-complementary base sequence. In this case, the average V_T shift was 0.03 mV with the standard deviation of 0.67 mV. Thus, the V_T shifts based on single-base extension were big enough to be detected with a sufficient signal to noise ratio. Moreover, the V_T shifted in the positive direction by the amount of 11.5 mV, when the field effect device was soaked again into the buffer solution containing dCTP. This is because four dCTPs with negative charges were incorporated into the probe-target duplex on the gate surface.

We evaluated the field effect devices in combination with single-base extension for DNA sequencing. We prepared four kinds of buffer solution containing both DNA polymerase and one of dCTP, dATP, dGTP or dTTP, respectively. The field effect devices hybridized with target DNA were immersed into the above-mentioned buffer solutions for single-base extension reaction and the V_T shift was measured in a 0.025 m phosphate buffer solution after washing the field effect devices. The cycle of single-base extension and measurement of the V_T was repeated iteratively to determine the base sequence of the target DNA. When the base sequence of R353Q region of the factor VII gene was used as a target DNA, the V_T shifted in the positive direction only after single-base extension with the specific deoxynucleotides which were complementary to the base sequence of the target DNA (Fig. 14.18(a)). The V_T change for three-base extension, GGG, was 6.9 mV, which was bigger than that for one-base extension, but was not three times as big as that expected from the number of intrinsic charges. Although the linear relationship between the base length synthesized by the extension reaction and the V_T shift was obtained in the range from 0 to 30 bases (Fig. 14.11), it is important to detect single-base extension quantitatively, in order to reduce base call error especially for continuous sequence of the same base. The density and orientation of the immobilized oligonucleotide probes have to be controlled during a series of extension reactions at 72 °C. Further improvement of precision of the base call is also expected by automation of extension reaction and V_T measurements.

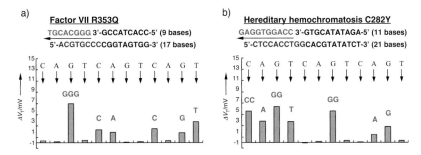

Figure 14.18. DNA sequencing using the FETs in combination with extension reaction. DNA polymerase extends the immobilized oligonucleotide probes with each deoxynucleotide in a template-dependent manner. Based on the proposed method, the base sequence of R353Q region of the factor VII gene (a) and C282Y region of hereditary hemochromatosis gene (b) were successfully determined.

C282Y region of hereditary hemochromatosis gene was used as another example of DNA sequencing using the field effect device and single base extension (Fig. 14.18(b)). The positive V_T shifts could be detected in accordance with the base sequence of the target DNA. In this case, the average V_T shifts for two-base incorporation was 5.8 mV with the standard deviation of 0.4 mV, while the average V_T shift for single-base extension was 3.2 mV as described previously. The V_T shifts for two-base extension was approximately twice as big as that for single-base extension. Thus, the results of iterative extension reaction and detection of the V_T indicated the ability of a direct, simple and potentially precise DNA sequencing analysis using the field effect devices. The number of bases which can be analyzed by the proposed method is about ten bases at present. The V_T shift for single base extension became gradually smaller as the number of bases increased more than ten bases. One of the reasons for this limitation would be the Debye length at the gate insulator/solution interface. Any charge density change induced outside the Debye length cannot be detected with the field effect devices. Lateral extension reaction in which DNA probes are extended in parallel with the gate surface would be effective for DNA sequencing with long bases. Another reason for the limitation would be peeling off the immobilized oligonucleotide probes from the surface of the gate insulator as the temperature stress of the extension reaction at 72 °C is applied repeatedly. The stronger immobilization method for oligonucleotide probes on the Si_3N_4 surface has to be adopted to analyze longer base sequence.

6. Conclusions and Future Perspectives

In this chapter, we have demonstrated the potentiometric detection of DNA molecular recognition events by use of the genetic field effect devices and that SNP genotyping and DNA sequencing can be realized based on intrinsic charges of DNA molecules without any labeling materials. It is possible to integrate multiple field effect devices and signal processing circuits in a single chip using advanced semiconductor technology. Simultaneous analyses of various base sequences including SNPs can be realized based on the field effect devices. Since the output of the field effect devices is an electrical signal, it is easy to standardize the results obtained with the field effect devices as compared with the fluorescent detection-based analyses. Therefore, the platform based on the field effect devices is suitable for a miniaturized and arrayed system for SNP genotyping and DNA sequencing in clinical research and diagnostics.

The principle and fundamental characteristics of the biologically coupled field effect transistor are described in this chapter. Since bio-FETs are fabricated using the advanced semiconductor technology, they have advantages such as miniaturization of the analytical system and integration of multiple sensors and signal-processing circuits on a single chip. Using the bio-FET technologies described in this chapter, a small instrument for genetic analysis would be realized and SNP typing and DNA sequencing of specific genes can be analyzed not only in the large scale hospitals but also in the small hospitals or physicians' offices for personalized medicine. In addition, we can take the small instrument out of the laboratory to detect nucleic acids of viruses or micro organisms on site for infectious disease testing. Thus, the bio-FET technology is very useful and effective to monitor security level rapidly in our environment. In the advanced countries such as Japan, highly aged society is progressing and medical needs are increasing in spite of the limited capacity of hospitals and medical doctors. We believe that one of the approaches to improve this situation is to realize the homecare system in which medical treatment and clinical diagnostics can be done near patients at home. In such a system, a simple, small and highly sensitive detection system is required as a point of care testing.

On the other hand, gene functional analyses have remarkably proceeded in the fields of molecular biology, pharmacogenomics and clinical research, based on the completion of the decoding of the human genome. Analysis of DNA sequencing and nucleotide variation has been becoming increasingly important for assembly of a high-resolution map of disease-related loci and for clinical diagnostics. Although a number of methods for genetic analysis have been developed, DNA sequencing techniques are still required to be improved in terms of cost, simplicity, high throughput in order to analyze not only human genome but also DNA sequence of other micro-organisms. Several DNA sequencing techniques based on new principles other than electrophoresis have been proposed for that purpose. Since the genetic FETs described in this chapter can be integrated and arrayed easily using the semiconductor technology, we think that integrated genetic FETs with high density is one of potential approaches to realize simple, cost effective and high throughput DNA sequencer.

References

Beattie WG, Meng L, Turner SL, Varma RS, Dao DD and Beattie KL (1995) Hybridization of DNA targets to glass-tethered oligonucleotide probes. Mol. Biotechnol. 4:213–225

Boger DL, Fink BE, Brunette SR, Tse WC and Hedrick MP (2001) A simple, high-resolution method for establishing DNA binding affinity and sequence selectivity. J. Am. Chem. Soc. 123:5878–5891

Chee M, Yang R, Hubbell E, Berno A, Huang XC, Stern D, Winkler J, Lockhart DJ, Morris M. S and Fodor SPA (1996) Accessing genetic information with high-density DNA arrays. Science 274:610–614

Chrisey LA, Lee GU and O'Ferrall CE (1996) Covalent attachment of synthetic DNA to self-assembled monolayer films. Nucl. Acids. Res. 24:3031–3039

Cooper MA, Dultsev FN, Minson T, Ostanin VP, Abell C and Klenerman D (2001) Direct and sensitive detection of a human virus by rupture event scanning. Nat. Biotechnol. 19:833–837

Crumbliss AL, Perine SC, Stonehuerner J, Tubergen KR, Zuhao J, Henkens RW and O'Daly JP (1992) Colloidal gold as a biocompatible immobilization matrix suitable for the fabrication of enzyme electrodes by electrodeposition. Biotechnol. Bioeng. 40:483–490

Dengler WA, Schulte J, Berger DP, Mertelsmann R and Fiebig HH (1995) Development of a propidium iodide fluorescence assay for proliferation and cytotoxicity assays. Anticancer Drugs 6:522–532

Disley DM, Cullen DC, You H-X and Lowe CR (1998) Covalent coupling of immunoglobulin G to self-assembled monolayers as a method for immobilizing the interfacial-recognition layer of a surface plasmon resonance immunosensor. Biosens, Bioelectron. 13:1213–1225

Duschl C, Sevin-Landais AF and Vogel H (1996) Surface engineering: optimization of antigen presentation in self-assembled monolayers. Biophys. J. 70:1985–1995

Eggers M, Hogan M, Reich RK, Lamture J, Ehrlich D, Hollis M, Kosicki B, Powdrill T, Beattie K and Smith S (1994) A microchip for quantitative detection of molecules utilizing luminescent and radioisotope reporter groups. Biotechniques 17:516–525

Fritz J, Cooper EB, Gaudet S, Sorger PK and Manails SR (2002) Proceedings of the.National Academy of Science, USA 99:14142–14146

Gilles PN, Wu DJ, Foster CB, Dillon PJ and Chanock SJ (1999) Single nucleotide polymorphic discrimination by an electronic dot blot assay on semiconductor microchip. Nat. Biotechnol. 17:365–370

Guo Z, Guilfoyle RA, Thiel AJ, Wang R and Smith LM (1994) Direct fluorescence analysis of genetic polymorphisms by hybridization with oligonucleotide arrays on glass supports. Nucleic Acids Res. 22:5456–5465

Guo Z, Gatterman MS, Hood L, Hansen JA and Petersdorf EW (2001) Oligonucleotide arrays for high-throughput SNPs detection in the MHC class I genes: HLA-B as a model system. Genome Res. 12:447–457

Haff L and Smirnov IP (1997) Single-nucleotide polymorphism identification assays using a thermostable DNA polymerase and delayed extraction MALDI-TOF mass spectrometry. Genome Res. 7:378:388

Hiller M, Kranz C, Huber J, Baeuerle P and Schuhmann W (1996) Amperometric biosensors produced by immobilization of redox enzymes at polythiophene-modified electrode surfaces. Adv. Mater 8:219–222

Howell WM, Jobs M, Gyllensten U and Brookes AJ (1999) Dynamic allele-specific hybridization. Nat. Biotechnol. 17:87–88

Huang E, Satjapipat M, Han S and Zhou F (2001) Surface structure and coverage of an oligonucleotide probe tethered onto a gold substrate and its hybridization efficiency for a polynucleotide target. Langmuir 17:1215–1224

Jobs M, Howell WM, Stromqvist L, Mayr T and Brookes AJ (2003) DASH-2: Flexible, Low-Cost, and High-Throughput SNP Genotyping by Dynamic Allele-Specific Hybridization on Membrane Arrays. Genome Res. 13:916–924

Kajiyama T, Miyahara Y, Kricka LJ, Wilding P, Graves DJ, Surrey S and Fortina P (2003) Genotyping on a thermal gradient DNA chip. Genome Res. 13:467–475

Kallury KMR, Krull UJ and Thompson M (1998) X-ray photoelectron spectroscopy of silica surfaces treated with polyfunctional silanes. Anl. Chem. 60:169–172

Kim D-S, Jeong Y-T, Park H-J, Shin J-K, Choi P, Lee J-H and Lim G (2004) An FET-type charge sensor for highly sensitive detection of DNA sequence. Biosensors and Bioelectronics 20:69–74

Kumar A, Larsson O, Parodi D and Liang Z (2000) Silanized nucleic acids: a general platform for DNA immobilization. Nucl. Acids. Res. 28:e71

Landegren U, Kaiser R, Sanders J and Hood L (1988) A ligase-mediated detection technique. Science 241:1077–1080

Landegren U, Nilsson M and Kwok P-Y (1998) Reading bits of genetic information: methods for single-nucleotide polymorphism analysis. Genome Res. 8:769–776

Livak KJ, Marmaro J and Todd JA (1995) Towards full automated genome-wide polymorphism screening. Nat. Genet. 9:341–342

Lockhart DJ, Dong H, Byrne MC, Follettie MT, Gallo MV, Chee MS, Mittmann M, Wang C, Kobayashi M, Horton H and Brown EL (1996) Expression monitoring by hybridization to high-density oligonucleotide arrays. Nat. Biotechnol. 14:1675–1680

Lu B, Xie J, Lu C, Wu C and Wei Y (1995) Oriented immobilization of Fab' fragments on silica surfaces. Anal. Chem. 67:83–87

Manz A, Miyahara Y, Miura J, Watanabe Y, Miyagi H and Sato K (1990) Design of an open-tubular column liquid chromatograph using silicon chip technology. Sensors and Actuators B 1:249–255

Miyahara Y, Moriizumi T and Ichimura K (1985) Integrated enzyme FETs for simultaneous detections of urea and glucose. Sensors and Actuators 7:1–10

Miyahara Y, Tsukada K and Miyagi H (1988) Field-effect transistor using a solid electrolyte as a new oxygen sensor. J. Appl. Phys. 63:2431–2434

Miyahara Y, Tsukada K, Miyagi H and Simon W (1991) Urea sensor based on an ammonium ion-sensitive field effect transistor. Sensors and Actuators B 3:287–293

Miyahara Y and Simon W (1991) Comparative studies between ion-selective field effect transistors and ion-selective electrodes with polymeric membranes. Electroanalysis 3:287–291

Miyahara Y, Tsukada K, Shibata Y and Watanabe Y (1994) Long-life planar oxygen sensor. Sensors and Actuators B 20:89–94

Mrksich M, Chen CS, Xia Y, Dike LE, Ingber DE and Whitesides GM (1996) Controlling cell attachment on contoured surfaces with self-assembled monolayers of alkanethiolates on gold. Proceedings of the National Academy of Science 93:10775–10778

Nakajima H, Esashi M and Matsuo T (1980) The pH-response of organic gate ISFETs and the influence of macromolecule adsorption. Nippon Kagaku Kaishi No.10:1499–1508

Orita M, Iwahana H, Kanazawa H, Hayashi K and Sekiya T (1989) Detection of polymorphisms of human DNA by gel electrophoresis as single-strand conformation polymorphisms. Proceedings of the National Academy of Science 86:2766–2770

Parker M, Patel N, Davies MC, Roberts CJ, Tendler SJB and Williams PM (1996) A novel organic solvent-based coupling method for the preparation of covalently immobilized proteins on gold. Protein Sci. 5:2329–2332

Parsons BL and Heflich RH (1997) Genotypic selection methods for the direct analysis of point mutations. Mutat. Res. 387:97–121

Pastinen T, Partanen J and Syvanen AC (1996) Multiplex fluorescent, solid-phase minisequencing for efficient screening of DNA sequence variation. Clin. Chem. 42:1391–1397

Pastinen T, Kurg A, Metspalu A, Peltonen L and Syvanen AC (1997) A specific tool for DNA analysis and diagnostics on oligonucleotide assays. Genome Res. 7:606–614

Pastinen T, Raitio M, Lindroos K, Tainola P, Peltonen L and Syvanen A-C (2000) A system for specific, high-throughput genotyping by allele-specific primer extension on microarray. Genome Res. 10:1031–1042

Pirrung MC (2002) How to make a DNA chip. Angew. Chem. Int. Ed. 41:1276–1289

Pouthas F, Gentil,C, Cote D and Bockelmann U (2004) DNA detection on transistor arrays following mutation-specific enzymatic amplification. Appl. Phys. Let. 84:1594–1596

Qian J, Liu Y, Liu H, Yu T and Deng J (1997) Immobilization of horseradish peroxidase with a regenerated silk fibroin membrane and its application to a tetrathiafulvalene-mediating H_2O_2 sensor. Biosens, Bioelectron 12:1213–1218

Rickert J, Weiss T and Gopel W (1996) Self-assembled monolayers for chemical sensors: molecular recognition by immobilized supramolecular structure. Sens. Actuators B 31:45–50

Ronagi M, Uhlen M and Nyren P (1998) A sequence method based on real-time pyrophosphate detection. Science 281:363–365

Rye HS, Yue S, Quesada MA, Haugland RP, Mathies RA and Glazer AN (1993) Picogram detection of stable dye-DNA intercalation complexes with two-color laser-excited confocal fluorescence gel scanner. Methods Enzymol. 217:414–431

Saiki RK, Walsh RS, Levenson CH and Erlich HA (1989) Genetic analysis of amplified DNA with immobilized sequence-specific oligonucleotide probes. Proc. Nat. Acad. Sci. USA 86:6230–6234

Sakahara H and Saga T (1999) Avidin–biotin system for delivery of diagnostic agents. Adv. Drug Delivery Rev. 37:89–101

Sakata T and Miyahara Y (2005a) Detection of DNA recognition events using multi-well field effect devices. Biosensors and Bioelectronics 21:827–832

Sakata T and Miyahara Y (2005b) Potentiometric detection of single nucleotide polymorphism by using a genetic field-effect transistor. ChemBioChem 6:703–710

Sakata T and Miyahara Y (2006) DNA sequencing based on intrinsic molecular charges. Angewandte Chemie International Edition 45:2225–2228

Sauer S, Lechner D, Berlin K, Lehrach H, Escary J-L, Fox N and Gut IG (2000) A novel procedure for efficient genotyping of single nucleotide polymorphisms. Nucl. Acids. Res. 28:e13

Schene M, Shalon D, Davis RW and Brown PO (1995) Quantitative monitoring of gene expression patterns with a complementary DNA microarray. Science 270:467–470

Shlyakhtenko LS, Gall AA, Weimer JJ, Hawn DD and Lyubchenko YL (1999) Atomic force microscopy imaging of DNA covalently immobilized on a functionalized mica substrate. Biophys. J. 77:568–576

Silberzan P, Leger L, Ausserre D and Bennattar JJ (1991) Silanation of silica surfaces. A new method of constructing pure or mixed monolayers. Langmuir 7:1647–1651

Sirkar K and Pishko MV (1998) Amperometric biosensors based on oxidoreductases immobilized in photopolymerized poly(ethylene glycol) redox polymer hydrogels. Anal. Chem. 70:2888–2894

Souteyrand E, Cloarec JP, Martin JR, Wilson C, Lawrence I, Mikkelsen S and Lawrence MF (1997) Direct detection of the hybridization of synthetic homo-oligomer DNA sequences by field effect. J. Phys. Chem. B 101:2980–2985

Sriram M, van der Marel GA, Roelen HLPF, van Boom JH and Wang AH-J (1992) Structural consequences of a carcinogenic alkylation lesion on DNA: effect of O^6-ethylguanine on the molecular structure of the d(CGC[e^{6}G]AATTCGCG) -netropsin complex. Biochemistry 31:11823–11834

Steel AB, Herne TM and Tarlov MJ (1998) Electrochemical quantitation of DNA immobilized on gold. Anal. Chem. 70:4670–4677

Strother T, Cai W, Zhao X, Hamers RJ and Smith LM (2000) Synthesis and characterization of DNA-modified silicon (111) surfaces. J. Am. Chem. Soc. 122:1205–1209

Strother T, Hamers RJ and Smith LM (2000) Covalent attachment of oligodeoxyribonucleotides to amine-modified Si (001) surfaces. Nucleic Acids Res. 28:3535–3541

Syvanen A-C (2001) Accessing genetic variation: genotyping single nucleotide polymorphisms. Nat. Rev. Genet. 2:930–942

Thiel AJ, Frutos AG, Jordan CE, Corn RM and Smith LM (1997) In situ surface plasmon resonance imaging detection of DNA hybridization to oligonucleotide arrays on gold surfaces. Anal. Chem. 69:4948–4956

Tiefenauer L and Ros R (2002) Biointerface analysis on a molecular level new tools for biosensor research. Colloids and Surfaces B 23: 95–114

Tsukada K, Miyahara Y, Shibata Y and Miyagi H (1990) An integrated chemical sensor with multiple ion and gas sensors. Sensors and Actuators B 2:291–295

Tyagi S, Bratu DP and Kramer FR (1998) Multicolor molecular beacons for allele discrimination. Nat. Biotechnol. 16:49–53

Tyagi S, Marras SAE and Kramer FR (2000) Wavelength-shifting molecular beacons. Nat. Biotechnol. 18:1191–1196

Uslu F, Ingebrandt S, Mayer D, Böcker-Meffert S, Odenthal M and Offenhäusser A (2004) Labelfree fully electronic nucleic acid detection system based on a field-effect transistor device. Biosensors and Bioelectronics 19:1723–1731

Vandenberg E, Elwing H, Askendal A and Lundstrom I (1991) Structure of 3-aminopropyl triethoxy silane on silicon oxide. J. Colloids Interface Sci. 147:103–118

Wikstrom P, Mandenium CF and Larsson P (1988) Phase Silylation, a rapid method for preparation of high-performance liquid chromatography supports. J. Chromatogr. 455:105–117

Wilson WD, Tanious FA, Barton HJ, Strekowski L and Boykin DW (1989) Binding of 4',6-diamidino-2-phenylindole (DAPI) to GC and mixed sequences in DNA: intercalation of a classical groove-binding molecule. J. Am. Chem. Soc. 111:5008–5010

Yon-Hin B, Smolander M, Crompton T and Lowe CR (1993) Covalent electropolymerization of glucose oxidase in polypyrrole. Evaluation of methods of pyrrole attachment to glucose oxidase on the performance of electropolymerized glucose sensors. Anal. Chem. 65: 2067–2071

15

Impedance-Based Biosensors for Pathogen Detection

Xavier Muñoz-Berbel, Neus Godino, Olivier Laczka, Eva Baldrich, Francesc Xavier Muñoz and Fco. Javier Del Campo

Abstract

Electrochemical impedance spectroscopy (EIS) is an important detection technique for biosensors. In the field of immunosensors, and particularly pathogen detection, it is one of the preferred electrochemical techniques because it does away with the use of enzyme labels or redox mediators. This chapter provides an introduction to the fundamentals of EIS and basic data analysis, with an emphasis on the most common features found in immunosensors and possible experimental limitations.

This chapter then discusses a series of functionalisation approaches that can be used in the development of an immunosensor for the detection of bacteria. This is followed by a selection of impedance-based immunosensor examples from the literature.

1. Introduction

As regulations in the areas of food safety, health, and environment become more strict, there is a pressing need for the development of novel and faster methods to detect pathogen agents. Traditional methods, namely plate count, ELISA, and PCR, are very selective and can be very sensitive, but at a high time cost. In fact, most of these methods may take up to a few days to yield an answer because of the multiple steps needed in the analytical process. Having said this, PCR techniques are evolving fast and probably represent the fastest among traditional methods. However, they are also subject to certain limitations, such as inhibition of the PCR, cross contamination between samples and or the ability to determine whether the detected pathogen was viable or not at the time of sampling, among others.

Biosensors have been recently defined (http://www.biosensors-congress.elsevier.com/about.htm) as "analytical devices incorporating a biological material (e.g., tissue, microorganisms, organelles, cell receptors, enzymes, antibodies, nucleic acids, natural products, etc.), a biologically derived material (e.g., recombinant antibodies, engineered proteins, aptamers, etc.), or a biomimic (e.g., synthetic catalysts, combinatorial ligands, imprinted polymers) intimately associated with or integrated within a physicochemical transducer or transducing microsystem, which may be optical, electrochemical, thermometric, piezoelectric, magnetic, or

Xavier Muñoz-Berbel, Neus Godino, Olivier Laczka, Eva Baldrich and Francesc Xavier Muñoz   •
Centro Nacional de Microelectrónica, IMB-CNM-CSIC, Esfera UAB, Campus Universidad Autónoma de Barcelona; 08193–Bellaterra; Barcelona, Spain. **Fco. Javier Del Campo**   •   Instituto de Biotecnología y Biomedicina, Departamento de Microbiología y Genética, Universidad Autónoma de Barcelona; 08193–Bellaterra; Barcelona, Spain.

M. Zourob et al. (eds.), *Principles of Bacterial Detection: Biosensors, Recognition Receptors and Microsystems*,
© Springer Science+Business Media, LLC 2008

micromechanical." They can thus be classified according to their signal transduction method or according to the biological recognition element used. Biosensors are postulated as an alternative technology for the fast detection of pathogen microorganisms. This chapter aims to give an overview of pathogen detection using electrochemical impedance spectroscopy, EIS. As will be discussed in more detail in the following section, EIS consists in the application of a small amplitude A.C. signal superimposed over a base constant potential or current. The resulting impedance of the system then provides information related to interfacial and bulk phenomena. Although most applications of EIS deal with the study of coatings and corrosion processes, the technique can be applied to other fields such as the characterisation of fuel cells and batteries or the study of reaction kinetics.

As will be shown later in this chapter, EIS-based biosensors consist mainly of an electrode or set of electrodes suitably modified with a biological receptor, so that when the process of interest takes place, an impedance change can be observed. As will be further shown, the impedance response of a system is an aggregate magnitude that includes the contribution of mass transport and charge transfer events. From the viewpoint of EIS-based biosensors, the system parameters most commonly measured are changes in interfacial capacitance, electron transfer resistance, and medium conductivity. Depending on the particular features of the system under study, the experimentalist may choose to observe these parameters individually or simply monitor the variation of the impedance modulus and phase. The following section gives an overview of the basic notions underpinning EIS biosensors. The chapter then describes the main surface modification methods that can be used to produce EIS biosensors. Last, a series of pathogen detection applications based on impedance spectroscopy are described and discussed.

2. Fundamentals of Electrochemical Impedance Spectroscopy

This section aims to provide a general overview of the principles underpinning EIS, paying special attention to its use in the field of biosensors. EIS represents a powerful method for the study of conducting materials and interfaces, and its basics can be found in the generalist electrochemistry literature (Brett and Oliveira Brett 1993; Christensen and Hamnet 1994; Bard and Faulkner 2001; Bagotsky 2006). A few fully devoted monographs can also be found (Sluyters-Rehbach 1994; Gabrielli 1995, 1998; Barsoukov and Macdonald 2005), and the reader is encouraged to take them up.

Electrochemical techniques study interfacial phenomena by looking at the relation between current and potential. A perturbation in either the current or the potential of the working electrode is imposed, and the response of the system to those perturbations is observed. The study of such dependence then gives access to a wealth of information about the system (kinetic, thermodynamic and mechanistic). In contrast to the most commonly used electrochemical techniques, which are based on the application of direct currents, such as the case of chronoamperometry or voltametry, EIS is based on the superimposition of a sine-wave potential of small amplitude over a polarisation potential of constant value:

$$E(t) = E_{polarisation} + \Delta E \sin(\omega t), \tag{15.1}$$

where $E_{polarisation}$ is the base potential of the working electrode measured against a suitable reference, ΔE is the amplitude of the sine wave and ω the frequency of the signal in rad s^{-1}. If the system is linear, the response is a sinusoidal current of the same frequency, but different in amplitude and phase from the voltage:

$$I(t) = \Delta I \sin(\omega t + \phi), \tag{15.2}$$

where $I(t)$ is the instant current value, ΔI is the current amplitude and ϕ is the phase shift angle.

The ratio

$$Z = \frac{E(t)}{I(t)} = \frac{\Delta E \sin(\omega t)}{\Delta I \sin(\omega t + \phi)}, \tag{15.3}$$

is a new phasor called impedance. Using Euler's notation for complex numbers, equation 15.3 transforms into:

$$Z = |Z|\, e^{j\phi} = |Z| \cos \phi + j\, |Z| \sin \phi, \tag{15.4}$$

where $|Z|$ is the impedance modulus (in Ω). This is represented in Fig. 15.1. When the current and the potential are in phase ($\phi = 0$), equation 15.3 is simply the expression of Ohm's law, and the impedance is a resistor. Unfortunately, electrochemical systems are hardly this simple and their impedance always presents an "imaginary" component, which may be identified with interfacial phenomena such as double-layer charging. This is because the charge separation found at the electrode-solution interface provides an electrical structure that very much resembles a capacitor. However, other phenomena such as mass transport and reaction kinetics also contribute to the imaginary part of the impedance. This is why most electrochemical problems are discussed in terms of combinations of capacitances and resistors. Although inductive terms may appear in electrochemical problems, often due to homogeneous reactions coupled to electrochemical processes, they very rarely appear in biosensor applications. If they do appear, it is probably a high frequency artefact due to excessively long cabling between the frequency response analyser and the sensor, and it can be easily overcome by using shorter leads.

Because impedance data may be found in Cartesian and polar coordinates alike, it is important to know how the various parameters relate to each other. Thus,

$$|Z| = \sqrt{(Z')^2 + (Z'')^2} \quad \text{and} \quad \phi = \arctan\left(\frac{Z''}{Z'}\right), \tag{15.5a-b}$$

and also

$$Z' = |Z| \cos \phi \quad \text{and} \quad Z'' = |Z| \sin \phi, \tag{15.6a-b}$$

where Z' and Z'' are the terms used to name the real and the imaginary parts of the impedance, respectively.

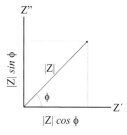

Figure 15.1. Generic complex plane plot diagram. The real part of the impedance is plotted in the abscissas and the imaginary part is represented in the ordinates axis.

The inverse of the impedance is called admittance, and it is commonly represented by the letter Y. In some cases, it may be more convenient to handle the data in terms of admittance rather than impedance. This is because networks of impedances and admittances can be calculated using Kirchhoff's laws. Hence, if several impedances are arranged in series, the total impedance is the sum of the individual impedances. However, if the impedances are distributed in parallel, then it is more convenient to use admittances, because the total admittance will be the sum of the individual admittances.

Electrochemical systems can be treated as networks of impedances arranged in series and parallel. In this context, the impedance measured in EIS experiments is the impedance of the whole electrochemical cell, Z_{cell}, and not simply that of the working electrode alone. In addition, every experimental setup contains features that can have an influence on the measured impedance: for example, the cables and connections used.

Moreover, electrochemical processes present nonlinear dependencies with potential, so the amplitude of the perturbation needs to be carefully chosen to ensure the linearity of the response. This is particularly important at low frequencies, and in general it is recommended to use amplitudes as small as possible (Gabrielli 1995). Typical potential amplitudes range from 5 up to 25 mV. The lower limit of the amplitude is determined by the parasitic noise of the experimental setup and the higher limit is marked by the onset of faradaic nonlinearities. It is also worth pointing out that in the absence of electron transfer processes, larger amplitudes may be used. Besides the frequency and amplitude of the perturbation signal, the polarisation potential is another key parameter to control in EIS experiments, because interfacial capacitance and electron transfer rates are potential dependent magnitudes (Bard and Faulkner 2001).

2.1. Data Analysis: Plotting

In order to analyse the data more conveniently, impedance may be plotted in a variety of ways depending on the information sought by the experimentalist. One very common way of plotting the data consists in representing Z' along the x axis and Z'' along the y axis. This representation is known as a complex plane plot and is often referred to as a *Nyquist* plot. Here each point corresponds to a frequency of measurement and represents a vector of length $|Z|$. As depicted in Fig. 15.1, the angle between the vector and the x axis is equal to the phase shift ϕ at that given frequency.

Bode plots are another convenient type of representation by which impedance-related parameters are plotted versus frequency on a logarithmic scale. Representations of the phase shift and the impedance modulus are the most usual kind of Bode plots, but representations of other magnitudes, such as admittance versus the logarithm of frequency, are also possible. Fig. 15.2 shows the typical response (complex plane plot and Bode plots) of a metal electrode immersed in an electrolyte solution. The response may be compared to a resistance (solution) and a capacitor (electrode-solution interface) in series.

2.2. Data Analysis: Interpretation

Given the complexity of real systems, impedance data are analysed according to simplified models. These models can be defined according to two main approaches, both of which require the understanding of the physical structure of the system, although expressed in different terms. From a scientific viewpoint, the best approach consists in the formulation of the system of equations describing the electrochemical phenomena occurring in the sample. However, the resulting system of partial differential equations is not trivial to solve, and numerical methods are required to do so. Experimental data may, for instance, be interpreted by fitting of the simulated frequency response after iteratively changing the set of key variables of interest, such as diffusion coefficients, concentrations, medium permittivity, or electrode geometry, amongst

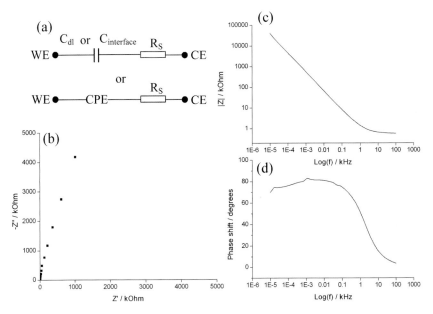

Figure 15.2. (a) In the absence of heterogeneous electron transfer processes, and assuming that the impedance of the counter electrode is negligible, the cell impedance may be compared to a resistance (electrolyte solution) connected in series to a capacitance or a CPE representing the electrode-solution interface. (b) Complex plane plot obtained at a 0.8 mm diameter platinum disk electrode at open circuit potential in 0.1M KCl. (c) Bode plot showing the logarithm of the impedance modulus *versus* the logarithm of the frequency in kHz. The value at high frequencies corresponds to the solution resistance. (d) Bode plot representing phase shift versus the logarithm of frequency. The fact that the low frequency angle does not reach 90 is an indication of electrode surface non-idealities (roughness).

others. This method is perhaps able to provide the best insights into the structure of the system under study, but it is much too tedious to be of practical use.

The second and more common way to analyse EIS data consists in defining the independent processes contributing to the system behaviour, and laying out the system's equivalent circuit in terms of electrical components, namely resistors and capacitors. In order to adequately build the equivalent circuit, one should identify each phenomenon with an equivalent element (Sluyters-Rehbach 1994). Once the equivalent circuit is defined, the experimental data is fitted to it using nonlinear least square fitting techniques. This second methodology is much simpler to implement than the previous one, but it can be extremely risky, as the same data can always be fitted to innumerable different circuits. Because of this, the worst thing to do is to try to fit electrochemical impedance data to random combinations of elements until a best fit is obtained. Another drawback of this approach is that the information is obtained in capacitance and resistance terms and their translation into actual physical system parameters is not always immediate.

The following section will address the most relevant parameters that have to be considered in impedimetric biosensing.

2.2.1. Non-Faradaic Parameters

2.2.1.1. Solution Resistance

The current between the working and the auxiliary or counter electrode flows through an electrolyte solution, the conductivity of which depends on its ionic composition. The resistance of such a solution can be treated as a resistor connected in series with the electrode, and its magnitude depends mainly on electrode and cell geometry. It can generally be estimated from Z' at very high frequencies ($\omega \to \infty$). In biosensor applications, solution resistance may appear in the literature as Rs or R. When a reference electrode is used, the measured solution resistance corresponds to that of the solution path between the working electrode and the tip of the reference electrode (Fig. 15.3). This may be a source of variability in the data unless the electrodes are always placed in the same relative position within the sample (Vanýsek 1997). In addition to electrode position, temperature is another variable that can affect solution resistance. Therefore, the use of thermostated systems whenever possible is also recommended.

2.2.1.2. Geometric Capacitance

Geometric capacitance appears between the working and auxiliary electrodes and in parallel to the electrode impedances and solution resistance (Sluyters-Rehbach 1994; Bagotsky 2006). This capacitance is due to the solution between the electrodes and depends on medium permittivity and the distance between the electrodes. Because of its small value, in the picofarad range, it can usually be neglected in the measurement frequencies used in biosensor applications. It is represented by C_g.

2.2.1.3. Double Layer and Interfacial Capacitance

Every time a metal electrode is immersed in an electrolyte solution, a charge separation develops across the interface. This provides the interface with an electrical structure analogous to a capacitor. The extent of this charge separation, and hence the capacity of the interface, is a function of potential. In biosensor applications, where the electrode surface is modified by the incorporation of a receptor biomolecule, the structure of the electrode-solution interface is modified and new capacitive terms may have to be considered in series with the double layer capacitance. Because of this, it makes more sense in biosensors to use the term interfacial capacitance, $C_{interface}$, rather than double-layer capacitance. As shown in Fig. 15.4a, the value of the $C_{interface}$ can be extracted, in the absence of other interfacial processes (such as electron transfer steps), from the slope of a graph representing $-1/Z''$ vs ω. For interdigitated electrodes, where both the working and auxiliary electrodes are equivalent (and the two interfaces are in series), the slope corresponds to half the capacitance of an individual set of electrodes.

2.2.1.4. The Constant Phase Element

In reality, no electrode-solution interfaces behave as true capacitors, except perhaps for mercury electrodes. The phase angle observed for real systems always deviates from $-\pi/2$, and it has been found that the smoother the electrode surface, the closer the phase angle is

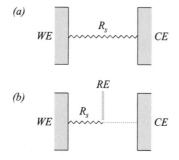

Figure 15.3. Diagrammatic representation of a working and counter electrode in solution, and the effect of using a reference electrode on the measured solution resistance.

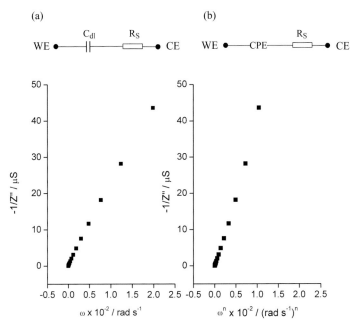

Figure 15.4. Data obtained at a 0.8 mm Pt disc electrode at open circuit potential in 0.1M KCl. (a) Interpretation of the interface in terms of a capacity. The resulting capacity is $0.253 \pm 0.135 \,\mu F$. (b) Interpretation of the same data in terms of a CPE where $K = 0.350 \pm 0.029$ and the index $n = 0.884 \pm 0.028$.

to this limit. Thus, electrochemists may use the constant phase element (CPE) to obtain an estimation of both interfacial capacitance and surface roughness. CPEs are nonintuitive circuit elements with a phase angle of constant value and independent of the frequency (Fricke 1932; Macdonald 1984; McAdams 1989). The impedance of the CPE can be expressed as:

$$Z = \frac{1}{K(jw)^n},$$
(15.7)

where K is the magnitude of the CPE which, despite being proportional to the interfacial capacitance, is not an actual capacitance itself (Sluyters-Rehbach 1994). On the other hand, n is a constant parameter which ranges from 1 for perfectly smooth surfaces to 0.5 for very rough surfaces. Thus, for planar surfaces, the CPE almost behaves as a pure capacitor. The phase angle of the CPE can be calculated (in radians) as shown in Macdonald (1984). Fig. 15.4.b shows a plot of subceptance $(1/Z'')$ versus ω^n, from which the value of K can be extracted.

$$\phi = -n\frac{\pi}{2}.$$
(15.8)

In biosensor applications the interfacial capacitance may be substituted by a CPE ($CPE_{interfacial}$) with a value of n between 0.8 and 1. Actually, metallic electrodes usually present values of n closer to 0.8 (McAdams 1989).

2.2.2. Faradaic Parameters

So far we have described only the circuit elements that appear when an electrode set is immersed in an inert electrolyte solution. It is possible to develop biosensor applications where only the above elements appear: for example, capacitive immunosensors, or DNA sensors. Here detection is provided by changes in interfacial capacitance or electron transfer resistance, changes brought about by the capture of an antigen by a suitable antibody immobilised over the electrode surface, or by the hybridisation between two single strands of DNA.

Provided that the impedance of the counter electrode can be neglected, the circuits previously presented in Fig. 15.3 and Fig. 15.4 represent the general starting point for most electrochemical systems. However, the presence of electroactive substances may introduce significant impedance changes, depending upon the measurement potential. The basic electrochemical reaction can be described as:

$$O + ne^- \Leftrightarrow R, \tag{15.9}$$

where n is the number of electrons exchanged at the electrode-solution interface, and its equilibrium is defined by the Nernst equation:

$$E = E_{O/R}^{0'} + \frac{RT}{nF} \log \frac{C_O}{C_R}, \tag{15.10}$$

where E is the electrode potential, $E_{O/R}^{0'}$ the formal potential for the couple, R is the gas constant, T the temperature, F the Faraday's constant, and C_O and C_R represent the molar concentrations of oxidised and reduced species, respectively.

2.2.2.1. Charge-Transfer Resistance

If the potential is such that these electroactive moieties present in the solution exchange electrons across the interface, a whole new set of phenomena, known as faradaic, appear. The first of these phenomena is charge transfer across the electrode-solution interface. This step is observed as a resistance in terms of equivalent circuits and occurs in parallel to the charge/discharge of the interfacial capacitance. It is commonly represented by R_{ct}, and it is an indicator of the rate at which oxidation or reduction processes occur at the electrode. The value of R_{ct} can be extracted from the complex plane plot and corresponds to the diameter of the semicircle that appears. When the measurement is not performed under strict kinetic control, mass transport is also involved and then numerical fitting is needed in order to estimate this diameter.

Electrochemical processes are commonly assumed to follow Butler-Volmer kinetics (Christensen and Hamnet 1994; Bard and Faulkner 2001):

$$i = i_0 \left[e^{\left(\frac{\alpha_A nF}{RT} \eta \right)} - e^{-\left(\frac{\alpha_C nF}{RT} \eta \right)} \right], \tag{15.11}$$

where i_0 is the exchanged current density, η is the overpotential ($\eta = E - E_0$), n is the number of electrons exchanged in the reaction, R is the constant of the ideal gases, F is Faraday's constant, T is the temperature, and α_A and α_C are the coefficient of anodic and cathodic transfer, respectively. This expression can be further simplified by considering a reversible system ($\alpha_A = \alpha_C = 0.5$) in equilibrium under a very small overpotential. Consider a first order Taylor's expansion, $e^x \approx 1 + x$.

The expression previously presented can be simplified to

$$i = i_0 \left[1 + \frac{nF}{2RT} \eta - 1 + \frac{nF}{2RT} \eta \right], \tag{15.12}$$

$$i = i_0 \frac{nF}{RT} \eta, \tag{15.13}$$

The resulting expression describes the charge-transfer resistance, R_{CT}:

$$\frac{\eta}{i} = R_{CT} = \frac{RT}{nFi_0}, \tag{15.14}$$

R_{CT} is a function of the electrode material properties and the structure of the interface (Bard and Faulkner 2001; Scholz 2002).

2.2.2.2. Warburg Impedance

When electrode kinetics are fast, electrochemical processes are usually limited by mass transport (Franceschetti and Macdonald 1979; Gabrielli 1995). Therefore, if the potential is such that electron transfer is sustained, eventually the supply of material to and from the solution bulk towards the electrode surface is limited by the diffusion of the electroactive species. Mass transport can take several forms, namely convection (mass transport is generated by the stirring of the solution), migration (mass transport is produced by a charge gradient), and diffusion (mass transport occurs via a concentration gradient). The experimental conditions are normally chosen so that the current is diffusion-limited. The thickness of the diffusion layer in the interface depends on the diffusion coefficient of the electroactive species, the stirring rate, and the temperature. This mass transport limitation brings about an additional resistive term, named the Warburg impedance, Z_W, that appears in series with the previously observed charge transfer resistance.

The derivation of the Warburg impedance is out of the scope of the present chapter. However, it is important to note that depending on electrode geometry and experimental conditions, the Warburg impedance component may show different behaviours. The first of them, presented in Fig. 15.5, can be observed at large electrodes, where the effect of radial

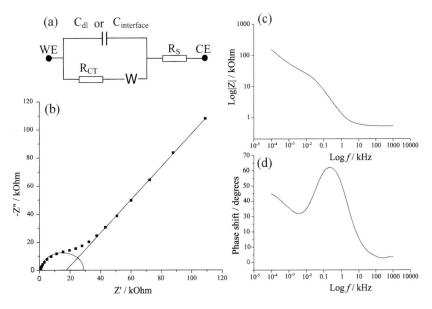

Figure 15.5. (a) Equivalent circuit for a system where electron transfer occurs between the electrode and species in solution. This is typical of mediated biosensors. (b) Complex plane plot obtained at a 0.8 mm diameter Pt above the open circuit potential disc electrode in 1mM ferrocyanide and 0.1 M KCl. The diameter of the semicircle corresponds to the charge transfer resistance, R_{CT}, and the low frequency tail presents the 45° angle typical of a diffusion-controlled process. This is also visible from the low frequency phase shift in Bode plot (d). Bode plots (c) and (d) enable the observation of the frequencies at which the various phenomena involved control the impedance.

diffusion around the edges is negligible, diffusion is planar, and the diffusion layer is assumed to be semi-infinite (Sluyters-Rehbach 1994; Gabrielli 1995, 1998). In these cases:

$$Z_W = R_{CT}\left(1 + \frac{\lambda}{\sqrt{j\omega}}\right), \tag{15.15}$$

where ω is the radial frequency and λ is the Warburg coefficient which can be defined as:

$$\lambda = \frac{k_f}{\sqrt{D_O}} + \frac{k_b}{\sqrt{D_R}}, \tag{15.16}$$

D_O and D_R are the diffusion coefficients of the oxidised and reduced species, as per equation 9 above, respectively. k_f and k_b are the forward and backward heterogeneous rate constants for electron transfer in equation 15.9. This Warburg impedance is observed on the complex plane plot at low frequencies as a straight line of slope 1 arising from the semicircle formed by the parallel R_{CT} and $C_{interface}$.

   At any given frequency, the relative weight of R_{CT} and Z_W in the measured impedance is a measure of the balance between kinetic and diffusion control.

   A special case of the Warburg impedance appears when microelectrodes are employed or when the diffusion layer is controlled by forced convection (Franceschetti and Macdonald 1979). In these cases the expression of the Warburg impedance is:

$$Z_W = R_{CT}\left(1 + \frac{k_f \tanh\left(\delta\sqrt{j\omega/D_O}\right)}{\sqrt{j\omega D_O}} + \frac{k_b \tanh\left(\delta\sqrt{j\omega/D_R}\right)}{\sqrt{j\omega D_R}}\right), \tag{15.17}$$

where δ is the length of the diffusion layer. The complex plane plot of this Warburg type no longer shows a straight line, but it has the appearance of a distorted semicircle whose intercept on the Z' axis corresponds to:

$$Z' = R_s + R_{CT}\left(1 + \frac{k_f \delta}{D_O}\right). \tag{15.18}$$

2.3. Measuring at Impedimetric Biosensors

   So far we have described the most common processes and their place in a possible equivalent circuit. It has already been mentioned that, in the case of EIS-based biosensors, the modification of the electrode brings about important interfacial changes. For a start, it may no longer make sense to talk about double-layer capacitance, but to leave it all in terms of an overall interfacial capacitance. The presence of a coating (be it a self-assembled monolayer or the biorecognition elements directly adsorbed) on the electrode surface changes both the interfacial capacitance and the electron transfer rate constant. However, depending on the detection strategy chosen for the biosensor, one of these parameters may either be easier to measure or show bigger changes upon detection. The following sections describe the most common experimental arrangements, along with a few considerations in terms of electrode design, materials, and cell configuration.

2.3.1. Measurement Modes

2.3.1.1. Two- and Three-Electrode Configurations

   As with any other electrochemical technique, EIS requires the use of at least two electrodes to operate. In such a two-electrode configuration the bias potential is commonly fixed with

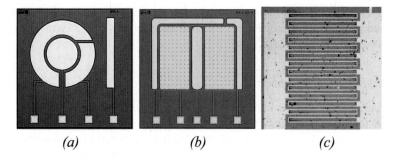

(a)   (b)   (c)

Figure 15.6. Three different electrode designs. (a) Disc macroelectrode 0.8 mm in diameter. (b) Chip containing two microelectrode arrays and (d) typical interdigitated structure.

respect to the auxiliary open circuit potential. This implies that the potential of the working electrode is almost certainly unknown. Two-electrode measurements are thus discouraged, except perhaps when interdigitated electrodes (Mamishev et al. 2004) (Fig. 15.6c) or other methods of achieving a reproducible electrode setup are used. Two-electrode electrochemical cells can be used to monitor solution conductivity changes—particularly for measurements in low conductivity media (Olthuis et al. 1995) or interfacial capacitance (Berggren et al. 2001)— brought about by the presence of bacteria in the sample.

The main drawback of this technique is its inability to control the electrode potential accurately, which can be critical in some applications: for instance, when redox mediators are used to facilitate the detection. This can be overcome by the introduction into the system of a reference electrode.

This three-electrode configuration requires the use of a potentiostat coupled to the frequency response analyser, as shown in Fig. 15.7. In this case, the potential of the working electrode is controlled with respect to the reference electrode, which needs to be placed as close as possible to it and along the current path between the working and auxiliary electrodes. In some cases, the use of a reference electrode may introduce artefacts (phase angles beyond $-\pi/2$), especially at high frequencies, due to the relative placement of the reference and the current path (Vanýsek 1997). One instance where this situation may arise is when using interdigitated electrodes in combination with an external reference electrode.

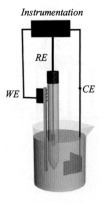

Figure 15.7. Diagrammatic representation of a three-electrode electrochemical cell. Ideally, the reference electrode (RE) should be placed in the current path and as close as possible to the working electrode and the counter electrode (CE) should have a much larger area than the working electrode (WE) so that its contribution to the overall impedance may be neglected.

On the other hand, by providing better control over electrode potential, the use of a reference electrode is recommended when charge transfer processes are involved in the system, e.g., biosensors using redox mediators.

2.3.1.2. Tetrapolar Configuration

In certain cases it may be interesting to add a second reference electrode to the system. This is called tetrapolar configuration, and its main merit is the elimination of an interfacial contribution to the measurement. In a four-electrode setup, such as the one presented in Fig. 15.8, the current flows between the working and auxiliary electrodes, but the impedance measurement is performed between the two reference electrodes, which are located in the current path between the working and the auxiliary electrodes (Ferris 1975; Gabrielli 1998). Since no current flows through these reference electrodes, the potential drop across them is measured without polarisation. The total impedance can thus be assumed to come from the solution because of the small influence of the interfacial impedance, which is critically linked with the electrode polarisation. Therefore, the use of a four-electrode configuration is recommended to monitor solution events, or transport across membranes. This kind of configuration is seldom used in biosensor applications.

2.3.1.3. Electrodes

Regardless of whether two, three, or four electrodes are used in the impedance measurements, it is important to choose an adequate and consistent electrode arrangement (Vanýsek 1997; Gibson 2001). Thus, and depending on the measurement of interest, the experimentalist needs to decide on geometrical factors such as electrode separation, size, and distribution. Fig. 15.6 shows three different types of electrodes that may be used in biosensor development. Because the geometry of the system has a very strong impact on biosensor performance, this section aims to provide some guidelines as to which geometrical parameters should be considered, depending on the type of measurement.

It has already been mentioned that the impedance measured is the total impedance of the cell and not just that of the working electrode. In order to simplify the analysis and be able to assume that the impedance corresponds to the working electrode alone, the auxiliary electrode needs to present much less impedance. This can be achieved by using an inert auxiliary electrode, the area of which is many times larger than that of the working electrode. In addition to this, the working electrode can be placed sufficiently far from the auxiliary electrode

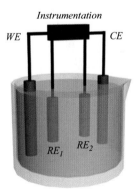

Figure 15.8. Diagram of a tetrapolar configuration. A current is passed between a working and a counter electrode, and the resulting potential drop is measured between two reference electrodes placed in the current path.

so that its surface can safely be assumed to be equipotential. Distance between electrodes is an important parameter that can influence the sensitivity of the impedimetric measurement, particularly in the case of conductivity measurements.

Because the field of pathogen detection is commonly concerned with the detection of very low levels of bacteria, then a conductivity biosensor may benefit from smaller interelectrode distances. This is particularly true in the case of small size interdigitated electrodes, IDE, where the electrode width and gap, typically of comparable sizes, are also of a similar order of magnitude to the target pathogen. Although IDEs are very well suited for conductivity measurements because of their well-defined geometry and the reproducibility of their fabrication techniques, it is best to use a tetrapolar configuration for conductivity measurements. On the other hand, the use of such devices presents a series of problems; for example, as the distance between electrodes becomes smaller, the chances of suffering short circuiting of the electrode set caused by the presence of debris or the bacteria itself in the sample also increase, which may result in a ruined measurement.

Another application of IDE is the capacitive biosensor (Berggren et al. 2001). In fact, the most commonly reported electrode design for EIS immunosensors is based on IDE structures. Although IDE may present very important advantages, there are a few further considerations to make when using them. First, IDE are usually produced by photolithographic techniques over silicon or glass wafers. Because silicon is a semiconductor, an oxide layer needs to be grown in order to avoid parasitic currents. Then a thin film of the electrode material is deposited and patterned. Therefore, in addition to the metal-solution interface, there is also a metal-oxide interface which also contributes, with an additional resistance and capacitance in parallel to the solution. The resistance is usually so large that it can be neglected, but the capacitance may need to be taken into account. It can easily be determined by performing a two-electrode measurement in the absence of solution (*e.g.*, in air). The capacitance measured may be in the order of the picofarads and can be subtracted from later measurements. Another disadvantage of interdigitated electrodes lies in the difficulty of using them in a three-electrode configuration, which can hardly be overcome by using an external reference electrode. In the latter case, artefacts appear due to the remoteness of the reference electrode from the current path between the working and auxiliary electrode structures.

2.3.1.4. Materials

The right choice of electrode materials is key to the successful performance of any biosensor. It is important to choose the right material, so that its functionalisation can be achieved with relative simplicity; but other criteria also, such as its durability and stability under the measurement conditions, the possibility of sterilising it, and its toxicity should be evaluated before working with bacteria (Gibson 2001). The most common material is gold, but biosensor applications using platinum or carbon electrodes have also been reported. As will be shown in later sections of this chapter, gold is a very good electrode material for biosensor applications because of its relatively fast electrode kinetics and the possibility of forming self-assembled monolayers where biorecognition elements can easily be immobilised.

2.4. Bacterial Parasitizing Effect on Electrode Surface

Bacteria in suspension tend to move and attach onto solid surfaces (Characklis 1990; Lappin-Scott and Costerton 1995). Thus, when an electrode is immersed and polarised in a bacterial sample, it is eventually covered by nonspecifically-bound bacteria (Fig. 15.9). This binding is initially reversible (Costerton et al. 1999), but microoganisms eventually start secreting certain exopolysaccharides which allow them to irreversibly adhere to the surface. If

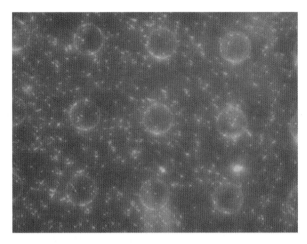

Figure 15.9. Photograph showing bacteria nonspecifically adsorbed over the surface of a microelectrode array composed of 20 micron diameter discs separated 50 microns between centers.

the exposure to this kind of bacterial environment is long enough, these spontaneous colonies develop into biofilms (Characklis 1990; Melo et al. 1992; Lappin-Scott and Costerton 1995; Costerton et al. 1999).

In the case of biosensor applications, there is usually not enough time for biofilms to grow. However, even after only a few minutes, bacteria reversibly attach to the electrode surface and modify the electrode impedance by affecting interfacial capacitance and electron transfer rate constants. These variations are almost impossible to control and filter out because they are not only time-dependent, they are also influenced by applied potential and bacterial concentration. This is the reason why it is important to protect the biosensor surface against nonspecific adsorption of material and to perform negative controls when developing any sort of biosensors. Suitable blocking of the electrode is described in a later section.

3. Development of an Immunosensor

3.1. Biological Recognition Elements in Biosensors for Pathogen Detection

The simplest impedimetric detection approach consists of the transient monitoring of the response at a bare electrode immersed in a sample containing bacteria. In "impedance microbiology," changes in impedance are correlated with the metabolic activity of live microorganisms which, for instance, transform complex molecules present in the sample into ionic metabolites (i.e., amino acids and organic acids from proteins and polysaccharides) (Cady 1975; Ur and Brown 1975). As in traditional methods, performing a 24 to 48-hour sample pre-enrichment prior to impedance measurements may enable the identification of different microorganism groups (ranging from total aerobic plate count to coliforms, yeast, mold, or lactic acid bacteria) and even specific pathogens (such as *E. coli, Salmonella*, or *pseudomonas*). This is the basis of existing commercial impedimetric systems (Bactometer® from Biomerieux; BacTrac® from Sy-Lab, Malthus® System from Malthus Instruments, RABIT® from Don Whitley Scientific). Nevertheless, the demand for faster, more sensitive, and more specific assays favours the development of alternative detection strategies, such as biosensors, based on the exploitation of recognition molecules specifically binding certain target-pathogen components. In this context, as antibodies (Ab) and nucleic acid probes continue to consolidate their role as biorecognition elements, alternative new molecules such as aptamers, phages, or peptides are starting to be used.

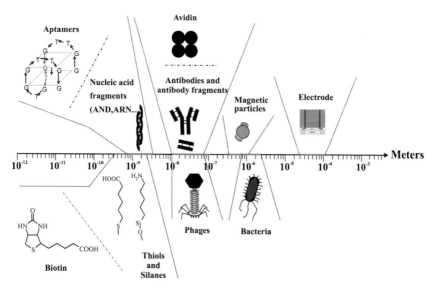

Figure 15.10. Relative sizes of various components involved in biosensors.

The following sections describe the use of the most common biorecognition elements, such as antibodies and DNA probes, and also of other new systems such as aptamers, peptides, and phages. Fig. 15.10 aims to give an idea of the relative sizes of the various elements involved in the construction of biosensors, ranging from the nanometric size of aptamers and DNA probes up to the micrometric size of the pathogens themselves and the transducers employed. See Part III of this book for more informations about the different recognition elements used in biosensors.

3.1.1. Antibodies

Antibodies (Ab), applied to detection for the first time in the 1950s (Yallow and Berson 1959), are the most widely used biorecognition elements, thanks to their proven sensitivity and specificity. They can be raised against the whole pathogen (their surface proteins) or against some of the pathogen components (lysate, enzymes, toxin, spore, pili). Depending on their production methodology, they can be polyclonal (PAb) or monoclonal (MAb). While monoclonal antibodies provide the highest specificity, they are also much more expensive. Thus, PAb are commonly used in approaches not demanding "excessively high" specificity, such as the simultaneous detection of different bacterial serotypes.

The easiest way to immobilise Abs is to unspecifically deposit or adsorb them on the sensing surface. Nonetheless, most authors maintain that directed immobilisation, by chemical conjugation or crosslinking of precise functional groups in the Ab (i.e., amines, carboxylates, carbohydrates), better preserves their integrity and functionality, while promoting more organised structures. An important number of Ab-based biosensors have been developed for pathogen detection, including a few impedimetric devices that will be further discussed later in the text. See chapter 21 for more information about antibodies and immunoassays.

3.1.2. Nucleic Acids

Single-chain nucleic acids are used as molecular probes to bind, by base pairing, complementary sequences present in the DNA/RNA of a target microorganism. In order to attain detection specificity, the target sequences (a) must be known, allowing design and synthesis of the complementary probe, and (b) must be highly specific for the pathogen under study. As the probe-target hybridisation event increases the interface negative charge and slows electron transfer rates, it can be directly monitored by impedance spectroscopy (Katz and Willner 2003). Additionally, oligonucleotides modified with SH_2 terminal groups can be directly self-assembled on gold surfaces. Otherwise, the incorporation of SH- or NH-terminal groups allows their chemical conjugation to other molecules preassembled or deposited on the electrode surface (SAMs, streptavidin-biotin, etc). Only few authors have described the successful unspecific adsorption of DNA/RNA on electrodes (Zhao et al. 1999) and the entrapment into membrane or polymer matrices (Sastry 2002). See chapter 22 for more information about Nucleic acid recognition elements.

3.1.3. Aptamers

Aptamers, described for the first time in 1990, are artificial nucleic acid ligands, specifically generated against the target under study. To date, aptamers towards targets as varied as small molecules, peptides, proteins, and even whole cells, viruses, and spores, have been reported. Their advantages over alternative approaches include a relatively simple isolation and production procedure (completely in vitro and thus possible against small, non-immunogenic or toxic targets), their small size, and their chemical simplicity (allowing regeneration as well as easy modification, immobilisation, and labelling). The main drawback, on the other hand, has always been the high sensitivity of nucleic acids to nuclease attack. The solution is being seen as spiegelmers (aptamers produced by a complex "mirror-image" SELEX procedure; Klussmann et al. 1996; Nolte et al. 1996) and chemically modified aptamers (containing modified nucleotides in the SELEX library; Kusser 2000), both species having demonstrated their survival for a few days in *in vivo* studies.

Aptamers can in principle be manipulated and immobilised as DNA probes, being more resistant to repeated thaw-freezing and drying-reconstituting cycles than Ab, but more sensitive to enzymatic degradation. To date, work on pathogen-binding aptamers has been geared mainly towards therapeutics as a less aggressive solution compared to traditional drugs, inducing minimal immune response, toxicity, and known secondary effects in treated animals. However, the increasing number of publications describing their integration in sensing devices and their recent production against structural proteins of various pathogens anticipates their potential role in diagnostics, and will be extensively discussed in Chapter 25.

3.1.4. Other Recognition Strategies

Some authors have described the existence of peptides that are much smaller in size than Ab but which specifically bind to certain pathogens (Barak et al. 2005; Rosenfeld and Shai 2006). Such peptides were initially derived from pathogen-binding proteins, but are now produced in vitro, exploiting phage displayed peptide libraries and a selection procedure similar to that used for aptamers. The selected sequences, however, can be not only chemically synthesised, but also amplified, culturing the displaying phage into bacterial broths. As an example, peptides binding to some glycoproteins partly responsible for the Newcastle disease virus infectivity and virulence have been reported; and, in fact, they seem to allow detection and discrimination between high, medium, and low virulence strains by ELISA and dot-blot assays (Lee et al. 2006). Also, fluorescently labelled peptides or phage-displayed peptides have been applied to fluorescence microscopy detection of both bacterial toxins and spores (Goldman et al. 2000; Williams et al. 2003).

Bacteriophages are bacterial viruses, whose life cycle includes binding to specific receptors present on the host cell membrane, injecting their nucleic acid, and parasitizing part of the cell metabolic and/or replicative machinery in order to replicate. The facts that coevolution of the two species has led towards an important level of phage-target specificity, and that a great variety of target-specific phages have been described, suggest that a number of phages may potentially be used as specific recognition elements in sensor development. In this respect, phages, adsorbed on chips, have been employed as capture agents to detect bacteria using an acoustic wave biosensor (Olsen et al. 2006). Alternatively, a few reports describe the use of a phage, labelled with a fluorophore or a quantum dot, as a reporter element to detect bacteria by fluorescence microscopy (Awais et al. 2006; Edgar et al. 2006). These results suggest that phages could be used as mass amplifiers for impedance detection if labelled with the appropriate structures. More information on phages can be found in Chapter 27 and 28.

At least one work reports on the impedimetric detection of a viral infection on a cell monolayer formed on gold electrodes (McCoy and Wang 2005). Electric cell-substrate impedance sensing (ECIS) takes advantage of the fact that most cells can successfully attach and grow on a gold surface, developing a biofilm that behaves as an insulating coating. Viral infection induces, among other things, changes in the cells' shape and ultimately desorption and/or cell disruption, and thus changes in impedance. Though this methodology was initially developed to study cell morphology/motility and cell-substrate interaction, it could be applied to detect viruses present in the samples under study by monitoring impedance over time, provided that the appropriate host cell line is available and used. The main limitation will probably be the time required for viruses to induce detectable changes on the biofilm.

A completely different approach consists of pathogen (or its antigenic components) immobilisation on the sensing surface, with the aim to detect anti-pathogen Abs present in potentially infected patients. Although this strategy has been extensively exploited in sandwich ELISA formats, a series of problems prevent its transfer to reagentless formats such as impedance biosensors. Among others, the small size of the detected Ab may not generate significant signal changes when present at low concentrations. Also, as with DNA detection, Ab may reflect past exposure to the pathogen and not necessarily its current presence. On the other hand, pathogen infection correlates with a "window period" in which the microorganism replicates, but Ab have not yet been produced in measurable amounts.

3.2. Surface Modification Methods

The development of an impedimetric biosensor requires the functionalisation of the electrode surface in order to integrate the selected biorecognition elements. This is one of the most critical steps in biosensor development because biosensor performance (sensitivity, dynamic range, reproducibility, and response time) hinges on how far the original properties of the bioreceptor are kept after its immobilisation. The existing immobilisation strategies include biomolecule physisorbtion, entrapment and encapsulation into polymers or membranes, silanisation, and SAM formation coupled to biomolecule cross-binding or covalent bonding. Whatever the chosen strategy, it has to ensure that the biorecognition element will be immobilised in a stable way, and that it will retain both its accessibility for the target molecule and its recognition ability. In addition, the modified surface has to be inert and biocompatible, so that it does not affect the sample composition or integrity in any way; and it should also guarantee a constant signal baseline. The following sections address the most common immobilisation techniques, as depicted in Fig. 15.11.

3.2.1. Adsorption

Nonspecific adsorption is the easiest way to immobilise molecules on a physical substrate, and consists of just depositing the molecules (Ab, antigen, etc.) on the surface and leaving them to interact in a completely random way. The driving forces may be initially hydrophobic (for hydrophobic surfaces such as polystyrene) or electrostatic (for more polar substrates such as silica), but protein adsorption will be further stabilised by a combination of hydrophobic

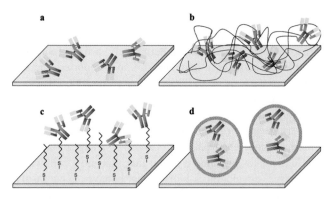

Figure 15.11. Different examples of biomolecule immobilisation strategies. (a) Unspecific adsorption, (b) entrapment, (c) cross-linking to a pre-assembled SAM, and (d) encapsulation.

interactions, hydrogen bonding, and/or Van der Waals forces, resulting in a behaviour highly dependent on each individual protein-surface involved, and a highly stable product (Parida et al. 2006). Adsorption on gold surfaces of proteins containing free SH_2 groups and/or S-S bonds, on the other hand, profits from the strong affinity of these functional groups for gold. Nevertheless, several reports demonstrate that Abs perform better if immobilisation is partly directed by reducing the interchain S-S bonds, generating free SH_2 groups, or using long-chain spacers prior to adsorption on gold (Park and Kim 1998; Karyakin et al. 2000).

As a consequence of adsorption, proteins may partly denature and thus lose structure and/or function (Heitz and Van Mau 2002). For example, activity losses up to 90% have been described for some Abs, presumably due to a combination of (a) loss of binding sites caused by denaturation, (b) binding site hiding in poorly oriented molecules, and (c) steric hindrance as a result of molecule crowding and aggregate formation (Butler 2000). However, thanks to the high number of molecules immobilised, the success rate is usually high enough to ensure a reasonable assay performance, with the exceptions of small molecules (such as peptides) that may require previous conjugation to a bigger carrier molecule, and MAbs which for unknown reasons show worse performance following unspecific adsorption than PAbs (Butler 2000). Although optimal conditions should be experimentally determined for each surface-molecule couple, the maximal coverage is often attained for most soluble proteins on noncharged surfaces at neutral pH and physiological ionic strength, regardless of the protein isoelectric point (pI). Protein concentrations between 5 and $20\mu g/ml$ are normally sufficient if incubations of 2–4 hours at 37°C or 12–17 hours at 4°C are to be carried out.

3.2.2. Self-assembled Monolayers

Self-assembled monolayers (SAM) are crystalline chemisorbed organic single layers formed by spontaneous organization of thiolated molecules, generally on metal surfaces (Love et al. 2005). They enable the formation of organic surfaces whose composition, structure, and properties can be varied rationally. Research in this area started in the 1980s (Nuzzo and Allara 1983; Porter et al. 1987; Bain and Whitesides 1988; Rubinstein et al. 1988; Troughton et al. 1988; Bain and Whitesides 1989; Hautman and Klein 1990), focusing on assemblies formed by the adsorption of organosulfur compounds from solution or the vapour phase onto metal substrates of gold and silver. Later, in the 1990s, the study of SAMs evolved into efforts to broaden the types of substrates and molecules used to form them. The most extensively studied class of SAMs is derived from the adsorption of n-alkanethiols on gold, silver, copper, palladium, platinum, and mercury. The formation of these well-defined organic surfaces with useful and highly alterable chemical functionalities is made possible by the high affinity of sulphur for noble and coinage metals. SAMs, the thickness of which is typically a few nanometers (Chaki and Vijayamohanan 2002; Ko et al. 2004), provide huge advantages for biosensor development as they (a) are easy to prepare and functionalize in an ordinary chemistry laboratory, (b) can form on surfaces of any size, and (c) allow linking molecular-level structures to macroscopic interfacial phenomena.

Gold became the standard substrate for SAM formation for various reasons: (a) it is easy to obtain, (b) it is easy to pattern using photolithography, and (c) it is bio-compatible.

For SAM generation, a thiolated molecule (or molecule mixture) diluted in a suitably polar solvent, depending on the thiolated compound (usually an ethanolic solution of thiol concentration in the high micromolar-low millimolar range), is left in contact with the substrate for a few minutes to up to several hours. In fact, a few minutes are generally sufficient for SAM formation, but in order to maximize the molecular density and enable a well ordered structure, particularly for long-chain alkanethiols, a time in the order of hours is required. The main factors that affect SAM formation are solvent, temperature (usually room temperature is sufficient; Gyepi-Garbrah and Silerova 2001), thiol concentration, purity and structure, substrate immersion time, concentration of oxygen in solution (less relevant for gold substrates), and cleanliness of the substrate.

SAM deposition is not always an irreversible reaction, and different removal techniques allow reutilization of the substrate. SAM desorption can be attained by chemical or physical methods. Amongst the chemical methods are the piranha treatment (Calvo et al. 2004), and an "exchange" method consisting of introducing the recovered substrate into another thiol solution. In this way the SAM can gradually (in minutes to hours) replace the thiols adsorbed onto the surface with the ones in solution (Schlenoff et al. 1995; Liu et al. 2004). Photooxidation (UV irradiation) is an efficient physical method for SAM removal (Huang and Hemminger 1993; Brewer et al. 2005). Electrochemical methods can also be used, since the application of a negative potential causes the reductive desorption of the thiols (Unwin and Bard 1992; Schneider and Buttry 1993; Everett and Fritschfaules 1995).

SAMs are often the basis for the subsequent immobilisation of the recognition elements described previously (antibodies, nucleic acids, aptamers, peptides, etc.). The possibilities are endless, since the functional groups provided by the SAM layer termination can be tailored to suit any particular requirements. These incorporation techniques will be described later in the text.

3.2.3. Silanisation

Silanes have the general formula $RSiX_3$, where R is an organofunctional group selected according to the desired surface properties; and X is a hydrolysable group, typically an alkoxy group (alkyl group linked to oxygen), which is capable of reacting with the substrate. Silanization of hydroxyl-terminated substrates is an effective and frequently used procedure for modification of chemical and physical properties of the substrate. Silanisation is a SAM substitute for substrates such as silica (Bhatia et al. 1989; Billard et al. 1991), silicon (Tlili et al. 2005), glass (Ruckenstein and Li 2005; Saal et al. 2006), cellulose (Pope et al. 1993), and metal oxide surfaces (Laureyn et al. 2000; Ruan C 2002). Silanes are normally hydrolysed at some stage in the coating process, allowing interaction with the substrate either via hydrogen or covalent bonds. They are used to immobilise proteins, peptides, antibodies (Piehler et al. 1996), or nucleic acids (Kumar A 2000) on micropatterned surfaces, because of their ability to improve the adhesion capacity of the surface. Two main deposition techniques are used for silanization: either by immersing the substrate in a solvent containing the silanes (Piehler et al. 1996), or by vapour deposition (de la Rica et al. 2006), which consists in keeping the sensor close to the silane in an oxygen-free sealed chamber. This technique prevents the formation of unnecessarily thick layers of silanes on the substrate surface.

3.2.4. Protein A and Protein G

Protein A and protein G are cell surface receptors produced by several strains of *Staphylococcus aureus* and *Streptococcus sp.*, respectively (Goding 1978). They both bind to Abs from several species with high efficiency (especially IgG), but in a reversible way, which has made them extremely useful in Ab purification (most of the products advertised by the providers as "affinity purified" have been recovered through protein A/G columns). Additionally, they have been used as capture elements in several immunochemical assay formats, including western blot, immunohistochemistry, and ELISA, as well as in immunoprecipitation assays (Goding 1978).

Among their characteristics, both proteins A and G interact towards the Ab COOH-extreme, not affecting the target binding site and promoting a highly ordered and directed binding. In this respect, each molecule of protein A (approximately 40KDa molecular weight) and protein G (around 20KDa molecular weight) can simultaneously bind 4 and 2 molecules of IgG. As protein G can also bind serum albumin, the commercial protein is usually a recombinant product from which this binding domain has been eliminated. In addition, while protein A is very resistant to heat and also to the presence of denaturing agents and nonionic detergents, the interaction with IgG is sensitive to extreme pH and salinity. This implies that the captured Ab can be efficiently eluted and suggests that these reagents could be used in the production of reusable sensing surfaces. Consequently, several authors have incorporated protein A/G into their reported biosensors (Anderson et al. 1997; Babacan et al. 2000; Lu et al. 2000). The composition of such surfaces, however, should be carefully optimised, as any protein A/G molecule not having captured an Ab will potentially bind any Ab present in the sample under study, or any labelled Ab subsequently used for signal amplification.

3.2.5. The Biotin-(Strept)Avidin System

This technique is based on the natural strong binding of avidin for the small molecule biotin (also called vitamin H). It is one of the most popular noncovalent conjugation methods and a very useful tool for targeting applications (Hermanson 1996). The dissociation constant of this complex is 1.3×10^{-15}M, which makes it one of the strongest noncovalent affinities known. The biospecificity of the interaction is similar to that of Ab-antigen or receptor-ligand recognition; and variation of buffer salts and pH, extremes of temperature, or denaturants and detergents do not prevent the interaction from occurring. A functional biotin group can be added to proteins, nucleic acids, and other molecules through well known reactions. Specific biotinylation agents exist for targeting amine, carboxylates, sulfhydryls, and carbohydrate groups. When none of these functional groups are present, the biotin can also be added using photoreactive biotinylation reagents (photobiotin; Pavlinkova et al. 2000; Pantano and Chin 2003; Chin and Pantano 2006).

The glycoprotein called avidin contains four identical subunits (tetramer) and has a total molecular weight of 66,000 Daltons. Each one of these subunits contains one binding site for biotin. Another very similar protein called streptavidin is often used, as it prevents nonspecific bindings that are quite elevated with avidin due to its high isoelectric point (10 for avidin versus 5–6 for streptavidin) and carbohydrate content (which is not so present in streptavidin that is not a glycoprotein; Bayer et al. 1990). Disappointingly, streptavidin exhibits a peptide that presents a high affinity for the integrins displayed on the cell surface, which are proteins partly responsible for cell adhesion (Alon et al. 1993). Consistently, Huang et al. reported high levels of bacteria-unspecific adsorption on biochips coated with streptavidin (Huang et al. 2003). Another factor to consider is that varieties of proteins naturally possess covalently bound biotin and, when present in the samples under study (e.g., some tissues), will contribute to increasing the background (Wood and Warnke 1981). Both streptavidin and avidin can be conjugated to other proteins or labelled with various detection reagents without loss of biotin binding activity.

NeutrAvidinTM and ExtrAvidin®, provided by Molecular Probes (Invitrogen) and Sigma, respectively, are modified forms of avidin that do not contain carbohydrates and thus retain the high affinity characteristic of avidin and the low level of background generated by streptavidin (Chung et al. 2006).

3.2.6. Chemical Conjugation

The most commonly used strategy consists in cross-binding between carboxylic and amine groups (-NH2) exploiting EDC/NHS chemistry (Fig. 15.12). The COOH group can be situated either on the termination of the SAM or on the (bio) component to be immobilised. The use of glutaraldehyde allows the reaction of two amino groups. Thiols can be coupled to amino groups using heterofunctional cross-linkers such as SMCC or SATA (Hermanson 1996; Nanda et al. 2002). The polysaccharides present in some proteins can be oxidised with sodium periodate (Hermanson 1996) and later conjugated to amine or hydrazyde groups in the SAM by reductive amination. Hydroxyl groups on a SAM can be treated the same way. In addition, the recognition element may also self-assemble; this is the case with nucleic acid immobilisation. A thiol-modified nucleic acid fragment (incorporating an –SH terminal group) will directly self-assemble on the metal surface. Another technique often used for biosensor fabrication is the modification of the SAM layer with polymers, which contributes to SAM durability and toughness. Polymers such as dextrans, polyethylene glycol, or polyacrylamide have been demonstrated to provide a good surface for the immobilisation of the recognition elements, while decreasing to a certain extent the level of unspecific binding of unrelated molecules.

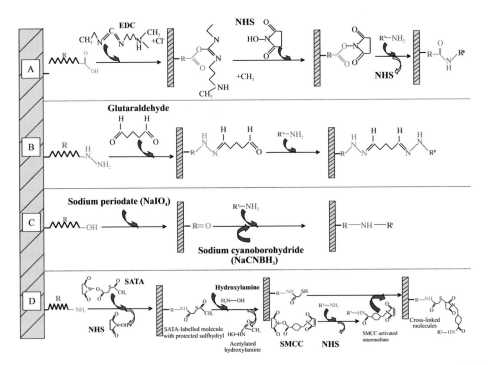

Figure 15.12. Description of chemical conjugations currently used to bind biomolecules to electrodes. (a) The EDC/NHS reaction, (b) homobifunctional cross linking with glutaraldehyde, (c) conjugation *via* aldehyde group obtained by sodium periodate oxidation, and (d) heterobifunctional cross linking with N-succinimidyl S-acetylthioacetate (SATA) and Succinimidyl 4-(N-maleimidomethyl)-cyclohexane-1-carboxylate (SMCC).

3.2.7. Entrapment

Biomolecules can be immobilised within organic or inorganic polymer matrices by entrapment during the matrix polymerisation, without suffering any chemical modification that could affect their integrity. Confinement of proteins into small inert spaces contributes to stabilising them by reducing unfolding and by shifting equilibrium between different configurations (Zhou and Dill 2001). For example, enzymes entrapped within the nanopores of silica beads profit from improved stability for longer times than enzymes free in the bulk solution, and provide a good matrix for biosensor development (Sotiropoulou et al. 2005). Polyacrylamide has been used for Ab entrapment and subsequent myoglobin detection (Hanbury et al. 1997). A more spread method consists in electrochemical entrapment. This technique uses a solution containing both the bioreceptor and an electroactive monomer. As a suitable potential is applied to the electrode and the electropolymer grows, bioreceptor molecules are trapped within its matrix (Cosnier 1999, 2003).

The main drawbacks associated with entrapment are the high concentrations of both monomer and biomolecule required, the biomolecule potential loss of structure and/or function following entrapment, the poor accessibility to certain target molecules, certain biomolecules' sensibility to the polymerisation conditions and/or polymer components, and the lack of reproducibility between sensing surfaces (Minett et al. 2002; Cosnier 2003). As an alternative, some authors describe the polymerisation of a mixture of monomers and monomer-derivatised biomolecules (i.e., Ab, biotin), or the chemical conjugation of biomolecules to the surface of a preformed polymer, in an attempt to reduce the amount of biomolecule required while preserving its integrity and access to the target.

Entrapment has been applied mainly to enzyme immobilisation followed by voltammetric or amperometric detection of its substrate in solution (Bartlett et al. 1998; Shan et al. 2006; Yidiz and Toppare 2006)). Other sensing molecules successfully entrapped include Abs, coenzymes, and cells. Only a few reports exist related to impedimetric detection, exploiting entrapped Ab for the detection of human serum albumin, luteinising hormone, and human IgG in solution (Sargent et al. 1999; Farace et al. 2002; Ouerghi et al. 2002).

3.2.8. Microencapsulation

Microencapsulation involves the entrapment of molecules within micro/nano-capsules of different composition (particles, spheres, tubes, fibers, vesicles; made of hydrogel, polymer, carbon, silica, lipids, etc.) and formed by different strategies (via template moulding, polymerisation, self-assembly, emulsification, etc.). Encapsulation, as entrapment, has been reported to protect proteins from unfolding and degradation, ensuring longer activity times. On the other hand, microencapsulation requires relatively high biocomponent concentrations and generates longer response times compared to the case when the biocomponent is free in solution.

Although microencapsulation has been exploited mainly for detection and signal amplification purposes, a few publications describe its utilisation when deposited or arrayed on sensing surfaces by electrospinning, photopatterning, and chemical or electrochemical procedures (Martin and Parthasarathy 1995; Koh et al. 2002; Patel et al. 2006). At least one example exists for glucose amperometric detection using liposome encapsulated enzyme deposited onto electrodes (Memoli et al. 2002), and biotin exposing liposomes have been immobilised on electrodes for target capture and second label potentiometric detection (Lee et al. 2005).

3.3. Blocking

The goal of blocking is to ensure that every ligand detected is being specifically bound to the immobilised receptor, and not unspecifically adsorbed to the surface. Blocking becomes especially important when reagentless strategies, such as impedimetric detection, are attempted. Even though a wide variety of reagents have been described to significantly reduce unspecific binding, including long-chain hydrophilic diamines, dextrans, and polyethylene glycol, amongst others, few of them have been extensively applied to pathogen detection. Furthermore, according to our own experience, they show limited performance when bacteria have to be sensed. In these cases, a blocking treatment, following immobilisation and prior to sample capture, may be required. Bovine serum albumin, casein, and skimmed milk are the most widely used blocking agents. They are able to fill any unoccupied gaps left on the surface. On the other hand, nonionic detergents such as Tween, Triton, and Nonidet P-40 have also been used, because of their transient effect in reducing and/or disrupting protein-protein as well as protein-surface hydrophobic interactions (Kenny and Dunsmoor 1987; Steinitz 2000). In any case, there are no silver bullets, and each case calls for its own blocking strategy which needs to be experimentally determined to avoid interferences and other biosensor performance-related problems.

3.4. Signal Amplification

One of the main virtues attributed to IES is that reagentless detection strategies are possible; that is, detection in the absence of labels or additional species. Unfortunately, this also poses its own problems and sometimes makes the technique less sensitive and specific than other approaches derived from the sandwich ELISA. Therefore, it demands better optimised detector surfaces that can ensure specific binding. However, sandwich-like strategies have been described in order to enhance sensitivity and selectivity and improve detection by EIS. This is achieved by incorporating, following the target capture, a second Ab or nucleic acid probe labelled with either an enzyme or a "mass/electrical enhancer," as shown in Fig. 15.13.

One interesting example is the use of an enzyme-labelled detector Ab or oligonucleotide, coupled with the addition of a substrate (i.e., 5-bromo-4-chloro-3-indolyl phosphate for alkaline phosphatase and 3-amino-9-ethyl-carbazole for horseradish peroxidase) that generates an insoluble product over any target molecules captured. Depending on the target concentration, such a product accumulates to a different extent, forming an insulating cover over the electrode and further amplifying the impedance change detected (Ruan et al. 2002; Lucarelli et al. 2005; Yu et al. 2006). A related strategy was applied to viral genome detection, coupled to in situ enzymatic target replication (Patolsky et al. 2001). Labelling the detector molecule with biotin instead permits signal amplification by sequentially incubating avidin/streptavidin and a biotinylated molecule, generating a bulky net over the electrode (Pei et al. 2001). The use of Ab or DNA probes labelled with gold nanoparticles has also been reported to amplify

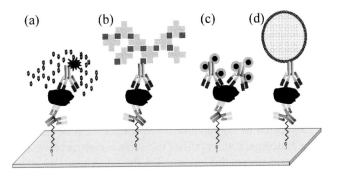

Figure 15.13. EIS amplification strategies exploiting the use of a second detector biomolecule labelled with: (a) an enzyme, coupled to the use of a precipitable substrate; (b) biotin, followed by incubation with streptavidin and biotin in solution; (c) gold nanoparticles submitted to silver reduction on their surface, and (d) negatively charged liposome encapsulating (or not) additional components. The different component dimensions do not correspond to the real ones.

impedance, capacitance, and conductivity changes induced by the target-probe binding event on either the electrode or the gap between microelectrodes (Park et al. 2002; Wang et al. 2006). The signal can be additionally amplified in this case by promoting Ag(I) reduction on the gold nanoparticles, that results in a decrease in the gap resistance (Park et al. 2002). Also for DNA detection, the use of Ab binding dsDNA has been applied for QCM signal amplification and could potentially be used in impedance (Bardea et al. 1999). The exploitation of Abs conjugated to magnetic beads, on the other hand, can play two independent roles. On the one hand, magnetic recovery of bound bacteria contributes to their physical concentration and/or separation from the sample matrix. On the other hand, the beads can behave as mass enhancers following capture of the complex on the electrode surface (Kim et al. 2003).

Liposomes appear as new promising tools for encapsulating strategies. They can be generated by phospholipid self-assembly in aqueous solvents via sonication, extrusion, or homogenisation, and often consist of spheres filled with solvent, which may be exploited for encapsulation purposes (Lasic 1998). Of special interest are temperature- and pH-dependent liposomes, which burst outside a specific condition range and liberate the encapsulated components. Resembling cell membranes in structure and composition, made from natural, biodegradable, nontoxic, and nonimmunogenic lipid molecules, liposomes have been extensively studied as drug- and gene-delivery vehicles. In the biosensor field, the conjugation or embedding of biorecognition elements has allowed the use of microencapsulated labels for fluorescent detection (Nichkova et al. 2005; Ho et al. 2006) and suggests that they could be used as signal enhancers in impedance sensing, a single microcapsule potentially containing multiple enzyme or substrate units. In this context, negatively charged liposomes have been successfully used as impedance mass enhancers for DNA detection, even leading to the identification of single base-pair mismatches (Patolsky et al. 2001). Liposomes will probably have a poor application to whole bacteria detection, as the membranes of both compartments are susceptible to fusion. Nevertheless, liposomes encapsulating a redox mediator have been successfully applied to the amperometric detection of haemolytic bacteria, including *Listeria* (Kim et al. 2006).

3.5. The Need for Negative Controls

As has been previously stated, impedance allows the direct detection of the target-probe interaction event, in the absence of any additional reactions or labels. Nonetheless, this methodology will detect any species binding onto the interface, without discriminating between specific and unspecific capture. Because of this, it is of extreme importance that the electrode surface is conveniently functionalised as to ensure that only the target is being bound and thus detected. The only way to guarantee this is to perform carefully chosen negative controls in parallel. Surprisingly, most publications in the field do not include such negative controls to confirm the specificity of the detected signals.

In this context, scientific papers reporting impedimetric sensors for pathogen detection should demonstrate at least that (a) the pathogen binds to the functionalised surface and not just the physical substrate and (b) the sensor specifically binds the target pathogen but not other related microorganisms. This implies that at least part of the experiments should be repeated, including as negative control a similar sensing surface, this time modified with a noncrossreacting binding molecule (i.e., an Ab or a DNA probe towards a completely different target). Even if the utilisation of nonfunctionalised surfaces can be useful in some cases, the fact is that biomolecule incorporation modifies the surface electrochemical properties and may change its behaviour versus unspecific binding. On the other hand, the experimentalist should demonstrate that the sensor is not binding other pathogens or components potentially present in the samples under study. The leap to real sample evaluation additionally requires that the sample matrix does not interfere with the target recognition or signal transduction either. These are all very difficult obstacles to overcome.

3.6. Development of Novel Strategies: Assessing Performance Using ELISA and Microscopy

The development of any novel impedance-based biosensor needs to be validated against long established techniques, and a series of performance-related parameters need to be assessed. For example, the structure of modified surfaces has been studied using techniques such as spectroscopy, diffraction, and microscopy. QCM, SPR, and impedance spectroscopy favour real-time surface coverage evaluation and determination of the saturation time and optimal reagent concentrations needed. Electrochemical study by sweep voltammetry, amperometry, or impedance may also help to determine the level of surface coverage by analysing electron transfer rates before, during, and after surface modification. Nucleic acid immobilisation can be traced additionally through studying the electrochemical reduction or oxidation of its nucleotide components by differential pulse, stripping, and cyclic voltammetry (Palecek et al. 1986; Palecek 1988; Palecek and Fojta 1994). More recently, the development of atomic force microscopy (AFM) and scanning tunnelling microscopy (STM) has allowed surface characterisation at an atomic and molecular level, allowing the visualisation of parameters such as surface homogeneity and roughness, as well as the existence of islands, granules, or defects of different height.

The incorporation of biomolecules during surface functionalisation can also be studied by scanning electrochemical microscopy (SECM), as well as by QCM and SPR. In the absence of the appropriate equipment, however, the simplest strategy consists of carrying out an ELISA-like sandwich assay. For instance, incubation with an enzyme-labelled Ab that binds the immobilised biocomponent (i.e., rabbit Ab raised against mouse IgG), followed by detection with an enzymatic substrate, allows tracing the amount of immobilised reagent, and also the level of surface blocking against unspecific adsorption. Besides this, the implementation of a model ELISA format may be useful to optimise and compare immobilisation strategies. Thanks to ELISA, the simultaneous evaluation of an Ab or enzyme performance following different alternative immobilisation procedures is possible, which greatly facilitates the optimisation process.

Once bacteria have been captured and detected, their presence/absence on the electrodes can be easily confirmed by optical microscopy after staining with a suitable dye. This is a simple way to assess the functionalised surface directly, and determine the conditions that result in optimum performance. In addition, ELISA also serves as a yardstick to measure the performance of a new biosensor method.

Although the number of electrochemical biosensor applications is steadily increasing, there are still very few works dedicated to the use of impedance spectroscopy in the detection of bacterial pathogens. The following sections comment on a series of biosensor applications for pathogen impedimetric detection where some of the fundamentals described above are used.

4. Current EIS Biosensors for Pathogen Detection

This section presents recent advances in impedimetric biosensors for pathogen detection. However, we have also decided to include here certain examples of impedimetric biosensors where the analyte is not necessarily a pathogen.

Amongst the various recognition elements discussed in the previous sections, only two are systematically used for pathogen detection using EIS. These are antibodies and nucleic acids (DNA).

As was mentioned in Section 2, the modification of the electrodes with coating films can change the value of the total impedance. Electrochemical biosensors are based on the functionalisation of the electrode surface with at least one biological receptor that can specifically recognize a target analyte, in this case a pathogen. In EIS biosensors, the biological receptor and the bound pathogen together can be considered as a coating film which is expected to change the measured impedance. It has already been said that impedimetric detection is based on monitoring impedance changes brought about by a biorecognition event. In order to classify the different applications, we have divided them into two groups according to the measured parameter. One group focuses on the interface, while the other is concerned with solution changes. There are two interfacial parameters that are mainly used in EIS biosensors (Ivnitski et al. 2000; Berggren et al. 2001; Katz and Willner 2003; K'Owino and Sadik 2005; Pejcic and De Marco 2006). The first one is interfacial capacitance, C_{int} (Rickert et al. 1996; Berggren and Johansson 1997); and the other one, which requires the presence of electroactive species, is charge-transfer resistance, R_{CT} (Bardea et al. 1999; Patolsky et al. 1999; Alfonta et al. 2001; Ruan et al. 2002; Yang et al. 2004). On the other hand, the group of applications monitoring changes produced in solution parameters generally measures conductivity by looking at solution resistance for a fixed electrode arrangement (Mubammad-Tahir and Alocilja 2003; Tahir et al. 2005). It is worth noting that in addition to the few works showing EIS biosensors for pathogen detection, even fewer works provide the necessary negative control data to demonstrate the specificity of the biosensor between target and nontarget pathogen.

4.1. Biosensors Based on Interfacial Capacitance Changes

In addition to EIS, other electrochemical techniques can also be used to measure interfacial capacitance. For example, Johansson and coworkers (Berggren and Johansson 1997) studied, through the use of controlled potential steps, capacitance changes produced by an Ab layer on gold substrates and their interactions with antigen (in this case the human chorionic gonadotropin hormone, HCG). The capacitance changes are measured from the transient current response when a potential step of 50 mV and 50 kHz is applied. The equivalent circuit is supposed to be a simple RC model (Fig. 15.14), and these parameters are fitted to the current response. Thiotic acid and a MAb resemble two capacitances, in series with the double-layer capacitance, C_{dl}. These two capacitances decrease the interfacial capacitance of the immunosensor. When HCG antigens bind to the Ab layer, there will be an additional layer that also contributes to the decrease of the capacitance. Water molecules may be displaced by less polar molecules near the electrode, which would imply a decrease in the dielectric constant, ε_r, while increasing the interface thickness. This is the fundamental detection mechanism by means of unlabelled antigens at the typical EIS immunosensor.

Heiduschka et al. (Rickert et al. 1996) studied nonfaradaic impedance changes via the biorecognition of a MAb for a thiol monolayer fuctionalised with a synthetic polypeptide of the foot-and-mouth disease virus. Although the impedance modulus changes are small, the

Figure 15.14. The total equivalent capacitance is composed of three single capacitances in series that stem from a thiotic acid layer, the bound antibody, and the double-layer capacitances, respectively.

authors were able to carry out a transient analysis of the interfacial capacitance and differentiate between specific and nonspecific interactions of the Ab.

4.2. Biosensors Based on Charge-Transfer Resistance Changes

Several authors have based the impedimetric detection of a pathogen on measuring variations in the charge-transfer resistance, R_{CT}, using redox couples in solution, typically $[Fe(CN)_6]^{3-/4-}$. In most of the cases, the detection also involves strategies for electro-chemical amplification.

Yanbin Li and coworkers (Yang et al. 2004) have reported on a label-free electrochemical impedance immunosensor for the detection of *Escherichia coli* O157:H7. The sensing interface consists of an indium tin oxide (ITO) interdigitated microelectrode array modified with anti-*E. coli* O157:H7 antibodies. They proposed an equivalent circuit consisting of C_{dl} (which, as we explained above, is not strictly correct; $C_{interface}$ should be used instead) in parallel with Z_w and R_{ct} for each electrode, and R_s in series between them (Fig. 15.15). The simulated values for each parameter are calculated for the bare electrode, after Ab immobilisation, and after *E. coli* capture (2.6×10^6 cells). Amongst the various parameters involved in the equivalent circuit, R_{ct} is the parameter showing the most important changes; after Ab immobilisation R_{CT} increases by 34% compared to the bare electrode. After *E. coli* binding, the resistance increases a further 28%. This increase can be explained by considering the bound cell as a barrier to electron transfer between the ferro/ferricyanide and the electrode. A linear relationship appears between charge-transfer resistance and the logarithm of the concentration of *E. coli* in a range 10^5–10^8 CFUmL^{-1}, with a detection limit of approximately 10^6 CFUmL^{-1}. This detection limit is not that good compared to ELISA, where detection limits one order of magnitude lower can be attained.

Willner and coworkers (Alfonta et al. 2001) have developed an ultrasensitive method for the detection of the cholera toxin, CT, which is also based on the monitoring of R_{ct} changes. The sensing interface consists of gold wire electrodes (0.5 mm diameter, ~ 0.2 cm^2 geometrical area, and ~ 1.2–1.5 rms roughness coefficient) modified with protein G and then linked to monoclonal anti-CT-Ab. The detection is based on the affinity binding of GM1 glansiosides to the CT and is operated in a "sandwich" fashion. The CT is first captured by the anti-CT Ab and then bound by an HRP/GM1 glanglioside-functionalised liposome (Fig. 15.16). HRP acts as a catalyst for a precipitation reaction whose product creates an insulating film on the electrode. This increases the charge-transfer resistance R_{ct} significantly. The increase in the resistance is directly related to the time interval used for the precipitation step. The authors also provide negative controls in order to demonstrate the role of each element in the detection and

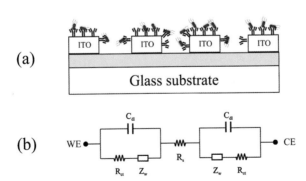

Figure 15.15. (a) Sketch of antibody immobilization and bacterial recognition on an interdigitated array of micro-electrodes. (b) Equivalent circuit for the electrochemical impedance measurement.

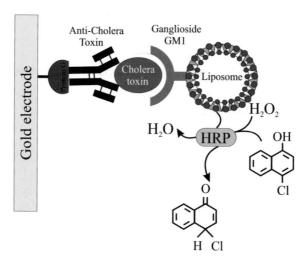

Figure 15.16. Immunodetection mechanism of the cholera toxin and signal amplification strategy based on HRP-functionalized liposomes.

amplification. The three different tests consist of a measurement without CT, a measurement using liposomes without GM1 glanglioside, and a measurement without anti-CT antibodies. In all of the latter cases, the observed R_{ct} increases are significantly smaller than for the positive control. Finally, after measuring the charge-transfer resistance increases in a range of CT concentrations (1×10^{-13}, 1×10^{-11}, 1×10^{-9}, 1×10^{-7} M), the authors propose a detection limit of 1×10^{-13} M.

A similar enzymatic amplification strategy has been presented by another author. Yanbin Li and coworkers (Ruan et al. 2002) have developed an immunobiosensor for the detection of *E. coli* O157:H7. The detection is also based in a "sandwich" assay format, but the amplification this time is accomplished via an alkaline phosphatase (AP) functionalised anti-*E. coli* Ab. For concentrations of *E. coli* between 6×10^4 and 6×10^7 cell mL^{-1} there is a linear relationship between charge-transfer resistance and the logarithm of the concentration, providing a detection limit of $\sim 10^5$ cell mL^{-1}, which is closer to the limits found for ELISA.

Amplification through the use of enzyme-labelled reagents can also serve to increase the sensitivity of DNA sensors (Bardea et al. 1999; Patolsky et al. 1999). In Patolsky, the hybridisation is amplified by affinity binding of an HRP-avidin to a biotinylated DNA probe (Fig. 15.17). The layer formed by oligonucleotide-DNA is negatively charged, and repulsion forces between the hybridised chain and the redox species $[Fe(CN)_6]^{3-/4-}$ appear, which increases the charge-transfer resistance. The enzyme acts as a catalyst of a precipitation reaction,

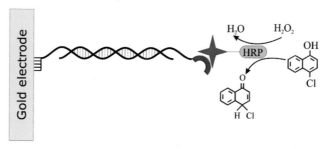

Figure 15.17. Signal amplification after recognition of an immobilised thiol-tagged DNA using a biotinylated DNA probe and HRP-avidin.

which insulates the electrode surface and further increases R_{ct}. The minimum concentration of the DNA analyte in the bulk is $20 \times 10^{-9}\,g\,mL^{-1}$, and a significant charge-transfer resistance R_{ct} change is measured after 40 min of precipitation of the insoluble product.

4.3. Biosensors Based on Conductivity Changes

A different detection strategy is based on monitoring changes produced in the conductivity of the solution, measured as changes in solution resistance. Alocilja et al. (Mubammad-Tahir and Alocilja 2003) have developed a conductimetric biosensor for the detection of foodborne pathogens. This is an integrated disposable microsystem consisting of four key parts for sample application, conjugate binding, immunocapture, and adsorption (Fig. 15.18). The liquid sample moves from one part to another driven by capillary forces towards the capture part, where detection occurs. It consists of two silver electrodes with immobilised PAb (*Salmonella*, *E.coli* O157:H7, or nonpathogenic *E. coli*) between them. In the conjugate part the bacteria (again, either *Salmonella*, *E. coli* O157:H7, or nonpathogenic *E. coli*) are bound by a PAb modified with polyaniline. This conductive polymer is responsible for the conductivity changes when the "sandwich" detection is done in the capture part. In order to measure the sensitivity of the biosensor, conductivity is measured for several bacterial concentrations (each bacteria with its specific biosensor). The relationship between concentration and conductivity is "*proportional for concentrations ranging between 10 and 10^4 CFU mL^{-1} and inversely proportional at higher concentrations*" [sic], and it can be explained on the grounds that conductivity increases with bacterial concentration as polyaniline units build up between the electrodes. However, when bacterial concentration in the sample is too high, the functionalised surface becomes "overcrowded" with bacteria. At this point the polyaniline units bound to the labelled Ab cannot come in contact with each other, and hence the conductivity drops. The detection limit is around 8×10^1 CFU mL^{-1} and the detection is qualitative only. The authors claim that statistical analysis of the data shows that the measurement errors found for the different concentration levels overlap, which prevents discrimination between one concentration level and the next. Negative controls are also given where the conductivity change due to the presence of nontarget organisms is close to zero. The same conductimetric biosensor is used for the detection of bovine viral diarrhea virus (BVDV) (Tahir et al. 2005), using monoclonal antibodies modified with protonated polyaniline and with a detection limit of 10^3 CCID mL^{-1}.

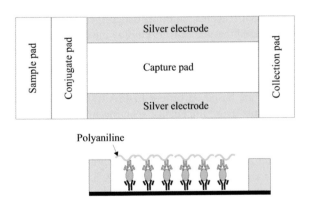

Figure 15.18. Diagrammatic representation of the conductimetric device. Sample and collection pads are cellulose membranes, and the conjugate pad is a fiberglass membrane and capture pad is a nitrocellulose membrane. (a) Corresponds to a top view, and (b) is a cross-section of the capture pad and silver electrodes.

4.4. Other Approaches

There are also several other interesting pathogen detection works based on EIS that cannot be considered biosensors because the biorecognition element is not integrated within the electrochemical transducer. Liju Yang et al. (Yang and Li 2006) have demonstrated the feasibility of the detection of *Salmonella* using interdigitated microelectrodes and immuno-magnetic separation. Magnetic beads are modified with anti-*Salmonella* antibodies in order to separate pathogens from the sample medium. The bacterial cells present in the medium are detected by changes in the impedance modulus that are directly related to changes in the interface capacitance C_{int}. The physical effect is a decrease in the capacitance thickness (d) produced by charged molecules associated with bacterial growth. The detection time is related to cell concentration. Thus, 8 hours are required for the detection of $10\,CFU\,mL^{-1}$, and only 1.5 hours for samples containing $10^6\,CFU\,mL^{-1}$. There is a linear relationship between the logarithm of bacterial concentration and the time of detection for a concentration range going from 10 to $10^6\,CFU\,mL^{-1}$. In Yang et al. 2004, *Salmonella* detection is also based on EIS using interdigitated microelectrodes, but in a selective medium.

Gomez-Sjöberg et al. 2005 have developed a microelectromechanical system to detect the metabolism of *Listeria* cells using EIS at a set of interdigitated platinum microelectrodes. The cells are diverted from the sample and driven to the detection chamber by dielectrophoresis. Before detection, there is a previous preconcentration step based on magnetic beads in order to achieve concentration factors between 10^4 and 10^5. The analysis time is at least 12h.

5. Conclusions and Future Perspectives

Impedance spectroscopy encompasses a whole family of versatile electroanalytical techniques. It is based on the application of an a.c. signal over a base potential or current. The resulting impedance can be monitored as a function of frequency and it conveys information relating to the properties of both the electrode-solution interface and the solution itself. Impedance spectroscopy has been traditionally applied to the study of corrosion and the characterisation of batteries and fuel cells. More recently, it has also been used to develop immunosensors, given its sensitivity and rapidity of response. Indeed, the wealth of information provided by this technique makes it extremely useful for the study and understanding of the chemical and biological interactions governing biosensor performance.

The detection of pathogen bacteria is an important application towards which immunosensors are being developed. An ideal biosensor should be able to perform in the absence of labels or additional reagents, and impedance spectroscopy makes this possible. The most sensitive impedance based immunosensors employ interdigitated microelectrode structures and measure changes in interfacial capacitance. There are two main drivers of performance in impedance based immunosensors. Probably the most important one is the functionalisation of the electrodes and, more specifically, the antibody used for the recognition of the target cell. Since most immunosensors are based on former ELISA tests, it should be possible to reach at least the same detection limits using impedance spectroscopy. However, these limits are often insufficient to make impedance spectroscopy sufficiently attractive, and different approaches have been attempted by several research groups to overcome this limitation. Most of these consist in the use of signal amplification strategies, such as enzyme labelled antibodies. This permits measuring greater changes in electron transfer resistance in the presence of reversible redox couples or on the precipitation of a product by means of suitably labelled antibodies. Other authors prefer to induce conductivity changes in the vicinity of the electrodes. While these strategies may improve the detection limits, they complicate the methodology and may introduce further variability. Certainly, these strategies are necessary in amperometric immunosensors, but there is another way for impedance-based immunosensors to improve their detection limits. The latter consists in the optimization of electrode geometry, so that the changes occurring within a given distance from the sensing surface bring about more significant changes to the measured impedance. The use of microfabrication techniques enables the production of electrode sets of dimensions in the micrometer range, and even in the nano-scale, in a cost effective and with a high degree of reproducibility.

This chapter presented the basic principles underpinning impedance spectroscopy, stressing the most important points to be considered from an experimental and data analysis viewpoint. This has been followed by a description and discussion of recent applications in the field of pathogen detection using electrochemical impedance spectroscopy, where the most important diagnostic parameters are interfacial capacitance and electron transfer resistance.

References

Alfonta L, Bardea A, Khersonsky O, Katz E and Willner I (2001) Chronopotentiometry and faradaic impedance spectroscopy as signal transduction methods for the biocatalytic precipitation of an insoluble product on electrode supports: Routes for enzyme sensors, immunosensors and DNA sensors. Biosensors & Bioelectronics 16:675–687

Alfonta L, Singh AK and Willner I (2001) Liposomes labeled with biotin and horseradish peroxidase: A probe for the enhanced amplification of antigen-antibody or oligonucleotide-DNA sensing processes by the precipitation of an insoluble product on electrodes. Analytical Chemistry 73:91–102

Alfonta L, Willner I, Throckmorton DJ and Singh AK (2001) Electrochemical and quartz crystal microbalance detection of the cholera toxin employing horseradish peroxidase and gm1-functionalized liposomes. Anal. Chem. 73:5287–5295

Alon R, Bayer EA and Wilchek M (1993) Cell-adhesion to streptavidin via rgd-dependent integrins. European Journal of Cell Biology 60:1–11

Anderson GP, Jacoby MA, Ligle FS, and King KD (1997) Effectiveness of protein a for antibody immobilization for a fiber optic biosensor. Biosensors and Bioelectronics 12: 329–336

Awais R, Fukudomi H, Miyanaga K, Unno H and Tanji Y (2006) A recombinant bacteriophage-based assay for the discriminative detection of culturable and viable but nonculturable escherichia coli o157 : H7. Biotechnology Progress 22:853–859

Babacan S, Pivarnik P, Letcher S and Rand AG (2000) Evaluation of antibody immobilization methods for piezoelectric biosensor application. Biosensors & Bioelectronics 15:615–621

Bagotsky VS (2006) Fundamentals of Electrochemistry. John Wiley & Sons, Hoboken

Bain CD and Whitesides GM (1988) Formation of 2-component surfaces by the spontaneous assembly of monolayers on gold from solutions containing mixtures of organic thiols. Journal of the American Chemical Society 110:6560–6561

Bain CD and Whitesides GM (1989) A study by contact-angle of the acid-base behavior of monolayers containing omega-mercaptocarboxylic acids adsorbed on gold - an example of reactive spreading. Langmuir 5:1370–1378

Barak O, Treat JR and James WD (2005) Antimicrobial peptides: Effectors of innate immunity in the skin. Adv Dermatol. 21:357–374

Bard AJ and Faulkner LR (2001) Electrochemical Methods: Fundamentals and Applications. John Wiley & Sons, New York

Bardea A, Dagan A and Willner I (1999) Amplified electronic transduction of oligonucleotide interactions: Novel routes for tay-sachs biosensors. Analytica Chimica Acta. 385:33–43

Bardea A, PatolskyF, Dagan A and Willner I (1999) Sensing and amplification of oligonucleoatide-DNA interactions by means of impedance spectrocospy: A route to a tay-sachs sensor. Chem. Commun. 21–22

Barsoukov E and Macdonald JR (eds) (2005) Impedance Spectroscopy: Theory, Experiment and Applications. John Wiley & Sons, Hoboken

Bartlett PN, Birkin PR, Wang JH, Palmisano F and De BenedettoG (1998) An enzyme switch employing direct electrochemical communication between horseradish peroxidase and a poly(aniline) film. Analytical Chemistry 70:3685–3694

Bayer EA, Benhur H and Wilchek M (1990) Isolation and properties of streptavidin. Methods in Enzymology 184:80–89

Berggren C and Johansson G (1997). Capacitance measurements of antibody-antigen interactions in a flow system. Anal. Chem. 69:3651–3657

Berggren C, Bjarnason B and Johansson G (2001) Capacitive biosensors. Electroanalysis 13:173–180

Bhatia SK, Shriverlake LC, Prior KJ, Georger JH, Calvert JM, Bredehorst R and Ligler FS (1989) Use of thiol-terminal silanes and heterobifunctional crosslinkers for immobilization of antibodies on silica surfaces. Analytical Biochemistry 178:408–413

Billard V, Martelet C, Binder P and Therasse J (1991) Toxin detection using capacitance measurements on immunospecies grafted onto a semiconductor substrate. Analytica Chimica Acta. 249:367–372

Brett CMA and Oliveira Brett AM (1993) Electrochemistry: Principles, Fundamentals and Applications. Oxford University Press, Oxford

Brewer NJ, Janusz S, Critchley K, Evans SD and Leggett GJ (2005). Photooxidation of self-assembled monolayers by exposure to light of wavelength 254 nm: A static sims study. Journal of Physical Chemistry B. 109:11247–11256

Butler JE (2000) Solid supports in enzyme-linked immunosorbent assay and other solid-phase immunoassays. Methods 22:4–23

Cady P. (1975) Rapid automated bacterial identification by impedance measurement. In: C.G. Heden, Editor, *New Approaches to the Identification of Mocroorganisms*, Wiley, New York, NY (1975), pp. 73–99.

Calvo EJ, Danilowicz C, Lagier CM, Manrique J and Otero M (2004) Characterization of self-assembled redox polymer and antibody molecules on thiolated gold electrodes. Biosensors & Bioelectronics 19:1219–1228

Chaki NK and Vijayamohanan K (2002) Self-assembled monolayers as a tunable platform for biosensor applications. Biosensors & Bioelectronics 17:1–12

Characklis WG and Marshall KC (ed) (1990) Biofilms. John Wiley, New York

Chen TH, Small DA, McDermott MK, Bentley WE and Payne GF (2003) Enzymatic methods for in situ cell entrapment and cell release. Biomacromolecules 4:1558–1563

Chin SF and Pantano P (2006) Antibody-modified microwell arrays and photobiotin patterning on hydrocarbon-free glass. Microchemical Journal 84:1–9

Christensen PA and Hamnet A (1994) Techniques and Mechanisms in Electrochemistry. Blackie Academic and Professional, London-Glasgow-New York

Cosnier S (1999) Biomolecule immobilisation on electrode surfaces by entrapment or attachment to electrochemically polymerized films. Biosensors and Bioelectronics 14:443–456

Cosnier S (2003) Biosensors based on electropolymerized films: New trends. Analytical and Bioanalytical Chemistry 377:507–520

Costerton JW, Stewart PS and Greenberg EP (1999) Bacterial biofilms: A common cause of persistent infections. Science. 284: 1318–1322.

de la Rica R, Fernandez-Sanchez C and Baldi A (2006) Polysilicon interdigitated electrodes as impedimetric sensors. Electrochemistry Communications 8:1239–1244

Ding SJ, Chang BW, Wu CC, Lai MF and Chang HC (2005) Electrochemical evaluation of avidin-biotin interaction on self-assembled gold electrodes. Electrochimica Acta. 50:3660–3666

Edgar R, McKinstry M, Hwang J, Oppenheim AB, Fekete RA, Giulian G, Merril C, Nagashima K and Adhya S (2006) High-sensitivity bacterial detection using biotin-tagged phage and quantum-dot nanocomplexes. Proceedings of the National Academy of Sciences of the United States of America 103:4841–4845

Ellington AD and Szostak JW (1990) In vitro selection of rna molecules that bind specific ligands. Nature 346:818–822

Elsholz B, Worl R, Blohm L, Albers J, Feucht H, Grunwald T, Jurgen B, Schweder T and Hintsche R (2006) Automated detection and quantitation of bacterial rna by using electrical microarrays. Analytical Chemistry 78:4794–4802

Everett WR and Fritschfaules I (1995) Factors that influence the stability of self-assembled organothiols on gold under electrochemical conditions. Analytica Chimica Acta. 307:253–268

Farabullini F, Lucarelli F, Palchetti I, Marrazza G and Mascini M (2006) Disposable electrochemical genosensor for the simultaneous analysis of different bacterial food contaminants. Biosensors and Bioelectronics 22:1544–1549

Farace G, Lillie G, Hianik T, Payne P and Vadgama P (2002) Reagentless biosensing using electrochemical impedance spectroscopy. Bioelectrochemistry 55:1–3

Ferris CD (1975) Introduction to bioelectrodes. Plenum Press, New York

Franceschetti DR and Macdonald JR (1979) Diffusion of neutral and charged species under small-signal ac conditions. Journal of Electroanalytical Chemistry 101:307–316

Fricke H (1932) The theory of electrolytic polarization. Philosophical Magazine 7:310–318

Gabrielli C (1995) Need name of article here. In: Rubinstein I (ed) Physical Electrochemistry Marcel Dekker, Inc., New York

Gabrielli C (1998) Identification of electrochemical processes by frequency response analysis. Solartron Analytical, Technical Note 004/83

Gibson DM (ed) (2001) Conductance/impedance techniques for microbial assay. CRP Press Inc., Boca Raton, Florida

Goding JW (1978) Use of staphylococcal protein-a as an immunological reagent. Journal of Immunological Methods 20:241–253

Goldman ER, Pazirandeh MP, Mauro JM, King KD, Frey JC and Anderson GP (2000) Phage-displayed peptides as biosensor reagents. Journal of Molecular Recognition 13:382–387

Gomez-Sjoberg R, Morisette DT and Bashir R (2005) Impedance microbiology-on-a-chip: Microfluidic bioprocessor for rapid detection of bacterial metabolism. Journal of Microelectromechanical Systems 14:829–838

Guiseppi-Elie A, Sheppard Jr. NF, Brahim S and Narinesingh D (2001) Enzyme microgels in packed-bed biore-actors with downstream amperometric detection using microfabricated interdigitated microsensor electrode arrays. Biotechnology and Bioengineering 75:475–484

Gyepi-Garbrah SH and Silerova R (2001) Probing temperature-dependent behaviour in self-assembled monolayers by ac-impedance spectroscopy. Physical Chemistry Chemical Physics 3:2117–2123

Hanbury CM, Miller WG and Harris RB (1997) Enzyme microgels in packed-bed bioreactors with downstream amperometric detection using microfabricated interdigitated microsensor electrode arrays. Clinical Chemistry 43:2128–2136

Hautman J and Klein ML (1990) Molecular-dynamics simulation of the effects of temperature on a dense monolayer of long-chain molecules. Journal of Chemical Physics 93:7483–7492

Heitz F and Van Mau N (2002) Protein structural changes induced by their uptake at interfaces. Biochimica et Biophysica Acta 1597:1–11

Hermanson GT (1996) Bioconjugate Techniques. Academic Press, London

Ho JAA, Zeng SC, Huang MR and Kuo HY (2006) Development of liposomal immunosensor for the measurement of insulin with femtomole detection. Analytica Chimica Acta. 556:127–132

Holliger P and Hudson PJ (2005) Engineered antibody fragments and the rise of single domains. Nature Biotechnology 23:1126–1136

Hoogenboom HR (2005) Selecting and screening recombinant antibody libraries. Nature Biotechnology 23:1105–1116

Huang JY and Hemminger JC (1993) Photooxidation of thiols in self-assembled monolayers on gold. Journal of the American Chemical Society 115:3342–3343

Huang TT, Sturgis J, Gomez R, Geng T, Bashir R, Bhunia AK, Robinson JP and Ladisch MR (2003) Composite surface for blocking bacterial adsorption on protein biochips. Biotechnology and Bioengineering 81:618–624

Ivnitski D, Abdel-Hamid I, Atanasov P, Wilkins E and Stricker S (2000) Application of electrochemical biosensors for detection of food pathogenic bacteria. Electroanalysis 12:317–325

K'Owino IO and Sadik OA (2005) Impedance spectroscopy: A powerful tool for rapid biomolecular screening and cell culture monitoring. Electroanalysis 17:2101–2113

Karyakin A, Presnova G, Rubtsova M and Egorov A (2000) Oriented immobilization of antibodies onto the gold surfaces via their native thiol groups. Anal Chem. 72:3805–3811

Katz E and Willner I (2003) Probing biomolecular interactions at conductive and semiconductive surfaces by impedance spectroscopy: Routes to impedimetric immunosensors, DNA-sensors, and enzyme biosensors. Electroanalysis 15:913–947

Kenny G and Dunsmoor C (1987) Effectiveness of detergents in blocking nonspecific binding of igg in the enzyme-linked immunosorbent assay (elisa) depends upon the type of polystyrene used. Israel Journal of Medical Sciences 23:732–734

Kim GH, Rand AG and Letcher SV (2003) Impedance characterization of a piezoelectric immunosensor part ii: Salmonella typhimurium detection using magnetic enhancement. Biosensors & Bioelectronics 18:91–99

Kim HJ, Bennetto HP, Halablab MA, Choi CH and Yoon S (2006) Performance of an electrochemical sensor with different types of liposomal mediators for the detection of hemolytic bacteria. Sensors and Actuators B-Chemical 119:143–149

Klussmann S, Nolte A, Bald R, Erdmann VA and Fürste JP (1996) Mirror-image rna that binds d-adenosine. Nature Biotecnology 14:1112–1115

Ko HY, Lee HW and Moon J (2004) Fabrication of colloidal self-assembled monolayer (SAM) using monodisperse silica and its use as a lithographic mask. Thin Solid Films 447:638–644

Koh W-G, Revzin A and Pishko MV (2002) Poly(ethylene glycol) hydrogel microstructures encapsulating living cells. Langmuir 18:2459–2462

Kohler G and Milstein C (1975) Continuous cultures of fused cells secreting antibody of predefined specificity (reprinted from Nature, vol 256, 1975). Nature 256:495–497

Kumar A, Larsson O, Parodi D, Liang Z (2000) Silanized nucleic acids: A general platform for DNA immobilization. Nucleic Acids Res. 28(14):e71

Kusser W (2000) Chemically modified nucleic acid aptamers for in vitro selections: Evolving evolution. Journal of Biotechnology 74:27–38

Lappin-Scott HM and Costerton JW (eds) (1995) Microbial biofilms. Cambridge University Press, Cambridge

Lasic DD (1998) Novel applications of liposomes. Trends in Biotechnology 16:307–321

Laureyn W, Nelis D, Van Gerwen P, Baert K, Hermans L, Magnee R, Pireaux JJ and Maes G (2000) Nanoscaled interdigitated titanium electrodes for impedimetric biosensing. Sensors and Actuators B-Chemical. 68:360–370

Lee HY, Jung HS, Fujikawa K, Park JW, Kim JM, Yukimasa T, Sugihara H and Kawai T (2005) New antibody immobilization method via functional liposome layer for specific protein assays. Biosensors & Bioelectronics 21:833–838

Lee TC, Yusoff K, Nathan S and Tana WS (2006) Detection of virulent newcastle disease virus using a phage-capturing dot blot assay. Journal of Virological Methods 136:224–229

Liu GY, Yang GH and Amro NA (2004) Molecular level approach to inhibit degradations of passivation layers on metal surfaces. Abstracts of Papers of the American Chemical Society 227:U1538-U1538

Liu RH, Yang JN, Lenigk R, Bonanno J and Grodzinski P (2004) Self-contained, fully integrated biochip for sample preparation, polymerase chain reaction amplification, and DNA microarray detection. Analytical Chemistry 76:1824–1831

Love JC, Estroff LA, Kriebel JK, Nuzzo RG and Whitesides GM (2005) Self-assembled monolayers of thiolates on metals as a form of nanotechnology. Chemical Reviews. 105:1103–1169

Lu HC, Chen HM, Lin YS and Lin JW (2000) A reusable and specific protein a-coated piezoelectric biosensor for flow injection immunoassay. Biotechnology Progress 16:116–124

Lucarelli F, Marrazza G and Mascini M (2005) Enzyme-based impedimetric detection of pcr products using oligonucleotide-modified screen-printed gold electrodes. Biosensors & Bioelectronics 20:2001–2009

Macdonald JR (1984) Note on the parameterization of the constant-phase admittance element. Solid State Ionics 13:147–149

Mamishev AV, Sundara-Rajan K, Yang F, Du YQ and Zahn M (2004) Interdigital sensors and transducers. Proceedings of the IEEE 92:808–845

Mao XL, Yang LJ, Su XL and Li YB (2006) A nanoparticle amplification based quartz crystal microbalance DNA sensor for detection of escherichia coli o157 : H7. Biosensors & Bioelectronics 21:1178–1185

Martin CR and Parthasarathy RV (1995) Polymeric microcapsule arrays. Advanced Materials 7:487–488

McAdams ET (1989) Effect of surface topography on the electrode-electrolyte interface impedance. Surface Topography 2:107–122

McCoy MH and Wang E (2005) Use of electric cell-substrate impedance sensing as a tool for quantifying cytopathic effect in influenza a virus infected mdck cells in real-time. Journal of Virological Methods 130:157–161

Melo LF, Bott TR, Fletcher M and Capdeville B (ed) (1992) Biofilms: Science and technology. Kluwer Academic Publishers, The Netherlands

Memoli A, Annesini MC, Mascini M, Papale S and Petralito S (2002) A comparison between different immobilised glucoseoxidase-based electrodes. Journal of Pharmaceutical and Biomedical Analysis 29:1045–1052

Minett AI, Barisci JN and Wallace GG (2002) Coupling conducting polymers and mediated electrochemical responses for the detection of listeria. Analytica Chimica Acta. 475:37–45

Mubammad-Tahir Z and Alocilja EC (2003) A conductometric biosensor for biosecurity. Biosensors & Bioelectronics 18:813–819

Nanda S, Muralidhar K and Kar SK (2002) Thermostable alpha-amylase conjugated antibodies as probes for immunodetection in elisa. Journal of Immunoassay & Immunochemistry 23:327–345

Nichkova M, Dosev D, Gee SJ, Hammock BD and Kennedy IM (2005) Microarray immunoassay for phenoxybenzoic acid using polymer encapsulated eu: Gd2o3 nanoparticles as fluorescent labels. Analytical Chemistry 77:6864–6873

Nolte A, Klussmann S, Bald R, Erdmann VA and Fürste JP (1996) Mirror-design of l-oligonucleotide ligands binding to l-arginine. Nature Biotechnology 14:1116–1119

Nuzzo RG and Allara DL (1983) Adsorption of bifunctional organic disulfides on gold surfaces. Journal of the American Chemical Society 105:4481–4483

Olsen EV, Sorokulova IB, Petrenko VA, Chen IH, Barbaree JM and Vodyanoy VJ (2006) Affinity-selected filamentous bacteriophage as a probe for acoustic wave biodetectors of salmonella typhimurium. Biosensors & Bioelectronics 21:1434–1442

Olthuis W, Streekstra W and Bergveld P (1995) Theoretical and experimental-determination of cell constants of planar-interdigitated electrolyte conductivity sensors. Sensors and Actuators B-Chemical 24:252–256

Ouerghi O, Touhami A, Jaffrezic-Renault N, Martelet C, Ouada H and Cosnier S (2002) Impedimetric immunosensor using avidin-biotin for antibody immobilization. Bioelectrochemistry 56:131–133

Palecek E, Jelen F and Trnkova L (1986) Cyclic voltammetry of DNA at a mercury electrode: An anodic peak specific for guanine. General Physiology and Biophysics 5:315–329

Palecek E (1988) Adsorptive transfer stripping voltammetry: Determination of nanogram quantities of DNA immobilized at the electrode surface. Anal Biochem. 170:421–431

Palecek E and Fojta M (1994) Differential pulse voltammetric determination of rna at the picomole level in the presence of DNA and nucleic acid components. Analytical Chemistry 66:1566–1571

Pantano P and Chin SF (2003) Direct photobiotin modification of glass surfaces for antibody patterning applications. Abstracts of Papers of the American Chemical Society 225:U125-U125

Parida SK, Dash S, Patel S and Mishra BK (2006) Adsorption of organic molecules on silica surface. Advances in Colloid and Interface Science 121:77–110

Park I-S and Kim N (1998) Thiolated salmonella antibody immobilization onto the gold surface of piezoelectric quartz crystal. Biosensors and Bioelectronics 13:1091–1097

Park S-J, Taton TA and Mirkin CA (2002) Array-based electrical detection of DNA with nanoparticle probes. Science 295:1503–1506

Patel AC, Li SX, Yuan JM and Wei Y (2006) In situ encapsulation of horseradish peroxidase in electrospun porous silica fibers for potential biosensor applications. Nano Letters 6:1042–1046

Patolsky F, Katz E, Bardea A and Willner I (1999) Enzyme-linked amplified electrochemical sensing of oligonucleotide-DNA interactions by means of the precipitation of an insoluble product and using impedance spectroscopy. Langmuir 15:3703–3706

Patolsky F, Lichtenstein A and Willner I (2001) Electronic transduction of DNA sensing processes on surfaces: Amplification of DNA detection and analysis of single-base mismatches by tagged liposomes. Journal of the Americal Chemical Society 123:5194–5205

Patolsky F, Lichtenstein A, Kotler M and Willner I (2001) Electronic transduction of polymerase or reverse transcriptase induced replication processes on surfaces: Highly sensitive and specific detection of viral genomes. Angew Chem Int Ed Engl. 40:2261–2265

Pavlinkova G Lou DY and Kohler H (2000) Site-specific photobiotinylation of antibodies, light chains, and immunoglobulin fragments. Methods 22:44–48

Pei R, Cheng Z, Wang E, Yang X (2001) Amplification of antigen-antibody interactions based on biotin labeled protein-streptavidin network complex using impedance spectroscopy. Biosensors & Bioelectronics. 16:355–361

Pejcic B and De Marco R (2006) Impedance spectroscopy: Over 35 years of electrochemical sensor optimization. Electrochimica Acta. 51:6217–6229

Piehler J, Brecht A, Geckeler KE and Gauglitz G (1996) Surface modification for direct immunoprobes. Biosensors & Bioelectronics 11:579–590

Pope NM, Kulcinski DL, Hardwick A and Chang YA (1993) New application of silane coupling agents for covalently binding-antibodies to glass and cellulose solid supports. Bioconjugate Chemistry 4:166–171

Porter MD, Bright TB, Allara DL and Chidsey CED (1987) Spontaneously organized molecular assemblies 4. Structural characterization of normal-alkyl thiol monolayers on gold by optical ellipsometry, infrared-spectroscopy, and electrochemistry. Journal of the American Chemical Society 109:3559–3568

Rickert J, Gopel W, Beck W, Jung G and Heiduschka P (1996) A 'mixed' self-assembled monolayer for an impedimetric immunosensor. Biosensors and Bioelectronics 11:757–768

Robertson DL and Joyce GF (1990) Selection in vitro of an rna enzyme that specifically cleaves single-stranded DNA. Nature 344:467–468

Rosenfeld Y and Shai Y (2006) Lipopolysaccharide (endotoxin)-host defense antibacterial peptides interactions: Role in bacterial resistance and prevention of sepsis. Biochimica Et Biophysica Acta-Biomembranes 1758:1513–1522

Ruan CM, Yang LJ and Li YB (2002) Immunobiosensor chips for detection of escherichia coli o157 : H7 using electrochemical impedance spectroscopy. Analytical Chemistry 74:4814–4820

Rubinstein I, Steinberg S, Tor Y, Shanzer A and Sagiv J (1988) Ionic recognition and selective response in self-assembling monolayer membranes on electrodes. Nature 332: 426–429

Ruckenstein E and Li ZF (2005) Surface modification and functionalization through the self-assembled monolayer and graft polymerization. Advances in Colloid and Interface Science 113:43–63

Saal K, Tatte T, Tulp I, Kink I, Kurg A, Maeorg U, Rinken A and Lohmus A (2006) Sol-gel films for DNA microarray applications. Materials Letters 60:1833–1838

Sargent A, Loi T, Gal S and Sadik OA (1999) The electrochemistry of antibody-modified conducting polymer electrodes. Journal of Electroanalytical Chemistry 470:144–156

Sastry M (2002) Entrapment of proteins and DNA in thermally evaporated lipid films. Trends in Biotechnology 20:185–188

Schlenoff JB, Li M and Ly H (1995) Stability and self-exchange in alkanethiol monolayers. Journal of the American Chemical Society 117:12528–12536

Schneider TW and Buttry DA (1993) Electrochemical quartz-crystal microbalance studies of adsorption and desorption of self-assembled monolayers of alkyl thiols on gold. Journal of the American Chemical Society 115:12391–12397

Scholz F (ed) (2002) Electroanalytical methods. Guide to experiments and applications. Springer-Verlag, Berlin-Heidelberg-New York

Shan D, He YY, Wang SX, Xue HG and Zheng H (2006) A porous poly(acrylonitrile-co-acrylic acid) film-based glucose biosensor constructed by electrochemical entrapment. Analytical Biochemistry 356: 215–221

Sluyters-Rehbach, M (1994) Impedances of Electrochemical Systems: Terminology, Nomenclature and Representation. Part I: Cells with Metal Electrodes and Liquid Solutions. Pure & Appl. Chem. 66: 1831–1891

Sotiropoulou S, Vamvakaki V and Chaniotakis NA (2005) Stabilization of enzymes in nanoporous materials for biosensor applications. Biosensors & Bioelectronics 20:1674–1679

Steinitz M (2000) Quantitation of the blocking effect of tween 20 and bovine serum albumin in elisa microwells. Analytical Biochemistry 282:232–238

Storri S, Santoni T, Minunni M and Mascini M (1998) Surface modifications for the development of piezoimmunosensors. Biosensors & Bioelectronics 13:347–357

Tahir ZM, Alocilja EC and Grooms DL (2005) Polyaniline synthesis and its biosensor application. Biosensors & Bioelectronics 20:1690–1695

Tlili A, Jarboui MA, Abdelghani A, Fathallah DM and Maaref MA (2005) A novel silicon nitride biosensor for specific antibody-antigen interaction. Materials Science & Engineering C-Biomimetic and Supramolecular Systems 25:490–495

Troughton EB, Bain CD, Whitesides GM, Nuzzo RG, Allara DL and Porter MD (1988) Monolayer films prepared by the spontaneous self-assembly of symmetrical and unsymmetrical dialkyl sulfides from solution onto gold substrates - structure, properties, and reactivity of constituent functional-groups. Langmuir 4:365–385

Tuerk C and Gold L (1990) Systematic evolution of ligands by exponential enrichment: Rna ligands to bacteriophage t4 DNA polymerase. Science 249:505–510

Unwin PR and Bard AJ (1992) Scanning electrochemical microscopy .14. Scanning electrochemical microscope induced desorption - a new technique for the measurement of adsorption desorption-kinetics and surface-diffusion rates at the solid liquid interface. Journal of Physical Chemistry 96:5035–5045

Ur A and Brown D (1975) Impedance monitoring of bacterial activity. J Med Microbiol. 8:19–28

Vanýsek P (1997) Impact of electrode geometry, depth of immersion, and size on impedance measurements. Can. J. Chem. 75:1635–1642

Wang JB, Profitt JA, Pugia MJ and Suni II (2006) An nanoparticle conjugation for impedance and capacitance signal amplification in biosensors. Analytical Chemistry. 78:1769–1773

Williams DD, Benedek O and Turnbough CL (2003) Species-specific peptide ligands for the detection of bacillus anthracis spores. Applied and Environmental Microbiology 69:6288–6293

Wood GS and Warnke R (1981) Suppression of endogenous avidin-binding activity in tissues and its relevance to biotin-avidin detection systems. Journal of Histochemistry & Cytochemistry 29:1196–1204

Yallow R and Berson S (1959) Assay of plasma insulin in human subjects by imunological methods. Nature 185:1648–1649

Yang LJ and Li YB (2006) Detection of viable salmonella using microelectrode-based capacitance measurement coupled with immunomagnetic separation. Journal of Microbiological Methods 64:9–16

Yang LJ and Li YB, Griffis CL and Johnson MG (2004) Interdigitated microelectrode (ime) impedance sensor for the detection of viable salmonella typhimurium. Biosensors & Bioelectronics 19:1139–1147

Yang LJ, Li YB, and Erf GF (2004) Interdigitated array microelectrode-based electrochemical impedance immunosensor for detection of escherichia coli o157 : H7. Analytical Chemistry 76:1107–1113

Yidiz HB and Toppare L (2006) Biosensing approach for alcohol determination using immobilized alcohol oxidase. Biosensors & Bioelectronics 21:2306–2310

Yu XB, Lv R, Ma ZQ, Liu ZH, Hao YH, Li QZ and Xu DK (2006) An impedance array biosensor for detection of multiple antibody-antigen interactions. Analyst 131:745–750

Zhao Y-D, Pang D-W, Hu S, Wang Z-L, Cheng J-K, Qi Y-P, Dai H-P, Mao B-W, Tian Z-Q, Luo J and Lin Z-H (1999) DNA-modified electrodes part 3.: Spectroscopic characterization of DNA-modified gold electrodes. Analytica Chimica Acta 388:93–101

Zhou H-X and Dill KA (2001) Stabilization of proteins in confined spaces. Biochemistry 40:11289–11293

16

Label-Free Microbial Biosensors Using Molecular Nanowire Transducers

Evangelyn Alocilja and Zarini Muhammad-Tahir

Abstract

There is an increasing public awareness and concern regarding the safety of our food supply. The complexity of the US food supply chain provides numerous entry points and routes in which pathogens and other disease-causing organisms can be introduced into the nation's food system. Foodborne illness and product recalls have been increasing in incidence. It is not easy to address the hundreds of microbial contaminants associated with microbial foodborne diseases. Conventional methods of identifying these pathogens require 2–7 days. Although these methods are highly sensitive and specific, they are elaborate and laborious. The use of biosensors as emerging technologies could revolutionize the study and detection of these foodborne microorganisms. The development of biosensors will further serve the food industry, agricultural sector, regulatory community, and public health. Biosensor techniques will play an extensive role in understanding the occurrence of contamination at the source during the next decade and help forecast the potential for risk and mitigation before foodborne outbreaks occur. This chapter describes emerging and novel biosensor technologies for rapid and sensitive detection of pathogens of concern to the food supply. Particularly, molecular nanowires as transducers in biosensor devices are covered. Antibody and DNA based biosensors are reviewed and two illustrations on immunosensors are presented.

1. Introduction

1.1. Rationale for Rapid Tests

The complexity of the U.S. food supply chain provides numerous entry points and routes through which contaminants and pathogens can be introduced into the nation's food system (Figure 16.1). The detection and identification of these foodborne pathogens in raw food materials, ready-to-eat food products, restaurants, processing and assembly lines, hospitals, ports of entry, and drinking water supplies continue to rely on conventional culturing techniques. Conventional methods involve enriching the sample and performing various media-based metabolic tests (agar plates or slants). Though these methods are highly sensitive and specific, they are elaborate, laborious, and typically require 2–7 days to obtain conclusive results (FDA 2005). In contrast, rapid tests are user-friendly and require less than 12 hours to attain a result.

An increased demand for high-throughput screening, especially in the clinical and pharmaceutical industries, has produced several technological developments for detecting biomolecules. Some of these emerging technologies include enzyme-linked immunosorbent

Evangelyn Alocilja and Zarini Muhammad-Tahir • Biosystems and Agricultural Engineering, Michigan State University, East Lansing, Michigan.

M. Zourob et al. (eds.), *Principles of Bacterial Detection: Biosensors, Recognition Receptors and Microsystems*,
© Springer Science+Business Media, LLC 2008

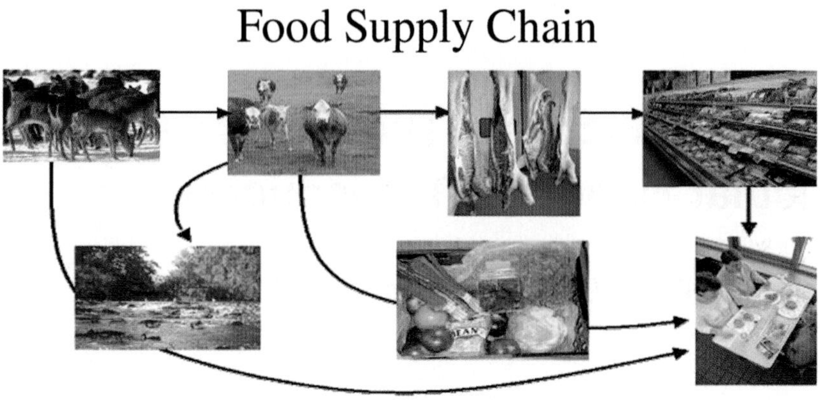

Food Supply Chain

Figure 16.1. Schematic of food a supply chain showing potential entry points and routes for foodborne pathogen contamination.

assay (ELISA), polymerase chain reaction (PCR) and hybridization, flow cytometry, molecular cantilevers, matrix-assisted laser desorption/ionization, immunomagnetics, artificial membranes, and spectroscopy. Pathogen detection utilizing ELISA methods for determining and quanti-fying pathogens in food has been well established (Cohn 1998).The PCR method is extremely sensitive but requires pure sample preparation and hours of processing, along with expertise in molecular biology (Meng et al. 1996). Flow cytometry is another highly effective means for rapid analysis of individual cells at rates up to 1000 cells/sec (McClelland and Pinder 1994); however, it has been used almost exclusively for eukaryotic cells. These detection methods are relevant for laboratory use but cannot adequately serve the needs of health practitioners and monitoring agencies in the field. Furthermore, these systems are costly, require specialized training, have complicated processing steps in order to culture or extract the pathogen from food samples, and are time consuming. In comparison, a field-ready biosensor is usually inexpensive, easy to use, portable, and provides results in minutes.

1.2. Target Microorganisms and Matrices

Pathogenic bacteria, viruses, and other microorganisms are ubiquitous in the environment. Bacterial pathogens are found in soil, animal intestinal tracts, and fecal-contaminated water. Human beings, on average, harbor more than 150 types of bacteria inside and outside the body (Madigan et al. 1997). Although many microorganisms are harmless, some are known to be the causative agents of many different infectious diseases including botulism, cholera, diarrhea, emesis, pneumonia, and typhoid fever (Doyle et al. 1997). The list of foodborne pathogens and bioterrorism-potential agents can be found from the Food and Drug Administration (FDA 2006) and the Centers for Disease Control and Prevention's websites (CDC 2006). More than 200 known diseases are transmitted through food and drink alone (Mead et al. 1999).

1.2.1. *Escherichia coli*

Escherichia coli are bacteria that naturally occur in the intestinal tracts of humans and warm-blooded animals to help the body synthesize vitamins. A particularly dangerous type is the enterohemorrhagic *Escherichia coli* O157:H7 or EHEC. In 2000, EHEC was the etiological agent in 69 confirmed outbreaks (twice the number in 1999) involving 1564 people in 26 states (CDC 2001a). Of the known transmission routes, 69% were attributed to food sources, 11% to animal contact, 11% to water exposures, and 8% to person-to-person transmission (CDC 2001a).

E. coli O157:H7 produces toxins that damage the lining of the intestine, cause anemia, stomach cramps and bloody diarrhea, and a serious complication called hemolytic uremic syndrome (HUS), and thrombotic thrombocytopenic purpura (TTP) (Doyle et al. 1997). In North America, HUS is the most common cause of acute kidney failure in children, who are particularly susceptible to this complication. TTP has a mortality rate of as high as 50% among the elderly (FDA 2006). Recent food safety data indicates that cases of *E. coli* O157:H7 are rising in both the US and other industrialized nations (WHO 2002).

Human infections with *E. coli* O157:H7 have been traced back to individuals having direct contact with food in situations involving food handling or food preparation. In addition to human contamination, *E. coli* O157:H7 may be introduced into food through meat grinders, knives, cutting blocks, and storage containers. *E. coli* O157:H7 has also been found in drinking water that has been contaminated by runoff from livestock farms as a result of heavy rains. Regardless of the source, *E. coli* O157:H7 has been traced to a number of food products, including meat and meat products, apple juice or cider, milk, alfalfa sprouts, unpasteurized fruit juices, dry-cured salami, lettuce, game meat, and cheese curds (Doyle et al. 1997; FDA 2005). Possible points of entry into the food supply chain include naturally occurring sources from wild animals and ecosystems, infected livestock, contaminated processing operations, and unsanitary food preparation practices, as illustrated in Figure 16.1.

1.2.2. *Salmonella*

Salmonella enterica serovar Typhimurium and *Salmonella enterica* serovar Enteritidis are the most common *Salmonella* serotypes found in the United States. According to CDC, salmonellosis is the most common foodborne illness (CDC 2002b). Over 40,000 actual cases are reported yearly in the U.S. (CDC 2002a). Approximately 500 (Mead et al. 1999) to 1,000 (CDC 2001b) persons die annually from *Salmonella* infections in the U.S. The estimated annual cost of human illness caused by *Salmonella* is $3 billion (Mead et al. 1999). *Salmonella* Enteritidis has frequently been observed as a contaminant in foods, such as fresh produce (De Roever 1998) and eggs and poultry products (Cohen et al. 1994). While various *Salmonella* species have been isolated from the outside of egg shells, the presence of *S*. Enteritidis inside the egg, in the yolk, is of great concern, as it suggests vertical transmission, i.e., deposition of the organism in the yolk by an infected layer hen prior to shell deposition (FDA 1992).

Human *Salmonella* infection can lead to enteric (typhoid) fever, enterocolitis, and systemic infections by nontyphoid microorganisms. Typhoid and paratyphoid strains are well adapted for invasion and survival within host tissues, causing enteric fever, a serious human disease. Nontyphoid *Salmonella* causes salmonellosis, which is manifested as gastroenteritis with diarrhea, fever, and abdominal cramps. Severe infection could lead to septicemia, urinary tract infection, and even death in at-risk populations (the young, the elderly, and immunocompromised individuals). Raw meats, poultry, eggs, milk and dairy products, fish, shrimp, frog legs, yeast, coconut, sauces and salad dressings, cake mixes, cream-filled desserts and toppings, dried gelatin, peanut butter, cocoa, and chocolate are some of the foods associated with *Salmonella* infection (D'Aoust 1997).

All known strains (about 2400) of *Salmonella* are pathogenic with a very low infectious dose, as observed in some of the foodborne outbreaks traced back to *Salmonella* contamination. Newborns, infants, the elderly, and immunocompromised individuals are more susceptible to *Salmonella* infections than healthy adults (D'Aoust 1997). The developing immune system in newborns and infants, the frequently weak and/or delayed immunological responses in the elderly and debilitated persons, and low gastric acid production in infants and seniors facilitate the intestinal proliferation and systemic infection of salmonellae in this susceptible population (Blaser and Newman 1982). Moreover, in recent years, concerns have been raised because many strains of *Salmonella* have become resistant to several of the antibiotics traditionally used to treat it, in both animals and humans, making *Salmonella* infections an important health concern in both developed and developing countries. The majority of the increased incidence of resistance can be attributed to *Salmonella* Typhimurium DT104. Treatment of this infectious disease is complicated by its ability to acquire resistance to multiple antibiotics (Carlson et al. 1999). Evidence suggesting that ingestion of only a few *Salmonella* cells can develop a variety of clinical conditions (including death) is a reminder for food producers, processors, and distributors that low levels of *Salmonella* in a finished food product can lead to serious public health consequences, and undermine the reputation and economic viability of the incriminated food manufacturer. Thus, early and rapid detection of *Salmonella* is very important to the food industry so that appropriate measures can be taken to eliminate the source of infection.

1.2.3. Bovine Viral Diarrhea Virus

Bovine viral diarrhea virus (BVDV), one of the most insidious and economically devastating viral pathogens in cattle (Brock 2003), is chosen as a model for potential bioterrorism agents, such as the viruses indicated on CDC's website (CDC 2006). Though BVDV is predominantly found in cattle, the ability of this virus to replicate in numerous wild ruminant species such as camels, deer, elk, and bison has also been documented (Nettleton 1990). Bovine viral diarrhea virus is classified in the genus *Pestivirus* within the family *Flaviviridae* with a single-stranded, enveloped RNA genome that is prone to high mutation rates. High mutation rates lead to heterogeneity, which helps BVDV and other pestiviruses to adapt and evade host immune systems (Ridpath 2003). Bovine viral diarrhea virus has two predominant genotypes, BVDV Type 1 and Type 2, based on their RNA makeup, the structure of their protein capsules, and the antibodies that are made in response to their infection (Pellerin et al. 1994). A specific virus may also be either cytopathic or noncytopathic, indicating its ability to cause visible damage to experimentally infected cells in the laboratory (Fulton et al. 2000). Although both biotypes can cause infections, the noncytopathic strain of BVDV is more common in the cattle population (Dubovi 1992). Studies also show that there are two different ways that cattle can be infected by BVDV: acute infection of animals that have not been previously exposed to the virus (immunocompetent cattle population), and fetal infection occurring in the early pregnancy stage (Houe 1995).

Most of the BVDV infections in an immunocompetent cattle population are subclinical, demonstrated by a small increase in body temperature and a decrease in milk production (Baker 1995). A bovine viral diarrhea disease is manifested when the BVDV infection becomes clinical. The virus incubation period is 5-7 days, and a transient viremia occurs 4–5 days post-infection and may continue up to 15 days (Duffell and Harkness 1985). At the time animals develop the clinical infection, they are usually starting to make neutralizing antibodies (Brock 2003). Clinical symptoms include depression, diarrhea, sometimes oral lesions characterized by ulceration, and a rapid respiratory rate, which can be mistakenly diagnosed as pneumonia (Perdrizet et al. 1987). The concentration of virus shed from these acutely infected animals is much lower than that of the persistently infected (PI) animals (Duffell and Harkness 1985). In Denmark alone, an annual incidence of acute infection of 34% is estimated, with total annual losses of $20 million per one million calves (Houe 1995).

Persistent infection with BVDV develops when a fetus is exposed to the virus between 50 to 150 days of gestation (Brock 2003). The ability to induce fetal persistent infections is a unique aspect of BVDV pathogenesis. Although PI animals may represent less than one percent of the cattle population, they shed the virus and initiate further virus replication and genetic variation (Brock 2003). Therefore, control programs must focus on the prevention of persistent infections, and identification and removal of PI animals. Breaking the cycle of exposure of pregnant animals in the first 125 to 150 days of gestation is the key to preventing persistent infections. The concentration level of virus in these PI animals is extremely high, up to 10^6 cell culture infective dose per milliliter (CCID/ml) in serum samples (Houe 1995). Cattle that are persistently infected with BVDV shed a high concentration of virus in their secretion and excretion, such as nasal discharge, saliva, blood serum, tissue, semen, urine and milk (Brock 1991). Direct contact with PI cattle is probably the most common method of transmission of the infection (Houe 1995).

Mucosal disease can occur when PI animals are exposed to a cytopathic strain of BVDV that shares close homology with the noncytopathic BVDV (Bolin 1995). Acute mucosal disease is characterized by pyrexia, depression, weakness, and anorexia (Baker 1995). Animals with chronic mucosal disease may develop high fever, continual diarrhea, weight loss, long-term erosive epithelial lesion, and may ultimately die from severe debilitation (Baker 1995). Generally, mucosal disease has a high morbidity rate and a low mortality rate (Bolin 1995).

1.3. Food Safety Applications

It is estimated that foodborne diseases cause approximately 76 million illnesses, including 325,000 hospitalizations and 5,000 deaths in the U.S. each year (Mead et al. 1999). Of these, known pathogens account for an estimated 14 million illnesses, 60,000 hospitalizations, and 1,800 deaths, indicating that these pathogens are a substantial source of infectious diseases (Mead et al. 1999). The causes of foodborne illness include viruses, bacteria, parasites, fungi, toxins, and metals, with the symptoms ranging from mild gastroenteritis to life-threatening neurological, hepatic, and renal problems. Researchers at the Economic Research Service (ERS) of the United States Department of Agriculture (USDA) estimate that the total annual medical cost associated with foodborne illness caused by pathogens is $6.5–9.4 billion (Buzby et al. 2000). The four major foodborne pathogens are *Campylobacter, Salmonella, Listeria monocytogenes*, and *E. coli* O157:H7 and are characterized in Table 16.1.

The National Institute of Allergy and Infectious Diseases (NIAID), an institute of The National Institutes of Health (NIH), and the Centers for Disease Control and Prevention (CDC) categorize biological pathogens as either Category A, B, or C (CDC 2004). Category A agents include organisms that pose a risk to national security because they can be easily disseminated or transmitted from person to person; result in high mortality rates and have the potential for major public health impact; might cause public panic and social disruption; and require special action for public health preparedness (CDC 2004). Category B agents include those that are moderately easy to disseminate; result in moderate morbidity rates and low mortality rates; and require specific enhancements of the nation's diagnostic capacity and enhanced disease surveillance (CDC 2004). Category C agents include emerging pathogens that could be engineered for mass dissemination in the future because of availability; ease of production and dissemination; and potential for high morbidity and mortality rates and major health impact

Table 16.1. Estimated illness from four major pathogens (Buzby et al. 2000)

Pathogen	Number of Cases	Hospitalizations	Deaths
Campylobacter	1,963,141	10,539	99
E. coli O157:H7	62,458	1,843	52
L. monocytogenes	2,498	2,298	499
Salmonella	1,342,532	16,102	556

(CDC 2004). NIAID and the CDC have identified foodborne pathogens such as *Salmonella* spp., *L. Monocytogenes*, and *E. coli* O157:H7 as Category B agents (CDC 2004; NIAID 2004).

2. Biosensor Formats

2.1. Definition

Biosensors are analytical instruments possessing a biomolecule as a reactive surface or receptor in close proximity to a transducer, which converts the binding of an analyte to the capturing or receptor biomolecule into a measurable signal (Turner and Newman 1998; D'Souza 2001) (Figure 16.2). They often operate in a reagentless process enabling the creation of user friendly and field ready devices. Biosensors are needed to quickly detect disease-causing agents in food and water in order to ensure the continued safety of the nation's food supply.

Biosensors make use of a variety of transducers, such as electrical, electrochemical, optical, piezoelectric crystal, and acoustic wave. A transducer should be highly sensitive for the analyte of interest and should have a moderately rapid response time. The transducer element also should be reliable, able to be miniaturized, and suitably designed for practical applications (Foulds and Lowe 1985). The sensing elements may be enzymes, antibodies, DNA (deoxyribonucleic acid), receptors, organelles, and microorganisms, as well as animal and plant cells. Some of the major attributes of a biosensor technology are its specificity, sensitivity, reliability, portability, real-time analysis, and simplicity of operation (D'Souza 2001). The most important features of a biosensor are its high sensitivity and specificity, and rapid detection time (Foulds and Lowe 1985).

2.2. Antibodies as Biological Sensing Element

This section illustrates the use of antibodies as the biological sensing elements in the biosensor design. Antibodies are frequently used in biosensor research due to their high specificity attributes (Sadana 2002). The specificity of antibodies is based on the immunological reaction involving the unique structure recognition on the antigen surface and the antibody binding site (Barbour and George 1997). The basic structure of antibody or immunoglobulin molecules consists of four polypeptides—two heavy chains and two light chains joined like a capital letter "Y," and linked by disulphide bonds (Sadana 2002) (Figure 16.3). The Fc

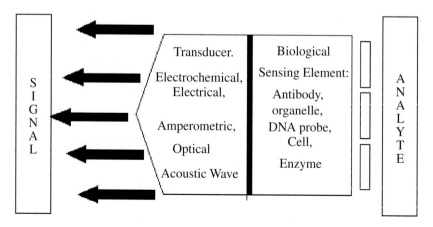

Figure 16.2. Schematic of a biosensor.

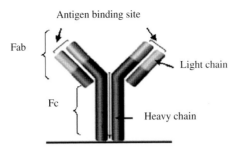

Figure 16.3. Antibody structure (adapted from Sadana 2002).

fragment comprises the effector functions, such as complement activation and cell membrane receptor interaction. The Fab fragment, on the other hand, comprises the antigen binding sites. The amino acid sequence on the tips of the "Y" (antigen binding sites) varies greatly among different antibodies. This antigen binding site is composed of 110–130 amino acids, giving the antibody its specificity for binding antigen (Barbour and George 1997). Typically, antibodies are prepared as monoclonal or polyclonal. Polyclonal antibodies are produced by multiple clones of antibody-producing molecules. Polyclonal antibodies recognize multiple epitopes or binding sites on the surface of the antigen, making them more tolerant to the variability in antigen structures. Monoclonal antibodies, on the other hand, are those that are derived from a single clone antibody-producing molecule and thus react only to a specific epitope on the antigen (CHEMICON International 2004). Because of their high specificity, monoclonal antibodies are excellent for use in immunoassay techniques. Studies also show that results from monoclonal antibodies are highly reproducible between experiments (CHEMICON International 2004). Monoclonal antibodies, however, are much more vulnerable to epitope loss due to chemical treatment compared to polyclonal antibodies.

Biosensors that use antibodies as receptors are often referred to as immunosensors. In immunosensor development, antibody immobilization is a vital step. The immobilization method must preserve the biological activity of the antibody and enable efficient binding. These methods can be grouped into three main categories: (1) adsorption; (2) immobilization via entrapment in acrylamide membranes (Ye et al. 1997; Vikholm 2005); and (3) immobilization via glutaraldehyde (GA) and other cross-linking agents (Bhatia et al. 1989; Narang et al. 1997; Nashat et al. 1998; Slavik et al. 2002; Radke and Alocilja 2005).

Attachment of antibodies on quartz or glass can be achieved by simple adsorption; however, the immobilized proteins suffer partial denaturation, and tend to leach or wash off the surface (Bhatia et al. 1989; Huang et al. 2004; Zhou and Muthuswamy 2004). Also, this approach does not provide permanent attachment because the complex is weakly bound to the solid support by adsorption.

The selection of an effective attachment method for a particular biosensor requires a detailed study of different immobilization methods. For example, when selecting a method for piezoelectric biosensors, the approach should address: (1) the best biological material for immobilization; (2) the optimum immobilization parameters for the highest yield and reproducibility of the attached layer on the crystal; and (3) the effect of the immobilization layers on the frequency and surface characteristics of the quartz crystal (Babacan et al. 2000). The results of these individual studies can then be combined to select the best method for the particular application.

Two methods have been reported to have the best potential for use in piezoelectric flow injection analysis (FIA) systems: (1) immobilization on a precoated crystal with polyethylenimine (PEI) (Ye et al. 1997; Lin and Tsai 2003; Tsai and Lin 2005); and (2) immobilization through Protein A coupling (Boltovets et al. 2002; Su and Li 2005). Protein A is a cell wall protein, produced by most strains of *Staphylococcus aureus*. Protein A is a directed immobilization method due to its natural affinity towards the Fc region of IgG molecules. This does not block the active sites of the antibodies for analyte binding (Babacan et al. 2000). In the PEI–glutaraldehyde (GA) method, immobilization of antibodies is achieved via surface aldehyde groups of GA on a quartz crystal pre-coated with PEI. Glutaraldehyde is a homobifunctional cross-linker. In this method, the antibodies are randomly oriented and bound to the active surface (Babacan et al. 2000). The self-assembled monolayer (SAM) technique offers one of the simplest ways to provide a reproducible, ultra thin, and well-ordered layer suitable for further modification with antibodies, which has potential in improving detection sensitivity, speed, and reproducibility (Chen et al. 2003; Su and Li 2004; Vikholm 2005).

Covalent binding is the preferred method of attaching an antibody to a surface due to the strong, stable linkage that is formed. Hydroxyl groups on the biosensor platform surface provide sites for covalent attachment of organic molecules. Several investigators have modified surface hydroxyl groups to provide a functionality that would react directly with antibodies (Bhatia et al. 1989; Shriver-Lake et al. 1997; Nashat et al. 1998). Coating the biosensor surface with a silane film provides a method for modifying the reactive hydroxyl groups on the surface to attach cross-linking agents. Silanes reportedly used include 4-aminobutyldimethylmethoxysilane, 4-aminobutyltriethoxysilane, 3-mercaptopropyltrimethoxysilane, mercaptomethyldimethylethoxysilane, and 3-aminopropyl-triethoxysilane with cross-linkers such as glutaraldehyde, N-N-maleimidobutyryloxy succinimide ester, N-succinimidyl-3-(2-pyridyldithio) propionate, N-succinimidyl-(4-iodoacetyl) aminobenzoate, and succinimidyl 4-(p-aleimidophenyl) butyrate (Bhatia et al. 1989; Shriver-Lake et al. 1997). Silanes utilized for the modification of the biosensor surface (gold, silicon, glass) introduce amino groups on it, which in turn provide reaction sites for covalently bonding to glutaraldehyde. Antibodies are then immobilized through a Schiff base to the cross-linker, glutaraldehyde. Glutaraldehyde has been known to form large polymers. These reactive polymers may bind many residues and form multiprotein complexes (Bhatia et al. 1989). These effects are likely to interfere with protein function, lowering the antibody activity on the biosensor surface. To obviate the problem, Sportsman and Wilson (1980) coated glass with glycidoxypropylsilane and oxidized the silane to produce aldehyde groups reacting directly with an antibody. This method needs fewer functionalization steps and less processing time, while providing a functionalized biosensor with a high activity of immobilized antibodies.

The optimization of biomolecule immobilization, however, is a critical issue in the performance of a biosensor. Most current immobilization methodologies are not at 100% efficiency due to the limited immobilization capacity, partial loss of bioactivity of the immobilized molecules (Bunde et al. 1998), cross-reactivity (Cheung et al. 1997), nonspecific binding (analyte binding occurs at places where it should not), and interference with the transducer elements (Scheller et al. 1991). For instance, during the antibody immobilization process on the conducting polymer film, the antibody may bind at the Fc region (Figure 16.4a), on both Fc and Fb regions, or one of the antigen binding sites (Figures 16.4b and 16.4c). Lu et al. (1996) emphasize that if immobilization occurs on the antigen-binding sites, the ability of the antibody to bind the antigen may be lost completely, or at least to a high degree, thus affecting the overall performance of the biosensor.

2.3. DNA as Biological Sensing Element

A DNA probe is a segment of nucleic acid that specifically recognizes, and hybridizes (binds) to, a nucleic acid target. The recognition is dependent upon the formation of stable hydrogen bonds between the two nucleic acid strands. This contrasts with interactions of antibody-antigen complex formation where hydrophobic, ionic, and hydrogen bonds play a

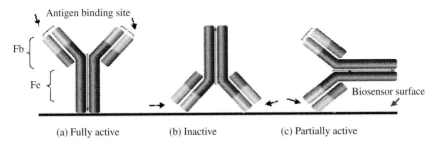

Figure 16.4. Potential antibody orientations after an immobilization process (adapted from Lu et al. 1996). Dotted arrows correspond to antigen binding sites.

role. The bonding between nucleic acids takes place at regular (nucleotide) intervals along the length of the nucleic acid duplex, whereas antibody-protein bonds occur only at a few specific sites (epitopes). The frequency of bonding is reflected in the higher association constant for a nucleic acid duplex in comparison with an antibody-protein complex, and thus indicates that highly specific and sensitive detection systems can be developed using nucleic acid probes (McGown et al. 1995; Skuridin et al. 1996). The specificity of nucleic acid probes relies on the ability of different nucleotides to form bonds only with an appropriate counterpart. Since the nucleic acid recognition layers are very stable, an important advantage of nucleic acid ligands as immobilized sensors is that they can easily be denatured to reverse binding and then regenerated simply by controlling buffer-ion concentrations (Graham et al. 1992). Most nucleic acid biosensors are based on this highly specific hybridization of complementary strands of DNA and also RNA molecules.

Besides the immobilization techniques described for antibodies, DNA probes also can be immobilized on a biosensor platform via avidin-biotin chemistry (Wang et al. 2004). Avidin is a basic glycoprotein with an isoelectric point (pI) of approximately 10, originally isolated from chicken egg white. It is also found in the tissues of birds. Biotin is a naturally occurring vitamin found in every living cell. The tissues with the highest amounts of biotin are the liver, kidney, and pancreas. Yeast and milk are also high in biotin content. Cancerous tumors have more biotin than normal tissue (Savage et al. 1994). The most commonly used method in surface plasmon resonance (SPR)-based optical sensing is the immobilization of biotinylated DNA probes or antibodies onto a layer of streptavidin. Streptavidin is a biotin-binding protein isolated from a culture broth of *Streptomyces avidinii*. Streptavidin binds four moles of biotin per mole of protein, and has a pI of approximately 5–6 (Savage et al. 1994). Streptavidin is covalently linked to carboxylated dextran fixed onto the gold surface of the SPR chip on a self-assembled monolayer of 11-mercaptoundecanol (Bianchi et al. 1997; Jordan et al. 1997; Silin and Plant 1997; Tombelli et al. 2000; Feriotto et al. 2002; Gao et al. 2004; Ruan et al. 2004). This method is very efficient in terms of sensitivity, selectivity, and stability of the realized sensor chip for DNA hybridization detection (Silin and Plant 1997; Kukanskis et al. 1999; Mariotti et al. 2002), but remains a time consuming procedure (5 days) if one starts from the gold surface.

2.4. DNA-Based Biosensors

The detection of bacteria and other pathogens in food, drinking water, and air, based on their nucleic acid sequences, has been explored using various detection systems. Biosensors have been developed to detect DNA hybridization at subpicomolar to micromolar levels using gravimetric detection systems (Sung Hoon et al. 2001). A quartz crystal nanobalance system could detect DNA hybridization at 0.3 nanogram levels using frequency shift nanogravimetric measurement (Nicolini et al. 1997). Tombelli et al. (2000) developed a DNA piezoelectric biosensor for the detection of bacterial toxicity based on the detection of a PCR-amplified *aer* gene of *Aeromonas hydrophila*. The biosensor was applied to vegetables, environmental water, and human specimens (Tombelli et al. 2000). The biosensor was able to successfully distinguish between samples containing the pathogen and those not contaminated. Zhao et al. (2001) developed a QCM biosensor using 50 nm gold nanoparticles as the amplification probe for DNA detection in the order of 10 fM of target, higher than ever before reported using the same method. The high sensitivity was explained by the weight of the larger particles, and the larger area occupied by those particles that needed less target DNA for their binding (Zhao et al. 2001). Another QCM biosensor applied to the detection of *E. coli* in water in combination with PCR amplification (of the *lac* gene) was able to detect a 10 fg of genomic *E. coli* DNA (few viable *E. coli* cells in 100 ml of water) (Mo et al. 2002). When used for detection of *Hepatitis B virus*, Zhou et al. (2002) observed that the QCM could detect frequency shifts of DNA hybridization as a linear relationship in the range 0.02–0.14 μg/ml with a detection limit of 0.1 μg/ml, similar to the QCM biosensor developed by He and Liu (2004) for *Pseudomonas aeruginosa*. While QCM biosensors have been demonstrated to successfully detect bacterial and viral targets, they rely on PCR amplification of the target, in addition to requiring 3–12 h for DNA hybridization and detection.

Optical biosensor systems developed for DNA detection exhibit higher sensitivity compared to the QCM biosensors. An automated optical biosensor system based on fluorescence excitation and detection in the evanescent field of a quartz fiber was used to detect 16-mer oligonucleotides in DNA hybridization assays. The detection limit for the hybridization with a complementary fluorescein-labeled oligonucleotide was 2×10^{-13} M (Abel et al. 1996). Another optical fiber evanescent wave DNA biosensor used a molecular beacon (MB) DNA probe that became fluorescent upon hybridization with target DNA. The detection limit of the evanescent wave biosensor with synthesized complementary DNA was 1.1 nM. Testing with environmental samples was not performed (Liu and Tan 1999). Liu et al. later developed MB-DNA biosensors with micrometer to submicrometer sizes for DNA/RNA analysis. The MB-DNA biosensor was highly selective with single base-pair mismatch identification capability, and could detect 0.3 nM and 10 nM of rat gamma-actin mRNA with a $105 - \mu$m biosensor and a submicrometer biosensor, respectively (Liu et al. 2000b).

Optical biosensors targeting RNA as the analyte offer an added advantage over traditional DNA-based detection methods, i.e., viable cell detection. Baeumner et al. (2003) detected as few as 40 *E. coli* cells/ml in samples using a simple optical dipstick-type biosensor coupled to nucleic acid sequence-based amplification (NASBA), emphasizing the fact that only viable cells were detected, and no false positive signals were obtained from dead cells present in the sample. The detection of viable cells is important in respect to safety, as well as food and environmental sample sterilization assessments. Similarly, a biosensor for the protozoan parasite *Cryptosporidium parvum* was developed (Esch et al. 2001). Hartley and Baeumner (2003) developed a simple membrane strip-based biosensor for the detection of viable *Bacillus anthracis* spores. The study combined the optical detection process with a spore germination procedure as well as a nucleic acid amplification reaction to identify as little as one viable *B. anthracis* spore in 12 h. A quantitative universal biosensor was developed on the basis of olignucleotide sandwich hybridization for the rapid (30 min total assay time) and highly sensitive (1 nM) detection of specific nucleic acid sequences (Baeumner et al. 2004). The biosensor consisted of a universal (polyethersulfone) membrane, a universal dye-entrapping liposomal nanovesicle, and two oligonucleotides—a reporter and a capture probe that could hybridize specifically with the target nucleic acid sequence. Limits of detection of 1 nM per assay and dynamic ranges of 1–750 nM were obtained. While the RNA-based biosensor can be an excellent tool for the detection of viable bacterial cells, the inherent disadvantages to the technique include the short life span of the mRNA target, and a high susceptibility to contaminants and inhibitors from environmental and food samples.

Other biosensors targeting DNA that have been developed include MEMS-based amperometric (Gau et al. 2001) and high throughput PCR biosensors (Nagai et al. 2001); carbon nanotube-based field effect transistor biosensors (Maehashi et al. 2004); microcantilever-based cyclic voltammetry biosensors (Zhang and Li 2005); and pulsed amperometry- (Ramanaviciene and Ramanavicius 2004), capacitance- (Berney et al. 2000; Lee et al. 2002), and absorbance-based biosensors (Mir and Katakis 2005).

2.5. Antibody-Based Biosensors

Antibody-based biosensors or immunosensors have also been developed for pathogen analysis, but only some of these biosensors exhibit excellent detection limits with relatively short analysis times. A quartz crystal microbalance (QCM)-based biosensor was used to detect *Salmonella* species in milk samples with detection limits around 10^6 CFU/ml (Park et al. 2000). *Listeria* and *Salmonella* spp. were detected with a similar detection limit by a surface plasmon resonance (SPR) biosensor (Koubova et al. 2001). QCM and SPR biosensors for different bacterial targets, such as *Pseudomonas aeruginosa*, *Bacillus cereus*, and *E. coli* O157:H7, were later developed by researchers (Vaughan et al. 2003; Kim et al. 2004; Su and Li 2004; Su and Li 2005), and showed no significant improvement in detection limits. In a dipstick-type assay, Park and Durst (2000) were able to detect *E. coli* O157:H7 in food matrices at a low detection limit of about 10^3 CFU/ml without any enrichment required.

An impedance biosensor chip for detection of *E. coli* O157:H7 was developed based on the surface immobilization of affinity-purified antibodies onto indium tin oxide (ITO) electrode chips, with a detection limit of 6×10^3 cells/ml (Ruan et al. 2002). Shah et al. (2003) developed an amperometric immunosensor with a graphite-coated nylon membrane serving as a support for antibody immobilization and as a working electrode. This approach was used for detection of *E. coli* with a low detection limit of 40 CFU/ml. An antibody-based conductometric biosensor developed by Muhammad-Tahir and Alocilja also showed a similar detection limit of 50 CFU/ml for bacteria and 10^3 cell culture infective dose (CCID)/ml for viruses within 10 min (Muhammad-Tahir and Alocilja 2003a; Muhammad-Tahir and Alocilja 2003b; Muhammad-Tahir et al. 2005b; Muhammad-Tahir et al. 2005a).

An antibody-based fiber-optic biosensor, to detect low levels of *L. monocytogenes* cells following an enrichment step, was developed using a cyanine 5-labeled antibody to generate a specific fluorescent signal. The sensitivity threshold was about 4.3×10^3 CFU/ml for a pure culture of *L. monocytogenes* grown at 37°C. Results could only be obtained after 2.5 h of sample processing. In less than 24 h, this method could detect *L. monocytogenes* in hot dog or bologna naturally contaminated or artificially inoculated with 10^1–10^3 CFU/g after enrichment in buffered Listeria enrichment broth (Geng et al. 2004). In another study, a microcapillary flow injection liposome immunoanalysis system (mFILIA) was developed for the detection of heat-killed *E. coli* O157:H7. Liposomes tagged with anti-*E. coli* O157:H7 and an encapsulating fluorescent dye were used to generate fluorescence signals measured by a fluorometer. The mFILIA system successfully detected as few as 360 cells/ml with a total assay time of 45 min (Ho et al. 2004).

The ability to detect small amounts of materials, especially bacterial organisms, was demonstrated using microelectromechanical systems (MEMS) for the qualitative detection of specific *Salmonella enterica* strains with a functionalized silicon nitride microcantilever. Detection was achieved due to a change in the stress on the cantilever surface in situ upon binding of a small number of bacteria with less than 25 adsorbed bacteria required for detection (Weeks et al. 2003). A MEMS fabricated high-density microelectrode array biosensor, developed for the detection of *E. coli* O157:H7 using a change in impedance caused by the bacteria measured over a frequency range of 100 Hz–10 MHz, was able to detect and discriminate 10^4–10^7 CFU/ml (Radke and Alocilja 2005).

The potential use of immunosensors is due to their general applicability, the specificity and selectivity of the antigen-antibody reaction, and the high sensitivity of the method, depending on the detection method used. The antigen-antibody complex may be utilized in all types of sensors. The physicochemical change induced by antigen-antibody binding does not generate an electrochemically detectable signal. Therefore, enzymes, fluorescent compounds, electrochemically active substrates, or avidin-biotin complexes are used to label either the antigen or the antibody to detect the biological recognition event. Immunosensors cannot be employed to specifically detect viable cells. The antibodies used are selective to the epitope on the antigen. If the epitope is present on a living or dead microorganism, the antibody will capture the antigen and register a positive signal. In the case of bacterial foodborne pathogens that must be ingested by the host to cause disease, a positive result from nonviable cells may raise a false alarm.

2.6. Biosensor Transducing Element: Conducting Polymer

Conducting polymers have become one material of choice in the recent advancement of biosensor technologies. Historically, the research in conducting polymers began in the 1960s when a Japanese scholar, Hideki Shirakawa, discovered a silvery film of polyacetylene by accidentally adding a thousandfold extra catalyst to a reaction vessel (Chiang et al. 1977). In another part of the world, MacDiarmid and Heeger were experimenting with a metallic-like film of the inorganic polymer sulphur nitride (Akhtar et al. 1977). The following collaboration between MacDiarmid, Shirakawa, and Alan Heeger led to the historic discovery of polyacetylene and they became the recipients of the Nobel Prize in Chemistry in 2000. This finding generated new interest in the scientific community leading to the discovery of new conducting polymer compounds.

Polyparaphenylene was first discovered by Ivory and his coworkers (Ivory et al. 1979). The discovery of polyparaphenylene is particularly interesting because of its processable properties, which open the door for commercially viable conducting plastics (Rabolt et al. 1980). Polypyrrole also is extensively investigated for its excellent conductive properties (Kanazawa et al. 1979). Polyaniline (Pani) is probably the most rapidly growing class of conducting polymers. Its interest lies in the fact that this polymer family can be doped by a variety of different dopants, either by chemical or electrochemical syntheses (MacDiarmid and Epstein 1990). Many other conducting polymer compounds, such as poly(3,4-ethylenedioxythiophene), polyfuran, polycarbazole, and polyindole have also been synthesized and extensively studied (Lagowski et al. 1998; Ivanov et al. 2001).

In becoming electrically conductive, a polymer has to imitate a metal, that is, its electrons are free to move. Conducting polymers have π electrons per carbon atom in their backbones, and these unpaired electrons are responsible for the unusual electronic properties of conducting polymers (Ivory et al. 1979; Macdiarmid et al. 1987; MacDiarmid and Epstein 1990). This extended π-conjugated system of conducting polymers has single and double bonds alternating along the polymer chain, e.g., the structure of polyaniline (Figure 16.5). These conjugated double bonds act collectively, "knowing" that the next nearest bond is also a double bond (Kroschwitz 1988).

Conjugation alone, however, is not sufficient to make the polymer material conductive. In addition to the conjugated double bond, charge carriers in the form of extra electrons or "holes" (a hole is an empty position where an electron is missing) have to be introduced into the material. When a hole is filled by an electron jumping in from a neighboring position, a new hole

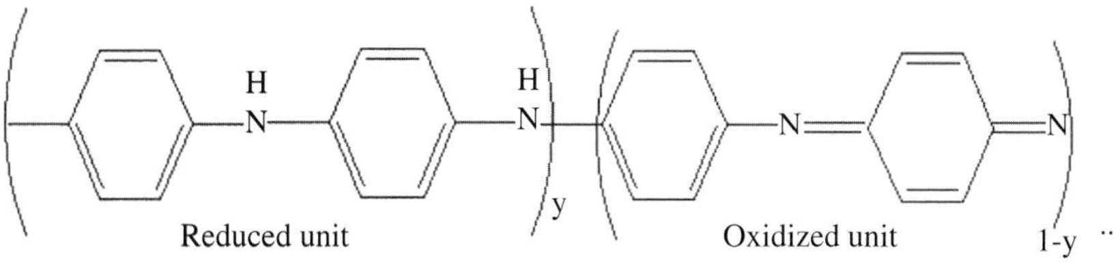

Figure 16.5. Polyaniline structure (adapted from Macdiarmid et al. 1987; Yen Wei 1989; Genies et al. 1990).

is created, which allows a charge transport (Winokur 1998). Photoexcitation, charge injection, and/or doping are employed to introduce the charge carriers necessary for electron transport (Winokur 1998). Of the three aforementioned mechanisms, only doping yields a permanent transition to the conductive state. By definition, doping is "an intentional introduction of a selected chemical impurity (dopant) into the crystal structure of a semiconductor to modify its electrical properties" (Streetman 1995). Doping, either by the addition of electrons (reduction reaction) or the removal of electrons (oxidation reaction) from the polymer is carried out via chemical or electrochemical methods (Salaneck and Lundstrom 1987). The conductivity of the polymer increases as the doping level increases (Winokur 1998).

Conducting polymer compounds are used extensively in the biosensor technology to enhance stability, speed, and sensitivity of detection (Deshpande and Hall 1990; Contractor et al. 1994; Guiseppi-Elie 1998; Castillo-Ortega et al. 2002; Gerard et al. 2002). Biomolecules are added, fabricated, or conjugated onto the conducting polymers by way of physical adsorption, covalent attachment based on ethyl-dimethylaminopropylcarbodiimide and N-hydroxy-succinimide coupling chemistry, electrostatic attachment, or electrochemical entrapment (Sadik and Emon 1996; Liu et al. 2000a, 2003; Tang et al. 2004). The existence of various methods of incorporating biomolecules into conducting polymer films permits the localization of biologically active materials on an electrode of any size and geometry (Unwin and Bard 1992).

Generally, conducting polymers act as transducers in a biosensor system to translate biochemical reactions into quantifying signals. Biosensors make use of a variety of transducers, such as electrical, electrochemical, optical, piezoelectric crystal, thermal, and acoustic wave (Liu et al. 2003; Muhammad-Tahir and Alocilja 2004; Ko and Grant 2006; Komarova et al. 2005; Radke and Alocilja 2005; Wilson et al. 2005; Lange et al. 2006). A suitable transducing system is adapted into a biosensor design depending on the nature of the biological/biochemical interaction between the biological element immobilized on the biosensor and the target of detection. The most common transducer being utilized in conducting polymer-biosensors is based on the electrochemical system. The electrochemical technique is concerned with the interplay between electricity and chemistry, such as the measurement of current, potential, or charge, and their relationship to chemical properties. The fundamental process in the electrochemical method is the transfer of electrons between the electrode surface and the molecules in the solution (electrolyte) adjacent to the electrode (Wang 2000).

An electrochemical-based biosensor may also be known as a conductometric, ampero-metric, impedimetric, or potentiometric biosensor, depending on the type of electrical properties being evaluated. Conductometric biosensors measure the change in conductance of the biological complex situated between electrodes (Gerard et al. 2002). Conductometric biosensors based on conducting polymer compounds have been developed for penicillin (Nishizawa et al. 1992), glucose, urea (Castillo-Ortega et al. 2002), lipids, hemoglobin (Contractor et al. 1994), bacteria, and viruses (Kim et al. 2000; Muhammad-Tahir et al. 2005b). Amperometric biosensors measure the current or charge produced between the electrode and the electrolyte by applying a constant potential value. The most important factor affecting the functioning of amperometric biosensors is the electron transfer between the catalytic molecule and the electrode surface, which most often involves a mediator or conducting polymer (Gerard et al. 2002; Minett et al. 2002, 2003). Various studies have been done in recent years on the application of electrochemically grown conducting polymer layers in amperometric biosensors. Enzymes or antibodies are either entrapped within conducting polymer layers or covalently bound to functional groups (Hammerle et al. 1992; Bartlett et al. 1996; Rodriguez and Alocilja 2005).

Few studies have been done on potentiometric biosensors with enzymes or antibodies immobilized within the polymer matrix (Gerard et al. 2002). This type of biosensor normally utilizes the pH sensitivity feature of conducting polymer compounds (Ratcliffe 1990) instead of the electroactive property of the polymer. For example, polypyrrole's sensitivity to NH_3 was used to produce a potentiometric biosensor for urea detection (Pandey and Mishra 1988; Trojanowicz and Krawczyski 1995).

Impedimetric biosensors, also known as capacitive biosensors, measure the changes in the capacitance layer between electrodes upon binding of a biological element to its receptor (Jian-Guo et al. 2004). Such biosensors are used for the detection of whole bacterial cells (Radke and Alocilja 2005), enzymes (Myler et al. 2005), viruses, and oligonucleotides (Davis et al. 2005). For whole cell detection, it has been demonstrated that an impedimetric biosensor based on a microelectrode array silicon chip can detect as few as 10^4 colony-forming units/ml of *Escherichia coli* O157:H7 grown in nutrient broth (Radke and Alocilja 2005).

2.6.1. Polyaniline

This section will focus mainly on Pani and Pani composite-based biosensors. Polyaniline is chosen as a transducer in biosensors because of its excellent stability in liquid form, promising electronic properties (Syed and Dinesan 1991), and strong biomolecular interactions (Lu et al. 1995; Ashok Mulchandani 1996; Imisides 1996; Gau et al. 2005). Also, Pani structure can be modified to attach to protein molecules, such as antibodies, by binding to its polymer backbone (Situmorang et al. 2000; Grennan et al. 2003). Additionally, Pani has the ability to efficiently transfer the electric charges produced by biochemical reactions to electronic circuits (De Taxis du Poet et al. 1990). Polyaniline also acts as an enzyme amplifier and conductivity modulating agent to provide signal amplification in the recognition process (Sergeeva 1996; Kim et al. 2000; Grennan et al. 2003; Morrin et al. 2003). The use of Pani as an enzyme switch, which yields "on" and "off" responses, was demonstrated by Iribe and Suzuki (2002). Additionally, the biological sensing elements entrapped within the polymer matrix have been shown to maintain their biological activity (Sadik and Emon 1996). Additionally, the concept of "electron wiring" was reported for the proper relay of electrons from the surface of an electrode to the biologically active sites (Alt et al. 2002).

Polyaniline is extensively researched for its electrical, optical, chemical, and electrochemical properties because of its simple synthesis method, its stability in air, and its myriad range of applications (Winokur 1998). Additionally, Pani is the best known semiflexible rod conducting polymer system (Genies et al. 1990). The chemical and structural flexibility surrounding the amine nitrogen linkages in Pani creates enormous diversity in its properties. Different forms of Pani exist based on the oxidation levels. Polyaniline is believed to be composed of the basic chemical units shown in Figure 16.5 (Macdiarmid et al. 1987; Yen Wei 1989; Genies et al. 1990).

When $0 < y < 1$, they are called poly (paraphenyleneamineimines) in which the oxidation state of the polymer increases continuously with decreasing value of y. The fully reduced form, also known as leucoemeraldine, has a y value equal to 1. The most oxidized form (y equal to 0) is called pernigraniline. Finally, the intermediate form, with y equal to 0.5, is called emeraldine. The terms leucoemeraldine, emeraldine, and pernigraniline refer to the different oxidation states of Pani (Macdiarmid et al. 1987). Between the three types of Pani, emeraldine is the most studied Pani compound (Sergeeva 1996). Polyaniline in the form of conductive emeraldine is synthesized via chemical oxidation by doping aniline in acidic media (e.g., 1 M hydrochloric acid). The doping mechanism is reported to increase the conductivity level of Pani from 10^{-10} S/cm to 1 S/cm (Macdiarmid et al. 1987).

2.6.2. Self-doped Polyaniline

The electrical property of Pani is pH-dependent with most studies conducted at a pH level lower than 4.0 (Shaolin and Jincui 1999). In a biosensor design, however, nearly neutral pH solution (pH 6–8) is used, since most biocatalyst and immunological reactions occur optimally at neutral pH. Thus it is very challenging to incorporate biological elements with the conventional pH-dependent Pani that requires an acidic environment to maintain its conductive property. Recently, significant progress has been made to improve the chemical and physical properties of Pani. One method has been to introduce sulfoacid residues to the emeraldine base to give rise to the so-called "self-doped" Pani. In contrast to the Pani discussed in the previous section, self-doped Pani has a negatively charged functional group bound to the polymer backbone to act as an inner anion dopant. Therefore, no anion or electron exchange between the polymer and the surrounding solution is required during the oxidation and reduction process (Lukachova et al. 2003). Such Pani has a conductivity of approximately 1 S/cm within a pH range of up to 7 (Lukachova et al. 2003).

A self-doped Pani is prepared by chemical or electrochemical synthesis. A distinct route for the preparation of self-doped Pani is by chemical modification of the emeraldine base with the sulfoacidic group, such as camphorsulfonic acid (Winokur et al. 2001). The discovery of polymer doped with camphorsulfonic acid attracts considerable attention for both fundamental research and technological applications such as biosensor development (Snejdarkova et al. 2004).

2.6.3. Carbon Nanotubes

Carbon nanotubes (CNT) represent an exciting nanomaterial with a wide range of applications in biotechnology and biomedical areas because of their excellent structural and electrical properties (Dresselhaus et al. 2001). Like Pani and other conducting polymer compounds, the molecular structure of CNT also consists of alternating single and double bonds; it therefore exhibits similar characteristics as conducting polymers (Rao 2001). These nanomaterial composites are made of graphitic carbon cylinders with one or several concentric tubules (Rao 2001). Carbon nanotubes play a role as catalysts to promote electron transfer in the electrochemical processes for many biological substances such as antibodies (Takeda et al. 2005; Yun et al. 2006), insulin (Wang and Musameh 2004), cytochrome c (Wang et al. 2002), etc. It is also reported that CNTs are made water-soluble via esterification of nanotubes with carboxylic acid (Varfolomeyev et al. 2002). Biological elements, such as bovine serum albumin, coupled with this water soluble CNT are found to remain biologically active.

The fabrication of carbon nanotubes (CNT) and Pani composites have received great interest, especially in the construction of biosensors. The incorporation of CNT has been shown to enhance the electroactive and structural properties of the conducting polymers and possibly the performance of the biosensor as well. For instance, a study by Mottaghitalab et al. (2005) showed that Pani fibers imbedded with CNT exhibited higher mechanical strength and electrical properties than conductive polymers alone (Mottaghitalab et al. 2005). It also has been reported that enzymes and antibodies may be absorbed on the surface of the CNT particles (Davis et al. 1998; Wang and Musameh 2005; Luo et al. 2006). For example, Luo et al. reported that CNTs can enhance the stability of their Pani based biosensor by effectively adsorbing the enzymes present in the reaction and also by increasing the electron transfer between the electrode and the enzyme (Luo et al. 2006). Additionally, Liu et al. discovered a way to shift the conductive property of Pani to a neutral pH by constructing Pani/CNT layers through layer-by-layer assembly. The resulting multilayered film was reported to be very stable and capable of detecting nicotinamide adenine dinucleotide at a very low potential (Liu et al. 2005). Ramanathan et al. also have fabricated a bioaffinity sensor using biologically functionalized conducting polymer nanowires with the sensitivity to detect a single molecule (Ramanathan et al. 2004, 2005).

2.7. Conducting Polymer-Based Biosensor for Microbial/Viral Detection

Microbial detection-based biosensors provide advantages over other types of biosensors especially in overcoming the tedious and costly process of DNA, RNA, and enzyme purification. Although there has been minimal study of the direct detection of bacteria or whole cells using conducting polymers, the electrochemical method of detection is the most common transducing system mentioned (Minett et al. 2002, 2003). Minett et al. reported that a toluidine blue-labeled antibody immobilized on a conducting polymer electrode can be used to reproducibly and selectively detect *Listeria monocytogenes* at a level of 10^5 cells per ml in 30 minutes. The method unfortunately was not able to detect other microorganisms, such as *E. coli* and *Salmonella* (Minett et al. 2002, 2003). Liu et al. also reported that conducting polymer-antibody complexes immobilized on a quartz crystal microbalance was successful in detecting pseudorabies virus (Liu et al. 2003). Because of the limited number of studies on direct microbial detection to date, the next sections will illustrate the fabrication and the performance of two different Pani-based biosensors for foodborne pathogen detection.

3. Illustration: Biosensor Using Self-doped and Non-self-doped Pani

The design, fabrication, performance, and illustrations presented in this section (section 3) are excerpted from Muhammad-Tahir 2003a,b; 2004; and Muhammad-Tahir et al. 2005a,b; 2007.

3.1. Pani Preparation

Self-doped Pani was prepared by doping emaraldine base with camphorsulfonic acid (Cao et al. 1993; Duic et al. 1994; Cao et al. 1995). A commercially available emaraldine base compound was doped with camphorsulfonic acid in a ratio of 1.0 p−phenyleneimine (C_6H_4N) unit per 0.5 mol doping acid. Emeraldine and camphorsulfonic acid were mixed and dissolved in chloroform to give a final concentration of 0.5 % by weight. To improve the processability of Pani, phenol was added in a ratio of 0.5 mol per p-phenyleneimine.

Non-self-doped Pani compounds were synthesized by polymerizing a polyaniline monomer, aniline, with different doping acids, such as phenylphosphonic acid, 4-hydroxy-benzenesulphonic acid, sulfobenzoic acid, hydrochloride acid, and percholoric acid, based on a standard procedure of oxidative polymerization in the presence of ammonium persulfate (Sergeeva 1996; Laska and Widlarz 2003a, 2003b).

3.2. Pani Characterization

3.2.1. Conductivity Measurement

The conductivity meter by Oakton (Vernon Hills, IL) was used to determine the conductivity of the polymer samples in 0.1 M phosphate buffer (pH 6.2–7). The dried Pani, on the other hand, was compressed into pellets and used to measure the conductivity of the polymer by using a four-point probe (Signatone model S-301, CA).

3.2.2. Biosensor Fabrication

This section describes the fabrication of an indium tin oxide (ITO)/Pani amperometric biosensor and a lateral flow-based conductometric biosensor. (Muhammad-Tahir 2003a,b; 2004; Muhammad-Tahir et al. 2005a,b; 2007.) Both biosensors were used to detect the presence of microbial/viral pathogens of concern to the food supply: BVDV, E. coli O157:H7, and Salmonella spp. Polyclonal and/or monoclonal antibodies were used as biological sensing elements in both of the biosensor designs.

3.2.3. Indium Tin Oxide/Pani Biosensor

Indium tin oxide glass ($1.27\,cm \times 2.54\,cm$) was prepared (Muhammad-Tahir et al. 2007) and used as the biosensor platform. Polyaniline solution was used to spin-coat a thin layer of polymer on the ITO platform at 500 rpm for 6 seconds. The ITO/Pani substrate was then ready to be functionalized with antibodies. Antibody functionalization was conducted using Liu's method (Liu et al. 2000a). Using the three-electrode cell and the potentiostat, a negative potential (-0.5 volts) was applied to the ITO/Pani substrate for 25 minutes. Forty-five ml of phosphate buffer (PB) was used as the electrolyte. After applying the potential for 25 minutes, the substrate was removed from the cell and then immersed in 1 ml of solution containing the antibody, 1% glutaraldehyde, and PB in a volume ratio of 2.5:0.5:1. The substrate was left in the antibody solution at room temperature for 30 minutes (Liu et al. 2000a).

The presence of antibody on the ITO/Pani substrate was confirmed by repeating the above immobilization procedures using fluorescein isothiocyanate-labeled antibodies. A confocal florescence microscope was used to visualize, measure the intensity of the emitted fluorescence, and confirm the antibody immobilization.

3.2.4. Lateral Flow Conductometric Biosensor

The conductometric biosensor consisted of two parts: an immunosensor and an electronic data collection system. The immunosensor was comprised of four one-time use, disposable membranes: sample application, conjugate, capture, and absorption membranes (Figure 16.6). The fabrication of each membrane is described as follows:

Conjugate membrane: A mixture of the monoclonal antibody ($150\mu g/ml$) and Pani ($0.1\,g/ml$) was left to react for 90 minutes at $21°C$. After inactivation of the nonreacted aldehyde group, the conjugate was then precipitated by centrifugation using a blocking reagent. The conjugated antibody was diluted in 0.1 M phosphate buffer (pH 7.2). The conjugate membrane was then immersed in the final product and left to air dry.

Capture membrane: Polyclonal antibodies were immobilized on the capture membrane by the following steps. First, the membrane was saturated in methanol in water for 45 minutes and left to dry. The membrane was then treated in glutaraldehyde for one hour. After drying, the polyclonal antibodies were pipetted on the membrane and incubated at $37°C$ for one hour. Inactivation of residual functional groups and blocking were carried out simultaneously by incubating the membrane with the blocking reagent. The membrane was then allowed to air dry. Silver electrodes were manually drawn on the capture membrane to electrically connect the immunosensor to the electronic data acquisition system, consisting of a copper wafer and an ohmmeter linked to a computer.

3.2.5. Signal Measurement

For the ITO/Pani biosensor, the amperometric measurement was conducted by first immersing the functionalized ITO/Pani substrate in 45 ml of phosphate buffer. Using the

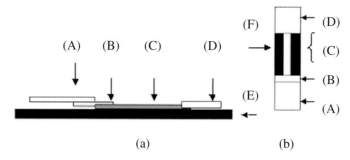

Figure 16.6. Schematic of the immunosensor: (a) cross section view and (b) top view; (A) sample application membrane (20 mm X 5 mm X 1 mm); (B) conjugate membrane for polyaniline-labeled antibody absorption (10 mm X 5 mm X 1 mm); (C) capture membrane (22 mm X 5 mm X 1 mm) coated with (F) silver electrodes on both sides; (D) absorption membrane (17 mm X 5 mm X 1 mm); (E) copper wafer platform (75 mm X 25 mm X 2 mm) (Reprinted from Muhammad-Tahir and Alocilja (2005b) IEEE SENSORS JOURNAL. (©2005 IEEE)).

three-electrode electrochemical cell setup, a fixed potential of 0.5 volts (V) was applied over 25 minutes. The theoretical value of the faradaic current between the working electrode and the electrolyte was obtained by performing a linear regression analysis on the charge-versus-time experimental data. Detection process was conducted by 1) calculating the amperometric response after antibody immobilization, 2) incubating the biosensor from (1) with 1 ml of microbial/viral culture for 30 minutes and 3) calculating the amperometric response of the biosensor from (2).

On the other hand, the signal of a conductometric biosensor was measured by connecting the biosensor to a BK multimeter Model AK-2880A (Worcester, MA) with the RS-232 interface and BK software. A resistance measurement was taken before and after sample application. After the addition of 0.1 ml of sample on the sample application membrane, the generated signal fluctuated for the first minute, while the sample flowed to the absorption membrane. The dispersion time from sample membrane to absorption membrane was about one minute. Measurements were taken three times at two-minute intervals. Three replications were performed for each experiment. The biosensor was calibrated with the blank. For data analysis, the measured signal (R^a) was normalized by calculating the drop in resistance between the resistance output of the sample (R^s) and that of the blank (R^b), that is, $R^a = R^s - R^b$. The resistance drop (normalized signal) was plotted against varying viral concentrations.

3.3. Properties of Pani

Sergeyeva et al. reported that the conductivity of the non-self-doped Pani (Pani chemically polymerized with hydrochloric acid) linearly increased with increasing molecular weight from 10 KDa up to 45 KDa. The conductivity of 100 KDa of the same Pani, however, was reported to be lower than that of 45 KDa (Sergeyeva et al. 1996). Following this result, three commercial emeraldine base compounds were self-doped with camphorsulfonic acid: two with molecular weights larger than 45 KDa (65 and 50 KDa) and one lower than 45 KDa (20KDa). Table 16.2 shows the conductivity of the self- and non-self-doped Pani compounds using a four-point probe meter. Among the self-doped Pani compounds, the Pani with the 65 KDa molecular weight had the highest conductivity at 1.5 Siemens (S)/cm, followed by the 50 (0.44 S/cm) and the 20 (0.36 S/cm) KDa Pani compounds. The conductivities of these self-doped Pani are within the range of conductivity levels found in a previous study (Lukachova et al. 2003) and are also reported to increase with increasing molecular weight.

Interestingly, the commercially available non-self-doped Pani has the highest conductivity level (6.7 S/cm) among all the tested Pani compounds, even though its molecular weight

Table 16.2. Conductively of the various types of Polyaniline and doping agents

Pani Type	Conductivity (S/cm)
Self doped	
Self doped-20KDa	0.36
Self doped-50KDa	0.44
Self doped-65KDa	1.5
Commercial non-self doped (15 KDa)	6.7
Synthesized non-self doped with:	
HBSA	10^{-2}
PPA	10^{-3}
SBA	10^{-2}
PA	10^{-3}
HCL	10^{-3}

($\sim$15 KDa) is smaller than the self-doped compounds (Table 16.2). Table 16.2 also shows the conductivity levels of the synthesized non-self-doped Pani in different doping acid. All samples have a conductivity level in the range of 10^{-2}–10^{-3} S/cm. The difference in conductivity levels found in Table 16.2 may be due to the different doping acids used in the Pani polymerization process (Duic et al. 1994; Stejskal et al. 1998). This proposition is further supported by the subsequent transmission electron microscope (TEM) analysis. Since the molecular weight of these synthesized non-self-doped Pani compounds was not investigated, a correlation between the polymer size and its conductivity level could not be concluded.

A transmission electron microscope (TEM) was used to study the morphology of Pani compounds (Figure 16.7). The TEM images show that the higher the molecular weight of the self-doped Pani, the larger is the polymer structure. For instance, the 65-KDa self-doped Pani is approximately 12μm in length (Figure 16.7c), while the 20-KDa Pani is only 2μm Figure 16.7a). Additionally, Figure 16.7 also shows that the self-doped Pani compounds have more globular shapes and are smaller than the commercial non-self-doped Pani. The commercial non-self-doped Pani is shown to have a nonuniform structure with an approximate length of 100μm.

Figure 16.8 shows the effect of protonating acids on the Pani structures. Protonation of the non-self-doped Pani with SBA is shown to produce the longest structure of Pani. Polymerization in the presence of PA, in contrast, is shown to produce a short and thick rod, consistent with the finding by Abe et al. (1989). In general, the non-self-doped Pani compounds synthesized with the selected doping acids were shown to have a rod-shaped structure. This difference in shapes and sizes among the Pani compounds may be due to the different doping acids used in the Pani polymerization process. Duic et al. and Stejskal et al. also observed differences in the size and shape of Pani when using different types of doping acids in their polymerization processes (Duic et al. 1994; Stejskal et al. 1998).

Thermogravimetric (TG) analysis was conducted to study the effect of temperature on the self-doped Pani weight. Figure 16.9 shows the changes in the weight loss of the self-doped Pani (65 KDa) after exposing the compound to temperatures ranging from 22°C to 300°C. A similar trend was also observed when testing the 50 and 20 KDa of Pani compounds. Figure 16.9 shows a steep rate of weight loss in regions A (22°C to 75°C) and B (210°C to 300°C), and a slightly increased rate of change in region C (75°C to 210°C). This finding shows that it is optimal to

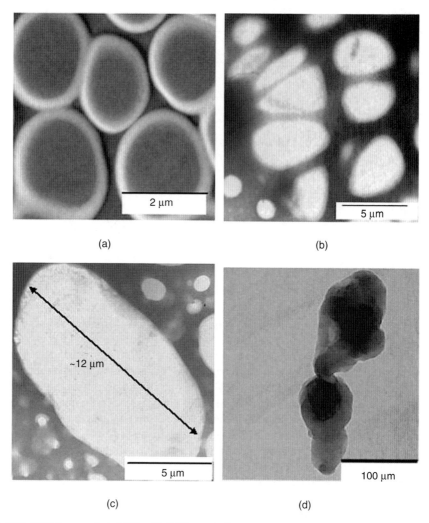

Figure 16.7. TEM images of a) 20 KDa, b) 50 KDa, and c) 65 KDa self doped Pani; and d) commercial non-self doped Pani (Reprinted from Muhammad-Tahir et al. (2007), with permission of MDPI publishing).

use the Pani between the temperature levels of 75°C and 210°C (region C), since a temperature fluctuation within this region leads to only a small change in Pani weight loss (Figure 16.9). Since it was concluded earlier that the weight of Pani affects the conductive property of the polymer, the use of Pani in region C ensures minimal changes in the polymer conductive property. Temperature levels at this region, however, are too high for any biological elements in a biosensor design, such as antibodies, to be functional. An antibody thermal stability study concluded that a heat treatment at 60°C resulted in the cleavage of the antibody heavy and light chains and promoted the denaturation of the protein (Alexander and Hughes 1995). For this reason, most biosensor operations are conducted at room temperature (~ 25°C) (Radke and Alocilja 2005; Rodriguez and Alocilja 2005). Since the Pani weight is sensitive to temperature changes between 22°C and 75°C (region A), a temperature-controlled mechanism needs to be introduced to the biosensor design to minimize temperature-dependent variations in the polymer properties. Though the TGA analysis of the non-self-doped Pani compounds has not been studied to date, a similar pattern of response is speculated there as well.

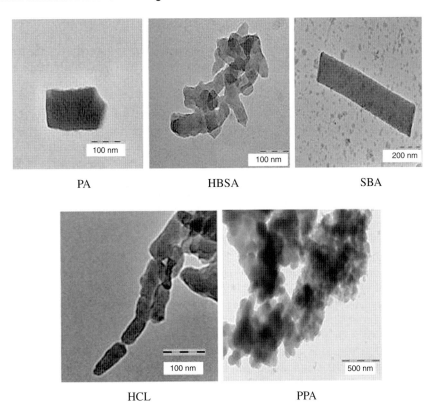

Figure 16.8. TEM images of the synthesized non-self doped Pani with their associated protonating acids (reprinted from Muhammad-Tahir et al. (2005a), with permission of Elsevier).

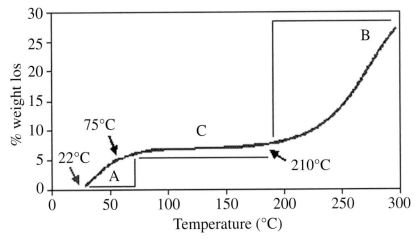

Figure 16.9. Percent of weight loss of self doped Pani in varying temperature levels (Reprinted from Muhammad-Tahir et al. (2007), with permission of MDPI publishing).

3.4. Detection Concept of the Biosensor

The ITO/Pani biosensor uses a direct antibody-antigen binding format with the self-doped Pani as the transducer. The authors also have found that the Pani is not only required as the biosensor transducing system but also as the mediator for the antibody binding on the biosensor surface.

By using the three-electrode electrochemical cell setup, the input signal is transferred from the auxiliary electrode to the working electrode by the ionic charges forming in the electrolyte solution. When a fixed potential is applied, electrons are allowed to flow freely from the auxiliary electrode to the ITO-Pani substrate due to the conductive property of the latter. When proteins (e.g., antibodies with a molecular weight of 150 KDa) are immobilized within the polymer backbones, electron flows are restricted. This phenomenon could be caused by the insulating protein membrane interfering with the transfer of electrons within the polymer π-backbone (Kim et al. 2000). The electron flow is restricted even more when a bigger antigen-antibody complex (molecular weight of BVDV at least 4 MDa) is present within the Pani backbone. It is here hypothesized that the bigger the protein complex present in the Pani backbone, the more restricted is the flow of electrons.

A potential of between 0.2 V and 0.8 V was demonstrated in previous studies to be a sufficient input signal, especially when dealing with whole cells or biological elements (Cattaneo et al. 1992; Darain et al. 2003; Tsiafoulis et al. 2004). Therefore, in this study, a constant potential of 0.5 V was chosen arbitrarily as the input signal.

The biosensor detection concept is based on the difference between the signal before (I^o) and the signal after (I^s) antibody-antigen binding (Figure 16.10). This current drop (ΔI) is expressed mathematically as follows:

$$\Delta I = I^o - I^s.$$

Theoretically, the higher the current drop between I^o and I^s, the more antibody-antigen complexes are formed on the biosensor surface, blocking the transfer of electrons. Therefore, the value of the current drop (ΔI) should increase with increasing antigen concentration.

The conductometric biosensor, in contrast, uses the lateral flow technique to enable the liquid sample to move from one pad to another by capillary action. Figure 16.11 illustrates the concept of the sandwich immunoassay. Before the sample is applied, the gap between the electrodes in the capture pad is open (Figure 16.11a). Immediately after the sample application, the solution carrying the antigen flows to the conjugate pad and dissolves the polyaniline-labeled

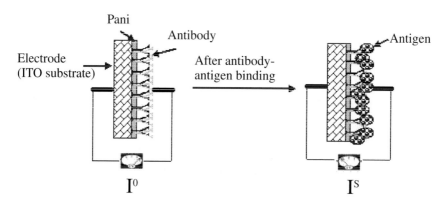

Figure 16.10. Schematic of an ITO-Pani biosensor before and after antibody-antigen binding (Reprinted from Muhammad-Tahir et al. (2007), with permission of MDPI publishing).

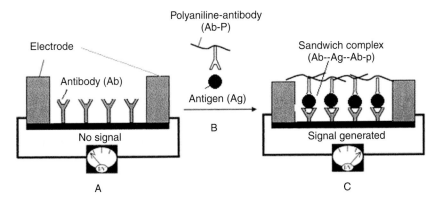

Figure 16.11. Cross section of a capture pad before (A) and after (C) analyte application (Reprinted from Muhammad-Tahir and Alocilja, 2003b, IEEE SENSORS JOURNAL. (© 2003 IEEE)).

antibody (Ab-P). The antibody-antigen binding occurs and forms a complex (Figure 16.11b). This complex is carried into the capture pad containing the immobilized antibody. A second antibody-antigen reaction occurs and forms a sandwich (Figure 16.11c). Polyaniline in the sandwich forms a molecular wire and bridges the two electrodes. The polymer structures extend out to bridge adjacent cells for signal generation. The unbound nontarget organisms are subsequently separated by capillary flow to the absorption membrane. In this biosensor design, the Pani acts solely as a transducing element.

3.5. Biosensor Properties

3.5.1. ITO-Pani Biosensor

The ITO/Pani biosensor consisted of two components: an immunosensor and an amperometric measuring device (Figure 16.12). The immunosensor was constructed from an ITO glass, and layered with Pani and antibodies. The amperometric measuring device is described in the previous section. Indium tin oxide glass is a common substrate used in an amperometric biosensor, due to its structural flexibility to bind directly with biological elements

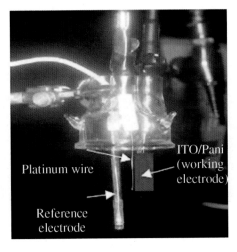

Figure 16.12. Picture of ITO-Pani biosensor set-up (Reprinted from Muhammad-Tahir et al. (2007), with permission of MDPI publishing).

(Fang et al. 2003), its stable electrical property with a high density of charge carrier (Marks et al. 2002), and its inexpensive production cost. In this study, an ITO glass substrate with dimensions of $1.27 \times 2.54 \times 0.1 \text{cm}^3$ and a resistance value ranging from 15 to 25 ohms was prepared for the biosensor fabrication.

To fabricate the biosensor, the ITO substrate was first cleaned and treated with a strong oxidizing agent, ammonium hydroxide, to enable the adherence of Pani to the ITO substrate (Ram et al. 1999). Then the self-doped Pani was layered onto the ITO substrate using a spin coating method. This method was chosen because of its ease of use, rapid processing time, reproducibility, and low cost compared to other types of polymer coating mechanisms, such as the Langmuir-Bloggert technique (Rubner 1991) or layer-by-layer molecular deposition (Ferreira and Rubner 1995). With a speed of 500 rpm for 6 seconds, a thickness of 324.62, 134.2, and 91.05 nm were observed on the ITO substrates coated with 65, 50, and 20 KDa Pani, respectively. After the coating procedure, the substrate was then functionalized into a biosensor by immobilizing antibodies onto the surface. A three-electrode electrochemical cell was used to charge the substrate by applying a small (0.5 V) negative potential. This step was essential to promote electrostatic bonding between the negatively charged substrate and the NH^+ site of the antibodies (Liu et al. 2000a). The use of a divalent crosslinker, glutaraldehyde, also helped facilitate the antibody binding mechanism (Irina et al. 1983; Diao et al. 2005).

The successful fabrication of the biosensor was evaluated using an atomic force microscope (AFM). Figures 16.13–16.18 show the AFM images of an ITO/Pani biosensor prepared with 20, 50, and 65 KDa Pani, functionalized with antibodies, and incubated with 10^4 and 10^6 CCID/ml of BVDV. An increase in height in the z direction (thickness) between plain ITO (Figure 16.13) and the ITO/Pani substrates (Figure 16.14) indicates a successful polymer coating. Figure 16.14 also shows that the higher molecular weight Pani formed a thicker layer on the ITO substrate. Similarly, an increase in thickness was observed between Figure 16.14 and Figure 16.16, where each of the ITO/Pani substrates was immobilized with antibodies, suggesting a successful antibody immobilization process. More importantly, Figures 16.16–16.18 show the AFM images of the biosensor surface after incubating them in the BVDV culture. A thicker substrate was observed when incubating the biosensors with the higher concentration of BVDV, supporting the logical conclusion that the higher the antigen concentration, the more antibody-antigen complex occurs. However, when the biosensors were tested with the same level of BVDV concentration (e.g., 10^6 CCID/ml), the biosensors coated with the higher molecular weight Pani (e.g., 65 KDa) were observed to be thicker than those coated

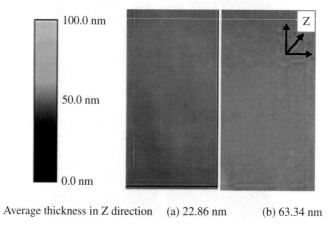

Average thickness in Z direction   (a) 22.86 nm   (b) 63.34 nm

Figure 16.13. Atomic force microscopy images of a) untreated ITO glass and b) ITO glass treated with ammonium hydroxide (NH₄OH) (Reprinted from Muhammad-Tahir et al. (2007), with permission of MDPI publishing).

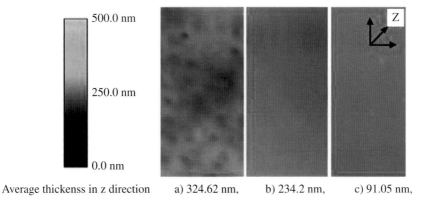

Average thickenss in z direction   a) 324.62 nm,   b) 234.2 nm,   c) 91.05 nm,

Figure 16.14. Atomic force microscopy images of a) ITO glass treated with (NH_4OH) and spin-coated with 65KDa; b) 50KDa; and c) 20KDa Pani (Reprinted from Muhammad-Tahir et al. (2007), with permission of MDPI publishing).

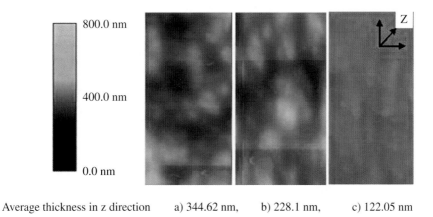

Average thickness in z direction   a) 344.62 nm,   b) 228.1 nm,   c) 122.05 nm

Figure 16.15. Atomic force microscopy images of a) ITO glass spin-coated with Pani (65KDa) + ab; b) Pani (50 KDa) + ab; and c) Pani (20 KDa) + ab. (Reprinted from Muhammad-Tahir et al. (2007), with permission of MDPI publishing).

with the Pani with a lower molecular weight (e.g., 20 KDa) (Figures 16.16a and 16.18a). This finding shows that the higher the molecular weight of Pani, the more antibody binding sites are available. This proposition is further evaluated in the next experiment.

To elucidate the use of Pani as a mediator or a "glue" for the binding of antibody on the biosensor platform, ITO glasses undergoing A) the fabrication process from sections in Method and Materials, the same fabrication process without the Pani spin-coating step, were tested for their amperometric responses (Figure 16.19). Depending on the fabrication process, the substrate is referred to as substrate A or substrate B thereafter. Figure 16.19a shows a significant difference between (I^o) and (I^s) for substrate A. The difference between (I^o) and (I^s) responses for substrate B, on the other hand, is statistically insignificant. It is suggested that the latter finding is caused by the absence of Pani on the ITO substrate, which may contribute to the lack of antibody binding, and subsequently to the insignificant differences between (I^o) and (I^s). To investigate this theory further, the presence of antibodies on both substrates was confirmed by repeating the above experiment with flourescent-tagged antibodies. The results showed that a much higher fluorescence emission level (250 out of 256-bit color mode) was observed from

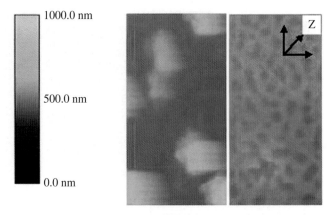

Average thickness in z direction   a) 524.62 nm,   b) 345.1 nm

Figure 16.16. Atomic force microscopy images of a) ITO-Pani biosensor (65KDa) tested with 10^6 CCID/ml; and b) 10^4 CCID/ml of BVDV (Reprinted from Muhammad-Tahir et al. (2007), with permission of MDPI publishing).

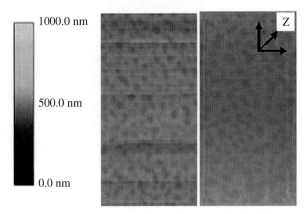

Average thickness in z direction a) 364.22 nm   b) 278.1 nm

Figure 16.17. Atomic force microscopy images of a) ITO-Pani biosensor (50KDa) tested with 10^6 CCID/ml; and b) 10^4 CCID/ml of BVDV (Reprinted from Muhammad-Tahir et al. (2007), with permission of MDPI publishing).

substrate A than from substrate B (20 out of 256-bit color mode). This implies that more antibodies are present from substrate A than substrate B. Therefore, it can be concluded that Pani is not only required as the biosensor transducing system but also as a mediator for the antibody binding. Kim et al. (2000) also demonstrated the use of Pani as a mediator between the antibodies and gold electrodes in their Pani-based biosensor.

The finding shown in Figure 16.19a also supports the detection concept of the ITO/Pani biosensor. An increase in current response was observed after an ITO substrate was coated with Pani. Then a decrease in current response was observed after antibody immobilization, suggesting that the reduction in the electron flow could be due to the insulating property of antibodies. A much higher drop in current response was observed after the antibody-antigen binding. This result supports the proposition made earlier that the bigger the protein molecules present on the surface of the biosensor, the lower the flow of electrons and thus the smaller the signal response.

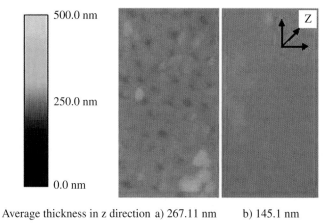

Average thickness in z direction a) 267.11 nm   b) 145.1 nm

Figure 16.18. Atomic force microscopy images of a) ITO-Pani biosensor (20KDa) tested with 10^6 CCID/ml; and b) 10^4 CCID/ml of BVDV (Reprinted from Muhammad-Tahir et al. (2007), with permission of MDPI publishing).

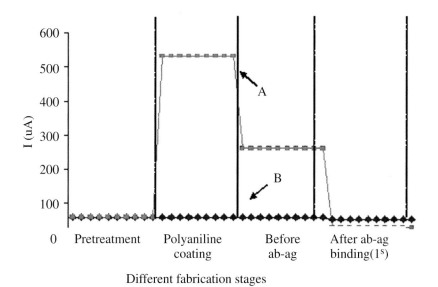

Figure 16.19. Amperometric responses of A) substrate A (ITO biosensors consisted of self-doped Pani; and B) substrate B (ITO biosensor without Pani coating) (Reprinted from Muhammad-Tahir et al. (2007), with permission of MDPI publishing).

3.6. Lateral Flow Conductometric Biosensor

For the conductometric biosensor, cellulose membrane was found to be the best material for both the sample application and absorption pads; fiberglass membrane grade G6 was best for the conjugate pad; and the nitrocellulose (NC) membrane, with a flow rate of 160 sec per 4 cm, was determined to be the best material for the capture pad. The electrodes were fabricated on the NC membrane to electrically connect the immunosensor to the electronic data acquisition system. Silver paste was found to be the easiest to handle and more conductive than the copper wire and thus was used in the subsequent development of the biosensor. When attaching the immunosensor to the platform, the etched copper wafer was found to generate a more stable signal than the glass and thus was used in subsequent experiments. Except for the platform, all other membranes were for one-time use only.

3.7. Biosensor Performance

3.7.1. ITO/Pani Biosensor

The response of the ITO/Pani biosensor was proportional to a virus concentration between 10^4 and 10^6 cell culture infective dose (CCID)/ml. Within this region of virus concentration, the biosensor response was significantly different than that of the blank and also between virus concentration levels (e.g., between 10^4 and 10^6 CCID/ml). The detection limit of the biosensor was found to be 10^4 CCID/ml. The biosensor did not respond to the presence of the IBR viruses, indicating the specificity of the BVDV antibody. While the detection time of the biosensor was found to be 55 minutes, this can be cut down to 30 minutes by minimizing the time for measuring the amperometric responses.

3.8. Conductometric Biosensor

The performance of the conductometric biosensor is summarized in Table 16.3 with a total detection time of 6 minutes. The sensitivity of the biosensor for *Salmonella, E. coli* O157:H7, and genetic *E. coli* in pure culture is 10^1 CFU/ml. The specificity study also shows that the biosensors were irresponsive to the nontarget antigen (e.g., a biosensor prepared with *Salmonella* specific antibodies did not respond to *E. coli* samples).

The performance of the biosensors were assessed in artificially inoculated strawberry, lettuce, and alfalfa sprout samples. For most of the samples tested, the resistance drop was proportional to the analyte concentration from 10^1 to 10^4 or 10^5 CFU/ml. The detection limits for the pathogenic and nonpathogenic *E. coli* in lettuce samples inoculated with the target bacteria were $5.5 \pm 0.04 \times 10^2$ and $6.3 \pm 0.1 \times 10^2$ CFU/ml, respectively. The detection limits for the pathogenic and nonpathogenic *E. coli* biosensors for the inoculated sprout samples were

Table 16.3.   Performance of the conductometric biosensor

Target Pathogen	Sensitivity	References
E. coli O157:H7, Generic *E. coli*, and *Salmonella spp* Pure culture	10^1-10^2 colony forming unit (CFU)/ml	(Muhammad-Tahir and Alocilja 2003b; Muhammad-Tahir and Alocilja 2003a)
Artificial inoculation in produce sample	10^2–10^3 CFU/ml	(Muhammad-Tahir and Alocilja 2004)
Bovine Viral Diarrhea Virus Pure culture	10^3 CCID/ml	(Muhammad-Tahir et al. 2005b; Muhammad-Tahir et al. 2005a)
Artificial inoculation in blood serum sample	10^3 CCID/ml	(Muhammad-Tahir et al. 2005a)

$7.8 \pm 0.1 \times 10^1$ and $8.2 \pm 0.4 \times 10^1$ CFU/ml, respectively. As for the strawberry samples, the detection limits for the pathogenic and nonpathogenic *E. coli* biosensors were $8.2 \pm 0.07 \times 10^2$ and $7.9 \pm 0.4 \times 10^3$ CFU/ml, respectively. The results also indicate that the pathogenic *E. coli* biosensors (prepared with antibodies specific only to *E. coli* O157:H7) responded only to *E. coli* O157:H7 contaminations and not to other organisms. The nonpathogenic *E. coli* biosensors (prepared with antibodies specific to all *E. coli* subspecies) responded to both *E. coli* and *E. coli* O157:H7 contaminations. The drop in resistance from the nontarget bacteria (for instance, the signal generated from pathogenic *E. coli* biosensors in produce samples inoculated with the nonpathogenic *E. coli*) was not significantly different from that of the control.

The conductometric biosensor was also adapted to detect viruses in both pure culture and serum sample. The detection limit of the biosensor with this characteristic in pure culture and blood serum was 10^3 CCID/ml. Though the conductometric biosensor had a low detection limit (Table 16.3), the response pattern showed a parabolic curve rather than the declining linear pattern that is expected in the semilog plot. For example, at bacterial concentrations above 10^3–10^4 CFU/ml, a crowding effect was observed (Muhammad-Tahir and Alocilja 2003a, 2003b). At this point, as the antigen concentration increased, resistance also increased. This phenomenon may be explained by the nature of sandwich immunoassays, which rely on the interaction between the labeled antibodies attached to the antigen and the limited number of antibodies fixed on the capture site (Hatch 1999). At high concentrations (i.e., above 10^3–10^4 CFU/ml), the binding site could be overoccupied with the antigen, thus obstructing the charge transfer within the conductive polymer structure. Scanning electron microscope (SEM) images (Figure 16.20) of the capture regions with different cell concentrations show differences in the density of bound cells between 10^1 (Figure 16.20a) and 10^7 CFU/ml of bacteria (Figure 16.20b).

Variation from sensor to sensor is a common obstacle in biosensor research. An example of this variation can be observed in the membrane-based conductometric biosensor designed for food pathogen and viral detection (Muhammad-Tahir and Alocilja 2003a, 2003b; Muhammad-Tahir et al. 2005b). Although the conductometric biosensor has a promising performance in detecting a low level of antigen concentration, its design has a disadvantage: its platform can only be used once because of the nonreversible properties of the nitrocellulose and cellulose

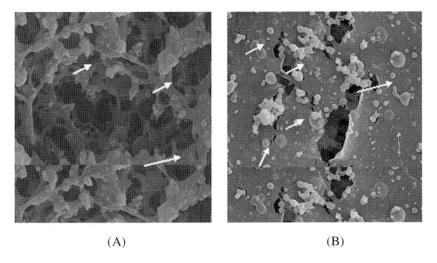

(A)   (B)

Figure 16.20. SEM images (3000X) of capture regions tested with (A) 10^1 and (B) 10^7 CFU/ml of bacterial cultures. Arrows indicate possible bacterial sites (reprinted from Muhammad-Tahir and Alocilja, (2003a), with permission of Elsevier).

membranes employed. Therefore, the calibration (control sample) and the sample testing cannot be performed on the same biosensor platform. The inability to calibrate and test samples using the same conductometric biosensor contributes to the variability observed between experiments (Muhammad-Tahir and Alocilja 2004). This has motivated the authors to redesign a biosensor that allows both the control (calibration) and the sample testing to be done on the same platform; hence the development of the ITO/Pani biosensor. The performance of the ITO/Pani biosensor in different foodborne pathogen samples, however, needs to be further investigated before more conclusive results can be obtained.

4. Conclusions and Future Perspectives

The microbial biosensor designs presented in this chapter are only a few of many novel devices and assays that are being researched on, developed, and commercialized for applications in biodefense, healthcare, food safety, environmental quality, animal welfare, and bioenergy. As demonstrated, biosensors have the potential of being highly useful in reducing the time for diagnosis, a major factor in survival, prevention, protection, and remediation. They are showing to have excellent sensitivity and specificity performance, allowing for the detection of contaminants at infective-dose levels. Furthermore, they can be multiplexed into tens or hundreds of arrays, making the device a critical tool for rapid screening. The biosensor chips can be miniaturized and are compatible with automated electronic data processing technologies as well as remote data transmission, thus they can be integrated into small portable field-based devices and can be used off site with data transmission into a centralized location. Because they have low manufacturing and production cost, they are very affordable. Biosensors will dominate the market in a very short time and could become an indispensable detection tool in the field and in the labs.

References

Abe A, Ohtani A, Umemoto Y, Akizuki S, Ezoe M, Higuchi H, Nakamoto K, Okuno A and Noda Y (1989) Soluble and high molecular weight polyaniline. Journal of Chemical Society-Chemical Communication 2:1736–1738

Abel AP, Weller MG, Duveneck GL, Ehra, M and Widmer HM (1996) Fiber-optic evanescent wave biosensor for the detection of oligonucleotides. Anal Chem 68:2905–12

Akhtar M, Kleppinger J, Macdiarmid AG, Milliken J, Moran MJ, Chiang CK, Cohen MJ, Heeger AJ and Peebles DL (1977) Metallic Derivative of Polymeric Sulfur Nitride - Poly (Thiazyl Bromide). Journal of the Chemical Society-Chemical Communications 13:473–474

Alexander AJ and Hughes DE (1995) Monitoring of IgG antibody stability by micellar electrokinetic capillary chromatography and matrix assisted laser desorption/ionization mass spectrophotometry. Analytical Chemistry 67:3636–3632

Alt F, Weber G, Messerschmidt J, Bohlen AV, Kastenholz B and Guenther K (2002) A simple assay for 2,4-dichloriphenoxyacetic acid using coated test strips. Analytical Letters 35:1341–1348

Ashok Mulchandani C-LW (1996) Bienzyme sensors based on poly(anilinomethylferrocene)-modified electrodes. Electroanalysis 8:414–419

Babacan S, Pivarnik P, Letcher S and Rand AG (2000) Evaluation of antibody immobilization methods for piezoelectric biosensor application. Biosensors and Bioelectronics 15:615–621

Baeumner AJ, Cohen RN, Miksic V and Min J (2003) RNA biosensor for the rapid detection of viable *Escherichia coli* in drinking water. Biosensors and Bioelectronics 18:405–413

Baeumner AJ, Pretz J and Fang S (2004) A universal nucleic acid sequence biosensor with nanomolar detection limits. Anal Chem 76: 888–94

Baker JC (1995) The clinical manifestation of bovine viral diarrhea infection. The Veterinary Clinics of North America: Food Animal Practice 11:425–445

Barbour WM and George T (1997) Genetic and immunologic techniques for detecting foodborne pathogens and toxins. In: Montville TJ (ed) Food Microbiology: Fundamentals and Frontiers. ASM Press, Washington D.C., pp 30–65

Bartlett PN, Wang JH and Wallace ENK (1996) A microelectrochemical switch responsive to "NADH." Journal of the Chemical Society-Chemical Communications 2:359–360

Berney H, West J, Haefele E, Alderman J, Lane W and Collins JK (2000) A DNA diagnostic biosensor: development, characterisation and performance. Sensors and Actuators B: Chemical 68:100–108

Bhatia SK, Shriver-Lake LC, Prior KJ, Georger JH, Calvert JM, Bredehorst R and Ligler FS (1989) Use of thiol-terminal silanes and heterobifunctional crosslinkers for immobilization of antibodies on silica surfaces. Anal Biochem 178:408–13

Bianchi N, Rutigliano C, Tomassetti M, Feriotto G, Zorzato F and Gambari R (1997) Biosensor technology and surface plasmon resonance for real-time detection of HIV-1 genomic sequences amplified by polymerase chain reaction. Clinical and Diagnostic Virology 8:199–208

Blaser MJ and Newman LS (1982) A review of human salmonellosis 1. Infective dose. Rev. Infect. Dis. 4:1096–1106

Bolin SR (1995) The pathogenesis of mucosal disease. The Veterinary Clinics of North America-Food Animal Practice 11: 489–500

Boltovets PM, Boyko VR, Kostikov IY, Dyachenko NS, Snopok BA and Shirshov YM (2002) Simple method for plant virus detection: effect of antibody immobilization technique. Journal of Virological Methods 105:141–146

Brock KV (1991) Detection of persistence bovine viral diarrhea virus infections by DNA hybridization and PCR assay. Archives of Virology 3:181–190

Brock KV (2003) The persistence of bovine viral diarrhea virus. Biologicals 31:133–135

Bunde RL, Jarvi EJ and Rosentrerer JJ (1998) Piezoelectric quartz crystal biosensor. Talanta 46:1223–1229

Buzby, J and Frenzen, P (2000) ERS updates Foodborne Illness Cost Food Safety v.23.

Cao Y, Qiu J and Smith P (1995) Effect of solvents and co-solvents on the processibility of polyaniline: Solubility and Conductivity studies. Synthetic Metals 69:187–190

Cao Y, Smit, P and Heeger AJ (1993) Counter-ion induced processibility of conducting polyaniline. Synthetic Metals 57:3514–3519

Carlson SA, Bolton LF, Briggs CE, Hurd HS, Sharma VK, Fedorka Cray PJ and Jones BD (1999) Detection of multiresistant *Salmonella typhimurium* DT104 using multiplex and fluorogenic PCR. Molecular and Cellular Probes 13:213–222

Castillo-Ortega MM, Rodriguez DE, Encinas JC, Plascencia M, Mendez-Velarde FA and Olayo R (2002) Conductometric uric acid and urea biosensor prepared from electroconductive polyaniline-poly(n-butyl methacrylate) composites. Sensors and Actuators B: Chemical 85:19–25

Cattaneo MV, Luong JHT and Mercille S (1992) Monitoring glutamine in mammalian cell cultures using an amperometric biosensor. Biosensors and Bioelectronics 7: 329–334

CDC (2001a) Outbreaks caused by Shiga toxin-producing Escherichia coli-Summary of 2000 Surveillance Data. Centers for Disease Control and Prevention. http://www.cdc.gov/foodborneoutbreaks/ecoli/2000_summaryLetter.pdf

CDC (2001b) Salmonellosis. http://www.cdc.gov/ncidod/dbmd/diseaseinfo/salmonellosis_g.htm

CDC (2002a) Notice to Readers: Final 2001 Reports of Notifiable Diseases. MMWR 51: 710

CDC (2002b) Preliminary FoodNet Data on the Incidence of Foodborne Illnesses - Selected Sites, United States, 2001. MMWR 51: 325–9

CDC (2002c) Report on the Decline of Foodborne Illness. Centers for Disease Control and Prevention. http://www.cdc.gov/foodborne/publications/201-nelson_2004.pdf

CDC (2004) Bioterrorism agents/diseases. Centers for Disease Control and Prevention. http://www.bt.cdc.gov/agent/agentlist-category.asp

CDC (2006) Bioterrorism Agents/Diseases. http://www.bt.cdc.gov/Agent/agentlist.asp

CHEMICON International (2004) Introduction to antibodies. CHEMICON International, Inc. http://www.chemicon.com/resource/ANT101/a1.asp

Chen ZZ, Wang KM, Yang XH, Huang SS, Huang HM, Li D and Wang Q (2003) Determination of hepatitis B surface antigen by surface plasmon resonance biosensor. Acta Chimica Sinica 61:137–140

Cheung JH, Stockton WB and Rubner MF (1997) Molecular-Level Processing of Conjugated Polymers: Layer-by-Layer Manipulation of Polyaniline via Electrostatic Interactions. Macromolecules 30:2712–2716

Chiang CK, Fincher CR, Park YW, Heeger AJ, Shirakawa H, Louise EJ, Gau SC and MacDiarmid AG (1977) Electrical Conductivity in Doped Polyacetylene. Physical Review Letters 39:1098–1101

Cohen ND, McGrudder ED, Neibergs HL, Bhele RW, Wallis DE and Hargis BM (1994) Detection of *Salmonella enteritidis* in feces from poultry using booster polymerase chain reaction and oligonucleotide primers specific for all number of the genus *Salmonella*. Poultry Science 73:354–357

Cohn GE (1998) Systems and Technologies for Clinical Diagnostics and Drug Discovery. SPIE Proceedings. 3259:11–17

Contractor AQ, Sureshkumar TN, Narayanan R, Sukeerthi S, Lal R and Srinivasa RS (1994) Conducting polymer-based biosensors. Electrochimica Acta 39:1321–1324

D'Aoust J-Y (1997) Salmonella Species. In: Montville TJ (ed) Food Microbiology: Fundamentals and Frontiers. ASM, Washington, D.C., pp 138–139

Darain F, Park S-U and Shim Y-B (2003) Disposable amperometric immunosensor system for rabbit IgG using a conducting polymer modified screen-printed electrode. Biosensors and Bioelectronics 18:773–780

Davis F, Nabok AV and Higson SPJ (2005) Species differentiation by DNA-modified carbon electrodes using an ac impedimetric approach. Biosensors and Bioelectronics Spec Iss 20:1531–1538

Davis JJ, Green MLH, Allen O, Hill H, Leung YC, Sadler PJ, Sloan J, Xavier AV and Chi Tsang S (1998) The immobilisation of proteins in carbon nanotubes. Inorganica Chimica Acta 272:261–266

De Roever C (1998) Microbiological safety evaluations and recommendations on fresh produce. Food Control 9:321–347

De Taxis du Poet P, Miyamoto S, Murakami T, Kimura J and Karube I (1990) Direct electron transfer with glucose oxidase immobilized in an electropolymerized poly-N-methylpyrrole film on a gold microelectrode. Analytica Chemica Acta 235:255–264

Deshpande MV and Hall EA H (1990) An electrochemically grown polymer as an immobilisation matrix for whole cells: Application in an amperometric dopamine Sensor. Biosensors and Bioelectronics 5:431–448

Diao J, Ren D, Engstrom JR and Lee KH (2005) A surface modification strategy on silicon nitride for developing biosensors. Analytical Biochemistry 343:322–328

Doyle MP, Zhao T, Meng J and Zhao S (1997) Escherichia coli O157:H7. In: Montville TJ (ed) Food Microbiology Fundamentals and Frontiers. American Society for Microbiology, Washington, D.C.

Dresselhaus MS, Dresselhaus G and Avouris P (2001) Carbon Nanotubes: Synthesis, Properties, and Applications. Springer-Verlag, Berlin

D'Souza SF (2001) Microbial Biosensors (Review). Biosensors and Bioelectronics 16: 337–353

Dubovi EJ (1992) Genetic diversity and BVD virus. Comparative Immunology, Microbiology & Infectious Diseases 15:155–162

Duffell S and Harkness J (1985) Bovine virus diarrhoea-mucosal disease infection in cattle. The Veterinary Record 117:240–245

Duic L, Mandic Z and Kovacicek F (1994) The effect of supporting electrolyte on the elctrochemical synthesis, morphology, and conductivity of polyaniline. Journal of Polymer Science: Part A: Polymer Chemistry 32:105–111

Esch MB, Locascio LE, Tarlov MJ and Durst RA (2001) Detection of viable Cryptosporidium parvum using DNA-modified liposomes in a microfluidic chip. Analytical Chemistry 73:2952–8

Fang A, Ng HT and Li SFY (2003) A high-performance glucose biosensor based on monomolecular layer of glucose oxidase covalently immobilised on indium-tin oxide surface. Biosensors and Bioelectronics 19:43–49

FDA (1992) Foodborne Pathogenic Microorganisms and Natural Toxins Handbook : Salmonella spp. http://www.cfsan.fda.gov/~mow/chap1.html

FDA (2005) Bacteriological Analytical Manual. Food and Drug Administration, Rockville, Maryland http://www.cfsan.fda.gov/~ebam/bam-toc.html

FDA (2006) Foodborne Pathogenic Microorganisms and Natural Toxins Handbook: The "Bad Bug Book." FDA-CFSAN. http://www.cfsan.fda.gov/~mow/intro.html

Feriotto G, Borgatti M, Mischiati C, Bianchi N and Gambari R (2002) Biosensor technology and surface plasmon resonance for real-time detection of genetically modified roundup ready soybean gene sequences. Journal of Agricultural and Food Chemistry 50:955–962

Ferreira M and Rubner MF (1995) Molecular -level processing of conjugated polymers: layer by layer manipulation of conjugated polyions. Macromolecules 28:7107–7114

Foulds NC and Lowe CR (1985) What's new: Biosensors: Current applications and future potential. BioEssays 3:129–132

Fulton RW, Saliki JT, Confer AW, Burge LJ, d'Offay JM, Helman RG, Bolin SR, Ridpath JF and Payton ME (2000) Bovine viral diarrhea virus cytopathic and noncytopathic biotypes and type 1 and 2 genotypes in diagnostic laboratory accessions: clinical and necropsy samples from cattle. Journal of Veterinary Diagnostic Investigation 12:33–38

Gao ZX, Fang YJ, Ren J, Ning B, Zhu HZ and He YH (2004) Studies on biotin-avidin indirect conjugated technology for a piezoelectric DNA sensor. International Journal of Environmental Analytical Chemistry 84:599–606

Gau J-J, Lan EH, Dunn B, Ho C-M and Woo JCS (2001) A MEMS based amperometric detector for E. coli bacteria using self-assembled monolayers. Biosensors and Bioelectronics 16:745–755

Gau V, Ma S-C, Wang H, Tsukuda J, Kibler J and Haake DA (2005) Electrochemical molecular analysis without nucleic acid amplification. Methods 37:73–83

Geng T, Morgan MT and Bhunia AK (2004) Detection of low levels of Listeria monocytogenes cells by using a fiber-optic immunosensor. Applied and Environmental Microbiology 70:6138–6146

Genies EM, Boyle A, Lapkowski M and Tsintavis C (1990) Polyaniline: A historical survey. Synthetic Metals 36:139–182

Gerard M, Chaubey A and Malhotra BD (2002) Application of conducting polymers to biosensors. Biosensors and Bioelectronics 17:345–359

Graham CR, Leslie D and Squirrell DJ (1992) Gene probe assays on a fiberoptic evanescent wave biosensor. Biosensors and Bioelectronics 7:487–493

Grennan K, Strachan G, Porter AJ, Killard AJ and Smyth M (2003a) Atrazine analysis using an amperometric immunosensor based on single-chain antibody fragments and regeneration-free multi-calibrant measurement. Analytica Chemica Acta 500: 287–298

Guiseppi-Elie A (1998) Chemical and Biological Sensors based on Electrically Conducting polymers. In: Reynolds JR (ed) Handbook of conducting polymers. Marcel-Dekker, Inc., New York, pp 963–992

Hammerle M, Schuhmann W and Schmidt H-L (1992) Amperometric polypyrrole enzyme electrodes: effect of permeability and enzyme location. Sensors and Actuators B: Chemical 6:106–112

Hartley HA and Baeumner AJ (2003) Biosensor for the specific detection of a single viable B. anthracis spore. Anal Bioanal Chem 376:319–27

Hatch A, Weigl B, Zebert D,Yager P (1999) Microfluidic approaches toward immunoassays. Microfluidic devices and systems II. 169–172

He FJ and Liu SQ (2004) Detection of *P. aeruginosa* using nano-structured electrode-separated piezoelectric DNA biosensor. Talanta 62:271–277

Ho JA, Hsu HW and Huang M-R (2004) Liposome-based microcapillary immunosensor for detection of *Escherichia coli* O157:H7. Analytical Biochemistry 330:342–349

Houe H (1995) Epidemiology of bovine viral diarrhea virus. The Veterinary Clinics of North America-Food Animal Practice 11:521–547

Huang TS, Tzeng Y, Liu YK, Chen YK, Walker KR, Guntupalli R and Liu C (2004) Immobilization of antibodies and bacterial binding on nanodiamond and carbon nanotubes for biosensor applications. Diamond and Related Materials 13:1098–1102

Imisides MD, John R, Wallace GG (1996) Microsensors based on conducting polymers. Chemtech 26 1:9–25

Iribe Y and Suzuki M (2002) Integrated enzyme switch as a novel biosensing device. Biosensors and Bioelectronics. Proceedings of the Seventh World Congress on Biosensors, Kyoto, Japan

Irina I-M, Sanchez Y, Fields HA and Dreesman GR (1983) Development of sensitive immunoassays for detection of antibodies against hepatitis B surface antigen. Journal of Virological Methods 6:41–52

Ivanov I, Gherman BF and Yaron D (2001) Comparison of the INDO band structures of polyacetylene, polythiophene, polyfuran, and polypyrrole. Synthetic Metals 116:111–114

Ivory DM, Miller GG, Sowa JM, Shacklette, L. W., Chance, R. R. and Baughman, R. H. 1979. Highly conducting charge-transfer complexes of poly(p-phenylene). The Journal of Chemical Physics 71:1506–1507

Jian-Guo G, Yu-Qing M and Qing-Jie Z (2004) Review: Impedimetric biosensors. Journal of Bioscience and Bioengineering 97:219–226

Jordan CE, Frutos AG, Thiel AJ and Corn RM (1997) Surface plasmon resonance imaging measurements of DNA hybridization adsorption and streptavidin/DNA multilayer formation at chemically modified gold surfaces. Analytical Chemistry 69:4939–4947

Kanazawa KK, Diaz AF, Geiss RH, Gill WD, Kwak JF, Logan JA, Rabolt JF and Street GB (1979) Organic metals: polypyrrole, a stable synthetic metallic polymer. Journal of Chemical Society-Chemical Communications 854–855

Kim JH, Cho JH and Cha GS (2000) Conductimetric membrane strip immunosensor with polyaniline-bound gold colloids as signal generator. Biosensors and Bioelectronics 14:907–915

Kim N, Park IS and Kim DK (2004) Characteristics of a label-free piezoelectric immunosensor detecting *Pseudomonas aeruginosa*. Sensors and Actuators B-Chemical 100:432–438

Ko S and Grant SA (2006) A novel FRET-based optical fiber biosensor for rapid detection of *Salmonella typhimurium*. Biosensors and Bioelectronics 21:1283–1290

Komarova E, Aldissi M and Bogomolova A (2005) Direct electrochemical sensor for fast reagent-free DNA detection. Biosensors and Bioelectronics 21:182–189

Koubova V, Brynda E, Karasova L, Skvor J, Homola J, Dostalek J, Tobiska P and Rosicky J (2001) Detection of foodborne pathogens using surface plasmon resonance biosensors. Sensors and Actuators B-Chemical 74:100–105

Kroschwitz JI (1988) Electrical and Electronic Properties of Polymers-State-of-the-Art Compendium. Wiley, New York

Kukanskis K, Elkind J, Melendez J, Murphy T, Miller G and Garner H (1999) Detection of DNA hybridization using the TISPR-1 surface plasmon resonance biosensor. Analytical Biochemistry 274:7–17

Lagowski JB, Salzner U, Pickup PG and Poirier RA (1998) Comparison of geometries and electronic structures of polyacetylene, polyborole, polycyclopentadiene, polypyrrole, polyfuran, polysilole, polyphosphole, polythiophene, polyselenophene and polytellurophene. Synthetic Metals 96:177–189

Lange K, Blaess G, Voigt A, Gotze, R and Rapp M (2006) Integration of a surface acoustic wave biosensor in a microfluidic polymer chip. Biosensors and Bioelectronics 22:227–232

Laska J and Widlarz J (2003a) One-step polymerization leading to conducting polyaniline. Synthetic Metal 135–136:263–264

Laska J and Widlarz J (2003b) Water soluble polyaniline. Synthetic Metal 135–136: 261–262

Lee JS, Choi Y-K, Pio M, Seo J and Lee LP (2002) Nanogap capacitors for label free DNA analysis. BioMEMS and Bionanotechnology. 729:185–190

Lin HC and Tsai WC (2003) Piezoelectric crystal immunosensor for the detection of staphylococcal enterotoxin B. Biosensors and Bioelectronics 18:1479–1483

Liu CH, Liao KT and Huang HJ (2000a) Amperometric immunosensors based on protein A coupled polyaniline-perfluorosulfonated ionomer composite electrodes. Analytical Chemistry 72:2925–2929

Liu J, Tian S and Knoll W (2005) Properties of Polyaniline/Carbon Nanotube Multilayer Films in Neutral Solution and Their Application for Stable Low-Potential Detection of Reduced Nicotinamide Adenine Dinucleotide. Langmuir 21:5596–5599

Liu X, Farmerie W, Schuster S and Tan W (2000b) Molecular beacons for DNA biosensors with micrometer to submicrometer dimensions. Analytical Biochemistry 283:56–63

Liu X and Tan W (1999) A fiber-optic evanescent wave DNA biosensor based on novel molecular beacons. Anal Chem 71:5054–9

Liu Y-C, Wang C-M, Hsiung K-P and Huang C (2003) Evaluation and application of conducting polymer entrapment on quartz crystal microbalance in flow injection immunoassay. Biosensors and Bioelectronics 18:937–942

Lu B, Smyth MR and O'Kennedy R (1996) Oriented immobilization of antibodies and its applications in immunoassays and immunosensors. Analyst 121:29R-32R

Lu W, Zhao H and Wallace GG (1995) Pulsed electrochemical detection of proteins using conducting polymer based sensors. Analytica Chimica Acta 315:27–32

Lukachova LV, Shkerin EA, Puganova EA, Karyakina EE, Kiseleva SG, Orlov AV, Karpacheva GP and Karyakin AA (2003) Electroactivity of chemically synthesized polyaniline in neutral and alkaline aqueous solutions: Role of self-doping and external doping. Journal of Electroanalytical Chemistry 544:59–63

Luo X, Killard AJ, Morrin A and Smyth MR (2006) Enhancement of a conducting polymer-based biosensor using carbon nanotube-doped polyaniline. Analytica Chimica Acta 575:39–44

Macdiarmid AG, Chiang JC and Richter AF (1987) Polyaniline: A new concept in conducting polymer. Synthetic Metals 18:285–290

MacDiarmid AG and Epstein AJ (1990) The polyanilines: Potential technology based on the new chemistry and new properties. In: Samuelsen EJ (ed) Science and Applications of Conducting Polymers. IOP Publishing Ltd., UK, pp 117–127

Madigan M, Martinko J and Parker J (1997) Biology of Microorganisms. Viacom, Upper Saddle River, NJ

Maehashi K, Matsumoto K, Kerman K, Takamura Y and Tamiya E (2004) Ultrasensitive detection of DNA hybridization using carbon nanotube field-effect transistors. Japanese Journal of Applied Physics Part 2-Letters & Express Letters 43:L1558-L1560

Manchester A, Clauson A (1995) Spending for food away from home. Food Review 18:12–15

Mariotti E, Minunni M and Mascini M (2002) Surface plasmon resonance biosensor for genetically modified organisms detection. Analytica Chimica Acta 453:165–172

Marks RS, Novoa A, Konry T, Krai, R and Cosnier S (2002) Indium tin oxide-coated optical fiber tips for affinity electropolymerization. Materials Science and Engineering: C-Biomimetic and Supramolecular Systems 21:189–194

McClelland R and Pinder A (1994) Detection of Salmonella typhimurium in dairy products with flow cytometry and monoclonal antibodies. Applied Environmental Microbiology A 60:4255–4262

McGown LB, Joseph MJ, Pitner JB, Vonk GP and Linn CP (1995) The nucleic-acid ligand - a new tool for molecular recognition. Analytical Chemistry 67:A663-A668

Mead PS, Slutsker L, Dietz V, McGaig L, Bresee J, Shapiro C, Griffin P and Tauxe R (1999) Food-Related Illnesses and Death in the United States. Emerging Infectious Disease 5:607–625

Meng J, Zhao S, Doyle M and Kresovich S (1996) Polymerase chain reaction for detection E. coli O157:H7. International Journal of Food Microbiology 32:103–113

Minett AI, Barisci JN and Wallace GG (2002) Immobilisation of anti-Listeria in a polypyrrole film. Reactive and Functional Polymers 53:217–227

Minett AI, Barisci JN and Wallace GG (2003) Coupling conducting polymers and mediated electrochemical responses for the detection of Listeria. Analytica Chimica Acta 475:37–45

Mir M and Katakis I (2005) Towards a fast-responding, label-free electrochemical DNA biosensor. Analytical and Bioanalytical Chemistry 381:1033–1035

Mo XT, Zhou YP, Lei H and Deng L (2002) Microbalance-DNA probe method for the detection of specific bacteria in water. Enzyme and Microbial Technology 30: 583–589

Morrin A, Guzman A, Killard A, Pingarron J and Smyth M (2003) Characterisation of horseradish peroxidase immobilisation on an electrochemical biosensor by colorimetric and amperometric techniques. Biosensors and Bioelectronics 18:715–720

Mottaghitalab V, Spinks GM and Wallace GG (2005) The influence of carbon nanotubes on mechanical and electrical properties of polyaniline fibers. Synthetic Metals 152:77–80

Muhammad-Tahir Z and Alocilja E (2004) A Disposable Biosensor for Pathogen Detection in Fresh Produce Samples. Biosystems Engineering 88:145–151

Muhammad-Tahir Z, Alocilja E and Grooms D (2007) Indium Tin Oxide-Polyaniline Biosensor: Fabrication and Characterization. Sensors 7:1123–1140

Muhammad-Tahir Z and Alocilja EC (2003a) A conductimetric biosensor for biosecurity. Biosensors and Bioelectronics 18:813–9

Muhammad-Tahir Z and Alocilja EC (2003b) Fabrication of a disposable biosensor for *Escherichia coli* O157:H7 detection. IEEE Sensors Journal 3:345–51

Muhammad-Tahir Z, Alocilja EC and Grooms DL (2005a) Polyaniline synthesis and its biosensor application. Biosensors and Bioelectronics 20:1690–1695

Muhammad-Tahir Z, Alocilja EC and Grooms DL (2005b) Rapid detection of *Bovine Viral Diarrhea Virus* as surrogate of bioterrorism agents. IEEE Sensors Journal 5:757–62

Myler S, Collyer SD, Davis F, Gornall DD and Higson SPJ (2005) Sonochemically fabricated microelectrode arrays for biosensors: Part III. AC impedimetric study of aerobic and anaerobic response of alcohol oxidase within polyaniline. Biosensors and Bioelectronics 21:666–671

Nagai H, Murakami Y, Yokoyama K and Tamiya E (2001) High-throughput PCR in silicon based microchamber array. Biosensors and Bioelectronics 16:1015–1019

Narang U, Anderson GP, Ligler FS and Buran, J (1997) Fiber optic-based biosensor for ricin. Biosensors and Bioelectronics 12:937–45

Nashat AH, Moronne M and Ferrari M (1998) Detection of functional groups and antibodies on microfabricated surfaces by confocal microscopy. Biotechnology and Bioengineering 60:137–146

Nettleton PF (1990) Pestivirus infections in ruminants other than cattle. Revue Scientifique Et Technique (International Office Of Epizootics) 9:131–150

NIAID (2004) NIAID Category A, B & C Priority Pathogens. National Institute of Allergy and Infectious Disease http://www.niaid.nih.gov/biodefense/bandc_priority.htm

Nicolini C, Erokhin V, Facci P, Guerzoni S, Ross A and Paschkevitsch P (1997) Quartz balance DNA sensor. Biosensors and Bioelectronics 12:613–618

Nishizawa M, Matsue T and Uchida I (1992) Penicillin sensor based on a microarray electrode coated with pH-responsive polypyrrole. Analytical Chemistry 64:2642–2644

Pandey PC and Mishra AP (1988) Conducting polymer-coated enzyme microsensor for urea. Analyst 113:329–331

Park IS, Kim WY and Kim N (2000) Operational characteristics of an antibody-immobilized QCM system detecting Salmonella spp. Biosensors and Bioelectronics 15:167–172

Park S and Durst R-A (2000) Immunoliposome sandwich assay for the detection of Escherichia coli O157:H7. Analytical Biochemistry 280:151–158

Pellerin C, Van Den Hurk J, Lecomte J and Tijssen P (1994) Identification of a New Group of Bovine Viral Diarrhea Virus Strains Associated with Severe Outbreaks and High Mortalities. Virology 203:260–268

Perdrizet J, Rebhun W, Dubovi E and Donis R (1987) Bovine virus diarrhea–clinical syndromes in dairy herds. Cornell Veterinarian 77:46–74

Rabolt JF, Clarke TC, Kanazawa KK, Reynolds JR and Street GB (1980) Organic metals: poly (p-phenylene sulphide) hexafluoroarsenate. Journal of Chemical Society-Chemical Communications 347–348

Radke SM and Alocilja EC (2005) A high density microelectrode array biosensor for detection of E. coli O157:H7. Biosensors and Bioelectronics SPEC ISS 20:1662–1667

Ram M, Salerno M, Manuela A, Faraci P and Nicolini C (1999) Physical properties of Polyaniline films: Assembled by the Layer-by-Layer Technique. Langmuir 15: 125–1259

Ramanathan K, Bangar MA, Yun M, Chen W, Mulchandani A and Myung NV (2004) Individually Addressable Conducting Polymer Nanowires Array. Nano Lett. 4:1237–1239

Ramanathan K, Bangar MA, Yun M, Chen W, Myung NV and Mulchandani A (2005) Bioaffinity Sensing Using Biologically Functionalized Conducting-Polymer Nanowire. J. Am. Chem. Soc. 127:496–497

Ramanaviciene A and Ramanavicius A (2004) Pulsed amperometric detection of DNA with an ssDNA/polypyrrole-modified electrode. Analytical and Bioanalytical Chemistry 379:287–293

Rao BCS, Govindaraj A, Nath M (2001) Nanotubes. ChemPhys Chem 2: 78–105

Ratcliffe NM (1990) Polypyrrole-based sensor for hydrazine and ammonia. Analytica Chimica Acta 239:257–262

Ridpath JF (2003) BVDV genotypes and biotypes: practical implications for diagnosis and control. Biologicals 31:127–131

Rodriguez M and Alocilja E (2005) Embedded DNA-polypyrrole biosensor for rapid detection of Escherichia coli. IEEE Sensors Journal 5:733–736

Ruan CM, Yang LJ and Li YB (2002) Immunobiosensor chips for detection of *Escherichia coli* O157 : H7 using electrochemical impedance spectroscopy. Analytical Chemistry 74:4814–4820

Ruan CM, Zeng KF, Varghese OK and Grimes CA (2004) A staphylococcal enterotoxin B magnetoelastic immunosensor. Biosensors and Bioelectronics 20:585–591

Rubner MF (1991) Conjugated polymers. Kluwer Academic Publisher, Boston

Sadana A (2002) Engineering Biosensors: Kinetic and Design Application. Academic Press, London

Sadik OA and Emon JMV (1996) Application of electrochemical immunosensors to environmental monitoring. Biosensors and Bioelectronics 11:i–xi

Salaneck WR and Lundstrom I (1987) Electronics properties of some polyaniline. Synthetic Metal 18:291–296

Savage MD, Mattson G, Desai S, Nielander GW, Morgensen S and Conklin EJ (1994) Avidin-Biotin Chemistry: A Handbook. Pierce Chemical Company, Rockford, Illinois

Scheller FW, Hintscher R, Pfeiffer P, Schubert F, Riedel K and Kindervater R (1991) Biosensors: Fundamentals, applications and trends. Sensors and Actuators B 4:197–206

Sergeeva TA, Pilletskii SA, Rachkov AE and Skaya AVE (1996) Synthesis and Examination of Polyaniline as labels in Immunosensor Analysis. Journal of Analytical Chemistry 51:394–396

Sergeyeva TA, Lavrik NV and Rachkov AE (1996) Polyaniline label-based conductimetric sensor for IgG detection. Sensors and Actuators B 34:283–288

Shah J, Chemburu S, Wilkins E and Abdel-Hamid I (2003) Rapid amperometric immunoassay for *Escherichia coli* based on graphite coated nylon membranes. Electroanalysis 15:1809–1814

Shaolin M and Jincui L (1999) Electrochemical activity and electrochromism of polyaniline in the buffer solutions. http://web.chemistrymag.org/cji/1999/011001pe.htm

Shriver-Lake LC, Donner B, Edelstein R, Breslin K, Bhatia SK and Ligler FS (1997) Antibody immobilization using heterobifunctional crosslinkers. Biosensors and Bioelectronics 12:1101–1106

Silin V and Plant A (1997) Biotechnological applications of surface plasmon resonance. Trends in Biotechnology 15:353–359

Situmorang M, Hilbert DB and Gooding JJ (2000) An experimental design study of interferences of clinical relevance of polytyramine immobilized enzyme biosensor. Electroanalysis 12:111–119

Skuridin SG, Yevdokimov YM, Efimov VS, Hall JM and Turner APF (1996) A new approach for creating double-stranded DNA biosensors. Biosensors and Bioelectronics 11:903–911

Slavik R, Homola J and Brynda E (2002) A miniature fiber optic surface plasmon resonance sensor for fast detection of Staphylococcal enterotoxin B. Biosensors and Bioelectronics 17:591–5

Snejdarkova M, Svobodova L, Evtugyn G, Budnikov H, Karyakin A, Nikolelis DP and Hianik T (2004) Acetyl-cholinesterase sensors based on gold electrodes modified with dendrimer and polyaniline: A comparative research. Analytica Chimica Acta 514:79–88

Sportsman JR and Wilson GS (1980) Chromatographic properties of silica-immobilized antibodies. Analytical Chemistry 52:2013–2018

Stejskal J, Riede A, Hlavatá D, Helmstedt M, Holler P and Prokeš J (1998) The effect of polymerization temperature on molecular weight, crystallinity, and electrical conductivity of polyaniline. Synthetic Metals 96:55–61

Streetman B (1995) Solid State Electronic Devices. Prentice Hall, Englewood Cliffs, New Jersey

Su XL and Li Y (2005) Surface plasmon resonance and quartz crystal microbalance immunosensors for detection of *Escherichia coli* O157 : H7. Transactions of the ASAE 48:405–413

Su XL and Li YB (2004). A self-assembled monolayer-based piezoelectric immunosensor for rapid detection of *Escherichia coli* O157 : H7. Biosensors and Bioelectronics 19:563–574

Sung Hoon R, In Seon P, Namsoo K and Woo Yeon K (2001) Hybridization of *Salmonella* spp.-specific nucleic acids immobilized on a quartz crystal microbalance. Food Science and Biotechnology 10:663–667

Syed AA and Dinesan MK (1991) Polyaniline -novel polymeric material. Talanta 38:815–837

Takeda S, Sbagyo A, Sakoda Y, Ishii A, Sawamura M, Sueoka K, Kida H, Mukasa K and Matsumoto K (2005) Application of carbon nanotubes for detecting anti-hemagglutinins based on antigen-antibody interaction. Biosensors and Bioelectronics 21:201–205

Tang DP, Yuan R, Chai YQ, Zhong X, Liu Y, Dai YJ and Zhang LY (2004) Novel potentiometric immunosensor for hepatitis B surface antigen using a gold nanoparticle-based biomolecule immobilization method. Analytical Biochemistry 333:345–350

Tombelli S, Mascini M, Sacco C and Turner APF (2000) A DNA piezoelectric biosensor assay coupled with a polymerase chain reaction for bacterial toxicity determination in environmental samples. Analytica Chimica Acta 418:1–9

Trojanowicz M and Krawczyski T (1995) Electrochemical biosensors based on enzymes immobilized in electropoly-merized films. Microchimica Acta 121:167–181

Tsai WC and Lin IC (2005) Development of a piezoelectric immunosensor for the detection of alpha-fetoprotein. Sensors and Actuators B-Chemical 106:455–460

Tsiafoulis CG, Prodromidis MI and Karayannis MI (2004) Development of an amperometric biosensing method for the determination of -fucose in pretreated urine. Biosensors and Bioelectronics 20:620–627

Turner AP and Newman JD (1998) An Introduction to Biosensor. In: Gateshead TW (ed) Biosensor for Food Analysis. Athaenaeum Press Ltd, United Kingdom, pp. 13–27

Unwin PR and Bard AJ (1992) Ultramicroelectrode voltammetry in a drop of solution: a new approach to the measurement of adsorption isotherms at the solid-liquid interface. Analytical Chemistry 64:113–119

Varfolomeyev S, Kurockhin I, Eremenko A and Efremenko E (2002) Chemical and biological safety. Biosensors and nanotechnological methods for the detection and monitoring of chemical and biological agents. Pure and Applied Chemistry 74:2311–2316

Vaughan RD, Carter RM, O'Sullivan CK and Guilbault GG (2003) A quartz crystal microbalance (QCM) sensor for the detection of *Bacillus cereus*. Analytical Letters 36:731–747

Vikholm I (2005) Self-assembly of antibody fragments and polymers onto gold for immunosensing. Sensors and Actuators B-Chemical 106:311–316

Wang J (2000) Analytical Electrochemistry. Wiley-VCH, location of publication

Wang J, Li M, Shi Z, Li N and Gu Z (2002) Direct electrochemistry of cytochrome c at a glassy carbon electrode modified with single-wall carbon nanotubes. Analytical Chemistry 74:1993–1997

Wang J and Musameh M (2004) Electrochemical detection of trace insulin at carbon-nanotube-modified electrodes. Analytica Chimica Acta 511:33–36

Wang J and Musameh M (2005) Carbon-nanotubes doped polypyrrole glucose biosensor. Analytica Chimica Acta 539:209–213

Wang RH, Tombelli S, Minunni M, Spiriti MM and Mascini M (2004) Immobilisation of DNA probes for the development of SPR-based sensing. Biosensors and Bioelectronics 20:967–974

Weeks BL, Camarero J, Noy A, Miller AE, Stanker L and De Yoreo JJ (2003) A microcantilever-based pathogen detector. Scanning 25:297–299

WHO (2002) Terrorist threats to food: guidance for establishing and strengthening prevention and response systems. World Health Organization Food Safety Dept., Geneva, Switzerland

Wilson PK, Jiang T, Minunni ME, Turner APF and Mascini M (2005) A novel optical biosensor format for the detection of clinically relevant TP53 mutations. Biosensors and Bioelectronics 20:2310–2313

Winokur MJ, Guo H and Kaner RB (2001) Structural study of chiral camphorsulfonic acid doped polyaniline. Synthetic Metals 119:403–404

Winokur WJ (1998) Structural studies of conducting polymers. In: Reynolds JR (ed) Handbook of Conducting Polymer. Marcel Dekker, Inc., New York, pp 707–726

Ye JM, Letcher SV and Rand AG. (1997) Piezoelectric biosensor for detection of *Salmonella* Typhimurium. Journal of Food Science 62:1067-&

Yen Wei XT, Yan Sun (1989) A study of the mechanism of aniline polymerization. Journal of Polymer Science: Part A: Polymer Chemistry 27:2385–2396

Yun Y, Bange A, Heineman WR, Halsall HB, Shanov VN, Dong Z, Pixley S, Behbehani M, Jazieh A and Tu Y (2007) A nanotube array immunosensor for direct electrochemical detection of antigen-antibody binding. Sensors and Actuators B: Chemical 123:177–182

Zhang ZX and Li MQ (2005) Electrostatic microcantilever array biosensor and its application in DNA detection. Progress in Biochemistry and Biophysics 32:314–317

Zhao HQ, Lin L, Li JR, Tang JA, Duan MX and Jiang L (2001) DNA biosensor with high sensitivity amplified by gold nanoparticles. Journal of Nanoparticle Research 3:321–323

Zhou AH and Muthuswam, J (2004) Acoustic biosensor for monitoring antibody immobilization and neurotransmitter GABA in real-time. Sensors and Actuators B-Chemical 101:8–19

Zhou XD, Liu LJ, Hu M, Wang LL and Hu JM (2002) Detection of *Hepatitis B virus* by piezoelectric biosensor. Journal of Pharmaceutical and Biomedical Analysis 27:341–345

17

Magnetic Techniques for Rapid Detection of Pathogens

Yousef Haik, Reyad Sawafta, Irina Ciubotaru, Ahmad Qablan, Ee Lim Tan and Keat Ghee Ong

Abstract

In situations of widespread infectious disease an action that might result, the rapid diagnosis of pathogenic states will assist first responders in implementing prompt treatments, in a huge reduction in the number of illnesses and deaths. Currently available detection/diagnostic procedures are either time-consuming (8–48 h) and require enrichment and culturing of bacteria before testing, or provide only qualitative results. Magnetic immunoassay technology appears to have particularly superior performance over other immunodetection methods. A typical magnetic immunoassay entails a capture part and a detection part, between which the target is immobilized. The capture part of the immunoassay consists of magnetic particles functionalized to capture the target from the sample. The immobilized target is then sandwiched between the capture and detection complexes and subjected to a detection process that will provide accurate and rapid results, most of the time in a matter of minutes. Another important advantage that a sensitive magnetic immunoassay confers is the reduced volume of samples and reagents needed. This chapter discusses the elements associated with a magnetic immunoassay specifically designed for the rapid detection of pathogens. The chapter presents a review of the different techniques used in the synthesis and encapsulation of magnetic particles, as well as strategies for the immobilization and detection of the targeted pathogen. Several magnetic separation strategies are also discussed.

1. Introduction

The need for a quick identification by first responders of the source and type of pathogens responsible for infectious disease or for contaminating surfaces otherwise pathogen-free has increased dramatically in recent years. Many of the conventional techniques that are based on the biochemical reaction developed by Pasteur and others, as well as on bacterial culture such as the disk diffusion method developed by Kirby-Bauer, are time-consuming and necessitate enrichment and culturing on specific media (Annas 2002; Bauer et al. 1966). The automation of biochemical tests has improved the detection time, but not sufficiently to diagnose illnesses in critical patients. Rapid isolation and detection of pathogens and sensitive pathogenic indicators are therefore demanded in multiple different application settings, such as buildings,

Yousef Haik • Department of Mechanical Engineering, United Arab Emirates University, Al Ain-UAE; Center of Research Excellence in Nanobioscience, University of North Carolina, Greensboro, NC, USA. **Reyad Sawafta and Irina Ciubotaru** • QuarTek Corporation, Greensboro, North Carolina. **Ahmad Qablan** • The Hashemite University, Zarqa, Jordan. **Ee Lim Tan and Keat Ghee Ong** • Department of Biomedical Engineering, Michigan Technological University, Houghton, Michigan.

M. Zourob et al. (eds.), *Principles of Bacterial Detection: Biosensors, Recognition Receptors and Microsystems*,
© Springer Science+Business Media, LLC 2008

the environment, and food and water sources. The early detection of pathogens will enable first responders to administer the necessary disinfection or decontamination treatment on-site. Early identification of food borne pathogens in production lines or in shelved produce will protect the consumer, reduce health care costs associated with infectious disease, and prevent further contamination and spoilage of produce.

Food borne pathogens have been identified as a major cause of illness for millions of people in the U.S. alone. Since the early 1990s, an increasing number of food borne illnesses have been associated with fresh and minimally processed produce, such as green onions (*Hepatitis A* virus), lettuce (*Escherichia coli* O157:H7), cantaloupes (*Salmonella* spp), and tomatoes (*Listeria monocytogenes*) (Frost et al. 1995). An increase in global trade, a longer food chain, exposure to exotic microflora, distribution to a larger population in more geographically dispersed areas, and an aging population may all play a role in the increased number of food borne illnesses that implicate fresh produce. For example, outbreaks of shigellosis in Norway, Sweden, and the U.K. in 1994 were mainly due to contaminated lettuce imported from southern Europe (Frost et al. 1995; Kapperud 1995); and cyclosporiasis in the U.S. was linked to consumption of contaminated raspberries imported from Guatemala (CDC 1996). In developing countries, the continual use of untreated waste water and manure as fertilizer for the production of fruits and vegetables is a major contributing factor to contamination that causes numerous food borne disease outbreaks.

Recently developed isolation and recognition techniques for pathogens use immunological methods (e.g., latex agglutination, lateral flow immunoassays, ELISA, magnetic immunoassays); genetic identification (e.g., real-time PCR, RFLP, FISH); and proteomics or mass spectroscopy, or a combination of these (Safarik and Safarikova 2004; Feng 1992; Fung 1995; Gehring, Patterson, and Tu 1998; Matsunaga et al. 1996; Sharma 2002; Tu et al. 2002).

Although real-time PCR is a relatively fast, sensitive, and accurate technology, it is expensive and requires specialized personnel as well as complex equipment. The immunodetection of biological molecules appears to be simpler and is expected to develop further with improvements in the affinity, specificity, and mass production of new antibodies (Mary 1997; Chatterjee, Haik, and Chen 2001). A common characteristic of most commercially available immunoassays is the use of plastic tubes, wells, or beads as the solid phase that immobilizes the antigen or antibody on their surface.

Highly sensitive immunodetection procedures are usually costly and require complex equipment and specialized personnel, being therefore inappropriate for field testing. One of the factors affecting the sensitivity of the assay is the complexity of the tested sample (Zborowski and Chalmers 2005; Jeníková, Pazlarová, and Demnerová 2000). Therefore, research has been focused on developing filters and other solid supports to improve the capture of the target molecule (Abdel-Hamid et al. 1999; Brewster and Mazenko 1998; Leonard et al. 2004). However, filtration presents problems in collecting the target from complex samples that may clog the filter, thereby limiting the sample volume that can be used for the test. Recently, the use of magnetic particles in immunological assays has increased considerably, as magnetic properties can improve the assay sensitivity and duration (Chatterjee, Haik, and Chen 2001; Zborowski and Chalmers 2005; Jeníková, Pazlarová, and Demnerová 2000; Abdel-Hamid et al. 1999; Brewster and Mazenko 1998; Leonard et al. 2004; Varshney 2005). Considerable progress in immunodetection has been also achieved by coupling sensitive biosensors with magnetic immunoseparation (Fukuda et al. 2005; Gehring et al. 2004; Yang and Li 2005; Chen, Lei, and Tong 2005; Haik et al. 2002; Chen, Haik, and Chatterjee 2003; Haik, Chatterjee, and Chen 2005).

In a magnetic immunoassay a superparamagnetic particle is linked with the target pathogen using different immobilization techniques. Among these techniques is the use of antibody-antigen complex. In this case the magnetic immunoassay is considered as a modified ELISA, utilizing superparamagnetic particles as the solid substrate to which the base antibody can attach. The remainder of the process closely resembles a traditional ELISA with a few minor adjustments. The magnetic particles are first coated with a protein envelope that is capable of binding a modified antibody. Upon mixing with the sample, this antibody/magnetic particle complex selectively captures the target pathogen from the sample. The presence of capture beads in the suspension confers a more efficient interaction between the capture antibodies and the pathogen target, as opposed to having the antibodies anchored onto the bottom of a microwell dish. As this heterogeneous solution is subjected to a magnetic field, the pathogen/magnetic particle complex is separated from the background media without the loss of the particle complex and with higher efficiency compared to separation through centrifugation. Different techniques have been proposed for the quantification process. Among these is the use of an enzymatic reaction producing a measurable colorimetric response (Haik et al. 2002; Chen, Haik, and Chatterjee 2003). A solution containing the antibody-enzyme complex is added to the particle-pathogen complex. Once the new solution has been added and mixed with the particles, and enough time has elapsed to insure the coupling of the antibody-enzyme to the particle-antibody-antigen complex, an external magnetic field is applied to isolate the newly formed sandwich from the background media. The isolated particle sandwich is then washed repeatedly to remove any excess enzymes. Upon completion, a solution containing a substrate is added to the particle sandwich and allowed to react for a given amount of time, after which a reading of the concentration can be made via a spectrophotometer. Different techniques, including the use of chemiluminescent or fluorescent dyes, and most recently quantum dots, have been reported (Feng 1992; Fung 1995; Gehring, Patterson, and Tu 1998; Matsunaga et al. 1996; Sharma 2002; Tu 2000).

In this chapter the current techniques reported for the magnetic isolation and detection of pathogens will be reviewed. This will include a discussion of the synthesis of inorganic magnetic particles suitable for the procedure, the coating and functionalization of these particles for a targeted pathogen in a heterogeneous solution, the design of a magnetic separation system capable of capturing the complex, and the utilization of an identification technique that is sensitive and reliable and requires minimum training of the user.

2. Synthesis of Magnetic Particles

Magnetic materials are distinguished by their ability to exhibit a long-range order of their atomic magnetic moments through an exchange interaction between the neighboring atoms below a characteristic ordering temperature, the so-called Curie temperature (T_c). Above this temperature the long-range order is destroyed and material is said to be in a paramagnetic state. Magnetic materials form domains to minimize their energy. All magnetic moments are aligned in one direction within a domain, and various domains adjust themselves such as to create a net zero or nearly zero moment for a bulk sample. Magnetic domains respond to external magnetic fields and start aligning in the direction of the applied field. Magnetic domains are typically spread over a range of a few microns to hundreds of microns in a particular dimension. As the sample size is reduced the number of domains decreases, until finally a single domain is formed. When the size is further reduced, the long-range order is destroyed. Such a state is referred to as superparamagnetism. This is the most relevant magnetic state for the purpose of magnetic separation.

Magnetic separation consists of applying a magnetic force to the magnetic particles capable of moving these particles in a preferred direction against frictional drag force or gravitational attraction force. The magnetic force is proportional to the volume of the magnetic particle, the magnetic susceptibility of the particle (magnetic properties), the strength of the applied field, and the magnetic field spatial gradient. An optimized magnetic separation requires:

1. The use of magnetic fields and magnetic field gradients. High magnetic fields such as those produced at the National High Magnetic Field Laboratory can be utilized for separation, counting on either the paramagnetic or the diamagnetic separation of components within the sample. Figure 17.1 shows the separation of red cells from whole blood in a 20-Tesla (T) field. However, this sort of separation is not commercially viable. Instead, enhancing the magnetic properties of the targeted compounds by tagging these compounds with magnetically responsive particles would allow for the use of low magnetic field separators.

2. Synthesis of magnetic particles with high magnetic moments. Today's commonly used biocompatible magnetic particles are primarily ferrite-based, as ferrites are easily stabilized in fluid solutions (Chatterjee, Haik, and Chen 2001; Haik et al. 2000); although they have a relatively small magnetization capability compared to other materials, e.g., pure iron, iron cobalt, and rare earth-based materials (Fu et al. 2002). To produce particles with superior magnetic moments, the particles need to be composed of ferromagnetic materials. To minimize the particle-particle interaction and ensure superior magnetic separation, the particles need to display a superparamagnetic behavior. To achieve this behavior, the particle size needs to be maintained below the domain size (i.e., nm scale).

Optimal magnetic separation requires particles with the largest magnetic core that still maintains the superparamagnetic behavior, and with the highest magnetic moment that enables the use of particles in conjunction with a commercially viable magnetic separator. This section discusses the influence of particle size on magnetic properties and different synthesis techniques for magnetic particles. The design of magnetic separators will be discussed in a later section.

2.1. Effect of Particle Size

The magnetic body of a particle has a multidomain structure, comprised of uniformly magnetized regions (domains) separated from each other by domain walls. This type of structural

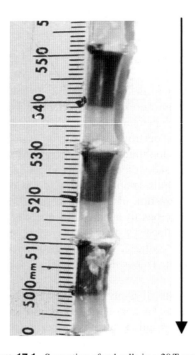

Figure 17.1. Separation of red cells in a 20 T magnet.

arrangement minimizes the magnetostatic energy and makes the system stable. Conversely, a complete balance between the magnetostatic energy, the anisotropy energy, and the exchange energy determines the domain structure and shape. When the dimension of the particle is reduced, the domain size is also reduced. Due to the energy requirement for the domain wall formation and the balance with the magnetostatic energy it limits the subdivision of the domains to an optimum value. In fact, there is a corresponding lower limit of the crystal size, below which a single domain structure exists (Kittel 1946). For a typical magnetic material the dimension limit is 20–800 nm (Dormam and Fiorani 1992). For a spherical crystal the characteristic radius is given by $R_{sd} = 9E_\sigma/\mu_0 M_s^2$, where M_s is the saturation magnetization, μ_0 is the vacuum permeability, and E_σ is the total domain energy per unit area—equivalent to $2(K/A)^{1/2}$, where K is the anisotropy energy constant, and A is a parameter representing exchange energy density. The change from a multidomain to a single domain structure is accompanied by a large increase in the coercive field (Hc $\approx K/3M_s$). With decreasing particle size, an increasing number of atoms lie near or on the surface and in the interfacial regions. This surface and interfacial electronic structure actually affects the magnetic properties. The large number of atoms located on the surfaces or interfaces, for example, whose local environments differ greatly from those of the interior atoms, leads to a distinction between the intrinsic and extrinsic properties. In small ferromagnetic materials one might expect intrinsic properties, such as M_s (saturation magnetization per unit volume), K (magnetocrystalline anisotropy constant per unit volume), T_c (Curie temperatures), M_r (remanent magnetization), and H_c (coercivity per unit volume) to differ from their bulk counterparts in a size dependent way. Another important property is that unlike the bulk ferromagnetic materials which have multiple domains, sufficiently small ferromagnetic particles (nanosized) are single-domained. A large surface to volume ratio in the nanomagnetic materials and the single domain behavior could lead to new and unusual magnetic properties.

Although the critical particle size for the single domain formation is dependent on some other factors, for most of the magnetic materials the critical size is in the range of 20–800nm (Dormam and Fiorani 1992; Morup 1993). The equilibrium magnetic properties of a large assembly of noninteracting uniaxial single domain particles, each of volume V, are largely determined by the relative magnitude of three characteristic energies: the thermal energy ($E_T = K_B T$, where K_B is the Boltzman constant); the anisotropic energy ($E_A = CV$, where C is the total anisotropy energy per unit particle volume; and the magnetostatic energy ($E_m = \mu H$, where H is the applied magnetic field and μ is the permeability of the medium). Thus, with decreasing particle size, the anisotropy energy decreases; and for a grain size lower than a characteristic value it may become so low as to be comparable to or lower than the thermal energy $K_B T$. This implies that the energy barrier for magnetization may be overcome and the total magnetic moment of the particle can fluctuate thermally like a single spin.

Thus the entire spin system may be rotated, and the spins within the single domain particles may be magnetically coupled. Superparamagnetism is exhibited by particles within a defined range of dimension. If the particles are too small, their magnetic and electrical properties change a lot compared to the bulk ones, and the superparamagnetic model cannot be applied. The lower dimensional limit has been stated as approximately 2 nm by Dorman et al. (1997), and the upper limit is given in principle by the characteristic single domain size and structure; though the characteristic grain size of a magnetic material for superparamagnetic relaxation depends on the anisotropy constant and the saturation magnetization values, and is usually less than 20 nm for spherical particles having a uniaxial anisotropy (Dorman, Fiorani, and Tronc 1997).

The size dependence of the magnetic properties is demonstrated in the following examples. Commercially available iron oxide particles from Nanophase Technologies Incorporated, along with a synthesized maghemite that was surface modified with cetyltrimethylammonium bromide

(CTAB), were subjected to magnetic measurements. The particles from Nanophase were fractionated into three different portions based on the particle diameter: fraction 1 with an average diameter of 100–150 nm, fraction 2 from 20–30 nm, and fraction 3 from 10–12 nm. The average diameter for the synthesized maghemite was 5–11 nm, with most particles at 9 nm.

Figure 17.2a shows the susceptibility vs. temperature plot for the whole unfractionated sample on warming after (a) cooling in a zero field and (b) cooling in an applied field of 50G. No blocking temperature within 300 K was observed. The blocking temperature is determined by measuring the peak position in zero-field cooled (ZFC) and field-cooled (FC) curves. Due to the presence of a wide distribution in sizes, there might be an overlap of blocking temperatures for particles with different sizes, and so they could not be determined. But when the same experiment is repeated for different fractions, the blocking temperatures for each fraction were observed as shown in Figure 17.2b. This figure shows the magnetic moment (emu/gm) vs. temperature plot for all the three fractions of maghemite, along with the synthesized one. Even though fractions 2 and 3 contain smaller particles, there is still a significant distribution of particle sizes (Chatterjee, Haik, and Chen 2003). For this reason, a very sharp blocking temperature was not observed.

The variation of temperature at which ZFC and FC separate (T sep) and the maximum (T_B) in each plot, with the sizes, are shown in Table 17.1.

Magnetization plots as a function of the magnetic field are shown in Figure 17.3 for sample fraction 3 (the fraction with the smallest size in the commercial sample), at both 5 K and 300 K. The blocking temperature of this fraction was about 100K, which was below room temperature. Below the blocking temperature, the particles showed ferrimagnetic behavior, with an increase in Mr and Hc. A field is required to bring the total sample moment to zero, since the thermal energy of the system ($K_B T$) is less than the energy barrier, and there is no other

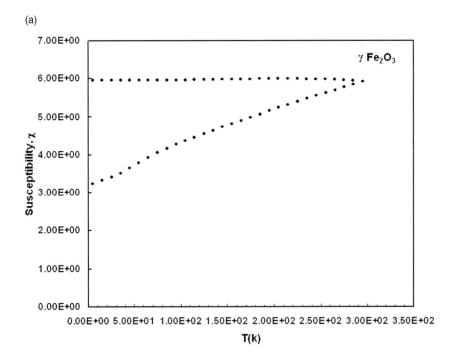

Figure 17.2a. Temperature-susceptibility plot for unfractionated commercial maghemite (reprinted from Chatterjee et al. (2003), with permission of Elsevier).

(b)

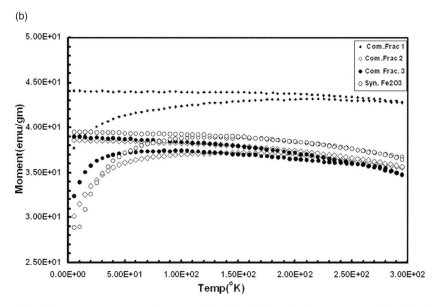

Figure 17.2b. Temperature-moment plot for synthesized and for different fractions of commercial maghemite as well as synthesized maghemite (reprinted from Chatterjee et al. (2003), with permission of Elsevier).

mechanism than the external field to rotate the particle/domain orientations and randomize the system. In the paramagnetic case, K_BT is greater than the energy barrier, and thus thermal energy can "randomize" the system and bring the moment to zero. It is totally expected that Hc and Mr are greater than zero for ferrimagnetic and zero for superparamagnetic systems (Chatterjee, Haik, and Chen 2003). The results showed that at 5 K, Hc was about 300 Oe, while at 300 K it was about 30 Oe. Similarly, Mr changed. The area under the M-H curve, which represents the work done by the magnetic field, has also been changed. In the superparamagnetic case, given the fact that K_BT is greater than the energy barrier, only thermal energy is required to reorient the 8 domains/particles, and diminishing hysteresis is observed as expected in the superparamagnetic behavior.

Rare earth magnetic materials have magnetic moments larger than those of conventionally characterized magnetic materials. Synthesizing magnetic particles that contain a rare earth high-energy magnetic material has gained interest in biorelated applications. These particles can remain small but maintain relatively high magnetic moments compared to iron oxides. This enables effective separation with low magnetic fields.

The composition of the metallic content in the synthesized particle dictates the stability and the magnetic moments. Fe-Nd-B particles were synthesized with three different compositions:

Table 17.1. Comparison between the commercially available and synthesized maghemite

Sample	Tmax, K	Tsep, K	Dia(nm)
Frac 1 (com)	205	280	100–150
Frac 2 (com)	130	270	20–30
Frac 3 (com)	100	125, 270	10–12
Syn. Maghemite	75	235	5–11

(a)

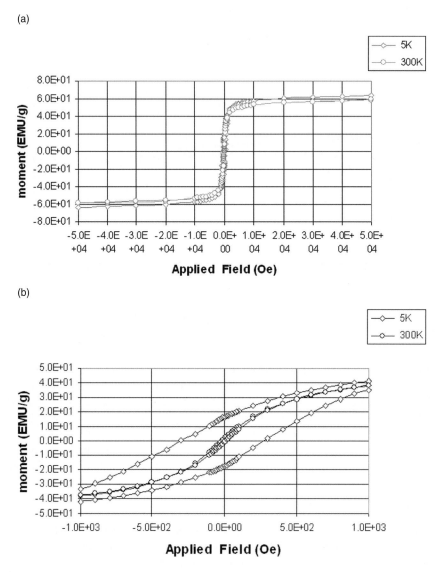

(b)

Figure 17.3. Applied field vs. magnetization plot for the third fraction (smallest size) of the commercial maghemite (reprinted from Chatterjee et al. (2003), with permission of Elsevier).

$Fe_{5.6}NdB_2$, $Fe_{5.4}NdB_9$, and $Fe_{5.2}NdB_{13.7}$. The particles were mostly in the size range of 50–200 nm. The hysterics plots (Figure 17.4) at 300 K for all samples showed a superparamagnetic nature as there is almost no remanent magnetization. The coercivity of these samples is negligible (6 Oe-20Oe), compared to the $Fe_2Nd_{14}B$ alloy particles (6 KOe) produced by the reduction diffusion method. The saturation magnetization is in the range of 6–24 emu/gm. The magnetic moment of the sample with higher borohydride concentration ($Fe_{5.6}NdB_2$) is much lower, compared to the other. The composition of the material certainly plays an important role in the magnetic property as observed from the hysterisis plot (Haik, Chatterjee, and Chen 2005).

The properties of magnetic particles are modulated depending on their application. For example, if particles are intended for use in pathogen immunomagnetic separation or detection, then their size and composition are important. Particles with sizes equal to or smaller than those of the pathogen to be detected are favored over larger particles. Also, high magnetic

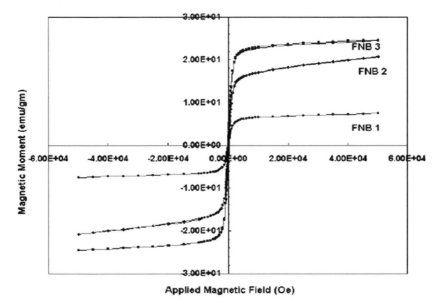

Figure 17.4. Applied magnetic field vs. magnetization plot for Fe-Nd-B particles. (Reprinted from Haik et al. (2005), with kind permission of Springer Science and Business Media).

moment particles are favored. The optimization targets the synthesis of small particles (less than 100 nm) with high magnetic moments.

2.2. Synthesis Techniques

The performance of magnetic particles depends on their properties. These properties in turn depend on the structural atomic composition of the particles, and on their interfaces and defects, which are all controlled by the thermodynamics and kinetics of chemical synthesis. There are two different methods to synthesize magnetic particles: chemical and physical.

Chemistry has played a major role in developing new materials with novel technologically important properties. The advantage of chemical synthesis is its versatility in designing and synthesizing new materials that can be refined into the final product. The primary advantage that chemical processes (e.g., borohydride reduction, chemical coprecipitation, and the polyol method) offer over other methods is a good chemical homogeneity, since chemical synthesis offers mixing at the molecular level, which enables the synthesis of dispersed, fine particles with a very narrow size distribution.

One of the physical methods of synthesizing magnetic nanoparticles (NPs) involves arc melting of the constituent metals in an inert furnace and ultrasonication for reducing the particle size to the nanometer range.

2.3. Encapsulation of Magnetic Particles

During the immunoseparation/immunodetection, metallic particulates may inappropriately interact with the pathogen target, causing a denaturation in its cellular structure or even cell death. If the metal composition of the capture beads is biocidal towards the targeted pathogen, the performance of the magnetic detection system is compromised. A solution to this problem is the encapsulation of metallic particles with materials (e.g., polymers, proteins) to protect the biological targeted against the potential toxic effects of the particles. Various studies have been conducted to identify coating polymers or proteins that act as a protective shell (Gref et al. 1995; Gupta and Wells 2004; Chastellain, Petri, and Hofmann 2004; Vila et al. 2004). Polyethylene glycol (PEG) (Gupta and Wells 2004; Chastellain, Petri, and Hofmann 2004), polyvinyl alcohol (PVA) (Vila et al. 2004), polylactic acid-polyethylene glycol (PLA-PEG) and PLA-PEG-PLA (Ruan and Feng 2003; Ruan, Feng, and Li 2002), and polylactic glycolide (PLGA) (Solaro 2002) have been reported as suitable coating materials. Albumin, a protein isolated from human serum (HSA) or bovine serum (BSA) has also been used for coating magnetic NPs (Chatterjee, Haik, and Chen 2001).

2.3.1. Methods of Preparing Polymer/Protein Coatings

Coating the particles using polymer/protein materials takes place by entrapping the particles inside the hollow nanospheres or microspheres formed by the coating material in suspension. Polymer-coated microparticles (MPs) and nanoparticles (NPs) belong to the class of multiphase systems in which one or more microphases are dispersed in a continuous matrix of different composition or physical state. The main characteristic of colloidal dispersions is the extremely large interface area between the dispersed phase and the continuous phase. Colloidal dispersions are metastable or unstable, since minimization of interface free energy between two different phases is dictated by thermodynamic constraints.

However, in some cases colloids display significant kinetic stability that prevents their aggregation. Hence, production of MPs and NPs relies essentially upon the chemical production of colloidal dispersions, their kinetic stabilization, and effective recovery of the final formulates. Polymeric materials are constituted by large molecules whose peculiar solution characteristics often allow for the preparation of stable and size-controlled colloidal dispersions, which in turn can be converted into MPs and NPs. In addition, several polymers can be used as stabilizers of colloidal dispersions, since they provide a surface coating of the metastable microphase, thus lowering its tendency to phase-aggregation. The common feature of all methods for the preparation of MPs and NPs is the externally-induced separation of at least two phases. This process is better known as coacervation, and it may be promoted by a number of different techniques.

2.3.1.1. Solvent Displacement Method

The solvent displacement method is a straightforward procedure in which the polymer, particles, and, if necessary, lipophilic stabilizers are dissolved in a semipolar water-miscible solvent, such as acetone or ethanol. The organic solution is then poured or injected into an aqueous solution containing a stabilizer under stirring. NPs are formed instantaneously by rapid solvent diffusion, and the organic solvent is then eliminated from the suspension under reduced pressure. Even though precipitation or nanoprecipitation is often used to define this method, it is important to stress that the formation of NPs is due to polymer aggregation in stabilized emulsion droplets.

The major limit to the application of this technique is the difficulty of finding a polymer/solvent/nonsolvent system in which NPs are formed and the particles efficiently entrapped (Vanderhoff, El-Aasser, and Ugelstad 1979).

2.3.1.2. Salting Out Technique

This method is based on the separation of a water-miscible solvent (acetone) phase from aqueous solutions, promoted by a salting-out effect. Water is added to the emulsion obtained by the addition of an acetone solution of polymer and particles emulsified in an aqueous gel containing the salting-out agent and the colloidal stabilizer. As a consequence of dilution, acetone diffuses into the water, resulting in NP formation. Solvent and salting-out agents are then eliminated by cross-flow filtration. This procedure allows for the incorporation of large amounts of particles with excellent yields, and procedure scale-up is fairly easy.

2.3.1.3. Emulsion Diffusion Method

The emulsion diffusion method is a slight modification of the salting-out technique. It differs mainly in that the organic solvent is only partially miscible with water; and it is previously saturated with water, in order to reach an initial thermodynamic equilibrium between the water and the organic phase. After the addition of water, solvent diffusion is observed, and a suspension of NPs is formed.

2.3.1.4. Solvent Evaporation Method

The solvent evaporation method is a well-known technique (Ruan and Feng 2003; Ruan, Feng, and Li 2002) that basically consists in the formation of a bi-phase (o/w) or tri-phase (w/o/w) emulsion. The inner phase is constituted by a polymer solution in organic solvent in the biphase procedure, and water in oil emulsion in the triphase method. In both cases, the continuous phase is an aqueous solution in which the polymer is insoluble. The resulting emulsion is then exposed to a high-energy mixer, such as an ultrasonic device, a homogenizer, a colloid mill, or a microfluidizer to reduce the globule size. The removal of the organic solvent by heat, vacuum, or both results in the formation of a fine aqueous dispersion of NPs which can be then collected and purified. This method is widely used for the preparation of MPs and NPs made of polysaccharides; aliphatic polyesters, such as PLA, PLGA, or PGA; and other synthetic polymers, such as PEG copolymers. The solvent evaporation method may present some drawbacks. In fact, toxic chlorinated solvents, such as chloroform and methylene chloride, are often used because of their water insolubility, easy emulsification, solubilizing properties, and low boiling point. Moreover, the evaporation step can result in the agglomeration of MPs. The HSA-coated NPs were prepared using a modified version of this method by chemical crosslinking.

2.3.1.5. Polymer Emulsion Process

The polymer emulsification process patented by Vanderhoff et al. (1979) is a modification of the solvent evaporation method. This process comprises intimately dispersing a liquefied polymer phase in an aqueous liquid medium phase containing at least one nonionic, anionic, or cationic oil-in-water functioning emulsifying agent, in the presence of a compound selected from the group consisting of those hydrocarbons and hydrocarbyl alcohols, ethers, alcohol esters, amines, halides, and carboxylic terminal aliphatic hydrocarbyl group of at least 8 carbon atoms, and mixtures thereof, and subjecting the resulting crude emulsion to the action of comminuting forces sufficient to enable the production of an aqueous emulsion containing polymer particles averaging less than about 0.5 μm in size. The solvent is preferably one which may be readily removed following the comminuting step. In most instances an emulsion is desired which is devoid of organic solvents that pose problems of flammability, pollution, toxicity, and/or odor and the like. To remove the solvent, a vacuum evaporation technique is generally employed. The solvent should be devoid of an aliphatic hydrocarbyl group of 8 or more carbon atoms, and inert in the emulsion. The oil-in-water functioning emulsifying agents employed in the process are generally surface active agents also useful in the detergents field. The emulsifying agent is operative in the aqueous medium in relatively low concentrations, being generally included in proportions of about 0.1–5% (and preferably about 0.2–3%) by weight of water in the aqueous phase. The higher aliphatic hydrocarbyl-containing additive compound, or mixture thereof, required in carrying out the process is generally employed in proportions of about 0.2–12% (and preferably about 0.4–6%) by weight of the polymer phase.

The resulting crude emulsion of coarse polymer phase droplets containing the magnetic NPs is then subjected to the action of comminuting forces sufficient to enable the production of an aqueous emulsion containing polymer particles with magnetic NPs. An ultrasonicator was used in this study to supply this force.

The crude emulsion is passed through the comminuting device a sufficient number of times, usually two, three, or more, until an emulsion is obtained containing the desired small size polymer phase particles.

2.3.2. Examples of Polymer/Protein Encapsulated Particles

In the examples presented below metallic particles were coated with various polymers or with the protein albumin. The morphology of these coated particles under TEM is shown in Figures 17.5–17.9. The dark shaded regions (metallic NPs) within the lighter shaded circle (polymer spheres) indicate that the NPs were well encapsulated within the polymeric shell. Also, the empty light grey circles are empty polymeric shells produced by the excess of polymer introduced in the solution compared to the metallic particles.

- *PVA encapsulation (Figure 17.5)*: Iron oxide NPs (obtained from Nanophase Technology Corporation) were coated using PVA (MW: 125,000, Polysciences Inc.).
- *PEG encapsulation (Figure 17.6)*: Iron oxide NPs (Nanophase Technology Corporation) were coated using PEG (MW: 1,540, Polysciences Inc.), by the polymer emulsion method. The other ingredients used with this method were a solvent (methylene chloride), water, emulsifying agent (sodium dodecyl sulfate), and an inhibitor compound (1-octanol). The magnetic particle to polymer ratio was approximately 1:40.
- *Ethyl-cellulose encapsulation (Figure 17.7)*: Iron oxide NPs (Nanophase Technology Corporation) were coated using ethyl cellulose (glass transition temperature of 42°C), by the polymer emulsion method. The other ingredients used with this method were a solvent (methylene chloride), water, emulsifying agent (sodium dodecyl sulphate), and an inhibitor compound (1-octanol). The magnetic particle to polymer ratio was approximately 1:40.
- *PEG encapsulation using the glutaraldehyde crosslinking method (Figure 17.8)*: Iron oxide NPs (Nanophase Technology Corporation) were suspended in methylene chloride, and PEG (MW: 125,000, Polysciences Inc.) was added to this suspension. Glutaraldehyde was then added as a crosslinking agent, the suspension stirred, and the particles washed in acetone and stored in PBS.
- *HSA-encapsulated Gd-Zn-Ferrite nanoparticles (Figure 17.9)*. Human serum albumin (HSA) is a single-chain polypeptide of 585 residues, which is synthesized in the liver and comprises about 60% of the plasma protein (Mohamadi-Nejad 2002). Gd-Zn-Ferrite NPs with Gd = 0.02 were used. Encapsulation of Gd-Zn-Ferrite particles with HSA required the preparation of a glutaraldehyde solution saturated with toluene, then the preparation of HSA–Gd-Zn-Ferrite particles in water, followed by the crosslinking of HSA microspheres/nanospheres with glutaraldehyde.

3. Immobilization Strategies

Magnetic NPs have controllable sizes ranging from a few up to tens of nanometres, which places them at dimensions that are smaller than or comparable to those of a cell (10–$100\,\mu$m), a virus (20–450 nm), a protein (5–50 nm), or a gene (2 nm wide and 10–100 nm long). This means that they can "get close" to a biological entity of interest. They can be coated with

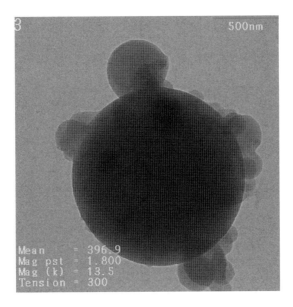

Figure 17.5. PVA-encapsulated iron oxide particles.

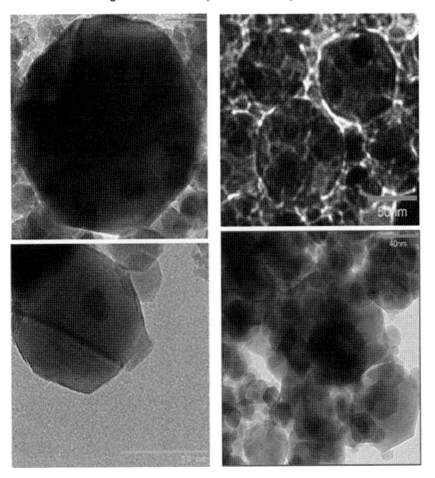

Figure 17.6. PEG-encapsulated iron oxide particles.

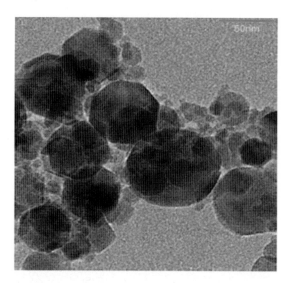

Figure 17.7. Ethyl-cellulose-encapsulated iron oxide particles.

biological molecules or synthetic materials that facilitate the binding and immobilization of biological targets on their surfaces. The particles used in the magnetic separation are usually made out of superparamagnetic materials. These materials are magnetically responsive only in the presence of an applied magnetic field; upon completing the separation and removing the applied fields, the superparamagnetic particles go back into suspension. The choice of magnetic particles is important, to avoid the agglomeration of particles because of magnetic interaction between them.

There are several strategies that have been reported in the literature for functionalizing the surface of magnetic particles with the aim of immobilizing specific targets:

1. Encapsulated magnetic core with a coating bearing an immobilized moiety with a specific affinity for biological entities. The immobilized structure can be a specific antibody, a

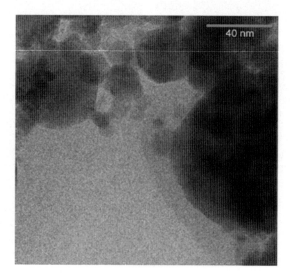

Figure 17.8. PEG-encapsulated iron oxide particles prepared by the glutaraldehyde crosslinking method.

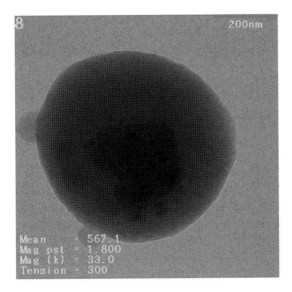

Figure 17.9. HSA-encapsulated Gd-Zn-Ferrite particles.

hydrophobic ligand, or an ion-exchange group. A commonly used system for specific separation of a biological entity in heterogeneous solutions uses a streptavidin-biotin linkage to attach the antibody to the magnetic particles. Also, imidized carbohydrate capsules enable binding the antibodies to the particle surface via an imido-amine linkage. Some other affinity ligands such as lectins, nitrilotriacetic acid, and gelatine, to name a few, have also been reported for the functionalization of the particles.

2. Encapsulated magnetic core with a surface structure having a nonspecific affinity for biological entities. Biopolymers with functionalized groups such as chitosan and agarose can be impregnated with a magnetic core for magnetic separation applications. Polyionic macromolecules obtained by polymerization of functional monomers provide cationic or anionic groups along the polymer chain. Examples of these include poly-4-vinyl-N-methylprifinium bromide and potassium polyacrylate.

Table 17.2. Functionalization of magnetic particles surface

Group on particle surface	Activating Reagent	Active Derivative
-COOH	R-N=C=R'	$-COO-C\begin{smallmatrix}NHR\\ \\N^+HR'\end{smallmatrix}$
-NH$_2$	OHC(CH$_2$)$_3$CHO	-N=CH(CH$_2$)$_3$CHO
-CONH$_2$	OHC(CH$_2$)$_3$CHO	-CON=CH(CH$_2$)$_3$CHO
-OH -OH	CNBr	$\begin{smallmatrix}-O\\ \\ -O\end{smallmatrix}C=NH$
-OH	$\begin{smallmatrix}O\\\|\\Cl-S-R\\\|\\O\end{smallmatrix}$	$\begin{smallmatrix}O\\\|\\-O-S-R\\\|\\O\end{smallmatrix}$
$-CH_2-\overset{O}{\overset{\diagup\diagdown}{CH}}-CH_2$	None	

3. Naked magnetic particles with a modified surface having an affinity to biological entities. The surface of an inorganic magnetic material such as maghemite can be modified by silanization to functionalize the surface with appropriate groups that enable the immobilization of ligands.
4. Adsorption of naked or encapsulated magnetic particles to the surface of biological structures. Carbonated magnetic NPs can be adsorbed onto the surface of biological entities.

3.1. Modification of Particle Surface with a Ligand

The formation of protein monolayers onto hydrophobic surfaces is desirable because of the monolayers' resistance to repeated washing. For this, covalent binding between specific groups on the hydrophobic material, and the NH_2 or SH groups on the protein, is required. Olsvik et al. (1994) lists (Table 17.2) the most common surface groups, activating agents, and active derivatives that are used in functionalizing the surface of particles.

4. Biological Targets

As discussed above, the biological target can be immobilized onto the surface of magnetic particles in a specific (e.g., antibody) or nonspecific (e.g., adsorption) way. Functionalized magnetic particles are used to immobilize the biological target and separate it from other components in a process known as immunomagnetic separation (IMS). This rapid method can specifically identify and efficiently collect even low numbers of biological targets that will be further subjected to detection. For example, bacteria bound to the surface of magnetic particles are usually viable and are further tested, with or without amplification, on appropriate media. Alternatively, the immobilized bacteria can be detected using a labeled antibody paired to the antibody used for capturing the target; this will emit a signal proportional to the amount of pathogen captured during IMS. This sandwich method is described in Section 5 below. Table 17.3 lists pathogens and examples of capture and detection antibodies that have been used as pairs in several studies (Olsvik et al. 1994). Various IMS systems have been developed; their efficiency in separating the targets varies widely, depending on the magnetic moment of the particles used, the binding efficiency on their surfaces, the complexity of the samples from which the target is separated, and also the magnitude of the external magnetic field to which the particles are subjected after the collection.

5. Magnetic Immunoassays

5.1. Direct Immunoassay Detection Using Magnetic Beads

Traditionally, magnetic beads are used mainly in preparing samples prior to measurement by other techniques. However, in recent years they have been increasingly used for direct quantification of bacteria by measuring their magnetic signals. When magnetic beads are exposed to a magnetic field generated by an inductive coil, they alter the magnetic field B; and the degree of alteration, in terms of the measured voltage V, can be measured based on Faraday's Law:

$$V = -NA\frac{dB}{dt}, \tag{17.1}$$

where N is the number of turns in the coil, dB/dt is the change of magnetic flux, and A is the area of the coil. Therefore, the number of magnetic beads can be directly determined from the measured voltage.

Table 17.3. List of Ligands

Pathogen	Ligand
Escherichia coli	Goat anti-E. coli (Gehring et al. 2004) Monoclonal mouse anti-K88 (Lund, Hellemann and Vartdal 1988; Lund, Wasteson and Olsvik 1991)
Listeria moncytogenes	MAb anti-L. monocyogenes flagella (Skjerve, Rorvik, and Olsvik 1990). PAb Anti-L monocyteogenes rabbit IgG (Hibi et al. 2006)
Pseudomonas putida	MAb anti-P-putida flagella (Morgan et al. 1991)
Salmonella	Goat-anti-Salmonella structural antigen (Blackburn, Patel and Gibbs 1991; Skjerve and Olsvilk 1991; Vermunt, Franken, and Beumer 1992; Yang and Li 2006)
Shigella dysenteriae and S. Flexneri	MAbs anti-S. dysenteria serotype I and S. flexneri serotypes 1 to 5 (Islam et al. 1993)
Staphylococcus aureus	Rabbit anti-S. aureus exopolysaccharide (Johne, Jarp and Haaheim 1989)
Vibrio parahaemolyticus	Rabit anticapsule (Olsvik et al. 1991)
Yersinia enterocolitica	Rabbit anti-O antigen (Kapperud et al. 1993)
Chlamydia trachomatis	Rabbit anti-chlamydia lipopolysaccharide (Hedrum et al. 1992)
Rickettsia conorii	MAb against endothelial cells (Drancourt et al. 1992)
HIV	MAb against CD4 antigen (Brinchmann, Albert and Vartdal 1991; Brinchmann 1989).
Cytomegalovirus	Biotinlayted amplicons (Brytting et al. 1992)

MAb: Monoclonal antibodies; PAb: Polyclonal antibodies

Quantification of analytes using magnetic markers is preferred to conventional techniques mainly because this approach is less expensive, more time efficient, and more user-friendly (e.g., remote control, query nature, mobile). Furthermore, in the case of paramagnetic beads, the specificity of the magnetic markers can be enhanced by controlling the applied magnetic field, a process known as "magnetic washing" (Baselt et al. 1998; Lee et al. 2000). The magnetic markers can easily differentiate specific and nonspecific analytes simply by varying the applied magnetic field.

5.1.1. Superconducting Quantum Interference Devices

Superconducting quantum interference devices (SQUIDs) remain the most sensitive magnetometers known. They are used to measure extremely small magnetic fields with a noise level as low as $3\,\text{fT} \cdot \text{Hz}^{-1/2}$. Due to their sensitivity, SQUIDs are widely used in biological studies which, most of the time, involve the emission of weak magnetic fields. The incorporation of SQUIDs and nanosized magnetic beads has been widely applied in clinical and research studies. Flynn and coworkers (Flynn et al. 2007) have reported a technique using SQUIDs and magnetic NPs to inspect transplant rejection. This technique allows the monitoring of transplanted organs without performing any invasive surgeries, thus reducing the incidence of infection and allowing long-term monitoring.

Many studies have been done to incorporate SQUIDs in microbial detection by measuring the magnetic beads, commonly magnetite particles, produced by various living microorganisms. The emissions of magnetic fields from these biogenic nanosized magnetic particles are extremely weak and can be detected only with highly sensitive magnetometers such as SQUIDs. To detect the magnetic beads using a SQUID, antibody-functionalized beads are first added into the sample solution. Magnetically-labeled analytes are extracted using an externally applied magnetic field. The extracted target analytes are then exposed to a magnetic field generated by a SQUID, and the magnetic flux produced by the magnetic beads is measured.

The incorporation of magnetic markers with SQUIDs offers certain advantages. For instance, the incorporation of magnetic beads in an ELISA assay has been shown to improve the detection limit. As a result, SQUIDs can be used along with the magnetic markers when an extremely low concentration of target analytes is used. Furthermore, SQUIDs have been used in the detection of weak magnetic fields produced by the motion of magnetotactic bacteria (Chemla et al. 1999). Magnetite particles produced by microorganisms are somewhat correlated to bacteria activities. One such study performed by Chemla and coworkers (Chemla et al. 1999) was to evaluate the motion of bacteria by determining the rotational drag coefficient, the magnetic moment, and the frequency and amplitude of the vibrational and rotational modes. Chemla's work has suggested that similar assays can be done when cells of interest are tagged with magnetic beads.

5.1.2. ABICAP Column

The measuring system based on the ABICAP column is one of the commonly used techniques for magnetic beads quantification. As illustrated in Figure 17.10, the main components of this measuring system are the excitation coil, the detection coil, and the ABICAP column. The ABICAP column consists of an ABICAP filter, which is commonly made of polyethylene that can be further modified to bind a specific antibody. When the sample solution is added into the column, target analytes bind to the antibody coated on the ABICAP filter. This is followed by adding secondary antibody-coated magnetic beads into the column. The subsequent binding of the secondary antibody to the bound antigens immobilizes the magnetic beads. The changes in magnetic flux due to the incoming magnetic beads can be measured with the detection coil, which indirectly quantifies the concentration of the targeted antigens. This technique has been used for the detection of microbial species including *Yersinia pestis* (Meyer et al. 2007a) and *Francisella tularensis* (Meyer et al. 2007b).

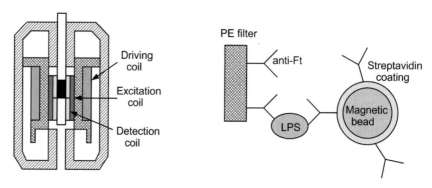

Figure 17.10. (a) Schematic diagram of an ABICAP column; (b) binding of target analytes to primary (capturing) antibody and secondary-antibody-coated magnetic beads.

A few studies have reported that the type of saline solution used in a sample has little or no effect on the magnetic signal. One such study conducted by Meyer and coworkers (Meyer et al. 2007a) was to determine the effect of phosphate buffered saline (PBS) and human blood serum in bacteria detection, in this case, YPF1. This study demonstrated a linear relationship between the magnetic signal and the antigen concentration. In addition, the study showed a high coefficient of determination regardless of the buffer solutions used, indicating that there was no nonspecific binding of antibody to antigen. The high linearity of this technique implies that this system is stable and reliable within a range of antigen concentration.

5.2. Indirect Immunoassay Detection Using Magnetic Beads

The principle of magnetic immunoassays is based on the identification, separation, and detection of a biological target using a capture moiety on the surface of magnetic particles and a labeled detection moiety that acts as an indicator agent (e.g., color change, fluorescence, electrical impedance). Figure 17.11 shows a schematic of a sandwich method.

For certain applications, magnetic beads are not directly used in bacteria quantification but simply act as a means to assist other techniques for bacteria detection. Magnetic beads are used as parts of the process to enhance the sensitivity of the detection techniques by specifically selecting the target analytes. In some assays requiring multiple washing steps, the utilization of magnetic beads is also able to shorten the washing steps, thus reducing the damage and degradation to the analytes.

5.2.1. ELISA

Enzyme-linked immunosorbent assay (ELISA) is a technique that involves the pairing of antigen and antibody to specifically isolate analytes prior to rendering them to detection. ELISA has found itself applicable in clinical immunology and food industries in detecting the presence of antigens such as HIV, hepatitis B and C, and hormone levels, as well as allergens in foods such as milk, peanuts, and eggs. Generally, there are two types of ELISA: sandwich ELISA and competitive ELISA. Both assays use the concept of an antigen-antibody complex formation. This technique is broadly used because of its high specificity. Figure 17.11 demonstrates a typical sandwich ELISA operating procedure.

As illustrated in the figure, a known sample of primary antibody is immobilized on a solid substrate and an unknown sample of antigen is added and bound to the primary antibody. This is followed by adding a secondary antibody that leads to subsequent binding to the already bound antigen. To quantify the analytes, an enzyme that is specifically bound to the secondary

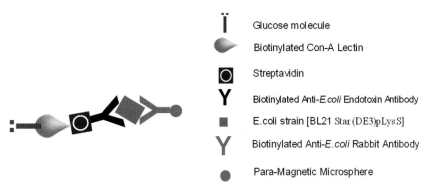

Figure 17.11. Schematic representation of a two-site magnetic immunoassay.

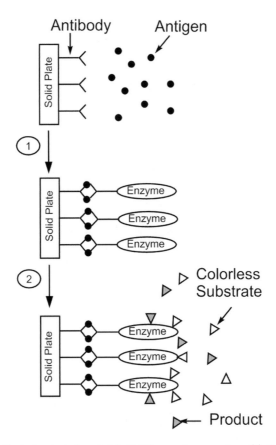

Figure 17.12. Diagram illustrating typical sandwich ELISA operating procedure: (1) Binding of target antigen to primary antibody; (2) Binding of enzyme-conjugated secondary antibody to bound antigen.

antibody is added, followed by an enzymatic substrate that leads to visible light generation. Irrespective of the detection system used for ELISA, the detected signal is typically proportional to the amount of analyte being measured.

The ability to encounter many more biological targets, as well as the tremendous increase in capture surface that the particles offer compared to the flat surface of an ELISA plate, is responsible for the superior sensitivity that magnetic immunoassays have. For example, ELISA kits for bacteria detection have sensitivities of 10^5–10^7 bacterial cells/ml (Kim et al. 1999) and require overnight enrichment and amplification of the sample prior to analysis (Feldsine et al. 1997), while the sensitivity of the magnetic immunoassay can be as low as 1–10 bacteria/cell.

Moreover, conventional ELISA has been known for its false positive and false negative detection because of its nonspecific binding to some antigens. Hence, magnetic beads are often used in magnetic ELISA to ensure efficient IMS and therefore to obtain good quality analyte isolation from complex samples. The incorporation of magnetic beads enhances the sensitivity of ELISA while maintaining its accuracy by specifically selecting targeted antigens before proceeding to the following steps (Nagasaki et al. 2007; Chou et al. 2001). This technique is particularly useful when a mixture of sample solution is being examined. Functionalized magnetic beads that are specifically bound to targeted antigens can be extracted by applying an external magnetic field. This narrows down the possibilities of nonspecific binding and shortens the duration of the assay evaluation (Gundersen et al. 1992; Gehring et al. 2004). The incorporation of magnetic markers with ELISA also improves the cost efficiency compared to other techniques used in similar detections. In addition, the incorporation of magnetic beads and ELISA has shown good stability in antigen detection (Chou et al. 2001). It is known that the stability of an assay decreases when it involves complex multiple steps. As in ELISA, these complex multiple steps are analogous to washing steps that remove any unbound particles surrounding the solid substrate. The use of magnetic markers essentially reduces or eliminates these complex steps, thus preserving the integrity of the assay.

The quality of the immunodetection depends not only on the efficiency of IMS but is greatly affected by the detection method. The detection markers that have been reported in the literature are described briefly below.

1. Colorimetric indicators: A photometric signal is given by a change in the detection side of the sandwich where the labeled moiety undergoes transformations, such as redox or ion exchange reactions, proportional to the amount of the target detected. The detection of such colorimetric change is performed using a spectrophotometer. The commonly used detection moiety labels are enzymes, such alkaline phosphatase (AP) or horseradish peroxidase (HRP) (Padhye and Doyle 1991; Wright, Chapman, and Siddons 1994; Yu and Bruno 1996; Johnson, Brooke, and Fritschel 1998; Bayliss 1999).

Recently we reported a technique that utilizes the glucose/glucose oxidase system as the detection system of an in-house developed magnetic immunoassay. In this system, glucose is used as a label attached to the detection antibody, and the amount of glucose in the sandwich is directly proportional to the amount of target immobilized in the sandwich. Glucose can be measured using electrical, colorimetric, and acidity detection methods. Due to sensitivity problems in both the electrical and the acidity approaches, the colorimetric approach is preferred. In a colorimetric assay, glucose is allowed to react with its enzyme glucose oxidase (GOX) to form gluconic acid and hydrogen peroxide (H_2O_2). Subsequently, H_2O_2 reacts with a reduced form of chromogenic compound and oxidizes it to a colored compound that is monitored by reflectance photometry (Vote et al. 2001).

$$\text{Glucose} + O_2 \xrightarrow{\text{GOX}} \text{Gluconic acid} + H_2O_2$$

$$H_2O_2 + \text{Reduced Chromogen} \xrightarrow[H_2O]{} + \text{Oxidized Chromgen}$$

$$\text{(Colorless)} \qquad\qquad\qquad \text{(Colored)}$$

Magnetic particles used in this assay were produced using established protocols (Chen, Haik, and Chatterjee 2003; Chatterjee, Haik, and Chen 2003). They were then coated with albumin, which was further coupled with avidin. This allowed a further anchorage of the capture antibody onto the particle surface via a (strept)avidin-biotin binding. Both monoclonal and polyclonal antibodies have been used with this assay depending on the biological target (e.g., proteins, bacteria). The detection antibody was labeled with glucose in a series of steps that entailed attaching a biotinylated lectin (Concanavalin-A) to the avidinated antibody, followed by the addition of the glucose to the system via its binding to Con-A. The Con-A, which is one of the most widely used and well-characterized lectins, has broad applicability primarily because it recognizes the commonly occurring sugar structures α-D-Mannose and α-D-Glucose through its four saccharide binding sites. At neutral and alkaline pH, Con-A exists as a tetramer of four identical subunits of approximately 26 kDa each. Below pH 5.6, Con-A dissociates into active dimers of 52 kDa. The nonspecific binding that occurred during the assay was prevented by using blocker buffer containing albumin and surfactant. The separation of the sandwich containing the biological target from the sample (e.g., blood, food, saline) was accomplished by applying an external magnetic field. After washing, the sandwich was incubated with a GOX reagent. The chemical reaction that occurred between the glucose oxidase enzyme and its substrate (glucose label) resulted in a pink-like color, which was detected spectrophotometrically at $A_{500\,nm}$. The determination of peripheral glucose molecules attached to the end of the whole conjugate is the indirect measurement of the amount of biological target in the sandwich. Figure 17.13 represents the relationship between various bacterial concentrations and their correlated glucose concentrations. The absorbencies of each bacterial dilution were matched with the standard absorbencies that were taken from the glucose standard curve, and the glucose concentration was calculated for each dilution using the curve standard equation $y = 41.046 \times^{-0.1945}$. In order to standardize glucose measurements, serial glucose concentrations prepared in distilled water were treated under the same experimental conditions and read spectrophotometrically at the same wavelength, A_{500}. The resulting transmittances were different and logarithmically correlated with the various glucose concentrations. The highest concentration had the lowest transmittance, as shown in Figure 17.14.

2. *Fluorescence indicators*: The classical principle of colorimetry brings sensitivity to the assay, but at the same time it is a laborious process that requires multiple washes and reagents. A more convenient method for the sensitive and simultaneous detection of multiple biomarkers uses fluorescence tagging. However, there are chemical and physical limitations associated with the use of classical organic fluorescent dyes. Dye photo-bleaching that occurs upon prolonged exposure to excitation light or storage is one of these limitations. Another drawback is the variation of the excitation wavelengths of different colored dyes. As a result, simultaneously using two or more fluorescent tags with different excitation wavelengths requires multiple excitation light sources.

This requirement thus adds to the cost and complexity of methods utilizing multiple fluorescent dyes (Yang and Li 2005; Zhao and Shippy 2004). Another drawback of organic dyes is the spectral overlap that exists from one dye to another. This is due in part to the relatively wide emission spectra of organic dyes and the overlap of the spectra near the tailing region. Few low molecular weight dyes have a combination of a large Stokes shift, which is defined as the separation of the absorption and emission maxima, and high fluorescence

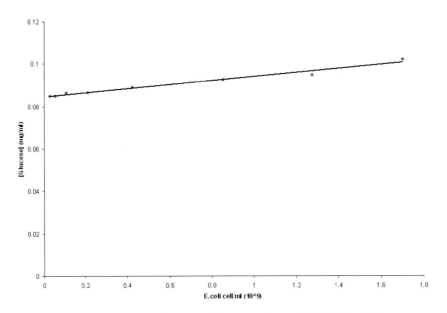

Figure 17.13. The relationship between various bacterial concentrations and their calculated glucose concentrations.

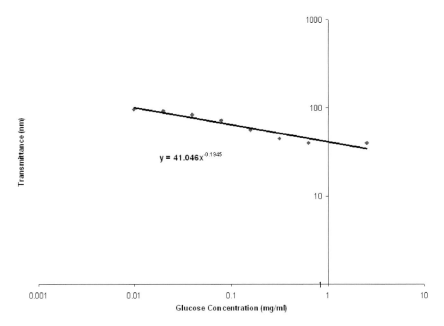

Figure 17.14. The relationship between each glucose concentration and its transmittance.

output. In addition, low molecular weight dyes may be impractical for some applications because they do not provide a bright enough fluorescent signal. The ideal fluorescent label should fulfill many requirements. Among the desired qualities are the following: (i) high fluorescent intensity (for detection in small quantities), (ii) a separation of at least 50 nm between the absorption and fluorescing frequencies, (iii) solubility in water, (iv) the ability to be readily linked to other molecules, (v) stability towards harsh conditions and high temperatures, (vi) a symmetric, nearly gaussian emission lineshape for easy deconvolution of multiple colors, and (vii) compatibility with automated analysis. At present, none of the conventional fluorescent labels satisfies all these requirements (Pinaud et al. 2006).

3. *Quantum dots*: Recently, fluorescent quantum dots have emerged as promising fluorescent labels that bring greater speed, less complexity, greater sensitivity, and greater selectivity in detecting the biomarkers of interest. Quantum dots are inorganic nanocrystals that are restricted in three dimensions to a somewhat spherical shape, typically with a diameter of 2–10 nm (on the order of 200–10,000 atoms) (Brewster and Mazenko 1998; Santra et al. 2005; Wang and Moffitt 2005; Ji et al. 2005). Several characteristics which distinguish quantum dots from the commonly used organic fluorophores are (i) a fluorescent wavelength tunable by size, (ii) a sharp and symmetrical fluorescent peak, (iii) strong and long-life emission, (iv) a wide excitation wavelength, and (v) photostability. Because of these advantages, quantum dots are superior alternatives of organic fluorophores in biodetection (Santra et al. 2005; Wang and Moffitt 2005; Ji et al. 2005; Nagasaki et al. 2005; Goldman et al. 2005; Mulvaney, Mattoussi, and Whitman 2004).

4. *Electrical Impedance*: Magnetically isolated bacteria are brought into or allowed to grow into an impedance measuring device such as a capacitor-meter. Detectable impedance change can be achieved using a low frequency or a DC integrated microelectrode probe (Yang and Li, 2006).

5. *Chemiluminescence*: A luminometer is used to quantify the signal of light emissions from a chemiluminescent reaction (Chen, Lei, and Tong 2005).

*6. **Magnetic indicators***: The magnetic sensing in this approach depends on the measurement of the difference in magnetic moment in a unit volume between magnetic particles that have bacteria bound to their surface and those that are bacteria-free. The sensitivity of detection correlates directly with the measurement of magnetic moments.

6. Handling Techniques

The free-floating magnetic beads can be washed out or remain in the sample solution. Figure 17.15 shows two alternative ways that can be used to analyze a magnetic signal. In Figure 17.15a, when the free-floating magnetic beads are washed out, the magnetic signal obtained is purely from antigen-bound magnetic beads. In contrast, when the free-floating magnetic beads are not removed, the magnetic signal obtained would be the sum of the signals from antigen-bound and unbound magnetic beads. However, because of the Brownian rotation of free-floating magnetic beads, magnetic signals resulting from free-floating magnetic beads are zeroed out (Figure 17.15b). Hence, theoretically, free-floating magnetic beads have little or no effect on the final result.

The disadvantage of using a magnetic marker technique is that only antigens that are bound to the functionalized magnetic beads are detected. In some cases, the antigen concentration is well above the antibody concentration. As a result, the actual concentration of the antigens becomes inconsequential. Apart from that, the concentration of primary (capture) and secondary (label) antibodies also determines the binding efficiency and hence affects the magnetic signal. An experiment was conducted by Meyer and coworkers (Meyer et al. 2007b) to determine the concentration of primary and secondary antibodies for an optimal magnetic signal. It was

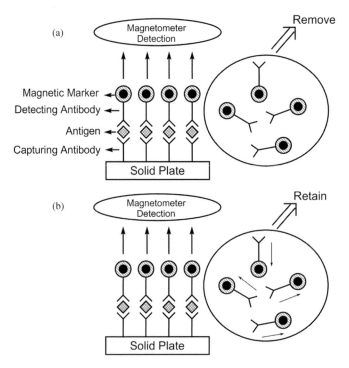

Figure 17.15. Detection of bound and unbound magnetic beads: (a) magnetic markers detection with separation of unbound magnetic markers; (b) due to Brownian rotation of free floating magnetic beads, their magnetic signals are zeroed out.

demonstrated that different antibody concentrations produced different magnetic signals, and that there was only one particularly optimal concentration.

In addition, the outcome of the assay is highly dependent on the type of substrate use, the technique of immobilization, and the immobilization conditions (pH, temperature, chemical reagent concentration). An experiment was conducted by Meyer and coworkers (Meyer et al. 2007a) to determine the relationship of the immobilization technique and the magnetic signal. In that study, three antibody immobilization techniques were used on a sintered polyethylene (PE) substrate. It was shown that adsorptive immobilization exhibited a better sensitivity and lower standard deviation compared to immobilization techniques such as GDA and NHS/DCC, covalent immobilization techniques.

7. Magnetic Separation

Magnetic separation occurs when a magnetic force is applied to magnetic elements and attracts the elements in a preferred direction. The magnetic force depends on the magnetic field and its spatial gradient. Most biological elements are diamagnetic in nature, causing them to repel an applied magnetic field. One of the exceptions to this rule is deoxygenated red blood cells, which have a paramagnetic behavior. The magnetic susceptibility of biological elements remains minute. To enable magnetic separation using commercially available magnetic separators, the magnetic susceptibility of the biological compound is enhanced by coupling the compound with magnetic particles, preferably with superparamagnetic properties. With the assumption that the magnetic particles and the biological compound are nonconducting, the magnetic field equation can be decoupled from the fluid motion equation and hence it can be solved separately.

In this section, magnetic field generation and numerical simulation of the magnetic field distribution are discussed. Several cases of different magnetic field arrangements that may produce a large magnetic force are investigated and analysis performed, under the assumption of a static, steady magnetic field.

7.1. Magnetic Force

The magnetic force per unit volume on the biological compound in a nonuniform field is given as:

$$\vec{F} = \left(\vec{M} \bullet \nabla \right) \vec{H},$$

where $\vec{M}$ is the intensity of magnetization per unit volume induced by the magnetic field $\vec{H}$.

The phenomenological relationship between the magnetization $\vec{M}$ and the magnetic field $\vec{H}$ is given as:

$$\vec{M} = \bar{\chi} \bullet \vec{H},$$

where $\bar{\chi}$ is the susceptibility tensor per unit volume. Because of the anisotropic rotational behavior of the biological compounds when subjected to a magnetic field, $\bar{\chi}$ is a tensor with two diagonal components, $\chi_{\perp}$ and $\chi_{//}$, Where $\chi_{\perp}$ is the susceptibility when the principal axis of the biological compound is perpendicular to the applied magnetic field, and $\chi_{//}$ is the susceptibility when the principal axis of the biological compound is parallel to the applied magnetic field. Then the magnetization can be written as:

$$\vec{M} = \chi_\perp \vec{H} + \chi_a \left(\vec{H} \bullet \vec{n} \right) \vec{n},$$

where $\vec{n}$ is a unit vector along the principal axis of the biological compound.

The magnetic force per unit volume when the fluid is static can be written as:

$$\vec{F} = \frac{\chi_\perp}{2} \nabla \left[\vec{H} \bullet \vec{H} \right] + \chi_a \left(\vec{H} \bullet \vec{n} \right) \nabla \left(\vec{n} \bullet \vec{H} \right),$$

where $\chi_a = \chi_{//} - \chi_\perp$ is the diamagnetic anisotropy.

In the case where the biological compound (or in this case the magnetic particle-biological compound complex) reaches equilibrium (i.e., 100% of the complex orients with the applied magnetic field) then the magnetization $\vec{M}$ can be further simplified and presented as a function of the principal susceptibility or the average susceptibility, χ, as:

$$\vec{M} = \chi \vec{H},$$

and the magnetic force per unit volume can be written as:

$$\vec{F} = \chi \left(\vec{H} \bullet \nabla \right) \vec{H} = \frac{1}{2} \nabla \left(\vec{H} \bullet \vec{H} \right) = \frac{1}{2} \nabla H^2,$$

If it is desirable to have as large a magnetic force as possible, then it is necessary to maximize ∇H^2.

Before we go any further, it is important to distinguish between the applied magnetic field and the induced magnetic field. The induced magnetic field in a material can be defined as:

$$\vec{B} = \mu_o \left(\vec{H} + \vec{M} \right),$$

where $\mu_o = 4\pi \times 10^{-7} H/m$ is the free space permeability. If $\vec{M} = \chi \vec{H}$, then the induced magnetic field can be written as:

$$\vec{B} = \mu_o \left(1 + \chi \right) \vec{H} = \mu \vec{H},$$

where μ is the magnetic material permeability. In free space the induced magnetic field is equal to $\vec{B} = \mu_o \vec{H}$; and for weak magnetic materials (susceptibility $\sim 10^{-6}$), as in the case of the biological compounds, the induced magnetic field is similar to the one in free space. In other words, the biological compound will not alter the applied magnetic field.

7.2. High-Field Electromagnets

In this section a brief discussion and analysis of the generation of a high and steady magnetic field by electrical current is given. The purpose is to consider the engineering aspects associated with the generation of the high field and to study the practical limitations of generating high fields. The discussion is focused on solenoid magnets. Due to the very low temperature requirement (liquid helium at 4 K), it is not practical to use a superconducting magnetic device for biomedical applications, since the freezing point for biological compounds is around 273 K.

Francis Bitter in 1938 was the first to build a resistive magnet of 10 T at room temperature. The magnet power requirement was 1.7 MW; the magnet bore diameter was 50 mm. His design employed an iron-free conductor in the form of a stack of annular plates, each separated by a thin sheet of insulation. Each plate has hundreds of holes so that the cooling water can pass through. The electrical current passes from one stack to another via small segments connected in a helical path. Around 10,000 A are required in the electrical resistive stack to produce a 10 T field.

The Bitter magnet in general requires less power than the uniform current solenoid. For example, at 2 T field, the uniform current solenoid will require 33 KW, while the Bitter magnet will need 3 KW. At 0.5 T, the power requirement is 2.0 KW for the uniform current solenoid, and 0.2 KW for the Bitter magnet. A permanent magnet of the same dimensions (1.5 mm, 10 mm) can produce up to a 1.2 T field without any power or cooling requirements.

Electromagnets made using present technologies can generate up to a 45 T field, while permanent magnets that are available today cannot provide a magnetic field higher than 2.2 T. Permanent magnets can be manufactured with almost any shape. Because of the high power requirement and the shape constraints of electromagnets, magnetic separation technology has concentrated on using permanent magnets.

7.3. Permanent Magnets

The strength of a magnet is described by its maximum energy product. The maximum energy product is defined as the maximum product of the magnetic induction and the external magnetic field, given in energy per unit volume. The strength of magnets increase largely through the discoveries of new magnetic materials: steel, alnico (aluminum, nickel, and cobalt), samarium, cobalt, and neodymium-iron-boron. Hard ferrites are commercially significant because they are inexpensive to produce. Recently the technology for producing high strength permanent magnets has experienced noticeable advances. Steel was used to produce permanent magnets until the 1930s. With recent advances in material science, permanent magnets have been improved to the point where they can compete with electromagnets, both functionally and economically. Rare earth magnets with high magnetic field strength (~ 2.2 T) are now available.

It should be mentioned here that permanent magnets have the advantage of being small in size compared to electromagnets. It is clear that in small devices the permanent magnets have a major advantage over electromagnetism. The disadvantage of the permanent magnets is that their maximum strength is still limited by the material property. Electromagnets with the proper design can reach a magnetic field of $40 +$ T, which is much higher than the field in the permanent magnet (maximum of ~ 2.2 T). The higher the field strength produced by the electromagnet, the higher the volume of wiring will be. For small design of devices, permanent magnets are more suitable than electromagnets. High strength permanent magnets are available commercially at a relatively low cost.

The details of producing permanent magnets may be found in Campbell (1994). Permanent magnets are characterized by their magnetization curve, as shown in Figure 17.13. The magnetization curve is a plot of B vs. H. Assume that the material is initially unmagnetized. Then as the H field increases, the B field initially rises, roughly proportional to the applied field. However, as the applied field increases, the slope decreases and then becomes roughly linear again at very high fields (the saturation region). After the sample is magnetized to saturation, the curve is no longer reversible, and it ceases to follow the initial magnetization curve. Instead it follows the indicated curve until the applied field goes to zero. At this point the sample has a residual magnetization B_r called the remanence. If the applied field is now applied in the opposite direction, the measured induction B can be driven to zero and back to saturation in the opposite direction. The field necessary to drive the induction to zero, H_c, is called the corrective force. If the material is driven back and forth from saturation to saturation, it follows the outer

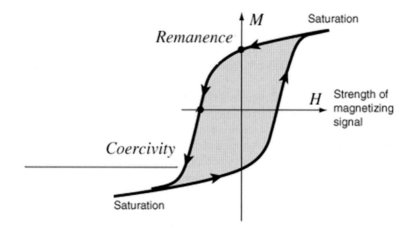

Figure 17.16. Magnetization curve for a permanent magnet.

curve called the saturation hysteresis curve. However, suppose the applied field is reversed at some other point than saturation. The material then follows a minor hysteresis curve as shown in Figure 17.16. The shape of the hysteresis curve depends on the material.

7.4. Numerical Analysis for Permanent Magnet Arrangements

This section discusses the magnetic field generated by different arrangements of magnets. The magnetic field generated by different permanent magnet arrangements has been calculated using ANSYS (ANSYS Co. 1997). The permanent magnets for this calculation are considered to be made out of rare earth material (neodymium-iron-boron).

The axisymmetric magnetic distribution is utilized in this demonstration and hence only two-dimensional models are solved. The elements used in the analysis are quadrilateral eight nodes elements. Because the models are two–dimensional, each node has only one vector potential, the z direction. The demagnetization data for the material are required by the program. The demagnetization data provide information about the relation between the magnetic field $\vec{H}$ and the magnetization $\vec{M}$ in the magnet. Outside the magnet, the governing equation reduces to the Laplace equation.

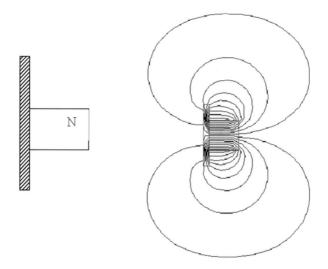

Figure 17.17. Case 1: Magnetic field for a single magnet attached to Premenorm 5000 H3 with $\mu = 5000$.

The magnetic force at a point in space depends on the magnetic susceptibility of the substance occupying the space and the magnetic field strength and its gradient. Therefore one attractive arrangement of magnets is one that provides the maximum force to cover a large surface area, since it then provides a large domain of magnetic influence for potential engineering applications.

In the following, different magnetic field arrangements that are calculated using ANSYS are presented. Each case studied is documented with a case number.

In Case 1, a magnet is attached to a soft magnetic material with linear relative permeability of 5000 (Permenorm 5000 H3).

Figure 17.17 shows the magnetic field distribution for Case 1. It is found that the magnetic material works as a sink for the magnetic field lines. It should be noted that the material provides a magnetic shield. The different permeability of the soft magnetic material does not change

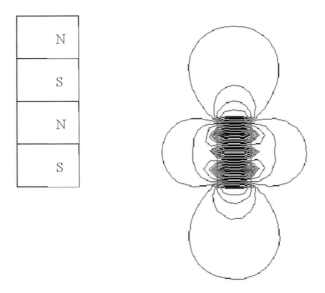

Figure 17.18. Case 2: Magnetic field distribution for four alternating magnets.

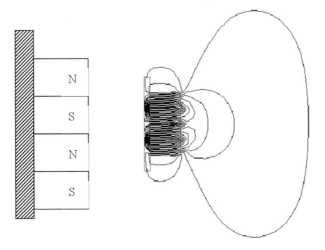

Figure 17.19. Case 3: Magnetic field for four alternating magnets attached to a plate with $\mu = 3000$.

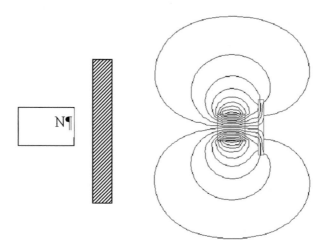

Figure 17.20. Case 4: Magnetic field for a single magnet facing a plate with $\mu = 3000$ and 1 cm away.

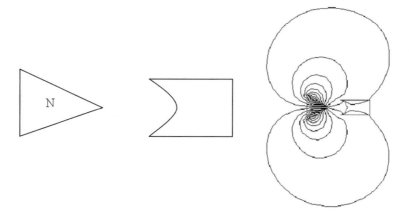

Figure 17.21. Case 5: Magnetic field for a triangular magnet that face a semi-circle with $\mu = 1000$.

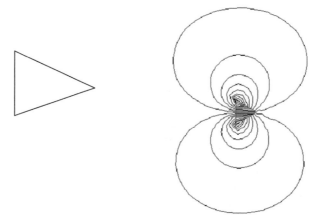

Figure 17.22. Case 6: Magnetic field for a triangular magnet.

Table 17.4. Force calculation for different cases at different distances form the surface of the magnet

Case No.	0 mm	0.5 mm	1.0 mm	1.5 mm	2.0 mm	2.5 mm	3.0 mm
1	143	120	95	72	57	45	35
2	564	450	350	225	160	56	20
3	650	470	324	250	130	65	10
4	50	60	65	72	80	60	30
5	2063	620	180	50	5	5	5
6	1142	570	250	114	5	3	0.5

the overall characteristics of the field, though the magnetic force is enhanced by introducing the soft magnetic material on one side. The magnetic force is calculated along a point in the middle of the magnet surface. The force is proportional to $B\frac{dB}{ds}$, where ds is along the normal to the surface.

In Case 2, an alternating arrangement of N-S-N-S of four equal-dimension magnets is computed. The magnetic field flux is shown in Figure 17.18. In Case 3, a soft magnetic material is placed in one side of the magnets (Figure 17.19). The magnetic material serves as a magnetic sink for the magnets. The force was enhanced by introducing the magnetic material. This arrangement is found to be one of the best arrangements for providing an overall enhancement of the magnetic force.

Figures 17.20–17.22 show the magnetic field flux of the different cases. The magnetic force was calculated at different locations along the normal for each case.

Table 17.4 shows the results for all cases. It should be pointed out that some of these cases have a very strong force at the surface of the magnet, but the magnetic force decays fast, as in Cases 5 and 6.

In order to evaluate the magnetic force for circular shaped magnetic arrangements, five different arrangements were considered. The simulation was conducted using Integrated Engineering Software's 3D magnetic field solver, which is based on the boundary element method. The magnets are made out of $Nd_2Fe_{14}B$ grade 35. Its residual induction B_r is 12.3 KG,

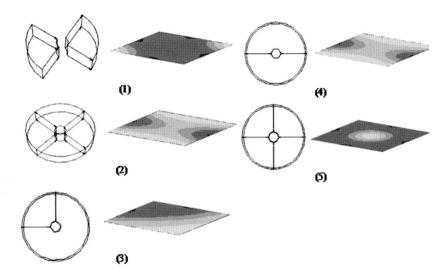

Figure 17.23. Simulation of magnetic field distribution for circular magnetic separators.

Figure 17.24. Sample of commercially available separators: 1) PickPen magnetic tool (Bio-Nobile), Finland; 2) MCB 1200 (Sigris Research), USA; 3) BioMag Solo-Sep (Polysciences) USA.

its coercive force H_c is 11.3 KOe, and its intrinsic coercive force H_{ci} is 14 KOe. The inner tube radius surrounded by the magnet is 1 mm, and the outer radius of the magnets is 6mm. From the field distribution shown in Figure 17.23, design 5 has the most uniform force distribution around the circular path.

Commercially available magnetic separators are usually made out of rare earth high energy magnets. These separators utilize magnetic configurations that are similar to those presented. Some commercially available separators use wire meshes or stakes of balls within the separation column to enhance the magnetic force by increasing the magnetic gradient at the wire or ball surface. In principle, they are trying to match the force shown in Case 6. However, one also needs to be aware that these meshes and balls may increase the shear stresses on the cellular compounds passing through the separation column. The increased shear stress may lyse the cells. Samples of commercially available magnetic separators are shown in Figure 17.24.

8. Giant Magnetoresistive (GMR) Devices for Bacterial Detection

GMR describes a phenomenon where a huge change in electrical resistance is observed within a material when it is exposed to an external magnetic field (Baibich et al. 1988; Binasch et al. 1989). A typical GMR structure is composed of alternating layers of magnetic and nonmagnetic materials (Coffey et al. 1995; Millen et al. 2005). The staging of magnetic and nonmagnetic materials forms a thin structure, the thickness of which is comparable to the mean free path of an electron in the material (Daughton et al. 1992). The distance between two magnetic materials is designed to produce a coupling of nuclear magnetic moments within the structure (Grünberg, Schreiber, and Pang 1986), resulting in the formation of an antiferromagnet, where the spins of electrons align in such a way that adjacent magnetic materials have opposite spin directions (Figure 17.25).

The resistance of a GMR structure is dependent on the orientation of the electron spins. A GMR structure is designed in such a way that at zero-field condition, it exhibits the highest resistance due to antiparallel electron spins (Millen et al. 2005). In other words, a GMR structure exhibits the highest degree of electron spin scattering when there is an absence of an external magnetic field. However, the electron spin orientation is altered when an external magnetic field is applied, changing the resistance of the GMR structure (Daughton et al. 1992). The degree of alteration is dependent on the intensity of the applied magnetic field. As shown in Figure 17.26, as the applied magnetic field increases or decreases from the zero-field state, the resistance experienced by the GMR structure decreases until the point where it gets saturated. Right before the resistance saturates, it undergoes a linear region, in which the change in resistance is proportionate to the change in the magnetic field.

The incorporation of GMR structures in bacteria sensing is illustrated in Figure 17.27 (Millen et al. 2005). Generally, the surface of the GMR sensing region is modified to allow the binding of capture antibody. When the GMR structure is exposed to a sample solution that contains target antigens, complex binding between the target antigen and antibody occurs. This

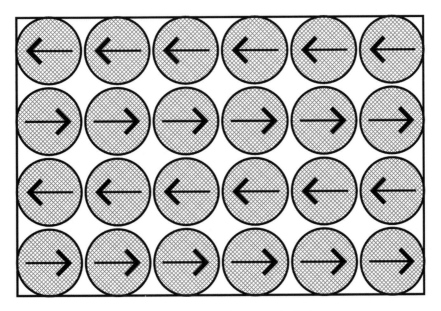

Figure 17.25. Alignment of anti-ferromagnetic dipole moments.

is followed by the addition of antibody-coated magnetic particles that subsequently label the target antigens and form a series of sandwich-like structures (Figure 17.27).

In order to detect the magnetic particles bound on a GMR structure surface, an external magnetic field is applied in the z-direction, as illustrated in Figure 17.28 (Rife et al. 2003). Bound magnetic particles that are exposed to a magnetic field will generate magnetic induction in the x-direction, as shown in the figure. Since the GMR structure detects only the x-component of the magnetic field, the external magnetic field in the z-direction does not have any effect on the detection.

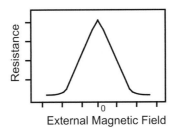

Figure 17.26. A graph illustrating the change of resistance as a function of applied magnetic fields.

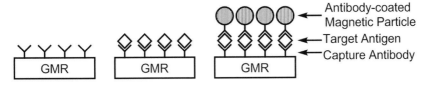

Figure 17.27. Bacteria sensing using a GMR structure.

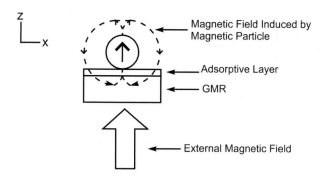

Figure 17.28. Detection of a magnetic particle on a GMR structure.

As indicated in Figure 17.26, increasing the magnetic field in either direction will reduce the resistance of a GMR material. Hence, the magnetic induction in both directions generated by magnetic particles will subsequently lower the resistance of the GMR structure. The induced magnetic field in the x-direction can be described as:

$$B_x = \mu_0 M \frac{a^3(a+t)d}{((a+t)^2 + d^2)^{5/2}},$$

(17.1)

where M is the external magnetic field, a is the radius of the magnetic particle, t is the distance of the magnetic particle to the GMR structure separated by the top layer, and d is the distance of B_x along the trace and relative to the center of the magnetic particle.

9. Bacteria Detection with Magnetic Relaxation Signal

Magnetic relaxation signals have been used for bacteria detection by distinguishing the bacteria-bound magnetic particles from the unbound ones. This technique is based on the distinct relaxation behavior of bacteria-bound magnetic particles when they are exposed to a magnetic field (Volkov et al. 2006). When an external magnetic field is applied to a sample solution containing antibody-conjugated magnetic particles, the magnetic dipole moments will tend to align in parallel to the magnetic field. If the external magnetic field is turned off, the particles will slowly rotate back to their initial state where the energy level is the lowest. The phenomenon in which the particles rotate to their lowest energy state is called Néel relaxation and can be described as $\tau_N = \tau_0 \exp(KV_M/k_BT)$, where τ_0 is $\approx 10^{-9}$ s, K is the magnetic anisotropy constant, V_M is the magnetic core volume, k_B is Boltzmann's constant, and T is the temperature. The values of τ_0 and K are highly dependent on the shape of the magnetic particles (Fannin and Charles 1994). The relaxation time of the particles can be measured by using a highly sensitive magnetometer, particularly the SQUID (Grossman et al. 2003; Volkov et al. 2006).

Figure 17.29 demonstrates the alignment of the magnetic dipole moments when an external magnetic field is applied to a sample solution, as well as the Néel relaxation effect when the magnetic field is turned off. As with the bacteria-bound particles, magnetic dipole moments in unbound particles also align in parallel to the magnetic field. However, when the magnetic field is turned off, the dipole moments of unbound particles return to their initial states faster than those of bound particles because of the unattached surface that leads to ease of movement. This phenomenon is known as Brownian relaxation and it can be described as $\tau_B = 3\eta V_H/k_BT$,

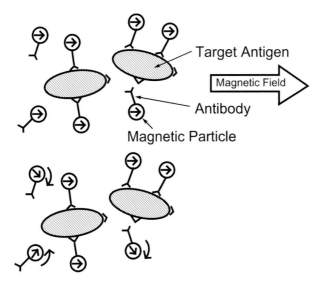

Figure 17.29. The alignment of magnetic dipole moments when an external magnetic field is applied on a sample solution and the Néel relaxation when the magnetic field is turned off.

where η is the viscosity of the medium, V_H is the hydrodynamic volume, k_B is Boltzmann's constant, and T is the temperature.

The advantage of using relaxation signals in the detection of bacteria is that no washing steps are required. For example, Grossman and coworkers (Grossman et al. 2003) have shown that the relaxation time of target bound and unbound particles can be easily distinguished in vivo with a highly sensitive magnetometer.

10. Magnetoelastic Sensors for Bacterial Detection

Magnetoelastic sensors are typically made of amorphous ferromagnetic ribbons. They are highly attractive for chemical, biological, and environmental monitoring not only because of their small size and low cost but also because of their passive and wireless nature. By applying a mass- and/or elasticity-changing biological-responsive layer, magnetoelastic sensors are ideal for measuring biological targets such as *E. coli* O157:H7 (Ruan et al. 2003). Magnetoelastic sensors track parameters of interest via changes in resonant behavior. They mechanically deform when subjected to a magnetic field, launching elastic waves within the sensor, the magnitude of which are greatest at the mechanical resonant frequency of the sensor. As illustrated in Figure 17.30, the mechanical deformations of the sensor launch magnetic flux that can be detected remotely by a sensing coil (Jain et al. 2001), as well as acoustic waves that can be monitored over a range of meters (Stoyanov and Grimes 2000). Furthermore, the sensors can be monitored optically through the amplitude modulation of a laser beam reflected from the surface of the sensor (Mungle 2001).

There are many techniques to interrogate and analyze the magnetoelastic sensor response. For example, the sensor can be interrogated by performing a frequency sweep to obtain the resonant spectrum (Zeng et al. 2002), or by capturing the time-domain transient response and then performing a fast Fourier transformation to obtain the frequency-domain resonance spectrum (Loiselle and Grimes 2000). Alternatively, the sensor can be interrogated using a transient frequency-counting operation (Zeng et al. 2002), which first excites the sensor with a series of sinusoidal bursts at a frequency near the resonant frequency of the sensor, and then

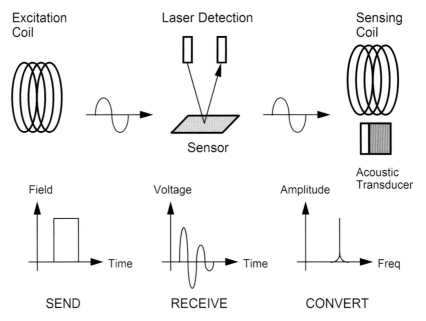

Figure 17.30. Schematic drawing illustrating the remote query nature of the passive, wireless magnetoelastic sensor.

captures its transient response, which is a damped sinusoidal waveform. The resonant frequency of the sensor is determined from the captured signal by measuring the frequency of the transient signal. The resonant frequency, f_0, of an uncoated sensor is related to its mechanical properties and physical length as (Grimes et al. 2002):

$$f_0 = \frac{1}{2L}\sqrt{\frac{E}{\rho}},$$
(17.2)

where L is the sensor length, ρ is the density, and E is the elasticity. Coating a sensor such as applying a mass load changes the resonant frequency. For small mass loads, the change in the resonant frequency is given by (Grimes et al. 2002):

$$\Delta f = f - f_0 = -f_0 \frac{\Delta m}{2M},$$
(17.3)

Based on Eq. (17.3), bacteria detection can be realized by using a bacteria-affinity coating that specifically increases the mass loading on the sensor in response to the targeted bacteria.

10.1. *E. coli* Detection

Alkaline phosphatase (AP) was used as a labeled enzyme to the anti-*E. coli* O157:H7 antibody, amplifying the mass change associated with the antibody-antigen binding reaction by the biocatalytic precipitation of 5-bromo-4-chloro-3-indolyl phosphate in a pH 10.0 PBS solution. The sensor was first coated with gold and then immersed in a 10 mM solution of 2-aminoethanethiol hydrochloride in ethanol overnight at room temperature. After the reaction, the sensor was rinsed with ethanol to remove aminoethanethiols, and then dried under a stream of nitrogen, resulting in an exposed active amino group on the surface of the sensor (Heleg-Shabtai et al. 1997). The sensor was subsequently immersed into a 5% glutaric dialdehyde solution for 1 hr, rinsed with water, and dried under a stream of nitrogen. Anti-*E. coli* O157:H7 antibodies were introduced onto one side of the sensor surface by dispersing $20 \mu l$ of 1 mg/ml anti-*E. coli* O157:H7 antibodies and allowing them to react for one hour, and then rinsing with deionized water. Two $100 \mu l$ samples of pH 7.2 PBS solution containing diluted heat-killed *E. coli* O157:H7 cells, ranging in concentration from 10^2 to 10^7 cells/ml, were sequentially dropped onto the sensor, and the sensor was incubated for 1 hr at 37°C. After the binding reaction between antibodies and *E. coli* O157:H7 antigens on the surface of the sensor, $20 \mu l$ of AP-conjugated antibody to *E. coli* O157:H7 (0.01 mg/ml) was dropped on the sensor surface, and allowed to react for 1 hr. The sensor was then washed with PBS buffer solution to remove any nonspecifically bound AP-conjugated antibodies. The fabrication process of the sensor is described in Ruan et al. (2003).

Figure 17.31 depicts the sandwich enzyme-linked immunosorbent assay procedure used with the *E. coli* sensor. The biocatalytical precipitates of 5-bromo-4-chloro-3-indolyl phosphate induced onto the sensor surface by AP in pH 10 PBS buffer increases the mass change, lowering the resonant frequency of the magnetoelastic sensor. Figure 17.31 shows that the precipitation mechanism includes hydrolysis of the phosphate moiety of the BCIP by the AP. The product of BCIP hydrolysis is subsequently oxidized by dissolved oxygen to produce an insoluble blue BCIP dimer that is strongly bound to the sensor surface.

Figure 17.32 shows the response of the magnetoelastic *E. coli* O157:H7 sensor, as a function of sensor immersion time, to the enzymatic catalytic reaction on the sensor surface, with *E. coli* O157:H7 concentrations ranging from 10^2 to 10^6 CFU/ml. The *E. coli* sensor shows a smooth decrease in frequency after only a small induction period (<60 s), which is a result of the time required for the concentration of the BCIP dimer to exceed its solubility product. A steady state response is generally achieved at 1 hr for a sensor exposed to *E. coli* concentrations of 10^3 CFU/ml or less. Sensors exposed to higher concentrations of *E. coli* cells require longer periods to reach steady state, since more sandwich conjugates are formed in high *E. coli* concentrations, and thus more alkaline phosphatase is accumulated on the sensor surface. This increases the enzymatic reaction products to cover the surface in order to terminate further reaction. It is clearly evident that the rate of resonant frequency change increases with increasing *E. coli* O157:H7 concentration. In these sandwich assays the conjugate enzyme is only present when *E. coli* O157:H7 is bound to the surface, and therefore the amount of precipitate directly reflects the number of *E. coli* O157:H7 cells. The background change in resonant frequency due to nonspecific binding of the AP-labeled anti-*E. coli* antibody is approximately 20 Hz after

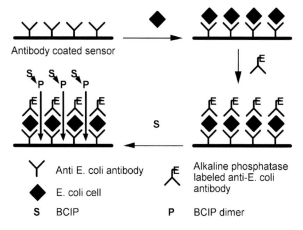

Figure 17.31. Schematic representation of the sandwich enzyme-linked immunosorbent assay procedure used with the magnetoelastic *E. coli* O157:H7 sensor (Reprinted from Ong et al. IEEE SENSORS JOURNAL. (© 2006 IEEE)).

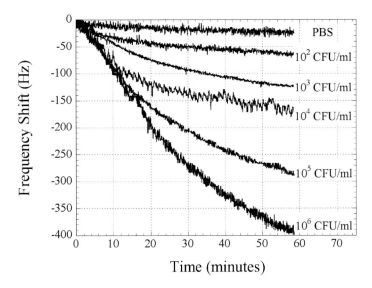

Figure 17.32. Real-time response of the *E. coli* sensor resonant frequency to *E. coli* O157:H7 concentration from 0 to 10^6 cells/ml in pH 10.0 PBS containing BCIP. (Reprinted from Ong et al. IEEE SENSORS JOURNAL. (© 2006 IEEE)).

50 minutes, as compared to the 60 Hz change for 10^2 CFU/ml of *E. coli* O157:H7 over the same period.

Figure 17.33 shows the relationship between the steady-state changes in resonant frequency versus the logarithmic value of the *E. coli* O157:H7 concentration (10^2–10^6 CFU/ml). Results indicate the changes in resonant frequency are linearly proportional to the logarithmic value of the *E. coli* O157:H7 concentration, with a detection limit of 10^2 CFU/ml. The results of the sensor are comparable to methods developed by other researchers for the specific detection

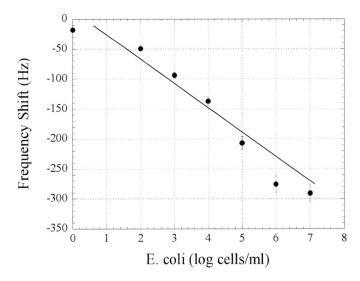

Figure 17.33. Shift in resonant frequency of *E. coli* O157:H7 sensor as a function of *E. Coli* O157:H7 concentration in test solution. (Reprinted from Ong et al. IEEE SENSORS JOURNAL. (© 2006 IEEE)).

of *E. coli* O157:H7. For example, an enzyme-linked immunomagnetic separation coupled with a bienzyme electrochemical biosensor has a minimum detectable level of 6×10^2 cells/ml (Ruan et al. 2002a), an electrochemical impedance immunobiosensor has a detection limit of 6×10^4 cells/ml (Ruan et al. 2002b), an enzyme-linked immunomagnetic electrochemical sensor using 1-naphthyl phosphate as an enzymatic substrate has a detection limit of 4.7×10^3 cells/ml (Gehring et al. 1999), a sensor based on immunoligand assay/light-addressable potentiometry has a detection limit of 7.1×10^2 cells/ml (Gehring et al. 1998), a sensor based on immunomagnetic separation/flow injection analysis/mediated amperometrics has a detection limit of 10^5 cell/ml (Perez et al. 1998), and a sensor based on immunomagnetic separation/flow cytometry has a detection limit of 10^3 cells/ml (Seo et al. 1998).

11. Conclusions and Future Perspectives

Every year in the United States, infections cause millions of illnesses and thousands of deaths; most infections go undiagnosed and unreported. It is estimated that $5–6 billion are spent annually in direct medical expenses and lost productivity as a consequence of food-borne diseases alone. Concerted efforts have been made by a number of U.S. agencies to increase the safety of the food and water supply as well as the prompt diagnosis of infections.

To accomplish this, there is a need for the rapid and accurate detection of pathogens. The currently available procedures are either time-consuming (8–48 h) and require enrichment and culturing of bacteria before testing, or provide only qualitative results. Also, in most cases, the current technologies available for rapid immunodetection lack sensitivity. On the other hand, highly sensitive procedures are usually costly and require complex equipment and specialized personnel, being therefore inappropriate for field testing. One of the factors affecting the sensitivity of the assay is the nature and complexity of the tested sample (e.g., whole blood). Therefore, research has been focused on developing filters and other solid supports to improve the capture of the target molecule. However, filtration presents problems in collecting the target from complex samples that may clog the filter, thereby limiting the sample volume that can be used for the test. The quality of the immunodetection is greatly affected by the detection method. The classical principle of colorimetry brings sensitivity to the assay, but at the same time it is a laborious process that requires multiple washes and reagents that are hard to accommodate into a portable device. A more convenient method for the sensitive and simultaneous detection of multiple biomarkers uses fluorescence tagging. However, there are chemical and physical limitations associated with the use of classical organic fluorescent dyes. Recently, fluorescent quantum dots have emerged as promising fluorescent labels that bring greater speed, less complexity, greater sensitivity, and greater selectivity in detecting the biomarkers of interest. Because of these advantages, quantum dots are superior alternatives to organic fluorophores in biodetection. Other methods, such as changes in pH or impedance, are greatly dependent on external factors and lack sensitivity.

Magnetic immunoassay technology appears to provide solutions to the problems that current pathogen detection technologies have. A typical magnetic immunoassay entails a capture part and a detection part, between which the target is immobilized. The capture part of the immunoassay consists of magnetic particles functionalized to capture the target from the sample. The immobilized target is then sandwiched between the capture and detection complexes and subjected to detection, a process that will provide accurate and rapid results in a matter of minutes. The detection part of the assay consists of moieties that are complementary to different sites on the target from those occupied during capture. Also, the detection part entails a signal producing moiety that allows the detection of the target in the sandwich. The signal produced in the presence of the specific target (e.g., color, fluorescence, impedance, pH) is proportional to the amount of target in the sandwich.

The advantages brought by magnetic immunoassays in pathogen detection are increased specificity, sensitivity, detection time, cost efficiency, and practicability. All these great features depend primarily on the quality of the capture (IMS). The sensitivity of the assay is given by its ability to detect low counts of pathogens without a need for enrichment. This property is the result of several factors, such as efficient capture even from complex samples (e.g., particles with high magnetic moment, binding surface, and functionalization with specific moieties), and a strong detection signal of an amplitude that is also dependent on the number of pathogens captured in the sandwich.

The performance of magnetic particles depends on their properties, which in turn depend on the particles' structural atomic composition, interfaces, and defects. Nanoparticles have been increasingly used in biological applications. A critical length scale of under 100 nm is material dependent and imprints novel properties (e.g., optical, magnetic, thermal, mechanical, biological) that differentiate NPs from other materials. For example, nanocrystalline iron oxide particles are superparamagnetic when the particle size is sufficiently small, and they behave as ferromagnetic when the size is in the micrometer range. Magnetic particles have been increasingly used in immunoseparation to efficiently isolate biological targets from complex sample matrices (e.g., blood, feces, food). For this purpose, the use of NPs is advantageous, since their magnetic properties and binding surface increase dramatically at nanoscale dimensions. Additionally, the surface properties of the nanomagnetic particles can be manipulated by selecting an appropriate encapsulating material to strongly improve the capture of the target from the sample with minimal particle loss. Functionalization may be accomplished by coating the surface of the particles with natural (e.g., albumin) or synthetic (e.g. PVA, PEG) materials that have moieties complementary to those found on the target.

A very important advantage that sensitive magnetic immunoassays confer is the reduced volume of samples and reagents needed. This is useful mainly for field testing, when a quick pathogen detection test, if possible integrated into a portable device, is needed. Although portable technologies are available, there are still challenges in designing cost effective, sensitive, hand-held devices for the quantitative detection of biological targets. Recently, biosensors that incorporate the use of a lab-on-a-chip, nanomagnetic particles, and quantum dots for biodetection have been successfully developed. This suggests that a portable device that uses high-efficiency IMS in conjunction with high-sensitivity immunorecognition may be the future for the rapid detection of biological targets.

In conclusion, immunomagnetic pathogen detection technology offers a route to cost reduction by using a rapid and sensitive detection system that uses uniquely manufactured nanomagnetic particles and minimal volumes of reagents. The instrument could be adapted for the testing of multiple pathogens, therefore constituting a platform for use in other areas such as medical diagnostics and environmental protection.

References

Abdel-Hamid I, Ivnitski D, Atanasov P and Wilkins E (1999) Flow-through immunofiltration assay system for rapid detection of *E. coli* O157:H7. Biosens Bioelectron 14(3):309–316

Annas GJ (2002) Bioterrorism, Public Health, and Civil Liberties. N. Engl J Med. 346 (17):1337–1342

ANSYS Co. (1997) Product Literature, New York

Baibich MN, Broto JM, Fert A, Nguyen Van Dau F, Petroff F (1988) Giant Magnetoresistance of (001)Fe/(001)Cr Magnetic Superlattices. Physical Review Letters 61:2472–2475

Baselt D, Lee GU, Natesan M, Metzger SW, Sheehan PE, Colton RJ (1998) A biosensor based on magnetoresistance technology. Biosens. Bioelectron. 13:731–739

Bauer AW, Kirby WMM, Scherris JC and Truck M (1966) Antibiotic susceptibility testing standardised single disk method.Am. J. Clin. Pathol. 45:493–496

Bayliss CL (1999) Detection and Separation of Pathogens and their Toxins. In: MAFF Research Program FS 12, MAFF UK, Center for Applied Microbiology and Research, Porton Down

Binasch G, Grünberg P, Saurenbach F and Zinn W (1989) Enhanced Magnetoresistance in Layered Magnetic Structures with Antiferromagnetic Interlayer Exchange. Physical Review B, Condensed Matter and Materials Physics 39:4828–4830

Blackburn C, Patel PD and Gibbs PA (1991) Separation and Detection of Salmonellae Using Immunomagnetic Particles. Biofouling 5:143–156

Brewster JD and Mazenko RS (1998) Filtration capture and immunoelectrochemical detection for rapid assay of *Escherichia coli* O157:H7. J Immunol Methods 211:1–8

Brinchmann JE, Gaudernack G, Thorsby E, Jonassen TO and Vartdal F (1989) Reliable isolation of human immunodeficiency virus from cultures of naturally infected CD4+ T cells. J. Virol. Methods 25:293–300

Brinchmann JE, Albert J and Vartdal F (1991) Few infected CD4+ T cells but a high proportion of replication-competent provirus copies in asymptomatic human immunodeficiency virus type 1 infection. J. Virol. 65:2019–2023

Brytting M, Wahlberg J, Lundberg J, Wahren B, Uhlen M, Sundqvist V-A (1992) Variations in the cytomegalovirus major immediate-early gene found by direct genome sequencing. J. Clin. Microbiol. 30:955–960

Campbell P (1996) Permanent Magnet Materials and their applications. Cambridge University Press, Cambridge, UK

Centers for Disease Control and Prevention (Date) Update: Outbreaks of Cyclospora cayetanaensis infection—U.S. and Canada 1996. Morbidity and Mortality Weekly Report 45: 611–612

Chastellain M, Petri A and Hofmann H (2004) Particle size investigations of a multistep synthesis of PVA coated superparamagnetic nanoparticles. J. Colloid. Interface Sci. 278:353–360

Chatterjee J, Haik Y and Chen CJ (2001a) Modification and characterization of polystyrene-based magnetic microsperes and comparison with albumin-based magnetic microspheres J Mag Mag Mat. 225:21–29

Chatterjee J, Haik Y and Chen C-J (2001b) Synthesis and characterization of heat-stabilized albumin magnetic microspheres Colloid Poly Sci. 279:1073–1081

Chatterjee J, Haik Y and Chen C-J (2003) Size dependent magnetic properties of iron oxide nanoparticles J Mag Mag Mat, 257:113–118

Chemla YR, Grossman HL, Lee TS, Clarke J, Adamkiewicz M and Buchanan BB (1999) A new study of bacterial motion: superconducting quantum interference device microscopy of magnetotactic bacteria. Biophysical Journal 76:3323–3330

Chen T, Lei JD and Tong AJ (2005) Immunosorbent assay microchip system for analysis of human immunoglobulin G on MagnaBind carboxyl derivatized beads. Luminescence 20(4–5):256–60

Chou C, Tsai Y, Liu J, Wei JCC, Liao T, Chen M and Liu L (2001). The detection of the HLA-B27 antigen by immunomagnetic separation and enzyme-linked immunosorbent assay—comparison with a flow cytometric procedure. Journal of Immunological Methods 255:15–22

Coffey KR, Hylton TL, Parker MA, Howard JK (1995) Thin Film Structures for Low Field Granular Giant Magnetoresistance. Scripta Metallurgica et Materialia 33:1593–1602

Daughton JM, Bade PA, Jenson ML, Rahmati MMM (1992) Giant Magnetoresistance in Narrow Stripes. IEEE Transactions on Magnetics 28:2488–2493

Dorman JL and Fiorani D (1992) Magnetic Properties of Fine Particles. Publisher, Amsterdam

Dorman JL, Fiorani D and Tronc E (1997) Magnetic relaxation in fine-particle systems. In: Prigogine I and Rice SA (eds) Advances in Chemical Physics, Vol. XCVIII. John Wiley and Sons, New York, 283–494

Drancourt M, George F, Brouqui P, Sampol J and Raoult D (1992). Diagnosis of Mediterranean spotted fever by indirect immunofluorescence of Rickettsia conorii in circulating endothelial cells isolated with monoclonal antibody-coated immunomagnetic beads. J. Infect. Dis. 166:660–663

Fannin PC, Charles SW (1994) On the Calculation of the Néel Relaxation Time in Uniaxial Single-Domain Ferromagnetic Particles. J. Phys. D Appl. Phys. 27:185–188

Feldsine PT, Forgey RL, Falbo-Nelson MT and Brunelle S (1997) *Escherichia coli* O157:H7 Visual Immunoprecipitation assay: a comparative validation study. J. AOAC 80:43–48

Feng PJ (1992) Commercial assay systems for detecting foodborne *Salmonella*: a review. Food Prot. 55:927–934

Flynn ER, Bryant HC, Bergemann C, Larson RS, Lovato D and Sergatskov DA (2007) Use of a SQUID array to detect T-cells with magnetic nanoparticles in determining transplant rejection. Journal of Magnetism and Magnetic Materials 311:429–435

Frost JA, McEvoy MB, Bentley CA and Andersson Y (1995) An outbreak of Shigella sonnei infection associated with consumption of iceberg lettuce. Emerg. Infect. Dis. 1(1): 26–29

Fu L, Dravid VP, Klug K, Liu X and Mirkin CA (2002) Synthesis and patterning of magnetic nanostructures. European Cells and Materials Journal 3:156–157

Fukuda S, Tatsumi H, Igimi S, Yamamot, S (2005) Improved bioluminescent enzyme immunoassay for the rapid detection of Salmonella in chicken meat samples. Lett Appl Microbiol. 41(5):379–384

Fung DYC (1995) What's needed in rapid detection of foodborne pathogens. Food Technol. 49:64–67

Gehring AG, Patterson DL and Tu SI (1998) Use of a light-addressable potentiometric sensor for the detection of *Escherichia coli* O157:H7. Anal. Biochem. 258:293–298

Gehring AG, Irwin PL, Reed SA, Tua S, Andreotti PE and Akhavan-Tafti HRS (2004) Enzyme-linked immunomagnetic chemiluminescent detection of Escherichia coli O157:H7. Immunol Methods 293(1–2):97–106

Goldman ER, Mattoussi H, Anderson GP, Medintz IL and Mauro JM (2005) Fluoroimmunoassays using antibody-conjugated quantum dots. Methods Mol Biol. 303:19–34

Gref R, Domb A, Quellec P, Blunk T, Muller RH, Verbavatz JM and Langer R (1995) The controlled intravenous delivery of drugs using PEG-coated sterically stabilized nanospheres. Advanced Drug Delivery Reviews 16 (2):215–233

Grimes CA, Mungle CS, Zeng K, Jain MK, Dreschel WR, Paulose M and Ong KG (2002) Wireless magnetoelastic resonance sensors: a critical review. Sensors 2:294–313

Grossman HL, Myers WR, Vreeland VJ, Bruehl R, Alper MD, Bertozzi CR, Clarke J (2003) Detection of Bacteria in Suspension by Using a Superconducting Quantum Interface Device. PNAS 101:129–134

Grünberg P, Schreiber R, Pang Y (1986) Layered Magnetic Structures: Evidence for Antiferromagnetic Coupling of Fe Layers across Cr Interlayers. Physical Review Letters 57:2442–2445

Gundersen SG, Haagensen I, Jonassen TO, Figenschau KJ, de Jonge N and Deelder AM (1992) Magnetic bead antigen capture enzyme-linked immunoassay in microtitre trays for rapid detection of schistosomal circulating anodic antigen. J. Immunol. Methods 148:1–8

Gupta AK and Wells S (2004) Surface-modified superparamagnetic nanoparticles for drug delivery: Preparation, characterization, and cytotoxicity studie. IEEE Trans. In Nanobioscience 3(1):66–73

Haik Y, Chen C-J, Chatterjee J and Kanuri S (2000) The use of biotinylated lectin for separating red cells from whole blood. Biomolecular Eng. 16(5):179

Haik Y, Cordovaz M, Chen C-J and Chatterjee J (2002) Magnetic Immunoassay for Rapid Assessment of Acute Myocardial Infarction. European Cells and Materials 3:41–44

Haik Y, Chatterjee J and Chen C-J (2005) Synthesis and stabilization of Fe-Nd-B nanoparticles by chemical method. J Nanoparticles Res. 7(6):675–679

Hedrum A, Lundeberg J, Pahlson C and Uhlen M (1992) Immunomagnetic recovery of Chlamydia trachomatis from urine with subsequent colorimetric DNA detection. PCR Methods & Applications 2:167–171

Heleg-Shabtai V, Katz E and Willner I (1997) Assembly of microperoxidase-11 and Co(II)-protoporphyrin IX reconstituted myoglobin monolayers on Au-electrodes: integrated bioelectrocatalytic interfaces. J Am. Chem. Soc. 119:8121–8122

Hibi K, Abe A, Ohashi E, Mitsubayashi K, Ushio H, Hayashi T, Ren H and Endo H (2006) Combination of immunomagnetic separation with flow cytometry for detection of Listeria monocytogenes. Analytica Chemica Acta 573:158–163

Islam D, Tzipori S, Islam M and Lindberg A (1993) Rapid detection of Shigella dysenteriae and Shigella flexneri in faeces by an immunomagnetic assay with monoclonal antibodies. A. Eur. J. Clin. Microbiol. Infect. Dis. 12:25–32

Jeníková ZG, Pazlarova J and Demnerova K (2000) Detection of Salmonella in food samples by the combination of immunomagnetic separation and PCR assay Int Microbiol. 3(4):225–229

Jain MK, Schmidt S and Grimes CA (2001) Magneto-acoustic sensors for measurement of liquid temperature, viscosity, and density. Appl. Acoustic 62:1001–1011

Ji X, Zheng J, Xu J, Rastogi VK, Cheng T-C, DeFrank JJ and Leblanc RM (2005) (CdSe)ZnS quantum dots and organophosphorus hydrolase bioconjugate as biosensors for detection of paraoxon J. Phys. Chem. B. 109 (9):3793–3799

Johne B, Jarp J and Haaheim LR (1989) *Staphylococcus aureus* exopolysaccharide in vivo demonstrated by immuno-magnetic separation and electron microscopy. J. Clin. Microbiol. 27:1631–1635

Johnson JL, Brooke CL and Fritschel SJ (1998) Comparison of the BAX for screening/E. coli O157:H7 method with conventional methods for detection of extremely low levels of *Escherichia coli* O157:H7 in ground beef. Appl. Environ. Microbiol. 64: 4390–4395

Kapperud, G, Varund T, Skjerve E, Hornes E and Michaelsen TE (1993) Detection of pathogenic *Yersinia enterocolitica* in food and water by immunomagnetic separation, nested polymerase chain reactions, and colorimetric detection of amplified DNA. Appl. Environ. Microbiol. 59:2938–2944

Kapperud G, Rorvik LM, Hasseltvedt V, Hoiby EA, Iverson BG, Staveland K, Johnson G, Leitao J, Herikstad H, Andersson Y, Langeland G, Gondrosen B and Lassen J (1995) Outbreak of Shigella sonnei infection traced to imported iceberg lettuce. J. Clin. Microbiol. 33: 609–614

Kim JW, Jin Cho LZ, Marquardat SH, Forhilch AA, Baidoo SK (1999) Use of chicken egg-yolk antibodies against K88+ fimbrial antigen for quantitative analysis of enterotoxigenic *Escherichia coli* (ETEC) K88+ by a sandwich ELISA J. Sci. Food Agric. 79: 1513–1518

Kittel C (1946) Physical theory of ferromagnetic domains. Phys. Rev. 70:965–971

Lee GU, Metzger S, Natesan M, Yanavich C, Dufrěne YF (2000) Implementation of force differentiation in the immunoassay. Analytical Biochemistry 287(2):261–271

Leonard P, Hearty S, Quinn J and O'Kennedy R (2004) A generic approach for the detection of whole Listeria monocytogenes cells in contaminated samples using surface plasmon resonance. Biosens Bioelectron. 19(10):1331–1335

Loiselle KT and Grimes CA (2000) Viscosity measurements of viscous liquids using magnetoelastic thick-film sensors. Rev. Sci. Instrum. 71:1441–1446

Lund A, Hellemann AL and Vartdal F (1988) Rapid isolation of K88+ Escherichia coli by using immunomagnetic particles. J. Clin. Microbiol. 26:2572–2575

Lund A, Wasteson Y and Olsvik O (1991) Immunomagnetic separation and DNA hybridization for detection of enterotoxigenic Escherichia coli in a piglet model. J. Clin Microbiol. 29:2259–2262

Mary M (1997) Applications of magnetic particles in immunoassays. In: Hafeli U, Schutt W, Teller J, Zborowski M (eds) Scientific and Clinical Applications of Magnetic Carriers. Plenum Press, New York

Matsunaga T, Kawasaki M, Tu X, Tsujimaura N and Nakamura N (1996) Chemiluminescence enzyme immunoassay using bacterial magnetic particles. Anal. Chem. 68: 3551–3554

Meyer MHF, Krause HJ, Hartmann M, Miethe P, Oster J and Keusgen M (2007a) Francisella tularensis detection using magnetic labels and a magnetic biosensor based on frequency mixing. Journal of Magnetism and Magnetic Materials 311:259–263

Meyer MHF, Stehr M, Bhuju S, Krause HJ, Hartmann M, Miethe P, Singh M and Keusgen M (2007b) Magnetic biosensor for the detection of yersinia pestis. Journal of Microbiological Methods 68:218–224

Millen RL, Kawaguchi T, Granger MC, Porter MD (2005) Giant Magnetoresistive Sensors and Superparamagnetic Nanoparticles: A Chip-Scale Detection Strategy for Immunosorbent Assays. Anal. Chem. 77:6581–6587

Mohamadi-Nejad A, Moosavi-Movahedi AA, Safarian S, Naderi-Manesh MH, Ranjbar B, Farzami B, Mostafavi H, Larijani MB and Hakimelah GH (2002) The Thermal Analysis of Nonezymatic Glycosylation of human serum albumin: differential scanning calorimetry and circular dichroism studies. Thermochimica acta 389:141–151

Morgan JAW, Winstanley C, Pickup RW and Saunders JR (1991) Rapid Immunocapture of Pseudomonas putida Cells from Lake Water by Using Bacterial Flagella. Appl. Environ. Microbiol. 57:503–509

Morup S (1993) Studies of Superparamagnetism in Samples of Ultrafine Particles. In: Hernando A (ed) Nanomagnetism. Kluwer Academic Publishers, Boston, pp 93–99

Mulvaney SP, Mattoussi HM and Whitman LJ (2004) Incorporating fluorescent dyes and quantum dots into magnetic microbeads for immunoassays. Biotechniques 36(4):602–6, 608–609

Mungle CS (2001) Optical detection of magnetoelastic sensors and the variable temperature response of the resonant frequency. Dissertation, University of Kentucky

Nagasaki Y, Ishii T, Sunaga Y, Watanabe Y, Otsuka H and Kataoka K (2004) Novel Molecular Recognition via Fluorescent Resonance Energy Transfer Using a Biotin-PEG/Polyamine Stabilized CdS Quantum Dot. Langmuir 20(15):6396–6400

Nagasaki Y, Kobayashi H, Katsuyama Y, Jomura T and Sakura T (2007) Enhanced immunoresponse of antiboy/mixed-PEG co-immobilized surface construction of high performance immunomagnetic ELISA system. Journal of Colloid and Interface Science 309:524–530

Olsvik O, Skjerve E, Hornes E et al. (1991) Magnetic separation techniques applied to cellular and molecular biology. In: Kemshead JT (ed) Clinical microbiology. Wordsmiths' Conference Publications, Somerset, England, pp 207–221

Olsvik O, Popovic T, Skjerve E, Cudjoe S, Hornes E, Ugelstad J and Uhlen M (1994) Magnetic separation techniques in diagnostic microbiology. Clinical Microbiol Rev. 7(1): 43–54

Padhye NV and Doyle MP (1991) Production and characterization of a monoclonal antibody specific for enterohemorrhagic Escherichia coli of serotypes O157:H7 and O26:H11. J. Clin. Microbiol. 29:99–103

Perez FG, Mascini M, Tothill EI and Turner AP (1998) Immunomagnetic separation with mediated flow injection analysis amperometric detection of viable Escherichia coli O157. Anal. Chem. 70:2380–2386

Pinaud F, Michalet X, Bentolila LA, Tsay JM, Doose S, Li JJ, Iyer G and Weiss S (2006) Advances in fluorescence imaging with quantum dot bio-probes. Biomaterials 27(9):1679–1678

Rife JC, Miller MM, Sheehan PE, Tamanaha CR, Tondra M, Whitman LJ (2003) Design and Performance of GMR Sensors for the Detection of Magnetic Microbeads in Biosensors. Sensors and Actuators A 107:209–218

Ruan G, Feng S and Li Q (2002) Effects of material hydrophobicity on physical properties of polymeric microspheres formed by double emulsion process. J of Controlled Release 84:151–160

Ruan C, Wang H and Li Y (2002a) A bienzyme electrochemical biosensor coupled with immunomagnetic separation for rapid detection of escherichia coli O157:H7 in food samples. Trans. ASAE 45:249–255

Ruan C, Yang L and Li Y (2002b) Immunobiosensor chips for detection of Escherichia coli O157:H7 using electrochemical impedance spectroscopy. Anal. Chem. 74:4814–4820

Ruan G and Feng S (2003) Preparation and characterization of poly(lactic acid)–poly(ethylene glycol)–poly(lactic acid) (PLA-PEG-PLA) microspheres for controlled release of paclitaxel. Biomaterials 24:5037–5044

Ruan C, Zeng K, Varghese OK and Grimes CA (2003) Magnetoelastic immunosensors: amplified mass immunosorbent assay for detection of escherichia coli O157:H7. Anal. Chem. 75:6494–6498

Safarik I and Safarikova M (2004) Magnetic techniques for the isolation and purification of proteins and peptides. BioMagn. Res. Technol. 2 (7):1–17

Santra S, Yang H, Holloway PH, Stanley JT and Mericle RA (2005) Synthesis of water-dispersible fluorescent, radio-opaque, and paramagnetic CdS:Mn/ZnS quantum dots: a multifunctional probe for bioimaging. J Am Chem Soc. 127(6):1656–1657

Seo KH, Brackett RE, Frank JF and Hilliard S (1998) Immunomagnetic separation and flow cytometry for rapid detection of E. coli O157:H7. J. Food Prot. 61:812–816

Sharma VK (2002) Detection and quantitation of enterohemorrhagic *Escherichia coli* O157, O111, and O26 in beef and bovine feces by real-time polymerase chain reaction. Food Prot. 65:1371–1380

Skjerve E, Rorvik ML and Olsvik O (1990) Detection of *Listeria monocytogenes* in foods by immunomagnetic separation. Appl. Environ Microbiol. 56:3478–3481

Skjerve E and Olsvilk O (1991) Immunomagnetic separation of Salmonella from foods. Int. J. Food Microbiol. 14:11–18

Solaro R (2002) Nanostructured Polymeric Systems in Targeted Release of Proteic Drugs and in Tissue Engineering. Proceedings of China-EU Forum on Nanosized Technology, pp 225–244

Stoyanov PG and Grimes CA (2000) A remote query magnetostrictive viscosity sensor. Sens. Actuators 80:8–14

Tu S-I, Uknalis J, Irwin P and Yu LSL (2000) The use of streptavidin coated magnetic beads for detecting pathogenic bacteria by light addressable potentiometric sensor (LAPS). J. Rapid Methods Autom. Microbiol. 8:95–109

Vanderhoff JW, El-Aasser MS and Ugelstad J (1979) US Patent: 4,177,177

Varshney M, Yang L, Su XL and Li Y (2005) Magnetic nanoparticle-antibody conjugates for the separation of Escherichia coli 0157:H7 in ground beef. J Food Prot. 68(9):1804–1811

Vermunt AE, Franken AA and Beumer RR (1992) Isolation of salmonellas by immunomagnetic separation. J. Appl. Bacteriol. 72:112–118

Vila A, Gill H, McCallion O and Alonso M (2004) Transport of PLA-PEG particles across the nasal mucosa: effect of particle size. J. of Controlled Release 98:231–244

Volkov I, Gudoshnikov S, Usov N, Volkov A, Moskvina M, Maresov A, Snigirev O, Tanaka S (2006) SQUID-measurements of Relaxation Time of Fe_3O_4 Superparamagnetic Nanoparticles Ensembles. Journal of Magnetism and Magnetic Materials 300: e294-e297

Vote D, Doar O, Moon RE and Toffaletti JG (2001) Blood glucose meter performance under hyperbaric oxygen conditions. Clinica Chimica Acta 305:81–87

Wang CW and Moffitt MG (2005) Use of Block Copolymer-Stabilized Cadmium Sulfide Quantum Dots as Novel Tracers for Laser Scanning Confocal Fluorescence Imaging of Blend Morphology in Polystyrene/Poly(methyl methacrylate) Films. Langmuir 21(6):2465–2473

Wright DJ, Chapman PA and Siddons CA (1994) Immunomagnetic separation as a sensitive method for isolating Escherichia coli O157from food samples. Epidemiol. Infec. 113: 31–39

Yang L, Li Y (2005) Quantum dots as fluorescent labels for quantitative detection of Salmonella typhimurium in chicken carcass wash water. J Food Prot. 68(6):1241–1245

Yang L and Li Y (2006) Detection of viable Salmonella using microelectrode-based capacitance measurement coupled with immunomagnetic separation. J Microbiol Methods 64: 9–16

Yu H and Bruno JG (1996) Immunomagnetic-electrochemiluminescent detection of Escherichia coli O157 and Salmonella typhimurium in foods and environmental water samples. Appl. Environ. Microbiol. 62:587–592

Zborowski M and Chalmers JJ (2005) Magnetic cell sorting. Methods Mol Biol. 295:291–300

Zeng K, Ong KG, Mungle CS and Grimes CA (2002) Time domain characterization of oscillating sensors: application of frequency counting for resonant frequency determination. Rev. Sci. Instrum. 73:4375–4380

Zhao X and Shippy SA (2004) Competitive Immunoassay for Microliter Protein Samples with Magnetic Beads and Near-Infrared Fluorescence Detection. Anal Chem. 76(7):1871–1876

Zhao L, Wu D, Wu L and Song T (2007) A simple and accurate method for quantification of magnetosomes in magnetotactic bacteria by common spectrophotometer. Journal of Biochem. Biophys. Methods 70:377–383

18

Cantilever Sensors for Pathogen Detection

Raj Mutharasan

Abstract

In this chapter we summarize briefly the use of cantilever sensors for pathogen detection. Both micro- and macro-cantilever sensors have been investigated for detecting pathogens in liquid samples. In this review we examine previous work and summarize progress on using piezoelectric-excited millimeter-sized cantilever (PEMC) sensors developed in the author's laboratory. PEMC sensors immobilized with an antibody specific to the target pathogen has been shown to be very highly sensitive for detecting one cell per mL in one liter samples and 10 cells per mL in 10 mL samples, both in buffers and at similar concentrations in food matrices. After a brief introduction, the physics of sensing is reviewed, followed by a characterization of PEMC sensors, and finally the results from detection experiments are described.

1. Introduction

A number of biosensors have been reported for the detection of pathogenic bacteria in a variety of applications such as food analysis, clinical diagnostics, and for environmental monitoring (Abdel-Hamid et al. 1999a, 1999b; Ivnitski et al. 1999). Current methods of detection and identification of pathogens include plate culture wherein the target bacteria is grown selectively, and then identified with labeled reagents, or DNA from the enriched bacteria is extracted and amplified using polymerase chain reaction (PCR) for confirming identity. If the bacteria are present in copious amounts, one can use directly enzyme-linked immunosorbent assays (ELISA). Alternatively, they can be visualized after a concentrating step using immunobeads. While these types of methods-based tests are well established and can accurately identify pathogens, they are laborious and time-consuming requiring 24–48 hours for identification (Prescott et al. 2005). In practical systems, however, there is a great need for biosensors that accurately and rapidly detect a few cells in real matrices. In this regard, cantilever biosensors have attracted considerable interest for label-free detection of proteins and pathogens because of their promise of very high sensitivity (Craighead 2003; Ilic et al. 2004a). The binding of an antigenic target (pathogen) to an antibody-immobilized cantilever causes a resonance frequency decrease due to the increase in mass and can be measured conveniently. Two excellent reviews on cantilever sensors are available and the reader is referred to them (Lavrik et al. 2004; Ziegler 2004). The cantilever sensors reported in the literature, by and large, use optics or piezoresistive transduction mechanism for measuring resonance frequency. In this paper, we describe a new class of cantilever sensors developed in the author's laboratory over the past five years. They are self-excited and self-sensing and have been labeled as piezoelectric-excited

Raj Mutharasan • Department of Chemical and Biological Engineering, Drexel University, Philadelphia, PA

M. Zourob et al. (eds.), *Principles of Bacterial Detection: Biosensors, Recognition Receptors and Microsystems*,
© Springer Science+Business Media, LLC 2008

millimeter-sized cantilever (PEMC) sensors. This new class of cantilever sensors consists of a composite of a piezoelectric layer bonded to a nonpiezoelectric base. The piezoelectric layer is used both for exciting the vibration of the cantilever, and for sensing its resonance. Electrical excitation of PZT causes it to expand and contract, which induces bending, twisting, and buckling oscillations of the composite cantilever. At mechanical resonance, stress levels are higher than at nonresonance frequencies, causing PZT's resistance value to increase, which is conveniently monitored by measuring phase angle or impedance. The PZT is used both to excite the cantilever oscillation and to sense its resonance.

2. Millimeter-Sized Cantilever Sensors

Miniaturization of cantilever sensors to micron- and nano-meter length scales provides smaller mass and often higher resonance frequencies in air. Since sensitivity is higher at higher frequencies with microsensors, and micron-sized cantilevers can be micro-fabricated in large quantities economically, there has been a strong drive in the scientific community to explore their use for pathogen and molecular detection. The sensitivity of a miniaturized cantilever is enough to detect a single virus particle (Gupta 2004; Ilic et al. 2004a). On the other hand, micro- and nano-sized cantilevers are highly damped under liquid due to fluid viscous forces; for example, see a recent analysis (Davila et al. 2007). The dynamics of a cantilever in a liquid medium are affected by the viscosity and density of the liquid (Boskovic et al. 2002; Inaba et al. 1993; Naik et al. 2003; Rijal and Mutharasan 2007). The surrounding liquid offers resistance to oscillations. The added mass of liquid increases the inertia of the cantilever proportional to the cantilever's acceleration, and the liquid viscosity causes dissipative losses proportional to the velocity of the cantilever's motion. Mass loading on the cantilever decreases the sensor's resonance frequency in liquid relative to that in air and vacuum. The net effect of damping is a decrease in resonance intensity, which broadens the resonance peak. For a detailed mathematical analysis of the frequency responses of cantilever sensors in a liquid medium, the reader is refered to Sader's article (Sader 1998). Cantilevers obeying the three conditions: (1) uniform cross section, (2) length much longer than thickness, and (3) amplitude of vibration much smaller than any length scale of the sensor, one can determine the resonance frequency in inviscid fluid from:

$$\frac{f_{fluid}}{f_{vac}} = \left(1 + \frac{\pi \rho b}{4 \rho_c h}\right)^{-1/2}, \tag{18.1}$$

where f_{fluid} and f_{vac} are the resonance frequencies in fluid and vacuum respectively, ρ_c is the density of the cantilever, b and h are the width and thickness of the cantilever, and ρ is the density of the surrounding fluid. The Reynolds number (Re) of the oscillating cantilever is defined as $2\pi f \rho b^2/(4\eta)$, where η is the viscosity of the liquid medium. For micrometer- and nanometer-size cantilevers, the Reynolds number (Re) is low (0.01 to 10), while for the millimeter-sized cantilevers, Re is in the order of 10^5 to 10^6. As a result, the hydrodynamics are dominated by interial forces, and the reduction in sharpness of the resonance peak is not significant (Rijal and Mutharasan 2007). In a separate study, the author's lab showed that PEMC sensors can be used under flow conditions (Campbell and Mutharasan 2006b) without a significant loss of Q-value. Before we examine the performance of PEMC sensors, it is useful to briefly describe a sample of previous work on cantilever sensors used in pathogen detection. We focus only on the resonance modes, and in the following discussion we have not included the copious literature on quratz crystal microbalance (QCM).

3. Reported Work on Detecting Cells Using Cantilever Sensors

A very comprehensive review of current technologies for pathogen detection was recently published (Lazcka et al. 2007). While there have been many publications (Arora et al. 2006; Balasubramanian et al. 2007; Hancock and McPhee 2005; Koubova et al. 2001; Oli et al. 2006; Rijal et al. 2005; Taylor et al. 2006; Taylor et al. 2005; Waswa et al. 2007; Zezza et al. 2006) using optical and other techniques (Su and Li 2005; Vaughan et al. 2003; Vaughan et al. 2001) for detecting pathogens, only a few research groups have examined the binding and growth of pathogens and cells on cantilever sensors (Davila et al. 2007; Detzel et al. 2006; Gfeller et al. 2005; Nugaeva et al. 2005). A fairly comprehensive summary of reported research on pathogen detection is presented in Table 18.1. In Table 18.2 a summary of the detection and growth of benign cells or spores is given. Inspection of the summary in Table 18.1 suggests that detection of *E. coli* 0157:H7 has been investigated by a number of research groups in various matrices, using both bending and resonating cantilever sensors. Recent interest in biothreat agents has spawned experiments with *Bacillus anthracis*, followed

Table 18.1. Summary of cantilever-based pathogen detection publications

Target Pathogen	Transduction Mechanism	Limit of Detection	Matrix	Reference
E. coli 0157:H7	Resonance cantilever	50 cells/mL	ground beef wash	(Campbell et al. 2007b)
E. coli 0157:H7	Resonance cantilever	10 cells/mL	ground beef wash	(Maraldo and Mutharasan 2007b)
E. coli 0157:H7	Resonance cantilever	10 cells/mL	buffer in presence of wild E coli strain	(Maraldo et al. 2007c)
E. coli 0157:H7	Resonance cantilever	1 cell	buffer; dry and measure	(Ilic et al. 2000; Ilic et al. 2001)
E. coli 0157:H7	Resonance cantilever	1 cells/mL	buffer; flow	(Campbell and Mutharasan 2007a)
E. coli 0157:H7	Bending cantilever	10^6 cells/mL	buffer	(Zhang and Ji 2004)
E. coli 0157:H7	Resonance cantilever	10 cells/mL	spinach, spring lettuce washes	(Maraldo and Mutharasan 2007d)
Bacillus anthracis (Sterne Strain)	Resonance cantilever	300 spores/mL	buffer, and in presence of 10^9 *Bacillus thuringiensis* spores/mL	(Campbell and Mutharasan 2006b; Campbell and Mutharasan 2006c)
Bacillus anthracis (Sterne Strain)	Resonance cantilever	330 spores/mL	buffer in presence of $\sim 10^5$ *Bacillus cereus* and *Bacillus thuringiensis*	(Campbell and Mutharasan 2007c)
Bacillus anthracis (Sterne Strain)	Resonance cantilever	50 spores	dispensed on sensor in buffer	(Davila et al. 2007)
Group A *Streptococcus pyogens*	Resonance cantilever	700 cells/mL	buffer	(Campbell and Mutharasan 2006e)
Salmonella typhimurium	Resonance cantilever	5,000 cells/mL	buffer	(Zhu et al. 2007b)
Salmonella typhimurium	Resonance cantilever	100,000 cells/mL	buffer	(Zhu et al. 2007a)
Salmonella enterica	Bending mode	25 cells	buffer	(Weeks et al. 2003)

Table 18.2. Summary of publications using cantilever sensors for benign cells

Cell Type	Measurement Mode	Limit of Detection	Matrix	Reference
Autographa californica nuclear polyhedrosis virus	Resonance cantilever	10^5 pfu/mL	buffer	(Ilic et al. 2004b)
Vaccinia virus	Resonance cantilever	~1–5	dispensed on sensor in buffer, pure	(Gupta et al. 2004; Johnson et al. 2006)
Bacillus subtilis	Resonance and bending cantilever	~1,000 spores	buffer	(Dhayal et al. 2006)
E. coli	Resonance cantilever	Measured growth rate	agar	(Detzel et al. 2006)
Yeast	Resonance cantilever	10^9/mL	buffer	(Yi et al. 2003)
E. coli	Resonance cantilever	Measured growth	agar	(Gfeller et al. 2005)
A. niger	Resonance cantilever	10^3–10^6 spores/mL	agar	(Nugaeva et al. 2005)

by the food-borne pathogen *Salmonella*. Most investigators have used buffer as the medium of choice that does not contain any interfering organisms or particles, leading to a good sensor response. Detection in a real matrix (ground beef washes, vegetable washes) or in the presence of other contaminating species is rare (see Table 18.1). It is also of interest to note that the limit of detection reported varies over a very large range, and the methods used are varied—from continuous measurement under sample flow conditions to batch measurement by dipping into a sample followed by drying prior to measuring resonance frequency change. In many cases the objective appears to be to demonstrate the sensor's sensitivity rather than to detect anything.

As noted in Table 18.2, a few investigators have examined nonpathogenic cells as model systems to explore the sensitivity of detection or the sensitivity of cantilever sensors. We have included three studies that followed the growth of cells on a cantilever sensor. Viruses have been a popular target, as they offer a much smaller mass than bacterial pathogens. By and large the detection limit noted in Table 18.2 is large except in one study.

In the following section we provide a brief summary of a few sample publications, followed by a review of results obtained with the PEMC sensors.

Gfeller (2005) and coworkers studied the growth of *Escherichia coli* using an array of eight microcantilever sensors which were agarose-coated and then inoculated with the bacteria. Several sensors were not inoculated and served as reference sensors. Using a position-sensitive detector (PSD), the resonance frequency was monitored in a controlled environment of $37° \pm 0.2°$C and $93\% \pm 2\%$ relative humidity. They reported that the resonance frequency decreased exponentially over the first 5 h, suggesting that the reduction was due to an increase in mass on the sensor as a result of bacterial growth. The authors estimated the sensitivity of the cantilever as 140 pg/Hz and used this value to determine the number of cells on the sensor.

In another study, a silicon-based microcantilever immobilized with antibody to *E. coli* O157:H7 was used in bending mode for in situ detection of *E. coli* (Zhang and Ji 2004). The sensor deflection was measured optically. The experiment, carried out in a fluid cell at $20° \pm 0.2°$C, was done at various *E. coli* concentrations. The authors showed bending results to a sample at 5×10^6 cfu/mL. The bending response did not reach a steady state in 5 h, and the authors suggested steric hindrance as the cause of it. In addition, when the *E. coli* sample was replaced with a buffer, the cantilever's deflection did not change. Furthermore, the reference sensor showed little or no deflection in the *E. coli* sample. The experimental data was fitted to a first-order Langmuir kinetic model and the reaction rate was calculated as $2.3 \times 10^{-4}\,s^{-1}$. The authors reported the detection limit of the sensor as 1×10^{6} cfu/mL.

In a more recent study, microcantilevers as resonance mass sensors for detection of *Bacillus anthracis(BA)* were used both in air and liquid (Davila et al. 2007). The authors relied on measuring the resonance frequency decrease driven by thermally induced oscillations. For the liquid phase detection of spores, the cantilevers were first functionalized with anti-BA antibodies followed by exposure to BA spores. The authors concluded that as few as 50 spores on the cantilever can be detected in water. They estimated that the measurement sensitivity was 0.1 fg/Hz in air and 10 fg/Hz in liquid.

Yi et al. showed detection of yeast cells using a PZT/stainless steel cantilever sensor (Yi et al. 2003). Although yeast is not considered a pathogen, this is an early study examining the macrocantilever response to cell binding, and is worthy of mention. The stainless steel layer was coated with 0.1 % polylysine and then exposed to yeast at 1 and 2 mg/mL. The basis of detection was the positively charged coating attracting negatively charged yeast cells. A fundamental resonance at 22.3 kHz was monitored for measuring mass change. The authors reported the sensitivity of the two cantilevers used as 0.4 and 2.3 µg/Hz.

In another interesting study, the binding of small well-defined phospholipid vesicles was monitored by the resonance frequency of cantilevers 125 µm long, 35 µm wide, and 4 µm thick (Ghatnekar-Nilsson et al. 2005b). These sensors, excited by a piezoelectric layer, exhibited resonance modes in air between 270 and 310 kHz. Upon immersion in liquid, the resonance frequency decreased $\sim$142 kHz and the Q-value decreased from 240 to 60. Upon introduction of phospholipid vesicles, the resonance frequency decreased in $\sim$1 minute, and stabilized in 8 minutes. The authors interpreted the frequency stabilization as due to sensor surface saturation. They estimated the adsorbed mass of vesicles as 450 pg, which was in agreement with the estimate of total monolayer coverage of 400 pg.

From the foregoing survey, we conclude that only a few studies have explored the use of cantilver sensors for pathogen detection. We now turn to the applications demonstrated using the PEMC sensor.

4. Physics of Cantilever Sensors

A material that is piezoelectric develops a voltage across it when mechanically stressed, which is referred to as the direct piezoelectric effect. Conversely, the application of a potential across the piezoelectric material generates a strain, which is the converse piezoelectric effect. The piezoelectric effect is caused by charge distribution within the material molecular structure, due to applied stress that results in an internal electric field. Therefore, the piezoelectric effect converts mechanical into electrical energy and vice versa. The piezoelectric ceramic used in the construction of PEMC sensors is made from oxides of lead, zirconium, and titanium and is referred to as lead zirconate titanate (PZT). The PZT gives a sensitive response to weak stresses due to the direct piezoelectric effect and generates high strain via the converse piezoelectric phenomenon.

PEMC sensors are fabricated to provide predominantly the bending mode vibration. Two designs, labeled A and B, obtained by bonding PZT to glass, are illustrated in Fig. 18.1. An electric voltage applied across the thickness of the PZT film will lengthen or shorten the film, depending on the polarity of the electric field. Such a change in dimension causes the cantilever to bend, twist, or buckle. If the applied field is alternated periodically, the composite cantilever will vibrate, and will resonate when the electric excitation frequency coincides with the mechanical or natural resonance frequency. The natural frequency of the cantilever depends on the flexural modulus and the mass density of the composite cantilever. At resonance,

Design A
SIDE VIEW

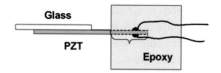

Design B
SIDE VIEW

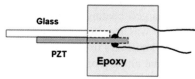

Figure 18.1. Two basic versions of PEMC sensor designs. Typical width is 1 mm, length of PZT and glass are a few millimeters. Design B showed good resonant modes below 150 kHz while Design A showed multiple resonant modes from 50 kHz – 2 MHz. Some of the high-order modes were found to be give sensitivity of ~1 fg/Hz.

the cantilever undergoes a significantly higher level of vibration and larger stresses. As a consequence, the PZT layer exhibits a sharp change in impedance, and is conveniently followed by measuring the phase angle (Campbell and Mutharasan 2005a).

The natural frequency of a cantilever with a flexural rigidity of EI, where E is the modulus of elasticity and I the moment of inertia, can be obtained by solving the general equation representing transverse vibration:

$$EI\frac{\partial^4 y}{\partial x^4} + (\rho wt)\frac{\partial^2 y}{\partial \tau^2} + (c_0)\frac{\partial y}{\partial \tau} = 0, \tag{18.2}$$

Here, y is the displacement along the thickness of the cantilever, x is the length along the cantilever, τ is time, and ρ is density. The term c_0 is the damping parameter intrinsic to the cantilever. The moment of inertia, I, of a rectangular cross section is $wt^3/12$, where w is the width and t is the thickness. A number of previous investigators have determined solutions to the above model. A practical expression for the resonance frequency for sensing purposes is obtained when one considers the distributed mass of the cantilever to be located at the tip, and is given as (Naik 2003):

$$f_n = \frac{\nu_n'^2}{2\pi}\sqrt{\frac{K}{M_e}}, \tag{18.3}$$

where $\nu_n^2 = \nu_n'^2\sqrt{3/0.236}$, with $\nu_n = 1.8751$, 4.6941, 7.8548, and 10.9956, corresponding to the first four eigenvalues for a rectangular cantilever. The parameter K is the effective spring constant which depends on the thickness, density, and modulus of the cantilever materials, both the nonpiezoelectric material and the PZT layer. The solution to Eq. (18.2) has been reported (Sader 1998).

The macrocantilevers investigated in the author's lab were fabricated by adhesive-bonding a piezoelectric layer to a nonpiezoelectric layer (borosilicate glass), and then anchoring either the PZT alone (Design A, Figure 18.1) (Campbell and Mutharasan 2007a,c; Maraldo et al. 2007) or PZT and glass (Design B, Fig. 18.1) (Campbell and Mutharasan 2005; Campbell and Mutharasan 2006a-e). As a result, the effective tip mass, M_e, for Design B of length L can be approximated as (Campbell and Mutharasan 2005):

$$M_e = 0.236 \left(\rho_p t_p + \rho_{np} t_{np} \right) w L_p + \rho_{np} t_{np} w \left(L - L_p \right). \tag{18.4}$$

Also, the effective spring constant, neglecting the nonpiezoelectric tip, is:

$$K = \frac{3w^2 \left(E_p^2 t_p^4 + E_{np}^2 t_{np}^4 + 2 E_p E_{np} t_p t_{np} \left(2t_p^2 + 2t_{np}^2 + 3t_p t_{np} \right) \right)}{12 L_p^3 \left(E_p t_p + E_{np} t_{np} \right)}, \tag{18.5}$$

where subscripts p and np refer to PZT and nonpiezoelectric layers, respectively. The Design A sensors have only the PZT layer anchored at one end, and at the other end a $1 \times 2\,\mathrm{mm}^2$ nonpiezoelectric layer is bonded. Therefore, the effective tip mass, M_e, can be approximated as:

$$M_e = \rho_p t_p w L_p + 0.236 \left(\rho_p t_p + \rho_{np} t_{np} \right) \left(L - L_p \right). \tag{18.6}$$

Also, the effective spring constant of Design B sensors, neglecting the nonpiezoelectric layer, can be expressed as:

$$K = \frac{3 E_p I w}{L^3}. \tag{18.7}$$

When a PEMC sensor is immersed in a liquid, the surrounding fluid offers resistance to its motion. Consequently, the sensor behaves as though an added mass of fluid is attached to it, which results in an additional inertial force in Eq. (18.2). This force is in phase with the cantilever motion. In addition, a dissipative force proportional to the velocity of the cantilever should be included to describe the behavior in the liquid. The fluid does not respond instantaneously to cantilever motion, and causes a phase shift between the cantilever motion and the fluid motion. Inclusion of these two effects in Eq. (18.2) results in:

$$EI \frac{\partial^4 y}{\partial x^4} + \left(\rho w t + M_a \right) \frac{\partial^2 y}{\partial \tau^2} + \left(c_0 + c_v \right) \frac{\partial y}{\partial \tau} = 0, \tag{18.8}$$

where M_a represents the added mass of liquid per unit length of cantilever and c_v represents the added damping coefficient due to fluid motion. Thus Eq. (18.3) can be rewritten as:

$$f_{nf} = \frac{v_n'^2}{2\pi} \sqrt{\frac{K}{M_e + M_a}}, \tag{18.9}$$

where f_{nf} is the resonance frequency of mode n in liquid. In detection applications, the added mass consists of two terms. The first is due to the fluid surrounding the cantilever and the second is due to the attachment of the analyte at the cantilever's tip. Thus Eq. (18.9) can be rewritten as:

$$f_{nf}' = \frac{v_n'^2}{2\pi} \sqrt{\frac{K}{M_e + m_{ae} + \Delta m}}, \tag{18.10}$$

where f'_{nf} is the resonance frequency of the nth mode in fluid when an analyte of mass Δm is attached at the cantilever's tip. In the above, m_{ae} is the effective added mass of fluid at the cantilever tip, similar to the effective cantilever mass M_e. From the above equation one gets:

$$f'_{nf} - f_{nf} = \frac{1}{2} f_{nf} \frac{\Delta m}{M_{ef}}, \qquad (18.11)$$

where $M_{ef} = M_e + m_{ae}$ and f_{nf} is the resonance frequency of the nth mode in fluid. That is, a change in resonance frequency represented by the left hand side of Eq. (18.11) is linearly dependent on the change in mass, for a particular resonance mode. The other parameters in Eq. (18.11) are constants for a given cantilever sensor. The effective mass in fluid, M_{ef}, can be determined experimentally from the resonance frequency under liquid immersion conditions. Thus, one can use Eq. (18.11) to calculate mass change due to target antigen attachment. For a small Δm, the change in resonance frequency (Eq. (18.11)) depends linearly on the attached mass. Eq. (18.15) can be rearranged to:

$$\sigma_{nf} = \frac{\Delta m}{f_{nf} - f'_{nf}} = \frac{2M_{ef}}{f_{nf}}, \qquad (18.12)$$

where the term (σ_{nf}) represents mass change sensitivity. The above indicates that for the same cantilever mass, the sensitivity is higher (meaning a lower numerical value) as resonance frequency (f_{nf}) increases. For a typical value of $M_{ef} = 1$ mg, and $f_{nf} = 60$ kHz, Eq. (18.12) predicts a sensitivity of $\sim 15\mu g/Hz$. However, experimentally, we find a considerably higher sensitivity of 10.6 ng/Hz (Campbell and Mutharasan 2006c) with PEMC sensors at 45–65 kHz. The higher sensitivity is thought to be due in part to sensor design and PZT, and in part to the higher mode. The frequency change is amplified because of the impedance change of the electrically active PZT. In light of the experimental determination, we modify Eq. (18.12) as:

$$\sigma_{nf-PZT} = A \frac{2M_{ef}}{f_{nf}}, \qquad (18.13)$$

where A is called the amplification factor and depends on the resonance mode (n). For the mode used in our early studies (Campbell and Mutharasan 2006c), A was $\sim 10^3$. In a recent study we determined the sensitivity of Design A in the range of 0.3 to 2 fg/Hz (Campbell and Mutharasan 2007a,c). In another study the high-order resonance frequency near 900 kHz exhibited a sensitivity (σ_{nf-PZT}) of 1.47 femtrogram/Hz (Campbell and Mutharasan 2007a). That is, for Design A, the amplification factor A is $\sim 10^8$ for the 900 kHz mode. The implication of such a high sensitivity is that the attachment of a single pathogen (mass of ~ 1 pg) would be detectable in near real time.

5. Resonance Modes

The cantilever resonates at various modes of vibration, as was noted earlier. For example, in Eq. (18.3) the eigenvalues that satisfy the governing equation are infinite, and thus a very large number of resonance modes exist. Generally, higher resonance modes are of such poor quality that they are not useful for measurement. In this section we show simulation results that provide a useful guidance to high modes, as many of the very sensitive measurements developed in the author's lab have relied on them.

A nominally-sized PEMC sensor (1 mm wide, 2 mm PZT, and 5 mm glass) was numerically simulated in a finite element modeling platform using FEMLAB®. The structural

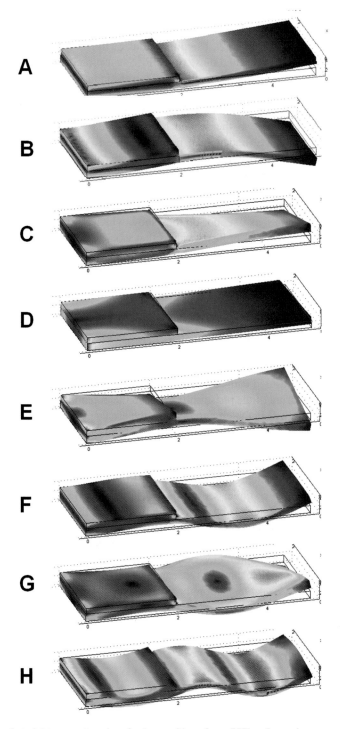

Figure 18.2. The first eight resonant modes of a 1 mm wide × 2 mm PZT × 5 mm glass sensor of Design B. The resonant modes, respectively, are: A) 10,893 Hz, 1st bending mode; B) 41,727 Hz, 2nd bending mode; C) 45,612 Hz, 1st torsional mode; D) 78,834 Hz, 1st buckling mode; E) 105.038 Hz, 2nd torsional mode; F) 105,874 Hz, 3rd bending mode; G) 186,187 Hz, 3rd torsional mode; H) 214,964 Hz, 4th bending mode.

mechanics plane stress model, using the appropriate subdomain property values and boundary conditions for the various vibrating modes, was determined. The results for PEMC sensor Design B are summarized in Fig. 18.2. The dimensions of the nonpiezoelectric layer (borosilicate glass) were $6 \times 2 \times 0.16\,mm^3$ and those of the PZT layer (on top) were $2 \times 2 \times 0.127\,mm^3$. Fig. 18.2 shows the first eight modes of oscillation, corresponding to the first eight eigenvalues. The fundamental mode is a bending mode; the cantilever oscillates in the Z-direction. The second mode is also a bending mode that resonates at approximately four times the frequency of the first. In close proximity to the second bending mode is the first tortional mode. At approximately seven times the fundamental frequency, the first buckling mode appears. The figure is a plot of stress level with position; red being higher than blue. Dark blue is the lowest stress point. It is clear from the figure that the largest stresses were generated in the bending modes, which suggest that the bending mode is likely to be the most sensitive of the three mode types. The simulated results in Fig. 18.2 give a sense of what modes are present. However, they do not show which mode is sensitive, nor do they show which is likely to give good Q-values. This is due to the inherent lack of property values that would enable us to calculate dissipative energy in the model. Thus one of the experimental efforts is to determine which modes are dominant, and which modes remain separate during a sensing episode, so that reliable sensing can be accomplished.

6. Characterization of PEMC Sensors

The mechanical resonance frequency of PEMC sensors can be experimentally determined from the phase angle, measured using an impedance analyzer (e.g., HP 4192A or HP 4294A). The phase angle is a measure of the phase shift between the potential applied to the PZT and the current generated (the signal) in the circuit; and it is the arc tangent of the ratio of the reactance to the resistance of the circuit. A typical measured spectrum of a PEMC sensor of Design B is shown in Fig. 18.3. One notes that three resonance modes are located at 23, 91, and 211 kHz. The modes successively become smaller or lower in intensity. Note also that the sharpness of the peak decreases with an increase in mode number. In Fig. 18.3, Panel B presents a close-up view of one of the modes of a similar sensor, in which the mode is located near 20 kHz. In the same graph the impedance change is presented as a function of the excitation frequency, which shows the classic resonance/antiresonance behavior of electrical circuits.

The spectrum shown in Fig. 18.3 is strongly influenced by the piezoelectric properties, the mechanical properties of the nonpiezoelectric layer, and the relative and absolute dimensions of the two layers. The phase angle change at resonance has been observed as high as 100 degrees, and the sharpness of the peak very strongly depends on the Q of the material used in the fabrication.

7. Mass Change Sensitivity

The mass change sensitivity (σ_{nf} in Eq. 18.12) can be determined experimentally by adding the known mass (nanograms to picograms) to the sensor while monitoring the resonance peak(s) of interest. A plot of the mass change versus the frequency change should give a straight line whose slope is the sensitivity in g/Hz. In a recent paper, we used paraffin wax dissolved into hexane (Campbell and Mutharasan 2007c). Nanoliter amounts of the wax solution of known concentration were dispensed on the sensor surface. After allowing the solvent to evaporate at room temperature in vacuum (50 mTorr), the resonance frequency was measured. The reader is cautioned that most commercial "pure" waxy hydrocarbons have a small percentage that will

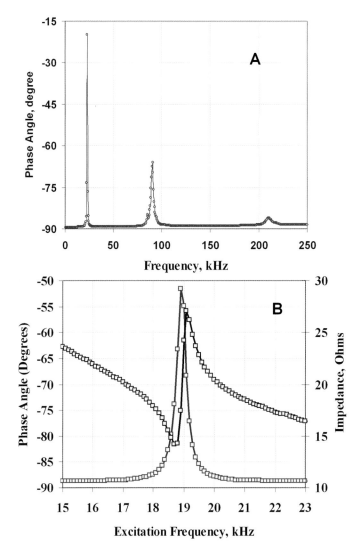

Figure 18.3. Panel A: Typical resonance spectrum of a PEMC sensor (Design B). Note that the first three modes are visible with progressively lower phase angles, and progressively wider peak base. Upon immersion in water, the Q-value decreases by 10–20 % and resonant frequency shifts down. Panel B: The mode near 19 kHz is expanded and the profile of impedance of the sensor is also shown. The latter shows a classical resonance anti-resonance characteristic.

volatilize at 50 mTorr, and a small percentage will be extracted by hexane. Thus, purification or removing volatile and extractable compounds of the paraffin is strongly suggested before using it for calibration.

8. Antibody Immobilization Methods

A PEMC sensor surface (glass) is immobilized with an antibody to the pathogen of interest. Several functionalization methods have been reported in the literature; an excellent practical book is available (Hermanson 1996). In adiditon, Pierce Chemicals provides a rich source of "how to" protocols for the surface immobilization of proteins in general. In our laboratory, two main functionalization schemes are used, but only one has been optimized

(Maraldo and Mutharasan 2007a). The three methods are: (1) derivatize the glass surface with an amine-terminal silane, (2) gold coat the glass surface, and (3) immobilize protein G.

The Ab-immobilization technique via amine-terminated silane involves three main steps: cleaning, silanization, and antibody immobilization. The sensing glass surface is cleaned sequentially with methanol-hydrochloric acid solution (1:1 v/v), concentrated sulfuric acid, hot sodium hydroxide, and finally boiling water. The surface is rinsed between each washing step with deionized water. The cleaning procedure produces reactive hydroxyl groups on the glass surface. After cleaning, the glass surface is silanylated with 0.4 % 3-aminopropyl-triethoxysilane (APTES; Sigma-Aldrich) in deionized water at pH 3.0 (adjusted by hydrochloric acid, 0.1 N) and 75°C for 2 hours. APTES reacts with glass, leaving a free amine terminal for further reaction with the carboxyl group to form a peptide bond. The carboxyl group present in the Fc region of the antibody is activated using the zero-length cross linker 1-ethyl-3-(3-dimethylaminopropyl)-carbodiimide (EDC; Sigma-Aldrich) and promoted by sulfo-N-hydroxysuccinimide (Sigma-Aldrich). EDC converts carboxyl groups into reactive unstable intermediates susceptible to hydrolysis. Sulfo-NHS replaces the EDC, producing a more stable reactive intermediate that is susceptible to attack by amines. Covalent coupling of the stable intermediate with the silanylated glass surface is carried out at room temperature for 2 hours. The glass surface with the immobilized antibody is used to detect the antigen of interest.

9. Detection in Batch and Stagnant Samples

In batch detection, the Ab-immobilized sensor was dipped in one mL of sample containing the pathogen, and the resonance frequency was measured continuously. A model pathogen *Group A Streptococcus pyogenes* (GAS) was detected in buffer solutions at 700 cells/mL (Campbell and Mutharasan 2006e). Two sensors of Design B were used (the exposed dimensions were 1 mm PZT and 4 mm glass with either 1 mm width or 2 mm width). The 1 mm wide sensor had resonance modes at 18 kHz ($Q = 70$) and 72.5 kHz ($Q = 38$). When the same sensor design was fabricated at 2 mm width, the resonance frequencies decreased to 17.5 kHz and 53 kHz with Q values of 25 and 38, respectively. The 1 mm showed superior sensitivity, 22 pg/Hz vs 0.72 ng/Hz for the 2-mm sensor, measured with known added mass. Thus, the authors used 1 mm sensor in their experiments. The resonance frequency decrease of the second mode at concentrations of 700, 7×10^3, 7×10^5, 7×10^6, 7×10^7, and 7×10^9 cells/mL resulted in, respectively, 136, 509, 690, 1130, 1260, and 1782 Hz (see Fig. 18.4). The control, consisting of either a blank sensor exposed to 7×10^6 cells/mL or an Ab-immobilized sensor submerged in PBS, gave essentially a zero response (3 ± 6 Hz). The authors report that the lowest concentration sample (700 cells/mL) showed the slowest binding kinetics, and the steady state was reached in 25 minutes. A kinetic model based on Langmuir kinetics for the pathogen attachment was proposed and verified. The observed binding rate constant was found to be in the range of 0.051 to 0.166 min^{-1}.

The steady state resonance frequency response of the PEMC sensor showed good correlation with the log of the GAS bulk concentration. That is, the results suggest that the calibration relationships for estimating the GAS concentration can be stated as:

$$\log(C_{b0}) = \frac{(-\Delta f)_{ss} + A}{B}, \tag{18.14}$$

where the parameters A and B are constants, and depend on cantilever dimensions, antibody type, and immobilization method. In the above, $(-\Delta f)_{ss}$ is the steady state resonance frequency change and C_{b0} is the pathogen concentration in the sample.

A similar approach was taken by the same authors in detecting *Bacillus anthracis* (Sterne strain; BA) in a batch stagnant sample (Campbell and Mutharasan 2006c). Their study indicates

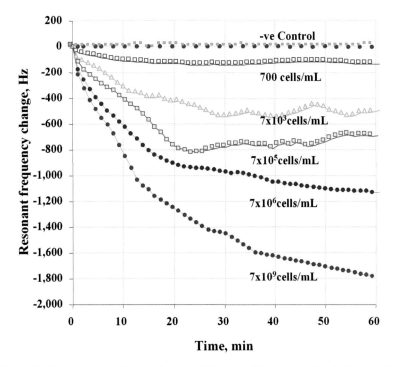

Figure 18.4. A PEMC sensor (Design B) operating at 64.5 kHz in liquid was immobilized with antibody to *Group A Streptococcus pyogenes*, and was then exposed to various pathogen concentrations in batch mode. Negative control was a clean sensor immersed in 7×10^6 cells/mL sample, and the response of an Ab-immobilized sensor in PBS (adapted from Campbell and Mutharasan 2006e).

that the PEMC sensor response is nonlinearly related to BA concentration in liquid. The second mode resonance frequency was monitored upon exposure to *Bacillus anthracis* spores at concentrations of 3×10^2, 3×10^3, 3×10^4, and 3×10^6 spores/mL (see Figs. 18.5 and 18.6). In all cases, the response showed a rapid decrease during the first 10 minutes, followed by a slower change, reaching a constant resonance frequency. For the highest spore sample (3×10^6 spores/mL), the rate of decrease was more rapid compared to the lower concentration (300 spores/mL) sample. This is the expected response, as the binding rate is proportional to the concentration. The total change was 2696 ± 6 Hz (n = 2) for the 3×10^6/mL sample and 92 ± 7 Hz (n = 3) for the 300 spores/mL sample. The sample containing 30 spores/mL gave a response of 31 Hz in one out of three trials, and is not shown. Several control experiments with antibody-functionalized sensors were conducted in PBS. The responses showed fluctuations in the resonance frequency of ± 5 Hz, which is considerably lower than the signal obtained with 300 spores/mL. Thus, it is clear that the steady state response to noise was in the order of 20 for the low spore count sample. For higher spore concentration samples, the steady state change in resonance frequency was significantly higher than measurement errors in resonance frequency. The observed resonance frequency fluctuation of ± 5 Hz for the final steady state value indicates that the cantilever resonance characteristic is quite stable under liquid immersion.

The same authors examined the selectivity of the PEMC sensor when a copious amount of *Bacillus thuringiensis* (BT) was present (Campbell and Mutharasan 2007c). The sensor selectivity to *Bacillus anthracis* (BA) spore was investigated in batch stagnant 1 mL samples by examining the response to mixed samples containing *Bacillus thuringiensis* (BT) spores. Four samples, A (BA: 3×10^6/mL, BT: 0 /mL), B (BA: 2.4×10^6/mL, BT: 3.0×10^8 /mL), C (BA: 2.0×10^6/mL, BT: 5.0×10^8 /mL), and D (BA: 1.5×10^6/mL, BT: 7.5×10^8 /mL) were

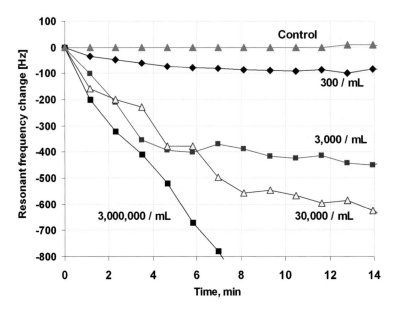

Figure 18.5. A PEMC sensor (Design B) response to *Bacillus anthracis* spores in batch mode detection. Sample containing 30 spores/mL gave a response of 31 Hz in one out of three trials, and is not shown. Negative control was a clean sensor in PBS. Sample volume was 1 mL. Note that the rapid binding of spores to the sensor surface. Note that a positive detection is indicated in less than four minutes (adapted from Campbell and Mutharasan 2006c).

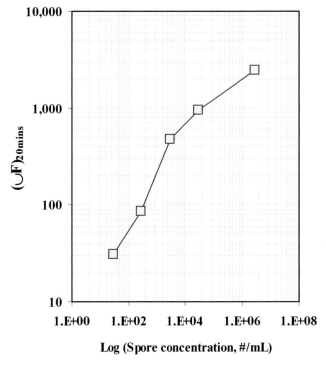

Figure 18.6. A PEMC sensor (Design B) response to *Bacillus anthracis* spore concentration in batch measurement. Lowest spore concentration that was reproducibly measured was 300 spores/mL. The authors (Campbell and Mutharasan 2006c) found 30 spores/mL (n = 1) is detectable using the same sensor in one out of three trials. The logarithmic response of the sensor is similar to that observed with *Group A Streptococcus pyogenes*. See Eq. (18.14).

prepared. Sample A gave the highest frequency change (2360 Hz; n = 1) in 1 h. As BT spores were added to the pure BA spore samples (samples B, C, and D) the resonance frequency reached a lower steady state value: 1980 ± 10, 1310 ± 10, and 670 ± 10 Hz for samples B, C, and D, respectively. The fluctuation in resonance frequency of the final steady state value was $\pm$ 10 Hz. It is to be noted that the pure BT spore sample (1.5×10^9/mL) gave only a 10 Hz change, suggesting that nonspecific binding, if any, was insignificant. Sample D contained 1.5×10^6 BA spores/mL and therefore one would expect a higher resonance frequency response in comparison to the pure *Bacillus anthracis* spore sample of 3×10^4 BA spores/mL. However, the opposite result was observed. That is, the pure BA sample (3×10^4 BA spores/mL) gave a much higher overall resonance frequency change (1030 Hz) than the sample D containing 7.5×10^8 /mL BT spores (670 ± 10 Hz). These results suggest that the presence of large amounts of BT spores hindered the attachment of BA spores, in batch mode. The authors hypothesized that the BT in these experiments "crowded" the sensor surface because of its very high relative concentration, which in effect reduced the accessibility of binding sites for the BA. The crowding effect can be overcome with a flow field imposed on the sensor and is discussed in the next section. From the results presented above, we conclude that the antibody-functionalized PEMC sensors are very sensitive to low pathogen concentrations, and are also selective to the *Bacillus anthracis* spores, but at a compromised sensitivity when high BT spores are present.

10. Detection in Flowing Samples

One of the disadvantages of stagnant or batch measurement of pathogens is that the target settles because of gravity, and only partial sampling of the sample solution is feasible. In batch stagnant samples, there is no convective transport of the target pathogen to the sensor surface. This deficiency is overcome by flowing the sample across the sensor. Based on both simulation and experiments, it was found that a flow parallel to the sensor's length introduced minimal change to the resonance frequency values, as long as the flow rate was kept constant (Campbell and Mutharasan 2006b). The authors used an apparatus, shown in Fig. 18.7, which consisted of a sensor flow cell that was connected to reagent reservoirs and a circulation flow circuit facilitated by a peristaltic pump. Typical flow rates used were 0.2 to 17 mL/min. The sensor flow cell was kept at a constant temperature. Two flow cell designs (parallel to sensor length and parallel to sensor width) were evaluated by measuring the flow-induced resonance frequency shifts. One of them that had a hold-up volume of 300 μL, showed small fluctuations ($\pm$ 20 Hz) around a common and constant resonance frequency response of 217 Hz in the flow rate range of 1 to 17 mL/min. The flow rate corresponded to a velocity in the sensor chamber of 0.05 to 0.85 cm/s. Of special note is the total resonance frequency change obtained for the binding of 300 spores/mL under flow, which was almost twice (162 ± 10 [n = 2]) the values measured in batch (90 ± 5 [n = 2]). The binding kinetics were modeled by the authors as:

$$(\Delta f) = (\Delta f_\infty)\left(1 - e^{-k_{obs}\tau}\right), \tag{18.15}$$

where (Δf) is the change in resonance frequency at time τ, and (Δf_∞) is the steady state resonance frequency change. The parameter k_{obs} is the observed binding rate constant and depends on the concentration. The characteristic sensor response rate constant, k_{obs}, was found to be higher under flow conditions: 0.195 ± 0.02 min^{-1} (stagnant) vs. 0.263 ± 0.05 min^{-1} (flow). That is, the binding rate of the pathogen was 35 % higher under flow than under stagnant conditions.

In another study (Campbell and Mutharasan 2007a), a PEMC sensor was investigated for detecting *E. coli* O157:H7 at 1 cell/mL in 1 mL and 1 L samples in batch mode and flow mode, respectively. Both sensor designs (A and B) were tested. Design

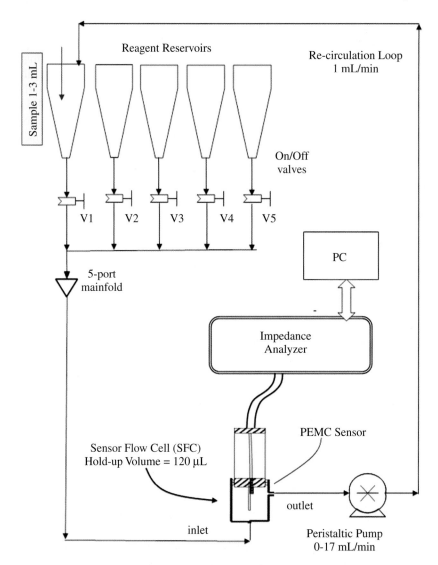

Figure 18.7. Continuous flow apparatus used in all pathogen detection experiments (adapted from Campbell and Mutharasan 2006b). The flow cell was maintained at a fixed temperature (30 C) using a circulating water bath (not shown). The sample circulating peristaltic pump was typically run at 0.5 to 3 mL/min flow rate. A LabView program managed running the impedance analyzer, with data collection and display.

B showed a sensitivity for detecting *E. coli* at 10 cells/mL, using the second bending mode at 85.5 kHz in air. Design A exhibited resonance frequencies at 186.5, 883.5, and 1,778.5 kHz in air and 162.5, 800.0, and 1,725.5 kHz in liquid flow conditions. A one-liter sample containing 1,000 *E. coli* cells was introduced at various flow rates, and the change in resonance frequency was monitored. The total frequency change observed for the mode at 800 kHz and sample flow rates of 1.5, 2.5, 3, and 17 mL/min were 2,230 ± 11, 3,069 ± 47, 4,686 ± 97, and 7,188 ± 52 Hz, respectively. Detection was confirmed by exposing the sensor to a low pH solution followed by a phosphate buffered saline (PBS) rinse, which caused the release of the attached *E. coli*. The final frequency change observed was nearly identical to the value prior to the *E. coli* attachment. Kinetic analysis showed that the observed binding rate constant at 1.5, 2.5, and 3 mL/min were 0.009, 0.015, and 0.021 min⁻¹, respectively. The significance of these results is that flow can considerably enhance the kinetics of attachment and gives a higher level of attachment.

11. Selectivity of Detection

In practical applications, the selectivity of the sensor to the target is important. Two types of experiments have been carried out in the author's laboratory. The first group of experiments consisted of the detection of pathogenic bacteria in the presence of nonpathogenic ones in clean buffers (Campbell and Mutharasan 2006d-a; Maraldo et al. 2006). The second type consisted of detection in complex food matrices (Campbell et al. 2007b; Maraldo and Mutharasan 2007b).

In Fig. 18.8, the transient resonance frequency responses of the PEMC sensor (Design B) to the binding of *Bacillus anthracis* (BA) spores at concentrations of 333 BA/mL are presented. In each experiment, the sensor responded with a rapid decrease in resonance frequency before reaching the same steady state value at different time periods. For the pure BA sample (333 spores/mL), the resonance frequency decreased most rapidly and reached steady state in 27 minutes. As the concentration of the nonantigenic *Bacillus* species (*Bacillus thuringiensis* [BT] and *Bacillus cereus* [BC]) increased, the rate of resonance frequency change decreased, and took a longer time to reach steady state. Steady states of $2,742 \pm 38$ (n = 3), $3,053 \pm 19$ (n = 2), $2,777 \pm 26$ (n = 2), $2,953 \pm 24$ (n = 2), and $3,105 \pm 27$ (n = 2) Hz were obtained for the 1:0, 1:1, 1:10, 1:100, and 1:1000 BA:BT+BC samples, respectively, in 27, 45, 63, 154, and

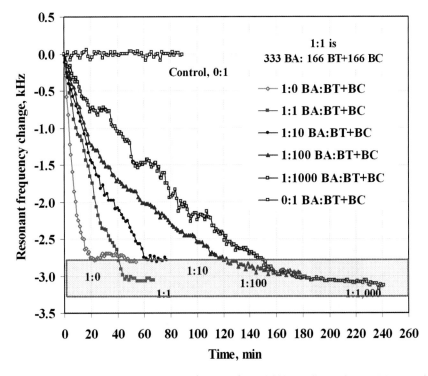

Figure 18.8. Transient response of PEMC sensor (Design A) to of 333 *Bacillus anthracis (BA)* spores/mL from samples containing various amounts of other Bacillus species. The control response shown is that of the anti-BA functionalized sensor exposed to a sample containing a mixture of BT and BC spores, at 166 BT/mL and 166 BC/mL. Note that the total sensor response was the same in all cases. However, in presence of a large number of non-pathogenic BC and BT, the kinetics of attachment decreased. Sample flow rate was 1 mL/min in the flow apparatus given in Fig 18.7 (adapted from Campbell and Mutharasan 2007c).

219 minutes. The authors noted that the steady state responses yielded an average frequency decrease of 2926 ± 162 (n = 11) Hz. The deviation of ± 162 Hz from eleven separate sensor preparations and detection experiments is small. For all practical purposes, the results indicate that the presence of nonantigenic components in the sample does not affect the steady state sensor response in flow conditions. Recall a similar experiment with BA and BT discussed in an earlier section, in which the steady state response was affected by the concentration of BT present. The authors concluded that nonantigenic *Bacillus* species (BT and BC) hindered the transport of the BA spores to the sensor surface, but never completely prevented attachment of the antigenic spores. The authors reported that flow in combinations with vibration of the sensor surface enabled the reduction of weak binding, if any. Corresponding to each detection experiment, a control consisting of exposing an anti-BA functionalized cantilever to a sample of only BT and BC at the same concentration and experimental conditions was carried out. The resonance frequency change fluctuated around zero for the control. The sensor response (14 ± 31 [n = 11] Hz) to the control containing BT+BC presented in Fig. 18.8 was typical.

The feasibility of detecting *Escherichia coli* O157:H7 in samples prepared in various ground beef matrices was investigated using a PEMC sensor (Design B; sensing area 4 mm²). The matrices included (Campbell et al. 2007b): (1) controls consisting of sterile broth, sterile broth plus raw ground beef, and sterile broth plus sterile ground beef without inoculation of *E. coli* O157:H7; (2) samples from 100 mL of sterile broth inoculated with 25 *E. coli* cells incubated at 37°C at t = 0, 2, 4, and 6 hours; (3) samples from 100 mL of sterile broth containing 25 grams of raw ground beef incubated at 37°C at t = 0, 2, 4, and 6 hours; and (4) samples from 100 mL of broth with 25 grams of sterile ground beef inoculated with 25 *E. coli* cells incubated at 37°C at t = 0, 2, 4, and 6 hours. The samples from ground beef containing preps were optically dense and contained blood, meat, and fat particles (see inset in Fig. 18.9). The total resonance

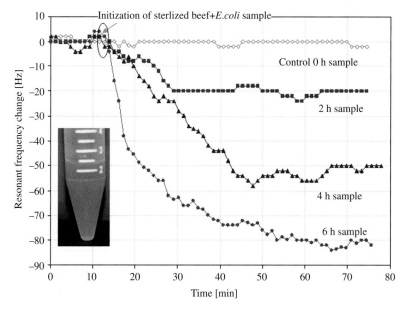

Figure 18.9. Resonant frequency change as a function of time for the binding of *E. coli* O157:H7 to antibody immobilized PEMC sensor (Design B) from samples flowing at 1 mL/min. The 25 g beef was sterilized, suspended in 100 mL of growth medium, and inoculated with 25 *E. coli* cells at t = 0 and incubated at 37 C. Three mL samples were drawn at 2h, 4h, and 6h. The response labeled control was from initial ground beef sample which contained zero *E. coli* O157:H7 cells. (Campbell et al. 2007b). Inset: typical sample obtained at t = 2 h. Blood, meat, and fat particles are visible.

frequency change obtained for the broth plus *E. coli* samples were 16 ± 2 ($n = 2$), 30 ($n = 1$), and 54 ± 2 ($n = 2$) Hz, corresponding to 2, 4, and 6 hours growth at 37°C, respectively. The responses to the broth plus 25 g of sterile ground beef plus *E. coli* cells were 21 ± 2 ($n = 2$), 37 ($n = 1$), and 70 ± 2 ($n = 2$) Hz, corresponding to 2, 4, and 6 hours, respectively (see Fig. 18.9). In all cases, the three different control samples yielded a frequency change of 0 ± 2 ($n = 6$) Hz. The *E. coli* O157:H7 concentration in each broth and beef sample was also quantified both by plating and by a pathogen modeling program. See the detailed evaluation given in Campbell et al. (2007b). The authors concluded that the PEMC sensor can selectively detect *E.coli* O157:H7 reliably at 50 to 100 cells/mL with a 3 mL complex matrix sample.

In a later study using PEMC sensors of Design A, the authors (Maraldo and Mutharasan 2007b) showed detection of 10 cells/mL of *E. coli* O157:H7 in spiked raw ground beef samples in 10 minutes. The composite PEMC sensors with a sensing area of 2 mm^2 were prepared by immobilizing polyclonal antibodies specific to *E. coli* O157:H7. Ground beef (2.5 g) was spiked with *E. coli* at 10–10,000 cells/mL in phosphate buffered saline. After a mixing step, and a wait period of ten minutes, one mL of supernatant was used to perform the detection experiments. The total resonance frequency change obtained for the inoculated samples was 138 ± 9, 735 ± 23, $2,603 \pm 51$, and $7,184 \pm 606$ Hz, corresponding to *E. coli* concentrations of 10, 100, 1,000, and 10,000 cells/mL, respectively. Positive detection of *E. coli* in the sample solution was observed within the first 10 minutes. The responses of the sensor to the three controls (cells present but no antibody on sensor, absence of cells with antibody-immoblized sensor, and buffer) were 36 ± 6, 27 ± 2, and 2 ± 7 Hz, respectively. Positive verification of *E. coli* O157:H7 attachment was confirmed by low-pH buffer (PBS/HCl pH 2.2) release, microscopy analysis, and second antibody binding post-E. Coli detection. These results and our previous studies suggest that it is feasible to detect *E. coli* O157:H7 at less than 10 cells/ml in 10 minutes without sample preparation, and with label-free reagents.

12. Conclusions

The main conclusion we draw from the studies conducted to date is that selective detection of small number of cells in complex matrices and in the presence of various organic and inorganic entities is feasible using cantilver sensors. We illustrated in this chapter the detection of *E. coli* O157:H7, Group A *Streptococcus pyogenes*, and *Bacillus anthracis*. Future work will include the detection of parasites (*Cryptospordium, Giardia*), food pathogens (*Salmonella, Listeria*), and food toxins (SEB).

A typical bacterial pathogen (about 1 μm in size) has a mass of one picogram. Given that the sensitivity of PEMC sensors is in the order of one femtogram per Hz, the attachment of a single bacterium would cause a sufficient signal that is higher than measurement noise. We showed that this is indeed the case through two examples: *Bacillus anthracis* in the presence of copious number of other *Bacillus* spp., and the case of *E. coli* O157:H7 in meat samples.

It is important to note that one must provide a flow that will bring the target antigen in the vicinity of the sensor surface for binding to occur. The sensor does not "attract" the pathogen, and thus a proper flow cell is paramount for low cell concentration detection. At the same time, a high flow rate does deteriorate sensor response. In the author's lab, the flow rate has been limited to less than 3 mL/min for continuous sensing. If flow-stop mode is employed, a much higher flow rate can be used. But the reader is cautioned that a flow rate without a suitable flow cell design is likely to segregate the target pathogen in low fluid renewal areas, resulting in a poor sensor response.

Finally, developing the cantilever sensors into anybody-can-use sensors would require significant engineering design and fabrication. Adding redundancy into the measurement, meaning measuring the same sample multiple times, and adding sensors as controls would reduce false readings. These practical considerations are surely the domain of product design.

References

Abdel-Hamid I, Ivnitski D, Atanasov P, Wilkins E (1999a) Highly sensitive flow-injection immunoassay system for rapid detection of bacteria. Analytica Chimica Acta 399(1–2):99–108

Abdel-Hamid I, Ivnitski D, Atanasov P, Wilkins E (1999b) Flow-through immunofiltration assay system for rapid detection of E. coli O157:H7. Biosensors and Bioelectronics 14(3):309–316

Arora K, Chand S, Malhotra BD (2006) Recent developments in bio-molecular electronics techniques for food pathogens. Analytica Chimica Acta 568(1–2):259–274

Balasubramanian S, Sorokulova IB, Vodyanoy VJ, Simonian AL (2007) Lytic phage as a specific and selective probe for detection of Staphylococcus aureus: A surface plasmon resonance spectroscopic study. Biosensors and Bioelectronics 22(6):948–955

Boskovic S, Chon JWM, Mulvaney P, Sader JE (2002) Rheological measurements using microcantilevers. Journal of Rheology 46(4):891–899

Campbell GA, Mutharasan R (2005) Sensing of liquid level at micron resolution using self-excited millimeter-sized PZT-cantilever. Sensors and Actuators A: Physical 122(2):326–334

Campbell GA, Mutharasan R (2006a) Use of piezoelectric-excited millimeter-sized cantilever sensors to measure albumin interaction with self-assembled monolayers of alkanethiols having different functional headgroups. Analytical Chemistry 78(7):2328–2334

Campbell GA, Mutharasan R (2006b) Detection of Bacillus anthracis spores and a model protein using PEMC sensors in a flow cell at 1 mL/min. Biosensors and Bioelectronics 22(1):78–85

Campbell GA, Mutharasan R (2006c) Piezoelectric-excited millimeter-sized cantilever (PEMC) sensors detect Bacillus anthracis at 300 spores/mL. Biosensors and Bioelectronics 21:1684–1692

Campbell GA, Mutharasan R (2006d) Use of Piezoelectric-Excited Millimeter-Sized Cantilever Sensors To Measure Albumin Interaction with Self-Assembled Monolayers of Alkanethiols Having Different Functional Headgroups. Anal. Chem. 78(7):2328–2334

Campbell GA, Mutharasan R (2006e) PEMC Sensor's Mass Change Sensitivity is 20 pg/Hz under Liquid Immersion. Biosensors and Bioelectronics 22(1):35–41

Campbell GA, Mutharasan R (2007a) A method of measuring Escherichia Coli O157:H7 at 1 Cell/mL in 1 liter sample using antibody functionalized Piezoelectric-Excited Millimeter-Sized Cantilever sensor. Environ. Sci. Technol. 41(5):1668–1674

Campbell GA, Mutharasan R (2007c) Method of measuring Bacillus anthracis spores in the presence of copious amounts of Bacillus thuringiensis and Bacillus cereus. Anal. Chem. 79(3):1145–1152

Campbell GA, Uknalis J, Tu S-I, Mutharasan R (2007b) Detection of Escherichia coli O157:H7 in ground beef samples using piezoelectric excited millimeter-sized cantilever (PEMC) sensors. Biosensors and Bioelectronics 22(7):1296–1302

Craighead HG (2003) Nanostructure science and technology: Impact and prospects for biology. Journal of Vacuum Science & Technology A21:S216–S221

Davila AP, Jang J, Gupta AK, Walter T, Aronson A, Bashir R (2007) Microresonator Mass Sensors for Detection of Bacillus anthracis Sterne Spores in Air and Water. Biosens. Bioelectron. 22:3028–3035

Detzel AJ, Campbell GA, Mutharasan R (2006) Rapid assessment of Escherichia coli by growth rate on piezoelectric-excited millimeter-sized cantilever (PEMC) sensors. Sensors and Actuators B: Chemical 117(1):58–64

Dhayal B, Henne WA, Doorneweerd DD, Reifenberger RG, Low PS (2006) Detection of Bacillus subtilis Spores Using Peptide-Functionalized Cantilever Arrays. J. Am. Chem. Soc. 128(11):3716–3721

Gfeller KY, Nugaeva N, Hegner M (2005) Micromechanical oscillators as rapid biosensor for the detection of active growth of Escherichia coli. Biosens. Bioelectron. 21(3):528–533

Ghatnekar-Nilsson S, Lindahl J, Dahlin A, Stjernholm T, Jeppesen S, Hook F, Montelius L (2005b) Phospholipid vesicle adsorption measured in situ with resonating cantilevers in a liquid cell. Nanotechnology 16(9):1512–1516

Gupta A, Akin D, Bashir R (2004) Single virus particle mass detection using microresonators with nanoscale thickness. Applied Physics Letters 84(11):1976–1978

Hancock REW, McPhee JB (2005) Salmonella's Sensor for Host Defense Molecules. Cell 122(3):320–322

Hermanson GT (1996) Bioconjugate Technique. Elsevier, San Diego, California

Ilic B, Craighead HG, Krylov S, Senaratne W, Ober C, Neuzil P (2004a) Attogram detection using nanoelectromechanical oscillators. Journal of Applied Physics 95:3694–3703

Ilic B, Czaplewski D, Craighead HG, Neuzil P, Campagnolo C, Batt C (2000) Mechanical resonant immunospecific biological detector. Applied Physics Letters 77(3):450–452

Ilic B, Czaplewski D, Zalalutdinov M, Craighead HG, Neuzil P, Campagnolo C, Batt C (2001) Single cell detection with micromechanical oscillators. Journal of Vacuum Science & Technology B 19(6):2825–2828

Ilic B, Yang Y, Craighead HG (2004b) Virus detection using nanoelectromechanical devices. Applied Physics Letters 85(13):2604–2606

Inaba S, Akaishi K, Mori T, Hane K (1993) Analysis of resonance characteristics of a cantilever vibrated photothermally in a liquid. Journal of Applied Physics 73(6):2654–2658

Ivnitski D, Abdel-Hamid I, Atanasov P, Wilkins E (1999) Biosensors for detection of pathogenic bacteria. Biosensors and Bioelectronics 14(7):599–624

Johnson L, Gupta ATK, Ghafoor A, Akin D, Bashir R (2006) Characterization of vaccinia virus particles using microscale silicon cantilever resonators and atomic force microscopy. Sens. Actuator B-Chem. 115(1): 189–197

Koubova V, Brynda E, Karasova L, Skvor J, Homola J, Dostalek J, Tobiska P, Rosicky J (2001) Detection of foodborne pathogens using surface plasmon resonance biosensors. Sensors and Actuators B: Chemical 74(1–3): 100–105

Lavrik NV, Sepaniak MJ, Datskos PG (2004) Cantilever transducers as a platform for chemical and biological sensors. Rev. Sci. Instrum. 75(6):2229–2253

Lazcka O, Campo FJD, Munoz FX (2007) Pathogen detection: A perspective of traditional methods and biosensors. Biosensors and Bioelectronics 22(7):1205–1217

Maraldo D, Mutharasan R (2007a) Optimization of antibody immobilization for sensing using piezoelectrically excited-millimeter-sized cantilever (PEMC) sensors. Sensors and Actuators B: Chemical 123 (1):474–479

Maraldo D, Mutharasan R (2007b) 10-minute assay for detecting *Escherichia coli O157:H7* in ground beef samples using Piezoelectrically-Excited Millimeter-Sized Cantilever (PEMC) sensors. Journal of Food Protection, 70(7): 1670–1677

Maraldo D, Mutharasan R (2007d) Preparation-free method for detecting *Escherichia coli O157:H7* in spinach, spring lettuce mix, and ground beef matrices. Journal of Food Protection, 70(11):2651–2655

Maraldo D, Rijal K, Campbell G, Mutharasan R (2007c) Method for Label-Free Detection of Femtogram Quantities of Biologics in Flowing Liquid Samples. Analytical Chemistry 79 (7):2762–2770

Naik T, Longmire EK, Mantell SC (2003) Dynamic response of a cantilever in liquid near a solid wall. Sensors and Actuators A: Physical 102(3):240–254

Nugaeva N, Gfeller KY, Backmann N, Lang HP, Duggelin M, Hegner M (2005) Micromechanical cantilever array sensors for selective fungal immobilization and fast growth detection. Biosens. Bioelectron. 21(6):849–856

Oli MW, McArthur WP, Brady LJ (2006) A whole cell BIAcore assay to evaluate P1-mediated adherence of Streptococcus mutans to human salivary agglutinin and inhibition by specific antibodies. Journal of Microbiological Methods 65(3):503–511

Prescott LM, Harley JP, Klein DA (2005) Microbiology, 6th ed. McGraw-Hill Education, Boston

Rijal K, Leung A, Shankar PM, Mutharasan R (2005) Detection of pathogen *Escherichia coli O157:H7* at 70 cells/mL using antibody-immobilized biconical tapered fiber sensors. Biosensors and Bioelectronics 21(6): 871–880

Rijal K, Mutharasan R (2007) Piezoelectric-excited millimeter-sized cantilever sensors detect density differences of a few micrograms/mL in liquid medium. Sensors and Actuators B: Chemical, 121(1):237–244

Sader JE (1998) Frequency Response of Cantilever Beams Immersed in viscous fluids with applications to the atomic force microscope. Journal of Applied Physics 84(1):64–76

Su XL, Li Y (2005) Surface plasmon resonance and quartz crystal microbalance immunosensors for detection of Escherichia coli O157: H7. Transactions of the Asae 48(1):405–413

Taylor AD, Ladd J, Yu Q, Chen S, Homola J, Jiang S (2006) Quantitative and simultaneous detection of four foodborne bacterial pathogens with a multi-channel SPR sensor. Biosensors and Bioelectronics 22(5):752–758

Taylor AD, Yu Q, Chen S, Homola J, Jiang S (2005) Comparison of *E. coli O157:H7* preparation methods used for detection with surface plasmon resonance sensor. Sensors and Actuators B: Chemical 107(1):202–208

Vaughan RD, Carter RM, O'Sullivan CK, Guilbault GG (2003) A quartz crystal microbalance (QCM) sensor for the detection of Bacillus cereus. Analytical Letters 36(4):731–747

Vaughan RD, O'Sullivan CK, Guilbault GG (2001) Development of a quartz crystal microbalance (QCM) immunosensor for the detection of Listeria monocytogenes. Enzyme and Microbial Technology 29(10): 635–638

Waswa J, Irudayaraj J, DebRoy C (2007) Direct detection of *E. coli O157:H7* in selected food systems by a surface plasmon resonance biosensor. LWT-Food Science and Technology 40(2):187–192

Weeks BL, Camarero J, Noy A, Miller AE, Stanker L, De Yoreo JJ (2003) A microcantilever-based pathogen detector. Scanning 25(6):297–299

Yi JW, Shih WY, Mutharasan R, Shih WH (2003) In situ cell detection using piezoelectric lead zirconate titanate-stainless steel cantilevers. Journal of Applied Physics 93(1):619–625

Zezza F, Pascale M, Mule G, Visconti A (2006) Detection of Fusarium culmorum in wheat by a surface plasmon resonance-based DNA sensor. Journal of Microbiological Methods 66(3):529–537

Zhang J, Ji HF (2004) An anti *E-coli O157 : H7* antibody-immobilized microcantilever for the detection of Escherichia coli (E-coli). Analytical Sciences 20(4):585–587

Zhu Q, Shih WY, Shih W-H (2007a) In-Situ, In-Liquid, All-Electrical Detection of Salmonella typhimurium Using Lead Titanate Zirconate/Gold-Coated Glass Cantilevers at any Dipping Depth. Biosensors and Bioelectronics, 22(12):3132–3138

Zhu Q, Shih WY, Shih W-H (2007b) Real-Time, Label-Free, All-Electrical Detection of Salmonella typhimurium Using Lead Titanate Zirconate/Gold-Coated Glass Cantilevers at any Relative Humidity. Sensors and Actuators B: Chemical, 125(2):379–388

Ziegler C (2004) Cantilever-based biosensors. Anal. Bioanal. Chem. 379(7–8):946–959

19

Detection and Viability Assessment of Endospore-Forming Pathogens

Adrian Ponce, Stephanie A. Connon and Pun To Yung

Abstract

In this chapter, we explore technology developments for the rapid detection, identification, and viability assessment of endospore-forming pathogens with a focus on *Bacillus anthracis*. First, we introduce various toxin-producing species and their role as bioinsecticides, probiotics, and bioweapons. We also review the role of endospores as biological indicators (i.e., dosimeters) for evaluating sterilization regimens, such as autoclaving and wastewater remediation. Monitoring the effectiveness of cleaning and sterilization regimens to maintain good hygiene is required in several major industries, including health care, food, and pharmaceutical industries. In the next section, we review recent developments in DNA-, immuno-, and dipicolinic acid assays, and their applications for detection and monitoring of *Bacillus anthracis* and other endospore-forming pathogens. Finally, we review viability assays capable of rapid validation of endospore inactivation after sterilization, including assays based on ATP synthesis during stage II germination, and DPA release during stage I germination.

1. Introduction

1.1. Historical Perspective

The era of modern microbiology began in the 1870s when the lifecycle of an endospore-forming pathogen, *Bacillus anthracis*, was elucidated using new methods for isolating pure cultures from single cell clones on solid growth media (Drews 2000). Robert Koch showed that pure cultures of *B. anthracis* isolated from diseased animals could infect healthy animals, which in turn yielded further *B. anthracis* isolates, thus identifying a specific microorganism as the causative agent for a specific disease for the first time (Koch 1876). At this stage, it became clear that microorganisms could be defined in terms of particular morphologic, metabolic, and pathogenic characteristics. Under the microscope, anthrax organisms appeared as long filaments, with some cells containing oval, translucent bodies (i.e., endospores) (Koch 1877). Koch found that the dried endospores could remain viable for years, even under harsh conditions. This accounted for the recurrence of anthrax in long unused pastures, where the dormant endospores persisted until infecting additional animals, allowing the endospores to germinate and develop into vegetative anthrax bacilli (Fig. 19.1).

To this day, isolation *via* culture and phenotypic analysis remain the gold standard for detection, identification, and viability determination of microorganisms. However, these

Adrian Ponce, Stephanie A. Connon and Pun To Yung • California Institute of Technology, Pasadena, CA 91125/Jet Propulsion Laboratory, Pasadena, CA 91109.

M. Zourob et al. (eds.), *Principles of Bacterial Detection: Biosensors, Recognition Receptors and Microsystems*,
© Springer Science+Business Media, LLC 2008

 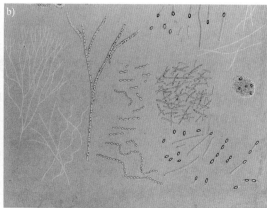

Figure 19.1. a) Robert Koch was one of the pioneer microbiologists researching bacterial spores. In his investigation of *B. anthracis*, he discovered the dormant-vegetative life cycle of endospore-formers. b) Koch developed staining methods to observe bacterial isolates under a microscope. Rod-shaped vegetative bacilli cells and phase-bright round-shaped endospores are depicted in his hand-drawn illustration, originally published in his paper in 1876 (Koch, 1876).

methods are labor intensive and require several days, which is often longer than the course of the disease. For example, inhalational anthrax progresses over 1–6 days with initial flu-like symptoms, and leads to high mortality rates within 36 hours after the onset of respiratory distress (Meselson et al. 1994; Weyant, Ezzell and Popovic 2001). In the case of another anthrax attack, rapid "detect-to-treat" technologies are critical for identifying exposed victims for prompt treatment.

1.2. Endospore Dormancy, Resistance and Longevity

Bacterial spores (i.e., endospores) were independently discovered by Cohn (Cohn 1876), Koch (Koch 1876), and Tyndall (Tyndall 1877), and identified as dormant structures of some bacteria in the Firmicutes phylum (Onyenwoke et al. 2004). In preparation for times of environmental stress or lack of nutrients, vegetative cells initiate the process of sporulation to produce resilient endospores. They remain dormant and exhibit nearly undetectable levels of endogenous metabolism (Desser and Broda 1965), with some reports citing no detectable metabolism (Foster and Johnstone 1990; Nicholson and Setlow 1990). When conditions become more favorable, the surviving endospore population can germinate to produce vegetative cells (Setlow 2003). The extraordinary resistance of endospores to environmental extremes is engendered by the many protective layers surrounding the spore core, and protection of the DNA by decreased water content in the spore core, synthesis of a special class of DNA-binding proteins, and DNA repair processes, which are activated during germination (Setlow and Kornberg 1970; Gest 1987; Lindahl 1993; Setlow 1995; Nicholson et al. 2000; Setlow 2001). The spore core is enveloped first by a thick layer of peptidoglycan forming the cortex that is required for heat resistance (Henriques et al. 2004), and then several protein layers coat and protect the cortex from chemical and enzymatic lysis (Fig. 19.2).

Rod-shaped, endospore-forming bacteria were first classified into aerobic *Bacillus* and anaerobic *Clostridum* genera (Allen, Emery and Lyerly 2003; Fritze 2004a). Other endospore-producing prokaryotes include *Desulfotomaculum*, *Sporolactobacillus*, and *Sporosarcina* genera (Dworkin 2006; Fritze 2004b). In general, endospore-forming bacteria are most commonly found in soils (Felske 2004). However, endospores exist almost everywhere—on the surface and

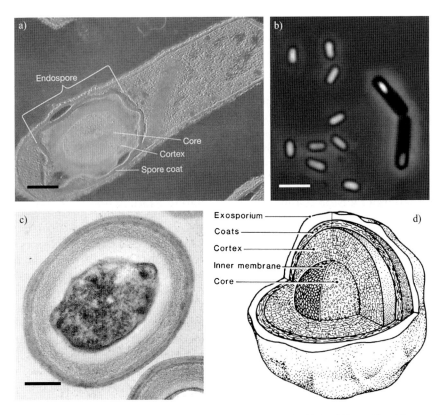

Figure 19.2. a) Transmission electron micrograph of an endospore embedded within a *Clostridium perfringens* vegetative cell. Bar, 0.3 μm. (*Reproduced from Figure 8B, page 121 in Principles of Microbiology 6th edition (2001) with permission from John Wiley and Sons, Inc.*) b) Endospores appear phase-bright under a phase contrast microscope and are readily distinguished from the phase-dark rod-shaped vegetative mother cells. Bar, 3 μm. c) TEM cross section of an endospore of *Geobacillus stearothermophilus*. Bar, 0.2 μm. (*Courtesy of Stuart Pankratz*). d) Schematic representation of the internal structure of a typical endospore. (*Reprinted from Molecular Biology, 4(1), Foster, SJ and Johnstone, K, Pulling the trigger: the mechanism of bacterial spore germination, 137-141 (1990), with permission from Wiley Blackwell*).

deep subsurface, in the oceans (Nicholson 2004), and in the atmosphere, where they are transported on particles of dust to populate all five continents (Keeton 1980; Dart 1996; Nicholson et al. 2000). While the structure and composition of endospores across the various phylogenetic groups are very similar, it is clear from DNA hybridization experiments, morphology, and metabolic characteristics that similar spores are formed by very different classes of organisms, suggesting the distinct possibilities of convergent evolution (Keynan and Sandler 1983) and/or horizontal gene transfer (Koonin, Makarova and Aravind 2001; Jain et al. 2003). For example, it has been routinely noted in the literature that endospore-formers are only found in the Firmicutes phylum (Onyenwoke et al. 2004). However, when an endospore-former from the Proteobacteria phylum—*Serratia marcescens* subsp *sakuensis*—was isolated and described (Ajithkumar et al. 2003), it was suggested that the endospore-forming genes were obtained by horizontal gene transfer since it was isolated from a wastewater treatment plant with high concentrations of *Bacillus*.

Endospores also exhibit remarkable longevity, with reports ranging from thousands (Sneath 1962; Gest 1987; Kennedy, Reader and Swierczynski 1994; Potts 1994; Nicholson et al. 2000) up to a quarter billion years (Dombrowski 1963; Cano and Borucki 1995; Vreeland, Rosenzweig and Powers 2000). If the longevity is truly hundreds of millions of years, it sets the precedent that life can be stored on a geological timescale. This, however, remains a matter of significant debate and uncertainty (McGenity et al. 2000; Nicholson 2004), largely because contamination by modern microbes cannot be ruled out. The lower limit of endospore longevity is in the thousands of years based on numerous accounts of viable endospore-forming organisms isolated from ancient samples (Gould 2005). In one account on microbial longevity, the number of cultivable cells per gram were measured in root samples from pressed plants in the Herbaria of the British Museum of Natural History and of Kew Gardens, which were obtained from botanical voyages dating from 1640–1962. Data from this investigation indicates that the microbial longevity is, at least, on the order of 1,000 years (Sneath 1962).

1.3. Endospores as Biodosimeters for Evaluating Sterilization Regimes

Endospores are also highly resistant to chemical, physical, and radiation sterilization processes (Nicholson et al. 2000). In fact, *B. subtilis* spores have survived for six years in space while exposed to high vacuum, temperature extremes, and intense solar and galactic radiation (Horneck, Bucker and Reitz 1994), which has lent support to the hypothesis that endospores are the most likely to survive an interplanetary lithopanspermic journey (Kennedy, Reader and Swierczynski 1994; Parsons 1996; Hoch and Losick 1997; Nicholson et al. 2000) (Fig. 19.3). Since bacterial spores are the last to remain viable during the sterilization processes that readily kill vegetative bacteria, they are employed as biological indicators for monitoring the effectiveness of sterilization processes, such as vaporized hydrogen peroxide treatments (Johnston, Lawson and Otter 2005; Kanemitsu et al. 2005), UV irradiation (Nicholson and Galeano 2003; Mamane-Gravetz and Linden 2004), and notably autoclaving, where *Geobacillus stearothermophilus* and *B. subtilis* spores are used to verify autoclave performance (Keeton 1980;

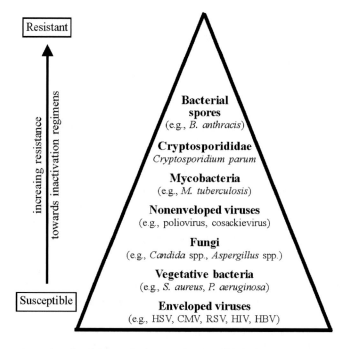

Figure 19.3. Increasing order of resistance of microorganisms to disinfection. Endospores are the most resilient organisms known to exist, requiring the most rigorous sterilization procedures, such as autoclaving and ethylene oxide treatment. (*Adapted from Figure 1, page 82 in the Manual of Clinical Microbiology, 8th edition*).

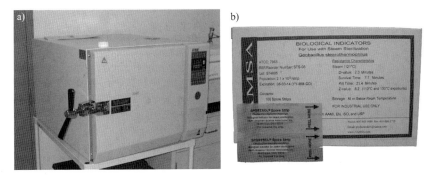

Figure 19.4. a) The autoclave is the gold standard sterilization instrument in microbiology. Samples are usually heated under pressure at 121°C. Flash sterilization (134°C for 3 min) is often employed in hospitals where a high turnover of medical instruments is needed. b) Autoclave performance is validated using endospore strip indicators (e.g., *Geobacillus stearothermophilus* spores).

Dart 1996; Nicholson et al. 2000) (Fig. 19.4). In another example, bacterial spores of *Clostridium perfringens* are used as an indicator species to estimate the sterilization efficiency of waterborne pathogens and parasites such as *Giardia* and *Cryptosporidium* in commercial water treatment facilities (Rice et al. 1996).

Bacterial spores are also used as a metric for bioburden analysis of NASA's interplanetary spacecraft (Office of Space Science 1999). NASA is required to validate that the bioburden on certain interplanetary spacecraft is below a threshold mandated by international law. For example, Class IVa planetary missions to Mars, comprising landers and probes without life-detection experiments, must meet a bioburden limit of 3×10^5 spores/vehicle, and less than 300 spores/m^2. Missions with life-detection experiments must undergo additional procedures to ensure that the total bioload does not exceed 30 spores (Board and Council 2000). Bacterial spores are ideal indicator species to validate bioburden reduction, because they are the most resistant life forms towards sterilization regimens. Thus, validation of bacterial spore inactivation to a certain concentration threshold guarantees that other, less resistant, microorganisms are reduced to lower concentrations.

1.4. Endospore-Forming Pathogens

Two endospore-forming genera of medical importance, *Bacillus* and *Clostridium*, contain the causative species for anthrax, tetanus, botulism, gas gangrene, and many cases of food poisoning (Allen, Emery and Lyerly 2003; Logan and Turnbull 2003; Lukasova, Vyhnalkova and Pacova 2001). Some of the common endospore-forming pathogens are summarized in Table 19.1. The fact that endospores are so durable is the primary reason for time consuming and expensive sterilization procedures employed in hospitals, canneries, and other food preparation facilities. Since spores are resistant to sterilization treatments, food that has been incorrectly or minimally processed can give rise to foodborne diseases. For example, *B. cereus* is a common aerobic foodborne pathogen, and *Clostridium botulinum* and *C. perfringens* are anaerobic foodborne pathogens commonly associated with food poisoning from canned foods. This health issue is exacerbated by the fact that minimally treated foods are becoming extremely popular (Zink 1997).

Similarly, endospore-forming bacteria can survive disinfectant treatments for medical equipment used in hospitals. The survival of pathogenic endospore-forming bacteria is one of the main causes of infections acquired in hospitals. For example, *C. difficile* is a major cause of infective hospital-acquired diarrhea, and has been associated with more than 16,000 cases per year in England and Wales alone (Wilcox and Fawley 2000). In the United States, it may

Table 19.1. Pathogenicity of endospores

Species	Target	Disease	Pathogenicity
Bacillus anthracis	Lungs	Inhalational anthrax	Initial symptoms resemble influenza and second phase is characterized by chest pain and hemorrhage. Highly fatal if not treated immediately.
	Wounds, mucous membrane	Cutaneous anthrax	Gelatinous edema develops at the wound, which may develop into papule, pustule, and necrotic ulcer if untreated.
	Gastrointestinal system	Gastrointestinal anthrax	Infects *via* intestinal mucosa and may spread to the lymphatic system.
	Livestock	Septicaemia	Ingested spores initiate infection in the pharyngeal and intestinal mucosa. Onset of septicaemia is followed by death.
B. cereus	Gastrointestinal system	Food poisoning	Categorized into emetic type and diarrheal type food poisoning, usually associated with farinaceous foods and dairy products.
B. thuringiensis	Lepidopterous insects	Death of insect	Release of lethal exo- and endo-toxins in insect gut upon ingestion. Germinated spores invade the haemocoel liberating enzymes to attack the host tissues and soon reduce them to blackened cadavers.
Paenibacillus popilliae	Japanese beetle and scarab beetle larvae	Milky spore disease	Spores germinate in the larval gut and multiply in the haemococel to kill some strains of insects, characterized by decolorized blood.
P. lentimorbus	Japanese beetle and scarab beetle larvae	Milky spore disease	Spores germinate in the larval gut and multiply in the haemococel to kill some strains of insects, characterized by decolorized blood.
P. alvei	Honeybee larvae	European Foulbrood	Spores germinate and multiply to cause infection and death to larvae before they are capped and sealed. Usually associated with other bacteria.
P. larvae	Bee larvae	American Foulbrood	Spores germinate and multiply until the larvae are killed after they are capped and sealed. More deadly than European foulbrood.
Clostridium botulinum	Gastrointestinal, nervous and muscular systems	Botulism	Ingestion of food contaminated with botulinic toxin causes foodborne botulism, characterized by lethal respiratory and musculoskeletal paralysis. Infant botulism is caused by ingestion of spores which germinate, multiply, and release toxin in the gut of infants, known as floppy baby syndrome.
C. perfringens	Gastrointestinal system, wounds and extremities	Food poisoning Gas gangrene	Ingestion of contaminated food leads to food poisoning and gastroenteritis, characterized by diarrheal and abdominal cramp. Wound and oral infections lead to necrosis and decay of tissues, known as gas gangrene.
C. tetani	Nervous and muscular systems	Tetanus	Germination and outgrowth of spores in contaminated wounds and broken skins release a lethal neurotoxin to cause muscle spasms and seizures.
C. difficile	Gastrointestinal system (colon)	Pseudomembraneous colitis	Spores surviving an antibiotic therapy in the gut germinate and proliferate to cause abdominal pain, diarrhea, fever and toxic megacolon.

be the cause of nearly half of all cases of nosocomial diarrhea in adult hospitalized patients (McFarland 1995). There is strong evidence suggesting that contact transmission from contaminated articles or the hands of staff is responsible for the spread of the disease. For example, it was noted that when culture-negative patients were placed in a hospital room currently or previously occupied by a person with *C. difficile* diarrhea, they were more likely to develop this type of diarrhea than patients placed in rooms where no patient had had *C. difficile* diarrhea (McFarland et al. 1989). This suggests the organism can persist on inanimate articles (e.g., lamps, door handles or bed rails) for some time unless rooms are thoroughly cleaned between patients.

The most well-known disease caused by an endospore-forming species is anthrax, which was first described 3,500 years ago, when a Greek poet and scientist gave a richly detailed account of this disease: "If anyone wore a garment made from the tainted wool, his limb was soon attacked by inflamed papules and a foul exudate, and if he delayed too long to remove the material a violent inflammation covered the parts it had touched (Dirckx 1981)." In the general population, anthrax is feared because *B. anthracis* spores were used as a biological weapon in the 2001 terrorist attack *via* the U.S. Postal Service (Malecki et al. 2001). However, anthrax is largely confined to the herbivore population, and before the vaccine became available in 1930, anthrax was the primary cause of domesticated herbivore mortality (Oggioni et al. 2004). Humans inhalational anthrax lethal doses range from 500 spores to 55,000 spores (Logan 2004). Anthrax can occur in cutaneous, gastrointestinal, and inhalational (i.e., pulmonary) clinical forms (Logan and Turnbull 2003; Edwards, Harriet and Antje 2006). While anthrax is not very contagious, if left untreated, all forms of anthrax can be fatal, with inhalational anthrax yielding near 100% mortality (Dixon et al. 1999; Logan and Turnbull 2003; Edwards, Harriet and Antje 2006). In 1979, in the Ural mountain city Svedlovsk, an accidental release of weaponized anthrax spores from a military laboratory resulted in 66 inhalational anthrax deaths (Meselson et al. 1994). Sick animals or their products may also transmit anthrax after entry of *B. anthracis* spores into the host body. While cutaneous anthrax is easily recognizable as a black eschar with associated edema that forms within days upon exposure, inhalational and gastrointenstinal forms are more difficult to diagnose, as no visual cues are available and the symptoms initially mimic bronchopneumonia and gastroenteritis, respectively (Logan and Turnbull 2003). Rapid progression of anthrax to a systemic form can render the victim unresponsive to treatment leading to high mortality due to sepsis and respiratory failure (Logan and Turnbull 2003; Logan 2004; Levine, Tang and Pei 2005; Edwards, Harriet and Antje 2006).

1.5. Bioweapons, Bioinsecticides and Probiotics

The genus *Bacillus* includes three species—*B. anthracis, B. thuringiensis*, and *B. cereus*—that are of primary interest because of their ability to produce toxins. Endospores are ideal delivery vehicles for the distribution of toxin-producing species into the environment because their extreme resilience enables them to survive in bioaerosols for extended periods. The most ominous use of endospore-formers is the release of *B. anthracis* as a bioweapon (Jernigan et al. 2002; Weis 2002; Sanderson et al. 2004) with devastating health, psychological, and economic impacts (Kaufmann, Meltzer and Schmid 1997). In 2001, the intentional release of weaponized anthrax spore powders *via* the mail system resulted in 19 infections, 5 deaths, and 10,000 prescriptions of antibiotics (Centers for Disease Control and Prevention 2002) (Fig. 19.5). While the 2001 anthrax attack was relatively limited, it has been estimated that a 100 kg release of *B. anthracis* spores in Washington, D.C., if undetected early on, would result in up to 3 million deaths (Office of Technology Assessment 1993).

B. cereus is a widespread cause of food poisoning due to its ability to produce endospores that can readily survive normal food cooking procedures. It is commonly found on vegetables,

Figure 19.5. a) The letter addressed to Senator Tom Daschle, one of the notorious envelopes that contained anthrax spore powder in 2001. Spores were aerosolized during processing and handling in U.S. Postal Service sorting rooms, leading to deadly inhalational anthrax and consequently a total of 22 deaths. b) Biohazard response teams dressed in protective gear performing quarantine and decontamination of anthrax in one of the contaminated postal facilities.

rice, pasta, meat, and milk. It produces both emetic and diarrheal type toxins and will germinate and multiply in temperatures above 12°C. *B. cereus* can cause diarrhea by producing heat-labile enterotoxins, including tripartite hemolysin BL, which has been shown to be the causative agent for diarrhea. Some strains of *B. cereus* also produce the more dangerous emetic heat-stable toxin, cereulide, which causes vomiting within a few hours of ingestion. For a review of the toxins produced by *B. cereus*, see Schoeni and Wong 2005. The highest levels of cereulide production were found in temperatures ranging from 12–15°C (Finlay 2000). This heat-stable toxin can directly cause illness even if the contaminated food is well-cooked prior to consumption. While symptoms such as vomiting and diarrhea are generally mild, *B. cereus*-caused food poisoning can also be fatal, with reports of death from the emetic toxin within as few as 13 hours after consumption of contaminated food (Mahler et al. 1997; Dierick 2005).

A number of toxin-producing *Bacillus* species are employed as biopesticides for agricultural pest control, including *B. thuringiensis*, which accounts for approximately 90 percent of the bioinsecticide market (Chattopadhyay, Bhatnagar and Bhatnagar 2004). *B. thuringiensis* specifically attack lepidoterous insect larvae, and the gene for the toxin has been introduced into several transgenic crops, such as corn and cotton, to protect the host from insect devastation without the use of more toxic pesticides (Keynan and Sandler 1983). Another important biopesticide is *Paenibacillus popilliae* (formerly *Bacillus popilliae*), which infects the larvae of Japanese beetles to cause milky spore disease. In both species, insect pathogenicity is linked to the sporulation process, during which the endospore and a crystalline parasporal body containing the toxin are formed. These toxins kill by causing autolysis of the insect gut (Keynan and Sandler 1983). Unlike chemical pesticides, these bioinsecticides are specific to the target insect group without developing resistance—they leave no residue in the environment, and do not become concentrated in the tissues of non-target species.

Live endospore-forming organisms have been used as food supplements to promote improved health (i.e., probiotics) (Ricca, Henriques and Cutting 2004). Experiments have shown that the normal gut microflora serves as an effective barrier against pathogenic microorganisms, such as *Salmonella typhimurium*, *Escherichia coli*, *C. perfringens*, and *C. difficile* (Senesi 2004). Supplementing the normal gut microflora with probiotics, such as *B. subtilis* spores, has been used to

- Prevent and treat many gastrointestinal disorders
- Reduce the risk of infection in patients not subjected to prophylactic antibiotic treatment
- Treat rheumatoid arthritis
- Lower cholesterol and improve lactose intolerance. It is believed that *Bacillus* probiotics help maintain or restore the intestinal microbial balance (Senesi 2004)

Live endospore-forming organisms are also being investigated for applications in vaccinations (Cutting 2004), tumor therapy (Pennington et al. 2004), and antifungal treatment (Azizbekyan 2004).

The purified *C. botulinum* neurotoxin is one of the most poisonous naturally occurring substances in the world, and has been tested as a chemical weapon (Ting and Freiman 2004). However, toxin degradation upon exposure to air makes it an inferior chemical warfare agent. The botulin toxin has recently become well known as a medical and cosmetic product, where it is used to treat muscle pain and for temporary relief of facial frown lines (Ting and Freiman 2004). In another industrial application, fermenting endospore-formers such as *C. acetobutylicum* have been employed to produce butanol, acetone, and butyric acid (Prescott and Dunn 1959; Keynan and Sandler 1983). Acetone produced in this manner was used in World War I for explosives, and butanol was needed for automobile lacuers. The sporoluation process is apparently essential in butanol product formation. As oil prices are increasing, the use of anaerobic endospore-formers for fermenting cellulose into fuels is being reconsidered.

Considering the health and economic impacts of pathogenic endospore-forming organisms, rapid detection, identification, and viability assessment are essential requirements for several commercial and defense applications—especially industrial hygiene in health care, food preparation, and biodefense arenas. Below we review the current state-of-the-art technology for the rapid detection and viability assessment of endospore-forming pathogens.

2. Detection of Endospore-Forming Pathogens and their Endospores

2.1. Phenotypic Identification

Phenotypic characteristics are the result of expressed gene products, and are observed using traditional protocols for microbial identification. Generally, these protocols require large cell populations obtained from pure culture, which entails long incubation times. The established system of bacterial classification based on colony characteristics, cellular morphology, differential staining, motility, physiology, biochemical reactions, and substrate utilization was systematized in Bergey's Manual of Determinative Bacteriology (1923). These tests can require the use of sometimes dozens of different media substrates and staining procedures to confirm identification. Streamlined substrate screening methods include API, Vitek, and Biolog microbial identification systems, which allow simultaneous screening of many different media at a time. While these traditional tests allow thorough testing and screening of unknown isolates for phenotypic characteristics and identification the turnaround is typically on the order of days or longer for identification. Phenotypic characterization is often used as a backup method for other more rapid detection methods such as molecular or immunological detection.

In response to the increased threat of bioterrorism, the Centers for Disease Control and Prevention established a Laboratory Response Network (LRN) tasked to provide the appropriate laboratory response to acts of bioterrorism and other public health threats and emergencies. Upon receiving a suspect specimen, a given LRN laboratory will attempt to culture and perform phenotypic analysis to identify the suspected pathogen. For timely response, a clinical microbiologist must be familiar with the culture conditions and phenotypic characteristics of these agents and the technologies available for their detection and identification. Detailed methods and flowcharts for identification of *Bacillus* and *Clostridium* species have been outlined in several major texts (Holt and Bergey 1994; Allen, Emery and Lyerly 2003; Logan 2004). Below we briefly review two specific examples—one for an aerobic and another for an anaerobic endospore-forming pathogen.

2.1.1. Phenotypic Identification of *Bacillus anthracis*

The '*Bacillus cereus* group' includes the closely related species *anthracis, cereus, thuringiensis, mycoides, pseudomycoides*, and *weihenstephanensis*. The '*Bacillus cereus* group' can generally be identified by Gram positive rods often in chains, endospores that are elliptical and subterminal, not swelling the sporangium as viewed by phase microscopy, facultative anaerobic growth, and lecithinase positive on egg yolk agar. For the case of differentiating *B. anthracis* from its closest relative, *B. cereus*, the commonly used tests include motility, hemolysis, susceptibility to gamma phage and penicillin susceptibility. *B. anthracis* is nonmotile, nonhemolytic on sheep blood agar, lysed by the gamma phage, and susceptible to penicillin, while *B. cereus* is motile, hemolytic, resistant to lysis by gamma phage and penicillin resistant. A selective medium, polymyxin lysozyme EDTA thallous-acetate (PLET) agar can be used to selectively grow *B. anthracis* when other *Bacillus* species may be possible contaminants. *B. anthracis* forms rough, mat, gray-white colonies with curly projections called comet-tail or medusa head on sheep blood agar, while *B. cereus* produces larger colonies with edges that are smoother. Colonies typically grow up within 24–48 hours. Virulent *B. anthracis* strains also have the capacity to produce capsular material that is involved in pathogenicity. Capsule production can be detected by the formation of mucoid colonies on sodium bicarbonate spiked nutrient agar when incubated in a CO_2 enriched atmosphere. Caution is warranted, because results may vary among different *Bacillus* species strains, decreasing the degree of confidence in the species identification. For example, several *B. anthracis* strains have been found to have resistance to gamma phage and some rare *B. cereus* strains are sensitive to gamma phage and penicillin and are nonhemolytic (Schuch, Nelson and Fischetti 2002; Marston et al. 2006) (Fig. 19.6). Once a presumptive identification has been made using the standard microbiological identification assays, serological- or DNA-based detection methods can be employed for confirmation (Papaparaskevas et al. 2004; Klee et al. 2006). The major distinguishing feature of *B. anthracis* is the presence of two large virulence plasmids—pXO1 and pXO2—that harbor the tripartite toxin complex and the genes responsible for the synthesis of a poly-c-D-glutamic acid capsule, respectively (Table 19.2).

2.1.2. Phenotypic Identification of *Clostridium perfringens*

Genus level identification for *Clostridium* can be achieved by demonstrating obligatory anaerobic growth, Gram positive, endospore-forming rods, an inability to form endospores in the presence of oxygen and an inability to carry out dissimilatory sulfate reduction. Some *Clostridium* species may be aerotolerant and show a Gram negative result, especially at later stages of growth. *C. perfringens* is one of the most common bacterial causes of foodborne illness

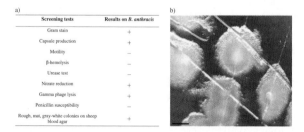

Figure 19.6. a) Examples of phenotypic tests for the presumptive identification of *Bacillus anthracis*. b) Rough, mat and gray-white anthrax colonies with curly projections called comet-tail or medusa head grow on sheep blood agar. Bar, 2 mm. Identification of *B. anthracis* using phenotypic screening tests based on agar-plate cultivation results is very time-consuming and labor intensive. (*Reproduced from Figure 4a, page 453 in the Manual of Clinical Microbiology, 8th edition with permission from ASM Press*).

Table 19.2. Characteristics of DNA and antigen markers for *B. anthracis*. (*Reproduced from Table 2, page 128 in Levine, Tang and Pei (2005) with permission of Lippincott Williams & Wilkins*)

Markers	Location	Copy Number	Specificity
DNA			
Chromosome	Intracellular	Single	*B. cereus* group
pXO1	Intracellular	Single or few	*B. anthracis / B. cereus* group
pXO2	Intracellular	Single or few	*B. anthracis / B. cereus* group
Antigens			
Exosporium proteins	Exosporium	Multiple	*B. anthracis, B. cereus*
PA	Spore coat	Multiple	*B. anthracis*
Poly-γ-D-glutamic acid	Capsule	Polymers	*B. anthracis, B. licheniformis*
SAP, EA1	Surface layer	Multiple	*B. anthracis* (epitopes)
Galactose-*N*-acetyl- D-glucosamine polysaccharide	Cell wall	Polymers	*B. anthracis, B. cereus*

in the United States and is one of several *Clostridium* species that can cause gas gangrene. The presumptive identification of *C. perfringens* requires further observations, including evidence of short squat boxcar shaped rods with oval subterminal endospores, lecithinase activity, nonmotility, reduction of nitrates, production of acid and gas from lactose, gelatin hydrolysis within 48 hours, and stormy fermentation within 5 hours in iron-milk medium and black colonies on tryptose-sulfite-cycloserine (TSC) agar. *C. perfringens* produces a number of toxins, identified as type A through E, that are important for pathogenicity. Confirmatory testing for the *C. perfringens* enterotoxins can be accomplished using an immunoassay such as a reversed passive latex agglutination (RPLA) test kit or commercial enzyme-linked immunosorbent assay (ELISA) kit.

2.2. Parameters of a Sensor

As in the era of Robert Koch, phenotypic analysis of cultured isolates remains the standard approach for pathogen detection, identification, and viability assessment. While these methods give rise to high sensitivity and a high degree of confidence in identification and viability assessment, they are not rapid enough to provide a diagnosis in time for treatment prior to the onset of disease. Rapid "detect-to-treat" sensor systems for endospore-forming pathogens, with a focus on *B. anthracis*, are reviewed below. The reader should be mindful that each method and instrument described will have pros and cons in terms of specific sensor parameters, such as sensitivity, selectivity (false positive rate), and false negative rates in relevant environments. Other important sensor parameters include response time, unit and operating costs, size, weight, power requirements, multianalyte detection, automation, reliability, ease of operation, and ability to distinguish live from dead organisms. Detailed information on these sensor performance parameters is beyond the scope of this text, and may be obtained from the primary literature, although caution is warranted as many reports highlight only the best feature(s) (e.g., sensitivity) while other performance parameters may not necessarily be disclosed (e.g., false positive rates). For example, in some technology reports, an assay may be advertised as the most sensitive method for a given analyte, but may be susceptible to false negatives when operated in the presence of environmental pollution.

Unfortunately, current technology development has not yet produced a "Tricorder" that can do it all with the push of button. Therefore, the specific application and operating environment will govern the choices and compromises among sensor parameters acceptable for a given application. For many applications, false positive rates are of primary concern, and although a sensor system may always respond correctly when presented the biological detection target, it may not be acceptable if it responds positively to nonpathogenic organisms. In some cases, portability and rapid response time requirements may force a compromise in sensitivity, specificity, or other parameters. This is analogous of situations where the tool set of a Swiss army knife may suffice, and the full compliments of a machine shop are not required. In other cases, a combination of instruments operating in series can provide high sensitivity, selectivity, and acceptable false negatives at greatly reduced operating costs, as in the case of the Anthrax Smoke Detector described below, which serves as an inexpensive front-end monitor for expensive-to-operate automated PCR-based systems (see anthrax smoke detector in section 2.5.2).

As new instrumentation advances in readiness levels, validation against characterized samples with known analyte and potential interferent contents in a realistic environment must be performed—preferably by a third party. Third party testing of this nature can be performed at proving grounds, which are facilities where the performance of new instrumentation can be tested under controlled and realistic conditions with fully characterized samples. Some well-known examples of proving grounds include Aberdeen Proving Ground where new military equipment is tested, and Dugway Proving Ground where chemical/biological detection hardware is tested.

2.3. Rapid Immunoassays

Immunological-based detection methods are adaptable to the detection of any biological agent that has the potential to elicit an immune response, including protein toxins that are not amenable to PCR-based methods. Antigens include structures that elicit an antibody response in an animal, and include viruses, toxins, proteins, vegetative cells, and endospores. Choosing the correct antibody is the most critical aspect of designing immunological assays which are often limited by low antibody affinity and/or specificity. The antibody must be tested against an extensive panel of molecules including compounds closely related to the target, potential interferents, and likely contaminants. The antibody is chosen based on its ability to bind the target antigen with a high affinity, while at the same time retaining a high specificity. The antigens of endospore-forming *Bacillus* species and the antigenic properties of endospores themselves have been studied since the early part of the last century. This early antigen and antibody work has been compiled in a review (Norris 1962). Antibodies continue to be developed for *B. anthracis* spores and have been used in a wide array of immuno-based tests. The assays presented here are focused on detection of *B. anthracis* spores on which extensive research has been done since the deliberate release of endospores as a toxic agent in the 2001 anthrax attacks.

2.3.1. Enzyme-Linked Immunosorbent Assays

Enzyme-linked immunosorbent assays (ELISA)—a basic and widely used immunoassay—can be used to detect endospores, capsules, or cell-wall components of endospore-forming bacteria. ELISA tests involve several steps, including a wash step to eliminate unbound compounds. First, an antibody is attached to a solid surface such as the surface of a well in a multi-well plate. The suspension of the antigen to be tested, which includes whole microorganisms or endospores, is then applied to the antibody-coated surface. The antibody captures the antigen, binding it to the surface. An antibody linked with an enzyme is bound to the antigen, and when the appropriate compound is added, the enzyme acts as a catalyst oxidizing it to produce a colored or fluorescent compound that can be detected by the appropriate instrument such as a spectrophotometer. Variations of the sandwiching of antibody and antigen exist and can include the binding of several antibodies before the final enzyme-linked antibody is bound. Typical ELISA detection limits for endospore and whole-cell detection are usually in the range of 10^4–10^6 bacteria (Speight et al. 1997; Rowe et al. 1999). The ELISA test typically has several drawbacks, including the time needed to run the assay, and reliance on reagents that have a short shelf life and can be prone to cross-reactivity.

Variations of the ELISA technique have been developed to increase its sensitivity and detection time. When cooled ultrasonic cavitation was applied to bacterial spores, detection increased 20 fold using an ELISA-based assay (Borthwick et al. 2005). A field portable detection device was developed using an ELISA-based immunoassay preformed inside a capillary and coupled with a laser-induced fluorescence detection system for the detection of a single endospore or bacterium (Song et al. 2005). A rapid and sensitive bead-based enzyme-labeled immunoassay was developed for the detection of *Bacillus* spores (Farrell, Halsall and Heineman 2005). Paramagnetic Dynal capture beads were prepared with biotinylated antibody to capture endospores in an immunoassay sandwich. Enzymatic conversion of fluorescein-diphosphate to fluorescein was measured in real time within 30 minutes, using luminescence spectrometry with a detection limit of 2.6×10^3 endospores per ml, or 78 total endospores.

2.3.2. Lateral-Flow Immunoassays

Lateral-flow immunoassays (LFIAs), commonly used for pregnancy tests, have been successfully adapted for species-specific endospore detection. In a lateral-flow test system, a membrane strip is embedded with the appropriate antibody at the capture line. The sample is suspended in a buffer and added to one end of the membrane strip and wicking action causes chromatographic separation and the removal of excess reactants. The test result is usually evaluated visually within minutes by two lines—an antigen capture line and a positive control line. Several LFIA-based test kits for endospore detection using the naked eye are on the market and include the SMART-II Anthrax Spore test kit, (New Horizons Diagnostics, Columbia, Md.), the Anthrax BioThreat Alert Kit (Tetracore, Gaithersburg, Md.), and the BioWarfare Agent Detection Devices (BADD) (Osborne Scientific, Lakeside, Ariz.). These test systems can be used by unskilled personnel and give results in 3–15 minutes, however the sensitivity of these kits are low, with a 10^5–10^6 endospore detection range (King et al. 2003). The RAMP System, developed by Response Biomedical Corp., is a LFIA system that is not readable by eye but with the use of a scanning fluorescence reader that is field portable and fits into a small briefcase. The sample is added to a test strip cartridge that is inserted into the reader and fluorescent-dyed latex particles coated with antigen-specific antibodies fluoresce when antigen is bound. This fluorescent-based detection is very sensitive, allowing detection of as few as 4000 *B. anthracis* spores in 15 minutes. The RAMP System has been approved by AOAC International for the detection of anthrax spores, and was used by the United Nations biological weapons inspectors in Iraq.

2.3.3. Immunomagnetic Electrochemiluminescence

Immunomagnetic electrochemiluminescence (ECL) is a sensitive antigen detection method that involves the use of capture antibodies conjugated to magnetic beads and detector antibodies conjugated to ruthenium. Ruthenium, which itself has little or no influence on antibody-antigen interactions, is involved in a voltage-dependent, cyclic oxidation-reduction reaction that triggers luminescence in the presence of tripropylamine. This method requires 30–90 min for detection. ECL has good sensitivity and has been shown to detect as few as 100 *B. anthracis* spores in a phosphate buffer (Gattomenking et al. 1995; Bruno and Yu 1996), with detection limits reduced by three orders of magnitude in soil suspensions (Bruno and Yu 1996). ECL has also been proven effective on *B. anthracis* protective antigen (PA) and botulinum toxin with a sensitivity that depends on the carrier liquid and ranges from 0.1 to 1.0 pg of antigen or toxin per ml volume (Higgins et al. 1999), as well as for the detection of *B. anthracis* lethal factor (LF) (Rivera et al. 2003). The need for a high antigen-antibody affinity in this method necessitates very specific antibodies that must be highly purified to have an efficient coupling to the magnetic beads and ruthenium. One advantage of this method is that these antibody reagents, once prepared, are highly stable and can be lyophilized for easy transport to the field for use.

2.3.4. Flow Cytometry

Flow cytometry has been successfully adapted as a detection method for immuno-based assays of endospores. Phillips was the first to use immuno-based flow cytometry to identify *Bacillus* spores, including *anthracis* (Phillips and Martin 1983). A simplified and rapid protocol was later developed to detect *Bacillus* spores in five minutes with fluorescein-labeled antibodies simply allowed to incubate with a sample of endospores before analysis on a flow cytometer (Stopa 2000). This method was able to detect both live and irradiated endospores equally with a detection sensitivity of approximately 10^3 endospores per ml. The frequency resonance energy transfer (FRET) effect was adapted to minimize background noise and increase selectivity of immuno-based flow cytometry detection assays for *B. anthracis* spores (Zahavy et al. 2003). The approach of combining FRET labeling and flow cytometry analysis improved the selectivity of the *B. anthracis* spores by a factor of 10–100 with respect to other *Bacillus* species. An immunoassay for the detection of *B. anthracis* spores was developed with the use of specific capture peptides conjugated with fluorescent quantum dots (Park et al. 2006). They reported the ability to distinguish *B. anthracis* spores from the spores of *B. thuringiensis* and *B. cereus* within one hour by three methods, including confocal laser scanning microscopy, spectrofluorometry, and flow cytometry.

The Luminex is a flow cytometry system that has been combined with the sequential injection analysis (SIA) for the automated detection of aerosolized *B. anthracis* and *Yersinia pestis* (McBride et al. 2003; Hindson et al. 2004). It incorporates liquid arrays and detects fluorescent antigen microsphere complexes. The microspheres have surfaces with carboxylated groups that can be coated with antibodies for immuno-based detection. Up to 100 different antigens can be detected on unique microspheres that each has a distinct fluorescent signature. When the sample is exposed to the antibody-tagged microspheres, a bioagent will bind to the appropriate bead. A second fluorescently labeled antibody is then added to the sample, which binds to the bioagent-microsphere complex, resulting in a highly fluorescent target for analysis by flow cytometry. Sample preparation and analysis takes less than 30 minutes. The Luminex system also serves as a component of the autonomous pathogen detection system (APDS) installed in U.S. Postal Service mail sorting rooms, which runs without user intervention and continuously monitors the air for pathogens to act as an early warning system in the event of an airborne release (McBride et al. 2003; Hindson et al. 2005a; Hindson et al. 2005b).

A unique membrane filter microchip-based flow cell system was developed to detect *B. globigii* spores as a model for *B. anthracis* (Floriano et al. 2005). Endospores were captured on a membrane surface within a flow cell chamber and labeled with fluorescently tagged anti-*B. globigii* antibody. This fluidics system is coupled to a flow cell positioned on a motorized stage of a microscope with fluorescent detection capabilities. This system was reported to have a sensitivity of 500 endospores and results within 5 min.

2.3.5. Vegetative Cells

Several immuno-based methods rely on detection of the vegetative cells rather than the spores of *B. anthracis*, which have a unique cell wall polysaccharide called galactose-N-acetyl-D-glucosamine (Gal-NAG) representing a major portion of the cell wall. However, this anthrax polysaccharide has also been identified on occasion in *B. cereus*, a closely related species (Ivanovics and Foldes 1958). Monoclonal antibodies were developed that targeted the *B. anthracis* cell wall component and were able to detect vegetative cells but not capsulated vegetative cells or endospores (Ezzell et al. 1990). Direct fluorescent antibody (DFA) testing on capsule antigens and cell wall polysaccharides can also be used to identify vegetative cells of *B. anthracis* using monoclonal antibodies against the cell wall polysaccharide antigen and capsule antigen (De et al. 2002). The DFA assay reportedly had a *B. anthracis* detection specificity of 99 percent and was capable of detecting concentrations as low as 10^4 cells/ml vegetative or encapsulated *B. anthracis* cells.

2.4. Rapid Nucleic Acid Assays

The polymerase chain reaction (PCR) is by far the most commonly used and widespread method for nucleic acid-based pathogen detection (Saiki et al. 1985; Saiki et al. 1988). Other nucleic acid-based assays include the ligase chain reaction (Barany 1991), the Q-beta replicase-based system (Blumenthal 1979; Dobkin et al. 1979), the NASBA system (Compton 1991), and strand displacement amplification (SDA) (Walker et al. 1992a; Walker et al. 1992b). The NASBA system and its application for the detection of viable endospores is discussed in section 3.2.2. In this section, we focus on the PCR detection methods as they specifically apply to the detection of *B. anthracis*. There has been a flurry of publications addressing this topic since the 2001 anthrax attacks, and while we are focusing on the detection of *B. anthracis*, these methods are also generally applicable to other endospore-forming microorganisms, as well as nonendospore-forming pathogens.

PCR is used to detect the nucleic acids present in both vegetative cells and endospores and can distinguish between different species and subspecies depending on the uniqueness of the gene or target sequence used for identification. One advantage of PCR identification and detection compared to traditional microbiological methods is the ability to identify a pathogen without necessarily isolating the microorganism into a culture first. This enables more rapid identification than traditional culture-based approaches. The main drawback of PCR-based techniques is the inability to determine the viability of the sample, which is important in determining the effectiveness of decontamination treatments (see endospore viability in section 3). PCR analysis traditionally involves three main steps:

1) Sample preparation and purification of the nucleic acids
2) PCR amplification of DNA
3) Detection of amplification products

However, in the increasingly popular real-time PCR, these last two steps are combined.

2.4.1. PCR Sample Preparation and Endospore Lysis

The DNA from cells and endospores can be purified using one of the many widely available DNA extraction kits on the market prior to adding to the PCR reaction mixture. Small amounts of cells or lysed endospores from a culture can also be used directly in the reaction mixture after a cell lysis step, although the detection sensitivity may be severely diminished due to PCR inhibitory compounds that are also present in the cellular contents. It may also be necessary to inactivate the cells or endospores before subsequent handling of pathogenic bacteria, especially the hardier aerosol prone endospore-formers that pose a greater risk to laboratory workers. A heat inactivation method has been developed that renders the endospores harmless and has little effect on the efficiency of subsequent PCR reactions

(Fasanella et al. 2003). Methods for lysing endospores both for inactivation and for subsequent PCR analysis include sonication (Belgrader et al. 1999b; Belgrader et al. 2000), plasma lysis that was shown to recover more intact DNA than sonication (Birmingham 2006), and Whatman's FTA cards, which also include a purification step and results in a reported PCR detection sensitivity of five endospores (Lampel et al. 2004).

2.4.2. The PCR Reaction

The PCR reaction requires a pair of short single-stranded oligonucleotide primers—usually 15–25 base pairs in length—that are each specific to a short region of sequence on opposing strands of the double-stranded genomic DNA. These primers are used in combination with a DNA polymerase (usually *Taq* polymerase), dNTPs, and reagent buffers to amplify the target DNA between and including the two primer target sequences. The target DNA, typically 100–2000 bp in length, is copied in an exponential fashion by cycling the reaction mixture through a range of temperatures. PCR can have a detection limit of femtograms of DNA and can, under ideal conditions, detect a single copy of a gene. However, practically the detection limits are much higher, ranging from 10 to 10^4 copies of a gene depending on other reaction variables. Variables that strongly affect the sensitivity of the PCR include the purity of the target DNA, the presence of inhibitors, the specificity of the primers to the target DNA, and the ratio of nontarget to target DNA in the sample. Care must be taken in the sample preparation which usually includes a lysis step to release the DNA followed by a purification step to remove PCR inhibitory compounds. PCR can also be prone to false positives without the adherence of strict laboratory cleanliness procedures to ensure that the DNA previously amplified in the laboratory does not contaminate subsequent reactions. Cross-reactivity of the oligonucleotide primers to nontarget DNA can also be a problem due to the similarity of the DNA sequence between closely related strains.

2.4.3. Specificity of PCR Primers for *Bacillus anthracis* Detection

The detection of *B. anthracis* is challenging, in part due to the close genetic similarity between *B. anthracis, B. thuringiensis*, and *B. cereus* (Helgason et al. 2000; Lan and Reeves 2001; Cherif et al. 2003a; Cherif et al. 2003b; Ivanova et al. 2003; Radnedge et al. 2003; Read et al. 2003; Zhang and Zhang 2003; Fritze 2004a; Fritze 2004b; Gohar et al. 2005; Rasko et al. 2005; Han et al. 2006). Generally, the main genetic differences between these three *Bacillus* species are the presence of plasmids coding for Bt toxin in *B. thuringiensis*, two plasmids (pXO1 and pXO2) that code for anthrax toxicity in *B. anthracis*, and the lack of these plasmids in *B. cereus*. However, these characteristics can not be held to rigidly, as variations exist including several *B. anthracis* species that have only one or neither plasmid (Hoffmaster et al. 2002a) and a *B. cereus* strain that was found to have a plasmid similar to pXO1 with similar disease-causing effects (Hoffmaster et al. 2004; Okinaka, Pearson and Keim 2006). Complicating the matter, both *B. cereus* and *B. thuringiensis* have stretches of genomic sequence similar to the plasmid sequence of the *B. anthracis*-specific plasmids pXO1 and pXO2 (Pannucci et al. 2002a; Pannucci et al. 2002b). Due to the close genetic similarity within the *Bacillus* genus, it is especially challenging to design species specific primers, which precludes the use of the highly conserved 16S rRNA gene to distinguish between them. The 16S rRNA gene has nevertheless been found useful for the broad detection of the 'Bacillus cereus group,' which includes *B. cereus, B. thuringiensis, B. anthracis, B. mycoides, B. pseudomycoides*, and *B. weihenstephanensis*— distinguishing members of this group from all other known bacterial species (Hansen, Leser and Hendriksen 2001; Rasko et al. 2005).

Primers that specifically detect *B. anthracis* have been designed for chromosomal genes and markers including the *gyrB* gene (Yamada et al. 1999), the *vrrA* gene (Andersen, Simchock and Wilson 1996; Jackson et al. 1997; Jackson et al. 1998), the SG-850 marker (Daffonchio et al. 1999), and the *rpoB* gene, including the Ba813 variable region (Patra et al. 1996; Patra et al. 1998; Ramisse et al. 1999; Qi et al. 2001; Drago et al. 2002; Ellerbrok et al. 2002; Oggioni et al. 2002; Patra et al. 2002; Ko et al. 2003; Elzi et al. 2005). However, due to the close genetic similarity, other *Bacillus* species are sometimes targeted. For example, primers developed by Qi et al. (2001) targeting *rpoB* were able to distinguish 144 *B. anthracis* strains from 175 other *Bacillus* species with one notable exception where the primers also targeted a non-*anthracis Bacillus* sp. strain Ba813_11. The Ba813 primers developed by Patra et al. (1996) to detect *B. anthracis* were later found to target a *B. cereus* isolated from an infected patient (Elzi et al. 2005).

While chromosomal genes have the advantage of being genetically more stable for PCR detection since plasmids can inherently be lost and not all *B. anthracis* strains carry plasmids, primers designed to target the *B. anthracis* specific plasmids, pXO1 and pXO2, are of great interest because these plasmids have been found to code for toxicity genes that are required for anthrax disease to occur (Okinaka et al. 1999). These plasmids are found in most, but not all, naturally occurring *B. anthracis* strains, and are both required for full pathogenicity (Turnbull et al. 1992; Qi et al. 2001). The plasmid pXO1 is especially important since it has the genes that produce the anthrax toxin proteins including *cya* (edema factor), *lef* (lethal factor), and *pagA* (protective antigen), while pXO2 carries genes required for capsule synthesis, including *capA, capB,* and *capC,* which confers resistance to phagocytosis. One plasmid-based detection method that was used in the anthrax attacks of 2001 was the amplification and sequencing of the *pagA* gene to confirm that the strain used in the attacks was indistinguishable from the Ames laboratory strain (Hoffmaster et al. 2002a).

A variation of PCR known as multiplex PCR uses multiple primer sets to detect several genes in one reaction, and has been successfully used to detect *B. anthracis* (Shangkuan et al. 2001; Levi et al. 2003; Wang et al. 2004). For example, a multiplex PCR designed for the detection of *B. anthracis* DNA included a primer set for the 16S rRNA gene, the Ba813 primers that target the chromosome, and a set of primers that targeted a region on the pXO2 plasmid (Levi et al. 2003). This method has the distinct advantage of a more comprehensive detection system allowing not only the identification of *B. anthracis* but also giving some indication of virulence since both plasmids are necessary to be fully infectious.

2.4.4. Rapid PCR Detection Methods High Throughput and real-time PCR

Traditionally, PCR products are detected by agarose gel electrophoresis, which separates the DNA fragments according to length. This is followed by visualization with a DNA interca-lating dye such as ethidium bromide or SYBR Green that allows the fragments of DNA to be visualized when illuminated with the appropriate excitation wavelength. The visualized band corresponds to the detection target, and multiple bands are seen for multiplexed PCR. This method is cumbersome and inefficient, requiring post-PCR handling of the sample and does not lend itself well to high throughput robotic processing.

Methods that can more rapidly and efficiently detect PCR products using high throughput technologies have been recently developed for the detection of *B. anthracis*. Flow cytometry was adapted to detect and distinguish PCR products from several species including *B. anthracis, Y. pestis, F. tularensis,* and *Bacillus melitensis* (Wilson et al. 2005). A high throughput robotic process for PCR amplification was coupled with flow cytometry detection of the amplified product DNA hybridized to fluorescent beads. This method included a 10-plexed PCR assay to screen 10 genetic targets at once and allowed the rapid screening of up to 384 environmental samples simultaneously. Microarray detection technology employs PCR amplification of the target gene and then exposes the products to a microarray of oligonucleotides specific for various species of *Bacillus* for both detection and identification. The intergenic transcribed spacer (ITS) region of the rrna operon has been targeted using this method (Nubel et al. 2004) with a sensitivity limit of 10^3 cells using target endospores added directly to the PCR without prior treatment. Microarray detection has also been applied to target genes from the plasmids, pXO1 and pXO2, and a chromosomal marker which allowed the unambiguous identification of *B. anthracis* from other *Bacillus* species as well as distinguishing between plasmid-containing and plasmid-free strains (Volokhov et al. 2004).

Real-time PCR is the most popular rapid PCR detection method because the products can be detected *in situ* while the reaction progresses, making post PCR processing unnecessary. Real-time PCR methods make use of fluorescence detection with a built-in fluorimeter. Products can be detected in as little as seven minutes when real-time PCR detection is combined with new PCR technologies that allow more rapid cycling of the reaction. Gene-specific fluorophore-labeled oligonucleotides such as Molecular Beacons, TaqMan® probes, Amplifluor®, and Scorpion® primers are used to measure fluorescence of the specific gene of interest as it is being amplified. Nonspecific detection methods can also be used that are based on a DNA-binding fluorogenic molecule such as SYBR Green, which fluoresces brightly when bound to double-stranded (ds)DNA. As the product increases in the vessel, so does the fluorescence. The drawback is that these fluorogenic dyes bind to all dsDNA and thus nonspecific amplification products and primer-dimers may also be detected during amplification—obscuring the results. Fluorescently labeled oligonucleotide probes and primers allow more specific detection of products. There are a number of variations of the fluorescent oligonucleotide detection techniques; for a review, see Mackay, Arden and Nitsche (2002). Variations of real-time PCR continue to be developed as users become more familiar with this rapid detection technique. One variation uses a combination of both SYBR Green and sequence-specific oligonucleotide probes to acquire several layers of specificity. Real-time PCR is also amenable to multiplexing for more comprehensive identification of one organism or the detection of multiple species in one reaction. The main drawbacks of real-time PCR, as compared to conventional PCR, are the high start-up expense for equipment, the high cost of reagents, and the inability to determine the size or total number of amplicons, unless a post analysis is preformed.

Real-time PCR has been used for the detection of a variety of genes for the identification of *B. anthracis* (Makino et al. 1993; Makino et al. 2001; Qi et al. 2001; Bell et al. 2002; Drago et al. 2002; Ellerbrok et al. 2002; Hoffmaster et al. 2002b; Oggioni et al. 2002; Patra et al. 2002; Makino and Cheun 2003; Rantakokko-Jalava and Viljanen 2003; Ryu et al. 2003; Bode, Hurtle and Norwood 2004; Priha et al. 2004; Kim et al. 2005; Klee et al. 2006). For example, it has been applied successfully by (Ryu et al. 2003) to detect *B. anthracis* concentrations of 10 spores per ml in pure suspension, 10^4 spores in a gram of soil, or 0.1 ng per µl of purified DNA in as little as three hours without the need for enrichment cultures and with minimal preprocessing using chemical, heat, and/or centrifugation methods. It has become a widely accepted and often used diagnostic PCR technique due to its rapid detection ability as compared to more traditional PCR (Mackay 2004; Espy et al. 2006).

2.4.5. Field Implementation of Rapid PCR for Analysis of Environmental Samples

PCR analysis traditionally involves three main steps, including sample preparation and purification, the PCR reaction, and detection of the amplified products. Most field-applicable technologies combine either two or all three of these steps for rapid automated identification of pathogens. The following field-compatible machines range from large automatic units that continuously monitor the air for pathogens to small handheld devices that identify an inserted sample, including bacterial endospores. The detection time—from sample collection to identification—usually takes 30 minutes to several hours depending on the nature of the necessary sample preparation. Real-time PCR is often incorporated into these devices because it can be multiplexed and already combines the amplification and detection steps.

In response to the October 2001 anthrax attacks, a laboratory trailer was set up near the contaminated sites equipped with a portable (49 lb) real-time PCR detection machine known as Ruggedized Advanced Pathogen Identification Device (RAPID), developed by Idaho Technology. The lethal factor on the pXO1 plasmid was detected by real-time PCR using RAPID after a DNA extraction protocol was implemented on collected swabbings from the mailrooms and other suspected facilities, with an overall sensitivity of 100 dry-swabbed endospores within five hours (Higgins et al. 2003a). Any sample that was found to be lethal factor positive with real-time PCR was then confirmed *via* traditional PCR using primers for the detection of *capA* on the pXO2 plasmid and the chromosomal *vrr* gene. This was done in conjunction with traditional culturing methods at the National Veterinary Service Laboratory in Ames, Iowa, which was found to have a better detection sensitivity of 10 endospores per swab but a longer detection time of 24 hours. The RAPID detection system requires pre-preparation of the DNA before setting up a PCR reaction that is then loaded onto the RAPID system requiring professionally trained and experienced personnel.

A Miniature Analytical Thermal Cycling Instrument (MATCI) was developed and tested at the Lawrence Livermore National Laboratory (LLNL) to perform rapid fluorogenic real-time PCR detection assays (Belgrader et al. 1998b; Ibrahim et al. 1998; Northrup et al. 1998). The complete MATCI setup includes a laptop computer and rechargeable battery power supply that fits into a briefcase for transport. The system can detect TaqMan type probes and intercalating fluorescent dyes with a 40-cycle PCR, taking less than 30 minutes in a single-reaction chamber. However, the DNA must first be purified and set up into an appropriate PCR reaction by skilled personnel because the MATCI does not have the capability to purify or prepare the sample.

The Advanced Nucleic Acid Analyzer (ANAA) is a redesign of MATCI that has an array of ten real-time PCR reaction chambers allowing rapid and parallel sample analysis (Belgrader et al. 1998a; Belgrader et al. 1999a). The portable ANAA fits in a suitcase that includes a laptop to run the reactions, and has been shown to detect five cells in as little as nine minutes or 500 cells in seven minutes. Some microorganisms take longer to detect, with higher limits up to 10^4 cells/ml. A commercial version of this instrument, called the Smartcycler, is now available from Cepheid in Sunnyvale, CA, with 96 separate individually programmable multiplex reaction chambers. Although the Smartcycler is not specifically field portable, Cepheid has designed a notebook-sized (3.3 kg) real-time PCR field portable unit that they tested for detection of *Bacillus subtilis* and *Bacillus thuringiensis* DNA and that uses independently programmable real-time multiplexed PCR chambers (Belgrader et al. 2001). The ANAA, like the MATCI, requires the DNA sample to be prepared prior to adding it to a real-time PCR reaction mix that is inserted into the machine, and can only be used by personnel experienced in PCR methods.

The handheld advanced nucleic acid analyzer (HANAA), also developed at the LLNL, was tested on preparations of *B. anthracis*, *Erwinia herbicola*, and *Eschericia coli* (Higgins et al. 2003b). It is a handheld device weighing less than 1 kg and has the same basic real-time PCR detection capabilities as the ANAA with only four reaction chambers, allowing it to maintain a small profile. It is the first highly mobile handheld PCR instrument and was used in 2003 by the United Nations inspectors in Iraq during their search for biological weapons. It is highly sensitive and able to detect concentrations as low as ten organisms (100 fg DNA) under ideal conditions, however using environmental samples that include dust and particulates, 100 organisms per milliliter can be detected within as little as 30 min. HANAA can also perform two independent identifications in each chamber due to the use of two LED excitation sources of 490 and 525 nm for multiplexing. One sample can therefore be tested for up to eight pathogens simultaneously in the four chambers. It rapidly detects amplified products within 7–30 minutes, however it requires a skilled user to prepare the sample and PCR reaction before it is inserted into the HANAA for amplification and detection.

IQuum's (Marlborough, MA) Liat Analyzer is a field portable nucleic acid testing system that can be used by nonspecialized personnel with minimal training (www.iquum.com) (Chen 2007). This unit integrates and fully automates sample preparation, nucleic acid extraction, and real-time PCR detection, thus only requiring the user to collect the sample into a Liat Tube and then insert the Liat Tube into the Liat Analyzer. Identification of the sample typically takes 30–60 minutes. It accepts most sample types, including blood, liquids, soils, and swabbings. The automation of the system is possible due to the Liat Tube, which is a multi-segment tube that contains all the reagents necessary for automated sample preparation, DNA extraction, and real-time PCR reaction setup. The system provides onsite and on-demand testing capabilities to a wide range of applications, including infectious disease diagnostics (e.g. avian influenza, Leishmaniasis, *Chlamydia trachomatis, N. gonorrhoeae*, Cytomegalovirus, Epstein-Barr Virus) and biodefense detection (e.g. anthrax, dengue). For example, the Liat Anthrax Assay is specifically designed to detect live *B. anthracis* spores using *in situ* spore germination, followed by automated DNA extraction and real-time PCR. With a sensitivity on the order of 10^2 endospores, the multiplex assay targets both plasmids (pXO1 and pXO2) and chromosomal DNA to achieve high specificity, and allows the differentiation of anthrax strains. The three anthrax targets are detected, along with an internal control using the four optical channels of the Liat Analyzer. IQuum is also developing the Liat Bioagent Autonomous Networked Detector (BAND) system with funding from the Department of Homeland Security (McBride 2007). The Liat BAND system uses the same lab-in-a-tube technology as the Liat Analyzer for fully automated around-the-clock detection of multiple aerosolized endospores, pathogens, toxins, and viruses.

2.4.6. Monitoring the Air for *Bacillus anthracis* Endospores by PCR

Biohazard Detection System (BDS), which detects *B. anthracis* spores from air samples, has been installed at U.S. Postal Service mail sorting centers (Knight 2002). In general, the BDS employs

- A hood and vacuum system to sample for hazardous materials,
- A device that separates small and large particles,
- A robotic arm that prepares and loads samples, and
- A PCR machine that amplifies the DNA of interest, which is subsequently detected with fluorescent probe-labeled DNA (i.e., TaqMan® detection scheme) (Holland et al. 1991)

They are fully automated, entirely self-contained units that can reportedly detect as few as 30 endospores per sample within a time frame of 30 min, and can process a sample once per hour continuously. This commercial technology was developed by Northrop Grumman and team members at Cepheid Inc. in Sunnyvale, CA, and incorporates the Cepheid GeneXpert® technology. The BDS concentrates airborne particles into a sterile water base. This creates a liquid sample that is injected into a cartridge (Belgrader et al. 2000). The system then uses sonication to lyse samples, followed by purification and concentration in microfluidic chambers prior to real-time PCR amplification of two key plasmid virulence genes—*pag* (pXO1) and *capB* (pX02)—along with internal controls. The unit only needs periodic replacement of sealed cartridges that contain all necessary reaction reagents. Current development is underway on a system that simultaneously detects *B. anthracis, Yersinia pestis*, and *Francisella tularensis*.

The autonomous pathogen detection system (APDS) is a podium-sized instrument developed at the Lawrence Livermore National Laboratory (LLNL) that can also continuously monitor the air for pathogens as an early warning system in the event of an airborne release (McBride et al. 2003; Hindson et al. 2005a; Hindson et al. 2005b). It combines an immunoassay detection step with a follow-up real-time PCR confirmation if an initial positive result is acquired to minimize the chance of false positives. It was shown that APDS could detect both aerosolized *B. anthracis* and *Yersinia pestis* (McBride et al. 2003). Like the BDS system installed in U.S. Postal Service mail sorting rooms, APDS runs without user intervention automatically collecting air samples followed by automatic sample preparation and detection of pathogens.

2.5. Rapid Detection of Endospores via Dipicolinic Acid Biomarker

Detection of endospores based on the chemical fingerprint, or distributions of molecular markers, have been investigated, including various chromatography, mass spectrometry, and spectroscopy methods. These include gas chromatography (GC) to analyze distribution patterns of long-chain fatty acid methyl esters (Lawrence, Heitefuss and Seifert 1991), the detection of the presence of various carbohydrates (Logan et al. 1985; Ezzell et al. 1990), and the analysis of dipicolinic acid (DPA) that is a large part of the bacterial spore. Dipicolinic acid (DPA, 2,6-pyridinedicarboxylic acid) was first identified in 1953 as a major component of all endospore-forming species. Each endospore contains a concentration of approximately 1 molar DPA, corresponding to about 15 percent of dry weight or more than 10^8 molecules of DPA per endospore (Powell 1953). DPA is synthesized and incorporated into the spore core as a 1:1 complex with Ca^{2+} ($K_{11} = 10^{4.4} \, M^{-1}$). Investigations with DPA-free mutants indicated that DPA is important for engendering heat resistance and facilitating germination (Hanson et al. 1972; Balassa 1979). DPA is not present in vegetative cells and is a unique constituent of all endospores, making it an ideal indicator molecule for the presence of bacterial spores. Chronologically, spectroscopic methods for DPA detection include UV absorbance (Perry and Foster 1955; Murty and Halvorson 1957), iron(II)-DPA colorimetry (Janssen, Lund and Anderson 1958), and most recently Tb^{3+}-DPA luminescence (Sacks 1990; Pellegrino et al. 1998; Hindle 1999; Rosen 1999; Jones and Vullev 2002b; Lester and Ponce 2002; Lester, Bearman and Ponce 2004; Li et al. 2004; Shafaat and Ponce 2006; Cable et al. 2007; Yung et al. 2007).

2.5.1. Terbium Dipicolinic Acid Luminescence Assay

DPA is released from the endospore body upon germination, chemical or physical rupture. High-affinity binding of DPA to Tb^{3+} triggers intense green luminescence under UV excitation, enabling a very sensitive assay for bacterial spore detection (Fig. 19.7) (Rosen, Sharpless and McGown 1997; Rosen 1998; Hindle 1999; Rosen 1999). Thus, the turn-on of green luminescence signals the presence of bacterial spores, and the intensity of the luminescence can be correlated to the number of endospores per milliliter. The mechanism of DPA-triggered Tb luminescence is based on the unique photophysical properties of lanthanide ions. The luminescence of lanthanide ions is characterized by long lifetimes (0.1 to 1 ms), small extinction coefficients (i.e., absorbtivity, $\sim 1 \, M^{-1} \, cm^{-1}$), and narrow emission bands. These characteristics arise because the valence f orbitals are shielded from the environment by the outer 5s and 5p electrons, and because the transition between the emitting excited state and ground state is highly forbidden (Horrocks and Sudnick 1981; Sabbatini, Guardigli and Lehn 1993). Thus, direct excitation of terbium ions leads to weak luminescence due to the small extinction coefficient (Sinha 1983). However, coordination of aromatic chromophores, like DPA, triggers intense terbium luminescence. The juxtaposition of DPA, which has an absorbtivity of $5000 \, M^{-1} \, cm^{-1}$ (Rosen 1999), serves as a light-harvesting center (e.g., antenna effect) (Horrocks and Sudnick 1981; Sinha 1983; Horrocks 1984; Balzani 1990; Balzani et al. 1990;

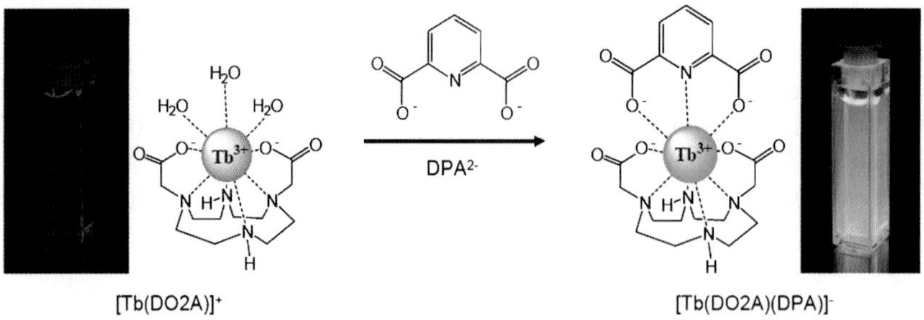

Figure 19.7. Dipicolinic acid (DPA) is a unique constituent of bacterial spores—a dormant form of *Bacillus* and *Clostridium*—which can be detected using DPA-triggered Tb^{3+} luminescence. The dramatic luminescence turn-on is depicted with the cuvettes under UV illumination before (left) and after (right) upon addition of DPA to $[Tb(DO2A)]^+$ solution. (*Reproduced with permission from J. Am. Chem. Soc. 2007, 129, 1474-1475, Copyright 2007 American Chemical Society*).

Sabbatini, Guardigli and Lehn 1993). Strong electronic coupling and downhill energetics allow the DPA-centered excitation energy to be efficiently transferred to the lanthanide ion that subsequently luminesces bright green.

Any fluorescence background from interferents can be eliminated by lifetime-gated detection of the Tb^{3+} luminescence. Lifetime gating takes advantage of the fact that terbium luminescence lifetimes are on the order of milliseconds, while fluorescence lifetimes from organic impurities generally are on the order of nanoseconds (Morgan and Mitchell 1996; Vereb 1998; Xiao and Selvin 1999; Beeby 2000; Connally, Veal and Piper 2002). Lifetime gating drastically reduces the chance of false negatives, which could arise if the terbium luminescence is masked by background fluorescence from impurities. Potential interferents or inhibitors such as sugars, nucleic acids, and amino acids are also present in much lower concentrations in endospores and vegetative cells and have binding constants for Tb^{3+} that are approximately six orders of magnitude less than that of DPA, making this method relatively immune to these interferents.

The detection limit for DPA in aqueous solution when using the Tb-DPA luminescence assay can be limited by instrumentation, such as excitation source (e.g., Xe-flash lamp versus UV LED (Li et al. 2004)) or type of detector (e.g., PMT versus CCD), but is ultimately governed by the binding constant between Tb^{3+} and DPA ($K_a = 10^{8.7}$) (Grenthe 1961; Cable et al. 2007). Indeed, below nanomolar concentration (i.e., $\sim 10^3$ endospores/ml), the percentage of DPA molecules bound to Tb^{3+} approaches nil. Because binding is required to trigger the luminescence turn-on, detection below those concentrations is not possible regardless of the excitation source or detection system.

In an effort to further improve the sensitivity, selectivity, and enable incorporation into waveguide or fiber optics, we are employing macrocyclic ligands to construct Tb^{3+}-receptor platforms that are specific for DPA (Cable et al. 2007). The macrocycle DO2A (*1,4,7,10*-tetraazacyclododecane-*1,7*-diacetate) meets our basic requirements for a receptor ligand in that

- The DO2A fraction bound at micromolar concentrations is near unity,
- DO2A binding keeps three adjacent coordination sites open and does not inhibit DPA binding,
- DO2A can be chemically modified with various covalent pendant groups to test receptor site constructs and enable polymer incorporation, and
- The $[Tb(DO2A)(DPA)]^-$ complex eliminates water quenching.

Binding experiments show that the affinity of $[Tb(DO2A)]^+$ for DPA^{2-} ($K_A = 10^{10.5} M^{-1}$) is much greater than that of $Tb^{3+}(aq)$ ($K_A = 10^{8.7} M^{-1}$). We conclude that the remarkable ternary $[Tb(DO2A)(DPA)]^-$ complex stability is due to additional binding interactions, such as the interligand hydrogen bonding between DO2A and DPA indicated in the crystal structure.

2.5.2. Anthrax Smoke Detector

While the 2001 anthrax attack led to 22 cases of life-threatening infections, including five deaths (Jernigan et al. 2002), it also taught us that even massive exposures to *B. anthracis* spores can be effectively treated with a simple antibiotic regimen (Tahernia 1967). However, for treatment to be effective it must be initiated soon after exposure. While rapid PCR methods for DNA analysis appear as an ideal solution, the cost of ownership in terms of reagents for continuous monitoring makes this option cost prohibitive (Knight 2002). Given the costs, the U.S. General Accounting Office (GAO) was prompted to investigate alternative technologies that have the potential to greatly reduce the operating costs for an anthrax alarm system. Among the desired characteristics for an anthrax alarm system are long-term, online operation with minimal maintenance, low operating costs, and low susceptibility to false alarms.

In general, an anthrax surveillance system requires a method of bioaerosol sampling (Griffiths and Decosemo 1994; Cox and Wathes 1995; Griffiths et al. 1997; Baron and Willeke 2001) coupled with an analysis method with high specificity and sensitivity for detecting *B. anthracis* spores. In an effort to reduce the operation costs of an anthrax surveillance system by approximately two orders of magnitude, we have developed and tested an automated anthrax smoke detector (ASD) that measures airborne bacterial spores using an air sampler coupled to the simple, robust, and inexpensive Tb-DPA luminescence assay (Fig. 19.8). The ASD is intended to serve as a front-end monitor for species-specific anthrax surveillance systems (e.g., DNA analysis *via* real time PCR) where, in the case of a bacterial spore event, the ASD triggers expensive-to-operate validation technology to confirm the result. A large increase of airborne bacterial spore concentration is a strong signature of an anthrax attack, because an anthrax attack must employ *B. anthracis* spores. The combination of the ASD as a front-end monitor with an anthrax surveillance system would drastically cut operating costs (~100 fold) while maintaining low false-positive rates of species-specific detection methods.

The ASD implementation of the terbium luminescence assay in conjunction with an aerosol capture device, thermal induced bacterial spore lysis unit, and a miniature time-gated spectrometer enables inexpensive and fully automated measurement of airborne bacterial spores every 15 minutes (Fig. 19.9). While in operation, the ASD samples 16.7 L of air per minute and collects micron-sized particles—including airborne bacterial spores if present—onto a meshed quartz fiber tape. After collection, the sample is automatically processed with the thermal lysis unit at ~ 250°C to release DPA from any captured endospores. Addition of the $TbCl_3$ reagent solution *via* syringe pump forms the luminescent Tb-DPA complex. Finally, the Tb-DPA luminescence intensity is measured under pulsed UV excitation using the time-gated spectrometer. A sharp increase in airborne bacterial spore concentration is a strong signature of an anthrax attack, because 1) bacterial spores, rather than the vegetative cells, are capable of surviving in the atmosphere (Nicholson et al. 2000) and are thus excellent vehicles for *B. anthracis* dispersal, and 2) natural background concentrations are very low, varying between 0.01 and 1 endospore/L (Pastuszka et al. 2000).

The ASD employs time-gated detection of long-lived terbium dipicolinate luminescence, which is a sensitive assay for the detection of trace amounts of endospores in real world environments such as office buildings, subway stations, and post offices. Such environments are filled with a plethora of fluorescent compounds that can confound fluorescence-based assays. Time-gated detection, however, essentially eliminates background fluorescence by taking advantage of the fact that fluorescent compounds have lifetimes in the nanosecond regime (Lakowicz 1983), whereas terbium complexes have lifetimes in the millisecond regime (Jones and Vullev 2002a).

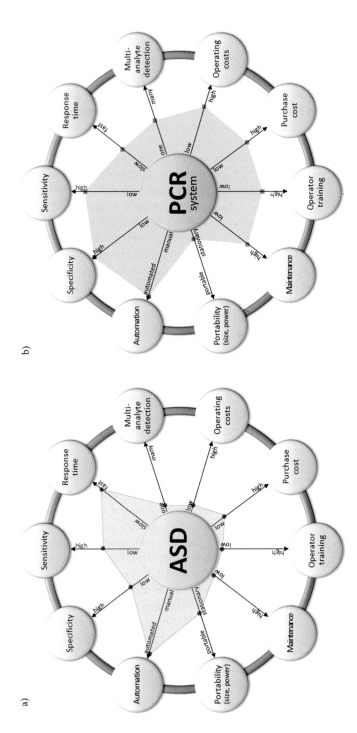

Figure 19.8. Spider charts illustrating instrument characteristics for a) anthrax smoke detector (ASD) and b) example of a polymerase chain reaction (PCR) system, such as the Biohazard Detection System. These spider charts serve to qualitatively illustrate the relation and potential synergies of characteristics between different instruments. Quantitative comparison will require categories such as portability and specificity to be broken out into measurable subcategories. For example, portability could be quantified in terms of size, mass, and power consumption, and specificity can be quantified in terms of false positive rate and detection confidence. (*Adapted from Figure 7, page 15 in Chemical and Biological Sensor Standards Study by Carrano et al.*).

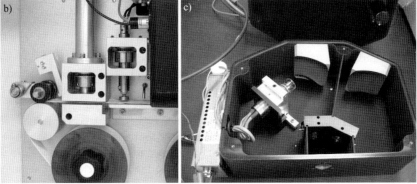

Figure 19.9. The Anthrax Smoke Detector (ASD) is a cost effective front-end monitor for anthrax surveillance systems. The principle of operation is based on measuring airborne endospore concentrations, where a sharp concentration increase signals an anthrax attack. a) The ASD consists of an air sampler, a thermal lysis unit, a syringe pump, a time-gated spectrometer, and endospore detection chemistry comprised of Tb^{3+}-DPA luminescence. b) Detailed view of the ASD components. c) Time-gated spectrometer.

The ability to detect 16 endospores/L in 15 minutes with full automation and inexpensive operation costs qualifies the ASD as a rapid front-end monitor for anthrax surveillance systems. Endospores are the ideal detection target for a front-end anthrax monitor, because 1) endospores can be detected with low operating cost instrumentation, 2) endospores are going to be the vehicle for anthrax delivery, and 3) false positives due to natural fluctuations are expected to be very rare and verifiable when coupled with species-specific detection technology.

3. Validation of Sterilization by Rapid Endospore Viability Assessment

3.1. Measuring Endospore Viability and Inactivation

Endospores can lay dormant for long periods and remain viable, even while exposed to harsh environmental conditions. When more favorable conditions are signaled by the presence of water, nutrients and germinants, endospores may break dormancy and germinate to become metabolically active and multiply. Figure 19.10 outlines the life cycle of an endospore-forming

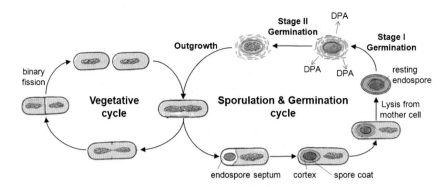

Figure 19.10. Life cycle of an endospore-forming bacterium. Sporulation produces resting endospores in preparation for environmental changes or extremes. In the presence of water and nutrients, endospores may germinate and re-engage in the vegetative cycle. DPA is released during stage I germination. (*Adapted from Figure 6, page 78 in Principles of Microbiology 6th edition (2001)*).

bacterium. As an endospore proceeds through germination towards cell division, there are various stages, including spore activation, stage I germination (during which DPA is released and water rehydrates the spore), stage II germination (during which cortex hydrolysis occurs and metabolism begins), and finally outgrowth (during which cell division occurs) (Setlow 2003). Each of these stages in the life cycle of endospore-forming organisms can be observed with the experimental techniques described below. Certainly outgrowth, manifested as visible colonies on growth media, is the most direct measurement of endospore viability. However, colony formation requires approximately 20 cycles of replication over several days of incubation before the colony becomes visible, and is therefore not amenable for rapid endospore viability analysis. In contrast, observation of water influx with phase contrast microscopy, DPA release with Tb-DPA luminescence assays, and ATP production with the luciferin-luciferase bioluminescence assays provide rapid measures of endospore viability, since these events occur much earlier in the germination to outgrowth pathway.

Before proceeding to review these rapid endospore viability assays, it is instructional to carefully consider the definition of viability and related terms in the context of the endospore life cycle. While the definition of viability has traditionally been synonymous with culturability, the realization that more than 99 percent of environmental microorganisms are viable-but-not(yet)-culturable (VBNC) has led to much debate about redefining the term viability (Roszak and Colwell 1987). The term VBNC has been most widely applied to a dormant state where normally culturable strains of bacteria such as *Vibrio* enter an unculturable state (Colwell and Grimes 2000). Many uncultured species that were once considered to be "VBNC," however, have been successfully grown in a laboratory environment with new culturing techniques (Connon and Giovannoni 2002; Janssen et al. 2002; Kaeberlein, Lewis and Epstein 2002; Rappe et al. 2002). Given the VBNC populations, we adopt the more appropriate definition of a viable microorganism as having the potential for replication. With this definition, however, the viable population will remain a theoretical, unmeasurable term, because measuring the *potential* for reproduction is beyond the reach of experimental methods or timescale. For example, it is not possible to explore in the laboratory all possible germination conditions for a superdormant endospore, which may lie in its resting state over many millennia before conditions for germination finally lead to reproduction. Measuring *actual* reproduction rates becomes more feasible *via* culturing, doubling rate measurements, or ATP production rates, although some fraction of an environmental microbial population may consist of cells that may have doubling times—even under ideal conditions—that are too slow to be measurable. Given these limitations in experimentally accessing viability, we define indicators of endospore

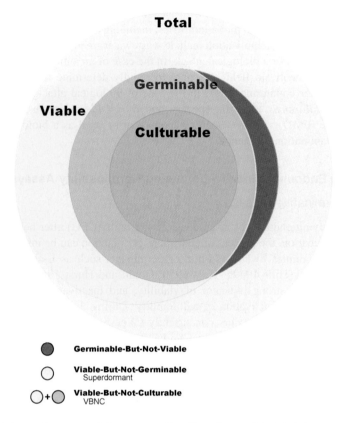

Figure 19.11. Various endospore populations expressed in a Venn diagram. Labeled circles represent *total* endospore population (e.g., as counted with phase contrast microscopy), *germinable* endospore population (e.g., as measured by EVA, or phase transition from bright to dark), *culturable* endospore population (e.g., as CFU counted on agar growth plate) and *viable* endospore population (i.e., defined as the proportion of endospores capable of metabolic activity). The indicated overlapping regions illustrate *germinable-but-not-viable*, *viable-but-not-germinable* (i.e., superdormant), and *viable-but-not-culturable* (VBNC) endospore populations. Relative area varies with species/environment and is not quantitative.

viability as observables in the germination-replication pathway of the endospore life cycle, including rapid indicators of DPA release (germination stage I), water uptake (germination stage I), and onset of ATP production (germination stage II).

Figure 19.11 shows the relationship of the various measurable populations of endospores with respect to the theoretical and unknown viable population for which no direct measurement yet exists. Given a total population of endospore bodies, a subset of the total population is germinable, and a subset of the germinable population is culturable. Within the germinable population, subsets may be germinable-but-not-culturable and germinable-but-not-viable. Within the viable endospore population, subsets may be viable-but-not-germinable (i.e., super-dormant), VBNC, and viable-and-germinable-but-not-culturable. While the plethora of spore populations defined above may be more confounding than helpful to the general reader, the most important point in evaluating the applicability of rapid endospore viability assays for monitoring endospore inactivation is the empirical relationship of inactivation/reduction of germinability versus culturability for a given sterilization or cleaning regimen.

The rapid viability assays described below aim to provide endospore inactivation information in near real-time. As the most resilient form of life, *Bacillus* spores are used as biological indicators for monitoring the efficiency of cleaning or sterilization processes (e.g., autoclave performance). Monitoring the effectiveness of cleaning and sterilization regimens to maintain good hygiene is required in several major industries, including health care, food, and pharmaceutical industries. Other areas of application include wastewater treatment facilities and validation of bioagent inactivation after a biological attack. In the case of an anthrax attack, rapid viability assessment technology will aid field personnel to rapidly determine the viability of anthrax spores before and after countermeasures. In the case of biological attacks with another agent (e.g., *Y. pestis*, *F. tularensis*, *Brucella* species, and *Burkholderia* species, Variola and Foot and Mouth Disease (FMD) viruses), *Bacillus* spores can be used as a biological indicator for monitoring decontamination efficiency.

3.2. Measuring Endospore Inactivation using Germinability Assays

3.2.1. Rapid Germinability Assays

While quantifying endospores as colony-forming units (CFU) after heat shock treatment requires days of incubation, the process of endospore germination can be initiated and observed on the timescale of minutes by the addition of germinants such as L-alanine, L-asparagine, or glucose (Hills 1949a; Hills 1949b; Hills 1950; Gould and Hurst 1969; Setlow 2003). The ability to germinate is a strong indicator of viability, and inactivation of germinability can readily be correlated with inactivation of culturability. During the first stage of germination, water uptake and DPA release occurs concurrently (Woese and Morowitz 1958; Gould and Hurst 1969; Sacks 1990). The water uptake can be observed by phase contrast microscopy as the phase-bright endospores transition into phase-dark germinated endospores. The DPA release can be observed with two novel endospore viability assays (EVA), based on bulk spectroscopic analysis (i.e., spectroEVA) and direct microscopic enumeration (i.e., microEVA). Rather than requiring full outgrowth before enumeration, these rapid germinability assays probe for viability much earlier—during stage I germination when DPA is released and water begins to enter the core.

For a given population of endospores, a subset will germinate, and a subset of that population capable of germination will form colonies. The actual viable endospore population will likely fall between the populations capable of germination and outgrowth, since culture-based assays often underestimate viable endospore bioburden due to viable-but-non-culturable (VBNC) populations (Colwell and Grimes 2000), while EVA will most likely overestimate the viable endospore bioburden because not every endospore that germinates will be capable of outgrowth. It is possible, however, that the germinable endospores population, measured on the time scale of the experiment, will underestimate the total population capable of germination if a significant superdormant populations is present. If superdormant endospores are not germinable or culturable under the conditions provided or the timescales of the experiment then, EVA would provide a closer approximation of the viable endospore population than culture-based assays.

3.2.2. Nucleic Acid-Based Amplification Methods for Detecting Germinable, Viable *Bacillus anthracis* Spores

Viable bacteria can be detected based on the production of mRNA. There are several nucleic amplification methods that are useful in detecting mRNA, including reverse transcriptase-strand displacement amplification (RT-SDA), nucleic acid sequence-based amplification (NASBA), and reverse transcriptase PCR (RT-PCR). mRNA is a highly labile molecule that is an intermediary for protein production from genes encoded in the DNA with a half-life on the order of seconds, making it a good marker for cell viability, while DNA itself makes a poor indicator of viability due to its ability to persist in the environment and clinical samples on the order of weeks (Masters, Shallcross and Mackey 1994; Deere et al. 1996; Baez et al. 1997; Herman 1997; Hellyer et al. 1999a; Hellyer et al. 1999b; McKillip, Jaykus and Drake 1999; Szabo and Mackey 1999; Rijpens et al. 2002). However, reports indicate that under some treatment regimes used to kill cells, mRNA persisted for hours after cells were killed and also depended on the storage of samples after they were killed (Sheridan et al. 1998; Yaron and Matthews 2002).

Nucleic acid sequence-based amplification (NASBA) can be used to amplify RNA for the detection of microorganisms and is fundamentally different from PCR in that it does not require temperature cycling but rather is run at a constant temperature of usually 41°C. NASBA of RNA occurs in two phases—noncyclic and cyclic—with the use of three enzymes: reverse transcriptase (RT), RNase H, and T7 RNA polymerase. The noncyclic phase includes three steps:

1. Extension of cDNA onto the target mRNA by RT with a primer that includes a T7 promoter sequence at the 5' end.
2. RNase H degradation of the initial RNA strand in the newly formed RNA-cDNA strand, leaving behind a single cDNA strand with the primer and its T7 promoter sequence still bound to it.
3. Formation of double-stranded DNA from the cDNA template with a second primer creating a now-active T7 promoter site.

The cyclic phase begins at this point with the transcription of the double-stranded DNA by T7 RNA polymerase into RNA. NASBA has been combined with standard lateral flow technology to detect viable *B. anthracis* spores (Hartley and Baeumner 2003; Baeumner et al. 2004). The endospore sample was first germinated, and the mRNA from a gene encoded on the pXO1 plasmid was purified and amplified by NASBA. A lateral-flow membrane strip was impregnated at the capture line with a DNA probe that had been coupled to liposomes encapsulating the dye sulforhodamine B. The captured target RNA could be read by eye or quantified with a hand-held reflectometer. Baeumner et al. reported the detection of as few as ten viable *B. anthracis* spores in four hours.

3.2.3. Germination Observed via Loss of Phase Brightness

During germination, highly refractile endospores change from bright to dark when viewed under a phase contrast microscope (Pulvertaft and Haynes 1951; Powell 1957) due to the influx of water (Ross and Billing 1957; Lewis, Snell and Burr 1960). Phase contrast microscopy can be carried out on a normal microscope slide-cover slip setup, or on solidified agar covered with a cover slip under an oil immersion objective lens. This technique has been applied to observe germination of both *Bacillus* and *Clostridium* spores (Levinson and Hyatt 1966; Hitchins, Kahn and Slepecky 1968; Rowley and Feeherry 1970).

The time course of refractility loss follows a biphasic kinetics (Vary and Halvorson 1965; Hashimoto, Frieben and Conti 1969a). In the first phase, endospores change into partial-phase dark and lose part of the heat resistance. The proteinaceous coat becomes more porous, leading to the hydrolysis of water and removal of calcium dipicolinate from the endospore core. The second phase marks the complete hydration of the spore core and degradation of the spore cortex, which render the endospore phase dark. Duration of germination depends on a number of factors, such as species, inoculum size, germinants, temperature, and the optics used for observation. The reported phase transition for individual bacterial spores ranges from 75 seconds to approximately an hour (Hashimoto, Frieben and Conti 1969b; Leuschner and Lillford 1999).

An inverse relationship has been reported between the endospore inoculum size and germination time (Caipo et al. 2002). When used in conjunction with other endospore detection assays, phase contrast microscopy proves to be a very useful validation test and provides total and germinable counts of endospore suspensions (Shafaat and Ponce 2006). There are, however, some limitations inherent with this technique. For example, lipid inclusions may be mistakenly assigned as endospores because they are phase bright and are about a micron in size. Also, it is nearly impossible to observe endospores with phase contrast microscopy in environmental samples, such as soils, without a rigorous separation.

3.2.4. Germination Observed via DPA release

Two related rapid endospore viability assays (EVA) based on the quantification of dipicolinic acid release during stage I germination have recently been demonstrated by Yung et al. (2007). SpectroEVA is a spectroscopy-based endospore viability assay that measures the germinable endospore concentrations in bulk suspension with Tb-DPA luminescence intensities tabulated against a calibration curve using *B. atrophaeus* spores, which contain an average 10^8 DPA molecules per endospore. The total endospore concentration may also be determined with spectroEVA by forcing DPA release by physical lysis of the total endospore population (e.g., with autoclaving), which enables the percentage of germinable endospores in a sample to be calculated. This was recently demonstrated and validated in comparison to phase contrast microscopy results, and successfully applied to environmental ice core samples from Greenland (Yung et al. 2007), which contained 295 ± 19 germinable endospores/ml and 369 ± 36 total endospores/ml (i.e., the percentage of germinable endospores is $79.9\% \pm 9.3\%$). Results from side-by-side comparison experiments using spectroEVA, phase contrast microscopy, and traditional heterotrophic plate counts on *B. atrophaeus* suspensions showed that of the total endospore population, $49\% \pm 4\%$ germinated as per spectroEVA, $54\% \pm 4\%$ germinated as per phase-bright to phase-dark transition, and $28\% \pm 7\%$ produced visible colonies. SpectroEVA was also applied to rapidly measure loss of endospore germinability as a function of UV exposure to determine the lethal dosage, and correlated this to loss in culturability. These results show that spectroEVA can be used as a rapid alternative over standard culture-based methods for monitoring the efficacy of sterilization processes.

In microEVA experiments, individual germinable endospores are counted in a microscope field of view after germinant addition. As the endospores germinate, $\sim 10^8$ molecules of DPA are released into the immediate area surrounding the spore. DPA combines with Tb^{3+} in the matrix to form the Tb^{3+}-DPA luminescence halos under UV excitation, and are enumerated in a microscope field of view. The germinating endospores manifest as bright spots in the field-of-view that grow in intensity over a period of 30 minutes. DPA release during germination resulted in bright luminescent spots due to Tb-DPA complex formation with the Tb^{3+} that was present in the surrounding agarose medium. While individual endospores—with sizes ranging between $0.8\,\mu m$ to $4\,\mu m$—cannot be spatially resolved with the current stereoscopic microscope, they can nonetheless be easily counted as result of the intense Tb-DPA luminescence emanating from the location of germinating endospores (Figure 19.12a,b).

Long luminescent lifetime ($\tau \sim 1$ ms) of Tb^{3+} enables the use of time gating to effectively remove background fluorescence (i.e., interferent fluorophores with nanosecond lifetimes), thus eliminating potential false positive features and rendering the image background dark. Elimination of this background enables a striking increase in image contrast and detection sensitivity. The characteristic germination time course allows unambiguous assignments of germinating endospores (Figure 19.12c,d). Time-lapse images are considered as 3D signals with the z-axis being the germination time. Quantitative image analysis is carried out on the image stack, and a characteristic temporal intensity change associated with formation of a bright spot is scored as a germinated endospore.

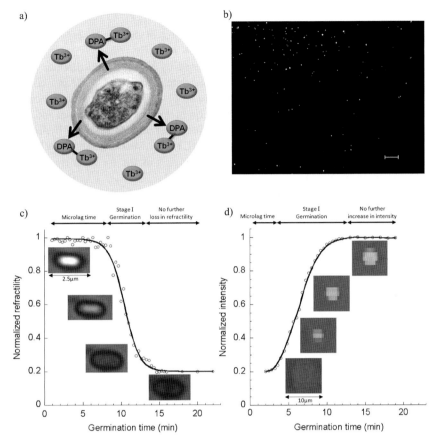

Figure 19.12. a) Principle of the micro EVA. Endospores are inoculated on agarose doped with germinants and terbium chloride. During germination, DPA released from endospores binds with terbium ions in the local vicinity of each individual germinating spore. b) Formation of the terbium dipicolinate complex manifests as bright spots under a time-gated microscope. Bar, 500 μm. Germination timecourses of *B. atrophaeus* spores based on c) phase transition using phase contrast microscopy, and d) release of DPA using micro EVA.

To validate this novel *Bacillus* spore assay, side-by-side comparison experiments were performed with both microEVA and traditional heterotrophic plate counts (Yung, Kempf and Ponce 2006). Both assays were used to assess the endospore population on 2.25" × 2.25" stainless steel coupons inoculated with *B. atrophaeus* spores. The comparison has been performed for test coupons inoculated with 1.3×10^5, 5.0×10^4, and 7.0×10^3 endospores. Using microEVA, 38.2% ± 7.2% of the inoculated endospore population was recovered and counted within 30 minutes. The NASA standard assay recovered and counted only 8.9% ± 5.8% of the inoculated endospores, but required almost three days before results became available. Remarkably, the microEVA requires less than five minutes to enumerate endospores on a given surface. The observed germination time course data are consistent with previously reported microgermination times (Vary and Halvorson 1965; Hashimoto, Frieben and Conti 1969a; Hashimoto, Frieben and Conti 1969b). MircoEVA was also successfully applied to environmental ice cores samples from Greenland. MicroEVA does not require full colony formation to obtain an endospore count, but only stage I germination during which DPA is released from the endospores. Thus, noncultivable endospore-formers can be counted as long as they can germinate under the experimental conditions provided.

3.3. Measuring Endospore Inactivation Using Metabolic Activity Assays

Adenosine 5'triphosphate (ATP) is the primary source of chemical energy and a ubiquitous energy currency in all living organisms. The use of a firefly (*Photuris pyralis*) enzyme to quantify ATP in biological systems was first proposed by McElroy and Strehler in the 1940s (McElroy 1947; McElroy and Strehler 1949). The detection is based on the conversion of chemical energy to light energy during the breakdown of ATP. Firefly luciferase catalyzes the ATP-dependent oxidative decarboxylation of luciferin in the presence of oxygen and magnesium ions into AMP and light. One photon of light is produced per molecule of ATP hydrolyzed when ATP is the limiting component in the reaction (Karl 1980). Measurement of ATP is a direct indication of cellular metabolism and is often reckoned as a metric for viability (Chappelle and Levin 1968; Thore et al. 1975).

First, nonmicrobial ATP is eliminated from the sample using a somatic cell releasing agent and a subsequent incubation in apyrase or ATPase. Bacterial cells are then disrupted using chemicals such as benzalkonium chloride. The ATP released is quantified using the luciferin-luciferase reaction. A differential filtration procedure is also reported to separate somatic from microbial cells (Cross 1992). Several different ATP reagents are available commercially and the protocol has been optimized over the years. For instance, a mutant luciferase resistant to benzalkonium chloride has been isolated to achieve maximum extraction of intracellular ATP from microbes and inactivation of the ATP-eliminating enzymes for removal of extracellular ATP (Hattori et al. 2003). Hattori *et al.* achieved a detection limit of 7.7 CFU/ml using vegetative cells of *B. subtilis*, while Promega Corporation reports a detection limit of 10 CFU/ml of vegetative *B. cereus*.

Challenges are encountered in the detection of endospores using the luciferin/luciferase system. While a vegetative bacterium contains approximately 10^{-17} mole of ATP per cell (Philip 1989), dormant spores of a number of *Bacillus* species have no detectable biosynthetic or metabolic activity and contain low levels of ATP (Church and Halvorson 1957; Setlow and Kornberg 1970; Setlow and Kornberg 1970). Kodake *et al.* reported that an endospore contains about 10^{-21} mole ATP per cell, four orders of magnitude lower than that of a vegetative bacterium (Kodaka et al. 1996). In addition, ATP cannot be sufficiently extracted from endospores, unlike their vegetative counterparts, due to a nonporous and hardy proteinaceous spore coat (Jenkinson, Kay and Mandelstam 1980; Makino et al. 1994; Henriques and Moran 2000). Theoretically speaking, if the current luminometers can detect approximately ten vegetative bacilli cells per milliliter, the limit of detection for endospores will be 10^5 endospores/ml.

There are several approaches for endospore detection using the firefly luciferase assay. One method is the screening for endospores by heat shock at 80°C for 15 minutes allowing only spores to survive, followed by incubation and detection of ATP in the subsequent outgrowth of vegetative cells. In this manner bacterial outgrowth from endospores on test strips was measured after five hours of incubation to validate the sterilization efficiency of autoclaves (Webster et al. 1998). This method is relatively faster than the traditional cultivation method, but can only provide semi-quantitative counts of the original endospore population.

Other approaches take advantage of the production of ATP from endospores during germination. Less than 1 percent of the adenine nucleotide pool in endospores is ATP, but it accounts for 80 percent in vegetative cells. Most of the nucleotides in endospores are stored in the form of 3-phosphoglyceric acid (Setlow and Kornberg 1970). Nevertheless, within the first minute of germination, the large depot of 3-phosphoglyceric acid is catabolized into ATP (Singh, Setlow and Setlow 1977). In addition, coat porosity increases after the onset of germination, which permits easy extraction of intracellular ATP (Santo and Doi 1974). Fujunami *et al.* measured a large increase in light intensity of *B. subtilis* spores after 30 minutes of incubation in nutrient broth supplemented with L-alanine in the presence of various white powders (Fujinami et al. 2004). Rapid accumulation of ATP upon nutrient-induced germination also held true for anaerobic *Clostridium* spores (Hausenbauer, Waites and Setlow 1977). Pressure-induced germination of *B. subtilis* at 100 MPa resulted in a rapid production of ATP, but no ATP was formed during germination at 600 MPa (Wuytack 1998) hydrogen peroxide-treated endospores can germinate, but accumulate very little ATP (Melly 2002). The mechanism of ATP accumulation during germination is still not very clear. Further study may shed light on the subject of endospore viability and the phenomenon of germinable-but-not-culturable.

Direct extraction of ATP from endospores is also another feasible method for detection. Venkateswaran used ATP as a biomarker of viable microorganisms in clean-room facilities. The use of benzalkonium chloride (Kikkoman International, Inc.) completely lyzed vegetative cells and endospores in surface swab samples to release ATP for detection. A low ATP-CFU ratio has been associated with endospores, ranging from 10^{-18} to 10^{-20} moles ATP per CFU (Venkateswaran et al. 2003).

Compared with other endospore detection techniques, ATP bioluminescence measurements offer many advantages, such as high sensitivity, large dynamic range, high specificity, and rapidity. The luciferase assay can easily be automated for high throughput processing. The instruments tend to be inexpensive, portable and can detect viable bacteria in relatively complex media, such as powder and milk. False positives are rare because the enzyme is highly specific for ATP, and ATP is lost rapidly upon cell death. Nevertheless, the assay suffers from drawbacks such as a low level of ATP in endospores and interference from extracellular and somatic ATP. Also the assay is not species-specific and some food samples have been shown to contain inhibitory substances that interfere with luciferase activity (Leach and Webster 1986).

At present, the ATP assay is mainly used to detect and enumerate vegetative cells of pathogenic bacteria for food quality control and hygiene testing (Griffiths 1996; Poulis et al. 1993). It has been used to detect as few as 10^4 CFU/ml of bacteria in milk in five to ten minutes (Griffiths 1993), a bacterial population of 10^5 CFU/ml in fruit juice (Ugarova, Brovko and Kutuzova 1993) and 5×10^4 CFU/g in meat (Basol and Gogus 1996). The application of the ATP assay on endospore detection in foodstuffs is still in its initial phase. Because endospores are the likely candidates for surviving pasteurization, steaming, and vacuum processes, the ATP assay could be expected to play a bigger role in the detection of pathogenic foodborne endospores in the future.

More recently, the ATP luciferase assay has shown promising results in detecting anthrax spores. The lysin *plyG*, isolated from a phage that infects *B. anthracis* was used to specifically target and lyse germinating *B. anthracis* spores causing a pronounced release of intracellular ATP. ATP could be detected within 60 minutes of the addition of germinants from as few as 100 spores using a phage sensitive *B. cereus* as a surrogate for *B. anthracis* (Schuch, Nelson and Fischetti 2002). The ATP assay was also used to detect airborne bacterial spores. A detection limit of 10^5 CFU/ml was reported using aerosolized *B. globigii* spores—a surrogate for *B. anthracis* (Stopa et al. 1999). In a nutshell, the luciferin-luciferase reaction could potentially be used for the detection of anthrax spores and validation of decontamination regimes after an anthrax attack.

4. Conclusions and Future Perspectives

Since *B. anthracis* was recognized as a human pathogen at the beginning of modern microbiology, many other endospore-forming pathogens have been identified and characterized. Detection and viability assessment of these pathogens is still performed using traditional phenotypic approaches. Since September 11, 2001 and the subsequent anthrax attacks, however, much technology development has focused on automating rapid biological detection to enable continuous monitoring for an attack. Rapid viability assessment technology has also been thriving over the past few years, aiming to aid in post-biological attack clean-up operations, and in evaluating sterilization and cleaning regimes to ensure hygiene. We have attempted to capture these technology developments. Clearly these developments do not yet approach the ultimate goal of an inexpensive, handheld, easy-to-use, Tricorder-like device that can specifically detect many pathogens at trace concentrations. This goal will remain the Holy Grail for biodetection for the foreseeable future. Nonetheless, recent advances have taken a large step towards this goal by providing automated technology for online monitoring of specific pathogens under environmental conditions. In the near future, we envision that several of these technologies will be combined into monitoring systems and operated in a triage mode, as in the anthrax smoke detector example that serves as a front-end for more expensive confirmatory technology. Combination of synergistic technologies like this will enable researchers to take advantage of the salient features of each technology, providing a higher degree of confidence in the results, which is reminiscent of the battery of phenotypic assay puzzle pieces that in combined analysis provide a clear picture for assigning the pathogen identity and viability.

References

Ajithkumar B, Ajithkumar VP, Iriye, et al. (2003) Spore-forming *Serratia marcescens* subsp *sakuensis* subsp nov., isolated from a domestic wastewater treatment tank. International Journal of Systematic and Evolutionary Microbiology 53:253–258

Allen SD, Emery CL and Lyerly DM (2003) *Clostridium*. In Murray P, Baron E, Pfaller M, Jorgensen J, and Yolken R (eds) Manual of Clinical Microbiology. ASM Press: Washington D.C., pp 835–856

Andersen GL, Simchock JM and Wilson KH (1996) Identification of a region of genetic variability among *Bacillus anthracis* strains and related species. Journal of Bacteriology 178(2):377–384

Azizbekyan RR (2004) The fungicidal activity of spore-forming bacteria. In Ricca E, Henriques AO, and Cutting SM (eds) Bacterial Spore Formers - Probiotics and Emerging Applications. Horizon Bioscience: Norfolk, pp 229–236

Baeumner AJ, Leonard B, McElwee J et al. (2004) A rapid biosensor for viable *B. anthracis* spores. Analytical and Bioanalytical Chemistry 380(1):15–23

Baez LA, Juneja VK, Thayer DW et al. (1997) Evaluation of PCR and DNA hybridization protocols for detection of viable enterotoxigenic *Clostridium perfringens* in irradiated beef. Journal of Food Safety 17(4):229–238

Balassa G, Milhaud P, Raulet E, Silva MT, Sousa JC (1979) A *Bacillus subtilis* mutant requiring dipicolinic acid for the development of heat-resistant spores. Journal of General Microbiology 110(2):365–379

Balzani V (1990) Supramolecular photochemistry. Pure and Applied Chemistry 62(6):1099–1102

Balzani V, Decola L, Prodi L et al. (1990) Photochemistry of supramolecular species. Pure and Applied Chemistry 62(8):1457–1466

Barany F (1991) The ligase chain reaction in a PCR world. PCR Methods and Applications 1(1):5–16

Baron PA and Willeke K (2001) Aerosol Measurement: Principles, Techniques, and Applications. Wiley, John & Sons, Inc.: New York

Basol MS and Gogus U (1996) Methods of antibiotic applications related to microbiological quality of lamb by PCA and bioluminescence. Journal of Food Science 61:348–349

Beeby A, Botchway SW, Clarkson IM, Faulkner S, Parker AW, Parker D, Williams JA (2000) Luminescence imaging microscopy and lifetime mapping using kinetically stable lanthanide (III) complexes. Journal of Photochemistry and Photobiology B 57(2-3):83–89

Belgrader P, Benett W, Hadley D et al. (1998a) Rapid pathogen detection using a microchip PCR array instrument. Clinical Chemistry 44(10):2191–2194

Belgrader P, Smith JK, Weedn VW et al. (1998b) Rapid PCR for identity testing using a battery-powered miniature thermal cycler. Journal of Forensic Sciences 43(2):315–319

Belgrader P, Benett W, Hadley D et al. (1999a) Infectious disease - PCR detection of bacteria in seven minutes. Science 284(5413):449–450

Belgrader P, Hansford D, Kovacs GTA et al. (1999b) A minisonicator to rapidly disrupt bacterial spores for DNA analysis. Analytical Chemistry 71(19):4232–4236

Belgrader P, Okuzumi M, Pourahmadi F et al. (2000) A microfluidic cartridge to prepare spores for PCR analysis. Biosensors & Bioelectronics 14(10-11):849–852

Belgrader P, Young S, Yuan B et al. (2001) A battery-powered notebook thermal cycler for rapid multiplex real time PCR analysis. Analytical Chemistry 73(2):286–289

Bell CA, Uhl JR, Hadfield TL et al. (2002) Detection of *Bacillus anthracis* DNA by LightCycler PCR. Journal of Clinical Microbiology 40(8):2897–2902

Birmingham JG (2006) Plasma lysis for identification of bacterial spores using ambient-pressure nonthermal discharges. IEEE Transactions on Plasma Science 34(4):1270–1274

Blumenthal T (1979) Qbeta RNA replicase and protein synthesis elongation factors EF-Tu and EF-Ts. Methods in Enzymology 60:628–638

Board SS and Council NR (2000) Preventing the Forward Contamination of Europa. National Academy Press: Washington, D.C.

Bode E, Hurtle W and Norwood D (2004) Real-time PCR assay for a unique chromosomal sequence of *Bacillus anthracis*. Journal of Clinical Microbiology 42(12):5825–5831

Borthwick KAJ, Love TE, McDonnell MB et al. (2005) Improvement of immunodetection of bacterial spore antigen by ultrasonic cavitation. Analytical Chemistry 77(22):7242–7245

Bruno JG and Yu H (1996) Immunomagnetic-electrochemiluminescent detection of *Bacillus anthracis* spores in soil matrices. Applied and Environmental Microbiology 62(9):3474–3476

Cable ML, Kirby JP, Sorasaenee K et al. (2007) Bacterial spore detection by [Tb^{3+}(macrocycle)(dipicolinate)] luminescence. Journal of the American Chemical Society 129:1474–1475

Caipo M, Duffy S, Zhao L et al. (2002) *Bacillus megaterium* spore germination is influenced by inoculum size. Journal of Applied Microbiology 92:879–884

Cano RJ and Borucki MK (1995) Revival and identification of bacterial spores in 25-million-year-old to 40-million-year-old Dominican amber. Science 268(5213):1060–1064

Chappelle EW and Levin GV (1968) Use of the firefly bioluminescence reaction for rapid detection and counting of bacteria. Biochemical Medicine 2:41–52

Chattopadhyay A, Bhatnagar NB and Bhatnagar R (2004) Bacterial insecticidal toxins. Critical Reviews in Microbiology 30(1):33–54

Chen S (2007) One hour molecular diagnostics at the bedside. Next Generation Pharmaceutical issue 7, http://www.ngpharma.com

Cherif A, Borin S, Rizzi A et al. (2003a) *Bacillus anthracis* diverges from related clades of the *Bacillus cereus* group in 16S-23S ribosomal DNA intergenic transcribed spacers containing tRNA genes. Applied and Environmental Microbiology 69(1):33–40

Cherif A, Brusetti L, Borin S et al. (2003b) Genetic relationship in the '*Bacillus cereus* group' by rep-PCR fingerprinting and sequencing of a *Bacillus anthracis*-specific rep-PCR fragment. Journal of Applied Microbiology 94(6):1108–1119

Cohn F (1876) Untersuchungen uber Bacterien. IV. Beitrage zur Biologie der Bacillen. Beitr. Biol. Pflanz. 2:249–276

Colwell R and Grimes D (2000) Non-Culturable Microorganisms in the Environment. ASM Press: Washington, D.C.

Compton J (1991) Nucleic-acid sequence-based amplification. Nature 350(6313):91–92

Connally R, Veal D and Piper J (2002) High resolution detection of fluorescently labeled microorganisms in environmental samples using time-resolved fluorescence microscopy. FEMS Microbiology Ecology 41(3):239–245

Connon SA and Giovannoni SJ (2002) High-throughput methods for culturing microorganisms in very-low-nutrient media yield diverse new marine isolates. Applied and Environmental Microbiology 68(8):3878–3885

Cox CS and Wathes CM (eds) (1995) Bioaerosols Handbook: Handbook of Samplers and Sampling. CRC Press: Boca Raton, FL

Cross J (1992) Harnessing the firefly. Food Manufacture 67:25

Cutting SM (2004) Spores as oral vaccines. In Ricca E, Henriques AO, Cutting SM (eds) Bacterial Spore Formers - Probiotics and Emerging Applications. Horizon Bioscience: Norfolk, pp 201–206

Daffonchio D, Borin S, Frova G et al. (1999) A randomly amplified polymorphic DNA marker specific for the *Bacillus cereus* group is diagnostic for *Bacillus anthracis*. Applied and Environmental Microbiology 65(3):1298–1303

Dart RK (1996) Microbiology for the Analytical Chemist. The Royal Society of Chemistry: Cambridge, UK

De BK, Bragg SL, Sanden GN et al. (2002) Two-component direct fluorescent-antibody assay for rapid identification *Bacillus anthracis*. Emerging Infectious Diseases 8(10):1060–1065

Deere D, Porter J, Pickup R et al. (1996) Direct analysis of starved *Aeromonas salmonicida*. Journal of Fish Diseases 19(6):459–467

Desser H and Broda E (1965) Radiochemical determination of the endogenous and exogenous respiration of bacterial spores. Nature 206(4990):1270–1271

Dierick K (2005) Fatal family outbreak of *Bacillus cereus*-associated food poisoning. Journal of Clinical Microbiology 43(8):4277

Dirckx JH (1981) Virgil on anthrax. American Journal of Dermatopathology 3(2):191–195

Dixon TC, Meselson M, Guillemin J et al. (1999) Anthrax. New England Journal of Medicine 341(11):815–826

Dobkin C, Mills DR, Kramer FR et al. (1979) RNA replication: required intermediates and the dissociation of template, product, and Q beta replicase. Biochemistry 18(10):2038–2044

Dombrowski H (1963) Bacteria from paleozoic salt deposits. Annals of the New York Academy of Sciences 108(2):453–460

Drago L, Lombardi A, De Veechi E et al. (2002) Real-time PCR assay for rapid detection of *Bacillus anthracis* spores in clinical samples. Journal of Clinical Microbiology 40(11):4399–4399

Drews G (2000) The roots of microbiology and the influence of Ferdinand Cohn on microbiology of the 19th century. FEMS Microbiology Reviews 24(3):225–249

Dworkin M, Falkow S, Rosenberg E, Schleifer K-H, Stackebrandt E, eds (2006) The Prokaryotes, A Handbook on the Biology of Bacteria, Vol. 4. Springer.

Edwards KA, Harriet AC and Antje JB (2006) *Bacillus anthracis*: toxicology, epidemiology and current rapid-detection methods. Analytical and Bioanalytical Chemistry 384(1):73–84

Ellerbrok H, Nattermann H, Ozel M et al. (2002) Rapid and sensitive identification of pathogenic and apathogenic *Bacillus anthracis* by real-time PCR. FEMS Microbiology Letters 214(1):51–59

Elzi MV, Mallard K, Droz S et al. (2005) Polyphasic approach for identifying *Bacillus* spp. Journal of Clinical Microbiology 43(2):1010–1010

Espy MJ, Uhl JR, Sloan LM et al. (2006) Real-time PCR in clinical microbiology: Applications for a routine laboratory testing. Clinical Microbiology Reviews 19(1):165–256

Ezzell JW, Abshire TG, Little SF et al. (1990) Identification of *Bacillus-anthracis* by using monoclonal-antibody to cell-wall galactose-N-acetylglucosamine polysaccharide. Journal of Clinical Microbiology 28(2):223–231

Farrell S, Halsall HB and Heineman WR (2005) Immunoassay for *B-globigii* spores as a model for detecting *B-anthracis* spores in finished water. Analyst 130(4):489–497

Fasanella A, Losito S, Adone R et al. (2003) PCR assay to detect *Bacillus anthracis* spores in heat-treated specimens. Journal of Clinical Microbiology 41(2):896–899

Felske ADM (2004) Ecology of *Bacillus* species in soil. In Ricca E, Henriques AO, Cutting SM (eds) Bacterial Spore Formers - Probiotics and Emerging Applications. Horizon Bioscience: Norfolk, pp 35–44

Finlay WJJ (2000) *Bacillus cereus* produces most emetic toxin at lower temperatures. Letters in Applied Microbiology 31(5):385

Floriano PN, Christodoulides N, Romanovicz D et al. (2005) Membrane-based on-line optical analysis system for rapid detection of bacteria and spores. Biosensors & Bioelectronics 20(10):2079–2088

Foster SJ and Johnstone K (1990) Pulling the trigger: the mechanism of bacterial spore germination. Molecular Microbiology 4(1):137–141

Fritze D (2004a) Taxonomy and systematics of the aerobic endospore forming bacteria: *Bacillus* and related genera. In Ricca E, Henriques AO, Cutting SM (eds) Bacterial Spore Formers - Probiotics and Emerging Applications. Horizon Bioscience: Norfolk, pp 17–34

Fritze D (2004b) Taxonomy of the genus *Bacillus* and related genera: The aerobic endospore-forming bacteria. Phytopathology 94(11):1245–1248

Fujinami Y, Kataoka M, Matsushita K et al. (2004) Sensitive detection of bacteria and spores using a portable bioluminescence ATP measurement assay system distinguishing from white powder materials. Journal of Health Science 50(2):126–132

Gattomenking DL, Yu H, Bruno JG et al. (1995) Sensitive detection of biotoxoids and bacterial-spores using an immunomagnetic electrochemiluminescence sensor. Biosensors & Bioelectronics 10(6-7):501–507

Gest H, Mandelstam J (1987) Longevity of microorganisms in natural environments. Microbiological Science 4(3):69–71

Gohar M, Gilois N, Graveline R et al. (2005) A comparative study of *Bacillus cereus, Bacillus thuringiensis* and *Bacillus anthracis* extracellular proteomes. Proteomics 5(14):3696–3711

Gould GW and Hurst A (1969) The Bacterial Spore. Academic Press, New York

Gould GW (2005) History of science - spores: Lewis B Perry memorial lecture 2005. Journal of Applied Microbiology 101(3):507–513

Grenthe I (1961) Stability relationships among the rare earth dipicolinates. Journal of the American Chemical Society 83:360–364

Griffiths MW (1993) Applications of bioluminescence in the dairy industry. Journal of Dairy Science 76(10):3118–3125

Griffiths MW (1996) The role of ATP bioluminescence in the food industry: New light on old problems. Food Technology 50(6):64–66

Griffiths WD and Decosemo GAL (1994) The assessment of bioaerosols - a critical-review. Journal of Aerosol Science 25(8):1425–1458

Griffiths WD, Stewart IW, Futter SJ et al. (1997) The development of sampling methods for the assessment of indoor bioaerosols. Journal of Aerosol Science 28(3):437–457

Han CS, Xie G, Challacombe JF et al. (2006) Pathogenomic sequence analysis of *Bacillus cereus and Bacillus thuringiensis* isolates closely related to *Bacillus anthracis*. Journal of Bacteriology 188(9):3382–3390

Hansen BM, Leser TD and Hendriksen NB (2001) Polymerase chain reaction assay for the detection of *Bacillus cereus* group cells. FEMS Microbiology Letters 202(2):209–213

Hanson RS, Halvorson HO, Curry MV et al. (1972) Mutants of *Bacillus-cereus* strain T that produce thermoresistant spores lacking dipicolinate and have low levels of calcium. Canadian Journal of Microbiology 18(7):1139–43

Hartley HA and Baeumner AJ (2003) Biosensor for the specific detection of a single viable *B-anthracis* spore. Analytical and Bioanalytical Chemistry 376(3):319–327

Hashimoto T, Frieben WR and Conti SF (1969a) Germination of single bacterial spores. Journal of Bacteriology 98:1011–1020

Hashimoto T, Frieben WR and Conti SF (1969b) Microgermination of *Bacillus cereus* spores. Journal of Bacteriology 100(3):1385–1392

Hattori NN, Sakakibara TT, Kajiyama NN et al. (2003) Enhanced microbial biomass assay using mutant luciferase resistant to benzalkonium chloride. Analytical Biochemistry 319(2):287–95

Hausenbauer JM, Waites WM and Setlow P (1977) Biochemical properties of *Clostridium bifermentans* spores. Journal of Bacteriology. 129(2):1148–1150

Helgason E, Okstad OA, Caugant DA et al. (2000) *Bacillus anthracis, Bacillus cereus*, and *Bacillus thuringiensis* - One species on the basis of genetic evidence. Applied and Environmental Microbiology 66(6):2627–2630

Hellyer TJ, DesJardin LE, Hehman GL et al. (1999a) Quantitative analysis of mRNA as a marker for viability of *Mycobacterium tuberculosis*. Journal of Clinical Microbiology 37(2):290–295

Hellyer TJ, DesJardin LE, Teixeira L et al. (1999b) Detection of viable *Mycobacterium tuberculosis* by reverse transcriptase-strand displacement amplification of mRNA. Journal of Clinical Microbiology 37(3):518–523

Henriques AO and Moran CP (2000) Structure and assembly of the bacterial endospore coat. Methods 20:95–110

Henriques AO, Costa TV, Martins LO et al. (2004) The functional architecture and assembly of the spore coat. In Ricca E, Henriques AO, Cutting SM (eds) Bacterial Spore Formers - Probiotics and Emerging Applications. Horizon Bioscience: Norfolk, pp 65–86

Herman L (1997) Detection of viable and dead Listeria monocytogenes by PCR. Food Microbiology 14:103–110

Higgins JA, Ibrahim MS, Knauert FK et al. (1999) Sensitive and rapid identification of biological threat agents. Food and Agricultural Security 894:130–148

Higgins JA, Cooper M, Schroeder-Tucker L et al. (2003a) A field investigation of *Bacillus anthracis* contamination of U.S. Department of Agriculture and other Washington, D.C., buildings during the anthrax attack of October 2001. Applied and Environmental Microbiology 69(1):593–599

Higgins JA, Nasarabadi S, Karns JS et al. (2003b) A handheld real time thermal cycler for bacterial pathogen detection. Biosensors & Bioelectronics 18(9):1115–1123

Hills GM (1949a) Chemical factors in the germination of spore-bearing aerobes - the effect of yeast extract on the germination of *Bacillus anthracis* and its replacement by adenosine. Biochemical Journal 45(3):353–362

Hills GM (1949b) Chemical factors in the germination of spore-bearing aerobes - the effects of amino-acids on the germination of *Bacillus anthracis*, with some observations on the relation of optical form to biological activity. Biochemical Journal 45(3):363–370

Hills GM (1950) Chemical factors in the germination of spore-bearing aerobes - observations on the influence of species, strain and conditions of growth. Journal of General Microbiology 4(1):38–47

Hindle AA, Hall EAH (1999) Dipicolinic acid (DPA) assay revisited and appraised for spore detection. Analyst 124(11):1599–1604

Hindson BJ, Brown SB, Marshall GD et al. (2004) Development of an automated sample preparation module for environmental monitoring of biowarfare agents. Analytical Chemistry 76(13):3492–3497

Hindson BJ, Makarewicz AJ, Setlur US et al. (2005a). APDS: the autonomous pathogen detection system. Biosensors & Bioelectronics 20(10):1925–1931

Hindson BJ, McBride MT, Makarewicz AJ et al. (2005b) Autonomous detection of aerosolized biological agents by multiplexed immunoassay with polymerase chain reaction confirmation. Analytical Chemistry 77(1):284–289

Hitchins AD, Kahn AJ and Slepecky RA (1968) Interference contrast and phase contrast microscopy of sporulation and germination of *Bacillus megaterium*. Journal of Bacteriology 96(5):1811–1817

Hoch JA and Losick R (1997) Genome sequencing - Panspermia, spores and the *Bacillus subtilis* genome. Nature 390(6657):237–238

Hoffmaster AR, Fitzgerald CC, Ribot E et al. (2002a) Molecular subtyping of *Bacillus anthracis* and the 2001 bioterrorism-associated anthrax outbreak, United States. Emerging Infectious Diseases 8(10):1111–1116

Hoffmaster AR, Meyer RF, Bowen MP et al. (2002b) Evaluation and validation of a real time polymerase chain reaction assay for rapid identification of *Bacillus anthracis*. Emerging Infectious Diseases 8(10):1178–1182

Hoffmaster AR, Ravel J, Rasko DA et al. (2004) Identification of anthrax toxin genes in a *Bacillus cereus* associated with an illness resembling inhalation anthrax. Proceedings of the National Academy of Sciences 101(22):8449–8454

Holland PM, Abramson RD, Watson R et al. (1991) Detection of specific polymerase chain-reaction product by utilizing the 5'- 3' exonuclease activity of *Thermus-aquaticus* DNA-polymerase. Proceedings of the National Academy of Sciences 88(16):7276–7280

Holt JG and Bergey DH (1994) Bergey's Manual of Determinative Bacteriology. Lippincott Williams & Wilkins: Baltimore

Horneck G, Bucker H and Reitz G (1994) Long-term survival of bacterial spores in space. Advances in Space Research: The Official Journal of the Committee on Space Research (COSPAR) 14(10):41–45

Horrocks Jr. WD and Sudnick D (1981) Lanthanide ion luminescence probes of the structure of biological macro-molecules. Accounts of Chemical Research 14:384–392

Horrocks Jr. WD (1984) Lanthanide ion luminescence in coordination chemistry and biochemistry. In Lippard SJ (ed) Progress in Inorganic Chemistry. John Wiley & Sons, Inc.: New York, pp 1–104

Hurst A and Gould GW (eds) (1983) The Bacterial Spore - Volume 2. Academic Press, Inc.: New York

Ibrahim MS, Lofts RS, Jahrling PB et al. (1998) Real-time microchip PCR for detecting single-base differences in viral and human DNA. Analytical Chemistry 70(9):2013–2017

Ivanova N, Sorokin A, Anderson I et al. (2003) Genome sequence of *Bacillus cereus* and comparative analysis with *Bacillus anthracis*. Nature 423(6935):87–91

Ivanovics G and Foldes J (1958) An immunospecific substance of *Bacillus-cereus* similar to polysaccharide obtained from *Bacillus-anthracis*. Naturwissenschaften 45(1):15–15

Jackson PJ, Walthers EA, Kalif AS et al. (1997) Characterization of the variable-number tandem repeats in *vrrA* from different *Bacillus anthracis* isolates. Applied and Environmental Microbiology 63(4):1400–1405

Jackson PJ, Hugh-Jones ME, Adair DM et al. (1998) PCR analysis of tissue samples from the 1979 Sverdlovsk anthrax victims: The presence of multiple *Bacillus anthracis* strains in different victims. Proceedings of the National Academy of Sciences 95(3):1224–1229

Jain R, Rivera MC, Moore JE et al. (2003) Non-clonal evolution of microbes. Biological Journal of the Linnean Society 79(1):27–32

Janssen FW, Lund AJ and Anderson LE (1958) Colorimetric assay for dipicolinic acid in bacterial spores. Science 127(3288):26–27

Janssen PH, Yates PS, Grinton BE et al. (2002) Improved culturability of soil bacteria and isolation in pure culture of novel members of the divisions *Acidobacteria, Actinobacteria, Proteobacteria*, and *Verrucomicrobia*. Applied and Environmental Microbiology 68(5):2391–2396

Jenkinson HF, Kay D and Mandelstam J (1980) Temporal dissociation of late events in *Bacillus subtilis* sporulation from expression of genes that determine them. Journal of Bacteriology 141(2):793–805

Jernigan DB, Raghunathan PL, Bell BP et al. (2002) Investigation of bioterrorism-related anthrax, United States, 2001: epidemiologic findings. Emerging Infectious Diseases 8(10):1019–1028

Johnston MD, Lawson S and Otter JA (2005) Evaluation of hydrogen peroxide vapour as a method for the decon-tamination of surfaces contaminated with *Clostridium botulinum* spores. Journal of Microbiological Methods 60(3):403–411

Jones G and Vullev VI (2002a) Medium effects on the photophysical properties of terbium(III) complexes with pyridine-2,6-dicarboxylate. Photochemical & Photobiological Sciences 1(12):925–933

Jones G and Vullev VI (2002b) Medium effects on the stability of terbium(III) complexes with pyridine-2,6-dicarboxylate. Journal of Physical Chemistry A 106(35):8213–8222

Kaeberlein T, Lewis K and Epstein SS (2002) Isolating "uncultivable" microorganisms in pure culture in a simulated natural environment. Science 296(5570):1127–1129

Kanemitsu K, Imasaka T, Ishikawa S et al. (2005) A comparative study of ethylene oxide gas, hydrogen peroxide gas plasma, and low-temperature steam formaldehyde sterilization. Infection Control and Hospital Epidemiology 26(5):486–489

Karl DDM (1980) Cellular nucleotide measurements and applications in microbial ecology. Microbiological Reviews 44(4):739–96

Kaufmann AF, Meltzer MI and Schmid GP (1997) The economic impact of a bioterrorist attack: Are prevention and postattack intervention programs justifiable? Emerging Infectious Diseases 3(2):83–94

Keeton WT (1980) Biological Science. W. W. Norton & Co.: New York

Kennedy MJ, Reader SL and Swierczynski LM (1994) Preservation records of microorganisms - evidence of the tenacity of life. Microbiology-UK 140:2513–2529

Keynan A and Sandler N (1983) Spore research in historical perspective. In Hurst A and Gould GW (eds) The Bacterial Spore, Volume 2. Academic Press, Inc.: New York, pp 8

Kim K, Seo J, Wheeler K et al. (2005) Rapid genotypic detection of *Bacillus anthracis* and the *Bacillus cereus* group by multiplex real-time PCR melting curve analysis. FEMS Immunology and Medical Microbiology 43(2):301–310

King D, Luna V, Cannons A et al. (2003) Performance assessment of three commercial assays for direct detection of *Bacillus anthracis* spores. Journal of Clinical Microbiology 41(7):3454–3455

Klee SR, Nattermann H, Becker S et al. (2006) Evaluation of different methods to discriminate *Bacillus anthracis* from other bacteria of the *Bacillus cereus* group. Journal of Applied Microbiology 100(4):673–681

Knight J (2002) US postal service puts anthrax detectors to the test. Nature 417(6889):579–579

Ko KS, Kim JM, Kim JW et al. (2003) Identification of *Bacillus anthracis* by *rpoB* sequence analysis and multiplex PCR. Journal of Clinical Microbiology 41(7):2908–2914

Koch R (1876) Untersuchungen uber Bakterien V. Die Atiologie der Milzbrandkrankheit, begrundet auf die Entwick-lungsgeschichte des *Bacillus anthracis*. Beitr. Biol. Pflanz. 2:277–310

Koch R (1877) Untersuchungen uber Bakterien VI. Verfahren zur Untersuchung, zum Conservieren und Photogra-phieren. Beitr. Biol. Pflanz. 2:399–434

Kodaka H, Fukuda K, Mizuochi S et al. (1996) Adenosine triphosphate content of microorganisms related with food spoilage. Japanese Journal of Food Microbiology 13:29–34

Koonin EV, Makarova KS and Aravind L (2001) Horizontal gene transfer in prokaryotes: Quantification and classification. Annual Review of Microbiology 55:709–742

Lakowicz JR (1983) Principles of Fluorescence Spectroscopy. Plenum Press: New York

Lampel KA, Dyer D, Kornegay L et al. (2004) Detection of *Bacillus* spores using PCR and FTA filters. Journal of Food Protection 67(5):1036–1038

Lan R and Reeves PR (2001) When does a clone deserve a name? A perspective on bacterial species based on population genetics. Trends in Microbiology 9(9):419–424

Lawrence D, Heitefuss S and Seifert HSH (1991) Differentiation of *Bacillus-anthracis* from *Bacillus-cereus* by gas-chromatographic whole-cell fatty-acid analysis. Journal of Clinical Microbiology 29(7):1508–1512

Leach FR and Webster JJ (1986) Commercially available firefly luciferase reagents. Methods in Enzymology 133:51–70

Lester ED and Ponce A (2002) An anthrax "smoke" detector: Online monitoring of aerosolized bacterial spores. IEEE Engineering in Medicine and Biology Magazine 21(5):38–42

Lester ED, Bearman G and Ponce A (2004) A second-generation anthrax "smoke detector". IEEE Engineering in Medicine and Biology Magazine 23(1):130–135

Leuschner RGK and Lillford PJ (1999) Effects of temperature and heat activation on germination of individual spores of *Bacillus subtilis*. Letters in Applied Microbiology 29:228–232

Levi K, Higham JL, Coates D et al. (2003) Molecular detection of anthrax spores on animal fibres. Letters in Applied Microbiology 36(6):418–422

Levine SM, Tang Y-W and Pei Z (2005) Recent advances in the rapid detection of *Bacillus anthracis*. Reviews in Medical Microbiology 16(4):125–133

Levinson HS and Hyatt MT (1966) Sequence of events during *Bacillus megaterium* spore germination. Journal of Bacteriology 91(5):1811–1818

Lewis JC, Snell NS and Burr HK (1960) Water permeability of bacterial spores and the concept of a contratile cortex. Science 132(3426):544–545

Li QY, Dasgupta PK, Temkin H et al. (2004) Mid-ultraviolet light-emitting diode detects dipicolinic acid. Applied Spectroscopy 58(11):1360–1363

Lindahl T (1993) Instability and decay of the primary structure of DNA. Nature 362:709–715

Logan NA (2004) Safety of aerobic endospore-forming bacteria. In Ricca E, Henriques AO, Cutting SM (eds) Bacterial Spore Formers - Probiotics and Emerging Applications. Horizon Bioscience: Norfolk, pp 93–106

Logan NA, Carman JA, Melling J et al. (1985) Identification of *Bacillus anthracis* by Api Tests. Journal of Medical Microbiology 20(1):75–85

Logan NA and Turnbull PCB (2003) *Bacillus* and other aerobic endospore-forming bacteria. In Murray P et al. (ed) Manual of Clinical Microbiology. ASM Press: Washington, D.C., pp 445–460

Lukasova J, Vyhnalkova J and Pacova Z (2001) *Bacillus* species in raw milk and in the farm environment. Milchwissenschaft-Milk Science International 56(11):609–611

Mackay IM, Arden KE and Nitsche A (2002) Real-time PCR in virology. Nucleic Acids Research 30(6):1292–1305

Mackay IM (2004) Real-time PCR in the microbiology laboratory. Clinical Microbiology and Infection 10(3):190–212

Mahler H, Pasi A, Kramer JM et al. (1997) Fulminant liver failure in association with the emetic toxin of *Bacillus cereus*. The New England Journal of Medicine 336(16):1142–1148

Makino S, Iinumaokada Y, Maruyama T et al. (1993) Direct detection of *Bacillus-anthracis* DNA in animals by polymerase chain-reaction. Journal of Clinical Microbiology 31(3):547–551

Makino S, Ito N, Inoue T et al. (1994) A spore-lytic enzyme released from *Bacillus cereus* spores during germination. Microbiology 140(6):1403–1410

Makino S and Cheun H (2003) Application of the real-time PCR for the detection of airborne microbial pathogens in reference to the anthrax spores. Journal of Microbiological Methods 53(2):141–147

Makino SI, Cheun HI, Watarai M et al. (2001) Detection of anthrax spores from the air by real-time PCR. Letters in Applied Microbiology 33(3):237–240

Malecki J, Wiersma S, Cahill K et al. (2001) Update: Investigation of bioterrorism-related anthrax and interim guidelines for exposure management and antimicrobial therapy, October 2001. The Journal of the American Medical Association 286(18):2226–2232

Mamane-Gravetz H and Linden KG (2004) UV disinfection of indigenous aerobic spores: implications for UV reactor validation in unfiltered waters. Water Research 38(12):2898–2906

Marston CK, Gee JE, Popovic T et al. (2006) Molecular approaches to identify and differentiate *Bacillus anthracis* from phenotypically similar *Bacillus* species isolates. BMC Microbiology 6:22

Masters CI, Shallcross JA and Mackey BM (1994) Effect of stress treatments on the detection of *Listeria* monocytogenes and enterotoxigenic *Escherichia coli* by the polymerase chain reaction. Journal of Applied Bacteriology 77(1):73–79

McBride MT, Masquelier D, Hindson BJ et al. (2003) Autonomous detection of aerosolized *Bacillus anthracis* and *Yersinia pestis*. Analytical Chemistry 75(20):5293–5299

McBride R (2007) Feds fund IQuum's bioterror test technology into Phase 3. Mass High Tech: The Journal of New England Technology, http:/masshightech.bizjournals.com

McElroy WD (1947) The energy source for bioluminescence in an isolated system. Proceedings of the National Academy of Sciences 33(11):342–345

McElroy WD and Strehler BL (1949) Factors influencing the response of the bioluminescent reaction to adenosine triphosphate. Archives of Biochemistry 22:420–433

McFarland LV, Mulligan ME, Kwok RYY et al. (1989) Nosocomial acquisition of *Clostridium-difficile* infection. New England Journal of Medicine 320(4):204–210

McFarland LV (1995) Epidemiology of infectious and iatrogenic nosocomial diarrhea in a cohort of general medicine patients. American Journal of Infection Control 23(5):295–305

McGenity TJ, Gemmell RT, Grant WD et al. (2000) Origins of halophilic microorganisms in ancient salt deposits. Environmental Microbiology 2(3):243–250

McKillip JL, Jaykus LA and Drake MM (1999) Nucleic acid persistence in heat-killed *Escherichia coli* O157:H7 from contaminated skim milk. Journal of Food Protection 62(8):839–844

Melly E, Cowan AE, Setlow P (2002) Studies on the mechanism of killing of *Bacillus subtilis* spores by hydrogen peroxide. Journal of Applied Microbiology 93(2):316–325

Meselson M, Guillemin J, Hugh-Jones M et al. (1994) The Sverdlovsk anthrax outbreak of 1979. Science 266(5188):1202–1208

Morgan CG and Mitchell AC (1996) Fluorescence lifetime imaging: An emerging technique in fluorescence microscopy. Chromosome Research 4(4):261–263

Murty GGK and Halvorson HO (1957) Effect of duration of heating, L-alanine and spore concentration on the oxidation of glucose by spores of *Bacillus cereus* var. *terminalis*. Journal of Bacteriology 73(2):235

Nicholson W and Setlow P (1990) Sporulation, germination, and outgrowth. In Cutting S (ed) Molecular Biology Methods for Bacillus. John Wiley and Sons: Sussex, England, pp 391–450

Nicholson WL, Munakata N, Horneck G et al. (2000) Resistance of *Bacillus* endospores to extreme terrestrial and extraterrestrial environments. Microbiology and Molecular Biology Reviews 64(3):548–572

Nicholson WL and Galeano B (2003) UV resistance of *Bacillus anthracis* spores revisited: Validation of *Bacillus subtilis* spores as UV surrogates for spores of *B. anthracis* sterne. Applied and Environmental Microbiology 69(2):1327–1330

Nicholson WL (2004) Ubiquity, longevity, and ecological roles of *Bacillus* spores. In Ricca E, Henriques AO, Cutting SM (eds) Bacterial Spore Formers - Probiotics and Emerging Applications. Horizon Bioscience: Norfolk, pp 1–16

Norris JR (1962) Bacterial spore antigens - a review. Journal of General Microbiology 28(3):393–408

Northrup MA, Benett B, Hadley D et al. (1998) A miniature analytical instrument for nucleic acids based on micromachined silicon reaction chambers. Analytical Chemistry 70(5):918–922

Nubel U, Schmidt PM, Reiss E et al. (2004) Oligonucleotide microarray for identification of *Bacillus anthracis* based on intergenic transcribed spacers in ribosomal DNA. FEMS Microbiology Letters 240(2):215–223

Office of Space Science, NASA (1999) Planetary Protection Provisions for Robotic Extraterrestrial Missions: Washington, D.C.

Office of Technology Assessment, U. C. (1993) Proliferation of Weapons of Mass Destruction. U. C. Office of Technology Assessment: Washington D.C., pp 53–55

Oggioni MR, Meacci F, Carattoli A et al. (2002) Protocol for real-time PCR identification of anthrax spores from nasal swabs after broth enrichment. Journal of Clinical Microbiology 40(11):3956–3963

Oggioni MR, Ciabattini A, Cassone M et al. (2004) Pathogenic bacilli: *Bacillus anthracis* and close relatives. In Ricca E, Henriques AO, Cutting SM (eds) Bacterial Spore Formers - Probiotics and Emerging Applications. Horizon Bioscience: Norfolk, pp 45–52

Okinaka R, Pearson T and Keim P (2006) Anthrax, but not *Bacillus anthracis*? PLOS Pathogens 2(11):1025–1027

Okinaka RT, Cloud K, Hampton O et al. (1999) Sequence and organization of pXO1, the large *Bacillus anthracis* plasmid harboring the anthrax toxin genes. Journal of Bacteriology 181(20):6509–6515

Onyenwoke RU, Brill JA, Farahi K et al. (2004) Sporulation genes in members of the low G+C Gram-type-positive phylogenetic branch (Firmicutes). Archives of Microbiology 182(2-3):182–192

Pannucci J, Okinaka RT, Sabin R et al. (2002a) *Bacillus anthracis* pXO1 plasmid sequence conservation among closely related bacterial species. Journal of Bacteriology 184(1):134–141

Pannucci J, Okinaka RT, Williams E et al. (2002b) DNA sequence conservation between the *Bacillus anthracis* pXO2 plasmid and genomic sequence from closely related bacteria. BMC Genomics 3:34

Papaparaskevas J, Houhoula DP, Papadimitriou M et al. (2004) Ruling out *Bacillus anthracis*. Emerging Infectious Diseases 10(4):732–735

Park TJ, Park JP, Seo GM et al. (2006) Rapid and accurate detection of *Bacillus anthracis* spores using peptide-quantum dot conjugates. Journal of Microbiology and Biotechnology 16(11):1713–1719

Parsons P (1996) Dusting off panspermia. Nature 383(6597):221–222

Pastuszka JS, Paw UKT, Lis DO et al. (2000) Bacterial and fungal aerosol in indoor environment in Upper Silesia, Poland. Atmospheric Environment 34(22):3833–3842

Patra G, Sylvestre P, Ramisse V et al. (1996) Isolation of a specific chromosomic DNA sequence of *Bacillus anthracis* and its possible use in diagnosis. FEMS Immunology and Medical Microbiology 15(4):223–231

Patra G, Vaissaire J, Weber-Levy M et al. (1998) Molecular characterization of *Bacillus* strains involved in outbreaks of anthrax in France in 1997. Journal of Clinical Microbiology 36(11):3412–3414

Patra G, Williams LE, Qi Y et al. (2002) Rapid genotyping of *Bacillus anthracis* strains by real-time polymerase chain reaction. In Domestic Animal/Wildlife Interface: Issue for Disease Control, Conservation, Sustainable Food Production, and Emerging Diseases. New York Academy of Sciences: New York, pp 106–111

Pellegrino PM, Fell NF, Rosen DL et al. (1998) Bacterial endospore detection using terbium dipicolinate photoluminescence in the presence of chemical and biological materials. Analytical Chemistry 70(9):1755–1760

Pennington OJ, Van Mellaert L, Theys J et al. (2004) Recombinant clostridial spores in tumor therapy. In Ricca E, Henriques AO, Cutting SM (eds) Bacterial Spore Formers - Probiotics and Emerging Applications. Horizon Bioscience: Norfolk, pp 207–216

Perry JJ and Foster JW (1955) Studies on the biosynthesis of dipicolinic acid in spores of *Bacillus cereus* var. *mycoides*. Journal of Bacteriology 69:337–346

Philip ES (1989) A review of bioluminescent ATP techniques in rapid microbiology. Journal of Bioluminescence and Chemiluminescence 4(1):375–380

Phillips AP and Martin KL (1983) Immunofluorescence analysis of *Bacillus* spores and vegetative cells by flow-cytometry. Cytometry 4(2):123–131

Potts M (1994) Desiccation tolerance of prokaryotes. Microbiological Reviews 58(4):755–805

Poulis JJA, de Pijper MM, Mossel DDA et al. (1993) Assessment of cleaning and disinfection in the food industry with the rapid ATP-bioluminescence technique combined with the tissue fluid contamination test and a conventional microbiological method. International Journal of Food Microbiology 20(2):109–16

Powell E (1957) The appearance of bacterial spores under phase-contrast illumination. Journal of Applied Bacteriology 3:342–348

Powell JF (1953) Isolation of dipicolinic acid (pyridine-2-6-dicarboxylic acid) from spores of *Bacillus megatherium*. Biochemical Journal 54(2):210–211

Prescott SC and Dunn CG (1959) Industrial Microbiology. McGraw Hill: New York, pp 250–284

Centers of Disease Control and Prevention (2002) Evaluation of postexposure antibiotic prophylaxis to prevent anthrax. (Reprinted from MMWR, vol 51, pg 59, 2002) Journal of the American Medical Association 287(6):710

Priha O, Hallamaa K, Saarela M et al. (2004) Detection of *Bacillus cereus* group bacteria from cardboard and paper with real-time PCR. Journal of Industrial Microbiology & Biotechnology 31(4):161–169

Pulvertaft RJV and Haynes JA (1951) Adenosine and spore germination: phase contrast studies. Journal of General Microbiology 5:657–663

Qi YA, Patra G, Liang XD et al. (2001) Utilization of the *rpoB* gene as a specific chromosomal marker for real-time PCR detection of *Bacillus anthracis*. Applied and Environmental Microbiology 67(8):3720–3727

Radnedge L, Agron PG, Hill KK et al. (2003) Genome differences that distinguish *Bacillus anthracis* from *Bacillus cereus* and *Bacillus thuringiensis*. Applied and Environmental Microbiology 69(5):2755–2764

Ramisse V, Patra G, Vaissaire J et al. (1999) The Ba813 chromosomal DNA sequence effectively traces the whole *Bacillus anthracis* community. Journal of Applied Microbiology 87(2):224–228

Rantakokko-Jalava K and Viljanen MK (2003) Application of *Bacillus anthracis* PCR to simulated clinical samples. Clinical Microbiology and Infection 9(10):1051–1056

Rappe MS, Connon SA, Vergin KL et al. (2002) Cultivation of the ubiquitous SAR11 marine bacterioplankton clade. Nature 418:630–633

Rasko DA, Altherr MR, Han CS et al. (2005) Genomics of the *Bacillus cereus* group of organisms. FEMS Microbiology Reviews 29(2):303–329

Read TD, Peterson SN, Tourasse N et al. (2003) The genome sequence of *Bacillus anthracis* Ames and comparison to closely related bacteria. Nature 423(6935):81–86

Ricca E, Henriques AO and Cutting SM (eds) (2004) Bacterial Spore Formers - Probiotics and Emerging Applications. Horizon Bioscience: Norfolk

Rice EW, Fox KR, Miltner RJ et al. (1996) Evaluating plant performance with endospores. Journal American Water Works Association 88(9):122–130

Rijpens NP, Nancy P, Herman LM et al. (2002) Molecular methods for identification and detection of bacterial food pathogens. Journal of AOAC International 85(4):984–995

Rivera VR, Merill GA, White JA et al. (2003) An enzymatic electrochemiluminescence assay for the lethal factor of anthrax. Analytical Biochemistry 321(1):125–130

Rosen DL, Sharpless C and McGown LB (1997) Bacterial spore detection and determination by use of terbium dipicolinate photoluminescence. Analytical Chemistry 69(6):1082–1085

Rosen DL (1998) Wavelength pair selection for bacterial endospore detection by use of terbium dipicolinate photoluminescence. Applied Optics 37(4):805–807

Rosen DL (1999) Bacterial endospore detection using photoluminescence from terbium dipicolinate. Reviews in Analytical Chemistry 18(1-2):1–21

Ross KFA and Billing E (1957) The water and solid content of living bacterial spores and vegetative cells as indicated by refractive index measurements. Journal of General Microbiology 16:418–425

Roszak DB and Colwell RR (1987) Survival strategies of bacteria in the natural environment. Microbiological Reviews 51(3):365–379

Rowe CA, Tender LM, Feldstein MJ et al. (1999) Array biosensor for simultaneous identification of bacterial, viral, and protein analytes. Analytical Chemistry 71(17):3846–3852

Rowley DB and Feeherry F (1970) Conditions affecting germination of *Clostridium botulinum* 62A spores in a chemically defined medium. Journal of Applied Bacteriology 104(3):1151–1157

Ryu C, Lee K, Yoo C et al. (2003) Sensitive and rapid quantitative detection of anthrax spores isolated from soil samples by real-time PCR. Microbiology and Immunology 47(10):693–699

Sabbatini N, Guardigli M and Lehn JM (1993) Luminescent lanthanide complexes as photochemical supramolecular devices. Coordination Chemistry Reviews 123(1-2):201–228

Sacks LE (1990) Chemical germination of native and cation-exchanged bacterial-spores with trifluoperazine. Applied and Environmental Microbiology 56(4):1185–1187

Saiki RK, Scharf S, Faloona F et al. (1985) Enzymatic amplification of beta-globin genomic sequences and restriction site analysis for diagnosis of sickle cell anemia. Science 230(4732):1350–1354

Saiki RK, Gelfand DH, Stoffel S et al. (1988) Primer-directed enzymatic amplification of DNA with a thermostable DNA polymerase. Science 239(4839):487–491

Sanderson WT, Stoddard RR, Echt AS et al. (2004) *Bacillus anthracis* contamination and inhalational anthrax in a mail processing and distribution center. Journal of Applied Microbiology 96(5):1048–1056

Santo LY and Doi RH (1974) Ultrastructural analysis during germination and outgrowth of *Bacillus subtilis* spores. Journal of Bacteriology 120(1):475–481

Schoeni JLJL and Wong ACACL (2005) *Bacillus cereus* food poisoning and its toxins. Journal of Food Protection 68(3):636–48

Schuch R, Nelson D and Fischetti VA (2002) A bacteriolytic agent that detects and kills *Bacillus anthracis*. Nature 418(6900):884–889

Senesi S (2004) Bacillus spores as probiotic products for human use. In Ricca E, Henriques AO, Cutting SM (eds) Bacterial Spore Formers - Probiotics and Emerging Applications. Horizon Bioscience: Norfolk, pp 131–142

Setlow P and Kornberg A (1970) Biochemical studies of bacterial sporulation and germination. XXII. Energy metabolism in early stages of germination of *Bacillus megaterium* spores. Journal of Biological Chemistry 245(14):3637–3644

Setlow P (1995) Mechanisms for the prevention of damage to DNA in spores of *Bacillus* species. Annual Review of Microbiology 49:29–54

Setlow P (2001) Resistance of spores of *Bacillus* species to ultraviolet light. Environmental and Molecular Mutagenesis 38(2-3):97–104

Setlow P (2003) Spore germination. Current Opinion in Microbiology 6(6):550–556

Shafaat HS and Ponce A (2006) Applications of a rapid endospore viability assay for monitoring UV inactivation and characterizing Arctic ice cores. Applied and Environmental Microbiology 72(10):6808–6814

Shangkuan YH, Chang YH, Yang JF et al. (2001) Molecular characterization of *Bacillus anthracis* using multiplex PCR, ERIC-PCR and RAPD. Letters in Applied Microbiology 32(3):139–145

Sheridan GE, Masters CI, Shallcross JA et al. (1998) Detection of mRNA by reverse transcription-PCR as an indicator of viability in *Escherichia coli* cells. Applied and Environmental Microbiology 64(4):1313–1318

Singh RP, Setlow B and Setlow P (1977) Levels of small molecules and enzymes in the mother cell compartment and the forespore of sporulating *Bacillus megaterium*. Journal of Bacteriology 130:1130–1138

Sinha S (1983) Systematics and the Properties of the Lanthanides (NATO Science Series C). Springer: Dordrecht, Holland

Sneath PHA (1962) Longevity of micro-organisms. Nature 195(4842):643–646

Song JM, Culha M, Kasili PA et al. (2005) A compact CMOS biochip immunosensor towards the detection of a single bacteria. Biosensors & Bioelectronics 20(11):2203–2209

Speight SE, Hallis BA, Bennett AM et al. (1997) Enzyme-linked immunosorbent assay for the detection of airborne microorganisms used in biotechnology. Journal of Aerosol Science 28(3):483–492

Stopa PJ, Tieman D, Coon PA et al. (1999) Detection of biological aerosols by luminescence techniques. Field Analytical Chemistry & Technology 3(4-5):283–290

Stopa PJ (2000) The flow cytometry of *Bacillus anthracis* spores revisited. Cytometry 41(4):237–244

Szabo EA and Mackey BM (1999) Detection of *Salmonella enteritidis* by reverse transcription-polymerase chain reaction (PCR). International Journal of Food Microbiology 51(2-3):113–122

Tahernia AC (1967) Treatment of anthrax in children. Archives of disease in childhood 42(222):181–182

Thore AA, Ansehn SS, Lundin AA et al. (1975) Detection of bacteriuria by luciferase assay of adenosine triphosphate. Journal of Clinical Microbiology 1(1):1–8

Ting PT and Freiman A (2004) The story of *Clostridium botulinum*: from food poisoning to Botox. Clinical Medicine 4(3):258–261

Turnbull PCB, Hutson RA, Ward MJ et al. (1992) *Bacillus anthracis* but not always anthrax. Journal of Applied Bacteriology 72(1):21–28

Tyndall J (1877) Further researches on the deportment and vital persistence of putrefactive and infective organisms from a physical point of view. Philosophical Transactions of the Royal Society of London 167:149–206

Ugarova NN, Brovko YL and Kutuzova GD (1993) Bioluminescence and bioluminescent analysis: recent development in the field. Biokhimiya 58:1351–1372

Vary JC and Halvorson HO (1965) Kinetics of germination of *Bacillus* spores. Journal of Bacteriology 89:1340–1347

Venkateswaran KK, Hattori NN, La Duc MTMT et al. (2003) ATP as a biomarker of viable microorganisms in clean-room facilities. Journal of Microbiological Methods 52(3):367–77

Vereb G, Jares-Erijman E, Selvin PR, Jovin TM (1998) Temporally and spectrally resolved imaging microscopy of lanthanide chelates. Biophysical Journal 74(5):2210–2222

Volokhov D, Pomerantsev A, Kivovich V et al. (2004) Identification of *Bacillus anthracis* by multiprobe microarray hybridization. Diagnostic Microbiology and Infectious Disease 49(3):163–171

Vreeland RH, Rosenzweig WD and Powers DW (2000) Isolation of a 250 million-year-old halotolerant bacterium from a primary salt crystal. Nature 407(6806):897–900

Walker GT, Fraiser MS, Schram JL et al. (1992a) Strand displacement amplification - an isothermal, *in vitro* DNA amplification technique. Nucleic Acid Research 20(7):1691–1696

Walker GT, Little MC, Nadeau JG et al. (1992b) Isothermal in vitro amplification of DNA by a restriction enzyme/DNA polymerase system. Proceedings of the National Academy of Sciences 89(1):392–396

Wang SH, Wen JK, Zhou YF et al. (2004) Identification and characterization of *Bacillus anthracis* by multiplex PCR on DNA chip. Biosensors & Bioelectronics 20(4):807–813

Webster JJ, Walker BG, Ford SR et al. (1998) Determination of sterilization effectiveness by measuring bacterial growth in a biological indicator through firefly luciferase determination of ATP. Journal of Bioluminescence and Chemiluminescence 2(3):129–133

Weis CP, Intrepido AJ, Miller AK, Cowin PG, Durno MA, Gebhardt JS, Bull R (2002) Secondary aerosolization of viable *Bacillus anthracis* spores in a contaminated US Senate Office. Journal of American Medical Association 288(22):2853–2858

Weyant R, Ezzell J and Popovic T (2001) Basic laboratory protocols for the presumptive identification of *Bacillus anthracis*. Centers for Disease Control and Prevention: Atlanta.

Wilcox MH and Fawley WN (2000) Hospital disinfectants and spore formation by *Clostridium difficile* The Lancet 356(9238):1324–1324

Wilson WJ, Erler AM, Nasarabadi SL et al. (2005) A multiplexed PCR-coupled liquid bead array for the simultaneous detection of four biothreat agents. Molecular and Cellular Probes 19(2):137–144

Woese C and Morowitz HJ (1958) Kinetics of the release of dipicolinic acid from spores of *Bacillus subtilis*. Journal of Bacteriology 76(1):81–83

Wuytack EY, Boven S, Michiels CW (1998) Comparative study of pressure-induced germination of *Bacillus subtilis* spores at low and high pressures. Applied and Environmental Microbiology 64(9):3220–3224

Xiao M and Selvin PR (1999) An improved instrument for measuring time-resolved lanthanide emission and resonance energy transfer. Review of Scientific Instruments 70(10):3877–3881

Yamada S, Ohashi E, Agata N et al. (1999) Cloning and nucleotide sequence analysis of *gyrB* of *Bacillus cereus, B-thuringinesis, B-mycoides*, and *B-anthracis* and their application to the detection of *B-cereus* in rice. Applied and Environmental Microbiology 65(4):1483–1490

Yaron S and Matthews KR (2002) A reverse transcriptase-polymerase chain reaction assay for detection of viable *Escherichia coli* O157:H7: investigation of specific target genes. Journal of Applied Microbiology 92(4):633–640

Yung PT, Kempf MJ and Ponce A (2006) A rapid single spore enumeration assay. IEEE Aerospace Conference, Big Sky, Montana

Yung PT, Lester ED, Bearman G et al. (2007) An automated front-end monitor for anthrax surveillance systems based on the rapid detection of airborne endospores. Biotechnology and Bioengineering 98(4):864–871

Yung PT, Shafaat HS, Connon SA et al. (2007) Quantification of viable endospores from a Greenland ice core. FEMS Microbiology Ecology 59(2):300–306

Zahavy E, Fisher M, Bromberg A et al. (2003) Detection of frequency resonance energy transfer pair on double-labeled microsphere and *Bacillus anthracis* spores by flow cytometry. Applied and Environmental Microbiology 69(4):2330–2339

Zhang R and Zhang CT (2003) Identification of genomic islands in the genome of *Bacillus cereus* by comparative analysis with *Bacillus anthracis*. Physiological Genomics 16(1):19–23

Zink DL (1997) The impact of consumer demands and trends on food processing. Emerging Infectious Diseases 3(4):467–469

20

Label-Free Fingerprinting of Pathogens by Raman Spectroscopy Techniques

Ann E. Grow

Abstract

Raman spectroscopy is a label-free technique for generating unique spectral fingerprints from intact microorganisms. Studies conducted for more than a decade have shown that these "whole-organism fingerprints" can be used to identify pathogens, including bacteria, yeasts, and spores, at the strain level, even when the microorganisms are so closely related that they are difficult to distinguish by conventional techniques. Emerging techniques such as Raman microscopy and surface-enhanced Raman scattering (SERS) can enhance the magnitude of the signal to the point that Raman fingerprinting can achieve single-cell sensitivity. More recently, Raman microscopy and SERS have been integrated with biomolecule capture to produce a new microarray technology, dubbed "microSERS," for rapid identification of pathogens and their toxins in complex samples, without any labels, pre-processing of the sample, or culturing. This chapter reviews the studies that have been done on Raman microscopy and SERS for pathogen identification, and innovative methods for sample collection, concentration, and manipulation that can be combined with fingerprinting techniques. It also presents recent progress on microSERS analysis for the identification of bacteria, spores, and toxins in complex samples; differentiation between viable and nonviable microorganisms; and evaluation of growth conditions on microbial phenotype and specificity/affinity for capture biomolecules.

1. Introduction

Vibrational spectroscopies such as infrared (IR) and Raman can be used for the detection and identification of pathogens, without the need for any extractions, amplifications, labeling, or staining steps.

The potential for using IR to identify bacteria had already been recognized by the early 1950s (Stevenson and Bolduan 1952; Thomas and Greenstreet 1954). This spectral analysis problem is basically the same as that widely used for the identification of chemicals, in which the spectrum is collected from an "unknown" chemical and compared against a reference database of spectra from diverse "known" chemicals. Because each compound produces its own unique spectrum, or "fingerprint," the unknown can be identified solely on the basis of its spectral fingerprint. Similarly, when IR is used to analyze intact bacteria, the resulting "whole-organism fingerprint" can be used for identification, by comparing the fingerprint from the "unknown" bacteria with a reference database of fingerprints collected from diverse known bacteria.

Ann E. Grow • Biopraxis, Inc., San Diego, California

M. Zourob et al. (eds.), *Principles of Bacterial Detection: Biosensors, Recognition Receptors and Microsystems*,
© Springer Science+Business Media, LLC 2008

However, bacterial identification is complicated by the fact that bacterial fingerprints are the integrated spectra of the total chemical composition of the cell. Owing to the multitude of cellular constituents, superimposed spectral bands are observed throughout the entire spectral range; i.e., the spectra show broad and complex contours rather than distinct peaks. This makes it much more difficult to compare fingerprints from unknowns against large reference databases. Accordingly, very little progress was made toward bacterial identification by IR fingerprinting until pattern recognition algorithms were developed, which proved to be a major breakthrough. By using search algorithms, cluster analysis, and multivariate statistics to compare the "unknown" whole-organism fingerprint against a reference database of IR fingerprints from "known" organisms, bacteria could be identified at the species, and usually the strain, levels, with a high degree of confidence (Helm et al. 1991; Naumann, Helm, and Labischinski 1991; Naumann et al. 1991; Holt et al. 1995). To identify the bacteria in complex media such as clinical samples, the sample was plated out, and microcolonies grown under standardized conditions. Pure microcolonies were selected by visual observation for analysis, transferred onto an IR-transparent sample holder, and dried; and the fingerprint was collected and compared against the reference database. IR fingerprinting was so specific, it could be more useful than biochemotyping, plasmid or lipopolysaccharide pattern analysis, multilocus enzyme electrophoresis, or outer membrane or whole cell protein pattern analysis, in identifying strains for epidemiology applications (Seltmann, Voigt, and Beer 1994).

IR spectroscopy measures the absorption of infrared light by the sample. Raman spectroscopy, on the other hand, measures the light that is inelastically scattered following excitation of the sample by a laser. The interaction between a laser photon and a molecule results in the excitation of vibrations in the molecule bonds. The many vibrational modes of a molecule give rise to a spectrum that contains information about chemical composition, bonding situation, symmetry, structures, and physical parameters (e.g., the length of bonds.) Different molecules have different vibrational modes, and so a Raman spectrum, like an IR spectrum, represents a chemical fingerprint of the molecular composition of a sample.

While IR showed considerable promise, Raman spectroscopy offered a number of potential advantages over IR. (1) Since Raman is a scattering phenomenon, spectra can be collected directly from an opaque surface (e.g., bacterial colonies growing on solid culture medium). Because IR is based on absorption, solid samples must be processed (e.g., smeared on an IR-transparent window) before they can be analyzed. (2) Since water is a poor Raman scatterer, hydrated and even aqueous samples can be analyzed by Raman fingerprinting. Because water absorbs in the IR range so strongly that its signal masks other interesting peaks in the spectrum, samples must be thoroughly dried for IR fingerprinting. And (3) Raman spectral bands are sharper and more readily distinguishable than those in an infrared spectrum, and can therefore often provide more information.

Like IR fingerprints of intact bacterial cells, Raman whole-organism fingerprints can be used for identification of pathogens at the strain level (Goodacre et al. 1998; Maquelin et al. 2003; Ibelings et al. 2005; Maquelin et al. 2006), even when the microorganisms are so closely related that they are difficult to distinguish by conventional techniques (Kirschner et al. 2001; Hutsebaut et al. 2006.) Raman spectra provide such a rich body of information that they have been used to characterize the variations in biochemical composition of different bacteria, such as *Listeria monocytogenes* strains with different susceptibilities to sakacin P (Oust et al. 2006.)

The main drawback of so-called "normal" Raman spectroscopy for microbial fingerprinting is an inherent lack of sensitivity, due to the fact that only ~1 in 10^8 incident photons is Raman scattered. However, several methods have recently been developed for enhancing the magnitude of the Raman signal, to the point that Raman fingerprinting can achieve single-cell sensitivity. These methods include Raman microscopy and surface-enhanced Raman scattering (SERS).

This chapter will discuss the use of Raman microscopy and SERS to generate unique spectral fingerprints for identifying pathogens; and the studies that have been done to integrate sensitive Raman fingerprinting techniques with biomolecule capture to produce a new microarray technology, dubbed "microSERS," for rapid identification of pathogens and their toxins in complex samples, without any labels, preprocessing of the sample, or culturing.

2. Raman Microscopy for Whole-Organism Fingerprinting

As noted earlier, Raman spectroscopy offers a number of advantages over IR. Raman microscopy offers yet another advantage over IR microscopy; i.e., Raman has a much better spatial resolution. IR is limited to a spatial resolution of ∼10 μm by the wavelength of light that is used, and so can achieve only single-cell sensitivity for the relatively large cells of higher organisms. Raman microscopes, on the other hand, can collect spectra from individual bacterial cells or spores, and even from organelles inside microbial cells, such as nuclei and mitochondria in yeasts.

In a Raman microscope, the light from the laser photon source is focused on the sample through the microscope objective. The resulting light scatter is then captured through that same objective, and processed by a spectrometer. A Raman microscope can be coupled with "laser tweezers" to isolate and manipulate individual microbial particles, identify them, and sort them.

Raman whole-organism fingerprinting was first evaluated as an adjunct to IR (Naumann et al. 1995; Goodacre et al. 1998; Kirschner et al. 2001; Maquelin et al. 2002b, 2003.) Since the results with Raman were as promising as those with IR, and since Raman offers many advantages over infrared, the more recent studies have tended to focus exclusively on Raman microscopy. Hutsebaut et al. (2006) found that Raman microscopy performed very well in distinguishing among eight species belonging to the "*Bacillus subtilis*" group, which are homogenous at the phenotypic and phylogenetic level. Maquelin et al. (2006) showed that it could be used for typing *Acinetobacter* species, using a collection of strains from five hospital outbreaks (Fig. 20.1). Raman microscopy was even capable of measuring the relative concentrations of *Streptococcus mutans* and *Streptococcus sanguis* in mixtures (Zhu, Quivey, and Berger 2004).

In many of these studies, the samples were cultured for 6–8 hours, and then microcolonies were loop-transferred to calcium fluoride slides for analyses. However, other studies demonstrated that Raman fingerprints could be collected directly from colonies growing on solid culture media (Maquelin et al. 2000, 2002a; Berger and Zhu 2003), simplifying the analyses and minimizing the handling steps even further. Fingerprints collected directly from microcolonies on solid culture media contain spectral contributions from the underlying culture medium, and the intensity of the culture medium signal can vary depending on the depth of the microcolony at which the spectrum is collected. A mathematical routine, involving vector algebra, was developed for the nonsubjective correction of spectra for variable signal contributions of the medium (Maquelin et al. 2000.)

Most researchers have used "unsupervised" pattern recognition methods to analyze the fingerprints, e.g., principal component analysis, hierarchical cluster analysis (Fig. 20.2), discriminant function analysis, and generalized discriminant analysis (Choo-Smith et al. 2001; Kirschner et al. 2001; Maquelin et al. 2002; Xie et al. 2005). Better results have been reported with "supervised" methods, such as artificial neural networks (Goodacre et al. 1998). Rösch and coworkers used the "support vector machine" technique (Rösch et al. 2005, 2006). The full spectral range was typically used in the analyses; and the results were, generally, quite good. However, better results were obtained when select spectral windows were used, rather than the entire Raman fingerprint (Kirschner et al. 2001; Berger and Zhu 2003). This is not surprising, since it is well-established that superior performance is achieved during spectral analyses when

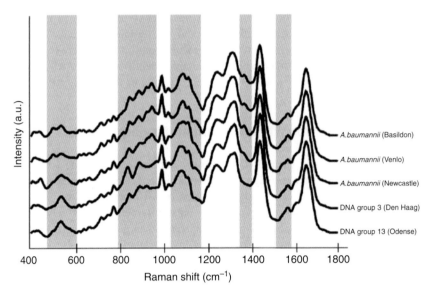

Figure 20.1. Whole organism fingerprints collected by Raman microscopy from five strains of *Acinetobacter* belonging to the *Acinetobacter calcoaceticus-Acinetobacter baumannii* (Acb) complex. Some visible spectral differences between the isolates of the respective outbreaks are highlighted in grey areas. Average of five 30-s spectra collected at 830 nm for each strain (reprinted from J. Microbiol. Methods 64(1); Maquelin K, Dijkshoorn L, van der Reijdenm TJ, and Puppels GJ; rapid epidemiological analysis of *Acinetobacter* strains by Raman spectroscopy, page 129, copyright 2006, with permission of Elsevier).

high value spectral regions are emphasized, and valueless regions (e.g., those that are inherently noisy or variable) are ignored (Gray et al. 1998).

In order to identify unknowns on the basis of their Raman fingerprints, spectral reference databases must be developed against which unknown fingerprints can be compared. Since Raman whole-organism fingerprints reflect the overall molecular composition of the cell, and since culture conditions influence the molecular makeup of a cell, culture conditions might be expected to affect the reproducibility of the Raman spectra. And, indeed, many laboratories have reported that the culture conditions must be carefully controlled for reliable identification. Parameters that can affect the fingerprints include the culture medium, the incubation temperature, and the growth time (Schuster, Urlaub, and Gapes 2000; Choo-Smith et al. 2001; Maquelin et al. 2003; Hutsebaut et al. 2004). In fact, Raman whole-organism fingerprints are so sensitive to the culture medium, they can be affected by isotopic labels. For example, when bacteria were grown in media containing incrementally varying ratios of $^{13}C_6$- to $^{12}C_6$-labeled glucose, ^{13}C incorporation shifted several characteristic Raman peaks to lower wavenumbers (Fig. 20.3) (Huang et al. 2004). In addition, the fingerprints collected from individual cells could differentiate between growth phases of a given species, although identification, at least at the species level, was still possible (Huang et al. 2004; Xie et al. 2005).

Laser tweezers (i.e., trapping individual particles in the focus of a laser beam) can manipulate bacterial cells and separate them from other sample particles. Laser tweezers have been combined with Raman spectroscopy (LTRS) for the analysis of single bacterial cells in aqueous samples. The same laser beam that was used to trap the bacteria was also used to excite Raman fingerprints for differentiation among bacterial species (Xie et al. 2005), or to separate bacteria in spoiled foods, such as milk, from food particles (Xie, Chen, and Li 2005). The ability to trap, sort, and identify individual bacterial cells is very impressive. However, LTRS has some limitations when it comes to the rapid identification of pathogens in complex samples. For example, a laser beam strong enough to trap microbial particles may also damage

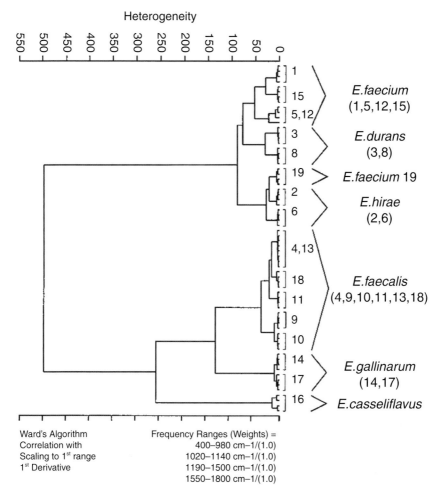

Figure 20.2. Typical dendrogram produced by hierarchical cluster analysis of Raman whole organism fingerprints—in this example, for six *Enterococcus* species. Average of ten spectra collected for 30-s at 830 nm. First derivative spectra were subjected to cluster analysis using Ward's algorithm and Euclidean distance measure (reproduced with permission from Kirschner et al. (2001)).

them, possibly to the point that identification is compromised; even microorganisms as robust as *Bacillus* spores can be damaged enough to release dipicolinate when exposed to the laser tweezers for more than 20 seconds (Chan et al. 2004.) The spectroscopic changes seen with laser trapping were similar to those reported for spores autoclaved at >100°C (Grow et al. 2003a.) The intensity of the signal from cells changes as the time in the laser beam lengthens, the reasons for which are still unknown (Singh et al. 2005). It is also difficult to trap a budding or dividing cell in a single-beam trap (Singh et al. 2006). Finally, the method as currently used is slow and labor-intensive, especially for complex samples containing particles that cannot be distinguished from bacteria by microscopic analysis, since all such particles must be fingerprinted to differentiate between the bacteria and the nonbacterial particles (Xie, Chen, and Li 2005).

The vast majority of Raman fingerprinting studies have been done on bacteria. However, a few studies have explored the potential of using this technique for yeasts. Maquelin et al. (2002a) successfully identified *Candida* species in blood cultures from hospital patients, by

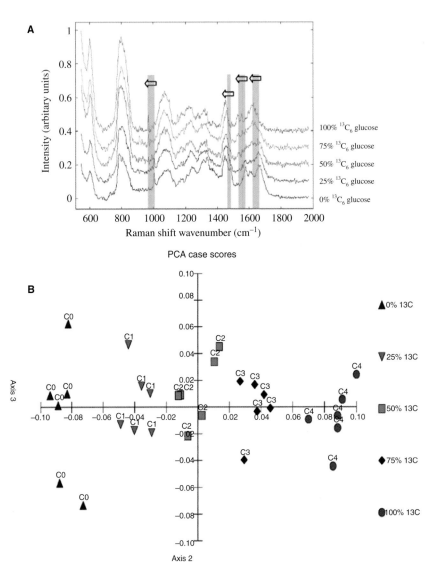

Figure 20.3. Effects of ^{13}C substitution in growth medium glucose on the Raman fingerprints of *Pseudomonas fluorescens* SBW25. (A) Substituting cellular biomass ^{12}C with ^{13}C causes visible red-shift in certain Raman spectral bands (arrows indicate the direction of the shift). Fingerprints collected for 90 s at 532 nm from individual cells. (B) PCA analysis of spectra of *P. fluorescens* SBW25 incubated with different percentages of $^{13}C_6$-glucose (reproduced with permission from Huang et al. (2004), copyright 2004 American Chemical Society).

fingerprinting microcolonies cultured for 6–8 hours. Since the spectra were collected from "bulk" samples, they comprised average spectra of many cells; and the identification process was the same as for bacteria. The identification of yeast at the single-cell level, on the other hand, was more challenging than the identification of individual bacteria. Bacteria are not compartmentalized (i.e., do not have a nucleus), and so their spectral fingerprints show spatial homogeneity (Rösch et al. 2005). Eukaryotic cells, on the other hand, store their DNA in a separate internal compartment. Owing to this molecular compartmentalization, different Raman fingerprints can be obtained from different locations in the eukaryotic cell. Nevertheless, Rösch et al. (2006) successfully identified yeasts by collecting ten Raman spectra from each cell and

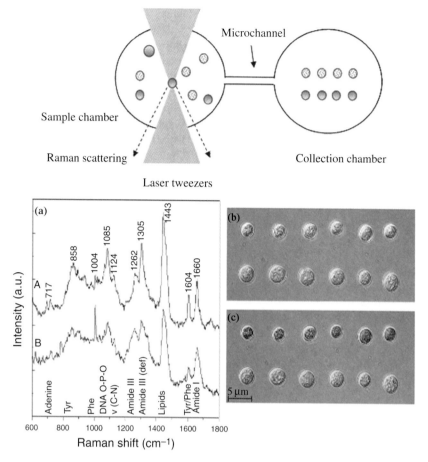

Figure 20.4. Schematic of LTRS sorting and identification. A particle in the sample chamber is captured with laser tweezers, identified by Raman microscopy, and then optically manipulated to a clean collection chamber. (a) Raman fingerprints of single live yeast cells (top) and dead yeast cells (bottom), collected for 20 s at 785 nm. Tyr, tyrosine; phe, phenylalanine; def, deformed. (b) Image of the sorted yeast cells in the collection chamber. Top row, dead yeast cells; bottom row, live yeast cells. (c) Image of the sorted yeast cells stained with 2 % eosin solution (reproduced with permission from Xie et al. 2005).

averaging them into a composite spectrum. Xie, Chen, and Li (2005) demonstrated the ability to use LTRS to distinguish between and separate live and dead yeast cells (Fig. 20.4).

3. Surface-Enhanced Raman Scattering (SERS) for Whole-Organism Fingerprinting

In surface-enhanced Raman scattering (SERS), the surface plasmon of a metal substrate (typically silver or gold, in the form of a roughened metal film, colloidal solution, or colloids immobilized on a surface) is in resonance with the excitation laser. When an analyte is in close proximity to the metal, the energy from the plasmon resonance may be coupled into bonds of the analyte, resulting in an enhancement of the Raman signal by many orders of magnitude. Under the right circumstances, SERS can achieve single-molecule sensitivity (Kneipp et al. 1997; Nie and Emory 1997; Kneipp et al. 1998a, 1998b, 1999; Xu et al. 1999).

Interest in SERS for use as a label in "surface-enhanced Raman immunoassay (SERIA)" is rapidly growing, since it has the potential to be more sensitive than fluorescence. For example, Ni et al. (1999) introduced a sandwich immunoassay in which a gold colloid is coated with Raman "reporter" molecules and then conjugated to a detection antibody. A waveband from the SERS spectrum of the reporter is monitored to detect antigen binding. Other researchers have applied this concept to the detection of pathogens, including bacteria (Lin et al. 2005), spores (Guicheteau and Christesen 2004), viruses (Driskell et al. 2005), and viral antigens (Xu et al. 2004). For an excellent review of SERIA, see Xu et al. (2005).

The idea of collecting SERS spectra directly from the microorganism appears to offer significant advantages in comparison with normal Raman and/or SERIA, including not only exquisite sensitivity and fewer sample processing steps, but also the ability to identify—not just detect—microorganisms on the basis of their fingerprints. However, despite three decades of research, much remains unknown about the SERS phenomenon and the surfaces that will be optimum for exciting strong signal enhancement in different types of analytes. The vast majority of studies on the SERS phenomenon have focused on small-molecule dye analytes; generating SERS signal enhancement from microorganisms presents an entirely different challenge. Accordingly, less progress has been made toward SERS fingerprinting than normal Raman fingerprinting of intact microorganisms. Nevertheless, SERS already shows exceptional promise.

Early studies to evaluate SERS for whole-organism fingerprinting were limited to a handful of model microorganisms, and often had the simple goal of demonstrating the ability to differentiate between Gram-positive and -negative organisms (Efrima, Bronk, and Czégé 1999; Spencer et al. 2002; Zeiri et al. 2004). It quickly became apparent that the spectra produced by SERS fingerprinting sometimes differed dramatically from those produced by normal Raman. SERS fingerprints did not contain nearly as many features as might be expected from the many constituents of the cell—certainly not nearly as many as seen in normal Raman fingerprints. In addition, since the SERS phenomenon is sensitive to distance, most researchers expected the SERS fingerprints to be dominated by cell-surface constituents and, therefore, expected dramatic differences between the fingerprints from Gram-positive and -negative bacteria. However, the fingerprints were surprisingly similar (Efrima, Bronk, and Czégé 1999; Jarvis, Brooker, and Goodacre 2004; Premasiri, Moir, and Ziegler 2005), although Gram-positive and -negative bacteria could still be distinguished.

Moreover, while each laboratory reported obtaining reproducible spectra from a given microorganism, the spectra produced from the same microorganism by different laboratories could differ substantially (Zeiri and Efrima 2005). These differences appeared to be related to differences in the experimental conditions used to induce the SERS phenomenon (Fig. 20.5). For example,

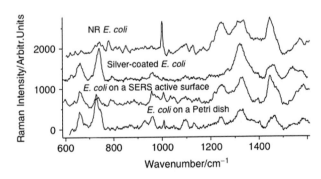

Figure 20.5. Normal Raman fingerprints of *E. coli* (NR *E. coli*); and SERS fingerprints of silver-coated *E. coli*, *E. coli* on a SERS-active substrate, and *E. coli* on a Petri dish, all excited at 633 nm (from Zeiri and Efrima 2005; copyright 2005; copyright John Wiley & Sons Ltd. Reproduced with permission).

some researchers reduced silver ions in the presence of bacteria, causing colloids to form on the bacterial surface (Efrima, Bronk, and Czégé 1999); some grew the bacteria in the presence of colloid (Sockalingum et al. 1999); others mixed the bacteria with preformed colloidal solutions and either analyzed the mixture in suspension (Sengupta, Laucks, and Davis 2005) or dried the mixture on support substrates (Jarvis and Goodacre 2004); and still others deposited the bacteria on a preformed SERS-active surface comprising a substrate coated with colloid clusters (Premasiri et al. 2005). Changing the properties of the colloidal particles and the location of the particles on the cell surface—by changing the protocols used to produce the colloids and bring them into contact with the bacteria—apparently changed the cell constituents whose spectra were SERS-enhanced (Efrima, Bronk, and Czégé 1999; Zeiri et al. 2004). Moreover, when bacteria were mixed with colloids and analyzed in suspension, the adsorption rate of the silver colloid particles on the bacteria was found to be strongly dependent on the pH, the density of the bacteria in solution, and even, to some extent, the type of bacteria (Sengupta, Laucks, and Davis 2005).

Nevertheless, SERS showed significant promise for pathogen identification. When a single protocol was consistently used to collect SERS fingerprints from different bacteria species and strains, bacteria could be differentiated at the strain level. Premasiri, Moir, and Ziegler (2005), for example, found that the SERS fingerprints from the Gram-negative *E. coli* and *S. typhimurium*, and the Gram-positive *B. subtilis*, *B. cereus*, *B. thuringiensis*, and *B. anthracis* Sterne, were not only species- and strain-specific, but exhibited greater differentiation than their corresponding normal Raman spectra. Jarvis and Goodacre (2004) analyzed a collection of clinical bacterial isolates associated with urinary tract infections, including multiple strains of *E. coli*, *Klebsiella oxytoca*, *Klebsiella pneumoniae*, *Citrobacter freundii*, *Enterococcus* spp, and *Proteus mirabilis*, demonstrating the ability to distinguish bacteria at the strain level, using multivariate statistical techniques and a relatively narrow window of the spectrum (i.e., ~400 – 980 cm^{-1}). SERS finger-printing coupled with principal components-discriminant function analysis showed that closely related bacteria belonging to the genus *Bacillus* could be identified at the strain level (Jarvis, Brooker, and Goodacre 2006). Vegetative *Bacillus* cells could be readily differentiated from their spores (Jarvis, Brooker, and Goodacre 2006; Guicheteau and Christesen 2007; Fig. 20.6);

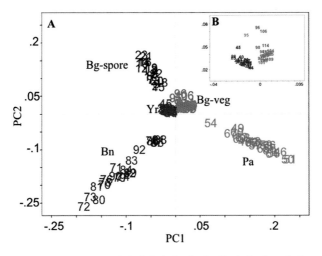

Figure 20.6. Typical principal components analysis (PCA) plot for the discrimination of microorganisms on the basis of their SERS fingerprints. (A) In this example, the plot shows the separation between *Pantoea agglomerans*, *Brucella neotomae*, *Yersinia rohdei*, *Bacillus globigii* spores, and *Bacillus globigii* vegetative cells. (B) Closer examination of the area between the vegetative cells of Bg and Yr shows further discrimination between the two species (reproduced, with permission from Guicheteau and Christesen 2007).

and the spores from two *Bacillus stearothermophilus* variants could also be distinguished (Alexander, Pellegrino, and Gillespie 2003).

Hou, Maheshwari, and Chang (2007) began addressing the problem of sampling, i.e., concentrating the bacteria in dilute clinical or environmental samples to the point that they can be detected. Their approach is a chip-scale device that relies on convection by a long-range converging vortex to concentrate the microorganisms and trap them on a gold spiral electrode. Silver nanoparticles added to the sample are trapped along with the microorganisms. The combination of the converging vortex and the electrode makes it possible to trap pathogens that vary widely in size and morphology; and the gold surface of the electrode enhances the SERS signal produced by the silver particles. Within 15 minutes, 150 μL samples containing as few as 10^4 colony-forming units per mL of *Saccharomyces cerevisiae*, *Escherichia coli*, or *Bacillus subtilis* produced whole-organism SERS fingerprints that were unique to the microorganisms.

4. MicroSERS for the Detection and Identification of Pathogens and Toxins

All of the studies discussed above provide evidence that Raman whole-organism finger-printing techniques offer significant potential for label-free analysis of pathogens. However, when complex samples, such as foods, clinical samples, or environmental media are to be analyzed, the microorganisms must be separated from the rest of the sample constituents before fingerprinting can take place. Most researchers have used conventional culturing methods; some have suggested laser tweezers. MicroSERS uses biomolecular recognition.

In microSERS (Grow 1999; Grow et al. 2003a), the analytes of interest are selectively isolated from the sample by capture biomolecules immobilized in a microarray. The captured analytes are then identified by Raman microscopy. To enable the rapid collection of strong, high-quality spectra from individual organisms, and even from bacterial toxins, the surface on which the microarray is printed is a roughened metal film capable of inducing SERS signal enhancement. Studies have shown that this simple approach can be used to detect analytes as small as cyanide, to those as large as *Giardia* cysts. MicroSERS can differentiate among species/strains of bacteria and spores that cross-react with a given biomolecule, and can even differentiate among closely related toxins that cross-react with a given biomolecule.

Perhaps most importantly, microSERS can produce information on the physiological state of the microorganism. For example, bacteria respond to environmental triggers, such as temperature, pH, and nutrient concentrations, by switching to different physiological states; and one state can be far more virulent than another (e.g., Miller and Mekalanos 1988; Litwin and Calderwood 1994; Mauchline et al. 1994; Maresca 1995; James et al. 1995, 1997; Eichenbaum, Green, and Scott 1996; Kapatral et al. 1996; Samoilova et al. 1996; Pettersson et al. 1997; Byrne and Swanson 1998). Moreover, Raman fingerprints reflect the growth history of microorganisms, which may be very helpful in, e.g., determining the source of a food poisoning outbreak (Huang et al. 2007). Finally, many bacteria can enter a dormant, "viable but nonculturable (VBNC)" state in low-nutrient environments. There is growing evidence that at least some VBNC pathogens retain their virulence (for a review, see Oliver 2005), posing a significant human health hazard. The changes that take place when a pathogen enters the VBNC state should be readily apparent in its SERS fingerprints; and since microSERS does not require culturing, it should be capable of detecting and identifying VBNC cells.

This section summarizes some of the work that has been done to develop the microSERS technology (Thompson et al. 2000a, 2000b; Grow 2001, 2002; Grow et al. 2002, 2003a, 2003b). The studies have evaluated the reproducibility of the SERS fingerprints of microorganisms and toxins; the ability to identify bacteria and spores at the species and subspecies levels on the basis of their SERS fingerprints; the impact of growth conditions on the microbial fingerprints; the ability to differentiate between viable and nonviable organisms; the ability to combine biomolecule capture with subsequent SERS fingerprinting; the ability to use biomolecules associated with virulence as capture biomolecules on the microSERS microarray; the ability to differentiate among structurally-similar toxins, even in mixtures; and the potential for analyzing complex food and environmental samples for pathogens and toxins. Tentative peak assignments in the analyses of the SERS whole-organism fingerprints are based on Notingher, Selvakumaran, and Hench (2004); De Gussem et al. (2005); Huang et al. (2005); and Krafft et al. (2006). Studies have also shown the ability to fingerprint *Cryptosporidium* oocysts, *Giardia* cysts, and viruses, but they are outside the scope of this chapter.

4.1. MicroSERS Detection of Bacteria

The Gram-positive *Listeria* and *Bacillus* and the Gram-negative *Legionella* and *Escherichia* were chosen as model systems for the initial microSERS studies. The bacteria included five species of *Listeria* (i.e., *Listeria monocytogenes*, *List. innocua*, *List. seeligeri*, *List. welshimeri*, and *List. ivanovii*, including three strains of *List. monocytogenes*, and two strains of *List. innocua*); seven species of *Bacillus* (i.e., *Bacillus brevis*, *B. cereus*, *B. coagulans*, *B. stearothermophilus*, *B. subtilis*, *B. megaterium*, and *B. thuringiensis*, including two strains each of *B. cereus*, *B. coagulans*, *B. subtilis*, and *B. thuringiensis*); six species of *Legionella* (i.e., *Legionella pneumophila*, *Leg. bozemanii*, *Leg. israelensis*, *Leg. micdadei*, *Leg. maceachernii*, and *Leg. dumoffii*); and one strain of *Escherichia coli*.

4.1.1. SERS Fingerprinting of Bacteria

As noted above, SERS fingerprinting of intact microorganisms is in its infancy. Therefore, the first experiments were designed simply to confirm that the SERS fingerprints would be reproducible if the bacteria were grown under a single set of culture conditions. Since microSERS will utilize biomolecules immobilized on a SERS-active surface to selectively capture bacteria from the sample and SERS fingerprints collected directly from the captured bacteria, a SERS-active surface suitable for a microarray assay configuration had to be used. Silver is the metal most commonly used to excite SERS, and has been, by far, the most commonly used for whole-organism fingerprinting. However, silver is very reactive; nanostructured silver oxidizes easily, losing its SERS activity; and silver might be expected to denature many proteins used as capture biomolecules. Gold, on the other hand, is much more robust and biocompatible, and under the right circumstances, can achieve single-molecule sensitivity. Accordingly, a preassembled gold SERS-active surface was selected for use.

Several dozen spectra were collected from each species/strain of the chosen model bacteria. As other researchers have reported, the fingerprints were reproducible, provided that the growth conditions were held constant. Interestingly, the *Listeria* fingerprints were the same whether grown in brain heart infusion (BHI) liquid media or on BHI agar slants, provided the temperature was the same. The analyses were very sensitive, with spectra collected for 60 seconds from a $\sim$1 μm diameter surface area (i.e., the size of a single bacterial cell).

As expected, the fingerprints from different species belonging to a given genus were readily differentiated. For example, *Leg. pneumophila* and *Leg. maceachernii* could be distinguished on the basis of peaks associated with nucleic acids, polysaccharides, protein backbone, and phospholipid conformation (Fig. 20.7.) Even species as closely related as *List. monocytogenes* and its "nonpathogenic variant" *List. innocua*, which are often misidentified by conventional tests (McLauchlin 1997), could be distinguished, as could the *Bacillus* species *B. anthracis*, *B. cereus*, and *B. thuringiensis*, which cannot be differentiated by multilocus enzyme electrophoresis or chromosomal gene sequencing (Helgason et al. 2000).

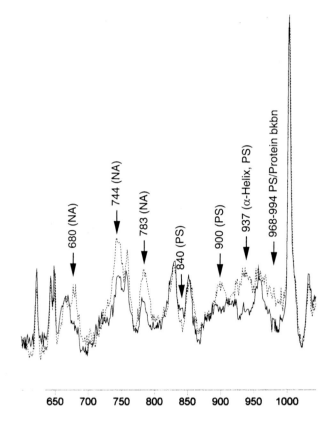

Figure 20.7. Mid-frequency window from the SERS fingerprints of *Legionella maceachernii* (- - - - -) and *Leg. pneumophila* (———) grown in BHI medium at 37°C. NA = nucleic acids, PS = polysaccharide, and bkbn = backbone. All microSERS fingerprints in this chapter collected at 632.8 nm; baseline corrected.

Preliminary results indicated that SERS analysis could often be used to identify bacteria at the subspecies level, as well. For example, SERS fingerprints could be used to differentiate among *B. cereus* ATCC11778 and ATCC25621, *B. coagulans* ATCC51232 and ATCC7050, *B. subtilis* ATCC6633 and ATCC9372, and *B. thuringiensis* ATCC35646 and ATCC10792. The three strains of *List. monocytogenes* belonging to serovars 1/2a, 3, and 4b also produced unique SERS fingerprints (Fig. 20.8); and all three could be readily differentiated from two strains of *List. innocua* belonging to serovars 6a and 6b. However, there were no striking differences between the SERS spectra of the two *List. innocua* strains; more sophisticated analyses using chemometric algorithms are needed to determine whether this nonpathogenic species can be identified at the subspecies level by SERS fingerprinting.

4.1.2. Impact of Growth Conditions on Bacterial Fingerprints

As noted earlier, bacteria respond to environmental triggers by switching to different physiological states. If such changes can be detected in the SERS fingerprints, then microSERS analysis can produce information that can be very useful in determining virulence, conducting epidemiological studies, or determining the source of a food poisoning outbreak. Accordingly, studies were initiated to determine whether growth conditions that were expected to affect virulence would change the bacterial fingerprints. For example, when grown on blood agar, virulent strains of *List. monocytogenes* cause b-hemolysis by secreting listeriolysin O (LLO).

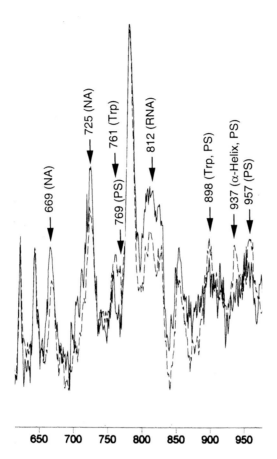

Figure 20.8. Mid-frequency window from the SERS fingerprints of *Listeria monocytogenes* strains from serovars 4b (———) and 3 (- - - -) grown in BHI medium at 37°C. NA = nucleic acids, PS = polysaccharide, and Trp = tryptophan.

List. monocytogenes 1/2a was grown on blood agar, and the fingerprints collected for comparison against those from cells grown on BHI. As expected, many bands in the *List. monocytogenes* fingerprints were affected by growth under conditions that induce the production and secretion of exotoxins. Notably, there were distinct changes in bands associated with protein, polysaccharides, and lipids; and bands associated with RNA were weaker. The differences were very distinct, and very reproducible.

The temperature at which a pathogen is grown can also have a dramatic effect on phenotype. For example, growth at higher temperatures can suppress the formation of flagella in many pathogenic bacteria, such as *Leg. pneumophila* (Mauchline et al. 1992) and *List. monocytogenes* (Peel, Donachie, and Shaw 1988). When the SERS fingerprints from *Leg. pneumophila* grown at 23°C and 36°C were compared, they showed many distinct differences; e.g., the fingerprints from cells grown at 23°C had increased peak intensities at ~722 (lipids) and 780 (RNA), and decreased intensities at ~1044–1060 (lipid and polysaccharide conformation), 1175 (lipids and aromatic amino acids), and 1368 (nucleic acid conformation) cm^{-1}, in comparison with those grown at 36°C.

Similarly, when the fingerprints of *List. monocytogenes* 4b grown at 25°C and at 37°C were compared, many differences were seen throughout the spectral range, including peaks associated with polysaccharides, lipids, and protein. Two new peaks were seen in the 25°C fingerprint that were not readily interpreted—i.e., a very strong peak at 740 cm^{-1} and a moderately strong peak at 1612 cm^{-1}. The 740 cm^{-1} peak varied in intensity from experiment to experiment, the only peak that was found to do so at either temperature.

One of the more worrisome aspects of *List. monocytogenes* contamination is the fact that this organism can grow—not just survive—at refrigerated temperatures, and can become exceptionally virulent (Durst 1975; Basher et al. 1984; Czuprynski, Brown, and Roll 1989; Picard-Bonnaud, Cottin, and Carbonnelle 1989; Stephens et al. 1991). The fingerprint of *List. monocytogenes* 4b grown at 4°C differed significantly from those of the same strain grown at 25°C and 37°C. Among the more striking changes were a dramatic increase in the intensity of the peaks at 780 (RNA), 1099 (nucleic acid, lipid, and/or polysaccharide conformation), and 1449 (CH$_2$ bend, probably associated with saturated lipids); strong increases in peaks at 898 (polysaccharide), 935 (protein and polysaccharide backbone) and 956 (unknown); and a distinct decrease in the intensity of the peak at 1247 cm^{-1} (protein amide III.) The sharp increase at 1449 cm^{-1} in the fingerprints of cells grown at 4°C was interesting, since cold adaptation in *List. monocytogenes* is known to involve an increase in the short-chain saturated fatty acids of the membrane and not, like some bacteria, increased unsaturation (Annous et al. 1997; Mastronicolis et al. 2006).

The ability to produce information on the physiological state of a microorganism offers many potential benefits. Nevertheless, it does have a potential "down side"—i.e., it means that a variety of different fingerprints, produced under a variety of growth conditions, must be developed for each pathogen, for inclusion in the spectral reference database. When identification is made solely on the basis of the whole-organism fingerprints, all the fingerprints must be included in a single reference database; and the identification algorithm must be capable of sorting through them all, to find a single, reliable, positive identification for the unknown. Other researchers have chosen to culture the pathogens under controlled conditions to avoid this complication.

However, it might be expected that, since changes in physiological state are dictated by genes, related bacteria will respond to many environmental triggers in the same way, and their fingerprints therefore be affected in a similar fashion. If the responses to environmental triggers of one strain can be used to predict those of other, closely related strains, the level of effort needed to develop the reference databases and identification algorithms will be minimized. Moreover, it must be remembered that microSERS will base identification on biomolecular recognition as well as whole-organism fingerprints. Accordingly, only the fingerprints from microorganisms capable of binding to a capture biomolecule need to be included in the spectral reference database associated with that biomolecule; and different identification algorithms can be individually tailored for each biomolecule in the microarray.

To begin evaluating the prediction that related microorganisms will respond to many environmental triggers in the same way, and their fingerprints therefore be affected in a similar fashion, studies were conducted on the impact of temperature on *Listeria* fingerprints. First, the *List. monocytogenes* 1/2a and 3 strains were grown at 25°C, and their SERS fingerprints compared with *List. monocytogenes* 4 grown at 25°C. The impact of the growth temperature produced changes in the fingerprints of each strain throughout the spectral range, which followed the same trends that were seen in *List. monocytogenes* 4b in the polysaccharide, lipid, and protein peaks that were affected. The four other *Listeria* species were also grown at 25°C, and the fingerprint peaks that were affected also followed many of the same trends. All species/strains even produced a new peak at 740 cm^{-1} that fluctuated in intensity when grown at 25°C, but not at 37°C. However, each species/strain still produced a unique fingerprint, and so it was still possible to individually identify all five species of *Listeria* and all three serotypes of *List. monocytogenes*, when grown at 25°C. Finally, the impact of growth at 4°C was evaluated. Of the five strains of *List. monocytogenes* and *List. innocua* under study, only three grew at 4°C; i.e., *List. monocytogenes* 1/2a and 4b, and *List. innocua* 6a. The fingerprints of *List. monocytogenes* and *List. innocua* grown at 4°C showed many of the same trends in the way their fingerprints were affected, in comparison with cells grown at 25°C and 37°C (Fig. 20.9).

Studies were also initiated to confirm that the impact of the growth medium on bacterial fingerprints would be similar for related strains. The *List. monocytogenes* 3 and 4b strains were grown on the blood agar, for comparison with the impact that this nutrient medium had on *List. monocytogenes* 1/2a. As predicted, the changes in the fingerprints were very similar. To confirm that the growth medium could have similar effects on the bacterial fingerprints of related species, studies were done with *List. ivanovii*; which is also a human pathogen and which produces large quantities of ivanolysin O, a substance similar, but not identical, to LLO. When *List. ivanovii* was grown on blood agar, inducing the secretion of large amounts of its hemolysin (which could be seen in significant bizonal hemolysis), many of its spectral features were affected in the same way as those from *List. monocytogenes*. There were, however, a few distinct differences: e.g., unlike the *List. monocytogenes* fingerprint, the *List. ivanovii* fingerprint bands at 667 (nucleic acid) and 1126 cm^{-1} (protein C–N and/or lipid) did not change when the cells were grown on blood agar; whereas the band at 1662 (protein amide I) did change, i.e., decreased significantly in intensity, narrowed, and shifted to a doublet at 1656/1659 cm^{-1}. The three *List. monocytogenes* could still be differentiated when grown on blood agar, and readily distinguished from *List. ivanovii*.

In short, although the database is still very small, it appears that related bacteria respond to environmental triggers in a similar fashion, and therefore a given environmental trigger will affect their whole-organism fingerprints in a similar way.

4.1.3. Viable vs. Nonviable Bacteria

The ability to discriminate between viable and nonviable organisms can be very important for many pathogen detection applications. Factors that contribute to cell death affect the structure and functionality of many biomolecules within the cell; it therefore seems likely that conditions which are severe enough to cause the death of bacteria will affect the spectral fingerprint. And, indeed, the normal Raman fingerprints of dead yeast cells differ significantly from those of viable cells (Singh et al. 2005; Xie, Chen, and Li 2005).

Heat killing (autoclaving) is often used for sterilization. Preliminary studies were conducted to evaluate the impact of heat killing on eleven strains of *Bacillus*, four of the five strains of *Legionella* (i.e., *Leg. maceachernii*, *Leg. israelensis*, *Leg. bozemanii*, and *Leg. pneumophila*), *List. monocytogenes* serotype 4b, and *E. coli* O157:H7. Aliquots of a bacterial suspension were either autoclaved at 121°C for 30 min and then deposited on a SERS-active surface for analysis, or deposited onto a SERS-active surface and then autoclaved and analyzed. The spectral changes caused by killing the bacteria were consistent from experiment to experiment, as long as the treatment conditions were the same. In all cases, the heat-killed preparations were easily distinguished from their viable counterparts on the basis of their SERS fingerprints, with differences observed over the entire spectral range (Fig. 20.10). The impact of the autoclaving tended to be similar from species to species within a genus. For example, all of the key peaks associated with RNA decreased in intensity in the autoclaved *Legionella* cells, as did peaks associated with lipid and polysaccharide conformation; the protein amide III regions changed; and the peak at ~1660 shifted to ~1670 cm^{-1} (amide I changed from a structured a-helix conformation to a random coil.) The heat-killed *Bacillus* showed similar, but not identical, changes in their spectra. Moist heat inactivates enzymes, causes changes in nucleic acids, alters the cytoplasmic membrane, and coagulates proteins. The changes in the SERS fingerprints were consistent with such effects.

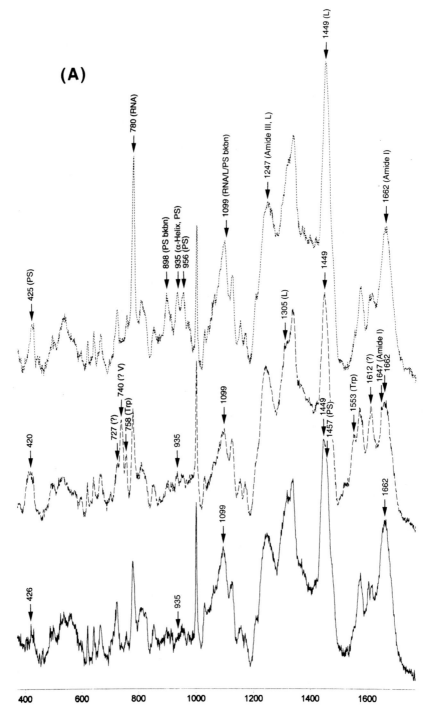

Figure 20.9. SERS fingerprints of (A) *Listeria monocytogenes* serovar 4b and

Continued

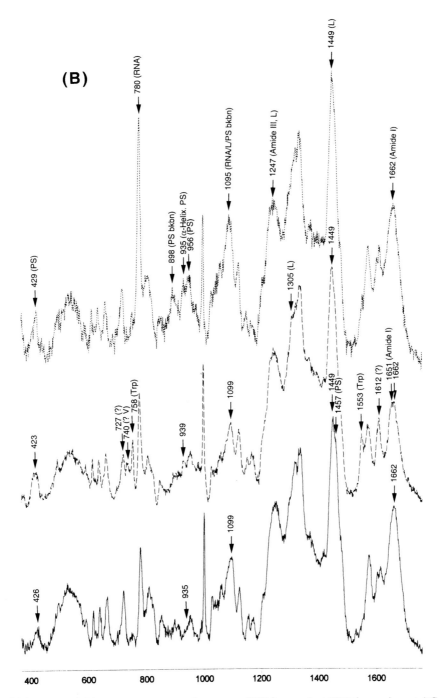

Figure 20.9. *continued* (B) *List. innocua* serovar 6a grown at 37°C (————), at 25°C (- - - - -), or at 4°C (·····) in BHI. Some of the major changes in response to temperature that are seen in the fingerprints of both organisms are marked. The differences may be in peak intensity, shape, and/or maximum. L = lipids, PS = polysaccharide, Trp = tryptophan, V = variable, and bkbn = backbone.

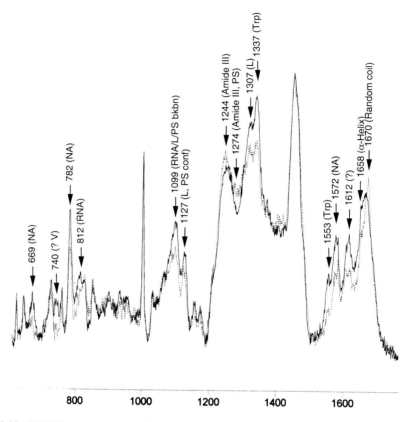

Figure 20.10. SERS fingerprints of viable (————) and heat-killed (- - - - -) *Listeria monocytogenes*. Some of the major changes associated with heat-killing are marked. L = lipids, NA = nucleic acids, PS = polysaccharide, Trp = tryptophan, V = variable, bkbn = backbone, and conf = conformation.

4.1.4. Integrated MicroSERS Detection and Identification of Bacteria

Having confirmed that the SERS fingerprints from bacteria were reproducible when growth and storage conditions were held constant, could be used to identify bacteria at the species/subspecies levels, and could differentiate between viable and heat-killed organisms, the next question was whether biomolecule capture would affect the SERS fingerprints. These experiments focused on *Listeria*, including the three representative strains of *List. monocytogenes* serotypes 4b, 3, and 1/2a, and *List. innocua* serotype 6a. *Escherichia coli* O157:H7 was used as a nontarget control.

It should be noted that the results reported here are preliminary. There is still much to learn about the optimum techniques for sampling pathogens by microarrays, as well as the optimum methods for immobilizing biomolecules on SERS-active surfaces.

Biochips prepared with polyclonal antibodies (PAbs) reportedly specific for *Listeria*, or for *List. monocytogenes*, were incubated in bacterial cell suspensions, briefly rinsed, and analyzed. The genus-specific PAb showed much higher affinities for *List. monocytogenes* 4b and 1/2a than for *List. monocytogenes* 3 or *List. innocua*. Surprisingly, genus-specific PAb biochips incubated in *E. coli* O157:H7 were sparsely but uniformly coated with *E. coli*, although the number of cells was very low in comparison with biochips incubated in any of the *Listeria*, including *List. innocua*. The species-specific PAb, on the other hand, had a higher affinity for *List. monocytogenes* 4b and 3 than for *List. monocytogenes* 1/2a, and captured only an isolated cell or two of *List. innocua* or *E. coli*.

In general, spectra collected from the antibody biochips were weaker than spectra collected from cells deposited on the surface, and contained features contributed by the antibodies themselves. Nevertheless, captured cells could be identified on the basis of their SERS fingerprints.

Other types of capture biomolecules produced much stronger fingerprints for *Listeria* or for spores (see below), without spectral contributions from the protein. The most likely reason for the difference is that the method that was used to immobilize the antibodies produced a thick, multilayered crust of protein that completely covered the surface, preventing the cells from coming into close proximity with the SERS-active surface. Since the SERS phenomenon is sensitive to distance, the signals from the captured cells would not be enhanced as strongly as those from cells deposited directly on the surface. Immobilization methods that produced a thin layer of capture biomolecules were much more effective at producing strong, high-quality spectra in 60 seconds from bacteria.

Almost all *List. monocytogenes* serovars present in food have clear virulent properties (Notermans et al. 1998). However, this does not necessarily mean that all *strains* present in food are virulent, or that the level of virulence is the same. Over 90 % of isolates responsible for human listeriosis belong to only three serovars; i.e., 4b, 1/2a, and 1/2b. Although humans are frequently exposed to *List. monocytogenes*, and high numbers may be ingested during consumption of certain foods, listeriosis is still a relatively rare disease. By using "virulence biomolecules" (e.g., cell surface receptors that pathogens use to gain entry into cells, or biomolecules associated with pathogen movement within host tissues) as capture biomolecules on the microSERS biochip, the biomolecule capture step can select for virulent strains, thereby identifying strains that pose a serious health threat—and, at the same time, providing a means to produce information on the virulence and modes of pathogenicity of the captured cells.

Preliminary studies were conducted to explore the "virulence biomolecule" concept. For example, *Listeria* virulence has been correlated with lectin binding profiles (Facinelli et al. 1998; Cowart et al. 1990), with *List. monocytogenes* serotype 4b exhibiting a pattern of lectin binding that is distinct from other listeriae (Thuan et al. 2000). Some *List. monocytogenes* strains are known to bind to fibronectin (Gilot, Andre, and Content 1999; Gilot, Jossin, and Content 2000), an extracellular matrix molecule that is often exploited by pathogens to gain entry into host cells.

Concanavalin A (ConA) from *Canavalia ensiformis* (jack bean), which binds a broad range of listeriae; wheat germ agglutinin (WGA) from *Triticum vulgaris*, which binds many virulent *Listeria* strains; and fibronectin were selected for preliminary studies. ConA biochips captured a dense, uniform coating of *List. monocytogenes* 3 and 1/2a from suspension, but a significantly lower percentage of 4b. With the WGA biochips, this trend was reversed, with a high capture rate for 4b and a significantly lower recovery of 3 and 1/2a. Fibronectin biochips captured *List. monocytogenes* serotypes 4b and 1/2a, but not serotype 3, from suspension. (It is interesting that 1/2a and 4b are the most dominant serotypes in human patients, whereas serogroup 3 is not often found.) None of the three types of biochip captured *List. innocua* 6a.

The cells could be identified on the basis of their SERS fingerprints when captured by lectins or fibronectin. The fingerprints collected from *Listeria* on the ConA biochips were much stronger than those collected from *Listeria* captured by the antibody biochips. The method used for immobilizing ConA may have produced a submonolayer of protein molecules, permitting the cells to come into direct contact with the SERS-active surface. (Since none of the bacteria adhered to the bare SERS-active unless dried on it, ConA was capturing those cells that were seen.) Alternatively, the high affinity ConA showed for these strains resulted in a much more dense, uniform coating of cells, which made it easier to focus directly on the cells.

Rinsing did not affect the number of *Listeria* cells remaining on any biochip after the capture step. The same was not true when it came to ConA biochips incubated in suspensions of *E. coli* O157:H7 that had been grown at 37°C, however. Without rinsing, ConA chips were heavily coated with *E. coli*, whereas a brief rinsing step removed almost all of the cells. When the brief rinsing step was used, the WGA biochips did not capture any *E. coli* O157:H7. A few *E. coli* cells were retained on the fibronectin biochips. The *E. coli* cells on the ConA and fibronectin biochips produced the distinctive *E. coli* fingerprint, which was readily distinguishable from any *Listeria* fingerprint; i.e., even when an unrelated pathogen unexpectedly bound itself to a given biomolecule, microSERS was still able to differentiate between the target and nontarget bacteria.

4.1.5. Impact of Growth Conditions on Biomolecule Capture

Since growth conditions can affect the expression of cell surface virulence factors, they might be expected to affect binding affinities for receptors. Tests with biochips incubated in suspensions of *Listeria* and *E. coli* O157:H7 grown at 25°C confirmed this. For example, both the WGA and species-specific PAb biochips exhibited a higher affinity for *List. monocytogenes* 4b grown at 25°C than for those grown at 37°C; whereas the ConA biochips exhibited a lower affinity for *List. monocytogenes* 3 grown at 25°C than for those grown at 37°C. Fibronectin biochips captured significantly more of the *E. coli* O157:H7 grown at 25°C than cells grown at 37°C.

Biochip affinity studies were also done with the *Listeria* strains that could grow at 4°C. The ConA, WGA, and species-specific PAb biochips exhibited lower affinities for *List. monocytogenes* 4b and 1/2a grown at 4°C in comparison with the same strains grown at 37°C. The fingerprints collected from the biochips were the same (aside from being weaker on the PAb chips) as those collected from bacteria grown at 4°C and deposited directly onto SERS-active surfaces.

In short, microSERS shows promise for being a powerful tool for screening the impact of growth conditions on pathogen binding to biomolecules associated with virulence, such as cell surface receptors, as well as an effective tool for detecting virulent organisms.

Studies have not yet been initiated to determine the degree to which heat killing will affect the biomolecule capture step of microSERS analysis. However, it is anticipated that many nonviable strains will be screened out during the capture step, since heat will denature many of the bacterial proteins involved in the biomolecular recognition process.

4.1.6. Analysis of Bacteria in Complex Samples

Milk is one of the foods most likely to be contaminated with *Listeria*, and methods for rapid detection of this pathogen are badly needed by the dairy industry. Moreover, lipids from whole milk were found, unexpectedly, to bind to a capture biomolecule for aflatoxins during toxin detection experiments (see the section on microSERS analysis of toxins). Accordingly, a few simple tests were conducted to determine whether milk will interfere with biomolecule capture or SERS fingerprint analysis of *List. monocytogenes*. ConA biochips were used in these experiments.

Prolonged incubation of the ConA biochips in fresh, "organically produced" milk did not affect the spectrum of the lectin, indicating that constituents in the milk did not bind to ConA, nor did incubation in milk denature the lectin. Next, ConA biochips were incubated in milk and then in saline suspensions of *List. monocytogenes*, to determine if constituents in the milk would block the cell binding sites on the lectin. The biochips captured the cells of the serotype 4b strain with roughly the same efficiency as chips that had not been incubated in milk, as expected from the fact that milk did not affect the ConA spectrum. The fingerprints collected from the captured *List. monocytogenes* 4b were the same as those captured by ConA that had not been incubated in milk; i.e., the milk did not affect the biomolecule capture nor the whole-organism fingerprinting step during the analysis for *List. monocytogenes* 4b.

However, ConA biochips showed a distinctly lower affinity for the serotype 3 strain, whether the chips were incubated first in milk and then in suspensions of the serotype 3 strain, or were incubated in samples of the serotype 3 strain that had been grown briefly in milk and then rinsed and resuspended in saline. When ConA biochips were incubated in milk that had already been contaminated with serotype 3, on the other hand, a significant number of cells were captured. These results were intriguing, but the reasons for lower affinity when the cells were grown in milk, rather than simply added to it, have not yet been determined. The milk was not, apparently, affecting the binding interaction per se; instead, growth in milk affected the ability of the cells to bind the lectin. It should be noted that the virulence of this particular strain is unknown; it may be less virulent than other *List. monocytogenes* strains, due, e.g., to a susceptibility to one or more of the various natural antimicrobials found in milk. Certainly, this strain behaved in a very different fashion from the serotype 1/2a and 4b strains, which belong to the serotypes predominant in human patients, in other studies; its binding profile was distinctly different, and it failed to grow at 4°C. Therefore, the reduced "sensitivity" for serotype 3 grown in milk may actually reflect microSERS' ability to evaluate virulence associated with biomolecular recognition binding profiles.

4.2. MicroSERS Detection of Spores

To date, the spores from twenty-two strains belonging to eleven species of *Bacillus* have been used in microSERS studies: *B. anthracis* (three strains), *B. brevis*, *B. cereus*, *B. coagulans* (two strains), *B. licheniformis*, *B. megaterium* (two strains), *B. nealsonii*, *B. pumilus* (six strains), *B. stearothermophilus*, *B. subtilis* (three strains), and *B. thuringiensis*, Some of the species were chosen because they exhibit a high degree of genetic homology: e.g., *B. anthracis*, *B. cereus*, *B. megaterium*, and *B. thuringiensis*. *B. subtilis* is often used as a simulant for anthrax, and as a standard for sterilization validation. *B. megaterium* is used for the development of sterilization methods on NASA's Planetary Protection Program. And *B. cereus* tops the list of pathogens for which food and beverage companies need a simple, rapid assay.

4.2.1. SERS Fingerprinting of Spores

As with bacteria, the first step in developing microSERS for spore detection and identification was to confirm that spore whole-organism fingerprints would be reproducible. Because so little biomass was needed for the analyses, a very simple method was developed for spore production: an agar slant was inoculated, incubated overnight, rinsed with distilled water, pelleted, and resuspended. This typically produced enough spores for several weeks' experiments.

For SERS fingerprinting, the spore preparations were diluted, and deposited directly on SERS-active surfaces for analysis. Fingerprints were collected from isolated, individual spores, typically for 60 seconds, although high-quality spectra could sometimes be collected from individual spores in as little as 10 seconds. The initial experiments were done with fourteen strains belonging to ten species: *B. anthracis* (three strains), *B. brevis*, *B. cereus* (two strains), *B. coagulans* (two strains), *B. megaterium*, *B. stearothermophilus*, *B. subtilis* (two strains), and *B. thuringiensis* (two strains). Several dozen spectra were collected from each strain, including spores in different aliquots from a given harvest, from different harvests, and from aliquots of the spores stored at 4°C and periodically removed for analysis over the course of ten weeks. With one notable exception, the fingerprints were reproducible from spore to spore, from aliquot to aliquot, from harvest to harvest, and throughout storage. Even though variations in colony morphology and color occurred when *B. subtilis* ATCC9372 (*B. globigii*) was grown on nutrient agar, with both white and orange colonies forming on the slants, the fingerprints of the spores were found to be reproducible, regardless of the type of colony that produced them.

B. cereus ATCC11778 was the one exception. Five peaks in its fingerprints varied in intensity from spore to spore, even within a given sample, with the peaks being very strong in roughly two-thirds of the spores in each batch, and significantly weaker, although still very distinct, in the remaining third. These peaks, at ~663, 825, 1015, 1396, and 1574 cm^{-1}, were seen in all spore spectra, and were assigned to calcium dipicolinate (Shibata et al. 1986). Given that the SERS phenomenon is sensitive to distance, it was surprising that dipicolinate could be seen in the fingerprints, since dipicolinate is found in the core of the spore. However, spores are known to release dipicolinate when they are autoclaved (Kozuka, Yasuda, and Tochikubo 1985); and when spores were autoclaved on the microSERS studies, these same five peaks disappeared from the spore SERS fingerprints (see below). Other researchers subsequently reported that they had also observed the dipicolinate signature in SERS spectra from spores (Alexander, Pellegrino, and Gillespie 2003; Jarvis, Brooker, and Goodacre 2006). Esposito et al. (2003) were unable to differentiate among the spores of *B. cereus*, *B. megaterium*, *B. subtilis*, and *B. thuringiensis* using normal Raman scattering, due to the overwhelming dipicolinate signature. SERS may be better able to discriminate among spores than normal Raman techniques, since spectral contributions from the spore coat are, apparently, more strongly enhanced than internal components.

Because all spore spectra contain the dipicolinate signature, these peaks do not appear to be particularly useful for identification at the species/subspecies levels; therefore, variability in their intensities, such as that seen in *B. cereus*, is unlikely to affect the ability to distinguish among the different species/strains. The dipicolinate signature does, however, appear to be very useful for rapidly differentiating between *Bacillus* spores and other types of organisms, and for differentiating between *Bacillus* spores and vegetative cells.

All of the spores that were studied could be individually and uniquely identified at the species level. As might be expected, some species that exhibit a high degree of genetic homology—e.g., *B. cereus* and *B. thuringiensis*—had very similar fingerprints. The SERS fingerprints could also be used to differentiate among spores at the strain level; although again, as might be expected, spores from different strains belonging to the same species sometimes had very similar fingerprints—e.g., the two *B. thuringiensis* strains and the two *B. cereus* strains. The spores could be identified on the basis of peaks associated with aromatic amino acids, cysteine disulfides, protein backbone, amide I, and lipids; many of the most useful peaks could be attributed to those associated with polysaccharide structure and composition.

Despite the fact that microSERS was at a very early stage of development, NASA decided to provide eleven samples of "unknowns" isolated from space flight hardware and NASA facilities, for use in a blind study. The basic question was whether SERS fingerprinting could determine which of the samples were identical. SERS spectra were collected from 6–18 spores from each sample. The 250–600 cm^{-1} window tended to be more variable than the other windows, and was therefore not included in the analysis. The samples were grouped into three clusters, listed in order of "relatedness," based on the similarities of their spectral patterns as determined by simple "eyeball" assessment of obvious peak patterns, rather than any known correlations between key peaks and biological moieties. (Pattern recognition algorithms might well uncover differences that were not readily apparent to the eye and/or assign spore fingerprints to different clusters.)

After the SERS results had been reported, NASA provided the identities of the samples, determined by 16S rDNA analysis. The eleven samples, all of which were different, contained spores from five different species, including six strains of *B. pumilus*, two of *B. subtilis*, and one each of *B. lichenformis*, *B. nealsonii*, and *B. megaterium*. Aside from the type strain of *B. pumilus* and a lab strain of *B. subtilis*, the samples had been isolated from Mars Odyssey or NASA facilities.

Despite the very simplistic method used to group the spectra, and despite the fact that six of the samples were different strains belonging to a single species, the SERS fingerprinting analyses not only determined that all the samples contained different strains, but the placement of nine of the eleven samples by SERS fingerprinting correlated with the placement by 16S rDNA analysis. SERS analysis determined that three samples had the most unusual spectral patterns; these same three samples, i.e., *B. licheniformis* and the two *B. subtilis* strains, were found by 16S rDNA to belong to their own cluster, which branched off very early from the other strains. Both SERS analysis and 16S rDNA found that three of the *B. pumilus* strains were very similar, and clustered with *B. megaterium*. Both techniques also assigned *B. nealsonii* and the *B. pumilus*-type strain to the same cluster; and both found *B. nealsonii* to be more closely related to the three-strain *B. pumilus* cluster than the *B. pumilus*-type strain. There were only two significant differences between the SERS and 16S rDNA analyses: 16S rDNA analysis placed *B. pumilus* 370 and 36 with *B. nealsonii* and *B. pumilus* 1607; while the SERS spectral patterns for *B. pumilus* 370 and 36 were relatively similar to those of *B. megaterium* and the three-strain *B. pumilus* cluster. And while strain 370 was recognized as being different from this group, it did not have the stronger spectral features that would place it in a different cluster. Given that SERS fingerprinting is still in its infancy, and the method used to assign the spectra to clusters was relatively arbitrary, the parallels with the 16S rDNA results were very intriguing. It will be interesting to see what parallels may exist between a larger SERS database analyzed by sophisticated pattern recognition algorithms in comparison with 16S rDNA or similar molecular biology techniques.

Despite the success at matching the NASA unknowns with their phylogenic clusters, however, the similarities in the SERS fingerprints of other strains did not always correspond to genetic homology. While some strains belonging to the same species produced similar fingerprints, others did not. For example, the spectra from the two strains of *B. subtilis* (i.e., ATCC6633 and ATCC9372 *B. subtilis* var *niger*, also known as *B. globigii*) were easily distinguishable, with strong differences throughout the ~400–1700 cm^{-1} spectral range. Other researchers later reported that the SERS fingerprints of spores in their studies also did not necessarily demonstrate any systematic trends correlated with evolutionary lineage (Premasiri, Moir, and Ziegler 2005).

This may sometimes prove to be a very distinct advantage. For example, *B. anthracis*, *B. cereus*, and *B. thuringiensis* are indistinguishable by multilocus enzyme electrophoresis and by sequence analysis of nine chromosomal genes (Helgason et al. 2000). However, the SERS fingerprints from *B. anthracis* spores were sharply different from those of *B. cereus* and *B. thuringiensis* (Fig. 20.11).

In any event, whether or not it is possible to "type" spores on the basis of their SERS fingerprints, all data produced to date, both in microSERS studies and by other groups, indicate that it will be possible to identify spores belonging to the strains in the spectral reference database. And, as above, it should be remembered that microSERS identification will be based on biomolecular recognition, as well as and in combination with, SERS fingerprinting.

4.2.2. Impact of Growth Conditions on Spore Fingerprints

Many researchers have reported that the conditions under which vegetative cells are grown can affect their Raman fingerprints. Spores are produced inside the mother cell where the "growth conditions" may be relatively consistent; and, as dormant structures, spores do not respond to external environmental conditions until germination is triggered. Therefore, spores might be expected to exhibit far less variability than actively growing cells. Little is known about the impact that changing the conditions under which the mother cell is grown may have on the spores' structure and makeup. However, factors have been identified that affect spore

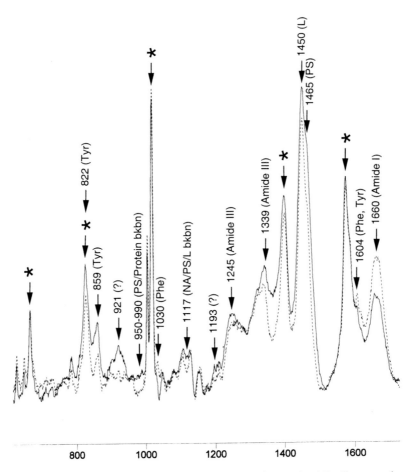

Figure 20.11. SERS fingerprints of single spores of *Bacillus anthracis* (————) and *Bacillus cereus* (- - - - -). Some of the major differences between the two fingerprints are marked. L = lipids, NA = nucleic acids, PS = polysaccharide, Phe = phenylalanine, Tyr = tyrosine, bkbn = backbone, and dipicolinate =*.

heat resistance, such as the temperature (Williams and Robertson 1954; Raso, Barbosa-Cánovas and Swanson 1998; Palop, Mañas and Condón 1999; González et al. 1999) or divalent metal cation environment (Marquis and Bender 1985; Palop, Mañas and Condón 1999; Cazemier, Wagenaars, and ter Steeg, 2001) in which the mother cell is grown. A change in heat resistance suggests that there might be differences in the spore composition.

Accordingly, studies were conducted to determine whether conditions known to affect spore heat resistance could affect their fingerprints. Two strains each of *B. thuringiensis*, *B. subtilis*, and *B. cereus* were tested. Spores were produced using the widest range of temperatures that each strain could tolerate; the lowest temperature that produced spores for any strain was 20°C and the highest was 52°C. Temperature had little effect on vegetative cell colony morphology, although some strains would not grow at the highest temperatures. The temperature at which the mother cells were grown had little or no impact on the spore fingerprints (Fig. 20.12).

In addition, two strains each of *B. thuringiensis* and *B. subtilis* were grown in nutrient agar supplemented with metals Ca^{2+}, Fe^{2+}, K^+, Mg^{2+}, and Mn^{2+}, which is the medium that was reported to enhance spore heat resistance (Palop, Sala, and Condón 1999; Cazemier, Wagenaars, and ter Steeg, 2001); or on nutrient agar and tryptic soy agar for comparison. The

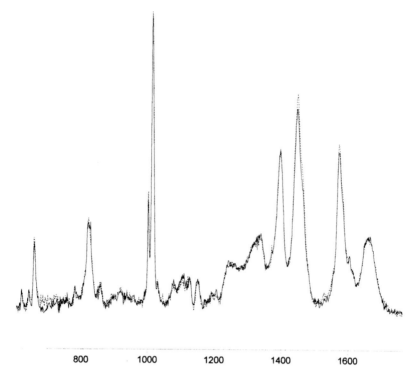

Figure 20.12. SERS fingerprints of single spores of *Bacillus thuringiensis* ATCC35646 from mother cells grown at 23°C (———) or at 42°C (- - - - -). As can be seen, the growth conditions had little impact on the spore fingerprints.

growth medium affected the rate at which the *B. thuringiensis* strains grew and sporulated, but not colony morphology. However, the colony morphologies for the *B. subtilis* strains were distinctly different for each of the three media, indicating that the growth conditions affected the vegetative cell constituents. Nevertheless, the impact of cationic environment on the spore spectra, if any, was extremely subtle.

The fact that neither temperature nor cationic environment had a significant effect on the spore fingerprints was not particularly surprising, since such factors apparently affect spore heat resistance by influencing the protoplast water content (Beaman et al. 1984; Beaman and Gerhardt 1986). Therefore, any impact on the spectrum would be associated with alterations in the packing and layering of core macromolecules, and, hence, with the intermolecular interactions within the core. With dipicolinate dominating the spectrum, the impact of such interactions would be difficult to see. In any event, since the spore spectra were very reproducible when the mother cells were cultured under a range of growth conditions, it appears that it will be a relatively simple matter to develop the spectral reference libraries and identification algorithms for spores.

If the captured spores are allowed to sit on the microSERS biochip for an extended period of time prior to analysis, it is conceivable that the spores will germinate. Dipicolinate is released from spores during germination (Bekhtereva et al. 1975; Scott and Ellar 1978; Alekseev et al. 1985; Kozuka, Yasuda, and Tochikubo 1985; Boschwitz et al. 1991). In addition, Shibata et al. (1986) noted that the calcium dipicolinate peaks seen in the normal Raman spectra of dormant spores were not seen in the spectra of germinated spores. As expected, spore germination was readily apparent in the microSERS studies as well, due to the loss of the dipicolinate signature as well as to other changes found throughout the microbial spectrum. Accordingly, microSERS analysis will be able to distinguish between spores and germinated spores.

4.2.3. Viable vs. Nonviable Spores

NASA funded some of the *Bacillus* spore studies under its Planetary Protection Program. One of the primary objectives was to determine whether microSERS would be able to differentiate between viable and nonviable spores.

The method typically used for sterilizing equipment is heat killing (autoclaving). In preliminary studies, *Bacillus* spores were heat treated at 121°C for 30 min, and their SERS fingerprints compared with those of viable organisms. As noted earlier, five peaks in the SERS spectra of *Bacillus* spores were tentatively assigned to dipicolinate; and Kozuka et al. (1985) found that dipicolinate is released when spores are autoclaved. When spores were autoclaved in microSERS experiments, the five dipicolinate peaks disappeared from the spore fingerprints. In addition, when spores were heat killed directly on a SERS surface, a puddle formed on the surface surrounding the spores; and the SERS spectra collected from the puddles contained a strong dipicolinate signature, along with what appeared to be contributions from small amounts of protein. The disappearance of the dipicolinate peaks, as well as many other changes throughout the SERS fingerprint, made it very easy to differentiate between viable and heat-killed spores (Fig. 20.13.) Some of the changes in the fingerprints caused by heat killing were unique to a given strain; others were seen in the fingerprints of all spores, such as changes in the amide III region and a shift in the maximum of the amide I band, which are consistent with protein denaturation.

NASA was evaluating a number of alternative methods for sterilizing spaceflight hardware, e.g., gamma irradiation, UV irradiation, and hydrogen peroxide treatment, and

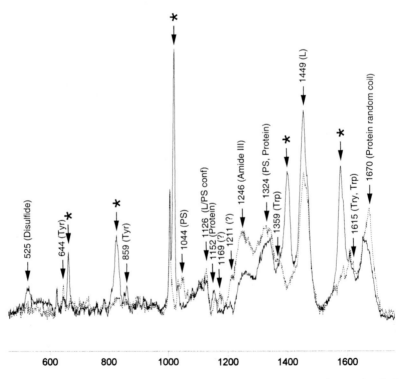

Figure 20.13. SERS fingerprints of *Bacillus thuringiensis* ATCC10792 spores before (———) and after (- - - - -) autoclaving. Note the disappearance of the five key dipicolinate peaks (*), and the significant changes in protein and polysaccharide conformation peaks, such as the protein amide I shift from structured at 1658 to random coil at 1670 cm^{-1}. L = lipids, PS = polysaccharide, Trp = tryptophan, and Tyr = tyrosine.

provided samples of *B. megaterium* spores that had been treated by these methods. There were clear differences in the spectra of spores that had been treated, in comparison with those of the untreated control spores. The fingerprints of all of the damaged spores showed changes in the amide I and III regions, plus an significant increase in the peak at $1054 \, cm^{-1}$, tentatively assigned to pyranose polysaccharide conformation. This $1054 \, cm^{-1}$ appeared to be correlated with the degree of damage that was incurred; in general, the stronger this peak, the stronger and more numerous the changes in the rest of the spectrum. Unlike autoclaving, the gamma irradiation, UV irradiation, and hydrogen peroxide treatments did not affect the dipicolinate content of the spores unless the spores were very severely damaged; and when the spores were damaged enough to begin losing dipicolinate, the peak at $1054 \, cm^{-1}$ was particularly strong. The different treatment methods appeared to affect the spores in different ways. Hydrogen peroxide tended to affect peaks associated with protein; gamma irradiation affected many peaks associated with carbohydrates as well as protein, producing, in particular, a sharp, narrow hallmark peak at $996 \, cm^{-1}$ (pyranose polysaccharide conformation); and UV irradiation affected several lipid peaks (1102, 1129, 1296, 1447, and 1738), as well as disulfide bonds and carbohydrates, notably producing a very strong, broad pyranose polysaccharide peak at $986 \, cm^{-1}$.

Spectra were collected from 45 spores in each sample, and the severity of the damage done to each spore was ranked according to the degree of change to its fingerprint. Six of the UV-irradiated spores appeared normal and another twenty had mild damage; very few appeared to be heavily damaged. Four of the gamma-irradiated spores appeared normal; most had moderate to severe damage, and eleven were losing their dipicolinate content. The hydrogen peroxide spores were the most heavily damaged, with none of the fingerprints appearing normal. Nutrient broth medium was inoculated with each sample. The UV-irradiated sample was showing signs of growth by the third day; the gamma-irradiated sample grew much more slowly, indicating heavier damage to a higher percentage of spores; and the hydrogen peroxide-treated sample did not show any signs of growth, even after 24 days' incubation;i.e., the spore fingerprints reflected the degree of damage that was done by the various treatment methods.

4.2.4. Integrated MicroSERS Detection and Identification of Spores

The next question was whether biomolecule capture would affect the SERS fingerprints from spores. Antibodies capable of binding *B. globigii* spores (anti-Bg) or *B. thuringiensis* (anti-Bt) were obtained, and several different methods for immobilizing them were developed. During the course of the experiments, it was found that the spores of some *Bacillus* species would adhere to the bare gold SERS-active surface. Bovine serum albumin (BSA) was therefore used as a blocking agent. Anti-Bg biochips were incubated in suspensions of *B. globigii* spores, and anti-Bt biochips in suspensions of *B. thuringiensis* spores, for 30 min and then rinsed in buffer. Chips coated with BSA-only were used as controls. The antibody chips captured a thick lawn of spores (Fig. 20.14), whereas only a very light scattering of spores was seen on the BSA control chips.

SERS fingerprints were collected from individual spores captured by the biochips, and compared with fingerprints from spore suspensions deposited directly on uncoated surfaces. When immobilization methods that produced a submonolayer of antibody were used, the fingerprints from the captured spores were not noticeably affected; they were strong, and appeared to be very similar, if not identical, to those from spores deposited directly on uncoated SERS surfaces (Fig. 20.15.) When antibodies were immobilized using a thiol conjugate, the spectra from the captured spores and those deposited directly on the SERS surface were very similar; the only difference was a low, broad peak in the $500-550 \, cm^{-1}$ window of the antibody-captured spore spectrum, presumably due to disulfide bonds from the thiol conjugate. Since this peak was consistent, it did not interfere with spore detection.

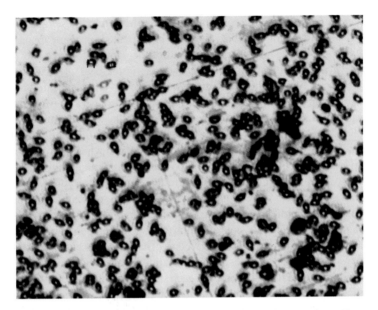

Figure 20.14. *Bacillus globigii* spores captured by a microSERS biochip coated with anti-Bg. The spores are ovoids, approximately 1.5 μm in length and 1.0 μm in diameter.

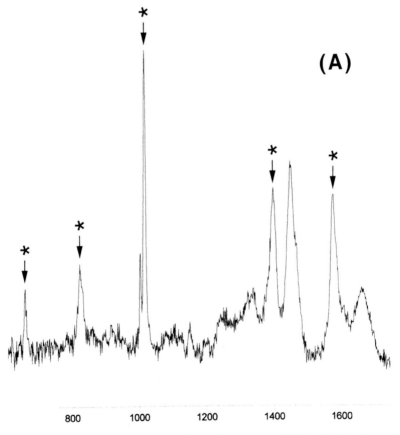

Figure 20.15. SERS fingerprints from individual *Bacillus thuringiensis* spores (A) deposited directly on a SERS-active surface or

Continued

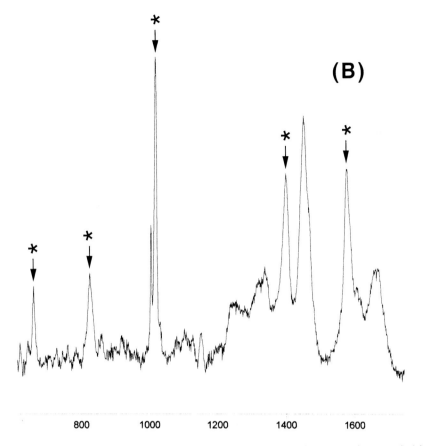

Figure 20.15. *continued* (B) captured by an anti-Bt biochip. The five dipicolinate peaks are marked (*).

Spores could be identified even when biochips were incubated in a mixture of two cross-reactive strains. Tests confirmed that *B. subtilis* ATCC6633 (Bs) is so closely related that its spores would bind to anti-Bg; however, the affinity was not as high. Antibody-coated chips were incubated in mixtures containing various ratios of *B. globigii* and *B. subtilis* spores; and the 975–1325 cm^{-1} window of the SERS fingerprints was used to identify the captured spores. The ratio of spores with the Bg fingerprint was consistently higher than the ratio in the mixtures. Given that the *B. globigii* spores bind more efficiently to the antibody, the results were what would be expected, and demonstrated that the microSERS approach can, indeed, be used to identify spores that cross-react with capture biomolecules.

Viable spores could also be differentiated from nonviable spores. Samples were prepared that contained viable *B. globigii* spores, heat-killed *B. globigii* spores, or mixtures containing different ratios of the two. All of the spores that were captured by biochips incubated in the suspension of viable *B. globigii* spores produced the distinctive dipicolinate signature; none of the spores captured by chips incubated in the suspension of heat-killed spores had any of the dipicolinate peaks in their spectra. When biochips were incubated in mixtures of viable/heat-killed spores, the percentage of spores exhibiting the dipicolinate signature in their fingerprints was significantly higher than the percentage of viable spores in the mixture. These results confirmed that viable spores could be differentiated from nonviable when both were captured by the biochip. It was also interesting to note that the heat-killing process affected the antigens recognized by the goat anti-Bg, so that nonviable spores were not captured as readily.

Finally, to confirm that microSERS can use a single biochip to detect and identify diverse pathogens, studies were done with mixtures of *Cryptosporidium parvum* oocysts and *Bacillus globigii* spores. Methods were developed for immobilizing a monoclonal antibody raised against *Cryptosporidium parvum* oocysts (anti-Cp). Biochips were prepared with four "pixels," namely, anti-Bg, anti-Bg plus BSA, anti-Cp plus BSA, and BSA-only. Biochips were incubated with samples containing a 100:1 or a 10:1 ratio of spores:oocysts. Results were very consistent. The anti-Bg and anti-Bg/BSA pixels all captured a dense lawn of spores from the 100:1 samples; fewer spores were seen on pixels incubated with the 10:1 samples, as expected, but coverage was still relatively high and relatively uniform in distribution. Oocysts were sprinkled across the surfaces of all of the anti-Cp/BSA pixels. No oocysts were seen on any of the BSA-only or anti-Bg-only pixels, or on seven of the anti-Bg/BSA pixels, indicating that the oocysts were selectively captured by the anti-Cp monoclonal. A handful of oocysts were seen on the remaining two BSA-only pixels, possibly due to defects in the surface or incomplete rinsing. A few spores were lightly sprinkled on the BSA-only and anti-Cp/BSA pixels.

Spectra were collected from some of the individual microorganisms on each type of pixel. Since oocysts are so much larger than spores, it was obvious which type of microorganism produced a given spectrum. The oocyst and spore fingerprints were typical of *C. parvum* and *B. globigii*, although some of the fingerprints collected late in the day showed that the captured oocysts were starting to suffer from having been dry for so long. The SERS fingerprints from spores found on the anti-Cp/BSA pixels were readily differentiated from the fingerprints of the oocysts. These findings helped show that even if a nontarget pathogen adheres to a capture biomolecule in the microSERS microarray, it will not generate a false response, since its fingerprint will be different from that of the target pathogen.

4.3. MicroSERS Detection of Bacterial Toxins

MicroSERS can be used to detect and identify the toxins that are produced by pathogens, as well as the pathogens themselves. The same basic approach is used—i.e., (1) capture biomolecules specific for the toxins of interest are immobilized in a microarray on a SERS-active surface; (2) the microarray is incubated in the sample; (3) SERS fingerprints are collected from each biomolecule spot in the microarray; and (4) the presence and identities of the toxins are determined by comparing the SERS fingerprints of the "unknowns" with spectral reference databases comprising the SERS fingerprints of compounds that are known to bind a given capture biomolecule.

Unlike microorganisms, the toxins of interest are often much smaller than the capture biomolecules in the microarray. Nevertheless, even toxins as small as cyanide and ammonia can produce strong, distinctive fingerprints, by exploiting Raman difference spectroscopy (RDS). In RDS, spectra are collected from two samples that are identical except for the change under investigation, e.g., the presence or absence of a ligand. By subtracting the spectrum of an uncomplexed biomolecule from the spectrum of a biomolecule-ligand complex, RDS produces the spectrum of the bound ligand as well as signals that result from changes in the biomolecule induced by ligand binding (Callender and Deng 1994; Callender, Deng, and Gilmanshin, 1998; Mulvaney and Keating 2000). Recently, RDS has been exploited for use in Raman crystallography, a label-free method for following protein-ligand interactions in protein crystals in hanging drops (Carey 2006). Other researchers use RDS to obtain information on the bonds involved in the interaction of a biomolecule and a ligand known to bind to it; microSERS uses RDS to determine whether binding took place, and the identity and quantity of the ligand that has been bound.

The microbially-produced toxins and antibiotics that have been studied to date have ranged in size from ammonium (14 Da) to *Staphylococcus* enterotoxin B (28,366 Da). A variety of different biomolecules that bind toxins via diverse mechanisms were used in their detection.

Toxins that comprised as little as 0.02 % by weight of the biomolecule-toxin complex produced strong, distinctive fingerprints when SERS spectra were collected from ~1 μm diameter areas of the capture biomolecule spots.

4.3.1. SERS Fingerprinting of Toxins

Cyanide and ammonium are among the smallest toxic compounds produced by human pathogens (Goldfarb and Margraf 1967; Hosseini, Ghaffariyeh, and Nikandish 2007). MicroSERS studies have shown that both of these toxics can be detected by biomolecule capture and SERS fingerprinting. For example, creatine kinase and glutathione reductase produced a simple fingerprint for cyanide, primarily comprising two strong peaks, one in the short wavelength region of the spectrum and the other in the long wavelength region (Fig. 20.16). These peaks correlate to the two peaks seen in the Raman spectrum of uncomplexed cyanide itself, at 313 and 2161 cm^{-1}. Four capture biomolecules were tested for ammonium detection, namely, urease, which is inhibited by ammonium; adenosine deaminase, which is stimulated by ammonium; glutamic dehydrogenase, which recognizes ammonium as a substrate as well as an inhibitor; and glutathione reductase, which is inactivated by ammonium. All four produced distinctive fingerprints for ammonium, including a peak contributed by the ammonium itself, as well as changes in protein bands caused by the binding interactions inducing structural changes in the proteins.

Biomolecules are rarely, if ever, totally specific for a single compound. MicroSERS studies have therefore also addressed the ability to differentiate among cross-reactive compounds. Adenosine deaminase, for example, is inhibited by coformycin (284 Da), a microbially-produced toxin that is used as a drug and that also has herbicidal activity; urease is inhibited by urea derivatives such as hydroxyurea; and glutamic dehydrogenase is inhibited by acetate. Coformycin, hydroxyurea, and acetate were readily detected, and produced fingerprints that differed significantly from those produced by ammonium.

Other studies confirmed that microSERS can differentiate among cross-reactive toxins that are more similar in structure. Trimethylamine (59 Da), cadaverine (102 Da), and histamine (111 Da) are metabolic products formed during food spoilage, and are valuable for assessing and predicting microbial quality in a variety of foods. Such biogenic amines typically cross-react with any biomolecule that binds one of them. For example, heme enzymes in general may represent common targets where bioamines interact to influence cell function. The heme proteins cytochrome P450 microsome and cytochrome *c*, as well as the non-heme diamine

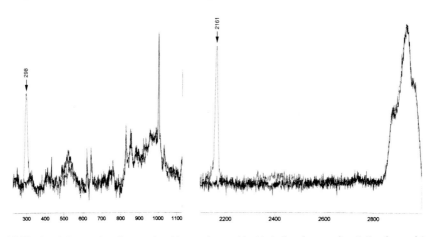

Figure 20.16. SERS fingerprints from a glutathione reductase biochip before (———) and after (- - - - -) incubation in cyanide solution. The very simple fingerprint reflects the simple structure of the very small toxin.

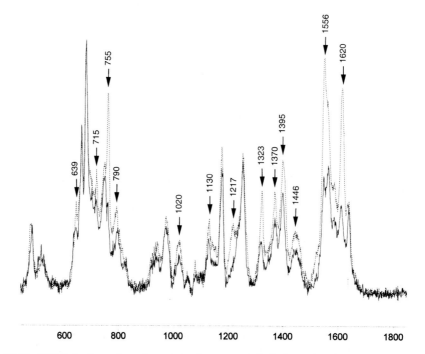

Figure 20.17. SERRS (surface-enhanced resonance Raman scattering) fingerprints from a cytochrome c biochip before (———) and after (- - - - -) incubation in trimethylamine solution. Because the heme moiety of the enzyme is resonantly enhanced at 632.8 nm irradiation, its spectral contributions dominate the spectra to the point that contributions from the amino acid moieties are undetectable. The fingerprint reflects the impact that the bioamine has on the heme vibrational modes, and not peaks that are contributed directly by the toxin.

oxidase, were used to prepare biochips. The heme moieties of cytochrome P450 and cytochrome c were resonantly enhanced at the wavelength that was used for analysis (i.e., 632.8 nm). Consequently, the spectral fingerprints were dominated by vibrational modes of the heme group; and the fingerprints produced by the bioamines were associated with their impact on the vibrational modes of the heme, rather than with bonds in the bioamines themselves (Fig. 20.17). Nevertheless, all three bioamines produced unique fingerprints when captured by any of the three types of protein.

Brevetoxins 2 (895 Da) and 3 (897 Da) could be detected and individually distinguished when captured by anti-brevetoxin polyclonal antibodies. *Staphylococcus* enterotoxin B (SEB) could be detected, and distinguished from *Staphylococcus* enterotoxins C_1 and/or A, in tests with a monoclonal antibody.

Toxins are often present in mixtures of closely related structures. Aflatoxins are a family of toxins often found in mixtures that exhibit a high degree of cross-reactivity. Of the naturally occurring aflatoxins produced by *Aspergillus flavus* and *A. parasiticus*, aflatoxin B_1 is the most prevalent and is also the most toxic. Aflatoxin B_1 and the most closely related aflatoxins, i.e., B_2, G_1, G_2, and M_1, which range in size from 312 to 330 Da, were selected for study. Four biomolecules that bind aflatoxins via diverse mechanisms were used in their detection. Glutathione transferase detoxifies aflatoxin; lipoxygenase activates aflatoxin to its more toxic epoxide metabolite; RNA polymerase is inhibited by covalent bonding with aflatoxin; and AT-B1 binds aflatoxin noncovalently. The biomolecules also ranged widely in molecular weight, from 24,530 Da (glutathione transferase) to almost twenty times that size (RNA polymerase, 440,000 Da.) The capture biomolecules all yielded fingerprints unique to aflatoxins B_1, B_2, G_1, G_2, and, when it would bind, M_1. Even RNA polymerase, which differs in molecular mass

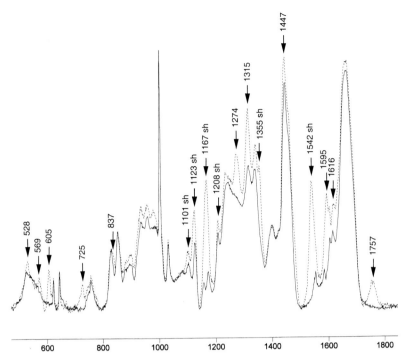

Figure 20.18. SERS fingerprints from an RNA polymerase biochip before (———) and after (- - - - -) incubation in aflatoxin G_1 solution. RNA polymerase differs in molecular mass from the aflatoxin by a factor of $\sim$1,410. New peaks that are contributed by the bound toxin are marked with arrows; peaks that are significantly shifted ($\geq 4\,cm^{-1}$) from the spectrum of the unbound toxin are marked "sh." Note that the fingerprint includes protein peaks that change in intensity, shape, and/or maxima as well.

from aflatoxins by a factor of $\sim$1,340–1,410, produced very strong, distinctive fingerprints (Fig. 20.18). The most toxic of the family, aflatoxin B_1 was tested in mixtures with aflatoxins B_2, G_1, G_2, or M_1, and was readily detected and identified in their presence. Aflatoxins B_1 and/or G_1 could be detected and identified in mixtures with nonligand toxins such as T-2 toxin.

Similar studies with antibody and/or enzyme biochips incubated in series of related antibiotics (e.g., sulfamethazine, sulfamerazine, sulfadimethoxine, and sulfamethizole) also showed that cross-reactive compounds could be detected and individually identified when microSERS biochips were incubated in mixtures.

Several types of evidence confirmed that fingerprint formation was due to specific binding between the toxin and the biomolecule. (1) Major peaks contributed by the toxin and/or biomolecule were often shifted, changed in relative intensity, or sometimes even missing altogether in comparison with uncomplexed species, indicating that a binding interaction had affected the chemistries. (2) Different biomolecules gave different fingerprints when complexed with the same toxin, reflecting differences in the binding mechanisms. For example, the short wavelength peak in the glutathione reductase biochip fingerprint for cyanide was sharp and narrow, at $298\,cm^{-1}$; whereas that in the creatine phosphokinase fingerprint for cyanide was broad, centered at 300, and had a shoulder at $323\,cm^{-1}$. (3) Different toxin concentrations were needed to produce a detectable fingerprint with different biomolecules; i.e., biomolecule affinity determined whether the toxin could be detected at a given concentration. (4) In general, the higher the concentration of the toxin in the sample, the stronger its fingerprints from the biochip. However, once the biomolecule active sites became saturated, the toxin fingerprint stopped increasing in intensity, even though more concentrated toxin samples were analyzed. (5) Nonligand compounds did not produce fingerprints, even after prolonged (60 min) incubation of the capture biomolecule in concentrated nonligand samples. For example, T-2 toxin and diisopropylfluorophosphate did not have any effect on the spectra of the anti-aflatoxin monoclonal antibody biochips. (6) Fingerprint formation was not due to nonspecific adsorption of the toxin directly onto the SERS-active surface. The concentrations used in the microSERS studies were too low for the toxin to produce a SERS spectrum by itself. Even a significantly higher (e.g., 100-fold) concentration usually did not generate a spectrum when deposited directly onto the SERS surface for analysis.

MicroSERS analysis was also shown to be quantitative. Every other microarray technology suffers from problems associated with uneven pixel printing, since there is no way to determine whether every pixel has been evenly coated with active biomolecule during microarray fabrication. The relatively crude methods that were used to prepare microSERS biochips for toxin analysis produced spots of protein that varied in size, shape, and thickness. Since spectra were collected from a very small percentage of the biomolecule spot, the amount of active protein contributing to the spectra could vary widely from chip to chip, and even from area to area of the same spot. Nevertheless, standard curves with an r^2 as high as 0.9999 were obtained, by using an internal standard to adjust for the amount of protein contributing to the spectrum. For example, the phenylalanine aromatic ring breathing mode at $1003 \, \text{cm}^{-1}$ is not affected when the local environment of the amino acid residue changes, and its intensity therefore correlates to the amount of protein contributing to the spectrum. Biochips were incubated in various concentrations of toxin, typically for 10 min, and their fingerprints were collected. Standard curves were constructed by normalizing the intensity of a major peak contributed by the toxin (e.g., the peaks at ~ 1168 and $\sim 1550 \, \text{cm}^{-1}$ in aflatoxin G_1 and B_1 fingerprints, respectively) against the intensity of the phenylalanine peak, and then plotting the peak intensity ratio vs. the ligand concentration. Even in the case of RNA polymerase biochips analyzing aflatoxins, the standard curves were linear over three orders of magnitude.

4.3.2. Analysis of Toxins in Complex Samples

Aflatoxins B_1 and G_1 were also used in studies on the microarray analysis of complex samples. The samples in the first series of experiments were controlled mixtures of known composition containing multiple inorganic and organic constituents (some of which are major components in urine), spiked with neither, one, or both aflatoxins. Diverse proteins were used in the microarrays; some of them had been shown to capture aflatoxins, whereas others were known to capture one or more of the other sample constituents. Two different types of microarrays and eight different samples were used. The resulting SERS fingerprint analyses showed that (a) different sample constituents were captured at different spots on the microarray, as predicted on the basis of the specificities and affinities of the biomolecules for those constituents; (b) when sample constituents were cross-reactive for a given biomolecule, each constituent could be identified by its contributions to the fingerprint produced at the cross-reactive capture biomolecule; and (c) microSERS could be used to analyze samples for inorganic and organic targets simultaneously.

Finally, complex "real world" samples of unknown composition were used. The "real world" samples included the fluid from liquid impingers used to scrub laboratory air for 24 hours, the extract from a kimwipe used to scrub a dirty lab bench, an opaque suspension from long-grain rice powdered and blended with boiling water, whole milk, and peanut butter dissolved in hot water. Two antibodies and the enzyme lipoxygenase were used as aflatoxin capture biomolecules. Prolonged incubation of biochips in any sample, except milk, had no discernable impact on the SERS spectra of the biomolecules. When aflatoxins B_1 and/or G_1 were spiked into these samples, however, a brief incubation produced the distinctive toxin fingerprints. A few bands, apparently from lipids, were observed in the spectra after biochips made with one of the antibodies were incubated in milk; however, they were in windows that did not interfere with aflatoxin detection.

Innovative software was used to begin developing spectral reference databases and identification algorithms. This software included a patched orthogonal (Gray et al. 1998), genetic evolutionary algorithms, and integrated multivariate patch and intelligent neural net codes (Wagner et al. 1996). A total of 56 "unknown" samples were analyzed by the identification algorithms. The test matrix included negative samples (clean water, liquid impinger fluid, kimwipe extract, and boiling water rice "extract"), and aliquots of these samples spiked with one or both of the aflatoxins alone, or mixed with T-2 toxin. There were no false positives or false negatives for either aflatoxin, and no false positives for T-2 toxin, for any sample analyzed by any of the three different capture biomolecules. This was particularly impressive in light of the fact that only clean samples (i.e., spiked and unspiked milliQ water) were used in developing the reference database and identification algorithms; i.e., the algorithms were not trained on "real world" samples before being used in the experiment.

5. Conclusion and Future Perspectives

Raman whole-organism fingerprinting shows exceptional potential for rapid identification of bacterial pathogens, no matter which Raman technique is used. The different techniques that are under investigation provide the flexibility to design experiments that yield the type of information that is needed for a given application. For example, normal Raman microscopy provides information on all of the constituents in the bacterial or spore sample, whereas metal colloid suspensions generate SERS fingerprints dominated by contributions from microbial surface structures. In addition to identity, microSERS can produce information on the physiological state of the microorganism, its ability to bind "virulence" biomolecules, and its ability to produce toxins.

Raman fingerprinting is still in its infancy. Many recent technological advances, which are outside the scope of this chapter, can be used to transform the basic Raman fingerprinting concept into user-friendly, automated systems for high-throughput analyses. They include, for example, hyperspectral imaging Raman microscopes that are capable of collecting hundreds of thousands of spectra from a surface, in minutes, with a spatial resolution of as fine as 250 nm. Some of these are already being investigated for use in detecting pathogens in complex samples (e.g., Escoriza et al. 2006). Innovative algorithms are being explored for processing and interpreting fingerprint data (Maquelin et al. 2003; Buttingsrud and Alsberg 2004; Jarvis and Goodacre 2005; Rösch et al. 2005; Preisner et al. 2007). New lab-on-a-chip methods are being developed for concentrating microbial samples, which, when coupled with SERS whole-organism fingerprinting, can rapidly analyze very dilute bacterial samples (Hou, Maheshwari, and Chang 2007). Approaches that rely on SERS fingerprinting may benefit from new, cost-effective methods for manufacturing robust, uniform, highly-reproducible SERS-active surfaces (Perney et al. 2006). And microSERS can utilize new microarray printing technologies (Lynch et al. 2004) to produce arrays of micron-sized biomolecule spots, for use in the detection of toxins and viruses. Preliminary studies have already confirmed that microSERS fingerprints can be collected from arrays of capture biomolecules printed in spots as small as 5 μm diameter, after incubation in submicroliter samples of small-molecule ligands.

Acknowledgements

Portions of the microSERS work were funded by the National Institutes of Health (NIEHS and NIGMS), the National Aeronautics and Space Administration, the U.S. Army Soldier and Biological Chemical Command, the National Oceanographic and Atmospheric Administration, the Army Research Office, and the Water Environment Research Foundation.

References

Alekseev AN, Karabanova LN, Krainova OA, Krasov ES and Kashparova EV (1985) Amino acid and mineral element content and the activity of various enzymes in germinating spores of *Bacillus thuringiensis*. Mikrobiologiia 54(2):181–185

Alexander TA, Pellegrino PM and Gillespie JB (2003) Near-infrared surface-enhanced-Raman-scattering-mediated detection of single optically trapped bacterial spores. Appl. Spectrosc. 57(11):1340–1345

Annous BA, Becker LA, Bayles DO, Labeda DP and Wilkinson BJ (1997) Critical role of anteiso-C15:0 fatty acid in the growth of *Listeria monocytogenes* at low temperatures. Appl. Environ. Microbiol. 63(10):3887–3894

Basher HA, Fowler DR, Rodgers FG, Seaman A and Woodbine M (1984) Role of haemolysin and temperature in the pathogenesis of *Listeria monocytogenes* in fertile hens' eggs. Zentralbl. Bakteriol. Mikrobiol. Hyg. [A] 258(2–3):223–231

Beaman TC and Gerhardt P (1986) Heat resistance of bacterial spores correlated with protoplast dehydration, mineralization, and thermal adaptation. Appl. Environ. Microbiol. 52:1242–1246

Beaman TC, Koshikawa T, Pankratz HS and Gerhardt P (1984) Dehydration partitioned within core protoplast accounts for heat resistance of bacterial spores. FEMS Microbiol. Lett. 24:47–51

Bekhtereva MN, Marchenko IV, Galanina LA and Loginova ON (1975) Change in *Bacillus anthracoides* spores and their content of dipicolinic acid during germination. Mikrobiologiia 44(2):233–236

Berger AJ and Zhu Q (2003) Identification of oral bacteria by Raman microspectroscopy. J. Modern Opt. 50(15-17):2375–2380

Boschwitz H, Gofshtein-Gandman L, Halvorson HO, Keynan A and Milner Y (1991) The possible involvement of trypsin-like enzymes in germination of spores of *Bacillus cereus* T and *Bacillus subtilis* 168. J. Gen. Microbiol. 137(Pt 5):1145–1153

Buttingsrud B and Alsberg BK (2004) A new maximum entropy-based method for deconvolution of spectra with heteroscedastic noise. J. Chemometrics 18(12):537–547

Byrne B and Swanson MS (1998) Expression of *Legionella pneumophila* virulence traits in response to growth conditions. Infect. Immun. 66(7):3029–3034

Callender R and Deng H (1994) Nonresonance Raman difference spectroscopy: a general probe of protein structure, ligand binding, enzymatic catalysis, and the structures of other biomacromolecules. Annu. Rev. Biophys. Biomol. Struct. 23:215–245

Callender R, Deng H and Gilmanshin R (1998) Raman difference studies of protein structure and folding, enzymatic catalysis and ligand binding. J. Raman Spectrosc. 29:15–21

Carey PR (2006) Raman crystallography and other biochemical applications of Raman microscopy. Annu. Rev. Phys. Chem. 57: 527–554

Cazemier AE, Wagenaars SFM and ter Steeg PF (2001) Effect of sporulation and recovery medium on the heat resistance and amount of injury of spores from spoilage bacilli. J. Appl. Microbiol. 90:761–770

Chan JW, Esposito AP, Talley CE, Hollars CW, Lane SM and Huser T (2004) Reagentless identification of single bacterial spores in aqueous solution by confocal laser tweezers Raman spectroscopy. Anal. Chem. 76(3):599–603

Choo-Smith LP, Maquelin K, van Vreeswijk T, Bruining HA, Puppels GJ, Ngo Thi NA, Kirschner C, Naumann D, Ami D, Villa AM, Orsini F, Doglia SM, Lamfarraj H, Sockalingum GD, Manfait M, Allouch P and Endtz HP (2001) Investigating microbial (micro)colony heterogeneity by vibrational spectroscopy. Appl. Environ. Microbiol. 67(4):1461–1469

Cowart RE, Lashmet J, McIntosh ME and Adams TJ (1990) Adherence of a virulent strain of *Listeria monocytogenes* to the surface of a hepatocarcinoma cell line via lectin-substrate interaction. Arch. Microbiol. 153(3):282–286

Czuprynski CJ, Brown JF and Roll JT (1989) Growth at reduced temperatures increases the virulence of *Listeria monocytogenes* for intravenously but not intragastrically inoculated mice. Microb. Pathog. 7(3):213–23

De Gussem K, Vandenabeele P, Verbeken A and Moens L (2005) Raman spectroscopic study of *Lactarius* spores (Russulales, Fungi). Spectrochim. Acta A 61:2896–2908

Driskell JD, Kwarta KM, Lipert RJ and Porter MD (2005) Low-level detection of viral pathogens by a surface-enhanced Raman scattering based immunoassay. Anal. Chem. 77:6147–6154

Durst J (1975) The role of temperature factors in the epidemiology of listeriosis. Zentralbl. Bakteriol. [Orig. A] 233(1):72–74

Efrima S, Bronk BV and Czégé J (1999) Surface enhanced Raman spectroscopy of bacteria coated by silver. Proc. SPIE 3602:164–171

Eichenbaum Z, Green BD and Scott JR (1996) Iron starvation causes release from the group A streptococcus of the ADP-ribosylating protein called plasmin receptor or surface glyceraldehyde-3-phosphate-dehydrogenase. Infect. Immun. 64(6):1956–1960

Escoriza MF, Van Briesen JM, Stewart S, Maier J and Treado PJ (2006) Raman spectroscopy and chemical imaging for quantification of filtered waterborne bacteria. J. Microbiol. Methods 66(1):63–72

Esposito AP, Talley CE, Huser T, Hollars CW, Schaldach CM and Lane SM (2003) Analysis of single bacterial spores by micro-Raman spectroscopy. Appl. Spectrosc. 57:868–871

Facinelli B, Giovanetti E, Magi G, Biavasco F and Varaldo PE (1998) Lectin reactivity and virulence among strains of *Listeria monocytogenes* determined in vitro using the enterocyte-like cell line Caco-2. Microbiol. 144(Pt 1): 109–118

Gilot P, Andre P and Content J (1999) *Listeria monocytogenes* possesses adhesins for fibronectin. Infect. Immun. 67(12):6698–6701

Gilot P, Jossin Y and Content J (2000) Cloning, sequencing and characterisation of a *Listeria monocytogenes* gene encoding a fibronectin-binding protein. J. Med. Microbiol. 49(10):887–896

Goldfarb WB and Margraf H (1967) Cyanide production by *Pseudomonas aeruginosa*. Ann. Surg. 165(1):104–110

González I, López M, Matnez S, Bernardo A and González J (1999) Thermal inactivation of *Bacillus cereus* spores formed at different temperatures. Int. J. Food Microbiol. 51:81–84

Goodacre R, Timmins EM, Burton R, Kaderbhai N, Woodward AM, Kell DB and Rooney PJ (1998) Rapid identification of urinary tract infection bacteria using hyperspectral whole-organism fingerprinting and artificial neural networks. Microbiology 144:1157–1170

Gray PC, Shokair I, Rosenthal S, Tisone GC, Wagner JS, Rigdon LD, Siragusa GR and Heinin RJ (1998) Distinguishability of biological material using ultraviolet multi-spectral fluorescence. Appl. Opt. 37:6037–6041

Grow AE (1999) Raman optrode processes and devices for detection of chemicals and microorganisms. U.S. Patent No. 5,866,430

Grow AE (2001) SBIR Phase I Final Report, Contract No. 50-DKNA-0-90046. U.S. Department of Commerce, NOAA

Grow AE (2002) SBIR Phase I Final Report, Grant No. 1R43ES11226-01. NIEHS

Grow AE, Wood L, Deal M, Claycomb J, Lee S and Thompson P (2002) SBIR Phase II Final Report, Contract No. NAS5-00222. NASA

Grow AE, Wood LL, Claycomb JL and Thompson PA (2003a) New biochip technology for label-free detection of pathogens and their toxins. J. Microbiol. Methods. 53(2):221–233

Grow AE, Deal MS, Thompson PA and Wood LL (2003b) Evaluation of the Doodlebug: A Biochip for Detecting Waterborne Pathogens. IWA Publishing, London

Guicheteau JA and Christesen SD (2004) Surface-enhanced Raman immunoassay (SERIA): detection of *Bacillus globigii* in ground water. Proc. SPIE 5585:113–121

Guicheteau J and Christesen SD (2007) Principal component analysis of bacteria using surface-enhanced Raman spectroscopy. Proc. SPIE 6218:62180G

Helgason E, Okstad OA, Caugant DA, Johansen HA, Fouet A, Mock M, Hegna I and Kolsto A-B (2000) *Bacillus anthracis*, *Bacillus cereus*, and *Bacillus thuringiensis*–one species on the basis of genetic evidence. Appl. Environ. Microbiol. 66:2627–2630

Helm D, Labischinski H, Schallehn G and Naumann D (1991) Classification and identification of bacteria by Fourier transform infrared spectroscopy. J. Gen. Microbiol. 137:69–79

Holt C, Hirst D, Sutherland A and MacDonald F (1995) Discrimination of species in the genus *Listeria* by Fourier transform infrared spectroscopy and canonical variate analysis. Appl. Environ. Microbiol. 61(1):377–378

Hosseini H, Ghaffariyeh A and Nikandish R (2007) Noxious compounds in exhaled air, a potential cause for ocular manifestations of *H. pylori* gastrointestinal infection. Med. Hypotheses 68(1):91–93

Hou D, Maheshwari S and Chang H-C (2007) Rapid bio-particle concentration and detection by combining a discharge driven vortex with surface-enhanced Raman scattering. Biomicrofluidics 1:014106–014118

Huang WE, Griffiths RI, Thompson IP, Bailey MJ and Whiteley AS (2004) Raman microscopic analysis of single microbial cells. Anal. Chem. 76:4452–4458

Huang YS, Karashima T, Yamamoto M and Hamaguchi HO (2005) Molecular-level investigation of the structure, transformation, and bioactivity of single living fission yeast cells by time- and space-resolved Raman spectroscopy. Biochemistry 44(30):10009–10019

Huang WE, Bailey MJ, Thompson IP, Whiteley AS and Spiers AJ (2007) Single-cell Raman spectral profiles of *Pseudomonas fluorescens* SBW25 reflects in vitro and in planta metabolic history. Microbial Ecology 53:414–425

Hutsebaut D, Maquelin K, De Vos P, Vandenabeele P, Moens L and Puppels GJ (2004) Effect of culture conditions on the achievable taxonomic resolution of Raman spectroscopy disclosed by three *Bacillus* species. Anal. Chem. 76(21):6274–6281

Hutsebaut D, Vandroemme J, Heyrman J, Dawyndt P, Vandenabeele P, Moens L and de Vos P (2006) Raman microspectroscopy as an identification tool within the phylogenetically homogeneous "*Bacillus subtilis*"-group. Syst. Appl. Microbiol. 29(8):650–660

Ibelings MS, Maquelin K, Endtz HP, Bruining HA and Puppels GJ (2005) Rapid identification of *Candida* spp. in peritonitis patients by Raman spectroscopy. Clin. Microbiol. Infect. 11(5):353–358

James BW, Mauchline WS, Fitzgeorge RB, Dennis PJ and Keevil CW (1995) Influence of iron-limited continuous culture on physiology and virulence of *Legionella pneumophila*. Infect. Immun. 63(11):4224–4230

James BW, Mauchline WS, Dennis PJ and Keevil CW (1997) A study of iron acquisition mechanisms of *Legionella pneumophila* grown in chemostat culture. Curr. Microbiol. 34(4):238–243

Jarvis RM and Goodacre R (2004) Discrimination of bacteria using surface-enhanced Raman spectroscopy. Anal. Chem. 76(1):40–47

Jarvis RM, Brooker A and Goodacre R (2004) Surface-enhanced Raman spectroscopy for bacterial discrimination utilizing a scanning electron microscope with a Raman spectroscopy interface. Anal. Chem. 76(17):5198–5202

Jarvis RM and Goodacre R (2005) Genetic algorithm optimization for pre-processing and variable selection of spectroscopic data. Bioinformatics 21(7):860–868

Jarvis RM, Brooker A and Goodacre R (2006) Surface-enhanced Raman scattering for the rapid discrimination of bacteria. Faraday Discus. 132:281–292

Kapatral V, Olson JW, Pepe JC, Miller VL and Minnich SA (1996) Temperature-dependent regulation of *Yersinia enterocolitica* Class III flagellar genes. Mol. Microbiol. 19(5):1061–1071

Kirschner C, Maquelin K, Pina P, Ngo Thi NA, Choo-Smith L-P, Sockalingum GD, Sandt C, Ami D, Orsini F, Doglia SM, Allouch P, Mainfait M, Puppels GJ and Naumann D (2001) Classification and identification of enterococci: a comparative phenotypic, genotypic, and vibrational spectroscopic study. J. Clin. Microbiol. 39(5):17631770

Kneipp K, Wang Y, Kneipp H, Perelman LT, Itzkan I, Dasari RR and Feld MS (1997) Single molecule detection using surface-enhanced Raman scattering. Phys. Rev. Lett. 78:1667–1670

Kneipp K, Kneipp H, Kartha VB, Manoharan R, Deinum G, Itzkan I, Dasari RR and Feld MS (1998a) Detection and identification of a single DNA base molecule using surface-enhanced Raman scattering (SERS). Phys. Rev. E 57, Rapid Comm., R6281

Kneipp K, Kneipp H, Manoharan R, Hanlon EB, Itzkan I, Dasari RR and Feld MS (1998b) Extremely large enhancement factors in surface-enhanced Raman scattering for molecules on colloidal gold clusters. Appl. Spectrosc. 52: 1493–1497

Kneipp K, Kneipp H, Itzkan I, Dasari RR and Feld MS (1999) Surface-enhanced non-linear Raman scattering at the single molecule level. Chem. Phys. 247:155–162

Kozuka S, Yasuda Y and Tochikubo K (1985) Ultrastructural localization of dipicolinic acid in dormant spores of *Bacillus subtilis* by immunoelectron microscopy with colloidal gold particles. J. Bacteriol. 162(3):1250–1254

Krafft C, Knetschke T, Funk RH and Salzer R (2006) Studies on stress-induced changes at the subcellular level by Raman microspectroscopic mapping. Anal. Chem. 78(13):4424–4429

Lin FY, Sabri M, Alirezaie J, Li D and Sherman PM (2005) Development of a nanoparticle-labeled microfluidic immunoassay for detection of pathogenic microorganisms. Clin. Diagn. Lab. Immunol. 12(3):418–425

Litwin CM and Calderwood SB (1994) Analysis of the complexity of gene regulation by fur in *Vibrio cholerae*. J. Bacteriol. 176(1):240–248

Lynch M, Mosher C, Huff J, Nettikadan S, Johnson J and Henderson E (2004) Functional protein nanoarrays for biomarker profiling. Proteomics 4(6):1695–1702

Maquelin K, Choo-Smith L-P, van Vreeswijk T, Endtz HP, Smith B, Bennett R, Bruining HA, and Puppels GJ (2000) Raman spectroscopic method for identification of clinically relevant microorganisms growing on solid culture medium. Anal. Chem. 72:12–19

Maquelin K, Choo-Smith LP, Endtz HP, Bruining HA and Puppels GJ (2002a) Rapid identification of *Candida* species by confocal Raman microspectroscopy. J. Clin. Microbiol. 40(2):594–600

Maquelin K, Kirschner C, Choo-Smith LP, van den Braak N, Endtz HP, Naumann D and Puppels GJ (2002b) Identification of medically relevant microorganisms by vibrational spectroscopy. J. Microbiol. Methods 51(3): 255–271

Maquelin K, Kirschner C, Choo-Smith LP, Ngo-Thi NA, van Vreeswijk T, Stammler M, Endtz HP, Bruining HA, Naumann D and Puppels GJ (2003) Prospective study of the performance of vibrational spectroscopies for rapid identification of bacterial and fungal pathogens recovered from blood cultures. J. Clin. Microbiol. 41(1):324–329

Maquelin K, Dijkshoorn L, van der Reijdenm TJ and Puppels GJ (2006) Rapid epidemiological analysis of *Acinetobacter* strains by Raman spectroscopy. J. Microbiol. Methods 64(1):126–131

Maresca B (1995) Unraveling the secrets of *Histoplasma capsulatum*. A model to study morphogenic adaptation during parasite host/host interaction. Verh. K. Acad. Geneeskd. Belg. 57(2):133–156

Marquis RE and Bender GR (1985) Mineralisation and heat resistance of bacterial spores. J. Bacteriol. 161:789–791

Mastronicolis SK, Boura A, Karaliota A, Magiatis P, Arvanitis N, Litos C, Tsakirakis A, Paraskevas P, Moustaka H and Heropoulos G (2006) Effect of cold temperature on the composition of different lipid classes of the foodborne pathogen *Listeria monocytogenes*: focus on neutral lipids. Food Microbiol. 23(2):184–194

Mauchline WS, Araujo R, Wait R, Dowsett AB, Dennis PJ and Keevil CW (1992) Physiology and morphology of *Legionella pneumophila* in continuous culture at low oxygen concentration. J. Gen. Microbiol. 138(Pt 11): 2371–2380

Mauchline WS, James BW, Fitzgeorge RB, Dennis PJ and Keevil CW (1994) Growth temperature reversibly modulates the virulence of *Legionella pneumophila*. Infect. Immun. 62(7):2995–2997

McLauchlin J (1997) The identification of *Listeria* species. Int. J. Food Microbiol. 38(1):77–81

Miller VL and Mekalanos JJ (1988) A novel suicide vector and its use in construction of insertion mutations: osmoregulation of outer membrane proteins and virulence determinants in *Vibrio cholerae* requires toxR. J. Bacteriol. 170(6):2575–2583

Mulvaney SP and Keating CD (2000) Raman spectroscopy. Anal. Chem. 72:145R–157R

Naumann D, Helm D and Labischinski H (1991) Microbiological characterizations by FT-IR spectroscopy. Nature 351:81–82

Naumann D, Helm D, Labischinski H and Giesbrecht P (1991) The characterization of microorganisms by Fourier-transform infrared spectroscopy (FT-IR). In: Nelsen WH (ed) Modern Techniques for Rapid Microbiological Analysis. VCH, New York, pp 43–96

Naumann D, Keller S, Helm D, Schultz NC and Schrader B (1995) FT-IR spectroscopy and Raman spectroscopy are powerful analytical tools for the non-invasive characterization of intact microbial cells. J. Mol. Struct. 347:399–406

Ni J, Lipert RJ, Dawson GB and Porter MD (1999) Immunoassay readout method using extrinsic Raman labels adsorbed on immunogold colloids. Anal. Chem. 71(21):4903–4908

Nie S and Emory SR (1997) Probing single molecules and single nanoparticles by surface-enhanced Raman scattering. Science 275:1102

Notermans S, Dufrenne J, Teunis P and Chackraborty T (1998) Studies on the risk assessment of *Listeria monocytogenes*. J. Food Prot. 61(2):244–248

Notingher I, Selvakumaran J and Hench LL (2004) New detection system for toxic agents based on continuous spectroscopic monitoring of living cells. Biosens. Bioelectron. 20(4):780–789

Oliver JD (2005) The viable but nonculturable state in bacteria. J. Microbiol. 43(5):93–100

Oust A, Moretro T, Naterstad K, Sockalingum GD, Adt I, Manfait M and Kohler A (2006) Fourier transform infrared and Raman spectroscopy for characterization of *Listeria monocytogenes* strains. Appl. Environ. Microbiol. 72(1):228–232

Palop A, Mañas P and Condón S (1999) Sporulation temperature and heat resistance of *Bacillus* spores: a review. J. Food Safety 19:57–72

Palop A, Sala FJ and Condón S (1999) Heat resistance of native and demineralised spores of *Bacillus subtilis* sporulated at different temperatures. Appl. Environ. Microbiol. 65:1316–1319

Peel M, Donachie W and Shaw A (1988) Temperature-dependent expression of flagella of *Listeria monocytogenes* studied by electron microscopy, SDS-PAGE and western blotting. J. Gen. Microbiol. 134(Pt 8):2171–2178

Perney NMB, Baumberg JJ, Zoorob ME, Charlton MDB, Mahnkopf S and Netti CM (2006) Tuning localized plasmons in nanostructured substrates for surface-enhanced Raman scattering. Opt. Express 14(2):847–857

Pettersson,A, Poolman JT, van der Ley P and Tommassen J (1997) Response of *Neisseria meningitidis* to iron limitation. Antonie Van Leeuwenhoek 71(1–2):129–136

Picard-Bonnaud F, Cottin J and Carbonnelle B (1989) Preservation of the virulence of *Listeria monocytogenes* in different sorts of soil. Acta Microbiol. Hung. 36(2–3):269–272

Preisner O, Lopes JA, Guiomar R, Machado J and Menezes JC (2007) Fourier transform infrared (FT-IR) spectroscopy in bacteriology: towards a reference method for bacteria discrimination. Anal. Bioanal. Chem. 387:1739–1748

Premasiri WR, Moir DT and Ziegler LD (2005) Vibrational fingerprinting of bacterial pathogens by surface enhanced Raman scattering (SERS). Proc. SPIE 5795:19–29

Premasiri WR, Moir DT, Klempner MS, Krieger N, Jones G 2nd and Ziegler LD (2005) Characterization of the surface enhanced Raman scattering (SERS) of bacteria. J. Phys. Chem. B 109(1):312–320

Raso J, Barbosa-Cánovas G and Swanson BG (1998) Sporulation temperature affects initiation of germination and inactivation by high hydrostatic pressure of *Bacillus cereus*. J. Appl. Microbiol. 85:17–24

Rösch P, Harz M, Schmitt M, Peschke KD, Ronneberger O, Burkhardt H, Motzkus HW, Lankers M, Hofer S, Thiele H and Popp J (2005) Chemotaxonomic identification of single bacteria by micro-Raman spectroscopy: application to clean-room-relevant biological contaminations. Appl. Environ. Microbiol. 71(3):1626–1637

Rösch P, Harz M, Peschke KD, Ronneberger O, Burkhardt H and Popp J (2006) Identification of single eukaryotic cells with micro-Raman spectroscopy. Biopolymers 82(4):312–316

Samoilova SV, Samoilova LV, Yezhov IN, Drozdov IG and Anisimov AP (1996) Virulence of pPst+ and pPst- strains of *Yersinia pestis* for guinea-pigs. J. Med. Microbiol. 45(6):440–444

Schuster KC, Urlaub E and Gapes JR (2000) Single-cell analysis of bacteria by Raman microscopy: spectral information on the chemical composition of cells and on the heterogeneity in a culture. J. Microbiol. Methods 42:29–38

Scott IR and Ellar DJ (1978) Study of calcium dipicolinate release during bacterial spore germination by using a new, sensitive assay for dipicolinate. J. Bacteriol. 135(1):133–137

Seltmann G, Voigt W and Beer W (1994) Application of physico-chemical typing methods for the epidemiological analysis of *Salmonella enteritidis* strains of phage type 25/17. Epidemiol. Infect. 113(3):411–424

Sengupta A, Laucks ML and Davis EJ (2005) Surface-enhanced Raman spectroscopy of bacteria and pollen. Appl. Spectrosc. 59(8):1016–1023

Shibata H, Yamashita S, Ohe M and Tani I (1986) Laser Raman spectroscopy of lyophilized bacterial spores. Microbiol. Immunol. 30(4):307–313

Singh GP, Creely CM, Volpe G, Grötsch H and Petrov D (2005) Real-time detection of hyperosmotic stress resonse in optically trapped single yeast cells using Raman microspectroscopy. Anal. Chem. 77:2564–2568

Singh GP, Volpe G, Creely CM, Grötsch H, Geli IM and Petrov D (2006) The lag phase and G1 phase of a single yeast cell monitored by Raman microspectroscopy. J. Raman Spectrosc. 37:858–864

Sockalingum GD, Lamfarraj H, Beljebbar A, Pina P, Delavenne D, Witthuhn F, Allouch P and Manfait M (1999) Vibrational spectroscopy as a probe to rapidly detect, identify, and characterize micro-organisms. Proc. SPIE 3608:185–194

Spencer KM, Sylvia JM, Clauson SL and Janni JA (2002) Surface-enhanced Raman as a water monitor for warfare agents. Proc. SPIE 4577:158–165

Stephens JC, Roberts IS, Jones D and Andrew PW (1991) Effect of growth temperature on virulence of strains of *Listeria monocytogenes* in the mouse: evidence for a dose dependence. J. Appl. Bacteriol. 70(3):239–244

Stevenson HJR and Bolduan OE (1952) Infrared spectrophotometry as a means for identification of bacteria. Science 116:111–113

Thomas LC and Greenstreet JES (1954) The identification of micro-organisms by infrared spectrophotometry. Spectrochim. Acta 6:302–319

Thompson PA, Guan Y, Wood LL and Grow AE (2000a) SBIR Phase I Final Report, Contract DAAD16-00-C-9217. U.S. Army Soldier & Biological Chemical Command

Thompson PA, Guan Y, Wood LL and Grow AE (2000b) STTR Phase II Final Report, Contract No. DAAG55-98-C-0004. U.S. Army Research Office

Thuan BP, Calderon de la Barca AM, Buck G, Galsworthy SB and Doyle RJ (2000) Interactions between listeriae and lectins. Roum. Arch. Microbiol. Immunol. 59(1–2):55–61

Wagner JS, Trahan MW, Nelson WE, Tisone GC and Prepernau BL (1996) How intelligent chemical recognition benefits from multivariate analysis and genetic optimization. Computers in Physics 10(2):113–118

Williams OB and Robertson WJ (1954) Studies on heat resistance. VI. Effect of temperature of incubation at which formed on heat resistance of aerobic thermophilic spores. J. Bacteriol. 67:377–378

Xie C, Chen D and Li YQ (2005) Raman sorting and identification of single living micro-organisms with optical tweezers. Opt. Lett. 30(14):1800–1802

Xie C, Mace J, Dinno MA, Li YQ, Tang W, Newton RJ and Gemperline PJ (2005) Identification of single bacterial cells in aqueous solution using confocal laser tweezers Raman spectroscopy. Anal. Chem. 77(14):4390–4397

Xu H, Bjerneld EJ, Kall M and Borjesson L (1999) Spectroscopy of single hemoglobin molecules by surface enhanced Raman scattering. Phys. Rev. Lett. 83(21):4357–4360

Xu S, Ji X, Xu W, Li X, Wang L, Bai Y, Zhao B and Ozaki Y (2004) Immunoassay using probe-labelling immunogold nanoparticles with silver staining enhancement via surface-enhanced Raman scattering. Analyst 129(1):63–68

Xu S, Ji X, Xu X, Zhao B, Dou X, Bai Y and Osaki Y (2005) Surface-enhanced Raman scattering studies on immunoassay. J. Biomed. Opt. 10(3):031112-1- 031112-12

Zeiri L, Bronk BV, Shabtai Y, Eichler J and Efrima S (2004) Surface-enhanced Raman spectroscopy as a tool for probing specific biochemical components in bacteria. Appl. Spectrosc. 58(1):33–40

Zeiri L and Efrima S (2005) Surface-enhanced Raman spectroscopy of bacteria: the effect of excitation wavelength and chemical modification of the colloidal milieu. J. Raman Spectrosc. 36(6–7):667–675

Zhu Q, Quivey RG and Berger AJ (2004) Measurement of bacterial concentration fractions in polymicrobial mixtures by Raman microspectroscopy. J. Biomed. Opt. 9(6):1182–1186

Recognition Receptors

21

Antibodies and Immunoassays for Detection of Bacterial Pathogens

Padmapriya P. Banada and Arun. K. Bhunia

Abstract

Antibody, also known as immunoglobulin, is normally made in the body in defense of foreign antigen or invading pathogen. Highly specific biorecognition property of antibody with antigen has made antibody as one of the most indispensable molecules for broad application, not only in the diagnosis or detection but also in prevention or curing of diseases. Animals are routinely used for production of both polyclonal and monoclonal antibodies; however, recombinant and phage display technologies are being adopted to improve antibody specificity and to cut cost for antibody production. Available genome sequence of pathogens is also allowing researchers to find and select suitable target antigens for production of antibody with improved specificity. In recent years, however, demand for antibody is even greater as novel biosensor or nanotechnology-based methods continue to utilize antibody for analyte capture and interrogation. Conventional immunoassay methods such as lateral flow and enzyme-linked immunoassays, though lack sensitivity, are available commercially and are widely used. While biosensor-based methods such as time-resolved fluorescence immunoassay, chemiluminescence assay, electrochemical immunoassay, surface plasmon resonance sensor, fiber optic sensor, and microfluidic biochip have, in some cases, demonstrated improved sensitivity, they require further optimization with real-world samples. Furthermore, environmental stress and the growth media are known to affect the physiological state of microorganism and antigen expression, often rendering unsatisfactory signal response from immunoassays. Thus, one must understand the microorganisms' response to these factors before designing an immunoassay to avoid false results. With the advent of microfluidics and nanotechnology, the adaptation of lab-on-chip concept in immunoassays will soon be a reality for near real-time detection of pathogens from food or clinical specimens.

1. Introduction

Our ability to produce specific antibodies against target analytes by conventional animal immunization methods or by cloning or recombinant DNA technology has revolutionized the field of immunoassay-based detection. Antibodies are among the most important molecules, with limitless applications in the field of biology, microbiology, medicine, and agriculture. For example, the most rapid diagnostic tests used in food or clinical laboratories are based on antigen-antibody reactions. Antibodies are used in the discovery of new bioactive molecules, the prevention of diseases, and the treatment of cancer. In recent years, however, antibodies have been found to be invaluable molecules in biosensor and nanotechnology applications for

Padmapriya P. Banada and Arun. K. Bhunia • Molecular Food Microbiology Laboratory, Department of Food Science, Purdue University, West Lafayette, Indiana

M. Zourob et al. (eds.), *Principles of Bacterial Detection: Biosensors, Recognition Receptors and Microsystems*,
© Springer Science+Business Media, LLC 2008

the interrogation of analytes in macro- to nano-scale, including pathogens (bacteria, viruses) and toxins. In this chapter we discuss the significant role antibodies play in bacterial pathogen detection. Bacteria are prokaryotes, and their distribution is ubiquitous, including humans, animals, and the environment. They are the oldest living organisms in the history of this planet and play a profound role in maintaining the ecosystem. Humans, animals, and plants harbor bacteria in large numbers, and mostly they have a symbiotic relationship. Only a small fraction of bacteria are harmful and can cause diseases. It is estimated that on the average, about 1–2 billion people around the world are infected by bacteria each year, with 70 % of the cases being foodborne (Allos et al. 2004). Annually, foodborne diseases are estimated to affect about 1 in 4 persons in the United States and 1 in 5 persons in England (Kendall et al. 2006). In recent years there has been a significant decline in the number of outbreaks and illnesses due to foodborne bacterial pathogens. However, during late 2006 the US experienced multiple outbreaks caused by *Escherichia coli* O157:H7 that was associated with spinach and lettuce. A total of 199 people were infected in 26 states by consuming tainted spinach, with 3 fatalities (CDC 2006). A *Salmonella* outbreak involving tomatoes and peanut butter resulted in several illnesses too. Peanut butter contamination resulted in 288 cases in 39 states in early 2007. Bacteria also pose a concern because of an increase in antibiotic resistance during clinical infections (for example, methicillin- and vancomycin-resistant staphylococci). Additionally, they cause stomach ulcers (eg., *Helicobacter pylori*), tuberculosis (*Mycobacterium tuberculosis*), meningitis (species of *Streptococci, Neisseria*), cholera (*Vibrio cholerae*), sexually transmitted diseases, and nosocomial or hospital-acquired infection. In recent years, several bacterial pathogens have been considered as potential threats for bioterrorism: *Bacillus anthracis* (anthrax bacteria); botulinum toxin, staphylococcal enterotoxin B (SEB), *Yersinia pestis*, etc. (Bhunia 2006).

Consequently, rapid detection and diagnostic tools are being developed as a measure to combat these pathogenic bacteria. Culturing methods continue to be the "gold" standard; followed by nucleic acid-based assays, ranked "silver;" and the immunoassays, ranked "bronze" (Gracias and McKillip 2004; Bhunia 2006). However, immunoassay is by far the most rapid compared to the other two methods. Culture methods require 24–48 h or more in order to get the bacterial colonies on a petri dish. Some of the bacteria like *M. tuberculosis* take 7–14 days to grow on the selected media (Cheng et al. 2005). A nucleic acid-based assay requires good technical expertise and a nucleic acid extraction step. Currently attempts are being made to automate the system for on-site application. In contrast, most of the antibody-based detection methods, such as lateral flow immunoassays, dipstick assays, and slide agglutination tests, can be done outside the convenience of a laboratory, with little technical knowledge; and the results can be obtained relatively quickly, in 10–15 minutes.

In this chapter, we also discuss the methods or strategies for developing a "good" antibody against bacterial pathogens for use in the capture and concentration of bacterial cells using immunomagnetic separation (IMS) technology, and their detection by various immunoassay procedures, including immunosensors and biosensors.

2. Antibodies

Antibodies (immunoglobulins) are glycoproteins belonging to the immunoglobulin (Ig) supergene family. An antibody (Ab) molecule has been viewed as a "Y" shaped molecule consisting of two pairs of identical polypeptide chains, called light and heavy chains, joined by disulphide bonds. The two variable domains on light (V_L) and heavy (V_H) chains make up the antigen (Ag) recognition and binding site (Fig. 21.1). The amino acid sequence of this region

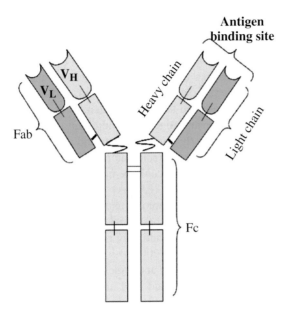

Figure 21.1. Structure of an immunoglobulin molecule.

is highly variable, and this contributes to the broad recognition power of the antibody to a wide range of target molecules. When digested with the enzyme papain, a mammalian antibody molecule yields two 50 kDa Fab (fragment antibody) fragments and one 50 kDa Fc (fragment crystalline) fragment. The Fab fragment binds to the antigen, while the Fc fragment binds to the Fc receptors located on many mammalian cells. Antibodies consist of different classes: IgA, IgG, IgM, IgE, and IgD (rare); and subclasses which slightly vary between humans and mice, primarily for IgG. Both human and mouse IgA consists of IgA1 and IgA2; while mouse IgG consists of IgG1, IgG2a, IgG2b, and IgG3, and human IgG consists of IgG1, IgG2, IgG3, and IgG4. The generation of a specific class or subclass of antibodies in a host depends largely on the nature of the antigens, the type of adjuvants, and the route of immunization, such as intramuscular vs. intradermal vs. subcutaneous. For all practical purposes, certain classes (or subclasses) are highly desirable for immunoassay applications because of their stability, binding affinity, and low cross reactions.

The antibody's ability to recognize and bind with high affinity to specific antigenic sites (epitopes), even in a complex mixture, is exploited for qualitative and quantitative measurement of the antigens. Thus antibody application is broad—it is not only used for detection and classification of the antigens, but also for understanding the microheterogeneity among proteins resulting from recombinant or somatic mutations.

The production and selection of a suitable antibody is imperative for the successful design of an immunoassay, which depends on the assay parameters: the choice of a polyclonal or monoclonal antibody; of purified or native sera; of fragmented, bispecific, or fusion proteins; and the relative cost (Liddell 2005).

Several reviews have described in depth the available polyclonal and monoclonal antibodies for various bacteria (Macario and Macario 1988; Bhunia 1997). Along with traditional polyclonal and monoclonal antibody production, in this chapter we have addressed different methodologies and selection approaches for the successful production of a polyclonal or monoclonal antibody for downstream applications in immunoassays.

2.1. Polyclonal Antibody

A polyclonal antibody (PAb) is a heterogeneous mixture of antibody molecules arising from a variety of constantly evolving B-lymphocytes, so that even successive bleeds from one animal are unique (Kane and Banks 2000). The assortment of antibodies present in a PAb preparation may consist of different classes and subclasses, and they may recognize multiple antigens or multiple epitopes located on the same antigen. In contrast, a monoclonal antibody recognizes only a specific epitope on an antigen. Polyclonal antibodies are produced for bacterial detection and were widely used by early immunologists and microbiologists for their ability to react with a variety of epitopes to characterize an antigen. Most of the commercial assays use polyclonal antibodies, and the assay format could be in the form of agglutination, precipitation, or an enzyme–linked immunosorbent assay (ELISA). Polyclonal antibodies are also found to be superior to monoclonal antibodies in capturing and concentrating target molecules and are used in immunomagnetic- or immunobead-based captures. Sheep, goats, and rabbits are the most common animals used for polyclonal antibody production, although chickens have been used occasionally (Kovacs-Nolan and Mine 2005).

Unlike mammalian antibodies, chicken antibodies are composed of three immunoglobulin subclasses: IgA, IgM, and IgY. The IgA and IgM are similar to mammalian IgA and IgM, while the IgY is equivalent to mammalian IgG. These antibodies are found in serum as well as in eggs. In eggs, IgA and IgM are primarily present in the albumen in trace amounts, while IgY is found in the yolk in large quantities (~25 mg/ml). Structurally, IgY (180 kDa) is larger than IgG (150 kDa) and can be readily harvested in large quantities from egg yolks (Kovacs-Nolan and Mine 2005).

Recently, several mammalian PAbs were used for the capture and detection of *Salmonella* (Kramer and Lim 2004; Hahm and Bhunia 2006), *E. coli* (Geng et al. 2006b), and *L. monocytogenes* (Geng et al. 2004; Gray and Bhunia 2005) on biosensor platforms, which will be discussed further in later sections.

2.2. Monoclonal Antibody

Development of monoclonal antibodies was first introduced by Kohler and Milstein in 1975 (Fig. 21.2), which revolutionized the understanding of the structure, nature, and distribution of antigens. Conventionally, during the production of polyclonal antibodies, a target-specific antibody is present in a pool of B-lymphocytes expressing unique antibodies. Separation of the target-specific lymphocytes would yield a highly specific antibody, which in practice is impossible because the lymphocytes cannot be regenerated in vitro. During MAb production, the B-lymphocytes are collected from the spleen of immunized mice, rats, or rabbits (in humans, from blood) and fused with myeloma cells (Sp2/0, NS1), producing the hybrid cells that in turn produce the antibodies and proliferate infinitely in a flask (Bhunia et al. 1991; Liddell 2005) (Fig. 21.2).

Target-specific antibodies are developed either for the whole-cell detection of bacteria such as *Salmonella, Listeria, E. coli, Campylobacter, Clostridium, Staphylococci, Pseudomonas,* etc., or their toxins. For whole-cell detection, the heat-killed cells (Bhunia 1997; Warschkau and Kiderlen 1999), or whole cellular antigens (Zhao and Liu 2005), or surface antigens like flagella (Kim et al. 2005), cell wall-associated lipopolysaccharides (Thirumalapura et al. 2005), cell surface antigens (Kunhe et al. 2004), and surface-expressed virulence proteins (Hearty et al. 2006) have been used as antigens for monoclonal antibody production. The epitope-masking method, using an antibody to block undesirable epitopes in an antigen, has been used to successfully develop a specific monoclonal antibody against *L. monocytogenes* (Bhunia and Johnson 1992).

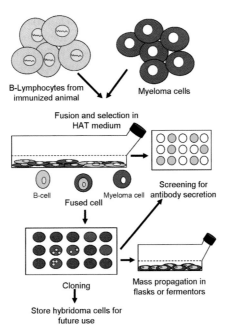

Figure 21.2. Flow diagram showing steps involved in monoclonal antibody production. HAT = hypoxanthine, aminopterin, and thymidine.

The type of antibodies (PAb vs. MAb) to be used depends on the specific application. Whenever possible, monoclonal antibodies are preferred. However, due to the high cost involved in the production of monoclonal antibodies, polyclonal antibodies still have a broader appeal. Irrespective of the type, the production of a target-specific, high performance antibody depends on the proper strategy in selecting and delivering the antigenic molecules such as small peptides or polypeptides as immunogens. Some proteins are not very easy to isolate in pure form in sufficient amounts from pathogens. In some situations, only a small fraction of the potential target proteins are isolated and characterized. Therefore, under these conditions the native proteins cannot be used to generate the desired antibodies. Thus several other strategies like the use of synthetic peptides, recombinant DNA technology, and phage display were introduced as alternative methods for antibody production.

2.3. Use of Synthetic Peptides for Antibody Production

The use of synthetic peptides as immunogens was first reported in 1982 by Young and Atassi and have since been widely used. Such use for the production of antibodies has been a popular choice for vaccine research and development (Tsurumi et al. 2003; Cauchard et al. 2006; Maruta et al. 2006). The major advantage of using this technique is to avoid the high degree of cross-reactivity which often is a major concern in many immunoassay experiments. Furthermore, peptides can be conveniently purified by using liquid chromatography. These peptides are then used for generating polyclonal or monoclonal antibodies. In order to design a good peptide, one should bear in mind that its optimal length should be 10–20 residues long, it should possess an antigenic region or domain, and it should be stable. Too long, too short, or unstable molecules result in steric or conformational changes in the molecule during conjugation or synthesis, resulting in cross-reactivity. Since peptides are too small to induce an immune response, a suitable carrier molecule such as bovine serum albumen (BSA) or keyhole limpet hemocyanine (KLH) is attached, or sometimes peptides are synthesized as multiple antigen peptides (MAPs) (Angeletti 1999). MAPs are 13–17 kDa protein, composed of 4–8 identical peptides linked by a poly L-lysine core, and are suitable for antibody production without the aid of a carrier protein. Maruta et al. (2006) used alum along with inactivated *Bordetella pertusis* for eliciting antipeptide antibodies against an array of peptides. They showed that this combination increased antibody production in mice. A novel approach that utilized nanoparticles as adjuvants elicited a strong immunologic response in foals against virulence associated protein A (Vap A) peptides of *Rhodococcus equi*, and the resulting antibody showed a strong reaction with the organism (Cauchard et al. 2006).

Several researchers have developed antibodies against peptides to neutralize the bacterial superantigens or toxins (Visvanathan et al. 2001; Dale et al. 2002) and against various epitopes on the cell surface proteins of *S. aureus*, Streptococci, *Helicobacter pylori*, and others (Huesca et al. 2000; Shin, Roe and Kim 2004; Bialek et al. 2006). IgY antibodies were produced in hens using the synthetic peptides of urease epitope from *Helicobacter pylori* (Shin et al. 2004). Subsequent characterization of antibodies using ELISA revealed that among the five peptides, UreB peptide containing 15- aminoacid residues was highly specific for IgY-*Hp* antibodies.

In our laboratory, we identified from the database several unique 20-mer amino acid sequence long peptides (epitopes) from surface-associated proteins of *L. monocytogenes* that are likely to be present on the surface of cells in their native configuration. The goal was that when the antibodies were developed, they should bind to the surface-exposed epitope on the bacterial cell for reliable detection of whole cells (Lathrop 2005; Lathrop et al. 2008). Sequence analysis of the published genome sequences of *L. monocytogenes* (Glaser et al. 2001) revealed 22 unique surface proteins of *L. monocytogenes*. Of these 22 proteins, 5 belonged to the internalin (invasion protein)-multigene family; one ActA (actin polymerization) protein and the remaining 16 were unknown (Lathrop 2005). Nine 20-mer peptides, representing each of nine proteins, were synthesized with an extra cysteine residue that was used to form a disulfide cross-link with the KLH. KLH-peptides were purified by HPLC and the identity of each was confirmed by mass spectrometry. The peptides were injected subcutaneously into rabbits for antibody production. In spite of careful and thorough analysis, not all peptides produced desirable antibodies. Some antibodies showed highly specific reactions against *L. monocytogenes* cells, while some showed reactions against the multiple species within the genus *Listeria* or other bacterial species (Lathrop 2005; Lathrop et al. 2008). One of the major problems that arose from this study was the presence of nonspecific antibodies in the pool of rabbit serum immunized with *Listeria*-specific peptides that were showing reactions with several nontarget pathogens. Further testing of preimmune serum revealed the presence of nonspecific antibodies against an array of bacterial pathogens. This indicated that the rabbits had previous exposure to nontarget microorganisms that induced immune responses. Therefore, for the production of peptide-specific PAbs in rabbits, thoroughly testing the preimmune serum of each animal for the presence of antibodies to the target pathogens as well as to common microorganisms is advised before initiating the immunization regimen.

In a test for the presence of antisera against 27 microorganisms involving 19 rabbits raised under conventional or pathogen-free environments, Lathrop et al. (2006) reported that 17 rabbits were positive for at least one pathogen, and 14 of 27 cultures showed positive reactions with 50 % or more preimmune sera. If these rabbits were used for the production of pathogen-specific antibodies for the purpose of diagnostic immunoassay, they would produce cross-reactive low quality antibodies, which might not be suitable for use in diagnostic immunoassays. To overcome this problem, monoclonal antibodies could be developed against the synthetic peptides. As the genome sequences of several microbial pathogens are currently available, the synthetic peptide approach continues to be an attractive and highly desirable method for the development of specific antibodies.

2.4. Recombinant DNA Technology

Orlandi et al. (1989) introduced a breakthrough method of antibody production when they demonstrated an innovative approach to clone antibody genes through a recombinant DNA technology. Recombinant antibodies are similar to monoclonal antibodies, except that they consist of only antigen-binding domains, without the Fc domain (Fig. 21.2). The mRNA is isolated from the lymphocytes of immunized animals or nonimmune donors and a cDNA library is constructed. These genes are amplified by polymerase chain reactions (PCR), using the antibody-specific primers designed for heavy and light chain fragments of the antibody. The size of the fragment depends on whether a whole Fab fragment or a single chain variable fragment (ScFv) is to be made (Emanuel et al. 2000; Liddell 2005). The PCR-amplified products are inserted into phagemid vectors and a combinatorial library is created by cloning into *E. coli* and infecting it with a helper bacteriophage, which helps express the antibody fragments on the surface (Fig. 21.3).

Several different platforms are used to select the recombinant antibodies: (1) a phage display, (2) a protein-mRNA link using a ribosome or mRNA display, and (3) a microbial cell display on yeast, bacteria, or retrovirus. Other selection methods may include a microbead display by in vitro compartmentalization, a display based on protein-DNA linkage, and an in vivo-based growth selection based on the protein fragment complementation (Hoogenboom 2005). However, the phage display, which is discussed more in this chapter, seems to be the most often applied approach for bacterial antibody production.

The major improvement of recombinant antibodies over polyclonal or monoclonal antibodies is that they are produced by bacteria, which offer a stable genetic source. Furthermore, screening is less time–consuming, and clones that are not specific or not producing antibodies against the target pathogen are discarded. The advantages are that the bacteria can be genetically manipulated and the recombinant antibodies can be produced in abundance in a relatively short period of time without the need for a cell culture system or the requirement of a surrogate animal (mice are often used to grow hybridoma in their peritoneal cavity) for monoclonal antibody production in large quantities (Emanuel et al. 2000).

2.4.1. Phage Display

Phage display was first reported in 1985 by George P. Smith, who demonstrated that peptide fragments can be expressed on the surface of bacteriophages. It was introduced as a method for antibody selection in 1990 by McCafferty and his colleagues, and it has since become a widely used method for recombinant antibody production. The method is based on the expression of functional antibody fragments (Fab) on the surface of a filamentous phage. A large number of antibodies can be quickly selected from libraries on the basis of the antigen-binding behavior of individual clones (Posner et al. 1993; Hoogenboom 2005).

Random peptide phage display has been successfully employed to identify bacteria (Williams et al. 2003; Paoli et al. 2004; Nanduri et al. 2007a, 2007b), new receptor molecules, receptor ligands (Lu et al. 2003) in epitope mapping and mimicking of protein antigens, antibodies (Khuebachova et al. 2002; Muhle et al. 2004), and drug discovery (Pan et al. 2006).

The most commonly used phages are the filamentous phages M13, f1, and fd. A phage-display library consists of a group of filamentous phages carrying genes encoding foreign proteins (antigens), which are expressed as fused proteins in their outer coat (Smith and Petrenko 1997). Briefly, the specific peptide represented by the DNA is fused to the coat protein on the surface of the virion, and each phage displays a single peptide. When the infective carrier phages are cloned individually into a suitable host, e.g., *E. coli*, they propagate indefinitely (Fig. 21.3).

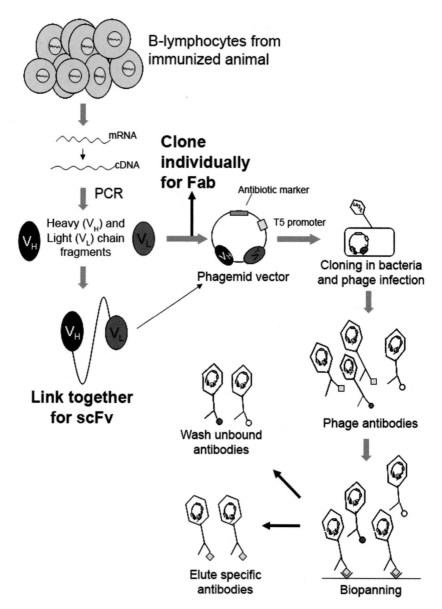

Figure 21.3. Recombinant antibody production through phage display (adapted from references: Liddell 2005; Smith and Petrenko 1997; Petrenko and Vodyanoy 2003; Emanuel et al. 2000).

The phage antibodies carry the sites for antigen recognition and binding domains. Through the affinity selection, a specific antibody is recognized and a particular phage is isolated and propagated by infecting a suitable host. Thus, a single-phage antibody library could be constructed and useful quantities of the antibody could be made (Petrenko and Vodyanoy 2003). The screening of the library for the selection of suitable phage-bearing peptides is called biopanning. Traditionally, 2–10 biopannings are done to select one phage antibody.

The development of several phage antibodies for numerous pathogens has been reported. Goldman et al. (2000) developed a phage antibody against staphylococcal enterotoxin B (SEB). They labeled the whole phage with the Cy5 fluorescent tag, and used it in a fiber-optic biosensor and a fluorescence microplate assay for the direct detection of SEB from samples. They were able to detect 1.4 ng of SEB. A phage display antibody was also developed to detect botulinum neurotoxin (Emanuel et al. 1996). Williams et al. (2003) screened commercial phage-display libraries for the presence of peptide ligand for *Bacillus* spores. They identified phage peptides that were specific for the spores of *Bacillus anthracis* and the spores of other *Bacillus* spp. Using those two peptides in tandem, they were able to detect *B. anthracis* spores. Kanitpun et al. (2004) developed single chain variable fragment (ScFv) molecules from hybridoma clones that produced immunoglobulins specific for the LPS and flagellar antigen of *E. coli* O157:H7, using phage display technology. Single chain variable fragment (ScFv) antibodies have been shown to be successful for the detection of *L. monocytogenes* (Paoli et al. 2004; Nanduri et al. 2007a) and *E. coli* (Nanduri et al. 2007b). Phage display methods for detection of bacterial pathogens are described in chapter 28.

3. Capture and Concentration of Cells by Immunomagnetic Separation

Paramagnetic beads coated with antibodies are currently used for the detection as well as for the concentration of bacterial cells from complex environmental samples (Fig. 21.4). Paramagnetic beads show a magnetic property only when placed under a magnetic field, and the magnetism disappears when the magnetic field is removed. This is an important property for their application in immunomagnetic separation (IMS), since magnetic beads would be free to interact with the target antigens (cells) in liquid suspension without being attracted to each other by inter-magnetic force.

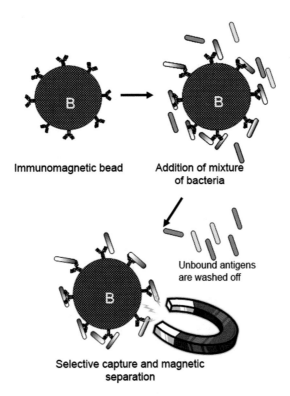

Immunomagnetic bead

Addition of mixture of bacteria

Unbound antigens are washed off

Selective capture and magnetic separation

Figure 21.4. Schematic diagram showing immunomagnetic separation technology for capture and concentration of target antigens. The magnetic beads are conjugated with the specific antibodies when added to a mixture of bacteria; the antibody binds specifically to the target antigen. Under the magnetic field, the paramagnetic bead concentrates in the area of magnetism and unbound antigens are washed off.

The IMS step is considered a selective enrichment step since it allows bacterial capture and concentration from liquid food matrices. Since the capture rate varies (10–70 %) depending on the type of microorganisms and the antibody used, the food suspension must carry a significant number of cells. Most often, IMS is suitable for use with food suspensions that have been subjected to a pre-enrichment or a selective enrichment step to bolster bacterial counts. Captured microbes (antigens) are then plated on nutrient/selective agar plates or further tested using other assays such as PCR, immunoassay, flow cytometry, and chemiluminescent- or pathogenicity-based assays (Feng 2001). Over the past two decades, IMS has been widely used to separate bacterial cells from complex materials including food, water, and clinical samples. Conventional methods of bacterial cell isolation require growing on a selective agar medium, which takes about 2–5 days, depending on the bacteria. IMS offers a rapid recovery of the cells and improves the sensitivity of the downstream microbiological analysis.

The Dynal Biotech company (Oslo, Norway) initially designed the paramagnetic beads called Dynabeads® and commercialized them by coating them with specific antibodies. Their Dynabeads® Anti-*Salmonella* and anti-*E. coli* antibodies have been shown to be superior to the conventional ISO or culturing methodologies for isolation of these pathogens. Culturing methods coupled with IMS increased the sensitivity of detection of many bacterial pathogens like *Salmonella* (Hara-Kudo et al. 2001; Jordan et al. 2004), *Staphylococcus aureus* (Yazdankhah et al. 1999), and *Listeria* (Bauwens et al. 2003).

IMS coupled with immunoassays or PCR assays was shown to significantly improve the speed and sensitivity of *E. coli* O157 detection to as low as 1 CFU in 25 g of the food sample (Chapman and Ashton 2003). IMS-PCR combinations for the detection of *Salmonella* (Soumet et al. 1999; Jenikova et al. 2000; Mercanoglu and Griffiths 2005), *L. monocytotenes* (Hudson et al. 2001; Amagliani et al. 2006; Ueda et al. 2006), and *E. coli* (Fu et al. 2005) have become popular in recent years. IMS coupled with multiplex PCR was used to simultaneously detect *L. monocytogenes* and *Salmonella* from food samples, and the detection limit was established to be about 1000 cells (Li et al. 2000; Hsih and Tsen 2001). They suggested that additional pre-enrichment with universal pre-enrichment broth would enhance the detection sensitivity. Hudson et al. (2001) showed that IMS-PCR could detect as little as 1.1 CFU/g of *L. monocytogenes* in a ham sample.

IMS has also found a wide application in flow cytometry (Jung et al. 2003; Splettstoesser et al. 2003) and fiber-optic biosensors (DeMarco and Lim 2002; Liu et al. 2003) for the detection of bacterial pathogens.

Nonmagnetic immunobeads, such as the protein A-conjugated sepharose bead (Sigma, St. Louis, MO) coated with an anti-*Listeria* monoclonal antibody (MAb-C11E9), were developed in our laboratory for the detection of *L. monocytogenes* (Gray and Bhunia 2005). First, the immunobeads and magnetic Dynabeads were used to capture bacteria from food matrices; and second, the captured bacteria were tested for cytotoxicity on a mammalian cell line to differentiate pathogenic from nonpathogenic *Listeria* species. This immunobead method was highly specific for *L. monocytogenes*, and performed better than the Dynabeads because the antibody (MAb-C11E9) on the immunobead is specific for *L. monocytogenes* and some strains of *L. innocua* (Bhunia et al. 1991; Lathrop et al. 2003; Gray and Bhunia 2005), while the antibody in Dynabeads reacts with all species of *Listeria* (Vytrasova et al. 2005). When tested in a cytotoxicity assay, the immunocaptured cells *L. monocytogenes* (captured with an immunobead assay) gave a positive cytotoxicity; while *L. monocytogenes, L. ivanovii*, and *L. seeligeri* (captured by Dynabeads) also showed a positive cytotoxicity (Gray and Bhunia 2005).

Recently, an improved IMS technique was described by Varshney et al. (2005), in which magnetic nanoparticle conjugates (MNCs) were coated with polyclonal anti-*E. coli* antibodies using a biotin and streptavidin system and then used to capture *E. coli* O157:H7. Capture efficiency values increased from 69 to 94.5 %, while bacterial concentrations decreased from

3.4×10^7 to $8.0\,CFU/ml$ in ground beef samples. This technology requires minimal sample preparation and results in higher capture efficiency compared to the traditional Dynal bead-based IMS.

Immunomagnetic beads were also used together with quantum dots for their ability to detect bacterial pathogens quantitatively (Su and Li 2004; Yang and Li 2005; Tully et al. 2006). Quantum dots naturally emit fluorescence and are used as the source of tracer fluorophor. Magnetic bead-captured bacterial cells can be quantitatively detected using antibody conjugated to quantum dots. Yang and Li (2005) used streptavidin-coated fluorescent semiconductor quantum dots to detect *Salmonella* cells at 10^3–10^7 CFU/ml bound to the magnetic beads conjugated with biotin-labeled anti-*Salmonella* antibody. The measure of fluorescence intensity was correlated to the number of cells present in the test sample.

3.1. Automated IMS Systems

Dynal Biotech Ltd. (Wirral, UK) automated the IMS procedure, and the commercial system is called BeadRetriever™. This system works on the inverse magnetic particle processing principle, in which the dynabeads are moved among tubes containing specific reagents with the aid of a magnetic bar, rather than by moving the liquids. The tube strips with their compartments contain reagents such as bacteria, secondary antibody, washing fluids, and the substrate. The system can process 15 pre-enriched samples in 20 minutes. BeadRetriever™ has been applied widely for the detection of *E. coli* (Chapman and Cudjoe 2001; Reinders et al. 2002; Fegan et al. 2004), *Salmonella* (Duncanson et al. 2003), and *L. monocytogenes* (Amagliani et al. 2006). Table 21.1 lists some of the commercialized and laboratory-developed IMS-integrated models.

4. Immunoassays for Pathogen Detection

Immunoassays are developed to measure the presence of an analyte (usually proteins) through antigen-antibody interaction. The detection signal can be radioactive, colorimetric, or fluorescent. The sensitivity and specificity of the immunoassays is highly dependent on the choice of antibodies.

4.1. Radioimmunoassay

The radioimmunoassay (RIA) was developed in 1960 by Yalow and Berson, who received the Nobel Prize in medicine in 1977 and demonstrated the application of RIA for the detection of insulin. In this method, a gamma-radioactive isotope (of iodine or tyrosine) was labeled to the antigen and then mixed with the antibody. A nonlabeled antigen was then added to this mixture at higher concentrations, to compete with the labeled antigen. Antibody preferentially binds to nonlabeled antigen, and the radioactive signal from the displaced labeled antigen is measured and the binding curve is plotted. Although this method is very sensitive and highly specific for bacterial cells or toxins, the cost and high risks associated with radioisotopes prohibited its widespread use. In recent years, this method has been replaced by enzyme immunoassays (EIA), where the detection is colorimetric or fluorescence-based.

4.2. Enzyme Immunoassays

Among the enzyme immunoassays, enzyme-linked immunosorbent assay (ELISA) has emerged as the most reliable quantitative detection method. ELISA is a rapid immuno-chemical method in which an antigen-antibody reaction is catalyzed by an enzyme producing a

Table 21.1.   Partial list of automated or semi-automated IMS integrated systems (AIMS)

Name of the Assay OR Commercial Name	Antibody Used	Integrated Method to AIMS	Detection	Detection Limit	Commercialized (yes or no) and Company Name	Reference
Pathatrix system*	Anti-E. coli, anti-Salmonella Anti-Listeria	Plating	Characteristic colonies on the plate, PCR	1CFU/25 g	Yes/ Matrix microsciene, CO	cfsan.fda.gov
Immunomagnetic particle- Enzyme Linked ImmunoSorbent Assay (IMP-ELISA) using BeadRetriever™	Dynal® Anti-Salmonella, E. coli and Listeria	ELISA	Colorimetric	10^5 CFU/mL	Yes/ Invitrogen Corporation, Carlsbad, CA	invitrogen.com
Enhanced Automated Immunomagnetic Separation (eAIMS)	Dynabeads anti-E.coli O157	Charge Switch® Technology (CST)	qPCR	10CFU/25 g	Yes/ Invitrogen, CA	invitrogen.com
SI-RSC (sequential injection renewable separation column) IMS	Anti-E. coli O157	TaqMan PCR and DNA Microarray	Real time amplification Fluorescence	$10–10^3$ CFU/mL without enrichment	No	Chandler et al. 2001
EIAFOSS™*	Anti-E. coli O157	ELISA	Colorimetric	10^5 CFU/mL	Yes/ Foss Electric A/S, Hillerød, Denmark	Reinders et al. 2002
Automated Immunomagnetic Separation-Enzyme Immuno Assay (AIMS-EIA)	Dynal® Anti-Salmonella	ELISA	Chemiluminiscence	ND	No	Duncanson, Wareing and Jones 2003
IMS/PCR Biodetection Enabling Analyte Delivery System (IMS/PCR BEADS)	Dynabeads anti-E.coli O157	PCR	PCR	10 cells/mL	No	Straub et al. 2005

*AOAC certified; ND- Not determined; Most or all of these have been applied in detection of *E. coli, Listeria* and *Salmonella* unless specified.

chromogenic or fluorescent signal. ELISA is performed in different formats depending on the location of antigen or the antibody on the solid surface (Fig. 21.5).

In competitive ELISA, the primary antibody is mixed in a separate tube with various dilutions of bacteria (or antigen) and added to the wells containing immobilized antigen. Only the free unbound antibody will bind the immobilized antigen. A secondary antibody-enzyme conjugate and substrate system is added for color development. The highest dilution of cells showing the minimum reaction or equivalent to a background control is considered positive (Fig. 21.5a).

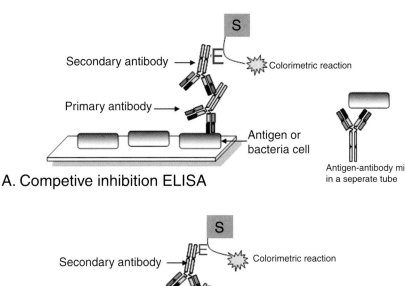

A. Competive inhibition ELISA

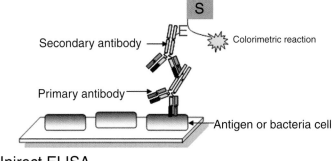

B. Inirect ELISA

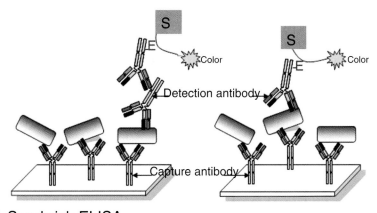

C. Sandwich ELISA

Figure 21.5. Diagrammatic representation of different ELISA formats for colorimetric detection of bacteria (E = enzyme; S = substrate).

In indirect ELISA (Fig. 21.5b), the antigen is first immobilized in the wells of a microtiter plate and is then added to the antibody solution. A secondary antibody conjugated with substrate modifying enzyme, e.g., horseradish peroxidase (HRP) or alkaline phosphatase (AP), is added to bind to the primary antibody, and the reaction is developed with a suitable substrate to give a color reaction. If a fluorescent molecule (e.g., FITC, Cy-5, Alexa-Flour) is used, the reaction is quantified by the amount of fluorescence emitted.

Sandwich ELISA works similarly to indirect ELISA, except that the microtiter plates are first coated with a capture antibody and then the antigen is added. The reaction is developed using the labeled detection or tracer antibody in one step, or using an additional secondary antibody conjugate in a two-step reaction (Fig. 21.5c).

ELISA has been a popular choice among microbiologists for the sensitive and quantitative detection of bacteria or their toxins in food, water, and environmental and clinical samples (Yeh et al. 2002; Gracias and McKillip 2004; Haggerty et al. 2005). Some of the commercially available immunoassay kits are listed in Table 21.2. ELISA assays available for some foodborne pathogens are discussed below.

4.2.1. *Escherichia coli*

Bacterial surface antigens, like proteins, lipopolysachharides (LPS), and flagella, have been used as targets for detection of whole cells present in water and food samples. Low molecular weight outer membrane proteins (<10,000 kDa) and LPS from *E. coli* O157:H7 have been targeted in ELISA using polyclonal antibodies. Blais et al. (2006) used polymyxin B, which binds LPS more efficiently, to immobilize the *E. coli* cells. They used commercially available anti-*E. coli* O111 or anti-*E. coli* O26 antisera to detect the respective strains in ground beef.

Most of the detection kits for *E. coli* O157 use sandwich ELISA with polyclonal antibodies that are raised against the whole cells. LMD-ELISA (LMD Laboratories, Carlsbad, CA) for *E. coli* O157 has been shown to detect *E. coli* cells within 1 h (Park et al. 1996). In the LMD-ELISA kit for *E. coli*, the microwell test strips were coated with polyclonal antibodies to *E. coli* O157. Test samples were prepared by mixing 0.2 ml of the sample with 0.1 ml of 10 % buffered formalin, and 0.1 ml of this suspension was then added to the appropriate wells. The plates were incubated at room temperature for 20 min, then washed, and two drops of enzyme conjugate (peroxidase-labeled anti-*E. coli* O157 antibodies) were added to each well. Plates were then incubated at room temperature for 10 min, washed again with the buffer, and rinsed with distilled water. One drop each of substrate A (tetramethylbenzidine) and substrate B (peroxide) were added, and the plates were incubated at room temperature for 5 min. The reaction was stopped using the stop solution (phosphoric acid), and the plates were read spectrophotometrically at 450 nm.

At least 12 other ELISA-based kits are available for the detection of *E. coli* cells (Feng 2001). Several ELISA-based assay kits have been developed for the detection of enterotoxins or Shiga-like toxins (Stx). *E. coli* O157:H7 produces either Stx1 or Stx2, or both. Of these, Stx2 is reported to be the most significant pathogenic factor. The Premier EHEC test (Meridian Diagnostics, Inc.) is a sandwich ELISA which uses monoclonal antibodies directed against Stx1 and Stx2 as capture antibodies and a polyclonal anti-Stx antibody conjugated with horseradish peroxidase for detection. It is easy to perform and is suitable for the routine analysis of food and stool samples (Nataro and Kaper 1998).

Competitive ELISA has been developed for the detection of enterotoxins, and as few as 3–20 pg of heat stable (ST) and heat labile (LT) enterotoxins from *E. coli* were detected using this method (Germani et al. 1994). Oxoid (Hampshire, UK) commercialized an ELISA-based kit for the detection of ST toxins. ELISA-based VEROTEST (MicroCarb) and Premier EHEC (Meridian) kits have also been developed for the detection of Shiga toxins in *E. coli* (Feng 2001).

Table 21.2. Partial list of commercially available immunoassay kits for bacterial pathogens (expanded from Feng 2001)

Target Bacteria/Toxin	The Commercial Name and the Company	Assay Principle
Helibacter pylori	**GAP-IgG,** BioRad	ELISA
	HM-CAP, Enteric products	
	Pylori test EIA, Orion	
	Premier Platinum HpSA ELISA, The Meridian (Cincinnati, OH)	
	Pylori test, Oxoid	LA
Salmonella	**Salmonella-Tek**, Organon Teknika Durham, NC	
	Equate, Binax	
	BacTrace, KPL	
	LOCATE, Rhone-Poulenc	ELISA
	Assurance, Biocontrol Inc, Belleview, Wash.	
	TRANSIA plate Salmonella Gold, Diffchamb	
	Bioline, Bioline	
	TECRA and **OPUS**, Tecra	
	VIDAS	
	VIDAS automatic ELFA, BioMerieux	ELFA
	Seobact, Remel	
	Welcolex, Laboratoire Wellcome	LA
	Bactigen, Wampole labs	
	Spectate, Rhone-Poulenc	
	Microscreen, Mercia	
	Salmonella Latex test, Oxoid	
	Singlepath Salmonella, EMD, Germany	
	Reveal, Neogen	
	Clearview, Unipath	Ab-ppt
	PATH-STIK, LUMAC	
Salmonella serotype typhi	Sanofi qualitative agglutination test kits (Bio-Rad, CA)	Tube agglutination
	TyphiDot™	Dot-ELISA
	Multi-Test Dip-S-Ticks, PANBIO INDX, Inc., Baltimore, Md.	Lateral flow
	TUBEX IDL Biotech, Sollentuna, Sweden	Agglutination
E. coli	**EHEC-TEK**, Organon Teknika	ELISA
	HEC O157ELISA, 3M Company (St. Paul, MN)	
	TRANSIA Card E. coli O157, DiffChamb	
	TECRA, TECRA	
	Assurance, Biocontrol	
	E. coli O157, LMD lab	
	Premier O157, The Meridian	
	E. coli O157:H7, Binax	
	E. coli Rapitest, Microgen	
	VIDAS (ELFA), BioMerieux	
	RIM, Remel	LA
	E. coli O157 latex test and Dryspot E. coli O157 Oxoid	
	Prolex, Prolab	
	Ecolex, Orion diagnostica	
	Welcolex, Murex	

E. coli	**VIP,** Biocontrol	Ab-ppt
	Reveal, Neogen	
	QuixRapid O157, Universal Health watch	
	ImmunoCardSTAT, The Meridian	
	Agilent 2100 Bioanalyzer (Agilent Technologies)	Microfluidic-flowcytometry
E. coli Shiga toxin (Verotoxin)	*Verotest*, Microcarb	ELISA
	Premier EHEC The Meridian	
	Verotox-F, Denka Seiken	RPLA
	VTEC-RPLA, Oxoid	
Enterotoxigenic *E. coli*		
E. coli heat stable toxin (ST)	**E. coli *ST* EIA,** Oxoid	ELISA
E. coli heat labile toxin (LT)	***Vet-RPLA,*** Oxoid,	RPLA
Listeria	**Listeria-*TEK*,** Organon Teknika	
	***TRANSIA Plate* Listeria,** DiffChamb	
	***TECRA*,** TECRA	ELISA
	***Assurance*,** Biocontrol	
	Pathalert, Merck	
	***EIAFOSS, VIDAS* Listeria express (ELFA),** BioMerieux	
	Microscreen, Listeria latex, Microgen	RPLA
	VIP, BioControl	Ab-ppt
	Clearview and ***Rapidtest***, Unipath	
Staphylococcus aureus	*S. aureus* VIA, TECRA	ELISA
	Enterotoxins:	
	SET-EIA, Toxin technology	
	TRANSIA Plate SE, DiffChamb	
	TECRA and *OPUS,* TECRA	
	VIDAS (ELFA), BioMerieux	
	RIDASCREEN, R-Biopharm	
	Staphyloslide, BD	LA
	Aureustest, Trisum	
	StaphLatex, DIFCO	
	Dry Spot Staphytect Plus, Oxoid	
	PBP2′ test (MRSA), Oxoid	
	SET-RPLA, Oxoid	RPLA
	TSST-1:	
	TST-RPLA, Oxoid	
Campylobacter	VIDAS, BioMerieux	
	EIAFOSS, Foss Electric A/S, Hillerød, Denmark	ELISA
	TECRA, TECRA	
	Campyslide, BD, Cockeysville, Md.	
	Meritechcampy, Meridian	LA
	Microscreen, Mercia	
Bacillus cereus diarrhoeal *toxin*	TECRA, TECRA	ELISA
	BceT, Oxoid	RPLA
Clostridium botulinum toxin	ELCA, ELCA Tech	ELISA
Vibrio cholerae	CholeraSMART, New Horizon	Ab-ppt
	BengalSMART	
	CholeraScreen	
	BengalScreen	Agglutination
	Enterotoxin	
	VET-RPLA	RPLA

Note: Analyte 2000 is a fiberoptic system. SPREETA-2000 (Texas Instruments) is a fiber optic SPR system. BIAcore SPR systems have been used for detection of various bacteria with some modifications.

Premier EHEC is an EIA which utilizes monoclonal anti-Stx antibodies adsorbed to microtitre wells to capture Stx, and a polyclonal anti-Stx antibody to detect bound toxin. Bennett-Wood et al. (2004) reported 100 % sensitivity and specificity for this kit after an overnight enrichment of stool samples to detect EHEC.

4.2.2. *Listeria monocytogenes*

Listeria monocytogenes is an intracellular bacterial pathogen which primarily affects immunocompromised patients, pregnant women, and children. A zero tolerance limit imposed by U.S. regulatory agencies on this bacterium has resulted in the development of numerous sensitive detection methods (Gasanov et al. 2005). The commercial kits for *Listeria* (*Listeria* Unique, TECRA International, Frenchs Forest, Australia; and VIDAS *Listeria* Express, bioMerieux, Marcy Etoile, France) claim equal sensitivity to traditional culture methods and enable a result within 30 h of sample receipt (Gasanov et al. 2005).

Bhunia (1997) has extensively reviewed different antibodies developed against *L. monocytogenes* surface antigens and their potential application. Kim et al. (2005) reported the development of five different monoclonal antibodies and a polyclonal IgY antibody against flagella from the 4b strain of *L. monocytogenes*. They showed that a combination of HRP-conjugated MAb7A3 and MAb 2B1 in a sandwich ELISA format was able to detect 10^5 cells/mL, compared to HRP-labeled IgY and MAb 2B1, which detected 10^6–10^7 cells/mL. Yu et al. (2004) designed a sandwich ELISA using the monoclonal antibodies against p60, a highly immunogenic murein hydrolase (essential for cell division), encoded by the *iap* (invasion associated protein) gene. Their two monoclonal antibodies, p6007 and p6017, were highly specific to *L. monocytogenes* and *Listeria* spp., respectively, in ELISA. Similar studies showed that monoclonal antibodies against Internalin A (MAb2B3; Hearty et al. 2006) and unknown proteins (Lin et al. 2006) reacted specifically with *L. monocytogenes* in ELISA. Polyclonal antibodies have been developed against Phospholipase C (Chaudhari et al. 2004), Listeriolysin O (Barbuddhe et al. 2002; Boerlin et al. 2003), Internalin B (Leonard et al. 2005) and Internalin A (Boerlin et al. 2003).

4.2.3. *Salmonella*

Salmonella Newport was detected with a commercially available ELISA kit (TECRA *Salmonella* Visual Immunoassays, International Bioproducts Inc., Vaughn, Ontario, Canada) (Bohaychuk et al. 2005). The VIDAS *Salmonella* test, a commercial ELISA kit, has been shown to be a good screening kit for the detection of *Salmonella* in fecal, tissue, feed, or meat samples (Uyttendaele et al. 2003; Sommerhauser and Failing 2006). However, the TRANSIA card showed a poor specificity for *Salmonella* serovars in food samples (Fratamico 2003). Veling et al. (2001) evaluated two ELISA methods in milk, using the antibodies raised against LPS and the flagellar antigen of *Salmonella*. The specificity for both the assays were 95–100 %; however, the sensitivity was very poor.

4.2.4. Staphylococcal Enterotoxins

Immunoassays have been the major choice for staphylococcal enterotoxin (SE) detection in food and clinical samples (Bhunia 2006). ELISA-based kits for enterotoxin detection are commercially available from various manufacturers like TECRA (*S. aureus* VIA, Tecra, OPUS), bioMerieux (VIDAS), Diffchamb AB (Transia Plate SE- Official Methods recommended by Ministere de l'Agriculture, France, Transia Tube SE), R-Biopharm (RIDASCREEN-A, B, C, D, E) and Toxin Technology (SET-EIA) (Feng 2001).

Since as little as 100–200 ng of the toxin can cause symptoms of staphylococcal intoxi-cation, the development of highly sensitive detection systems is obligatory. Studies have indicated that VIDAS SET2 has a greater sensitivity ($<$ 0.5 ng/g of toxins A and B; and $<$ 1 ng/g of toxins C, D, and E) and specificity (100 %) than VIDAS SET and TRNASIA PLATE SE (Vernozy-Rozand et al. 2004) and RIDASCREEN could detect SEs A through E as low as 0.35 ng/mL of food extract or 0.5–0.75 ng/g of foods within 3 h (Park et al. 1994). Later, Schotte et al. (2002) demonstrated that a rapid immunochromatographic-based handheld assay can detect as little as 50 pg/g of SEB within 15 min. Staphylococcal superantigens (SEA, SEB, and SEC), toxic shock syndrome toxin-1 (TSST-1), and streptococcal pyrogenic toxin A (SPEA) were detected by ELISA and were quantifiable at picogram levels within 2.5 h (Miwa et al. 2000).

4.2.5. *Clostridium botulinum* Toxins

ELISA formats have been the most common method for the sensitive detection of botulinum neurotoxins (Ferreira et al. 2003; Lindstrom and Korkeala 2006). Using a sandwich ELISA format with a polyclonal antibody (PAb) as the capture antibody and a monoclonal antibody (BA93) as the detection antibody, Type A neurotoxin was detected at a concentration of 4–8 pg/mL or 1–2 mouse lethal dose (MLD)/mL (Ekong et al. 1995). An antibotulinum polyclonal antibody in ELISA was able to detect very low concentrations of botulinum neuro-toxins A, B, E, and F within 8 h (Lindstrom and Korkeala 2006). Ferreira et al. (2004) developed ELISA-based detection systems for botulinum and have successfully detected botulinum toxin from foods that were responsible for botulism outbreaks. Both polyclonal and monoclonal antibodies for *C. botulinum* toxin and toxoids A through F are commercially available for assay development from KPL Inc. (Gaithersburg, MD).

4.3. Lateral Flow Immunoassay

The lateral flow immunoassay (LFI), or immunochromatographic strip (ICS), or dipstick test is one of the most attractive and widely used popular immunoassay methods in food and clinical diagnostic work today. In this assay the capture antibody is immobilized on the nitrocellulose membrane in a predefined position. The detection antibody, coupled with colloidal gold or latex particles, is placed in an area near the sample application port. The absorbing blot membrane located at the opposite end of the sample application serves as the wick and facilitates fluid movement on the membrane. When a sample of bacteria, toxin, or antigen suspended in liquid is applied to the sample port of the device, it binds to the detection antibody conjugated to a gold or latex particle. The antigen-antibody complex migrates laterally on the membrane by capillary action to the opposite end of the strip and through a porous membrane that contains two capture zones, one specific for the bacterial pathogen and another specific for unbound antibodies coupled to the gold or latex (control line). The presence of only one (control) line on the membrane indicates a negative sample, and the presence of two lines indicates a positive result, which is visualized within 5–10 min (Chapman and Ashton 2003; Ray and Bhunia 2008). One drawback to this method is that it is less sensitive than the ELISA-based assays and requires a bacterial cell concentration of about 10^7–10^9 cells for a positive reaction. In recent years, however, attempts have been made to improve the sensitivity of the assay by introducing an automatic reader, which avoids the ambiguity in reading the positive reactive bands with human eyes; or by introducing chemiluminescent-based detection of the antigen-antibody complex on the membrane.

A large number of LFI kits have been introduced in recent years, especially for the detection of *E. coli* O157:H7, *Salmonella*, and *Listeria* from enriched food samples. Commer-cially available LFI kits include, VIP (Biocontrol); Immunocard STAT (Meridian); Singlepath®

and Duopath® (Merck); DuPont LFI strips (Dupont); Tecra UNIQUE (De Paula et al. 2002; Briggs et al. 2004); RapidChek® (Strategic Diagnostics Inc. Newark, DE); and REVEAL (Neogen Corporation, Lansing, MI) (Table 21.2).

The performance of LFI kits (REVEAL, Neogen Corp.) was compared with that of ELISA and of PCR for the detection of *Salmonella* Newport, *L. monocytogenes*, and *E. coli*, and the study demonstrated that the LFI produced no false negatives and thus would be a better choice when screening for these pathogens in meat or poultry (Bohaychuk et al. 2005). However, LFI strips take about 2–4 days for the recommended pre-enrichment of the bacterial test samples before analyses.

4.4. Other Immunoassays

4.4.1. Latex Agglutination (LA) and Reverse Passive Latex Agglutination (RPLA) Tests

LA- and RPLA-based commercial detection kits are the most rapid methods used for bacterial or toxin detection. Generally these methods require large amounts of antigen to show a positive reaction. In these methods, antigen-specific antibodies are immobilized on latex particles and mixed with a sample in wells of microtiter plates. If the specific antigen (toxin) is present in the sample in LA, a coagulated precipitate is observed. In RPLA, a diffuse pattern will appear in the bottom; in its absence, a ring or button will appear and the latex does not play a role here—thus it is called a "passive latex agglutination test." These methods have been successfully applied to detect somatic or flagellar antigens or toxins of several foodborne pathogens, including *Staphylococcus aureus*, *Clostridium perfringens*, *Bacillus cereus*, *Vibrio cholerae*, *E. coli*, and *Campylobacter* (On 1996; Feng 2001; Gasanov et al. 2005).

4.4.2. Enzyme-Linked Fluorescent Assay

Fluorescence-based detection by ELISA, called enzyme linked fluorescent assay (ELFA), became popular because of its improved sensitivity and quick results (Vernozy-Rozand et al. 2004; Ray and Bhunia 2008). In this assay, the enzyme (i.e., alkaline phosphatase) conjugated to the detection antibody breaks down the substrate (4-methyl umbellliferyl phosphate, MUP) to produce a fluorescent end product (methyl umbelliferyl), which can be sensitively detected by a spectrofluorometer. In a different format, a fluorophore molecule, instead of an enzyme, is attached to the detection antibody for direct interrogation of antigens or pathogens. The commonly used fluorescent molecules are rhodamine B, fluorecein isocyanate, and fluorescein isothiocyanate (FITC). Decory et al. (2005) used liposome nanovesicles encapsulating fluorescent dyes to increase the fluorescence signal, thus reducing the detection time and limit. They were able to detect < 1 CFU/mL of *E. coli* within 8 h by combining IMS and fluorescence detection.

4.4.3. Time-Resolved Fluorescence Immunoassay

Time-resolved fluorescence immunoassay (TRFIA) is commercially marketed as a dissociation enhanced lanthamide fluorescent immunoassay (DELFIA) by Perkin-Elmer Life Sciences (Akron, OH). A lanthamide chelate (europium, samarium, terbium, or dysprosium) is used as a label in the detection antibody. The method works similarly to ELISA in a microtiter plate, in which an antibody first captures an antigen which is then detected by using the lanthamide-labeled antibody. A low pH enhancement solution is added to dissociate the label from the antibody after the reaction, and these free molecules rapidly form a stable new fluorescent chelate which can be read by a fluorescent reader. Unlike other fluorescent labels, lanthamide has a long fluorescence decay time and an exceptionally large Stokes' shift, thus can be read after the background noise has reduced. Europium (Eu^{3+}) is the label that has been commonly used in TRFIA (Peruski and Peruski 2003; Tu et al. 2004; Lim et al. 2005; Bhunia 2006). TRFIA in combination with IMS (Tu et al. 2002; Yu et al. 2002) and PCR (Watanabe et al. 2002) has been shown to increase the detection sensitivity for *E. coli* and *Salmonella enterica*. It has also been used to detect toxins of *S. aureus* and *C. botulinum* at 4–20 pg ranges (Peruski et al. 2002).

4.4.4. Chemiluminescent Immunoassay

Chemiluminescent immunoassay (CLIA) has been developed to detect pathogens from food samples. Detection antibodies conjugated to the chemiluminescent dyes such as 3(2′ spiroadamantane) 4 methoxy 4(3″ phosphoryloxy) phenyl 1,2 dioxetane (AMPPD), APS-5 (Gehring et al. 2004), and luminol (3-aminophthalhydrazide) enhanced with 4-iodophenol (Zamora and Hartung 2002) are shown to elicit a sensitive signal upon binding to target bacteria. CLIA has proved to be an efficient method for detecting botulinum and staphylococcal toxins (Kijek et al. 2000; Cadieux et al. 2005). However, for *Salmonella* detection, CLIA showed no improvement over the conventional ELISA in specificity and sensitivity (Zamora and Hartung 2002).

4.4.5. Capillary Microbead (Spheres) Immunoassay

Capillary microbead (spheres) immunoassay for staphylococcal enterotoxins (SEs) was developed individually by two different groups during 1997–1998 (Strachan et al. 1997; Giletto and Fyffe 1998). Strachan et al. (1997) used sandwich ELISA, in which microbeads coated with antibody were used to capture antigens in a capillary tube and the SEB was detected with fluorescently labeled anti-SEB antibody. They automated the immunosensor and showed that it was capable of detecting as little as 5 ng/g of toxin within 10 min. Giletto and Fyffe (1998) used a similar approach, except that the detection was chromogenic, yielding a localized coloration upon addition of the substrate mimicking the lateral flow strip reaction. They could detect 0.5 ng/g of SEB in certain foods, compared to 2–10 ng/g by TECRA and SET-RPLA.

4.4.6. Electrochemical-Immunoassay

Electrochemical-immunoassay was adopted for bacterial detection as an alternative to a label-free assay system, in which the antigen-antibody reaction is measured amperometrically using the redox electrodes. In practice, electrochemical immunosensors are an extension of conventional antibody-based enzyme immunoassays (ELISA), in which catalysis of substrates by an enzyme conjugated to an antibody causes pH change, produces ions, or allows oxygen consumption that generates electrical signals on a transducer (Warsinke et al. 2000). Amperometric, potentiometric, and capacitive transducers have been used for such applications. In amperometric detection, for example, alkaline phosphatase conjugated to an antibody hydrolyzes *p*-nitrophenyl phosphate to phenol, which is detected by voltammetry. In light-addressable potentiometric sensors (LAPS), urease-conjugated antibody hydrolyzes urea, resulting in the production of carbon dioxide and ammonia that changes the pH of the solution. A silicon chip coated with a pH-sensitive insulator and an electrochemical circuit measures the alternating photocurrent as a light emitting photodiode shines on the silicon chip. These sensors are very sensitive and have been used for detection of *Salmonella* and *E. coli* O157:H7 rapidly in 0.5 to 1.5 h (Ghering et al. 1998). Croci et al. (2004) developed an electrochemical immunoassay coupled with fluorescence to detect 10^3 CFU of *Salmonella* in 25g of the meat sample.

4.5. Optical Biosensors

4.5.1. Surface Plasmon Resonance

Surface plasmon resonance (SPR) measures the changes in the refractive index resulting from binding of the antigen molecule to an immobilized antibody on the surface of metal films (Au or Ag) (Hsieh et al. 1998; Geng and Bhunia 2007). The binding kinetics could be measured in as soon as a few seconds to a maximum of 15 min (Dmitriev et al. 2002; Lathrop et al. 2003). Label-free, quantitative detection is the major advantage of SPR. Several commercial SPR instruments are currently available: Biacore (Biacore International SA, Switzerland), SPR-670M (Nippon Laser and Electronic Lab, Nagoya, Japan), Spreeta™ (Texas Instruments, Dallas, TX) (Homola et al. 1999; Rich and Myszka 2006). The sensitivity of most of the SPR-based methods is equivalent to an ELISA (10^5–10^8 CFU/mL); however, the major advantage is that it is a label-free detection assay. SPR has been demonstrated to be an efficient optical detection system for *E. coli* O157:H7 (Fratamico et al. 1998), *L. monocytogenes* (Lathrop et al. 2003; Bergwerff and Van Knapen 2006; Hearty et al. 2006), and staphylococcal enterotoxins (Rasooly and Herold 2006). Hsieh et al. (1998) used monoclonal antibodies against *Clostridium botulinum* neurotoxin in a BIAcore instrument to detect the toxin within 20 min. A polyclonal antibody against Internalin B was used in an SPR to detect about 10^5 CFU/mL of *L. monocytogenes*, with coefficients of variation between 2.5 % and 7.7 % (Leonard et al. 2005). Thomas et al. (2006) developed an SPR assay to detect the lipopolysaccharide antigen of *S. enterica* serovar Enteritidis using IgY, and the sensitivity and specificity was determined to be 84 % and 100 %, respectively.

Surface plasmon resonance was used to detect staphylococcal enterotoxins (Homola et al. 2002) and surface proteins of *L. monocytogenes* at very low concentrations (Lathrop et al. 2003). Another SPR-based instrument, Spreeta™, has been evaluated for specific and sensitive detection of *E. coli* O157:H7 in near real time, with the total assay taking about 35 min (Meeusen et al. 2005). However, the detection limit was about 10^6–10^7 CFU/mL. Balasubramanian et al. (2007) applied Spreeta™ for label-free detection of *Staphylococcus aureus*, using lytic bacteriophages, with a detection limit of 10^4 CFU/mL. See chapter 5 for more information about the use of SPR for bacterial detection.

4.5.2. Fiber-Optic Biosensors

Fiber-optic biosensors utilize the total internal reflection (TIR) property of light when it travels through the waveguide and generates a boundary of evanescent waves on the surface of the waveguide. Antibody- or immunoassay-based fiber-optic biosensors provide increased sensitivity, selectivity, and speed compared to the conventional immunoassay techniques (Geng and Bhunia 2007; Bhunia et al. 2007).

Antibody-coupled fluorescence wave guide biosensors are commonly employed in bacterial detection. In principle, a specific antibody is first covalently linked to the optic fiber that captures the bacteria of interest and a fluorescently labeled (e.g., Cy-5 or Alexa-Fluor 647) detection antibody binds specifically to the bacteria. When a 635-nm laser light is launched at the proximal end of the waveguide, fluorescence molecules are excited and generate an evanescent wave. Part of the emitted light energy is transmitted through the fiber and detected by a photodetector at wavelengths of 670–710 nm (Fig. 21.6). Portable sensors, e.g., Analyte 2000 and RAPTOR, manufactured by Research International (Monroe, WA), are widely used for such applications (Anderson et al. 2000; Geng et al. 2004; Nanduri et al. 2006). These systems allow a qualitative detection of the target entity, and the signal is proportional to the amount of antigen or hapten present in the sample (Taitt et al. 2005).

In recent years there has been an increase in the use of fiber-optic sensors for bacterial pathogens and their toxins (Lim 2003). This sensor has been successfully used for the detection

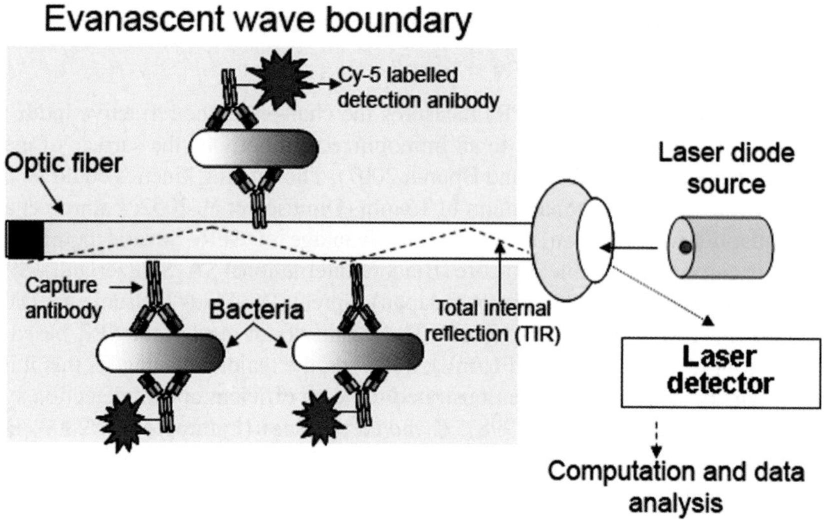

Evanascent wave boundary

Figure 21.6. Schematic diagram showing bacteria detection using a fiber optic biosensor.

of *E. coli* (DeMarco and Lim 2002; Geng et al. 2006), *Streptococci* (Kishen et al. 2003), *Salmonella* (Kramer and Lim 2004), and *L. monocytogenes* (Geng et al. 2004; Nanduri et al. 2006). Geng et al. (2006) were able to detect *E. coli* O157:H7 with initial inoculation of 1 CFU/g of ground beef after only 4 h of enrichment. Using an automated fiber-optic–based detector, RAPTOR™, *Salmonella typhimurium* was detected in sprout rinse water at a concentration of 5×10^5 CFU/mL (Kramer and Lim 2004). Geng et al. (2004) reported a fiber-optic–based detection assay for *L. monocytogenes* with a detection limit of 10^3–10^4 CFU/mL in hot dogs and bologna, and it was later confirmed using the automated RAPTOR™ (Nanduri et al. 2006).

The combination of a fiber-optic biosensor with PCR has been shown to increase the sensitivity of detection of bacteria and dramatically improve the speed of detection from 10 h, needed for conventional fiber-optic sensors, to 2 h by conjugating with PCR (Simpson and Lim 2005). Chapter 6 covers the optical fibre biosensors for bacterial detection.

4.5.3. Antibody-Based Microfluidic Sensors

Research on microfluidics has burgeoned in the past decade from a fascinating concept to applications in clinical, molecular, biochemical, and medical diagnostics. The common materials used in the manufacture of microfluidic systems are silicon, glass, and polymers. The polymer, poly(dimethylsiloxane) (PDMS) has been widely used for designing immunoassays on chips (Bange et al. 2005).

A microfluidic immunosensor promises to improve analytical performance by reducing the assay time and reagent consumption, increasing sensitivity and reliability through automation, and integrating multiple processes in a single device (Bange et al. 2005; Lim and Zhang 2006). A number of researchers have demonstrated the ability of these micro- and nano-biochips to detect the presence of bacteria in food or clinical samples using optical or electrical methods (Stokes et al. 2001; Lin et al. 2005; Li and Su 2006).

Stokes et al. (2001) demonstrated the use of a microfluidic biochip with an integrated 2-dimensional photosensor array for the detection of *E. coli*. The chip contained an array of integrated, independently operating photodiodes, along with amplifiers, discriminators, and logic circuitry on a single platform. They used cellulosic membrane as the platform to capture *E. coli* cells with an antibody on an integrated circuit (IC) biochip and then detected the bacteria using a Cy5-labeled polyclonal antibody in a sandwich immunoassay format. They demonstrated that the biochip had a linear dynamic range of three orders of magnitude greater than that observed for conventional assays, and can detect as few as 20 *E. coli* cells.

Gold nanoparticles, which are about 40–120 nm in diameter, have been exploited in the detection of protein analytes because of their basic ability to scatter a white light to yield a monochromatic light (Thanh and Rosenzweig 2002; Lin et al. 2005). Lin et al. (2005) coated the microchannels created on the PDMS with bacterial cell lysates, which reacted with biotin-labeled polyclonal antibodies raised against *Helicobacter pylori* and *E. coli* as the primary antibodies. The nanoparticles immobilized with antibiotin antibodies were used as a secondary antibody, and the positive reaction was recorded, as the amount of light emitted, under a dark-field stereo microscope. They were able to achieve a detection limit of 10–1000 ng in a 1.5 μL chamber, which is similar to the sensitivity achieved through a conventional ELISA. However, this technology permits small volumes and less time for detection.

Huang et al. (2003) showed that the specificity of the protein biochips could be enhanced using blocking agents such as biotinylated bovine serum albumin (BSA) on a C_{18}-derivatized SiO_2 surface. They used the monoclonal antibody C11E9 (Bhunia et al. 1999) to demonstrate the specific detection of *L. monocytogenes*, and minimized the nonspecific binding by other bacteria.

Alternatively, several studies have used antibodies to enhance the capture efficiency and quantification of bacteria on microfluidic biochips (Sakamoto et al. 2005; Yang et al. 2006). Yang et al. (2006) demonstrated a coupling of immunocapture with dielectrophoresis (DEP) on a microfluidic system to concentrate and detect *L. monocytogenes* cells. The microfluidic biochip consisted of an array of interdigitated microelectrodes on a flat oxidized silicon substrate, and an array of microelectrodes was made on a PDMS cover. A biotinylated anti-*L. monocytogenes* monoclonal antibody, C11E9, was immobilized on the surface of the SiO_2, and positive DEP (at 20Vpp and 1 MHz) was applied to concentrate the bacterial cells from the flowing sample in the chamber. DEP could capture 90 % of the cells during a continuous flow of the sample at the flow rate of 0.2 μL/min. DEP-concentrated cells were captured by the immobilized antibodies on the channel surface with an efficiency of 18–27 % when cells were present at 10^1–10^3/mL.

Accelr8 is a commercial microfluidic lab-on-chip device that allows immunochemical microscopic identification of bacterial cells (http://www.accelr8.com/index.php). The product BACcelr8r is in developmental stages, in which the bacterial cells from biological samples are concentrated on the surface of a chip by applying an electronic potential and are detected by an indirect ELISA format, using specific antibodies. The cells are detected fluorescently using microscopy and image analysis.

4.5.4. Serodiagnosis

Serodiagnosis refers to the serological identification of antibodies against a particular antigen in the serum or blood of the patients. Several kits have been developed based on the principle of ELISA, latex agglutination (LA), and reverse passive latex agglutination (RPLA) for the detection of antibodies against *M. tuberculosis* and *Salmonella typhi* (Perkins et al. 2003; Imaz et al. 2004; Jesudason and Sivakumar 2006). Typically, the ELISA plates or the latex particles are coated with the antigen and when reacted with the serum, the specific antibodies bind to the antigen and give an agglutination reaction.

Imaz et al. (2004) evaluated four different commercial kits (Omega Diagnostics Ltd., Alloa, Scotland) for the serological diagnosis of pulmonary tuberculosis (TB). The Panthozyme-TB Complex Plus detects IgG antibodies against recombinant forms of two antigens from the *M. tuberculosis* complex: r38 kDa (PstS-1, PhoS, antigen 5, antigen 78) and r16 kDa, a member of the alpha-crystalline family of low-molecular-weight heat shock proteins. The other tests, Pathozyme-Myco G (Myco G), Myco M, and Myco A, which utilize the r38 kDa antigen and the lipoarabinomannan (LAM—a common lipoglycan component of the mycobacterial cell wall), detect human IgG, IgM, or IgA, respectively, in the serum of individuals infected with the tuberculosis bacilli (Imaz et al. 2004).

Typhidot® (Malaysian Biodiagnostics Research Sdn. Bhd, Selangor Darul Ehsan, Malaysia) uses an ELISA format, which detects IgG and IgM antibodies against *S. typhi* in the blood (Jesudason and Sivakumar 2006). Olsen et al. (2004) evaluated three different kits for *S. typhi*, Multi-Test Dip-S-Ticks, TyphiDot, and TUBEX, to detect immunoglobulin G (IgG), IgG and IgM, and IgM, respectively, and compared the results with those of the Widal test using the commercial kit from Bio-Rad, CA (Sanofi qualitative agglutination test kit). The Widal test is done as a tube agglutination test to detect the H and O antigens present on the surface of the typhoid bacteria. However, Olsen et al. (2004) observed that the Widal test was insensitive and displayed interoperator variability; and that the two rapid kits, TyphiDot and TUBEX, demonstrated promising results.

5. Recent Developments in Immunoassays

Although the technologies described above yield sensitive and rapid detection of bacteria and their toxins, the need for portability in the instrumentation and even faster results, along with increased specificity and sensitivity, has led to the design of more sophisticated technologies. Here we review some of the novel immunobased assays which show promise in bacterial detection.

5.1. Protein/Antibody Microarrays

Microarrays were originally developed as a tool for genotyping and gene expression analyses. The success of DNA and mRNA microarrays led to more promising ventures into targeting proteins, either toxins or cells using an antibody array format. Unlike the nucleic acid arrays, in which the lysis of cells to release the DNA or RNA is required, protein microarrays have the advantage of detecting cells or toxins in one step. Although the protein microarray is a commercial success in medicine, it still remains in its infancy in respect to bacterial detection.

In recent years, patterned protein (Cai et al. 2005; Steller et al. 2005) or antibody microarrays (Morhard et al. 2000; Howell et al. 2003; Gehring et al. 2006) have been developed for the identification and detection of different bacteria, viz., *Salmonella* (Cai et al. 2005), *E. coli* O157:H7 (Gehring et al. 2006), *Renibacterium salmoninarum* (Howell et al. 2003), and *Neisseria meningitidis* (Steller et al. 2005).

In a protein microarray, the antigens or the recombinant proteins are spotted onto a coated glass slide using an automated spotter and then reacted with fluorescent-labeled antibodies, and the spots are read in the microarray reader. In an antibody microarray, antibodies are patterned, using microcontact printing or soft lithography, onto the substrates (such as glass or silicon), without the loss of biological activity (Inerowicz et al. 2002; Howell et al. 2003; Pavlickova et al. 2004). Binding of antigen to the captured antibodies is detected using fluorescent-labeled antibodies. Phage display technology has been used for antibody production against large arrays of antigens for the purpose of using them on a microarray platform (Pavlickova et al. 2004). Most commonly, different substrates like glass (Howell et al. 2003), gold (Morhard et al. 2000), and silicon (Nijdam et al. 2007) have been used to create patterns for bacteria.

Cai et al. (2005) developed antibody microarrays for serotyping of *Salmonella enterica* strains. With their model, they were able to identify 86 target strains, partially identify 30 more, and differentiate 73 nontarget strains from the targets. Gehring et al. (2006) developed antibody microarrays for detection of *E. coli* O157:H7 in a sandwich format. A linear signal was obtained when cell concentrations were between 3×10^6 and 9×10^7 cells/mL.

Microarrays were successfully used for the simultaneous detection of a multitude of different biowarfare agents: *Staphylococcus* enterotoxin B, ricin, Venezuelan equine encephalitis virus, St. Louis encephalitis virus, West Nile virus, yellow fever virus, orthopox virus, *Francisella tularensis, Yersinia pestis, Brucella melitensis, Burkholderia mallei*, and *Escherichia coli* EHEC O157:H7. A chip was developed for the ArrayTube platform, with classical sandwich ELISA and streptavidin HRP-based detection, which could be accomplished within 1–1.5 h (Huelseweh et al. 2006).

The most common problems associated with fluorescence or direct labeling are lower sensitivity and impaired protein solubility. Thus label-free detection systems like scattering and mass spectrometry are considered alternative methods of choice for the interrogation of antigens. Chapter 26 describes the protein microarray technologies for detection and identification of bacterial analytes.

5.2. Mass Spectrometric Immunodetection

Matrix-assisted laser desorption/ionization (MALDI) is the most common technique used for mass spectrometric analysis of proteins using laser pulses. MALDI coupled with time-of-flight (TOF) measures the mass of intact peptides. In recent years, MALDI-TOF has been applied to protein biochips, to study the interaction of recombinant antibody-antigen, and for direct detection of bacterial cells (Pavlickova et al. 2004). MALDI-TOF has been also shown to be an effective and rapid method for identification of whole bacterial cells (Madonna et al. 2001, 2003) and toxins (Nedelkov and Nelson 2003). *Salmonella* was detected at a concentration of $\sim 10^5$ cells/mL within 1 h using MALDI-TOF (Madonna et al. 2001). In order to increase the sensitivity of detection, IMS and phage-typing were integrated into MALDI-TOF, to rapidly and specifically identify very low numbers of bacterial cells (Madonna et al. 2003). Paramagnetic beads coated with anti-*E. coli* polyclonal antibodies were used to isolate and concentrate *E. coli* from a complex mixture. Captured cells were then infected with a specific lytic bacteriophage (MS2 phage), which produced a large number of phage progeny. Using MALDI-TOF MS, the capsid protein of the MS2 phage was detected, which correlated with the presence of the target bacteria. This integrated approach improved the detection limit twofold, to 10^4 cells/mL.

SPR in combination with MALDI-TOF was found to be highly sensitive for the detection of bacterial lysates and toxins, and was able to detect <1 ng/mL of staphylococcal enterotoxin B, corresponding to the mass analysis of $\sim$500 amol of SEB (Nedelkov and Nelson 2003). Chapter 36 describes the use of mass spectrometric techniques for bacterial detection.

5.3. μSERS Biochip Technology

μSERS is a novel label-free detection technology which uses surface-enhanced Raman scattering (SERS) microscopy. The chip comprises pixels of capture antibodies on a SERS active metal surface, which selectively binds the target bacteria in a sample. Using the Raman microscope, the SERS fingerprints are collected from the pixels on the chip. At each pixel, the bacteria are identified in the spectral domain by matching the unique SERS fingerprint against the library of known fingerprints (Grow et al. 2003). An array of microorganisms consisting of *Listeria, Legionella, Bacillus* spores, and *Cryptosporidium* oocysts were identified and differentiated in a mixed sample (Grow et al. 2003). See chapter 20 for more information about μSERS.

6. Limitations and Challenges

6.1. Specificity and Sensitivity

Universal questions about the immunoassays have been about how specific and sensitive the reaction is. The specificity of any immunoassay-based detection depends on the specificity of the antibodies used. Monoclonal antibodies provide a high degree of specificity because they are specific for an epitope and the antibody-producing clone is selected to provide a limitless supply of homogeneous antibody. However, if that epitope is shared by pathogenic as well as nonpathogenic microorganisms, the assay may not be specific; thus detailed knowledge about an antibody is a prerequisite before it can be incorporated into an assay. On the other hand, polyclonal antibodies could be the source of nonspecific reactions in an immunoassay, since they are comprised of antibodies that react with various epitopes on the same antigen or multiple antigens, and the chances of epitope sharing among different microorganisms are very high. In addition, the animals that are used for antibody development may be the source of undesirable/ unintended antibodies, because they may harbor background antibodies against common microorganisms. In a survey, Lathrop et al. (2006) observed that more than 50 % of the pre-immune sera from 19 different rabbits cross-reacted with 14 different bacterial antigens, suggesting that the presence of cross-reactive antibodies in pre-immune serum is a common problem in the production of specific antibodies. Raising a highly specific antibody is still a challenge, and thus adopting improved methods in the selection and screening of antibodies using recombinant technologies would enhance the specificity of the immunoassays. Conversely, highly specific antibodies can also result in a number of false-negatives. Thus a thorough analysis of the antibodies and evaluation of the assay should be done, if the target is to detect all the pathogenic serotypes and strains in a particular bacterial species.

The sensitivity of the immunoassay still remains a limitation. Although it has shown to be highly sensitive in detecting pico- and femtogram quantities of bacterial toxins, the detection of whole cells below 1,000 cells/mL is still a challenge. With the USDA/FSIS-imposed regulations on "zero tolerance" for many bacterial pathogens (*Clostridium perfringens, Listeria monocytogenes, Salmonella*, and *E. coli*) in various foods, the demand for sensitivity in detection technologies is elevated. Thus, a short enrichment of the samples before use in immunoassays has become a common practice for sensitive detection of bacterial pathogens. Additionally, automated instruments, such as those developed for microtitre plates or biochips, may be a potential source of errors. Bak et al. (2006) noticed significant errors in his assay instrument, even when several data points were collected from over a short period of time.

6.2. Effect of Physical and Chemical Stresses on the Expression Profile of Antigens in Bacteria

6.2.1. Effect of Media Composition on the Expression of Proteins in Bacteria

Antibody-based immunoassays are rapid, but most commercially available assays to detect bacterial species are performed after the test samples are enriched for 24–48 h (Hitchins 1998; Donnelly 2002). Conventionally, USDA/FSIS procedure recommends enrichment of the food sample for isolation and detection using specific selective enrichment broths/media. However, these media may or may not support the expression of the proteins that are used as targets for immunoassays, since each component in the media can interfere with the regulation of gene expression in bacterial pathogens.

Selective enrichment broths have previously been shown to severely affect the expression of anti-*Listeria* monoclonal and polyclonal antibodies (Geng et al. 2003, 2006b; Nannapaneni et al. 1998a, 1998b). Studies in our lab have demonstrated that *L. monocytogenes* expression is differentially regulated under different media and stress conditions (Jaradat and Bhunia 2002; Geng et al. 2003; Lathrop 2005; Geng et al. 2006b; Hahm and Bhunia 2006; Lathrop et al. 2008). Virulence factors like Internalin A, Internalin B, Listeriolysin O, actin polymerization protein, and phospholipases are obvious target antigens for antibody development (Bhunia 1997) because they are associated with the pathogenicity of *L. monocytogenes*.

Glucose, through its active metabolism by *L. monocytogenes*, may be indirectly responsible for the repression of many pathogenic factors by lowering the pH of the media. Glucose-induced, low pH-mediated suppression of the virulence protein, LLO (Milenbachs et al. 1997) and expression of LAP (Jaradat and Bhunia 2002) have been reported earlier. Similar observations have been made by earlier researchers in *E. coli* strains, in which the carbohydrates and pH have resulted in differential expression of various proteins in *E. coli* (Stancik et al. 2002; Vanmaele and Armstrong 1997).

Differential protein expression among strains of *L. monocytogenes* under different growth and environmental conditions has been extensively studied (Sokolovic et al. 1996; Geng et al. 2003; Milohanic et al. 2003; Chatterjee et al. 2006). Geng et al. (2006b) showed that commonly used bacteriological media for *Listeria* isolation and recovery, such as brain-heart infusion broth (BHI), buffered *Listeria* enrichment broth (BLEB), *Listeria* repair broth (LRB), University of Vermont medium (UVM), and Fraser broth (FB) have affected the stress recovery and immunodetection of *L. monocytogenes*. Indirect ELISA using a monoclonal antibody, C11E9, and a polyclonal anti-*Listeria* PAb revealed that BLEB and LRB favorably supported increased expression of antigens and proved to be superior to UVM and FB for the immunodetection of stressed *L. monocytogenes* cells. An earlier study by Sokolovic et al. (1996) also demonstrated that the ActA expression was significantly higher in *L. monocytogenes* serogroup 4 cells grown in a mammalian cell culture medium, MEM (minimum essential medium), than in the nutrient-rich BHI broth. Interestingly, other virulence genes, *prfA*, *plcA*, and *hly*, behaved similarly in BHI (Sokolovic et al. 1996). According to Marr et al. (2006), *L. monocytogenes* growing in the presence of glucose showed that higher PrfA expression resulted in lower carbohydrate intake and slower growth. This study indicates that overexpression of regulatory genes hinders the expression of metabolic genes in *L. monocytogenes*. Potassium nitrate and anaerobiosis has been shown to significantly affect the expression of genes in *Pseudomonas aeruginosa* (Filiatrault et al. 2005; Wu et al. 2005).

6.2.2. Effect of Stress on the Expression of Proteins in Bacteria

Bacteria are exposed to various stress conditions in the host cell, the environment, or during processing and storage. In response to the changes in the environment, the physiological response changes, so that the bacteria are able to survive and cope under the new conditions. These bacteria respond to stress conditions by activating small or large groups of genes under the control of common regulatory proteins. Stress conditions result in the accumulation of these regulatory proteins; and the subsequent transcription of many genes allows cells to cope with specific stress situations, conferring stress tolerance and survival (Chung et al. 2006). Studies have shown that environmental or growth factors such as nutrient concentration, acidity, temperature, carbon sources, and osmotic and oxidative stresses, could down-regulate certain antigen expression and thus their detection (Jaradat and Bhunia 2002; Milohanic et al. 2003; Shetron-Rama et al. 2003; Lemes-Marques and Yano 2004; Sue et al. 2004).

E. coli O157:H7, *Salmonella Enteritidis*, and *L. monocytogenes* were subjected to various stress conditions and were tested by ELISA for their reaction to specific antibodies (Hahm and Bhunia 2006). In general, the study demonstrated that under all stress conditions, including temperature (4° and 45°C), NaCl (5.5 %), oxidative stress (15 mmol^{-1} H_2O_2), acidic pH (5.5), and ethanol (5 %) for 3 h (short-term stress) or for 5 days (long-term stress), the bacteria differentially expressed the antibody reactive antigens. Reaction with *Listeria* PAb showed an up-regulation with most stresses; whereas anti-*E. coli* and anti-*Salmonella* antibodies demonstrated reduced expression levels. Overall, these stress conditions caused an 18–59 % reduction in immunoreaction (Hahm and Bhunia 2006). Likewise, Geng et al. (2003) observed a similar reduced reaction of monoclonal antibodies C11E9 and EM-7G1 to stress-exposed

L. monocytogenes in ELISA and Western immunoblot, due to the reduced surface expression of antibody reactive proteins. Banada et al. (2006) evaluated the virulence protein (InlB and ActA) expression profile in *L. monocytogenes* using specific antibodies after growth in a low conductive growth medium, which was designed for microfluidic biochip applications. This medium, though a minimal medium, did not affect the expression of those proteins, thus suggesting a favorable application on protein biochips for antibody-mediated capture and growth-based sensitive detection of this pathogen (Yang et al. 2006).

A 55 kDa protein of *Salmonella* serovar Typhi was expressed at high intensity when the cells were grown under inorganic acid stress (pH 5.5, 5.0, and 4.5), organic acid stress (2, 4, and 6% of 100-mM stock of acetate, propionate, and butyrate), and heat stress (42°, 45°, and 50°C for 30 min). However, there was no or reduced expression of the same protein when exposed to the same temperatures for less than 30 min (Chander et al. 2004).

During stress exposure, *E. coli* and *Vibrio cholerae* can convert into a viable but noncultarable (VBNC) state or a dormant state. Generally, it is difficult to culture and detect those cells. Desnues et al. (2003) showed that protein expression in *E. coli* under a VBNC state results in increased and nonreversible oxidative damage, which affects various bacterial compartments and proteins. They further showed that the VBNC state is due to stochastic deterioration, rather than an adaptive program, and they pinpoint oxidation management as the "Achilles' heel" of these cells.

These studies suggest that not only antibodies, but also the bacterial physiological status, their response to stress, the growth media, selective enrichment broths, and media compositions can substantially affect the antibody-based detection of pathogens. Therefore, while designing an immunoassay, it is essential to understand the antibody-reactive protein expression profile in a pathogen during exposure to stress or selective enrichment media in order to achieve an optimal immune reaction. This is particularly important in immunosensor applications in which the system is highly sensitive and a slight decline in the immune reaction may lead to a false-negative result.

7. Conclusions and Future Perspectives

An antibody is one of the most important molecules used in biorecognition and detection. Its use continues to grow as we introduce more novel detection technologies, such as those that are bionanosensor-based. There is always a great demand for highly specific and good quality antibodies for assay development. Conventional animal-based antibody production strategies are still useful and valuable, but recombinant and phage display technologies have been shown to be promising in developing a new class of antibodies. In order to obtain the optimum signal from an antibody reaction, one must understand the microorganism's response to environmental stress, the growth media, and its physiological state while designing an immunoassay. When these conditions are thoroughly understood, one can expect improvement in specificity and sensitivity. As we continue to improve our assay systems, there is great demand for a portable device for on-site application. Furthermore, the assay should be inexpensive, easy to read/interpret, automated, and produce a minimum of false results. With the rapid advancement in microfluidics and nanotechnology, the adaptation of the lab-on-chip concept in immunoassays will soon be a reality for commercial use. Bioanalytical immunosensors for real-time monitoring of pathogenic bacterial presence in the food and pharmaceutical industries will greatly reduce the contamination risks to consumers and the call-back risks to the industries. On-field handheld immunodiagnostic instruments will be best suited for farmers and for soldiers to combat the possible spread of pathogenic and bioterrorism-related microbial agents.

References

Allos BM, Moore MR, Griffin PM and Tauxe RV (2004) Surveillance for sporadic foodborne disease in the 21st century: The foodnet perspective. Clin. Infect. Dis. 38: S115–S120

Amagliani G, Omiccioli E, del Campo A, Bruce IJ, Brandi and Magnani M (2006) Development of a magnetic capture hybridization-PCR assay for *Listeria monocytogenes* direct detection in milk samples. J. Appl. Microbiol. 100: 375–383

Anderson GP, King KD, Gaffney KL and Johnson LH (2000) Multi-analyte interrogation using the fiber optic biosensor. Biosens. Bioelectron. 14: 771–777

Angeletti RH (1999) Design of useful peptide antigens. J. Biomol. Tech. 10: 2–10

Bak H, Ekeroth L and Houe H (2007) Quality control using a multilevel logistic model for the danish pig salmonella surveillance antibody-ELISA programme. Prev. Vet. Med. 78: 130–41

Balasubramanian S, Sorokulova IB, Vodyanoy VJ and Simonian AL (2007) Lytic phage as a specific and selective probe for detection of staphylococcus aureus - a surface plasmon resonance spectroscopic study. Biosens. Bioelectron. 22: 948–955

Banada PP, Liu YS, Yang LJ, Bashir R and Bhunia AK (2006) Performance evaluation of a low conductive growth medium (LCGM) for growth of healthy and stressed *Listeria monocytogenes* and other common bacterial species. Intl. J. Food Microbiol. 111: 12–20

Bange A, Halsall HB and Heineman WR (2005) Microfluidic immunosensor systems. Biosens. Bioelectron. 20: 2488–2503

Barbuddhe SB, Chaudhari SP and Malik SVS (2002) The occurrence of pathogenic *Listeria monocytogenes* and antibodies against Listeriolysin-o in buffaloes. J. Vet. Med. B-Infect. Dis. Vet. Pub. Health. 49:181–184

Bauwens L, Vercammen F and Hertsens A (2003) Detection of pathogenic *Listeria* spp. In zoo animal faeces: Use of immunomagnetic separation and a chromogenic isolation medium. Vet. Microbiol. 91:115–123

Bennett-Wood VR, Russell J, Bordun A-M, Johnson PDR and Robins-Browne RM (2004) Detection of enterohaemorrhagic *Escherichia coli* in patients attending hospital in Melbourne, Australia. Pathology 36: 345–51

Bergwerff AA and Van Knapen F (2006) Surface plasmon resonance biosensors for detection of pathogenic microorganisms: Strategies to secure food and environmental safety. J. AOAC Int. 89:826–831

Bhunia AK (1997) Antibodies to *Listeria monocytogenes*. Crit. Rev. Microbiol. 23: 77–107

Bhunia AK (2006) Detection of significant bacterial pathogens and toxins of interest in homeland security. In: Amass SF, Bhunia AK, Chaturvedi AR, Dolk DR, Peeta S, and Atallah MJ (eds) The Science of Homeland Security. Purdue University Press, West Lafayette, Indiana

Bhunia AK and Johnson MG (1992) Monoclonal antibody specific for *Listeria monocytogenes* associated with 66-kDa cell surface antigen. Appl. Environ.Microbiol. 58:1924–1929

Bhunia AK, Ball PH, Fuad AT, Kurz BW, Emerson JW and Johnson MG (1991) Development and characterization of a monoclonal-antibody specific for *Listeria monocytogenes* and *Listeria innocua*. Infect. Immun. 59:3176–3184

Bhunia AK, Banada PP, Banerjee P, Valadez A, Hirleman ED (2007) Light scattering, fiber optic and cell-based sensors for sensitive detection of foodborne pathogens. J. Rapid Methods Automat. Microbiol. 15:121–145

Bialek M, Grabowski S, Kaminski Z and Kaca W (2006) Synthetic peptides mimicking antigenic epitope of helicobacter pylori urease. Acta Biochimica. Polonica. 53:83–86

Blais BW, Bosley J, Martinez-Perez A and Popela M (2006) Polymyxin-based enzyme-linked immunosorbent assay for the detection of *Escherichia coli* O111 and O26. J. Microbiol. Methods 65: 468–475

Boerlin P, Boerlin-Petzold F and Jemmi T (2003) Use of Listeriolysin O and internalin A in a seroepidemiological study of Listeriosis in swiss dairy cows. J. Clin. Microbiol. 41:1055–1061

Bohaychuk VM, Gensler GE, King RK, Wu JT and McMullen LM (2005) Evaluation of detection methods for screening meat and poultry products for the presence of foodborne pathogens. J. Food Prot. 68: 2637–2647

Briggs J, Dailianis A, Hughes D and Garthwaite I (2004) Validation study to demonstrate the equivalence of a minor modification (TECRA®, ULTIMA™ protocol) to AOAC method 998.09 (TECRA® *Salmonella* visual immunoassay) with the cultural reference method. J. AOAC Int. 87:374–379

Cadieux B, Blanchfield B, Smith JP and Austin JW (2005) A rapid chemiluminescent slot blot immunoassay for the detection and quantification of *Clostridium botulinum* neurotoxin type E, in cultures. Int. J Food Microbiol. 101: 9–16

Cai HY, Lu L, Muckle CA, Prescott JF and Chen S (2005) Development of a novel protein microarray method for serotyping *Salmonella enterica* strains. J. Clin. Microbiol. 43: 3427–3430

Cauchard J, Taouji S, Sevin C, Duquesne F, Bernabe M, Laugier C and Ballet JJ (2006) Immunogenicity of synthetic rhodococcus equi virulence-associated protein peptides in neonate foals. Int. J. Med. Microbiol. 296:389–396

CDC (2006) Update on multi-state outbreak of *E. coli* O157:H7 infections from fresh spinach, October 6, 2006. In: Department of Health and Human Services (ed) *E. coli* O157:H7 Outbreak in Spinach. Center for Disease control and Prevention (CDC), Atlanta, Georgia, http://www.cdc.gov/foodborne/ecolispinach/100606.htm

Chander H, Majumdar S, Sapru S and Rishi P (2004) Reactivity of typhoid patients sera with stress induced 55 kda phenotype in *Salmonella* enterica serovar Typhi. Mol. Cell. Biochem. 267: 75–82

Chandler DP, Brown J, Call DR, Wunschel S, Grate JW, Holman DA, Olson L, Stottlemyre MS and Bruckner-Lea CJ (2001) Automated immunomagnetic separation and microarray detection of *E. coli* O157: H7 from poultry carcass rinse. Int. J. Food Microbiol. 70:143–154

Chapman PA and Ashton R (2003) An evaluation of rapid methods for detecting *Escherichia coli* O157 on beef carcasses. Int. J. Food Microbiol. 87, 279–285

Chapman PA and Cudjoe KS (2001) Evaluation of BeadRetriever™, an automated system for concentration of *Escherichia coli* O157 from enrichment cultures by immunomagnetic separation. J. Rapid Methods Automat. Microbiol. 9: 203–214

Chatterjee SS, Hossain H, Otten S, Kuenne C, Kuchmina K, Machata S, Domann E, Chakraborty T and Hain T (2006) Intracellular gene expression profile of *Listeria monocytogenes*. Infect. Immun. 74:1323–1338

Chaudhari SP, Malik SVS, Chatlod LR and Barbuddhe SB (2004) Isolation of pathogenic *Listeria monocytogenes* and detection of antibodies against phosphatidylinositol-specific phospholipase C in buffaloes. Comp. Immunol. Microbiol. Infect. Dis. 27:141–148

Cheng VCC, Yew WW and Yuen KY (2005) Molecular diagnostics in tuberculosis. Eur. J. Clin. Microbiol. Infect. Dis. 24:711–720

Chung HJ, Bang W and Drake MA (2006) Stress response of *Escherichia coli*. Compr. Rev. Food Sci. Food Safety 5:52–64

Croci L, Delibato E,Volpe G, De Medici D and Palleschi G (2004) Comparison of PCR, electrochemical enzyme-linked immunosorbent assays, and the standard culture method for detecting salmonella in meat products. Appl. Environ. Microbiol. 70:1393–1396

Dale JB, Chiang EY, Hasty DL and Courtney HS (2002) Antibodies against a synthetic peptide of saga neutralize the cytolytic activity of streptolysins from group A streptococci. Infect. Immun. 70:2166–2170

De Paula AMR, Gelli DS, Landgraf M, Destro MT and Franco B (2002) Detection of *Salmonella* in foods using TECRA *Salmonella* VIA and TECRA *Salmonella* UNIQUE rapid immunoassays and a cultural procedure. J. Food Prot. 65: 552–555

DeCory TR, Durst RA, Zimmerman SJ, Garringer LA, Paluca G, DeCory HH and Montagna RA (2005) Development of an immunomagnetic bead-immunoliposome fluorescence assay for rapid detection of *Escherichia coli* O157: H7 in aqueous samples and comparison of the assay with a standard microbiological method. Appl. Environ. Microbiol. 71: 1856–1864

DeMarco DR and Lim DV (2002) Detection of *Escherichia coli* O157: H7 in 10-and 25-gram ground beef samples with an evanescent-wave biosensor with silica and polystyrene waveguides. J. Food Prot. 65: 596–602

Desnues B, Cuny C, Gregori G, Dukan S, Aguilaniu H and Nystrom T (2003) Differential oxidative damage and expression of stress defence regulons in culturable and non-culturable *Escherichia coli* cells. EMBO Reports 4: 400–404

Dmitriev DA, Massino YS, Segal OL, Smirnova MB, Pavlova EV, Gurevich KG, Gnedenko OV, Ivanov YD, Kolyaskina GI, Archakov AI, Osipov AP, Dmitriev AD and Egorov AM (2002) Analysis of the binding of bispecific monoclonal antibodies with immobilized antigens (human IgG and horseradish peroxidase) using a resonant mirror biosensor. J. Immunol. Methods 261: 103–118

Donnelly CW (2002) Detection and isolation of *Listeria monocytogenes* from food samples: Implications of sub lethal injury. J. AOAC Int. 85: 495–500

Duncanson P, Wareing DRA and Jones O (2003) Application of an automated immunomagnetic separation-enzyme immunoassay for the detection of *Salmonella* spp. During an outbreak associated with a retail premises. Lett. Appl. Microbiol. 37:144–148

Ekong TAN, McLellan K and Sesardic D (1995) Immunological detection of *Clostridium botulinum* toxin type A in therapeutic preparations. J. Immun. Method. 180:181–191

Emanuel P, Obrien T, Burans J, DasGupta BR, Valdes JJ and Eldefrawi M (1996) Directing antigen specificity towards botulinum neurotoxin with combinatorial phage display libraries. J. Immunol. Methods 193: 189–197

Emanuel PA, Dang J, Gebhardt JS, Aldrich J, Garber EAE, Kulaga H, Stopa P, Valdes JJ and Dion-Schultz A (2000). Recombinant antibodies: A new reagent for biological agent detection. Biosens. Bioelectron. 14:751–759

Fegan N, Higgs G, Vanderlinde P and Desmarchelier P (2004) Enumeration of *Escherichia coli* O157 in cattle faeces using most probable number technique and automated immunomagnetic separation. Lett. Appl. Microbiol. 38: 56–59

Feng P (2001) Rapid methods for detecting foodborne pathogens. In Bacteriological Analytical Manual, Ed. 8 (revised: Jan 25, 2001)

Ferreira JL, Eliasberg SJ, Edmonds P and Harrison MA (2004) Comparison of the mouse bioassay and enzyme-linked immunosorbent assay procedures for the detection of type a botulinal toxin in food. J. Food Prot. 67:203–206

Filiatrault MJ, Wagner VE, Bushnell D, Haidaris CG, Iglewski BH and Passador L (2005) Effect of anaerobiosis and nitrate on gene expression in *Pseudomonas aeruginosa*. Infect. Immun. 73: 3764–3772

Fratamico PM (2003) Comparison of culture, polymerase chain reaction (PCR), Taqman *Salmonella*, and TRANSIA card *Salmonella* assays for detection of *Salmonella* spp. in naturally-contaminated ground chicken, ground turkey, and ground beef. Mol. Cell. Probe. 17: 215–221

Fratamico PM, Strobaugh TP, Medina MB and Gehring AG (1998) Detection of *Escherichia coli* O157: H7 using a surface plasmon resonance biosensor. Biotechnol. Tech. 12: 571–576

Fu Z, Rogelj S and Kieft TL (2005) Rapid detection of *Escherichia coli* O157:H7 by immunomagnetic separation and real-time PCR. Int. J. Food Microbiol. 99: 47–57

Gasanov U, Hughes D and Hansbro PM (2005) Methods for the isolation and identification of *Listeria* spp. and *Listeria monocytogenes*: A review. FEMS Microbiol. Rev. 29: 851–875

Gehring AG, Patterson DL and Tu SI (1998) Use of a light-addressable potentiometric sensor for the detection of *Escherichia coli* O157: H7. Anal. Biochem. 258: 293–298

Gehring AG, Irwin PL, Reed SA, Tu SI, Andreotti PE, Akhavan-Tafti H and Handley RS (2004) Enzyme-linked immunomagnetic chemiluminescent detection of *Escherichia coli* O157:H7. J. Immunol. Methods 293: 97–106

Gehring AG, Albin DM, Bhunia AK, Reed SA, Tu S-I and Uknalis J (2006) Antibody microarray detection of *Escherichia coli* O157:H7: Quantification, assay limitations, and capture efficiency. Anal. Chem. 78:6601–6607

Geng T and Bhunia AK (2007) Optical biosensors in foodborne pathogen detection. In: Knopf GK and Bassi AS (eds) Smart Biosensor Technology. Taylor and Francis, Boca Raton, Florida, pp 503–519

Geng T, Kim KP, Gomez R, Sherman DM, Bashir R, Ladisch MR and Bhunia AK (2003) Expression of cellular antigens of *Listeria monocytogenes* that react with monoclonal antibodies C11E9 and EM-7G1 under acid-, salt- or temperature-induced stress environments. J. Appl. Microbiol. 95: 762–772

Geng T, Morgan MT and Bhunia AK (2004) Detection of low levels of *Listeria monocytogenes* cells by using a fiber-optic immunosensor. Appl. Environ. Microbiol. 70: 6138–6146

Geng T, Uknalis J, Tu SI and Bhunia AK (2006) Fiber-optic biosensor employing Alexa-fluor conjugated antibody for detection of *Escherichia coli* O157: H7 from ground beef in four hours. Sensors 6: 796–807

Geng T, Hahm BK and Bhunia AK (2006b) Selective enrichment media affect the antibody-based detection of stress-exposed *Listeria monocytogenes* due to differential expression of anti body-reactive antigens identified by protein sequencing. J. Food Prot. 69: 1879–1886

Germani Y, Deroquigny H and Begaud E (1994) *Escherichia coli* heat-stable enterotoxin (stA)-biotin enzyme-linked-immunosorbent-assay (stA-biotin ELISA). J. Immunol. Methods 173: 1–5

Giletto A and Fyffe JG (1998) A novel ELISA format for the rapid and sensitive detection of staphylococcal enterotoxin A. Biosci. Biotechnol. Biochem. 62: 2217–2222

Glaser P, Frangeul L, Buchrieser C, Rusniok C, Amend A, Baquero F et al. (2001) Comparative genomics of *Listeria* species. Science 294: 849–852

Goldman ER, Pazirandeh MP, Mauro JM, King KD, Frey JC and Anderson G P (2000) Phage-displayed peptides as biosensor reagents. J. Mol. Recognit. 13: 382–387

Gracias KS and McKillip JL (2004) A review of conventional detection and enumeration methods for pathogenic bacteria in food. Can. J. Microbiol. 50: 883–890

Gray KM and Bhunia AK (2005) Specific detection of cytopathogenic *Listeria monocytogenes* using a two-step method of immunoseparation and cytotoxicity analysis. J. Microbiol. Methods 60: 259–268

Grow AE, Wood LL, Claycomb JL and Thompson PA (2003) New biochip technology for label-free detection of pathogens and their toxins. J. Microbiol. Methods 53: 221–233

Haggerty TD, Perry S, Sanchez L, Perez-Perez G and Parsonnet J (2005) Significance of transiently positive enzyme-linked immunosorbent assay results in detection of helicobacter pylori in stool samples from children. J. Clin. Microbiol. 43:2220–2223

Hahm BK and Bhunia AK (2006) Effect of environmental stresses on antibody-based detection of *Escherichia coli* O157: H7, *Salmonella* enterica serotype Enteritidis and *Listeria monocytogenes*. J. Appl. Microbiol.100:1017–1027

Hara-Kudo Y, Kumagai S, Masuda T, Goto K, Ohtsuka K, Masaki H, Tanaka H, Tanno K, Miyahara M and Konuma H (2001) Detection of *Salmonella* enteritidis in shell and liquid eggs using enrichment and plating. Int. J. Food Microbiol. 64:395–399

Hearty S, Leonard P, Quinn J and O'Kennedy R (2006) Production, characterisation and potential application of a novel monoclonal antibody for rapid identification of virulent *Listeria monocytogenes*. J. Microbiol. Methods 66: 294–312

Hitchins AD (1998) Chapter 10, *Listeria monocytogenes*, FDA Bacteriological Analytical Manual. AOAC Int., Maryland

Homola J, Dostalek J, Chen SF, Rasooly A, Jiang SY and Yee SS (2002) Spectral surface plasmon resonance biosensor for detection of staphylococcal enterotoxin B in milk. Int. J. Food Microbiol. 75:61–69

Homola J, Yee SS and Gauglitz G (1999) Surface plasmon resonance sensors: Review. Sens. Actuat. B-Chem. 54:3–15

Hoogenboom HR (2005) Selecting and screening recombinant antibody libraries. Nature Biotechnol. 23:1105–1116

Howell SW, Inerowicz HD, Regnier FE and Reifenberger R (2003) Patterned protein microarrays for bacterial detection. Langmuir 19:436–439

Hsieh HV, Stewart B, Hauer P, Haaland P and Campbell R (1998) Measurement of *Clostridium perfringens* beta-toxin production by surface plasmon resonance immunoassay. Vaccine 16: 997–1003

Hsih HY and Tsen HY (2001) Combination of immunomagnetic separation and polymerase chain reaction for the simultaneous detection of *Listeria monocytogenes* and *Salmonella* spp. in food samples. J. Food Prot. 64: 1744–1750

Huang TT, Sturgis J, Gomez R, Geng T, Bashir R, Bhunia AK, Robinson JP and Ladisch MR (2003) Composite surface for blocking bacterial adsorption on protein biochips. Biotechnol. Bioeng. 81:618–624

Hudson JA, Lake RJ, Savill MG, Scholes P and McCormick RE (2001) Rapid detection of *Listeria monocytogenes* in ham samples using immunomagnetic separation followed by polymerase chain reaction. J. Appl. Microbiol. 90: 614–621

Huelseweh B, Ehricht R and Marschall HJ (2006) A simple and rapid protein array based method for the simultaneous detection of biowarfare agents. Proteomics 6: 2972–2981

Huesca M, Sun Q, Peralta R, Shivji GM, Sauder DN and McGavin MJ (2000) Synthetic peptide immunogens elicit polyclonal and monoclonal antibodies specific for linear epitopes in the d motifs of *Staphylococcus aureus* fibronectin-binding protein, which are composed of amino acids that are essential for fibronectin binding. Infect. Immun. 68:1156–1163

Imaz MS, Comini MA, Zerbini E, Sequeira MD, Latini O, Claus JD and Singh M (2004) Evaluation of commercial enzyme-linked immunosorbent assay kits for detection of tuberculosis in Argentinean population. J. Clin. Microbiol. 42: 884–887

Inerowicz HD, Howell S, Regnier FE and Reifenberger R (2002) Multiprotein immunoassay arrays fabricated by microcontact printing. Langmuir 18: 5263–5268

Jaradat ZW and Bhunia AK (2002) Glucose and nutrient concentrations affect the expression of a 104-kilodalton listeria adhesion protein in *Listeria monocytogenes*. Appl. Environ. Microbiol. 68: 4876–4883

Jenikova G, Pazlarova J and Demnerova K (2000) Detection of *Salmonella* in food samples by the combination of immunomagnetic separation and PCR assay. Int. Microbiol. 3: 225–9

Jesudason MV and Sivakumar S (2006) Prospective evaluation of a rapid diagnostic test TyphiDot® for typhoid fever. Ind. J. Med. Res. 123:513–516

Jordan D, Vancov T, Chowdhury A, Andersen LM, Jury K, Stevenson AE and Morris SG. (2004) The relationship between concentration of a dual marker strain of *Salmonella typhimurium* in bovine faeces and its probability of detection by immunomagnetic separation and culture. J. Appl. Microbiol. 97:1054–1062

Jung YS, Frank JF and Brackett RE (2003) Evaluation of antibodies for immunomagnetic separation combined with flow cytometry detection of *Listeria monocytogenes*. J. Food Prot. 66:1283–1287

Kane MM and Banks JN (2000) Raising antibodies. In: Goosling J (ed) Immunoassays: a practical approach. Oxford University Press, Oxford

Kendall PA, Hillers VV and Medeiros LC (2006) Food safety guidance for older adults. Clin. Infect. Dis. 42:1298–1304

Khuebachova M, Verzillo V, Skrabana R, Ovecka M, Vaccaro P, Panni S, Bradbury A and Novak M (2002) Mapping the c terminal epitope of the Alzheimer's disease specific antibody MN423. J. Immunol. Methods 262: 205–215

Kim S-H, Park M-K, Kim J-Y, Chuong PD, Lee Y-S, Yoon B-S, Hwang K-K and Lim Y-K (2005) Development of a sandwich ELISA for the detection of *Listeria* spp. using specific flagella antibodies. J. Vet. Sci. 6:41–6

Kishen A, John MS, Lim CS and Asundi A (2003) A fiber optic biosensor (FOBS) to monitor mutants of streptococci in human saliva. Biosens. Bioelectron. 18:1371–1378

Kovacs-Nolan J, Marshall P and Mine Y (2005) Advances in the value of egg and egg components for human health., J. Agric. Food Chem. 53: 8421–8431

Kramer MF and Lim DV (2004) A rapid and automated fiber optic-based biosensor assay for the detection of *Salmonella* in spent irrigation water used in the sprouting of sprout seeds. J. Food Prot. 67:46–52

Lathrop AA (2005) Development of *Listeria monocytogenes* specific antibodies using a proteomics/genomics approach and expression of antibody-specific antigens inlB and actA under different environments. Food Science. Purdue University, West Lafayette

Lathrop AA, Banada PP and Bhunia AK (2008) Differential expression of InlB and ActA in *Listeria monocytogenes* in selective and nonselective enrichment broths J. Appl. Microbiol. 104:627–639

Lathrop AA, Huff K and Bhunia AK (2006) Prevalence of antibodies reactive to pathogenic and nonpathogenic bacteria in preimmune serum of New Zealand white rabbits. J. Immunoass. Immunochem. 27:351–361

Lathrop AA, Jaradat ZW, Haley T and Bhunia AK (2003) Characterization and application of a *Listeria monocytogenes* reactive monoclonal antibody C11E9 in a resonant mirror biosensor. J. Immun. Methods 281:119–128

Lemes-Marques EG and Yano T (2004) Influence of environmental conditions on the expression of virulence factors by *Listeria monocytogenes* and their use in species identification. FEMS Microbiol. Lett. 239:63–70

Leonard P, Hearty S, Wyatt G, Quinn J and O'Kennedy R (2005) Development of a surface plasmon resonance - based immunoassay for *Listeria monocytogenes*. J. Food Prot. 68: 728–735

Leung WK, Ng EKW, Chan FKL, Chung SCS and Sung JJY (1999) Evaluation of three commercial enzyme-linked immunosorbent assay kits for diagnosis of *Helicobacter pylori* in Chinese patients. Diag. Microbiol. Infect. Dis. 34: 13–17

Li XM, Boudjellab N and Zhao X (2000) Combined PCR and slot blot assay for detection of *Salmonella* and *Listeria monocytogenes*. Int. J. Food Microbiol. 56:167–177

Li YB and Su XL (2006) Microfluidics-based optical biosensing method for rapid detection of *Escherichia coli* O157: H7. J. Rapid Methods Automat. Microbiol. 14:96–109

Liddell E (2005) Antibodies. In: Wild D (ed) The Immunoassay Handbook, 3rd ed. Elsevier Ltd., New York

Lim CT and Y Zhang (2007) Bead-based microfluidic immunoassays: The next generation. Biosens. Bioelectron. 22:1197–1204

Lim DV (2003) Detection of microorganisms and toxins with evanescent wave fiber-optic biosensors. Proc. IEEE. 91: 902–907

Lim DV, Simpson JM, Kearns EA and Kramer MF (2005) Current and developing technologies for monitoring agents of bioterrorism and biowarfare. Clin. Microbiol. Rev. 18: 583–607

Lin FYH, Sabri M, Alirezaie J, Li DQ and Sherman PM (2005) Development of a nanoparticle-labeled microfluidic immunoassay for detection of pathogenic microorganisms. Clin. Diagn. Lab. Immunol. 12: 418–425

Lin M, Todoric D, Mallory M, Luo BS, Trottier E and Dan HH (2006) Monoclonal antibodies binding to the cell surface of *Listeria monocytogenes* serotype 4b. J. Med. Microbiol. 55: 291–299

Lindstrom M and Korkeala H (2006) Laboratory diagnosis of botulism. Clin. Microbiol. Rev. 19: 298–314

Liu YC, Ye JM and Li YB (2003) Rapid detection of *Escherichia coli* O157: H7 inoculated in ground beef, chicken carcass, and lettuce samples with an immunomagnetic chemiluminescence fiber-optic biosensor. J. Food Prot. 66: 512–517

Lu D, Shen JQ, Vil MD, Zhang HF, Jimenez X, Bohlen P, Witte L and Zhu ZP (2003) Tailoring in vitro selection for a picomolar affinity human antibody directed against vascular endothelial growth factor receptor 2 for enhanced neutralizing activity. J. Biol. Chem. 278: 43496–43507

Macario AJL and De Macario EC (1988) Monoclonal-antibodies against bacteria. Biotechnol. Adv. 6:135–150

Madonna AJ, Basile F, Furlong E and Voorhees KJ (2001) Detection of bacteria from biological mixtures using immuno-magnetic separation combined with matrix-assisted laser desorption/ionization time-of-flight mass spectrometry. Rapid Commun. Mass Spectrom. 15: 1068–1074

Madonna AJ, Van Cuyk S and Voorhees KJ (2003) Detection of *Escherichia coli* using immunomagnetic separation and bacteriophage amplification coupled with matrix-assisted laser desorption/ionization time-of-flight mass spectrometry. Rapid Commun. Mass Spectrom.17: 257–263

Marr AK, Josephj B, Mertins S, Ecke R, Muller-Altrock S and Goebel W (2006) Overexpression of PrfA leads to growth inhibition of *Listeria monocytogenes* in glucose-containing culture media by interfering with glucose uptake. J. Bacteriol. 188: 3887–3901

Maruta T, Oshima M, Deitiker PR, Ohtani M and Atassi MZ (2006). Use of alum and inactive *Bordetella pertussis* for generation of antibodies against synthetic peptides in mice. Immunol. Invest. 35: 137–148

McCafferty J, Griffiths AD, Winter G and Chiswell DJ (1990) Phage antibodies: Filamentous phage displaying antibody variable domains. Nature 348:552–4

Meeusen CA, Alocilja EC and Osburn WN (2005) Detection of *E. coli* O157:H7 using a miniaturized surface plasmon resonance biosensor. Trans ASAE 48, 2409–2416

Mercanoglu B and Griffiths MW (2005) Combination of immunomagnetic separation with real-time pcr for rapid detection of *Salmonella* in milk, ground beef, and alfalfa sprouts. J. Food Prot. 68: 557–561

Milohanic E, Glaser P, Coppee JY, Frangeul L, Vega Y, Vazquez-Boland JA, Kunst F, Cossart P and Buchrieser C 2003. Transcriptome analysis of *Listeria monocytogenes* identifies three groups of genes differently regulated by *prfA*. Mol. Microbiol. 47:1613–1625

Miwa K, Fukuyama M, Sakai R, Shimizu S, Ida N, Endo M and Igarashi H (2000) Sensitive enzyme-linked immunosorbent assays for the detection of bacterial superantigens and antibodies against them in human plasma. Microbiol. Immunol. 44:519–523

Morhard F, Pipper J, Dahint R and Grunze M (2000) Immobilization of antibodies in micropatterns for cell detection by optical diffraction. Sensor Actuat. B-Chem. 70:232–242

Muhle C, Schulz-Drost S, Khrenov AV, Saenko EL, Klinge J and Schneider H (2004) Epitope mapping of polyclonal clotting factor VIII-inhibitory antibodies using phage display. Thromb. Haemostasis. 91: 619–625

Nanduri V, Kim G, Morgan MT, Ess D, Hahm BK, Kothapalli A, Valadez A, Geng T and Bhunia AK (2006) Antibody immobilization on waveguides using a flow-through system shows improved *Listeria monocytogenes* detection in an automated fiber optic biosensor: RAPTOR™. Sensors 6: 808–822

Nanduri V, Bhunia AK, Tu S-I, Paoli GC and Brewster JD (2007a) SPR biosensor for the detection of *L. monocytogenes* using phage-displayed antibody. Biosens. Bioelectron. 23:248–252.

Nanduri V, Sorokulova IB, Somoylov AM, Simonian AL, Petrenko VA and Vodyanoy V (2007b) Phage as a molecular recognition element in biosensors immobilized by physical adsorption. Biosens. Bioelectron. 22: 986–992.

Nannapaneni R, Story R, Bhunia AK and Johnson MG (1998a) Reactivities of genus-specific monoclonal antibody EM-6E11 against *Listeria* species and serotypes of *Listeria monocytogenes* grown in nonselectvie and selective enrichment broth media. J. Food Prot. 61:1195–1198.

Nannapaneni R, Story R, Bhunia AK and Johnson MG (1998b) Unstable expression and thermal instability of a species-specific cell surface epitope associated with a 66-kilodalton antigen recognized by monoclonal antibody

EM-7G1 within serotypes of Listeria monocytogenes grown in nonselective and selective broths. Appl. Environ. Microbiol. 64:3070–3074.

Nataro JP and Kaper JB (1998) Diarrheagenic *Escherichia coli*. Clin. Microbiol. Rev. 11:142–201

Nedelkov D and Nelson RW (2003) Detection of staphylococcal enterotoxin B via biomolecular interaction analysis mass spectrometry. Appl. Environ. Microbiol. 69: 5212–5215

Nijdam AJ, Ming-Cheng Cheng M, Geho DH, Fedele R, Herrmann P, Killian K, Espina V, Petricoin EF 3rd, Liotta LA and Ferrari M (2007) Physicochemically modified silicon as a substrate for protein microarrays. Biomaterials 28:550–8

Olsen SJ, Pruckler J, Bibb W, Thanh NTM, Trinh TM, Minh NT, Sivapalasingam S, Gupta A, Phuong PT, Chinh NT, Chau NV, Cam PD and Mintz ED (2004) Evaluation of rapid diagnostic tests for typhoid fever. J. Clin. Microbiol. 42: 1885–1889

On SLW (1996) Identification methods for campylobacters, helicobacters, and related organisms. Clin. Microbiol. Rev. 9:405–

Orlandi R, Gussow DH, Jones PT and Winter G (1989) Cloning immunoglobulin variable domains for expression by the polymerase chain-reaction. Proc. Nat. Acad. Sci. USA 86:3833–3837

Padhye NV and Doyle MP (1991) Rapid procedure for detecting enterohemorrhagic *Escherichia coli* O157:H7 in food. Appl. Environ. Microbiol. 57:2693–2698

Pan K, Wang H, Zhang H-B, Liu H-W, Lei H-T, Huang L and Sun Y-M (2006) Production and characterization of single chain Fv directed against beta 2-agonist clenbuterol. J. Agric. Food Chem. 54: 6654–9

Paoli GC, Chen CY and Brewster JD (2004) Single-chain Fv antibody with specificity for *Listeria monocytogenes*. J. Immunol. Methods 289:147–155

Park CE, Akhtar M and Rayman MK (1994) Evaluation of a commercial enzyme-immunoassay kit (RIDASCREEN) for detection of staphylococcal enterotoxins A, B, C, D and E in foods. Appl. Environ. Microbiol. 60: 677–681

Park CH, Vandel NM and Hixon DL (1996) Rapid immunoassay for detection of *Escherichia coli* O157 directly from stool specimens. J. Clin. Microbiol. 34:988–990

Pavlickova P, Schneider EM and Hug H (2004) Advances in recombinant antibody microarrays. Clinica. Chimica. Acta. 343:17–35

Perkins MD, Conde MB, Martins A and Kritski AL (2003) Serologic diagnosis of tuberculosis using a simple commercial multiantigen assay. Chest 123: 107–112

Peruski AH, Johnson LH and Peruski LF (2002) Rapid and sensitive detection of biological warfare agents using time-resolved fluorescence assays. J. Immunol. Methods 263: 35–41

Peruski AH and Peruski LF (2003) Immunological methods for detection and identification of infectious disease and biological warfare agents. Clin. Diagn. Lab. Immun. 10: 506–513

Petrenko VA and Vodyanoy VJ (2003) Phage display for detection of biological threat agents. J. Microbiol. Methods 53, 253–262

Posner B, Lee I, Itoh T, Pyati J, Graff R, Thorton GB, Lapolla R and Benkovic SJ (1993) A revised strategy for cloning antibody gene fragments in bacteria. Gene. 128:111–117

Rasooly A and Herold KE (2006) Biosensors for the analysis of food- and waterborne pathogens and their toxins. J. AOAC Int. 89: 873–883

Ray B and Bhunia AK (2008) Fundamental Food Microbiology, Chapter 41. CRC Press, Boca Raton, Florida

Reinders RD, Barna A, Lipman LJA and Bijker PGH (2002) Comparison of the sensitivity of manual and automated immunomagnetic separation methods for detection of shiga toxin-producing *Escherichia coli* O157: H7 in milk. J. Appl. Microbiol. 92:1015–1020

Rich RL and Myszka DG (2006) Survey of the year 2005 commercial optical biosensor literature. J. Mol. Recognit. 19:478–534

Sakamoto C, Yamaguchi N and Nasu M (2005) Rapid and simple quantification of bacterial cells by using a microfluidic device. Appl. Environ. Microbiol. 71:1117–1121

Schotte U, Langfeldt N, Peruski AH and Meyer H (2002) Detection of staphylococcal enterotoxin B (SEB) by enzyme-linked immunosorbent assay and by a rapid hand-held assay. Clin. Lab. 48:395–400

Shetron-Rama LM, Mueller K, Bravo JM, Bouwer HGA, Way SS and Freitag NE (2003) Isolation of *Listeria monocytogenes* mutants with high-level in vitro expression of host cytosol-induced gene products. Mol. Microbiol. 48: 1537–1551

Shin JH, Roe IH and Kim HG (2004) Production of anti-helicobacter pylori urease-specific immunoglobulin in egg yolk using an antigenic epitope of *H. pylori* urease. J. Med. Microbiol. 53: 31–34

Simpson JM and Lim DV (2005) Rapid PCR confirmation of *E. coli* O157: H7 after evanescent wave fiber optic biosensor detection. Biosens. Bioelectron. 21:881–887

Smith GP and Petrenko VA (1997) Phage display. Chem. Rev. 97: 391–410

Sokolovic Z, Schuller S, Bohne J, Baur A, Rdest U, Dickneite C, Nichterlein T and Goebel, W (1996) Differences in virulence and in expression of *prfA* and *prfA*-regulated virulence genes of *Listeria monocytogenes* strains belonging to serogroup 4. Infect. Immun. 64: 4008–4019

Sommerhauser J and Failing K (2006) Detection of *Salmonella* in faecal, tissue, and feed samples by conventional culture methods and VIDAS *Salmonella* test. Berl. Munch. Tierarztl. Wochenschr. 119:22–7

Splettstoesser WD, Grunow R, Rahalison L, Brooks TJ, Chanteau S and Neubauer H (2003) Serodiagnosis of human plague by a combination of immunomagnetic separation and flow cytometry. Cytometry A. 53:88–96

Stancik LM, Stancik DM, Schmidt B, Barnhart DM, Yoncheva YN and Slonczewski JL (2002) pH-dependent expression of periplasmic proteins and amino acid catabolism in *Escherichia coli*. J. Bacteriol. 184:4246–4258

Steller S, Angenendt P, Cahill DJ, Heuberger S, Lehrach H and Kreutzberger J (2005) Bacterial protein microarrays for identification of new potential diagnostic markers for *Neisseria meningitidis* infections. Proteomics 5:2048–2055

Stokes DL, Griffin GD and Tuan VD (2001) Detection of *E. coli* using a microfluidics-based antibody biochip detection system. Fresen. J. Anal. Chem. 369: 295–301

Strachan NJC, John PG and Millar IG (1997) Application of a rapid automated immunosensor for the detection of *Staphylococcus aureus* enterotoxin B in cream. Int. J. Food Microbiol. 35: 293–297

Straub TM, Dockendorff BP, Quinonez-Diaz MD, Valdez CO, Shutthanandan J I, Tarasevich BJ, Grate JW and Bruckner-Lea CJ (2005) Automated methods for multiplexed pathogen detection. J. Microbiol. Methods 62: 303–316

Su X-L and Li Y (2004) Quantum dot biolabeling coupled with immunomagnetic separation for detection of *Escherichia coli* O157:H7. Anal. Chem. 76: 4806–10

Sue D, Fink D, Wiedmann M and Boor KJ (2004) SigmaB-dependent gene induction and expression in *Listeria monocytogenes* during osmotic and acid stress conditions simulating the intestinal environment. Microbiology-UK 150:3843–3855

Taitt CR, Anderson GP and Ligler S (2005) Evanescent wave fluorescence biosensors. Biosens. Bioelectron. 20: 2470–2487

Thanh NTK and Rosenzweig Z (2002) Development of an aggregation-based immunoassay for anti-protein a using gold nanoparticles. Anal. Chem. 74: 1624–1628

Thirumalapura NR, Morton RJ, Ramachandran A and Malayer JR (2005) Lipopolysaccharide microarrays for the detection of antibodies. J. Immunol. Methods 298: 73–81

Thomas E, Bouma A, van Eerden E,Landman WJM, van Knapen F, Stegeman A and Bergwerff AA (2006) Detection of egg yolk antibodies reflecting salmonella enteritidis infections using a surface plasmon resonance biosensor. J. Immunol. Methods 315: 68–74

Tsurumi Y, Hayakawa M, Shibata Y and Abiko Y (2003) Production of antibody against a synthetic peptide of *Porphyromonas gingivalis* 40-kDa outer membrane protein. J. Oral Sci. 45:111–116

Tu S-I, Golden M, Andreotti P, Irwin P (2002) The use of time-resolved fluoroimmunoassay to simultaneously detect *Escherichia coli* O157:H7, *Salmonella enterica* serovar Typhimurium and *Salmonella enterica* serovar Enteriditis in foods. J. Rapid Methods Automat. Microbiol. 10:37–48

Tu S-I,Golden M, Paoli G, Gore M and Gehring A (2004) Time-resolved fluorescence detection of shiga-like toxins produced by *Escherichia coli* O157 and non-O157 in ground beef. J. Rapid Methods Automat. Microbiol. 12: 247–258

Tully E, Hearty S, Leonard P and O'Kennedy R (2006) The development of rapid fluorescence-based immunoassays, using quantum dot-labelled antibodies for the detection of *Listeria monocytogenes* cell surface proteins. Int. J. Biol. Macromol. 39: 127–134

Ueda S, Maruyama T and Kuwabara Y (2006) Detection of *Listeria monocytogenes* from food samples by PCR after IMS-plating. Biocontrol Sci. 11:129–34

Uyttendaele M. Van Hoorde I and Debevere J (2000) The use of immuno-magnetic separation (IMS) as a tool in a sample preparation method for direct detection of *L. monocytogenes* in cheese. Int. J. Food Microbiol. 54: 205–212

Uyttendaele M, Vanwildemeersch K and Debevere J (2003) Evaluation of real-time PCR vs automated ELISA and a conventional culture method using a semi-solid medium for detection of *Salmonella*. Lett. Appl. Microbiol. 37: 386–391

Vanmaele RP and Armstrong GD (1997) Effect of carbon source on localized adherence of enteropathogenic *Escherichia coli*. Infect. Immun. 65: 1408–1413

Varshney M,Yang L, Su X-L and Li Y (2005) Magnetic nanoparticle-antibody conjugates for the separation of *Escherichia coli* O157:H7 in ground beef. J. Food Prot. 68:1804–11

Veling J, van Zijderveld FG, Bemmel A, Schukken YH and Barkema HW (2001) Evaluation of two enzyme-linked immunosorbent assays for detecting salmonella enterica subsp enterica serovar dublin antibodies in bulk milk. Clin. Diagn. Lab. Immunol. 8:1049–1055

Vernozy-Rozand C, Mazuy-Cruchaudet C, Bavai C and Richard Y (2004) Comparison of three immunological methods for detecting staphylococcal enterotoxins from food. Lett. Appl. Microbiol. 39: 490–494

Visvanathan K, Charles A, Bannan J, Pugach P, Kashfi K and Zabriskie JB (2001) Inhibition of bacterial superantigens by peptides and antibodies. Infect. Immun. 69:875–884

Vytrasova J, Zachova I, Cervenka L, Stepankova J and Pejchalova M (2005) Non-specific reactions during immuno-magnetic separation of *Listeria*. Food Technol. Biotechnol. 43:397–401

Warschkau H and Kiderlen AF (1999) A monoclonal antibody directed against the murine macrophage surface molecule F4/80 modulates natural immune response to *Listeria monocytogenes*. J. Immunol. 163: 3409–3416

Warsinke A, Benkert A and Scheller FW (2000) Electrochemical immunoassays. Fresenius J. Anal. Chem. 366:622–634

Watanabe K, Arakawa H and Maeda M (2002) Simultaneous detection of two verotoxin genes using dual-label time-resolved fluorescence immunoassay with duplex PCR. Luminescence 17:123–129

Williams DD, Benedek O and Turnbough Jr. CL (2003) Species-specific peptide ligands for the detection of *Bacillus anthracis* spores. Appl. Environ. Microbiol. 69:6288–6293

Wu MH, Guina T, Brittnacher M, Nguyen H, Eng J and Miller SI (2005) The *Pseudomonas aeruginosa* proteome during anaerobic growth. J. Bacteriol. 187: 8185–8190

Yang LJ, Banada PP, Chatni MR, Lim KS, Bhunia AK, Ladisch M and Bashir R (2006) A multifunctional microfluidic system for dielectrophoretic concentration coupled with immuno-capture of low numbers of *Listeria monocytogenes*. Lab Chip. 6:896–905

Yang LJ and Li YB (2005) Quantum dots as fluorescent labels for quantitative detection of *Salmonella* Typhimurium in chicken carcass wash water. J. Food Prot. 68:1241–1245

Yazdankhah SP, Solverod L, Simonsen S and Olsen E (1999) Development and evaluation of an immunomagnetic separation-ELISA for the detection of *Staphylococcus aureus* thermostable nuclease in composite milk. Vet. Microbiol. 67:113–125

Yeh KS, Tsai CE, Chen SP and Liao CW (2002) Comparison between VIDAS automatic enzyme-linked fluorescent immunoassay and culture method for *Salmonella* recovery from pork carcass sponge samples. J. Food Prot. 65:1656–1659

Yu KY, Noh Y, Chung MS, Park HJ, Lee N, Youn M, Jung BY and Youn BS (2004) Use of monoclonal antibodies that recognize p60 for identification of *Listeria monocytogenes*. Clin. Diagn. Lab. Immunol. 11:446–451

Yu LSL, Reed SA and Golden MH (2002) Time-resolved fluorescence immunoassay (TRFIA) for the detection of *Escherichia coli* O157: H7 in apple cider. J. Microbiol. Methods 49:63–68

Zamora BM and Hartung M (2002) Chemiluminescent immunoassay as a microtiter system for the detection of *Salmonella* antibodies in the meat juice of slaughter pigs. J. Vet. Med. B. 49: 338–345

Zhao ZJ and Liu XM (2005) Preparation of monoclonal antibody and development of enzyme-linked immunosorbent assay specific for *Escherichia coli* O157 in foods. Biomed. Environ. Sci. 18:254–259

22

Rapid Nucleic Acid-Based Diagnostics Methods for the Detection of Bacterial Pathogens

Barry Glynn

Abstract

The ultimate goal in microbial testing is the ability to accurately and sensitively detect pathogens in real-time or as quickly as possible. Nucleic acid diagnostics (NAD) offer many advantages over traditional microbiological and immunological methods for the detection of infections micro-organisms. These include faster processing time as well as greater potential for intra-species identification and identification of antibiotic susceptibility and strain typing based upon unique sequences. The original techniques of PCR and gel electrophoresis are being superseded by real-time PCR while the development of integrated sample preparation and amplification devices with a simplified user interface will allow for true point-of-care disease detection and suitably tailored treatments. This chapter describes the principles of nucleic acid diagnostics including an overview of the technology's history as well as the general properties of an ideal nucleic acid diagnostics target. Special emphasis is placed upon the detection of pathogens relevant to the food industry. While traditional culture-based methods will retain the lead position as bioanalytical test methods for food safety for the foreseeable future, rapid NAD methods will increasingly compliment or provide alternatives to these methods to meet the ever-evolving challenges in food safety. Ongoing developments in molecular detection platforms including microarrays and biosensors provide potential for new test methods that will enable multi-parameter testing and at-line monitoring for microbial contaminants.

1. Introduction

Molecular diagnostics is revolutionising the clinical management of infectious disease (Yang and Rothman 2004) in a wide range of areas, including pathogen detection, evaluation of emerging novel infections, surveillance, early detection of biothreat agents, and anti-microbial resistance profiling. Figures reporting on the United States alone suggest that as many as 5 million cases of infectious disease-related illnesses are recognised annually. A far greater number remain unrecognised, both among inpatients and in the general community, resulting in substantial morbidity and mortality (Sands et al. 1997).

This chapter will describe some of the technologies employed in the detection of bacterial pathogens, with particular examples described for the detection of food-borne pathogens. Culturing still has a place in pathogen detection, and improvements continue to be made in the

Barry Glynn • National University of Ireland, Galway, The National Diagnostics Centre, Galway, Ireland

M. Zourob et al. (eds.), *Principles of Bacterial Detection: Biosensors, Recognition Receptors and Microsystems*,
© Springer Science+Business Media, LLC 2008

areas of assay automation and the reduction of necessary sample volume. The main focus of this chapter, however, will be in the area of nucleic acid diagnostics (NAD) assays for food-borne pathogens. Areas covered will include direct probe-based detection of target genes and in vitro amplification methodologies, focusing on conventional as well as real-time polymerase chain reaction (PCR) assays. Diagnostics targets for NAD assays will be described and the ideal properties of such a target outlined. Of special interest to the food microbiologist is the problem of separating sufficient target material from a typically complex food matrix. Culture based pre-enrichment remains a commonplace necessity for these assays, but alternative pre-enrichment practices such as immuno-magnetic separation will be discussed. The tantalising prospect of improving assay sensitivity by using "high-copy" number target RNAs is only starting to be realised; this approach also has the potential to differentiate live from dead cells. Finally, as covered in detail elsewhere in this book, improvements in technology continue to be made in the areas of signal amplification, direct detection on biosensors and microarrays, and the application of these emerging technologies to the detection of food-borne pathogens; the detection and enumeration of genetically modified organisms (GMOs) will also be briefly discussed.

The aim of all diagnostic protocols is the identification of a contaminating pathogen in a timely manner, facilitating the rapid administration of appropriate treatment or preventative measures. In food microbiology settings the aim is the "real-time" and "at line" detection of small numbers of contaminants before the produce reaches the food chain. Improvements in diagnostic technologies have not altered these aims; rather, they have redefined the accepted time scale for a positive/negative assay result and lowered the threshold required for an acceptable assay to less than one viable cell per gram of foodstuff.

1.1. Detection of Pathogenic Bacteria from Clinical Samples

Rapid accurate detection of infectious agents is crucial for the timely administration of appropriate treatments. Unfortunately, conventional laboratory practices in many hospitals still employ slow culture-based assays, which often means that accurate diagnoses arrive post-treatment. These problems are only amplified in the case of fastidious pathogens (several weeks culture time is not uncommon) while critical characteristics of a microbial pathogen, such as strain type, presence of virulence factors, and an anti-microbial resistance profile often further delay the delivery of critical information to the bedside. Samples from hospitalised patients often are further complicated by the routine administration of antibiotics upon presentation of any signs of bacterial infection. Such empiric antibiotic therapy has amongst its consequences increased rates of antimicrobial resistance within hospital-acquired strains of bacteria (Cosgrove 2003). A rapid diagnostics assay would permit the application of more informed therapeutic intervention by critical care clinicians (Yang and Rothman 2004).

PCR is particularly useful for the identification of organisms that cannot be cultured, or where culturing conditions are insensitive or require prolonged incubation times. Thus PCR has opened up new possibilities for the detection of slow-growing pathogens, intracellular bacteria as well as viable but non-culturable pathogens. While outside the area covered by this chapter, the potential to detect viruses (and retroviruses) has also enormously expanded.

1.2. NAD Assays for the Detection of Respiratory Infection, Sepsis and Sexually Transmitted Infection

Nucleic acid diagnostics and in particular PCR have been applied to the detection of respiratory disease, sepsis, and sexually transmitted infections.

One of the earliest clinical applications of PCR was for the detection of the respiratory pathogen *Mycobacterium tuberculosis* (D'Amato et al. 1995). Standard culture methods for this pathogen typically included week- to month-long incubations. Public health considerations related to the potential transmission of infection during this prediagnosis period prompted the application of non-culture-based methods. Currently the FDA-approved use of this diagnostics assay is as an adjunct to conventional culture. However, improvements in patient management, clinical outcome, and cost efficacy have resulted in cases where these assays have been deployed (Yang and Rothman 2004).

Detection of the STI organisms *Chlamydia trachomatis* and *Neisseria gonorrhoeae* from both genital swabs and urine samples has been demonstrated using direct hybridisation methods (Gen-probe) as well as PCR and other in vitro amplification technologies (Table 22.1). These have proven particularly useful in the case of *Chlamydia*, where culturing is difficult and limited by the requirements for specialised facilities. Development of this assay has facilitated the routine screening of patients considered at-risk for STIs upon presentation at clinics and hospitals. Early indications are that the application of this technology has resulted in up to threefold greater rates of disease detection and treatment compared to standard practice.

Sepsis is a clinical term used to describe a patient who has symptomatic bacteremia, with or without organ dysfunction; while septicemia refers to the active multiplication of bacteria in the bloodstream that results in an overwhelming infection. Septic shock is the most common cause of mortality in the intensive care unit. Bacterial infections are the most common cause of septic shock. Almost any bacterium can cause bacteremia. Bacteremia is not necessary for the development of septic shock. Only 30–50 percent of patients with sepsis have positive blood culture results. *S. pneumoniae, Neisseria meningitidis*, or *S. aureus* usually cause sepsis in the child. Sepsis due to *H. influenzae* was very common; however, since the introduction of the Hib vaccine, invasive *H. influenzae* infections have virtually disappeared. Other causes include *E. coli, S. agalactiae* (Group B Strep), *Klebsiella* spp., and *Enterobacter* spp. Sepsis is the 7th leading cause of death in children 1–4 years of age and is the 9th leading cause of death in children 5–14 years of age (World Health Organization, 2003).

In Europe and the United States, GBS is a leading cause of infant mortality and serious neonatal infections such as sepsis, pneumonia, and meningitis (Ke et al. 2000). Identification of GBS-colonised women is critical for the prevention of neonatal GBS infections, as it can be passed from mother to child during birth. Culture methods require up to 48 h to yield results and predict only 87 % of women likely to be colonized by GBS at delivery (Yancey et al. 1996a; Yancey et al. 1996b). Rapid real-time NAD assays targeting the *S. agalactiae cfb* gene have

Table 22.1. Commercial nucleic acid-based diagnostics assays for the detection of bacterial pathogens from clinical samples (Yang and Rothman 2004)

Organism	Product	Manufacturer	Method
Chlamydia trachomatis	Amplicor	Roche	PCR
	LCX	Abbott	LCR
	AMP	Gen-Probe	TMA
	PACE 2	Gen-Probe	Hybridisation
	BD Probe Tec	Becton Dickinson	SDA
Neisseria gonorrhoeae	Amplicor	Roche	PCR
	LCX	Abbott	LCR
	Hybrid Capture II CT/GC	Digene	Hybrid capture
	BD Probe Tec	Becton Dickinson	SDA
	PACE 2	Gen-Probe	Hybridisation
M. tuberculosis	TB Amplicor	Roche	PCR
	E-MTD	Gen-Probe	TMA
Group B streptococcus	Xpert GBS	Cepheid	Real-Time PCR

successfully been designed for the rapid (<75 min) detection of this organism (Ke, Menard et al. 2000) and are marketed by Cephid as the Xpert GBS system.

1.3. Profiling of Multi-drug Resistance

Due to the increasing rates of multi-drug resistance pathogens, early antimicrobial resistance profiling is crucial for the timely treatment of patients as well as for broader public health considerations. Culturing typically takes 48–72 hours and is subject to many sources of variability (innoculum size, culture conditions, etc.).

PCR assays have been described for detection of the *mecA* gene in *Staphylococcus aureus* and are becoming established as the recognised method of identifying meticillin-resistant *S. aureus* (MRSA) (Tenover et al. 1999). Rifampicin resistance in *M. tuberculosis* is conferred by mutations within a short section of the *rpoB* gene. Detection based upon PCR combined with line-probe diction technologies is available for the detection of these mutations. The possibility of multiple mutations conferring resistance adds a level of complexity to the easy detection of rifampicin resistance; this may be solved by the application of multiples or microarray technologies (Traore et al. 2006).

1.4. Bioterrorism

The limitations of conventional culture methods make them totally unsuitable for application in the area of biodefence. In more recent times, the advent of bioterrorism has highlighted the need for rapid, simple, and robust diagnostics assays to detect select agents (Ivnitski et al. 2003). Common targets in bioterrorism detection systems are *B. mallei, B. anthracis, Y. pestis*, and *F. tularensis*.

2. Detection of Bacterial Food-Borne Pathogens

2.1. Recent Outbreaks

Food-borne illness poses a significant economic burden for nations, damages consumer confidence, and impacts the international trading of food products. Worldwide, the number of cases of gastroenteritis associated with food is estimated to be between 68 million and 275 million per year (Naravaneni and Jamil 2005). In Ireland, during 2003, there were approximately 2000 cases of bacterial enteritis caused by the three most common food-borne pathogens, Salmonella, Campylobacter, and *E. coli* O157:H7; while *Listeria monocytogenes* was responsible for 7 cases of illness (Health Protection Surveillance Centre 2003). During the past decade there has been heightened awareness among food producers and legislators regarding food safety. Quality assurance systems, including Hazard Analysis of Critical Control Points (HACCP), have been introduced by food producers to meet legislative requirements and this has led to a significant increase in food microbiology testing.

Recent serious and high profile cases of outbreaks of food poisoning within Ireland, the United Kingdom, and the United States has led to heightened public awareness and concern over food-borne illness (Glynn et al. 2006). This increased public concern has, in turn, resulted in consumer demands for assurances from the food industry as to the microbiological quality and safety of food and food products. These outbreaks have also highlighted the need for new improved, user friendly, and cost effective diagnostics assays to enable the effective monitoring of food during its preparation and processing for human consumption (Hodgson 1998).

Microbiological quality testing of foods has relied almost entirely on conventional microbiological methods, involving the isolation and enumeration of pathogenic bacteria from food using specialised microbiological media, which yields results only after several days and repeated culture enrichment steps. Culturing techniques are often complicated as well as time-consuming; for example, there are about 9 different plating media and 7 different broths described for the isolation of Campylobacter from foods (Corry et al. 1995). Campylobacter also have fastidious growth requirements and conventional detection and identification usually takes from 4–6 days (Giesendorf and Quint 1995).

2.2. Benefits and Limitations of Conventional Methods

Traditional methods remain the mainstay of food microbiology testing. Many of these methods were developed early in the last century and involve the isolation and enumeration of pathogenic bacteria from food using specialised microbiological media. These methods can involve multiple enrichment and culturing steps, followed by morphological and biochemical analysis. Results are usually obtained after 3 to 7 days (de Boer and Beumer 1999). Recent improvements to traditional methods include improved isolation media, alternative plating techniques, standardised equipment for the preparation of plates and application of test samples, and rapid and automated pathogen monitoring and identification systems. Examples include VITEK (BioMerieux, Marcy-Etoile, France), and the MicroLog™ system (BiOLOG, CA, USA). Novel methods to measure bacterial growth have included impedance measurements and ATP bioluminescence for bacterial detection in food (Fung 2002; Gracias 2004). These methods are nonspecific, as they measure the accumulation of by-products of microbial growth without identifying the particular organism present. Applications of these technologies in the food industry include the measurement of surface contamination on meats and the post-processing evaluation of pasteurisation processes. The merits of these approaches to the detection and enumeration of food pathogens have been reviewed by Gracias and McKillip (2004). Commercially available assays based on bacterial ATP detection applicable to food include Profile-1 (New Horizons Diagnostic, MD, USA) and Biotrace (BioTrace International, Bridgend, Wales).

2.3. Development of Rapid Diagnostics Methods

Developments in the fields of immunology and molecular biology offer the potential to develop rapid, high throughput tests that will allow the food industry to make timely assessments on the microbiological safety of its food products, prior to their release to the consumer market (Meng and Doyle 1997). Current research is focused on the development of improved nucleic acid-based diagnostics tests to enable more sensitive and rapid detection of pathogens in foods; while future developments, e.g., real-time fluorescent-PCR, automated closed systems, and microarrays or "DNA chips," will provide the next generation of rapid diagnostic technologies.

3. Rapid Nucleic Acid Diagnostics for Bacterial Food-Borne Pathogens

3.1. In Vitro Nucleic Acid Amplification-Based Detection of Food-Borne Pathogens

The application of a test for food-borne pathogen identification and detection that includes an in vitro amplification step has the potential to increase the speed and sensitivity of food quality testing, facilitating a more expedient release of products from the industry. In vitro amplification technologies are designed to amplify a target nucleic acid fragment that contains unique sequences, which are specific to the pathogen of interest. A number of in vitro amplification technologies have been developed which are applicable to food-borne pathogen diagnosis

(Scheu, Berghof, and Stahl 1998). The polymerase chain reaction (PCR) is at present one of the most widely utilised molecular techniques to complement classical microbiological methods for the detection of pathogenic microorganisms in foods.

The basis of any NAD assay is a specific nucleic acid target sequence, unique to the bacterial pathogen of interest. Even within a given genus of bacteria, there are unique sequences that are specific to each species, and these can be exploited to determine the presence of that bacterial species in a sample. DNA is a very stable molecule, which makes it particularly suitable as a molecular target for NAD assays, as it can be isolated relatively simply from a variety of complex biological samples (Fitzmaurice et al. 2004; Grennan et al. 2001; Maher et al. 2003; O'Connor et al. 2000; O'Sullivan et al. 2000). Nucleic acid-based diagnostic tests have several advantages, including faster turnaround time and improved sensitivity and specificity compared to conventional technologies for the detection of bacterial food-borne pathogens. NAD systems are available in a variety of formats, ranging from simple nucleic acid probe hybridisation systems to tests incorporating the amplification of a specific genomic target by means of one of many available in vitro amplification technologies. Recently, there has been a move towards more sophisticated NAD platforms for bacterial pathogen identification, including real-time in vitro amplification systems, biosensors, and microarray-based platforms.

3.2. Requirements for a NAD-Based Food Assay

Ideally, a candidate NAD target should be present in the cell at relatively high copy numbers, while being sufficiently heterologous at the sequence level to allow for differentiation of the pathogen at both genus and species levels. While NAD tests have utilised a wide variety of genomic targets, multicopy ribosomal RNA (rRNA) genes, genes encoding toxins or virulence factors, and genes involved in cellular metabolism are popularly used for the identification of food-borne bacterial pathogens. At the National University of Ireland, Galway (NUI, Galway) researchers have successfully employed the 16S/23S rRNA intergenic spacer region in bacteria as a molecular target in NAD assays for common food-borne pathogens (Grennan, O'Sullivan et al. 2001; O'Connor, Joy et al. 2000; O'Sullivan, Fallon et al. 2000). Table 22.2 lists examples of gene targets applied in NAD assays for the identification of food-borne pathogens.

3.3. Polymerase Chain Reaction (PCR)

PCR is an in vitro technique used to enzymatically amplify, in an exponential manner, a specific fragment of DNA, through a series of repetitive reaction cycles. During each cycle the number of copies of the target sequence doubles; and since newly synthesised copies also serve as templates for subsequent rounds of synthesis, the amount of DNA generated increases exponentially. Detection of amplification products can be carried out in a variety of ways,

Table 22.2. Genomic targets utilised in PCR-based diagnostic assays for common bacterial food-borne pathogens [1]

Organism	Food Type	Target(s)
E. coli O157, STEC	Beef, milk	*stx*1, *stx*2, *eae*, *rfb,vt1*, *vt2*, *elyA*, *hlyA*
C. jejuni	Chicken, chicken fecal samples, other poultry	16S rRNA, *flaA*, *flab*, 16S/23S rRNA intergenic spacer
L. monocytogenes	Beef, seafoods, cooked ground beef, pork, milk	*hlyA*, *hly* O, *iap*, *inl*A, *inl*B, 16S/23S rRNA intergenic spacer
Salmonella	Chicken, raw meat, pooled raw egg	*invA, sefA*

e.g., visually using gel electrophoresis, ethidium bromide staining, and examination of the gel using UV light (Eyigor et al. 1999; Woodward and Kirwan 1996). Southern blot-based hybridisation with a specific DNA probe following gel electrophoresis of the PCR products allows confirmation of the identity of the PCR products (Gonzalez et al. 1997). Colourimetric or fluorimetric detection of amplification products can also be carried out following hybridisation to specific labelled DNA probes, allowing rapid and simplified detection (Mandrell and Wachtel 1999; Metherell, Logan and Stanley 1999). DNA probe hybridisation to amplified PCR products increases the sensitivity of the test by up to a hundredfold (Lin and Tsen 1996), and adds a second level of specificity to the assay if a species-specific DNA probe is combined with organism-specific PCR amplification. The sensitivity of the PCR/DNA probe assays depends on the PCR reaction conditions used, the food matrix, and the post-PCR detection method employed.

PCR/DNA probe assays have been developed targeting the 16S/23S rRNA intergenic spacer region for *L. monocytogenes*, *Campylobacter jejuni*, *C. coli*, and Salmonella. These assays are compiled as colorimetric membrane-based detection tests with DNA probes cross linked to a specialised membrane solid support. Biotinylated PCR products are heat denatured, hybridised to the capture probe by reverse hybridisation kinetics, and detected by an alkaline phosphatase mediated colorimetric reaction. As many as 20 different probes can be applied per test strip for multiparameter detection, with the potential to simultaneously screen a sample for the presence of a range of bacterial pathogens (O'Connor, Joy et al. 2000). A number of these food-borne pathogen diagnostic assays have also been adapted for microtitre plate-based detection using the Nucleolink technology (Nunc A/S, Denmark).

3.4. Application of PCR-Based Tests to Pathogen Detection in Food Samples

Currently, the strategy for application of PCR-based assays for the detection of food-borne pathogens requires enrichment of the food sample prior to PCR amplification and PCR product detection. A potential limitation of PCR is that it is capable of detecting DNA released from nonviable organisms. This was demonstrated in a study in which heat-killed *Campylobacter jejuni* cells were added to raw milk and detected by PCR for up to 5 weeks without loss of sensitivity (Allmann et al. 1995). However, the enrichment of the food sample prior to PCR amplification significantly reduces the likelihood of detecting nonviable organisms and offers the advantages of increasing the number of live target cells and diluting the inhibitory effects of the food matrix.

Alternatively, the application of assays that target the in vitro amplification and detection of RNA will ensure the detection of viable organisms (McKillip, Jaykus, and Drake 1998). The isolation of RNA is technically more difficult than that of DNA, as RNA is considerably less stable than DNA. However, as sample preparation and handling improves, this limitation will be overcome, and using high copy number RNA targets will enable the development of amplification-based diagnostics tests for pathogens in food.

One of the major problems associated with the use of PCR assays for food samples is the presence of PCR inhibitors in the food sample. False negative results can occur for a variety of reasons, including (1) the degradation of the target nucleic acid sequences in the sample, and (2) the presence of substances that inhibit the PCR reaction. The degree of inhibition varies greatly with food type. Certain foods are more problematic than others. Soft cheeses, for example, can completely inhibit PCR assays (Rossen et al. 1992). Calcium ions in milk have also been identified as a source of PCR inhibition (Bickley et al. 1996). Studies have shown that high levels of oil, salt, carbohydrate, and amino acids have no inhibitory effect; while casein hydrolysate, calcium ions, and certain components of some enrichment broths are inhibitory for PCR. The removal of inhibitory substances from DNA to be amplified is an important prerequisite to successful PCR amplification. Several methods of sample preparation have been reported, including filtration, centrifugation, the use of detergents and organic solvents, enzyme treatment, sample dilution, and immunomagnetic separation (IMS).

IMS is a technique commonly used to specifically capture and extract intact bacterial cells from complex food matrices (Dziadkowiec, Mansfield, and Forsythe 1995; Fluit et al. 1993). Magnetic particles coated with antibodies to the organism of interest can be used to capture organisms from the food sample for inclusion directly in the PCR. A point to note in relation to IMS is that specificity will be determined by the antibody used for coating the magnetic particles. A potential problem with IMS is that certain food components can interfere with the antibody-organism interaction.

The complexity of food samples dictates that no one sample preparation method will be suitable for all food types. In addition, the analysis of foods for pathogens demands highly sensitive tests because of the need to detect very low numbers of organisms. Investigation is ongoing into the simplification of food sample preparation methods, the amplification process, and the final detection system.

In summary, most of the (PCR) amplification-based assays currently applied to food samples include a pre-enrichment step of 18 hours or more to increase cell numbers while diluting potential PCR inhibitors present in the food matrix being sampled. More recently, there have been reports of successful application of PCR-based assays to samples enriched for 6 hours (Gouws, Visser, and Brozel 1998; Nastasi, Mammina, and Mioni 1999). A number of commercial PCR-based kits are available; e.g., the BAX system developed by Qualicon recommends a short culture-based enrichment of the food sample and PCR amplification with gel-based detection of the PCR products (Stewart and Gendel 1998).

3.5. Use of RNA as an Alternative Nucleic Acid Diagnostic Target

RNA, an alternative nucleic acid target for NAD assays, is a labile molecule that is quickly and easily degraded, particularly once the organism is killed. This property makes handling RNA much more difficult than DNA but means RNA has the advantage of enabling viable organisms to be distinguished from nonviable rRNA remains the "gold standard" target for bacterial pathogen identification and has been exploited in a range of NAD assays for identification of food pathogens (Pfaller 2001). Gen-Probe Incorporated (CA, USA) have patented the use of rRNA as a target technology and have exploited this target in a range of assays for bacterial pathogens (www.gen-probe.com).

The possibility of using functional, high copy number RNAs other than rRNA for the rapid, sensitive, and specific detection of bacterial pathogens has also been investigated. The bacterial *ssrA* gene and its encoded transcript, tmRNA, present in all bacterial phyla, has many of the desired properties of a NAD target (Schonhuber et al. 2001). tmRNA is present in all bacterial species at relatively high transcript numbers (Keiler and Shapiro 2003). *E. coli* has previously been shown to have approximately 5×10^2 tmRNA molecules per cell (Lee, Bailey, and Apirion 1978), while we have applied quantitative real-time PCR and in vitro transcribed cRNA standards to show that tmRNA is present from 3×10^2 to 2×10^3 copies per cell in a variety of bacterial pathogens (Glynn, unpublished observation). The high copy number of tmRNA is advantageous in developing NAD assays while the use of an RNA target has the potential to enable viable pathogens to be distinguished from nonviable organisms. Additionally, the $5'$ and $3'$ ends of the *ssrA* gene and its tmRNA transcript are highly conserved between different species and across bacterial genera, while the intervening regions of tmRNA contain sufficient sequence heterogeneity between species to permit the development of specific DNA probes and pathogen-specific NAD tests (Schonhuber, Le Bourhis et al. 2001). Reviews of the tRNA/mRNA functionality of the tmRNA molecule are provided by Muto (Muto, Ushida, and Himeno 1998) and on the tmRNA website (http://www.indiana.edu/~tmrna/).

3.6. Sample Preparation for NAD from Clinical Sample Types

NAD assays have found application in the detection of microorganisms in environmental and clinical sample types as well as from food matrices. While this chapter deals mainly with the area of food-borne bacterial detection, it is also useful to give an overview of the preparation steps common to all sample types. Key to any NAD assay are the steps for the conversion of heterogeneous samples (biological materials, foods, soil, or water samples) into a homogeneous PCR-compatible sample. Once this has been achieved, amplification protocols become somewhat standard irrespective of the starting sample material or microbial cell type. The principles of nucleic acid diagnostics assay design can be easily transferred from food to clinical to environmental samples once initial pre-treatment, cell concentration, cell lysis, and nucleic acid recovery have been accomplished.

An issue of particular relevance to clinical samples is the interpretation of a positive result. Some clinicians have expressed concern that current amplification-based methods are too sensitive, as in some cases the presence of a single organism may not be clinically relevant. The persistence of free DNA in samples from dead cells or the isolation and detection of target material from nonviable cells can also lead to false positives. A potential solution to this problem is the use of RNA targets which should select only for the presence of viable cells. RNA is known to be rapidly degraded following cell death. The technical obstacle to be overcome prior to the whole scale application of RNA-based assays is also related to the labile nature of the target material, and requires additional steps during sample preparation to stabilise the RNA as well as to ensure that all equipment and reagents are free of RNase enzymes.

3.7. Limitations of NAD in Clinical Settings

Due to the amplification power of PCR, even minute amounts of carryover contamination will lead to false positive results. This can become a particular problem in the clinical laboratory, where large numbers of samples are handled. Good laboratory practice in the reduction of this problem often includes physical separation (separate rooms or even buildings), some sample preparation, amplification, and analysis; while the use of enzymatic agents such as uracil N-glycosylase can further reduce contamination due to carryover of amplification products from previous reactions. Future technologies designed to integrate cell lysis, target amplification, and detection into lab-on-a-chip systems will also reduce this problem by sealing the entire process with a single microdevice and reduce opportunities for the introduction of contaminants.

False negatives are also a cause for concern. Problems arise from the relative small volume of sample material that can be included in an amplification, as well as issues arising from PCR processing. A simple solution to the first problem involves steps to concentrate the target material from a larger volume of biological samples. Regardless of the biological sample type (blood, cerebrospinal fluid, urine, sputum), pretreatments are required to remove eukaryotic cells or to liquefy viscous material before the concentration step to increase bacterial yield. Such methods can be nonspecific, such as centrifugation or filtration; or specific, such as capture antibodies coated onto magnetic beads.

Following target concentration the next obstacle to overcome is from inhibitory material that co-purifies during sample processing. The three most common sample processing issues related to false negative results are: (1) the failure to remove PCR inhibitors form the sample; for example in the case of blood samples, the *Taq* enzyme used in most PCRs is rendered totally inactive by as little as 0.004 % (vol/vol) blood in the PCR mix (Hoorfar, Wolffs, and Radstrom 2004); (2) ineffective lysis of bacterial cells and the failure to release sufficient nucleic acid for detection, which is particularly relevant when dealing with gram positive bacteria as well as spores; and (3) poor DNA recovery after extraction and purification. Methods to overcome these obstacles need to be tailored to individual combinations of sample type, target micro-organism, and assay sensitivity. Because of these issues the importance of a co-purified positive control target nucleic should be stressed. Target DNA is added to the biological sample before sample preparation, and failure to detect this material following the intervening steps is indicative of failure to remove PCR inhibitors, or of the loss of nucleic acid during processing.

4. Formats of NAD Assays for Food Pathogen Detection

4.1. Nucleic Acid-Based Diagnostics Based on In Vitro Amplification Technologies

The polymerase chain reaction (PCR) has been the most popular platform applied in NAD assays for food-borne pathogens. Post-PCR detection methods are many and varied, and include gel electrophoresis, Southern blot hybridisation analysis, and detection formats incorporating the use of specific nucleic acid probes to enable the detection and identification of food-borne pathogens (Eyigor and Carli 2003; Gonzalez, Grant et al. 1997; Lin and Tsen 1996; Woodward and Kirwan 1996). Fluorometric or colorimetric detection of PCR products can be achieved using labelled probes in a range of detection formats, including membrane and microwell formats (Grennan, O'Sullivan et al. 2001; Mandrell 1999; O'Connor, Joy et al. 2000; O'Sullivan, Fallon et al. 2000, 2001). PCR-based NAD assays dominate the in vitro amplification-based publications for common food-borne bacterial pathogens.

4.2. PCR-ELISA and PCR-DNA Probe Membrane Based Assays for *Campylobacter* and *Salmonella*

PCR-ELISA assays usually involve the specific capture of a non-radioactively labelled PCR product by a specific probe immobilised on a microwell, yielding a colorimetric endpoint that can be detected visually or by a microwell plate reader. Alternatively, the PCR primer can be immobilised on the microwell plate and PCR can be performed in the plate (www.nuncbrand.com).

A PCR-ELISA described by Sails et al. (2001) targeting the ORF-C region in *C. jejuni* and *C. coli* and utilising a commercially available PCR-ELISA format determined a detection limit of 1 cell-equivalent for *C. jejuni*. The group subsequently applied the assay to detect *Campylobacter* in enriched meat, milk, and shellfish samples with a sensitivity of 97% (Bolton et al. 2002). Waller and Ogata (2000) developed an immunocapture PCR-ELISA assay and applied it to detect *Campylobacter* at a level of 1 cfu/ml in artificially contaminated milk and chicken rinse samples. Another study of a PCR-ELISA assay designed to detect the *ceuE* gene in *Campylobacter* was successfully applied to detect *Campylobacter* in 31 of 32 culture-positive chicken rinse samples. This PCR-ELISA was rapid, was amenable to automation, and at a cost of $3.00 per test may have potential for application to the detection of *Campylobacter* in foods. This group also developed a similar PCR-ELISA based on the *invA* gene for the identification of *Salmonella* in foods (Hong et al. 2003). Grennan et al. (Grennan, O'Sullivan et al. 2001) developed a PCR-ELISA assay based on amplification and probe-based detection of the 16S/23S rRNA intergenic spacer region in *C. jejuni*, *C. coli*, and the *Campylobacter* genus; and demonstrated that the assay could be applied to detect *Campylobacter* in enriched poultry-meat specimens. Perelle et al. developed a PCR-ELISA assay for *Salmonella* exploiting the previously standardised *invA* gene assay for *Salmonella* (Malorny et al. 2003). The PCR-ELISA had a detection limit of 5 cfu/25g food after enrichment and had 100% concordance with the ISO method.

4.3. Specific Examples of Nucleic Acid Diagnostics Assays for the Detection of Bacterial Food-Borne Pathogens

An extensive range of PCR-based assays relying on the amplification of specific genomic targets in food-borne pathogenic bacteria, followed by the detection of the PCR product by gel electrophoresis analysis, have been developed. Recent studies have employed amplification of a conserved region of the *flaA-flaB* genetic locus for the detection of *C. jejuni*, *C. coli*, and *C. lari* in environmental water samples (Moore, Caldwell, and Millar 2001), and the *ceuE* gene in a multiplex assay to identify *C. jejuni* and *C. coli* in Thailand (Houng et al. 2001). Bang et al. (2002) developed a rapid molecular test based on nested PCR amplification of 16S rRNA and hippuricase O genes for the identification of *Campylobacter* spp. in environmental samples collected from Danish broiler farms. The group applied these assays to the detection of *C. jejuni* and *C. coli* in environmental samples and reported a detection rate that was 10 times higher than they achieved by conventional culture-based isolation. They concluded that the PCR-based assays were more sensitive at detecting sub-lethally injured cells, viable but non-culturable cells, and the low cell numbers likely to be present in environmental samples (Bang, Wedderkopp et al. 2002). In 2003, On and Jordan compared the performance of 11 PCR-based assays for *Campylobacter* for their sensitivity and specificity on a panel of 111 well-characterised reference, type, and field strains of *Campylobacter*. The assays for *C. coli* exhibited 100 % sensitivity and specificity for the detection and identification of *C. coli* strains. The assays designed for the identification of *C. jejuni* exhibited sensitivities of 84 %–100 % and specificities varying from 88 %–100 %. The authors concluded that *C. jejuni* is a more heterogeneous species than *C. coli* and that it may be difficult to design a PCR-based assay to detect 100 % of *C. jejuni* strains.

PCR-based assays developed for Campylobacter *spp.* in food include assays targeting the *flaA*, *cadF*, *ceuE* and *cdt* genes and the 16S rRNA region (Bang et al. 2002, Hong, Berrang et al. 2003, Waller and Ogata 2000). A PCR-based assay used in conjunction with an immunomagnetic separation (IMS) technique identified a single cell of *C. jejuni* ml^{-1} in spiked milk and chicken wash samples, in an assay time of 8 hours without sample enrichment (Waller 2000). A PCR-ELISA has also been described to detect *C. jejuni* in poultry samples. This assay targeting the *ceuE* gene was applied directly without culture-based enrichment to chicken carcass rinse samples and a detection limit of 40 CFU ml^{-1} was reported (Hong, Berrang et al. 2003).

PCR assays have been developed for the detection and identification of *Salmonella* spp. in foods. An *S. typhimurium* species-specific assay based on the amplification of the *ogdH* gene with a detection limit of 100 CFU from pure cultures and 200 CFU from positive chicken meat samples has been described (Jin et al. 2004). A more sensitive PCR assay for the detection of *Salmonella* spp. in salami with a detection limit of 1–10 CFU has been optimised. The total assay time of 12 hours includes a 6-hour culture enrichment step, a DNA purification step, PCR amplification, and post-PCR analysis. The results obtained by the PCR assay targeting the *invA* gene of *Salmonella* spp. exhibited very good correlation with conventional methods when evaluated in naturally and artificially contaminated samples, but in a much shorter time frame (Ferretti et al. 2001).

PCR-based assays have been developed for the identification of Listeria and *L. monocytogenes* in food, milk, and dairy samples (Aznar and Alarcon 2003; Gouws, Visser et al. 1998). A specific PCR assay for the detection of the *hly* gene of *L. monocytogenes* and conventional culture methods for the identification of *Listeria* spp. were compared for their accuracy of detection of *L. monocytogenes* in a variety of food products (Gouws and Liedemann 2005). Of the 27 food products tested, 74 % were presumptively positive for *Listeria* on Oxford agar (Oxoid), and 44 % were identified as *L. monocytogenes* on RAPID'L. mono agar (Biorad). PCR was used as the confirmatory test and identified *L. monocytogenes* in 37 % of the samples tested. The authors concluded that PCR was able to eliminate false positive results found by culture-based methods (Gouws and Liedemann 2005).

A second PCR procedure was established for the routine detection of *L. monocytogenes* in food, with initial optimisation of the procedure performed on 60 beef samples artificially contaminated with *L. monocytogenes* (Aznar and Alarcon 2003). This PCR assay, performed after 48 hours of culture enrichment, and targeting the *hlyA* gene in *L. monocytogenes*, had a detection limit of $1 \, CFU \, g^{-1}$ of beef. Two hundred and seventeen naturally contaminated beef and other food samples were tested by the optimised PCR assay and by conventional culture-based methods. The culture method identified 17 samples as *L. monocytogenes* positive, while the PCR assay detected *L. monocytogenes* in 56 of the samples (Aznar and Alarcon 2003).

A further PCR assay based on the detection of the *prfA* gene in *L. monocytogenes* was designed (Holko et al. 2002). One hundred samples of milk and dairy products were analysed by a standard culture-based method and by the specific PCR-based method. Both methods identified 18 *L. monocytogenes* positive samples, but the PCR method yielded results in 2 days compared to 5 or more days with the traditional culture method (Holko, Urbanova et al. 2002).

PCR assays have been developed for *E. coli* serogroups O174 and O177 based on *wzx* and *wzy* genes. Detection limits for both strains were $0.1 \, CFU \, g^{-1}$ of inoculated culture-enriched pork samples (Beutin et al. 2005). Multiplex PCR assays have been developed for the detection of *E. coli* O157:H7 (Johnston et al. 2005; Paton and Paton 1998). Application of the PCR assay to spiked raw seed sprouts and irrigation water samples without prior culture enrichment proved to be a rapid detection method for *E. coli* O157:H7. The PCR assay targeting the *stx* genes in *E. coli* O157:H7 reported detection limits of $10 \, CFU \, g^{-1}$ of raw seed sprouts and $100 \, CFU \, L^{-1}$ of water (Johnston, Elhanafi et al. 2005). A previously described multiplex PCR assay for *vt1*, *vt2*, *eaeA,* and *hlyA* genes in *E. coli* O157:H7 was evaluated in combination with IMS for the detection of *E. coli* O157:H7 in inoculated, culture-enriched ground beef samples. Following 16 hours of culture enrichment, it was possible to detect 40 *E. coli* O157:H7 cfu g^{-1} (Fitzmaurice et al. 2004).

Multiplex PCR assays have also been developed for the simultaneous detection of two or more food-borne pathogens. A multiplex PCR assay for the simultaneous detection of *Salmonella* spp., *L. monocytogenes*, and *E. coli* O157:H7 following culture-based enrichment had a detection limit of $1 \, CFU \, g^{-1}$ of pork in a total assay time of 30 hours (Kawasaki et al. 2005). The simultaneous detection of *E. coli* O157:H7, *Salmonella*, and *Shigella* in apple cider following a 24-hour enrichment step has been demonstrated in a multiplex assay with a detection limit of $1 \, CFU \, ml^{-1}$ (Li and Mustapha 2004).

4.3.1. Commercially Available Conventional NAD Assays for Food-Borne Bacterial Pathogens

Examples of some commercially available NAD assays based on PCR for bacterial food-borne pathogen identification are included in Table 22.3. Commercially available conventional assays include the BAX® (DuPont Qualicon, DE, USA) and the Dr. Food™ (Dr. Chip Biotech Inc., Miao-Li, Taiwan) kits. BAX® kits are available for the identification of *Salmonella*, *E. coli* O157:H7, *Listeria*, *L. monocytogenes*, *C. jejuni*, and *C. coli*. BAX® tests are also available for other bacteria, including *S. aureus* and *E. sakazakii*. BAX® tests involve culture-based enrichment of the food sample and cell lysis to release the DNA, followed by PCR amplification with gel-based detection of the PCR products (Stewart 1998). Dr. Food™ kits provide rapid PCR-based methods for the identification of *E. coli*, *L. monocytogenes, Campylobacter* spp., *Salmonella* spp., and other food-borne pathogens such as *S. aureus*, *Y. enterocolitica*, *B. cereus*, *Clostridium* spp., *Shigella* spp., and *Vibrio* spp. in culture enriched foods. Identification of the pathogen is achieved following PCR amplification of extracted DNA and post-PCR hybridisation to an oligonucleotide probe in a colorimetric reaction.

Table 22.3. Examples of commercially available PCR and real-time PCR kits for common food-borne bacterial pathogens (Glynn, Lahiff et al. 2006)

Kit Name	Format	Target Species	Manufacturer
BAX® system	Conventional	Salmonella, *E. coli* O157:H7, Listeria and *L. monocytogenes* and *Campylobacter jejuni* and *C. coli*	Qualicon
Dr Food™ kit	Conventional	*E. coli, L. monocytogenes,* Campylobacter *spp.* and Salmonella *spp.*,	Dr. ChipBiotech Inc
PCR Diagnosis-Bacteria identification kit	Conventional	*E. coli*	BioChain
LightCycler® foodproof Listeria genus detection kit	Real-time	Listeria	Roche
LightCycler® foodproof Salmonella detection kit	Real-time	Salmonella	Roche
LightCycler® foodproof *E. coli* O157 detection kit	Real-time	*E. coli*	Roche
LightCycler® foodproof *L.monocytogenes* detection kit	Real-time	*L. monocytogenes*	Roche
LightCycler® foodproof campylobacter detection kit	Real-time	Campylobacter	Roche
RealArt™ L. monocytogenes PCR kits*	Real-time	*L. monocytogenes*	Qiagen
RealArt™ Campylobacter PCR Kits*	Real-time	*C. jejuni, C. lari, C. coli*	Qiagen
RealArt™ Salmonella PCR kits*	Real-time	Salmonella	Qiagen
Artus *L.monocytogenes* PCR kit	Real-time	*L. monocytogenes*	Artus
Artus Salmonella PCR kit	Real-time	Salmonella	Artus
Artus Campylobacter PCR kit	Real-time	*Camylobacter*	Artus
TaqMan® *Listeria monocytogenes* detection kit	Real-time	*L. monocytogenes*	Applied BioSystems
TaqMan® *Campylobacter jejuni* detection kit	Real-time	*Campylobacter jejuni*	Applied BioSystems
TaqMan® *Escherichia coli* O157:H7 detection kit	Real-time	*E. coli*	Applied BioSystems
TaqMan® *Salmonella enterica* detection kit	Real-time	*Salmonella enterica*	Applied BioSystems
SureFood® Pathogen Salmonella	Real-time	Salmonella *spp.*	Congen
SureFood® Pathogen Campylobacter	Real-time	*Campylobacter jejuni, C. lari, C. coli*	Congen
SureFood® Pathogen Listeria	Real-time	*L. monocytogenes*	Congen

4.3.2. Alternative In Vitro Amplification Technologies

A wide variety of in vitro amplification-based technologies have been described, including ligase chain reaction (LCR), QB replicase, branched DNA amplification (bDNA), nucleic acid sequence-based amplification (NASBA), strand displacement amplification (SDA), transcription mediated amplification (TMA), and rolling circle amplification (RCA). While many of these platforms have been applied to detect pathogens in clinical microbiology, their application in food testing has been limited to date. TMA, a proprietary technology of Gen-Probe Inc., is an isothermal RNA-based amplification system that combines the use of two enzymes to amplify either a DNA or an RNA target. NASBA, another isothermal amplification technology, uses three enzymes to amplify RNA and has been applied previously to detect *Campylobacter* and *Salmonella* in enriched food samples. As NASBA is an RNA-based amplification technology, it is particularly suited for applications in which the assessment of the viability status of a pathogen is a requirement (Cook 2003; Rodriguez-Lazaro et al. 2004; Simpkins et al. 2000; Uyttendaele, Bastiaansen and Debevere 1997). A combination NASBA-ECL assay for *Salmonella* determined a detection limit of approximately 100 cells in a variety of food types after 18 hrs culture enrichment (D'Souza and Jaykus 2003). Strand displacement amplification (SDA) is another isothermal amplification system that has been developed for bacterial pathogen identification. SDA enables amplification of 10^{10} copies of DNA in as little as 15 minutes. SDA has been applied to the identification of pathogenic *E. coli* in water samples (Ge et al. 2002).

Most of these platforms enable pathogen detection based on in vitro amplification of a specific DNA and RNA target; however, some technologies, such as branched DNA (bDNA), enable detection and identification through probe-mediated signal amplification. Other proprietary in vitro amplification technologies include Transcription Mediated Amplification Technology (Gen-Probe Inc, CA, USA), Ligase Chain Reaction (Abbott Diagnostics, IL, USA), Strand Displacement Amplification (Becton Dickinson, NJ, USA) and Hybrid Capture (Digene, MD, USA).

4.4. Standardisation of In Vitro Amplification-Based NAD Assays and Inter-Laboratory Validation Studies

As part of an EU programme (European Food-PCR Project), PCR assays have been evaluated and optimised as a step towards the development of standard PCR-based methods for the detection of *Campylobacter* spp., *Salmonella* spp., and *L. monocytogenes* in foods (D'Agostino et al. 2004; Lubeck et al. 2003; Malorny et al. 2004). The aim of these projects is to develop validated PCR-based assays that are suitable for routine application in food testing laboratories.

Researchers have developed assays for the identification of *Salmonella* based on the *invA* gene (Arnold et al. 2004), the 16S gene (Lin et al. 2004), and the 16S/23S rRNA intergenic spacer region (Chiu et al. 2005). Arnold et al. (Arnold et al. 2004) demonstrated that their molecular assay for *Salmonella* had comparable sensitivity to the ISO microbiological method for the detection of *Salmonella* in pigs. Recently Ziemer (Ziemer and Steadham 2003) evaluated the specificity of nine sets of primers for the specific detection of *Salmonella* and found that only three primer sets were specific for *Salmonella*, highlighting the need for good internal laboratory validation of the specificity of candidate species-specific primers. As part of an EU project (European Food-PCR project: www.pcr.dk), researchers have been evaluating and optimising PCR-based assays for the detection and identification of *Campylobacter* and *Salmonella*. For *Campylobacter*, Lübeck et al. (Lubeck, Wolffs et al. 2003) reported the evaluation of a range of 16S and 23S PCR primers for *Campylobacter* against a panel of 150 target and nontarget organisms. Following the evaluation, they selected a single 16S primer set for the detection of *C. jejuni*, *C. coli*, and *C. lari*. Malorny et al. (Malorny, Hoorfar et al. 2003) evaluated 4 primer sets for the PCR-based detection of *Salmonella* and selected a PCR primer set for the *invA* gene as being the most specific for *Salmonella* detection. Both groups optimised assays including a robust *Taq* polymerase and an internal standard control with a gel-based end-point detection (Lubeck, Wolffs et al. 2003; Malorny, Hoorfar et al. 2003). Subsequently, the assays were evaluated in participating partner laboratories. The studies suggested that these methods might have potential as standard molecular methods for the identification of *Campylobacter* and *Salmonella* pathogens. Researchers are also beginning to report the application of multiplex PCR assays for the simultaneous detection of *Salmonella* and *Campylobacter* in food and water samples (Gilbert et al. 2003; Morin, Gong and Li 2004).

4.5. Real-Time In Vitro Amplification-Based Nucleic Acid Diagnostics

Real-time PCR monitors the accumulation of PCR product in a reaction while it is taking place, compared to endpoint detection of the PCR product in conventional PCR. These technologies provide sensitive, quantitative detection of PCR products in a fast turnaround time in a closed tube format, thereby significantly reducing the risk of contamination (Csordas et al. 2004; Raoult, Fournier, and Drancourt 2004). Fluorescent technologies employed are either nonspecific, using dyes such as SYBR Green I or SYBR Gold, which are minor groove-binding dyes and intercalate into the PCR product during amplification; or specific, using probes to detect specific sequence amplification in the PCR. A number of different fluorescent probe chemistries have been employed in real-time PCR assays, including TaqMan (5′exonuclease) probes, Fluorescent Resonance Energy Transfer (FRET) probes, Molecular beacons, and Scorpion probes. While the mechanism of fluorescent signal generation is different for each of the probe chemistries, the fluorescent signal generated by the probes or minor groove-binding dyes is directly proportional to the amount of PCR product generated (Bustin 2002, McKillip, and Drake 2004). Real-time PCR is quantitative, with a broader dynamic range than conventional PCR. Informative reviews of the mechanisms of real-time PCR are available, including Bustin and Mueller (2005) and Dorak (dorakmt.tripod.com/genetics/realtime.html). Over the past five years, there have been an increasing number of reports in the literature describing the design and application of real-time PCR-based NAD tests for common bacterial food-borne pathogens. Examples of recent real-time PCR-based NAD assays for these pathogens are included here and listed along with commercially available NAD assays in Table 22.3.

4.5.1. Specific Examples of Real-Time PCR Assays for the Detection of Bacterial Food-Borne Pathogens

SYBR Green real-time PCR assays have been described for the detection of *Salmonella*, with assay times of approximately two hours when applied directly to samples without pre-enrichment (Fukushima et al. 2003; Jothikumar, Wang, and Griffiths 2003). Detection limits of 1–5 CFU were determined for other real-time assays incorporating SYBR Green applied to enriched milk and meat samples (Wang, Jothikumar, and Griffiths 2004b). Real-time PCR assays for *Salmonella* have also been developed using molecular beacon technology targeting the *invA* gene (Chen, Martinez, and Mulchandani 2000; Liming and Bhagwat 2004; Wan et al. 2004). Real-time PCR assays incorporating TaqMan or 5′exonuclease probe technologies for culture confirmation of *Salmonella* (Klerks, Zijlstra, and van Bruggen 2004) and for identification of *Salmonella* in enriched food samples have also been developed (Malorny et al. 2004; Piknova et al. 2005; Rodriguez-Lazaro et al. 2003; Seo et al. 2004). Single CFU detection from 600 grams of pooled egg sample was reported by Seo et al. (2004) with a real-time PCR including a 5′ nuclease probe for the *sefA* gene in *Salmonella*. This test yielded results in two days compared to five days for conventional culture.

Real-time PCR tests for *Campylobacter* spp. incorporating SYBR Green as the reporter and the *cadF*, *mapA*, 16S rRNA, and *gyrA* genes have recently been described (Cheng and Griffiths 2003; Fukushima et al. 2003; Inglis and Kalischuk 2004). *Campylobacter* DNA from 1–25 cells have been detected by real-time PCR assays based on 5′ nuclease technology in enriched food samples (Sails et al. 2003), cell suspensions (Wilson et al. 2000), and spiked fecal samples (Rudi et al. 2004). Detection limits of 10 CFU have been reported for *Salmonella* and *Campylobacter* in chicken faeces and carcass rinse samples using real-time PCR assays incorporating FRET hybridisation probes (Ellingson et al. 2004; Eyigor and Carli 2003; Perelle et al. 2004). Campylobacter assays with detection limits of 10 CFU ml^{-1} carcass rinse (Perelle, Josefsen et al. 2004) and 100–150 CFU ml^{-1} of fecal suspension (Lund and Madsen 2006) have been shown to perform comparably with conventional methods and to have potential for flock screening.

A SYBR Green-based real-time PCR test enabling simultaneous detection of *E. coli* O157:H7, *L. monocytogenes* and *Salmonella* strains was designed and evaluated using artificially contaminated fresh produce. This test successfully detected all three pathogens when fresh produce was washed with artificially contaminated water containing *E. coli* O157:H7, *S. typhimurium* (1–10 cells ml^{-1}), and *L. monocytogenes* (1000 cells ml^{-1}) (Bhagwat 2003). A quantitative real-time PCR assay incorporating TaqMan probe technology for the direct detection and enumeration of *L. monocytogenes* and *L. innocua* was developed and applied to artificially contaminated milk samples targeting the *iap* gene (Hein et al. 2001). The assay detected six copies of the *iap* gene using purified DNA as a template. When applied to the direct detection and quantification of *L. monocytogenes* in milk, the real-time quantitative PCR assay was as sensitive as the traditional plate count method (Hein, Klein et al. 2001).

A detection limit of from 6–60 CFU was achieved using a TaqMan PCR targeting the *L. monocytogenes hlyA* gene. This method employed a magnetic bead-based sample preparation strategy to extract bacterial cells from skim milk and unpasteurised raw milk, and the extraction and PCR detection method could be completed in three hours (Nogva et al. 2000).

Multiplex real-time PCR assays to simultaneously detect *Salmonella* serovars and *L. monocytogenes* using the SYBR Green detection format were shown to have a detection limit of 2.5 *Salmonella* cells and 1 *L. monocytogenes* cell from liquid culture targeting the *fimI* gene of *Salmonella* and the *hly* gene of *L. monocytogenes* (Jothikumar, Wang et al. 2003). A multiplex SYBR Green-based PCR detection assay targeting the *invA* gene of *Salmonella* and the *hlyA* gene of *L. monocytogenes* detected 4 *Salmonella* cells and 3 *L. monocytogenes* cells g^{-1} of raw meat sample with a total assay time of 10 hours, including 6–8 hours of pre-enrichment (Wang, Jothikumar and Griffiths 2004b).

E. coli O157:H7 toxin genes *vt1* and *vt2* were specifically detected from minced beef samples using FRET probes (Fitzmaurice, Glennon et al. 2004). Following inoculation, culture-based enrichment for 16 hours, and sample purification by IMS, a detection limit of 3.5 *E. coli* O157:H7 cells ml^{-1} was established. The use of IMS to concentrate *E. coli* O157:H7 cells from minced beef samples without culture-based enrichment has also been investigated (Fu, Rogelj, and Kieft 2005). A TaqMan assay targeting the *eaeA* gene was shown to be specific for *E. coli* O157:H7 with a detection limit of 1.3×10^4 cells g^{-1} minced beef. When used to capture *E. coli* cells from cell cultures, this IMS/TaqMan approach had a detection limit of less than 5×10^2 cells ml^{-1}.

4.5.2. Alternative Real-Time In Vitro Amplification-Based Diagnostics Technologies

While PCR has been the most popular platform for real-time NAD assays for food-borne pathogens studied to date, some alternative in vitro amplification technologies have been investigated for pathogen detection in food. *Mycobacterium avium* subsp. *paratuberculosis* (MAP) has been detected in artificially spiked milk samples by a real-time NASBA assay (Rodriguez-Lazaro, Lloyd et al. 2004). The sensitivity of this assay was poor, requiring more than 5×10^3 cells, while the reaction also failed to differentiate RNA from DNA, thereby reducing the primary advantage of NASBA for the detection of live cells only.

There are an increasing number of commercially available real-time PCR kits for the detection of food-borne pathogens (Table 22.3). Roche Diagnostics "foodproof" real-time PCR assays are available for *Campylobacter*, *Salmonella*, *Listeria* genus, *L. monocytogenes*, and *E. coli* O157. Other companies with kits on the market include Qiagen (*RealArt*™ kits for *L. monocytogenes, Salmonella*, and *Campylobacter*); Applied Biosystems (TaqMan® detection kits for *C. jejuni, L. monocytogenes, E. coli* O157, and *S. enterica*); Artus (*L. monocytogenes, Salmonella* and *Campylobacter* PCR kits); and Congen (SureFood® Pathogen Kits for *Salmonella, Campylobacter,* and *Listeria*).

4.6. Limitations and Other Considerations for In Vitro Amplification NAD Tests

The general acceptance and application of NAD tests as standard methods in the routine detection of food-borne pathogens has been limited due to the lack of standardisation and validation of diagnostic PCR assay protocols (Malorny, Hoorfar et al. 2003). An integrated approach to the development of guidelines for the standardisation and validation of NAD assays includes three aspects: first, sample-specific method development and validation that takes into account the effect of sampling, sample preparation, and preparation of the amplification mixture on test performance; second, the establishment of an internal quality assurance scheme in the form of an internal amplification control that checks for false negative results caused by inhibitory substances in the reaction mixture; and third, participation in external quality assurance programs implemented as interlaboratory trials to validate PCR-based pathogen detection protocols (Hoorfar, Wolffs et al. 2004).

The generation of false positive results by in vitro amplification of DNA originating from dead organisms in the food sample is a potential limitation of NAD-based systems. This problem can be circumvented by applying a culture enrichment step prior to PCR analysis (Scheu, Berghof et al. 1998). Alternatively, the application of assays that target the in vitro amplification and detection of RNA will ensure the detection of viable organisms (McKillip, Jaykus et al. 1998). The isolation of RNA is more difficult than that of DNA because of the labile nature of RNA; and its application as a NAD target may lead to reduced analytical detection limits (Scheu, Berghof et al. 1998).

Another of the major problems associated with the application of PCR assays for the identification of food-borne pathogens is the presence of PCR inhibitors in the food sample. False negative results can occur for a variety of reasons, including (1) the degradation of target nucleic acid sequences in the sample, and (2) the presence of substances that inhibit the PCR reaction. The degree of inhibition varies greatly with the food type. Certain foods, such as dairy products, are more problematic than others (Bickley et al. 1996; Rossen, Norskov et al. 1992). Hence the removal of inhibitory substances from the DNA to be amplified is an important prerequisite to successful PCR amplification. Several methods of sample preparation have been reported, including filtration, centrifugation, the use of detergents and organic solvent treatments, enzyme treatment, the addition of PCR additives, and sample dilution before PCR (Abolmaaty and Levin 2002; Lantz et al. 1994; Wilson 1997).

Immunomagnetic separation (IMS) has recently been applied to speed up the selective enrichment step prior to PCR, as well as to eliminate PCR inhibitors from the food matrix. The method uses paramagnetic beads coated with antibody to selectively concentrate target organisms from samples. Pre-enriched samples are incubated with antibody-coated beads that form a complex with the target organism. Antibody complexes are captured magnetically and can be plated for culturing or further tested using NAD assays (Hudson et al. 2001; Lamoureux et al. 1997; Mercanoglu 2005; Rijpens et al. 1999). IMS protocols are relatively simple to use and take less than one hour to complete. The method generally increases the sensitivity and specificity of the analysis and saves processing and identification time. Two of the companies that commercially produce antibody-coated magnetic particles for use in food diagnostics are Dynal Biotech (Oslo, Norway), which supplies anti-*Salmonella*, anti-*Listeria*, and anti-*E. coli* O157 Dynabeads® (Favrin, Jassim, and Griffiths 2003; Hsih and Tsen 2001; Shaw, Blais, and Nundy 1998); and Vicam (Watertown, MA), which offers Verify™ tests for the isolation of *Salmonella* and ListerTest® for the isolation of *Listeria* (Mitchell et al. 1994).

Despite their demonstration as rapid, sensitive, and specific detection methods, in vitro amplification NAD assays still await general acceptance and official approval as standard methods for the identification of food-borne pathogens. This is largely because of the lack of universal validation and standardisation of these methods, which therefore requires an international initiative to focus on the development of internal and external quality assurance programs, suitable sample processing guidelines, and the establishment of proficiency ring trials to facilitate the validation and standardisation of NAD assays, and thereby enable their routine application (Hoorfar, Wolffs et al. 2004).

4.7. Non-Amplified Direct DNA Probe-Based Nucleic Acid Diagnostics

Direct nucleic acid probe hybridisation tests have been applied in food testing laboratories for confirmation of the identity of organisms following culture-based isolation of the food-borne pathogen of concern (Fung 2002). Direct detection systems are available in multiple formats, they generally do not require sophisticated equipment, they are simple to perform, and they have detection limits of 10^4–10^5 bacterial cells (Curiale et al. 1990; Peterkin, Idziak, and Sharpe 1991; Todd et al. 1999).

Commercial NAD assays based on direct nucleic acid probe technology include Accuprobe® (Gen-Probe, CA, USA), available for detection of *Campylobacter* spp. and *L. monocytogenes*; and Gene-Trak® (Neogen, MI, USA) for *E. coli, Salmonella*, and *Listeria* spp. detection. Both these NAD systems are based on the hybridisation of specific DNA probes to 16S rRNA, taking advantage of the fact that the higher copy number of bacterial rRNA provides a naturally amplified target and affords greater assay sensitivity. Both test systems require that the food samples are culture-enriched before sample analysis. This is followed by a simple lysis procedure to release the target nucleic acid material before the probe analysis is performed. The Gene-Trak system comprises a pathogen-specific capture probe and a detection probe with an attached chromogenic substrate. Accuprobe® uses a single DNA probe to hybridise to rRNA, the unbound probe is degraded chemically, and the chromogenic label on the bound probe is detected using a luminometer.

4.8. DNA-Probe Based Detection Methods

The direct DNA probe hybridisation assays described for the identification of food-borne pathogens are usually performed for culture confirmation. In one assay type, a colony hybridisation is performed following the alkaline lysis of colonies on a nitrocellulose membrane. The DNA probe, labelled nonradioactively with biotin or digoxigenin, is hybridised to the denatured DNA or rRNA at a predetermined temperature; and a colorimetric signal is generated in positive samples following the binding of an enzyme-antibody conjugate. Alternatively, the probe can be bound to a membrane or microtitre well surface as a capture probe to which genomic DNA or rRNA binds; and the binding of a second DNA probe labelled with a specific reporter is responsible for the detection (Smith 2000). Peptide nucleic acid (PNA) probes are DNA mimics with a pseudopeptide backbone that can form very stable duplexes with dsDNA. PNA probes have been used successfully to detect bacterial contamination in a hygiene monitoring context (Stender et al. 2001). Commercially, available systems based on direct probe hybridisation technologies include the Accuprobe® line for culture confirmation from Gen-Probe Inc. (www.genprobe.com). Neogen Corporation has developed the GENE-TRAK® system, in which probes labelled with horseradish peroxidase allow the colorimetric detection of genomic targets from selected food pathogens, including *Salmonella*, following culture enrichment (www.neogen.com).

5. Conclusions and Future Perspectives

The ultimate goal in rapid microbial testing for the food industry is the ability to accurately and sensitively detect food-borne pathogens on the production line and in real time. Nucleic acid diagnostics have superior sensitivity and specificity in comparison to other types of diagnostics tests for microbial detection. In addition, they serve to identify the source of the contamination, since they can be used to distinguish between different populations of the same pathogen, using molecular typing methods. This gives food producers and processors the ability to detect the source of contamination quickly and alter their HACCP protocols appropriately. DNA probe technology has the potential to significantly shorten the sample analysis time in food-borne pathogen detection. However, the continued requirement for a culture enrichment period will reduce the effectiveness of the technology, since it is the rate-limiting step in the process. In the short to medium term, the most effective way forward will be to reduce the length of the enrichment period through improved growth media. Much research and development is ongoing to develop accelerated bacterial growth media, which will lead to shortened enrichment times prior to the nucleic acid detection assay. However, in the long term, to meet user expectations, tests will be required which are accurate, sensitive, specific, easy to use, and cost-effective, but which also can be completed (from food sample to result) within a working day without the need for any culture enrichment. This will require major technological and methodological improvements in the efficiency of extraction of specific intact bacterial cells and/or nucleic acids from food matrices, to enable the direct amplification of the bacterial genetic material from foods.

5.1. Emerging Nucleic Acid Diagnostic Technologies for Food-Borne Pathogen Detection

5.1.1. Biosensors

Diagnostic biosensors are devices and technologies that use a biologically derived material immobilised on a detection platform to measure the presence of one or more analytes (Mascini, Tombelli, and Palchetti 2005). For applications in food microbiology analysis, an ideal biosensor would be a self-contained, automated system capable of pathogen detection directly from a food matrix without pre-enrichment, and would also be capable of differentiating live from dead cells (Ivnitski et al. 2000). Biosensor technologies developed for the specific detection of pathogenic bacteria in food samples have included metabolism-based, antibody-based, and DNA-based systems. A thorough review of biosensor technology for environmental pollutants and food contaminants is provided by Baeumner (2003), covering assay formats and target recognition elements, including nucleic acids, enzymes, antibody, whole-cell, and bioelectric biosensors. Nucleic acid-based biosensors that have found application in the food industry include Quartz Crystal Microbalances (QCM), and optical detection systems.

QCM, in combination with PCR of the *lac* gene, has been used to detect 1–10 *E. coli* cells from 100 ml of water (Mo et al. 2002). Optical-based biosensors have been developed by a number of companies, including Biacore International and Texas Instruments. The current generation of optical biosensors use surface plasmon resonance (SPR) to monitor biomolecular interactions on a surface in real time (Rand et al. 2002). An advantage of SPR-based systems is that no labelling of the target molecule is required. The Biacore instrument and other SPR detectors have typically been used for monitoring antibody-antigen interactions. Targets have included whole-cell detection of *L. monocytogenes* (Leonard et al. 2004). Monitoring of surface plasmon resonance to measure DNA-DNA and DNA-RNA hybridisations has been demonstrated, and procedures are well established for the immobilisation of oligonucleotide probes to device surfaces and hybridisation of target nucleic acid to the bound probes. However, the applications of this technology for food-borne pathogen identification have yet to be fully realised (Mannelli et al. 1005; Wang et al. 2004a). To date, the potential of SPR-based biosensors in the food industry has been demonstrated by the detection of genetically modified (GM) soybeans and maize following PCR amplification of transgenic and wild-type sequences (Feriotto et al. 2002).

Novel biosensors with advanced visualisation and signal amplification technologies have created the possibility of monitoring single molecular interactions in real time (Yao et al. 2003). The application of gold nanoparticles for the detection of nucleic acid targets on a biosensor platform has been demonstrated to have a sensitivity of 3×10^6 target molecules when evaluated with dilutions of PCR products (Storhoff et al. 2004). A recent study combining biosensor-based nucleic acid detection, with an initial step to concentrate the target nucleic acid using DNA probes bound to supramagnetic nanoparticles on the biosensor prior to detection, achieved detection limits of as little as 2 molecules of HCV cDNA target, even when included in a 2.5 millionfold excess of nonhomologous DNA background (Fuentes et al. 2006). The next generation of biosensors will have applications in all sectors of the molecular diagnostics market.

5.1.2. Microarrays

DNA microarrays consist of large numbers of probes (either oligonucleotides or cDNAs) immobilised on a solid surface. Hybridisations are performed by the application of a labelled nucleic acid target in a liquid state to the microarray surface. Following appropriate hybridisation and washing steps, the target nucleic acid bound to probes on the array surface is visualised using a microarray scanner. For a full exploration of the nature and variety of microarray platforms and technologies, readers are directed to chapters 23 and 26 in this book. Some progress has been made with the identification of food pathogens from genomic DNA using microarrays under laboratory conditions (Ahn and Walt 2005; Borucki et al. 2005). Microarrays have been demonstrated for the molecular identification of *E. coli* O157:H7 (Wu et al. 2003) and *Campylobacter* spp. (Volokhov et al. 2003) from cultures, following PCR amplification of the target genes. As microarray technology matures, these planar arrays are being supplemented by further evolutions including microbead and suspension microarray formats. Suspension microarrays are a modification of the original planar microarrays, with the differentiating probes being immobilised onto the surface of polystyrene microbeads containing internal fluorescent dyes (Ahn and Walt 2005; Borucki et al. 2005; Straub et al. 2005). A microarray-based assay incorporating signal amplification and suspension microarray technologies has been reported for the identification and subtyping of *L. monocytogenes* from genomic DNA (Borucki et al. 2005). Microbead-type arrays have been developed for the identification of *Salmonella* spp. Detection sensitivities of 1×10^3 CFU ml^{-1} were reported when the *Salmonella invA2* and *spvB* genes were targeted (Ahn and Walt 2005). A multiplex PCR for *E. coli* O157:H7 (*eaeA, hlyA, stx1,* and *stx2*) and *Salmonella* (*invA*), combined with a suspension microarray detection system, had a similar sensitivity for each species (Straub, Dockendorff et al. 2005).

DNA microarray or "biochip" products are already available for the detection of GMO material in foods. GeneScan Europe offers a GMO chip to detect plant and virus species, construction elements used in genetic engineering, or specific genetically engineered modifications.

For the application of microarray and biosensor technology in bacterial food safety, several important issues remain. While biosensors are designed for the optimum detection of minute quantities of target nucleic acid, it is important that the sample material be free of contaminants before application to the biosensor surface (Fung 2002). This is of particular relevance to the food industry, where regulatory standards require the detection of 1 viable target cell in 25 g of foodstuff such as ground beef. Current biosensor technologies are unable to detect such a low bacterial load from a food matrix without either sample amplification (using culture-based enrichment or target amplification by PCR) or extensive sample purification techniques. These major hurdles have to be overcome before biosensors and microarrays will provide "real-time" detection of pathogens in food samples.

References

Abolmaaty A and Levin RE (2002) Development of a microslide agglutination assay with the aid of an inexpensive projection microscope. J. Microbiol. Methods 51: 421–423

Ahn S and Walt DR (2005) Detection of Salmonella spp. Using microsphere-based, fiber-optic DNA microarrays. Anal. Chem. 77: 5041–5047

Allmann M, Hofelein C, Koppel E, Luthy J, Meyer R, Niederhauser C, Wegmuller B, and Candrian U (1995) Polymerase Chain Reaction (PCR) for detection of pathogenic micro-organisms in bacteriological monitoring of dairy products. Res. Microbiol. 146: 85–97

Arnold T, Scholz HC, Marg H, Rosler U and Hensel A (2004) Impact of INVA-PCR and culture detection methods on occurrence and survival of Salmonella in the flesh, internal organs and lymphoid tissues of experimentally infected pigs. J. Vet. Med. B. Infect. Dis. Vet. Public Health 51: 459–463

Aznar R and Alarcon B (2003) PCR detection of *Listeria monocytogenes*: A study of multiple factors affecting sensitivity. J. Appl. Microbiol. 95: 958–966

Baeumner AJ (2003) Biosensors for environmental pollutants and food contaminants. Anal. Bioanal. Chem. 377: 434–445

Bang DD, Wedderkopp A, Pedersen K and Madsen M (2002) Rapid PCR using nested primers of the 16s rRNA and the hippuricase (*hip o*) genes to detect *Campylobacter jejuni* and *Campylobacter coli* in environmental samples. Mol. Cell. Probes 16: 359–369

Beutin L, Kong Q, Feng L, Wang Q, Krause G, Leomil L, Jin Q and Wang L (2005) Development of PCR assays targeting the genes involved in synthesis and assembly of the new *Escherichia coli* O174 and O177 O antigens. J. Clin. Microbiol. 43: 5143–5149

Bhagwat AA (2003) Simultaneous detection of *Escherichia coli* o157:H7, *Listeria monocytogenes* and Salmonella strains by real-time PCR. Int. J. Food Microbiol. 84: 217–224

Bickley J, Short JK, McDowell DG and Parkes HC (1996) Polymerase Chain Reaction (PCR) detection of *Listeria monocytogenes* in diluted milk and reversal of PCR inhibition caused by calcium ions. Lett. Appl. Microbiol. 22: 153–158

Bolton FJ, Sails AD, Fox AJ, Wareing DR and Greenway DL (2002) Detection of *Campylobacter jejuni* and *Campylobacter coli* in foods by enrichment culture and polymerase chain reaction enzyme-linked immunosorbent assay. J. Food Prot. 65: 760–767

Borucki MK, Reynolds J, Call DR, Ward TJ, Page B and Kadushin J (2005) Suspension microarray with dendrimer signal amplification allows direct and high-throughput subtyping of *Listeria monocytogenes* from genomic DNA. J. Clin. Microbiol. 43: 3255–3259

Bustin SA (2002) Quantification of mRNA using real-time reverse transcription PCR (RT-PCR): Trends and problems. J. Mol. Endocrinol. 29:23–39

Bustin SA and Mueller R (2005) Real-time reverse transcription PCR (qRT-PCR) and its potential use in clinical diagnosis. Clin. Sci. (Lond.) 109: 365–379

Chen W, Martinez G and Mulchandani A (2000) Molecular beacons: A real-time polymerase chain reaction assay for detecting salmonella. Anal. Biochem. 280:166–172

Cheng Z and Griffiths MW (2003) Rapid detection of *Campylobacter jejuni* in chicken rinse water by melting-peak analysis of amplicons in real-time polymerase chain reaction. J. Food Prot. 66:1343–1352

Chiu TH, Chen TR, Hwang WZ and Tsen HY (2005) Sequencing of an internal transcribed spacer region of 16s-23s rRNA gene and designing of PCR primers for the detection of Salmonella spp. in food. Int. J. Food Microbiol. 97:259–265

Cook N (2003) The use of NASBA for the detection of microbial pathogens in food and environmental samples. J. Microbiol. Methods 53:165–174

Corry JE, Post DE, Colin P and Laisney MJ (1995) Culture media for the isolation of Campylobacters. Int. J. Food Microbiol. 26:43–76

Cosgrove SE and Carmeli Y (2003) The impact of antimicrobial resistance on health and economic outcomes. Clin Infect Dis 36:1433–1437

Csordas AT, Barak JD and Delwiche MJ (2004) Comparison of primers for the detection of *Salmonella enterica* serovars using real-time PCR. Lett. Appl. Microbiol. 39:187–193

Curiale MS, Klatt MJ and Mozola MA (1990) Colorimetric deoxyribonucleic acid hybridization assay for rapid screening of Salmonella in foods: Collaborative study. J. Assoc. Off. Anal. Chem. 73:248–256

D'Agostino M, Wagner M, Vazquez-Boland JA, Kuchta, T, Karpiskova R, Hoorfar J, Novella S, Scortti M, Ellison J, Murray A, Fernandes I, Kuhn M, Pazlarova J, Heuvelink A and Cook N (2004) A validated PCR-based method to detect *Listeria monocytogenes* using raw milk as a food model–towards an international standard. J. Food Prot. 67:1646–1655

D'Amato RF, Wallman AA, Hochstein LH, Colaninno PM, Scardamaglia M, Ardila E, Ghouri M, Kim K, Patel RC and Miller A (1995) Rapid diagnosis of pulmonary tuberculosis by using roche amplicor *Mycobacterium tuberculosis* PCR test. J Clin Microbiol 33: 1832–1834

D'Souza DH and Jaykus LA (2003) Nucleic acid sequence based amplification for the rapid and sensitive detection of *Salmonella enterica* from foods. J. Appl. Microbiol. 95: 1343–1350

de Boer E and Beumer RR (1999) Methodology for detection and typing of foodborne microorganisms. Int. J. Food Microbiol. 50: 119–130

Dziadkowiec D, Mansfield LP and Forsythe SJ (1995) The detection of Salmonella in skimmed milk powder enrichments using conventional methods and immunomagnetic separation. Lett. Appl. Microbiol. 20:361–364

Ellingson JL, Anderson JL, Carlson SA and Sharma VK (2004) Twelve hour real-time PCR technique for the sensitive and specific detection of Salmonella in raw and ready-to-eat meat products. Mol. Cell. Probes 18:51–57

Eyigor A and Carli KT (2003) Rapid detection of Salmonella from poultry by real-time polymerase chain reaction with fluorescent hybridization probes. Avian Dis. 47:380–386

Eyigor A, Dawson KA, Langlois BE and Pickett CL (1999) Cytolethal distending toxin genes in *Campylobacter jejuni* and *Campylobacter coli* isolates: Detection and analysis by PCR. J. Clin. Microbiol. 37: 1646–1650

Favrin SJ, Jassim SA and Griffiths MW (2003) Application of a novel immunomagnetic separation-bacteriophage assay for the detection of *Salmonella enteritidis* and *Escherichia coli* o157:H7 in food. Int. J. Food Microbiol. 85: 63–71

Feriotto G, Borgatti M, Mischiati C, Bianchi N and Gambari R (2002) Biosensor technology and surface plasmon resonance for real-time detection of genetically modified roundup ready soybean gene sequences. J. Agric. Food Chem. 50: 955–962

Ferretti R, Mannazzu I, Cocolin L, Comi G and Clementi F (2001) Twelve-hour PCR-based method for detection of Salmonella spp. in food. Appl. Environ. Microbiol. 67: 977–978

Fitzmaurice J, Duffy G, Kilbride B, Sheridan JJ, Carroll C and Maher M (2004) Comparison of a membrane surface adhesion recovery method with an ims method for use in a polymerase chain reaction method to detect *Escherichia coli* o157:H7 in minced beef. J. Microbiol. Methods 59: 243–252

Fitzmaurice J, Glennon M, Duffy G, Sheridan JJ, Carroll C and Maher M (2004) Application of real-time PCR and RT-PCR assays for the detection and quantitation of vt 1 and vt 2 toxin genes in *E. Coli* o157:H7. Mol. Cell. Probes 18:123–132

Fluit AC, Torensma R, Visser MJ, Aarsman CJ, Poppelier MJ, Keller BH, Klapwijk P and Verhoef J (1993) Detection of *Listeria monocytogenes* in cheese with the magnetic immuno-polymerase chain reaction assay. Appl. Environ. Microbiol. 59:1289–1293

Fu Z, Rogelj S and Kieft TL (2005) Rapid detection of *Escherichia coli* o157:H7 by immunomagnetic separation and real-time PCR. Int. J. Food Microbiol. 99:47–57

Fuentes M, Mateo C, Rodriguez A, Casqueiro M, Tercero JC, Riese HH, Fernandez-Lafuente R and Guisan JM (2006) Detecting minimal traces of DNA using DNA covalently attached to superparamagnetic nanoparticles and direct PCR-ELISA. Biosens. Bioelectron. 21:1574–1580

Fukushima H, Tsunomori Y and Seki R (2003) Duplex real-time sybr green PCR assays for detection of 17 species of food- or waterborne pathogens in stools. J. Clin. Microbiol. 41: 5134–5146

Fung DY (2002) Predictions for rapid methods and automation in food microbiology. J. AOAC Int. 85:1000–1002

Ge B, Larkin C, Ahn S, Jolley M, Nasir M, Meng J and Hall RH (2002) Identification of *Escherichia coli* o157:H7 and other enterohemorrhagic serotypes by ehec- hlya targeting, strand displacement amplification, and fluorescence polarization. Mol. Cell. Probes 16: 85–92

Giesendorf BA and Quint WG (1995) Detection and identification of Campylobacter spp. Using the polymerase chain reaction. Cell Mol. Biol. (Noisy-le-grand) 41:625–638

Gilbert C, Winters D, O'Leary A and Slavik M (2003) Development of a triplex PCR assay for the specific detection of *Campylobacter jejuni*, Salmonella spp., and *Escherichia coli* o157:H7. Mol. Cell. Probes 17: 135–138

Glynn B, Lahiff S, Wernecke M, Barry T, Smith TJ and Maher M (2006) Current and emerging molecular diagnostic technologies applicable to bacterial food safety. Int. J. Dairy Tech. 59: 126–139

Gonzalez I, Grant KA, Richardson PT, Park SF and Collins MD (1997) Specific identification of the enteropathogens *Campylobacter jejuni* and *Campylobacter coli* by using a PCR test based on the *ceue* gene encoding a putative virulence determinant. J. Clin. Microbiol. 35:759–763

Gouws PA and Liedemann I (2005) Evaluation of diagnostic PCR for the detection of *Listeria monocytogenes* in food products. Food Tech. Biotech. 43:201–205

Gouws PA, Visser M and Brozel VS (1998) A polymerase chain reaction procedure for the detection of Salmonella spp. within 24 hours. J. Food Prot. 61: 1039–1042

Gracias KS and McKillip JL (2004) A review of conventional detection and enumeration methods for pathogenic bacteria in food. Can. J. Microbiol. 50: 883–890

Grennan B, O'Sullivan NA, Fallon R, Carroll C, Smith T, Glennon M and Maher M. (2001) PCR-ELISAS for the detection of *Campylobacter jejuni* and *Campylobacter coli* in poultry samples. Biotechniques 30:602–606, 608–610

Hein I, Klein D, Lehner A, Bubert A, Brandl E and Wagner M (2001) Detection and quantification of the *iap* gene of *Listeria monocytogenes* and *Listeria innocua* by a new real-time quantitative PCR assay. Res. Microbiol. 152:37–46

Hodgson J (1998) Shrinking DNA diagnostics to fill the markets of the future. Nat. Biotechnol. 16:725–727

Holko I, Urbanova J, Kantikova M, Pastorova K, and Kmet V (2002) Pcr detection of *Listeria monocytogenes* in milk and milk products and differentiation of suspect isolates. Acta. Veterinaria. Brno. 71: 125–131.

Hong Y, Berrang ME, Liu T, Hofacre CL, Sanchez S, Wang L and Maurer JJ (2003) Rapid detection of *Campylobacter coli, c. Jejui,* and *Salmonella enterica* on poultry carcasses by using PCR-enzyme-linked immunosorbent assay. Appl. Environ. Microbiol. 69:3492–3499

Hoorfar J, Wolffs P and Radstrom P (2004) Diagnostic PCR: Validation and sample preparation are two sides of the same coin. Apmis. 112:808–814

Houng HS, Sethabutr O, Nirdnoy W, Katz DE and Pang LW (2001) Development of a *ceue*-based multiplex polymerase chain reaction (PCR) assay for direct detection and differentiation of *Campylobacter jejuni* and *Campylobacter coli* in thailand. Diagn. Microbiol. Infect. Dis. 40:11–19

Hsih HY and Tsen HY (2001) Combination of immunomagnetic separation and polymerase chain reaction for the simultaneous detection of *Listeria monocytogenes* and Salmonella spp. in food samples. J. Food Prot. 64: 1744–1750

Hudson JA, Lake RJ, Savill MG, Scholes P and McCormick RE (2001) Rapid detection of *Listeria monocytogenes* in ham samples using immunomagnetic separation followed by polymerase chain reaction. J. Appl. Microbiol. 90:614–621

Inglis GD and Kalischuk LD (2004) Direct quantification of *Campylobacter jejuni* and *Campylobacter lanienae* in feces of cattle by real-time quantitative PCR. Appl. Environ. Microbiol. 70:2296–2306

Ivnitski D, Abdel-Hamid I, Atanasov P, Wilkins E and Stricker S (2000) Application of electrochemical biosensors for detection of food pathogenic bacteria. Electroanalysis 12:317–325

Ivnitski D, O'Neil DJ, Gattuso A, Schlicht R, Calidonna M and Fisher R (2003) Nucleic acid approaches for detection and identification of biological warfare and infectious disease agents. Biotechniques 35:862–869

Jin UH, Cho SH, Kim MG, Ha SD, Kim KS, Lee KH, Kim KY, Chung DH, Lee YC and Kim CH (2004) PCR method based on the *ogdh* gene for the detection of Salmonella spp. From chicken meat samples. J. Microbiol. 42:216–222

Johnston LM, Elhanafi D, Drake M and Jaykus LA (2005) A simple method for the direct detection of Salmonella and *Escherichia coli* o157:H7 from raw alfalfa sprouts and spent irrigation water using PCR. J. Food. Prot. 68:2256–2263

Jothikumar N, Wang X and Griffiths MW (2003) Real-time multiplex sybr green i-based PCR assay for simultaneous detection of Salmonella serovars and *Listeria monocytogenes*. J. Food. Prot. 66: 2141–2145

Kawasaki S, Horikoshi N, Okada Y, Takeshita K, Sameshima T and Kawamoto S (2005) Multiplex PCR for simultaneous detection of Salmonella spp., *Listeria monocytogenes*, and *Escherichia coli* o157:H7 in meat samples. J. Food. Prot. 68: 551–556

Ke D, Menard C, Picard FJ, Boissinot M, Ouellette M, Roy PH and Bergeron MG (2000) Development of conventional and real-time PCR assays for the rapid detection of group b streptococci Clin. Chem 46: 324–331

Keiler KC and Shapiro L (2003) tmRNA in *Caulobacter crescentus* is cell cycle regulated by temporally controlled transcription and RNA degradation. J. Bacteriol. 185:1825–1830

Klerks MM, Zijlstra C and van Bruggen AH (2004) Comparison of real-time PCR methods for detection of *Salmonella enterica* and *Escherichia coli* o157:H7, and introduction of a general internal amplification control. J. Microbiol. Methods 59:337–349

Lamoureux M, MacKay A, Messier S, Fliss I, Blais BW, Holley RA and Simard RE (1997) Detection of *Campylobacter jejuni* in food and poultry viscera using immunomagnetic separation and microtitre hybridization. J. Appl. Microbiol. 83:641–651

Lantz PG, Tjerneld F, Borch E, Hahn-Hagerdal B and Radstrom P (1994) Enhanced sensitivity in PCR detection of *Listeria monocytogenes* in soft cheese through use of an aqueous two-phase system as a sample preparation method. Appl. Environ. Microbiol. 60:3416–3418

Lee SY, Bailey SC and Apirion D (1978) Small stable RNAs from *Escherichia coli*: Evidence for the existence of new molecules and for a new ribonucleoprotein particle containing 6s RNA. J. Bacteriol. 133:1015–1023

Leonard P, Hearty S, Quinn J and O'Kennedy R (2004) A generic approach for the detection of whole *Listeria monocytogenes* cells in contaminated samples using surface plasmon resonance. Biosens. Bioelectron. 19: 1331–1335

Li Y and Mustapha A (2004) Simultaneous detection of *Escherichia coli* o157:H7, Salmonella, and shigella in apple cider and produce by a multiplex PCR. J. Food. Prot. 67:27–33

Liming SH and Bhagwat AA (2004) Application of a molecular beacon-real-time PCR technology to detect Salmonella species contaminating fruits and vegetables. Int. J. Food Microbiol. 95:177–187

Lin CK, Hung CL, Hsu SC, Tsai CC and Tsen HY (2004) An improved PCR primer pair based on 16s DNA for the specific detection of Salmonella serovars in food samples. J. Food Prot. 67: 1335–1343

Lin CK and Tsen HY (1996) Use of two 16s DNA targeted oligonucleotides as PCR primers for the specific detection of Salmonella in foods. J. Appl. Bacteriol. 80:659–666

Lubeck PS, Wolffs P, On SL, Ahrens P, Radstrom P and Hoorfar J (2003) Toward an international standard for PCR-based detection of food-borne thermotolerant Campylobacters: Assay development and analytical validation. Appl. Environ. Microbiol. 69:5664–5669

Lund M and Madsen M (2006) Strategies for the inclusion of an internal amplification control in conventional and real time PCR detection of Campylobacter spp. In chicken fecal samples. Mol. Cell. Probes 20:92–99

Maher M, Finnegan C, Collins E, Ward B, Carroll C and Cormican M (2003) Evaluation of culture methods and a DNA probe-based PCR assay for detection of Campylobacter species in clinical specimens of feces. J. Clin. Microbiol. 41:2980–2986

Malorny B, Cook N, D'Agostino M, De Medici D, Croci L, Abdulmawjood A, Fach P, Karpiskova R, Aymerich T, Kwaitek K, Hoorfar J and Malorny B (2004) Multicenter validation of PCR-based method for detection of Salmonella in chicken and pig samples. J. AOAC Int. 87: 861–866

Malorny B, Hoorfar J, Bunge C and Helmuth R (2003) Multicenter validation of the analytical accuracy of Salmonella PCR: Towards an international standard. Appl. Environ. Microbiol. 69:290–296

Malorny B, Paccassoni E, Fach P, Bunge C, Martin A and Helmuth R (2004) Diagnostic real-time PCR for detection of Salmonella in food. Appl. Environ. Microbiol. 70:7046–7052

Mandrell RE and Wachtel MR (1999) Novel detection techniques for human pathogens that contaminate poultry. Curr. Opin. Biotechnol. 10: 273–278

Mannelli I, Minunni M, Tombelli S, Wang R, Michela Spiriti M and Mascini M (2005) Direct immobilisation of DNA probes for the development of affinity biosensors. Bioelectrochemistry 66:129–138

Mascini M, Tombelli S and Palchetti I (2005) New trends in nucleic acid based biosensors: University of florence (Italy) Bioelectrochemistry 67:131–133

McKillip JL and Drake M (2004) Real-time nucleic acid-based detection methods for pathogenic bacteria in food. J. Food Prot. 67: 823–832

McKillip JL, Jaykus LA and Drake M (1998) RNA stability in heat-killed and uv-irradiated enterotoxigenic *Staphylococcus aureus* and *Escherichia coli* o157:H7. Appl. Environ. Microbiol. 64: 4264–4268

Meng J and Doyle MP (1997) Emerging issues in microbiological food safety. Annu. Rev. Nutr. 17:255–275

Mercanoglu B and Griffiths MW (2005) Combination of immunomagnetic separation with real-time PCR for rapid detection of Salmonella in milk, ground beef, and alfalfa sprouts. J. Food Prot. 68: 557–561

Metherell LA, Logan JM and Stanley J (1999) PCR-enzyme-linked immunosorbent assay for detection and identification of Campylobacter species: Application to isolates and stool samples. J. Clin. Microbiol. 37: 433–435

Mo XT, Zhou YP, Lei H and Deng L (2002) Microbalance-DNA probe method for the detection of specific bacteria in water. Enz. Micro. Technol. 30:583–589

Moore J, Caldwell P and Millar B (2001) Molecular detection of Campylobacter spp. In drinking, recreational and environmental water supplies. Int. J. Hyg. Environ. Health 204: 185–189

Morin NJ, Gong Z and Li XF (2004) Reverse transcription-multiplex PCR assay for simultaneous detection of *Escherichia coli* o157:H7, *Vibrio cholerae* o1, and *Salmonella typhi*. Clin. Chem. 50:2037–2044

Muto A, Ushida C and Himeno H (1998) A bacterial RNA that functions as both a tRNA and an mRNA. Trends Biochem. Sci. 23: 25–29

Naravaneni R and Jamil K (2005) Rapid detection of food-borne pathogens by using molecular techniques. J. Med. Microbiol. 54:51–54

Nastasi A, Mammina C, and Mioni R (1999) Detection of Salmonella spp. in food by a rapid PCR-hybridization procedure. New Microbiol. 22: 195–202

Nogva HK, Rudi K, Naterstad K, Holck A and Lillehaug D (2000) Application of 5'-nuclease PCR for quantitative detection of *Listeria monocytogenes* in pure cultures, water, skim milk, and unpasteurized whole milk. Appl. Environ. Microbiol. 66: 4266–4271

O'Connor L, Joy J, Kane M, Smith T and Maher M (2000) Rapid polymerase chain reaction/DNA probe membrane-based assay for the detection of Listeria and *Listeria monocytogenes* in food. J. Food Prot. 63: 337–342

O'Sullivan NA, Fallon R, Carroll C, Smith T and Maher M (2000) Detection and differentiation of *Campylobacter jejuni* and *Campylobacter coli* in broiler chicken samples using a PCR/DNA probe membrane based colorimetric detection assay. Mol. Cell. Probes 14:7–16

On SL and Jordan PJ (2003) Evaluation of 11 PCR assays for species-level identification of *Campylobacter jejuni* and *Campylobacter coli*. J. Clin. Microbiol. 41:330–336

Paton AW and Paton JC (1998) Detection and characterization of shiga toxigenic *Escherichia coli* by using multiplex PCR assays for *stx1, stx2, eaea*, enterohemorrhagic *E. coli hlya, rfbo111*, and *rfbo157*. J. Clin. Microbiol. 36: 598–602

Perelle S, Dilasser F, Malorny B, Grout J, Hoorfar J and Fach P (2004) Comparison of PCR-ELISA and lightcycler real-time PCR assays for detecting Salmonella spp. In milk and meat samples. Mol. Cell. Probes 18:409–420

Perelle S, Josefsen M, Hoorfar J, Dilasser F, Grout J and Fach P (2004) A lightcycler real-time PCR hybridization probe assay for detecting food-borne thermophilic Campylobacter. Mol. Cell. Probes 18: 321–327

Peterkin PI, Idziak ES and Sharpe AN (1991) Detection of *Listeria monocytogenes* by direct colony hybridization on hydrophobic grid-membrane filters by using a chromogen-labeled DNA probe. Appl. Environ. Microbiol. 57: 586–591

Pfaller MA (2001) Molecular approaches to diagnosing and managing infectious diseases: Practicality and costs. Emerg. Infect. Dis. 7:312–318

Piknova L, Kaclikova E, Pangallo D, Polek B and Kuchta T (2005) Quantification of Salmonella by 5′-nuclease real-time polymerase chain reaction targeted to *fimc* gene. Curr. Microbiol. 50:38–42

Rand AG, Ye JM, Brown CW and Letcher SV (2002) Optical biosensors for food pathogen detection. Food Technology. 56:32–39

Raoult D, Fournier PE and Drancourt M (2004) What does the future hold for clinical microbiology? Nat. Rev. Microbiol. 2:151–159

Rijpens N, Herman L, Vereecken F, Jannes G, De Smedt J and De Zutter L (1999) Rapid detection of stressed Salmonella spp. In dairy and egg products using immunomagnetic separation and PCR. Int. J. Food Microbiol. 46:37–44

Rodriguez-Lazaro D, Hernandez M, Esteve T, Hoorfar J and Pla M (2003) A rapid and direct real time PCR-based method for identification of Salmonella spp. J. Microbiol. Methods 54:381–390

Rodriguez-Lazaro D. Lloyd J, Herrewegh A, Ikonomopoulos J, D'Agostino M, Pla M, and Cook N (2004) A molecular beacon-based real-time NASBA assay for detection of *Mycobacterium avium* subsp. paratuberculosis in water and milk. FEMS Microbiol. Lett. 237:119–126

Rossen L, Norskov P, Holmstrom K and Rasmussen OF (1992) Inhibition of PCR by components of food samples, microbial diagnostic assays and DNA-extraction solutions. Int. J. Food Microbiol. 17:37–45

Rudi K, Hoidal HK, Katla T, Johansen BK, Nordal J and Jakobsen KS (2004) Direct real-time PCR quantification of *Campylobacter jejuni* in chicken fecal and cecal samples by integrated cell concentration and DNA purification. Appl. Environ. Microbiol. 70:790–797

Sails AD, Fox AJ, Bolton FJ, Wareing DR and Greenway DL (2003) A real-time PCR assay for the detection of *Campylobacter jejuni* in foods after enrichment culture. Appl. Environ. Microbiol. 69:1383–1390

Sails AD, Fox AJ, Bolton FJ, Wareing DR, Greenway DL and Borrow R (2001) Development of a PCR ELISA assay for the identification of *Campylobacter jejuni* and *Campylobacter coli*. Mol. Cell. Probes 15:291–300

Sands KE, Bates DW, Lanken PN, Graman PS, Hibberd PL, Kahn KL, Parsonnet J, Panzer R, Orav EJ, Snydman DR, Black E, Schwartz JS, Moore R, Johnson Jr. BL and Platt R (1997) Epidemiology of sepsis syndrome in 8 academic medical centers. Jama 278:234–240

Scheu PM, Berghof K and Stahl U (1998) Detection of pathogenic and spoilage micro-organisms in food with the polymerase chain reaction. Food Micro. 15:13–31

Schonhuber W, Le Bourhis G, Tremblay J, Amann R and Kulakauskas S (2001) Utilization of tmRNA sequences for bacterial identification. BMC Microbiol. 1:20

Seo KH, Valentin-Bon IE, Brackett RE and Holt PS (2004) Rapid, specific detection of *Salmonella enteritidis* in pooled eggs by real-time PCR. J. Food Prot. 67: 864–869

Shaw SJ, Blais BW and Nundy DC (1998) Performance of the dynabeads anti-Salmonella system in the detection of Salmonella species in foods, animal feeds, and environmental samples. J. Food Prot. 61: 1507–1510

Simpkins SA, Chan AB, Hays J, Popping B and Cook N (2000) An RNA transcription-based amplification technique (NASBA) for the detection of viable *Salmonella enterica*. Lett. Appl. Microbiol. 30: 75–79

Smith TA (2000) Occupational skin conditions in the food industry. Occup. Med. 50: 597–598

Stender H, Sage A, Oliveira K, Broomer AJ, Young B and Coull J (2001) Combination of atp-bioluminescence and PNA probes allows rapid total counts and identification of specific microorganisms in mixed populations. J. Microbiol. Methods 46:69–75

Stewart D and Gendel SM (1998) Specificity of the bax polymerase chain reaction system for detection of the foodborne pathogen *Listeria monocytogenes*. J. AOAC Int. 81: 817–822

Storhoff JJ, Marla SS, Bao P, Hagenow S, Mehta H, Lucas A, Garimella V, Patno T, Buckingham W, Cork W and Muller UR (2004) Gold nanoparticle-based detection of genomic DNA targets on microarrays using a novel optical detection system. Biosens. Bioelectron. 19: 875–883

Straub TM, Dockendorff BP, Quinonez-Diaz MD, Valdez CO, Shutthanandan JI, Tarasevich BJ, Grate JW and Bruckner-Lea CJ (2005) Automated methods for multiplexed pathogen detection. J. Microbiol. Methods 62: 303–316

Tenover FC, Jones RN, Swenson JM, Zimmer B, McAllister S and Jorgensen JH (1999) Methods for improved detection of oxacillin resistance in coagulase-negative staphylococci: Results of a multicenter study. J Clin Microbiol 37: 4051–4058

Todd EC, Szabo RA, MacKenzie JM, Martin A, Rahn K, Gyles C, Gao A, Alves D and Yee AJ (1999) Application of a DNA hybridization-hydrophobic-grid membrane filter method for detection and isolation of verotoxigenic *Escherichia coli*. Appl. Environ. Microbiol. 65: 4775–4780

Traore H, van Deun A, Shamputa IC, Rigouts L and Portaels F (2006) Direct detection of *Mycobacterium tuberculosis* complex DNA and rifampin resistance in clinical specimens from tuberculosis patients by line probe assay. J Clin Microbiol 44:4384–4388

Uyttendaele M, Bastiaansen A and Debevere J (1997) Evaluation of the NASBA nucleic acid amplification system for assessment of the viability of *Campylobacter jejuni*. Int. J. Food Microbiol. 37:13–20

Volokhov D, Chizhikov V, Chumakov K and Rasooly A (2003) Microarray-based identification of thermophilic *Campylobacter jejuni, C. coli, C. lari*, and *C. upsaliensis*. J. Clin. Microbiol. 41:4071–4080

Waller DF and Ogata SA (2000) Quantitative immunocapture PCR assay for detection of *Campylobacter jejuni* in foods. Appl. Environ. Microbiol. 66:4115–4118

Wan CS, Li JA and Luo J (2004) [detection of the Salmonella invasion gene inva using molecular beacon probe.]. Di Yi Jun Yi Da Xue Xue Bao 24:1257–1259

Wang R, Tombelli S, Minunni M, Spiriti MM and Mascini M (2004a) Immobilisation of DNA probes for the development of SPR-based sensing. Biosens. Bioelectron. 20:967–974

Wang X, Jothikumar N and Griffiths MW (2004b) Enrichment and DNA extraction protocols for the simultaneous detection of Salmonella and *Listeria monocytogenes* in raw sausage meat with multiplex real-time PCR. J. Food Prot. 67:189–192

Wilson DL, Abner SR, Newman TC, Mansfield LS and Linz JE (2000) Identification of ciprofloxacin-resistant *Campylobacter jejuni* by use of a fluorogenic PCR assay. J. Clin. Microbiol. 38:3971–3978

Wilson IG (1997) Inhibition and facilitation of nucleic acid amplification. Appl. Environ. Microbiol. 63: 3741–3751

Woodward MJ and Kirwan SE (1996) Detection of *Salmonella enteritidis* in eggs by the polymerase chain reaction. Vet. Rec. 138: 411–413

Wu CF, Valdes JJ, Bentley WE and Sekowski JW (2003) DNA microarray for discrimination between pathogenic 0157:H7 edl933 and non-pathogenic *Escherichia coli* strains. Biosens. Bioelectron. 19:1–8

Yancey MK, Duff P, Kubilis P, Clark P and Frentzen BH (1996a) Risk factors for neonatal sepsis. Obstet Gynecol 87:188–194

Yancey MK, Schuchat A, Brown LK, Ventura VL and Markenson GR (1996b) The accuracy of late antenatal screening cultures in predicting genital group b streptococcal colonization at delivery. Obstet Gynecol 88: 811–815

Yang S and Rothman RE (2004) PCR-based diagnostics for infectious diseases: Uses, limitations, and future applications in acute-care settings. Lancet Infect Dis 4: 337–348

Yao G, Fang X, Yokota H, Yanagida T and Tan W (2003) Monitoring molecular beacon DNA probe hybridization at the single-molecule level. Chemistry 9:5686–5692

Ziemer CJ and Steadham SR (2003) Evaluation of the specificity of Salmonella PCR primers using various intestinal bacterial species. Lett. Appl. Microbiol. 37:463–469

23

Oligonucleotide and DNA Microarrays: Versatile Tools for Rapid Bacterial Diagnostics

Tanja Kostic, Patrice Francois, Levente Bodrossy and Jacques Schrenzel

Abstract

The rapid and unambiguous detection and identification of microorganisms, historically a major challenge of clinical microbiology, gained additional importance in the fields of public health and biodefence. These requirements cannot be well addressed by classical culture-based approaches. Therefore, a wide range of molecular approaches has been suggested. Microarrays are molecular tools that can be used for simultaneous identification of microorganisms in clinical and environmental samples. Main advantages of microarrays are high throughput, parallelism and miniaturization of the detection system. Furthermore, they allow for both high specificity and high sensitivity of the detection.

Microarrays consist of set of probes immobilized on a solid surface. Even though the first application of the microarrays can be seen as relatively recent (Schena et al. 1995), the technology developed rapidly reaching the milestone of 5,000 published papers in 2004 (Holzman and Kolker 2004). This development encompasses both the successful transfer of various technological aspects as well as the expansion of the application scope. The most important technological elements of custom-made platforms as well as the characteristics of the commercially available formats are reviewed in this chapter. Furthermore, application potential is presented together with considerations about quality control.

1. Introduction

For almost two centuries, clinical microbiology laboratories have relied on culture methods. The essence of this discipline was to provide growth evidence of a suspected pathogen, and enable physicians to correlate this observation with a given clinical presentation. Koch's postulates, derived from his work on infectious diseases such as anthrax and tuberculosis, established a standard for causation in infectious disease. The standard was intended to convince skeptics that microbes can cause disease and to push microbiologists to use more rigorous criteria before claiming a causal relationship between a microbe and a disease (Fredericks and Relman 1996).

Tanja Kostic • Department of Bioresources, Austrian Research Centres GmbH - ARC, A-2444 Seibersdorf, Austria. **Patrice Francois** • Genomic Research Laboratory, Division of Infectious Diseases, University of Geneva Hospital, Geneva, Switzerland. **Levente Bodrossy** • Department of Bioresources, Austrian Research Centres GmbH - ARC, A-2444 Seibersdorf, Austria. **Jacques Schrenzel** • Genomic Research Laboratory, and Clinical Microbiology Laboratory, Service of Infectious Diseases, University Hospital of Geneva, Department of Internal Medicine, Geneva, Switzerland

M. Zourob et al. (eds.), *Principles of Bacterial Detection: Biosensors, Recognition Receptors and Microsystems*,
© Springer Science+Business Media, LLC 2008

Addressing Koch's postulates implies good knowledge of bacterial physiology. Current diagnostic methods have to care about dozens of living species that require various growth culture media, growth temperatures, and sometimes specific incubation atmospheres. Bacteria could therefore be considered as "challenging analytes."

Generally speaking, results generated by a clinical microbiology laboratory fall within the following categories: detection, identification (or speciation), antimicrobial susceptibility testing, and typing. Rapid reporting time is a constant requirement. Results have to be delivered to clinicians in due time, i.e., early enough to meaningfully influence patient work-up and therapy.

We have to admit that culture-based approaches have reached their golden age: their performance can no longer be significantly improved. In particular, time for bacterial growth represents a barrier that prevents more rapid, and clinically relevant, delivery of results to physicians. Simply speaking, we should bypass all culture steps that are, by essence, limited by microbial physiology. Culture-based approaches also display limitations towards groups of clinically important microorganisms (e.g., fastidious or hardly cultivable organisms), not to mention those that cannot be cultivated (e.g., *Treponema pallidum*) or that have been killed during handling (e.g., improper transport of sample containing *Neisseria meningitidis*) or by exposure to antimicrobials.

Thinking of molecular approaches, real-time PCR (qPCR) sounds rather appealing. However, due to limited multiplexing capabilities, qPCR will likely not be the appropriate molecular method to replace cultures. Miniaturization of conventional assays is a general trend in diagnostics as well as in biomedical research. Working with smaller volumes leads to reduced reagent consumption and faster reaction kinetics due to increased sample concentration. Parallel determination of numerous reactions is also highly desirable to save time and reduce costs, not to mention the possibility of large-scale comparison and intra-experimental quality control. Probes bound to a solid surface were therefore developed to allow spatial discrimination of numerous reactions performed in parallel, and culminated in the recent development of microarrays (Schena et al. 1995; Fodor et al. 1991, 1993).

Microarrays consist of an orderly arrangement of probes (oligonucleotides, DNA fragments, proteins, sugars, or lectins) attached to a solid surface. The main advantages of microarray technology are: high-throughput, parallelism, miniaturization, speed, and automation. Despite the fact that microarray analysis is a relatively novel technology, described only a decade ago (Schena et al. 1995), microarrays are now broadly applied. The scientific and technological background discussed here will be limited to DNA microarrays, excluding the new evolving field of protein microarrays and that of glycomics (Raman et al. 2005).

2. Microarray Technology

With the publication of the first microarray studies in 1995 (Schena et al. 1995; Lipshutz et al. 1995), and the milestone of nearly 5,000 published microarray papers in 2004 (Holzman and Kolker 2004), there is no doubt that this technology has rapidly spread into basic and applied research.

A microarray refers to a checkerboard-like ordering of molecules on a surface. It is the molecular equivalent of a spreadsheet, where each cell or address contains a specific probe designed to detect a given target (Southern, Mir and Shchepinov 1999). By analogy to antigen-antibody interactions on immunoarrays, DNA microarrays rely on sequence complementarity of the two strands. They put into practice the fundamentals of complementary base-pairing (hybridization) that were first described by Ed Southern (Southern 1975). In general, the strategy of microarray hybridization is reversed to that of a standard dot-blot, leading to recurring confusion in the nomenclature. Therefore, the suggestion has been made to describe tethered nucleic acid as the probe and the free nucleic acid as the target (Phimister 1999).

Earlier studies on duplex melting and reformation, carried out on DNA solutions, have provided the basic knowledge (the reaction kinetics as well as the computational determination of the melting point as a function of nucleic acid composition and salt concentration) (SantaLucia 1998; SantaLucia, Allawi and Seneviratne 1996; Allawi and SantaLucia 1998). Much of the pioneering work can be linked to the use of nitrocellulose membranes (Gillespie and Spiegelman 1965), dot-blots (Kafatos, Jones and Efstratiadis 1979), and Southern blots (Southern 1975). Development of cDNA or oligonucleotide arrays was made possible by combined innovations in microengineering, molecular biology (Case-Green and Southern 1994; Maskos and Southern 1993b; Mir and Southern 1999; Shchepinov, Case-Green and Southern 1997; Southern et al. 1994; Southern, Mir and Shchepinov 1999), and bioinformatics (Fodor et al. 1991). The real breakthrough in microarray technology was initiated by two key innovations: the use of nonporous solid supports (such as glass and silicon) and the development of methods for high density synthesis of oligonucleotides directly onto the microarray (Fodor et al. 1991).

Powerful fundamental or applied projects using miniaturized, high-density microarrays have been demonstrated for a broad variety of applications such as cell differentiation (Tamayo et al. 1999), whole-genome expression analysis (Gasch et al. 2000; Schena et al. 1995), cancer research (Alizadeh et al. 2000; Perou et al. 2000; Golub et al. 1999), comparative genome hybridization (CGH) (Salama et al. 2000), drug discovery (Debouck and Goodfellow 1999), vaccine development (Grifantini et al. 2002), and single nucleotide polymorphism (SNP) analysis (Fan et al. 2000).

Technology transfer to diagnostic applications is therefore very appealing (Lucchini, Thompson and Hinton 2001; Ye et al. 2001; Aitman 2001). High-throughput technologies, such as DNA microarrays, have significant potential for identifying organisms in many areas of biomedical science, including health care, biological defense, and environmental monitoring. Using data from increasing numbers of whole microbial genomes, thousands of sequences can be selected to probe numerous genes of interest in cultures, clinical specimens, environmental samples, or host tissues. Reports have already shown that oligoarray hybridization can provide bacterial detection of a conserved bacterial gene (Small et al. 2001), species identification (Wilson et al. 2002), and genotyping of bacterial pathogens, by using large sets of discriminative epidemiological markers (Gingeras et al. 1998; Yue et al. 2001; Call, Borucki and Loge 2003; Volokhov et al. 2002; Chizhikov et al. 2001). Detection of genetically-encoded virulence or antimicrobial resistance determinants (Lucchini, Thompson and Hinton 2001; Troesch et al. 1999; Chizhikov et al. 2001; Bekal et al. 2003) may also afford a major benefit for the selection of an adequate chemotherapy.

Current limitations for the routine implementation of microarrays to detect DNA signatures are, for example, their high manufacturing costs and the requirement of large amounts of nucleic acids. Availability of high amounts of nucleic acid targets requires either a large volume of bacterial culture (biological amplification), or target amplification (Mikulowska-Mennis et al. 2002; Puskas et al. 2002). Additional technical achievements in signal amplification methods (Francois et al. 2003) and novel optical techniques (Pawlak et al. 2002; Ferguson, Steemers and Walt 2000) have already improved target detection to the femtomole range.

Development of a microarray-based bacterial identification starting directly from the biological sample, without any enzymatic target amplification, is an important objective (Straub and Chandler 2003). This procedure would significantly reduce the turnaround time and overcome enzymatic-induced signal alterations or biases (Yershov et al. 1996; Bavykin et al. 2001).

3. Technical Aspects of Microarray Technology

The development of a new microarray platform requires consideration of many different features, most of them being co-dependent. Different approaches have been reported, each of them exhibiting certain advantages and limitations. A summary of the most important technological features will be presented to introduce the specific platforms in association with their intended experimental applications.

3.1. Probes

The nature of the probe used is related to the experimental question. In general a distinction can be made between genome fragments, PCR products, and oligonucleotide probes. The application potential as well as the advantages and limitations of each probe type will be described briefly.

3.1.1. Genome Fragments

The use of entire bacterial or community genomes (suitably fragmented) as probes was first employed for the reverse sample genome probing (RSGP) technique (Voordouw et al. 1991). The same principle was later applied for community genome arrays (Murray et al. 2001; Bae et al. 2005). The major problem related to such microarray platforms is the huge complexity of the system.

3.1.2. PCR Products

PCR products used as probes for microarray fabrication are mostly amplified inserts of the clone libraries. Different types of clone libraries can be used as a template for microarray fabrication (cDNA libraries, SSH libraries, shotgun libraries). Microarrays utilizing PCR products are used for gene expression analysis (Schena et al. 1995; Gill et al. 2002; Zaigler, Schuster and Soppa 2003; Stowe-Evans, Ford and Kehoe 2004; Murray et al. 2001; Lindroos et al. 2005). Furthermore, PCR products can be used as probes for functional gene arrays (Wu et al. 2001; Call et al. 2003). The inappropriate labeling of a substantial fraction of the PCR products (1 to 5 %) can lead to poorly controlled microarrays, even when originating from prestigious research centers or commercial entities (Knight 2001). The IMAGE consortium (Integrated Molecular Analysis of Genomes and their Expression) revealed that only 62 % of 1,189 cDNAs were pure and correct, after resequencing.

3.1.3. Oligonucleotide Probes

Compared to cDNA microarrays, oligoarrays provide a flexible design and are considered more reliable in terms of sensitivity and specificity (Blanchard and Friend 1999; Barczak et al. 2003; Kothapalli et al. 2002; Li, Pankratz and Johnson 2002). Differing from the previously described PCR probes, oligonucleotide probes are typically designed with a predefined specificity. The development of generic or universal microarrays will be described in a separate section (see Section 23.5.3).

Two main features influence probe specificity: probe length and the degree of conservation of the marker gene. In general, probe design is carried out *in silico* using different software tools, e.g., ARB (Ludwig et al. 2004) or OligoCheck (Charbonnier et al. 2005), and is based on a sequence database of the targeted marker gene. The extent and quality of the sequence database has a major effect on the probe quality.

The general criteria that need to be considered during probe design are: (1) the required probe specificity, and (2) the uniformity of the probe set regarding hybridization behavior. *In silico* approaches allow for partial prediction of the hybridization behavior of the designed probes. However, it has become clear that the simple notion that short oligonucleotides with a mismatch (MM) should hybridize less efficiently than perfect match (PM) probes is not always applicable (Pozhitkov et al. 2006). It has been demonstrated that the hybridization intensity of MM probes can depend on the nucleotide type (i.e., A, C, G, or T) and the position of the MM relative to the termini (Urakawa et al. 2002); and that some MM probes yield even higher signal intensities with the target than those of corresponding PM probes (Naef and Magnasco 2003). Even well-designed probes can display differences in maximal hybridization capacity of 2 orders of magnitude under different hybridization conditions (Bodrossy and Sessitsch 2004); and thus, it is difficult to find one set of conditions that is optimal for all probes on an array (Kajiyama et al. 2003; Urakawa et al. 2003). Factors affecting duplex formation on DNA microarrays include: probe density, microarray surface composition, spacer length, and the stabilities of the oligonucleotide-target duplexes, intra- and intermolecular self-structures, and RNA secondary structures (Matveeva et al. 2003; Peterson, Heaton and Georgiadis 2001; Halperin, Buhot and Zhulina 2005).

More generally, there is a lack of a simple relationship between hybridizations of probe-target duplexes as inferred from signal intensity values and *in silico* predictions based on Gibbs free energies (Pozhitkov et al. 2006). This does not apply as strictly for high-density microarrays, where the high level of redundancy accounts for the specificity of the signals; nor for long oligonucleotide microarrays, where inter-allele distinction is not required. In any case, a thorough wet-lab validation with a set of reference strains or clones is warranted, as with the implementation of any other molecular tool (Charbonnier et al. 2005).

Original approaches were recently published to address the issue of cross-hybridization while maintaining target sensitivity. Using *E. cuniculi* as a biological model and conventional probe design, Rimour et al. could determine specific 50-mer oligonucleotides for only approximately 40 % of the genome. When relying on their new probe design strategy (GoArrays), based on the determination of two specific subsequences, they were able to design specific probes for each CDS of the genome (Rimour et al. 2005).

3.1.3.1. Long Oligonucleotides

The main advantages of long oligonucleotides (over 50 nucleotides in length) are high target binding capacity and irreversible hybridization kinetics. These features allow for enhanced detection sensitivity. However, the threshold for the differentiation is at 85 to 90 % sequence similarity, resulting in reduced specificity. This can be compensated by the host specificity of the targeted genes. Due to their high sensitivity, long oligonucleotide microarrays are typically used in combination with universal amplification techniques or without any amplification, allowing the researcher to target an unlimited number of different genes (Tiquia et al. 2004). Long oligonucleotide microarrays have been used for the detection of viruses (Wang et al. 2002) and pathogens (Vora et al. 2004).

3.1.3.2. Short Oligonucleotides

Short oligonucleotide (15 to 30-mer) microarrays are more precise in the detection of shorter nucleotide polymorphisms, including single nucleotide differences under optimized hybridization conditions. On the downside, short oligoarrays frequently require a larger number of probes for reliable diagnostics. Reversible hybridization kinetics and lower target binding capacity (in comparison to long oligonucleotides) are responsible for somewhat limited detection

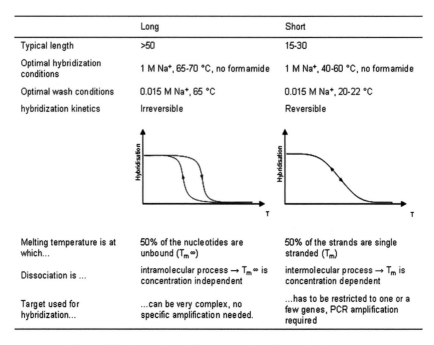

	Long	Short
Typical length	>50	15-30
Optimal hybridization conditions	1 M Na⁺, 65-70 °C, no formamide	1 M Na⁺, 40-60 °C, no formamide
Optimal wash conditions	0.015 M Na⁺, 65 °C	0.015 M Na⁺, 20-22 °C
hybridization kinetics	Irreversible	Reversible
Melting temperature is at which...	50% of the nucleotides are unbound ($T_m \infty$)	50% of the strands are single stranded (T_m)
Dissociation is ...	intramolecular process → $T_m \infty$ is concentration independent	intermolecular process → T_m is concentration dependent
Target used for hybridization...	...can be very complex, no specific amplification needed.	...has to be restricted to one or a few genes, PCR amplification required

Figure 23.1. Comparison of long *versus* short oligonucleotide probes.

sensitivity (Fig. 23.1). Short oligonucleotides are widely used for both environmental and diagnostic microbial diagnostic microarrays (MDMs) (Bodrossy et al. 2003; Hashsham et al. 2004; Sergeev et al. 2004; Loy et al. 2005).

3.2. Substrates for Printing

The choice of the substrate for microarray printing depends primarily on the nature of the probes. An important factor to be taken into consideration is the effect of steric hindrance on the hybridization efficiency (Bodrossy et al. 2003). This may be a considerable problem in the case of short oligonucleotide probes, and therefore such probes are generally appended to spacer molecules. Further important parameters to be considered during fabrication of the microarray are the probe concentration, spotting buffer, and surface blocking strategies. Most of these features have been discussed thoroughly in the literature (Zammatteo et al. 2000; Lindroos et al. 2001; Taylor et al. 2003; Hessner et al. 2004). The most widely used microarray format is a planar glass slide (1 × 3 in). Slides for microarray printing are usually coated with different active surfaces that facilitate deposition of nucleic acids. An overview of the more commonly used substrates and their applications will be given hereafter.

3.2.1. Slides with Poly-L-Lysine Coating

The binding of DNA fragments to poly-L-lysine involves charge interactions that can be converted to covalent bonding by baking or UV-crosslinking. Advantages of poly-L-lysine surfaces are their low background and good signal intensities. The main disadvantage is low temperature resistance that can lead to the damage of the surface during denaturation or hybridization at higher temperatures. Poly-L-lysine coated slides have been successfully used for the binding of PCR products (Schena et al. 1995; Diehl et al. 2001; Zaigler, Schuster and Soppa 2003) and short oligonucleotides (Taylor et al. 2003).

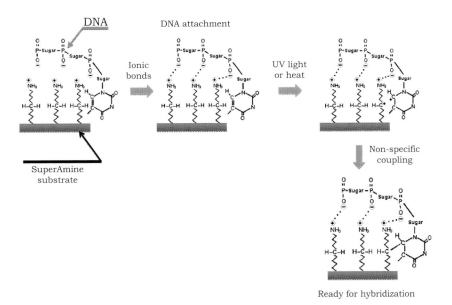

Figure 23.2. Binding chemistry of amino silane-coated slides (Reprinted with permission from TeleChem International, Inc., www.arrayit.com).

3.2.2. Slides with Amino Silane Coating

Amino silane surface chemistry allows for electrostatic interactions between the amino groups of the silane (positively charged at neutral pH) and the negatively charged phosphodiester backbone of the DNA. This interaction can be additionally stabilized by UV-crosslinking (Fig. 23.2). Amino silane coating demonstrates enhanced resistance towards high temperatures. However, somewhat higher background signals may occur when low quality coatings are used. Substrates based on amino silane surface chemistry are widely used for deposition of PCR products (Gill et al. 2002; Stowe-Evans, Ford and Kehoe 2004; Denef et al. 2003; Rhee et al. 2004; Rudi et al. 2003; Treimo et al. 2006).

3.2.3. Slides with Aldehyde Coating

Covalent binding between aldehyde groups and DNA fragments (Fig. 23.3) can be facilitated either through a 5'-amino linker on chemically modified DNA fragments or through aromatic amines of nucleotides. Probes with the 5' amino group bind more efficiently than native DNA fragments. Furthermore, coupling via the 5' amino group is directional, allowing for the defined orientation of the probes on the microarray. In general, slides with aldehyde coating are characterized by high binding capacity and low background. Aldehyde-coated slides are mostly used for short oligonucleotide microarrays (Bodrossy et al. 2003; Loy et al. 2002).

3.2.4. Slides with Epoxy Coating

Similarly to aldehyde coated slides, epoxy coated substrates (Fig. 23.4) also allow for covalent binding utilizing amino groups of the DNA fragments. Epoxy coated slides have been used for the deposition of PCR products (Call et al. 2003), and short (Bailly et al. 2006) and long (Wang et al. 2004) oligonucleotides.

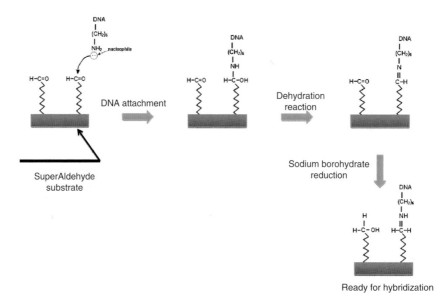

Figure 23.3. Binding chemistry of aldehyde-coated slides (Reprinted with permission from TeleChem International, Inc., www.arrayit.com).

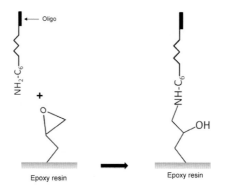

Figure 23.4. Binding chemistry of epoxy-coated slides.

3.2.5. Proprietary Surface Chemistries

Finally, it is worth mentioning that the array surface of various commercial microarrays can display very different properties, e.g., a highly hydrophobic surface on the Agilent SurePrint technology (Agilent, Palo Alto, CA) because of proprietary surface chemistry. This leads to the use of specific hybridization and washing buffers adapted to the surface chemistry.

3.2.6. Probe Spacers

To reduce the effect of steric interference (steric hindrance and surface electrostatic forces) (Vainrub and Pettitt 2002; Shchepinov, Case-Green and Southern 1997) on hybridization of targets to planar surfaces (e.g., glass and silicon), spacer molecules with a length of more than 50 Å can be used to physically separate the probes from the microchip surface (Peplies, Glockner and Amann 2003; Shchepinov, Case-Green and Southern 1997). These are typically

C_6-C_{12} alkane spacers and/or 5–15 thymidine or adenine residues added to the tethered end of the oligoprobe (Anthony, Brown and French 2000; Bodrossy et al. 2003; Zhang, Hurek and Reinhold-Hurek 2006; Francois et al. 2005).

3.3. Targets for Microarray Analysis

Targets used for microarray analysis are typically fluorescently labeled nucleic acid derivatives. Two basic types of targets can be distinguished: those derived from RNA, and those derived from DNA templates. The selection of nucleic acids used as a template for target preparation is primarily dependent on the experimental question. Gene expression studies use mRNA-based targets, whereas microbial diagnostic arrays employ primarily DNA-based targets, or rRNA targets that are more abundant. The parallel analysis of mRNA- and DNA-based targets provides a complex picture correlating presence and activity (Bodrossy et al. 2006). See Section 23.5 for further discussion of possible applications.

3.3.1. Target Amplifications and Sensitivity Issues

In general, targets used for short oligonucleotide diagnostic microarrays have been previously PCR-amplified. PCR amplification ensures enrichment of the targeted gene(s) and therefore increases the sensitivity of microarray detection, but this introduces an inherent PCR bias (Vora et al. 2004). Long oligonucleotide probes exhibit higher target binding capacity and therefore allow hybridization with highly complex target mixes (i.e., unamplified environmental DNA, a native mixture of mRNA from an organism, or the products of universal, whole genome amplification methods). The main advantage of the latter targets is that they represent the entire gene pool to be studied, without reduction of its complexity.

Thus, the potential of DNA microarray-based microbial screening and diagnostic technologies is currently limited by front-end target-specific nucleic acid detection. The presence of a ubiquitous poly-adenylated tail at the 3'-end of eukaryotic messenger RNAs offers the possibility of converting minute amounts of RNA to micrograms of labeled material, with minimal effects on the respective abundance of the mRNA mixture (Aoyagi et al. 2003; Puskas et al. 2002). Prokaryotic RNAs are not poly-adenylated and thus are more challenging to work with when starting material is scarce. In such cases, the use of generic primers able to amplify parts of the 16s rRNA gene is often required (Anthony, Brown and French 2000; Rudi et al. 2002), but the universality of such primers is questionable: false negative signals are not rare (Anthony, Brown and French 2000). Other options include intact or even degraded RNA amplification using T3-coupled random primers (Xiang et al. 2003) or a limited set of genome-derived cognate primers (Talaat et al. 2000).

The interested reader is referred to two recent publications that have analyzed and validated different target amplification strategies before array hybridization (Francois et al. 2006; Vora et al. 2004). Vora et al. investigated four front-end amplification strategies: random primed, isothermal Klenow fragment-based, Phi 29 DNA polymerase-based, and multiplex PCR. Their results underscore the feasibility of using random amplification approaches and begin to systematically address the versatility of these approaches for unbiased pathogen detection from environmental sources (Stenger et al. 2002). Francois et al. (2006) compared commercially available amplification methods, such as MessageAmp and GenomiPhi. They showed that this type of enzyme represents an interesting alternative of moderate cost for transcriptomic studies. Such amplifications permitted them to obtain significant amounts of nucleic acids, sufficient to perform microarray studies even when starting with a few tens of ng of material. Importantly, these methods showed exquisite reproducibility, even considering the data before normalization, which is the major requirement for their utilization in transcriptomic studies (Wilson et al.

2004). Finally, these nucleic amplification methods can be coupled to signal amplification; see for example Borucki et al. (2005) and/or array-based methods for improving detection sensitivity, as discussed in Section 23.3.5.

3.3.2. Labeling of the Targets

Fluorescently labeled targets are in general prepared using one of the many commercially available kits (Lynch et al. 2006) or following standardized labeling protocols (Bodrossy et al. 2003). Incorporation of the fluorescently labeled nucleotides occurs during enzymatic amplification of the nucleic acids (e.g., PCR amplification, in vitro transcription, reverse transcription, random DNA amplification). Alternatively, modified nucleotides (i.e., amino-allyl nucleotides) can be incorporated in the target followed by subsequent coupling with fluorescent dye esters.

3.3.3. Hybridization and Wash Conditions

Hybridization specificity is of paramount importance, especially when one has to differentiate targets from nontargets or to discriminate closely related DNA or RNA sequences that may possibly differ by only one base pair. Probes on the microarray are subjected to the same washing procedures (e.g., buffers, salt concentrations, and temperature). Strategies to overcome problems arising therefrom include the acquisition of melting curves for every individual probe (Liu, Mirzabekov and Stahl 2001); the careful design of probes with similar predicted hybridization properties (usually combined with the application of 2–3 probes per targeted group) (Bodrossy et al. 2003; Sanguin et al. 2006; Zhang, Hurek and Reinhold-Hurek 2006); the addition of tetramethylammonium chloride that equalizes the melting temperature of different probes by stabilizing the AT base pairs composition (Maskos and Southern 1993b); or the use of highly redundant probe sets with multiple probes to target each specific group of microorganisms (Wilson et al. 2002).

Secondary structure formation within the targets can reduce the binding constant of a specific probe by as much as 10^5 to 10^6 times (Lima et al. 1992), leading to an increase in false negative signals and a decrease in hybridization specificity (Armitage 2003; Southern, Mir and Shchepinov 1999). Several methods have been suggested to alleviate this problem, such as the use of helper oligonucleotides (Peplies, Glockner and Amann 2003), a two-probe proximal chaperon detection system (Small et al. 2001), an appropriate labeling method (Franke-Whittle et al. 2006), and a protocol to achieve optimized target lengths (Nguyen and Southern 2000; Southern, Mir and Shchepinov 1999; Yershov et al. 1996). Since long targets can form secondary and tertiary structures that hinder efficient probe-target duplex formation, the sizes of the target molecule and its amplicon are often reduced via chemical, enzymatic, or thermal fragmentation methods (Kelly et al. 2002; Liu, Mirzabekov and Stahl 2001; Nguyen and Southern 2000; Proudnikov and Mirzabekov 1996; Small et al. 2001; Bodrossy et al. 2003). Liu et al. have recently elegantly reviewed these issues and experimentally demonstrated that microarray hybridizations with short rRNA fragments were more dependent on target sequence than on the competition between probe-target interaction and RNA self-folding (Liu, Guo and Wu 2007).

Hybridization with short gene fragments increases the potential for the accumulation of background signal from nonspecific hybridization events. In order to circumvent this negative target effect, an alternative protocol, sequence-specific end labeling of oligonucleotides, was developed (Rudi et al. 2003; Kostic et al. 2007). In this approach, targets are complementary to oligonucleotide probes on the array, and the labeling is performed by incorporating single-labeled ddNTP in the presence of the targeted PCR product. This method ensures both high specificity and sensitivity; however, it is still affected by PCR bias.

3.4. Classical Commercially-Available Microarray Formats

All platforms share the common attribute that a sensor detects a signal from target sequences hybridized to immobilized nucleotidic probes. The intensity of this signal provides a measure of the amount of bound nucleic acid from a sample (Pozhitkov et al. 2006). Schematically, we have divided this section into discussions of spotted and in situ synthesized arrays. The next section focuses on alternative platforms that provide improved detection sensitivities.

3.4.1. Spotting Approaches

Currently, up to 50,000 gene fragments or oligonucleotides can be spotted onto a single microscope slide using robotic technology. The advantages of this technology are: flexibility in the design of the array; the relative ease of production; and its relatively low cost. Multiple identical microarrays can be robotically printed in batches of over a hundred in a single run. Most of the cost of printing such arrays is associated with the synthesis of oligonucleotide probes or primer pairs required for the amplification of the probe gene fragments (Dorrell, Hinchliffe and Wren 2005). We review here briefly various commercially-available microarray formats.

3.4.1.1. Operon

The Qiagen Operon format (www.operon.com) uses optimized 70-mer oligonucleotides to represent each gene in a given genome. Each 70-mer probe is designed to have optimal specificity for its target gene and is melting-temperature normalized. This approach provides a reduction in cross-hybridization and an increase in the differentiation of overlapping genes or highly homologous regions. Theoretically, mutant alleles could be detected using such oligonucleotide microarrays, owing to the shorter probe size compared to PCR product-based microarrays (Dorrell, Hinchliffe and Wren 2005).

3.4.2. In Situ Synthesis

In situ synthesis allows higher yields and lower chip-to-chip variation, as well as higher probe densities. These methods also allow the manufacture of true "random access" arrays, meaning that each oligonucleotide in any position can have any chosen sequence (Southern, Mir and Shchepinov 1999). Manufacturing techniques include photolithographic masks to control chemical activation by photodeprotection steps (Fodor et al. 1991; Lipshutz et al. 1999), ink-jet deposition (Stimpson et al. 1998; Hughes et al. 2001), and physical barriers to sequential flooding of precursors (Maskos and Southern 1993b).

3.4.2.1. Affymetrix

With Affymetrix microarrays (www.affymetrix.com), oligonucleotide probes are not deposited but directly synthesized on the surface. The company has coupled photochemical deprotection to solid-phase DNA synthesis by adapting techniques from the semiconductor industry (Pease et al. 1994; Lipshutz et al. 1999; Lockhart et al. 1996). The main advantage of this approach is a very high probe density (over 500,000 probes can be deposited on a surface of $1.6\,cm^2$). The limitations are a high price, low flexibility, and lack of properly validated probe sets. Therefore, in order to ensure the specificity of the detection, applications of the Affymetrix platform require multiple probes to monitor a single target, relying on empirical algorithms (Brodie et al. 2006).

3.4.2.2. NimbleGen

Recent technical developments, such as NimbleGen's micromirror device (www.nimblegen.com), facilitate maskless photoreactive synthesis of oligonucleotide probes, and currently permit the simultaneous deposition and analysis of as many as nearly 800,000 probes on one array platform (Albert et al. 2003). Such probe density now permits detailed comparative genome hybridizations (CGH) for detecting small deletion changes in the studied genome. However, insertions of genes compared with the sequenced reference strain cannot be detected by CGH DNA microarray analyses. This problem can be alleviated by adding nonredundant amplified sequences from several closely related bacteria to the array, once new genetic information is available (Borucki et al. 2005; Borucki et al. 2003; Porwollik et al. 2003; Cassat et al. 2005).

3.4.2.3. Agilent

A more versatile, but still essentially mechanical, method for producing DNA arrays is to use the print heads out of commercial piezoelectric ink jet printers to deliver reagents to individual spots on the array (Southern 1989; Brennan and Heinecker 1995; Baldeschwieler, Gamble and Thierault 1995; Blanchard, Kaiser and Hood 1996). A piezoelectric ink jet head consists of a small reservoir with an inlet port and a nozzle at the other end. When a voltage is applied to the crystal, it contracts laterally, thus deflecting the diaphragm and ejecting a small drop of fluid from the nozzle. Such devices are inexpensive and can deliver drops with volumes of tens of picoliters at rates of thousands of drops per second. In conjunction with a computer-controlled XY stepping stage to position the array with respect to the ink jet nozzles, it is possible to deliver different reagents to different spots on the array. Arrays of approx 250,000 spots can be addressed in a few minutes, with each spot receiving one drop of reagent. Agilent (www.agilent.com) has developed a flexible method for microarray production, centered around an in situ oligonucleotide synthesis method in which the ink jet printing process is modified to accommodate the delivery of phosphoramidites to directed locations on a glass surface (Blanchard, Kaiser and Hood 1996). Achieving high density with the ink jet approach requires one more trick. Two drops of liquid applied too closely together on a surface will tend to spread into each other and mix. For 40 picoliter drops the minimal center-to-center spacing is about 600 microns. This limits the array density achievable with the ink jet method. One way around this is to engineer patterns in the surface chemistry of the array to produce spots of a relatively hydrophilic character surrounded by hydrophobic barriers [Southern (1989) PCT WO 89/10977] [Brennan (1995) US Patent 5,474,796] (Blanchard, Kaiser and Hood 1996). Design flexibility and high densities constitute the two major advantages of this technique that can generate arrays at moderate costs.

3.4.2.4. CombiMatrix

CombiMatrix's technology (www.combimatrix.com) is a specially modified semiconductor adapted for biological applications. These integrated circuits contain arrays of microelectrodes that are individually addressable using embedded logic circuitry on the chip. Placed in a specially designed fluidic chamber, the chip digitally directs the molecular assembly of biopolymers in response to a digital command.

Under a controlled process, each microelectrode is addressed to selectively generate chemical reagents by means of an electrochemical reaction. These chemical reagents facilitate the in situ synthesis of complex molecules such as DNA oligonucleotides. The parallel process drastically reduces the cost and time of synthesizing hundreds or thousands of different molecules. Currently, this technology is able to produce arrays with approximately 45,000 features.

3.5. Alternative Methods for Improving Microarray-Based Detection Sensitivity

Most microarray applications are limited by the starting amounts of the nucleic acids to be studied. In other words, detection sensitivity is a major limitation of microarray-based approaches that has to be compensated for by several enzymatic steps for target amplification and/or labeling (Call, Borucki and Loge 2003; Loge, Thompson and Call 2002; Borucki et al. 2003), as discussed in Section 23.3.3.1.

The next section illustrates various array-based methods that can also improve detection sensitivity, independently from any target or signal amplification.

3.5.1. Resonance-Light Scattering (RLS)

New optical techniques are now available for microarray detection (Ferguson, Steemers and Walt 2000; Pawlak et al. 2002) which provide sensitivities high enough to detect femtomolar amounts of targets. Francois et al. (2003) nicely illustrated the improvements in detection sensitivity that can be achieved with different optical detection methods when using direct nonenzymatic labeling of bacterial nucleic acids. Microarrays detected by resonance light scattering (www.genicon.com) offer short turn-around times and exquisite sensitivity. Interestingly, the labeling and detection schemes offer an alternative at a reasonable cost to the expensive fluorescence-based methods. The principle of RLS is the following: when a suspension of nano-sized gold or silver particles is illuminated with a fine beam of white light, the scattered light has a clear (not cloudy) color that depends on its composition and particle size. This scattered light can be used as the signal for ultrasensitive analyte detection (Pasternack and Collings 1995).

3.5.2. Planar-Waveguide Technology (PWT)

Fig. 23.5 depicts PWT-based microarrays. A 150 to 300 nm thin metallic oxide film (green) with high refractive index (e.g., Ta_2O_5 or TiO_2) is deposited on a transparent support (grey) with a lower refractive index (e.g., glass or polymer). A parallel laser light beam (red) is coupled into the wave-guiding film by a diffractive grating that is etched into the substrate. The light propagates within this film and creates a strong evanescent field perpendicular to the direction of laser propagation into the adjacent medium (Duveneck et al. 2003). The field strength decays exponentially with the distance from the waveguide surface, and its penetration depth is limited to about 400 nm (large orange arrow). This effect results in the selective excitation of fluorophore molecules located at or near the surface of the waveguide (red circles). For microarray applications, specific capture probes or recognition elements are immobilized on the waveguide surface. Upon fluorescence excitation by the evanescent field, the excitation and detection of fluorophores by a CCD camera is restricted to the sensing surface, whilst signals from unbound molecules in the bulk solution (blue) are not detected. This yields a significant increase in the signal/noise ratio compared to conventional optical detection methods (Francois et al. 2005).

3.5.3. Liquid Arrays

The Luminex (www.luminex.com) suspension array is simply a transfer of the microarray format from a glass slide to a high-throughput and efficient bead format ("suspension microarray"). With this type of assay, the DNA probes (e.g., oligonucleotides) are attached to 5.6-nm polystyrene microspheres ("beads") containing an internal fluorescent dye. Each probe is assigned to a particular bead set containing a unique mixture of fluorescent dyes, or "spectral address." Bead sets coupled to the probes of interest are then mixed together in the wells

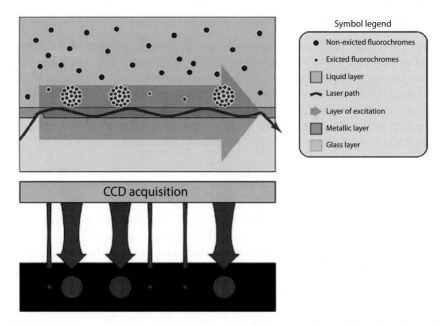

Figure 23.5. Schematic representation of PWT technology-based microarrays (reprinted from Francois et al. (2005), with permission of Elsevier).

of a 96-well microtiter plate, allowing many different probes to be analyzed simultaneously. Target DNA molecules are labeled with a different and spectrally distinct fluorescent dye and hybridized to the probes on the beads. Beads with the hybridized targets are then separated and quantified using a two-laser flow cytometer. The unique internal color of the bead is read by one laser and serves to identify which probe is present on the bead. The second laser measures the fluorescent signal of the reporter dye present on the labeled target DNA and allows one to assess the strength of the hybridization between the target DNA and the probe. Because this technology allows up to 100 different probes to be analyzed in a single well of a 96-well plate, it promises to make microarray subtyping faster and less expensive. The established suspension array protocol requires that relatively short PCR products be used as targets (Dunbar et al. 2003).

Microsphere-based fiber-optic arrays (www.illumina.com) provide many advantages over other array-based methods (Ahn, La and Forney 2006): higher sensor-packing density, smaller assay sample volumes, increased array reusability, flexible array design, and reduced false positives and false negatives (Epstein et al. 2003b). Previous work has demonstrated that the microsphere-based fiber-optic array can detect as few as 600 target DNA molecules and is sensitive enough to discriminate a single-base mismatch from a perfect match (Epstein et al. 2003a; Epstein and Walt 2003).

Finally, other bead-based arrays have been suggested for high-throughput sequencing approaches (Brenner et al. 2000). Such approaches are discussed under Section 23.5.4.

3.5.4. Three-Dimensional Microarray Formats

Three-dimensional microarray formats offer the option of recording hybridization and dissociation events in real time. This enables rapid establishment of the melting curves for all probes on the microarray, facilitating the development of validated probe sets. Three-dimensional microarray systems include gel-pads (Guschin et al. 1997; Pozhitkov et al. 2005;

Liu, Mirzabekov and Stahl 2001), flow-through systems such as PamGene (www.pamgene.com) (Wu et al. 2004) or MetriGenix (www.metrigenix.com) (Kessler et al. 2004).

3.6. Marker Genes Used on Microbial Diagnostic Microarrays (MDMs)

Microarrays employing long oligonucleotide probes or gene fragments can target an unlimited number of different genes (see Section 23.3.1.3.1). Short oligoarrays, on the other hand, depend on PCR amplification to reduce the target complexity to a level compatible with the sensitivity of the probes (see Section 23.3.1.3.2). They are thus usually limited to a small number of marker genes (typically between one and ten). Marker genes used for phylogenetic analysis and the development of short oligo-MDMs need to fulfill several criteria: (1) widespread distribution throughout the targeted organism group, (2) a high degree of conservation allowing for universal PCR amplification, (3) the existence of variable regions allowing for the design of discriminating probes, and (4) no (or low rate of) horizontal gene transfer.

The most commonly employed phylogenetic marker for the detection of microorganisms is the 16S rRNA gene. Ribosomal RNAs (rRNA) are particularly suitable for species identification procedures because they occur universally, contain conserved as well as divergent regions, and are highly abundant in cells. A further advantage of the 16S rRNA gene consists in the availability of large sequence and probe databases (http://www.arb-home.de, http://rdp.cme.msu.edu, http://greengenes.lbl.gov, http://www.microbial-ecology.net/probebase). Technical challenges faced by short oligo-MDMs are related to detection sensitivity and hybridization specificity. The former usually refers to the minimum amount of target that can be reproducibly detected by individual probes in a given complex sample (see also Section 23.3.3.1); and the latter refers to the ability of the DNA microarray technique to differentiate targets from nontargets or to discriminate closely related DNA or RNA sequences that may possibly differ by only one base pair (Liu, Mirzabekov and Stahl 2001; Urakawa et al. 2002; Liu, Guo and Wu 2007). See Section 23.3.3.3 for further details. The main limitation of the 16S rRNA gene lies in its extremely high degree of conservation. In many cases (a notable example being various genera of *Enterobacteriaceae*) it is not possible to design even species-specific probes based on it.

Therefore, a range of alternative phylogenetic and functional marker genes has been suggested (Santos and Ochman 2004; Loy and Bodrossy 2006). These include the 23S rRNA gene, the rRNA intragenic spacer region, so-called house-keeping genes (e.g., *gyrB, rpoB, recA, atpD, groEL*), virulence genes, antibiotic resistance genes, and functional genes (e.g., *pmoA, amoA, nifH, nirK, nirS*). Many of these have been successfully applied on various microarray platforms (Wu et al. 2001; Bodrossy et al. 2003; Kakinuma, Fukushima and Kawaguchi 2003; Taroncher-Oldenburg et al. 2003). The major limitation of these "alternative" marker genes is the limited organism coverage of published sequence databases.

4. Analysis and QC Aspects

Each step of microarray experiments needs to be optimized and validated, from the array design and manufacture to data collection and analysis. Among the critical technical parameters that need to be controlled are the microarray surface chemistry, the probe sequence, the probe deposition process, and the hybridization conditions. The MicroArray Quality Control (MAQC) Consortium, an unprecedented, community-wide effort, spearheaded by FDA scientists, recently addressed experimentally the key issues surrounding the reliability of DNA microarray data. They assessed the performance of seven microarray platforms in profiling the expression of two commercially available RNA sample types. The results were compared not only at different locations and between different microarray formats, but also in relation to three more traditional quantitative gene expression assays. MAQC's main conclusions confirm that, with careful experimental design and appropriate data transformation and analysis, microarray data can indeed be reproducible and comparable among different formats and laboratories, irrespective of sample labeling format. The data also demonstrate that fold change results from microarray experiments correlate closely with results from assays like quantitative reverse transcription PCR (Shi et al. 2006).

Most diagnostic microarray datasets can achieve optimal classification with no more than 5–50 discriminative genes (Bo and Jonassen 2002; Li 2005). This opens new possibilities for the design of small diagnostic microarrays used for gene expression-based diagnosis. Array-to-array normalization is crucial for microarray analysis (Yang and Speed 2002; Kroll and Wolfl 2002; Smyth and Speed 2003). Various methods for normalization have been suggested. One approach is to determine a set of invariant genes for normalization (Schadt et al. 2001; Tseng et al. 2001). Another approach recommends replicating genes on the array and using this within-array replication for normalization (Fan et al. 2005; Fan et al. 2004). Standard normalization protocols rely on the assumption that the majority of genes on the microarray are not differentially expressed between samples (Yang 2002). Jaeger et al. suggest including additional normalization genes on the small diagnostic microarrays, and they propose two strategies for selecting them from genome-wide microarray studies. The first is a data driven univariate selection of normalization genes. The second is multivariate and based on finding a balanced diagnostic signature (Jaeger and Spang 2006).

When shifting from expression arrays to detection/identification arrays, QC issues persist but they definitely require other validation approaches. These microarrays are typically hybridized with a single target (i.e., one-color hybridization). Signals on short oligonucleotide arrays are then usually normalized against positive controls. These positive controls are designed for conserved regions of the targeted gene, for the PCR primers used to amplify the targeted gene, or against exogenous spiked DNA. Long oligonucleotide arrays can be normalized against general probes, targeting conserved regions of the universal genes present in all bacteria, i.e., universal 16S probe(s) or other housekeeping genes. Normalized signals are compared to arbitrary threshold values, and the targeted microbes are rated as present or absent. For low density short oligonucleotide arrays, the threshold values are ideally individual values, reflecting the hybridization potential of the individual probes. For long oligoarrays and high density arrays with a highly redundant set of short oligos, it is possible to devise universal rules for calling a signal or a set of signals present or absent.

5. Applications of Microarray Technology in Microbial Diagnostics

Microarray technology offers a great potential for answering many different experimental questions. The nature of the experimental question at hand is the main issue that has to be taken into consideration when developing a new microarray platform. Depending on this question, there is an initial decision on the nature of the probes and on the method for target labeling. This subsequently influences the selection of substrates and hybridization strategies. Some of the most common experimental questions for which microarray technology is used will be reviewed here.

5.1. Gene Expression Studies

The most widely used application of DNA microarrays is the study of transcriptional responses. Consequently, targets are derived from mRNA. Initial studies were limited to an organism or tissue of interest and provided insights into particular aspects of the organism's physiology (Schena et al. 1995; Gill et al. 2002; Zaigler, Schuster and Soppa 2003; Stowe-Evans, Ford and Kehoe 2004). For these studies, probes representing the genetic profile of the organism or tissue of interest were used (e.g., clone libraries containing either gDNA or cDNA fragments). Recent developments of microarray technology enabled the environmental monitoring of gene expression. Even though these studies are still limited to a few genes of interest, they can provide valuable information regarding the functionality of the whole microbial community (Zhou and Thompson 2002).

More recently, and benefiting from advanced target amplification methods (see Section 23.3.3.1), Garzoni et al. (2007) could monitor genome-wide bacterial transcription changes after *S. aureus* was ingested by nonprofessional phagocyte cells. Similar approaches (Lucchini et al. 2005), sometimes coupled to proteomics (Scherl et al. 2006), have proven instrumental for deciphering genes and/or pathways involved in physiologic adaptation—including the establishment of chronic infections or the development of antimicrobial resistance mechanisms. These techniques are valuable for developing targeted diagnostic tools and discovering focused therapeutic interventions.

Finally, Relman et al. proposed an original diagnostic approach focusing exclusively on the host innate response. Instead of identifying the pathogen itself, the authors aim at detecting pathogens through the elicited immune response, by studying the gene expression profiles of circulating monocytes (Boldrick et al. 2002). This approach, however, remains plagued by challenges and complexities that have yet to be adequately addressed. The rapidly changing nature over time of acute infectious diseases in a host, and the genetic diversity of microbial pathogens present unique problems for the design and interpretation of functional-genomic studies in this field. In addition, there are the more common problems related to heterogeneity within clinical samples, the complex, nonstandardized confounding variables associated with human subjects, and the complexities posed by the analysis and validation of highly parallel data (Liu et al. 2006).

5.2. Comparative Genome Hybridizations (CGH)

Traditional phylogenetic classification of bacteria to study evolutionary relatedness is based on the characterization of a limited number of genes, rRNA, or signature sequences. However, owing to the acquisition of DNA through lateral gene transfer, the differences between closely related bacterial strains can be vast (Dorrell, Hinchliffe and Wren 2005). By contrast, whole-genome sequencing comparisons allow a multitude of genes to be compared. Unfortunately, whole-genome sequencing is currently too expensive to allow the comparison of a large number of isolates of a species in a high-throughput scenario, as the global surveillance of infectious diseases requires. Therefore, since microbial genotyping is increasingly being used to track infectious diseases as they spread in human populations, another usage of microarrays has emerged. Comparative genome hybridization (CGH) permits assessing the genetic similarities and differences between closely related organisms. This approach is an adaptation of array methods used in gene expression studies, but applied to total genomic DNA (Murray et al. 2001; Lindroos et al. 2005). CGH enables a "bird's-eye view" of all the genes absent or present in a given genome compared to the reference genome on the microarray. Whole-genome comparisons typically identify sets of "core genes," which are shared by all strains in a species; and "accessory genes," which are present in one or more strains in a species, and often result from gene acquisition. It is these differences that can often be used to identify genes and/or genetic islands related to "gain-of-function traits" in pathogenic strains (Dorrell, Hinchliffe and Wren 2005).

CGH approaches can be applied to further characterize strains and to identify novel marker genes and chromosomal regions specific for given groups of isolates, thus providing better discrimination and additional information compared to classical genotyping methods (Koessler et al. 2006). However, ambiguities in the interpretation of the ratios of hybridization and cross-hybridization to paralogous genes remain important limitations of the technique (van Bakel and Holstege 2004). Solid statistical criteria for the absence or presence of ORFs are still lacking as a result of the diversity of the microarray design approaches, affecting the meta-analysis of the data obtained by different investigators (Joyce et al. 2002).

Garaizar et al. (2006) concluded that the construction or purchase of DNA microarrays and the performance of strain-to-strain hybridization experiments are still prohibitively expensive for routine application. The future use of arrays in epidemiology is likely to depend on the development of more cost-effective protocols, more robust and simplified formats, and the adequate evaluation of their performance (efficacy) and convenience (efficiency), compared with other genotyping methods. Indeed, more focused assays are finding broad application in routine bacterial epidemiology. Using commercially-available low-density microarrays fitted within microtubes (www.clondiag.com), papers have reported the development and validation of assays for bacterial genotyping and virulence gene detection (Korczak 2005) as well as for extensive detection of antimicrobial resistance determinants in Gram-positive bacteria (Perreten et al. 2005). Different low-density array formats permitted the recent release of the first commercially available microarray-based assay for the genotyping of human papilloma viruses, HPV (www.genomica.es) (www.greinerbioone.com).

5.3. Generic or Universal Microarrays

Combinatorial strategies refer to methods developed to make microarrays containing all sequences of a given length (also referred to as "generic arrays"). Combinatorial arrays have been promoted mostly to study large-scale hybridization behavior (Southern et al. 1994) or for solid-state nucleic acid sequencing (Drmanac and Drmanac 1999; Drmanac et al. 1993; Drmanac et al. 1998; Strezoska et al. 1991; Macevicz 1991; Lipshutz et al. 1999).

Generic arrays have been proposed as an inexpensive alternative to sequencing. Using all possible combinations of an n-mer allows "walking" at every position along a nucleotide sequence. This approach is currently limited by the complexity of the algorithm required to generate contigs (conversion of a listing of hybridized n-mers into a meaningful sequence). Also, if the sequence undergoing analysis contains a repeat region (the same sequence appearing more than once within the target molecule), the reconstruction diagram will have to display a corresponding number of branching points, leading to an ambiguous sequence. This type of array is the only one capable of detecting sequences that are lacking in large electronic libraries. In contrast, dedicated arrays are used for repetitive sequencing (resequencing) of the same target for detection of nucleotide polymorphisms or functional mutations (see Section 23.5.4).

Universal arrays refer to strategies that can provide target identification without any a priori sequence knowledge. This approach has been named the Non-Cognate Hybridization System (NCHS) (Schrenzel and Hibbs 2003). By synthesizing all probes of a given length that can be generated by a combination of four nucleotides, microarrays could detect any single organism. Unfortunately, for generating realistic probe lengths (i.e., permitting unique sequence identification as well as adequate hybridization behavior), the total number of permutations of the four nucleotides would yield very large numbers of possible probes (i.e., $4^{13} = 67,108,864$ probes for 13-mers nucleotides). The synthesis of such large microarrays is currently technically impossible. Furthermore, bioinformatics tools that would be necessary for the microarray analysis of billons of probes are not yet available. Thus, reducing the probe set and the complexity of the analytical approach is warranted. Random reduction of the probe set to a smaller number of probes might result in potentially missing sequences characteristic for given pathogens. An alternative would be the pruning of low informative probes, i.e., probes that have no targets in a collection of selected micro-organisms. However, this approach yields ultimately to another cognate design, with the inherent limitations described above. Another way to reduce the probe set is to consider permutations of only two nucleotides. This divides the probe set by $2^{(n-2)}$ where n is the length of the probes. This microarray design is truly noncognate and guarantees that no organism is a priori favored or missed. By analogy, Roth et al. have designed a universal microarray system combined with an enzymatic manipulation step that is capable of generating expression profiles from any organism without requiring a priori sequence-specific knowledge of transcript sequences (Roth et al. 1998).

Finally, one should mention here "multipurpose" arrays that contain probes for detecting a series of molecular barcodes. These molecular barcodes can be used as tags in various experimental conditions, for probing different targets under various experimental formats. These multipurpose arrays serve only to quantitatively detect the presence of barcoded probes, whose specificity is determined by the user. Molecular inversion probe (MIP) technology was initially developed for the detection of single nucleotide polymorphisms (SNPs) in human genes (Hardenbol et al. 2003). MIP technology has been shown to work well for multiplexing, i.e., massive parallel processing (12,000 MIPs in the same reaction tube) (Hardenbol et al. 2005). The power and versatility of MIP technology makes it perfectly suited for the identification and quantification of microbes. MIP's high sensitivity and specificity in detecting large numbers of SNPs (Hardenbol et al. 2003; Hardenbol et al. 2005) should allow one to harness this technology to detect a large number of pathogens and to identify multiple infections in an individual sample. A molecular inversion probe is comprised of genomic recognition sequences, common amplification sequences, and a molecular barcode for each genotype assigned to a specific gene. This probe is a linear oligonucleotide with target-complementary sequences at the ends and a noncomplementary linking segment in between (Thiyagarajan et al. 2006).

5.4. Microarrays for Sequence Analysis

Microarray technology has been used successfully for sequence analysis. These applications are technologically very diverse and will not be discussed in detail here. Usually, on-chip sequence analysis involves fabrication of high-density short oligonucleotide microarrays. These arrays contain either all possible oligonucleotides of a given length (Bains and Smith 1988) or they display a range of oligonucleotides covering a DNA sequence of interest, employing the so-called "tiling strategy" (Hacia 1999). Sequence analysis is performed by comparison of the hybridization patterns of the reference vs. the test sample. Such microarrays have been used for sequencing (Yershov et al. 1996; Hacia 1999), for the detection of single nucleotide polymorphisms (Hacia 1999; Lindroos et al. 2001), and for the analysis of secondary structures (Sohail, Akhtar and Southern 1999). Alternative detection methods have been suggested, such as using a labeled common oligonucleotide primer that is extended to the site of the match or mismatch (Lodes et al. 2006). Publications have already started addressing the issue of automated interpretation of resequencing on microarrays (Malanoski et al. 1995).

Recent advances in high-density oligonucleotide arrays have enabled the development of high-throughput resequencing techniques. Resequencing arrays are designed to cover the entire genome by overlapping oligonucleotides. Multiple versions of each oligonucleotide are spotted on the array to represent the four possible base combinations (A, T, G, and C) for each nucleotide position. To date, this technique has been applied to *Bacillus anthracis*, with 56 strains being resequenced using a custom-designed resequencing array (Zwick et al. 2005). The same technique has also been used to track the evolution of the severe acute respiratory syndrome coronavirus (Wong et al. 2004). Conversely, similar tiling-based resequencing has been employed for *Staphylococcus* speciation (Couzinet et al. 2005a), genotyping (van

Leeuween et al. 2003), and the detection of mutations conferring resistance to quinolone antimicrobials (Couzinet et al. 2005b).

Finally, other bead-based arrays have been suggested for high-throughput sequencing-based approaches (Brenner et al. 2000). The latter approach resulted in a powerful high-throughput sequencing platform (Illumina-Solexa) that currently competes against the pyrosequencing method described by Margulies et al. (Margulies et al. 2005). One can certainly conclude that advances in sequencing technologies currently allow the complete decoding of an entire microbial genome in a few hours (Margulies et al. 2005), clearly out-competing arrays for the detection of SNPs, albeit at a substantially higher cost. In any event, more accessible, technically robust, and, above all, cheaper formats are needed before the broad application of any of these technologies in a clinical and epidemiological surveillance setting.

5.5. Microbial Diagnostic Microarrays

Microbial diagnostic microarrays (MDMs) are used for the simultaneous identification of microorganisms in clinical or environmental samples. Probes used for MDMs are usually oligonucleotides designed to be specific for a given strain, subspecies, species, genus, or higher taxon. Classification and nomenclature of MDMs throughout the literature is not unanimous. According to their intended use, environmental MDMs (Bodrossy et al. 2003; Loy et al. 2005) and detection/identification MDMs can be distinguished (Hashsham et al. 2004; Sergeev et al. 2004). The main difference between these two MDM types relies in their detection requirements.

Environmental MDMs are generally used to assess the whole microbial community structure or a subset of the microbial community in a particular environment. Therefore, reliable parallel detection of many different microorganisms and the potential for some level of quantification are required. Detection/identification MDMs are primarily used in clinical (medical, veterinary), food, and biodefense microbiology. For this purpose, highly sensitive and specific detection of a few microorganisms in a complex community is required. According to the nature of the marker gene, one can further distinguish phylogenetic (Rudi et al. 2003; Loy et al. 2005; Brodie et al. 2006) and functional MDMs, also referred to as functional gene arrays (Wu et al. 2001; Bodrossy et al. 2003; Taroncher-Oldenburg et al. 2003).

There are two concepts for the sensitivity of MDMs, both of them potentially posing a bottleneck to the detection of the targeted microbes. Absolute sensitivity refers to the amount of target DNA or the number of target microbes required for successful detection. Absolute sensitivity reflects the hybridization capacity and detection sensitivity of the microarray platform used. Relative sensitivity, on the other hand, refers to the ratio of the targeted microbe within the entire microbial community analyzed. It is primarily due to low level nonspecific background signal accumulation and to the fact that the amount of target DNA applicable in microarray hybridizations is limited.

A promising approach to increase the sensitivity of a microarray assay is tyramide signal amplification (TSA) (Denef et al. 2003). Upon hybridization, this method relies on enzymatic amplification of the signal by employing the horseradish peroxidase-mediated deposition of fluorochrome-labeled tyramides at the location of the probe. The relative sensitivity can be improved by limiting the labeling to very short, specific regions (Kostic et al. 2007; Ballmer et al. 2007; Rudi et al. 2003). High density microarrays employing multiple perfect match/mismatch probe sets for each targeted microbe also enable a significant improvement in relative sensitivity (Brodie et al. 2006). A novel method to analyze microarray data holds promise for a significant improvement in terms of the relative sensitivity of MDMs (Marcelino et al. 2006).

The ultimate specificity of microarray technology depends on the discrimination between a fully complementary target and a nontarget differing in only one single nucleotide. Various enzyme-assisted hybridization strategies, also used in single nucleotide polymorphism and resequencing assays (Ericsson et al. 2003; Lindroos et al. 2002), are being applied because of their promise in strongly discriminating single mismatches located near the 3′ end of microarray probes (Rudi et al. 2003; Cassat et al. 2005; Halperin, Buhot and Zhulina 2005; Lipshutz et al. 1995; Maskos and Southern 1993a; Porwollik et al. 2003; Proudnikov and Mirzabekov 1996).

Isotope microarrays represent further development of the traditional phylogenetic MDMs, enabling linking phylogeny (community structure) to function. This approach employs "double-labeled" targets, where the first radioactive labeling is substrate-mediated, and the second labeling is performed according to standard microarray protocols (Adamczyk et al. 2003).

6. Conclusions

The development of microarray-based bacterial identification systems starting directly from the biological sample, without any enzymatic target amplification, would be most welcome (Straub and Chandler 2003). During a single hybridization, arrays can integrate probes that provide microbial identification and also enclose large sets of discriminative epidemiological markers (Call et al. 2003; Gingeras et al. 1998; van Leeuwen et al. 2003; Volokhov et al. 2002; Yue et al. 2001), or contain probes to detect virulence or antimicrobial resistance determinants (Bekal et al. 2003; Chizhikov et al. 2001; Couzinet et al. 2005b; Korczak et al. 2005; Troesch et al. 1999). The high parallelism of microarray systems appears particularity adapted for this, provided the systems' design includes the following: (1) targeted universal gene(s), (2) simplified coupling and labeling protocols, (3) exquisite sensitivity, and (4) an adapted analysis strategy.

Clinical bacteriology is witnessing a revolution. We are shifting away from Koch's postulates that required evidence of bacterial growth to the routine use of advanced molecular biology tools. Revolutions bring important changes in the way we think. The extended use of molecular biology will help discover new species that could not be cultivated. Quantitative aspects of bacterial loads, easily addressed on petri dishes, will need to be transposed and validated using molecular biology tools. This revolution should now be supported also by a better appreciation of the remarkable diversity of the bacterial world; the understanding that adequate taxonomy is needed; and the knowledge that host-pathogen interactions will have to be addressed perhaps more straightforward without last part.

References

Adamczyk J, Hesselsoe M, Iversen N, Horn M, Lehner A, Nielsen PH, Schloter M, Roslev P, Wagner M (2003) The isotope array, a new tool that employs substrate-mediated labeling of rRNA for determination of microbial community structure and function. Appl Environ Microbiol 69:6875–6887

Ahn HJ, La HJ, Forney LJ (2006) System for determining the relative fitness of multiple bacterial populations without using selective markers. Appl Environ Microbiol 72:7383–7385

Aitman TJ (2001) DNA microarrays in medical practice. BMJ 323:611–615

Albert TJ, Norton J, Ott M, Richmond T, Nuwaysir K, Nuwaysir EF, Stengele KP, Green RD (2003) Light-directed 5′ → 3′ synthesis of complex oligonucleotide microarrays. Nucleic Acids Res 31:e35

Alizadeh AA, Eisen MB, Davis RE, Ma C, Lossos IS, Rosenwald A, Boldrick JC, Sabet H, Tran T, Yu X, Powell JI, Yang L, Marti GE, Moore T, Hudson J Jr, Lu L, Lewis DB, Tibshirani R, Sherlock G, Chan WC, Greiner TC, Weisenburger DD, Armitage JO, Warnke R, Staudt LM (2000) Distinct types of diffuse large B-cell lymphoma identified by gene expression profiling. Nature 403:503–511

Allawi HT, SantaLucia J Jr (1998) Nearest-neighbor thermodynamics of internal AC mismatches in DNA: sequence dependence and pH effects. Biochemistry 37:9435–9444

Anthony RM, Brown TJ, French GL (2000) Rapid diagnosis of bacteremia by universal amplification of 23S ribosomal DNA followed by hybridization to an oligonucleotide array. J Clin Microbiol 38:781–788

Aoyagi K, Tatsuta T, Nishigaki M, Akimoto S, Tanabe C, Omoto Y, Hayashi S, Sakamoto H, Sakamoto M, Yoshida T, Terada M, Sasaki H (2003) A faithful method for PCR-mediated global mRNA amplification and its integration into microarray analysis on laser-captured cells. Biochem Biophys Res Commun 300:915–920

Armitage BA (2003) The impact of nucleic acid secondary structure on PNA hybridization. Drug Discov Today 8:222–228

Bae JW, Rhee SK, Nam YD, Park YH (2005) Generation of subspecies level-specific microbial diagnostic microarrays using genes amplified from subtractive suppression hybridization as microarray probes. Nucleic Acids Res 33:e113

Bailly X, Bena G, Lenief V, de Lajudie P, Avarre JC (2006) Development of a lab-made microarray for analyzing the genetic diversity of nitrogen fixing symbionts *Sinorhizobium meliloti* and *Sinorhizobium medicae*. J Microbiol Methods 67:114–124

Bains W, Smith GC (1988) A novel method for nucleic acid sequence determination. J Theor Biol 135:303–307

Baldeschwieler JD, Gamble RC, Thierault TP (1995) Method and apparatus for performing multiple sequential reactions on a matrix. WO9525116, USA, Ref Type: Patent

Ballmer K, Korczak BM, Kuhnert P, Slickers P, Ehricht R, Hachler H (2007) Fast DNA-serotyping of *Escherichia coli* by oligonucleotide microarray. J Clin Microbiol 45:370–379

Barczak A, Rodriguez MW, Hanspers K, Koth LL, Tai YC, Bolstad BM, Speed TP, Erle DJ (2003) Spotted long oligonucleotide arrays for human gene expression analysis. Genome Res 13:1775–1785

Bavykin SG, Akowski JP, Zakhariev VM, Barsky VE, Perov AN, Mirzabekov AD (2001) Portable system for microbial sample preparation and oligonucleotide microarray analysis. Appl Environ Microbiol 67:922–928

Bekal S, Brousseau R, Masson L, Prefontaine G, Fairbrother J, Harel J (2003) Rapid identification of *Escherichia coli* pathotypes by virulence gene detection with DNA microarrays. J Clin Microbiol 41:2113–2125

Blanchard AP, Friend SH (1999) Cheap DNA arrays—it's not all smoke and mirrors. Nat Biotechnol 17:953

Blanchard AP, Kaiser RJ, Hood LE (1996) High-density oligonucleotide arrays. Biosensors & Bioelectronics 11:687–690

Bo T, Jonassen I (2002) New feature subset selection procedures for classification of expression profiles. Genome Biol 3:RESEARCH0017

Bodrossy L (2003) Diagnostic oligonucleotide microarrays for microbiology. In: Blalock E (ed) A Beginner's Guide to Microarrays, Kluwer Academic Publishers, New York, pp 43–92

Bodrossy L, Sessitsch A (2004) Oligonucleotide microarrays in microbial diagnostics. Curr Opin Microbiol 7:245–254

Bodrossy L, Stralis-Pavese N, Konrad-Koszler M, Weilharter A, Reichenauer TG, Schofer D, Sessitsch A (2006) mRNA-based parallel detection of active methanotroph populations by use of a diagnostic microarray. Appl Environ Microbiol 72:1672–1676

Bodrossy L, Stralis-Pavese N, Murrell JC, Radajewski S, Weilharter A, Sessitsch A (2003) Development and validation of a diagnostic microbial microarray for methanotrophs. Environ Microbiol 5:566–582

Boldrick JC, Alizadeh AA, Diehn M, Dudoit S, Liu CL, Belcher CE, Botstein D, Staudt L, Brown PO, Relman DA (2002) Stereotyped and specific gene expression programs in human innate immune responses to bacteria. Proc Natl Acad Sci U S A 99:972–977

Borucki MK, Krug MJ, Muraoka WT, Call DR (2003) Discrimination among *Listeria monocytogenes* isolates using a mixed genome DNA microarray. Vet Microbiol 92:351–362

Borucki MK, Reynolds J, Call DR, Ward TJ, Page B, Kadushin J (2005) Suspension microarray with dendrimer signal amplification allows direct and high-throughput subtyping of *Listeria monocytogenes* from genomic DNA. J Clin Microbiol 43:3255–3259

Brennan TM, Heinecker H (1995) Methods and compositions for determining the sequence of nucleic acids. 5′474′796, USA, Ref Type: Patent

Brenner S, Johnson M, Bridgham J, Golda G, Lloyd DH, Johnson D, Luo S, McCurdy S, Foy M, Ewan M, Roth R, George D, Eletr S, Albrecht G, Vermaas E, Williams SR, Moon K, Burcham T, Pallas M, DuBridge RB, Kirchner J, Fearon K, Mao J, Corcoran K (2000) Gene expression analysis by massively parallel signature sequencing (MPSS) on microbead arrays. Nat Biotechnol 18:630–634

Brodie EL, DeSantis TZ, Joyner DC, Baek SM, Larsen JT, Andersen GL, Hazen TC, Richardson PM, Herman DJ, Tokunaga TK, Wan JM, Firestone MK (2006) Application of a high-density oligonucleotide microarray approach to study bacterial population dynamics during uranium reduction and reoxidation. Appl Environ Microbiol 72:6288–6298

Call DR, Bakko MK, Krug MJ, Roberts MC (2003) Identifying antimicrobial resistance genes with DNA microarrays. Antimicrob Agents Chemother 47:3290–3295

Call DR, Borucki MK, Loge FJ (2003) Detection of bacterial pathogens in environmental samples using DNA microarrays. J Microbiol Methods 53:235–243

Case-Green SC, Southern EM (1994) Studies on the base pairing properties of deoxyinosine by solid phase hybridisation to oligonucleotides. Nucleic Acids Res 22:131–136

Cassat JE, Dunman PM, McAleese F, Murphy E, Projan SJ, Smeltzer MS (2005) Comparative genomics of *Staphylococcus aureus* musculoskeletal isolates. J Bacteriol 187:576–592

Charbonnier Y, Gettler BM, Francois P, Bento M, Renzoni A, Vaudaux P, Schlegel W, Schrenzel J (2005) A generic approach for the design of whole-genome oligoarrays, validated for genomotyping, deletion mapping and gene expression analysis on *Staphylococcus aureus*. BMC Genomics 6:95

Chizhikov V, Rasooly A, Chumakov K, Levy DD (2001) Microarray analysis of microbial virulence factors. Appl Environ Microbiol 67:3258–3263

Couzinet S, Jay C, Barras C, Vachon R, Vernet G, Ninet B, Jan I, Minazio MA, Francois P, Lew D, Troesch A, Schrenzel J (2005a) High-density DNA probe arrays for identification of staphylococci to the species level. J Microbiol Methods 61:201–208

Couzinet S, Yugueros J, Barras C, Visomblin N, Francois P, Lacroix B, Vernet G, Lew D, Troesch A, Schrenzel J, Jay C (2005b) Evaluation of a high-density oligonucleotide array for characterization of *grlA*, *grlB*, *gyrA* and *gyrB* mutations in fluoroquinolone resistant *Staphylococcus aureus* isolates. J Microbiol Methods 60:275–279

Debouck C, Goodfellow PN 1999 DNA microarrays in drug discovery and development. Nat Genet 21:48–50

Denef VJ, Park J, Rodrigues JL, Tsoi TV, Hashsham SA, Tiedje JM (2003) Validation of a more sensitive method for using spotted oligonucleotide DNA microarrays for functional genomics studies on bacterial communities. Environ Microbiol 5:933–943

Diehl F, Grahlmann S, Beier M, Hoheisel JD (2001) Manufacturing DNA microarrays of high spot homogeneity and reduced background signal. Nucleic Acids Res 29:E38

Dorrell N, Hinchliffe SJ, Wren BW (2005) Comparative phylogenomics of pathogenic bacteria by microarray analysis. Curr Opin Microbiol 8:620–626

Drmanac R, Crkvenjakov R (1993) Method of sequencing of genomes by hybridization of oligonucleotide probes. US1991000723712 USA, Ref Type: Patent

Drmanac R, Drmanac S (1999) cDNA screening by array hybridization. Methods Enzymol 303:165–78

Drmanac R, Drmanac S, Strezoska Z, Paunesku T, Labat I, Zeremski M, Snoddy J, Funkhouser WK, Koop B, Hood L (1993) DNA sequence determination by hybridization: a strategy for efficient large-scale sequencing. Science 260:1649–1652

Drmanac S, Kita D, Labat I, Hauser B, Schmidt C, Burczak JD, Drmanac R (1998) Accurate sequencing by hybridization for DNA diagnostics and individual genomics. Nat Biotechnol 16:54–58

Dunbar SA, Vander Zee CA, Oliver KG, Karem KL, Jacobson JW (2003) Quantitative, multiplexed detection of bacterial pathogens: DNA and protein applications of the Luminex LabMAP system. J Microbiol Methods 53:245–252

Duveneck GL, Bopp MA, Ehrat M, Balet LP, Haiml M, Keller U, Marowsky G, Soria S (2003) Two-photon fluorescence excitation of macroscopic areas on planar waveguides. Biosens Bioelectron 18:503–510

Epstein JR, Ferguson JA, Lee KH, Walt DR (2003a) Combinatorial decoding: an approach for universal DNA array fabrication. J Am Chem Soc 125:13753–13759

Epstein JR, Leung AP, Lee KH, Walt DR (2003b) High-density, microsphere-based fiber optic DNA microarrays. Biosens Bioelectron 18:541–546

Epstein JR, Walt DR (2003) Fluorescence-based fibre optic arrays: a universal platform for sensing. Chem Soc Rev 32:203–214

Ericsson O, Sivertsson A, Lundeberg J, Ahmadian A (2003) Microarray-based resequencing by apyrase-mediated allele-specific extension. Electrophoresis 24:3330–3338

Fan J, Chen Y, Chan HM, Tam PK, Ren Y (2005) Removing intensity effects and identifying significant genes for Affymetrix arrays in macrophage migration inhibitory factor-suppressed neuroblastoma cells. Proc Natl Acad Sci U S A 102:17751–17756

Fan J, Tam P, Woude GV, Ren Y (2004) Normalization and analysis of cDNA microarrays using within-array replications applied to neuroblastoma cell response to a cytokine. Proc Natl Acad Sci U S A 101:1135–1140

Fan JB, Chen X, Halushka MK, Berno A, Huang X, Ryder T, Lipshutz RJ, Lockhart DJ, Chakravarti A (2000) Parallel genotyping of human SNPs using generic high-density oligonucleotide tag arrays. Genome Res 10:853–860

Ferguson JA, Steemers FJ, Walt DR (2000) High-density fiber-optic DNA random microsphere array. Anal Chem 72:5618–5624

Fodor SP, Rava RP, Huang XC, Pease AC, Holmes CP, Adams CL (1993) Multiplexed biochemical assays with biological chips. Nature 364:555–556

Fodor SP, Read JL, Pirrung MC, Stryer L, Lu AT, Solas D (1991) Light-directed, spatially addressable parallel chemical synthesis. Science 251:767–773

Francois P, Bento M, Vaudaux P, Schrenzel J (2003) Comparison of fluorescence and resonance light scattering for highly sensitive microarray detection of bacterial pathogens. J Microbiol Methods 55:755–762

Francois P, Charbonnier Y, Jaquet J, Utinger D, Bento M, Lew DP, Kresbach G, Schlegel W, Schrenzel J (2005) Rapid bacterial identification using evanescent waveguide oligonucleotide microarray classification. J Microbiol Methods 65:390–403

Francois P, Garzoni C, Bento M, Schrenzel J (2007) Comparison of amplification methods for transcriptomic analysis of low abundance prokaryotic RNA sources. J Microbiol Methods 68(2):385–91

Franke-Whittle IH, Klammer SH, Mayrhofer S, Insam H (2006) Comparison of different labeling methods for the production of labeled target DNA for microarray hybridization. J Microbiol Methods 65:117–126

Fredericks DN, Relman DA (1996) Sequence-based identification of microbial pathogens: a reconsideration of Koch's postulates. Clin Microbiol Rev 9:18–33

Garaizar J, Rementeria A, Porwollik S (2006) DNA microarray technology: a new tool for the epidemiological typing of bacterial pathogens? FEMS Immunol Med Microbiol 47:178–189

Garzoni C, Francois P, Couzinet S, Tapparel C, Charbonnier Y, Huyghe A, Renzoni A, Lucchini S, Lew DP, Vaudaux P, Kelley WL, Schrenzel J (2007) A global view of the *Staphylococcus aureus* whole genome expression upon internalization in human epithelial cells, BMC Genomics 14:171

Gasch AP, Spellman PT, Kao CM, Carmel-Harel O, Eisen MB, Storz G, Botstein D, Brown PO (2000) Genomic expression programs in the response of yeast cells to environmental changes. Mol Biol Cell 11:4241–4257

Gill RT, Katsoulakis E, Schmitt W, Taroncher-Oldenburg G, Misra J, Stephanopoulos G (2002) Genome-wide dynamic transcriptional profiling of the light-to-dark transition in *Synechocystis* sp strain PCC 6803. J Bacteriol 184: 3671–3681

Gillepsie D, Spiegelman SA (1965) A quantitative assay for DNA-RNA hybrids with DNA immobilized on a membrane. J Mol Biol 12:829–842

Gingeras TR, Ghandour G, Wang E, Berno A, Small PM, Drobniewski F, Alland D, Desmond E, Holodniy M, Drenkow J (1998) Simultaneous genotyping and species identification using hybridization pattern recognition analysis of generic *Mycobacterium* DNA arrays. Genome Res 8:435–448

Golub TR, Slonim DK, Tamayo P, Huard C, Gaasenbeek M, Mesirov JP, Coller H, Loh ML, Downing JR, Caligiuri MA, Bloomfield CD, Lander ES (1999) Molecular classification of cancer: class discovery and class prediction by gene expression monitoring. Science 286:531–537

Grifantini R, Bartolini E, Muzzi A, Draghi M, Frigimelica E, Berger J, Ratti G, Petracca R, Galli G, Agnusdei M, Giuliani MM, Santini L, Brunelli B, Tettelin H, Rappuoli R, Randazzo F, Grandi G (2002) Previously unrecognized vaccine candidates against group B *meningococcus* identified by DNA microarrays. Nat Biotechnol 20:914–921

Guschin D, Yershov G, Zaslavsky A, Gemmell A, Shick V, Proudnikov D, Arenkov P, Mirzabekov A (1997) Manual manufacturing of oligonucleotide, DNA, and protein microchips. Anal Biochem 250:203–211

Hacia JG (1999) Resequencing and mutational analysis using oligonucleotide microarrays. Nat Genet 21:42–47

Halperin A, Buhot A, Zhulina EB (2005) Brush effects on DNA chips: thermodynamics, kinetics, and design guidelines. Biophys J 89:796–811

Hardenbol P, Baner J, Jain M, Nilsson M, Namsaraev EA, Karlin-Neumann GA, Fakhrai-Rad H, Ronaghi M, Willis TD, Landegren U, Davis RW (2003) Multiplexed genotyping with sequence-tagged molecular inversion probes. Nat Biotechnol 21:673–678

Hardenbol P, Yu F, Belmont J, Mackenzie J, Bruckner C, Brundage T, Boudreau A, Chow S, Eberle J, Erbilgin A, Falkowski M, Fitzgerald R, Ghose S, Iartchouk O, Jain M, Karlin-Neumann G, Lu X, Miao X, Moore B, Moorhead M, Namsaraev E, Pasternak S, Prakash E, Tran K, Wang Z, Jones HB, Davis RW, Willis TD, Gibbs RA (2005) Highly multiplexed molecular inversion probe genotyping: over 10,000 targeted SNPs genotyped in a single tube assay. Genome Res 15:269–275

Hashsham SA, Wick LM, Rouillard JM, Gulari E, Tiedje JM (2004) Potential of DNA microarrays for developing parallel detection tools (PDTs) for microorganisms relevant to biodefense and related research needs. Biosens Bioelectron 20:668–683

Hessner MJ, Meyer L, Tackes J, Muheisen S, Wang X (2004) Immobilized probe and glass surface chemistry as variables in microarray fabrication. BMC Genomics 5:53

Holzman T, Kolker E (2004) Statistical analysis of global gene expression data: some practical considerations. Curr Opin Biotechnol 15:52–57

Hughes TR, Mao M, Jones AR, Burchard J, Marton MJ, Shannon KW, Ziman M, Meyer MR, Kobayashi S, Dai H, He YD, Stephaniants SB, Cavet G, Walker WL, West A, Coffey E, Shoemaker DD, Stoughton R, Blanchard AP, Friend SH, Linsley PS (2001) Expression profiling using microarrays fabricated by an ink-jet oligonucleotide synthesizer. Nat Biotechnol 19:342–347

Jaeger J, Spang R (2006) Selecting normalization genes for small diagnostic microarrays. BMC Bioinformatics 7:388

Joyce EA, Chan K, Salama NR, Falkow S (2002) Redefining bacterial populations: a post-genomic reformation. Nat Rev Genet 3:462–473

Kafatos FC, Jones CW, Efstratiadis A (1979) Determination of nucleic acid sequence homologies and relative concentrations by a dot hybridization procedure. Nucleic Acids Res 7:1541–1552

Kajiyama T, Miyahara Y, Kricka LJ, Wilding P, Graves DJ, Surrey S, Fortina P (2003) Genotyping on a thermal gradient DNA chip. Genome Res 13:467–475

Kakinuma K, Fukushima M, Kawaguchi R (2003) Detection and identification of *Escherichia coli*, *Shigella*, and *Salmonella* by microarrays using the *gyrB* gene. Biotechnol Bioeng 83:721–728

Kelly JJ, Chernov BK, Tovstanovsky I, Mirzabekov AD, Bavykin SG (2002) Radical-generating coordination complexes as tools for rapid and effective fragmentation and fluorescent labeling of nucleic acids for microchip hybridization. Anal Biochem 311:103–118

Kessler N, Ferraris O, Palmer K, Marsh W, Steel A (2004) Use of the DNA flow-thru chip, a three-dimensional biochip, for typing and subtyping of influenza viruses. J Clin Microbiol 42:2173–2185

Knight J (2001) When the chips are down. Nature 410:860–861

Koessler T, Francois P, Charbonnier Y, Huyghe A, Bento M, Dharan S, Renzi G, Lew D, Harbarth S, Pittet D, Schrenzel J (2006) Use of oligoarrays for characterization of community-onset methicillin-resistant *Staphylococcus aureus*. J Clin Microbiol 44:1040–1048

Korczak B, Frey J, Schrenzel J, Pluschke G, Pfister R, Ehricht R, Kuhnert P (2005) Use of diagnostic microarrays for determination of virulence gene patterns of *Escherichia coli* K1, a major cause of neonatal meningitis. J Clin Microbiol 43:1024–1031

Kostic T, Weilharter A, Rubino S, Delogu G, Rudi K, Sessitsch A, Bodrossy L (2007) A microbial diagnostic microarray technique for the detection and identification of pathogenic bacteria in a background of non-pathogens. Anal Biochem 360:244–254

Kothapalli R, Yoder SJ, Mane S, Loughran TP Jr (2002) Microarray results: how accurate are they? BMC Bioinformatics 3:22

Kroll TC, Wolfl S (2002) Ranking: a closer look on globalisation methods for normalisation of gene expression arrays. Nucleic Acids Res 30:e50

Li J, Pankratz M, Johnson JA (2002) Differential gene expression patterns revealed by oligonucleotide versus long cDNA arrays. Toxicol Sci 69:383–390

Li W (2005) How many genes are needed for early detection of breast cancer, based on gene expression patterns in peripheral blood cells? Breast Cancer Res 7:E5

Lima WF, Monia BP, Ecker DJ, Freier SM (1992) Implication of RNA structure on antisense oligonucleotide hybridization kinetics. Biochemistry 31:12055–12061

Lindroos HL, Mira A, Repsilber D, Vinnere O, Naslund K, Dehio M, Dehio C, Andersson SG (2005) Characterization of the genome composition of *Bartonella koehlerae* by microarray comparative genomic hybridization profiling. J Bacteriol 187:6155–6165

Lindroos K, Liljedahl U, Raitio M, Syvanen AC (2001) Minisequencing on oligonucleotide microarrays: comparison of immobilisation chemistries. Nucleic Acids Res 29:E69

Lindroos K, Sigurdsson S, Johansson K, Ronnblom L, Syvanen AC (2002) Multiplex SNP genotyping in pooled DNA samples by a four-colour microarray system. Nucleic Acids Res 30:e70

Lipshutz RJ, Fodor SP, Gingeras TR, Lockhart DJ (1999) High density synthetic oligonucleotide arrays. Nat Genet 21:20–24

Lipshutz RJ, Morris D, Chee M, Hubbell E, Kozal MJ, Shah N, Shen N, Yang R, Fodor SP (1995) Using oligonucleotide probe arrays to access genetic diversity. Biotechniques 19:442–447

Liu M, Popper SJ, Rubins KH, Relman DA (2006) Early days: genomics and human responses to infection. Curr Opin Microbiol 9:312–319

Liu WT, Guo H, Wu JH (2007) Effects of target length on the hybridization efficiency and specificity of rRNA-based oligonucleotide microarrays. Appl Environ Microbiol 73:73–82

Liu WT, Mirzabekov AD, Stahl DA (2001) Optimization of an oligonucleotide microchip for microbial identification studies: a non-equilibrium dissociation approach. Environ Microbiol 3:619–629

Lockhart DJ, Dong H, Byrne MC, Follettie MT, Gallo MV, Chee MS, Mittmann M, Wang C, Kobayashi M, Horton H, Brown EL (1996) Expression monitoring by hybridization to high-density oligonucleotide arrays. Nat Biotechnol 14:1675–1680

Lodes MJ, Suciu D, Elliott M, Stover AG, Ross M, Caraballo M, Dix K, Crye J, Webby RJ, Lyon WJ, Danley DL, McShea A (2006) Use of semiconductor-based oligonucleotide microarrays for influenza A virus subtype identification and sequencing. J Clin Microbiol 44:1209–1218

Loge FJ, Thompson DE, Call DR (2002) PCR detection of specific pathogens in water: a risk-based analysis. Environ Sci Technol 36:2754–2759

Loy A, Bodrossy L (2006) Highly parallel microbial diagnostics using oligonucleotide microarrays. Clin Chim Acta 363:106–119

Loy A, Lehner A, Lee N, Adamczyk J, Meier H, Ernst J, Schleifer KH, Wagner M (2002) Oligonucleotide microarray for 16S rRNA gene-based detection of all recognized lineages of sulfate-reducing prokaryotes in the environment. Appl Environ Microbiol 68:5064–5081

Loy A, Schulz C, Lucker S, Schopfer-Wendels A, Stoecker K, Baranyi C, Lehner A, Wagner M (2005) 16S rRNA gene-based oligonucleotide microarray for environmental monitoring of the betaproteobacterial order "Rhodocyclales." Appl Environ Microbiol 71:1373–1386

Lucchini S, Liu H, Jin Q, Hinton JC, Yu J (2005) Transcriptional adaptation of *Shigella flexneri* during infection of macrophages and epithelial cells: insights into the strategies of a cytosolic bacterial pathogen. Infect Immun 73:88–102

Lucchini S, Thompson A, Hinton JC (2001) Microarrays for microbiologists. Microbiology 147:1403–1414

Ludwig W, Strunk O, Westram R, Richter L, Meier H, Yadhukumar, Buchner A, Lai T, Steppi S, Jobb G, Forster W, Brettske I, Gerber S, Ginhart AW, Gross O, Grumann S, Hermann S, Jost R, Konig A, Liss T, Lussmann R, May M, Nonhoff B, Reichel B, Strehlow R, Stamatakis A, Stuckmann N, Vilbig A, Lenke M, Ludwig T, Bode A, Schleifer KH (2004) ARB: a software environment for sequence data. Nucleic Acids Res 32:1363–1371

Lynch JL, deSilva CJ, Peeva VK, Swanson NR (2006) Comparison of commercial probe labeling kits for microarray: towards quality assurance and consistency of reactions. Anal Biochem 355:224–231

Macevicz SC (1991) Nucleic acid sequence determination by multiple mixed oligonucleotide probes. US1988000261702 USA Ref Type: Patent

Malanoski GJ, Samore MH, Pefanis A, Karchmer AW (1995) *Staphylococcus aureus* catheter-associated bacteremia. Minimal effective therapy and unusual infectious complications associated with arterial sheath catheters. Arch Intern Med 155:1161–1166

Marcelino LA, Backman V, Donaldson A, Steadman C, Thompson JR, Preheim SP, Lien C, Lim E, Veneziano D, Polz MF (2006) Accurately quantifying low-abundant targets amid similar sequences by revealing hidden correlations in oligonucleotide microarray data. Proc Natl Acad Sci U S A 103:13629–13634

Margulies M, Egholm M, Altman WE, Attiya S, Bader JS, Bemben LA, Berka J, Braverman MS, Chen YJ, Chen Z, Dewell SB, Du L, Fierro JM, Gomes XV, Godwin BC, He W, Helgesen S, Ho CH, Irzyk GP, Jando SC, Alenquer ML, Jarvie TP, Jirage KB, Kim JB, Knight JR, Lanza JR, Leamon JH, Lefkowitz SM, Lei M, Li J, Lohman KL, Lu H, Makhijani VB, McDade KE, McKenna MP, Myers EW, Nickerson E, Nobile JR, Plant R, Puc BP, Ronan MT, Roth GT, Sarkis GJ, Simons JF, Simpson JW, Srinivasan M, Tartaro KR, Tomasz A, Vogt KA, Volkmer GA, Wang SH, Wang Y, Weiner MP, Yu P, Begley RF, Rothberg JM (2005) Genome sequencing in microfabricated high-density picolitre reactors. Nature 437:376–380

Maskos U, Southern EM (1993a) A novel method for the parallel analysis of multiple mutations in multiple samples. Nucleic Acids Res 21:2269–2270

Maskos U, Southern EM (1993b) A study of oligonucleotide reassociation using large arrays of oligonucleotides synthesised on a glass support. Nucleic Acids Res 21:4663–4669

Matveeva OV, Shabalina SA, Nemtsov VA, Tsodikov AD, Gesteland RF, Atkins JF (2003) Thermodynamic calculations and statistical correlations for oligo-probes design. Nucleic Acids Res 31:4211–4217

Mikulowska-Mennis A, Taylor TB, Vishnu P, Michie SA, Raja R, Horner N, Kunitake ST (2002) High-quality RNA from cells isolated by laser capture microdissection. Biotechniques 33:176–179

Mir KU, Southern EM (1999) Determining the influence of structure on hybridization using oligonucleotide arrays. Nat Biotechnol 17:788–792

Murray AE, Lies D, Li G, Nealson K, Zhou J, Tiedje JM (2001) DNA/DNA hybridization to microarrays reveals gene-specific differences between closely related microbial genomes. Proc Natl Acad Sci USA 98:9853–9858

Naef F, Magnasco MO (2003) Solving the riddle of the bright mismatches: labeling and effective binding in oligonucleotide arrays. Phys Rev E Stat Nonlin Soft Matter Phys 68:011906

Nguyen HK, Southern EM (2000) Minimising the secondary structure of DNA targets by incorporation of a modified deoxynucleoside: implications for nucleic acid analysis by hybridisation. Nucleic Acids Res 28:3904–3909

Pasternack RF, Collings PJ (1995) Resonance light scattering: a new technique for studying chromophore aggregation. Science 269:935–939

Pawlak M, Schick E, Bopp MA, Schneider MJ, Oroszlan P, Ehrat M (2002) Zeptosens' protein microarrays: a novel high performance microarray platform for low abundance protein analysis. Proteomics 2:383–393

Pease AC, Solas D, Sullivan EJ, Cronin MT, Holmes CP, Fodor SP (1994) Light-generated oligonucleotide arrays for rapid DNA sequence analysis. Proc Natl Acad Sci U S A 91:5022–5026

Peplies J, Glockner FO, Amann R (2003) Optimization strategies for DNA microarray-based detection of bacteria with 16S rRNA-targeting oligonucleotide probes. Appl Environ Microbiol 69:1397–1407

Perou CM, Sorlie T, Eisen MB, van de Rijn M, Jeffrey SS, Rees CA, Pollack JR, Ross DT, Johnsen H, Akslen LA, Fluge O, Pergamenschikov A, Williams C, Zhu SX, Lonning PE, Borresen-Dale AL, Brown PO, Botstein D (2000) Molecular portraits of human breast tumours. Nature 406:747–752

Perreten V, Vorlet-Fawer L, Slickers P, Ehricht R, Kuhnert P, Frey J (2005) Microarray-based detection of 90 antibiotic resistance genes of gram-positive bacteria. J Clin Microbiol 43:2291–2302

Peterson AW, Heaton RJ, Georgiadis RM (2001) The effect of surface probe density on DNA hybridization. Nucleic Acids Res 29:5163–5168

Phimister B (1999) Chipping forecast: going global. Nat Genet 21 Suppl:1

Porwollik S, Frye J, Florea LD, Blackmer F, McClelland M (2003) A non-redundant microarray of genes for two related bacteria. Nucleic Acids Res 31:1869–1876

Pozhitkov A, Chernov B, Yershov G, Noble PA (2005) Evaluation of gel-pad oligonucleotide microarray technology by using artificial neural networks. Appl Environ Microbiol 71:8663–8676

Pozhitkov A, Noble PA, Domazet-Loso T, Nolte AW, Sonnenberg R, Staehler P, Beier M, Tautz D (2006) Tests of rRNA hybridization to microarrays suggest that hybridization characteristics of oligonucleotide probes for species discrimination cannot be predicted. Nucleic Acids Res 34:e66

Proudnikov D, Mirzabekov A (1996) Chemical methods of DNA and RNA fluorescent labeling. Nucleic Acids Res 24:4535–4542

Puskas LG, Zvara A, Hackler L Jr, Van Hummelen P (2002) RNA amplification results in reproducible microarray data with slight ratio bias. Biotechniques 32:1330–4, 1336, 1338, 1340

Raman R, Raguram S, Venkataraman G, Paulson JC, Sasisekharan R (2005) Glycomics: an integrated systems approach to structure-function relationships of glycans. Nat Methods 2:817–824

Rhee SK, Liu X, Wu L, Chong SC, Wan X, Zhou J (2004) Detection of genes involved in biodegradation and biotransformation in microbial communities by using 50-mer oligonucleotide microarrays. Appl Environ Microbiol 70:4303–4317

Rimour S, Hill D, Militon C, Peyret P (2005) GoArrays: highly dynamic and efficient microarray probe design. Bioinformatics 21:1094–1103

Roth FP, Hughes JD, Estep PW, Church GM (1998) Finding DNA regulatory motifs within unaligned noncoding sequences clustered by whole-genome mRNA quantitation. Nat Biotechnol 16:939–945

Rudi K, Flateland SL, Hanssen JF, Bengtsson G, Nissen H (2002) Development and evaluation of a 16S ribosomal DNA array-based approach for describing complex microbial communities in ready-to-eat vegetable salads packed in a modified atmosphere. Appl Environ Microbiol 68:1146–1156

Rudi K, Treimo J, Nissen H, Vegarud G (2003) Protocols for 16S rDNA array analyses of microbial communities by sequence-specific labeling of DNA probes. ScientificWorldJournal 3:578–584

Salama N, Guillemin K, McDaniel TK, Sherlock G, Tompkins L, Falkow S (2000) A whole-genome microarray reveals genetic diversity among *Helicobacter pylori* strains. Proc Natl Acad Sci U S A 97:14668–14673

Sanguin H, Herrera A, Oger-Desfeux C, Dechesne A, Simonet P, Navarro E, Vogel T M, Moenne-Loccoz Y, Nesme X, Grundmann GL (2006) Development and validation of a prototype 16S rRNA-based taxonomic microarray for Alphaproteobacteria. Environ Microbiol 8:289–307

SantaLucia J Jr (1998) A unified view of polymer, dumbbell, and oligonucleotide DNA nearest-neighbor thermodynamics. Proc Natl Acad Sci U S A 95:1460–1465

SantaLucia JJ, Allawi HT, Seneviratne PA (1996) Improved nearest-neighbor parameters for predicting DNA duplex stability. Biochemistry 35:3555–3562

Santos SR, Ochman H (2004) Identification and phylogenetic sorting of bacterial lineages with universally conserved genes and proteins. Environ Microbiol 6:754–759

Schadt EE, Li C, Ellis B, Wong WH (2001) Feature extraction and normalization algorithms for high-density oligonucleotide gene expression array data. J Cell Biochem Suppl Suppl 37:120–125

Schena M, Shalon D, Davis RW, Brown PO (1995) Quantitative monitoring of gene expression patterns with a complementary DNA microarray. Science 270:467–470

Scherl A, Francois P, Charbonnier Y, Deshusses JM, Koessler T, Huyghe A, Bento M, Stahl-Zeng J, Fischer A, Masselot A, Gallé F, Renzoni A, Vaudaux P, Lew D, Zimmermann-Ivol CG, Binz PA, Sanchez JC, Hochstrasser DF, Schrenzel J (2006) Exploring glycopeptide resistance in *Staphylococcus aureus*: a combined proteomics and transcriptomics approach for the identification of resistance related markers. BMC Genomics 7:-296

Schrenzel J, Hibbs J (2003) Non-cognate hybridization system (NCHS). 00/75377 A2 Ref Type: Patent

Sergeev N, Distler M, Courtney S, Al Khaldi SF, Volokhov D, Chizhikov V, Rasooly A (2004) Multipathogen oligonucleotide microarray for environmental and biodefense applications. Biosens Bioelectron 20:684–698

Shchepinov MS, Case-Green SC, Southern EM (1997) Steric factors influencing hybridisation of nucleic acids to oligonucleotide arrays. Nucleic Acids Res 25:1155–1161

Shi L, Reid LH, Jones WD, Shippy R, Warrington JA, Baker SC, Collins PJ, De Longueville F, Kawasaki ES, Lee KY, Luo Y, Sun YA, Willey JM, Setterquist RA, Fischer GM, Tong W, Dragan YP, Dix DJ, Frueh FW, Goodsaid FM, Herman D, Jensen RV, Johnson CD, Lobenhofer EK, Puri RK, Schrf U, Thierry-Mieg J, Wang C, Wilson M, Wolber PK, Zhang L, Slikker W Jr, Shi L, Reid LH (2006) The MicroArray Quality Control (MAQC) project shows inter- and intraplatform reproducibility of gene expression measurements. Nat Biotechnol 24:1151–1161

Small J, Call DR, Brockman FJ, Straub TM, Chandler DP (2001) Direct detection of 16S rRNA in soil extracts by using oligonucleotide microarrays. Appl Environ Microbiol 67:4708–4716

Smyth GK, Speed T (2003) Normalization of cDNA microarray data. Methods 31:265–273

Sohail M, Akhtar S, Southern EM (1999) The folding of large RNAs studied by hybridization to arrays of complementary oligonucleotides. RNA 5:646–655

Southern EM (1989) Analysing polynucleotide sequences. 19891100 GB Ref Type: Patent

Southern EM (1975) Detection of specific sequences among DNA fragments separated by gel electrophoresis. J Mol Biol 98:503–517

Southern EM, Case-Green SC, Elder JK, Johnson M, Mir KU, Wang L, Williams JC (1994) Arrays of complementary oligonucleotides for analysing the hybridisation behaviour of nucleic acids. Nucleic Acids Res 22:1368–1373

Southern EM, Mir K, Shchepinov M (1999) Molecular interactions on microarrays. Nat Genet 21:5–9

Stenger DA, Andreadis JD, Vora GJ, Pancrazio JJ (2002) Potential applications of DNA microarrays in biodefense-related diagnostics. Curr Opin Biotechnol 13:208–212

Stimpson DI, Cooley PW, Knepper SM, Wallace DB (1998) Parallel production of oligonucleotide arrays using membranes and reagent jet printing. Biotechniques 25:886–890

Stowe-Evans EL, Ford J, Kehoe DM (2004) Genomic DNA microarray analysis: identification of new genes regulated by light color in the cyanobacterium *Fremyella diplosiphon*. J Bacteriol 186:4338–4349

Straub TM, Chandler DP (2003) Towards a unified system for detecting waterborne pathogens. J Microbiol Methods 53:185–197

Strezoska Z, Paunesku T, Radosavljevic D, Labat I, Drmanac R, Crkvenjakov R (1991) DNA sequencing by hybridization: bases read by a non-gel-based method. Proc Natl Acad Sci U S A 88:10089–10093

Tamayo P, Slonim D, Mesirov J, Zhu Q, Kitareewan S, Dmitrovsky E, Lander ES, Golub TR (1999) Interpreting patterns of gene expression with self-organizing maps: methods and application to hematopoietic differentiation. Proc Natl Acad Sci U S A 96:2907–2912

Taroncher-Oldenburg G, Griner EM, Francis CA, Ward BB (2003) Oligonucleotide microarray for the study of functional gene diversity in the nitrogen cycle in the environment. Appl Environ Microbiol 69:1159–1171

Taylor S, Smith S, Windle B, Guiseppi-Elie A (2003) Impact of surface chemistry and blocking strategies on DNA microarrays. Nucleic Acids Res 31:e87

Thiyagarajan S, Karhanek M, Akhras M, Davis RW, Pourmand N (2006) Pathogen MIPer: a tool for the design of molecular inversion probes to detect multiple pathogens. BMC Bioinformatics 7:500

Tiquia SM, Wu L, Chong SC, Passovets S, Xu D, Xu Y, Zhou J (2004) Evaluation of 50-mer oligonucleotide arrays for detecting microbial populations in environmental samples. Biotechniques 36:664–665

Treimo J, Vegarud G, Langsrud T, Marki S, Rudi K (2006) Total bacterial and species-specific 16S rDNA micro-array quantification of complex samples. J Appl Microbiol 100:985–998

Troesch A, Nguyen H, Miyada CG, Desvarenne S, Gingeras TR, Kaplan PM, Cros P, Mabilat C (1999) *Mycobacterium* species identification and rifampin resistance testing with high-density DNA probe arrays. J Clin Microbiol 37:49–55

Tseng GC, Oh MK, Rohlin L, Liao JC, Wong WH (2001) Issues in cDNA microarray analysis: quality filtering, channel normalization, models of variations and assessment of gene effects. Nucleic Acids Res 29:2549–2557

Urakawa H, El Fantroussi S, Smidt H, Smoot JC, Tribou EH, Kelly JJ, Noble PA, Stahl DA (2003) Optimization of single-base-pair mismatch discrimination in oligonucleotide microarrays. Appl Environ Microbiol 69:2848–2856

Urakawa H, Noble PA, El Fantroussi S, Kelly JJ, Stahl DA (2002) Single-base-pair discrimination of terminal mismatches by using oligonucleotide microarrays and neural network analyses. Appl Environ Microbiol 68: 235–244

Vainrub A, Pettitt BM (2002) Coulomb blockage of hybridization in two-dimensional DNA arrays. Phys Rev E Stat Nonlin Soft Matter Phys 66:041905

van Bakel H, Holstege FC (2004) In control: systematic assessment of microarray performance. EMBO Rep 5:964–969

van Leeuwen WB, Jay C, Snijders S, Durin N, Lacroix B, Verbrugh HA, Enright MC, Troesch A, Van Belkum A (2003) Multilocus sequence typing of *Staphylococcus aureus* with DNA array technology. J Clin Microbiol 41:3323–3326

Volokhov D, Rasooly A, Chumakov K, Chizhikov V (2002) Identification of listeria species by microarray-based assay. J Clin Microbiol 40:4720–4728

Voordouw G, Voordouw JK, Karkhoff-Schweizer RR, Fedorak PM, Westlake DW (1991) Reverse sample genome probing, a new technique for identification of bacteria in environmental samples by DNA hybridization, and its application to the identification of sulfate-reducing bacteria in oil field samples. Appl Environ Microbiol 57:3070–3078

Vora GJ, Meador CE, Stenger DA, Andreadis JD (2004) Nucleic acid amplification strategies for DNA microarray-based pathogen detection. Appl Environ Microbiol 70:3047–3054

Wang D, Coscoy L, Zylberberg M, Avila PC, Boushey HA, Ganem D, DeRisi JL (2002) Microarray-based detection and genotyping of viral pathogens. Proc Natl Acad Sci U S A 99:15687–15692

Wang RF, Beggs ML, Erickson BD, Cerniglia CE (2004) DNA microarray analysis of predominant human intestinal bacteria in fecal samples. Mol Cell Probes 18:223–234

Wilson CL, Pepper SD, Hey Y, Miller CJ (2004) Amplification protocols introduce systematic but reproducible errors into gene expression studies. Biotechniques 36:498–506

Wilson KH, Wilson WJ, Radosevich JL, DeSantis TZ, Viswanathan VS, Kuczmarski TA, Andersen GL (2002) High-density microarray of small-subunit ribosomal DNA probes. Appl Environ Microbiol 68:2535–2541

Wong CW, Albert TJ, Vega VB, Norton JE, Cutler DJ, Richmond TA, Stanton LW, Liu ET, Miller LD (2004) Tracking the evolution of the SARS coronavirus using high-throughput, high-density resequencing arrays. Genome Res 14:398–405

Wu L, Thompson DK, Li G, Hurt RA, Tiedje JM, Zhou J (2001) Development and evaluation of functional gene arrays for detection of selected genes in the environment. Appl Environ Microbiol 67:5780–5790

Wu Y, de Kievit P, Vahlkamp L, Pijnenburg D, Smit M, Dankers M, Melchers D, Stax M, Boender PJ, Ingham C, Bastiaensen N, de Wijn R, van Alewijk D, van Damme H, Raap AK, Chan AB, van Beuningen R (2004) Quantitative assessment of a novel flow-through porous microarray for the rapid analysis of gene expression profiles. Nucleic Acids Res 32:e123

Xiang CC, Chen M, Ma L, Phan QN, Inman JM, Kozhich OA, Brownstein MJ (2003) A new strategy to amplify degraded RNA from small tissue samples for microarray studies. Nucleic Acids Res 31:e53

Yang YH, Dudoit S, Luu P, Lin DM, Peng V, Ngai J, Speed TP (2002) Normalization for cDNA microarray data: a robust composite method addressing single and multiple slide systematic variation. Nucleic Acids Res 30:e15

Yang YH, Speed T (2002) Design issues for cDNA microarray experiments. Nat Rev Genet 3:579–588

Ye RW, Wang T, Bedzyk L, Croker KM (2001) Applications of DNA microarrays in microbial systems. J Microbiol Methods 47:257–272

Yershov G, Barsky V, Belgovskiy A, Kirillov E, Kreindlin E, Ivanov I, Parinov S, Guschin D, Drobishev A, Dubiley S, Mirzabekov A (1996) DNA analysis and diagnostics on oligonucleotide microchips. Proc Natl Acad Sci U S A 93:4913–4918

Yue H, Eastman PS, Wang BB, Minor J, Doctolero MH, Nuttall RL, Stack R, Becker JW, Montgomery JR, Vainer M, Johnston R (2001) An evaluation of the performance of cDNA microarrays for detecting changes in global mRNA expression. Nucleic Acids Res 29:E41

Zaigler A, Schuster SC, Soppa J (2003) Construction and usage of a onefold-coverage shotgun DNA microarray to characterize the metabolism of the archaeon *Haloferax volcanii*. Mol Microbiol 48:1089–1105

Zammatteo N, Jeanmart L, Hamels S, Courtois S, Louette P, Hevesi L, Remacle J (2000) Comparison between different strategies of covalent attachment of DNA to glass surfaces to build DNA microarrays. Anal Biochem 280:143–150

Zhang L, Hurek T, Reinhold-Hurek B (2006) A nifH-based oligonucleotide microarray for functional diagnostics of nitrogen-fixing microorganisms. Microb Ecol

Zhou J, Thompson DK (2002) Challenges in applying microarrays to environmental studies. Curr Opin Biotechnol 13:204–207

Zwick ME, Mcafee F, Cutler DJ, Read TD, Ravel J, Bowman GR, Galloway DR, Mateczun A (2005) Microarray-based resequencing of multiple *Bacillus anthracis* isolates. Genome Biol 6:R10

24

Pathogenic Bacterial Sensors Based on Carbohydrates as Sensing Elements

Haiying Liu

Abstract

Protein–carbohydrate interactions are involved in a wide variety of cellular recognition processes including cell growth regulation, differentiation, adhesion, cancer cell metastasis, cellular trafficking, the immune response, and viral or bacterial infections. These specific interactions occur through glycoproteins, glycolipids, polysaccharides found on cell surfaces, and proteins with carbohydrate-binding domains called lectins through cooperative multiple interactions since it is known that individual carbohydrate–protein interactions are generally weak. A common way for bacteria to accomplish adhesion is through their cellular lectins, also called fimbriae or pili, which bind to complementary carbohydrates on the surface of the host tissues. Lectin-deficient mutant bacteria often fail to initiate infection. Carbohydrate-based detection of bacterial pathogens presents an exciting alternative to standard methods for screening and detecting bacterial targets in food industry, water and environment quality control, and clinical diagnosis. Conjugated fluorescent glycopolymers such as conjugated glycopoly(p-phenylene-ethynylene)s, glycopolythiophenes, glycopoly(p-phenylene)s, and carbohydrate-bearing polydiacetylenes have been prepared for quick detection of *Escherichia coli* (*E. coli*) through cooperative multivalent interactions between the polymeric carbohydrates and the bacterial pili because they combine fluorescent scaffolding and carbohydrate reporting functions into one package and possess intrinsic fluorescence and high sensitivity to minor external stimuli. Glyconanoparticles and galactose-functionalized carbon nanotubes (Gal-SWNTs) have been used as three-dimensional systems to study their specific multivalent interactions with *E. coli*. In addition, Gal-SWNTs have also been employed to detect *Bacillus anthracis* spores through divalent cation-mediated multivalent carbohydrate–carbohydrate interactions. Carbohydrate microarrays combine the benefits of immobilized format assays with the capability of detecting thousands of analytes simultaneously and can offer a general and powerful platform for whole-cell applications because their multivalent display of carbohydrates can mimic multivalent interactions at cell–cell interfaces. A simple but very effective diagnostic carbohydrate microarray has been reported for the quick detection of *E. coli* in complex biological mixtures with detection limit of 10^5–10^6 cells. Wang et al. reported another direct and unique approach to detect pathogenic bacteria by using a carbohydrate microarray of 48 carbohydrate-containing antigenic macromolecules for recognition of carbohydrate-binding antibodies from 20 human serum specimens.

Pathogenic bacteria possess a cell surface capsular polysaccharide (CPS) or lipopolysaccharide (LPS) shell, or both, which helps the pathogen initiate an infection. Lectin microarrays have been utilized as a very important and powerful tool to detect *E. coli*, profile diverse glycan structures of *E. coli* and discover dynamic changes in surface glycosylation of bacteria in response to environmental stimuli through specific multivalent interactions of lectins and the bacterial LPSs.

Haiying Liu • Department of Chemistry, Michigan Technological University, Houghton, Michigan.

M. Zourob et al. (eds.), *Principles of Bacterial Detection: Biosensors, Recognition Receptors and Microsystems*,
© Springer Science+Business Media, LLC 2008

1. Introduction

Detection of pathogenic bacteria is of the utmost importance in the food industry, water and environmental quality control, as well as in clinical diagnosis for health and safety reasons. A variety of bacteria have been identified as important food and waterborne pathogens (Doyl et al. 2001). The yearly incidence of foodborne illness in the United States is estimated at between 6 and 80 million cases, with bacterial foodborne outbreaks accounting for 91 % of the total outbreaks, resulting in approximately 500 to 9,000 deaths (Doyl et al. 2001; McCabe-Sellers and Beattie 2004). The consequences can be severe, even though most bacterial foodborne illnesses are mild. Many foodborne bacterial pathogens can cause invasive disease in addition to acute gastrointestinal symptoms, such as diarrhea and vomiting. For example, most people infected with *Salmonella* bacteria develop nausea, vomiting, diarrhea, fever, and abdominal cramps. In some patients, the *Salmonella* infection may spread from the intestines to the blood stream and then to other body sites, and can cause death unless they are treated promptly with antibiotics. The elderly, infants and those with impaired immune systems are more liable to have a severe illness (Ukuku 2006). Infection with some foodborne pathogens can be followed by chronic sequelae or disability. *Yersinia enterocolitica* infection can cause reactive arthritis (Fredriksson–Ahomaa et al. 2006; Viboud and Bliska 2005; Zhang and Bliska 2005), and *E. coli* O157:H7 strain causes severe damage to intestinal epithelial cells, leading to hemorrhaging, hemolytic uremic syndrome (HUS), permanent kidney damage or even death. (Sheng et al. 2006). Since *E. coli* can easily contaminate ground beef, chicken, fish, raw milk, rice and vegetables, it is extremely important to carefully control this pathogenic microorganism, especially in the fields of food production. The 2006 North American *E. coli* outbreak started in September 2006 with spinach contaminated with *E. coli* O157:H7. It caused 199 infections, three deaths, and thirty-one kidney failures by October 6, 2006 according to U.S. FDA News (October 2006). A number of pathogenic bacteria have the potential to be used as biological warfare agents. Outbreaks of diseases from pathogenic bacteria can pose a severe threat to global security since these bacteria are resistant to environmental conditions, most of the human population is easily vulnerable and the microbial diseases can cause a high fatality rate (Dando 1994). Current microbial disease prevention practices are to carefully control pathogenic bacteria in clinical medicine, food safety and environmental monitoring.

The detection of pathogenic bacteria is a very important step to prevent and identify health and safety problems. The effective detection of pathogenic bacteria requires analytical methods to meet several challenging criteria such as high selectivity and sensitivity with very low detection limits. Bacterial detection methods must be rapid and extremely sensitive because even a single pathogenic bacterium in the body or food can function as an infectious dose. In addition, bacterial detection methods should display extremely high selectivity because a very small number of pathogenic bacteria often coexist with many other nonpathogenic organisms in complex biological samples. For example, the strain of *E. coli* O157:H7 requires an infectious dose of as low as ten cells to cause disease, and the coliform standard for *E. coli* in water is four cells per 100 ml (Greenberg 1995; Katamay 1990; Pontius 2000).

The culture and colony counting methods remain the standard detection methods even though they are the oldest bacterial detection techniques. They usually involve an evaluation of bacterial morphology, testing of the bacteria for their abilities to grow in various media under different conditions, and colony counting of coliform bacteria (Kaspar and Tartera 1990). Although standard microbiological techniques can offer the detection of a single bacterium, they are excessively time-consuming, taking days to weeks to complete (Brooks et al. 2004; Hobson et al. 1996; Kaspar and Tartera 1990). Some new technologies are very sensitive but still require hours. For example, the polymerase chain reaction (PCR) can be used to amplify small quantities of genetic material to detect bacteria. The PCR method is highly sensitive and can provide conclusive and unambiguous results since it is based on the isolation, amplification and quantification of a short DNA sequence of the targeted bacteria (Gillespie 1998; Hill 1996; Louws et al. 1999; Mohammadi et al. 2006; Olsen et al. 1995; Seelig et al. 1991; Toze 1999). However, it takes from 5 to 24 hours to generate a detection result. More recently, faster methods have been reported using pathogen recognition by fluorescently labeled antibodies (Shriver–Lake et al. 2004; Zhao et al. 2004), DNA probes (Ikeda et al. 2006; Keum et al. 2006) or bacteriophages (Marks and Sharp 2000; Takikawa et al. 2002).

Carbohydrates are the key components of glycolipid glycoprotein and cell surface molecules, and play critical roles in cell–cell recognition, cell adhesion, differentiation, trafficking, signaling between cells, cellular metastasis and viral or bacterial infections (Jelinek and Kolusheva 2004; Lis and Sharon 1998). Identification of the specific carbohydrates involved in these processes is important to better understand cellular recognition at the molecular level and to aid the design of therapeutics and diagnostic tools (Sharon 2006). Carbohydrate-based detection of bacterial pathogens presents an exciting alternative to standard methods for screening and detecting bacterial targets for clinical applications and defense purposes (Jelinek and Kolusheva 2004). Recently many researchers have made great strides toward developing rapid and promising methods for detection of bacterial pathogens. However, much research and development work is still needed before new approaches become real and trustworthy alternatives. This book chapter offers an overview of trends in the area of detection of pathogenic bacteria based on carbohydrates as sensing elements.

2. Bacterial Surface Lectins

Bacterial adhesion is a first crucial step in the infection process as it is frequently associated with the ability of pathogenic bacteria to colonize host tissue (Telford et al. 2006). A common way for bacteria to accomplish adhesion is through lectins present on the surface of the infectious bacteria since the cellular lectins bind to complementary carbohydrates on the surface of the host tissues (Mammen et al. 1998; Telford et al. 2006). Lectin-deficient mutants often fail to initiate infection. Numerous bacterial strains produce surface lectins, also called fimbriae or pili, commonly in the form of submicroscopic multi-subunit proteins (Mammen et al. 1998). The mannose-specific type 1 pili, the galabiose-specific P pili and the N-acetylglucosamine-specific pili of E. coli have been well characterized (Lis and Sharon 1998; Mammen et al. 1998). Type 1 pili are found in most E. coli strains, including both uropathogenic and commensal strains, as well as throughout the Enterobacteriaceae family, and have been shown to mediate binding to mannose receptors on host cells through the FimH adhesion (Bouckaert et al. 2006). Type 1 pili of E. coli, encoded by the fim gene cluster (fimA–fimH), possess a short tip fibrillum containing the FimH adhesin, which is joined to the distal end of the FimA pilus rod (Jones et al. 1995; Sauer et al. 2000; Telford et al. 2006). The FimH adhesin binds specifically to mannosylated glycoproteins present in the bladder epithelium. FimC is the type 1 pilus chaperone, and FimD is the OM usher (Connell et al. 1996). The FimH adhesin of E. coli type 1 pili possesses an extended combining site and displays a considerably higher affinity for oligomannose or hybrid types such as Manα3Manβ4GlcNAc or Manα6(Manα3)Manα6(Manα3)Man than for mannose (Bouckaert et al. 2006). Type 1 fimbriated E. coli binds much more strongly to hydrophobic aromatic α-mannosides than to methyl α-D-mannopyranoside (MeαMan) because its type 1 fimbriae also possess a hydrophobic region next to the carbohydrate-binding site of FimH (Firon et al. 1987). The combining sites of type 1 fimbriae of Salmonella enteritidis and of other enteric bacteria are smaller than those of E. coli and Klebsiella pneumoniae, and do not possess an adjoining hydrophobic region (Kisiela et al. 2006).

Since type 1 fimbriated strains of E. coli were identified in 1980's, different specific strains of E. coli have been discovered (Table 24.1). (Sauer et al. 2000; Sharon, 2006;

Table 24.1.   Carbohydrates as binding sites for bacterial pathogens on animal tissues (adapted from Sharon (2006), with permission of Elsevier)

Organism	Target Tissue	Carbohydrate	Form
Bordetella pertussis	Respiratory	lactose	Glycolipids (Prasad et al. 1993)
Campylobacter Jejuni	Intestinal	Fucα2Galβ4GlcNAc	Glycoproteins (Ruiz-Palacios et al. 2003)
Clostridium perfringens	Gastrointestinal tract	Galβ4GlcNAc	glycoside hydrolases (Ficko-Blean and Boraston, 2006)
E. coli Type 1	Bladder	Manα3Manα6Man	Glycoproteins (Bouckaert et al. 2006)
E. coli P type	Kidney	Galα4Gal	Glycolipids (Kuehn et al. 1992; Tewari et al., 1994; Winberg et al. 1995)
E. coli S	Neural	NeuAc (α2–3)Gal	Glycolipids and glycoproteins (Hanisch et al. 1993; Korhonen et al. 1984; Stins et al. 1994)
E. coli CFA/I	Intestinal	NeuAc (α2–8)–	Glycoproteins (Wenneras et al. 1990)
E. coli F1C	Kidney and Bladder	GalNAcβ4Galβ	Glycolipids (Khan et al. 2000)
E. coli F17	Urinary	GlcNAc	Glycolipids (Le Bouguenec and Bertin 1999)
E. coli K99	Intestinal	NeuAc(α2–3)Galβ4Glc	Glycolipids (Ono et al. 1989; Teneberg et al. 1993)
Haemophilus influenzae	Respiratory	NeuAc(α2–3) Galβ4GlcNAcβ3Galβ4GlcNAc	Glycolipids (Berenson et al. 2005)
Helicobacter pylori	Stomach duodenum	NeuAcα2,3Galβ4GlcNAcβ3Galβ4 NeuAc (α2–3)–	Glycolipids (Roche et al. 2004) glycoproteins (Gustafsson et al. 2006; Mahdavi et al. 2002; Teneberg et al. 2005; Unemo et al. 2005)
Klebsiella pneumoniae	Respiratory	Man	Glycoproteins (Duncan et al. 2005)
Neisseria gonorrhoea	Genital	Galβ4GlcNAc	Glycolipids (Nassif et al. 1999)
Neisseria meningitidis	Respiratory	NeuAc(α2–3) Galβ4GlcNAcβ3Galβ4GlcNAc	Glycolipids (Schweizer et al. 1998; Smedley et al. 2005)
Pseudomonas aeruginosa	Respiratory	GalNAcβ4Galβ	Glycolipids (Krivan et al. 1988; Schweizer et al. 1998; Sheth et al. 1994; Smedley et al. 2005)
Salmonella typhimurium	Intestinal	Man	Glycoproteins (Kisiela et al. 2006; Tinker and Clegg, 2000)
Salmonella enterica serovar Enteritidis	Intestinal	Man	Glycoproteins (Kisiela et al. 2006; Kisiela et al. 2005)
Streptococcus pneumoniae	Respiratory	NeuAc(α2-3)Galβ4GlcNAcβ3Galβ4Glc	Glycolipids (Barthelson et al. 1998)
Staphylococcus saprophyticus	Urinary	GalNAc	Glycoproteins (Beuth et al. 1992)
Streptococcus suis	Respiratory	Galα4Gal	Glycolipids (Tikkanen et al. 1995)
Vibrio cholerae	Intestinal	Man	Glycoproteins (Marsh and Taylor 1999)
Yersinia enterocolitica	Intestinal	Gal, GalNAc, Lac	Glycolipids (Lotter et al. 2004; Mantle and Husar 1994)

Telford et al. 2006). P pili of *E. coli,* encoded by the 11 genes of the *pap* (*papA–papK*) gene cluster, consist of six structural proteins that interact to form a fiber composed of two distinct subassemblies: a 6.8-nm-thick helical rod comprising mainly PapA and a 2-nm-diameter linear tip pilus comprising mainly PapE (Kuehn et al. 1992; Sauer et al. 2000). The PapG adhesion is located at the distal end of the tip pilus and binds to Galα4Gal moieties present in kidney glycolipids (Kuehn et al. 1992). PapD and PapC are the chaperone and usher for P pili, respectively (Kuehn et al. 1992; Tewari et al. 1994; Winberg et al. 1995). S-fimbriated *E. coli* causes sepsis and meningitis in newborns (Schroten et al. 1998). S pili bind specifically to the terminal sialyl-(α2-3)galactoside on many epithelial surfaces as a structural component of glycoproteins or glycolipids (Hanisch et al. 1993; Korhonen et al. 1984; Stins et al. 1994). They also bind most strongly to sialylated β-galactosides carrying *N*-glycolylneuraminic acid or the (α2-8)-linked dimer of *N*-acetylneuraminic acid on gangliosides of the b series or fetal glycopeptides (Hanisch et al. 1993). Enterotoxigenic *E. coli* strains responsible for traveler's severe diarrhea mediate their specific adherence to the human intestinal mucosa by the fimbrial colonization factor antigen CFA/I consisting of ∼100 identical adhesive subunits. CFA/I has been shown to bind to free sialic acid (Evans et al. 1979), sialoglycoproteins (Wenneras et al. 1990) and GM2 (Faris et al. 1980). *E. coli* harboring F1C pili have been reported to bind to epithelial cells in the distal tubules and collecting ducts, as well as to endothelial cells of the human kidney and bladder (Backhed et al. 2002; Roos et al. 2006). F1C pili have a high affinity to bind to the GalNAcβ1-4Galβ sequence of glycolipids (Backhed et al. 2002; Klemm et al. 1994). F17 pili are 3-nm-wide, flexible, wire-like organelles of enterotoxigenic *E. coli* built up of the major pilin F17-A and exposing the F17-G adhesion at their tips (Le Bouguenec and Bertin 1999). The F17-G adhesion at the tip of flexible F17 fimbriae mediates binding to *N*-acetylβ-D-glucosamine-presenting receptors on the microvilli of the intestinal epithelium of ruminants, causing diarrhea or septicemia in ruminants (Le Bouguenec and Bertin 1999). Enterotoxlgenic *E. coli* possessing K99 pili causes diarrhea in neonatal calves, pigs and lambs (Ono et al. 1989). The first stage of the bacterial infection is adhesion by the pili on the small intestinal mucosa, and the adhesion is followed by colony formation. K99 pili bind specifically to iV-glycolylneuraminyl-lactosyl-ceramide (Ono et al. 1989; Teneberg et al. 1993).

In addition, other pathogenic bacteria with specific affinities for carbohydrates have been reported. *Campylobacter jejuni* is a major human intestinal bacterial pathogen and one of the most common causes of diarrhea worldwide (Tauxe, 1997). Symptoms of *C. jejuni* infection often present are fever, stomach pain, nausea, headache, muscle pain, cramps and vomiting (Allos and Blaser, 1995). The binding of *C. jejuni* to intestinal epithelial cells is achieved through fucosylated carbohydrate epitopes (Ruiz-Palacios et al. 2003). *Hemophilus influenzae* is an opportunistic pathogen that is most often spread by the inhalation of respiratory droplets from a colonized individual (Swords et al. 2003). Hif pili, encoded by the *hif* gene cluster (*hifA–hifE*), found in pathogenic *Haemophilus influenzae* strains, have rods with 6–7 nm in diameter and a short, thin tip differentiation. The rods possess a cross-over repeat with a double-stranded right-handed helical architecture. HifA is the major pilin subunit, whereas HifD and HifE are minor pilus subunits, located at the pilus tip. HifB is a periplasmic chaperone, and HifC functions as an outer membrane usher (Vanham et al. 1994). HifE protein mediates attachment to human epithelial cells (Gilsdorf et al. 1997). Nontypeable *Haemophilus influenzae* binds with high affinity and specificity to minor gangliosides of human respiratory (HEp-2) cells and macrophages (Berenson et al. 2005). *Helicobacter pylori* is a human- and primate-specific pathogen found in the gastric mucus layer or attached to the gastric epithelium. *H. pylori* infection affects about half the world population, results in chronic active gastritis and is a risk factor for the development of peptic ulcer disease, gastric adenocarcinoma, and gastric lymphoma (Montecucco and Rappuoli 2001). Particularly, *H. pylori* exhibits a quite unusual complexity in its specific interactions with sialylated oligosaccharides, gangliotetraosylceramide, Lewis b antigen, monohexosylceramide, lactosyl-ceramide, lactotetraosylceramide, sulfatide and heparan sulfate because it contains multiple adhesions recognizing all terminal α2,3-linked sialic acid (Mahdavi et al. 2002; Teneberg et al. 2005; Unemo et al. 2005) and NeuAcα2,3Galβ1,4GlcNAcβ1,3Galβ1,4GlcNAc structures (Roche et al. 2004). Neisseria gonorrhea, a genital pathogen, recognizes N-acetyllactosamine (Galβ4GlcNAc, LacNAc) (Nassif et al. 1999). The opportunistic pathogen *Pseudomonas aeruginosa* is a leading cause of nosocomial pneumonia, which is of major concern to immuno-suppressed or immunocompromised patients (Barasch et al. 2003). The *P. aeruginosa* pilus (*N*-methylphenylalanine type) consists of monomeric subunits, termed pilin, each conferring the receptor binding function. The pili PAK of *P. aeruginosa* binds to glycolipids contained within epithelial cell membranes (Krivan et al. 1988) and shows a specificity toward the carbohydrate sequence GalNAcβ4Galβ disaccharide found in the glycosphingolipids asialo-GM1 and asialo-GM2 (Schweizer et al. 1998; Sheth et al. 1994; Smedley et al. 2005). *Streptococcus pneumoniae* is a human pathogen of major importance, causing invasive diseases such as pneumonia, meningitis and bacteraemia, as well as otitis media and sinusitis (Levine et al. 2006). *S. pneumoniae* binds specifically to the pentasaccharide NeuAc(α2-3)Galβ4GlcNAcβ3Galβ4Glc, and the corresponding internal tetra and trisaccharides Galβ4GlcNAcβ3Galβ4Glc and GlcNAcβ3Galβ4Glc, respectively (Barthelson et al. 1998). *Streptococcus suis* is an important Gram-positive pathogen which causes meningitis, sepsis, and other serious infections in piglets and meningitis in humans who have been in contact with pigs (Staats et al. 1997). Two adhesion variants (PN and PO) of *S. suis* possess binding specificity to Galα4Gal-containing glycolipids, which are expressed in many pig tissues (Tikkanen et al. 1995).

Characteristic features of individual protein–carbohydrate interactions are highly specific and have low affinity in the millimolar range (Mammen et al. 1998). An increase of several orders of magnitude in the affinity to bacterial pili has been achieved to capitalize on the cluster glycoside effect by systems bearing spherical or linear arrays of carbohydrates including glycodendrimers (Roy 2003), glycopolymers (Appeldoorn et al. 2005; Baek et al. 2000; Disney et al. 2004; Joralemon et al. 2004; Li et al. 1999; Lin et al. 2005; Xue et al. 2006b), self-assembled monolayers of carbohydrates (Nilsson and Mandenius 1994; Qian et al. 2002), carbohydrate-functionalized polydiacetylene monolayers and vesicles (Li et al. 2002; Ma et al. 2003, 2000, 1998; Su et al. 2005; Sun et al. 2004a; Zhang et al. 2002, 2005, 2004), glycan microarrays (Disney and Seeberger 2004), glyconanoparticles (Lin et al. 2002), and carbohydrate-functionalized carbon nanotubes (Gu et al. 2005).

3. Surface Carbohydrate Structures of Pathogenic Bacteria

Most bacteria heavily expose carbohydrate structures on their surface, which can consist of a capsular polysaccharide (CPS) and/or a lipopolysaccharide (LPS) (Raetz 1993). These polymers are serotype-specific and give rise to the K- and O-antigens, respectively. CPS forms a slimy capsule around the bacteria, hides cell surface components of the bacterium from the immune system of the host, prevents complement activation by cell surface proteins and inhibits phagocytosis (Roberts 1996). CPS is built up by repeating units and shows a large structural variation. All Gram-negative bacteria contain in their outer membrane a unique glycolipid structure, called LPS, which consists of amphiphilic complex molecules with a molecular mass of about 10 kDa and displays wide variation in chemical composition both between and within bacterial species (Caroff and Karibian 2003). LPS typically consists of three distinct structural components: species-specific repeating units of a polysaccharide (O-antigen); a nonrepeating core oligosaccharide, the lipid A proximal inner core and the O-polysaccahride-proximal outer core; and a hydrophobic domain with a ketosidic linkage to the core polysaccharide known as

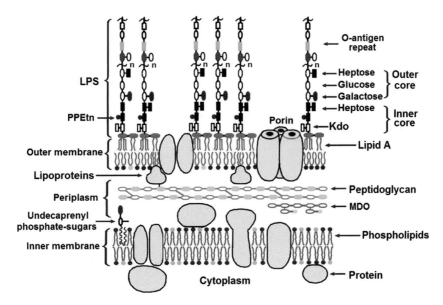

Figure 24.1. Molecular representation of the envelope of a Gram-negative bacterium (Raetz 1993). Ovals and rectangles stand for sugar residues, whereas circles depict the polar head groups of various lipids (phosphatidylethanolamine and phosphatidylglycerol). The core region shown is that of *E. coli* K-12 (Raetz 1990), a strain that does not normally contain an O-antigen repeat unless transformed with an appropriate plasmid (Raetz 1990). Kdo is 3-deoxy-D-manno-octulosonic acid. MDO and PPEtn stand for membrane-derived oligosaccharides, 2-aminoethyl diphosphate, respectively (Raetz 1993). (Adapted with permission, from Raetz and Whitfield (2002) and the Annual Review of Biochemistry, 2002 by Annual Reviews www.annualreviews.org).

lipid A (Fig. 24.1) (Caroff and Karibian 2003; Erridge et al. 2002; Raetz and Whitfield 2002). LPS offers a permeability barrier to large, negatively charged and/or hydrophobic molecules. The O-antigen, the outermost part of the LPS, contributes to the variation of the Gram-negative bacterial cell wall. Many pathogens, such as *E. coli*, *Salmonella enterica*, and *Vibrio cholerae*, are classified into different serological groups based on the variability of their O-antigens and capsular polysaccharides. In fact, *E. coli* is known to synthesize at least 170 different O-antigens and more than 80 different K-antigens (Comstock and Kasper 2006; Orskov et al. 1977; Whitfield and Roberts 1999). Because of their wide surface position and structural variation, the CPS and LPS structures are important bacterial antigens, which are used in the serological typing of bacteria, as well as being potential vaccine candidates (Ada and Isaacs 2003).

Lipid A, the hydrophobic anchor of LPS, is a unique and distinctive glucosamine-based phosphoglycolipid that makes up the outer monolayer of the outer membranes of most Gram-negative bacteria, and is responsible for toxicity of Gram-negative bacteria (Raetz and Whitfield 2002). There are $\sim 10^6$ lipid A residues and $\sim 10^7$ glycerophospholipids in a single cell of *E. coli* (Raetz 1990). The structure of lipid A is highly conserved among Gram-negative bacteria. Lipid A is virtually constant among Enterobacteriaceae. *E. coli* lipid A consists of $\beta(1\text{-}6)$-linked disaccharide of D-glucosamine, which is further acylated at its 2-amine and 3-hydroxyl groups, involving six fatty acid chains (Fig. 24.2) (Alexander and Rietschel 2001; Raetz and Whitfield 2002). This acylated disaccharide is also phosphorylated at positions 1 and 4' (Alexander and Rietschel 2001; Raetz 1990). However, there are significant variations in lipid A structure in some Gram-negative bacteria (Raetz and Whitfield 2002). These variations (compared with *E. coli* and *S. typhimurium*) involve chain length, number, and position of the acyl groups as in *P. aeruginosa* (Fig. 24.2B), the degree of phosphorylation, the absence of phosphate substituents in bacteria such as *Rhizobium etli*, *Rhizobium leguminosarum* and *Aquifex pyrophilus* (Fig. 24.2C

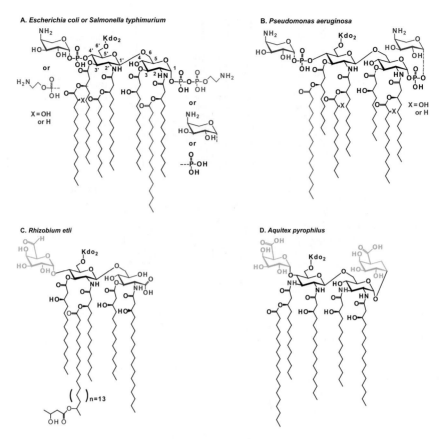

Figure 24.2. A selection of lipid A structures (adapted with permission, from Raetz and Whitfield (2002) and the Annual Review of Biochemistry, 2002 by Annual Reviews www.annualreviews.org).

and 24.2D) (Price et al. 1995; Que et al. 2000a,b, the presence of galacturonic acid residues (Fig. 24.2C and 24.2D) (Que et al. 2000a,b), and subtle modifications of one or both glucosamine units (Que et al. 2000a,b; Raetz and Whitfield 2002). Reviews dedicated to lipid A, LPS and the processes of LPS biosynthesis have reported extensive data on the structure of lipid A (Alexander and Rietschel 2001; Raetz and Whitfield 2002).

The LPS oligosaccharide core consists of a short chain of carbohydrates connecting the lipid A anchor to the O-antigen, is usually divided into an outer and an inner core, and also partially responsible for the toxicity of gram-negative bacteria (Fig. 24.1) (Raetz and Whitfield 2002). The inner core proximal to lipid A often contains unusual carbohydrates and shows little structural variability within a bacterial species. Two frequent components of the inner core are 3-deoxy-D-*manno*-2-octulosonic acid (Kdo) and L-*glycero*-D-*manno*-heptose (Hep) (Wilkinson 1996). Kdo residue attaches the core oligosaccharide to lipid A through an acid-labile ketosidic linkage, and three to six additional monosaccharides contribute to the rest of the inner core region (Raetz and Whitfield 2002). As a result, Kdo is unique to LPS and invariably present. The inner core may also have charged entities such as phosphate, pyrophosphate, 2-aminoethylphosphate and 2-aminoethylpyrophosphate (Wilkinson 1996). The outer core, serving as an attachment site for the O-antigen, consists of an oligosaccharide (up to six sugar units) (1→3)-linked to Hep, and shows a lot of variations (Raetz and Whitfield 2002). As a result, the core structures often display various degrees of heterogeneity, which is especially apparent in a number of bacteria. These bacteria never generate an O-antigen, and their LPSs are truncated and more oligosaccharides than polysaccharides. In these bacteria, because of the lack of O-antigen, the core structures are the outermost immunodominant part and of biological interest as potential bases for typing and as vaccine candidates. Examples of such bacteria are *Haemophilus*, *Neisseria*, *Chlamydia*, and *Moraxella* (Hansson and Oscarson 2000). The minimal LPS required for the growth of *E. coli* and most other Gram-negative bacteria consists of the lipid A and Kdo domains (Fig. 24.1) (Raetz, 1990). In wild-type strains, additional core and O-antigen sugars may be present (Fig. 24.1). (Raetz and Whitfield 2002). Reviews dedicated to LPS and LPS endotoxins have reported extensive data on the chemical structures of the LPS core (Hansson and Oscarson 2000; Raetz and Whitfield 2002; Wilkinson 1996).

The immunogenic O-antigen extends out from the core into the environment and directly contacts the host during infection. The structural diversity of O-antigens is remarkable, as more than 60 monosaccharides and 30 different noncarbohydrate components have been documented (Diks et al. 2001; Skurnik and Bengoechea 2003). The O-antigen constitutes a polymer of up to 40 repeating oligosaccharide units containing up to eight different or identical glycosyl residues. The repeating unit structures can differ in the monomeric glycosidic type, ring form, substitution, sequence, and the position and stereochemistry of the glycosidic linkages (Galanos and Freudenberg 1993; Pier, 2003; Samuel and Reeves 2003). The repeating units from different structures may comprise different numbers of monosacchrides as linear or branched homo-oligosaccharides or hetero-oligosaccharides (Skurnik and Bengoechea 2003). As a result, the repeating unit differs from strain to strain within a serotype, and thus shows an enormous structural variability and determines the serological specificity of the LPS, a kind of fingerprint for bacteria. During biosynthesis, repeating units are polymerized into blocks with different lengths and then added to the core. The individual O-antigen chains can vary in length ranging up to 40 repeat units. The O-antigen is much longer than the core oligosaccharide and functions as the hydrophilic domain of the LPS (Rocchetta et al. 1999; Samuel and Reeves 2003). The structure of the O-polysaccharide defines the O-antigen serological specificity in an organism. However, the numbers of unique O-antigens within a species vary significantly. For example, *E. coli* produces at least 170 O serotypes, and there are 46 serogroups in *S. enterica* (Comstock and Kasper 2006; Orskov et al. 1977). On the other hand, modifications to these basal O-repeat unit structures form many additional "O factors." As a result, the O-antigens of most Gram-negative bacteria are highly variable, strain-specific surface antigens. The nature of the O-antigen may, in some cases, directly contribute to the pathogenic effects of the bacteria. In addition, the O-antigen helps the bacterium escape the lytic action of the complement complex by a shielding process, and protects the bacteria from the effect of numerous antibiotics since the nature of the O side chain can easily be modified by Gram-negative bacteria to avoid detection (Banemann et al. 1998; Joiner et al. 1984). Reviews dedicated to bacterial LPS, LPS endotoxins, biosynthesis of O-antigens, and vaccine antigens have reported extensive data on the chemical structures of the O-antigens (Galanos and Freudenberg 1993; Pier, 2003; Raetz and Whitfield 2002; Rocchetta et al. 1999; Samuel and Reeves 2003; Whitfield 1995; Wilkinson 1996).

The surfaces of many bacteria are covered by polysaccharides consisting of a capsular polysaccharide (CPS) and/or a lipopolysaccharide (LPS), which helps the pathogen to establish an infection. In Gram-negative bacteria, LPS covers about 40% of the bacterial surface. *E. coli* isolates produce two serotype-specific surface polysaccharides: LPS O-antigen and CPS K-antigen. Variations in structures of these polysaccharides result in ~170 different O-antigens and ~80 K-antigens (Comstock and Kasper 2006; Orskov et al. 1977; Whitfield 2006; Whitfield and Roberts 1999). The CPSs constitute a homopolymer of monosaccharides like α-$(2{\rightarrow}8)$-linked sialic acid in *Neisseria meningitidis* groups B and C, and *E. coli* K1, or a homopolymer with repeating units consisting of two to six glycosyl residues (Lee 1987). The CPSs may be present in both Gram-negative bacteria such as *N. meningitidis*, *Haemophilus influenzae*, *E. coli* or *Salmonella typhi* and in Gram-positive bacteria such as *Streptococci* and *Staphylococci*

(O'Riordan and Lee 2004; Ovodov 2006b; Shinefield and Black 2005; Swiatlo and Ware 2003; Whitfield and Paiment 2003). The CPSs hide cell surface components of the bacterium from the immune system of the host by preventing complement activation by cell surface proteins and inhibiting phagocytosis. The CPSs present on the surface of pathogenic bacteria are essential virulence factors and protective antigens (Ovodov 2006a), and have been used as the basis of vaccines against several diseases caused by capsular microorganisms (Jones 2005; Ovodov 2006b; Sood et al. 1996). For example, E. coli K1 antigen (the capsular polysialic acid), a linear homopolymer of alpha-2,8-linked N-acetylneuraminic acid residues, plays an essential role in pathogenesis by protecting the invasive bacteria from host innate immunity (Ovodov 2006a; Whitfield 2006; Whitfield and Roberts 1999). Expression the of polysialic acid capsule appears to be essential in order for K1 strains to cause infection, since mutants lacking polysialic acid are not virulent (Cross 1990; Moxon and Kroll 1990). Reviews dedicated to bacterial CPSs, CPS-protein vaccines, and biosynthesis of CPSs in E. coli, have reported extensive data on the chemical structures of the bacterial CPSs (Cobb and Kasper 2005; Jones 2005; Ovodov 2006a,b; Sood et al. 1996; Whitfield 2006; Whitfield and Roberts 1999).

4. Carbohydrate Microarrays for Detection of Bacteria

For over a decade, microarray-based technologies have been extensively developed as low-cost, high-throughput analytical tools to facilitate fast, quantitative, and simultaneous analyses of a large number of biomolecular interactions (Angenendt 2005; Kersten et al. 2005; Stoughton 2005). DNA microarrays have been constructed in situ on glass slides by photolithographic methods or by using ink-jet technology (Stoughton 2005). The polymerase chain reaction (PCR) has provided quick amplification of libraries of complementary DNA to analyze mutation of genes, investigate change of patterns of gene expression in diseases, and track the activities of many genes simultaneously (Stoughton 2005). Protein microarrays on glass slides, microwells and three-dimensionally modified gel-pad chips have been applied for the high-throughput studies of protein–protein, protein–DNA and protein–ligand interactions, and profiling of protein expression in normal and diseased states (Angenendt 2005; Kersten et al. 2005; Winssinger and Harrisß 2005).

Recent progress in carbohydrate microarrays has significantly changed our ability to study carbohydrate–protein interactions at the molecular level (Campbell and Yarema 2005; de Paz and Seeberger 2006; Khan et al. 2004; Wang 2003). Carbohydrate microarrays combine the benefits of immobilized format assays with the capability of detecting thousands of analytes simultaneously. These microarrays can offer a general and powerful platform for whole-cell applications because their multivalent display of carbohydrates can mimic multivalent inter-actions at cell-cell interfaces. Interactions of proteins, cells or bacteria with a multivalent carbohydrate microarrays are much more avid and specific than those with the monovalent counterparts as individual carbohydrate-protein interactions are generally weak (Klefel and von Itzstein 2002; Lis and Sharon 1998; Mammen et al. 1998; Wong 1999). The advantages of carbohydrate microarrays also include assay miniaturization, simplification of the isolation, simultaneous detection of analytes, and high throughput (Seeberger 2006). A series of carbohydrate microarrays have been prepared to rapidly determine the binding profile of carbohydrate-binding proteins (Bryan et al. 2004; Feizi et al. 2003; Hanson et al. 2004; Hirabayashi 2003; Houseman et al. 2003; Houseman and Mrksich 2002; Love and Seeberger 2002; Mrksich 2004; Park et al. 2004; Park and Shin 2002; Ratner et al. 2004a,b; Smith et al. 2003; Wang 2003; Weis and Drickamer 1996; Wong 2005), detect pathogens and specific antibodies for the diagnosis of diseases (Wang et al. 2002), characterize carbohydrate-cell recognition events (Disney and Seeberger 2004), identify new inhibitors of protein–carbohydrate interactions (Fukui et al.

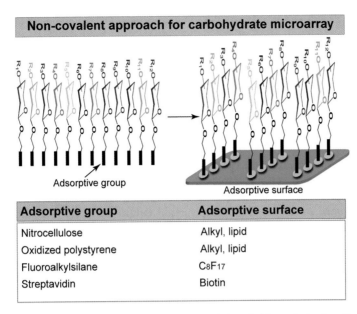

Figure 24.3. Noncovalent approach for carbohydrate microarrays (adapted with permission from Paulson et al. (2006), Copyright 2006 Nature Publishing Group).

2002) and examine the adhesion properties of hepatocytes and leukocytes (Palma et al. 2006) by using both naturally derived and synthetic glycans on media ranging from multiwell plates to modified glass slides. Noncovalent and covalent approaches have been used to prepare carbohydrate microarrays (Figs. 24.3 and 24.4), and there are a few excellent reviews about fabrication techniques and advances in carbohydrate microarrays (Campbell and Yarema 2005;

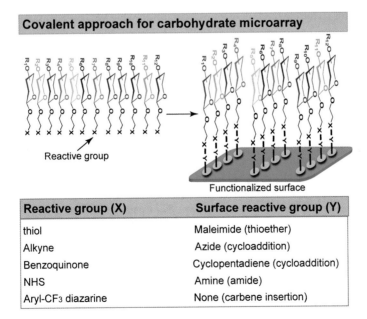

Figure 24.4. Covalent approach for carbohydrate microarrays (adapted with permission from Paulson et al. (2006), Copyright 2006 Nature Publishing Group).

Culf et al. 2006; de Paz and Seeberger 2006; Dyukova et al. 2006; Feizi et al. 2003; Khan et al. 2004; Monzo and Guttman 2006; Ratner et al. 2004a; Shin et al. 2005; Stevens et al. 2006; Wang 2003; Wang et al. 2006b).

Disney and Seeberger reported a simple but very effective diagnostic carbohydrate microarray for the quick detection of bacteria in complex biological mixtures (Disney and Seeberger 2004). They constructed a carbohydrate microarray through the covalent approach of amide bond formation by printing solutions of five different amine-functionalized monosaccharide derivatives onto glass slides that were coated with a hydrophilic polymer bearing N-hydroxysuccinimide esters. Specific binding of the fluorescently labeled *E. coli* strain ORN 178 to mannose was observed on the microarray. The microarray was also able to detect strain-specific binding differences between a strain ORN 178 and a mutant strain ORN 209 of *E. coli* with a reduced affinity for mannose. Moreover, identification of bacterial strains could be further confirmed by the more time-consuming culture and colony counting methods or PCR methods because bacteria can be collected from the carbohydrate microarray and further cultured. The detection limit of the microarray system for bacteria is 10^5–10^6 cells. The most significant aspect of this microarray is the effective detection of bacteria in a more complex sample containing sheep erythrocytes and serum even though the detection limits are worsened in the contaminated sample. The detection limits for bacteria in the complex samples could be further improved by using new fabrication strategies to prepare carbohydrate microarrays with excellent selectivity and quantitative performance. The ideal fabrication approaches can facilitate preparation of carbohydrate microarrays with desirable features that can prevent nonspecific adsorption, present carbohydrates in a homogenous environment with the same activity toward proteins and bacteria, and precisely control density and orientation of carbohydrates for optimal multivalent carbohydrate–protein interactions (Houseman and Mrksich 2002). Mrksich and colleagues have prepared a carbohydrate microarray of ten monosaccharides on gold-coated glass slides by covalently attaching the monosaccharides to mixed self-assembled monolayers terminated with benzoquinone and penta (ethylene glycol) groups on the gold-coated glass slides through the Diels–Alder cycloaddition reaction (Houseman and Mrksich 2002). The microarray was evaluated by profiling the binding specificities of five plant lectins, and specific monosaccharide binding was observed for the five lectins. This fabrication approach provides carbohydrate microarrays with many good features such as excellent control of carbohydrate density and efficient prevention of nonspecific adsorption of sticking protein fibrinogen (Houseman and Mrksich 2002). As a result, this fabrication approach could be employed to further enhance detection limits for bacteria in more complex samples by reducing background levels and preventing nonspecific adsorption.

Wang et al. reported another direct and unique approach to detect pathogenic bacteria by using a carbohydrate microarray with 48 microbiol polysaccharide probes for specific recognition of carbohydrate-binding antibodies from 20 human serum specimens (Wang et al. 2002). They screened serum samples from individuals and successfully detected antibodies that bound to polysaccharides of pathogenic strains of *E. coli* and *Pneumococcus*. This approach can open the way for rapid detection of pathogenic bacteria through specific interactions of antibodies with unique cell-surface polysaccharides of pathogenic bacteria.

5. Lectin Microarrays for Detection of Bacteria

Pathogenic bacteria possess a cell surface capsular polysaccharide (CPS) or lipopolysaccharide (LPS) shell, or both, which helps the pathogen initiate an infection (Weintraub 2003). The CPSs hide cell surface components of the bacterium from the immune system of the host, prevent complement activation by cell surface proteins and inhibit phagocytosis (Roberts

1996). LPS consists of a core lipid-anchored saccharide attached to a repeating sugar unit referred to as the O-antigen (Caroff and Karibian 2003). Bacteria can quickly modify their glycan structures to respond to growth conditions and other factors. Rapid profiling of glycan structures of pathogenic bacteria can provide insights into the role of bacterial glycans in pathogenesis, and help develop new approaches for detection and prevention of pathogenic bacteria. The capability of profiling diverse glycan structures of cells and pathogenic bacteria has been achieved by using naturally occurring carbohydrate-binding lectins (Carlsson et al. 2005; Ebe et al. 2006; Hsu et al. 2006; Mahal et al. 2004; Zweigner et al. 2006). Lectin microarrays have been prepared by immobilizing the lectins on glass slides, and then incubating them with fluorescently labeled cells and pathogenic bacteria for profiling of the diverse glycan structures (Hsu et al. 2006; Mahal et al. 2004; Zheng et al. 2005). The pattern of bound cells can dynamically profile specific glycan structures present on the bacterial surfaces according to the specific lectin-LPS interactions (Fig. 24.5).

Hsu *et al.* report a lectin microarray consisting of 21 lectins to profile diverse bacterial surface glycans by printing the lectins onto activated glass slides (Hsu et al. 2006). After the lectin microarrays were incubated with fluorescently labeled bacteria, the researchers analyzed each strain of bacteria by reading the patterns of small fluorescent lectin spots and determining which glycans were present in the bacterial surfaces according to specific binding of bacterial glycans to the immobilized lectins. They found out that the bacteria reproducibly and specifically bound only to certain lectins. Moreover, the lectin microarrays could be used to differentiate between strains that could not be distinguished by traditional hemagglutination assays since weak binding of LPS to some lectins could be detected on the microarray slides through cooperative multivalent carbohydrate-protein interactions. They further confirmed the binding specificity of the bacteria to the microarrays by adding a carbohydrate to the lectin microarrays to block the binding of the bacteria to the expected lectins. The microarrays readily monitored growth-related changes in surface glycosylation of a neonatal meningitis strain of *E. coli*. Understanding LPS structural variation in bacterial pathogens is important because O-antigen modifications play an important role in the infection process such as the adherence and the ability to protect the bacteria from the effect of numerous antibiotics (Banemann et al. 1998; Joiner et al. 1984). The lectin microarrays can provide a very important and powerful tool to detect bacteria according to their surface polysaccharides and discover dynamic changes in surface glycosylation of bacteria in response to environmental stimuli. However, they have some limitations. They are unable to detect all bacterial glycans in complex samples as they are limited by the number of commercially available lectins that can bind specifically to bacterial glycans.

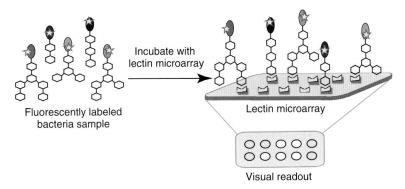

Figure 24.5. Lectin microarrays for bacterial screening (adapted from Prescher and Bertozzi (2006), with permission of Elsevier).

Ertl et al. reported lectin microarrays for electrochemical detection of *E. coli* subspecies and other five bacteria by taking advantage of the specific binding between lectins and bacterial LPSs (Ertl and Mikkelsen 2001; Ertl et al. 2003). They immobilized up to ten lectins on a porous membrane surface and confirmed that the lectins can specifically bind to bacterial LPSs. Electrochemical detection was achieved by measuring the current changes of redox ferricyanide within the bacterial respiratory chains during the bacterial LPS-lectin recognition process. Although carbohydrate-lectin recognition provides an indirect approach for bacterial detection, the specific multivalent interactions between bacterial LPSs and lectin microarrays could offer high sensitivity and selectivity of the electrochemical microarrays in the absence of other biological redox species such as ascorbate acid. The ascorbate acid could cause potential interference since it could undergo not only electrocatalytic oxidation by ferricyanide but also direct oxidation at Pt electrode under potential of 0.5 V versus Ag/AgCl (Zen, 2003).

6. Conjugated Fluorescent Glycopolymers for Detection of Bacteria

Fluorescent conjugated glycopolymers, combining fluorescent scaffolding and carbohydrate reporting functions into one package, are very attractive for bacterial biosensing applications because of their intrinsic fluorescence and high sensitivity to minor external stimuli due to signal amplification by a cooperative system response (Bunz 2000; Kim et al. 2005; McQuade et al. 2000).

Baek et al. reported glycopolythiophenes for colorimetric detection of the influenza virus and *E. coli* bacteria (Baek et al. 2000). A series of glycopolythiophenes were prepared by copolymerization of carbohydrate-functionalized thiophene monomers with methyl (3-thienyl) acetate as a comonomer. Tethered spacer-lengths between the polymer backbone and carbohydrate residues were adjusted to optimize carbohydrate–protein interactions. Colorimetric detection of the influenza virus and *E. coli* was achieved by incubating glycopolythiophenes bearing mannose and sialic acid residues with influenza virus A/B and *E. coli* strain HB101, respectively, and examining binding interactions by the naked eye and UV-visible absorption spectroscopy. The binding interactions cause an unusual red-shift in the visible absorption of the polymers, which was observed with the naked eye as a color of the glycopolymer solution changed to dark red. The red-shift in the visible absorption of the polymers may be due to enhanced conjugation of the polymer backbone and a more extended π-conjugated system caused by highly cooperative multivalent interactions between the carbohydrate residues and the receptor proteins of *E. coli* or influenza virus. The glycopolythiophenes offer a new platform for colorimetric assay of pathogenic bacteria.

Disney et al. reported fluorescent detection of *E. coli* bacteria by using fluorescent conjugated glycopoly(p-phenylene-ethynylene)s (Fig. 24.6) (Disney et al. 2004). They prepared the glycopolymers by postpolymerization functionalization of poly(p-phenylene-ethynylene) bearing carboxylic acid groups with amine-functionalized mannose or galactose through amide bridges, which provided a versatile scaffold for the rapid functionalization of conjugated polymers with a variety of different carbohydrates to detect a wide range of different bacteria. Fluorescent detection of *E. coli* was achieved by incubating *E. coli* with the glycopolymers and examining the binding interactions by spectrofluorometer and fluorescent confocal microscopy. The bacteria strongly bind to the glycopolymer bearing mannose residues, forming fluorescent clusters of about a thousand bacteria through multivalent interactions, and resulting in a red shift in the fluorescent spectra of the glycopolymer upon incubating the glycopolymer with *E coli* (Fig. 24.7b, c and d). The specific binding of the glycopolymer bearing mannose residues was further confirmed by using mutant *E coli* which failed to bind to the glycopolymer (Fig. 24.7a). The brightly fluorescent bacteria clusters could easily be detected by irradiating the solution. This method can detect *E. coli* within 10 to 15 minutes with a detection limit of 10^4 bacteria.

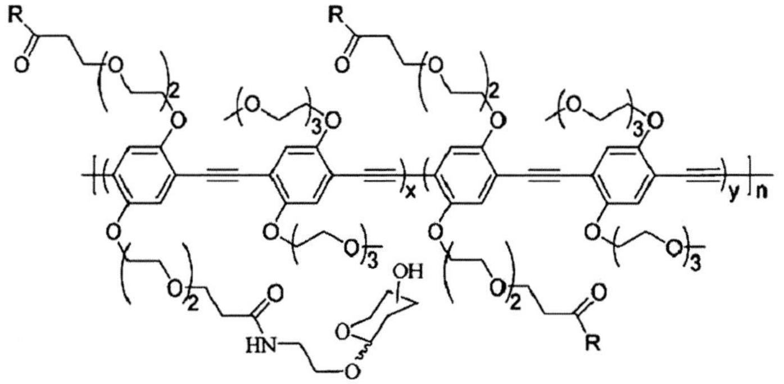

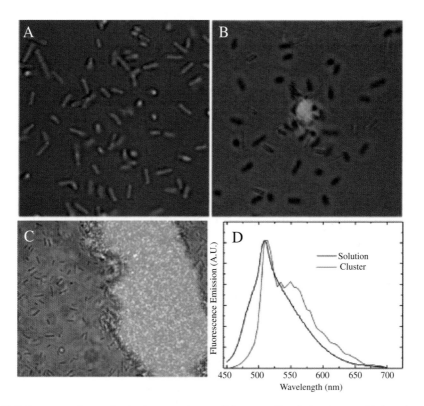

1 R=OH; x : y = 0 : 1
2a R = OH or NH(CH$_2$)$_2$OH; x : y = 1 : 1
    sugar = mannose
2b R = OH or NH(CH$_2$)$_2$OH; x : y = 1 : 1
    sugar = galactose

Figure 24.6. Chemical structures of glycopoly(p-phenylene-ethynylene)s (Reprinted with permission from Disney et al. (2004), Copyright 2004 American Chemical Society).

Figure 24.7. Laser scanning confocal microscopy image of **A** mutant *E. coli* that does not bind to glycopolymer bearing mannose. Individual cells observed without aggregation. **B** A fluorescent bacterial aggregate due to multivalent interactions between the mannose-binding bacterial pili and the glycopolymer bearing mannose. **C** Fluorescence microscopy image of a large fluorescent bacterial cluster. **D** Conventional fluorescence spectra of glycopolymer bearing mannose in PBS (black) and normalized fluorescence spectra of a bacterial cluster obtained using confocal microscopy (red) (Reprinted with permission from Disney et al. (2004), Copyright 2004 American Chemical Society).

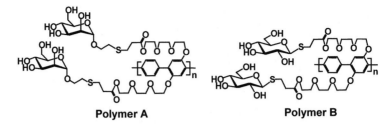

Polymer A   **Polymer B**

Figure 24.8. Chemical structures of glycopolymers bearing α-mannose (polymer A) and β-glucose (polymer B). (Reprinted with permission from Xue et al. (2006), Copyright 2006 American Chemical Society).

Xue et al. reported simple, convenient and versatile prepolymerization and postpolymerization functionalization approaches to prepare highly water-soluble fluorescent conjugated glycopoly(p-phenylene)s for detection of bacteria (Fig. 24.8) (Xue et al. 2006b). They prepared the glycopolymers by postpolymerization functionalization of bromo-bearing poly(p-phenylene) with thiol-functionalized mannose or glucose through formation of thioether bonds in a basic condition. This approach offers a simple and versatile means to quickly attach a variety of different carbohydrates to conjugated polymers for well-defined fluorescent conjugated glycopolymers due to almost 100% degree functionalization of the precursor polymers (Xue et al. 2006a). Glycopoly(p-phenylene)s are highly water-soluble and fluorescent presumably due to relatively weaker π-π stacking interactions between poly(p-phenylene) backbone than those between neutral glycopoly(p-phenylene-ethynylene)s, which show a relatively low water-solubility (Kim et al. 2004). Specific binding to *E. coli* was confirmed by incubating the glycopolymers with mannose-binding ORN178 strain and mutant ORN208 strain (which is deficient in the fimH gene and expresses abnormal type 1 pili that fail to mediate D-mannose-specific binding). Incubation of the mannose-bearing polymer with the ORN178 strain resulted in the formation of fluorescently stained bacterial clusters under irradiation (Fig. 24.9). The mannose-bearing polymer selectively binds to the wild-type type 1 pili of the ORN178 strain but not to abnormal type 1 pili of the mutant ORN208 strain, demonstrating specific binding of the glycopolymer to FimH protein.

Since the first report of colorimetric detection of the influenza virus by using a polydiacetylene film (Charych et al. 1993; Nagy et al. 1993; Reichert et al. 1995; Spevak et al. 1993),

Figure 24.9. Visualization of mannose-binding *E. coli* ORN178 strain (left) and mutant *E. coli* ORN208 strain (right) after incubation with mannose-bearing polymer A. (Reprinted with permission from Xue et al. (2006), Copyright 2006 American Chemical Society).

a variety of carbohydrate-functionalized polydiacetylene monolayers and vesicles have been prepared for colorimetric detection of bacteria since they are unique in terms of the method of preparation, molecular structure, and output signal (Li et al. 2002; Ma et al. 2003, 2000, 1998; Su et al. 2005; Sun et al. 2004a; Zhang et al. 2002, 2005, 2004). Unlike other conjugated polymers, functionalized polydiacetylenes (PDAs) are prepared by using photo-polymerization of self-assembled diacetylene monomers. Closely packed and properly ordered diacetylene lipids are polymerized through 1,4-addition reaction to form alternating ene-yne polymer chains upon irradiation (Figs. 24.10 and 24.11). PDAs display an intense blue color when they are generated under optimized photochemical conditions. The nanostructured polydiacetylenes for bacterial detection have been prepared as vesicles in aqueous solutions (Ma et al. 2000, 1998), Langmuir-Blodgett (LB)/Langmuir-Schaefer (LS) films (Zhang et al. 2004), or immobilized vesicles on solid supports (Figs. 24.8 and 24.9) (Zhang et al. 2005). The unique property of nanostructured PDAs for bacterial sensing applications is a color change from blue to red in the polymers in response to interactions of the polymeric carbohydrates with bacterial pili, which cause the changes in effective conjugated length of the delocalized π-conjugated polymer backbone. Ma et al. reported a simple approach to prepare glucose-functionalized PDAs by inserting easily-obtained dioctadecyl glyceryl ether-β-glucoside into PDAs (Ma et al. 1998). Incubation of *E. coli* with the PDA vesicles caused the polymer color to change from deeply

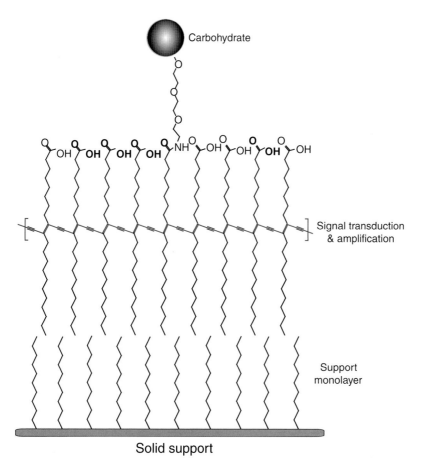

Figure 24.10. Polydiacetylene thin films on solid support (adapted from Charych et al. (1996), with permission of Elsevier).

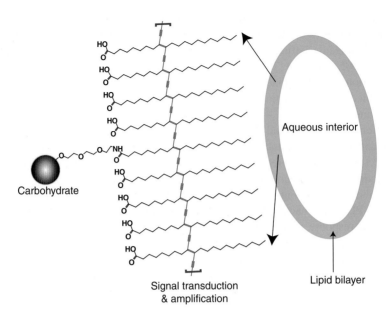

HO
O
HO
O
HO
O
HO
O
NH
O
HO
O
HO
O
HO
O
HO
O

Carbohydrate

Aqueous interior

Signal transduction
& amplification

Lipid bilayer

Figure 24.11. Polydiacetylene liposomes or vesicles (adapted from Charych et al. (1996), with permission of Elsevier).

blue to red within several seconds, which resulted from cooperative multivalent interactions of the polymeric carbohydrates with the bacterial pili. They further evaluated the effect of spacer length of glycolipid molecules and the different diacetylenic lipid matrixes in color changeable polydiacetylenic vesicles on the detection of *E coli*, and showed that the glycolipids with longer spacers enhance the sensitivity of the system (Ma et al. 2000). The simplicity of design and synthesis of a variety of carbohydrate-functionalized PDAs make them very promising prototype platforms for detection of pathogenic bacteria.

7. Glyconanoparticles for Detection of Bacteria

Glyconanoparticles and glyconanorods can provide ideal multivalent, 3-dimensional systems to study carbohydrate-protein interactions involving cellular recognition. They can display multivalent carbohydrates in a globular or red shape on the surface with chemically well-defined composition, and possess unique physical properties to study carbohydrate–protein interactions because of the quantum size effect (Daniel and Astruc 2004; de la Fuente and Penades 2004). As a result, glyconanoparticles and glyconanorods are expected to have a wide variety of potential applications including diagnostic kits, vaccines, and targeted drug delivery agents against cancer, and bacterial or viral infections. Glyconanoparticles stabilized with thiol-functionalized monosaccharide, disaccharide, oligosaccharide and glycopolymers have been prepared to study carbohydrate–carbohydrate, carbohydrate–protein, and carbohydrate–cell interactions by measuring color changes associated with the aggregation of metal glyconanoparticles or fluorescent changes of semiconductor glyconanoparticles (de la Fuente et al. 2005; Reynolds et al. 2006; Robinson et al. 2005).

The deep red color of gold nanoparticles in water is due to the surface plasmon absorption band in the visible region around 520 nm when an interparticle distance is greater than the average particle diameter (Schofield et al. 2006). The maximum absorption band is dependent on the sizes and shapes of gold nanoparticles. Gold nanoparticles with mean diameter of 9, 15, 22, 48, and 99 nm, display the maximum absorption band of the surface plasmon absorption at 517,

520, 521, 533, and 575 nm in an aqueous solution, respectively, while gold nanoparticles with core diameters less than 2 nm do not show the absorption (Daniel and Astruc 2004). Formation of aggregated gold nanoparticles triggers a reversible change in color in gold nanoparticles from red to violet since coupling interactions cause the surface plasmon absorption band to shift to longer wavelengths (Daniel and Astruc 2004). Monodisperse silver nanoparticles typically have a surface plasmon absorption band around 400 nm. Silver nanoparticles, well dispersed in aqueous solution, display a yellow color and the aggregated nanoparticles in aqueous solution appear orange (Schofield et al. 2006). The color changes associated with metal glyconanoparticle aggregation have been successfully exploited to develop colorimetric assays for the detection of carbohydrate–carbohydrate and carbohydrate–protein interactions as the multivalent carbohydrate-carbohydrate or carbohydrate-protein interactions trigger aggregation of metal glyconanoparticles (Reynolds et al. 2006; Schofield et al. 2006).

α-Mannose-functionalized gold glyconanoparticles have been shown to specifically bind to FimH adhesion of type 1 pili in *E. coli* by transmission electron microscopy (TEM) and exhibit stronger interactions with FimH on the basis of the glycoside cluster effect than free mannose in the competition assay since up to 200 mannoses were attached to each glyconanoparticle (Lin et al. 2002). Lin et al. verified the specific binding of the glyconanoparticles to FimH by incubating the glyconanoparticles with *E. coli* ORN178 and ORN208 strains and used TEM to examine the specific binding of the glyconanoparticles to the type 1 pili of the ORN178 strain but not to those of the ORN208 strain. Since the ORN178 strain expresses wild-type type 1 pili, and the mutant ORN208 strain deficient in the *fimH* gene expresses abnormal type 1 pili that fail to mediate D-mannose specific binding (Fig. 24.12). Glyconanoparticles serve as a multivalent nanoscaffold to achieve strong and selective binding to bacterial type 1 pili through cooperative multiple interactions, which offers a new method to label specific proteins on the cell surface with glyconanoparticles.

Quantum dots (QDs) have many important advantages over conventional fluorescent dyes (Akamatsu et al. 2005; Popescu and Toms 2006; Sharrna et al. 2006). The first emission of QDs is narrow, symmetric and independent of excitation wavelengths although their absorption band is broad (Akamatsu et al. 2005). Second, the emission wavelength of QDs is easily tuned by changing the material composition and size of the cores for simultaneous detection of analytes (Akamatsu et al. 2005). Third, QDs can be prepared to possess long-term photostability and high fluorescent quantum yield. Glyco-quantum dots combining fluorescent property and carbohydrate reporting groups into one package can offer powerful systems to study carbohydrate–protein interactions. Although glycol-quantum dots stabilized by monosaccharide

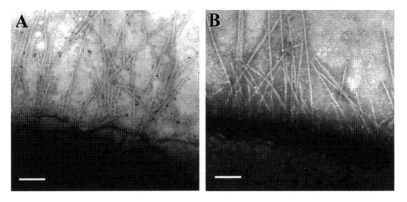

Figure 24.12. Typical TEM images of the *E. coli* ORN178 strain bound with the glyconanoparticles (A), and the *E. coli* ORN208 strain deficient of the *fimH* gene failing to bind to the glyconanoparticles (B). Scale bar = 100 nm (Reprinted with permission from Lin et al. (2002), Copyright 2002 American Chemical Society).

and polysaccharides have been successfully prepared to study carbohydrate–lectin interactions (Chen et al. 2003; Robinson et al. 2005; Sun et al. 2004b), there are not any reports of glyco-quantum dots for fluorescent detection of bacteria.

8. Carbohydrate-Functionalized Carbon Nanotubes for Detection of Bacteria

Carbon nanotubes (CNTs) have attracted enormous interest because of their unique structural, mechanical and electronic properties (Chen et al. 2004; Li et al. 2003; Sherigara et al. 2003). They are especially attractive candidates for nanoscale biosensing systems because of their exceptional chemical and biochemical stability (Chen et al. 2004; Li et al. 2003; Sherigara et al. 2003). The most important aspect of the one-dimensional direct narrow band gap of semi-conducting carbon nanotubes is their ability to fluoresce brightly in the 800- to 1600-nm wavelength range of the near infrared, a region important in bioimaging (Jeng et al. 2006; O'Connell et al. 2002).

Gu et al. (2005) used galactosylated single-walled carbon nanotubes (Gal-SWNTs) as polyvalent ligands for specifically capturing pathogenic *E. coli* (Fig. 24.13). They prepared water-soluble Gal-SWNTs through amidation of an amino group of 2′-aminoethyl-β-D-galactopyranoside with the nanotube-bound carboxylic acids. They confirmed the specific binding of Gal-SWNTs to *E. coli* P pili by incubating Gal-SWNTs with *E. coli* O157 : H7 strain C7927 (which expresses P-type *E. coli* pili) and examining the specific binding by scanning electron microscopy (SEM). SEM results showed that Gal-SWNTs exhibit strong cell adhesion, resulting in efficient agglutination of pathogenic *E. coli* through the cooperative multivalent effect (Fig. 24.14). In addition, they found that there is no apparent binding when Gal-SWNTs were replaced by SWNTs functionalized with either α-D-mannose (Man-SWNTs) or bovine serum albumin protein under essentially the same experimental conditions.

Wang et al. also demonstrated that Gal-SWNTs or Man-SWNTs are unique multivalent scaffolds that bind effectively to *Bacillus anthracis* spores through divalent cation-mediated multivalent carbohydrate–carbohydrate interactions, forming significant spore aggregation and the associated substantial reduction in colony forming units (Wang et al. 2006a). They confirmed effective binding of Gal-SWNTs and Man-SWNTs to *B. anthracis* spores through carbohydrate–carbohydrate interactions by mixing Gal-SWNTs and Man-SWNTs with *B anthracis* spore suspension to form apparently homogeneous mixtures, respectively, and sequentially adding

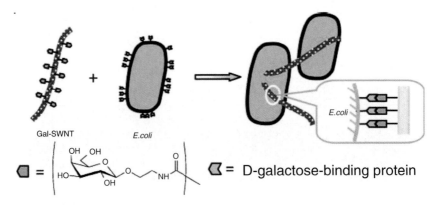

Figure 24.13. Multivalent interactions of Gal-SWNTs with *E. coli*. (Reprinted from Gu et al. (2005), with permission of The Royal society of Chemistry).

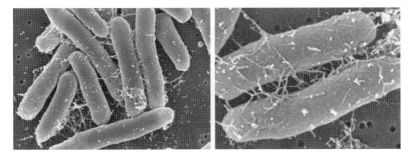

Figure 24.14. SEM images for Gal-SWNTs capturing pathogenic *E. coli* cells. (Reprinted from Gu et al. (2005), with permission of The Royal society of Chemistry).

divalent cation Ca^{2+} ions, which result in substantial aggregation of the *B. anthracis* spores. The aggregated *B anthracis* spores were easily observed by the naked eye, optical microscopy and scanning electron microscopy (Figs. 24.15 and 24.16). Interestingly, the binding to aggregate *B. anthracis* spores is unique to Gal-SWNTs or Man-SWNTs, and is not available to polymeric nanoparticles functionalized with mannoses or galactoses. This unique binding of Man-SWNTs or Gal-SWNTs to B. *anthracis* spores may be due to specific arrangements of the carbohydrate ligands required for multivalent interactions with the spore surface. They further confirmed divalent cation-mediated multivalent carbohydrate–carbohydrate interactions of *B. anthracis* spores with Man-SWNTs by adding EDTA (a strong chelating agent of Ca^{2+} ion) to a solution containing the aggregated *B. anthracis* spores, which cause the disappearance of the aggregated *B. anthracis* spores by disrupting the divalent cation-mediated multivalent carbohydrate–carbohydrate interactions. This approach may find potential applications in the detection and decontamination of *B. anthracis* spores.

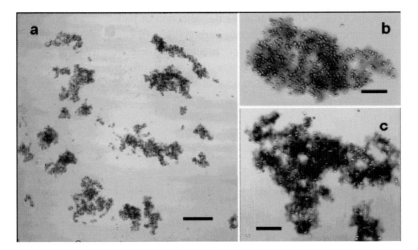

Figure 24.15. Optical micrographs showing the aggregation of *B. anthracis* spores induced by the Ca^{2+}-mediated binding with Man-SWNT (a and b) and Gal-SWNT (c). Scale bars: $100\,\mu m$ (a) and $20\,\mu m$ (b,c). (Reprinted with permission from Wang et al. (2006), Copyright 2006 American Chemical Society).

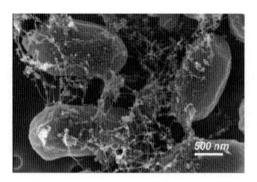

Figure 24.16. A high-resolution SEM image showing the interactions of Man-SWNTs with *B. anthracis spores.* (Reprinted with permission from Wang et al. (2006), Copyright 2006 American Chemical Society).

9. Conclusions and Future Perspectives

Traditional pathogen detection methods are often too slow to be of any practical use although they are highly sensitive. As a result, new rapid, highly sensitive and selective methods are still needed. Carbohydrate-based detection of bacterial pathogens represents an exciting alternative to standard methods. In order to become a real and trustworthy alternative, new methods based on carbohydrates as sensing elements should detect bacterial pathogens with high speed and selectivity, and possess at least the same detection limits as traditional techniques (between 10 and 100 CFU mL^{-1}).

With further advances in the generation of diverse glycan libraries, a series of arrays of water-soluble fluorescent conjugated glycopolymers or glyco-quantum dots will be prepared by tuning them to have different wavelength emissions and bear diverse glycan residues for simultaneous detection of different pathogenic bacteria present in complex biological samples in the near future. The field of biosensors for the detection of pathogenic bacteria will continue to grow with the further development of new fabrication strategies, new microarray surfaces and printing methods, miniaturization of microarrays, and enhanced sensitivities of detection. The carbohydrate microarrays consisting of vast numbers of natural and synthetic oligosaccharides, and also glycoconjugates, will revolutionize the detection of pathogenic bacteria. Future availability of diverse natural lectins, engineering carbohydrate-binding proteins and antibodies will facilitate lectin microarray construction for simultaneous detection of a variety of pathogenic bacteria. The unique near-infrared band-gap fluorescent properties of single-walled carbon nanotubes functionalized with a wide variety of carbohydrates will be explored for detection of bacteria in the near future.

Acknowledgements

The author gratefully acknowledges the Research Excellence Fund of Michigan Technological University, 21st Century Jobs Fund of Michigan (contract number: 06-1-P1-0283) for support of the work. This project was also supported by National Research Initiative Grant no. 2007-35603-17740 from the USDA Cooperative State Research, Education, and Extension Service - Nanoscale Science and Engineering for Agriculture and Food Systems. The author would like thank Mrs. Haihua Li in J. Robert Van Pelt library, Michigan Technological University, for helping to draw the adapted illustrations.

References

2006 North American *E. coli* outbreak (2006). U.S.FDA News, October 6

Ada G and Isaacs D (2003) Carbohydrate-protein conjugate vaccines. Clin Microbiol Infect 9:79–85

Akamatsu K, Tsuruoka T and Nawafune H (2005) Band gap engineering of CdTe nanocrystals through chemical surface modification. J Am Chem Soc 127:1634–1635

Alexander C and Rietschel ET (2001) Bacterial lipopolysaccharides and innate immunity. J Endoxtin Res 7:167-202

Allos BM and Blaser MJ (1995) Campylobacter-Jejuni and the Expanding Spectrum of Related Infections. Clin Infect Dis 20:1092–1099

Angenendt P (2005) Progress in protein and antibody microarray technology. Drug Discov Today 10:503–511

Appeldoorn CCM, Joosten JAF, el Maate FA, Dobrindt U, Hacker J, Liskamp RMJ, Khan AS and Pieters, RJ (2005) Novel multivalent mannose compounds and their inhibition of the adhesion of type 1 fimbriated uropathogenic *E. coli*. Tetrahedron-Asymmetr 16:361–372

Backhed F, Alsen B, Roche N, Angstrom J, von Euler A, Breimer ME, Westerlund-Wikstrom B, Teneberg S and Richter-Dahlfors A (2002) Identification of target tissue glycosphingolipid receptors for uropathogenic, F1C-fimbriated Escherichia coli and its role in mucosal inflammation. J Biol Chem 277:18198–18205

Baek MG, Stevens RC and Charych DH (2000) Design and synthesis of novel glycopolythiophene assemblies for colorimetric detection of influenza virus and E-coli. Bioconjug Chem 11:777–788

Banemann A, Deppisch H and Gross R (1998) The lipopolysaccharide of Bordetella bronchiseptica acts as a protective shield against antimicrobial peptides. Infect Immun 66:5607–5612

Barasch A, Gordon S, Geist RY and Geist JR (2003) Necrotizing stomatitis: Report of 3 Pseudomonas aeruginosa-positive patients. Oral Surgery Oral Medicine Oral Pathology Oral Radiology and Endodontics 96:136–140

Barthelson R, Mobasseri A, Zopf D and Simon P (1998) Adherence of Streptococcus pneumoniae to respiratory epithelial cells is inhibited by sialylated oligosaccharides. Infect Immun 66:1439–1444

Berenson CS, Sayles KB, Huang J, Reinhold VN, Garlipp MA and Yohe HC (2005) Nontypeable Haemophilus influenzae-binding gangliosides of human respiratory (HEp-2) cells have a requisite lacto/neolacto core structure. FEMS Immunol Med Microbiol 45:171–182

Beuth J, Ko HL, Tunggal L and Pulverer G (1992) Urinary-Tract Infection Caused by Staphylococcus Saprophyticus - Increased Incidence Depending on Blood-Group. Dtsch Med Wochenschr 117:687–691

Bouckaert J, Mackenzie J, de Paz JL, Chipwaza B, Choudhury D, Zavialov A, Mannerstedt K, Anderson J, Pierard D, Wyns L et al. (2006) The affinity of the FimH fimbrial adhesin is receptor-driven and quasi-independent of Escherichia coli pathotypes. Mol Microbiol 61:1556–1568

Brooks BW, Devenish J, Lutze-Wallace CL, Milnes D, Robertson RH and Berlie-Surujballi G (2004) Evaluation of a monoclonal antibody-based enzyme-linked immunosorbent assay for detection of Campylobacter fetus in bovine preputial washing and vaginal mucus samples. Vet Microbiol 103:77–84

Bryan MC, Fazio F, Lee HK, Huang CY, Chang A, Best MD, Calarese DA, Blixt C, Paulson JC, Burton D et al. (2004) Covalent display of oligosaccharide arrays in microtiter plates. J Am Chem Soc 126:8640–8641

Bunz UHF (2000) Poly(aryleneethynylene)s: Syntheses, properties, structures, and applications. Chem Rev 100: 1605–1644

Campbell CT and Yarema KJ (2005) Large-scale approaches for glycobiology. Genome Biol 6:236.1–236.8

Carlsson J, Mecklenburg M, Lundstrom I, Danielsson B and Winquist F (2005) Investigation of sera from various species by using lectin affinity arrays and scanning ellipsometry. Anal Chim Acta 530:167–171

Caroff M and Karibian D (2003) Structure of bacterial lipopolysaccharides. Carbohydr Res 338:2431–2447

Charych D, Cheng Q, Reichert A, Kuziemko G, Stroh M, Nagy JO, Spevak W and Stevens RC (1996) A 'litmus test' for molecular recognition using artficial membranes. Chem Biol 3:113–120

Charych DH, Nagy JO, Spevak W and Bednarski MD (1993) Direct Colorimetric Detection of a Receptor-Ligand Interaction by a Polymerized Bilayer Assembly. Science 261:585–588

Chen JR, Miao YQ, He NY, Wu XH and Li SJ (2004) Nanotechnology and biosensors. Biotechnology Advances 22:505–518

Chen YF, Ji TH and Rosenzweig Z (2003) Synthesis of glyconanospheres containing luminescent CdSe-ZnS quantum dots. Nano Letters 3:581–584

Cobb BA and Kasper DL (2005). Zwitterionic capsular polysaccharides: the new MHCII-dependent antigens. Cellular Microbiology 7:1398–1403

Comstock LE and Kasper DL (2006) Bacterial glycans: Key mediators of diverse host immune responses. Cell 126:847–850

Connell H, Agace W, Klemm P, Schembri M, Marild S and Svanborg C (1996) Type 1 fimbrial expression enhances Escherichia coli virulence for the urinary tract. Proceedings of the National Academy of Sciences 93:9827–9832

Cross AS (1990) The Biologic Significance of Bacterial Encapsulation. Curr Top Microbiol Immunol 150:87–95

Culf AS, Cuperlovic-Culf M and Ouellette RJ (2006) Carbohydrate microarrays: Survey of fabrication techniques. Omics 10:289–310

Dando M (1994) Biological Warfare in the 21st Century: Biotechnology and the Proliferation of Biological Weapons, 1st ed. Brassey's UK Ltd, London

Daniel MC and Astruc D (2004) Gold nanoparticles: Assembly, supramolecular chemistry, quantum-size-related properties, and applications toward biology, catalysis, and nanotechnology. Chem Rev 104:293–346

de la Fuente JM, Eaton P, Barrientos AG, Menendez M and Penades S (2005) Thermodynamic evidence for Ca^{2+}-mediated self-aggregation of Lewis X gold glyconanoparticles. A model for cell adhesion via carbohydrate-carbohydrate interaction. J Am Chem Soc 127:6192–6197

de la Fuente JM and Penades S (2004) Understanding carbohydrate-carbohydrate interactions by means of glyconanotechnology. Glycoconj J 21:149–163

De la Fuente JM and Penades S (2006) Glyconanoparticles: Types, synthesis and applications in glycoscience, biomedicine and material science. Biochimica Et Biophysica Acta-General Subjects 1760:636–651

de Paz JL and Seeberger PH (2006) Recent advances in carbohydrate microarrays. Qsar & Combinatorial Science 25:1027–1032

Diks SH, van Deventer SJH and Peppelenbosch MP (2001) Lipopolysaccharide recognition, internalisation, signalling and other cellular effects. J Endoxtin Res 7:335–348

Disney MD and Seeberger PH (2004) The use of carbohydrate microarrays to study carbohydrate-cell interactions and to detect pathogens. Chem Biol 11:1701–1707

Disney MD, Zheng J, Swager TM and Seeberger PH (2004) Detection of bacteria with carbohydrate-functionalized fluorescent polymers. J Am Chem Soc 126:13343–13346

Doyl MP, Beuchat LR and Montville TJ (2001) Food Microbiology: Fundamentals and Frontiers, 2nd ed, ASM Press, Washington, D.C.

Duncan MJ, Mann EL, Cohen MS, Ofek I, Sharon N and Abraham SN (2005) The distinct binding specificities exhibited by enterobacterial type 1 fimbriae are determined by their fimbrial shafts. J Biol Chem 280:37707–37716

Dyukova VI, Shilova NV, Galanina OE, Rubina AY and Bovin NV (2006) Design of carbohydrate multiarrays. Biochimica Et Biophysica Acta-General Subjects 1760:603–609

Ebe Y, Kuno A, Uchiyama N, Koseki-Kuno S, Yamada M, Sato T, Narimatsu H and Hirabayashi J (2006) Application of lectin microarray to crude samples: Differential glycan profiling of Lec mutants. J Biochem (Tokyo) 139:323–327

Erridge C, Bennett-Guerrero E and Poxton IR (2002) Structure and function of lipopolysaccharides. Microbes Infect 4:837–851

Ertl P and Mikkelsen SR (2001) Electrochemical biosensor array for the identification of microorganisms based on lectin-lipopolysaccharide recognition. Anal Chem 73:4241–4248

Ertl P, Wagner M, Corton E and Mikkelsen SR (2003) Rapid identification of viable Escherichia coli subspecies with an electrochemical screen-printed biosensor array. Biosens Bioelectron 18:907–916

Evans DG, Evans DJ, Clegg S and Pauley JA (1979) Purification and Characterization of the Cfa-I Antigen of Enterotoxigenic Escherichia-Coli. Infect Immun 25:738–748

Faris A, Lindahl M and Wadstrom T (1980) Gm2-Like Glycoconjugate as Possible Erythrocyte Receptor for the Cfa-I and K-99 Hemagglutinins of Entero-Toxigenic Escherichia-Coli. FEMS Microbiol Lett 7:265–269

Feizi T, Fazio F, Chai WC and Wong CH (2003) Carbohydrate microarrays - a new set of technologies at the frontiers of glycomics. Curr Opin Struct Biol 13:637–645

Ficko-Blean, E., and Boraston, A. B. (2006). The interaction of a carbohydrate-binding module from a Clostridium perfringens N-acetyl-beta-hexosaminidase with its carbohydrate receptor. J Biol Chem 281, 37748–37757.

Firon N, Ashkenazi S, Mirelman D, Ofek I and Sharon N (1987) Aromatic Alpha-Glycosides of Mannose Are Powerful Inhibitors of the Adherence of Type-1 Fimbriated Escherichia-Coli to Yeast and Intestinal Epithelial-Cells. Infect Immun 55:472–476

Fredriksson-Ahomaa M, Stolle A and Korkeala H (2006) Molecular epidemiology of Yersinia enterocolitica infections. FEMS Immunol Med Microbiol 47:315–329

Fukui S, Feizi T, Galustian C, Lawson AM and Chai WG (2002) Oligosaccharide microarrays for high-throughput detection and specificity assignments of carbohydrate-protein interactions. Nat Biotechnol 20:1011–1017

Galanos C and Freudenberg MA (1993) Bacterial-Endotoxins - Biological Properties and Mechanisms of Action. Mediators Inflamm 2:S11–S16

Gillespie SH (1998) New polymerase chain reaction-based diagnostic techniques for bacterial respiratory infection. Curr Opin Infect Dis 11:133–138

Gilsdorf JR, McCrea KW and Marrs CF (1997) Role of pili in Haemophilus influenzae adherence and colonization. Infect Immun 65:2997–3002

Greenberg AE (1995) Standard Methods for the Examination of Water and Wastewater, 19th ed. American Public Health Association, Washington, D.C.

Gu LR, Elkin T, Jiang XP, Li HP, Lin Y, Qu LW, Tzeng TRJ, Joseph R and Sun YP (2005) Single-walled carbon nanotubes displaying multivalent ligands for capturing pathogens. Chem Commun. 874–876

Gustafsson, A., Hultberg, A., Sjostrom, R., Kacskovics, I., Breimer, M. E., Boren, T., Hammarstrom, L., and Holgersson, J. (2006). Carbohydrate-dependent inhibition of Helicobacter pylori colonization using porcine milk. Glycobiology 16, 1–10.

Hanisch FG, Hacker J and Schroten H (1993) Specificity of S-Fimbriae on Recombinant Escherichia-Coli - Preferential Binding to Gangliosides Expressing Neugc-Alpha(2-3)Gal and Neuac-Alpha(2-8)Neuac. Infect Immun 61: 2108–2115

Hanson S, Best M, Bryan MC and Wong CH (2004) Chemoenzymatic synthesis of oligosaccharides and glycoproteins. Trends Biochem Sci 29:656–663

Hansson J and Oscarson S (2000) Complex bacterial carbohydrate surface antigen structures: Syntheses of Kdo- and heptose-containing lipopolysaccharide core structures and anomerically phosphodiester-linked oligosaccharide structures. Current Organic Chemistry 4:535–564

Hill WE (1996) The polymerase chain reaction: Applications for the detection of foodborne pathogens. Crit Rev Food Sci Nutr 36:123–173

Hirabayashi J (2003) Oligosaccharide microarrays for glycomics. Trends Biotechnol 21:141–143

Hobson NS, Tothill I and Turner APF (1996) Microbial detection. Biosens Bioelectron 11:455–477

Houseman BT and Mrksich M (2002) Carbohydrate arrays for the evaluation of protein binding and enzymatic modification. Chem Biol 9:443–454

Houseman BT, Gawalt ES and Mrksich M (2003) Maleimide-functionalized self-assembled monolayers for the preparation of peptide and carbohydrate biochips. Langmuir 19:1522–1531

Hsu KL, Pilobello KT and Mahal LK (2006) Analyzing the dynamic bacterial glycome with a lectin microarray approach. Nat Chem Biol 2:153–157

Ikeda M, Yamaguchi N, Tani K and Nasu M (2006) Rapid and simple detection of food poisoning bacteria by bead assay with a microfluidic chip-based system. J Microbiol Methods 67:241–247

Jelinek R and Kolusheva S (2004) Carbohydrate biosensors. Chem Rev 104:5987–6015

Jeng ES, Moll AE, Roy AC, Gastala JB and Strano MS (2006) Detection of DNA hybridization using the near-infrared band-gap fluorescence of single-walled carbon nanotubes. Nano Letters 6:371–375

Joiner KA, Schmetz MA, Goldman RC, Leive L and Frank MM (1984) Mechanism of Bacterial-Resistance to Complement-Mediated Killing - Inserted C5b-9 Correlates with Killing for Escherichia-Coli O111b4 Varying in O-Antigen Capsule and O-Polysaccharide Coverage of Lipid-a Core Oligosaccharide. Infect Immun 45:113–117

Jones C (2005) Vaccines based on the cell surface carbohydrates of pathogenic bacteria. An Acad Bras Cienc 77: 293–324

Jones CH, Pinkner JS, Roth R, Heuser J, Nicholes AV, Abraham SN and Hultgren SJ (1995) Fimh Adhesin of Type-1 Pili Is Assembled into a Fibrillar Tip Structure in the Enterobacteriaceae. Proc Natl Acad Sci USA 92:2081–2085

Joralemon MJ, Murthy KS, Remsen EE, Becker ML and Wooley KL (2004) Synthesis, characterization, and bioavailability of mannosylated shell cross-linked nanoparticles. Biomacromolecules 5:903–913

Kaspar CW and Tartera C (1990) Methods for Detecting Microbial Pathogens in Food and Water. Methods Mircrobiol 22:497–531

Katamay MM (1990) Assessing Defined-Substrate Technology for Meeting Monitoring Requirements of the Total Coliform Rule. J Am Water Work Assoc 82:83–87

Kersten B, Wanker EE, Hoheisel JD and Angenendt P (2005) Multiplex approaches in protein microarray technology. Expert Rev Proteomics 2:499–510

Keum KC, Yoo SM, Lee SY, Chang KH, Yoo NC, Yoo WM, Kim JM, Choi JY, Kim JS and Lee G (2006) DNA microarray-based detection of nosocomial pathogenic Pseudomonas aeruginosa and Acinetobacter baumannii. Mol Cell Probes 20:42–50

Khan AS, Kniep B, Oelschlaeger TA, Van Die I, Korhonen T and Hacker J (2000) Receptor structure for F1C fimbriae of uropathogenic Escherichia coli. Infect Immun 68:3541–3547

Khan I, Desai DV and Kumar A (2004) Carbochips: a new energy for old biobuilders. J Biosci Bioeng 98:331–337

Kim IB, Erdogan B, Wilson JN and Bunz UHF (2004) Sugar-poly(para-phenylene ethynylene) conjugates as sensory materials: Efficient quenching by Hg^{2+} and Pb^{2+} ions. Chemistry-a European Journal 10:6247–6254

Kim IB, Wilson JN and Bunz UHF (2005) Mannose-substituted PPEs detect lectins: A model for Ricin sensing. Chem Commun. 1273–1275

Kisiela D, Laskowska A, Sapeta A, Kuczkowski M, Wieliczko A and Ugorski M (2006) Functional characterization of the FimH adhesin from Salmonella enterica serovar Enteritidis. Microbiol-Sgm 152:1337–1346

Kisiela D, Sapeta A, Kuczkowski M, Stefaniak T, Wieliczko A, and Ugorski M (2005). Characterization of FimH adhesins expressed by Salmonella enterica serovar Gallinarum biovars Gallinarum and Pullorum: Reconstitution of mannose-binding properties by single amino acid substitution. Infect Immun 73, 6187–6190.

Klefel MJ and von Itzstein M (2002) Recent advances in the synthesis of sialic acid derivatives and sialylmimetics as biological probes. Chemical Reviews 102:471–490

Klemm P, Christiansen G, Kreft B, Marre R and Bergmans H (1994) Reciprocal Exchange of Minor Components of Type-1 and F1c Fimbriae Results in Hybrid Organelles with Changed Receptor Specificities. J Bacteriol 176:2227–2234

Korhonen TK, Vaisanenrhen V, Rhen M, Pere A, Parkkinen J and Finne J (1984) Escherichia-Coli Fimbriae Recognizing Sialyl Galactosides. J Bacteriol 159:762–766

Krivan HC, Roberts DD and Ginsburg V (1988) Many Pulmonary Pathogenic Bacteria Bind Specifically to the Carbohydrate Sequence Galnac-Beta-1-4gal Found in Some Glycolipids. Proc Natl Acad Sci USA 85:6157–6161

Kuehn, M. J., Heuser, J., Normark, S., and Hultgren, S. J. (1992). P Pili in Uropathogenic Escherichia-Coli Are Composite Fibers with Distinct Fibrillar Adhesive Tips. Nature 356, 252–255.

Le Bouguenec C and Bertin Y (1999) AFA and F17 adhesins produced by pathogenic Escherichia coli strains in domestic animals. Vet Res 30:317–342

Lee CJ (1987) Bacterial Capsular Polysaccharides - Biochemistry, Immunity and Vaccine. Mol Immunol 24:1005–1019

Levine OS, O'Brien KL, Knoll M, Adegbola RA, Black S, Cherian T, Dagan R, Goldblatt D, Grange A, Greenwood B et al. (2006) Pneumococcal vaccination in developing countries. Lancet 367:1880–1882

Li J, Zacharek S, Chen X, Wang JQ, Zhang W, Janczuk A and Wang PG (1999) Bacteria targeted by human natural antibodies using alpha-Gal conjugated receptor-specific glycopolymers. Bioorg Med Chem 7:1549–1558

Li NQ, Wang JX and Li MX (2003) Electrochemistry at carbon nanotube electrodes. Reviews in Analytical Chemistry 22:19–33

Li YJ, Ma BL, Fan Y, Kong XG and Li JH (2002) Electrochemical and Raman studies of the biointeraction between Escherichia coli and mannose in polydiacetylene derivative supported on the self-assembled monolayers of octadecanethiol on a gold electrode. Anal Chem 74:6349–6354

Lin CC, Yeh YC, Yang CY, Chen CL, Chen GF, Chen CC and Wu YC (2002) Selective binding of mannose-encapsulated gold nanoparticles to type 1 pili in Escherichia coli. J Am Chem Soc 124:3508–3509

Lin FYH, Sabri M, Alirezaie J, Li DQ and Sherman PM (2005) Development of a nanoparticle-labeled microfluidic immunoassay for detection of pathogenic microorganisms. Clin Diagn Lab Immunol 12:418–425

Lis H and Sharon N (1998) Lectins: Carbohydrate-specific proteins that mediate cellular recognition. Chem Rev 98:637–674

Lotter H, Russmann H, Heesemann J and Tannich E (2004) Oral vaccination with recombinant Yersinia enterocolitica expressing hybrid type III proteins protects gerbils from amebic liver abscess. Infect Immun 72:7318–7321

Louws FJ, Rademaker JLW and de Bruijn FJ (1999) The three Ds of PCR-based genomic analysis of phytobacteria: Diversity, detection, and disease diagnosis. Annu Rev Phytopathol 37:81–125

Love KR and Seeberger PH (2002) Carbohydrate arrays as tools for glycomics. Angew Chem Int Ed 41:3583–3586

Ma BL, Fan Y, Zhang LG, Kong XG, Li YJ and Li JH (2003) Direct colorimetric study on the interaction of Escherichia coli with mannose in polydiacetylene Langmuir-Blodgett films. Colloids and Surfaces B-Biointerfaces 27:209–213

Ma ZF, Li JR, Jiang L, Cao J and Boullanger P (2000) Influence of the spacer length of glycolipid receptors in polydiacetylene vesicles on the colorimetric detection of Escherichia coli. Langmuir 16:7801–7804

Ma ZF, Li JR, Liu MH, Cao J, Zou ZY, Tu J and Jiang L (1998) Colorimetric detection of Escherichia coli by polydiacetylene vesicles functionalized with glycolipid. J Am Chem Soc 120:12678–12679

Mahal LK, Pilobello K and Krishnamoorthy L (2004) Development of a lectin microarray for the glycomic profiling of cells. Glycobiology 14:1203–1203

Mahdavi J, Sonden B, Hurtig M, Olfat FO, Forsberg L, Roche N, Angstrom J, Larsson T, Teneberg S, Karlsson KA et al. (2002) Helicobacter pylori SabA adhesin in persistent infection and chronic inflammation. Science 297:573–578

Mammen M, Choi SK and Whitesides GM (1998) Polyvalent interactions in biological systems: Implications for design and use of multivalent ligands and inhibitors. Angew Chem Int Ed 37:2755–2794

Mantle M and Husar SD (1994) Binding of Yersinia-Enterocolitica to Purified, Native Small-Intestinal Mucins from Rabbits and Humans Involves Interactions with the Mucin Carbohydrate Moiety. Infect Immun 62:1219–1227

Marks T and Sharp R (2000) Bacteriophages and biotechnology: A review. J Chem Technol Biotechnol 75:6–17

Marsh JW and Taylor RK (1999) Genetic and transcriptional analyses of the Vibrio cholerae mannose-sensitive hemagglutinin type 4 pilus gene locus. J Bacteriol 181:1110–1117

McCabe-Sellers BJ and Beattie SE (2004) Food safety: Emerging trends of foodborne illness surveillance and prevention. J Am Diet Assoc 104:1708–1717

McQuade DT, Pullen AE and Swager TM (2000) Conjugated polymer-based chemical sensors. Chem Rev 100: 2537–2574

Mohammadi T, Savelkoul PHM, Pietersz RNI and Reesink HW (2006) Applications of real-time PCR in the screening of platelet concentrates for bacterial contamination. Expert Rev Mol Diagn 6:865–872

Montecucco C and Rappuoli R (2001) Living dangerously: How Helicobacter pylori survives in the human stomach. Nat Rev Mol Cell Biol 2:457–466

Monzo A and Guttman A (2006) Immobilization techniques for mono- and oligosaccharide microarrays. Qsar & Combinatorial Science 25:1033–1038

Moxon ER and Kroll JS (1990) The Role of Bacterial Polysaccharide Capsules as Virulence Factors. Curr Top Microbiol Immunol 150:65–85

Mrksich M (2004) An early taste of functional glycomics. Chem Biol 11:739–740

Nagy JO, Spevak W, Charych DH, Schaefer ME, Gilbert JH and Bednarski MD (1993) Polymerized Liposomes Containing C-Glycosides of Sialic-Acid Are Potent Inhibitors of Influenza-Virus Hemagglutination and Invitro Infectivity. J Cell Biochem 382–382

Nassif X, Pujol C, Morand P and Eugene E (1999) Interactions of pathogenic Neisseria with host cells. Is it possible to assemble the puzzle? Mol Microbiol 32:1124–1132

Nilsson KGI and Mandenius CF (1994) A Carbohydrate Biosensor Surface for the Detection of Uropathogenic Bacteria. Bio-Technology 12:1376–1378

O'Connell MJ, Bachilo SM, Huffman CB, Moore VC, Strano MS, Haroz EH, Rialon KL, Boul PJ, Noon WH, Kittrell C et al. (2002) Band gap fluorescence from individual single-walled carbon nanotubes. Science 297:593–596

O'Riordan K and Lee JC (2004) Staphylococcus aureus capsular polysaccharides. Clin Microbiol Rev 17:218–234

Olsen JE, Aabo S, Hill W, Notermans S, Wernars K, Granum PE, Popovic T, Rasmussen HN and Olsvik O (1995) Probes and Polymerase Chain-Reaction for Detection of Food-Borne Bacterial Pathogens. Int J Food Microbiol 28:1–78

Ono E, Abe K, Nakazawa M and Naiki M (1989) Ganglioside Epitope Recognized by K99-Fimbriae from Entero-Toxigenic Escherichia-Coli. Infect Immun 57:907–911

Orskov I, Orskov F, Jann B and Jann K (1977) Serology, Chemistry, and Genetics of O and K Antigens of Escherichia-Coli. Bacteriol Rev 41:667–710

Ovodov YS (2006a) Bacterial capsular antigens. Structural patterns of capsular antigens. Biochemistry-Moscow 71: 937–954

Ovodov YS (2006b) Capsular antigens of bacteria. Capsular antigens as the basis of vaccines against pathogenic bacteria. Biochemistry-Moscow 71:955–961

Palma AS, Feizi T, Zhang YB, Stoll MS, Lawson AM, Diaz-Rodriguez E, Campanero-Rhodes MA, Costa J, Gordon S, Brown GD and Chai WG (2006) Ligands for the beta-glucan receptor, Dectin-1, assigned using "designer" microarrays of oligosaccharide probes (neoglycolipids) generated from glucan polysaccharides. J Biol Chem 281:5771–5779

Park S, Lee MR, Pyo SJ and Shin I (2004) Carbohydrate chips for studying high-throughput carbohydrate-protein interactions. J Am Chem Soc 126:4812–4819

Park SJ and Shin IJ (2002) Fabrication of carbohydrate chips for studying protein-carbohydrate interactions. Angew Chem Int Ed 41:3180–3182

Paulson, J. C., Blixt, O., and Collins, B. E. 2006. Sweet spots in functional glycomics. Nat. Chem. Biol. 2: 238–248.

Pier GB (2003) Promises and pitfalls of Pseudomonas aeruginosa lipopolysaccharide as a vaccine antigen. Carbohydr Res 338:2549–2556

Pontius FW (2000) Reconsidering the Total Coliform Rule. J Am Water Work Assoc 92:14

Popescu MA and Toms SA (2006) In vivo optical imaging using quantum dots for the management of brain tumors. Expert Rev Mol Diagn 6:879–890

Prasad, S. M., Yin, Y. B., Rodzinski, E., Tuomanen, E. I., and Masure, H. R. (1993). Identification of a Carbohydrate-Recognition Domain in Filamentous Hemagglutinin from Bordetella-Pertussis. Infect Immun *61*, 2780–2785.

Prescher JA and Bertozzi CR (2006) Chemical technologies for probing glycans. Cell 126:851–854.

Price NPJ, Jeyaretnam B, Carlson RW, Kadrmas JL, Raetz CRH and Brozek KA (1995) Lipid-a Biosynthesis in Rhizobium-Leguminosarum - Role of a 2-Keto-3-Deoxyoctulosonate-Activated 4'-Phosphatase. Proc Natl Acad Sci USA 92:7352–7356

Qian XP, Metallo SJ, Choi IS, Wu HK, Liang MN and Whitesides GM (2002) Arrays of self-assembled monolayers for studying inhibition of bacterial adhesion. Anal Chem 74:1805–1810

Que NLS Lin SH, Cotter RJ and Raetz CRH (2000a) Purification and mass spectrometry of six lipid A species from the bacterial endosymbiont Rhizobium etli - Demonstration of a conserved distal unit and a variable proximal portion. J Biol Chem 275:28006–28016

Que NLS, Ribeiro AA and Raetz CRH (2000b) Two-dimensional NMR spectroscopy and structures of six lipid A species from Rhizobium etli CE3 - Detection of an acyloxyacyl residue in each component and origin of the aminogluconate moiety. J Biol Chem 275:28017–28027

Raetz CRH (1990) Biochemistry of Endotoxins. Annu Rev Biochem 59:129–170

Raetz CRH (1993) Bacterial-Endotoxins - Extraordinary Lipids That Activate Eukaryotic Signal-Transduction. J Bacteriol 175:5745–5753

Raetz CRH and Whitfield C (2002) Lipopolysaccharide endotoxins. Annu Rev Biochem 71:635–700

Ratner DM, Adams EW, Disney MD and Seeberger PH (2004a) Tools for glycomics: Mapping interactions of carbohydrates in biological systems. Chembiochem 5:1375–1383

Ratner DM, Adams EW, Su J, O'Keefe BR, Mrksich M and Seeberger PH (2004b) Probing protein-carbohydrate interactions with microarrays of synthetic oligosaccharides. Chembiochem 5:379–382

Reichert A, Nagy JO, Spevak W and Charych D (1995) Polydiacetylene Liposomes Functionalized with Sialic-Acid Bind and Colorimetrically Detect Influenza-Virus. J Am Chem Soc 117:829–830

Reynolds AJ, Haines AH and Russell DA (2006) Gold glyconanoparticles for mimics and measurement of metal ion-mediated carbohydrate-carbohydrate interactions. Langmuir 22:1156–1163

Roberts IS (1996) The biochemistry and genetics of capsular polysaccharide production in bacteria. Annu Rev Microbiol 50:285–315

Robinson A, Fang JM, Chou PT, Liao KW, Chu RM and Lee SJ (2005) Probing lectin and sperm with carbohydrate-modified quantum dots. Chembiochem 6:1899–1905

Rocchetta HL, Burrows LL and Lam JS (1999) Genetics of O-antigen biosynthesis in Pseudomonas aeruginosa. Microbiol Mol Biol Rev 63:523–553

Roche N, Angstrom J, Hurtig M, Larsson T, Boren T and Teneberg S (2004) Helicobacter pylori and complex gangliosides. Infect Immun 72:1519–1529

Roos V, Schembri MA, Ulett GC and Klemm P (2006) Asymptomatic bacteriuria Escherichia coli strain 83972 carries mutations in the foc locus and is unable to express F1C fimbriae. Microbiol-Sgm 152:1799–1806

Roy R (2003) A decade of glycodendrimer chemistry. Trends Glycosci Glyc 15:291–310

Ruiz-Palacios GM, Cervantes LE, Ramos P, Chavez-Munguia B and Newburg DS (2003) Campylobacter jejuni binds intestinal H(O) antigen (Fuc alpha 1, 2Gal beta 1, 4GlcNAc), and fucosyloligosaccharides of human milk inhibit its binding and infection. J Biol Chem 278:14112–14120

Samuel G and Reeves P (2003) Biosynthesis of O-antigens: genes and pathways involved in nucleotide sugar precursor synthesis and O-antigen assembly. Carbohydr Res 338:2503–2519

Sauer, F. G., Barnhart, M., Choudhury, D., Knights, S. D., Waksman, G., and Hultgren, S. J. (2000). Chaperone-assisted pilus assembly and bacterial attachment. Curr Opin Struct Biol 10, 548–556.

Schofield CL, Haines AH, Field RA and Russell DA (2006) Silver and gold glyconanoparticles for colorimetric bioassays. Langmuir 22:6707–6711

Schroten H, Stapper C, Plogmann R, Kohler H, Hacker J and Hanisch FG (1998) Fab-independent antiadhesion effects of secretory immunoglobulin A on S-fimbriated Escherichia coli are mediated by sialyloligosaccharides. Infect Immun 66:3971–3973

Schweizer F, Jiao HL, Hindsgaul O, Wong WY and Irvin RT (1998) Interaction between the pili of Pseudomonas aeruginosa PAK and its carbohydrate receptor beta-D-GalNAc(1 -> 4)beta-D-Gal analogs. Can J Microbiol 44:307–311

Seelig R, Renz M, Bottner C, Stockinger K, Czichos J, Schulz V and Seelig HP (1991) Rapid Diagnosis of Tuberculosis by Polymerase Chain-Reaction (Pcr). Immun Infekt 19:179–185

Sharon N (2006) Carbohydrates as future anti-adhesion drugs for infectious diseases. Biochimica Et Biophysica Acta-General Subjects 1760:527–537

Sharrna P, Brown S, Walter G, Santra S and Moudgil B (2006) Nanoparticles for bioimaging. Adv Colloid Interface Sci 123:471–485

Sheng HQ, Lim JY, Knecht HJ, Li J and Hovde CJ (2006) Role of Escherichia coli O157: H7 virulence factors in colonization at the bovine terminal rectal mucosa. Infect Immun 74:4685–4693

Sherigara BS, Kutner W and D'Souza F (2003) Electrocatalytic properties and sensor applications of fullerenes and carbon nanotubes. Electroanalysis 15:753–772

Sheth HB, Lee KK, Wong WY, Srivastava G, Hindsgaul O, Hodges RS, Paranchych W and Irvin RT (1994) The Pili of Pseudomonas-Aeruginosa Strains Pak and Pao Bind Specifically to the Carbohydrate Sequence Beta-Galnac(1-4)Beta-Gal Found in Glycosphingolipids Asialo-GM(1) and Asialo-GM(2). Mol Microbiol 11:715–723

Shin I, Park S and Lee MR (2005) Carbohydrate microarrays: An advanced technology for functional studies of glycans. Chemistry-a European Journal 11:2894–2901

Shinefield HR and Black S (2005) Prevention of Staphylococcus aureus infections: advances in vaccine development. Expert Rev Vaccines 4:669–676

Shriver-Lake LC, Taitt CR and Ligler FS (2004) Applications of array biosensor for detection of food allergens. J AOAC Int 87:1498–1502

Skurnik M and Bengoechea JA (2003) Biosynthesis and biological role of lipopolysaccharide O-antigens of pathogenic Yersiniae. Carbohydr Res 338:2521–2529

Smedley JG, Jewell E, Roguskie J, Horzempa J, Syboldt A, Stolz DB and Castric P (2005) Influence of pilin glycosylation on Pseudomonas aeruginosa 1244 pilus function. Infect Immun 73:7922–7931

Smith EA, Thomas WD, Kiessling LL and Corn RM (2003) Surface plasmon resonance imaging studies of protein-carbohydrate interactions. J Am Chem Soc 125:6140–6148

Sood RK, Fattom A, Pavliak V and Naso RB (1996) Capsular polysaccharide-protein conjugate vaccines. Drug Discov Today 1:381–387

Spevak W, Nagy JO, Charych DH, Schaefer ME, Gilbert JH and Bednarski MD (1993) Polymerized Liposomes Containing C-Glycosides of Sialic-Acid - Potent Inhibitors of Influenza-Virus Invitro Infectivity. J Am Chem Soc 115:1146–1147

Staats JJ, Feder I, Okwumabua O and Chengappa MM (1997) Streptococcus suis: Past and present. Vet Res Commun 21:381–407

Stins MF, Prasadarao NV, Ibric L, Wass CA, Luckett P and Kim KS (1994) Binding Characteristics of S-Fimbriated Escherichia-Coli to Isolated Brain Microvascular Endothelial-Cells. Am J Pathol 145:1228–1236

Stevens J, Blixt O, Paulson JC and Wilson IA (2006) Glycan microarray technologies: tools to survey host specificity of influenza viruses. Nature Reviews Microbiology 4;857–864

Stoughton RB (2005) Applications of DNA microarrays in biology. Annu Rev Biochem 74:53–82

Su YL, Li JR, Jiang L and Cao J (2005) Biosensor signal amplification of vesicles functionalized with glycolipid for colorimetric detection of Escherichia coli. J Colloid Interface Sci 284:114–119

Sun CY, Zhang YJ, Fan Y, Li YJ and Li JH (2004a) Mannose-Escherichia coli interaction in the presence of metal cations studied in vitro by colorimetric polydiacetylene/glycolipid liposomes. J Inorg Biochem 98:925–930

Sun XL, Cui WX, Haller C and Chaikof EL (2004b) Site-specific multivalent carbohydrate labeling of quantum dots and magnetic beads. Chembiochem 5:1593–1596

Swiatlo E and Ware D (2003) Novel vaccine strategies with protein antigens of Streptococcus pneumoniae. FEMS Immunol Med Microbiol 38:1–7

Swords WE, Jones PA and Apicella MA (2003) The lipo-oligosaccharides of Haemophilus influenzae: an interesting array of characters. J Endoxtin Res 9:131–144

Takikawa Y, Mori H, Otsu Y, Matsuda Y, Nonomura T, Kakutani K, Tosa Y, Mayama S and Toyoda H (2002) Rapid detection of phylloplane bacterium Enterobacter cloacae based on chitinase gene transformation and lytic infection by specific bacteriophages. J Appl Microbiol 93:1042–1050

Tauxe RV (1997) Emerging foodborne diseases: An evolving public health challenge. Emerg Infect Dis 3:425–434

Telford, J. L., Barocchi, M. A., Margarit, I., Rappuoli, R., and Grandi, G. (2006). Pili in Gram-positive pathogens. Nature Reviews Microbiology 4, 509–519.

Teneberg S, Unemo M, Aspholm-Hurtig M, Boren T and Danielsson D (2005) The sialic acid-binding SabA adhesin of Helicobacter pylori is essential for non-opsonic activation of human neutrophils. Helicobacter 10:495–495

Teneberg S, Willemsen PTJ, Degraaf FK and Karlsson KA (1993) Calf Small-Intestine Receptors for K99 Fimbriated Enterotoxigenic Escherichia-Coli. FEMS Microbiol Lett 109:107–112

Tewari R, Ikeda T, Malaviya R, Macgregor JI, Little JR, Hultgren SJ and Abraham SN (1994) The Papg Tip Adhesin of P-Fimbriae Protects Escherichia-Coli from Neutrophil Bactericidal Activity. Infect Immun 62:5296–5304

Tikkanen K, Haataja S, Francoisgerard C and Finne J (1995) Purification of a Galactosyl-Alpha-1-4-Galactose-Binding Adhesin from the Gram-Positive Meningitis-Associated Bacterium Streptococcus-Suis. J Biol Chem 270: 28874–28878

Tinker, J. K., and Clegg, S. (2000). Characterization of FimY as a coactivator of type 1 fimbrial expression in Salmonella enterica serovar typhimurium. Infect Immun 68, 3305–3313.

Toze S (1999) PCR and the detection of microbial pathogens in water and wastewater. Water Research 33:3545–3556

Ukuku DO (2006) Effect of sanitizing treatments on removal of bacteria from cantaloupe surface, and re-contamination with Salmonella. Food Microbiol 23:289–293

Unemo M, Aspholm-Hurtig M, Ilver D, Bergstrom J, Boren T, Danielsson D and Teneberg S (2005) The sialic acid binding SabA adhesin of Helicobacter pylori is essential for nonopsonic activation of human neutrophils. J Biol Chem 280:15390–15397

Vanham SM, Vanalphen L, Mooi FR and Vanputten JPM (1994) The Fimbrial Gene-Cluster of Haemophilus-Influenzae Type-B. Mol Microbiol 13:673–684

Viboud GI and Bliska JB (2005) Yersinia outer proteins: Role in modulation of host cell signaling responses and pathogenesis. Annu Rev Microbiol 59:69–89

Wang DN (2003) Carbohydrate microarrays. Proteomics 3:2167–2175

Wang DN, Liu SY, Trummer BJ, Deng C and Wang AL (2002) Carbohydrate microarrays for the recognition of cross-reactive molecular markers of microbes and host cells. Nat Biotechnol 20:275–281

Wang HF, Gu LR, Lin Y, Lu FS, Meziani MJ, Luo PGJ, Wang W, Cao L and Sun YP (2006a) Unique aggregation of anthrax (Bacillus anthracis) spores by sugar-coated single-walled carbon nanotubes. J Am Chem Soc 128: 13364–13365

Wang J, Uttamehandani M, Sun HY and Yao SQ (2006b) Small molecule microarrays: Applications using specially tagged chemical libraries. Qsar & Combinatorial Science 25:1009–1019

Weintraub A (2003) Immunology of bacterial polysaccharide antigens. Carbohydr Res 338:2539–2547

Weis WI and Drickamer K (1996) Structural basis of lectin-carbohydrate recognition. Annu Rev Biochem 65:441–473

Wenneras C, Holmgren J and Svennerholm AM (1990) The Binding of Colonization Factor Antigens of Entero-Toxigenic Escherichia-Coli to Intestinal-Cell Membrane-Proteins. FEMS Microbiol Lett 66:107–112

Whitfield C (1995) Biosynthesis of Lipopolysaccharide O-Antigens. Trends Microbiol 3:178–185

Whitfield C (2006) Biosynthesis and assembly of capsular polysaccharides in Escherichia coli. Annu Rev Biochem 75:39–68

Whitfield C and Paiment A (2003) Biosynthesis and assembly of Group 1 capsular polysaccharides in Escherichia coli and related extracellular polysaccharides in other bacteria. Carbohydr Res 338:2491–2502

Whitfield C and Roberts IS (1999) Structure, assembly and regulation of expression of capsules in Escherichia coli. Mol Microbiol 31:1307–1319

Wilkinson SG (1996) Bacterial lipopolysaccharides - Themes and variations. Prog Lipid Res 35:283–343

Winberg J, Mollby R, Bergstrom J, Karlsson KA, Leonardsson I, Milh MA, Teneberg S, Haslam D, Marklund BI and Normark S (1995) The Papg-Adhesin at the Tip of P-Fimbriae Provides Escherichia-Coli with a Competitive Edge in Experimental Bladder Infections of Cynomolgus Monkeys. J Exp Med 182:1695–1702

Winssinger N and Harris JL (2005) Microarray-based functional protein profiling using peptide nucleic acid-encoded libraries. Expert Rev Proteomics 2:937–947

Wong CH (1999) Mimics of complex carbohydrates recognized by receptors. Accounts Chem Res 32:376–385

Wong CH (2005) Protein glycosylation: New challenges and opportunities. J Org Chem 70:4219–4225

Xue CH, Donuru VRR and Liu HY (2006a) Facile, Versatile Pre-polymerization and Post-polymerization Functionalization Approaches for Well-defined Fluorescent Conjugated Fluorene-based Glycopolymers. Macromolecules 39:5747–5752

Xue CH, Jog SP, Murthy P and Liu HY (2006b) Synthesis of highly water-soluble conjugated fluorescent glycopoly(p-phenylene)s for lectin and *escherichia coli*. Biomacromolecules 7:2470–2474

Zen JM (2003) Flow injection analysis of ascorbic acid in real samples using a highly stable chemically modified screen-printed electrode. Electroanalysis 15:1171–1176

Zhang LG, Fan Y, Ma BL, Xu XY, Kong XG, Shen DZ and Li YJ (2002) Colorimetric transition process of polydiacetylene/mannoside derivative Langmuir-Schaefer film by binding Escherichia jm 109. Thin Solid Films 419:194–198

Zhang Y and Bliska JB (2005) Role of macrophage apoptosis in the pathogenesis of Yersinia. In: Editor Name (ed) Role of Apoptosis in Infection. Springer-Verlag, Berlin, pp 151–173

Zhang YJ, Fan Y, Sun CY, Shen DZ, Li YJ and Li JH (2005) Functionalized polydiacetylene-glycolipid vesicles interacted with Escherichia coli under the TiO2 colloid. Colloids and Surfaces B-Biointerfaces 40:137–142

Zhang YJ, Ma BL, Li YJ and Li JH (2004) Enhanced affinochromism of polydiacetylene monolayer in response to bacteria by incorporating US nano-crystallites. Colloids and Surfaces B-Biointerfaces 35:41–44

Zhao XJ, Hilliard LR, Mechery SJ, Wang YP, Bagwe RP, Jin SG and Tan WH (2004) A rapid bioassay for single bacterial cell quantitation using bioconjugated nanoparticles. Proceedings of the National Academy of Science 101:15027–15032

Zheng T, Peelen D and Smith LM (2005) Lectin arrays for profiling cell surface carbohydrate expression. J Am Chem Soc 127:9982–9983

Zweigner J, Schumann RR and Weber JR (2006) The role of lipopolysaccharide-binding protein in modulating the innate immune response. Microbes Infect 8:946–952

25

Aptamers and Their Potential as Recognition Elements for the Detection of Bacteria

Casey C. Fowler, Naveen K. Navani, Eric D. Brown and Yingfu Li

Abstract

DNA and RNA are well-known polymers that are central to the existence of every known form of life. Once thought to be strictly passive templates containing genetic information, it has since become clear that these nucleic acids are capable of much more. Synthetic biologists are working to exploit this potential, creating a wide spectrum of "functional nucleic acids". These molecules can be divided into two broad categories: catalysts (deoxyribozymes and ribozymes) and receptors (aptamers). This chapter begins by providing a background on the field of functional nucleic acids with an emphasis on aptamer technology. One major application of aptamers is their use as recognition elements in sensors of interesting molecules and cell types. Some common designs of these sensors are profiled, explaining how the aptamer-target binding is converted into a detectable signal. The chapter concludes with a discussion of aptamer-based sensors of bacteria, including some of the relevant targets, the progress to date and the future prospects.

1. Functional Nucleic Acids

Nucleic acids have long been celebrated for their crucial role in the existence of every life form on earth as the carriers of genetic information. RNA's seemingly endless repertoire of activities in gene expression and regulation (Bartel 2004; Eddy 1999; Erdmann et al. 2001; Storz 2002; Winkler and Breaker 2005)—a repertoire that seems to grow with each passing year—provided the first clues that nucleic acids are capable of much more than acting as a passive template of genetic information. Over the past decade and a half, this potential has begun to be exploited by scientists in wide-ranging research areas, creating the field of functional nucleic acids.

The term *functional nucleic acid*, as it will be applied here, refers to a single-stranded DNA (ssDNA) or RNA molecule that has a function beyond its stereotypical role in information storage. Much like proteins, ssDNA and RNA are capable of folding into complex three-dimensional shapes with chemical functionalities that are precisely positioned for carrying out a given task (Adams et al. 2004; Jiang et al. 1996; Lin and Patel 1997; Montange and Batey 2006; Rupert et al. 2002; Serganov et al. 2006; Yang et al. 1996). The two most common classes of functional nucleic acids are *nucleic acid enzymes* (Fiammengo and Jaschke 2005) (RNA enzymes are known as *ribozymes* and their DNA equivalents as *deoxyribozymes*) and *aptamers* (Bunka and Stockley 2006). Ribozymes and deoxyribozymes catalyze a specific chemical reaction in a synonymous manner as their protein enzyme counterparts. Aptamers, the

Casey C. Fowler, Naveen K. Navani, Eric D. Brown and Yingfu Li • Department of Biochemistry and Biomedical Sciences and Department of Chemistry, McMaster University, Hamilton, Canada.

M. Zourob et al. (eds.), *Principles of Bacterial Detection: Biosensors, Recognition Receptors and Microsystems*,
© Springer Science+Business Media, LLC 2008

name comes from the Latin word *aptus* (meaning "to fit") (Ellington and Szostak 1990; Green et al. 1991), are oligonucleotides that are selected to bind tightly and specifically to a particular ligand. This chapter will provide an introduction to the field of functional nucleic acids, with a focus on aptamers, leading to a discussion of how aptamer technology is being applied to the detection of hazardous bacteria.

1.1. Properties of Nucleic Acids

Nucleic acids are polymers of nucleotide subunits, each consisting of a common sugar-phosphate backbone and one of four distinct nucleobases (Fig. 25.1). The two natural forms of nucleic acids, DNA and RNA, have only two differences that distinguish them: (1) RNA contains a hydroxyl group at the 2′ position of the ribose ring rather than a hydrogen; (2) DNA uses the nucleobase thymine whereas RNA uses uracil – these bases vary by a single methyl group.

While these differences seem quite minor, they have a significant impact on the functional properties and stability of the molecule. This is particularly true for the addition of the 2′-hydroxyl group, which is typically viewed as the reason that RNA molecules have generally shown a greater capacity for carrying out complex tasks (Cate et al. 1996). The price to be paid for this additional functionality, however, is that this hydroxyl group also contributes greatly to the spontaneous degradation of RNA. The 2′ position of the ribose lies in close proximity to the phosphodiester linkage that connects the nucleotides within an RNA chain. A hydroxyl group at this position is capable of carrying out a nucleophilic attack on the phosphodiester bond, which results in the cleavage of the RNA backbone. Under physiological conditions, this makes a typical DNA linkage approximately 100,000 times more stable than the corresponding RNA linkage (Li and Breaker 1999a). Nucleases (ubiquitous enzymes catalyzing the degradation of nucleic acids) that target RNA are also far more prevalent than the DNA equivalents, further decreasing the stability of RNA molecules, particularly in an *in vivo* setting.

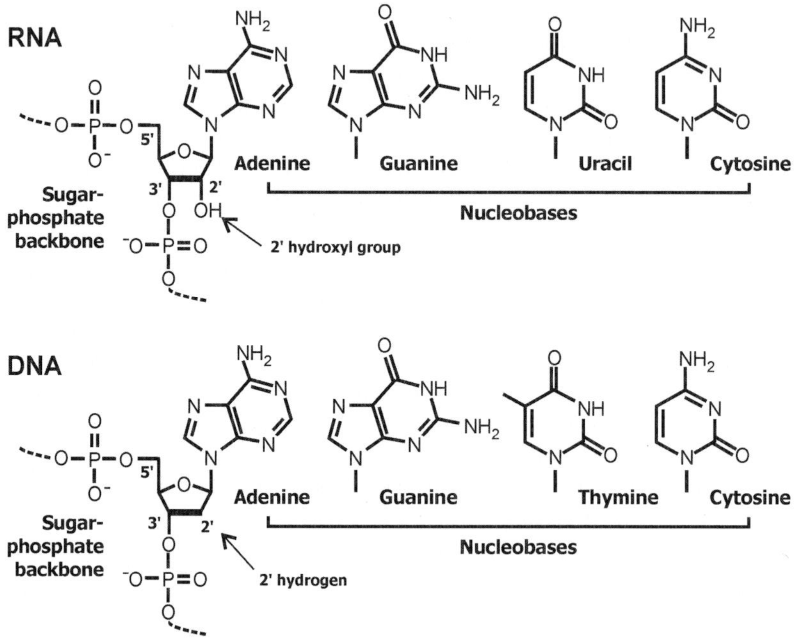

Figure 25.1. Chemical structures of RNA and DNA.

At the core of nucleic acid structure is the specific hydrogen-bonding pattern (dubbed Watson–Crick interactions or base pairing) between a nucleobase and its complement: Guanine forms three hydrogen bonds with cytosine (or *vice versa*), whereas adenine forms two hydrogen bonds with thymine (or *vice versa*). Nucleic acids are nonsymmetrical polymers and are therefore directional: The backbone of each strand contains a 5′- hydroxyl group at one end and a 3′-hydroxyl group at the other. DNA is well known for its double-stranded formation in which two antiparallel strands with a complementary base composition adopt a highly stable helical structure through Watson–Crick base pairings. These same nucleobase interactions can be exploited by single stranded DNA or RNA to form helical segments between complementary regions within the same molecule. This is considered to be the secondary structure of an oligonucleotide, and results in the formation of structures referred to as stem-loops or hairpins (Fig. 25.2a-c).

The crossroads of two or more neighboring stems (junctions; Fig. 25.2a-c) as well as sections of stems that contain unmatched nucleotides (bulges; Fig. 25.2a-b) are also often structurally important regions of a functional nucleic acid. The spatial orientation of various stem-loops, bulges and junctions can produce remarkable three-dimensional folds (Adams et al. 2004; Jiang et al. 1996; Lin and Patel 1997; Montange and Batey 2006; Rupert et al. 2002; Serganov et al. 2006; Yang et al. 1996). Even more complex structures can arise through the formation of higher-order structural motifs. One such motif involves the base pairing of the loop section of a stem-loop with another segment of the molecule to form a pseudoknot (Fig. 25.2c). G-quartets represent another highly stable nucleic acid structure. A G-quartet results from favorable hydrogen bonding and stacking interactions between properly aligned

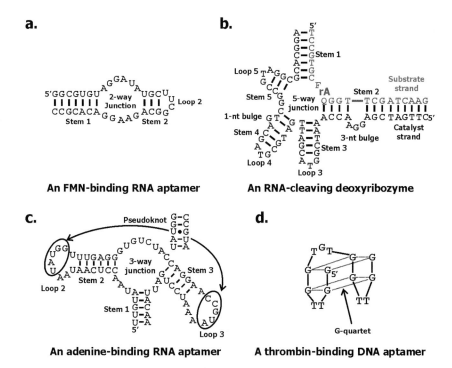

Figure 25.2. Secondary structures of representative functional nucleic acids. (a) An RNA aptamer that binds flavin mononucleotide (FMN) (Burgstaller and Famulok 1994; Fan et al. 1996). (b) An RNA-cleaving deoxyribozyme with fluorescence-quenching reporter functionality (Chiuman and Li 2006). F: fluorescein-dT, Q: DABCYL-dT, rA: adenosine ribonucleotide. (c) A natural RNA aptamer that interacts with adenine (Mandal and Breaker 2004). (d) A DNA aptamer that binds human thrombin (Bock et al. 1992; Padmanabhan et al. 1993).

guanosine residues, such as the four GG elements in a thrombin-binding DNA aptamer shown in Fig. 25.2d.

1.2. Synthesizing, Sequencing and Modifying Nucleic Acids

One of the primary reasons for the use of nucleic acids in biological research and biotechnological applications is the fact that they can be synthesized, amplified and modified in a fast, inexpensive and efficient manner. The chemical synthesis of oligonucleotides has become a very standard and inexpensive practice. Two hundred nanomoles of a custom-designed DNA oligonucleotide (2×10^{16} molecules) can be purchased at about US $100 and will generally be received within a matter of days. Driven by the human genome project, DNA sequencing technology has also advanced considerably. Small-scale sequencing reactions are offered commercially at a price of US $10 or less per reaction (approximately 500–1000 bp of quality sequence) and can be carried out in a larger 96-well plate scale for less than US $500 per plate. Sequencing results are also usually available in less than a week.

The wealth of readily available DNA and RNA synthesizing and modifying enzymes has made the most important contribution to development of functional nucleic acids. These enzymes can be purified in house from a recombinant source or can be purchased commercially. Described below are the most important of these enzymes and the relevant enzymatic reactions (Sambrook and Russell).

1.2.1. DNA Polymerase and Polymerase Chain Reaction

DNA polymerase is the enzyme responsible for the replication of DNA in all species (Sambrook and Russell). It catalyzes the synthesis of the complementary strand of a given DNA template from a small oligonucleotide primer and deoxyribonucleoside 5′-triphosphate (dNTP) subunits. DNA polymerase moves along the DNA template, beginning at the location where the primer binds, adding the complementary nucleotide to the growing primer until the end of the template is reached. The polymerase chain reaction (PCR) uses a heat-stable DNA polymerase for the repetitive replication of a DNA template from an excess of suitable primers and dNTPs (Sambrook and Russell). This is accomplished through iterative cycles of the following three steps (Fig. 25.3): (1) heating the reaction to melt double-stranded DNA molecules (denaturation), (2) cooling to allow for annealing the primers to the template, and (3) incubating at DNA polymerase's active temperature to allow for primer extension. This technique allows the generation of micrograms of DNA from a few copies of a given DNA template within a few hours.

1.2.2. RNA Polymerase and In Vitro Transcription

RNA polymerase, the key element of transcription, is the RNA-producing equivalent of DNA polymerase (Sambrook and Russell). RNA polymerase uses nucleoside 5′-triphosphate (NTP) subunits to produce an RNA copy of the coding strand of a double stranded DNA template. RNA polymerase isolated from the T7 phage, a relatively simple version of this enzyme, is commonly used to create RNA from a purified DNA template in a process known as in vitro transcription. T7 RNA polymerase requires a simple promoter sequence (to guide the initiation of transcription) that can readily be built into DNA templates. It is a robust polymerase that can be easily purified by recombinant techniques. In vitro transcription can be used to quickly produce large amounts of RNA for in vitro experiments (Sambrook and Russell).

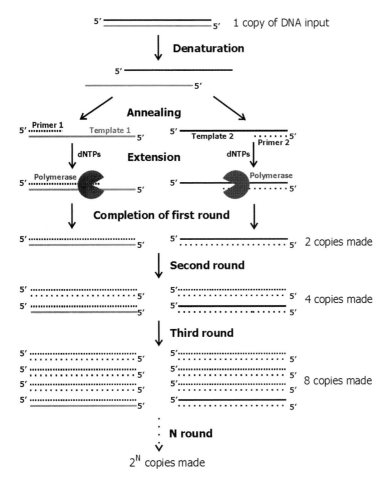

Figure 25.3. Schematic of polymerase chain reaction (PCR).

1.2.3. Reverse Transcription

Just as DNA can be converted to RNA through transcription using RNA polymerase, RNA can be converted to DNA using the enzyme reverse transcriptase (Sambrook and Russell). Reverse transcription (RT) is often combined with PCR in a technique known as RT-PCR. This allows for small quantities of RNA to be converted to a much larger quantity of the corresponding double-stranded DNA.

1.2.4. Other Modifications

Many other DNA modifying enzymes are also commonly used. DNA phosphorylation using polynucleotide kinase is particularly relevant for tracking DNA (Sambrook and Russell). Radioactively labeled phosphate groups can be added to the 5′ end of DNA using this enzyme, making it possible to track and quantify even small amounts of DNA. Robust enzymes that act on DNA and RNA through cleaving specific linkages, dephosphorylation, methylation, degradation and ligation (joining together) are also important tools for molecular biologists (Sambrook and Russell).

2. Isolation of Functional Nucleic Acids

2.1. Introduction to SELEX

The aptamer field first emerged in 1990 with the invention of the technique known as SELEX (selective evolution of ligands by exponential enrichment) (Tuerk and Gold 1990). This process, shown schematically in Fig. 25.4, is used to identify an oligonucleotide with a desired activity from an extremely large number of distinct DNA or RNA molecules. A typical DNA-based SELEX experiment (Fig. 25.4a) uses an ssDNA library consisting of somewhere between 10^{13}–10^{16} sequences, approximately 100-nt in length (nt: nucleotide), made up of a random-sequence region flanked on either side by constant sequences that are later used as primer binding sites for PCR. This library is subjected to a selection step that separates desirable (active) sequences from those that are undesirable (inactive). The surviving members of the library are then amplified by PCR, providing a greater number of the enriched sequences. This process is repeated for several rounds until active molecules begin to dominate the pool, at which point members of the library are cloned, sequenced, and studied on an individual basis.

For RNA selections (Fig. 25.4b), the starting point is again a DNA library; however, a T7 promoter is embedded in the 5′ end of each sequence. For these experiments, each round of selection begins with an in vitro transcription step that converts the DNA library to RNA. Following the selection step the RNA is converted back to DNA using reverse transcription, followed by PCR amplification to regenerate the pool for the next round of selection.

2.2. Selection Methods

The selection step used to isolate a functional nucleic acid varies with the activity being sought. This is particularly true for nucleic acid enzymes where unique selection strategies are usually required for each type of catalysis. A variety of nucleic acid enzymes have been isolated to catalyze a self-modification reaction, such as RNA cleavage (Breaker and Joyce 1994), RNA ligation (Bartel and Szostak 1993), DNA phosphorylation (Li and Breaker 1999b) and RNA alkylation (Wilson and Szostak 1995), using specific in vitro selection strategies. However, the catalytic repertoire of DNA and RNA is not limited to self-modification reactions: Many nucleic

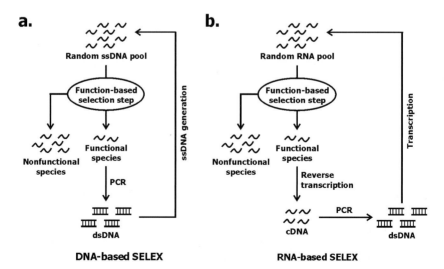

Figure 25.4. Schematic of SELEX for the isolation of (a) functional DNAs from a synthetic ssDNA pool or (b) functional RNAs from a RNA pool transcribed from a DNA library.

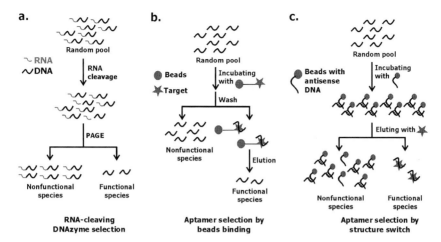

Figure 25.5. Representative selection strategies. (a) PAGE-based method for selecting RNA-cleaving deoxyribozymes (DNAzymes). (b) Aptamer selection using beads coated with a target of interest. (c) Aptamer selection using beads that contain a hybridized library.

acid enzymes that catalyze other types of chemical reactions (such as porphyrin metallation (Li and Sen 1996), peptide bond formation (Zhang and Cech 1997) and carbon–carbon bond formation (Tarasow et al. 1997)) have also been reported (for a recent review, see Fiammengo and Jaschke 2005). Selection for these types of catalysts can be more difficult because the reactant (or reactants) must be tethered to the initial DNA (or RNA) library as well as to the enriched pools during each selection cycle.

The selection step for a catalyst typically relies on active molecules experiencing a predictable change in physical properties that can be exploited by some means of separation. An example of such a selection is for deoxyribozymes that are capable of cleaving an attached RNA substrate (Fig. 25.5a) (Mei et al. 2003; Santoro and Joyce 1997). In this case, active molecules can be isolated using polyacrylamide gels that separate oligonucleotides on the basis of size. Molecules that catalyze the cleavage of the attached RNA substrate will become smaller, and will therefore travel faster through a polyacrylamide gel than their inactive counterparts. These shortened sequences are purified from the gel, amplified using PCR, and used for the next round of selection following a suitable procedure for the generation of ssDNA molecules modified again with the RNA substrate.

Selection strategies for aptamers have the advantage that they are generally more versatile and offer many more options for any given type of selection. The downside, however, is that it is often much more difficult to optimize the selection conditions for aptamers than for catalytic molecules. Although very few sequences can stimulate most types of catalysis, nonspecific interactions with target molecules are much more common. Most aptamer selections are based on target-binding sequences either co-separating with the target molecule or being eluted from some other type of interaction by the target molecule. Because binding interactions are transient, tight-binding molecules will be free in solution a certain percentage of the time and can therefore be lost. Similarly, less desirable molecules can form weak interactions that allow them to pass through a given round of selection. SELEX works on the principle that, on average, tighter binding molecules are more likely to proceed through the selection process, and after a sufficient number of rounds, these desirable candidates should dominate the pool. However the extreme purification that is required to go from a few active sequences in billions to almost entirely active sequences demands that the selection step be highly efficient. Summarized below are some examples of common aptamer selection strategies.

2.2.1. Bead and Column Based Selections

These methods rely on selectively retaining aptamers on beads or columns that are coated with the target molecule (Ellington and Szostak 1990; Huizenga and Szostak 1995). In a column-based selection, target-binding sequences will be retained in the column following incubation, while non-interacting molecules will pass through. The desirable oligonucleotides can then be collected from the column using an elution buffer containing an excess of the target molecule. Selections using beads work in a similar manner, except that target-coated beads that can be pulled out of solution are used in place of columns. Magnetic beads or beads that can be pelleted by centrifugation with surfaces that are covered with a variety of functional groups for target-coating are commercially available. These methods are illustrated in Fig. 25.5b.

A variation on this method is to coat the solid support with oligonucleotides that are complementary to a constant sequence element in the library (Nutiu and Li 2005b). The library will therefore bind the beads through standard Watson-Crick interactions with the complementary oligonucleotide. A solution containing the target molecule is then used to elute the desirable sequences (Fig. 25.5c).

2.2.2. Polyacrylamide Gel Electrophoresis (PAGE) Based Selections

PAGE based selections work on the principle that molecules interacting with the target will travel slower through a polyacrylamide gel. This is an inexpensive and relatively straightforward method of selection (Fig. 25.6). The library is incubated with the target for some amount of time and then loaded onto a nondenaturing gel. After running the gel for a suitable amount of time, the radioactively labeled library will be visible using autoradiography. Target-bound molecules will travel slower than the bulk of the library, and can be purified from the gel and amplified for the next round of selection. This strategy is mostly useful when selecting for large target molecules such as proteins that will significantly affect the movement of an oligonucleotide through the gel (Goodman et al. 1999).

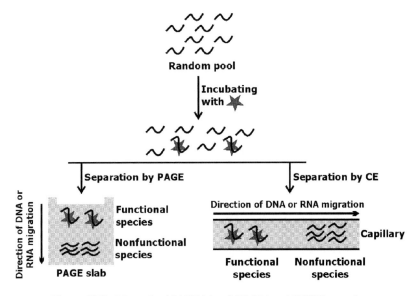

Figure 25.6. Schematic of PAGE (a) and CE (b) based SELEX strategies.

2.2.3. Capillary Electrophoresis (CE) Based Selections

CE is a separation technique wherein molecules pass through extremely thin tubes under the influence of an electric field (Landers 1996). Much like polyacrylamide gel electrophoresis described above, CE separates molecules based on their charge–mass ratio. The oligonucleotide library, pre-incubated with target, is passed through CE column and tracked through UV absorbance measurements. Sequences that bind the target will travel at a different rate than unbound molecules, normally slower, and will thus pass through the capillary at a later time (Fig. 25.6). These later fractions can be collected, amplified by PCR, and moved along to the next round of selection (Berezovski et al. 2005; Mendonsa and Bowser 2004).

2.3. Optimizing Functional Nucleic Acids

One of the trickiest aspects of SELEX is to avoid losing active sequences in early rounds when they are extremely rare, but to set conditions stringent enough to efficiently remove inactive molecules. A common practice to address this issue is to set the selection conditions to be relatively less stringent in earlier rounds and then to adjust them to be more stringent as the fraction of active sequences increases. This is often done by decreasing reaction times for catalytic selections, or by washing away undesirable molecules more thoroughly for aptamer selections. Another way to isolate only the most specific and most active sequences is to introduce a counter-selection step into the SELEX process, which is usually carried out in an identical manner to the selection step, but without a target molecule or with a different target molecule (Huizenga and Szostak 1995; Jenison et al. 1994). The oligonucleotides that do not interact with the mock-target in the counter-selection step are collected and applied for the selection with the real target.

Often times the initial SELEX experiment is not the end of a selection process. To improve on the initially isolated candidates, researchers can attempt to isolate their mutants that are more active. Mutagenic PCR (also known as error-prone PCR) is one way of achieving this (Beaudry and Joyce 1992). By adjusting the conditions of the PCR reaction, it is possible to increase the number of spontaneous mutations introduced by DNA polymerase. The selection can therefore be continued using mutagenic PCR, and as mutations manifest, those that are beneficial for activity will begin to dominate the population, whereas detrimental mutations will be lost. Another way of accomplishing this is to perform a reselection experiment. This involves repeating the selection process with a new synthetic library that is biased toward the sequence of a functional nucleic acid isolated in an original selection experiment. The sequence variants in the reselection library will have a high probability of obtaining the "wild type" nucleotide at each position (e.g., $\sim 76\%$) and one of the three remaining nucleotides at a reduced probability (e.g., $\sim 8\%$ each) (Li and Breaker 1999b). As a result, a pool that is extremely rich in variants of the original functional nucleic acid is produced and the reselection from this new pool will often produce variants with a significantly improved activity.

3. Aptamers: Properties and Targets

Aptamers that bind a large number of very diverse target molecules have been isolated (Bunka and Stockley 2006; Famulok and Mayer 1999; Osborne et al. 1997; Wilson and Szostak 1999). Cellular cofactors, organic dyes, amino acids, antibiotics and other drugs, peptides, and a large number of proteins are among the types of molecules have been successfully targeted by aptamers with good affinity and specificity (Lee et al. 2004). Aptamers have also been developed for the recognition of specific types of cells (Cerchia et al. 2005; Daniels et al. 2003; Shangguan et al. 2006). The following section will provide a brief overview of the arsenal of aptamers isolated to date as well as a discussion of the types of interactions and structures that they form to recognize their cognate ligands. The section will conclude with a comparison of aptamers to competing types of recognition elements.

3.1. The Growing Aptamer Catalogue

In 1990, the laboratories of Larry Gold at the University of Boulder Colorado and of Jack Szostak at Massachusetts General Hospital simultaneously published the first proof-of-principle papers demonstrating SELEX-type protocols (Ellington and Szostak 1990; Tuerk and Gold 1990). The publication by Tuerk and Gold (1990), where the term SELEX was first coined, used a partially random library based on an RNA sequence known to bind T4 DNA polymerase. They were able to uncover a novel sequence that bound the protein with the same affinity as the wild type sequence. The publication by Szostak's group reported the use of a completely random library to yield aptamers that bound several different organic dyes (Ellington and Szostak 1990). They used affinity chromatography with columns coated with the different dye molecules to isolate the aptamers that generally proved to be specific for their target and had K_d values in the high micromolar range. The Szostak lab later built on this success by repeating this same type of experiment using single-stranded DNA libraries. These efforts yielded the first DNA aptamers (Ellington and Szostak 1992).

Since this time, much effort has gone into developing and fine tuning novel aptamer selection strategies, building and applying libraries containing chemically modified nucleotides, and generally demonstrating the flexibility and potential of aptamers. A summary of some extraordinary aptamers is given in Table 25.1. This table does not represent a complete list of all published aptamers, but rather describes some of the properties of an assortment of the most remarkable and relevant aptamers uncovered to date.

3.2. Aptamer Specificity

The primary concern facing any expedition to engineer a system that binds a particular molecule is the prospect of isolating a promiscuous receptor. Indeed, without suitable specificity, any molecule isolated from a SELEX experiment will be worthless. Stringent selection conditions, counter-selections and using multiple selection strategies for a single experiment can help weed out such undesirable molecules. Taking proper precautions, it has been shown repeatedly that aptamers can display an extraordinary ability to distinguish between similar molecules. A milestone experiment to demonstrate this was published in 1994 by a group from NeXagen Inc. and the University of Colorado at Boulder (Jenison et al. 1994). The experiment was designed to isolate an aptamer that could tightly bind the small molecule theophylline, but would have poor affinity for caffeine, a molecule that differs from theophylline by a single methyl group (Fig. 25.7). They carried out five rounds of selection using a column based method, and then continued the selection for three more rounds with the addition of a counter-selection step aimed to remove caffeine-binding sequences. The best aptamer (Fig. 25.7a) obtained in this experiment bound theophylline with an apparent K_d of approximately 300 nM. What was truly remarkable about this study, however, was that this aptamer showed virtually no affinity for caffeine. An approximate K_d of 3.5 mM was measured for caffeine, indicating more than a 10,000-fold difference in affinity for ligands that differ only by the presence of a 15-Dalton, neutral functional group.

The theophylline aptamer is by no means the only example of outstanding recognition properties by an oligonucleotide receptor. Many aptamers have specificities as good as or even better than analogous proteins and antibodies. An aptamer targeting the human factor IX protein binds its target with greater than a 1000-fold preference over the bovine homologue of

Table 25.1. A summary of some particularly interesting aptamers

Target (and Type)	K_d (μM)	Comment	Reference
Cibacron Blue 3G-A dye (organic molecule)	100	Amongst the first RNA aptamers isolated	Ellington and Szostak 1990
Reactive green 19 dye (organic molecule)	30	Amongst the first DNA aptamers isolated	Ellington and Szostak 1992
Theophylline (metabolite)	0.3	An aptamer that is highly specific against structural analogues	Jenison et al. 1994
L-arginine (amino acid)	0.3	An aptamer that is highly enantioselective	Geiger et al. 1996
Neomycin (antibiotic)	0.1	Aptamer provided insight into activity of an RNA-binding antibiotic	Wallis et al. 1995
ATP (cofactor)	10	Widely adopted as a model small-molecule binding aptamer for bioanalytical applications	Huizenga and Szostak 1995
Keratinocyte growth factor (protein)	0.0003	2′-substituted RNA aptamer with very high affinity	Pagratis et al. 1997
Thrombin (protein)	0.05	"Toggle SELEX" used to create aptamers that recognize the target from multiple species	White et al. 2001
Vascular endothelial growth factor (VEGF; protein)	0.05	2′-fluoro substituted RNA aptamers that show in vivo inhibition of VEGF; one of the aptamers was eventually developed as a drug known as "Macugen" (also known as "pegaptanib") for ocular vascular disease	Ruckman et al. 1998 Ng et al. 2006
MUC1 (protein)	0.1	A DNA aptamer for a cancer marker	Ferreira et al. 2006
Thrombin (protein)	0.02	Widely adopted as a model protein binding aptamer for bioanalytical applications	Bock et al. 1992
RET-receptor kinase (protein)	0.04	An aptamer that blocks signaling of its cognate target in cell-based assays	Cerchia et al. 2005
Human influenza virus (virus)	0.0002	An aptamer that distinguishes between closely related viral strains	Gopinath et al. 2006a
African trypanosomes (protozoa)	0.07	An aptamer that recognizes a region on the surface of a dangerous blood parasite	Homann et al. 2006
Mesenchymal stem cells (mammalian cell)	Not determined	Developed for use in selectively purifying mesenchymal stem cells from bone marrow	Guo et al. 2006

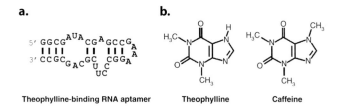

Figure 25.7. A small RNA aptamer that can distinguish theophylline and caffeine.

the same protein (Gopinath et al. 2006a). An aptamer for the amino acid arginine has shown 12,000-fold preference for the L-isoform over the D-isoform (Geiger et al. 1996). Aptamers that can discriminate between the oxidized and reduced forms of nicotinamide (Lauhon and Szostak 1995), phosphorylated versus unphosphorylated versions of a protein (Seiwert et al. 2000), and closely related viruses (Gopinath et al. 2006b) further illustrate that binding specificity is not a limitation of aptamers, but is instead one of their most significant strengths.

3.3. Aptamer–Ligand Interactions

Several high-resolution NMR and X-ray crystallographic structures of aptamer–ligand complexes have been published in the past decade or so. These reports have provided a great deal of insight into the molecular details of how aptamers work (Feigon et al. 1996; Hermann and Patel 2000; Patel and Suri 2000). A number of common features emerged in these studies that helped explain how molecules with such limited chemical diversity achieve such remarkable binding characteristics.

Unlike many proteins that possess prefolded binding pockets and thus experience minimal structural rearrangements upon ligand binding, most aptamers studied to date fold in a highly ligand-dependent manner. An unbound aptamer is in a less structured state, and through inter-actions with its target molecule it adopts a more complex architecture of which the ligand is an essential part. In the bound structure, sections of the ligand are often buried deeply within the surrounding oligonucleotide, allowing aptamers to achieve very tight binding and a good degree of selectivity without the diverse array of building blocks that is available to proteins. Adoption of ligand in the folding of the aptamer, referred to as *adaptive recognition*, appears to be an excellent strategy for nucleic acids to enhance their ability to recognize a broad range of potential targets.

Like their protein counterparts, aptamers adopt complex three-dimensional structures with well defined pockets occupied by the target molecule (Hermann and Patel 2000). Compared to proteins, aptamers tend to rely less on intermolecular hydrogen bonding and more on stacking interactions. Other than stacking of aromatic groups, aptamers also form favorable contacts with ligands through electrostatic interactions, hydrogen bonding and shape complementarity.

3.4. Aptamers vs. Other Recognition Elements

Oligonucleotides are only one of the many different types of molecules that can be engineered to bind a target molecule of your choosing. Other types of synthetic polymer libraries such as polypeptides can be applied in a somewhat similar manner to oligonucleotide SELEX experiments. Large libraries of small organic molecules, either chemically synthesized or purified from natural sources, have been screened with great success in a high throughput manner to uncover those that bind a chosen target. Perhaps the most significant competitors of aptamers, however, are antibodies isolated from animals that have been stimulated by the target antigen.

The main disadvantage of oligonucleotides compared to each of these other recognition elements is the lack of chemical diversity within DNA and RNA molecules. Although this can be enhanced by carrying out selections with modified nucleotides, it remains an obstacle for the aptamer field. The chemical functionality of aptamers is more compatible with some types of molecules than others. Molecules lacking functional groups commonly targeted by aptamers, such as aromatic rings and positively charged atoms, can be problematic for SELEX experiments. Although some success has been observed with targets that do not fit the profile of a good aptamer target (Jeong et al. 2001; Lozupone et al. 2003; Majerfeld and Yarus 1994; Majerfeld and Yarus 1998), this remains a legitimate concern. Proteins and peptides (that are composed of 20 building blocks compared to the 4 used by oligonucleotides) and small molecule libraries have a broader range of interactions and contacts than can theoretically be exploited.

A second disadvantage of nucleic acid receptors, particularly those composed of RNA, is their short half-life *in vivo*. This issue has been successfully addressed on several occasions by altering the originally selected aptamer with the use of chemically modified bases that provide greater stability and nuclease resistance (Brody and Gold 2000; Green et al. 1995; Jellinek et al. 1995; Latham et al. 1994; Pagratis et al. 1997).

There are also several advantages of aptamers that have made them an attractive option to many researchers. The chemical synthesis of nucleic acids is inexpensive and very well established—leading to low cost and very little variation in the concentration or activity from one batch to the next. Oligonucleotides can be easily modified with chemical groups that are useful for chromatography applications (such as amine modifications and biotinylation) and for signaling experiments (such as fluorophores). Aptamers can generally be forced to release their target molecule using excess antisense oligonucleotides. This provides a convenient antidote to help reverse any harmful effects of treatment for *in vivo* applications (Rusconi et al. 2002). This can also be useful for purposes of eluting the target in chromatography experiments. Aptamers have a long shelf life, are less prone to irreversible denaturation than proteins, and are isolated from libraries that are far less expensive to create, store, replenish and assay than are small molecule inhibitors.

4. Applications of Aptamers

There are many academic labs around the world as well as multiple pharmaceutical companies that focus on creating aptamers for a variety of purposes. In the beginning of aptamer research, targets were typically selected out of convenience or based on the perceived probability of success. These proof-of-principle type experiments were aimed at showing the versatility of aptamers and demonstrating new selection strategies. Recently, although new selection methods are still being developed, most novel aptamers are intended to have a specific practical function downstream. The major applications of aptamers can be lumped into three categories: therapeutics, tools for separation/purification, and reagents for molecular detection. This section will describe the advances in each of these fields with a particular emphasis on molecular detection applications.

4.1. Aptamers for Purification

Aptamers are uncovered from extremely large libraries of primarily noninteracting oligonucleotides on the basis of their interaction with the target molecule. Once an aptamer has been characterized and optimized, researchers have been able to turn this around and purify the target molecule from complex mixtures using aptamer-based purification. The protocols used in these experiments are often analogous to the selection strategies used in aptamer development (discussed in Section 25.2). Affinity chromatography has been used to purify a desired protein that has been over-expressed in cells. One group reported a 15,000-fold purification with 83 % recovery using such a method (Romig et al. 1999). Using aptamers for protein purification not only has the potential to be much more efficient than classical purification methods, but does not require the addition of affinity tags that can affect the function of the protein. Aptamers have also been used as the stationary phase in a range of chromatographic techniques to effectively purify a single enantiomer from a racemic mixture (Michaud et al. 2003) or to purify a small molecule from a mixture of cellular metabolites (Deng et al. 2001; Deng et al. 2003).

A similar concept that is gaining momentum is using aptamers to selectively purify a particular cell type from a heterogeneous population (Guo et al. 2006). This can be done by generating aptamers that bind a particular population of cells of the chosen type. Specificity for the desired cell type can be achieved by introducing counter-selection steps that remove sequences that bind other cell types (Shangguan et al. 2006). It is thought that an oligonucleotide isolated from this type of selection will normally bind a single molecule on the surface of the cell. Since no prior knowledge of the specific target is needed for this type of experiment, it is also possible to isolate novel cell type-specific biomarkers using this strategy (Shangguan et al. 2006). Biomarkers that are indicative of disease states such as cancers can be extremely valuable to help characterize the disorder and for directed delivery of therapeutics (Kallioniemi et al. 2001).

4.2. Aptamers with Therapeutic Potential

The most common strategy that has been used in drug development has been to identify factors (typically proteins) that are contributing to the ailment in question and then to try to block their activity. Ideally, preventing such a factor (referred to as a drug target) from performing its function will disrupt either the source or the symptoms of the sickness and provide therapeutic benefit to the patient. The screening of extremely large libraries of small molecules has been highly successful in identifying potent inhibitors of drug targets and has led to many drugs in common use today.

Aptamer technology offers an exciting alternative to standard small molecule screens. The potential of such endeavors is beginning to become evident (Lee et al. 2006; Nimjee et al. 2005). The first aptamer drug appeared on the market in 2005 after successfully completing clinical trials for the treatment of age related macular degeneration. The drug, dubbed *Macugen*, binds the human vascular endothelial growth factor and inhibits its ability to promote unwanted growth of new blood vessels in the eyes (Gragoudas et al. 2004; Lee et al. 2005; Ng et al. 2006; Siddiqui and Keating 2005). Macugen is undoubtedly going to be joined by other disease-treating aptamers in the clinic in the near future. Promising candidates for preventing blood clotting are currently in the late stages of clinical trials (Dyke et al. 2006; Rusconi et al. 2004), as are aptamers that inhibit the function of proteins that are known to play an important role in specific forms of cancer (Cerchia et al. 2002; Laber et al. 2005).

Another therapeutic area being tapped by aptamers is in the treatment of viral infections. An advantage held by aptamers over small molecule treatments of such infections is their larger size. The largest obstacle to overcome in the treatment of a virus is its ability to rapidly accumulate mutations to avoid the drug's effects. As opposed to small molecules that typically only interact with a small crevice of the target, aptamers can make extensive contacts with a more substantial surface. It is thought that having a larger contact area with the viral target will make it far less likely that one or a few mutations will be sufficient to negate the drug's effects. Aptamers that tightly bind several established drug targets for human immunodeficiency virus (HIV) and hepatitis C virus (HCV) have been developed in recent years, although their clinical efficacy is yet to be proven (Allen et al. 1995; Bellecave et al. 2003; Gomez et al. 2004; Tuerk et al. 1992).

4.3. Aptamers as Sensing Elements

Due to their superb binding characteristics, stability and easy modification, aptamers have all the necessary qualities to be used as the basis of detection elements. This facet of aptamer applications has been exploited with great success. Through their recognition capabilities, aptamers innately possess one of the two qualities of a detection element. In order to become a

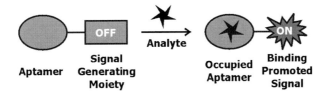

Figure 25.8. General concept of exploring aptamers as biosensors.

fully functional sensor, the binding of the ligand must be linked to the production of a recordable signal (Fig. 25.8). Aptamers can be chemically modified to include a signal-producing moiety or tethered to any number of surfaces, providing aptamer sensor engineers with many options for their designs. This section will illustrate several of these systems that have proven successful.

The list of different detection methods that have been employed by aptamer detectors is extensive and is not limited to those described below. Some detection methods not discussed in this chapter include proximity ligation assays (Fredriksson et al. 2002; Gustafsdottir et al. 2005), photoSELEX based detection (Golden et al. 2000), cantilever-aptamer sensing (Savran et al. 2004), and a variety of optical detection strategies (Navani and Li 2006).

4.3.1. Conformation-Dependent Fluorescent Sensors

A series of detection methods has been established that relies on conformational changes in the aptamer upon target binding and that alters the signal output of a covalently attached fluorophore (Navani and Li 2006; Nutiu and Li 2004; Nutiu and Li 2005a). To get a suitable signal enhancement, these methods employ a fluorescence-modifying moiety that is brought closer or farther from the fluorophore when the aptamer is in its ligand-bound structure. Three examples of detection schemes following this basic premise are shown in Fig. 25.9 (Li et al. 2002; Nutiu and Li 2003; Stojanovic et al. 2001). Each example employs a fluorophore–quencher pair with a suitable excitation–emission spectra. The fluorescence quenchers used in these experiments are specially designed molecules that decrease the quantum yield of a corresponding fluorophore in a highly distance-dependent manner. Therefore when the fluorophore–quencher pair is in very close proximity, the fluorescence will be very low. When separated by greater distances, the quencher is ineffective, and a greater fluorescent signal is observed.

Fluorescent aptamer sensors can also be designed with the use of two fluorophores to exploit a phenomenon known as fluorescent resonance energy transfer (FRET). Fluorophores are excited with a certain wavelength of light and re-emit photons at a longer, lower energy wavelength. If the emitted light from one fluorophore is suitable for the excitation of a second fluorophore, the energy can be transferred and the emission observed will be that of the second fluorophore (Lakowicz 1999). Much like fluorescent quenchers, FRET is a highly distance-dependent phenomenon and will only occur when the two fluorophores are in close proximity. Therefore the same types of designs often apply to both fluorophore–quencher and FRET designs. The major advantage of FRET is that the measurement taken is not simply a loss or gain of signal strength, but rather the creation of an otherwise absent signal. This helps eliminate background noise and erroneous results (Lakowicz 1999).

4.3.2. Quantum Dot Sensors

Quantum dots (QDs) are inorganic crystals of Cadmium selenide (CdSe) coated with Zinc Sulfide (ZnS). They can achieve ultra-sensitive detection limits due to high fluorescence capabilities, resistance to photo bleaching and long fluorescence lifetimes. QDs can be coated with oligonucleotides and converted into aptamer-based sensors using a second fluorophore for

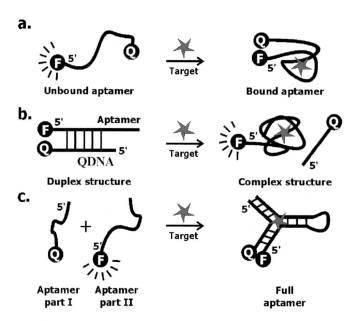

Figure 25.9. Representative strategies for designing conformation-dependent fluorescent sensors. (a) One-piece folding aptamer (Li et al. 2002). This method exploits the fact that many aptamers fold to place the two ends of the aptamer chain next to one another in the ligand-bound structure. (b) Structure-switching aptamer sensor (Nutiu and Li 2003). This approach takes advantage of the fact that an aptamer can form both a tertiary structure in the presence of its cognate target and a duplex structure with an oligonucleotide having a complementary sequence. (c) Two-piece aptamer (Stojanovic et al. 2001). An aptamer is broken into two separate molecules and target binding promotes their assembly.

FRET or fluorescence quenchers as described in the section above. Fig. 25.10 demonstrates an example of a sensor that relies on target binding to aptamers covalently attached to a quantum dot displacing a short quencher-labeled oligonucleotide (Levy et al. 2005). The fluorescence of the aptamer-modified QDs is quenched by the antisense DNA carrying a quencher (Q). The target binding causes the release of the Q-DNA strands, generating a fluorescence signal.

4.3.3. Target Detection by Fluorescence Anisotropy

When a fluorophore is excited by a photon there is a certain amount of time before the lower energy photon is emitted. If the fluorescent body is excited with plane-polarized light, the extent of the polarization of the emitted signal will depend on the rate at which the molecule is

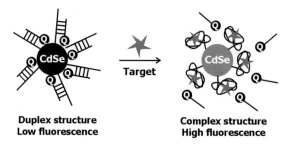

Figure 25.10. A QD-aptamer sensor. Aptamer-modified QDs are mixed with an antisense DNA strand (Q-DNA) carrying a fluorescence quencher. The target binding causes the release of the Q-DNA, generating a fluorescence signal.

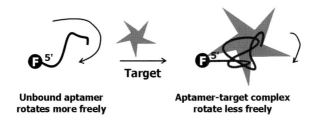

Figure 25.11. Target detection by fluorescence anisotropy, which requires the use of relatively small, labeled aptamers that bind a much larger target. The bound aptamer with the significantly increased size will rotate less freely in solution than the free aptamer, generating a signal that retains its polarity.

tumbling freely in solution. Small molecules that tumble quickly in solution will have rotated considerably during the time between excitation and emission, and therefore the emitted light will no longer be polarized. By contrast, larger complexes will generally rotate less freely in solution and will emit a signal that retains its polarity to a greater extent (Lakowicz 1999). A small, fluorophore-labeled aptamer that is unbound will therefore produce a lower anisotropy measurement than if it is bound to a large ligand (Fig. 25.11). Several aptamer sensors have been designed using this approach (Fang et al. 2001; Gokulrangan et al. 2005; Potyrailo et al. 1998).

4.3.4. Enzyme Linked Aptamer Assays

ELISA (enzyme linked immunosorbent assay) is a widely used, antibody based detection method. A standard "sandwich" ELISA uses assay plates that have been coated with an antibody (known as the *capture* antibody). The test sample is added to the plate and allowed to incubate for a period of time before it is removed and the plates are washed to remove unbound products. A second antibody that is coupled with an enzyme (known as the *detection* antibody) is then added, allowed to incubate, and washed. To determine the amount of target present in the sample, an assay is carried out to measure the activity of the enzyme coupled to the detection antibody (Fig. 25.12). The aptamer equivalent of ELISA is known as either *ELONA* (enzyme linked oligonucleotide assay) (Drolet et al. 1996) or *ALISA* (aptamer linked immunosorbent assay) (Vivekananda and Kiel 2006). ALISAs use aptamers rather than antibodies for the capture element, the detection element or for both (Fig. 25.12) (Guthrie et al. 2006; McCauley et al. 2006; Vivekananda and Kiel 2006). The great success of ELISA experiments, combined with the advantages that aptamers hold over antibodies with respect to stability and ease of modification, make ALISA technology an exciting possibility for the design of medical diagnostics.

4.3.5. Acoustic Sensors

This method uses materials such as quartz crystals that are said to be piezoelectric. Piezoelectric materials can generate a voltage in response to a mechanical force, and in acoustic sensors this voltage is converted to an acoustic wave. Aptamer-based acoustic sensors rely on the detection of small changes in amplitude or velocity of this wave as it passes through surfaces coated with an aptamer. The wave is very sensitive to changes occurring at the surface, such that when an aptamer binds its target, the characteristics of the wave are altered and can be recorded (Fig. 25.13). This principle has been used to create extremely sensitive detectors for protein targets and has compared favorably to synonymous antibody-based systems targeting the same protein (Gronewold et al. 2005).

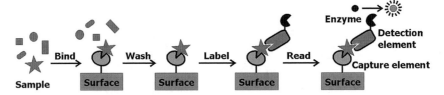

Figure 25.12. Enzyme-linked sandwich assays. The capture and detection elements can be either antibody or aptamer.

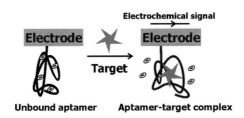

Figure 25.13. Target detection by an acoustic sensor. Target binding induces a change in amplitude or velocity of an acoustic wave as it passes through the surface of piezoelectric materials (such as quartz) coated with an aptamer.

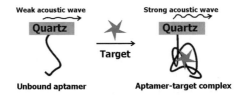

Figure 25.14. Aptamer-based electrochemical sensor. The binding of the aptamer causes the release of the electrochemical markers from the aptamer, resulting in an electrochemical signal.

4.3.6. Electrochemical Sensors

Aptamer-based electrochemical sensors have attracted considerable attention in the field of bioanalytical chemistry (Bang et al. 2005; Lai et al. 2007; Le et al. 2006; Xu et al. 2005). Some of the reported sensors are made of an electrode modified with an aptamer that has different affinities to an electrochemical marker in the absence and presence of the target. The addition of the target to the detection solution will therefore cause the release of the marker from the aptamer (or vice versa), resulting in an electrochemical signal (Fig. 25.14). Other designs employ an electrode attached with an aptamer that is covalently modified with an electrochemical marker. The aptamer experiences a large conformational change upon target binding, which brings the tethered marker either closer to or farther away from the electrode, leading to a detectable electric current.

5. Aptamers for Detection of Pathogenic Bacteria

The threat of bioterrorism and the emergence of new pathogenic species and strains that are resistant to or unaffected by traditional treatments are among the most pressing health concerns of our time. The bird flu virus epidemic and the outbreak of *Escherichia coli* and *Salmonella* contaminated lettuce in North America are the latest examples to drive this point home. These outbreaks can be a huge economic burden and, much more importantly, pose a tremendous risk to human health. Early and accurate identification of dangerous pathogens is a crucial step in preventing large-scale outbreaks. This is particularly relevant today as globalization of commerce and decreased travel times have significantly increased the rate and breadth of the spread of infectious agents. The development of fast, sensitive, accurate and inexpensive detectors of known biological threats is an area of intense research. It is also significant to have reliable methods in place for quickly creating new sensors and tests to respond to unanticipated outbreaks.

As described in Section 25.4.3, aptamers have emerged as exciting new tools for biosensing. Some initial steps have already been taken to apply this potential to the creation of detection systems for pathogenic bacteria. This section describes some of the bacterial species most in need of efficient detection, and some strategies to employ aptamers to achieve this task.

5.1. Categories of Microbial Agents to be Detected

Bacteria are generally classified as being either Gram-positive or Gram-negative. The origin of this grouping is a straightforward test that distinguishes bacteria that have an outer membrane from those that do not. Gram-positive organisms lack an outer membrane but have a thick cell wall, composed largely of peptidoglycan and teichoic acid, that resides outside the cytoplasmic membrane. Gram-negative organisms, by contrast, have a much thinner cell wall lacking teichoic acid that is sandwiched between the cytoplasmic membrane and an outer membrane. The outer leaflet of the latter is composed largely of lipopolysaccharide. These distinctions are relevant for engineering sensors to identify bacteria since it is typically the outermost layer of a bacterium that is targeted for detection. Introduced below are dangerous organisms from both of these phylogenies that are in need of fast and reliable detection methods.

5.1.1. Gram-Positive Bacteria

The major biothreat species belonging to this class include *Bacillus anthracis* and *Yersinia Pestis*. *B. anthracis*, the organism that causes anthrax, assumes special importance for its ability to form spores and have an aerobic metabolism. The biothreat agent is a spore that is resistant to degradation, dessication and other inclement conditions. If inhaled, the spore germinates and the bacterium multiplies, resulting in the formation of coal-like black patches on the skin (hence anthracis from the Greek word *anthrokis* meaning coal). There are at least 87 different strains of *B. anthracis*, several of which are pathogenic and can be particularly dangerous due to their ability to secrete potent toxins (Edwards et al. 2006). The largest concern with respect to anthrax is the use of specially designed strains that are optimized for inhalation as biological weapons. Inhalation anthrax is a very deadly form of infection that presents as a nondescript illness with symptoms such as a mild fever, cough, and chest and abdominal pains for the first several days. Left untreated, this phase can pass into a rapidly progressing second phase with much more severe symptoms often ending in death within 2–3 days (Swartz 2001).

The detection of spores is of particular importance since it provides an opportunity to identify the threat of anthrax prior to infection. One of the major components of the spore is dipicolinic acid; its detection can serve as an indication that spores are present in that environment. Alternatively, antigens or epitopes on the surface of the spore that are specific to anthrax represent viable targets for detection (Kobiler et al. 2006).

Clostridium botulinum is a sporulating motile rod that causes botulism through its production of highly potent neurotoxins. There are two families of *C. botulinum* that are pathogenic to humans, each of which is capable of producing several distinct neurotoxins. The classical form of the illness, food-borne botulism, is not actually an infection but instead results from the ingestion of food contaminated with preformed botulism toxin. Infection with *C. botulinum* spores can also occur through open wounds or in infants by ingestion of spores.

Infection leads to the production of neurotoxins that are absorbed into the blood and cause similar symptons to the foodborne version of the illness. These symptoms are the result of the toxin's interference with neurotransmitters, leading to paralysis—first of cranial muscles and eventually of limbs and respiratory musculature. Death from botulism usually results from respiratory dysfunction (Sobel 2005). As biological weapons, botulinum toxins may be aerosolized or used to sabotage food supplies.

The main targets for detection of botulism are the seven distinct neurotoxins produced by the various strains that have been described (Sobel 2005). A mouse bioassay has been traditionally used to detect the presence of the toxins in samples from stool, vomitus, gastric contents and serum. Recently, immunological methods have shown great promise for faster and more sensitive detection, however finding a reliable source of high-quality antibodies remains a problem (Lindstrom and Korkeala 2006).

5.1.2. Gram-Negative Bacteria

A particularly important Gram-negative species is *Francisella tularensis*, a highly virulent pathogen that can infect a host with as few as 25 colony forming units (McLendon et al. 2006). *F. tularensis* is a small aerobic bacillus with two main serotypes: Jellison Type A and Type B. Type A is the more virulent form. Infection can mature to various forms—tularemia (fever, chills, headache and other flulike symptoms), ulceroglandular tularemia (above symptoms with skin ulcers), oculoglandular tularemia (conjunctivitis), pulmonary tularemia (cough, chest pain and pulmonary infiltrates) and typhoidal tularemia (fatigue, bloody diarrhea, splenomegaly) (Bossi et al. 2006). In spite of long-standing recognition of the pathogenicity of *Francisella*, its extreme virulence is still not completely understood. It can be contracted either by skin contact, ingestion of contaminated food or water, or by inhalation of aerosolized organisms. The ability of *Francisella* to survive tough conditions, its high infectivity and ability to cause a very lethal disease make it an attractive target as an agent of biological warfare (Bossi et al. 2006).

Other gram-negative pathogens of interest belong to families of Enterobacteriaceae, Pseudomonas, Acinetobacter and Stenotrophomonas. These organisms differ in virulence, but all have intrinsic resistance to some commonly used antibiotics and the capacity to develop rapidly acquired resistance, which makes treatment of infections very difficult.

Two species of particular interest are *Yersinia pestis* and *Brucella*. *Yersinia pestis*, a member of the Enterobacteriaceae, is the causal organism for human plague that has been the cause of three pandemics and is linked to the deaths of millions of people. There are three clinical forms: bubonic plague, pneumonic plague, usually secondary to bubonic plague, and a septicemic form. Bipolar staining is typical of *Yersinia*, although it is not enough to conclusively identify the organism. Currently the best target for identification is a temperature-regulated antigen (dubbed F1) that is expressed when *Y. pestis* is incubated at 37°C (Huang et al. 2006).

Brucella species are small, faintly staining, single-appearing coccobacilli or short rods, arranged in pairs and short chains. They are nonmotile, nonsporulating strict aerobes. The genus *Brucella* has six species—among them *B. abortus*, *B. melitensis*, *B. suis* and *B. canis* are associated with human disease, with *Brucella melitensis* being the most virulent (Pappas et al. 2006).

5.2. Traditional Pathogen Detection Methods

Traditionally, microbial pathogen detection has been accomplished using classical microbiological techniques coupled with biochemical analysis. These techniques are slow and laborious because of defined media and differentiation tests to be done for specific microbes. It takes a good deal of time to arrive at a definitive conclusion—time that can be very precious when assessing a patient with a deteriorating condition or when trying to identify and control a large-scale outbreak. Immunological techniques have proven to be a significant advancement over microbiological methods. Such tests have typically matched or exceeded the sensitivity of classical methods but can be carried out much more quickly. Advanced versions of immunological sensing such as immunomagnetic-electrochemiluminescence have been employed with reasonable success for detection of biowarfare agents and show promise for future designs and improvements (Cote et al. 2005; Lim et al. 2005). Aptamers have many of the same benefits that antibodies do with respect to sensitivity and speed of detection, but offer a more stable and less expensive alternative with a more reliable and renewable synthesis.

5.3. Aptamers in Pathogen Detection

Nucleic acids are already in common use for detection of bacteria in PCR-based methods (Lim et al. 2005). These techniques play a vital role as a means to gain insight into the genetic makeup of an invading organism; however, they are limited in their applications. These methods require a clean sample and are unable to directly investigate nonDNA targets such as intact spores, surface antigens and toxins. Nucleic acid biosensing methods that use aptamers can overcome these drawbacks. While this technology is still very much in its infancy, the potential for future developments is great.

Depending on the detection needs for a particular organism or strain, there are many different approaches that can be taken by sensor engineers to develop aptamers against a suitable target. For the detection of intact bacteria, cell-based SELEX is an option that does not require an in-depth understanding of the organism, its genome and its surface markers. Using closely related organisms for counter-selection steps, researchers can aim to isolate aptamers that recognize a target on the cell's surface that is specific to the strain of interest.

More standard SELEX approaches can be used when more is known about the organism in question. Toxins, confirmed biomarkers and specific antigens can be purified and used as the target molecule for SELEX experiments. These selections can be customized to isolate aptamers that work in conditions of a researcher's choosing. Furthermore, one of any number of designs can be used to convert the aptamer into a functional sensor (see Section 25.4.3).

One of the initial attempts to use aptamers for pathogen detection was carried out using DNA aptamers isolated by Bruno and Kiel through a SELEX experiment against a heat-inactivated strain of anthrax spores (Bruno and Kiel 1999). Using two DNA/anthrax spore ratios, these investigators found that it was feasible to isolate aptamers for the anthrax spores with varying affinities. The aptamer-spore recognition was carried out using an aptamer-magnetic bead-ECL sandwich assay (ECL: enhanced chemiluminescent labeling) (Bruno and Kiel 1999).

DNA aptamers for detecting *Francisella tularensis* have also been generated (Vivekananda and Kiel 2006). The SELEX experiment was conducted using a library of 10^{14}–10^{16} different sequences and the *F. tularensis* subspecies *japonica* antigen as the target. The experiment yielded 25 different aptamer sequences. Since polyclonal antibodies against the tularemia antigen are also available, the researchers carried out a comparison study of ALISA vs. ELISA (see Section 25.4.3.4 for the discussion on these two assays). The results showed that ALISA (aptamer based) was more sensitive than ELISA (antibody based) in detecting *F. tularensis*. An antigen detection limit of 25 ng by ALISA compared to 100 ng by ELISA. More importantly, the antigen recognition by aptamers was found to be as specific as that by the antibody (Vivekananda and Kiel 2006). Though there has been progress in detecting *F. tularensis* in biological samples using capture-ELISA and other related immunodiagnostic techniques (Grunow et al. 2000), there is much room for improvement. The aptamer-based technique shows promise and represents a solid precedent for aptamers out-performing antibodies in an assay format that is still overwhelmingly dominated by antibodies.

Another example of aptamers breaking into this field made use of quantum dots (QDs; see Section 25.4.3). A DNA aptamer against the E.coli 0111:B4 surface antigen lipopolysaccharide

(LPS) was generated by SELEX, and then conjugated to QDs (Dwarakanath et al. 2004). Interestingly, it was found that the binding of the aptamer-QDs to live E. coli 0111:B4 cells promoted a large emission wavelength shift (which could reach 140 nm) toward the blue spectrum. The researchers speculated that the blue shift was likely caused by the change of the chemical environment experienced by the aptamer coupled QDs when they bound to the bacterial cell surface. This preliminary work opens an interesting possibility of exploring aptamer conjugated QDs for sensitive detection of bacterial pathogens.

6. Conclusions

Aptamers technology is being applied in many ways in diverse fields. Cancer imaging and biomarker discovery, purification and biotechnology, drug discovery and delivery, and biosensing all represent research areas that are being exploited by the aptamer community and its collaborators.

All signs indicate that the detection and differentiation of pathogens and biowarfare agents is on its way to being another success story for aptamers. Tremendous strides made in the field of genomics have opened up the genome sequences of microbes to researchers, and it is now possible to clearly identify the differences in the genomes. A combination of comparative bioinformatics of bacterial surface markers and in vitro selection can provide robust aptamer probes that can specifically detect the pathogen of interest. The amenability of DNA to be decorated with different fluorophores can make detection of multiple microorganisms possible. Even when the differences in the surface characteristics of related species or strains elude the researchers, cleverly designed in vitro selection approaches may still be able to generate aptamers specific to each strain.

A related application of the aptamers that can be envisaged is the use of aptamers as tools to neutralize pathogens. Aptamers can be generated against a pathogen or biowarfare agent so as to block disease causing factors. These factors can be, for example, pili for cell invasion, surface receptors on bacteria that bind human cells, or harmful secreted toxins. Generation of aptamers against these targets will not only warn of their presence but could also help attenuate or control the pathogens.

It is clear that aptamers have tremendous potential to be developed as reliable tools for the detection of pathogen/biowarfare agents. It is also true that, like any other technological development, the aptamer field faces challenges to make the transition from laboratory research to real-world applications. Further advances in aptamer generation methodology coupled with increased knowledge of microbial genomics and proteomics and improvements in signaling technologies will all be key to this transition. Considering the advantages with respect to cost, stability, and ease and speed of synthesis, aptamer technology has the potential to one day surpass immunological techniques and become the gold standard for the development of inexpensive, fast, accurate and sensitive detectors of dangerous bacterial species.

References

Adams PL, Stahley MR, Kosek AB, Wang J and Strobel SA (2004) Crystal structure of a self-splicing group I intron with both exons. Nature 430:45–50

Allen P, Worland S and Gold L (1995) Isolation of high-affinity RNA ligands to HIV-1 integrase from a random pool. Virology 209:327–336

Bang GS, Cho S and Kim BG (2005) A novel electrochemical detection method for aptamer biosensors. Biosens Bioelectron 21:863–870

Bartel DP (2004) MicroRNAs: genomics, biogenesis, mechanism, and function. Cell 116:281–297

Bartel DP and Szostak JW (1993) Isolation of new ribozymes from a large pool of random sequences. Science 261:1411–1418

Beaudry AA, and Joyce, G. F. 1992. Directed evolution of an RNA enzyme. Science 257:635–641

Bellecave P, Andreola ML, Ventura M, Tarrago-Litvak L, Litvak S and Astier-Gin T (2003) Selection of DNA aptamers that bind the RNA-dependent RNA polymerase of hepatitis C virus and inhibit viral RNA synthesis in vitro. Oligonucleotides 13:455–463

Berezovski M, Drabovich A, Krylova SM, Musheev M, Okhonin V, Petrov A and Krylov SN (2005) Nonequilibrium capillary electrophoresis of equilibrium mixtures: a universal tool for development of aptamers. J Am Chem Soc 127:3165–3171

Bock LC, Griffin LC, Latham JA, Vermaas EH and Toole JJ (1992) Selection of single-stranded DNA molecules that bind and inhibit human thrombin. Nature 355:564–566

Bossi P, Garin D, Guihot A, Gay F, Crance JM, Debord T, Autran B and Bricaire F (2006) Bioterrorism: management of major biological agents. Cell Mol Life Sci 63:2196–2212

Breaker RR and Joyce GF (1994) A DNA enzyme that cleaves RNA. Chem Biol 1:223–229

Brody EN and Gold L (2000) Aptamers as therapeutic and diagnostic agents. J Biotechnol 74:5–13

Bruno JG and Kiel JL (1999) In vitro selection of DNA aptamers to anthrax spores with electrochemiluminescence detection. Biosens Bioelectron 14:457–464

Bunka DH and Stockley PG (2006) Aptamers come of age - at last. Nat Rev Microbiol 4:588–596

Burgstaller P and Famulok M (1994) Isolation of RNA aptamers for biological cofactors by in vitro selection. Angew. Chem. Int. Ed. Engl. 33:1084–1087

Cate JH, Gooding AR, Podell E, Zhou K, Golden BL, Kundrot CE, Cech TR and Doudna JA (1996) Crystal structure of a group I ribozyme domain: principles of RNA packing. Science 273:1678–1685

Cerchia L, Duconge F, Pestourie C, Boulay J, Aissouni Y, Gombert K, Tavitian B, de Franciscis V and Libri D (2005) Neutralizing aptamers from whole-cell SELEX inhibit the RET receptor tyrosine kinase. PLoS Biol 3:e123

Cerchia L, Hamm J, Libri D, Tavitian B and de Franciscis V (2002) Nucleic acid aptamers in cancer medicine. FEBS Lett 528:12–16

Chiuman W and Li Y (2006) Evolution of high-branching deoxyribozymes from a catalytic DNA with a three-way junction. Chem Biol 13:1061–1069

Cote CK, Rossi CA, Kang AS, Morrow PR, Lee JS and Welkos SL (2005) The detection of protective antigen (PA) associated with spores of Bacillus anthracis and the effects of anti-PA antibodies on spore germination and macrophage interactions. Microb Pathog 38:209–225

Daniels DA, Chen H, Hicke BJ, Swiderek KM and Gold L (2003) A tenascin-C aptamer identified by tumor cell SELEX: systematic evolution of ligands by exponential enrichment. Proceedings of the National Academy of Sciences 100:15416–15421

Deng Q, German I, Buchanan D and Kennedy RT (2001) Retention and separation of adenosine and analogues by affinity chromatography with an aptamer stationary phase. Anal Chem 73:5415–5421

Deng Q, Watson CJ and Kennedy RT (2003) Aptamer affinity chromatography for rapid assay of adenosine in microdialysis samples collected in vivo. J Chromatogr A 1005:123–130

Drolet DW, Moon-McDermott L and Romig TS (1996) An enzyme-linked oligonucleotide assay. Nat Biotechnol 14:1021–1025

Dwarakanath S, Bruno JG, Shastry A, Phillips T, John AA, Kumar A and Stephenson LD (2004) Quantum dot-antibody and aptamer conjugates shift fluorescence upon binding bacteria. Biochem Biophys Res Commun 325:739–743

Dyke CK, Steinhubl SR, Kleiman NS, Cannon RO, Aberle LG, Lin M, Myles SK, Melloni C, Harrington RA, Alexander JH, Becker RC and Rusconi CP (2006) First-in-human experience of an antidote-controlled anticoagulant using RNA aptamer technology: a phase 1a pharmacodynamic evaluation of a drug-antidote pair for the controlled regulation of factor IXa activity. Circulation 114:2490–2497

Eddy SR (1999) Noncoding RNA genes. Curr Opin Genet Dev 9:695–699

Edwards KA, Clancy HA and Baeumner AJ (2006) Bacillus anthracis: toxicology, epidemiology and current rapid-detection methods. Anal Bioanal Chem 384:73–84

Ellington AD and Szostak JW (1990) In vitro selection of RNA molecules that bind specific ligands. Nature 346:818–822

Ellington AD and Szostak JW (1992) Selection in vitro of single-stranded DNA molecules that fold into specific ligand-binding structures. Nature 355:850–852

Erdmann VA, Barciszewska MZ, Hochberg A, de Groot N and Barciszewski J (2001) Regulatory RNAs. Cell Mol Life Sci 58:960–977

Famulok M and Mayer G (1999) Aptamers as tools in molecular biology and immunology. Curr Top Microbiol Immunol 243:123–136

Fan P, Suri AK, Fiala R, Live D and Patel DJ (1996) Molecular recognition in the FMN-RNA aptamer complex. J Mol Biol 258:480–500

Fang X, Cao Z, Beck T and Tan W (2001) Molecular aptamer for real-time oncoprotein platelet-derived growth factor monitoring by fluorescence anisotropy. Anal Chem 73:5752–5757

Feigon J, Dieckmann T and Smith FW (1996) Aptamer structures from A to zeta. Chem Biol 3:611–617

Ferreira CS, Matthews CS and Missailidis S (2006) DNA aptamers that bind to MUC1 tumour marker: design and characterization of MUC1-binding single-stranded DNA aptamers. Tumour Biol 27:289–301

Fiammengo R and Jaschke A (2005) Nucleic acid enzymes. Curr Opin Biotechnol 16:614–621

Fredriksson S, Gullberg M, Jarvius J, Olsson C, Pietras K, Gustafsdottir SM, Ostman A and Landegren U (2002) Protein detection using proximity-dependent DNA ligation assays. Nat Biotechnol 20:473–477

Geiger A, Burgstaller P, von der Eltz H, Roeder A and Famulok M (1996) RNA aptamers that bind L-arginine with sub-micromolar dissociation constants and high enantioselectivity. Nucleic Acids Res 24:1029–1036

Gokulrangan G, Unruh JR, Holub DF, Ingram B, Johnson CK and Wilson GS (2005) DNA aptamer-based bioanalysis of IgE by fluorescence anisotropy. Anal Chem 77:1963–1970

Golden MC, Collins BD, Willis MC and Koch TH (2000). Diagnostic potential of PhotoSELEX-evolved ssDNA aptamers. J Biotechnol 81:167–178

Gomez J, Nadal A, Sabariegos R, Beguiristain N, Martell M and Piron M (2004) Three properties of the hepatitis C virus RNA genome related to antiviral strategies based on RNA-therapeutics: variability, structural conformation and tRNA mimicry. Curr Pharm Des 10:3741–3756

Goodman SD, Velten NJ, Gao Q, Robinson S and Segall AM (1999) In vitro selection of integration host factor binding sites. J Bacteriol 181:3246–3255

Gopinath SC, Balasundaresan D, Akitomi J and Mizuno H (2006a) An RNA aptamer that discriminates bovine factor IX from human factor IX. J Biochem (Tokyo) 140:667–676

Gopinath SC, Misono TS, Kawasaki K, Mizuno T, Imai M, Odagiri T and Kumar PK (2006b) An RNA aptamer that distinguishes between closely related human influenza viruses and inhibits haemagglutinin-mediated membrane fusion. J Gen Virol 87:479–487

Gragoudas ES, Adamis AP, Cunningham Jr. ET, Feinsod M and Guyer DR (2004) Pegaptanib for neovascular age-related macular degeneration. N Engl J Med 351:2805–2816

Green LS, Jellinek D, Bell C, Beebe LA, Feistner BD, Gill SC, Jucker FM and Janjic N (1995) Nuclease-resistant nucleic acid ligands to vascular permeability factor/vascular endothelial growth factor. Chem Biol 2:683–695

Green R, Ellington AD, Bartel DP and Szostak JW (1991) In vitro genetic analysis: Selection and amplification of rare functional nucleic acids. Methods 2:75–86

Gronewold TM, Glass S, Quandt E and Famulok M (2005) Monitoring complex formation in the blood-coagulation cascade using aptamer-coated SAW sensors. Biosens Bioelectron 20:2044–2052

Grunow R, Splettstoesser W, McDonald S, Otterbein C, O'Brien T, Morgan C, Aldrich J, Hofer E, Finke EJ and Meyer H (2000) Detection of Francisella tularensis in biological specimens using a capture enzyme-linked immunosorbent assay, an immunochromatographic handheld assay, and a PCR. Clin Diagn Lab Immunol 7:86–90

Guo KT, SchAfer R, Paul A, Gerber A, Ziemer G and Wendel HP (2006) A new technique for the isolation and surface immobilization of mesenchymal stem cells from whole bone marrow using high-specific DNA aptamers. Stem Cells 24:2220–2231

Gustafsdottir SM, Schallmeiner E, Fredriksson S, Gullberg M, Soderberg O, Jarvius M, Jarvius J, Howell M and Landegren U (2005) Proximity ligation assays for sensitive and specific protein analyses. Anal Biochem 345:2–9

Guthrie JW, Hamula CL, Zhang H and Le XC (2006) Assays for cytokines using aptamers. Methods 38:324–330

Hermann T and Patel DJ (2000) Adaptive recognition by nucleic acid aptamers. Science 287:820–825

Homann M, Lorger M, Engstler M, Zacharias M and Goringer HU (2006) Serum-stable RNA aptamers to an invariant surface domain of live African trypanosomes. Comb Chem High Throughput Screen 9:491–499

Huang XZ, Nikolich MP and Lindler LE (2006) Current trends in plague research: from genomics to virulence. Clin Med Res 4:189–199

Huizenga DE and Szostak JW (1995) A DNA aptamer that binds adenosine and ATP. Biochemistry 34:656–665

Jellinek D, Green LS, Bell C, Lynott CK, Gill N, Vargeese C, Kirschenheuter G, McGee DP, Abesinghe P, Pieken WA and et al. (1995) Potent 2′-amino-2′-deoxypyrimidine RNA inhibitors of basic fibroblast growth factor. Biochemistry 34:11363–11372

Jenison RD, Gill SC, Pardi A and Polisky B (1994) High-resolution molecular discrimination by RNA. Science 263:1425–1429

Jeong S, Eom T, Kim S, Lee S and Yu J (2001) In vitro selection of the RNA aptamer against the Sialyl Lewis X and its inhibition of the cell adhesion. Biochem Biophys Res Commun 281:237–243

Jiang F, Kumar RA, Jones RA and Patel DJ (1996) Structural basis of RNA folding and recognition in an AMP-RNA aptamer complex. Nature 382:183–186

Kallioniemi OP, Wagner U, Kononen J and Sauter G (2001) Tissue microarray technology for high-throughput molecular profiling of cancer. Hum Mol Genet 10:657–662

Kobiler D, Weiss S, Levy H, Fisher M, Mechaly A, Pass A and Altboum Z (2006) Protective antigen as a correlative marker for anthrax in animal models. Infect Immun 74:5871–5876

Laber DA, Sharma VR, Bhupalam L, Taft B, Hendler FJ and Barnhart KM (2005) Update on the first phase I study of AGRO100 in advanced cancer. J Clin Oncol 23 (June 1 Supplement, 2005 ASCO Annual Meeting Proceedings):3064

Lai RY, Plaxco KW and Heeger AJ (2007) Aptamer-based electrochemical detection of picomolar platelet-derived growth factor directly in blood serum. Anal Chem 79:229–233

Lakowicz JR (1999) Principles of Fluorescence Spectroscopy, 2nd Ed. Kluwer Academic/Plenum Press, New York

Landers JP (1996) Handbook of Capillary Electrophoresis, 2nd Ed. CRC Press Inc., Boca Raton, Florida

Latham JA, Johnson R and Toole JJ (1994) The application of a modified nucleotide in aptamer selection: novel thrombin aptamers containing 5-(1-pentynyl)-2′-deoxyuridine. Nucleic Acids Res 22:2817–2822

Lauhon CT and Szostak JW (1995) RNA aptamers that bind flavin and nicotinamide redox cofactors. J Am Chem Soc 117:1246–1257

Le Floch F, Ho HA and Leclerc M (2006) Label-free electrochemical detection of protein based on a ferrocene-bearing cationic polythiophene and aptamer. Anal Chem 78:4727–4731

Lee JF, Hesselberth JR, Meyers LA and Ellington AD (2004) Aptamer database. Nucleic Acids Res 32:D95–100

Lee JF, Stovall GM and Ellington AD (2006) Aptamer therapeutics advance. Curr Opin Chem Biol 10:282–289

Lee JH, Canny MD, De Erkenez A, Krilleke D, Ng YS, Shima DT, Pardi A and Jucker F (2005) A therapeutic aptamer inhibits angiogenesis by specifically targeting the heparin binding domain of VEGF165. Proceedings of the National Academy of Science 102:18902–18907

Levy M, Cater SF and Ellington AD (2005) Quantum-dot aptamer beacons for the detection of proteins. Chembiochem 6:2163–2166

Li JJ, Fang X and Tan W (2002) Molecular aptamer beacons for real-time protein recognition. Biochem Biophys Res Commun 292:31–40

Li Y and Breaker RR (1999a) Kinetics of RNA degradation by specific base catalysis of transesterification involving the 2′-hydroxyl group. J Am Chem Soc 121:5364–5372

Li Y and Breaker RR (1999b) Phosphorylating DNA with DNA. Proceedings of the National Academy of Science 96:2746–2751

Li Y and Sen D (1996) A catalytic DNA for porphyrin metallation. Nat Struct Biol 3:743–747

Lim DV, Simpson JM, Kearns EA and Kramer MF (2005) Current and developing technologies for monitoring agents of bioterrorism and biowarfare. Clin Microbiol Rev 18:583–607

Lin CH and Patel DJ (1997) Structural basis of DNA folding and recognition in an AMP-DNA aptamer complex: distinct architectures but common recognition motifs for DNA and RNA aptamers complexed to AMP. Chem Biol 4:817–832

Lindstrom M and Korkeala H (2006) Laboratory diagnostics of botulism. Clin Microbiol Rev 19:298–314

Lozupone C, Changayil S, Majerfeld I and Yarus M (2003) Selection of the simplest RNA that binds isoleucine. Rna 9:1315–1322

Majerfeld I and Yarus M (1994) An RNA pocket for an aliphatic hydrophobe. Nat Struct Biol 1:287–292

Majerfeld I and Yarus M (1998) Isoleucine:RNA sites with associated coding sequences. RNA 4:471–478

Mandal M and Breaker RR (2004) Adenine riboswitches and gene activation by disruption of a transcription terminator. Nat Struct Mol Biol 11:29–35

McCauley TG, Kurz JC, Merlino PG, Lewis SD, Gilbert M, Epstein DM and Marsh HN (2006) Pharmacologic and pharmacokinetic assessment of anti-TGFbeta2 aptamers in rabbit plasma and aqueous humor. Pharm Res 23:303–311

McLendon MK, Apicella MA and Allen LA (2006) Francisella tularensis: taxonomy, genetics, and Immunopathogenesis of a potential agent of biowarfare. Annu Rev Microbiol 60:167–185

Mei SH, Liu Z, Brennan JD and Li Y (2003) An efficient RNA-cleaving DNA enzyme that synchronizes catalysis with fluorescence signaling. J Am Chem Soc 125:412–420

Mendonsa SD and Bowser MT (2004) In vitro evolution of functional DNA using capillary electrophoresis. J Am Chem Soc 126:20–21

Michaud M, Jourdan E, Villet A, Ravel A, Grosset C and Peyrin E (2003) A DNA aptamer as a new target-specific chiral selector for HPLC. J Am Chem Soc 125:8672–8679

Montange RK and Batey RT (2006) Structure of the S-adenosylmethionine riboswitch regulatory mRNA element. Nature 441:1172–1175

Navani NK and Li Y (2006) Nucleic acid aptamers and enzymes as sensors. Curr Opin Chem Biol 10:272–281

Ng EW, Shima DT, Calias P, Cunningham Jr. ET, Guyer DR and Adamis AP (2006) Pegaptanib, a targeted anti-VEGF aptamer for ocular vascular disease. Nat Rev Drug Discov 5:123–132

Nimjee SM, Rusconi CP and Sullenger BA (2005) Aptamers: an emerging class of therapeutics. Annu Rev Med 56:555–583

Nutiu R and Li Y (2003) Structure-switching signaling aptamers. J Am Chem Soc 125:4771–4778

Nutiu R and Li Y (2004) Structure-switching signaling aptamers: transducing molecular recognition into fluorescence signaling. Chemistry 10:1868–1876

Nutiu R and Li Y (2005a) Aptamers with fluorescence-signaling properties. Methods 37:16–25

Nutiu R and Li Y (2005b) In vitro selection of structure-switching signaling aptamers. Angew Chem Int Ed Engl 44:1061–1065

Osborne SE, Matsumura I and Ellington AD (1997) Aptamers as therapeutic and diagnostic reagents: problems and prospects. Curr Opin Chem Biol 1:5–9

Padmanabhan K, Padmanabhan KP, Ferrara JD, Sadler JE and Tulinsky A (1993) The structure of alpha-thrombin inhibited by a 15-mer single-stranded DNA aptamer. J Biol Chem 268:17651–17654

Pagratis NC, Bell C, Chang YF, Jennings S, Fitzwater T, Jellinek D and Dang C (1997) Potent 2′-amino-, and 2′-fluoro-2′-deoxyribonucleotide RNA inhibitors of keratinocyte growth factor. Nat Biotechnol 15:68–73

Pappas G, Panagopoulou P, Christou L and Akritidis N (2006) Brucella as a biological weapon. Cell Mol Life Sci 63:2229–2236

Patel DJ and Suri AK (2000) Structure, recognition and discrimination in RNA aptamer complexes with cofactors, amino acids, drugs and aminoglycoside antibiotics. J Biotechnol 74:39–60

Potyrailo RA, Conrad RC, Ellington AD and Hieftje GM (1998) Adapting selected nucleic acid ligands (aptamers) to biosensors. Anal Chem 70:3419–3425

Romig TS, Bell C and Drolet DW (1999) Aptamer affinity chromatography: combinatorial chemistry applied to protein purification. J Chromatogr B Biomed Sci Appl 731:275–284

Ruckman J, Green LS, Beeson J, Waugh S, Gillette WL, Henninger DD, Claesson-Welsh L and Janjic N (1998) 2′-Fluoropyrimidine RNA-based aptamers to the 165-amino acid form of vascular endothelial growth factor (VEGF165). Inhibition of receptor binding and VEGF-induced vascular permeability through interactions requiring the exon 7-encoded domain. J Biol Chem 273:20556–20567

Rupert PB, Massey AP, Sigurdsson ST and Ferre-D'Amare AR (2002) Transition state stabilization by a catalytic RNA. Science 298:1421–1424

Rusconi CP, Roberts JD, Pitoc GA, Nimjee SM, White RR, Quick Jr. G, Scardino E, Fay WP and Sullenger BA (2004) Antidote-mediated control of an anticoagulant aptamer in vivo. Nat Biotechnol 22:1423–1428

Rusconi CP, Scardino E, Layzer J, Pitoc GA, Ortel TL, Monroe D and Sullenger BA (2002) RNA aptamers as reversible antagonists of coagulation factor IXa. Nature 419:90–94

Sambrook J and Russell DW (2001) Molecular Cloning - A Laboratory Manual, 3rd ed. Cold Spring Harbor Laboratory Press, New York

Santoro SW and Joyce GF (1997). A general purpose RNA-cleaving DNA enzyme. Proceedings of the National Academy of Science 94:4262–4266

Savran CA, Knudsen SM, Ellington AD and Manalis SR (2004) Micromechanical detection of proteins using aptamer-based receptor molecules. Anal Chem 76:3194–3198

Seiwert SD, Stines Nahreini T, Aigner S, Ahn NG and Uhlenbeck OC (2000) RNA aptamers as pathway-specific MAP kinase inhibitors. Chem Biol 7:833–843

Serganov A, Polonskaia A, Phan AT, Breaker RR and Patel DJ (2006) Structural basis for gene regulation by a thiamine pyrophosphate-sensing riboswitch. Nature 441:1167–1171

Shangguan D, Li Y, Tang Z, Cao ZC, Chen HW, Mallikaratchy P, Sefah K, Yang CJ and Tan W (2006) Aptamers evolved from live cells as effective molecular probes for cancer study. Proceedings of the National Academy of Science 103:11838–11843

Siddiqui MA and Keating GM (2005) Pegaptanib: in exudative age-related macular degeneration. Drugs 65:1571–1577; discussion 1578–1579

Sobel J (2005) Botulism. Clin Infect Dis 41:1167–1173

Stojanovic MN, de Prada P and Landry DW (2001) Aptamer-based folding fluorescent sensor for cocaine. J Am Chem Soc 123:4928–4931

Storz G (2002) An expanding universe of noncoding RNAs. Science 296:1260–1263

Swartz MN (2001) Recognition and management of anthrax–an update. N Engl J Med 345:1621–1626

Tarasow TM, Tarasow SL and Eaton BE (1997) RNA-catalysed carbon-carbon bond formation. Nature 389:54–57

Tuerk C and Gold L (1990) Systematic evolution of ligands by exponential enrichment: RNA ligands to bacteriophage T4 DNA polymerase. Science 249:505–510

Tuerk C, MacDougal S and Gold L (1992) RNA pseudoknots that inhibit human immunodeficiency virus type 1 reverse transcriptase. Proceedings of the National Academy of Science 89:6988–6992

Vivekananda J and Kiel JL (2006) Anti-Francisella tularensis DNA aptamers detect tularemia antigen from different subspecies by Aptamer-Linked Immobilized Sorbent Assay. Lab Invest 86:610–618

Wallis MG, von Ahsen U, Schroeder R and Famulok M (1995) A novel RNA motif for neomycin recognition. Chem Biol 2:543–552

White R, Rusconi C, Scardino E, Wolberg A, Lawson J, Hoffman M and Sullenger B (2001) Generation of species cross-reactive aptamers using "toggle" SELEX. Mol Ther 4:567–573

Wilson C and Szostak JW (1995) In vitro evolution of a self-alkylating ribozyme. Nature 374:777–782

Wilson DS and Szostak JW (1999) In vitro selection of functional nucleic acids. Annu Rev Biochem 68:611–647

Winkler WC and Breaker RR (2005) Regulation of bacterial gene expression by riboswitches. Annu Rev Microbiol 59:487–517

Xu D, Yu X, Liu Z, He W and Ma Z (2005) Label-free electrochemical detection for aptamer-based array electrodes. Anal Chem 77:5107–5113

Yang Y, Kochoyan M, Burgstaller P, Westhof E and Famulok M (1996) Structural basis of ligand discrimination by two related RNA aptamers resolved by NMR spectroscopy. Science 272:1343–1347

Zhang B and Cech TR (1997) Peptide bond formation by in vitro selected ribozymes. Nature 390:96–100

26

Protein Microarray Technologies for Detection and Identification of Bacterial and Protein Analytes

Christer Wingren and Carl AK Borrebaeck

Abstract

Protein-based microarrays is a novel, rapidly evolving proteomic technology with great potential for analysis of complex biological samples. The technology will provide miniaturized set-ups enabling us to perform multiplexed profiling of minute amounts of biological samples in a highly specific, selective, and sensitive manner. In this review, we describe the potential and specific use of protein microarray technology, including both functional protein microarrays and affinity protein microarrays, for the detection and identification of bacteria, bacterial proteins as well as bacterial diseases. To date, the first generations of a variety of set-ups, ranging from small-scale focused biosensors to large-scale semi-dense array layouts for multiplex profiling have been designed. This work has clearly outlined the potential of the technology for a broad range of applications, such as serotyping of bacteria, detection of bacteria and/or toxins, and detection of tentative diagnostic biomarkers. The use of the protein microarray technology for detection and identification of bacterial and protein analytes is likely to increase significantly in the coming years.

1. Introduction

Entering the post-genomic era, proteomics—the large-scale analysis of proteins—has become a key discipline (Phizicky et al. 2003; Zhu et al. 2003). To this end, the need for (novel) technologies, allowing us to perform rapid and multiplexed analysis of biological samples in a selective, specific, and sensitive manner in various applications, ranging from focused assays to proteome-scale analysis, is tremendous (Yanagida 2002; Hanash 2003; Phizicky et al. 2003; Zhu et al. 2003; Wingren and Borrebaeck 2004). The protein-based microarray is a promising and rapidly evolving technology that may provide us with the unique means to perform high-throughput proteomics (Haab 2001; Zhu and Snyder 2003; Wingren and Borrebaeck 2004; Kingsmore 2006). In this review, we will describe the potential and specific use of protein microarray technology for the detection and identification of bacteria, bacterial proteins, and bacterial diseases. The number of such applications is still low, but is likely to increase significantly as the microarray technology develops.

Christer Wingren • Dept. of Immunotechnology, Lund University, Lund, Sweden. **Carl AK Borrebaeck** • CREATE Health, Lund University, Lund, Sweden.

M. Zourob et al. (eds.), *Principles of Bacterial Detection: Biosensors, Recognition Receptors and Microsystems*, © Springer Science+Business Media, LLC 2008

1.1. Definition and Classification of Protein Microarrays

The concept of protein microarrays is based on the arraying of a small amount (pL to nL scale) of protein in discrete positions in an ordered pattern, a microarray, onto a solid support where they will act as probes, or catcher molecules (Fig. 26.1) (Haab 2001; MacBeath 2002; Wingren and Borrebaeck 2004; Angenendt 2005; Kingsmore 2006). A minute quantity (μL scale) of the biological sample, e.g., serum, is then incubated on the array and any specifically bound analytes can be detected using mainly fluorescence as a read-out system. Adopting a high-performing setup, assay sensitivities in the pM to fM range can be observed, allowing low-abundance analytes to be readily targeted even in complex samples (Pawlak et al. 2002; Wingren et al. 2005; Wingren et al. 2006). Depending on the assay setup at hand, the observed microarray binding pattern can then be converted into, for example, a protein-ligand interaction profile, a (differential) protein expression profile, or even a proteomic map revealing the detailed composition of the proteome at the molecular level (MacBeath 2002; Wingren and Borrebaeck 2004; Kingsmore 2006).

Protein microarrays are frequently divided into two conceptual classes of array approaches, functional protein microarrays (functional proteomics) and affinity protein microarrays (quantitative proteomics) (MacBeath 2002; Poetz et al. 2005). While functional protein microarrays examine the biochemical activity, such as ligand binding properties and reactivity, of a set of immobilized target proteins (MacBeath and Schreiber 2000; Zhu et al. 2000, 2001; Phizicky et al. 2003); affinity microarrays utilize affinity reagents as probes to detect and measure the abundance of multiple proteins in a (semi-) quantitative way (Haab 2003; Wingren and Borrebaeck 2004, 2006a). As will be described in this review, both classes of protein microarrays have been used for detection and identification of bacteria and bacterial protein analytes, and the different protein array technology platforms have already been shown to display great promise within biomedical and biotechnological applications (Table 26.1). In this context, it might be of interest to note that protein-based microarray applications have recently emerged in a similar manner also for the detection of viruses, viral proteins, and viral diseases (Perrin et al. 2003; Yuk et al. 2004; Livingston et al. 2005; Lu et al. 2005; Zhu et al. 2006).

1.2. Functional Protein Microarrays

In general terms, functional protein microarrays have been designed and developed to investigate the biochemical properties, e.g., immunoreactivity (Robinson et al. 2003; Hueber

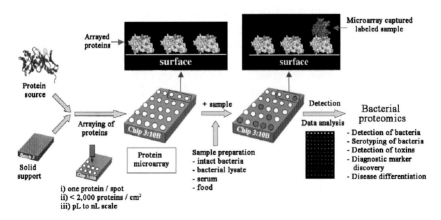

Figure 26.1. Schematic illustration of the protein microarray technology.

Table 26.1. Protein microarray technologies for detection and identification of bacterial and protein analytes

Application Type of Microarray	Probe Source	Support	Mode of Detection	Sample	References
1) Serotyping of bacteria					
Antibody microarray	Pabs (antisera)	SuperEpoxy	Fluorescence	Various *Salmonella* strains	Cai et al. 2005
		Epoxy-modified	TMB / red light	Several *E. coli* strains	Anjum et al. 2006
2) Detection of bacteria					
Antibody microarray	Mabs	C_{18} derivatized SiO_2	Impedance spectroscopy	*Listeria monocyotgenes* cells	Huang et al. 2003
	Abs	Streptavidin modified	Fluorescence	*E. coli*	Gehring et al. 2006
Reversed protein array (cell-based array)	Genetically engineered *E. coli* cells	gold electrodes	fluorescence	Streptavidin	Oh et al. 2006
	E. coli cells	cellulosic membrane	Fluorescence	*E. coli*	Stokes et al. 2001
3) Detection of toxins					
Antibody microarray	Pabs and Mabs	Avidin-coated borosilicate	Fluorescence (waveguide)	Several toxins / toxoids	Ligler et al. 2003
	Mabs	Epoxy-slides	Fluorescence	Several native toxins	Rucker et al. 2005
		Hydrogel-based	Fluorescence, MS chemiluminescence	Several plant and bacterial toxins	Rubina et al. 2005
	Abs	Glass-slides	Fluorescence	Toxic agents	Wadkins et al. 1998
4) Simultaneous detection of bacteria and toxins					
Antibody microarray	Pabs	Neutravidin derivatized	Fluorescence (waveguide)	Bacterial, viral, and protein analytes	Rowe et al. 1999
	Mabs and Pabs	Neutravidin derivatized	Fluorescence (waveguide)	Bacterial analytes and toxins	Rowe-Taitt et al. 2000; Taitt et al. 2002
		Neutravidin coated	Fluorescence	Bacteria and bacterial proteins	Delehanty and Ligler 2002
	Abs	SERS-activated metal	SERS microscopy	Bacteria and bacterial analytes	Grow et al. 2003

Table 26.1. (continued)

Application / Type of Microarray	Probe Source	Support	Mode of Detection	Sample	References
5) Identification of diagnostic markers					
Protein microarray	*Y. pestis* proteins	Silylated glass	Fluorescence	Serum from immunized rabbits	Li et al. 2005
	N. meningitides proteins	FAST-slides	Fluorescence	Human sera	Steller et al. 2005
6) Disease state differentiation					
Protein microarray	*M. tuberculosis* proteins	FAST-slides	Fluorescence	Human sera	Sartain et al. 2006
	M. tuberculosis antigens	Epoxy slides	Fluorescence	Human sera	Tong et al. 2005
7) Identification of toxin modulator / regulator					
Protein microarray	Yeast/human proteins	Nitrocellulose	Fluorescence	cytotoxic enterotoxin	Galindo et al. 2006
8) Detection of protein signatures associated with bacterial infection					
Antibody microarray	Recombinant scFv Ab library	Black polymer Maxisorb	Fluorescence	Human stomach tissue	Ellmark et al. 2006b
9) Carbohydrate fingerprinting of bacteria					
Protein microarray	Lectins	Nexterion H slides	Fluorescence	*E. coli strains*	Hsu et al. 2006
		Expoxysilane modified glass slide	Fluorescence (wave guide)	Glycoproteins	Uchiyama et al. 2006
Reversed protein array	Carbohydrates	CodeLink slides	Fluorescence	Various model bacteria systems	Disney and Seeberger 2004

Abs = Antibodies, not specified whether polyclonal and/or monoclonals; Mabs = Monoclonal antibodies; MS = mass spectrometry; Pabs = Polyclonal antibodies; SERS = Surface-Enhanced Raman Scattering; TMB = 3,3′, 5,5′-tetramethylbenzidine (TMB) peroxidase substrate et al. 2005; Lueking et al. 2005), and the functional properties, e.g., protein-protein interactions, of the arrayed proteins (MacBeath and Schreiber 2000; MacBeath 2002; LaBaer and Ramachandran 2005; Kingsmore 2006). In particular, the use of cDNA expression libraries as the probe source has been very rewarding (Zhu et al. 2000, 2001; Horn et al. 2006). For example, microarrays composed of 119 of 122 yeast protein kinases (Zhu et al. 2000), 5,800 of 6,200 yeast proteins (Zhu et al. 2001), or 37,200 redundant recombinant fetal brain proteins (Horn et al. 2006) have been designed, fabricated, and successfully applied to perform global protein-protein interaction studies. To date, mainly water-soluble protein analytes have been targeted, but the first microarray design also targeting membrane proteins in the format of intact mammalian cells has recently been published (Deviren et al. 2007).

Functional protein arrays have frequently been applied within the academic research community, but several commercial ventures have also developed such microarrays that are now available on the market. For example, focused microarrays targeting cell signaling proteins and tentative cancer-associated proteins have been launched by Sigma (http://www.sigma.com). In addition, comprehensive microarrays based on over 8,000 human recombinant proteins have recently been released by both Protagen (http://protagen.de) and Invitrogen (http://invitrogen.com).

In the case of bacteria and bacterial protein analytes, various designs of functional protein microarrays have been successfully applied (Table 26.1). In these first examples, microarrays based on a set of human or yeast proteins (Galindo et al. 2006), recombinant bacterial proteins (Li et al. 2005; Steller et al. 2005; Tong et al. 2005), biochemically isolated and fractionated bacterial proteins (Tong et al. 2005; Sartain et al. 2006), neoglycoproteins (oligosaccharides bound to bovine serum albumin) (Tong et al. 2005), and lipopolysaccharides and saccharides (Tong et al. 2005) have been developed and fabricated. In these studies, the first generation(s) of applications for the identification of toxin modulators/regulators (Galindo et al. 2006), disease state differentiation (Tong et al. 2005; Galindo et al. 2006; Sartain et al. 2006), and identification of diagnostic markers (Li et al. 2005; Steller et al. 2005) have been outlined.

1.3. Affinity Protein Microarrays

Affinity protein microarrays have been developed for multiplex protein expression profiling, specifically detecting whether the targeted analytes are expressed and at what (relative) levels (MacBeath 2002; Zhu and Snyder 2003; Angenendt 2005; Wingren and Borrebaeck 2006a). To date, antibodies are the most commonly used probe source for affinity protein microarrays (Wingren and Borrebaeck 2004; Haab 2006; Kingsmore 2006; Wingren and Borrebaeck 2006a). In more detail, microarrays based on polyclonal and monoclonal antibodies (Sreekumar et al. 2001; Miller et al. 2003; Gao et al. 2005; Sanchez-Carbayo et al. 2006), as well as recombinant antibody fragments (Wingren et al. 2003; Pavlickova et al. 2004; Wingren et al. 2005; Ellmark et al. 2006b), have been successfully designed and applied for (disease) proteomics. For example, tentative protein expression profiles associated with disease and clinical parameters have been identified (Gao et al. 2005; Borrebaeck 2006; Ellmark et al. 2006b; Sanchez-Carbayo et al. 2006).

Intense efforts are currently under way to develop the affinity protein microarray technology even further; for a review see Wingren and Borrebaeck (2004); Kingsmore (2006); Wingren and Borrebaeck (2006b, 2006a). In parallel with all the academic efforts, several commercial products have been launched, mainly for focused assays (protein expression profiling) and point-of-care applications (Wingren and Borrebaeck 2004); although larger arrays have also started to emerge on the market (e.g., http://www.clontech.com, http://www.raybiotech.com, http://www.protneteomix.com, http://www.whatman.com, http://www.sigmaaldrich.com). As for functional protein arrays, mainly water-soluble analytes have been targeted (Wingren and Borrebaeck 2004), but array designs targeting membrane proteins in the format of intact mammalian cells have recently been published as well (Belov et al. 2001, 2003; Ko et al. 2005b, 2005a; Campbell et al. 2006; Ellmark et al. 2006a) (Dexlin, Borrebaeck, and Wingren unpublished).

In Table 26.1, different applications in which various antibody-based microarray designs have been used to target bacteria and protein analytes thereof are listed. So far, the main applications are the detection of bacteria (Stokes et al. 2001; Huang et al. 2003; Gehring et al. 2006; Oh et al. 2006), toxins (Wadkins et al. 1998; Ligler et al. 2003; Rubina et al. 2005; Rucker et al. 2005), or both (Rowe et al. 1999; Rowe-Taitt et al. 2000; Delehanty and Ligler 2002; Taitt et al. 2002; Grow et al. 2003); the serotyping of bacteria (Cai et al. 2005; Anjum et al. 2006); and the detection of protein expression signatures associated with bacterial infections (Ellmark et al. 2006b). To date, the degree of multiplexity differs, with array designs based on a few single probes ranging up to a few hundred, depending on the actual assay at hand. In more detail, a low degree of multiplexity ($\leq$ 50 probes) may be sufficient to screen and detect toxins, etc., while a high degree of multiplexity (> 100 probes) may, at least initially, be required to detect, for example, disease-associated protein expression signatures.

1.4. Alternative Microarray Setups

Entering the post-genomic era, not only the proteome but also the glycome has gained significant biomedical interest (Shriver et al. 2004; Miyamoto 2006). In recent work this has been explored to single out pathogenic bacteria (Pohl 2006), as well as to detect pathogens (Disney and Seeberger 2004). In these efforts, work has been made to design both lectin microarrays (Hsu et al. 2006; Uchiyama et al. 2006) and carbohydrate microarrays (Disney and Seeberger 2004) (Table 26.1). Cell surface carbohydrates are in fact critical for many seminal interactions that define bacteria as pathogens and symbiotes. By adopting multiplex technologies for carbohydrate profiling, series of bacterial strains could be fingerprinted, based on their carbohydrate patterns, which might be used for identification, etc.

Lectins are sugar-binding proteins of nonimmune origin that contain at least two sugar-binding sites and are commonly used in agglutination tests to screen the bacterial glycome. However, these assays often suffer from inadequate sensitivity and subjective visual read-out. In comparison, MS, NMR, and HPLC-based analysis are alternative but time-consuming assays commonly applied for bacterial glycan analysis. In a recent paper, Hsu et al. presented a lectin-based microarray approach for analyzing the dynamic bacterial glycome (Hsu et al. 2006). The platform was based on 21 lectins on Nexterion H slides, and the arrays were imaged by fluorescence, since the bacteria were directly labeled with SYT 85. The results showed that (1) closely related *E. coli* strains could be distinguished based on their glycosylation pattern, i.e., enabling fingerprinting; and (2) dynamic alterations in the bacteria glycome could be observed. The range of specificities displayed by lectins is currently a bottleneck, which is why other carbohydrate binding probes, e.g., antibodies, may provide an alternative route.

Recently, proof-of-concept was reported for an alternative lectin microarray platform by Uchiyama and coworkers (Uchiyama et al. 2006). They have developed a novel setup that allows observation of lectin-glycoprotein interactions under equlibrium conditions, based on an evanescent-field fluorescence-assisted detection principle. This enables the assay to be performed without washing procedures, a clear advantage considering the relatively weak lectin-glycan interactions.

To examine carbohydrate-cell interactions and to detect pathogens, a carbohydrate microarray has been developed (Disney and Seeberger 2004). In this context, it may be of interest to note that cell-surface carbohydrates are exploited by many pathogens for tissue adherence and entry into host cells. The carbohydrates (e.g., mannose and fucose) were dispensed and immobilized onto CodeLink slides, and bound bacteria (directly labeled) were detected by fluorescence. Proof-of-principle was shown for this carbohydrate microarray as a means to detect bacteria, as illustrated by different *E. coli* strains. In addition, these nondestructive arrays allow the bacteria to be harvested and tested for, e.g., antibacterial susceptibility.

2. Detection of Bacteria and Bacterial Protein Analytes

2.1. Serotyping of Bacteria

Bacteria can be serotyped by determining their somatic (O) and flagellar (H) cell surface antigens. Serotyping is of clinical importance as, for example, many O bacterial serotypes are linked with a number of diseases (syndromes), in that subsets of serotypes, or pathotypes, can cause meningitis, systematic disease, diarrhea, etc. Despite being a primary diagnostic tool, the current methods available, e.g., agglutination tests, suffer from limited throughput, no or low multiplexity, requirement of large sample volumes, and are costly. In two recent publications, the possibility of using antibody-based microarrays for serotyping of *E. coli* (Anjum et al. 2006) and *S. entrecia* (Cai et al. 2005) strains have been explored and exploited (Table 26.1).

Anjum and coworkers adopted the ArrayTube platform to develop miniaturized antibody arrays on an epoxy-modified glass surface (Anjum et al. 2006). The authors employed 17 rabbit antisera raised against the most common *E. coli* pathogens (e.g., O157 and O26) associated with disease syndromes in both humans and animals. After adding the *E. coli* cell samples, bound cells were detected by secondary antibodies and signal amplification reagents, and the arrays were imaged by monitoring specific changes in red light transmission. This feasibility study showed that 88–100 % of the tested *E. coli* isolates could be correctly classified. In fact, the observed discrepancy was related to poor sample quality rather than to an inadequate identification. Hence, the results implied that the antibody array setup performed well for O serotyping, providing multiplexed, cost-effective, efficient, and accurate typing.

In comparison, Cai et al. developed an antibody array platform based on 35 polyclonal antibodies (antisera) against 20 common *Salmonella* serovars using SuperEpoxy slides (Cai et al. 2005). Numerous fluorescently labeled *Salmonella enterica* strains were analysed and the arrays were imaged using fluorescence as a read-out system. The results showed that the array setup enabled complete serovar identification of 86 of 117 target strains, and partial identification of 30 of 117. Further, all of the 73 analysed nontarget strains (negative controls) were successfully excluded. Hence, an array platform providing a rapid and cost-effective alternative to the traditional agglutination method for *Salmonella* serotyping has been developed.

2.2. Detection of Pathogenic Organisms

Foodborne pathogenic bacteria, such as *E. coli* O157:H7, are responsible for about 80 million illnesses in the United states each year, with thousands resulting in death. So far, the analytical approaches applied for the detection of bacteria have included plate culture, ELISA, and PCR. Hence there is a tremendous need for multiplexed technologies enabling combinations of pathogens to be screened and detected in a simple manner, and efforts to address this issue are underway (Table 26.1).

In recent years, two independent antibody microarray or biochip setups for the detection of *E. coli*, using O157:H as a test system, have been developed (Stokes et al. 2001; Gehring et al. 2006). In the first example, a microfluidics-based antibody biochip-based system was developed for the detection of *E. coli* (Stokes et al. 2001). This reversed affinity protein microarray is based on the exposure of a cellulosic membrane to a sample potentially containing *E. coli*. The bacteria is then immobilized (bound) to the membrane and detected by fluorescently labeled secondary antibodies. The setup was found to display rapid and selective detection of bacteria with at least three orders of magnitude linear dynamic range. Of note, the assay sensitivity was found to be as low as 20 organisms (in this case *E. coli* O157:H7). In comparison, Gerhing and coworkers have developed a sandwich fluorescent immunoassay in the microarray format (Gehring et al. 2006). Biotinylated capture antibodies were immobilized onto streptavidin-modified Superfrost Gold slides, and the setup was evaluated targeting *E. coli* O157:H7 samples. The applicability of this, so far, low-density setup was outlined and a limit of detection in the 3×10^6 cells/ml range was observed.

Similar to the setup by Gerhing et al. (2006), Oh and colleagues have developed a reversed affinity protein array setup for the detection of bacteria (Oh et al. 2006). Using a microfluidic device, the cells were electrokinetically immobilized onto gold electrodes and imaged by fluorescence after adding appropriate secondary reagents (antibodies). It should, however, be noted that this setup was not primarily developed for the detection of bacteria, but rather as a new tool for taking advantage of the bacteria to display a capture protein, e.g., a membrane protein, in its natural environment, and thereby increasing its on-chip functionality. In this context, it should be noted that designing and fabricating membrane protein microarrays is in general a major challenge that remains to be fully resolved (Fang et al. 2002a, 2002b; Wingren and Borrebaeck 2004). Hence, this setup might open new avenues not only for the detection of bacteria, but also for designing membrane protein-based microarrays.

In recent work, an antibody microarray setup based on biotinylated monoclonal antibodies was developed and optimized with respect to nonspecific adsorption of bacteria and proteins thereof. The antibodies were immobilized via streptavidin, which in turn was bound to biotinylated bovine serum albumin (BSA) adsorbed onto a C_{18}-derivatized SiO_2 surface (Huang et al. 2003). The results showed that the dual action of BSA, acting both as a surface blocker and as a probe immobilized, was successful. Directed immobilization and low nonspecific binding were reported.

2.3. Detection of Multiple Toxins

Rapid, sensitive, and multiplex detection of biological toxins in clinical samples, food, drinking water, and environmental samples is of great importance in revealing possible infections and contaminations, as well as potential bioterrorist threats (Table 26.1). In the long-term, small, simple, and portable devices (biosensors) would be an attractive format for the instrumentation behind such key applications. In this context, the sample format will also be critical, and assay designs allowing the sample to be directly applied without any significant pretreatment, e.g., fluorescent labeling, would clearly be advantageous.

Four studies have been published, in which the efforts at developing antibody-based microarray biosensors for the detection mainly of toxins have been successfully described (Wadkins et al. 1998; Ligler et al. 2003; Rubina et al. 2005; Rucker et al. 2005). In an early study by Wadkins et al. a planar array immunosensor for the detection of multiple toxic agents was fabricated (Wadkins et al. 1998). Polyclonal antibodies were covalently coupled to derivatized glass slides, and bound toxins, e.g., ricin and staphylococcal enterotoxin B, were monitored using fluorescently labeled secondary antibodies. Assay sensitivities in the 5–25 ng/ml range were observed.

In comparison, an array biosensor, based on monoclonal and polyclonal antibodies, capable of detecting multiple targets on the borosilicate glass surface of a single waveguide, was more recently designed (Ligler et al. 2003). Both competitive and sandwich fluoroimmuno setups were developed to enable small, as well as large, molecular weight toxins (e.g., ricin, botulinum toxoids, and trinitrotoluene) to be detected in complex samples, such as food or clinical specimens. Notably, the setup was capable of addressing up to 12 samples at the same time. The results showed that specific and sensitive ($\geq$0.5 ng/ml) detection of target analytes was accomplished. With additional development of the sensor instrument, this may in the end provide a rapid, fieldable, and low-tech assay for the detection of toxins.

Similarly, Rucker and colleagues have developed competitive and noncompetitive antibody microarray setups for native toxin detection (e.g., diphtheria toxin and anthrax lethal factor) in serum samples (Rucker et al. 2005). In this case, monoclonal antibodies were immobilized on epoxy-slides, and the arrays were imaged using fluorescence. While the competitive assay setup was favored for not having to label the sample, the direct assay benefited from superior sensitivity (low ng/ml vs. high ng/ml). In the end, the choice of setup may be dependent on whether the assay is run in field trials, where a simple assay is desired, or in the laboratory, where more complex assay principles providing higher sensitivity can be applied.

Further, a hydrogel-based monoclonal antibody microchip was recently designed and fabricated by Rubina et al. (2005). The platform was developed with the aim of performing a quantitative immunoassay of a series of plant toxins (e.g., ricin and viscumin) and bacterial toxins (e.g., diphtheria toxin and tetanus toxin). Direct, competitive, and sandwich assay setups were successfully developed and found to be compatible with the platform. In contrast to the previous studies, this system was interfaced with either a fluorescent-, chemiluminescent-, or MS-based read-out system, providing high flexibility. In all cases, the assay sensitivities were found to be in the low ng/ml range, i.e., within the range of sensitivity required in order to be able to perform clinical applications.

2.4. Simultaneous Detection and Identification of Bacterial Proteins and Bacteria

Similar to the work described in Section 26.2.3, antibody-based microarray biosensors have also been used for the simultaneous detection of bacteria and bacterial proteins (e.g., toxins) (Table 26.1), where again technologies for rapid and multiplexed detection will play a key role.

An antibody-based array biosensor composed of three parts, the antibody array (recognition element), an image capture and processing part, and an automated fluidics unit, has been developed by Ligler et al. (Rowe et al. 1999; Rowe-Taitt et al. 2000; Taitt et al. 2002). The capture polyclonal and/or monoclonal antibodies were biotinylated and immobilized on neutravidin-derivatized waveguides. Bound analytes (proteins, glycoproteins, Gram-negative, and Gram-positive bacteria) were detected using labeled secondary (tracer) antibodies. The results showed that assay sensitivities in the mid ng/ml range were readily observed targeting, for example, cholera toxin and *B. globigii*. Moreover, the assay was demonstrated to be rapid ($<$15 min) and easy to execute. Taken together, these studies have demonstrated proof-of-concept for an inexpensive and multiplex device for simple detection of bacteria and bacterial analytes. In addition, the setup is in a format amenable to automation and portability. In these first studies, the capture antibody spots were in the 2.5 mm^2 size range and generated by physically isolated patterning using polymer flow cells. In recent work, the spot size of the biotinylated capture antibodies has been considerably reduced (0.04 mm^2) by adopting a noncontact piezo-based dispenser to fabricate the arrays (Delehanty and Ligler 2002). Using confocal microscopy for detection, an assay sensitivity in the low ng/ml range was still obtained. Hence the latter study outlined a way of fabricating high-density arrays for bacterial detection, while maintaining assay sensitivity.

In a recent review by Grow et al. a new biochip technology for label-free detection of pathogens and their toxins was presented and discussed (Grow et al. 2003). The biochip is composed of spots of capture probes (e.g., antibodies) immobilized on a surface-enhanced Raman scattering (SERS) active metal surface. Once the chip has been subjected to sampling and target analytes have been bound, a Raman microscope is applied to collect SERS fingerprints from the spots (pixels) on the chip. This interesting technology has been named μSERS, as it couples SERS with microscopy. The identification is based on SERS finger-prints, and the authors demonstrated that both Gram-positive and Gram-negative bacteria often could be detected at the strain/subspecies level based on their SERS fingerprints. Further, the SERS fingerprints could also be used to differentiate viable vs. nonviable, e.g., heat- or UV-killed, microorganisms; different physiological states of the bacteria cells, e.g., when cultured under conditions known to affect virulence; and to detect toxins in a specific and sensitive manner. Work is currently under way to develop the Raman microscope instrumen-tation even further, to enable the simultaneous collection of hundreds or thousands of spectra from discrete positions on the chip with a spatial resolution of 250 nm to 1.5 μm. Future experiments will unravel the potential of this read-out system for protein microarray-based applications.

3. Detection of Diagnostic Markers, Toxin Regulators and Associated Protein Expression Profiles

3.1. Identification of Potential Diagnostic Markers and/or Vaccine Candidates

In two recent publications (Li et al. 2005; Steller et al. 2005), the possibility of using protein microarrays to identify novel potential diagnostic markers and/or vaccine candidates was explored and outlined (Table 26.1). Again, the array format was critical in order to enable sufficient multiplexity and throughput to gain success.

In the first study, a recombinant bacterial protein microarray was fabricated on FAST-slides and used for identification of new potential diagnostic markers for *Neisseria meningitides* (Steller et al. 2005). The authors succeeded in expressing 67 of 102 known phase-variable genes from *N. meningitides* serogroup B strain MC58 as recombinant proteins in *E. coli*. Subsequently, these proteins were used as probes in the array format and applied to screen sera from healthy controls vs. patients suffering from meningitis. The results showed that 47 of these proteins were immunogenic, i.e., that an antibody response had been mounted. Nine proteins were found to be immunogenic in at least 3 of 20 meningitis sera tested, while 1 protein showed a response in 11 of 20 sera. The potential of these *N. meningitis* proteins for diagnostic purposes remains to be elucidated, but this study clearly outlines the potential of the approach.

Yersina pestis causes plague, which is one of the most feared diseases. Work is ongoing to identify novel vaccine candidates to improve the current plague vaccines. In these efforts, Li et al. have developed a 149-recombinant *Yersina pestis* protein microarray to profile the antibody response in immunized rabbits, providing a new tool in the search for vaccine candidates and/or diagnostic antigens (Li et al. 2005). The authors found that an antibody response had been elicited against about 50 of the arrayed *Y. pestis* proteins. Among these 50, 11 proteins to which the predominant antibody response was directed were identified. Taken together, these 11 new proteins show promise for further evaluation as candidates for vaccines and/or diagnostic antigens.

3.2. Disease State Differentiation and Identification of Diagnostic Markers

The evaluation of serological reactivity from healthy vs. nonhealthy patients, in order to allow disease state differentiation and the identification of tentative diagnostic markers, has gained significant attention within the field of disease proteomics (Hanash 2003; Wingren and Borrebaeck 2004; Borrebaeck 2006). Focusing on bacterial related diseases, two independent protein microarray setups, focusing on tuberculosis, have been developed (Tong et al. 2005; Sartain et al. 2006) and applied to perform serological tuberculosis assays (Table 26.1).

Tuberculosis can be diagnosed by microscopy and culture of mycobacteria of the *Mycobacterium tuberculosis* complex from clinical samples. Still, these approaches are associated with limitations and technical hurdles. To be proven valuable, a serodiagnostic approach should (1) display a specificity > 90 % (i.e., comparable to microscopy and bacterial cultures), and (2) be able to differentiate/detect multiple disease states.

Interestingly, Tong and coworkers have developed a protein microarray setup based on 54 *M. tuberculosis* antigens on epoxy-slides, and the arrays were imaged by fluorescence (Tong et al. 2005). The probe antigens were obtained from five sources, including biochemical fractionation of *M. tuberculosis* cells/culture fluids, oligosaccharides bound to BSA, purified lipopolysaccharides, purified polysaccharides, and recombinant antigens. The clinical serum samples from healthy controls (non-TB) and tuberculosis (TB) patients were screened for IgG antibodies specific for any of these antigens, e.g., for a serum-specific IgG profile. Based on the analysis of 20 TB sera and 80 non-TB sera, combinations of TB antigens were ranked with respect to specificity and sensitivity of TB detection. The results showed that the highest-ranking TB antigen combination displayed a receiver operator curve (ROC) with an area under the curve (AUC) of 0.95. Of note, a single antigen, Ara_6-BSA, was found to give an AUC value of 0.90. The authors concluded that the TB antigen microarray provided a rapid and efficient means of finding TB antigens that could be used to discriminate between TB and non-TB patients.

In comparison, Sartain et al. fabricated a TB antigen microarray based on 960 unique fractions, obtained from *M. tuberculosis* cytosol and culture filtrates, by multidimensional protein fractionation (Sartain et al. 2006). TB antigen arrays were fabricated on FAST slides and interfaced with a fluorescent read-out system. Next, serum samples from 12 healthy individuals, 9 noncavitary TB patients, 11 cavitary TB patients, 10 HIV-positive TB patients, and 6 HIV-posititve TB-negative patients were analysed. The authors demonstrated that the TB antigen microarray setup provided them with a novel means of assessing antigen recognition profiles (e.g., specific IgG profiles) discriminating between different disease states. In more detail, the different sera were found to display partly overlapping reactivity patterns, e.g., containing antibodies specific for material in the arrayed subfractions, but also distinctly unique patterns. Hence, the results indicated that the setup could be useful for differentiating the different disease states, thus demonstrating the potential of array-based serodiagnostics for tuberculosis.

3.3. Identification of Potential Toxin Modulators/Regulators

The use of high-density protein microarrays to examine the protein-protein interaction patterns for bacterial toxins in a multiplex high-throughput manner, is very appealing (Table 26.1). In the end, this may allow for identification of novel toxin modulators and/or regulators.

In a recent paper by Galindo and colleagues, the potential of cytotoxic enterotoxin (Act) of *Aeromonas hydrophila* to bind to human and yeast proteins was investigated by adopting a protein microarray approach (Galindo et al. 2006). To this end, the human and yeast ProtoArrays composed of 1869 human proteins and 4319 yeast proteins, respectively, on nitrocellulose coated slides were used. The study showed that Act was capable of binding nine human proteins and 4 yeast proteins. For three of the interactions, a confirming Western blot analysis was performed. Next, a set of experiments, including small interfering RNA, was performed in order to explore the relevance of the observed interactions. These efforts indicated a potential involvement of galectin-3 and SNAP23 in *A. hydrophila* cytotoxic enterotoxin-induced host cell apoptosis. Hence, by adopting a high-density protein microarray screening approach, the authors were able to present the first report of tentative protein binding partners for Act, as well as potential mediators/regulators for Act-induced apoptosis.

3.4. Screening of Protein Expression Signatures Associated with Bacterial Infection

To date, a major focus has been placed upon using protein microarrays, and in particular antibody-based microarrays for oncoproteomics, with the aim of finding disease-specific (serum or tissue) protein signatures for diagnostics and biomarker discovery, etc. (Wingren and Borrebaeck 2004; Borrebaeck 2006; Kingsmore 2006). In a similar fashion, protein microarrays could be used to find protein expression signatures associated with bacterial infections and conditions (Table 26.1).

In a recent study by Ellmark et al. the authors examined stomach tissue samples from gastric adenoma carcinoma patients using a 127-human recombinant scFv antibody microarray on black polymer Maxisorb slides, interfaced with a fluorescent read-out system (Ellmark et al. 2006b). Of note, these cancer patients are often associated with *Helicobacter pylori* infections. The proteins were extracted from the tissue samples, biotinylated, and analysed on the recombinant antibody microarrays. The platform was found to display an assay sensitivity in the low pg/ml range. Further, the results showed that a 14-protein expression signature associated with *H. pylori* infection could be identified, where 10 analytes were distinctly different from the corresponding protein signature found to be associated with adenoma carcinoma. Taken together, these studies clearly demonstrate the use and potential of antibody (protein) microarray technology for rapid, sensitive, and multiplexed expression profiling of complex samples in order to identify disease-associated protein signatures.

4. Conclusions and Future Perspectives

Taken together, the first generations of protein-based microarray technology platforms for detection and identification of bacteria, bacterial proteins and bacterial diseases, have in recent years been developed. These miniaturized assay platforms include functional protein microarrays as well as affinity protein microarrays, and the designs range from small-scale focused biosensors targeting a few analytes to large-scale semi-dense microarray set-ups for multiplex screening. A broad range of applications, such as serotyping of bacteria, detection of bacteria, identification of toxins, disease state differentiation, and discovery of disease-associated biomarkers, have so far been demonstrated, clearly outlining the potential of the technology. Still, the number of applications is low, but is likely to increase significantly as the microarray technology progress and develops into a robust proteomic technology. In future, protein microarray based applications are likely to play an important role for detection and identification of bacterial and protein analytes.

Acknowledgements

This study was supported by grants from the Swedish National Science Council (VR-NT), the SSF Strategic Center for Translational Cancer Research (CREATE Health), the Alfred Österlund Foundation, and the Great and Johan Kock Foundation.

References

Angenendt P (2005) Progress in protein and antibody microarray technology. Drug Discovery Today 10:503–511

Anjum MF, Tucker JD, Sprigings KA, Woodward MJ, Ehricht R (2006) Use of miniaturized protein arrays for *Escherichia coli* O serotyping. Clin Vaccine Immunol 13:561–7

Belov L, de la Vega O, dos Remedios CG, Mulligan SP, Christopherson RI (2001) Immunophenotyping of leukemias using a cluster of differentiation antibody microarray. Cancer Res 61: 4483–9

Belov L, Huang P, Barber N, Mulligan SP, Christopherson RI (2003) Identification of repertoires of surface antigens on leukemias using an antibody microarray. Proteomics 3:2147–54

Borrebaeck CA (2006) Antibody microarray-based oncoproteomics. Expert Opin Biol Ther 6:833–8

Cai HY, Lu L, Muckle CA, Prescott JF, Chen S (2005) Development of a novel protein microarray method for serotyping *Salmonella enterica* strains. J Clin Microbiol 43:3427–30

Campbell CJ, O'Looney N, Chong Kwan M, Robb JS, Ross AJ, Beattie JS, Petrik J, Ghazal P (2006) Cell interaction microarray for blood phenotyping. Anal Chem 78:1930–8

Delehanty JB, Ligler FS (2002) A microarray immunoassay for simultaneous detection of proteins and bacteria. Anal Chem 74:5681–7

Deviren G, Gupta K, Paulaitis ME, Schneck JP (2007) Detection of antigen-specific T cells on p/MHC microarrays. J Mol Recognit 20:32–8

Disney MD, Seeberger PH (2004) The use of carbohydrate microarrays to study carbohydrate-cell interactions and to detect pathogens. Chem Biol 11:1701–7

Ellmark P, Belov L, Huang P, Lee CS, Solomon MJ, Morgan DK, Christopherson RI (2006a) Multiplex detection of surface molecules on colorectal cancers. Proteomics 6:1791–802

Ellmark P, Ingvarsson J, Carlsson A, Lundin SB, Wingren C, Borrebaeck CA (2006b) Identification of protein expression signatures associated with *H. pylori* infection and gastric adenocarcinoma using recombinant antibody microarrays. Mol Cell Proteomics 5:1638–46

Fang Y, Frutos AG, Lahiri J (2002a) Membrane protein microarrays. J Am Chem Soc 124:2394–5

Fang Y, Frutos AG, Webb B, Hong Y, Ferrie A, Lai F, Lahiri J (2002b) Membrane biochips. Biotechniques Dec Suppl:62–5

Galindo CL, Gutierrez C Jr, Chopra AK (2006) Potential involvement of galectin-3 and SNAP23 in *Aeromonas hydrophila* cytotoxic enterotoxin-induced host cell apoptosis. Microb Pathog 40:56–68

Gao WM, Kuick R, Orchekowski RP, Misek DE, Qiu J, Greenberg AK, Rom WN, Brenner DE, Omenn GS, Haab BB, Hanash SM (2005) Distinctive serum protein profiles involving abundant proteins in lung cancer patients based upon antibody microarray analysis. BMC Cancer 5:110

Gehring AG, Albin DM, Bhunia AK, Reed SA, Tu SI, Uknalis J (2006) Antibody microarray detection of *Escherichia coli* O157:H7: quantification, assay limitations, and capture efficiency. Anal Chem 78:6601–7

Grow AE, Wood LL, Claycomb JL, Thompson PA (2003) New biochip technology for label-free detection of pathogens and their toxins. J Microbiol Methods 53:221–33

Haab BB (2001) Advances in protein microarray technology for protein expression and interaction profiling. Curr Opin Drug Discov Devel 4:116–23

Haab BB (2003) Methods and applications of antibody microarrays in cancer research. Proteomics 3:2116–22

Haab BB (2006) Applications of antibody array platforms. Curr Opin Biotechnol 17:415–21

Hanash S (2003) Disease proteomics. Nature 422:226–32

Horn S, Lueking A, Murphy D, Staudt A, Gutjahr C, Schulte K, Konig A, Landsberger M, Lehrach H, Felix SeB, Cahill DeJ (2006) Profiling humoral autoimmune repertoire of dilated cardiomyopathy (DCM) patients and development of a disease-associated protein chip. Proteomics 6:605–613

Hsu KL, Pilobello KT, Mahal LK (2006) Analyzing the dynamic bacterial glycome with a lectin microarray approach. Nat Chem Biol 2:153–7

Huang TT, Sturgis J, Gomez R, Geng T, Bashir R, Bhunia AK, Robinson JP, Ladisch MR (2003) Composite surface for blocking bacterial adsorption on protein biochips. Biotechnol Bioeng 81:618–24

Hueber W, Kidd BA, Tomooka BH, Lee BJ, Bruce B, Fries JF, Sondersrup G, Monach P, Drijfhout JW, van Venrooij WJ, Utz PJ, Genovese MC, Robinson WH (2005) Antigen microarray profiling of autoantibodies in rheumatoid arthritis. Arthritis Rheum 52:2645–55

Kingsmore SF (2006) Multiplexed protein measurement: technologies and applications of protein and antibody arrays. Nat Rev Drug Discov 5:310–20

Ko IK, Kato K, Iwata H (2005a) Antibody microarray for correlating cell phenotype with surface marker. Biomaterials 26:687–96

Ko IK, Kato K, Iwata H (2005b) Parallel analysis of multiple surface markers expressed on rat neural stem cells using antibody microarrays. Biomaterials 26:4882–91

LaBaer J, Ramachandran N (2005) Protein microarrays as tools for functional proteomics. Curr Opin Chem Biol 9:14–9

Li B, Jiang L, Song Q, Yang J, Chen Z, Guo Z, Zhou D, Du Z, Song Y, Wang J, Wang H, Yu S, Wang J, Yang R (2005) Protein microarray for profiling antibody responses to *Yersinia pestis* live vaccine. Infect Immun 73:3734–9

Ligler FS, Taitt CR, Shriver-Lake LC, Sapsford KE, Shubin Y, Golden, J.P. (2003) Array biosensor for detection of toxins. Anal Bioanal Chem 377:469–77

Livingston AD, Campbell CJ, Wagner EK, Ghazal P (2005) Biochip sensors for the rapid and sensitive detection of viral disease. Genome Biol 6:112

Lu DD, Chen SH, Zhang SM, Zhang ML, Zhang W, Bo XC, Wang SQ (2005) Screening of specific antigens for SARS clinical diagnosis using a protein microarray. Analyst 130:474–82

Lueking A, Cahill DJ, Müllner S (2005) Protein biochips: A new and versatile platform technology for molecular medicine. Drug Discovery Today 10:789–94

MacBeath G (2002) Protein microarrays and proteomics. Nat Genet 32(Suppl):526–32

MacBeath G, Schreiber SL (2000) Printing proteins as microarrays for high-throughput function determination. Science 289:1760–3

Miller JC, Zhou H, Kwekel J, Cavallo R, Burke J, Butler EB, Teh BS, Haab BB (2003) Antibody microarray profiling of human prostate cancer sera: antibody screening and identification of potential biomarkers. Proteomics 3:56–63

Miyamoto S (2006) Clinical applications of glycomic approaches for the detection of cancer and other diseases. Curr Opin Mol Ther 8:507–13

Oh SH, Lee SH, Kenrick SA, Daugherty PS, Soh HT (2006) Microfluidic protein detection through genetically engineered bacterial cells. J Proteome Res 5:3433–7

Pawlak M, Schick E, Bopp MA, Schneider MJ, Oroszlan P, Ehrat M (2002) Zeptosens protein microarrays: a novel high performance microarray platform for low abundance protein analysis. Proteomics 2:383–93

Pavlickova P, Schneider EM, Hug H (2004) Advances in recombinant antibody microarrays. Clin Chim Acta 343:17–35

Perrin A, Duracher D, Perret M, Cleuziat P, Mandrand B (2003) A combined oligonucleotide and protein microarray for the codetection of nucleic acids and antibodies associated with human immunodeficiency virus, hepatitis B virus, and hepatitis C virus infections. Anal Biochem 322:148–55

Phizicky E, Bastiaens PI, Zhu H, Snyder M, Fields S (2003) Protein analysis on a proteomic scale. Nature 422:208–15

Poetz O, Schwenk JM, Kramer S, Stoll D, Templin MF, Joos TO (2005) Protein microarrays: catching the proteome. Mech Ageing Dev 126:161–70

Pohl NL (2006) Array methodology singles out pathogenic bacteria. Nat Chem Biol 2:125–6

Robinson WH, Steinman L, Utz PJ (2003) Protein arrays for autoantibody profiling and fine-specificity mapping. Proteomics 3:2077–84

Rowe CA, Tender LM, Feldstein MJ, Golden JP, Scruggs SB, MacCraith BD, Cras JJ, Ligler FS (1999) Array biosensor for simultaneous identification of bacterial, viral, and protein analytes. Anal Chem 71:3846–52

Rowe-Taitt CA, Golden JP, Feldstein MJ, Cras JJ, Hoffman KE, Ligler FS (2000) Array biosensor for detection of biohazards. Biosens Bioelectron 14:785–94

Rubina AY, Dyukova VI, Dementieva EI, Stomakhin AA, Nesmeyanov VA, Grishin EV, Zasedatelev AS (2005) Quantitative immunoassay of biotoxins on hydrogel-based protein microchips. Anal Biochem 340:317–29

Rucker VC, Havenstrite KL, Herr AE (2005) Antibody microarrays for native toxin detection. Anal Biochem 339:262–70

Sanchez-Carbayo M, Socci ND, Lozano JJ, Haab BB, Cordon-Cardo C (2006) Profiling bladder cancer using targeted antibody arrays. Am J Pathol 168:93–103

Sartain MJ, Slayden RA, Singh KK, Laal S, Belisle JT (2006) Disease state differentiation and identification of tuberculosis biomarkers via native antigen array profiling. Mol Cell Proteomics 5:2102–13

Shriver Z, Raguram S, Sasisekharan R (2004) Glycomics: a pathway to a class of new and improved therapeutics. Nat Rev Drug Discov 3:863–73

Sreekumar A, Nyati MK, Varambally S, Barrette TR, Ghosh D, Lawrence TS, Chinnaiyan AM (2001) Profiling of cancer cells using protein microarrays: discovery of novel radiation-regulated proteins. Cancer Res 61:7585–93

Steller S, Angenendt P, Cahill DJ, Heuberger S, Lehrach H, Kreutzberger J (2005) Bacterial protein microarrays for identification of new potential diagnostic markers for Neisseria meningitidis infections. Proteomics 5:2048–55

Stokes DL, Griffin GD, Vo-Dinh T (2001) Detection of E. coli using a microfluidics-based antibody biochip detection system. Fresenius J Anal Chem 369:295–301

Taitt CR, Anderson GP, Lingerfelt BM, Feldstein MJ, Ligler FS (2002) Nine-analyte detection using an array-based biosensor. Anal Chem 74:6114–20

Tong M, Jacobi CE, van de Rijke FM, Kuijper S, van de Werken S, Lowary TL, Hokke CH, Appelmelk BJ, Nagelkerke NJ, Tanke HJ, van Gijlswijk RP, Veuskens J, Kolk AH, Raap AK (2005) A multiplexed and miniaturized serological tuberculosis assay identifies antigens that discriminate maximally between TB and non-TB sera. J Immunol Methods 301:154–63

Uchiyama N, Kuno A, Koseki-Kuno S, Ebe Y, Horio K, Yamada M, Hirabayashi J (2006) Development of a lectin microarray based on an evanescent-field fluorescence principle. Methods Enzymol 415:341–51

Wadkins RM, Golden JP, Pritsiolas LM, Ligler FS (1998) Detection of multiple toxic agents using a planar array immunosensor. Biosens Bioelectron 13:407–15

Wingren C, Borrebaeck C (2006a) Antibody microarrays—current status and key technological advances. OMICS 10:411–427

Wingren C, Borrebaeck C (2006b) Recombinant antibody microarrays. Screening. Trends in Drug Discovery 2:13–15

Wingren C, Borrebaeck CA (2004) High-throughput proteomics using antibody microarrays. Expert Rev Proteomics 1:355–64

Wingren C, Ingvarsson J, Dexlin L, Szul D, Borrebaeck CA (2006) Design of recombinant antibody microarrays for complex proteome analysis: choice of sample labeling-tag and solid support. (submitted)

Wingren C, Ingvarsson J, Lindstedt M, Borrebaeck CA (2003) Recombinant antibody microarrays—a viable option? Nat Biotechnol 21:223

Wingren C, Steinhauer C, Ingvarsson J, Persson E, Larsson K, Borrebaeck CA (2005) Microarrays based on affinity-tagged single-chain Fv antibodies: sensitive detection of analyte in complex proteomes. Proteomics 5:1281–91

Yanagida M (2002) Functional proteomics: current achievements. J Chromatogr B Analyt Technol Biomed Life Sci 771:89–106

Yuk CS, Lee HK, Kim HT, Choi YK, Lee BC, Chun BH, Chung N (2004) Development and evaluation of a protein microarray chip for diagnosis of hepatitis C virus. Biotechnol Lett 26:1563–8

Zhu H, Bilgin M, Bangham R, Hall D, Casamayor A, Bertone P, Lan N, Jansen R, Bidlingmaier S, Houfek T, Mitchell T, Miller P, Dean RA, Gerstein M, Snyder M (2001) Global analysis of protein activities using proteome chips. Science 293:2101–5

Zhu H, Bilgin M, Snyder M (2003) Proteomics. Annu Rev Biochem 72:783–812

Zhu H, Hu S, Jona G, Zhu X, Kreiswirth N, Willey BM, Mazzulli T, Liu G, Song Q, Chen P, Cameron M, Tyler A, Wang J, Wen J, Chen W, Compton S, Snyder M (2006) Severe acute respiratory syndrome diagnostics using a coronavirus protein microarray. Proc Natl Acad Sci U S A 103:4011–6

Zhu H, Klemic JF, Chang S (2000) Analysis of yeast protein kinases using protein chips. Nature Genetics 26:283–290

Zhu H, Snyder M (2003) Protein chip technology. Curr Opin Chem Biol 7:55–63

27

Bacteriophage: Powerful Tools for the Detection of Bacterial Pathogens

Mathias Schmelcher and Martin J. Loessner

Abstract

Methods for detection of bacterial pathogens have to be rapid, sensitive, specific, inexpensive, easy to perform, and robust. Traditional culture-based plating techniques are hampered by time-consuming enrichment steps. This and other problems are tackled by culture-independent detection methods. The use of bacteriophage or parts thereof for bacterial detection attracts increasing attention, as reflected by a multitude of different phage-based techniques recently reported. Bacterial viruses have been optimized by evolution to specifically target their host organisms and are therefore ideal tools for detection of these microbes. For this purpose, every stage in the replication cycle of phages, from adsorption to host cell lysis, has been exploited.

Phage amplification assays are among the easiest methods to harness the host specificity of phages; the use unmodified phage particles to infect target organisms, followed by amplification of the infection as a signal by addition of helper cells. The capacity of phages to infect and lyse their host cells is utilized in assays which detect the release of cytoplasmic molecules. Cell wall recognition, phage adsorption, and injection of phage DNA into the host bacterium have also been exploited by detection methods, including capture of cells by immobilized phages and labeling of target organisms by fluorescently tagged phage particles. Phage encoded high affinity molecules such as tail fiber proteins or cell wall binding domains of phage endolysins have proven to be suitable for this purpose, especially when coupled with magnetic separation of captured bacteria. For bacterial detection, genetically modified reporter phages introduce reporter genes into their hosts. Upon infection, these gene products are produced in the target cells and can be detected with high sensitivity. Other detection methods employ phages without making use of their own host specificity. Phage display is a popular technique that can be used for production of non-phage derived high affinity molecules for recognition of pathogenic bacteria. Filamentous phages present entire libraries of randomized peptides on their surfaces, and the most suitable ones can be selected by repeated rounds of screening. In conclusion, the use of phages for detection of pathogenic bacteria offers interesting alternatives and advantages compared to traditional analytical methods.

1. Introduction

When assessing methods for detection of bacterial pathogens in food, clinical or environmental samples, the key points are rapidity, sensitivity and specificity. Such tests should be inexpensive, simple to perform, and robust, i.e., they should be feasible under various conditions and with diverse test materials. Conventional culture methods are most widespread and are often considered to be the "gold standard" for bacterial detection (Rees and Dodd 2006). In

Mathias Schmelcher and Martin J. Loessner • Institute for Food Science and Nutrition, Zurich, Switzerland.

M. Zourob et al. (eds.), *Principles of Bacterial Detection: Biosensors, Recognition Receptors and Microsystems*,
© Springer Science+Business Media, LLC 2008

general, these techniques exhibit high sensitivity but are mostly laborious and slow, due to the requirement for time-consuming enrichment steps and development of visible colonies on agar plates. In certain fields such as food processing, countermeasures against contamination have to be taken as quickly as possible, and conventional culture methods are often not sufficient to yield results in the desired time periods.

There are various culture-independent methods that have been developed to increase speed and sensitivity in detection. They include mass spectrometry (MS) and biochemical detection systems (Mosier-Boss et al. 2003). MS systems deliver mass spectra that can indicate the presence of certain bacteria in a sample by highly specific biomarkers. However, the drawbacks are high cost, difficult handling, and data that is often difficult to interpret. Biochemical or molecular approaches comprise PCR techniques and immunoassays, such as ELISA. The DNA of target organisms can be detected by PCR-based approaches in very low copy numbers, which makes this method highly sensitive. However, as it merely indicates the presence of nucleic acids, discrimination between viable and dead cells is not possible. Coupling of PCR with reverse transcription (RT-PCR) for detection of mRNA, which is more or less rapidly degraded after cell death, may solve this problem, but because of technical issues and cost limits, RT-PCR is not suitable for routine use as a diagnostic method (Rees and Dodd 2006). Immunoassays are rapid and relatively simple to perform; however, their sensitivity may be low and often fails to fulfill the legal requirements for the detection of pathogens in foods.

Although the basic idea to use phages for identification of bacterial cells is not new, it recently gained renewed attraction because of several advantages offered by harnessing the interaction of bacteria and their viruses. Bacteriophages have naturally co-evolved and been optimized to specifically target their host cells, which renders them ideal tools for detection of these organisms. Besides their specificity, their biggest advantages are easy production, robustness, and the capability of distinguishing between live and dead cells. Moreover, the reservoir of phages with various host ranges is virtually unlimited; they are available from the same environments as their bacterial hosts. The host specificity of bacteriophages has for many years been utilized in phage typing—i.e., discrimination of different bacterial isolates by challenging them with a set of phages with different lytic spectra. More recently, a multitude of phage-based methods for detection of microorganisms has been developed, including the use of native as well as genetically modified phages, and exploiting every stage of the phage's replication cycle, from adsorption to the cell surface to lysis of the host cell. This chapter will address these different techniques, trying to give an overview of the state-of-the-art in phage-based detection of bacterial pathogens.

2. Detection by Phage Amplification

In one of the first reports, Cherry et al. (1954) reported the detection of *Salmonella enterica* by means of the broad host range phage Felix-O1. The replication of phages in a plated bacterial culture results in the formation of plaques, i.e., clear zones of lysis within the lawn of bacteria (Ellis and Delbrück 1939). Every plaque traces back to one initial phage particle, so that the number of plaques corresponds to the number of phages infecting the culture. Therefore, the detection of plaques is doubtlessly the most self-evident approach when phages are to be used for detection of bacterial pathogens. Because the number of target microorganisms present in a sample to be examined is often too low to result in a bacterial lawn and the formation of plaques, the addition of a sufficient number of cells of a propagating strain for the phage used (helper cells) can serve as a means of signal amplification (Rees and Voorhees 2005). This is the underlying principle of the so-called *phage amplification assay* (Fig. 27.1), which was developed and first described about 15 years ago (Stewart et al. 1992; Wilson et al.

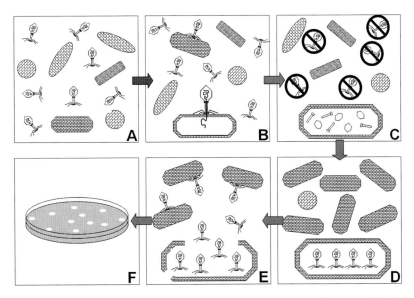

Figure 27.1. Phage Amplification Assay. A: Phage are mixed with the sample containing target cells and background flora. B: Only target cells present in the sample are infected by the phage; other bacteria are not affected. C: A virucide is added, which destroys all free phage. All phage that have successfully infected a target cell are protected inside the cell and can replicate. D: After neutralization of the virucide, helper cells are added to the sample and the mixture is spread on agar plates. E: At the end of the phage replication cycle, target cells are lysed and the liberated phage progeny infect adjacent helper cells. F: Every infected target cell initially present is indicated by formation of a plaque.

1997; McNerney et al. 1998; Stewart et al. 1998). In this assay, which is also termed *PhaB* (phage amplified biologically assay) (Wilson et al. 1997), the sample to be tested for a certain pathogen is first mixed with a phage specific for this pathogen. If the bacterium is present, it will be infected by the phage, while other microorganisms in the sample will not be affected. In a second step, a virucide is added, which should inactivate all phages that have not infected a target cell, while all phage genomes that have successfully entered a host are not harmed. After neutralizing the virucide and mixing the sample with helper bacteria (also termed *sensor cells* or *signal amplifying cells*) (Favrin, Jassim and Griffiths 2001), the whole sample is plated in a soft agar overlay and incubated. When the phage-infected cells complete the lytic replication cycle, the phage progeny is released and can now infect the bacteria of the propagating strain, which finally results in the formation of visible plaques. The number of plaques corresponds to the number of target cells initially present and infected by the phage in the sample. Obviously, the crucial point in the assay is the efficient inactivation of free phage that have not successfully infected a target cell, as all surviving viruses will give false positive results. The virucidal compound commonly used in phage amplification assays is ferrous ammonium sulphate (FAS) (McNerney et al. 1998; Park et al. 2003), but the use of essential oils and plant extracts has also been reported (Rees and Voorhees 2005). Stewart et al. (1998) achieved a reduction in phage titer of about 11 logs within minutes by using a combination of FAS and tannin-containing pomegranate rind extracts. A recent report mentioned the use of various tea infusions for phage inactivation (de Siqueira, Dodd and Rees 2006).

The phage amplification assay uses non-modified native phage particles, avoiding any legal concerns regarding GMO, and assuring easy, rapid, and cost-effective development of tests for different pathogenic bacteria. It has been successfully applied to various pathogens, including *Mycobacterium tuberculosis*, *Pseudomonas aeruginosa*, *Salmonella typhimurium*, *Staphylococcus aureus*, *Escherichia coli*, *Listeria*, and *Campylobacter* (Stewart et al. 1998; Rees and Voorhees 2005; Rees and Dodd 2006). Another benefit of this assay is that it can distinguish between live and dead bacteria, as phage replication can only take place in viable cells. Furthermore, host specificity of the phage ensures specificity of the method. On the other hand, for some applications, the use of broad host-range phages can be advantageous, as in the case of *Mycobacterium tuberculosis*. Detection of this organism in human sputum via phage amplification has been the subject of several reports (Stewart et al. 1992; Wilson et al. 1997; McNerney et al. 1998), and commercial tests (FAST*Plaque*TB and FAST*Plaque*TB-*RIF* for detection of rifampicin resistant mycobacteria) are available (Mole and Maskell 2001). In these assays, the slow-growing *Mycobacterium tuberculosis* is identified by use of a broad host-range phage that can also infect other mycobacterial species, such as the fast-growing *Mycobacterium smegmatis*. The latter serves as the sensor bacterium, which makes it possible to visualize plaques after only two days, while conventional viable culture procedures for detection of *Mycobacterium tuberculosis* take several weeks. In this case, the specificity of the test is ensured not by the phage, but by the nature of the sample, human sputum, which can only contain pathogenic and, therefore, no non-pathogenic mycobacteria (Rees and Dodd 2006). Besides its main application, the FAST *Plaque*TB test has also been successfully used for detection of *Mycobacterium bovis* and *Mycobacterium paratuberculosis* in milk samples (Rees and Dodd 2006).

Difficulties in the phage amplification assay may arise from a competitive microflora in the sample that can overgrow the lawn of helper bacteria. Usually this problem can be eliminated by dilution of the initial sample and the addition of a high number of helper cells, as long as the number of competitive organisms is below 10^6 cfu per ml (Rees and Voorhees 2005). At higher levels, specific decontamination of the sample prior to the test might be required (Mole and Maskell 2001). Another problem can be posed by the matrix of the sample itself, which may have inhibitory effects on phage infection, but decontamination steps can be helpful. With respect to sputum samples, the use of NaOH and SDS have been reported as decontamination agents that do not affect viability of the target cells (Park et al. 2003). A recent review that evaluated performance of bacteriophage-based tests for the detection of *Mycobacterium tuberculosis,* concluded that phage amplification methods have high specificity but lower and variable sensitivity, rendering them as yet unsuitable to replace conventional diagnostic tests such as microscopy and culture (Kalantri et al. 2005). However, in the case of slow-growing pathogens, the advantage compared to culture methods is evident, and very good results (sensitivity of 95 %–100 %) were obtained for the detection of rifampicin-resistant mycobacteria (Wilson et al. 1997; Mole and Maskell 2001; Simboli et al. 2005).

A variant of the phage amplification assay utilizes immunomagnetic separation of target bacteria to concentrate the target cells prior to the assay (Favrin, Jassim and Griffiths 2001), minimizing problems associated with the competitive microflora and inhibitory effects due to the sample matrix. Another variation of the method aims to couple immunomagnetic separation and phage amplification to MALDI-TOF MS (matrix-assisted laser desorption-ionization time-of-flight mass spectrometry), using the amplified phage structural proteins as specific biomarkers (Madonna, Van Cuyk and Voorhees 2003; Rees and Voorhees 2005). With this technique, simultaneous detection of two or more different pathogens in one sample is—theoretically, at least—possible.

3. Detection Through Phage-Mediated Cell Lysis

The lytic replication cycle of a bacteriophage results in lysis of the host bacterium and, thereby, release of the phage progeny. In most cases, destruction of the rigid cell wall from within the infected cell is mediated by two proteins: a holin, which forms small pores in the cytoplasmic membrane, and an endolysin, which gains access to the peptidoglycan via these pores and subsequently degrades its substrate (Young 1992). Through the cell lysis process, not only the phage but also a variety of detectable intracellular markers are released, which should make it possible to measure this lysis event. Again, the use of host specific phages obviously conveys high specificity to this detection method.

3.1. Measurement of ATP Release

The measurement of bacterial intracellular adenosine triphosphate (ATP) is an easy and rapid method to enumerate viable bacteria in almost any given sample (Stanley 1989). The amount of ATP in an average, actively metabolizing bacterial cell is quite consistent (approximately 10^{-15} g per cell). After death of the cell, ATP levels decrease rapidly (Rees and Dodd 2006). For determination of total viable cell numbers in a sample, the intracellular ATP can be released from the bacteria by use of detergent-based extractants (Blasco et al. 1998) and detected through a bioluminescent assay employing ATP-dependent firefly luciferase (Stanley 1989). With luciferin as the required substrate for this enzyme, detectable amounts of photons will be emitted at femtomolar levels of ATP, according to the following reaction (Squirrel et al. 2002):

$$ATP + luciferin + O_2 \xrightarrow[\text{Mg}^{2+}]{\text{Firefly luciferase}} AMP + PP_i + oxyluciferin + CO_2 + h\nu \text{ (light)}$$

The amount of light emitted can be determined by small hand-held luminometers and is proportional to the amount of ATP and thus the number of viable microbes. This approach can be applied to a wide range of samples, including food and clinical specimens. Several commercial kits based on this reaction are available (Stanley 1989; Rees and Dodd 2006). However, by using general detergent-based agents for cell disruption, detection of specific pathogens is not possible. This specificity can only be provided by specific cell lysis, which may again be achieved by phages that exclusively infect the target bacteria. As a proof of concept for this approach, detection of *Listeria* in mixed samples was described (Sanders 1995). However, a major drawback of the method is the insufficient detection limit (more than 10^4 cells required) caused by high unspecific ATP background levels in food and environmental samples. The detection limit can be reduced by a modification of the ATP assay, as described by Blasco et al. (1998). In this study, adenylate kinase (AK) released from the lysed cells was used to amplify the signal (Fig. 27.2). AK (E.C.2.7.4.3) is an intracellular enzyme of low molecular weight (20-30 kDa), which equilibrates concentrations of adenine nucleotides (AMP, ADP, ATP) in the cell by the following reaction (Corbitt, Bennion and Forsythe 2000):

$$ADP + ADP \xrightarrow[\text{Mg}^{2+}]{\text{Adenylate kinase}} ATP + AMP$$

As for ATP, the levels of bacterial AK within cells are also quite constant. Because ATP is generated by using ADP in a linear manner, the overall amount of fluorescent light emitted from a firefly-luciferase mediated reaction is roughly proportional to the number of viable cells in the sample (Rees and Dodd 2006). When adding additional ADP to the sample as free AK reaction substrate, the sensitivity of the assay can be increased by 10-100-fold. Under optimal conditions, the thresholds for detection of *E. coli* and *Salmonella* could be lowered to fewer than 10^3 cells within 1–2 hours (Blasco et al. 1998; Wu, Brovko and Griffiths 2001), which is in the same range as other frequently used rapid methods. As described above, further improvement of the phage-mediated bioluminescent AK assay was accomplished by combining it with an immunomagnetic separation (IMS) step, in which target cells are first captured on magnetic

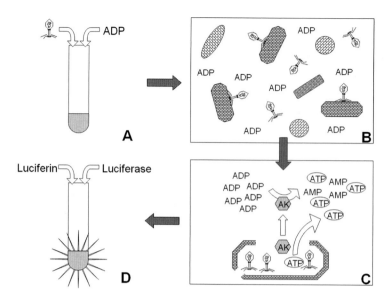

Figure 27.2. AK/ATP Assay. A: Phage specific for the target pathogen and ADP are added to the sample. B: Phage infect only target cells while all other microorganisms in the sample remain untouched. C: Phage-mediated cell lysis of target bacteria liberates intracellular ATP and adenylate kinase (AK). AK converts ADP into AMP and ATP. D: After addition of luciferin (substrate) and firefly luciferase (enzyme), detectable light is emitted by the ATP-driven reaction.

beads coated with specific antibodies, and thereby concentrated and semi-purified prior to the assay (Squirrel, Price and Murphy 2002).

All these phage lysis-based assays are relatively straightforward and of general applicability for various target organisms, since all that is needed is purified, native phages. In addition to intact phage particles, purified recombinant phage endolysins can also be employed for specific lysis from without. Detection of *Listeria monocytogenes* through endolysin Ply118 mediated lysis with subsequent bioluminescence-based ATP assay has been demonstrated (Stewart, Loessner and Scherer 1996). Similarly, detection of *Bacillus anthracis* was possible using recombinant PlyG (Schuch, Nelson and Fischetti 2002). Besides the vegetative cells, spores could also be identified after application of an aqueous germinant solution. Approximately 2.5×10^3 spores yielded a light signal after 10 minutes, and 60 min after addition of PlyG, as few as 100 spores could be detected.

3.2. Detection of Other Cytoplasmic Markers

ATP and adenylate kinase are not the only markers suitable for monitoring lysis events; other intracellular markers can also be utilized. Neufeld et al. (2003) introduced a method for rapid identification and quantification of pathogenic and polluting bacteria by amperometric measurement of enzymatic activity caused by intracellular enzymes released after phage - induced lysis (Neufeld et al. 2003). In their model (using *E. coli* strain MG1655) they monitored activity of the bacterial β-*D*-galactosidase, converting p-aminophenyl-β-*D*-galactopyranoside (PAPG) into p-aminophenol (PAP). Filtration and pre-incubation of the bacteria before infection enhanced the amperometric detection of the end product, and *E. coli* cell counts as low as 1 cfu per 100 ml could be identified within 6–8 hours. Again, due to the use of unmodified phage and the availability of numerous cytoplasmic markers in different organisms, this electrochemical method may be applied to other pathogens for which appropriate phage and enzyme markers exist.

A different approach is to first introduce a reporter gene encoding a detectable cytoplasmic marker into the target bacterium: Takikawa et al. (2002) cloned a chitinase gene into an *Enterobacter cloacae* strain isolated from tomato leaves, and assayed the presence of chitinase after infection with a bacteriophage by fluorometric or visible detection of the turnover of 4-methylumbelliferyl-β-D-N,N′,N″-triacetylchitotrioside into 4-methylumbelliferon (Takikawa et al. 2002). Only 30 minutes after infection, transformed bacteria could be identified by fluorescence of the product formed. This system aimed to track transgenic *E. cloacae* strains, used as biological control agents for plant diseases.

3.3. Measurement of Impedance

Direct impedance measurement is a rapid method that enables detection of microbial growth by measuring changes in the electrical conductivity of the growth medium due to the metabolism of the bacteria (Silley and Forsythe 1996; Wawerla et al. 1999). The concept was first introduced in 1899 by G.N. Stewart, who followed the electrical response caused by putrefication of blood and serum over a period of 30 days (Stewart 1899). Today, although the principle is the same, results can be obtained much more quickly, and the sensitivity is relatively high. Impedance can be defined as the resistance to flow of an alternating current as it passes through a conducting material, whereas any increase in conductance results in a decrease of impedance (Silley and Forsythe 1996). When microorganisms grow in a culture medium, they usually cause an increase in conductivity as they transform previously uncharged substrates (mainly carbohydrates) into highly charged end products (acids). For instance, glucose (non-ionic substrate) can be converted into two molecules of lactic acid, resulting in a higher conductance. Once an impedance system is established, changes in conductivity can be measured at regular intervals until a pre-determined threshold is exceeded, at which the change is big enough to be reliably measured. The time to reach this point is called the detection time (DT), and it depends on medium properties as well as growth kinetics and initial number of microorganisms. This near-real-time measurement of microbial activity makes it possible to detect less than 10 viable cells per milliliter (Silley and Forsythe 1996).

In its basic form, impedance enables us to monitor only the total microbial load in a sample. Specificity can be added by the use of suitable selective enrichment media, which contain substrates that can be exclusively or primarily metabolized by the target organisms (Easter and Gibson 1985). Another experimental approach is the use of phages specific for the target bacteria, as demonstrated by Chang et al. (2002), who were able to detect strains of *E. coli* O157:H7 by using phage AR1 (Chang, Ding and Chen 2002). Here, presence of target organisms can be monitored by changes in conductivity in phage-treated samples. Phages will infect and lyse target bacteria, resulting in reduced microbial growth and, therefore, increased DT in the phage-treated samples.

Limitations of direct impediometric detection are posed by the fact that the media have to be optimized for electrical measurements, which is time consuming and may be difficult, depending on the nature of the sample (Wawerla et al. 1999). Moreover, not all organisms are suitable for detection through direct impediometry, as some bacteria do not produce sufficient ionized metabolites. An alternative method to circumvent these problems is indirect impedance measurement (Silley and Forsythe 1996). In this technique, the production of CO_2 by the target organisms is monitored. The measurement does not take place in the culture medium itself but in a potassium hydroxide solution, which is separated from the medium and absorbs all CO_2 produced, resulting in a decrease of impedance. This indirect procedure has been successfully applied for detection of *Staphylococcus aureus*, *Listeria monocytogenes*, *Enterococcus faecalis*, *Bacillus subtilis*, *Escherichia coli*, *Pseudomonas aeruginosa*, *Aeromonas hydrophila*, *Salmonella*, and *Campylobacter* (Bolton 1990; Falahee, Park and Adams 2003).

4. Detection Through Cell Wall Recognition, Phage Adsorption and DNA Injection

The first step in host cell infection by phage is the adsorption of the phage particle to the surface of the bacterium. This occurs through specific recognition and binding to receptor molecules on the bacterial cell envelope, which can be manifold: Flagella or pili frequently play a role in this process, but membrane proteins and carbohydrates are also often involved (Beumer, Hannecart-Pokorni and Godard 1984). In case of tailed phages, which make up the majority of all bacterial viruses, short or long tail fiber proteins are responsible for the interaction with the receptors. In many cases, two stages of phage adsorption can be distinguished: reversible binding of the tail fibers to the receptors, followed by an irreversible docking of the base plate to the cell envelope. After that, the viral nucleic acid is injected into the host cell. The specificity of the adsorption process, which determines the host range of the phage, clearly represents an ideal feature that can be exploited for detection of bacterial pathogens.

4.1. Immobilized Phage

In 1997, Bennett et al. reported a method to separate and concentrate *Salmonella* from food materials by using a biosorbent consisting of unmodified *Salmonella*-specific phages passively immobilized to a solid phase (Bennett et al. 1997). In their approach, the lytic phage "Sapphire" was immobilized onto two different polystyrene surfaces, microplates and dipsticks, simply by soaking the surfaces with phage suspensions with a minimum titer of 5×10^{10} pfu per milliliter, followed by washing in order to remove unbound phage, and blocking of the remaining adsorption sites. The obtained biosorbents were then incubated with *Salmonella* and mixed bacterial cultures, and after washing, the specific recovery of *Salmonella* cells was assessed either by PCR or by epifluorescence microscopy using acridine orange as dye for labeling the bacterial nucleic acids. In both cases, *Salmonella* could be separated from mixed bacterial cultures, and nine out of eleven *Salmonella* strains gave positive signals. One of the strains not detected by the test, *Salmonella enterica* serovar Arizonae CRA 1568, is known to synthesize an incomplete lipopolysaccharide molecule, which results in an inability of the phage to adsorb to this strain. However, the detection limit for this method was rather high: With a PCR detection step, a concentration of 10^5 cfu per milliliter was required to generate a positive signal, and a concentration of 10^7 cfu/ml was necessary in the initial culture to ensure that this amount of cells was captured by the biosorbent. This represents a capture efficiency of 1 %, which is clearly not sufficient.

An improvement of the method was reported by Sun et al. (2001), who described specific immobilization of phages, exploiting the high affinity of biotin to streptavidin, as opposed to passive adsorption (Sun, Brovko and Griffiths 2001). In this work, *Salmonella* phage SJ2 was treated with sulfosuccinimidobiotin, which reacts with primary amino groups of the phage coat proteins, resulting in biotinylation of the phage particles. Afterwards, these were coated onto streptavidin-labeled magnetic beads. This phage-based biosorbent was applied to capture target cells of *S. enteritidis* by magnetic separation, using magnetic beads coated with nonbiotinylated phage as a negative control. The capture efficiency in this study was determined by using a recombinant bioluminescent *Salmonella* strain as target organism and measuring relative light output (RLU) after capture. With this technique, approximately 20 % of the target cells could be recovered when a culture of 2×10^6 cfu/ml was used, a significant improvement compared to the passive immobilization method.

4.2. Detection Through Phage-Encoded Affinity Molecules

Employing native or modified replicating phages for identification of pathogens has been demonstrated to be useful and reliable by many different approaches and techniques. Another approach is the utilization of some of those phage components that confer specificity to the virus, for instance the recognition proteins involved in the adsorption process or the host-specific lysis. Particularly interesting in this respect are phage-encoded peptidoglycan hydrolases (endolysins), which mostly show a modular organization consisting of an N-terminal domain that harbors the enzymatic activity, and a C-terminal domain, which is responsible for specific substrate recognition. The corresponding characteristics were described for several of these enzymes from phages infecting different bacteria (Garcia et al. 1990; Loessner, Wendlinger and Scherer 1995; Low et al. 2005). Regarding *Listeria*, the carboxy-terminal cell wall-binding domains (CBDs) of endolysins from phages A118 and A500 were genetically fused to GFP, yielding recombinant fusion proteins that enabled specific fluorescent in vitro and *in vivo* labeling of *Listeria* cells (Loessner et al. 2002; Lenz et al. 2003; Henry et al. 2006). With one exception, both constructs generally recognized only cells of the genus *Listeria*, making it possible to identify these pathogens within mixed bacterial cultures by using fluorescence microscopy. With respect to the different serovar groups, CBD118 and CBD500 display exclusive but complementary binding specificities. While the GFP-CBD118 marker protein labels all strains of serovars 1/2, 3, and "7," predominantly at the poles and septal regions, GFP-CBD500 binds to strains of serovars 4, 5, and 6, and is distributed evenly over the entire cell (Fig. 27.3). When used in combination, the proteins enable specific labeling and detection of all serovars of *Listeria*. Affinity measurements revealed very strong binding of the CBDs to their ligands in the bacterial cell wall (equilibrium constants in the nanomolar range), which occurs in a rapid, saturation-dependent manner. Only 30 seconds after addition of the GFP-CBD proteins to a sample containing *Listeria* cells, the target bacteria were completely labeled (Loessner et al. 2002). Moreover, the specific binding was also shown to occur in complex environments, such as infected eukaryotic cells (Henry et al. 2006).

The high affinity and specificity of the cell wall binding domains was also exploited for immobilization and separation of *Listeria*, using paramagnetic beads coated with GFP-CBD fusion proteins (Kretzer et al. 2007). The *Listeria* cells attach to the surface of the beads (Fig. 27.4) and can be captured by magnetic forces. In contrast to the poor recovery obtained by immunomagnetic separation, CBD-based magnetic separation (CBD-MS) technique yielded recoveries of more than 90 % from cell suspensions within 20 to 40 minutes (Kretzer et al. 2007). For their application for detection of *Listeria* cells, contaminated food samples were homogenized, incubated with CBD-coated beads for 40 minutes, and the beads subsequently magnetically separated and plated or used for PCR-based detection protocols. After

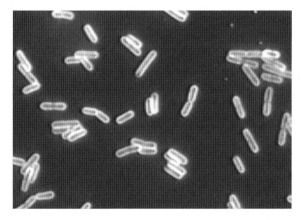

Figure 27.3. CBD-based cell wall decoration. Recombinant fusion protein, consisting of GFP and the *Listeria* phage-encoded cell wall binding domain CBD500, recognizes and binds to the cell wall of *L. monocytogenes* serovar 4b cells with high affinity. The even decoration of the cells is visualized by fluorescence microscopy.

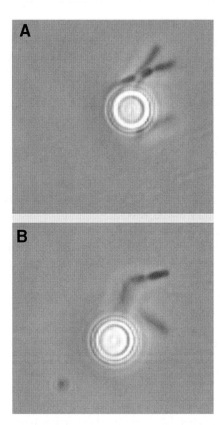

Figure 27.4. Use of CBD affinity molecules for capture of bacteria. Paramagnetic polystyrene beads coated with recombinant CBD proteins from the endolysins of *Listeria* bacteriophages can immobilize live *Listeria* cells on their surface. A: CBD118 and *Listeria monocytogenes* serovar 1/2c. B: CBD500 and *Listeria monocytogenes* serovar 4b.

enrichment of only 6 hours detection limits between 1 and 100 cfu/g were achieved. Extending the enrichment step to 24 hours, extremely low cell counts of less than one cfu per gram could be detected. Rapidity is the biggest advantage of the CBD-MS method compared to the standard plating procedure for detection of *Listeria* in food, which requires 96 hours. Moreover, the CBD-based approach could also be applied for detection of other pathogens such as *Bacillus cereus* and *Clostridium perfringens* (Kretzer et al. 2007). However, the utilization of CBDs from bacteriophage endolysins is primarily suited for Gram-positive bacteria, since Gram-negative cells feature an outer membrane that prevents direct access of the cell wall peptidoglycan from without.

For Gram-negative bacteria, another class of proteins encoded by phage can be utilized in a similar way. More specifically, the short tail fiber proteins of phages able to infect Gram-negative cells can specifically recognize and bind to lipopolysaccharide components on the cell surface. One example for the application of recombinant tail fiber proteins is the EndoTrap system (Profos AG, Regensburg, Germany), which enables elimination of endotoxic LPS from medical and biological fluids. These proteins are also useful for the labeling and/or immobilization of entire cells featuring the corresponding LPS type.

4.3. Fluorescently Labeled Phage

Correlation of light emission with presence of target organisms is an easy and rapid method for microbial detection. In the previous section, the detection of a recombinant bioluminescent bacterium after specific phage-mediated capture was described. A somewhat different and even more elegant way is to combine binding specificity and signaling ability in the detection agent itself, making it possible to detect bacteria not capable of light emission. A first approach was reported in 1965 by the use of a "phage-fluorescent antiphage staining system" for identification of *Listeria monocytogenes* (Watson and Eveland 1965). This technique was based upon the production of fluorescently labeled antibodies directed against the coat proteins of phage specific for this pathogen. Thirty years later, fluorescently labeled virus probes (FLVPs) for labeling, identification, and enumeration of specific bacteria in seawater samples were developed (Hennes, Suttle and Chan 1995). Here, fluorescent dyes were used to pre-label the genetic material within purified intact phage particles specific for the target bacteria. The stained viruses were added to seawater samples containing the target cells, which could then be identified and enumerated using epifluorescence microscopy.

Goodridge and co-workers modified this method by coupling it with immunomagnetic separation for detection of *E. coli* O157:H7 in broth (Goodridge, Chen and Griffiths 1999a). In this fluorescent–bacteriophage assay (FBA), target bacteria are first captured by magnetic beads coated with anti-O157 antibodies, and then labeled with dye-tagged bacteriophage LG1. This phage was found to be specific for all *E. coli* O157 strains tested, but also for some other strains and bacteria of other species and genera. However, by combination of two specific selection steps (IMS and phage specificity), the overall specificity of the FBA could be increased. After incubation with the fluorescent phage, target cells could be identified by epifluorescence microscopy due to their "halolike" appearance, and flow cytometry enabled reliable detection of the pathogens at a level of 10^4 cells per milliliter. The FBA was also tested for its ability to detect *E. coli* O157:H7 in food samples, namely ground beef and raw milk (Goodridge, Chen and Griffiths 1999b). For this purpose the food samples were first spiked with different concentrations of target bacteria. After an enrichment step, FBA signals were measured by flow cytometry, and the assay detected approximately 3 cfu/g in ground beef after 6 h enrichment, and 10^1-10^2 cfu/ml in raw milk after 10 h enrichment.

Several recent studies further developed the principle of using fluorescent tags for labeling phage particles. Examples are SYBR green (Lee et al. 2006) and SYBR gold, which was used to stain phage P22 for specific detection of *Salmonella typhimurium* LT2 (Mosier-Boss et al. 2003). Compared to other dyes such as YOYO-1, Ethidium bromide or DAPI, SYBR gold was found to be superior (enhancement of fluorescence of more than 1000 fold, high quantum yields, two different excitation maxima, sufficient difference between excitation and emission wavelengths). Interestingly, the latter study indicated that the labeled phage DNA was apparently injected into the *Salmonella* target cells, as they appeared as defined rod-shaped fluorescing objects. The authors claimed that the dye is "protected" inside the phage capsid until the target organism is bound (Mosier-Boss et al. 2003).

Although fluorescent phage assays must still be further developed and optimized (especially in terms of sensitivity), we conclude that the use of bacteriophages or their components as specific detection agents offers many advantages compared to antibodies: They are robust; less sensitive to changes in temperature, pH, and ionic strength; easy and inexpensive to produce and to purify; self replicating; and have a long shelf life.

5. Detection by Reporter Phage

Besides the use of native phage particles for the detection of pathogens, also possible is the engineering of recombinant phages that can transduce a reporter gene into the target cell, enabling the subsequent identification of infected hosts by monitoring the product encoded by the reporter gene (Fig. 27.5). The first such reporter phage was made by introducing

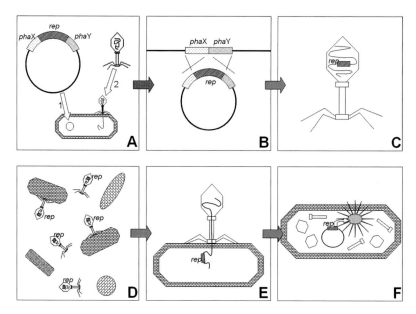

Figure 27.5. Construction of a reporter phage and its application. A: A reporter gene (*rep*) is cloned into a plasmid, flanked by regions of phage genomic DNA (*pha*X, *pha*Y). A cell of a suitable propagating strain is transformed with the plasmid (1), and subsequently infected with the phage (2). B: Within the cells, double-crossover homologous recombination between plasmid and phage DNA can take place, leading to insertion of the reporter gene into the phage genome. C: The genome, including the reporter gene, is packaged and the phage particles assembled. After plating, plaques containing the reporter phage can be identified by activity of the reporter gene product. D: For detection of target bacteria, reporter phage are mixed with the sample and allowed to infect target cells. E: Phage DNA, including the reporter gene, is injected (transduced) into the host cells. F: Expression of the reporter gene in the host cell results in a detectable signal.

a promoterless β-galactosidase gene *lacZ* into the bacteriophage Mu (Castilho, Olfson and Casadaban 1984). Upon infection of host cells, random insertion of recombinant Mu-lac into the bacterial genome occurred. However, only when *lacZ* was coincidentally inserted somewhere downstream of a host promoter, was the β-galactosidase produced, and the infected cells could be identified by a color-forming reaction.

   Until present, several different reporter genes have been employed for construction of recombinant phages, which are reviewed below. There exist a number of different ways in which a reporter gene can be introduced into the phage. The most commonly used are direct cloning, transposition, and homologous recombination (Ulitzur and Kuhn 2000). In most cases, however, direct cloning is not an option, as this method only works with relatively small phage genomes and requires established genetic manipulation, as well as systems and a way to package the modified nucleic acids into phage virions. As an alternative, transposition is efficient and easy to perform. Through the (random) insertion into non-essential or intergenic regions of the phage genome, each experiment results in different reporter phage particles, which are then amplified, selected, and purified on the basis of the desired properties. The third approach is based upon cloning of the reporter gene on a plasmid, flanked by phage DNA fragments; introduction of the plasmid into a suitable phage host; and homologous recombination through double crossover between plasmid and phage within the infected cell (Figure 27.5, A–C). (From our own experience and the reports of others, the difficulty is not to generate a recombinant phage, but to subsequently identify the needle in the haystack....)

   One drawback inherent to the use of recombinant phages for bacterial detection is the fact that genetically modified organisms (GMOs) are created by infection of the host. This may result in some resistance within the population against this detection technique, and may also create legal issues in many countries (Rees and Voorhees 2005). Moreover, the increasing requirements for certification and registration of laboratories in order to be able to work with GMOs may be time-consuming and expensive.

5.1. Luciferase Reporter Phage (LRP)

The basic principle of luciferase applications for phage-mediated detection of pathogens was described above, in the context of ATP measurement following cell lysis. Interestingly, luciferase genes are the most frequently used reporter genes in recombinant phage. In 1987, the first luciferase reporter phage (LRP) was constructed by insertion of bacterial bioluminescence genes (i.e., the entire *lux* operon from *Vibrio fischeri*) into a *lambda*-based cloning vector (Ulitzur and Kuhn 1987). In contrast to the insect luciferase described above, bacterial luciferases require long chain aldehydes as substrate. In presence of reduced monoflavin FMNH$_2$ and molecular oxygen, light is emitted according to the following reaction:

$$FMNH_2 + RCOH + O_2 \xrightarrow{\text{Bacterial luciferase}} FMN + RCOOH + H_2O + light$$

The *lux* operon consists of several genes, of which *lux*AB encodes the two subunits of the luciferase holoenzyme, while the *lux*CDE gene products are responsible for the reduction of fatty acids into the aldehyde substrate. *lux*I and *lux*R have important quorum sensing regulatory functions. The LuxI protein generates a specific acyl-homoserine lactone (AHL), a diffusible signaling molecule referred to as autoinducer, which interacts with LuxR to stimulate transcription of *lux*CDABE and of *lux*I itself (Ripp et al. 2006; Hagens and Loessner 2007). By inserting the complete *lux* operon, Ulitzur and Kuhn were able to detect as few as 10 *E. coli* cells within one hour in milk samples by using a luminometer (Ulitzur and Kuhn 1989). These researchers also made some other *E. coli* LRPs, based on *lux*AB only. They have also reported the application of some of these phages for detection of antibiotic sensitivity. The decrease of light emission in the presence of specific antibiotics indicated the respective sensitivity of target bacteria (Ulitzur and Kuhn 2000). An LRP constructed by random mini Tn*10::lux*AB mutagenesis was used for detection of enteric indicator bacteria in slaughterhouse and carcass samples (Kodikara, Crew and Stewart 1991), with a sensitivity of 10 to 10^4 cfu per gram with and without enrichment, respectively. A similar LRP (Waddell and Poppe 2000) carrying only the *lux*AB genes from *Vibrio harveyi* (Hill, Swift and Stewart 1991), permitted detection of *E. coli* O157:H7 in one hour. Recently, another approach for phage-mediated identification of *E. coli,* which is based on quorum sensing, was described (Ripp et al. 2006). It uses a binary reporter system consisting of an LRP, which carries the *lux*I gene only, and *E. coli* as luminescent bioreporter, harboring the *lux*CDABE and *lux*R genes (Ripp et al. 2006). Both phage and bioreporter cells are added to test samples, and infection of target cells present in the sample results in production of AHL. These diffusible autoinducers interact with the LuxR regulatory protein within the *E. coli* reporters, which triggers *lux*CDABE transcription and eventually generates bioluminescence. This system responded to target cell concentrations as low as 1 cfu/ml pure culture. An inherent drawback of this approach is that the *lux* genes of the bioreporters can potentially be induced by any "compatible" autoinducer from other bacteria present in the sample, resulting in false positive results. On the other hand, an advantage with respect to the limited capacity of the phage genome for additional genetic materials is the small size of *lux*I, compared to *lux*AB genes or even the complete operon. Moreover, because the bioreporter cells are not lysed themselves, FMNH$_2$ is continuously regenerated, resulting in longer signal emission and detection periods (Hagens and Loessner 2007).

To date, numerous LRPs are available for detection not only of *E. coli*, but also many other bacterial pathogens. Regarding *Salmonella*, Chen and Griffiths (1996) described several LRPs generated by homologous recombination, which detected as few as 10 cells per milliliter sample when a cocktail of these phages was used (Chen and Griffiths 1996). Noteworthy in terms of GMO concerns is a study by Kuhn et al. (2002). These authors constructed an LRP based on the *Salmonella* phage Felix-O1, which is genetically "locked," i.e., it can only infect, but not multiply on wild-type host cells (Kuhn et al. 2002). This was accomplished by replacing two non-essential and one essential gene of the phage by the *lux*AB genes. Only in a laboratory strain, where the missing essential gene was supplied *in trans*, could this defective phage be propagated. In samples containing wild-type *Salmonella,* the phage will adsorb to the target cells, inject its DNA, and cause the generation of bioluminescence, but it is not able to produce viable phage particles, so that a spread of genetically modified organisms in the environment is impossible. Felix-O1 is an ideal candidate for this purpose, as it has a very broad host range, infecting up to 96 % of all clinically relevant strains (Fey et al. 1978). Furthermore, the Felix-O1 based LRP generated bioluminescence also in a number of strains on which the native phage cannot propagate, probably because the host range regarding the initial stages of infection appears broader than what is required for completion of the propagation cycle (Kuhn et al. 2002).

With respect to *Listeria monocytogenes*, the causative agent of foodborne Listeriosis, the unusual A511 phage, characterized by an extremely broad host range within the genus, was an ideal candidate for creating an LRP (Loessner et al. 1996). A511 is a virulent myovirus that infects approximately 95 % of all *L. monocytogenes* strains of serovars 1/2 and 4, which are most often involved in outbreaks of Listeriosis (Farber and Peterkin 1991). The late gene region of A511 has been identified (Loessner et al. 1994; Loessner, Wendlinger and Scherer 1995), and a *lux*AB fusion gene from *Vibrio harveyi* was introduced into the phage genome by homologous recombination directly downstream of the major capsid protein (*cps*) gene, featuring a very strong promoter. Upon addition of the substrate nonanal to cells infected with the A511::*lux*AB reporter phage, emission of light corresponded to the expression of late genes, with a maximum immediately before lysis occurred. The use of A511::*lux*AB for detection of *L. monocytogenes* in a range of artificially and naturally contaminated food samples was evaluated (Loessner, Rudolf and Scherer 1997). In many of the spiked foods, very low initial contamination rates could be detected after an enrichment period of 20 hours and a total assay time of less than 24 hours. Clearly, speed is the main advantage of this method, compared to the standard plating method (takes around four days). Due to the high specificity of the phage, detection is possible in samples with high background flora without successive rounds of enrichment and selective plating (Rees and Dodd 2006).

By far the largest number of studies about LRPs that have been published dealt with the detection of *Mycobacterium tuberculosis*. As discussed above with respect to the phage amplification assay the main advantage offered by the phage-based techniques compared to conventional culture methods is the required time until detection. The first reported mycobacterial LRP utilized the firefly luciferase gene *luc* as reporter (Jacobs et al. 1993). Since then, a multitude of different *luc* reporter phages were constructed for detection of mycobacteria and determination of drug susceptibility, all of them based on phages TM4, L5, and D29 (Hagens and Loessner 2007). Phage TM4 is a strictly lytic broad host-range phage, infecting strains of *M. tuberculosis*, *M. smegmatis*, *M. bovis*, and *M. avium*. However, a drawback is that the TM4 lytic cycle is very rapid, which means that within the limited time only small amounts of luciferase can be produced, leading to low sensitivity, i.e., high detection limits of 10^4 cells (Jacobs et al. 1993). The temperate phage L5 was shown to express an inserted luciferase gene under control of a *Mycobacterium* promoter after insertion as a prophage, decreasing the detection limit considerably (Sarkis, Jacobs and Hatfull 1995). But as L5 has a very limited host range, infecting only fast-growing mycobacteria, L5-based LRPs are not capable of detecting *Mycobacterium tuberculosis*. An improved mycobacterial LRP based on TM4 was made by introduction of temperature sensitive mutations (Carriere et al. 1997). The reporter phage phAE88 can only replicate at 30°C, but not at the infection temperature of 37°C. When phage infected cells are incubated at the higher temperature, the result is in an extended luciferase expression period and a correspondingly higher light output. This modification allowed bioluminescence-based detection of as few as 120 cells within 12 hours. Phage D29, although highly related to L5, features the deletion of a putative repressor, which renders the phage strictly lytic and provides a broad host range. Unlike the L5 LRP, the D29 reporter phage can readily infect *M. tuberculosis* and *M. bovis* with equal detection sensitivity (Pearson et al. 1996). The broad host ranges of TM4 and D29 make it possible to propagate the respective LRPs in fast-growing strains like *Mycobacterium smegmatis*. However, a disadvantage of this decreased specificity is the possibility of false-positive results in *Mycobacterium tuberculosis* detection. This problem can be circumvented by using ρ-nitro-α-acetyl-amino-β-hydroxy-propiophenone (NAP) (Laszlo and Eidus 1978), a chemical that selectively inhibits the growth of *Mycobacterium tuberculosis*. When incorporating this test into an LRP assay as an additional control, samples containing only *M. tuberculosis* do not emit light in the sample containing NAP (Riska et al. 1997). A field test with this method was conducted in Mexico with more than 500 sputum samples, in comparison with cultivation methods (Banaiee et al. 2001). A total of 76 % of the samples were identified to contain mycobacteria with the LRP method, and 94 % of the bacteria were correctly assigned to the *Mycobacterium tuberculosis* complex, employing the NAP control test. In the same study (and numerous others), the usefulness of LRPs for antibiotic susceptibility testing was described (Jacobs et al. 1993; Sarkis, Jacobs and Hatfull 1995; Riska and Jacobs 1998; Banaiee et al. 2003). In order to obtain any detectable signal in an LRP-based assay, target cells must be actively metabolizing, i.e., growing. Growth inhibition inevitably results in inability of the phage to replicate the lack of intracellular metabolites required for the luciferase reaction. This is exploited in the drug susceptibility test, where the decrease or absence of a light signal upon addition of a tested antibiotic indicates susceptibility of the bacteria against the drug, and vice versa.

In LRP assays for detection of pathogens or in drug susceptibility testing, a luminometer is required for signal detection. However, because such equipment may be expensive and, therefore, not affordable for laboratories in poor countries, an inexpensive "low tech" alternative has been developed that uses a Polaroid film camera (Riska et al. 1999; Hazbon et al. 2003). In comparison with luminometer measurements, the co-called Bronx Box yielded the same sensitivity in drug susceptibility tests, but the time required to obtain results was approximately two times longer (Hazbon et al. 2003), which is still adequate for most situations.

5.2. Fluorescent Protein Reporter Phage

The green fluorescent protein (GFP) from the jellyfish *Aquorea victoria* has become the most commonly used marker in molecular biology, medicine and cell biology. Reasons for its popularity are its high stability and cellular compatibility, as well as the fact that its chromophore is formed in an autocatalytic cyclization and does not require cofactors (Zimmer 2002). In contrast to luciferases, fluorescent proteins do not need a substrate because they have properties very similar to fluorescent dyes. Excitation at a certain wavelength results in emission of light at a different wavelength, which can be conveniently measured. These properties seem to make GFP and related proteins promising candidates for construction of reporter phages. However, it was not until 2002 that Funatsu et al. modified the *E. coli* bacteriophage λ by direct cloning to express the GFP reporter gene (Funatsu et al. 2002). Four to six hours after infection of *E. coli* cells the fluorescent bacteria could be detected by fluorescence microscopy. Two years later, virulent phage PP01 specific for *E. coli* O157:H7 (Morita et al. 2002) was the basis for development of a GFP reporter phage (Oda et al. 2004). In this case, the GFP gene was cloned into the phage by homologous recombination, and was genetically fused to the N- or C-terminal end of the phage's small outer capsid protein (SOC). It was shown that fusion of GFP to SOC in both cases did not change the host range of PP01, but rather enhanced its affinity to the host cells. Visualization of infected target cells by fluorescence microscopy was possible after only 10 minutes. The test worked with viable cells, viable but nonculturable (VBNC) cells, and even dead cells. In mixes of target cells with an *E. coli* K12 cells, only O157:H7 target cells were labeled. The same researchers also constructed a T4-based reporter phage by fusing GFP to the SOC protein (Tanji et al. 2004). As this T4 mutant does not make the T4 lysozyme at the end of its replication cycle, the resulting reporter phage was defective and "locked" within the infected cells (as described above for Felix-O1). Visualization and distinction of target cells from T4-insensitive bacteria was simple and effective, especially as the accumulation of GFP inside the cells increased the intensity of fluorescence. The GFP-modified PP01 phage was also modified to be lysozyme-deficient (Awais et al. 2006), and exhibited increased intensity of fluorescence upon infection of target cells. In combination with nutrition uptake analysis, this reporter phage enabled discrimination between culturable, VBNC, and dead cells: While growing cells exhibited intense green fluorescence upon infection with PP01e⁻/GFP, VBNC and dead cells allowed adsorption but not proliferation and signal amplification, resulting in only low fluorescence.

5.3. Other Reporter Phages

Besides the above discussed luciferases and fluorescent proteins, several other reporter genes have been introduced into phages for detection of pathogens. Wolber and Green (1990) used the *ina* gene encoding an ice nucleation protein from *Pseudomonas* to construct a reporter phage based on *Salmonella* phage P22. Upon infection of target cells by *ina*-recombinant phage, the ice nucleation protein was produced and presented on the surface of the organism by incorporation into its outer membrane. When samples are cooled down, those that contain *Salmonella* presenting the INP would freeze at slightly elevated temperatures, which was indicated by a phase-sensitive fluorescent dye. Using phage-based ice nucleation (termed the BIND assay), very low levels of *Salmonella* could be detected (Wolber 1993). In buffer and in raw eggs, identification of only 2 cells per milliliter was possible within 3 hours, and even at high levels of background flora, 10 cells per milliliter were detectable. Combining the BIND assay with *Salmonella*-specific immunomagnetic separation, the minimum detection limit in food samples could be lowered to approximately 5 cells per ml (Irwin et al. 2000).

Using the *lacZ* gene product, another reporter phage specific for *E. coli* O157:H7 has been constructed. The test used cotton swabs, followed by selective enrichment and immunomagnetic separation and was termed Phast Swab (Goodridge 2006). Using chemiluminescence-based enzyme substrates, a detection limit of 10^2 to 10^3 cells within 12 hours was obtained.

Another recent study reports high-sensitivity bacterial detection by using biotin-tagged phages and quantum-dot based detection (QD) (Edgar et al. 2006). QDs are colloidal fluorescent semiconductor crystals (e.g., CdSe), with diameters in the nanometer range. They feature a broadband adsorption but narrow emission spectra. An outer shell of a few atomic layers, made from, e.g., ZnS), ensures high quantum yields and enhanced photostability, compared to organic fluorophores such as GFP. In their work, *E. coli* phage T7 was modified to express a 15-aa biotinylation peptide fused to the major capsid protein. Newly assembled phage T7-bio virions now display the peptide on their capsid surface. This enables biotinylation of phage particles *in vivo*, followed by binding of streptavidin-coated QDs, exploiting the high-affinity interaction of biotin and streptavidin. Using these QD reporter phages, as few as 10 bacteria per ml could be clearly distinguished from background noise, within only 1 hour. Besides the rapidity and sensitivity of the assay, another advantage seems its wide applicability. The small size of the reporter gene, which only serves to provide an anchor for biotin-tagging, minimizes problems in genetic engineering of the phage, especially concerning limits of packaging due to increased genome size. Furthermore, the availability of QDs of various emission wavelengths could support multiplex imaging of different pathogens in a single sample, using phages of different specificity. Tools for detection can include fluorescent microscopy, spectroscopy, flow cytometry and mass spectrometry.

6. Other Detection Methods Using Phage

The majority of phage-based assays for detection of bacterial pathogens exploits the natural specific relationship between phage and host to confer specificity to these tests. However, there are different methods in which phages do not serve as the detection agents themselves, but are utilized, e.g., for production of highly specific binding molecules, or are used to signal the successful recognition of a target organism by an antibody. These technologies are discussed in this section.

6.1. Phage Display for Production of Highly Specific Binding Molecules

The very popular phage display technique, developed more than 20 years ago (Smith 1985), allows the expression of complete libraries of peptides or proteins as fusions with phage coat proteins and their presentation on the surface of phage virions. This is then followed by selection of molecules with highest affinity to a target antigen (see below), such as specific ligands on the cell surfaces of bacterial pathogens (see Fig. 27.6). Display systems based on various phages, such as T4, T7 and λ, have been reported, but the technology is best established for the filamentous phage strains M13, fd and f1 (Smith and Petrenko 1997). These viruses contain small genomes consisting of single stranded DNA of approximately 6400 nucleotides, surrounded by a rod-shaped tube composed mainly of helically arranged molecules of the major coat protein pVIII. The ends (tips) of the particles contain a few copies of different minor coat proteins. Foreign gene sequences can be introduced by fusion with either the major coat protein gene VIII, or the minor coat protein genes III or VI, resulting in display of the cloned protein in thousands of copies over the complete surface, or only at one of the tips, respectively. The choice of the cloning strategy depends on the molecule to be introduced. For example, peptides exceeding a certain size might impair the integrity of the phage when fused to every single copy of pVIII. Large random libraries can be created by means of combinatorial chemistry, and typically consist of billions of phage clones, each of them displaying a different peptide variant (Petrenko and Sorokulova 2004) (Fig. 27.6a). Moreover, instead of random peptides, the variable domains of animal or human antibodies can be displayed in all variants by phage libraries (Rader and Barbas 1997; Hoogenboom 2002). Whatever the nature of the phage-borne molecules may be, the "fittest" among the complete library in terms of binding to the target ligand can be found by repeated rounds of selection and amplification, a procedure termed "panning" (Rader and Barbas 1997; Smith and Petrenko 1997; Petrenko and Sorokulova 2004) (Fig. 27.6d–e). For this purpose, the target ligand is immobilized on a solid support and exposed to the complete phage display library. Phage particles, whose display molecules can bind to the target ligand, are captured, while all others can be washed away. The bound subpopulation is then eluted, propagated in suitable host cells, and subjected to another round of panning. Selection stringency can be increased in each round, and the whole process ideally results in the identification of a protein variant with the desired properties in terms of specificity and

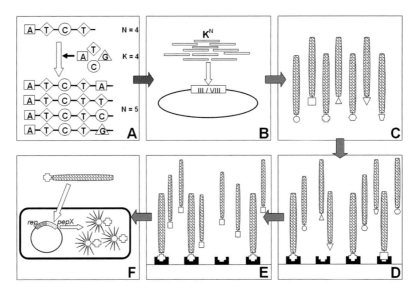

Figure 27.6. Selection of high affinity binders by phage display. A: Random oligonucleotides are synthesized by combinatorial chemistry. For every step, a mixture of all possible nucleotides is added to the growing chain. N is the number of positions in the oligonucleotide, K the number of possible nucleotides per position. B: The total number of variants is K^N. These are genetically fused to one of the coat protein genes III, VI, or VIII of a filamentous phage. C: The resulting library contains all possible different random peptides presented on the surface of phage particles. D: By panning with immobilized target antigens, a subpopulation of the library with affinity to the antigen is captured, while the rest of the phage can be removed by washing steps. E: Repeated rounds of panning with increased stringency allow for enrichment and isolation of phage displaying peptides with highest affinity to the target molecule. F: The gene coding for the peptide (*pep*X) is fused to a reporter gene (*rep*) and transferred to a high-level expression system. Fusion proteins consisting of a reporter domain and a high affinity binding domain for the desired target can then be produced in large amounts.

affinity to a given target ligand. The corresponding gene can then easily be transferred to any desired high-level expression system in order to produce the binding molecule in large amounts (Fig. 27.6f). Labeling of such evolved binding proteins by genetic fusion with reporter proteins enables the product to be used in detection assays (Rees and Voorhees 2005). Reporters that have been used include fluorescent proteins (Casey et al. 2000; Morino et al. 2001), alkaline phosphatase (Kerschbaumer et al. 1997; Muller et al. 1999), and streptavidin (Pearce et al. 1997). One major advantage of the phage display technology, compared to the production of antibodies in immunized animals, is the fact that selection is solely based on affinity. This makes it possible to also find high affinity binding molecules for target ligands, which are either not immunoreactive or are toxic for a given host (Petrenko and Vodyanoy 2003; Rees and Voorhees 2005).

High affinity molecules derived from phage displays have been successfully used for detection of various pathogenic bacteria, viruses, and bacterial toxins (Smith and Petrenko 1997; Petrenko and Vodyanoy 2003; Petrenko and Sorokulova 2004). Zhou et al. (2002) used a human single-chain Fv antibody (scFv) phage display library for isolating variants with high affinity against spores of *Bacillus subtilis* (Zhou, Wirsching and Janda 2002). Cross-reactivity with *B. licheniformis* spores was removed by deploying a subtractive panning strategy. Shortly thereafter, the generation of short peptides derived from phage display libraries for the binding of spores from different *Bacillus* species, including *B. anthracis*, *B. cereus*, *B. subtilis*, *B. globigii*, and *B. amyloliquefaciens* was reported (Knurr et al. 2003; Turnbough 2003). At least one of the peptide families was shown to selectively bind to *B. anthracis* spores.

Affinity molecule-based magnetic separation and PCR detection (IMS-PCR) for detection of *M. avium* subsp. *paratuberculosis* (Grant et al. 2000) was reported by Stratmann et al. (2002), who used a 12-mer peptide isolated from a phage display library (Stratmann et al. 2002). The short peptide could then be chemically synthesized and coated onto paramagnetic beads, enabling specific binding and capture of *M. paratuberculosis* from milk. Coupling with a suitable PCR protocol allowed the detection of target cells in a concentration of 10 cfu/ml.

Several other studies report the application of antibody fragments derived from phage display libraries, e.g., for immuno-electrochemical detection of *Listeria monocytogenes* (Benhar et al. 2001), differentiation of pathogenic and nonpathogenic strains of *Streptococcus suis* (de Greeff, van Alphen and Smith 2000) and specific detection of *Brucella abortus* (Hayhurst et al. 2003). For the latter case, it is noteworthy that the phage antibodies isolated did not show any cross reaction with *Yersinia pseudotuberculosis*, while many anti-*Brucella* sera do so because the O:9 antigenic epitope is shared by both organisms.

Besides the use of purified antibody fragments and peptides, complete filamentous phage particles displaying the desired molecules can also serve as probes in detection assays. As the viruses expose thousands of reactive amino groups on their surface, they can be conjugated with various fluorescent dyes such as Cy5 or Alexa (Petrenko and Sorokulova 2004). Goldman et al. (2000) used a Cy-5 labeled filamentous phage displaying a 12-mer peptide specifically binding to staphylococcal enterotoxin B (SEB) for detection of this toxin down to a concentration of 1.4 ng/well in a plate-based fluorescent immunoassay (Goldman et al. 2000). Recently, the construction of acoustic wave biosensors for detection of *Salmonella typhimurium* using affinity-selected filamentous phages as probes was reported (Olsen et al. 2006). In these quartz crystal microbalance (QCM) sensors acoustic waves are excited by application of alternating voltage to a piezoelectric crystal. Deposition of a certain mass (e.g., a phage or a bacterial cell) on the gold surface of the QCM electrode results in a change of resonance frequency, which can be detected. Filamentous phages specific for *S. typhimurium* displaying an 8-mer peptide were derived from a phage library and adsorbed to the sensor surface, resulting in a biosensor that allowed the detection of 10^2 target cells/ml in a near real-time response of less than 180 seconds. Advantages of employing phage particles for this technology, compared to the use of antibodies, include low-cost production, durability, reusability and stability, while achieving equivalent specificity and sensitivity (Petrenko and Smith 2000).

6.2. Dual Phage Technology

Yet another phage-based detection method (however, not exploiting specificity of the phage for a certain receptor) is dual phage technology, which can be applied for detection of almost any kind of target molecule (Wilson 1999). Specificity of this assay (at least in its original form) is provided by antibodies binding to the target antigen, while the phage serves for signaling the successful binding event. The technique is termed "dual phage assay," as it employs two different transducing phages covalently bound to the antibodies, conveying resistance against two different antibiotics. The use of two different antibodies that bind to two distinct epitopes of a single antigen molecule ensures a very high specificity of the method. The two different phage variants, each conjugated to a (monoclonal) antibody species, are mixed with the test sample. If the target antigen is present, the antibodies will attach to it, and the two different phages will be in direct proximity. In a second step, suitable phage host cells are added, and the mixture is plated on a medium containing both antibiotics. Uninfected cells or cells that are infected by only one of the phages will not receive the two antibiotic resistance genes, and will not be able to grow and form colonies. However, when both viruses are physically linked through the antibody-antigen complex, they can simultaneously transfer resistance against both antibiotics to a target bacterium, resulting in the formation of a colony of the selective medium. A crucial issue of the dual phage assay is the concentration of phages used. If the multiplicity of infection is too high, the chance of double infection by free phage particles increases, leading to false positive results. Besides specificity, another advantage of the dual phage technology is its high sensitivity, as every signal (linking of antigens) is amplified by growth of one bacterium into a colony of more than 10^8 cells (Rees and Voorhees 2005), which could be detected by many other means besides colony formation. The assay is quite universal, as every antigen for which antibodies are available can be targeted.

7. Conclusions and Future Perspectives

In this chapter we have tried to provide an overview of bacteriophage-based techniques for detection of bacterial pathogens. Many of these methods offer advantages compared to conventional culture methods, especially regarding time requirement and specificity. Furthermore, some of the techniques mentioned feature high sensitivity, with detection limits of less than one bacterial cell per gram sample. The application of genetically modified phages is currently limited by legal issues, but generally speaking, the use of bacteriophages in general is less costly than most other advanced detection methods, e.g. real time-PCR based techniques.

With estimated 10^{31} particles worldwide, bacteriophages are by far the most abundant self-replicating units on earth. Therefore, they represent a practically unlimited reservoir of tools of different specificities that can be utilized for various purposes. Not only complete virions, but also isolated phage-derived molecules such as cell wall hydrolases receive increasing attention, e.g. as antimicrobials, but also as means for bacterial detection. Overall, the application of bacteriophages certainly has a promising future in biotechnology, food science, and medicine.

References

Awais R, Fukudomi H, Miyanaga K, Unno, H and Tanji Y (2006) A recombinant bacteriophage-based assay for the discriminative detection of culturable and viable but nonculturable *Escherichia coli* O157:H7. Biotechnol Prog 22:853–9

Banaiee N, Bobadilla-Del-Valle M, Bardarov Jr. S, Riska PF, Small PM, Ponce-De-Leon A, Jacobs Jr. WR, Hatfull GF and Sifuentes-Osornio J (2001) Luciferase reporter mycobacteriophages for detection, identification, and antibiotic susceptibility testing of *Mycobacterium tuberculosis* in Mexico. J Clin Microbiol 39:3883–8

Banaiee N, Bobadilla-del-Valle M, Riska PF, Bardarov Jr. S, Small PM, Ponce-de-Leon A, Jacobs Jr. WR, Hatfull GF and Sifuentes-Osornio J (2003) Rapid identification and susceptibility testing of *Mycobacterium tuberculosis* from MGIT cultures with luciferase reporter mycobacteriophages. J Med Microbiol 52:557–61

Benhar I, Eshkenazi I, Neufeld T, Opatowsky J, Shaky S and Rishpon J (2001) Recombinant single chain antibodies in bioelectrochemical sensors. Talanta 55:899–907

Bennett AR, Davids FG, Vlahodimou S, Banks JG and Betts RP (1997) The use of bacteriophage-based systems for the separation and concentration of *Salmonella*. J Appl Microbiol 83:259–65

Beumer J, Hannecart-Pokorni E and Godard C (1984) Bacteriophage receptors. Bull. Inst. Pasteur 82:173–253

Blasco R, Murphy MJ, Sanders MF and Squirrell DJ (1998) Specific assays for bacteria using phage mediated release of adenylate kinase. J Appl Microbiol 84:661–6

Bolton FJ (1990) An investigation of indirect conductimetry for detection of some food-borne bacteria. J Appl Bacteriol 69:655–61

Carriere C, Riska PF, Zimhony O, Kriakov J, Bardarov S, Burns J, Chan J and Jacobs Jr. WR, (1997) Conditionally replicating luciferase reporter phages: improved sensitivity for rapid detection and assessment of drug susceptibility of *Mycobacterium tuberculosis*. J Clin Microbiol 35:3232–9

Casey JL, Coley AM, Tilley LM and Foley M (2000) Green fluorescent antibodies: novel in vitro tools. Protein Eng 13:445–52

Castilho BA, Olfson P and Casadaban MJ (1984) Plasmid insertion mutagenesis and lac gene fusion with mini-mu bacteriophage transposons. J Bacteriol 158:488–95

Chang TC, Ding HC and Chen S (2002) A conductance method for the identification of *Escherichia coli* O157:H7 using bacteriophage AR1. J Food Prot 65:12–7

Chen J and Griffiths MW (1996) *Salmonella* detection in eggs using Lux(+) bacteriophages. Journal of Food Protection 59:908–914

Cherry WB, Davis BR, Edwards PR and Hogan RB (1954) A simple procedure for the identification of the genus *Salmonella* by means of a specific bacteriophage. J Lab Clin Med 44:51–5

Corbitt AJ, Bennion N and Forsythe SJ (2000) Adenylate kinase amplification of ATP bioluminescence for hygiene monitoring in the food and beverage industry. Lett Appl Microbiol 30:443–7

de Greeff A, van Alphen L and Smith HE (2000) Selection of recombinant antibodies specific for pathogenic *Streptococcus suis* by subtractive phage display. Infect Immun 68:3949–55

de Siqueira RS, Dodd CE and Rees CE (2006) Evaluation of the natural virucidal activity of teas for use in the phage amplification assay. Int J Food Microbiol 111:259–62

Easter MC and Gibson DM (1985) Rapid and automated detection of *Salmonella* by electrical measurements. J Hyg (Lond) 94:245–62

Edgar R, McKinstry M, Hwang J, Oppenheim AB, Fekete RA, Giulian G, Merril C, Nagashima K and Adhya S (2006) High-sensitivity bacterial detection using biotin-tagged phage and quantum-dot nanocomplexes. Proc Natl Acad Sci U S A 103:4841–5

Ellis EL and Delbrück M (1939) The growth of bacteriophage. Journal of General Physiology 22:365–384

Falahee MB, Park SF and Adams MR (2003) Detection and enumeration of *Campylobacter jejuni* and *Campylobacter coli* by indirect impedimetry with an oxygen scavenging system. J Food Prot 66:1724–6

Farber JM and Peterkin PI (1991) *Listeria monocytogenes*, a food-borne pathogen. Microbiol Rev 55:476–511

Favrin SJ, Jassim SA and Griffiths MW (2001) Development and optimization of a novel immunomagnetic separation-bacteriophage assay for detection of *Salmonella enterica* serovar enteritidis in broth. Appl Environ Microbiol 67:217–24

Fey H, Burgi E, Margadant A and Boller E (1978) An economic and rapid diagnostic procedure for the detection of *Salmonella/Shigella* using the polyvalent *Salmonella* phage O-1. Zentralbl Bakteriol [Orig A] 240:7–15

Funatsu T, Taniyama T, Tajima T, Tadakuma H and Namiki H (2002) Rapid and sensitive detection method of a bacterium by using a GFP reporter phage. Microbiol Immunol 46:365–9

Garcia P, Garcia JL, Garcia E, Sanchez-Puelles JM and Lopez R (1990) Modular organization of the lytic enzymes of *Streptococcus pneumoniae* and its bacteriophages. Gene 86:81–8

Goldman ER, Pazirandeh MP, Mauro JM, King KD, Frey JC and Anderson GP (2000) Phage-displayed peptides as biosensor reagents. J Mol Recognit 13:382–7

Goodridge L (2006) Template reporter bacteriophage platform and multiple bacterial detection assays based thereon. US Patent Application 20060210968

Goodridge L, Chen J and Griffiths M (1999a) Development and characterization of a fluorescent-bacteriophage assay for detection of *Escherichia coli* O157:H7. Appl Environ Microbiol 65:1397–404

Goodridge L, Chen J and Griffiths M (1999b) The use of a fluorescent bacteriophage assay for detection of *Escherichia coli* O157:H7 in inoculated ground beef and raw milk. Int J Food Microbiol 47:43–50

Grant IR, Pope CM, O'Riordan LM, Ball HJ and Rowe MT (2000) Improved detection of *Mycobacterium avium* subsp. *paratuberculosis* In milk by immunomagnetic PCR. Vet Microbiol 77:369–78

Hagens S and Loessner MJ (2007) Luciferase Reporter Phages (LRPs). In: Marks RS, Lowe CR, Cullen DC, Weetal HH, and Karube I (eds) Handbook of Biosensors and Biochips John Wiley & Sons, Ltd., Chichester

Hayhurst A, Happe S, Mabry R, Koch Z, Iverson BL and Georgiou G (2003) Isolation and expression of recombinant antibody fragments to the biological warfare pathogen *Brucella melitensis*. J Immunol Methods 276:185–96

Hazbon MH, Guarin N, Ferro BE, Rodriguez AL, Labrada LA, Tovar R, Riska PF and Jacobs Jr. WR (2003) Photographic and luminometric detection of luciferase reporter phages for drug susceptibility testing of clinical *Mycobacterium tuberculosis* isolates. J Clin Microbiol 41:4865–9

Hennes KP, Suttle CA and Chan AM (1995) Fluorescently Labeled Virus Probes Show that Natural Virus Populations Can Control the Structure of Marine Microbial Communities. Appl Environ Microbiol 61:3623–3627

Henry R, Shaughnessy L, Loessner MJ, Alberti-Segui C, Higgins DE and Swanson JA (2006) Cytolysin-dependent delay of vacuole maturation in macrophages infected with *Listeria monocytogenes*. Cell Microbiol 8:107–19

Hill PJ, Swift S and Stewart GS (1991) PCR based gene engineering of the *Vibrio harveyi lux* operon and the *Escherichia coli trp* operon provides for biochemically functional native and fused gene products. Mol Gen Genet 226:41–8

Hoogenboom HR (2002) Overview of antibody phage-display technology and its applications. Methods Mol Biol 178:1–37

Irwin P, Gehring A, Tu SI, Brewster J, Fanelli J and Ehrenfeld E (2000) Minimum detectable level of Salmonellae using a binomial-based bacterial ice nucleation detection assay (BIND). J AOAC Int 83:1087–95

Jacobs Jr. WR, Barletta RG, Udani R, Chan J, Kalkut G, Sosne G, Kieser T, Sarkis GJ, Hatfull GF and Bloom BR (1993) Rapid assessment of drug susceptibilities of *Mycobacterium tuberculosis* by means of luciferase reporter phages. Science 260:819–22

Kalantri S, Pai M, Pascopella L, Riley L and Reingold A (2005) Bacteriophage- based tests for the detection of *Mycobacterium tuberculosis* in clinical specimens: a systematic review and meta- analysis. BMC Infect Dis 5:59

Kerschbaumer RJ, Hirschl S, Kaufmann A, Ibl M, Koenig R and Himmler G (1997) Single-chain Fv fusion proteins suitable as coating and detecting reagents in a double antibody sandwich enzyme-linked immunosorbent assay. Anal Biochem 249:219–27

Knurr J, Benedek O, Heslop J, Vinson RB, Boydston JA, McAndrew J, Kearney JF and Turnbough Jr. CL, (2003) Peptide ligands that bind selectively to spores of *Bacillus subtilis* and closely related species. Appl Environ Microbiol 69:6841–7

Kodikara CP, Crew HH and Stewart GS (1991) Near on-line detection of enteric bacteria using *lux* recombinant bacteriophage. FEMS Microbiol Lett 83:261–5

Kretzer JW, Lehmann R, Schmelcher M, Banz M, Kim K, Korn C and Loessner MJ (2007) High affinity cell wall-binding domains of bacteriophage endolysins for immobilization and separation of bacterial cells. Appl Environ Microbiol 73:1992–2000

Kuhn J, Suissa M, Wyse J, Cohen I, Weiser I, Reznick S, Lubinsky-Mink S, Stewart G and Ulitzur S (2002) Detection of bacteria using foreign DNA: the development of a bacteriophage reagent for *Salmonella*. Int J Food Microbiol 74:229–38

Laszlo A and Eidus L (1978) Test for differentiation of *M. tuberculosis* and *M. bovis* from other mycobacteria. Can J Microbiol 24:754–6

Lee SH, Onuki M, Satoh H and Mino T (2006) Isolation, characterization of bacteriophages specific to *Microlunatus phosphovorus* and their application for rapid host detection. Lett Appl Microbiol 42:259–64

Lenz LL, Mohammadi S, Geissler A and Portnoy DA (2003) SecA2-dependent secretion of autolytic enzymes promotes *Listeria monocytogenes* pathogenesis. Proceedings of the National Academy of Sciences, US 100:12432–7

Loessner MJ, Kramer K, Ebel F and Scherer S (2002) C-terminal domains of *Listeria monocytogenes* bacteriophage murein hydrolases determine specific recognition and high-affinity binding to bacterial cell wall carbohydrates. Mol Microbiol 44:335–49

Loessner MJ, Krause IB, Henle T and Scherer S (1994) Structural proteins and DNA characteristics of 14 *Listeria* typing bacteriophages. J Gen Virol 75 (Pt 4):701–10

Loessner MJ, Rudolf M and Scherer S (1997) Evaluation of luciferase reporter bacteriophage A511::*lux*AB for detection of *Listeria monocytogenes* in contaminated foods. Appl Environ Microbiol 63:2961–5

Loessner MJ and Scherer S (1995) Organization and transcriptional analysis of the *Listeria* phage A511 late gene region comprising the major capsid and tail sheath protein genes cps and tsh. J Bacteriol 177:6601–9

Loessner MJ, Wendlinger G and Scherer S (1995) Heterogeneous endolysins in *Listeria monocytogenes* bacteriophages: a new class of enzymes and evidence for conserved holin genes within the siphoviral lysis cassettes. Mol Microbiol 16:1231–41

Low LY, Yang C, Perego M, Osterman A and Liddington RC (2005) Structure and lytic activity of a *Bacillus anthracis* prophage endolysin. J Biol Chem 280:35433–35439

Madonna AJ, Van Cuyk S and Voorhees KJ (2003) Detection of *Escherichia coli* using immunomagnetic separation and bacteriophage amplification coupled with matrix-assisted laser desorption/ionization time-of-flight mass spectrometry. Rapid Commun Mass Spectrom 17:257–63

McNerney R, Wilson SM, Sidhu AM, Harley VS, al Suwaidi Z, Nye PM, Parish T and Stoker NG (1998) Inactivation of mycobacteriophage D29 using ferrous ammonium sulphate as a tool for the detection of viable *Mycobacterium smegmatis* and *M. tuberculosis*. Res Microbiol 149:487–95

Mole RJ and Maskell TW (2001) Phage as a diagnostic - the use of phage in TB diagnosis. J Chem Technol Biotechnol 76:683–688

Morino K, Katsumi H, Akahori Y, Iba Y, Shinohara M, Ukai Y, Kohara Y and Kurosawa Y (2001) Antibody fusions with fluorescent proteins: a versatile reagent for profiling protein expression. J Immunol Methods 257:175–84

Morita M, Tanji Y, Mizoguchi K, Akitsu T, Kijima N and Unno H (2002) Characterization of a virulent bacteriophage specific for *Escherichia coli* O157:H7 and analysis of its cellular receptor and two tail fiber genes. FEMS Microbiol Lett 211:77–83

Mosier-Boss PA, Lieberman SH, Andrews JM, Rohwer FL, Wegley LE and Breitbart M (2003) Use of fluorescently labeled phage in the detection and identification of bacterial species. Appl Spectrosc 57:1138–44

Muller BH, Chevrier D, Boulain JC and Guesdon JL (1999) Recombinant single-chain Fv antibody fragment-alkaline phosphatase conjugate for one-step immunodetection in molecular hybridization. J Immunol Methods 227:177–85

Neufeld T, Schwartz-Mittelmann A, Biran D, Ron EZ and Rishpon J (2003) Combined phage typing and amperometric detection of released enzymatic activity for the specific identification and quantification of bacteria. Anal Chem 75:580–5

Oda M, Morita M, Unno H and Tanji Y (2004) Rapid detection of *Escherichia coli* O157:H7 by using green fluorescent protein-labeled PP01 bacteriophage. Appl Environ Microbiol 70:527–34

Olsen EV, Sorokulova IB, Petrenko VA, Chen IH, Barbaree JM and Vodyanoy VJ (2006) Affinity-selected filamentous bacteriophage as a probe for acoustic wave biodetectors of *Salmonella typhimurium*. Biosens Bioelectron 21:1434–42

Park DJ, Drobniewski FA, Meyer A and Wilson SM (2003) Use of a phage-based assay for phenotypic detection of mycobacteria directly from sputum. J Clin Microbiol 41:680–8

Pearce LA, Oddie GW, Coia G, Kortt AA, Hudson PJ and Lilley GG (1997) Linear gene fusions of antibody fragments with streptavidin can be linked to biotin labelled secondary molecules to form bispecific reagents. Biochem Mol Biol Int 42:1179–88

Pearson RE, Jurgensen S, Sarkis GJ, Hatfull GF and Jacobs Jr. WR (1996) Construction of D29 shuttle phasmids and luciferase reporter phages for detection of mycobacteria. Gene 183:129–36

Petrenko VA and Smith GP (2000) Phages from landscape libraries as substitute antibodies. Protein Eng 13:589–92

Petrenko VA and Sorokulova IB (2004) Detection of biological threats. A challenge for directed molecular evolution. J Microbiol Methods 58:147–68

Petrenko VA and Vodyanoy VJ (2003) Phage display for detection of biological threat agents. J Microbiol Methods 53:253–62

Rader C and Barbas CF 3rd (1997) Phage display of combinatorial antibody libraries. Curr Opin Biotechnol 8:503–8

Rees CE and Dodd CE (2006) Phage for rapid detection and control of bacterial pathogens in food. Adv Appl Microbiol 59:159–86

Rees CED and Loessner MJ (2005) Phage for the Detection of Pathogenic Bacteria. In: Kutter E and Sulakvelidze A (eds) Bacteriophages: Biology and Applications. CRC Press, Boca Raton, pp 267–284

Rees JC and Voorhees KJ (2005) Simultaneous detection of two bacterial pathogens using bacteriophage amplification coupled with matrix-assisted laser desorption/ionization time-of-flight mass spectrometry. Rapid Commun Mass Spectrom 19:2757–61

Ripp S, Jegier P, Birmele M, Johnson CM, Daumer KA, Garland JL and Sayler GS (2006) Linking bacteriophage infection to quorum sensing signaling and bioluminescent bioreporter monitoring for direct detection of bacterial agents. J Appl Microbiol 100:488–99

Riska PF and Jacobs Jr. WR (1998) The use of luciferase-reporter phage for antibiotic-susceptibility testing of mycobacteria. Methods Mol Biol 101:431–55

Riska PF, Jacobs Jr. WR, Bloom BR, McKitrick J and Chan J (1997) Specific identification of *Mycobacterium tuberculosis* with the luciferase reporter mycobacteriophage: use of p-nitro-alpha-acetylamino-beta-hydroxy propiophenone. J Clin Microbiol 35:3225–31

Riska PF, Su Y, Bardarov S, Freundlich L, Sarkis G, Hatfull G, Carriere C, Kumar V, Chan J and Jacobs Jr. WR (1999) Rapid film-based determination of antibiotic susceptibilities of *Mycobacterium tuberculosis* strains by using a luciferase reporter phage and the Bronx Box. J Clin Microbiol 37:1144–9

Sanders MF (1995) A rapid bioluminescent technique for the detection and identification of *Listeria monocytogenes* in the presence of *Listeria innocua*. In: Campbell AK, Kricka LJ and Stanley PE (eds) Bioluminescence and Chemiluminescence: Fundamental and Applied Aspects. John Wiley & Sons, Chichester, U.K., pp 454–457

Sarkis GJ, Jacobs Jr. WR and Hatfull GF (1995) L5 luciferase reporter mycobacteriophage: a sensitive tool for the detection and assay of live mycobacteria. Mol Microbiol 15:1055–67

Schuch R, Nelson D and Fischetti VA (2002) A bacteriolytic agent that detects and kills *Bacillus anthracis*. Nature 418:884–9

Silley P and Forsythe S (1996) Impedance microbiology–a rapid change for microbiologists. J Appl Bacteriol 80:233–43

Simboli N, Takiff H, McNerney R, Lopez B, Martin A, Palomino JC, Barrera L and Ritacco V (2005) In-house phage amplification assay is a sound alternative for detecting rifampin-resistant *Mycobacterium tuberculosis* in low-resource settings. Antimicrob Agents Chemother 49:425–7

Smith GP (1985) Filamentous fusion phage: novel expression vectors that display cloned antigens on the virion surface. Science 228:1315–7

Smith GP and Petrenko VA (1997) Phage display. Chem Rev 97:391–410

Squirrel DJ, Price RL and Murphy MJ (2002) Rapid and specific detection of bacteria using bioluminescence. Anal Chim Acta 457:109–114

Stanley PE (1989) A review of bioluminescent ATP techniques in rapid microbiology. J Biolumin Chemilumin 4:375–80

Stewart GN (1899) The changes produced by the growth of bacteria in the molecular concentration and electrical conductivity of culture media. J Exp Med 4:235–243

Stewart GS, Jassim SA, Denyer SP, Newby P, Linley K and Dhir VK (1998) The specific and sensitive detection of bacterial pathogens within 4 h using bacteriophage amplification. J Appl Microbiol 84:777–83

Stewart GSAB, Jassim SAA, Denyer SP, Park S, Rostas-Mulligan K and Rees CED (1992) Methods for rapid microbial detection. Patent WO 92/02633

Stewart GSAB, Loessner MJ and Scherer S (1996) The bacterial *lux* gene bioluminescent biosensor revisited. Asm News 62:297–301

Stratmann J, Strommenger B, Stevenson K and Gerlach GF (2002) Development of a peptide-mediated capture PCR for detection of *Mycobacterium avium* subsp. *paratuberculosis* in milk. J Clin Microbiol 40:4244–50

Sun W, Brovko L and Griffiths M (2001) Use of bioluminescent *Salmonella* for assessing the efficiency of constructed phage-based biosorbent. J Ind Microbiol Biotechnol 27:126–8

Takikawa Y, Mori H, Otsu Y, Matsuda Y, Nonomura T, Kakutani K, Tosa Y, Mayama S and Toyoda H (2002) Rapid detection of phylloplane bacterium *Enterobacter cloacae* based on chitinase gene transformation and lytic infection by specific bacteriophages. J Appl Microbiol 93:1042–50

Tanji Y, Furukawa C, Na SH, Hijikata T, Miyanaga K and Unno H (2004) *Escherichia coli* detection by GFP-labeled lysozyme-inactivated T4 bacteriophage. J Biotechnol 114:11–20

Turnbough Jr. CL (2003) Discovery of phage display peptide ligands for species-specific detection of *Bacillus* spores. J Microbiol Methods 53:263–71

Ulitzur S and Kuhn J (1987) Introduction of *lux* genes into bacteria; a new approach for specific determination of bacteria and their antibiotic susceptibility. In: Slomerich R, Andreesen R, Kapp A, Ernst M and Woods WG (eds) Bioluminescence and Chemiluminescence: New Perspectives. John Wiley & Sons, New York, pp 463–472.

Ulitzur S and Kuhn J (1989) Detection and/or identification of microorganisms in a test sample using bioluminescence or other exogenous genetically introduced marker. U.S. Patent 4,861,709

Ulitzur S and Kuhn J (2000) Construction of *lux* bacteriophages and the determination of specific bacteria and their antibiotic sensitivities. Bioluminescence and Chemiluminescence, Pt C 305:543–557

Watson BB and Eveland WC (1965) The application of the phage-fluorescent antiphage staining system in the specific identification of *Listeria monocytogenes*. I. Species specificity and immunofluorescent sensitivity of *Listeria monocytogenes* phage observed in smear preparations. J Infect Dis 115:363–9

Wawerla M, Stolle A, Schalch B and Eisgruber H (1999) Impedance microbiology: applications in food hygiene. J Food Prot 62:1488–96

Wilson SM (1999) Analytical method using multiple virus labeling. PCT Patent WO99/63348

Wilson SM, AlSuwaidi Z, McNerney R, Porter J and Drobniewski F (1997) Evaluation of a new rapid bacteriophage-based method for the drug susceptibility testing of *Mycobacterium tuberculosis*. Nature Medicine 3:465–468

Wolber PK (1993) Bacterial ice nucleation. Adv Microb Physiol 34:203–37

Wu Y, Brovko L and Griffiths MW (2001) Influence of phage population on the phage-mediated bioluminescent adenylate kinase (AK) assay for detection of bacteria. Lett Appl Microbiol 33:311–5

Young R (1992) Bacteriophage lysis: mechanism and regulation. Microbiol Rev 56:430–81

Zhou B, Wirsching P and Janda KD (2002) Human antibodies against spores of the genus *Bacillus*: A model study for detection of and protection against anthrax and the bioterrorist threat. Proceedings of the National Academy of Sciences, US 99:5241–6

Zimmer M (2002) Green fluorescent protein (GFP): applications, structure, and related photophysical behavior. Chem Rev 102:759–81

28

Phage Display Methods for Detection of Bacterial Pathogens

Paul A. Gulig, Julio L. Martin, Harald G. Messer, Beverly L. Deffense and Crystal J. Harpley

Abstract

There exists a great need for detecting bacterial pathogens in food, water, environmental, and patient samples. Although significant developments are being made in nucleic acid-based detection, e.g., real time-PCR, there are advantages to detecting microbial antigens using a variety of assays such as enzyme-linked immunosorbent assay (ELISA), antibody arrays, fiber optics, and surface plasmon resonance. Immunological reagents used to detect microbial antigens have mostly consisted of antisera or monoclonal antibodies. However, over the past two decades new methods have been developed that use bacteriophages (phages), viruses of bacteria, as tools to express antibody fragments or random peptides that can detect microbes. The most widely used antibody fragment is the single chain F variable (scFv) portion that includes the antigen-binding regions of the heavy and light chains. The genes encoding antibodies can be obtained either from unimmunized animals or from animals immunized with the target antigen or microbe. scFv libraries are usually constructed in specialized plasmids called phagemids that create fusions of the antibody fragment to a phage coat protein. Phagemids must be packaged into phage particles by the use of helper phages during infection of *E. coli* hosts. Random peptides are usually constructed in phage genomes directly; hence, they do not require helper phages. DNA sequences encoding random peptides and antibodies are most often fused to the *gIII* gene of a filamentous phage such as M13 so that a hybrid pIII protein is expressed on the phage particle. Phage particles that display a fusion peptide that recognizes the desired antigen are selected from the library by a process called panning. Panning involves binding phages to the antigen, washing away unbound phages, eluting the specific phages, and amplification by infection of an *E. coli* host. A major advantage of phage display is that panning and screening of clones can be accomplished in only a few days, as opposed to months for antisera and monoclonal antibodies. Once a phage displaying an antibody fragment or random peptide of desired specificity is isolated, the gene encoding the binding peptide can be genetically manipulated. With scFvs, it is possible to have *E. coli* cells secrete the protein, rather than relying on the use of the phage particle displaying the scFv protein. This chapter reviews the literature on the use of phage display to detect bacteria in a variety of assays. There has been considerable success in the use of commercially available random peptide phage display libraries. Most reported success with scFv phage display has been from using libraries constructed from immunized animals. In addition to antibodies, other target-binding tools are under development, such as affibodies, anticalins, ankyrins, and trinectins. The ease, economy, rapidity, and genetic manipulability of phage display make it an effective tool for developing reagents to detect bacterial pathogens.

Paul A. Gulig, Julio L. Martin, Harald G. Messer, Beverly L. Deffense, and Crystal J. Harpley •
Department of Molecular Genetics and Microbiology, University of Florida College of Medicine, Gainesville, Florida.

M. Zourob et al. (eds.), *Principles of Bacterial Detection: Biosensors, Recognition Receptors and Microsystems*,
© Springer Science+Business Media, LLC 2008

1. Introduction

The purpose of this chapter is to help investigators interested in using phage display to develop peptide/protein reagents that recognize bacteria and their products primarily for purposes of detection. This chapter does not cover construction of the reagents themselves. Rather, the focus is on readily available tools such as the New England Biolabs Ph.D. peptide libraries and Griffin.1/Tomlinson scFv libraries.

1.1. Why Detect Bacteria and What Tools Are Available?

Before launching into the detail of phage display methods, one should consider the use of the tools. There are three basic reasons for detecting bacterial pathogens: 1) preventing disease by detecting organisms in food, water or air, with particular emphasis on detecting agents of bioterrorism; 2) diagnosing disease by detecting organisms or their products in patient samples so that appropriate therapy can be administered; and 3) determining which organisms are causing what diseases through epidemiology to aid in both preventive measures and detecting ongoing outbreaks.

There is a variety of methods, both old and new, used to detect bacterial pathogens in numerous situations (reviewed in Lim et al. 2005). The most longstanding and widely used method involves culture of the organisms coupled with biochemical or metabolic tests to enable definitive identification of the organisms. Originally these methods were time-consuming, requiring growth of the organisms on a variety of selective or differential media. More recently, automated systems as the VITEK 2 (bioMérieux) and OmniLog (Biolog) have become popular time-saving alternatives. Additionally, analysis of bacterial fatty acid and lipid profiles has become a useful method for rapid identification of microorganisms (Busse et al. 1996). Immunological tests for specific identification are also commonplace, either in conjunction with cultured organisms or on patient or environmental samples (Lim et al. 2005). Because immunological tests have clearly been useful in the development of phage display, we will first discuss these tools in detail.

1.2. Immunological Tools

In the realm of immunological tools, there is a limited set of possible reagents. The oldest and most basic immunological reagents consisted of antisera raised in animals. These antisera could be made more specific by absorption with cross-reactive antigens. However, the major problem with antisera is that they are often difficult to standardize, as production of antibodies, even from a single animal, can fluctuate over time, in addition to variability among different animals. See Table 28.1 for a comparison of benefits and detriments of different detection reagent tools. The creation of monoclonal antibodies in 1975 ushered in a new era of immunological analysis because these antibodies demonstrate exquisite specificity and can be made in essentially limitless supply (Kohler and Milstein 1975). The monoclonal antibody produced today will be the same as that produced ten years from now, as long as the hybridoma cell line is properly maintained. However, monoclonal antibodies are expensive and often difficult to isolate. Just because a mouse has been immunized with an appropriate antigen does not guarantee that useful monoclonal antibodies will be generated, even after several months and thousands of dollars have been invested in the process. Relatively few laboratories have the expertise and resources for immunizing animals and performing cell fusions. However, most research institutions have core laboratories to perform such procedures, and hybridomas may be produced commercially. Additionally, just as for raising antisera in animals, there are limits to the nature of the antigen that can be used for generating monoclonal

Table 28.1. Methods of producing specific reagents for detection

Reagent Type	Procedure	Advantages	Disadvantages	Time Frame
Conventional antisera	Immunize animals, collect sera	Technically simple	Cannot use toxic antigens, variability from animal to animal or different bleeds, contaminating antibodies	Months
Monoclonal antibodies	Immunize animals, purify lymphocytes, perform fusions, identify specific hybridomas	Highly specific reagents, invariant with time, unlimited supply	Cannot use toxic antigens, not guaranteed to obtain desired specificity, expensive	Months
Phage Display (Random, premade library)	Obtain library, pan on antigen, screen clones	Rapid, highly specific reagents, invariant with time, unlimited supply, can use toxic antigens, can be genetically manipulated	Not guaranteed to obtain desired specificity	Days to weeks
Phage display (immunized library)	Immunize animals, clone antibody sequences into vector, pan library on antigen, screen clones	Highly specific reagents, invariant with time, higher probability of success than unimmunized library, unlimited supply, can be genetically manipulated	Cannot use toxic antigens, not guaranteed to obtain desired specificity	Months
Ribosome display, mRNA display	Obtain libraries, pan on antigen, screen clones	Rapid, highly specific reagents, invariant with time, unlimited supply, can use toxic antigens, can be genetically manipulated	Technically difficult	Days

antibodies. Animals cannot be treated with toxins or infected with pathogenic microbes because their health will be compromised, and it is difficult, if not impossible, to immunize with self antigens.

Antibodies have been used in a variety of assays throughout the decades. Some of the oldest immunological tests involved immunodiffusion of antibodies and antigens in gel matrices so that lines of identity or partial identity could be observed. The development of enzyme-linked immunosorbent assays (ELISAs) enabled the analysis of numerous samples, either antigens or antibodies, under standardized and quantitative conditions. Similarly, Western blots have proven to be useful tools for identifying the antigens recognized by antibodies, as long as the antigenic determinant is stable under the denaturing conditions of SDS-PAGE. Antibodies can also be used to detect microbial antigens directly in patient tissue samples using immunohisto-chemistry. Another immunological test related to infectious disease is examining patient sera for antibodies to specific pathogens or toxins to determine if the patient has been exposed to the pathogen or toxin. Serology is still the assay of choice for diagnosing syphilis, AIDS and other diseases.

1.3. Nucleic Acid-Based Tools

Although this chapter deals with phage display reagents for detecting bacterial pathogens, we would be remiss if we did not mention the quantum leaps being made in a completely different realm, nucleic-acid-based detection. Of primary importance is the polymerase chain reaction (PCR) and related techniques, including real time (RT)-PCR and quantitative RT-PCR (q-PCR). Such tests are extremely rapid and sensitive, requiring as little as 13 minutes to identify as few as 10 cells of a pathogen (Lim et al. 2005). Although there may appear to be competition between nucleic-acid-based and antigen-based detection systems, each has its own strengths and weaknesses, and each system plays an important role in detection of microbial pathogens.

2. What Types of Antigen Detection Methods Are Being Developed?

Before attempting to detect bacteria via their structural components other than nucleic acids, one must be familiar with the available methods to ensure that reagents with appropriate qualities are used for the chosen methods. Standard techniques, such as ELISA and Western blot, will not be discussed further. Rather, this section will focus on newer methods that permit simultaneous screening of numerous types of samples for a variety of organisms without focusing on a single agent or limited set of likely targets (see Table 28.2). For example, similar to DNA-based microarrays, antibody arrays coupled with immunofluorescence are being developed (reviewed in Lim et al. 2005). Specific capture antibodies are arrayed on a solid matrix— a glass slide, a micro-bead housed in a chip, a capillary tube, or a fiber optic waveguide. The sample is passed over the capture matrix, and a second labeled detection antibody is reacted with the matrix. If the antigen is bound by the capture antibody and then reacts with the detection antibody, the specific locus on the array or specific capillary or waveguide will yield a signal. We and our collaborator, Daniel Lim at the University of South Florida, are developing tools to be used in conjunction with a fiber optic waveguide (DeMarco et al. 1999; Lim et al. 2005). The system works as a capture assay utilizing the principles of a sandwich immunoassay. The waveguide is coated with a primary capture reagent. When a sample is passed over the waveguide, antigen is captured onto the surface. Captured antigen is then reacted with a secondary detection reagent normally coupled to a fluorophore. A laser is passed through the waveguide, and if it encounters the fluorophore on the detection reagent that has been captured onto the waveguide, a fluorescent signal is generated that is detected via the waveguide. Such a system has been produced by the U.S. Naval Research Laboratory (Washington, D.C.) in a field-ready portable unit called the RAPTOR (Anderson et al. 2000).

Table 28.2. Antigen detection assay systems

Method	Signal
Conventional: ELISA, fluorescence immunosorbent assay (FISA)	Enzyme – color, chemiluminescence; fluorescence
Antibody arrays	Fluorescence
Fiber optic	Fluorescence
Surface plasmon resonance	Change in angle of reflected light
Piezoelectric	Change in resonant frequency of a piezoelectric cantilever
Electrical impedance/conductance	Change in conductivity
Surface acoustic wave and Love wave	Change in frequency of elastic acoustic wave

There are inherent limitations to the hardware used for such detection devices, but the antibodies or similar reagents used for capture and detection are of critical importance. The ability to modify the antibodies to improve their usefulness is also worth considering. With standard antisera and monoclonal antibodies, the proteins must be chemically labeled with the appropriate tags or ligands, such as fluorophores, enzymes, or chemicals such as biotin.

Technologies are being developed that do not rely on labeled reagents generating detectable signals or sandwich assays involving capture and detection reagents. Instead, these technologies depend on a signal generated by the interaction of the probe and its target, which can be detected as changes in either mass, frequency of oscillation, capacitance, resistance, surface plasmon resonance (SPR) or interference of acoustic waves (Table 28.2). SPR detects very subtle changes in electromagnetic characteristics of metal/dielectric interfaces (Lim et al. 2005). This method involves coating a metal surface with a bait ligand and then passing a solution with the prey ligand, so that binding of the prey to the bait can be detected by changes in reflection of light at the surface. For example, Rasooly et al. (2001) used SPR to detect as little as 10 ng staphylococcal enterotoxin B per gram of food. Likewise, Subramanian et al. (2006) used SPR to detect *E. coli* 0157:H7. A similar technique involving resonant mirrors couples aspects of SPR with waveguides for sensitive detection (Watts et al. 1994). This system was initially used to detect *Staphylococcus aureus* (Watts et al. 1994). Piezoelectric cantilever biosensors are based on the ability of certain crystals to generate electromagnetic activity when they are deformed (Kim et al. 2003). In some systems, subtle changes in the biosensor upon binding of the target to the ligand are detected as changes in resonant frequency in the crystal complex. A piezoelectric system was originally used to detect *Salmonella enterica* serovar Typhimurium (Kim et al. 2003) and subsequently used to detect *E. coli* O157:H7 in ground beef and *Bacillus anthracis* spores (Campbell et al. 2006; Campbell, Uknalis, Tu and Mutharasan 2006). Electrical impedance/conductance biosensors are based on bridging a gap in an electrical circuit using a sandwich assay system. A capture reagent is coated in the gap. If antigen is captured, it is detected with a second reagent, usually an antibody, that is coupled with a conducting material. Sufficient binding of target antigen by the conducting detection reagent generates an electrical signal. Muhammad-Tahir and Alocilja (2003) used such a system to detect *E. coli* O157:H7 and *S. enterica* serovar Typhimurium. Surface acoustic wave and Love wave immunosensors are based on interfering with a guided horizontal acoustic wave (Moll et al. 2006). The Love wave sensor was initially used to detect *E. coli* cells (Moll et al. 2006).

All of the assay systems for detecting bacterial antigens mentioned above have one thing in common—the need for specific reagents to recognize the antigen and generate a signal. This review discusses phage display technologies to produce and isolate such reagents, either as antibody fragments or random peptide sequences.

3. Phage Display

A solid understanding of phage display technology requires a firm grasp of phage biology. Bacteriophages (phages) are viruses that infect bacteria. As is the case with animal viruses, phages come in a variety of shapes, sizes, and genomic forms. For a review of phages, see Campbell (2003). Similarly to animal viruses, phages are limited in their host range, usually infecting a single species of bacteria and often only a subset of strains of the species. This fact has practical considerations for phage display in that phage display systems require working with specific bacterial strains. For example, the phage display systems discussed in this chapter work with certain strains of *E. coli*.

Phages bind to a specific receptor on the surface of the host bacterial cell, and the phage genome is introduced into the host. Some phages (lytic phages) immediately reproduce and lyse the host cell, while others (temperate phages) may enter into a quiescent state in which the phage genome is either integrated into the host genome or exists as a plasmid and is maintained throughout the growth and replication of the host cell with minimal phage gene expression in a process called lysogeny. Phages in the lysogenic state may convert to a lytic phase under certain conditions, such as damage to host cell DNA. Although any phage can theoretically be used for phage display, the most commonly used are filamentous phages. These phages do not lyse the host cell but instead convert the cell into a phage-producing factory with inhibited bacterial growth.

3.1. Phage M13

Phage M13 belongs to the Ff family of filamentous phages, which also includes f1 and fd. For reviews of M13 see (Webster 2001; Kehoe and Kay 2005). These phages are specific to *E. coli* strains that harbor the conjugative F plasmid because the F plasmid-encoded pilus is the receptor for the phages. This fact becomes important when choosing host strains for use with phage display. As a filamentous phage, the structure of M13 is somewhat different from other phages in that the phage particle lacks a head and tail motif. Instead the phage consists of a circular single-stranded DNA genome complexed with a major coat protein and a few minor structural proteins at either end (Fig. 28.1). The 6.4-kb genome encodes 11 genes, five of which contribute to the particle structure. There are nearly 3,000 copies of the major coat protein, pVIII, bound to the DNA genome. The pIII protein found at one end of the particle is present in about five copies and is responsible for binding of the phage to the F pilus receptor. It also has a role in release of the phage particles from infected *E. coli* cells. The cap is composed of a few copies of the pVII and pIX proteins.

Certain characteristics of M13 make it useful for phage display, among other molecular genetic manipulations. Ff phage replication and assembly is tolerated by the infected bacterial cell. Infected cells continue to grow but at a lowered rate of division. Therefore, infected *E. coli* cells producing M13 phages appear as turbid plaques in confluent lawns of bacteria growing on agar plates. In contrast, clear or turbid plaques resulting from lytic or temperate phages, respectively, are formed when infected bacteria die. The nonlytic nature of the M13 phage–host interaction means that infected *E. coli* strains can be grown and propagated for some time. Second, the replicative form of the M13 phage genome is a double-stranded circular DNA molecule, essentially a plasmid. Therefore, it is possible to purify the replicative form from infected cells and perform DNA manipulations that one typically performs on a plasmid. In fact, some of the earliest cloning vectors were derivatives of M13 (Messing 1991). As noted below, particularly for random peptide reagents, the ability to manipulate the phage genome is a critical aspect of the system. Finally, because the phage genome is simply bound by coat protein, as opposed to filling a phage head structure, there is not a strict limit on the size of the packaged DNA. This enables leeway in the size of recombinant tools.

3.2. Principles of Phage Display

Phage display was created by George Smith in 1985 (Smith 1985). Wilson and Finlay (1998) have written an excellent review of the detailed history of phage display, and several more recent reviews have been published (Pini and Bracci 2000; Azzazy and Highsmith 2002; Carmen and Jermutus 2002; Kehoe and Kay 2005). Phage display advantages are summarized in Table 28.1. Phage display is based on the idea that a binding ligand can be expressed on the surface of a phage particle and that the DNA encoding the ligand is part of the phage genome. Hence, selecting for a phage particle expressing a particular binding ligand essentially selects for the gene encoding the ligand. Having the gene for the binding ligand enables manipulation

Structure

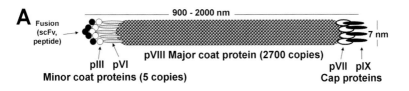

Genetics

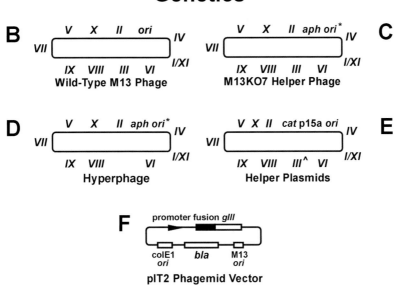

Figure 28.1. Structure and genetics of phage M13. Part A depicts the basic structure of phage M13. Part B: The 11 genes and origin of replication (*ori*) are shown for the wild-type M13 phage. Part C: In the M13KO7 helper phage the M13 *ori* is interrupted by the p15a plasmid origin of replication (*ori**) and an *aph* gene encoding kanamycin resistance. Part D: Hyperphage consists of the M13KO7 helper phage with the *gIII* gene deleted. Part E: The helper plasmids consist of the M13 genome with a p15a plasmid *ori* and *cat* gene encoding chloramphenicol resistance replacing the M13 *ori*. The *gIII* gene can be intact, truncated, or completely deleted (*III*^). Part F: The pIT2 phagemid consists of a colE1 plasmid *ori* with a *bla* gene encoding ampicillin resistance, the M13 *ori*, a *lac* promoter driving expression of scFv sequences fused to the *gIII* gene.

of the gene, for example, fusing the gene to a reporter cassette or performing in vitro evolution to increase the affinity of the ligand. To produce large quantities of phages for analysis of binding characteristics, one simply infects *E. coli* with the phage and harvests phages from the supernatant. Consequently, the self-propagating nature of phages is an important quality. Bacterial culture is simple and inexpensive compared with mammalian cell culture required for producing monoclonal antibodies. Another major advantage in generating phage display antibodies is that they do not involve the use of animals, if utilizing existing libraries. In contrast, production of monoclonal antibodies requires immunization of animals to stimulate the clonal selection and amplification of B cells to produce specific antibodies. Generation of immune responses to toxic substances or highly lethal infectious agents requires the use of inactivated toxins or killed microbes, and these treatments could adversely affect the experimental results. Phage display can be a more efficient method because clonal selection for target-binding members of the library is performed in vitro from randomly generated amino acid sequences or randomly assorted antibody fragments; hence, highly toxic substances or virulent infectious agents can be used directly to obtain the desired reagents.

How are binding ligands generated on the surface of the phage particle? Quite simply, the genes for major or minor coat proteins are fused with DNA sequences encoding the binding ligands. We will discuss two widely used systems: random peptides and antibody fragments. M13 phage has several major and minor coat proteins, all of which have been used as fusion partners for phage display; however, the F pilus-binding pIII is the most widely used today (Gao et al. 1999). Because the pVIII major coat protein comprises the vast majority of the phage structure, the deleterious effects of altering the peptide sequence and structure are more profound on phage stability and structure (Iannolo et al. 1995). It is believed that up to eight amino acid residues can be fused to pVIII, sometimes called landscape phage display, without seriously compromising its structure-function. Another disadvantage to using pVIII as the fusion partner is that M13 phage particles are often detected using antibodies to the pVIII major coat protein, and genetically altering pVIII could interfere with detection of the phages. The major advantage to using pVIII as the fusion partner is that the fusion will be present in very high copy with correspondingly high avidity. However, having the fusion represented nearly 3,000 times on the surface of the particle could, in theory, make selection of high versus low affinity sequences less effective.

The pIII protein offers numerous advantages and fewer disadvantages compared with pVIII. Of critical importance is the fact that relatively large amino acid sequences can be added to the N-terminus of the protein and the protein will retain its F pilus-binding activity. If this were not the case, constructing pIII fusions would render the phage uninfective and therefore nonpropagatable. Because pIII is less important structurally to the phage particle, it can tolerate larger fusion sequences as well. Finally, having five copies of the protein on the tip of the phage particle enables multivalency of binding for sensitivity balanced with the ability to select for high affinity ligands.

3.3. Phages Versus Phagemids

Before delving into the details of phage display, it is important to note that there are two very different systems widely in use that produce recombinant pIII proteins on the surface of M13 phage particles—recombinant phage libraries and recombinant phagemid libraries. The phage libraries, exemplified by the random peptide-pIII fusions available commercially (Ph.D. System, New England Biolabs), consist of a library of M13 phages with some modifications to the phage genome. The most important modifications are random sequences of seven or twelve amino acids fused to the N-terminus of pIII. Additionally, the *lacZα* gene fragment was inserted to enable screening for recombinant phages versus other phage contaminants when examining plaques (the recombinant phages form blue plaques under suitable conditions with the appropriate *E. coli* host strain). Therefore, this system almost entirely involves using phages from the library in rounds of panning (selecting phages with appropriate binding characteristics) and screening phages with desired specificities coupled with propagation of the phages in appropriate *E. coli* hosts.

In contrast, the single chain F variable (scFv) human antibody fragment Griffin.1 and Tomlinson libraries (Griffiths et al. 1994; de Wildt et al. 2000) are constructed in plasmids called phagemids because they replicate in bacteria as plasmids but can also be packaged into phage particles (Marks et al. 1991) (Fig. 28.1). These plasmids are essentially cloning vectors that contain the M13 *gIII* gene encoding the pIII protein with N-terminal scFv fusions, as discussed below, and the origin of replication of the M13 phage. Therefore, to create phage particles displaying the scFv-pIII fusions, *E. coli* possessing the phagemids must be infected with a helper M13 phage that provides all of the phage genes and their functions required for production of M13 phages.

The requirement for use of helper-type phages to produce the actual recombinant M13 phages when using phagemids raises several practical considerations. One of the most widely used helper phages is M13KO7. It consists of M13 phage with a kanamycin resistance gene, *aph*, added to the genome (Fig. 28.1). Therefore, to produce phages from an *E. coli* strain containing scFv-pIII fusion-encoding phagemids, such as with the Griffin.1 and Tomlinson libraries, the *E. coli* cells are infected with M13KO7, and the culture is grown in the presence of kanamycin. The M13KO7 helper phage will produce all of the structural and accessory proteins required for phage production, including more M13KO7 helper phage. The phagemid will produce the scFv-pIII fusion protein, which will be incorporated into the progeny phage. Finally, the M13 origin of replication on the phagemid will result in the phagemid being packaged in a phage particle. Therefore, the ideal outcome is the production of phage particles bearing scFv-pIII fusion proteins on the tip of the phage with the scFv-pIII-encoding phagemid packaged inside. Hence, when the phage particle of interest is isolated by virtue of its binding to the target antigen, the gene encoding the binding ligand has been captured as well.

Unfortunately, this represents an ideal that cannot be met. As noted, the M13KO7 helper phage reproduces itself, hence energy is wasted on production of helper phages. Production of helper phages along with the phage display phages is not as serious a problem as it may seem because when phages with appropriate binding activity are isolated, they are infected into *E. coli* that is grown in the presence of ampicillin, which selects for phagemids only. However, because the helper phage encodes its own wild-type *gIII gene*, the progeny phages from the helper phage superinfection consist of both scFv-pIII fusions encoded by phagemid and wild-type pIII encoded by the helper phage. Thus, a mixture of phage particles exists that can contain anywhere from zero to five recombinant pIII proteins. This detracts from the inherent advantages of having several copies of the pIII protein on each phage particle, because some copies of the protein will be wasted as wild-type pIII.

To overcome this problem, various forms of M13 helper systems have been created (Fig. 28.1). One system which is routinely used in our laboratory is Hyperphage (Broders et al. 2003) marketed by Progen (Heidelberg, Germany). Hyperphage is simply the M13KO7 helper phage with the *gIII* gene deleted. Therefore, when Hyperphage is used to produce M13 phages using pIII fusion phagemids, the only *gIII* gene available for pIII production is the recombinant *gIII* of the phagemid. Hyperphage is produced by pseudotyping the phages with pIII encoded by a *gIII* gene integrated into the chromosome of a specialized *E. coli* production strain. We and other researchers have found, however, that such pseudotyped helper M13 phages can be problematic in that they often yield lower levels of functional phages than desired. This is an important issue because libraries of recombinant phagemids can consist of 10^8 to 10^{10} different clones, and amplification of the libraries requires high quantities of helper phages to ensure that every clone retains its representation in the library. Finally, because Hyperphage is also a packageable M13 genome, Hyperphage pseudotyped with the recombinant pIII fusion will also be produced from cultures whose goal was the production of only packaged phagemids. As for M13KO7 helper phage, such phages are selected against when phagemids encoding desired specificity are infected into *E. coli* and grown with ampicillin.

To circumvent problems with pseudotyped, *gIII*-deleted helper phages, Baek et al. (2002) produced a helper phage, named *Ex-phage*, that contains an amber stop codon engineered into the *gIII* gene, so that the helper phage can only be produced in *E. coli* possessing appropriate suppressor mutations. In suppressor-free *E. coli* strains in which phagemids will be packaged, the helper phage pIII will not be translated, hence only recombinant pIII proteins will appear on the progeny phage. The Ex-phage system has a serious drawback in that the widely used Griffin.1 and Tomlinson scFv-pIII libraries contain an amber stop codon engineered in between the scFv peptide and the pIII-encoding sequences so that scFv protein can be produced out of the context of the pIII protein by expressing the gene in a suppressor-free *E. coli* strain. However, this means that to produce pIII fusions in progeny phages for panning and propagation, the phagemid must be placed in an amber suppressor-containing *E. coli* host, and the Ex-phage will therefore produce the wild-type pIII protein to compete with the recombinant pIII protein. Therefore, Ex-phage could be very useful for phagemids that do not rely on amber stop codons between the binding peptide sequences and the *gIII* sequences. Most recently, a third solution to the helper phage problem was reported by Chasteen et al. (2006). Instead of using helper phage to package phagemids, they created a series of M13-derived helper plasmids that encode either full length, truncated, or no pIII (Fig. 28.1). The use of a truncated pIII protein offered a high yield of recombinant helper display particles with antigen-binding qualities. Unlike helper phage and Hyperphage, the plasmids are not packaged.

3.4. Phage Display Formats

Having described the essentials of phage display biology and genetics, we will now discuss some widely used pIII fusions. All of these involve the insertion of DNA sequences encoding target-binding peptide sequences to the *gIII* gene of M13. All of the fusions that we are aware of are at the N-terminus. The most widely used fusions consist of either random amino acids or various fragments of antibodies. It should also be noted that the target-binding phage, itself, is not usually the ultimate product of phage display. Instead, it is usually the peptide sequence or antibody fragment expressed independently out of the context of the phage that is the ultimate tool. Therefore, when a peptide with desired binding characteristics is identified, the DNA sequence encoding the peptide is constructed into appropriate expression vectors, or the peptide is chemically synthesized. Similarly, when a useful antibody fragment is identified, it is expressed out of the context of the phage as an independent protein or the DNA sequence is cloned into a suitable expression vector for increased utility. That is why the possession of the DNA sequence encoding the peptide or antibody fragment is one of the major advantages and powers of phage display.

3.4.1. Random Peptides

Because of their simplicity, we will first discuss random peptides. Random peptide libraries are commercially available from New England Biolabs as the Ph.D. (Phage Display) systems. These are phage libraries, as opposed to phagemid libraries. The two simplest libraries consist of either seven or twelve amino acids with a short spacer fused to the N-terminus of the mature pIII protein of the M13 phage genome (Table 28.3). The fusion *gIII* gene is expressed from its native promoter. A third library consists of seven amino acids containing two cysteine residues to enable a disulfide loop to form in the random amino acid sequence. This loop is to enable conformational, as opposed to simple linear, determinants to form. When target-binding peptides are identified, they are either chemically synthesized for use by themselves or coupled with other binding (e.g., biotin) or signal-generating (e.g., Cy3) components for use. As noted below, it is rare that the phage particle bearing the target-binding peptide is used as the ultimate reagent.

3.4.2. Antibody Fragments

A common motif in designing binding proteins in any context is a constant peptide framework with a variable binding domain or domains. Antibody molecules fit this motif perfectly. Antibodies consist of constant and variable regions present on two different peptides, the heavy and light chains, with a combined mass of 150 kDa (Fig. 28.2). The antigen-binding domains are spread among sequences on the variable portion of the heavy and light chains. The variable sequences are called the complementarity determining regions (CDR) and are created

Table 28.3. Phage display detection motifs

Format	Frame	Comments
Random peptide	None	7-mer, 12-mer, C-C loop commercially available
Antibody Fab	Immunoglobulin	Intact antigen binding domain in native conformation
Antibody scFv	Immunoglobulin	Partial antigen binding domain in engineered conformation
Affibody	*Staphylococcus aureus* protein A, Z domain fragment	Flat surface with randomized sequences
Anticalin	Lipocalin protein	Cavity with randomized sequences
Ankyrin repeat	Ankyrin family of stacked proteins	Loop and helix with randomized sequences
Trinectins	Fibronectin, tenth type three domain (Fn3)	Loop with randomized sequences

by several different genetic events, including combinations of different gene sequences and mutation. There are theoretically several different portions of antibody molecules that can be expressed in phage display. There are practical considerations, however, that limit the portions that can be used effectively. First, because the antibody molecule consists of two peptides, both of which are required for function in antigen binding, getting the two peptides expressed in the same *E. coli* cell and then combined on the phage particle would be extremely difficult. The very large size of the antibody molecule also presents a problem. As noted above, different phage proteins are capable of tolerating only limited sizes of fusion peptides. Because most investigators are interested in only the antigen binding activity of the antibody molecule, there is no need to include the two C-terminal constant regions of the heavy chain. The next least complex antibody fragment for consideration is the Fab, which consists of the entire light chain and the variable and constant region 1 of the heavy chain linked by a disulfide bond (Fig. 28.2). Although size is less of an issue with Fab, the requirement for producing and linking two

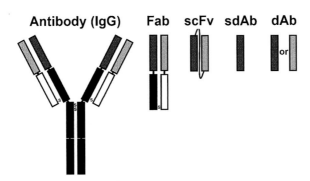

Figure 28.2. Antibody molecule structures: different portions and constructs of an immunoglobulin molecule. On the whole antibody (IgG) molecule, the heavy chain is black and the light chain is white. The constant regions are solid, and the variable regions are stippled. Interchain disulfide bonds are shown as *S*. The Fab molecule is composed of the N-terminal variable and constant domains of the heavy and light chains. The scFv (single chain F variable) molecule is a recombinant construct composed of the variable regions of the heavy and light chains linked by a synthetic linker peptide. Single domain antibodies (sdAbs) are variable heavy fragments from animals whose immunoglobulins naturally consist of only heavy chains. Domain antibodies (dAbs) are either the variable heavy or variable light fragment of immunoglobulins that normally consist of both a heavy and light chain.

peptides to form the active protein is problematic in phage display. There have been attempts at producing Fab in *E. coli*, but there are technical limitations and such libraries are not widely available (Skerra and Pluckthun 1991; Arndt et al. 2001).

The most widely used antibody fragment in phage display is the single chain F variable (scFv), which is composed of the F variable regions of both the heavy and light chains joined by a flexible glycine-serine amino acid linker (Griffiths et al. 1994) (Fig. 28.2). This peptide is relatively compact at about 27-kDa and has the advantage that it possesses both binding domains, hence there is no need for assembly of the antibody fragment post-translationally. It is not the intention of this chapter to go into the detail of how these libraries are made. Suffice it to say that Fv sequences of the heavy and light chains (as appropriate) are reverse transcribed from B cell mRNA, PCR amplified, and then cloned into phagemids constructed to enable the linkage of the Fv light and Fv heavy sequences with the pIII protein. By introducing increased genetic variability in the CDR sequences of the amplified immunoglobulin sequences, the diversity of the resulting libraries can be increased significantly in so-called semi-synthetic or synthetic libraries (Pini and Bracci 2000; Azzazy and Highsmith 2002; Carmen and Jermutus 2002). Phage display libraries have also been made using the Fv peptides that are not joined by a linker, but their inherent instability makes them less useful (Glockshuber et al. 1990). More recently, single domain antibodies (sdAbs) (Goldman et al. 2006; Liu et al. 2007) and domain antibodies (dAbs) (Holt et al. 2003) have been developed into phage display tools. As discussed below in the review of the literature, sdAbs are antibodies that consist of only a heavy chain naturally produced by some animals. dAbs consist of only the variable heavy chain fragment of antibodies that normally consist of both heavy and light chains; hence, they are essentially half of an scFv (Holt et al. 2003). As noted below, the ultimate tool is usually not a phage particle, but the scFv protein. To facilitate production of the scFv protein, the Griffin.1 and Tomlinson libraries encode an amber stop codon between the scFv sequence and the *gIII* sequence. In an amber suppressor *E. coli* strain such as TG1, enough fusion protein is made to enable the production of functional M13 phages with recombinant scFv-pIII at their tips. In contrast, in a suppressor-free strain such as HB2151, the amber stop codon is not suppressed and the scFv protein is produced by itself and secreted into the *E. coli* periplasm and/or culture supernatant. The N-terminus of the phagemid-encoded scFv-pIII fusion contains the PelB leader sequence, which enables the secretion of the peptide. The C-terminus of the scFv sequence also contains a poly-His tag for affinity purification on nickel matrices, a c-myc tag for detection with appropriate anti-c-myc antibodies, and for the Tomlinson libraries a trypsin-cleavable site between scFv and pIII sequences. The use of this protease site in working with scFv systems is discussed below.

A critical issue in working with antibody phage display is the source of the antibody sequences cloned for use. The Griffin.1 and Tomlinson scFv libraries were obtained using B cells from humans who were not specifically immunized with the intention of producing antibodies with a given specificity (Griffiths et al. 1994). Based on extensive analysis of the human immunoglobulin gene families, a synthetic human combinatorial antibody Fab library, HuCAL, was constructed (Knappik et al. 2000). These libraries have broad potential specificity, but the probability of finding reactive clones with high affinity will be lower. In contrast, if one immunizes an animal or human before obtaining B cells, the probability of obtaining specific clones with higher binding affinity is greater. However, to make an immunized library means that one is now dealing with animals, lengthy immunization protocols, and limitations from using self or toxic antigens, all of which are detriments in performing typical antibody generation (Table 28.1). Alternatively, obtaining lymphocytes from immune individuals due to natural infection bypasses issues with immunization. Additionally, creating immunized libraries requires going through all of the molecular genetic steps in library construction. Most investigators would rather not go through this tedious process and would rather simply purchase or acquire a pre-existing library. However, as is evident below, in terms of the publication record, most of the success in working with scFv libraries has come from immunized libraries.

3.5. The Phages Themselves Are Not the Ultimate Tool

It is very unlikely that the phages will be of ultimate use in detection systems. Although phages are among the most stable viruses, they are complex particulate entities that are subject to degeneration and degradation. Their large size also limits their use because it is difficult to obtain large molar quantities of phage, as opposed to smaller soluble proteins such as antibodies. However, if one is simply looking for a research tool for the laboratory, phages displaying binding ligands could serve that purpose. We have found this to be the case in our own work. As detailed below, our ultimate goal and purpose in working with phage display reagents are to be able to genetically engineer the tools to optimize their use, such as fusing reporter or binding domains. This would be difficult, if not impossible, to perform when working with whole phages as tools.

Unfortunately, getting scFv peptides expressed and secreted out of the context of the pIII fusion and then to retain antigen binding is very often unsuccessful. The failure of soluble scFvs to recognize an antigen with high affinity is a common problem (Cloutier et al. 2000; Wang et al. 2004a; Suzuki et al. 2005). Some have found that growing the scFv-expression cultures in the presence of L-arginine can increase the yield of refolded proteins by decreasing the quantity of aggregated forms, apparently a common problem with scFv expression (Das et al. 2004). Many evolutionary engineering or antibody reshaping methods have been successfully applied to increase stability and affinity of scFv proteins by using DNA chain-shuffling (Wang et al. 2004b), error prone PCR (Suzuki et al. 2005), or affinity maturation (Irving et al. 1996; Bose et al. 2003; Yau et al. 2005). In contrast, Wang et al. (2004a) applied a simple method to improve binding by mixing low affinity scFv fragments fused with c-myc tags with an anti-c-myc monoclonal antibody to form dimeric antibody fragment-anti-tag monoclonal antibody complexes. Relatively low affinity scFvs can be further manipulated by tetramerizing their biotinylated forms using tetravalent streptavidin forming high avidity streptavidin antibodies, stAbs, that are functional at 100 times lower concentration than unmodified scFv (Cloutier et al. 2000).

3.6. Using Phage Display

Having acquired a suitable phage display library, one finds the clones of relevant reactivity by panning, also called biopanning (Fig. 28.3). The typical basic steps in panning are binding, washing, elution, and optional amplification carried out in repetitive rounds. For binding, the library with considerable redundancy in representation, usually 100-fold, is incubated with the desired antigen. There are four possible methods of binding and washing. The antigen of choice can be bound to a solid support matrix, such as a polystyrene tube or microplate. Some have panned on cultured eukaryotic cells growing in tissue culture plates. The phage library is added to the tube/plate and incubated. Unbound phages are aspirated away, and the tube/plate is rinsed and washed. If the antigen is of a particulate nature, such as whole bacterial cells, the phage display library may be incubated with the cells in suspension. Unbound phages can be removed by pelleting the bacterial cells by centrifugation, and washing is carried out by suspending the cell pellet in wash buffer. Additionally, some have bound antigens to magnetic or epoxy beads. Some have used soluble antigens in solution. This has the advantage that the antigen is not altered by being bound to a solid matrix; hence, more natural conformations might be available to bind with phages. The difficulty in working with antigens in solution is washing. This can be accomplished by including a tag on the antigen, e.g., poly-His tag, so that the antigen with any bound phages can be collected on an affinity matrix, such as a nickel column. Of course, including the tag could compromise the natural structure of the antigen if that was the goal in working with a soluble antigen in the first place. Elution of phages bound to the antigen is effected in a variety of ways, most of which temporarily disrupt the antigen-phage binding using chemical means. The most widely used methods are acidic glycine and alkaline triethylamine. We have found that acidic elution is better when working with gram-negative bacteria as antigens because triethylamine solubilizes the bacteria, resulting in a viscous mixture that might trap the eluted phages. If phages are eluted with acidic or alkaline conditions, they must be neutralized quickly, within ten minutes. Phages can also be eluted by adding excess soluble antigen or hapten to compete the phage off of the immobilized antigen. Elution by pH shock can, in theory, elute both specifically and nonspecifically bound phages. In contrast, antigen or hapten elution should only elute specifically bound phages. One advantage of the Tomlinson I and J libraries is the trypsin-cleavable peptide sequence between the scFv and pIII sequences. Phages bound to the immobilized antigen can be eluted by cleavage with trypsin; however, because the scFv peptide has been removed, the eluted phages can only be used for infecting an *E. coli* host to recover and propagate the phagemid.

An important issue in panning is if the eluted phages/phagemids will be amplified by infection of *E. coli* before the next round of panning or if the eluted phages will be used immediately for the next round of panning without amplification (Fig. 28.3). As noted immediately above, if the phages have been eluted with trypsin, there is no choice but to infect *E. coli*. However, if sufficient numbers of phages were eluted from a round of panning, it is possible to use them directly in the next round of panning. This saves at least a day or two; however, it is balanced by the possibility that useful phages will be lost because they are not represented in the eluted population in sufficient numbers to be recovered from the next round of panning. By monitoring the numbers of eluted phages at each round of panning by infecting *E. coli* and counting plaques for phage libraries or colonies for phagemid libraries, one can determine if

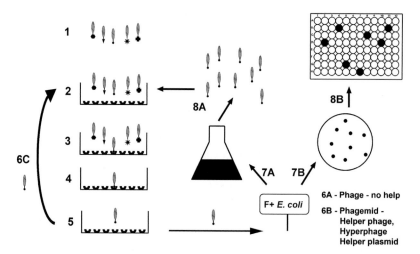

Figure 28.3. General panning and phage manipulation. Step 1 - phage display library. The different binding specificities of the phages are represented by symbols at their tips. Typical libraries consist of scFv, Fab, or random peptides fused to the pIII tip protein of M13 phage. Step 2 - panning the library. In this case, a well coated with antigen is represented as semicircles. Step 3 - one phage that recognizes the antigen binds. Step 4 - After washing, phages recognizing the coated antigen are enriched. Step 5 - bound phages are eluted chemically (acidic glycine or basic triethylamine, among others), or by adding excess soluble antigen to compete for binding of the phages to the antigen on the plate, or by cleavage of the pIII fusion peptide with trypsin. Step 6 - manipulation of eluted phages: 6A - if the fusion is encoded in the M13 phage genome, the eluted phages can be infected into *E. coli* with no need for helper phage; 6B - if the fusion is encoded on a phagemid, the *E. coli* will have to provide phage replication and packaging functions by being infected with helper phage or Hyperphage or having a helper plasmid; and 6C - alternatively, the eluted phages could be used directly in the next round of panning. Step 7-amplification or production of phage for further use. If additional panning is desired (Step 8A), phages are amplified in batch (Step 7A). If individual clones are to be examined for activity (Step 8B), colonies or plaques are picked and amplified individually (Step 7B).

amplification is required or not. We have evolved our panning to always include amplification after the first round of panning, but then optional amplification after subsequent rounds. For amplifying phagemid libraries the eluted phages are used to infect a suitable *E. coli* host strain, and cells acquiring the phagemid are selected by plating on an appropriate antibiotic (ampicillin for Griffin.1 and Tomlinson libraries). To generate phages from the phagemid clones, a helper phage must be used to infect the clones. The considerations in helper phage systems are discussed above. If phage libraries such as the New England Biolabs Ph.D. random peptide libraries are used, the eluted phages are infected into a suitable *E. coli* host, and the phages will amplify themselves. This simplicity is a benefit of phage libraries. In either case, the titer of the amplified phage is determined before further use.

Most procedures suggest three rounds of panning before one begins examining individual clones for their binding characteristics. One can analyze if panning is, in fact, enriching for phages with binding activity by either enumerating the number of eluted phages after each round (assuming that a constant number of phages was used at each panning step) or measuring the collective binding activity of the amplified phages in a suitable assay such as an ELISA. The number of eluted phages or the binding activity of the phages should increase with each round. The goal of the panning is to enrich the population of eluted phage clones with the desired specificity that can be identified by picking and testing individual clones. If a high throughput screening system is available, one can easily screen hundreds of clones to identify individual reactive clones. The number of rounds of panning required for any given project will be determined by the initial frequency of reactive clones (out of the hands of the investigator unless an immunized library is used) and the stringency of the binding, washing, and eluting steps. If one is fortunate enough to not need amplification, three rounds of panning could be accomplished in a single day. Each binding step can be 30 minutes or less, with washes in between binding of about 30 minutes duration, and with elution and neutralization not taking significant time. However, to perform a single step of amplification for a phagemid requires overnight incubation to infect *E. coli* and select for carriage of the phagemid followed by several hours for helper phage infection and growth for production of progeny phage.

Another important consideration in panning is the use of negative or subtractive pannings to increase the specificity of the resulting clones. This is most important when the target antigen is bound to a solid support such as a polystyrene tube or microtiter plate. Because the plastic matrix is usually designed to bind antigens, it must be blocked to prevent nonspecific binding of the phages, as if they were antigen. It is important to note that if blocking is used the blocking agent should be included with the phage suspension during panning to prevent the recovery of phages bound to the blocking reagent on the matrix. Even with blocking, it is highly probable that phages specific for the polystyrene will be obtained. Therefore, it is useful to include in such panning strategies a step in which the phages are incubated in an empty plastic vessel to remove plastic-specific phages. Additionally, if one desires the isolation of reagents specific to a select set of bacterial strains, for example those that express particular virulence attributes, one can pan on the virulent strains followed by subtractive panning on closely related strains lacking the desired attribute. Ideally, an isogenic mutant would enable the subtraction of every clone except those recognizing the desired target.

4. Review of Literature on Phage Display Against Bacterial Pathogens

There is a considerable body of literature from investigators using phage display methods to create reagents recognizing a variety of antigens for a variety of purposes. We highlight those that used phage display specifically to detect antigens on or produced by bacteria and that contribute a novel aspect to the use of phage display.

4.1. Random Peptide Phage Display

Of the numerous publications highlighted here, it will become immediately clear that the most widely used method for random peptide phage display is the Ph.D. system by New England Biolabs.

Kim et al. (2005) used the New England Biolabs 7-mer Ph.D. system to isolate peptides that recognized lipopolysaccharide (LPS) from a variety of gram-negative bacteria. They bound either lipid A, the most conserved portion of the LPS molecule, or whole LPS to epoxy beads for panning. After five rounds of panning, they chose isolated phages, tested their reactivity to LPS in a bead-based ELISA, and then sequenced the positive clones. They obtained two different peptide sequences that bound to LPS with high affinity. Both peptides recognized "polysaccharide" determinants, as opposed to lipid A. Because the reactivity crossed serovar and species lines, the peptides likely recognized the core oligosaccharide. However, when one of the chemically synthesized peptides was conjugated to beads, it was able to bind whole salmonella cells. Because the core oligosaccharide is less accessible on the bacterial cell surface, being covered by the LPS O antigen, this is an interesting result. The authors felt that the use of epoxy beads, as opposed to polystyrene binding matrices, in binding LPS for panning was a key to their success.

In an interesting twist to identifying peptides that recognize a bacterial pathogen, Gasanov et al. (2006), used the New England Biolabs random 12-mer library to pan on whole *Listeria monocytogenes* cells. *L. monocytogenes* is a gram-positive food-borne bacterial pathogen that causes lethal disease in immunocompromised people and pregnant women and their fetuses. Because these investigators were interested specifically in *L. monocytogenes*, they performed subtractive panning of panned clones on *L. innocua* and *L. ivanovii*. These investigators panned on cells attached to microtiter wells. When they examined the amino acid sequences of reactive clones, they observed that they had homology to the insulin-like growth factor II receptor (IGFIIR) on human cells, suggesting that this host protein might serve as a receptor for adherence by the bacteria. They then showed that the bacteria could bind to this protein. Furthermore, binding and invasion of Listeria cells to host cells could be inhibited by either the synthesized peptide or mannose-6-phosphate, which also binds IGFIIR. Although they had performed subtractive panning with *Listeria* species other than *L. monocytogenes*, the 12-mer peptide still bound to other Listerias; however, the peptide did not bind other gram-positive or gram-negative bacteria. It is interesting that even the nonpathogenic *Listeria* species apparently recognize the same receptor on host cells. It should also be noted that *L. monocytogenes* uses other receptors, such as E-cadherin and a methionine receptor tyrosine kinase, recognized by the *L. monocytogenes* virulence factors InlA and InlB (Mengaud et al. 1996; Shen et al. 2000), respectively. This study highlights one of the advantages of using peptide phage display for analysis of bacteria in that determining the amino acid sequence of the binding peptides led to an important biological discovery. Had antibodies or phage display with antibody fragments been used, the identification of the host cell receptor likely would not have occurred.

Ide et al. (2003) used the New England Biolabs 12-mer library to identify peptides that recognize the H7 flagella of *E. coli* O157:H7, otherwise known as Enterohemorrhagic *E. coli* (EHEC). EHEC is the pathogenic *E. coli* that causes hemorrhagic colitis and fatal kidney damage. Obtaining H7-specific reagents is significant because this antigen is shared among most EHEC strains. The investigators panned the library on flagella that had been immobilized to immunotubes. It is interesting that when they eluted with acid, a common procedure, they isolated phages that apparently were specific to polystyrene present in the immunotubes, as noted above. However, when they changed to elution with antigen, the H7 flagellin protein, they obtained specific clones. Sequence analysis of four clones failed to reveal a consensus, other than numerous proline residues. The best peptide demonstrated specific binding to the H7 flagella and could bind to intact *E. coli* cells expressing the H7 flagella.

Stratmann et al. (2002) used the New England Biolabs 12-mer library to isolate peptides that recognized *Mycobacterium avium* subspecies *paratuberculosis* (*M. paratuberculosis*), an acid fast pathogen found in milk that is associated with a chronic inflammatory bowel disease of animals called Johne's disease, which might be related to the human disease called Crohn's disease. They panned on whole bacterial cells bound to microtiter wells. One phage clone demonstrated high affinity to whole *M. paratuberculosis* cells. These investigators bound the peptide to magnetic beads and used the complex to capture *M. paratuberculosis* cells from both experimentally and naturally contaminated milk followed by PCR to detect the bacteria. This system demonstrated specificity and great sensitivity, as low as 10 CFU/mL of milk.

Turnbough and colleagues (Williams et al. 2003; Knurr et al. 2003) used the New England Biolabs 7- and 12-mer libraries to isolate phage clones that recognize spores of a variety of species of *Bacillus*, most importantly *B. anthracis,* of relevance for bioterrorism. In the first study, the 7-mer library was panned on whole *B. anthracis* spores in suspension, and numerous phages with a consensus sequence at the amino terminus were identified. One peptide was synthesized and conjugated to a fluorophore, and this conjugate demonstrated reasonable specificity to *B. anthracis* spores. In a follow-up study, this group isolated additional phage display peptides that recognized related *B. subtilis* spores, but not *B. anthracis* spores. This is important because if anti-spore reagents are to be used to detect acts of bioterrorism, they must be able to discriminate between the pathogenic *B. anthracis* species and other nonpathogenic and easily available species.

Mourez et al. (2001) used the New England Biolabs 12-mer library to isolate phages that recognized and even neutralized anthrax toxin. Although not technically used for detection of bacterial pathogens, this study demonstrates the usefulness of phage display technology in infectious diseases. These investigators panned the library on the cleaved heptamer of protective antigen (PA), which is responsible for binding and internalization of the two different forms of anthrax toxin, lethal toxin and edema toxin. They subtractively panned on intact PA to avoid peptides that recognize the unprocessed protein. One of the resulting two clones inhibited the binding of lethal factor to the PA heptamer, thus preventing the formation of the active toxin on the surface of host cells. A multivalent peptide derivative could prevent lethality to rats injected with lethal toxin. The ability of the peptide to inhibit formation of edema toxin from PA and edema factor was not reported. It will be interesting to see if these peptides ultimately have clinical use.

Goldman et al. (2000) used the New England Biolabs 12-mer library to isolate peptides with affinity for SEB, a toxin associated with food poisoning. After three rounds of panning on immobilized SEB, two distinct, yet highly similar, peptides were identified from nine phage clones selected for analysis. These phage clones demonstrated affinity for immobilized SEB by ELISA, and the synthesized 12-mer peptides possessed binding activity, which is not always the case when the peptide that was initially identified as part of a phage is synthesized independently. The clones possessed significant cross-reactivity with Staphylococcal enterotoxin C (SEC), which was not unexpected since SEC has 60 %–70 % homology with SEB. Interestingly, the authors fluorescently labeled the phage particles expressing the 12-mer of one of the anti-SEB clones and demonstrated its usefulness in both a microtiter immunofluorescence assay and using the RAPTOR fiber optic biosensor.

Houimel et al. (1999) created 25-mer and 6-mer peptide libraries in a phagemid vector and isolated two peptide sequences that bound to and neutralized the urease of *Helicobacter pylori*, the causative agent of gastric ulcers. Urease is an enzyme that is important for pathogenesis as well as detection of *H. pylori* infection. It will be interesting to see if such urease-neutralizing peptides achieve clinical use.

Bishop-Hurley et al. (2005) used a 15-mer random peptide phage display library made by G.P. Smith to isolate phages recognizing nontypeable *Haemophilus influenzae* (NTHI), a gram-negative bacterium that causes respiratory disease. The phages demonstrated a unique characteristic among phage display reagents discussed here—they were bactericidal. The phage library was subtractively panned against a nonpathogenic NTHI strain before being panned against a pathogenic NTHI strain. As opposed to the New England Biolabs libraries based on phage M13, this noncommercial library was based on another filamentous phage, fd. Their initial screen after three rounds of panning was to examine the ability of individual phages to inhibit growth of NTHI. The synthesized 15-mer peptides did not possess bactericidal activity. This could have been due to conformational changes in the peptides under the different conditions of fusion versus free peptide.

Sorokulova et al. (2005) panned against *S*. Typhimurium whole cells using an fd phage random 8-mer landscape phage peptide library. As noted above, landscape phages have fusions to the major coat protein, pVIII, as opposed to pIII protein located on the tip. As a result, fusions are more highly represented on each phage particle. Different panning methods were used with regard to immobilized versus suspended Salmonella whole cells, concentration of Tween-20, number of rounds of panning, and subtractive selection steps. One phage demonstrated high affinity to and specificity toward Salmonella whole cells. In a follow-up study, Olsen et al. (2006) used that phage in an acoustic wave biosensor and obtained a limit of detection of 100 Salmonella cells/ml, which is very low.

This same group, Brigati et al. (2004), again used the landscape phage library to pan against anthrax spores. After four rounds of panning, 16 clones were examined further. Eleven different sequences were identified and placed into three families based on homology. Some of the phages were capable of capturing *B. anthracis* spores with reasonable specificity. Analysis of the specific peptides was not reported.

Ma et al. (2006) constructed several cyclic peptide libraries, i.e., the random amino acids were flanked by cysteine residues to form disulfide loops, constructed in the pIX protein and panned against *Clostridium botulinum* neurotoxin. After four rounds of panning on botulinum neurotoxin A, 48 clones were examined for reactivity by ELISA, and a single clone was further analyzed. The clone was specific to neurotoxin A, not recognizing other botulinum neurotoxins. The synthesized peptide was not effective in surface plasmon resonance assays, so it was conjugated to a polymer matrix to increase valency. After confirming the usefulness of the polymer matrix peptide conjugate in surface plasmon resonance, it was used in a sandwich ELISA with the minimal detection level being 1 pg/ml in buffer and 1 ng/ml in body fluids and foods.

4.2. scFv Libraries

There are more publications involving scFv phage display libraries compared to random peptides. Because these libraries are based on antibody fragments, it is possible to enrich for DNA sequences encoding appropriate antigen-binding domains by cloning sequences from immunized animals or humans. In fact, some libraries have been constructed from monoclonal antibody-secreting hybridoma cell lines. Others have used the available semisynthetic Griffin.1 or Tomlinson I and J libraries.

Berger et al. (2006) cloned DNA sequences from sheep with Johne's disease, caused by *M. paratuberculosis*, to create an enriched scFv library fused to the *gIII* gene of phage M13. After two sets of extensive panning, including subtractive panning with subspecies not of interest, and screening of scFv clones, two promising candidates demonstrated binding specificity for *M. paratuberculosis* and *M. avium* subspecies *avium*. The clones recognized a 34-kDa protein. As was the case for the anti-*M. paratuberculosis* peptides discussed above, the scFvs were useful in magnetic bead enrichment procedures coupled with PCR.

Griep et al. (1998) used a combinatorial human nonimmunized scFv library (Vaughan et al. 1996) to isolate scFvs that recognize the LPS of *Ralstonia solanacearum* race 3, which is a plant pathogen. Panning on LPS bound to immunotubes resulted in isolating scFvs that bound to LPS and whole bacterial cells. These investigators performed an extensive series of cross reactivity studies, and the scFvs demonstrated better detection activity than the currently used polyclonal antiserum because of less cross reactivity with irrelevant strains. The same group then genetically fused the scFv DNA sequences to the gene encoding modified green fluorescent protein to yield a single reagent to detect bacteria by immunofluorescent staining or flow cytometry (Griep et al. 1999).

Sabarth et al. (2005) investigated *H. pylori* for surface antigens. To aid in their investigations, they created their own scFv library by immunizing mice with *H. pylori* antigens and cloning immunoglobulin gene fragments into a phagemid vector. They panned on whole *H. pylori* cells in suspension and ultimately ended up with a single clone that recognized whole cells. To identify the antigen recognized by the clone, they coupled the scFv protein to an affinity column matrix and affinity-purified a protein that was identified as HopQ, an outer membrane protein.

Reiche et al. (2002) took a novel approach to isolate scFvs against *H. pylori* by creating an antibody library from human patients with *H. pylori* infections. The library was panned on bacterial cell lysates and Helicobacter urease. Two clones to uncharacterized cellular antigens and a single clone to urease were identified. Houimel et al. (2001) had previously isolated scFvs to the *H. pylori* urease from an unimmunized human library. Their scFvs also demonstrated enzyme-inhibiting activity.

An even more focused and unique approach was taken by Kuepper et al. (2005), who desired to create a humanized version of a murine monoclonal antibody that recognized *Streptococcus mutans*, the causative agent of dental caries. Their ultimate goal was the pursuit of immunotherapy for caries, and the murine monoclonal antibody Guy's 13 had efficacy in preventing *S. mutans*-mediated caries in a mouse model. Therefore, these investigators cloned the variable light chain (V_L) DNA sequences from the murine hybridoma into a phagemid vector that contained a random human variable heavy (V_H) library and panned on streptococcal antigens. In this manner, a hybrid scFv was created that possessed both human and murine (Guy's 13) sequences. Several representative clones were chosen to donate the isolated human V_H sequence back into a phagemid vector that contained random human V_L sequences. These now fully human libraries were again panned on streptococcal antigens, and three different clones were isolated. The conservation of specificity relative to the original Guy's 13 murine monoclonal antibody was confirmed in an ELISA using the monoclonal antibody to inhibit the binding of the scFvs to the streptococcal antigens. Because the investigators desired to have bivalent, as opposed to monovalent antibodies, they cloned the V_L and V_H sequences into a new vector that changed the spacer between these sequences so that the only way that functional antibodies could be made was by two peptides annealing to each other, so called *diabodies*. The diabodies demonstrated ELISA activity against streptococcal antigens and agglutinated streptococcal cells, confirming their bivalent nature.

Paoli et al. (2004) used the Griffin.1 library to isolate a set of phages that bound to whole *L. monocytogenes* cells. As was done by others using peptide libraries (see above), they performed subtractive panning on non-*L. monocytogenes* cells. A single scFv clone was isolated that demonstrated reactivity against only *L. monocytogenes*. The target of the phage-scFv was not reported and neither was the breadth of reactivity against the different serotypes of *L. monocytogenes*.

Hayhurst et al. (2003) made an scFv library from immunized mice to isolate scFv clones to *Brucella melitensis*, a zoonotic pathogen that could be used for bioterrorism. The library was panned on irradiated whole cells bound to polystyrene plates. As has frequently been our experience, the scFv protein, when expressed out of the context of the pIII fusion, failed to function as well as the phage-scFv because of solubility and protein stability issues. They overcame these problems by genetically fusing a human kappa light chain constant domain to the scFv, resulting in a single chain antibody (scAb) fragment. They also expressed a periplasmic chaperone protein, Skp, with the anti-*Brucella* scAb to aid with solubility issues during production by *E. coli*. Both the scAb and chaperone alleviated problems of solubility and stability; however, their reactivity was still not as good as the phages.

Lindquist et al. (2002) panned a nonimmune human scFv library (Sheets et al. 1998) against *Chlamydia trachomatis* elementary bodies. *C. trachomatis* is an obligate intracellular bacterial pathogen that is the leading cause of bacterial sexually transmitted disease in the United States. It exists in two different cellular forms: Elementary bodies are infective but non-replicative, and reticulate bodies are the non-infectious but replicative form. These investigators bound elementary bodies to polystyrene plates and panned with the library. After panning, they screened over 350 individual phage clones by ELISA, several of which had useful specificity. Most were genus-specific, while others recognized only *C. trachomatis*. As would be expected in panning on bacteria that had to be recovered from host cells, some clones recognized host antigens tightly bound to the elementary bodies.

de Greeff et al. (2000) used the Griffin.1 scFv library to isolate scFv phages that recognized *Streptococcus suis*, a pathogen of swine causing a variety of infections in young animals. Panning was performed on whole cells or an extracellular factor, EF, associated with pathogenic strains. Whole cell-reactive clones appeared to recognize conformational determinants of surface proteins. Interestingly, when subtractive panning was performed using a nonpathogenic strain, a single clone that was identical to one obtained by panning on EF was isolated. This result confirmed the association of EF with virulence.

Boel et al. (1998) desired to identify antigens on *Moraxella catarrhalis* strains that exhibited increased virulence because they were resistant to complement. *M. catarrhalis* is a gram-negative coccus that causes otitis media and sinusitis, as well as pneumonia. To prefer-entially isolate scFvs that recognized complement resistance-associated antigens, they panned a human scFv phage library (de Kruif et al. 1995) on complement-resistant *M. catarrhalis* cells bound to a polystyrene matrix in the presence of cells of a complement-sensitive strain. Western blot analysis revealed that most of these phage scFvs recognized a high molecular weight protein antigen, HMW-OMP, which had been associated with complement resistance.

Wang et al. (2004a) produced scFvs against PA of *B. anthracis* by constructing their own library using amplified DNA sequences obtained from mice immunized with PA. Panning was performed with PA bound to polystyrene. A single high binding clone was isolated and was subjected to a variety of genetic manipulations to improve its utility. First, the scFv gene was subcloned into a new vector that added useful peptide tags for affinity purification (His tag) and detection (c-myc). To improve upon the binding qualities of the scFv, they changed the linker between the V_H and V_L sequences, resulting in diabodies and triabodies. The investigators also created fusions of the scFv sequence with *phoA* encoding alkaline phosphatase, and they chemically labeled the diabody protein with Cy3. This study highlights several useful genetic manipulations that make scFv phage display technology so useful.

Kühne et al. (2004) used an scFv library that was constructed by immunizing rabbits with the target antigens—the EspA and intimin proteins of EHEC. EspA is a component of the type III secretion apparatus that injects the Tir protein into host cells. The bacteria then use their intimin protein to bind to the Tir protein that acts as a receptor for bacterial attachment on host cells. scFv phages and then soluble secreted scFv proteins were isolated with reactivity to either of these proteins. Taking advantage of having the DNA sequence for the scFvs, alkaline phosphatase fusions were created for one step use reagents that function in a variety of assays.

Zhou et al. (2002) constructed an unimmunized human scFv library and panned against *Bacillus subtilis* spores as proof of principle in preparation for subsequent studies on anthrax spores. After panning, twelve of 48 clones examined reacted with *B. subtilis* spores. An interesting aspect of this study was that the group altered reactivity of the clones by creating a chain-shuffled Fab library from nine positive scFv phage clones and using subtractive panning to eliminate clones with undesired cross-reactivity. Phage particles expressing scFv proteins were fluorescently labeled and could be used to indentify spores by fluorescence microscopy.

To isolate scFv antibodies against *Clostridium difficile*, an important cause of nosocomial antibiotic-associated colitis and diarrhea, Deng et al. (2003) created an scFv library from a murine hybridoma cell line that produced a monoclonal antibody against toxin B. Interestingly, the best scFv demonstrated equivalent affinity and greater sensitivity toward toxin B than the original monoclonal antibody.

4.3. Single Domain Antibodies (sdAbs) and Domain Antibodies (dAbs)

One problem associated with recombinant antibodies based on the widely used human and murine antibodies is that the antigen-binding domains are present on different peptides—the heavy and light chains; hence, to construct a complete antibody molecule requires either linking the variable heavy and variable light sequences in an artificial manner (scFvs) or encouraging noncovalent binding of the two fragments (Fabs). Recently, the laboratory of Ellen Goldman performed provocative studies developing new recombinant antibody systems based on immunoglobulins from two animals that normally produce single chain antibodies—llamas and sharks (Goldman et al. 2006: Liu et al. 2007). In addition to being relatively easy to work with because they are single chain proteins, recombinant sdAbs apparently do not suffer from being half antibody molecules, presumably because they have evolved to be functional in their animal hosts (Hamers–Casterman et al. 1993). Furthermore, these sdAbs are also very stable to temperature and detergents. We believe that sdAbs represent an exciting and promising development in phage display technology. Additionally, the use of single chain variable heavy or variable light chains, called domain antibodies (dAbs) has recently been developed (Holt et al. 2003). We have found no published articles describing the use of phage-displayed dAbs to detect bacteria as of yet. It is interesting that, although dAbs represent essentially half antigen binding domains, they retain sufficient function to be useful.

Goldman et al. (2006) constructed a semisynthetic library using cloned llama antibody sequences by introducing random amino acid residues into the antigen binding regions to increase diversity (greater than 10^9 members). They panned the library on several antigens, including vaccinia virus (as a surrogate for variola, the causative agent of small pox), cholera toxin, SEB, and ricin toxin. The sdAbs demonstrated very good affinity and specificity, and the sdAbs possessed significantly more thermal stability than conventional antibodies. The sdAbs were useful as capture and reporter reagents in sandwich assays.

Liu et al. (2007) from the Goldman laboratory performed similar experiments using shark antibodies. Sharks produce antibodies composed of only heavy chains that are called new antigen receptor (NAR) or immunoglobulin isotype NAR (IgNAR). These investigators constructed several libraries from different shark species and panned one library against cholera toxin. After three rounds of panning, sixteen nearly identical clones were isolated. The clones exhibited very good sensitivity; however, affinities were lower than those observed for conventional anti-CT antibodies. As was observed for llama sdAbs, the shark sdAbs possessed excellent thermal stability.

5. Summary of Our Results Using and Developing Phage Display scFv and Peptides

We are attempting to isolate binding reagents to enable the detection of bacteria and viruses in real time systems, such as the fiber optic and antibody arrays mentioned above. Although we have been working with scFv phage display and random 12-mer peptide phage display for several years, we are constantly developing improved methods of panning and manipulation. Most of such effort has been devoted to panning, which we believe is the most critical step. However, we are also interested in modification of the binding reagents genetically or chemically to improve their utility in the detection systems.

5.1. Panning Methods

We initially attempted to pan scFv libraries on whole bacterial cells. Our first targets were gram-negative bacteria such as *S.* Typhimurium, *E. coli* O157:H7, and *Vibrio cholerae*. We initially obtained the Griffin.1 library followed by the Tomlinson I and J libraries. We considered several options in identifying useful phages. We could pan on whole cells and sort out the phages that reacted with relevant target antigens, e.g., LPS, after their initial isolation. We could purify the desired antigens, e.g., flagella or LPS, and pan on them in immunotubes. Finally, we could clone the genes for relevant protein targets and pan on the purified proteins. We initially attempted whole cells. Since our experience with whole cell ELISAs showed that binding of bacteria to the polystyrene plates was extremely inefficient, we used panning on whole bacterial cells in suspension. We used approximately 10^9 cells with 10^{11} phages from the Griffin.1 or Tomlinson scFv libraries. We quickly found that alkaline elution of these bacteria resulted in a viscous mixture from which it would be difficult to isolate eluted phages. In contrast, acidic glycine proved to be better, causing the bacteria to remain intact and be easily pelleted. The whole-cell panning enabled us to isolate several scFv-bearing phages; unfortunately, none of them demonstrated the desired specificity in terms of strains, and we could not definitively identify the specific antigen. For example, to detect *E. coli* O157:H7, we aimed to obtain scFvs reacting with either the O157 LPS or the H7 flagellar antigens. However, we never obtained such clones. Similarly, for *V. cholerae*, we hoped to obtain scFvs recognizing the O1 LPS antigenic determinants or those specific for the Inaba or Ogawa serotype antigens. However, we were unsuccessful in this. In this regard, it has been reported that scFvs have poor binding activity to carbohydrates, as opposed to proteins (Ravn et al. 2004). Our most productive panning for bacterial antigens was on semi-purified flagella bound to polystyrene tubes. For *S.* Typhimurium we obtained two scFv phages that recognized the major flagellin protein. Unfortunately, the phages and the scFv proteins had minimal reactivity against whole cells. We isolated several phages panned against semi-purified *L. monocytogenes* flagella that recognized the major flagellin protein by Western blot. Some of these scFvs recognized a carbohydrate modification on the flagellin. A final attempt at isolating anti-Listeria phages was by using a cloned and expressed fragment of the recently described Auto protein (Cabanes et al. 2004). We obtained several scFv phages that reacted with the recombinant Auto protein and even whole Listeria cells; however, when we attempted to express the scFv proteins out of the context of the M13 pIII fusion by expressing the clone in a suppressor-free *E. coli* host, the resulting scFv proteins lost their reactivity. This has been the biggest problem in our experience with phage display scFv. Finding a phage expressing an scFv-pIII fusion that has the desired characteristics often does not translate into a functional scFv protein.

Part of our work for detection of infectious pathogens involves using vaccinia virus as a surrogate for variola virus, which causes small pox. We initially cloned and expressed a fragment of the vaccinia A27L protein and panned on the purified protein using immunotubes. We obtained numerous reactive phages, many of which produced soluble scFv. Unfortunately, we experienced a common problem in working with recombinant proteins—the resulting reagents failed to bind to the wild-type protein. We therefore included a final panning step involving intact vaccinia virus. This resulted in the isolation of a secreted scFv protein that recognizes intact vaccinia virus.

5.2. Screening Methods

After we panned for scFv-phages, we needed an efficient method to screen for reactive clones. Since we had eluted the phages as the last step, we had to infect *E. coli* and then superinfect them with helper phage to produce large enough quantities of phages for screening. As noted above, this required at least one additional day. We picked individual colonies of *E. coli* TG1 containing phagemids, grew them in a 24-well culture plate, and infected them with either M13KO7 helper phage or Hyperphage. The resulting supernatants were then screened in an ELISA with antigen-coated and negative wells. For screening phages, we have found that an anti-M13 pVIII monoclonal antibody available from GE Healthcare is very effective. For screening scFv proteins, our favorite reagent is protein L (Sigma-Aldrich). Even though the scFv proteins contain a c-myc epitope and His tag, we have found these latter tags not nearly as useful as the protein L binding site, which seems to have greater reactivity.

The issue of producing target-binding phages that do not yield functional soluble scFv protein is a dilemma that arises all too frequently. We therefore recently attempted to initially screen clones for secretion of reactive, soluble scFv proteins, skipping the initial screening of phages. Screening for scFv-producing clones requires examining as many as ten-fold more initial clones because of the inefficiency in converting phage clones to scFv proteins. To screen phagemid clones in a more high-throughput manner, we have made attempts at a colony blot assay. *E. coli* containing phagemids are grown overnight on agar media with appropriate antibiotics. The next day the colonies are lifted or picked onto nitrocellulose that was soaked with the target antigen. The idea is that secreted scFv proteins will adhere to the antigen on the nitrocellulose. The nitrocellulose filters are blocked with casein to prevent nonspecific binding of secreted scFv protein and then reacted with a secondary reagent that recognizes the scFv, e.g., protein L-horseradish peroxidase. We are also working with high throughput growth of individual phagemid clones in 96 well plates and screening the supernatants by ELISA, which has less background.

5.3. Genetic Modification of Phagemid Clones

Our most useful scFv clone recognizes the A27L protein of vaccinia virus. We will therefore briefly describe how we have worked with this protein in assay development. The ultimate goal is to use these reagents in a sandwich-type assay—either using a fiber optic waveguide or antibody array; therefore, we require a capture reagent. Since the current matrices for these assays are polystyrene, we are attempting to increase the efficiency of binding of the capture scFv to the polystyrene. One method that has been promising is using nickel-treated ELISA plates (HisGrab Nickel Coated Plates, Pierce) that bind the His-tagged scFv proteins with higher efficiency than would plain polystyrene. To use the same scFv protein as a detection antibody, we must remove the His tag, otherwise it will simply bind to the nickel-treated plate and generate background. We therefore cloned the scFv-encoding sequence minus the C-terminal His tag and c-myc tag into a plasmid vector designed to create fusions with a peptide sequence that is biotinylated by *E. coli*—the AviTag system (Avidity). Advantages of the AviTag system are that chemical biotinylation of the protein is unnecessary, and the location of the biotinylation is limited to the 17 amino acid AviTag sequence. A streptavidin-conjugated reporter such as horseradish peroxidase or FITC is used in conjunction with the biotinylated detection scFv. The sandwich ELISA for vaccinia virus using the His-tagged anti-A27L scFv for capture and the AviTag-fused anti-A27L scFv for detection yielded promising results.

5.4. Random Peptide Phage Libraries

We have also worked with the New England Biolabs Ph.D. 12-mer library to isolate peptides that bind to bacterial targets or for assay development. We have yet to obtain useful clones. We also aimed to obtain peptide sequences that bound to polystyrene so that such peptides could be genetically fused to capture scFv reagents to improve their binding to the detection matrix. By panning on polystyrene we have isolated numerous polystyrene-binding phage clones. We are in the process of cloning the peptide sequences into scFv clones and examining if their binding to polystyrene has been improved.

6. New Directions

We have discussed the principles and theories of the most widely used phage display systems, and we have detailed the published experience of numerous investigators who have successfully used these procedures to isolate reagents for detecting bacteria and their products. We will now address some promising new technologies involving phage display that should become available in the near future. These are affibodies, anticalins, ankyrins, and trinectins (Table 28.3). Finally, we will discuss alternatives to phage display for obtaining antigen-binding peptides—ribosome display and mRNA display. The common theme in the new phage display systems is a stable, invariant backbone into which are cloned somewhat random peptide sequences to recognize the target antigen.

6.1. Proteins Based on Phage Display

We have discussed the two major forms of phage display reagents used today—random peptides and scFv antibodies. However, there are other target-binding tools that have been engineered to increase randomness of binding, and thus are useful as detection reagents. For a comprehensive review of non-immunoglobulin binding proteins see (Binz et al. 2005).

6.1.1. Affibodies

Affibodies are small 6-kDa proteins derived from *Staphylococcus aureus* protein A, which binds to the Fc domain of IgG (Renberg et al. 2005). The Z domain of protein A possesses a 58-amino-acid three-helix-bundle scaffold that supports 13 amino acid residues exposed at the surface of the protein. By randomizing these 13 amino acids, a great breadth in binding specificity can be obtained. Highly specific binding affibodies with micromolar and nanomolar affinities to protein targets have been obtained using phage display technology. Because affibodies are so small, they can be synthesized by solid phase peptide synthesis, which allows the addition of unnatural amino acids, reporter groups such as fluorophores, and tags for affinity purification.

6.1.2. Anticalins

Anticalins are based on lipocalin proteins, which transport or store molecules that are poorly soluble or are easily degraded (Beste et al. 1999). Prototypic lipocalins include retinol binding protein and bilin binding protein (BBP). They possess a one-domain fold containing a β-barrel composed of eight anti-parallel strands. One end of the structure forms a pocket that is suitable for ligand binding. By altering the DNA sequence encoding the pocket, the ligand-binding pocket can be manipulated to recognize a variety of targets. Anticalins with high affinity for a variety of targets have been produced.

6.1.3. Ankyrins

Ankyrins are proteins that have repeating structural units that stack together to form elongated protein domains with a continuous target binding surface (Binz et al. 2004). These proteins are heat stable and can be produced with high yields in bacterial culture. Because they are modular proteins, the size of the potential binding area can be altered. Changing the number of modules forming the binding protein increases the diversity of the library. Binding affinities can be evolved by shuffling different modules, inserting modules, or deleting modules.

6.1.4. Trinectins

Trinectins are small proteins of about 10 kDa mass that are based on the tenth type three domain (Fn3) of human fibronectin, which is an immunoglobulin-like fold lacking cysteines (Parker et al. 2005). Trinectins are very stable, especially to heat. Binding libraries are created by randomizing loop sequences at the end of two anti-parallel ß-sheets.

6.2. Alternatives to Phage Display

Although this chapter focused on the use of systems based on phage display to isolate proteins that recognize targeted molecules on bacteria, we mention here three other related systems that do not involve bacteriophage but can be used for the same purpose.

6.2.1. Aptamers

Aptamers are nucleic acid tools (either DNA or RNA) that have been selected for their ability to bind specific targets (reviewed in Bunka and Stockley 2006; Ngundi et al. 2006). The usefulness of aptamers is based on the fact that nucleic acids, especially single stranded, can assume conformations that confer specificity of binding. As is the case for phage display, the power of aptamers is that when the reagents are selected for specific binding characteristics, the genetic code for the binding reagent is included with the reagent. The novelty of aptamers is that the reagent itself is the gene for the reagent. Aptamers are isolated from libraries through an iterative "panning" process called *SELEX* (systematic evolution of ligands by exponential enrichment). SELEX differs from typical panning for phage display in that reagents with desired binding qualities are amplified by PCR instead of phage propagation. For RNA aptamers, reverse transcriptase-PCR and in vitro transcription are used to amplify and generate the RNA molecules. Also, SELEX typically involves 10 to 20 cycles, as opposed to as few as three for phage display. Ngundi et al. (2006) also discuss additional alternative methods for detection in their review, such as molecularly imprinted polymers.

6.2.2. Ribosome Display

Ribosome display is used to capture a nascent peptide as it is being translated from mRNA by a ribosome in vitro (reviewed in Lipovsek and Pluckthun 2004). The phenotype and genotype are linked through complexes of mRNA, ribosome, and encoded protein. A DNA library encoding polypeptides is transcribed and translated in vitro. The DNA fragments are ligated to a C-terminal spacer that fills the ribosomal tunnel, provides flexibility for the nascent peptide, and allows the peptide to fold as an independent unit while it is still connected to the ribosome. The mRNA lacks a stop codon, so the mRNA-ribosome-peptide complex does not dissociate since the release factors have no sequence to recognize. Therefore, if the ternary complex is incubated with the target antigen and the protein binds to the antigen, the mRNA encoding the protein will be bound as well. By collecting the complexes and performing reverse transcription, the gene encoding the binding peptide can be captured. Because ribosome display is a generic procedure involving in vitro transcription/translation of stop codon-deficient genes, it can be adapted to include peptides engineered into other binding scaffolds or antibody mimics.

6.2.3. mRNA Display

mRNA display is very similar to ribosomal display in advantages and procedure (reviewed in Lipovsek and Pluckthun 2004). Rather than depending on the lack of a stop codon to keep the peptide tethered to the ribosome and mRNA, the peptide is bound to the mRNA by addition of puromycin to the transcription/translation mix. Puromycin causes termination of translation without the dissociation of the ribosome-mRNA-peptide complex. Rather than using the ribosome-mRNA-peptide complex for panning, the mRNA-protein is purified from ribosomes and used for panning using puromycin as a bridge. As is the case for ribosome display, the use of in vitro technology results in high library diversity with 10^{12} to 10^{13} unique sequences. Using mRNA display, high affinity binding peptides have been selected from libraries using various scaffolds, such as scFv.

7. Conclusions

The goal of this chapter was to introduce the concepts and issues related to a variety of phage display technologies for use in detecting bacteria and their products. Because of inherent limitations and problems with phage display, this technology is not the answer to everyone's needs. There will always be a place for antisera and monoclonal antibodies. Phage display can be used for detecting essentially any form of antigen, regardless of its origin, so principles discussed here are applicable to a variety of goals. Additionally, there is a considerable body of literature detailing the use of phage display peptides and antibodies in treating diseases ranging from cancer to infection; however, this was beyond the scope of this work. We are confident that with the development of new sensitive detection assays and new display genetic systems mentioned above, phage display will be increasingly useful in creating binding reagents for numerous purposes.

Acknowledgements

We thank Ann Griswold for excellent review and suggestions with the manuscript. Our work on phage display has been funded by W911SR-05-C-0020 and 1209-168-LO-A from the United States Army through the University of South Florida.

References

Anderson GP, King KD, Gaffney KL and Johnson LH (2000) Multi-analyte interrogation using the fiber optic biosensor. Biosens. Bioelectron. 14:771–777

Arndt KM, Muller KM and Pluckthun A (2001) Helix-stabilized Fv (hsFv) antibody fragments: substituting the constant domains of a Fab fragment for a heterodimeric coiled-coil domain. J. Mol. Biol. 312:221–228

Azzazy HM and Highsmith Jr. WE (2002) Phage display technology: clinical applications and recent innovations. Clin. Biochem. 35:425–445

Baek H, Suk KH, Kim YH and Cha S (2002) An improved helper phage system for efficient isolation of specific antibody molecules in phage display. Nucleic Acids Res. 30:e18

Berger S, Hinz D, Bannantine JP and Griffin JF (2006) Isolation of high-affinity single-chain antibodies against *Mycobacterium avium* subsp. *paratuberculosis* surface proteins from sheep with Johne's disease. Clin. Vaccine Immunol. 13:1022–1029

Beste G, Schmidt FS, Stibora T and Skerra A (1999) Small antibody-like proteins with prescribed ligand specificities derived from the lipocalin fold. Proc. Natl. Acad. Sci. U. S. A 96:1898–1903

Binz HK, Amstutz P, Kohl A, Stumpp MT, Briand C, Forrer P, Grutter MG and Pluckthun A (2004) High-affinity binders selected from designed ankyrin repeat protein libraries. Nat. Biotechnol. 22:575–582

Binz HK, Amstutz P and Pluckthun A (2005) Engineering novel binding proteins from nonimmunoglobulin domains. Nat. Biotechnol. 23:1257–1268

Bishop-Hurley SL, Schmidt FJ, Erwin AL and Smith AL (2005) Peptides selected for binding to a virulent strain of *Haemophilus influenzae* by phage display are bactericidal. Antimicrob. Agents Chemother. 49:2972–2978

Boel E, Bootsma H, de Kruif J, Jansze M, Klingman KL, van Dijk H and Logtenberg T (1998) Phage antibodies obtained by competitive selection on complement-resistant *Moraxella* (*Branhamella*) *catarrhalis* recognize the high-molecular-weight outer membrane protein. Infect. Immun. 66:83–88

Bose B, Chugh DA, Kala M, Acharya SK, Khanna N and Sinha S (2003) Characterization and molecular modeling of a highly stable anti-Hepatitis B surface antigen scFv. Molecular Immunology 40:617–631

Brigati J, Williams DD, Sorokulova IB, Nanduri V, Chen IH, Turnbough Jr. CL and Petrenko VA (2004) Diagnostic probes for *Bacillus anthracis* spores selected from a landscape phage library. Clin. Chem. 50:1899–1906

Broders O, Breitling F and Dubel S (2003) Hyperphage. Improving antibody presentation in phage display. Methods Mol. Biol. 205:295–302

Bunka DH and Stockley PG (2006) Aptamers come of age - at last. Nat. Rev. Microbiol. 4:588–596

Busse HJ, Denner EB and Lubitz W (1996) Classification and identification of bacteria: current approaches to an old problem. Overview of methods used in bacterial systematics. J. Biotechnol. 47:3–38

Cabanes D, Dussurget O, Dehoux P and Cossart P (2004) Auto, a surface associated autolysin of *Listeria monocytogenes* required for entry into eukaryotic cells and virulence. Mol. Microbiol. 51:1601–1614

Campbell A (2003) The future of bacteriophage biology. Nat. Rev. Genet. 4:471–477

Campbell GA and Mutharasan R (2006) Piezoelectric-excited millimeter-sized cantilever (PEMC) sensors detect *Bacillus anthracis* at 300 spores/mL. Biosens. Bioelectron. 21:1684–1692

Campbell GA, Uknalis J, Tu SI and Mutharasan R (2006) Detect of *Escherichia coli* O157:H7 in ground beef samples using piezoelectric excited millimeter-sized cantilever (PEMC) sensors. Biosens. Bioelectron. 22:1296–1302

Carmen S and Jermutus L (2002) Concepts in antibody phage display. Brief. Funct. Genomic. Proteomic. 1:189–203

Chasteen L, Ayriss J, Pavlik P and Bradbury AR (2006) Eliminating helper phage from phage display. Nucleic Acids Res. 34:e145

Cloutier SM, Couty S, Terskikh A, Marguerat L, Crivelli V, Pugnieres M, Mani JC, Leisinger HJ, Mach JP and Deperthes D (2000) Streptabody, a high avidity molecule made by tetramerization of in vivo biotinylated, phage display-selected scFv fragments on streptavidin. Mol. Immunol. 37:1067–1077

Das D, Kriangkum J, Nagata LP, Fulton RE and Suresh MR (2004) Development of a biotin mimic tagged ScFv antibody against western equine encephalitis virus: bacterial expression and refolding. J Virological Methods 117:169–177

de Greeff A, van Alphen L and Smith HE (2000) Selection of recombinant antibodies specific for pathogenic *Streptococcus suis* by subtractive phage display. Infect. Immun. 68:3949–3955

de Kruif J, Boel E and Logtenberg T (1995) Selection and application of human single chain Fv antibody fragments from a semi-synthetic phage antibody display library with designed CDR3 regions. J. Mol. Biol. 248:97–105

de Wildt RM, Mundy CR, Gorick BD and Tomlinson IM (2000) Antibody arrays for high-throughput screening of antibody-antigen interactions. Nat. Biotechnol. 18:989–994

DeMarco DR, Saaski EW, McCrae DA and Lim DV (1999) Rapid detection of *Escherichia coli* O157:H7 in ground beef using a fiber-optic biosensor. J. Food Prot. 62:711–716

Deng XK, Nesbit LA and Morrow Jr. KJ (2003) Recombinant single-chain variable fragment antibodies directed against *Clostridium difficile* toxin B produced by use of an optimized phage display system. Clin. Diagn. Lab Immunol. 10:587–595

Gao C, Mao S, Lo CH, Wirsching P, Lerner RA and Janda KD (1999) Making artificial antibodies: a format for phage display of combinatorial heterodimeric arrays. Proceedings of the National Academy of Sciences, US 96:6025–6030

Gasanov U, Koina C, Beagley KW, Aitken RJ and Hansbro PM (2006) Identification of the insulin-like growth factor II receptor as a novel receptor for binding and invasion by *Listeria monocytogenes*. Infect. Immun. 74:566–577

Glockshuber R, Malia M, Pfitzinger I and Pluckthun A (1990) A comparison of strategies to stabilize immunoglobulin Fv-fragments. Biochemistry 29:1362–1367

Goldman ER, Anderson GP, Liu JL, Delehanty JB, Sherwood LJ, Osborn LE, Cummins LB and Hayhurst A (2006) Facile generation of heat-stable antiviral and antitoxin single domain antibodies from a semisynthetic llama library. Anal. Chem. 78:8245–8255

Goldman ER, Pazirandeh MP, Mauro JM, King KD, Frey JC and Anderson GP (2000) Phage-displayed peptides as biosensor reagents. J. Mol. Recognit. 13:382–387

Griep RA, van Twisk C, van Beckhoven JRCM, van der Wolf JM and Schots A (1998) Development of specific recombinant monoclonal antibodies against the lipopolysaccharide of *Ralstonia solanacearum* race 3. Phytopathology 88:795–803

Griep RA, van Twisk C, van der Wolf JM and Schots A (1999) Fluobodies: green fluorescent single-chain Fv fusion proteins. J. Immunol. Methods 230:121–130

Griffiths AD, Williams SC, Hartley O, Tomlinson IM, Waterhouse P, Crosby WL, Kontermann RE, Jones PT, Low NM, Allison TJ et al. (1994) Isolation of high affinity human antibodies directly from large synthetic repertoires. EMBO J. 13:3245–3260

Hamers-Casterman C, Atarhouch T, Muyldermans S, Robinson G, Hamers C, Songa EB, Bendahman N and Hamers R (1993) Naturally occurring antibodies devoid of light chains. Nature 363:446–448

Hayhurst A, Happe S, Mabry R, Koch Z, Iverson BL and Georgiou G (2003) Isolation and expression of recombinant antibody fragments to the biological warfare pathogen *Brucella melitensis*. J. Immunol. Methods 276:185–196

Holt LJ, Herring C, Jespers LS, Woolven BP and Tomlinson IM (2003) Domain antibodies: proteins for therapy. Trends Biotechnol. 21:484–490

Houimel M, Corthesy-Theulaz I, Fisch I, Wong C, Corthesy B, Mach J and Finnern R (2001) Selection of human single chain Fv antibody fragments binding and inhibiting *Helicobacter pylori* urease. Tumour. Biol. 22:36–44

Houimel M, Mach JP, Corthesy-Theulaz I, Corthesy B and Fisch I (1999) New inhibitors of *Helicobacter pylori* urease holoenzyme selected from phage-displayed peptide libraries. Eur. J. Biochem. 262:774–780

Iannolo G, Minenkova O, Petruzzelli R and Cesareni G (1995) Modifying filamentous phage capsid: limits in the size of the major capsid protein. J. Mol. Biol. 248:835–844

Ide T, Baik SH, Matsuba T and Harayama S (2003) Identification by the phage-display technique of peptides that bind to H7 flagellin of *Escherichia coli*. Biosci. Biotechnol. Biochem. 67:1335–1341

Irving RA, Kortt AA and Hudson PJ (1996) Affinity maturation of recombinant antibodies using *E. coli* mutator cells. Immunotechnology 2:127–143

Kehoe JW and Kay BK (2005) Filamentous phage display in the new millennium. Chem. Rev. 105:4056–4072

Kim GH, Rand AG and Letcher SV (2003) Impedance characterization of a piezoelectric immunosensor part II: *Salmonella typhimurium* detection using magnetic enhancement. Biosens. Bioelectron. 18:91–99

Kim YG, Lee CS, Chung WJ, Kim EM, Shin DS, Rhim JH, Lee YS, Kim BG and Chung J (2005) Screening of LPS-specific peptides from a phage display library using epoxy beads. Biochem. Biophys. Res. Commun. 329:312–317

Knappik A, Ge L, Honegger A, Pack P, Fischer M, Wellnhofer G, Hoess A, Wolle J, Pluckthun A and Virnekas B (2000) Fully synthetic human combinatorial antibody libraries (HuCAL) based on modular consensus frameworks and CDRs randomized with trinucleotides. J. Mol. Biol. 296:57–86

Knurr J, Benedek O, Heslop J, Vinson RB, Boydston JA, McAndrew J, Kearney JF and Turnbough Jr. CL (2003) Peptide ligands that bind selectively to spores of *Bacillus subtilis* and closely related species. Appl. Environ. Microbiol. 69:6841–6847

Kohler G and Milstein C (1975) Continuous cultures of fused cells secreting antibody of predefined specificity. Nature 256:495–497

Kuepper MB, Huhn M, Spiegel H, Ma JK, Barth S, Fischer R and Finnern R (2005) Generation of human antibody fragments against *Streptococcus mutans* using a phage display chain shuffling approach. BMC. Biotechnol. 5:4

Kühne SA, Hawes WS, La Ragione RM, Woodward MJ, Whitelam GC and Gough KC (2004) Isolation of recombinant antibodies against EspA and intimin of *Escherichia coli* O157:H7. J. Clin. Microbiol. 42:2966–2976

Lim DV, Simpson JM, Kearns EA and Kramer MF (2005) Current and developing technologies for monitoring agents of bioterrorism and biowarfare. Clin. Microbiol. Rev. 18:583–607

Lindquist EA, Marks JD, Kleba BJ and Stephens RS (2002) Phage-display antibody detection of *Chlamydia trachomatis*-associated antigens. Microbiology 148:443–451

Lipovsek D and Pluckthun A (2004) In-vitro protein evolution by ribosome display and mRNA display. J. Immunol. Methods 290:51–67

Liu JL, Anderson GP, Delehanty JB, Baumann R, Hayhurst A and Goldman ER (2007) Selection of cholera toxin specific IgNAR single-domain antibodies from a naive shark library. Mol. Immunol. 44:1775–1783

Ma H, Zhou B, Kim Y and Janda KD (2006) A cyclic peptide-polymer probe for the detection of *Clostridium botulinum* neurotoxin serotype A. Toxicon 47:901–908

Marks JD, Hoogenboom HR, Bonnert TP, McCafferty J, Griffiths AD and Winter G (1991) By-passing immunization. Human antibodies from V-gene libraries displayed on phage. J. Mol. Biol. 222:581–597

Mengaud J, Ohayon H, Gounon P, Mege R-M and Cossart P (1996) E-cadherin is the receptor for internalin, a surface protein required for entry of *L. monocytogenes* into epithelial cells. Cell 84:923–932

Messing J (1991) Cloning in M13 phage or how to use biology at its best. Gene 100:3–12

Moll N, Pascal E, Dinh DH, Pillot JP, Bennetau B, Rebiere D, Moynet D, Mas Y, Mossalayi D, Pistre J and Dejous C (2006) A Love wave immunosensor for whole *E. coli* bacteria detection using an innovative two-step immobilisation approach. Biosens. Bioelectron. 22:9–10

Mourez M, Kane RS, Mogridge J, Metallo S, Deschatelets P, Sellman BR, Whitesides GM and Collier RJ (2001) Designing a polyvalent inhibitor of anthrax toxin. Nat. Biotechnol. 19:958–961

Muhammad-Tahir Z and Alocilja EC (2003) A conductometric biosensor for biosecurity. Biosens. Bioelectron. 18: 813–819

Ngundi MM, Kulagina NV, Anderson GP and Taitt CR (2006) Nonantibody-based recognition: alternative molecules for detection of pathogens. Expert. Rev. Proteomics. 3:511–524

Olsen EV, Sorokulova IB, Petrenko VA, Chen IH, Barbaree JM and Vodyanoy VJ (2006) Affinity-selected filamentous bacteriophage as a probe for acoustic wave biodetectors of *Salmonella typhimurium*. Biosens. Bioelectron. 21:1434–1442

Paoli GC, Chen CY and Brewster JD (2004) Single-chain Fv antibody with specificity for *Listeria monocytogenes*. J. Immunol. Methods 289:147–155

Parker MH, Chen Y, Danehy F, Dufu K, Ekstrom,J, Getmanova E, Gokemeijer J, Xu L and Lipovsek D (2005) Antibody mimics based on human fibronectin type three domain engineered for thermostability and high-affinity binding to vascular endothelial growth factor receptor two. Protein Eng. Des. Sel. 18:435–444

Pini A and Bracci L (2000) Phage display of antibody fragments. Curr. Protein Pept. Sci. 1:155–169

Rasooly A (2001) Surface plasmon resonance analysis of staphylococcal enterotoxin B in food. J. Food Prot. 64:37–43

Ravn P, Danielczyk A, Jensen KB, Kristensen P, Christensen PA, Larsen M, Karsten U and Goletz S (2004) Multivalent scFv display of phagemid repertoires for the selection of carbohydrate-specific antibodies and its application to the Thomsen-Friedenreich antigen. J. Mol. Biol. 343:985–996

Reiche N, Jung A, Brableiz T, Vater T, Kirchner T and Faller G (2002) Generation and characterization of human monoclonal scFv antibodies against *Helicobacter pylori* antigens. Infect. Immun. 70:4158–4164

Renberg B, Shiroyama I, Engfeldt T, Nygren PK and Karlstrom AE (2005) Affibody protein capture microarrays: synthesis and evaluation of random and directed immobilization of affibody molecules. Anal. Biochem. 341: 334–343

Sabarth N, Hurvitz R, Schmidt M, Zimny-Arndt U, Jungblut PR, Meyer TF and Bumann D (2005) Identification of *Helicobacter pylori* surface proteins by selective proteinase K digestion and antibody phage display. J. Microbiol. Methods 62:345–349

Sheets MD, Amersdorfer P, Finnern R, Sargent P, Lindquist E, Schier R, Hemingsen G, Wong C, Gerhart JC and Marks JD (1998) Efficient construction of a large nonimmune phage antibody library: the production of high-affinity human single-chain antibodies to protein antigens. Proceedings of the National Academy of Sciences, US 95:6157–6162

Shen Y, Naujokas M, Park M and Ireton K (2000) InIB-dependent internalization of *Listeria* is mediated by the Met receptor tyrosine kinase. Cell 103:501–510

Skerra A and Pluckthun A (1991) Secretion and in vivo folding of the Fab fragment of the antibody McPC603 in *Escherichia coli*: influence of disulphides and cis-prolines. Protein Eng 4:971–979

Smith GP (1985) Filamentous fusion phage: novel expression vectors that display cloned antigens on the virion surface. Science 228:1315–1317

Sorokulova IB, Olsen EV, Chen IH, Fiebor B, Barbaree JM, Vodyanoy VJ, Chin BA, and Petrenko VA (2005) Landscape phage probes for *Salmonella typhimurium*. J. Microbiol. Methods 63:55–72

Stratmann J, Strommenger B, Stevenson K and Gerlach GF (2002) Development of a peptide-mediated capture PCR for detection of *Mycobacterium avium* subsp. *paratuberculosis* in milk. J. Clin. Microbiol. 40:4244–4250

Subramanian A, Irudayaraj J and Ryan T (2006) A mixed self-assembled monolayer-based surface plasmon immunosensor for detection of *E. coli* O157:H7. Biosens. Bioelectron. 21:998–1006

Suzuki Y, Ito S, Otsuka K, Iwasawa E, Nakajima M and Yamaguchi I (2005) Preparation of functional single-chain antibodies against bioactive gibberellins by utilizing randomly mutagenized phage-display libraries. Biosci. Biotechnol. Biochem. 69:610–619

Vaughan TJ, Williams AJ, Pritchard K, Osbourn JK, Pope AR, Earnshaw JC, McCafferty J, Hodits RA, Wilton J and Johnson KS (1996) Human antibodies with sub-nanomolar affinities isolated from a large non-immunized phage display library. Nat. Biotechnol. 14:309–314

Wang X, Campoli M, Ko E, Luo W and Ferrone S (2004a) Enhancement of scFv fragment reactivity with target antigens in binding assays following mixing with anti-tag monoclonal antibodies. J. Immunol. Methods 294:23–35

Wang XB, Zhou B, Yin CC, Lin Q and Huang HL (2004b) A new approach for rapidly reshaping single-chain antibody in vitro by combining DNA shuffling with ribosome display. J Biochem (Tokyo) 136:19–28

Watts HJ, Lowe CR and Pollard-Knight DV (1994) Optical biosensor for monitoring microbial cells. Anal. Chem. 66:2465–2470

Webster R (2001) Filamentous phage biology. In: Barbas CFI (ed) Phage display: a laboratory manual. Cold Spring Harbor Laboratory Press, Cold Spring Harbor, New York

Williams DD, Benedek O and Turnbough Jr. CL (2003) Species-specific peptide ligands for the detection of *Bacillus anthracis* spores. Appl. Environ. Microbiol. 69:6288–6293

Wilson DR and Finlay BB (1998) Phage display: applications, innovations, and issues in phage and host biology. Can. J. Microbiol. 44:313–329

Yau KY, Dubuc G, Li S, Hirama T, MacKenzie CR, Jermutus L, Hall JC and Tanha J (2005) Affinity maturation of a V(H)H by mutational hotspot randomization. J. Immunol. Methods 297:213–224

Zhou B, Wirsching P and Janda KD (2002) Human antibodies against spores of the genus Bacillus: a model study for detection of and protection against anthrax and the bioterrorist threat. Proceedings of the National Academy of Sciences, US 99:5241–5246

29

Molecular Imprinted Polymers for Biorecognition of Bioagents

Keith Warriner, Edward P.C. Lai, Azadeh Namvar, Daniel M. Hawkins and Subrayal M. Reddy

Abstract

There is a trend in biohazard diagnostics to develop integrated systems to extract, concentration and detection from sample matrices. Although biological recognition agents, such as antibodies, can be applied for concentration and detection, there are several limitations. Specifically, biological recognition agents are hard to produce in large quantities, expensive and inherently unstable. Due to such limitations there has been a sustained interest in developing artificial or plastic antibodies that can be readily mass produced, highly stable and cheap. One of the most promising approaches to date has been in the area of Molecular Imprinted Polymers (MIP's). In basic theory behind MIP's is to form a polymer matrix around a template (analyte or structural surrogate) which is subsequently removed to leave voids with high affinity for the target analyte. To date, the majority of MIP research has focused on concentrating or detecting low molecular weight analytes in analytical chemistry. However, there has been interest in applying MIP's to separate, concentrate or detect bioagents such as microbial metabolites, toxins, enzymes and even microbial cells. In the following chapter an overview on the principles of MIP will be outlined. The application of MIP's as solid phase extraction matrices for separating and concentrating biological agents will be reviewed and recent advances described. The utility of MIP's as biorecognition elements in biosensor devices will be covered. Finally, future directions in MIP research will be discussed and the main technological barriers to overcome identified.

1. Introduction

Molecularly imprinted polymers (MIPs) represent a new class of smart materials that have artificially created receptor structures to serve as plastic antibodies. Researchers can imprint a plastic material to create molecularly sized and precisely shaped cavities with specific chemistry that enables them to bind target analyte molecules. Since their invention in 1972, MIPs have attracted considerable interest from scientists and engineers involved with the development of chromatographic adsorbents, membranes, sensors, enzyme mimics, and sorbents for solid phase extraction (SPE). To date, the majority of MIP research has focused on concentrating or detecting low molecular weight analytes in analytical chemistry. However, there has been

Keith Warriner and Azadeh Namvar • Department of Food Science, University of Guelph, Guelph, Ontario, Canada. **Edward P.C. Lai** • Ottawa–Carleton Chemistry Institute, Department of Chemistry, Carleton University, Ottawa, Ontario, Canada. **Daniel M. Hawkins and Subrayal M. Reddy** • University of Surrey, School of Biomedical and Molecular Sciences, Guildford, Surrey, UK.

M. Zourob et al. (eds.), *Principles of Bacterial Detection: Biosensors, Recognition Receptors and Microsystems*,
© Springer Science+Business Media, LLC 2008

interest in applying MIPs to separate, concentrate or detect bioagents such as microbial metabolites, toxins, enzymes and even microbial cells. In the following chapter an overview of the principles of MIP will be outlined. The application of MIPs as solid phase extraction matrices for separating and concentrating biological agents will be reviewed and recent advances described. The utility of MIPs as biorecognition elements in biosensor devices will be covered. Finally, future directions in MIP research will be discussed and the main technological barriers to overcome identified.

2. Principles of Molecular Imprinting

Molecular Imprinting (MI) is a generic term used to describe a process that typically involves a template molecule acting as a substrate or antigen analogue that is associated with a number of so called "functional monomers" in a solvent (porogen), prior to the addition of a cross-linker and a polymerization initiator. After polymerization, the template is extracted from the three-dimensional polymer network, leaving an imprint in the polymer material that bears a steric arrangement of interactive groups defined by the structure of the template molecule. This process introduces a molecular memory to the newly formed polymer material, which is then capable of selectively rebinding the template molecule of interest.

Currently, two basic approaches to MI exist, which are typically performed in organic solvents, and known as "traditional" or "classical" MI (Allender et al. 1999). In the pre-organized or covalent approach (Wulff et al. 1972), the template is chemically derivatized with molecules containing polymerizable groups using reversible covalent bond forming techniques. The aim of this method is to produce an "exact fit" recognition site in which the same chemical bonds in the initial template-monomer complex reform during any subsequent binding of the imprinted polymer cast. Removing the template from the resulting polymer requires only mild chemical cleavage, leaving a polymer cavity with the spatial arrangement necessary to rebind the template. Traditionally, covalent MI has been performed using condensation reactions including boronate esters (Wulff et al. 1972, 1977), Schiff's bases (Wulff et al. 1984, 1989) and ketals (Shea et al. 1986, 1989).

The self-assembly approach involves the pre-arrangement between the template molecule and the functional monomers, formed by noncovalent interactions (Batra et al. 2003) (Fig. 29.1). This approach is much faster and simpler than the covalent approach and is a more flexible technique, considering the vast choice of functional monomers, possible target molecules and the use of imprinted materials. It is believed that the functional monomers become spatially fixed within the matrix, which is both sterically and chemically complimentary to the template. As no covalent bonds are formed between the template molecule and functional monomers, template removal simply involves washing repeatedly in a suitable solvent. With this method, the pre-polymerization complex is an equilibrium system, the stability of which depends on the affinity constants between imprint molecule and functional monomers. On a per-bond basis, noncovalent bonds are 1–3 orders of magnitude weaker than covalent bonds, and consequently

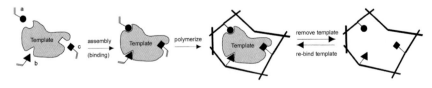

Figure 29.1. Schematic representation of a (noncovalent) molecular imprinting process. (Reprinted from Batra et al. (2003), with permission of Elsevier).

a greater degree of noncovalent bonding (with matching structural orientation) is required for the self-assembly approach (Elemans et al. 2003).

In addition to these two fundamental imprinting techniques, a hybrid of the two mechanisms has been suggested where the polymerization is performed in the presence of a template covalently linked with the functional monomer, followed by the cleavage of the template via decarboxylation, leaving imprints able to interact noncovalently with the template (Whitcombe et al. 1995). This method has effectively led to the development of a methodology termed "sacrificial spacing." Furthermore, a "semicovalent" procedure has been developed that combines both covalent and noncovalent imprinting principles (Sellergren et al. 1990; Bystrom et al. 1993). In all cases, MIPs are generally prepared as dense polymer monoliths that must either be ground or crushed prior to template removal, which is performed in organic solvents.

The advantages and disadvantages of the two main approaches differ considerably and should be considered in relation to the requirements of final imprinted polymer. In general, covalent approaches tend to exhibit low degrees of non-specific interactions that arise due to the stoichiometric nature of the imprinting process. The distinct disadvantage of this method is the unavoidable need for synthetic chemistry to be performed. In contrast, the noncovalent approach requires little or no synthetic chemistry, and a substantial range of chemical functionalities can be targeted (Mayes et al. 2005). However, noncovalently imprinted polymers do tend to display receptor sites that are heterogeneous in nature and of a low affinity. From this point, all references to MI will be made in reference to the noncovalent approach unless otherwise stated.

2.1. Imprinting Considerations

2.1.1. Versatility

MIPs provide a combination of mechanical and chemical robustness with highly selective molecular recognition properties. They are extremely practical compounds as they can be stored in a dry state at ambient temperatures for several years without loss of recognition capabilities and are prepared simply and inexpensively. Additionally, the generation of molecular imprints does not involve the use of laboratory animals or any material of biological origin (other than the potential use of a biological imprint molecule), as is necessary in the production of antibodies. There is also the potential to produce MIPs for analytes that traditionally cause a major problem when attempting to raise antibodies for them, such as highly lipophilic analytes, short peptides and highly toxic compounds. A summary of the advantages and disadvantages of synthetic receptors based on MIPs is given in Table 29.1.

Table 29.1. Summary of the advantages and disadvantages of synthetic receptors based on non-covalently prepared molecularly imprinted polymers for analytical applications (reprinted from Mahony et al. (2005), with permission of Elsevier)

Advantages	Disadvantages
Cost-effective alternative to biomolecule-based recognition	Lower catalytic capabilities than biological counterparts
Ease of preparation; enhanced thermal and chemical stability vs. antibodies	Binding site heterogeneity provides a distribution of binding site affinities
Can be prepared in different formats (bead/block/thin film) depending upon application	Template bleeding requires suitable template analogue for imprinting step and affects quantitative applications
Can be stored for years without loss of affinity for target analyte	Grinding and sieving of bulk polymer is labor intensive and inefficient in material yield

2.1.2. Template Molecule

Molecular imprints have been demonstrated for many classes of molecules including drugs (Vlatakis et al. 1993), pesticides (Siemann et al. 1996), amino acids (Kriz et al. 1995), peptides (Ramstrom et al. 1994), nucleotides (Norrlow et al. 1987), nucleotide bases (Shea et al. 1993), steroids (Ramstrom et al. 1996), sugars (Mayes et al. 1994) and hormones (Andersson et al. 1995), with the imprinting of small organic molecules now well established and considered routine.

Metals and other ions have also been used as templates to induce the specific arrangement of functional groups in the imprinting matrix (Kato et al. 1981; Chen et al. 1997; Saunders et al. 2000) Larger organic compounds such as peptides have also been imprinted via similar techniques, but generally the imprinting of large biological structures remains a challenge. Biological molecules require an aqueous environment to maintain their structure, yet traditional approaches to MI rely upon organic solvents to promote the intermolecular forces that are integral to the formation of host-guest complexes.

2.1.3. Functional Monomer

The role of the functional monomer in MI is to weakly interact with the template molecule with noncovalent interactions such as van der Waals forces, electrostatic bonding and hydrogen bonding (Nicholls et al. 1999). This occurs at one end of the monomer molecule, leaving the free unbound end of the monomer chain to covalently bond to the cross-linking molecule. The functional monomers are sometimes considered analogous to the twenty common amino acids that constitute the building blocks of all proteins.

Methacrylic acid remains the most commonly employed functional monomer, due to its ability to participate in ion–ion, ion–dipole and dipole–dipole complexing (Vlatakis et al. 1993; Titirici et al. 2004), and as a result, there has been reliance upon its use. This can be said of several functional monomers and occurs due to the demonstration of a reproducible imprinting effect, rather than a choice governed by predicted template-monomer chemistries that may occur. In saying that, a large variety of acidic (Andersson 1996; Suedee et al. 1998), basic (Kempe et al. 1994; Piletsky et al. 1995) and neutral functional monomers (Yu et al. 1997; Pap et al. 2004) have been successfully employed in MI (Fig. 29.2).

Combinatorial approaches to MI attempt to perform more a systematic evaluation of polymer compositions, rather than relying solely upon experience of the technique to govern

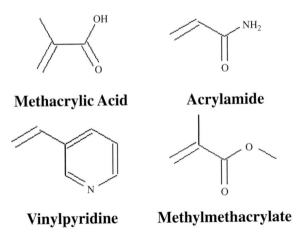

Figure 29.2. Examples of functional monomers commonly employed in molecular imprinting protocols.

the choice of imprinting parameters. Computational assessment of MI utilizes vast libraries of functional monomers to predict the manner in which the monomers will interact with the template (Piletska et al. 2004a; Piletsky et al. 2004b). Multiple co-monomer complexes have also been extensively employed in the design of imprinting strategies (Lubke et al. 2000; Zheng et al. 2002) as the combination of specific monomers for a particular target of interest present the possibility of producing truly "tailor-made" monomer complexes. The success of such an approach depends upon the use of monomers with strong interactions with the target molecule but with weak non-productive interactions with each other (Lubke et al. 2000). It is also possible to prepare custom functional monomers, with many excellent examples reported (Spivak et al. 1999; Piletsky et al. 2006). However, this approach tends to lean back toward covalent methodologies, as a substantial amount of synthetic chemistry is invariably involved in producing such monomers.

2.1.4. Cross-Linking

The cross-linking agent is also a key variable in MI as it provides the necessary structural scaffolding on which the MIP is formed, and is therefore the major fraction of the physical constitution of the polymer (typically 85 %–90 % by weight) (Sellergren 1989). A high degree of cross-linking is necessary to achieve the structural integrity that results in template specificity, but high proportions of cross-linkers result in rigid structures that are difficult to physically manipulate, which can be experimentally problematic. The percentage composition of the cross-linker in relation to the total functional monomer composition determines the total percentage of cross-linking. Fig. 29.3 demonstrates the cross-linking density theory. As percentage cross-linking increases, so does the density of the polymer matrix, therefore increasing the number of potential imprinted sites due to an improved structural architecture. As a direct result, however, physical manipulation of the polymer becomes difficult, and the degree of template removal decreases. As with functional monomers, a great number of cross-linkers are available that offer variations in structural characteristics such as rigidity, stability, porosity and hydrophobicity (Fig. 29.4). Originally, the cross-linkers of choice were isomers of divinylbenzene, but later it was found that acrylic- or methacyrlic-acid-based systems such as ethylene glycol dimethacrylate (EGDMA) were favorable (Sellergren 1989). Subsequently, it has been shown that novel cross-linkers such as trimethylolpropane trimethacrylate (TRIM) (Kempe 1995) and pentaerythritol triacrylate (PETA) (Kempe 1996) are superior to EGDMA, which only serves to highlight the variations available and considerations to be made when considering the experimental components of a MIP.

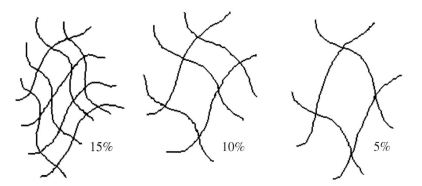

Figure 29.3. Schematic describing the percentage cross linking density theory of a molecular imprinting strategy.

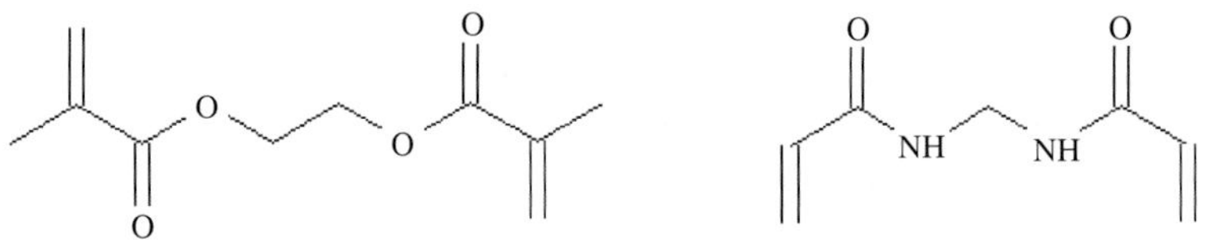

Ethyleneglycol Dimethacrylate   **N,N'-Methylene-bis-acrulamide**

Figure 29.4. Examples of cross linkers commonly employed in molecular imprinting protocols.

2.1.5. Polymerization

When conducting an MI protocol, the template–monomer association and cross-linking of the structure all occur in solution. In order for the polymer matrix to form, a polymerization reaction must occur, which creates the molecular scaffold in which the template is entrapped. This can occur in several ways, but most commonly occurs by free radical polymerization (FRP). FRP has three distinct stages: initiation, propagation and termination, as shown in Fig. 29.5.

FRP is a useful reaction that is commonly exploited in the manufacture of commercially important plastics. The feature that makes FRP important in so many fields is the amazing versatility of the technique; it can be performed easily in bulk or in solution at ambient temperatures and at atmospheric pressure, and it is tolerant of impurities. This makes the technique ideally suited to preparing MIPs. Free radicals display great chemical reactivity toward carbon–carbon double bonds and are completely inert to many other organic groups. Therefore, there are a huge variety of monomers that can be polymerized by this technique, which has subsequently led to many applications.

FRP can also occur spontaneously, due to the nature of the initiating precursor. This is true of polymer hydrogels, where polyacrylamide polymerization is initiated by ammonium persulfate (APS) (in the presence of the cross-linker N, N′-methylenebisacrylamide (bisacrylamide) and the catalyst N,N,N,N-tetramethyl-ethylenediamine (TEMED)).

2.1.6. Solvent

The solvent, or porogen, plays one of the most important roles in dictating the success of an MI process, as the strength of noncovalent interactions, along with the immediate influence upon polymer morphology, is governed by the solvent. The choice of solvent is regulated by

Initiation	$I—I \rightarrow 2\,I^*$	(equation 1)
	$I^* + M \rightarrow I—M^*$	(equation 2)
Propagation	$I—M^* + nM \rightarrow I—M^*_{n+1}$	(equation 3)
Termination	$2\,I—M^*_{n+1} \rightarrow I—M_n—I$	(equation 4)

The three stages of FRP, where I is the initiator, I* is the active radical, Mis the functional monomer and I-M* is the initial monomer radical.

Figure 29.5. The three stages of FRP, where I is the initiator, I* is the active radical, M is the functional monomer, and I-M* is the initial monomer radical.

three characteristics: the ability to solubilize the constituents, the effects upon porosity and surface area, and the effects upon template: monomer complexion.

The best imprinting porogens for accentuating the binding strengths are solvents of a very low dielectric constant such as toluene and dichloromethane. Apolar organic solvents are used to maximize the strengths of the noncovalent interactions involved in cavity formation. Conversely, binding strengths have been shown to be significantly reduced in aqueous solutions due to the high polarity of water, resulting in poorer recognition (Ramstrom et al. 1996). The use of more polar solvents will weaken the interactive forces formed between the print species and the functional monomers resulting in poorer recognition.

However, relatively polar solvents such as acetonitrile (Martin-Esteban et al. 2001) are commonly used, as the benefit of their effect upon the macroporous polymer pore size outweighs the drawbacks of weakened attractive forces. Subsequent specific recognition of the imprint molecule by the imprint polymer is strongest under conditions that most closely resemble the "cocktail" used for polymer synthesis. This avoids problems associated with the matrix swelling, which in turn leads to three dimensional configuration changes of the functional groups involved in recognition.

Many medical and environmental analytes are present in forms and concentrations that do not lend themselves well to an organic solvent environment. Therefore, it is often necessary to perform preconcentration and clean up steps, most commonly liquid–liquid extraction (Senholdt et al. 1997).

For many applications an extraction step into an organic solvent is laborious, and assays would be simpler and more widely accepted if they could be performed directly in the sample matrix, which whether for medical or environmental applications, is usually aqueous. Therefore, from the emergence of the technique, there has been a driving force to adapt molecular imprinting assays to aqueous solvents or organic/aqueous co-solvent mixtures. Although progress has been made to transfer the technique to polar protic solvents, a truly general method for aqueous phase MI using the noncovalent approach is yet to be developed.

2.2. Aqueous Phase MIP

Molecular imprinting has the potential to play an important role in many diverse scientific fields in years to come. The key areas that are retarding the natural progression of the technique include reliance upon organic solvents, difficulty to imprint large molecules, and a general inability to easily perform reproducible polymer chemistry that yields a standardized preparation procedure. However, the simple fact remains that combining the advantages of synthetic plastics such as low cost, durability and robustness, with the recognition properties of natural receptors, highlights the potential worth of furthering the field of aqueous phase MI and developing imprinting protocols for large biological structures.

In most applications, traditional MIPs have been difficult to adapt to aqueous conditions. This is firstly and in part to do with the specific polar interactions between good imprinted sites and the analyte that become weakened in an aqueous environment, and secondly due to the non-specific (hydrophobic) interactions between other small molecules and the polymer matrix that become strengthened. However, as long as the best recognition sites retain their selectivity, many MIP-based assays have been applied to aqueous conditions. Also, many assays have first been developed in organic solvent with secondary studies performed to find optimal conditions for aqueous analysis (Andersson 1996; Mayes et al. 1996; Ramstrom et al. 1996). This so called "solvent switching" strategy, although effective in facilitating analysis in the aqueous phase does not fully address the problems associated with true aqueous phase molecular imprinting. This is almost entirely due to the heavy reliance upon organic solvents, as the sample is only "loaded" in an aqueous medium, with subsequent detection performed in the original porogen. However, although the vast majority of the literature to date describes the interaction of polar functional groups, the polymer matrix within which the template is entrapped does contribute toward differing degrees of hydrophobic polymer–template interactions. There has been a significant number of publications that report of noncovalently imprinted polymers that recognize their respective templates in the absence of hydrogen bonding in an aqueous media (Andersson 1996; Karlsson et al. 2001). This clearly suggests that although still a considerable challenge, imprinting within an aqueous media is distinctly possible.

Another driving force behind the development of MI technologies, along with the move toward aqueous phase imprinting, is the molecular imprinting of large biological molecules. Historically, MI has been reserved for low molecular weight molecules and is now considered commonplace and routine.

The benefits of MI as an analytical technique are such that if developments in technology allowed the formation of MIPs specific to larger molecules such as proteins and enzymes, then the impact upon diagnostic clinical strategies would be immense, as the rapid screening, diagnosis and therapeutic monitoring of many pathophysiological conditions would become possible.

The majority of proteins are incompatible within the organic solvents that are traditionally employed in MI. Since the non-polar side chains (which are proximal upon the external protein structure) are more soluble in organic solvents than in water, the hydrophobic interactions that maintain the highly folded quaternary structure of proteins are weakened considerably. This results in a loss of structure and specificity leading to precipitation and denaturation.

Additionally, high ratios of cross-linker to functional monomer are traditionally used (to minimize macromolecular chain relaxation and swelling phenomena), resulting in a highly cross-linked polymer matrix with a low molecular mass between cross-links. This invariably makes template removal (especially with large molecules) extremely difficult, particularly when considering that organic solvents would not be a suitable tool for use as either the imprinting porogen or template eluant.

2.2.1. Hydrogels

Hydrogels are insoluble, cross-linked polymer network structures composed of hydrophobic homo or hetero copolymers that possess the ability to absorb significant amounts of water. The importance of hydrogels in biomedical applications was first realized toward the end of the 1950s when Wichterle and co-workers developed poly (2-hydroxyethyl methacrylate) (PHEMA) and employed it as a soft contact lens material (Wichterle et al. 1960).

Traditionally, hydrogels have found applications in controlled release systems (Peppas et al. 1987), as the hydrophilic and lipophilic balance of a gel carrier can be altered to provide significant contributions that present different solvent diffusion characteristics, which in turn influence the diffusive release of the drug contained within the gel matrix. Today, hydrogels are used in numerous biomedical applications including biomembranes, biosensors and ophthalmological devices (Lowman et al. 1999).

In recent years, it has become apparent that the principles and characteristics of hydrogels could be combined with those of molecular imprinting to produce intelligent imprinted gels that can memorize specific binding conformations. This would not only further the development of MI in the aqueous phase, but also involve molecular imprinting in new technology fields such as targeted drug delivery devices. Gel properties have been compared to that of proteins (Annaka 1992, 2000). Proteins may be in their folded, compact or expanded random coil conformations, depending upon the conditions of the surrounding environment. The similarity between polymer gels and proteins suggests that the production of synthetic gels with molecular recognition capabilities is viable.

2.2.2. MIP Within Hydrogels

A polymer network structure depends upon the type of monomer chemistry, the association interaction between monomers and pendant groups on the solvent, and the relative amounts of co-monomers in the feed from which the structure is formed. As good recognition between template and polymer matrix requires three-dimensional orientation, most techniques limit the movement of the memory site via chain relaxation, swelling phenomena and other processes by using high ratios of cross-linking agent to functional monomers.

An increase in cross-linking monomer leads to a decrease in the average molecular mass between cross-links, resulting in a more rigid system. In less cross-linked systems, movement of the macromolecular chains, specifically the spacing of functional groups, will change as the network expands or contracts, depending on the chosen rebinding solvent, application and environment.

Imprinting success depends on the relative amount of cross interaction between the solvent and the intended noncovalent interactions. If the solvent interferes or competes with any of these interactions, less effective recognition occurs. Proper tuning of noncovalent interactions, such as increasing macromolecular chain hydrophobicity (Yu et al. 1997a) or including stronger hydrogen-bond donors and acceptors has been shown to enhance binding and selective recognition in aqueous solutions (Yu et al. 1997b)

The theory and understanding of hydrogels is well established, yet the development of MI within hydrogels is very much in its infancy. It appears as if one of the most promising and straightforward approaches to gel imprinting mimics traditional approaches to MI, and involves the inclusion of the biomolecule of interest in the design of the polymer matrix (Hoffman 1992; Hubbell 1999). Another approach details the inclusion of a post cross-linking reaction between either an excess of functional monomers on opposite macromolecular chains or via an excess of additional monomers which are introduced to the network after the gel is formed and imprint the rebounded molecule (Enoki et al. 2000; Alvarez-Lorenzo et al. 2001) In addition, polymer gels that have included intelligent envirosensitive mechanisms capable of turning an active site on or off, have been reported for polymers with enzyme–conjugates (Yang et al. 1995), antibody fragment–conjugates (Lu et al. 1999) and lectin–conjugates (Kofufuta et al. 1991; Miyata et al. 1996; Obaidat et al. 1996).

In recent years, the use of hydrogels, specifically molecularly imprinted hydrogels, in controlled drug delivery systems has been discussed (Byrne et al. 2002; Chien 2002; Alvarez-Lorenzo et al. 2004). The hydrophilic and hydrophobic balance of a gel carrier can be altered to provide controllable solvent diffusion characteristics, which in turn influence the diffusive release of a drug contained within the gel matrix (Langer et al. 1981).

2.2.3. Polyacrylamide Gels—HydroMIPs

Polyacrylamide is a type of vinyl polymer that has been widely employed as a support matrix in molecular biology. Polyacrylamide gels provide a means of separating molecules by size, effectively acting as a sieve by retarding or completely obstructing the movement of large macromolecules while still allowing smaller molecules to migrate freely.

These gels have presented themselves as excellent candidates for HydroMIPs, as they possess or can be engineered to possess the structural parameters (polymer volume fraction in swollen state, average molecular weight between cross-links, and the network pore size) necessary to successfully produce an analyte-specific HydroMIP (Peppas et al. 2000).

Polyacrylamide gels are produced with great ease and are formed by copolymerization of acrylamide and bisacrylamide (Fig. 29.6). The reaction is a vinyl addition polymerization initiated by a free radical-generating system (Chrambach et al. 1971; Rodbard et al. 1971a,b). Polymerization is initiated by APD and TEMED: TEMED accelerates the rate of formation

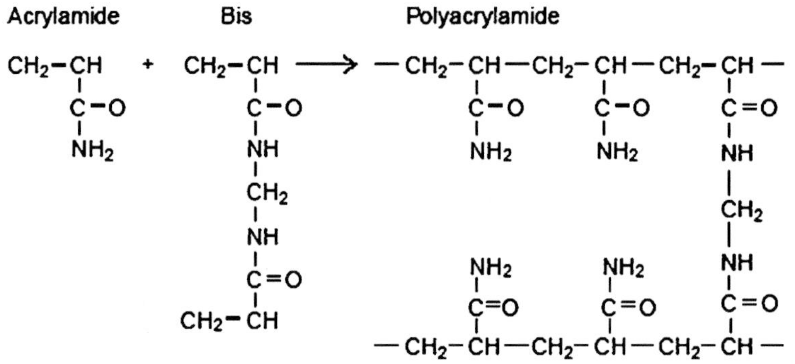

Figure 29.6. Schematic detailing the FRP of acrylamide and bis-acrylamide to form polyacrylamide.

of free radicals from persulfate, and these in turn catalyse polymerization. The persulfate free radicals convert acrylamide monomers to free radicals that react with unactivated monomers to begin the polymerization chain reaction (Shi et al. 1998). The elongating polymer chains are randomly cross-linked by bisacrylamide, resulting in a gel with a characteristic porosity that depends on the polymerization conditions and monomer concentrations.

TEMED is subject to oxidation, which causes the gradual loss of catalytic activity. This process is greatly accelerated by contaminating oxidizing agents. TEMED that contains oxidation products is characterized by a yellow colour. The practical consequences of the oxidative process are the requirement for greater amounts of TEMED to achieve adequate polymerization, and a gradual loss of TEMED reactivity with time. TEMED is also very hygroscopic and will gradually accumulate water, which will accelerate oxidative decomposition.

APS is also very hygroscopic. This property is particularly important, since the structure begins to break down almost immediately when dissolved in water. Therefore, the accumulation of water in APS results in a rapid loss of reactivity. This is why ammonium persulfate solutions should be prepared fresh daily. Persulfate is consumed in the polymerization reaction. Excess persulfate can cause oxidation of proteins and nucleic acids. This oxidation problem can be avoided if inhibitor-free gel-forming reagents are used, and APS is used at the recommended levels.

In physically preparing a HydroMIP, the template molecule is introduced to the pre-polymerization polyacrylamide solution, with polymerization subsequently occurring in the presence of the template. This allows the formation of a network of noncovalent bonds between the functional monomer, acrylamide, and the template molecule. Upon removal of the template, 3-dimensional molecular cavities displaying a structural and chemical memory for the template remain, which in turn are capable of selectively rebinding the template.

Although the use of HydroMIPs as molecular recognition elements is still very much a recent approach, there have been several excellent reports upon the use of polyacrylamide gels used in conjunction with molecular imprinting. Hjerten and co-workers pioneered the use of polyacrylamide gels as an imprinting matrix for a wide range of proteins (Liao et al. 1996; Hjerten et al. 1997; Tong et al. 2001).

The optimization of polymerization parameters was described and the molecularly imprinted gels were applied as artificial antibodies for the selective adsorption of proteins in affinity chromatography with excellent chromatographic resolution achieved. Regardless of the chromatographic resolution achieved, the nature of the polyacrylamide packed column resulted in a particularly poor mobile phase flow rate. The work of Zhang and co-workers (Shi et al. 1998; Guo et al. 2004, 2005a,b) report of the use of polyacrylamide modified chitosan beads for application as stationary chromatography columns. These protein-imprinted beads significantly increased the rate of flow through a packed column without compromising either the selectivity or specificity of the imprinted material.

The work of Ou et al. (2004) described the molecular imprinting of lysozyme in a polyacrylamide gel. In this study, template removal was performed by elution of protein in NaCl, and an increase in template removal was observed when "traditional" functional monomers such as methacrylic acid were incorporated in the prepolymerization polyacrylamide solution.

Historically, MI has found extensive application in the field of capillary chromatography (Nilsson et al. 1994; Lin et al. 1997; Schweitz et al. 1997; Valleno et al. 2000). Application in this field has continued with the advent of aqueous phase imprinting, with polyacrylamide being utilized as the imprinting matrix for a range of molecules (Gubitz et al. 2000; Zou et al. 2002).

The research of Pang et al. has described a method termed "inverse-phase seed suspension polymerization" (IPSSP) for the preparation of bovine serum albumin (BSA) imprinted polyacrylamide gels (Pang et al. 2005), as well as the synthesis of polyacrylamide beads also for BSA (Lu et al. 2006; Pang et al. 2006). IPSSP is a technique in which soft external gels are cast around a highly cross-linked inner gel core. It is proposed that this technique produces a shallow surface imprinting effect, and maximizes the use of the template molecule by avoiding the creation of hard-to-access imprinted sites embedded deep within the polymer matrix. In addition, it is also proposed that this technique produces imprinted particles that are highly homogenous in size and imprint distribution. Polyacrylamide gels have been utilized for their excellent temperature and pH sensitivity for the imprinting of BSA (Demuel et al. 2005) and copper (Tokuyama et al. 2006) and have also been reported for the successful bio-imprinting of tumor specific glycoproteins (Miyata et al. 2006), electrosynthesised membranes (Liao et al. 2004), copolymer membranes (Sallacan et al. 2002) and bacterial proteins (Zhao et al. 2006). Diverse applications of other molecularly imprinted hydrogels notably include the repair of brain lesions (Hou et al. 2005) and the imprinting of baculovirus (Bolisay et al. 2004).

3. Solid Phase Extraction Based on MIPs for Concentrating Bioagents

3.1. Antibiotics

In recent years there has been a strong interest in developing simple and rapid extraction of antibiotics in complex biological and food sample matrices (Fernández-González et al. 2006). MIPs have specifically received interest due to the low cost, robustness, versatility and compatibility with sensitive detection platforms such as HPLC-UV and LC-MS analysis (Lai et al. 2003; Wu et al. 2006)

Zhao and co-workers prepared a monolithic MIP for sulfamethoxazole (SMO) by *in situ* polymerization method as a HPLC stationary phase (Liu et al. 2006). By optimizing the polymerization conditions, the monolithic MIP showed highly specific recognition for the template SMO over its three structurally related analogs. As shown by SEM and the pore size distribution profile, the resultant MIP monolith showed a main pore diameter of 594 nm and a large specific surface area of $124 \, m^2 \, g^{-1}$, which allowed the mobile phase to flow through the monolithic column with low backpressure. Furthermore, the recognition abilities of the monolithic MIP in aqueous and organic media were studied. The results exhibited that the monolithic MIP possessed excellent recognition ability in aqueous media. Hydrophobic interactions, in addition to shape recognition, were the dominant effect for recognition in the mobile phase with high water content. Moreover, the binding sites and the dissociation constant were also determined by frontal chromatography as $122 \, \mu\text{mol} \, g^{-1}$ and $1.88 \times 10^{-5} \, \text{mol} \, l^{-1}$, respectively, which demonstrated that the obtained SMO-MIP monolith had a high binding capacity and strong affinity ability to the template molecule. Finally, the resultant SMO-MIP monolith was used as an HPLC column directly to determine the SMO contents in three kinds of pharmaceutical tablets with the optimized aqueous mobile phase.

Marce and co-workers synthesized a new MIP using enrofloxacin as a template molecule (Caro et al. 2006). The imprinting effect of the polymer was verified by chromatographic evaluation and, interestingly, this evaluation also revealed that the imprinted polymer showed a high degree of cross-reactivity for ciprofloxacin, the major metabolite of enrofloxacin. The MIP was then applied as a selective sorbent in a two-step solid-phase extraction (SPE) method focusing upon complex biological matrices, specifically human urine and pig liver. This two-step SPE protocol, in which a commercial Oasis HLB cartridge and a molecularly imprinted solid-phase extraction (MISPE) cartridge were combined, allowed enrofloxacin and ciprofloxacin to be determined by liquid chromatography coupled to a UV detector at levels below the maximum residue limits established by the European Union. The quantification and detection limits in tissue samples of enrofloxacin and ciprofloxacin were established at $50\,\mu g\ kg^{-1}$ and $30\,\mu g\ kg^{-1}$, respectively.

Syu et al. (2006) successfully synthesized a molecularly imprinted polymer (MIP) capable of detecting bilirubin. Bilirubin template was imprinted in poly(methacrylic acid-co-ethylene glycol dimethylacrylate) [poly(MAA-co-EGDMA)]. MAA and EGDMA were used as the monomer and the cross-linker, respectively. A solvent system based on ethylenediamine tetraacetic acid (EDTA) and ascorbic acid was compared with respect to the stability of bilirubin. Both pH and bilirubin concentration were investigated for the bilirubin stability. The cross-linking effect was further confirmed by the thermogravimetric analysis (TGA). The effect of salts, such as NaCl and KCl, on the binding capacity of the molecularly imprinted polymer was also discussed. Further, the rat serum and bile samples were applied and the binding of the MIPs for bilirubin was thus confirmed.

Pingarron and co-workers reported an electrochemical method for the determination of sulfamethazine at a low concentration level ($25\,\mu g\ L^{-1}$) in milk (Guzmán-Vázquez et al. 2006). The method involved sample cleanup and selective preconcentration of sulfamethazine with an MIP, and a further electrode surface preconcentration of the analyte at a Nafion-coated glassy carbon electrode (GCE). Square wave (SW) oxidative voltammetry of accumulated sulfamethazine was employed for its quantification. Sulfamethazine electrode preconcentration was carried out in 0.1 M Britton–Robinson buffer of pH 1.5, and by applying 5 min of accumulation at open circuit. A linear calibration graph was obtained for sulfamethazine at the Nafion-modified GCE over the 0.01 to 1.0 μM concentration range, with a detection limit of 0.007 μM ($1.9\,\mu g\ l^{-1}$). This detection limit is significantly better than those reported previously in the literature using electroanalytical techniques. Although the detection limit achieved was sufficient to allow the direct determination of sulfamethazine at the concentration level required in milk, a sample cleanup was shown to be necessary to obtain analytically useful SW voltammograms. This was accomplished by processing the deproteinized milk through a cartridge containing an MIP for sulfamethazine, also allowing for a selective preconcentration of the analyte. Elution of the analyte from the MIP cartridge was carried out with 2 mL of a methanol/acetic acid (9:1 v/v) mixture. Determination of sulfamethazine in milk samples was accomplished by interpolation into a calibration graph constructed with sulfamethazine standard solutions that were subjected to the same procedure as the deproteinized milk samples. Results obtained for five samples, spiked at the 25 $\mu g\ l^{-1}$ level, showed a mean recovery of 100 ± 3 %.

Tarley et al. (2006) coupled a sorbent flow preconcentration system to an amperometric detector for the chloroguaiacol (4-chloro-2-methoxyphenol) determination at submicromolar levels. Satisfactory selectivity was attained by using a chloroguaiacol-imprinted polymer, which was synthesized by bulk polymerization. Flow and chemical parameters associated with the preconcentration system, such as sample pH, preconcentration and elution flow rates, concentration of the carrier solution (KCl), and eluent volume were investigated through multivariate analysis. The flow preconcentration of chloroguaiacol was not affected by equimolar presence of structurally similar phenolic compounds including catechol, 4-chloro-3-methylphenol, 4-aminophenol and 2-cresol, thus showing the good performance of the MIP. Under the best experimental conditions, a preconcentration factor of 110-fold was obtained, as well as low detection and quantification limits of 0.03 and 0.08 μM respectively. The analytical curve covered a wide linear range from 0.05 up to 5.0 μM ($R^2 > 0.999$). Satisfactory precision was evaluated as 5.5 % and 4.2 % RSD ($n = 8$) for solutions of 1.0 and 5.0 μM chloroguaiacol. Recoveries varying from 93 % up to 112 % for water samples (tap water and river water) spiked with chloroguaiacol concentration were achieved, thus assuring the accuracy of the proposed flow preconcentration system.

Room temperature phosphorescence (RTP) has gained significance as a very useful mode of detection for optical sensing applications. Sanchez-Barragan et al. (2006) have presented an overview of the recent uses of, and future prospects for, RTP in this field. In particular, an iodinated MIP was developed for the selective optosensing of fluoranthene in a flow-through sensor measurement mode.

Soft contact lenses are receiving increased attention not only for correcting mild ametropia but also as drug delivery devices. To provide polyhydroxyethylmethacrylate (PHEMA) lenses with the ability to load norfloxacin (NRF) and to control its release, Alvarez-Lorenzo et al. (2006) carefully chose functional monomers and then spatially ordered them to apply the molecular imprinting technology. Isothermal titration calorimetry (ITC) studies revealed that maximum binding interaction between NRF and acrylic acid (AA) occurred at a 1:1, and that the process saturated at a 1:4 molar ratio. Hydrogels were synthesized using different NRF:AA molar ratios (1:2 to 1:16), at two fix AA total concentrations (100 and 200 mM), and using molds of different thicknesses (0.4 and 0.9 mm). The cross-linker molar concentration was 1.6 times that of AA. Control (non-imprinted) hydrogels were prepared similarly but with the omission of NRF. All hydrogels showed a similar degree of swelling (55 %) and, once hydrated, presented adequate optical and viscoelastic properties. After immersion in 0.025, 0.050 and 0.10 mM drug solutions, imprinted hydrogels loaded greater amounts of NRF than the non-imprinted ones. Imprinted hydrogels synthesized using NRF:AA 1:3 and 1:4 molar ratios showed the greatest ability to control the release process, sustaining it for more than 24 h. These results prove that ITC is a useful tool for the optimization of the structure of the imprinted cavities in order to obtain efficient therapeutic soft contact lenses. Byrne and co-workers have applied the principles of biomimesis by incorporating a natural receptor-based rational design strategy in the synthesis of novel recognitive soft contact lenses (Venkatesh et al. 2007). They demonstrated the potential of biomimetic carriers to load significant amounts of ocular medication such as H_1-antihistamines, as well as to release a therapeutic dosage of drug in vitro in a controlled fashion for 5 days, with an even further extension in the presence of protein. Gels of multiple complexation points with varying functionalities outperformed gels formed with less diverse functional monomers and showed superior loading with a six-fold difference over control gels and a three-fold difference over less biomimetic gels. Moreover, mechanical and optical properties of these hydrogels agreed with conventional lenses, and increased loading was reflected in a reduced propagation of polymer chains. This approach can be extended to a wider biological spectrum in the design of novel, controlled and modulated delivery devices to alleviate ocular disorders and provide an alternative to topical therapy.

Hu and co-workers achieved the spontaneous formation of an ordered nano-TiO$_2$/*p-tert*-butylcalix[4]arene hybrid thin film imprinted by parathion using a self-assembled technique (Li et al. 2006). A sensor based on the imprinted film was constructed for the selective determination of parathion. A linear response to parathion in the concentration range of 0.05 to 10 μM was observed with a good correlation coefficient ($R^2 = 0.992$). The linear regression equation was $I_p(\mu A) = 0.202C(\mu M) + 0.797$ and the detection limit of the sensor was 0.01 μM (S/N = 3).

The imprinted film sensor has been applied to the determination of parathion in spiked vegetable samples, and the recoveries were varied from 93 % to 103 % at 0.3 and 2.0 μM.

3.2. Mycotoxins

Mycotoxins (such as ochratoxin A, patulin and DON) present a significant food safety and public health risk. Ochratoxin A (OTA) is a nephrotoxic compound that was first detected in wine in the nineties. Since then, several grape-derived products have been found to be potentially contaminated. Ochratoxin contamination of grapes takes place in the field and is caused mainly by *Aspergillus* spp, especially *A. carbonarius*. Several factors affect ochratoxin production by *Aspergillus* on grapes. Attempts have been made to lower fungal contamination and thereby ochratoxin contamination of grapes, with varying success. Varga and Kozakiewicz (2006) have given an overview of recent knowledge regarding the occurrence and detection of, and legislation for ochratoxins in grapes and grape-derived products. They discussed the potential sources of ochratoxin contamination, and possible strategies for control of ochratoxins in wines and other grape products. Analytical methods for the detection of OTA were reviewed, including the application of a molecularly imprinted SPE method from Maier et al. (2004).

Aresta et al. (2006) have applied solid-phase microextraction (SPME), using a polydimethylsiloxane/divinylbenzene (PDMS/DVB) fiber, interfaced with liquid chromatography–fluorescence detection (LC–FD), to the determination of OTA in wine samples. Compared to the most widely adopted extraction/clean-up procedure based on immunoaffinity columns (IAC), the solvent-less extraction is simpler and more cost-effective, requiring the simple immersion of the fiber in 1.5 ml of diluted wine samples (1:20 v/v with 0.03 % HCl) for 60 min under magnetic stirring. The effect of dilution was two-fold: (i) the decrease in the ethanol content of the sample from 12 %–14 % to 0.6 %–0.7 % was beneficial since it is well known that the presence of an organic solvent in the aqueous extracting solution reduces the extraction efficiency of the fiber, and (ii) for mixed coatings with a porous solid as the primary extraction phase, the prevailing extraction mechanism is adsorption. A simplification of the matrix complexity can prevent (or mitigate) analyte displacement from binding sites by competitive effects and/or sorbent saturation. Furthermore, a fast LC separation is achieved under isocratic conditions (OTA retention time = 13.1 min). Data indicated that, after 1 h extraction at ambient temperature, ~10 % of the total OTA amount present in the sample solution was extracted. For operation under non-equilibrium conditions, this represented a good compromise between extraction time and sensitivity. The linear range investigated in wine was 0.25–8 ng/ml. At fortification levels of 0.5 and 2 ng/ml within-day intra-laboratory precision (repeatability) values, expressed as %RSD, were 5.9 and 5.1, respectively, whereas between days ($n = 4$) precision was 8.5 and 7.1 %, respectively. The limit of detection (LOD) at a signal-to-noise (S/N) ratio of 3 was 0.07 ng/ml; the limit of quantification (LOQ) calculated at S/N = 10 was 0.22 ng/ml, well below the European regulatory level of 2 ng/ml. The potential of the method has been demonstrated by the analysis of a number of different wine samples. Interestingly, the extraction efficiency was independent of the particular wine variety (red, rose or white). Furthermore no significant difference was observed between amounts extracted from wine sample and from standard solution, as ascertained from a "paired *t*-test" (95 % confidence level) performed at OTA concentration levels of 5, 2.5, 1.25 and 0.5 ng/ml. All the findings clearly indicated the absence of significant matrix effects. Compared to the most accredited clean/up step by IAC, SPME has a less favorable LOD (0.07 versus 0.01 ng/mL) but is completely solvent free, operationally simpler and definitely cheaper (one fiber can be used for more than 100 samples).

To demonstrate the broad applicability of MIP science, one more new MIP was developed for the determination of OTA in wheat extracts by Lai and co-workers (Zhou et al. 2004a,b). For better compatibility with a sensor system based on the optical phenomenon of surface plasmon resonance (SPR), a new preparation method was evaluated by electrochemical deposition of molecularly imprinted polypyrrole (MIPPy) on the sensor surface. Polypyrroles are excellent for electrochemical preparation of MIPs due to their stability under mild preparation conditions of room temperature and $+0.8$ V versus Ag/AgCl (even when a cross-linker is added). Conducting PPy is a biocompatible polymer matrix wherein biochemical molecules can be incorporated by way of doping. Functionalization with a wide range of bioorganic compounds (as templates) is possible. This novel application of MIPPy was demonstrated for real-time detection of OTA in biosensing (Yu et al. 2004, 2005). Real-time SPR mycotoxin sensing requires the rapid diffusion of the analyte into (and out of) the sensor element. This can be accomplished by imprinting into pseudo two-dimensional ultra-thin PPy films. A MIP sensor film recently reported by Tokareva et al. (2006) was only 32 ± 4 nm thick in the absence of analytes. Moreover, 100 % of the incorporated target molecule can be released by reversing the applied potential (Geetha et al. 2006). This can be exploited as a highly efficient electrochemically-controlled pulsed elution (ECPE) technique in the development of SPR sensors that can be regenerated for multiple analyses. Note that conductive MIPPy films, in 0.1 M Na_2SO_4 solution, allow for amperometric measurement of the total surface area (including CNTs and QDs) when the applied potential is pulsed between -0.2 V and $+0.4$ V versus Ag/AgCl (Gooding et al. 2004). MIPPy was next electrodeposited in macroporous stainless-steel frits for on-line micro solid phase preconcentration (SPP) of OTA prior to HPLC analysis with fluorescence detection (FD) (Yu et al. 2005a,b). Desirable properties such as a high surface area can be achieved on these macroporous stainless steel frits.

MIPs with selective recognition properties for zearalenone (ZON), an estrogenic mycotoxin, and structurally related compounds have been prepared using the noncovalent imprinting approach by Moreno-Bondi and co-workers (Urraca et al. 2006). Cyclododecyl 2,4-dihydroxybenzoate (CDHB), which exhibits resemblance to ZON in terms of size, shape and functionality, was synthesized and used as template for MIP preparation instead of the natural toxin. Several functional monomers were evaluated to maximize the interactions with the template molecule during the polymerization process. The polymer material prepared with 1-allylpiperazine (1-ALPP) as functional monomer, trimethyl trimethacrylate (TRIM) as cross-linker, and acetonitrile as porogen (in a 1:4:20 molar ratio) displayed binding capacities superior to any of the other MIPs tested. Selectivity of this material for ZON and structurally related and non-related compounds was evaluated using it as a stationary phase in liquid chromatography. The results demonstrated that the imprinted polymer showed significant affinity in the porogenic solvent for the template mimic (CDHB) as well as for ZON and other related target metabolites in food samples, dramatically improving the performance of previously reported MIPs for ZON recognition. Therefore, MIPs can be an excellent alternative for cleanup and preconcentration of the mycotoxin in contaminated food samples.

3.3. Nano-Sized Structures

Since 2001, analytical chemistry has undergone unprecedented developments in terms of automation capability, miniaturization, resolution power, sensitivity, and overall efficiency of new techniques. As researchers progress toward molecular level analysis, the design and fabrication of smart materials (when incorporated as interfacial recognition or transducer elements) will be of paramount importance. Applications of materials in analytical chemistry have recently been reviewed by He and Toh (2006), with focus on sensors, separations and extraction techniques. Materials with novel and interesting properties are developed through variation of their chemical composition, physical dimensions, and inclusion of biological components from nature. These interesting materials include hybrids, nanomaterials and biomolecular materials.

Nano-sized structures (with physical features less than 100 nm in one or more dimensions) may be in the form of particles, pores, wires or tubes. Structural dimension of the molecular recognition layer can influence partition or sensor kinetics to a large extent. Biomolecular materials are referred to as materials comprising biomolecules such as proteins and nucleic acids.

Carbon nanotubes (CNTs) were used by Lai et al. (2006) as structural fibers to build up a 3-D network of nanoporosity for improving the % recovery during SPP. These MIPPy/CNTs-modified stainless steel frits offered a high binding capacity and strong affinity for OTA (Yu et al. 2006c). A novel SPP device was also developed by electrochemically depositing MIPPy over CNTs packed inside a 22-gauge syringe needle. When applied to red wine analysis, HPLC-FD results demonstrated a significant enrichment of OTA at sub-ppb levels in the presence of wine matrix components. Using 0.5 ml of red wine, OTA could be determined down to 0.04 ppb. The total MIPPy/CNTs-μSPP-HPLC-FD analysis took only 40 min, including a μSPP time of 30 min (Wei et al. 2007). Alternatively, a syringe needle can be packed with MIPPy-encapsulated CNTs by electrodeposition. The MIPPy/CNTs-modified needle will be mounted in an autosampler for sequential HPLC-FD analysis of red wine samples. Operator-free processing of red wine samples >0.5-ml containing OTA at <0.04 ppb levels will be tested. Other mycotoxins (patulin, DON, etc.) are important templates for the making of new MIPPy/CNTs-modified syringe needles to further evaluate their autosampler compatibility.

Construction of simple and integrated sampling/extraction/elution/injection devices will be optimized by considering new ways to immobilize various sorbents in the stainless steel needle (Wang et al. 2005) and by programming different elution strategies on the HPLC autosampler. Knowing that conductive polymers are technically semiconductors, MIPPy can be effectively used as macromolecular glue to immobilize semiconductor QDs on the CNT's surface. The expected benefits would be a significantly larger total surface area and shorter diffusion distances for quantitative SPP of ultratrace mycotoxins.

Acrylate-based MIP nanospheres for the recognition of mycotoxins (patulin, DON) have been developed (Ciardelli et al. 2006). MIP nanospheres can be prepared by precipitation polymerization to exhibit a high degree of monodispersity (Wei et al. 2006). These nanospheres are glued by PPy electrodeposition onto a gold film surface. After template removal, the rebinding selectivity of MIP nanospheres will be tested on an SPR biosensor using patulin, DON, structural analogs, and metabolites. MIP micromembranes will be fabricated inside the 0.5-μm pores of a stainless steel frit for an improved recognition of biomolecules in physiological samples. Imprinted P(MMA-co-AA) copolymer membranes can be prepared from THF solution, via non-solvent induced phase separation (NIPS), by adopting the latest method of Silvestri et al. (2006). Specific recognition sites are introduced during membrane formation in the inversion bath. In addition, MIP nanospheres can be loaded onto the P(MMA-co-AA) micro-membrane (up to 100 mg/g) inside a macroporous stainless steel frit. After repetitive incubation in physiological solution (or acetonitrile), transmission electron microscopy (TEM) images will confirm the successful loading of particles as a stable layer on the membrane surface inside the macroporous frit structure. This approach promises interesting application in the development of composite membranes for analytical sample processing. Together with the typical selectivity of membranes (relying on their permeability properties and chemical nature), a significantly higher degree of selectivity is introduced with the specific cavities built in the MIP nanospheres.

3.4. Peptides and Proteins

It is generally accepted that misfolding of the cellular prion protein (PrP^C) leads to the accumulation of an insoluble, toxic PrP^{Sc} isoform (also known as scrapie PrP) in the brain. Abnormal prion proteins are associated with transmissible spongiform encephalopathies

(TSEs). The concentration of infectious PrP^{Sc} in blood is far too small to be detected by conventional methods, but it is still enough to spread the disease. A recent paper by Castilla et al. (2006) reported the detection of PrP^{Sc} in the blood of scrapie-infected hamsters using a biochemical amplification method that is conceptually analogous to the polymerase chain reaction. The method, protein misfolding cyclic amplification (PMCA), uses sound waves to vastly accelerate the process that prions use to convert normal proteins to infectious forms. This technological advance raises the possibility that prion diseases in humans and livestock could soon be diagnosed using blood samples if several technical limitations can be overcome (Supattapone et al. 2006). Lai et al. (unpublished results) have prepared new MIPs using two peptide sequences (DYEDRYYRENM and YPNQVYYRPMD) as templates. Each sequence contains the tri-peptide motif YYR, which is known to be partly responsible for the abnormal behavior of PrP^{Sc}. The molecular recognition properties of these MIPs are under investigation, in contrast to a control, in biomedical applications for the early diagnosis of animal diseases.

For presymptomatic detection of ultratrace PrP^{Sc} in TSE-infected animals during the preclinical period, development of a new class of MIP-encapsulated QDs for analyte preconcentration will be essential. The determination of any PrP^{Sc} at ultratrace levels is most promising by selective preconcentration using the molecular recognition property of a tailor-made MIP. Bovine recombinant prion protein (brecPrP) is used for all method development experiments because it is of a normal prion protein that does not represent a hazard to human health. In surface imprinting, brecPrP serves as a template around which a binding pocket is assembled that is complementary in shape and chemistry. Highly-selective recognition elements can be achieved through molecular imprinting on such self-assembled monolayer films (Li et al. 2006). These films, when prepared *in-situ* as a coating on QDs, can direct the target protein to adopt a given imprinted conformation through tailor-designed binding affinities. They can restrict random processes that occur readily in bulk solution to a surface-specific molecular recognition event. Both the shape recognition and the hydrophobic interaction between PrP^{Sc} and the MIP coating would constitute the predominant recognition forces in aqueous media (Liu et al. 2006).

For the determination of several abnormal proteins simultaneously after selective preconcentration on MIPs, capillary electrophoresis and microfluidics can offer many analytical separation merits in conjunction with the sensitive detection method of laser-induced fluorescence (LIF) (Chen et al. 2006). A unifying hypothesis is that MIP nanospheres, as well as modified QDs, can be characterized by their electrophoretic mobility (= observed rate of migration divided by electric field strength in given buffer medium) in CE based on their charge, mass, size, shape and surface functionality. Microfluidics can be used to make the CE immunoassay more rapid and robust. Its high speed and throughput are especially needed for screening live animals. Uniform MIP nanospheres will be prepared via adaptation of controlled suspension polymerization in a spiral-shaped microchannel using a perfluorocarbon liquid as the continuous phase. (Zourob et al. 2006). Monodisperse droplets containing the monomers, template, initiator, and porogenic solvent are introduced into the microchannel. The droplet size can be varied by changing the flow conditions in the microfluidic device. A surfactant may be required to stabilize the droplets. Nanospheres of uniform size are produced by subsequent UV polymerization, quickly and without wasting polymer materials or the brecPrP template. The specific binding sites that are created during the imprinting process can be tested via binding analysis.

3.5. Viruses

MIPs targeted for tobacco mosaic virus (TMV) recognition have been synthesized by Kofinas and coworkers (Bolisay et al. 2006). Batch equilibrium studies using imprinted and non-imprinted polymer hydrogels in TMV and TNV solutions were conducted to determine virus-binding capacities. TMV-imprinted hydrogels showed increased binding to TMV (8.8 mg TMV/g polymer) compared to non-imprinted hydrogels (4.2 mg TMV/g polymer), while non-imprinted hydrogels bound similar amounts of TMV or TNV. This research demonstrated that molecular imprinting of viruses can be used to selectively induce binding of target viruses based on shape differences of their virions.

3.6. Bacterial Cells and Endospores

Typically, in conventional microbiological analysis there is a need for a pre-enrichment step to increase the numbers of target pathogen in the sample. This is then followed by selective enrichment and subsequent differentiation on selective agar followed by confirmatory (genetic or physiological testing) tests. Depending on the pathogen screened, this can take on the order of 4–8 days to complete in addition to the time required to send the sample to the laboratory and interpret the results (Lazcka et al. 2006). The key to rapid microbiological diagnostics is to eliminate one or more of these steps, thereby reducing the time required for analysis (Lazcka et al. 2006). One of the most commercially successful approaches has been the application of immunomagnetic separation (IMS) that can reduce the pre-enrichment time and negate the need for selective pre-enrichment (Olsvik et al. 1994). IMS is based on capturing and concentrating the pathogen of interest using appropriate antibodies immobilized on the surface of 100μm diameter paramagnetic beads. When a magnetic field is applied the beads become magnetized and, hence, can be separated from the sample matrix, thereby concentrating the pathogen of interest and removing background microflora that could interfere with selective plating, as well as interferents that detrimentally affect detection, using techniques such as Real-Time PCR (Olsvik et al. 1994). IMS has become a standard method for screening food samples, such as ground beef, for *E. coli* O157:H7 (O'Brien et al. 2005).

Although antibody coated Dynabeads have contributed significantly to improving rapid detection of pathogens, there are disadvantages to the approach, such as limited shelf-life, cost and availability. Molecular imprinted beads provide an alternative to antibodies for capturing bacterial cells, and proof-of-concept has been demonstrated by Harvey et al. (2006). The approach involves introducing the bacteria of interest onto beads coated with a polymerizing polymer such as polyamine. The polymerization process is activated by illumination with UV and by cell templates subsequently released by refluxing in HCl:methanol solutions (Fig. 29.7). The imprints are then coated with lectins to enhance binding of target cells. When the imprinted

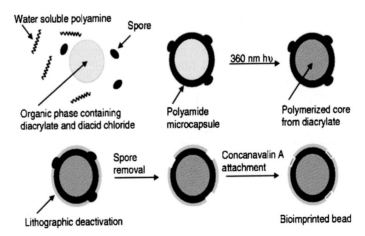

Figure 29.7. Schematic diagram for sysnthezing spore imprinted beads. (Adapted from Harvey et al. 2006 with kind permission of Springer Science and Business Media).

beads were suspended in suspensions of the target cells, recovery yields of 44 %–55 % were obtained. This compares to 40 %–65 % typically obtained using antibody based capture systems. The approach of microbial imprinted beads has been demonstrated for both spores (Harvey et al. 2006) and vegetative cells (Perez et al. 2001). Regarding the latter, imprints were prepared using *Listeria monocytogenes* alone or in combination with *Staphylococcus aureus*. Although imprinted films demonstrated enhanced capture of cells compared to non-imprinted control beads, selectivity was relatively poor. This may have been expected given that the recognition is based on the morphology of the cells as opposed to epitopes for antibody binding. As a consequence it is unlikely that microbial imprinted beads could be used as an alternative to immunobased systems for selectively capturing pathogens. Nevertheless, given the ease of production and low cost of microbial imprinted beads, such a technique could find utility where the diagnostic platform can compensate for poor selectivity.

4. Biosensors Based on MIPs

The application of imprinted films as recognition agents in biosensors has been considered from the early days of MIP research. For example, Vlatakis et al. (1993) developed a pseudo-immunosensor based on MIPs to detect morphine or diazepam. The approach was based on a competitive assay format using radiolabeled analytes. Here, the amount of nonbound radiolabeled substrate could be correlated to the concentration of analyte in the sample. Radiolabels are ideal for sensors based on MIPs since the probe and target are identical, in addition to the low levels of detection that can be achieved. However, radiolabels are not compatible with routine analysis, and alternatives are needed. Unlike immunoassays that can use sandwich formats to detect bound analytes, sensors based on MIPs can only be based on competitive assays or via detecting binding directly (Ansell 2002).

The main approaches used to detect analyte binding to MIPs can be broadly divided into four approaches:

1) Tag the probe/reporter with a chromagenic, fluorescent or enzyme label in a competitive assay. The main limitation of this approach is that the structure of the probe is altered and, as a consequence, the affinity of the MIP;
2) The target analyte is labeled and detected upon binding to the surface. The main problem with this approach is that the imprints may have higher affinity for the label compared to the analyte, thereby resulting in false-positive reactions;
3) The probe that is not structurally related to the analyte of interest. Here, in the absence of the analyte, the probe will bind weakly to the imprints on the polymer surface, thereby generating a high signal. However, in the presence of the target analyte the binding of the probe will be restricted due to the higher affinity of the imprints to the target. The choice of probe in this format has to be carefully selected in terms of chemical characteristics and size.
4) Directly detect analyte binding through monitoring changes in properties of the supporting polymer film. Here, the bound analyte can be detected through changes in mass/refractive index changes monitored using surface plasmon resonance or quartz-crystal microbalance (QCM). Alternatively, the binding of the analyte can be detected via optical (florescence) or electrochemical changes in the supporting polymer.

Examples of all the above reporting strategies will be provided in the following sections. The described sensors are restricted to those that are related to detecting microbial products (metabolites, toxins) or whole cells. More comprehensive reviews on MIP-based sensors can be found in several reviews published in the literature (Ansell 1996; Anderson 2000; Sellergren 2000; Ansell 2002).

4.1. MIP-based Sensors for Detection of Amino Acids

The majority of the early work with MIPs focused on fabricating imprinted films using amino acids as templates. Amino acids represent excellent model systems for MIPs due to their diverse structures and provide a means of assessing specificity, especially in terms of differentiating between enantiomers. In the context of the current chapter, amino acids have no direct relevance for detecting microbial activity. However, the work performed on amino acid MIPs highlights the potential for fabricating imprints with high affinity for microbial metabolic products and peptides.

A sensor for the detection of homocysteine (marker of vascular dysfunctions) based on fluorescence probes and MIPs was reported by Chow et al. (2002). Here, MIPs were fabricated (MAA, trimethylolpropane trimetharcylate and 1-azobiscyclohex- anecarbonitrile polymer) using DL-homocysteine derivatized with N-(1-pyrenyl)maleimide, to form N-(1-pyrenyl)maleimide-DL-homocysteine (PM-H). By using PM-H as the target analyte the lower limit of detection was reported to be 750 nM. Similar approaches have been used for fabricating sensors for D-fructose, D-glucose (Marvin 1997) and maltose (Gilardi and Zhou 1994). Although the use of conjugated fluorescent probes enhances sensitivity, the inclusion of a derivatization step prior to analysis can be viewed as a disadvantage in terms of sensor applications.

Liu et al. (2006) developed an MIP for L-tryptophan using QCM to detect binding of the analyte. Noncovalent molecular imprinting polymers were synthesized using acrylamide (AM) and trimethylolpropane trimethacrylate (TRIM) as a functional monomer and a cross-linking agent, respectively. The sensor was able to discriminate between the L- and D-tryptophan enantiomers with a lower detection limit of 8.8 mM.

A further example of a QCM-based sensor was fabricated by Stanley et al. (2003) for the detection of L-serine. Here, the imprinted polymer was directly cast onto the surface of the QCM electrode using L-serine as the template. The sensor was demonstrated to discriminate between the L- and D-forms with a lower detection limit of 0.2 ppb (Stanley et al. 2003).

In a similar approach, a piezoelectric sensor coated with a thin molecularly imprinted sol-gel film for determination of L-histidine in aqueous phase has also been reported (Zhang et al. 2005). L-histidine was imprinted directly into silica sol-gel films that consisted of a hybrid mixture of functionalized organosilicon precursors (phenyltrimethoxysilane and methyltrimethoxysolane). The developed sensor had a working linear range between 50nM to 0.1mM with a lower detection limit of 2.5nM and exhibited higher stereo selectivity.

An MIP-based sensor to detect the neurotoxic amino acid domoic acid, derived from algae, has been reported (Lotierzo et al. 2004). Surface plasmon resonance (SPR) was used to detect the binding of the target domoic acid to the imprinted film. The films were prepared on gold surfaces by initially treating the chip with 4,4'-azo-bis (cyanovaleric acid) photo-initiator, carbodiimmide and 1-hydroxybenzotriazole. The polymerization solution used to form the MIP comprised 2-(diethylamino) ethyl methacrylate, N,N'-methylenebisacrylamide and a domoic acid template. After UV polymerization was completed the template was removed by soaking in methanol. Due to the relatively low molecular weight of domoic acid (FW 311) the analyte was conjugated to horseradish peroxidase. The sensor had a lower detection limit of $5 \mu g \, l^{-1}$, which compares to $1.8 \mu g \, l^{-1}$ when monoclonal antibodies are used as the biorecognition agent (Lotierzo et al. 2004).

An L-glutamate MIP has been fabricated using titanium oxide deposited into a gold-coated quartz crystal microbalance. Template removal was achieved by submerging the sensor in water with the resultant voids exhibiting a high sensitivity toward L-glutamate with a lower detection limit of $10 \mu M$ (Feng et al. 2004). The main benefit derived from using TiO_2 is the ability to form films in the aqueous phase and enhanced imprinting efficacy of carboxylic acid-functionalized templates such as amino acids (Lee et al. 1999).

Traditional polymer materials used to fabricate MIPs are inert and consequently do not exhibit changes upon binding of the target analyte. This is the primary reason optical- and microbalance-based methods are applied to detect bound analyte. However, there is interest in preparing MIPs from more reactive polymers that can amplify the analyte binding event. In this respect, the potential of using conducting polymers as MIPs is gaining interest (Ramanaviciene and Ramanavicius 2004; Yu et al. 2005). *Conducting polymer* is a broad term to describe a group of conjugated polymers that possess electronic properties (Gerard et al. 2002). The conductivity of conducting polymers (for example, polypyrrole) can be modulated by external stimuli such as ion exchange, pH, applied bias potential (oxidative state) and polymer chain elongation.

An MIP-based conducting polymer sensor for detecting tyrosine has been reported (Liang et al. 2005). Imprinted films were prepared by spin coating a Ni electrode with a pyrrole solution containing tyrosine and polymerization process initiated by heating at 90°C for 1 min. The tyrosine template was released from the MIP by soaking in methanol for 90 min. The sensor could discriminate between L and D-tyrosine with a lower detection limit of 5mM. The researchers hypothesized that the change in film charge by interaction with L-tyrosine was due to preferential doping effect by virtue of the imprints within the film (Liang et al. 2005).

A similar approach has been undertaken to fabricate monosaccharide-imprinted films formed from polyaniline on electrode surfaces. Here, aniline was electropolymerized on the surface of electrodes in the presence of boronic acid, fluoride and monosaccharide template. The formation of a saccharide-aminophenylboronic acid complex in the presence of fluoride acted as dopant to counter the cation charges on the polyaniline film. The resultant films exhibited an enhanced affinity of D-fructose over D-glucose when the films were cycled between reductive and conductive states (Deore and Freund 2003).

In a different approach, imprinted films derived from polypyrrole have been prepared using colloidal polymerization. Here the polypyrrole is chemically polymerized in the presence of polyvinylpyrrolidone (steric stabilizer), peroxodisulfate (oxidant) and l-lactate (dopant). When the film was overoxidized the formed voids had a high affinity (as determined using QCM) for L-alanine and L-cysteine but not for the larger phenylalanine. However, films with enhanced uptake of phenylalanine could be fabricated by substituting L-lactate with L-phenyllactate (Chen et al. 2000; Shiigi et al. 2003).

4.2. Molecular-Imprinted Films for Toxins

With the advent of an increased threat of bioterrorism there has been interest in developing sensors to detect toxins of microbial or plant origin. Numerous papers have been published with respect to immuno or chromatography based methods (Bergwerff and Van Knapen 2005), although relatively little work has been performed with MIPs.

Nevertheless, an SPR based ochratoxin A sensor has been reported (Yu and Lai 2005). The imprinted film was formed by electropolymerization of pyrrole in the presence of ochratoxin A. Analyte binding was observed by monitoring changes in the refractive index with a lower detection limit of 0.05 ppb. Regeneration of the film was possible using acetic acid: methanol solutions with no interference being noted when assays were performed in wheat or wine extracts (Yu and Lai 2005).

A piezoelectric sensor has been reported by Chianella et al. (2003) for the analysis of microsystin LR, a highly toxic compound produced by freshwater cyanobacteria, present in aqueous samples. The polymer was synthesised using microcystin-LR template with AMPSA, UAEE, EGDMA (cross-linker) and (cyclohexane-carbonitrile) (initiator). Particles (38–63μm) were prepared from the film and template removed by repeated washing. The formed polymer was spin coated into a QCM electrode using PVC as the supporting medium. The lower detection limit of the sensor was 0.35 nM. The same imprinted film could also be used as an SPE, achieving up to 1000-fold concentration of the toxin (Chianella et al. 2003).

Toxins of microbial origins are invariably proteins and no sensors based on MIPs have yet been reported in the literature. This in part could be explained by the difficulty in handling toxins and also the relative high quantities (hence cost) to prepare imprinted films. Nevertheless, Rick and Chou (2005) described a protein (lysozyme and cytochrome C) sensor based on MIP formed from 3-aminophenylboronic acid (APBA) monomer using ammonium persulphate as the initiator. The MIPs were formed on the surface of gold QCM electrodes in aqueous solutions. The formed films were selective for the respective enzyme with a lower detection limit of 1.39 nM (Rick and Chou 2005). It is likely that further examples of sensors based on protein-imprinted films will appear in the future as the technology develops (Bossi et al. 2007).

4.3. Microbial Imprinted Polymers

The imprinting of microorganisms or mammalian cells has been hitherto considered to be unfeasible due to restricted diffusion of such bulky templates/analytes from and into the film. Nevertheless, sensors have been reported whereby imprints of microbes have been formed onto polymer films. Dickert has been the pioneer of microbial imprinted films, and his research has demonstrated the proof-of-principle of imprinting viruses through to erythrocytes (Dickert and Hayden 2002; Dickert et al. 2004a,b; Hayden et al. 2006). Films were formed by using a "stamping" technique whereby the immobilized cell template was pressed onto the surface of a sol-gel polymer layer. By using the stamping technique it is possible to form uniform imprints homogenously across the polymer surface, and problems relating to template release are negated.

The underlying principle of the stamping technique is provided in Fig. 29.8. Self-assembled layers of the cell or virus template are prepared on smooth, flat, surfaces such as glass slides. To facilitate release of the templates, following the polymerization process, blockers (for example, glucose, 4-aminophenol, or 4-aminobenzoic acid) are coated onto the template surface and excess removed. The prepolymerized polymer mixture (e.g., acrylic acid and ethylene glycol dimethacrylate) is deposited on the transducer of interest, for example QCM electrode. Polymerization is initiated by UV illumination and the stamp pressed onto the

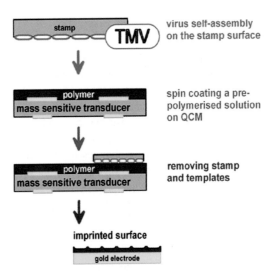

Figure 29.8. Stamping technique for preparing virus imprinted films (Adapted from Dickert et al. 2004 with kind permission of Springer Science and Business Media).

forming film. Upon completion of the polymerization process, the template is removed, leaving voids with high affinity for the target cell/virus particle. Films imprinted with tobacco mosaic virus (TMV) could detect the virus in the range of 100 ngml^{-1} – 1mgml^{-1} with an assay time less than 15 minutes (Hayden et al. 2006). The robustness of the sensor was demonstrated by the detection of TMV in plant sap derived from infected plants (Dickert et al. 2004). The sensor was also selective, being able to differentiate between TMV and human rhinovirus serotype 2 (Hayden et al. 2006). Conversely human rhinovirus imprinted films exhibited higher selectivity for the virus with negligible sensor responses being obtained for TMV (Hayden et al. 2006).

TMV is an example of a nonenveloped virus that is highly tolerant to environmental stress, which is not only important for survival but also is conducive to the imprinting process. However, many of the viruses important to animal and human health are enveloped, containing only a lipid–protein membrane to protect nucleic acid. Enveloped viruses are unstable outside the host and can easily be destroyed even by mild heating or exposure to disinfectants. With this background it would be expected that forming imprints of enveloped viruses would be almost impossible. However, imprints have been fabricated using parapox ovis virus using a process termed *soft lithography* (Hayden et al. 2006). A soft silicone rubber (poly(dimethylsiloxane)) is used as the base polymer. Again, the sensor could sensitively detect parapox ovis virus but exhibited a lower response to damaged virus particles (Lieberzeit et al. 2005). Given that viruses are difficult if not impossible to cultivate in the laboratory, it can be envisaged that virus imprinted films hold strong potential for future sensing devices.

Yeast-cell imprinted films have also been prepared using polyurethane as the base substrate and QCM as the transduction strategy (Dickert et al. 2003). The sensor could be used to measure up to 21 g l^{-1} in complex growth media and could be readily regenerated (Dickert et al. 2003). Another example is a blood group-selective sensor prepared by templating with erythrocyte ghosts. Both the blood-group A and B imprinted material selectively distinguish between blood groups A, B and O, whereas no difference in sensor signal has been observed for AB, where both blood group antigen types are present on the cell surface (Dickert and Hayden 2002; Dickert et al. 2004a, b).

Virus imprinted films have also been fabricated using polypyrrole (Ramanaviciene and Ramanavicius 2004). Polyprrole was electropolymerized onto the surface of platinum black electrodes in the presence of bovine leukemia virus (BLV) glycoprotein *gp*51. The template was removed from the polypyrrole by applying reducing potentials. When the polypyrrole was oxidized, gp51 protein was re-introduced (doped) onto the film, resulting in an increase in current that was detected using pulsed amperometric detection. The MIP could be regenerated although imprint integrity was progressively decreased following each de-doping/doping cycle (Ramanaviciene and Ramanavicius 2004).

Polypyrrole has also been applied as a base polymer to fabricate *Bacillus* endospore imprinted films (Namvar and Warriner 2007). In this example, *Bacillus subtilis* spores were deposited on the surface of glassy carbon electrodes onto which pyrrole was electrochemically polymerized. The polypyrrole film had a characteristic globular, open structure with no clear imprints of spores being observed. Therefore, a secondary poly(3-mthylthiophene) layer was deposited onto the surface of the polypyrrole. The selection of poly(3-methylthiophene) was based on its smoother and more compact structure. Release of the spore template was achieved by soaking the electrode in DMSO (Namvar and Warriner 2007). The interaction of spores with the imprinted film could be detected by changes in film Y' (susceptance). However, the sensitivity of the sensor was relatively poor, with a lower detection limit of 4 log cfu/ml. The sensitivity of the sensor was enhanced by stimulating germination of the attached/captured spores. Here, germination of captured spores was stimulated by incubating the sensor at 70°C for 10 min to prime the spores. The electrode was then submerged in germination solution (KCl containing L-alinine and glucose) and change in film charge monitored over time. By

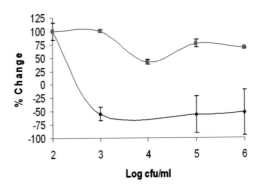

Figure 29.9. Relative change in imprinted film charge as a function of *Bacillus subtilis* spore density. *B. subtilis* endospore imprinted (■) or nonimprinted (•) polypyrrole/EGTA/poly(3-methylthiophene) was submerged in spore suspension for 10 min and subsequently held at 70°C for 10 min. The electrodes were then transferred to germination solution, and cyclic voltammograms performed between −0.3 to 0.4V vs Ag/AgCl at 50mV/s. The accumulated charge measured at t = 15 min was subtracted from that obtained at t = 0 min to represent the response. Values represent the change in charge at different spore densities relative to the response using 10^2 cfu/ml *B. subtilis* endospores. (Reprinted from Namvar and Warriner (2007), with permission of Elsevier).

using this approach the lower detection limit of the sensor was 2 log cfu/ml. In comparison, negligible changes in sensor response were observed for non-imprinted films (Fig. 29.9). The sensor response to attached spores was thought to be due to the release of dipicolinic acid (DPA) from the spore core during the germination process that under went ion-exchange with the conducting polymer matrix (Namvar and Warriner 2007).

29.In a similar approach, an endospore sensor has been fabricated based on using DAP imprinted polymers. The MIPs were formed from vinylic monomers polymerized in a dimethyl-formamide solution containing DPA as a template. Microbeads (10–20μm) were formed from the imprinted polymer by grinding and submerged in a dialysate (containing DPA) of spent culture medium derived from sporulating *Bacillus* spores. The bound DPA was detected using steadystate and time-resolved fluorescence. The sensor had a lower detection limit of 0.15 mM DPA, which corresponds to 5 log cfu/ml (Anderson et al. 2004).

5. Conclusions and Future Perspectives

MIPs hold great potential for developing devices for both concentrating and detecting microbes or their metabolic products. Although research is at an early stage, the versatility of MIPs to detect small analytes, proteins and even microbial cells holds strong potential for the future. It can be envisaged that MIPs will be applied for monitoring the formation of bioproducts in the biotechnology industry or presence of spoilage metabolites in industrial fermentations. Microbial imprinted films will find utility in concentrating and detecting a diverse range of microbes—especially those that are difficult to cultivate within the laboratory environment, such as enteric viruses and protozoans. Sensors based on MIPs to detect bacterial pathogens such as *Salmonella*, *Campylobacter* or *Escherichia coli* O157:H7 may find application for onsite sensors, although advantages over current (immuno) techniques remain to be demonstrated. A more likely application of MIPs will be in the concentration of pathogens to enhance the sensitivity of downstream diagnostic tests, such as Real-Time PCR. The most promising future for MIPs will be in the development of systems that can simultaneously detect and concentrate toxins—especially those linked to bioterrorism. There are several groups driving toward developing such integrated technologies, and it can be predicted that the first commercial MIP-based sensors will appear in the very near future.

References

Allender CJ,Brain KR and Heard CM1 (1999) Pharmaceutical applications for molecularly imprinted polymers Binding cross-reactivity of Boc-phenylalanine enantiomers on molecularly imprinted polymers. Chirality 9:233–237

Alvarez-Lorenzo C, Guney O, Oya T, Sakai M, Kobayashi M, Enoki T, Takeoka Y, Ishibashi K, Kuroda K, Tanaka K, Wang GQ, Grosberg AY, Masamune S and Tanaka T (2001) Reversible adsorption of calcium ions by imprinted temperature sensitive gels. J. Chem. Phys. 114:2812–16

Alvarez-Lorenzo C and Concheiro A (2004) Molecularly imprinted polymers for drug delivery J. Chrom B. 804:231–45

Alvarez-Lorenzo C, Yañez F, Barreiro-Iglesias R and Concheiro A (2006) Imprinted soft contact lenses as norfloxacin delivery systems J. Controlled Release. 113:236–244

Andersson L, Muller R,Vlatakis G and Mosbach K (1995) Mimics of the binding sites of opioid receptors obtained by molecular imprinting of enkephaline and morphine. Proc. Natl. Acad. Sci. 92:4788–4792

Andersson.L (1996) Application of molecular imprinting to the development of aqueous buffer and organic solvent based radioligand binding assays for (S)-propranolol. Anal. Chem. 68:111–17

Andersson LI (2000) Molecular imprinting for drug bioanalysis - A review on the application of imprinted polymers to solid-phase extraction and binding assay J. Chromatogr. B. 739:163–168

Anderson J, Nelson J, Charles R, Ringelberg D, Tepper G and Dmitry P (2004) Steady-state and frequency-domain lifetime measurements of an activated molecular imprinted polymer imprinted to dipicolinic acid. J Fluores. 14:269–274

Annaka M and Tanaka T (1992) Multiple phases of polymer gels. Nature. 355:430–32

Annaka M, Tokita M, Tanaka T, Tanaka S and Nakahira TJ (2000b) Multiple-phase behavior and its microscopic implication for 4-acrylamidosalicylic acid gel Chem. Phys. 112:471–77

Annaka M, Hara K, Du R, Chuang J, Wasserman K, Grosberg AY, Masamune S and Tanaka T (2000a) Frustrations in polymer conformation in gels and their minimization through molecular imprinting Phys. Rev. Lett. 85:5000–5003

Ansell RJ (1996) Towards artificial antibodies prepared by molecular imprinting Clin. Chem. 42:1506–1510

Ansell RJ (2002) MIP-ligand binding assays (pseudo-immunoassays). Bioseperations 10:365–377

Aresta A, Vatinno R, Palmisano F and Zambonin CG (2006) Determination of Ochratoxin A in wine at sub ng/mL levels by solid-phase microextraction coupled to liquid chromatography with fluorescence detection. J. Chromatogr. A. 115:196–201

Batra D, Shea KJ (2003) Combinatorial methods in molecular imprinting Curr. Opin. Chem. Biol. 7:434–442

Bergwerff AA and Van Knapen F (2006) Surface plasmon resonance biosensors for detection of pathogenic microorganisms: Strategies to secure food and environmental safety. J AOAC Int. 89:826–831

Bolisay LDV, March JF, Bentley WE and Kofinas P (2005) Virus recognition using molecularly imprinted polymer hydrogels. Abs Am Chem Soc. 603:4273–4273

Bolisay LD, Culver JN and Kofinas P (2006) Molecularly imprinted polymers for tobacco mosaic virus recognition. Biomaterials. 27:4165–4168

Bossi A, Bonini F, Turner APF et al. (2007) Molecularly imprinted polymers for the recognition of proteins: The state of the art. Biosens Bioelectron. 22:1131–1137

Byrne ME, Park K and Peppas NA (2002) Molecular imprinting within hydrogels. Adv Drug Delivery Rev. 54:149–61

Caro E, Marcé RM, Cormack PAG, Sherrington DC and Borrull F (2006) On-line solid-phase extraction with molecularly imprinted polymers to selectively extract substituted 4-chlorophenols and 4-nitrophenol from water Analytica Chimica Acta 562:145–151

Chen H, Olmstead MM, Albright RL, Devenyi J and Fish RH (1997) Measurement of enantiomeric excess using molecularly imprinted polymers. Angewandte Chemie Int. 36:642–45

Chen ZD, Takei Y, Deore BA and Nagaoka T (2000) Enantioselective duptake of amino acid with overoxidized polypyrrole colloid templated with L-lactate Analyst 125:2249–2254

Chen S, Liu BF, Fu L, Xiong T, Liu T, Zhang Z, Huang ZL, Lu Q, Zhao YD and Luo Q (2006) Continuous wave-based multiphoton excitation fluorescence for capillary electrophoresis J. Chromatogr A. 119:160–167

Chianella I, Piletsky SA, Tothill IE, Chen B and Turner APF (2003) MIP-based solid phase extraction cartridges combined with MIP-based sensors for the detection of microcystin-LR. Biosens Bioelectron 18:119–127

Chien YW, Lin S (2002) Optimisation of treatment by applying programmable rate-controlled drug delivery technology. Clinical Pharmacokinetics 41:1267–99

Chow CF, Lam MHW and Leung MKP (2002) Fluorescent sensing of homocysteine by molecular imprinting Anal. Chim Acta. 466:17–30

Chrambach A and Rodbard D (1971) Polyacrylamide Gel Electrophoresis Science.172:440–51

Ciardelli G, Borrelli C, Silvestri D, Cristallini C, Barbani N and Giusti P (2006) MIP-based solid phase extraction cartridges combined with MIP-based sensors for the detection of microcystin-LR. Biosens. Bioelectron. 2: 2329–2334

Demuel G, Ozcetin G, Turan E and Caykara T (2005) pH/temperature-sensitive imprinted ionic poly (N-tertbutylacrylamide-co-acrylamide/maleic acid) hydrogels for bovine serum albumin. Macromolecular Bioscience. 5:1032–1035

Deore B and Freund MS (2003) Saccharide imprinting of poly(aniline boronic acid) in the presence of fluoride. Analyst 128:803–806

Dickert FL and Hayden G (2002) Bioimprinting of polymers and sol-gel phases. Selective detection of yeasts with imprinted polymers. Anal Chem 74:1302–1306

Dickert FL, Hayden O, Lieberzeit P, Haderspoeck C, Bindeus R, Palfinger C, Wirl B (2003) Nano- and micro-structuring of sensor materials - from molecule to cell detection. Synth Met 138:65–69

Dickert FL, Hayden O, Bindeus R, Mann KJ, Blaas D, Waigmann E (2004a) Bioimprinted QCM sensors for virus detection-screening of plant sap. Anal Bioanal Chem 378:1929–1934

Dickert FL, Lieberzeit P, Miarecka SG, Mann KJ, Hayden O, Palfinger C (2004b) Synthetic receptors for chemical sensors–subnano- and micrometre patterning by imprinting techniques. Biosens Bioelectron 20:1040–1044

Elemans JAAW, Rowan AE and Nolte RJM (2003) Mastering molecular matter. Supramolecular architectures by hierarchical self-assembly J. Mater. Chem. 13:2661–2670

Enoki T, Tanaka K, Watanabe T et al. (2000) Frustrations in polymer conformation in gels and their minimization through molecular imprinting. Phy Rev Lett 85:5000–5003

Feng LA, Liu YJ and Hu JM (2004a) Molecularly imprinted solid-phase extraction for the screening of antihyper-glycemic biguanides. Langmuir 20:1786–1790

Feng SY, Lai EPC, Dabek-Zlotorzynska E and Sadeghi S (2004b) Molecularly imprinted TiO_2 thin film by liquid phase deposition for the determination of L-glutamic acid. J. Chromatogr. A. 1027:155–160

Fernández-González A, Guardia L, Badía-Laíño R and Díaz-García ME (2006) Mimicking molecular receptors for antibiotics - analytical implications.Trend Anal Chem 10:949–957

Gao S, Wang W and Wang B (2001) Building fluorescent sensors for carbohydrates using template-directed polymer-izations. Bioorg Chem 29:308–320

Gerard M, Chaubey A, Malhotra BD (2002) Application of conducting polymers to biosensors. Biosens Bioelectron 17:345–359

Geetha S, Rao CRK, Vijayan M and Trivedi DC (2006) Biosensing and drug delivery by polypyrrole. Anal. Chim. Acta. 568:119–122

Gilardi G, Zhou Q, Hibbert L et al. (1994) Engineering the maltose-maltose protein for reagentless fluorescent sensing. Anal Chem 66:3840–3847

Gooding JJ, Wasiowych C, Barnett D, Hibbert DB, Barisci JN and Wallace GG (2004) Electrochemical modulation of antigen-antibody binding Biosens. Bioelectron. 20:260–268

Gubitz G, Schmid MG (2000) Recent progress in chiral separation principles in capillary electrophoresis. Electrophoresis 21:4112–4135

Guo TY, Xia YQ, Hao GJ, Song MD and Zhang BH (2004) Study of the binding characteristics of molecular imprinted polymer selective for cefalexin in aqueous media. Biomaterials. 25:5905–5912

Guo TY, Xia YQ, Hao GJ, Zhang BH, Fu GQ, Yuan Z, He BL and Kennedy JF (2005a) Chemically modified chitosan beads as matrices for adsorptive separation of proteins by molecularly imprinted polymer. Carbohydr. Polym. 62:214–21

Guo TY, Xia YQ, Wang J, Song MD and Zhang BH (2005b) Chitosan beads as molecularly imprinted polymer matrix for selective separation of proteins. Biomaterials 26:5737–5745

Guzmán-Vázquez de Prada A, Reviejo AJ and Pingarrón JM (2006) A method for the quantification of low concentration sulfamethazine residues in milk based on molecularly imprinted clean-up and surface preconcentration at a Nafion-modified glassy carbon electrode. J. Pharm. Biomed. Anal. 40281–286

Harvey SD, Mong GM, Ozanich RM, Mclean JS, Goodwin SM, Valentine NB and Fredrickson JK (2006) Preparation and evaluation of spore-specific affinity-augmented bio-imprinted beads. Bioanlyt Chem. 386:211–219

Hawkins DM, Stevenson D and Reddy SM (2005) Investigation of Protein Imprinting in Hydrogel-based Molecularly Imprinted Polymers (HydroMIPs). Analytica Chimica Acta 542:61–65

Hawkins DM, Trache A, Stevenson D, Holzenburg A, Meininger G and Reddy SM (2006) Quantification and Confocal Imaging of Protein Specific Molecularly Imprinted Polymers (HydroMIPs). Biomacromolecules 7:2560–2564

Hayden O, Lieberzeit PA, Blaas D et al. (2006) Artificial antibodies for bioanalyte detection-sensing viruses and proteins. Advan Funct. Met 16:1269–1278

Hubbell JA (1999) Bioactive biomaterials. Curr. Opin. Biotechnol. 10:123–29

Karlsson JG, Andersson L and Nicholls IA (2001) Probing the molecular basis for ligand-selective recognition in molecularly imprinted polymers selective for the local anaesthetic bupivacaine Anal Chimica Acta 435:57–64

Kato M, Nishide H, Tsuchida E and Sasaki T (1981) Complexation of metal ion with poly(1-vinylimidazole) resin prepared by radiation induced polymerization with template metal ion. J Polymer Sci Chem 19:1803–09

Kempe M and Mosbach K (1994) Direct resolution of naproxen on a noncovalently molecular imprinted chiral stationary phase. J. Chromatogr. A 664:276–79

Kempe M and Mosbach K (1995) Receptor binding mimetics-a novel molecular imprinted polymer. Tetrahedron Lett 36:3563–66

Kempe M (1996) Antibody-Mimicking polymers as chiral stationary phases in HPLC Anal. Chem. 68:1948–1953

Kofufuta E, Zhang Y-Q and Tanaka T (1991) Saccharide sensitive phase transition of a lectin loaded gel. Nature 35:302–04

Kriz D, Ramstrom O, Svensson A and K Mosbach (1995) Introducing biomimectic sensors based on molecular imprinted polymers as recognition elements. Anal. Chem. 67:2142–2144

He L and Toh CS (2006) Recent advances in analytical chemistry - A material approach. Anal. Chim. Acta 556:156–165

Lai EPC and Wu SG (2003) Molecularly imprinted solid phase extraction for rapid screening of cephalexin in human plasma and serum Anal. Chim. Acta 481:165–167

Lai EPC and Feng SY (2003) Molecularly imprinted solid phase extraction for rapid screening of metformin. Microchem. J. 75:159–168

Langer RS and Peppas NA (1981) Present and future applications of biomaterials in controlled drug delivery systems. Biomaterials 2:201–214

Lee SW, Ichinose I, Kunitake T (1998) Molecular imprinting of azobenzene carboxylic acid on a TiO_2 ultrathin film by the surface sol-gel process. Langmuir 14:2857–2863

Lee M, Chae L and Lee KC (1999) Carbohydrate recognition by porphyrin-based molecularly imprinted polymers. Nano Mat 11:195–201

Liang HJ, Ling TR, Rick JF, Chou TC (2005) Molecularly imprinted electrochemical sensor able to enantroselectivly recognize D and L-tyrosine. Anal. Chim. Acta. 542:83–89

Liao JL, Wang Y and Hjerten S (1996) Novel support with artificially created recognition for the selective removal of proteins and for affinity chromatography. Chromatographia. 42:259–62

Liao HP, Zhang ZH, Nie LH and Yao SZ (2004) Preparation of the molecularly imprinted polymers-based capacitive sensor specific for tegafur and its characterization by electrochemical impedance and piezoelectric quartz crystal microbalance. J. Biochem. Bioph. Methods 59:75–87

Lin JM, Nakagama T, Uchiyama K and Hobo T (1997) Temperature effect on chiral recognition of some amino acids with molecularly imprinted polymer filled capillary electrochromatography. J. Pharm. Biomed. Anal. 15: 1351–1358

Liu F, Liu X, Ng SC and Chan HSO (2006a) Enantioselective molecular imprinting polymer coated QCM for the recognition of L-tryptophan. B-Chemical 113:234–240

Liu X, Ouyang C, Zhao R, Shangguan D, Chen Y and Liu G (2006b) Enantioselective molecular imprinting polymer coated QCM for the recognition of L-tryptophan. Anal. Chim. Acta. 571:23–241

Lieberzeit PA, Gazda-Miarecka S, Halikias K et al. (2005a) Imprinting as a versatile platform for sensitive materials - nanopatterning of the polymer bulk and surfaces. B-Chemical. 111:259–263

Lieberzeit PA, Glanznig G, Jenik M, Gazda-Miarecka S, Dickert FL and Leidl A (2005b) Softlithography in chemical sensing - Analytes from molecules to cells. Sensors 5:509–518

Lotierzo M, Henry OYF, Piletsky S, Tothill I, Cullen D, Kania M, Hock B and Turner APF (2004) Surface plasmon resonance sensor for domoic acid based on grafted imprinted polymer. Biosens Bioelectron 23:145–152

Lowman AM and Peppas NA (1999) Encyclopedia of Controlled Drug Delivery. John Wiley and Sons, New York

Lu Z-R, Kopeckova P, Kopecek (1999) Polymerizable Fab' antibody fragments for targeting of anticancer drugs. J. Nature Biotechnol. 17:1101–1104

Lu SL, Cheng GX and Pang X (2006) Study on preparation of protein-imprinted soft-wet gel composite microspheres with magnetic susceptibility and their characteristics. I. Preparation and particle morphology. J. Appl. Polym. Sci. 99:2401–2407

Lubke C, Lubke M, Whitcombe MJ and Vulfson EN (2000) Imprinted polymers prepared with stoichiometric template-monomer complexes: Efficient binding of ampicillin from aqueous solutions. Macromolecules 33:5098–5105

Maier NM, Buttinger G, Welhartizki S, Gavioli E and LindnerW (2004) Molecularly imprinted polymer-assisted sample clean-up of ochratoxin A from red wine: merits and limitations. J. Chromatogr. B. 804:103–111

Mahony JO, Nolan K, Smyth MR and Mizaikoff B (2005) Molecularly imprinted polymers-potential and challenges in analytical chemistry Analytica Chimica Acta. 534:31–39

Martin-Esteban A,Turiel E and Stevenson D (2001) Effect of template size on the selectivity of molecularly imprinted polymers for phenylurea herbicides. Chromatographia 53:434–37

Marvin JS, Corcoran EE and Hattangadi NA et al. (1997) The rational design of allosteric interactions in a monomeric protein and its applications to the construction of biosensors. Proc Natl Acad Sci 94:4366–4371

Mayes AG, Andersson L, Mosbach K (1994) Sugar binding polymers showing anomeric and epimeric discrimination obtained by noncovalent molecular imprinting. Anal Biochem 222:483–488

Mayes AG and Mosbach K (1996) Molecularly imprinted polymer beads: Suspension polymerization using a liquid perfluorocarbon as the dispersing phase Anal. Chem. 68:3769–3774

Mayes AG and Whitcombe MJ (2005) Synthetic strategies for the generation of molecularly imprinted organic polymers Adv. Drug Delivery Rev. 57:1742–1778

Miyata T, Jikihara A, Nakamae K and Hoffman AS (1996) Preparation of poly(2-glucosyloxyethyl methacrylate)concanavalin A complex hydrogel and its glucose-sensitivity Macromol. Chem. Phys.197:1147–1157

Miyata T, Jiye M, Nakaminami T and Uragama T (2006) Tumor marker-responsive behavior of gels prepared by biomolecular imprinting Proc. Natl. Acad. Sci., USA 103:1190–1193

Namvar A and Warriner K (2007) Microbial Imprinted Polypyrrole and poly3-methylthiophene Composite Films for the Detection of *Bacillus* Endospores. Biosens Bioelectron 22:2018–2024

Nicholls IA, Anderson H (1996) Molecularly imprinted polymers: Man made mimics of antibodies and their application in analytical chemistry. J. Mol Rec 9:652–657

Nilsson KGI, Lindell J, Norrlow O and Sellergren B (1994) Imprinted polymers as antibody mimetics and new affinity gels for selective separations in capillary electrophresis. J. Chromatogr. A .680. 57–61.

Norrlow O, Mansson M-O and Mosbach K 1(987) Improved chromatography prearranged distances between boronate groups by the molecular imprinting approach. J. Chromatogr. A 396:374–377

Obaidat AA and Park K (1996) Characterization of glucose dependent gel-sol phase transition of the polymeric glucose-concanavalin A hydrogel system Pharm. Res. 13. 989–995

Ou SH, Wu MC, Chou TC and Liu CC (2004) Polyacrylamide gels with electrostatic functional groups for the molecular imprinting of lysozyme Analytica Chimica Acta. 504. 163–66

Owega S and Lai EPC (1999) Silver cationization of thia fatty acids and esters in laser desorption/ionization time-of-flight mass spectrometry. J. Mass Spectrom. 34:872–879

Pang X, Cheng GX, Li R, Lu S and Zhang Y (2005) Bovine serum albumin-imprinted polyacrylamide gel beads prepared via inverse-phase seed suspension polymerization. Analytica Chimica Acta.550:13–17

Pang X, Cheng GX, Lu SL and Tang EJ (2006) Synthesis of polyacrylamide gel beads with electrostatic functional groups for the molecular imprinting of bovine serum albumin. Analytical Bioanal. Chem. 384:225–230

Pap T, Horvai G (2004) Characterization of the selectivity of a phenytoin imprinted polymer. J. Chromatogr. A 1034:99–107

Perez N, Alexander C and Vulfson EN (2001) Need title of article or chapter here. In: Sellergren B (ed) Molecularly imprinted polymers: man-made mimics of antibodies and their applications in analytical. chemistry. Elsevier, Amsterdam, The Netherlands, pp 295–303

Peppas NA (1987) Hydrogels in Medicine and Pharmacy, Volume 2: Polymers. CRC Press, Boca Raton, Florida

Peppas NA, Huang Y, Torres-Lugo M, Ward JH and Zhang J (2000) Physicochemical, foundations and structural design of hydrogels in medicine and biology. Ann Rev Biomed Engin.2:9–29

Piletsky SA, Piletska EV, Elgersma AV, Yano K, Karube I, Parhometz,YP and Elskaya AV (1995) Atrazine sensing by molecular imprinted membranes. Biosens. Bioelectron. 10:959–64

Piletska EV, Piletsky SA, Karim K, Terpetschnig E and Turner APF (2004a) Custom synthesis of molecular imprinted polymers for biotechnological application - Preparation of a polymer selective for tylosin. Analytica Chimica Acta 504:179–83

Piletsky SA, Piletska EV, Karim K, Foster G, Legge C and Turner APF (2004b) Biotin-specific synthetic receptors prepared using molecular imprinting. Analytica Chimica Acta 504:123–130

Piletsky SA, Andersson HS and Nicholls IA (2006) Polymer cookery. 2. Influence of polymerization pressure and polymer swelling on the performance of molecularly imprinted polymers. Macromolecules 32:633–636

Pison U, WelteT, Giersig M and Groneberg DA (2006) Nanomedicine for respiratory diseases. Eur. J. Pharmacol. 533:341–350

Ramanaviciene A and Ramanavicius A (2004) Molecularly imprinted polypyrrole-based synthetic receptor for direct detection of bovine leukemia virus glycoproteins. Biosens. Bioelectron. 20:1076–1082

Ramstrom O,Ye L and Mosbach K (1996) Artificial antibodies to corticosteroids prepared by molecular imprinting. Chem Biol 3:471–77

Ramstrvm O, Nicholls IA and Mosbach K (1994) Synthetic peptide receptor mimics - Highly steroselective recognition in noncovalent molecular imprinted polymers. Tetrahedron Asymmetry 5:649–656

Rick J and Chou TC (2005) Imprinting unique motifs formed from protein-protein associations. *Anal Chim Acta* 542:26–31

Rick J and Chou TC (2006) Using protein templates to direct the formation of thin-film polymer surfaces. Biosens Bioelectron 22:544–549

Sallacan N, Zayats M, Bourneko T and Kharitonov AB (2002) Imprinting of nucleotide and monosaccharide recognition sites in acrylamidephenylboronic acid-acrylamide copolymer membranes associated with electronic transducers. Anal. Chem. 74:702–712

Sanchez-Barragan I, Costa-Fernandez JM, Sanz-Medel A, Valledor M and Campo JC (2006) Room-temperature phosphorescence (RTP) for optical sensing. Tends in Anal Chem 25:958–967

Saunders GD, Foxon SP, Walton PH, Joyce MJ and Port SN (2000) A selective uranium extraction agent prepared by polymer imprinting. Chem. Commun. 4:273–274

Sellergren B (1989) Molecular imprinting by noncovalent interactions enantiospecificity and binding capacity of polymers prepared under condit ions favoring the formation of template complexes. Makromolekulare Chemie 190:2703–2711

Sellergren B and Andersson L (1990) Molecular recognition in microporous polymers prepared by a substrate-analog imprinting strategy. J. Org. Chem. 55:3381–3383

Sellergren B (2000) Application of imprinted synthetic polymers in binding assay development. Meth-A. Compan. Meth. Enzymol. 22:92–95

Senholdt MM, Siemann M and Mosbach K (1997) Determination of cyclosporin A and metabolites total concentration using a molecularly imprinted polymer based radioligand binding assay. Anal. Lett. 30:1809–1821

Schweitz L, Andersson L, Nilsson S (1997) Capillary electrochromatography with predetermined selectivity obtained through molecular imprinting. Anal. Chem.69:1179–1183

Shea KJ and Dougherty TK (1986) Molecular recognition on synthetic amporphous surfaces-The influence of functional group positioning in the effectiveness of molecular recognition. J. Am. Chem. Soc. 108:1091–1093

Shea KJ and Sasaki DY (1989) On the control of microenvironment shape of functionalized network polymers prepared by template polymerization. J. Am. Chem. Soc. 111:3442–3444

Shea KJ, Spivak DA and Sellergren B (1993) Polymer complements to nucleotide bases-selective binding of adenine-derivatives to imprinted polymers. J. Am. Chem. Soc. 115:3368–3369

Shi Q and Jackowski G (1998) One-dimensional polyacrylamide gel electrophoresis. In: Hames BD, (ed) Gel electrophoresis of proteins: A practical approach, 3rd ed. Oxford University Press, Oxford, pp 1–52

Shiigi H, Yakabe H, Kishimoto M et al. (2003) Molecularly imprinted overoxidized polypyrrole colloids: Promising materials for molecular recognition. Microchimica Acta 143:55–162

Siemann M, Andersson LI and Mosbach K (1996) Selective recognition of the herbicide atrazine by noncovalent molecularly imprinted polymers. J. Agric. Fd. Chem. 44:141–145

Silvestri D, Barbani N, Cristallini C, Giusti P and Ciardelli G (2006) Molecularly imprinted membranes for an improved recognition of biomolecules in aqueous medium. J. Membrane Science 282–284

Spivak DA and Shea KJ (1999) Molecular imprinting of carboxylic acids employing novel functional macroporous polymers. J. Org. Chem. 64:4627–34

Stanley S, Percival CJ, Morel T, Braithwaite A, Newton MI, McHale G and Hayes W (2003) Enantioselective detection of L-serine. Sensors Actuators B-Chemical 89:103–106

Suedee R, Songkram C, Petmoreekul A, Sangkunakup S, Sankasa S and Kongyarit N (1998) Thin-layer chromatography using synthetic polymers imprinted with quinine as chiral stationary phase. J. Pharm. Biomed. Anal 11:272–76

Supattapone S, Geoghegan JC and Rees JR (2006) On the horizon: a blood test for prions. Trends in Microbiol 14:149–151

Syu MJ, Nian YM, Chang YS, Lin XZ, Shiesh SC and Chou TC (2006) Ionic effect on the binding of bilirubin to the imprinted poly(methacrylic acid-co-ethylene glycol dimethacrylate). J. Chromatogr. A.1122:54–62

Tarley CRT, Segatelli MG and Kubota LT (2006) Amperometric determination of chloroguaiacol at submicromolar levels after on-line preconcentration with molecularly imprinted polymers. Talanta 69:259–266

Titirici MM, Sellergren B (2004) Peptide recognition via hierarchical imprinting. Anal BioAnal. Chem. 378:1913–21

Tokareva I, Tokarev I, Minko S, Hutter E and Fendler JH (2006) Ultrathin molecularly imprinted polymer sensors employing enhanced transmission surface plasmon resonance spectroscopy. Chem. Commun. 31:3343–3349

Tokuyama H, Fujioka M and Sakohara S (2006) Temperature swing adsorption of heavy metals on novel phosphate-type adsorbents using thermosensitive gels and/or polymers. J. Chem. Eng. Jpn. 38:633–640

Tong D, Hetenyi C, Bikadi Z, Gao JP and Hjerten S (2001) Some studies of the chromatographic properties of gels ('artificial antibodies/receptors') for selective adsorption of proteins. Chromatographia 54:7–14

Urraca JL, Marazuela MD, Merino ER, Orellana G and Moreno-Bondi MC (2006) Molecularly imprinted polymers applied to the clean-up of zearalenone and alpha-zearalenol from cereal and swine feed sample extracts. J. Chromatogr A. 1116:127–134

Varga J and Kozakiewicz Z (2006) Ochratoxin A in grapes and grape-derived products. Trends Food Sci. Technol. 17:72–81

Vallano PT, Remcho VT (2000) Affinity screening by packed capillary high-performance liquid chromatography using molecular imprinted sorbents. I. Demonstration of feasibility J.Chromatogr. A. 888:23–34

Venkatesh S, Sizemore SP and Byrne ME (2007) Biomimetic hydrogels for enhanced loading and extended release of ocular therapeutics. Biomaterials 28:717–724

Vlatakis G, Andersson LI Miller R and Mosbach K (1993) Drug assay using antibody mimics made by molecular imprinting. Nature 361 645–647

Wang A, Fang F and Pawliszyn P (2005) Sampling and determination of volatile organic compounds with needle trap devices. J. Chromatogr. A. 1072:127–132

Wei S, Molinelli A, Mizaikoff B (2006) Molecularly imprinted micro and nanospheres for the selective recognition of 17 beta-estradiol. Biosens. Bioelectron. 21:943–951

Whitcombe MJ, Rodriguez ME, Villar P and Vulfson EN (1995) A new method for the introduction of recognition site functionality into polymers prepared by molecular imprinting-Synthesis and characterization of polymeric receptors for cholesterol. J. Am. Chem. Soc. 117:7105–7111

Wichterle O and Lim D (1960) Hydrophilic Gels for Biological Use. Nature. 185:117–118

Wu SG, Lai EPC and Mayer PMJ (2006) Molecularly imprinted solid phase extraction–pulsed elution–mass spectrometry for determination of cephalexin and α-aminocephalosporin antibiotics in human serum. Pharmaceut. Biomed. Anal. 36:483–490

Wulff G and Sarhan A (1972) Über die Polykondensation von Diäthylphosphit mit aliphatischen Diolen. Angewandte Chemie. 84:364–365

Wulff G, Best W and Akelah A (1984) Enzyme-analogue built polymers. 17 investigations of the reacemic-resolution of amino acids. React. Polym. 2:167–174

Wulff G and Vietmeier J (1989) Enzyme-analogue built polymers. 25 synthesis of miroporous copolymers from alpha-amino-acid based vinyl compounds. Makromol. Chem. 190:1727–1735

Yang Z, Mesiano A, Venkatasunramanian S, Gross SH, Harris JM and Russell AJ (1995) Activity and stability of enzymes incorporated into acrylic polymers. J. Am. Chem. Soc. 117:4843–4850

Yu C and Mosbach K(1997) Molecular imprinting utilizing an amide functional group for hydrogen bonding leading to highly efficient polymers. J.Org.Chem. 62:4057–4064

Yu C, Ramstrom O and Mosbach K (1997) Enantiomeric recognition by molecularly imprinted polymers using hydrophobic interactions. Anal. Lett. 30:2123–2140

Yu JCC and Lai EPC (2004) Polypyrrole film on miniaturized surface plasmon resonance sensor for ochratoxin A detection. Synthet. Metals 143:253–258

Yu JCC and Lai EPC (2005a) Interaction of ochratoxin A with molecularly imprinted polypyrrole film on surface plasmon resonance sensor. Reactive Func Poly..63:171–176

Yu JCC and Lai EPC (2005b) Polypyrrole modified stainless steel frits for on-line micro solid phase extraction of ochratoxin A Anal. Bioanal. Chem. 381:948–952

Yu JCC, Krushkova S, Lai EPC and Dabek-Zlotorzynska E (2005c) Molecularly-imprinted polypyrrole-modified stainless steel frits for selective solid phase preconcentration of ochratoxin A. Anal. Bioanal. Chem. 382:1534–1540

Yu JCC and Lai EPC (2006) Molecularly imprinted polypyrrole modified carbon nanotubes on stainless steel frit for selective micro solid phase pre-concentration of ochratoxin A. React. Funct. Polymers. 66:702–711

Zhang ZH, Liao HP, Li H, Nie LH and Yao SZ (2005) Stereoselective histidine sensor based on molecularly imprinted sol-gel films. Anal. Biochem. 336:108–116

Zhao Z, Wang CH, Guo MJ, Shi LQ, Fan YG, Long Y and Mi HF (2006) Molecular imprinted polymer with cloned bacterial protein template enriches authentic target in cell extract. FEBS Lett. 580:2750–2754

Zheng N, Li Y, Chang W, Wang Z and Li T (2002) Sulfonamide imprinted polymers using co-functional monomers. Anal Chimica Acta. 452:277–283

Zhou SN and Lai EPC (2004a) N-phenylacrylamide functional polymer with high affinity for ochratoxin A. React. Funct. Polymers. 58:35–42

Zhou SN, Lai EPC and Miller JD (2004b) Analysis of wheat extracts for ochratoxin A by molecularly imprinted solid-phase extraction and pulsed elution. Anal. Bioanal. Chemistry. 378:1903–1906

Zou HF, Huang XD, Ye ML and Luo QZ (2002) Monolithic stationary phases for liquid chromatography and capillary electrochromatography. J.Chromatogr.A. 954:25–32

Zourob M, Mohr S, Mayes AG, Macaskill A, Perez-Moral N, Fielden PR and Goddard NJ (2006) A micro-reactor for preparing uniform molecularly imprinted polymer beads. Lab Chip 6:296

IV

Microsystems

30

Microfluidics-Based Lysis of Bacteria and Spores for Detection and Analysis

Ning Bao and Chang Lu

Abstract

The disruption of the membrane/coat, or lysis, of bacteria and spores is often a critical step for analyzing the intracellular molecules such as proteins and nucleic acids. In this chapter, we review recent advances in the application of microfluidic devices for lysis of bacteria and spores. We divide existent devices and methods into five categories: mechanical, chemical, thermal, laser, and electrical. We also point out future directions in this field.

1. Introduction

Rapid detection and analysis of pathogenic bacteria/spores has become increasingly important for applications ranging from public health and food safety to biological weapons defense. Traditional methods for the detection of bacteria on foods rely on culturing and plating of the bacteria followed by identification using biochemical or serological assays. These cultural methods are time consuming—taking three days to determine a total viable count and five to seven days to detect specific pathogenic bacteria (Kaspar and Tartera 1990). Recent advances in the development of rapid detection tools have dramatically improved both the speed and sensitivity (Swaminathan and Feng 1994; Ivnitski et al. 1999). Most of such assays are based on the detection and analysis of proteins and nucleic acids. For example, enzyme-linked immunosorbent assay (ELISA) has been used to identify bacteria by detecting their binding to the surface immobilized antibody (Bhunia 1997; Gehring et al. 1998, 2004). Gene probes are also used to recognize and bind to nucleic acid targets. Polymerase chain reaction (PCR) based DNA hybridization methods have shown superb sensitivity and specificity (Tietjen and Fung 1995; Fratamico and Strobaugh 1998; Fratamico 2003).

Although the size of bacteria cells is much smaller than those of animal and plant cells, they have almost the same structure from a morphological point of view. Depending on the type of the bacteria cells, generally, a barrier called a capsule (or outer membrane, cell wall, plasma membrane) composed of polysaccharide and/or proteins protects the bacteria from the extracellular environment. In this barrier the membrane proteins are doped to form one or more sites all over the hydrophobic surface. Most of the proteins and DNAs, which are of interest for detection, exist in the internal structures of the bacterial cells, such as nucleoi, ribosome, storage

Ning Bao and Chang Lu • Department of Agricultural and Biological Engineering, School of Chemical Engineering, Birck Nanotechnology Center, Bindley Bioscience Center, Purdue University, West Lafayette, Indiana.

M. Zourob et al. (eds.), *Principles of Bacterial Detection: Biosensors, Recognition Receptors and Microsystems*,
© Springer Science+Business Media, LLC 2008

granules, and sometimes, endospores. Cell/spore lysis can disrupt the capsule and release the majority of the intracellular materials for detection and analysis. As an added benefit, such lysis also leads to inevitable cell death, which is desired for sterilization purposes.

A number of methods have been developed for lysis of bacteria/spores during the past decades for assays at bench scale (Harrison 1991). Recent developments in microfluidics have opened up possibilities for new, revolutionary approaches to performing biological and chemical assays at micrometer scale (Verpoorte 2002). Microfluidic chips offer reduction in sample amounts (consuming nanoliter to picoliter volume), portability for field applications, high level of integration and automation, and high throughput. The concept of integrating different steps of the biological assay onto one microchip makes it possible to construct portable devices for point-of-care pathogenic detection (Manz et al. 1992). Effective coupling of lysis with other steps is challenging on a microfluidic platform. As we will review in this chapter, some conventional lysis methods have been adopted in microfluidic systems. Moreover, novel lysis approaches have also been demonstrated taking advantage of versatility offered by microscale device design. Successful detection of pathogenic bacteria on a microfluidic platform relies on a series of procedures including culturing and selection of the bacteria cells, lysis of bacteria cells, as well as separation and detection of intracellular materials using biochemical assays (El-Ali et al. 2006).

The choice of the lysis method largely depends on the target intracellular species and the particular assay following lysis. Although varying on a case-by-case basis, several general considerations often apply. First, one would want to avoid potential interference with the subsequent assay from the lysis step. When intracellular proteins (antigens) are targeted, the physical and chemical conditions (such as temperature, pH value and chemical composition of the buffer) can have dramatic effects on the functions of the assayed molecules. For example, the results of immunoassays can potentially be influenced by the surfactants, such as sodium dodecyl sulfate (SDS), as well as high temperature, which are often applied for lysis. Second, the time required for lysis needs to be considered, especially when the assay involves different steps in series. Rapid lysis decreases the total time for the detection assay. Third, another important factor is the throughput of the lysis method. Depending on the sensitivity of the subsequent assay, a large number of bacterial cells/spores are typically required for detection and analysis. The ability to handle cells/spores with high throughput in a microscale device is often important.

In this chapter we will mainly discuss the recent advances of bacteria/spores lysis based on microfluidics. We will point out the characteristics and potential applications of different mechanisms, such as mechanical, thermal, chemical, laser-based, and electrical lysis. We will also discuss the future directions that this field may take.

2. Bench Scale Methods for Bacteria/Spore Lysis

Cell lysis has been a routine operation and many commercialized techniques and protocols have been developed over the year. Here we will briefly review the conventional methods for bacteria/spores lysis because many of the microfluidics-based assays are based on the same or similar mechanisms as those of the large-scale ones. The conventional methods can be roughly classified into physical and chemical/biological methods (Geciova et al. 2002).

Physical methods dominated the field in the early days. They include manual grinding, freeze/thaw cycles, mechanical disruption, sonication and liquid homogenization.

Manual grinding is relatively simple and commonly used for the disruption of plant cells because the wall of the plant cells could be effectively destroyed in this way (Tonshoff and Raschke 1977). With this approach, the cells are usually frozen in liquid nitrogen, and the combination of a mortar and a pestle is necessary in the disruption procedure. Similar results can be achieved by adding glass beads to a cell sample (Ghuysen and Strominger 1963). The lysis of cells is accelerated due to their collision with the beads. This approach is often used for the lysis of yeast cells.

Cell lysis can be also achieved by repeatedly freezing and suspending cells at low temperatures no greater than 0°C and then thawing them at room or higher temperatures (Zhou et al. 1991). The cell membrane is destroyed due to the formation of large crystals during the gradually freezing procedure. Although it is time and labor consuming, this approach is especially effective for the release of recombinant cytoplasm proteins when used for bacteria lysis.

Waring blenders and Polytrons have been applied to mechanically destroy cell membranes with the rotating blades (Caprioli and Rittenbe 1969; Sherman et al. 1976). They are typically applied for the lysis of cells with volumes larger than 1 ml. Another tool, sonication (Kim et al. 2006) is also widely used for lysis of bacterial cells. The high frequency sound waves delivered by the vibrating probe can efficiently destroy the bacterial spores or cells.

Homogenization has been extensively used in combination with other lysis approaches because it is often necessary to reduce the viscosity of the lysate for subsequent biological assays. In this method, the cell membrane is sheared by water when cells are forced to pass through a narrow space. This equipment for homogenization has three common types: Dounce homogenizer (Penman et al. 1963), Potter-Elvehjem homogenizer (Miller et al. 1965) and French press (Garen and Echols 1962). It should be emphasized that the French press is often applied to lyse bacterial cells. In a French press, the suspension of cells is forced by a piston to pass through a very small hole in the press so that high pressure is applied to the cells. Homogenization requires expensive equipment.

Compared with the above physical approaches, presently chemical/biological lysis approaches and protocols are becoming increasingly popular due to several advantages. First, after the reactions between bacterial spores/cells and chemical detergents or biological enzymes, cell membrane is destroyed and intracellular materials are released (Malamy and Horecker 1964, Scandell and Kornberg 1971). Normally these reactions can take place under mild conditions, such as room temperature. In comparison with the above-mentioned physical lysis approaches, undesirable local heating is often inevitable. Such heating can easily denature or aggregate intracellular proteins. The device often needs to be pre-chilled and the samples placed on ice in order to avoid this problem. Second, chemical/biological lysis does not require expensive instruments, which are necessary for some physical lysis methods. Third, chemical/biological lysis allows quantitative addition of different chemical/biological reagents to cell/spore suspensions. Because of this, chemical/biological lysis often yields more reproducibility than physical lysis. Finally, chemical/biological lysis methods are target-oriented and these protocols have been optimized for obtaining a specific target molecule or category such as genomic DNA or a certain protein antigen.

A number of commercial kits have been developed for bacterial cell lysis. For obtaining intracellular proteins, Geno Technology developed the Bacterial–PE LBTM kit to lysis bacteria using the lysozome lysis approach. During the lysis procedure, nucleic acids, such as DNA and RNA, are cleaved and removed with nuclease treatment. In order to enhance extraction and improve stability of intracellular proteins, organic agents, mild non-ionic detergents and various salts are applied in this kit. The proteins prepared with this kit can be used for subsequent chromatographic and electrophoretic separation and protein folding. There are also several commercial kits for DNA extraction from bacteria/spores for subsequent polymerase chain reaction (PCR). Sigma-Aldrich offers the GenElute Bacterial Genomic DNA Kit for extraction of genomic DNA from a variety of cultured bacteria. This kit consists of the steps of bacteria lysis, binding of DNA on a membrane, removal of impurities with a centrifuge, and elution of DNA from the membrane. This kit is able to handle bacterial suspensions with volumes of less than 1.5 ml with a final yield of 15 ~ 20 μg DNA. The time required is less than 2 hours. Invitrogen offers the ChargeSwitch gDNA Mini Bacteria Kit for the same application. Their protocol consists of similar procedures with magnetic beads applied for DNA adsorption. The yield of DNA can reach 12 μg, and time for the purification of genomic DNA from bacterial culture is less than 15 minutes.

3. Bacteria/Spore Lysis Based on Microfluidic Systems

Microfluidics offers the manipulation of tiny amounts of the liquid samples (typically at pico to nanoliter scale) and the integration of a number of biological assays on a microchip. The integration of sample preparation and testing is often a prerequisite for reduction in sample amounts since the tiny amount of target materials obtained by a microfluidic device can only be tested using other microfluidics-based assays (Sanders and Manz 2000). Therefore, the implementation of on-chip lysis of bacteria cells is a critical problem to address before micro total analysis systems (μTAS) can be applied to bacteria/spores detection. We will see below that some of the bench scale lysis methods can be miniaturized on a microfluidic chip. However, there are also a number of novel lysis techniques that take advantage of what microscale structures offer. We will comment on how researchers develop bacteria/spores lysis in the context of integrating it with other procedures in microfluidic systems for detection and analysis purposes.

3.1. Mechanical Lysis

Due to the small size of microchannels, it is extremely difficult to duplicate conventional mechanical lysis methods in microfluidic systems. Having high-speed moving parts or taking advantage of inertia is often not practical in microfluidic systems. The small size of bacterial cells poses another difficulty. For example, mechanical lysis of mammalian cells using nanobars has been implemented on a microfluidic platform (Di Carlo et al. 2003). However, the same device has not been demonstrated for bacteria lysis. The bacterial cells may be able to slip through the gap in the structure and clog the device more easily.

Madou's research group developed a mechanical lysis device for bacterial cells by creating a rapid-shear-flow carrying microspheres in a compact-disk-based microfluidic system (Kim et al. 2004). They performed theoretical investigation into the collision and friction between the microspheres in the rotating annular chambers of a CD and demonstrated that the important parameters were the size and density of microspheres, the rotation rate, the angular speed and the solid volume fraction (SVF). The key step in their experiments was the intro-duction of rimming flow in the microfluidic chamber through rotating the CD at a high speed and changing directions periodically. *E. coli* cells were lysed in this CD-based system. Their results demonstrated that most DNA was extracted when the volumetric ratio of cell suspension to beads was set to be 2:1.

The mechanism of sonication for bacteria lysis is based on the gaseous cavitation generated by the sonication probe at high power. Therefore, in order to sonicate bacteria/spores in microfluidic systems, the sonication probe needs to be miniaturized. In order to rapidly lyse *Bacillus anthracis*, which could become the poor person's atomic bomb, Belgrader et al. (1999) developed a microsystem consisting of a minisonicator and a plastic spore lysis cartridge. A chamber for the lysis of bacterial spores was constructed between the minisonicator and the plastic cartridge with a flexible interface. The minisonicator was composed of stacked piezoelectric disks and a titanium horn interfaced with a 3-mm polypropylene membrane. The

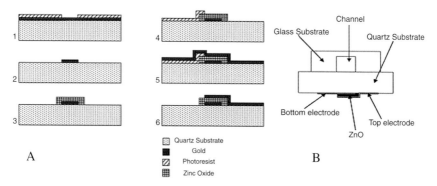

Figure 30.1. Microfluidic sonicator for real-time disruption of eukaryotic cells and bacterial spores. (A) Fabrication steps for the PZT element; (B) Cross section of the microfluidic channel (Reproduced from Marentis et al. (2005), Copyright 2005 World Federation for Ultrasound in Medicine & Biology).

size of the minisonicator was only one third of that of a commercial one. With the addition of glass beads, *Bacillus* spores were successfully lysed within only 30 seconds; normally this process might take 1 hour. Combined with the downstream microchip PCR analysis, the total time to perform the lysis and detection of the bacteria was greatly decreased to less than 15 minutes. In their follow-up work, further improvement was made in order to transfer more ultrasonic energy to the cell suspension (Taylor et al. 2001). They applied an analytical model on a horn-interface-liquid interaction to investigate the influence of pressure on the transfer of ultrasonic power. Their study indicated that the liquid should be pressurized to promise its tight contact with the horn tip and the flexible interface. The lysis of bacterial spores with this integrated system was further confirmed by SEM images of the lysed spores.

In another report, Marentis et al. (2005) developed a minisonicator integrated with microfluidic channels for the lysis of *Bacillus subtilis* spores. As shown in Fig. 30.1, in their system, the glass substrate with channel structure was combined with the quartz substrate to form the microchannel. On the other side of the quartz substrate, transducers consisting of a ZnO layer fabricated in between two golden layers were constructed to generate the ultrasonic power. An amplified constant sinusoidal signal was applied to the transducers to generate the ultrasonic power for cell lysis in the microchannel. Effective lysis of the bacterial spores could be realized by manipulating the flow rate and the density of the spore suspension. Results of PCR analysis demonstrated that 50 % of *Bacillus subtilis* spores were lysed with 30 seconds in this system. Their minisonicator design presented high compatibility with the microfluidic platform.

3.2. Chemical Lysis

Chemical lysis of bacterial cells on a microchip largely involves the mixing between cells and chemical reagents so that chemical lysis typically takes seconds or longer to finish. In addition, when applying chemical lysis, one needs to be extremely sensitive to whether the chemical reagents can affect the subsequent assays.

Laminar flow at the low Reynolds numbers in microchannels has been applied for the mixing necessary for chemical lysis. Yager's research group designed an "H" channel to perform the chemical lysis of *E. coli* and integrated protein analysis (Schilling et al. 2002). In their system, laminar flow was utilized for the mixing and separation of bacteria cells and lysis detergent on the basis of the different diffusion rates of the cells and lysis agent. It can be observed from Fig. 30.2 that there were three input ports, two output ports and two main channels (lysis channel and detection channel) in their microfluidic system. The flow rates at

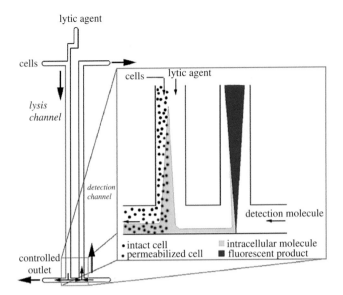

Figure 30.2. Schematic of microfluidic device for cell lysis and fractionation/detection of intracellular components. Pump rates are controlled at all inlets and one outlet. Lytic agent diffuses into the cell suspension, lysing the cells. Intracellular components then diffuse away from the cell stream, and some are brought around the corner into the detection channel, where their presence can be detected by the production of a fluorescent species from a fluorogenic substrate (Reproduced from Schilling et al. (2002). Copyright 1998 American Chemical Society).

all five ports were accurately controlled so that the expected flow patterns in two main channels could be maintained. When bacterial suspension and detergent solution were input into the lysis channel, their mixing occurred based on diffusion. The larger diffusion coefficient of smaller chemical detergent molecules made them diffuse rapidly into the bacteria suspension. As a result, the membranes of bacterial cells were disrupted and the intracellular materials were extracted in the lysis channel. In the detection channel, the intracellular materials were then brought together with a fluorogenic reagent for labeling. The authors applied the above system for the detection and quantification of β-galactosidase extracted from *E. coli*.

Heo et al. (2003) performed chemical lysis of bacterial cells in hydrogel patches inside microfluidic channels. *E. coli* cells were firstly entrapped in the hydrogel patches that were photopolymerized in the microchannel by introducing the hydrogel precursor solution containing the bacterial cells into the system. It should be pointed out that the application of UV light for the formation of hydrogel did not seriously compromise the bacterial membrane. The authors demonstrated that small molecules (including the dyes and chemical detergents) could diffuse into the hydrogel so that the chemical lysis could be carried out. They applied sodium dodecyl sulfate (SDS) for the chemical lysis of *E. coli* cells. The lysis was revealed by staining dead cells with propidium iodide (PI). It was suggested that the entrapped bacterial cell system could potentially work as biosensors.

For chemical lysis, the lysis reagents should be carefully selected because sometimes they may compromise or contaminate the intracellular materials that are needed for the downstream assays. Therefore, it is interesting that deionized (DI) water was applied as the lysis reagent for *E. coli* cells in the report of Prinz et al. (2002). The authors firstly applied lysozyme on the *E. coli* cells to remove the peptidoglycan layer between the inner and outer membranes of *E. coli* because this layer is able to protect cells from osmotic pressure from the surrounding solution. After this treatment, the shape of *E. coli* cells changed from rod shape to spherical. The obtained *E. coli* spheroplasts could be stored in a buffer solution with a high concentration of sucrose.

The authors applied a T-shape channel for the mixing of bacterial suspension and DI water. *E. coli* spheroplasts were effectively lysed within a second. The extracted *E. coli* chromosome was then trapped in the dielectrophoretic region on the same chip under the applied AC electric field. Although some off-chip procedure was needed, this approach provided a simple solution to performing chemical lysis of bacteria using water.

Stephen Quake's research group developed a two-layered microfluidic system to integrate parallel sequential operations including isolation and lysis of bacterial cells as well as extraction of DNA or RNA on a single microfluidic chip (Hong et al. 2004). They applied multilayer soft lithography to construct the pneumatic valves and pumps to realize microfluidic structures with complicated functions. The chemical lysis was carried out in closed chambers so that there was no contamination or leakage due to diffusion. In their approach, shown in Fig. 30.3, *E. coli* cell suspension, lysis buffer and dilution buffer were mixed for the lysis of bacterial cells. They designed an annular microchannel equipped with three sequentially operated pneumatic valves for the effectively diffusive mixing of the cell suspension, lysis buffer and dilution buffer so that the required time for effective mixing was greatly decreased from several hours to several minutes. After that, the bacterial lysate was carried by the washing buffer to go through a section of the channel with packed microscale beads (2.8 μm in the diameter). Here the beads adsorbed nucleic acid molecules. As a last step in this approach, the elution buffer flushed nucleic acids from the beads to the collection port. Three parallel units could work in parallel on a single microchip. Analysis of mRNA based on 2–10 cells and genomic DNA based on 28 cells was demonstrated in this study. This implementation adequately demonstrated the power and strength of microfluidic integrated systems for the analysis of tiny quantities of bacterial cells.

Although DNA and proteins are the relevant intracellular materials in most biological assays, some other intracellular materials can be of significance too when attempting to understand the biological activities inside cells. Liu et al. (2005) designed a microfluidic system integrating bacteria lysis, electrophoretic separation, and bioluminescence detection for the metabolic analysis of intracellular ATP. In their experiment, *E. coli* cells and the lysis detergent Triton X-100 were mixed in the buffer reservoir. After a three-minute extraction, the lysate was injected using gated mode for separation and detection. Their experimental results showed that this device was able to successfully monitor intracellular ATP of *E. coli*.

3.3. Thermal Lysis

Thermal lysis is performed based on disrupting the bacterial membrane under high temperature, so it is not suitable for the extraction of intercellular proteins due to their tendency to be denatured and aggregated at high temperatures. However, thermal lysis has been successfully applied to the extraction of DNA by using the same integrated heating device for PCR amplification. Therefore, thermal lysis has been extensively integrated with the downstream PCR procedure in microfluidic systems.

In 1998, Ramsey's research group reported a device to integrate thermal lysis of *E. coli* with PCR amplification as well as on-chip electrophoresis (Waters et al. 1998). They fabricated their chip based on glass using standard photolithograph, wet chemical etching, and thermal bonding techniques. They directly placed the glass chip containing the bacterial sample in a commercial thermal cycler. The microchip was first heated at 94°C for 4 minutes for the thermal lysis of *E. coli*. After that, in order to amplify DNA, the PCR cycle was repeated 24 times at 94°C for 2 minutes, 50°C for 3 minutes and 72°C for 4 minutes in each cycle. To finish the procedure of chain extension, the temperature was set at 72°C for 7 minutes. Finally, the temperature for the microchip with the sample was set at 5°C to finish the total program. Their downstream experimental results showed that with the above procedure, sufficient DNA was

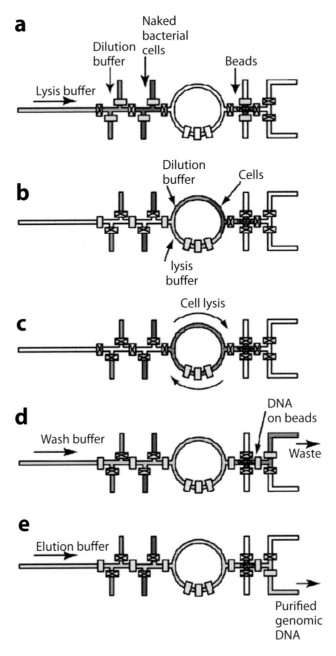

Figure 30.3. DNA purification chip and schematic diagram of one instance of the DNA isolation process (open valve, rectangle; closed valve, x in rectangle). (a) Bacterial cell culture (red) is introduced into the microfluidic chip through the "cell in" port located in the uppermost part of the chip. Buffer (green) for dilution of the cell sample is introduced through the "buffer in" port located next to the "cell in" port. Lysis buffer (yellow) is introduced from the left side of the chip. (b) The cell sample, dilution buffer and lysis buffer slugs are introduced into the rotary mixer. (c) The three different liquids are mixed thoroughly and consequently bacterial cells are lysed completely. (d) The lysate is flushed over a DNA affinity column and drained to the "waste port." (e) Purified DNA is recovered from the chip by introducing elution buffer from the left side of the chip and can be used for further analysis or manipulation (reproduced from Hong et al. 2004. Copyright 2004 Nature Publishing Group).

replicated for electrophoretic separation and detection. This early report unveiled the simplicity and feasibility of bacteria lysis integrated with PCR amplification on a single microchip.

Another device with built-in thermal lysis of spores was reported by Peterson et al. (2005). The heaters were fabricated with low-temperature co-fired ceramics (LTCC). The spore lysis by this system was demonstrated by comparing the turbidity of the harvested bacterial suspensions treated at 180°C to that at room temperature. With this LTCC-fabricated microfluidic system, the spores of *Bacillus subtilius*, which is much more resistant to outside forces than typical bacterial cells, were successfully lysed.

Liu et al. (2004) reported a highly integrated microfluidic system to perform the total cell-based assay including steps such as thermal cell lysis, PCR amplification and DNA microarray detection. Their microfluidic system (Fig. 30.4) consisted of a variety of elements including magnetic beads, electrochemical micropumps, microvalves, PCR reactor, heaters and sensors. It needs to be noted that the paper reported a new kind of planar close–open paraffin microvalve, which was based on temperature and pressure control for fluidic separation and transport. Although the response time of this microvalve (around 10 seconds) was slower than that of the pneumatic microvalves, this planar paraffin-based microvalve was extremely simple because auxiliary equipment was not necessary. The authors performed pathogenic bacteria detection with their system according to the following steps: First, the whole citrated rabbit blood containing *E. coli* cells was mixed with biotinylated polyclonal rabbit-anti-*E. coli* antibody and streptavidin-labeled Dynabeads in the sample storage chamber using cavitation microstreaming. The mixture was then pumped into the PCR chamber. Here the beads-antibody-cell complexes were retained with the help of magnetic force, and the other materials were pumped into the waste chamber. When the complex was isolated, the procedure of thermal lysis and PCR amplification was executed through controlling the integrated heater. Following this step, the solution in the PCR chamber was transferred into the eSensor microarray chamber. This system demonstrated a high level of integration, which was necessary for bacteria detection based on raw biological samples.

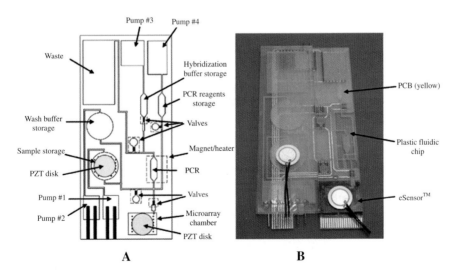

Figure 30.4. (A) Schematic of the plastic fluidic chip. Pumps 1-3 are electrochemical pumps, and pump 4 is a thermopneumatic pump. (B) Photograph of the integrated device that consists of a plastic fluidic chip, a printed circuit board (PCB), and a Motorola eSensor microarray chip (Reproduced From Liu et al. 2004, copyright 2004 American Chemical Society).

3.4. Laser-Based Lysis

Laser has been applied to exert energy on the cells/spores for the disruption of their membrane. Although the mechanism is not completely understood, such lysis is in general believed to be primarily due to thermal effects from concentrated energy.

Manz's research group developed a PDMS-based microfluidic system with UV laser light used for the lysis of *Bacillus globigii* spores (Hofmann et al. 2005). They designed a nanovial (0.8 nL) target plate in order to achieve higher efficiency of laser-induced lysis of bacterial spores. In addition, the spore sample was firstly mixed with a solution of 3-hydroxypicolinic acid (3-HPA), which was used for the enhancement of the laser adsorption. A large drop of mixture was put above the target plate. Because of the surface tension, the sample mixture was drawn into the nanovial, and then the target plate was connected to a microfluidic flow cell. For lysis, the irradiation of UV laser with a wavelength of 337 nm was applied on the sample zone via an optical fiber. The lysate was then flushed out of the nanovial for later PCR analysis. Their experimental results showed that the irradiation of laser light for around 1 second with an energy density of $40\,\mu J/cm^2$ was enough for the extraction of DNA from the spores. Furthermore, the SEM photos of the bacterial spores before and after irradiation by laser were compared for confirmation of the efficiency of this laser-based lysis method. Although this approach would benefit from further improvement in integration and automation, it presented the potential of laser-based lysis and its integration with microfluidics.

More recently, Lee et al. (2006) developed a portable system with laser-based lysis and PCR amplification of bacterial cells for rapid identification of pathogens. Because high-power lasers and magnetic beads were involved, this system was named the laser-irradiated magnetic bead system (LIMBS). In this system, shown in Fig. 30.5, a diode laser with a wavelength

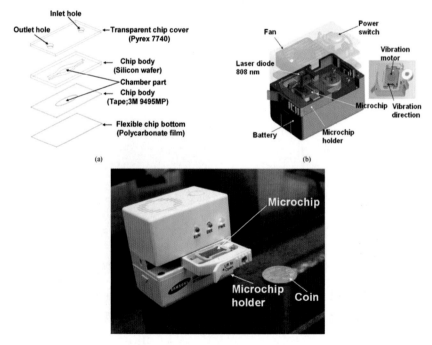

Figure 30.5. Portable device for LIMBS and the single-chamber real-time detection of pathogens. (a) Schematic diagram of a microchip for temperature measurement. (b) Schematic diagram of a hand-held type LIMBS using small laser diode and microchip holder with vibration motor. (c) Image of hand-held type sample preparation device ($58.3 \times 57.9 \times 37.0$ mm, 148 g) with a small laser diode (Reproduced from Lee et al. (2006). Copyright 2006 The Royal Society of Chemistry).

of 808 nm was chosen on the basis of its low adsorption in water, high availability of the laser source and minimal effect on the quality of DNA. In order to enhance the lysis effect, the microchip was shaken to induce the movement of magnetic beads during the period of laser illumination. Experimental results revealed that the mechanism of bacteria lysis might be due to the combination of laser heating and the collision of magnetic beads with bacterial cells in the compartment. Using the same system, DNA extraction was also carried out and optimized. In this procedure, magnetic beads were shown to remove denatured proteins and improve detection limits for DNA, as well as decrease the time for DNA separation. Other parameters were further optimized for the integrated system, such as the power of the laser and the vibration rate of the microchip, as well as the concentration of beads.

3.5. Electrical Lysis

Electrical lysis is based on electroporation (Zimmermann et al. 1981). Electroporation occurs when cells are placed in an electrical field with its intensity higher than a certain threshold. It is generally accepted that nanoscale pores are formed during electroporation. Such pores can be reversible or irreversible, depending on the strength and the duration of the applied electric field. Irreversible electroporation, or electrical lysis, disrupts the cell membrane. When electrical lysis was applied, heat was inevitably generated due to the current introduced. Such heat can be managed by controlling the ionic strength of the cell suspension when it is not desired. A conventionally pulsed electric field was applied for killing bacterial cells as a sterilization measure. For example, Geveke and Brunkhorst (2004) demonstrated a strategy to destroy *E. coli* in apple juice under the applied radio frequency electric field (RFEF). To avoid overheating, they applied high-intensity RFEF on a narrow channel of the chamber that allowed the apple juice to pass through. Theoretical investigation with finite element analysis showed that the electric field in the narrow channel was uniform so as to promise the same treatment of total juice. They studied the influences of RFEF strength and frequency, temperature, bacteria concentration and treatment stage on the lysis effect. Experimental results demonstrated that under optimal conditions, only one thousandth of *E. coli* in apple juice was left. Compared to pasteurization, which had been extensively applied for apple juice sterilization, electrical treatment with pulsed field avoids high temperatures that would destroy the nutrition components in the juices and influence its quality.

On microfluidic platforms, electrical lysis is gaining popularity due to its extreme rapid lysis without the introduction of chemical or biological reagents. In 1998, Cheng and co-workers developed a lab-on-a-chip platform integrating the isolation of *E. coli* from blood cells and electrical lysis of the bacterium (Cheng et al. 1998). They designed and fabricated a checkerboard array of five-by-five platinum microelectrodes with a distance of 200 μm in the biochip for the separation and lysis. With these electrodes, *E. coli* were dielectrophoretically separated from the blood cells in the biochip chamber. In the following step, these cells were *in situ* electrically lysed under a series of pulsed electric fields with a peak voltage of 500 V and peak width of 50 μs for up to 400 cycles. Then the lysate solution in the chamber was heated with proteinase K for 20 minutes at a temperature of 50°C to remove the contaminating proteins. On a separate biochip, the lysate was further analyzed based on electrically enhanced hybridization. Their experimental results with gel electrophoresis demonstrated that genomic DNA, RNA and plasmids were successfully extracted from *E. coli* cells.

In 1999, Lee and Tai systematically investigated electrical lysis of a variety of bacterial cells and yeast on a microfluidic platform (Lee and Tai 1999). They constructed a microelectrode array that had the counter electrodes with a microscale gap in between (5 μm) for bacteria lysis. This effectively lowered the total voltage required to generate a high field, which was sufficient for electrical lysis. *E. coli* cells were flown through the gap between microelectrodes while a square pulsed electric field was applied with a peak of 3.5 V and a width of 500 μs. Here it is worth noting that although the voltage was very low, the electric field intensity could still reach as high as 7 kV/cm due to the small gap between the electrodes. They used plate count to indicate the lysis of the bacterial cells.

In the above reports electrical lysis was applied by having pulsed or AC fields. Furthermore, the introduction of a microelectrode array entailed fairly involved fabrication, and the electrodes could be easily contaminated by the lysate since they were in contact during the whole process. Recently, we have demonstrated a simple device based on a common DC power supply for electrical lysis on a microfluidic platform (Wang et al. 2006). Specifically, we used a simple microfluidic channel that was composed of alternating narrow and wide sections (the depth was uniform in the channel) with a constant DC voltage established across the channel (as shown in Fig. 30.6). According to Ohm's law, the strength of the electric field distributed in the narrow channel would be much larger than that in the wide ones due to the difference in the cross sectional area. The following relationship holds:

$$E_2/E_1 = W_1/W_2, \tag{30.1}$$

We were able to adjust the total voltage across the channel so that only the field intensity in the narrow section was higher than the electrical lysis threshold. When *E. coli* cells flowed through the channel, the lysis exclusively occurred in the narrow section. The intensity of the lysis field in the narrow section was calculated or simulated based on basic laws of physics, and the duration of the lysis field was determined by the velocity of the cells and the length of the narrow section. As a result, electroporation or electric lysis of cells could be conveniently carried out in this simple device in a "flow-through" fashion without the need for an electrical pulse generator. The total voltage applied on the microchannel was greatly decreased compared to a device without the geometric variation due to the confinement of the high field in the narrow section. Using this approach, we investigated electric lysis of *E. coli* cells that expressed green fluorescent proteins. The performance of electric lysis in devices with different geometries was compared. Our experimental results indicated that once the applied electric field was over 1500 V/cm, 100 % of *E. coli* cells were lysed with less than one second in the lysis field. The application of constant DC voltage not only lowers the requirement for expensive equipment but also allows easy integration with other analytical tools such as electrophoresis. In principle, our device can be scaled up or down in size without significant effects on the performance.

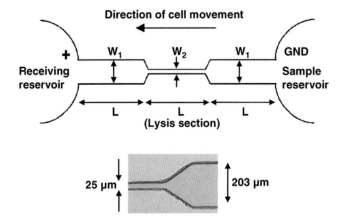

Figure 30.6. The design of the microfluidic device for cell lysis. Cells were loaded in the sample reservoir and transported to the receiving reservoir in a DC electric field. Cell lysis occurred in the narrow section when the field strength was sufficiently high. A microscope image shows the width reduction in a typical channel. (Reproduced from Wang et al. (2006), with permission of Elsevier).

4. Conclusions and Future Perspectives

As we described above, there have been a large number of tools for bacteria/spore lysis that are available to researchers working with micro total analysis systems. Based on specific application, one needs carefully choose an approach that would be compatible with other assays on the chip. As the issue of pathogen detection gains more importance and publicity, microfluidic systems requiring bacteria/spore lysis will become commonplace. The successful incorporation of the lysis assay with the rest of the chip or other devices remains a problem. Some of the existing designs for integrated systems are probably too complicated to be practical. A robust microfluidic system requires a simple and effective solution and seamless connection from one step to the next. In addition, the accumulation of cell lysate after lysis remains a challenge—especially when proteins are of interest. Most of the biological assays require a fairly large number of bacterial cells to generate significant signal. Other than having a closed chamber during lysis, such as was demonstrated in some chemical lysis or thermal lysis devices, there were very few schemes for gathering the protein components from cell lysate. However, the use of chambers often makes subsequent assays, such as electrophoresis, cumbersome. Future development in these directions will tremendously benefit the field.

Acknowledgements

The authors acknowledge support by a cooperative agreement with the Agricultural Research Service of the US Department of Agriculture, Project Number 1935-42000-035, through the Center for Food Safety Engineering at Purdue University.

References

Belgrader P, Hansford D, Kovacs GTA, Venkateswaran K, Mariella R, Milanovich F, Nasarabadi S, Okuzumi M, Pourahmadi F and Northrup MA (1999) A minisonicator to rapidly disrupt bacterial spores for DNA analysis. Anal. Chem. 71:4232–4236

Bhunia AK (1997) Antibodies to *Listeria monocytogenes*. Crit. Rev. Microbiol. 23:77–107

Caprioli R and Rittenbe D (1969) Pentose Synthesis In Escherichia Coli. Biochemistry 8:3375

Cheng J, Sheldon EL, Wu L, Uribe A, Gerrue LO, Carrino J, Heller MJ and O'Connell JP (1998) Preparation and hybridization analysis of DNA/RNA from E-coli on microfabricated bioelectronic chips. Nature Biotechnol. 16:541–546

Di Carlo D, Jeong KH and Lee LP (2003) Reagentless mechanical cell lysis by nanoscale barbs in microchannels for sample preparation. Lab Chip 3:287–291

El-Ali J, Sorger PK and Jensen KF (2006) Cells on chips. Nature 442:403–411

Fratamico PM (2003) Comparison of culture, polymerase chain reaction (PCR), TaqMan Salmonella, and Transia Card Salmonella assays for detection of Salmonella spp. in naturally-contaminated ground chicken, ground turkey, and ground beef. Mol. Cell. Probes 17:215–221

Fratamico PM and Strobaugh TP (1998) Simultaneous detection of Salmonella spp. and Escherichia coli O157:H7 by multiplex PCR. J. Ind. Microbiol. Biotechnol. 21:92–98

Garen A and Echols H (1962) Properties Of 2 Regulating Genes For Alkaline Phosphatase. J. Bacteriol. 83:297

Geciova J, Bury D and Jelen P (2002) Methods for disruption of microbial cells for potential use in the dairy industry - a review. Int. Dairy J. 12:541–553

Gehring AG, Patterson DL and Tu S (1998) Use of a light-addressable potentiometric sensor for the detection of Escherichia coli O157:H7. Anal. Biochem. 258:293–298

Gehring AG, Irwin PL, Reed SA, Tu S, Andreotti PE, Akhavan-Tafti H and Handley RS (2004) Enzyme-linked immunomagnetic chemiluminescent detection of Escherichia coli O157:H7. J. Immunol. Methods 293:97–106

Geveke DJ and Brunkhorst C (2004) Inactivation of Escherichia coli in apple juice by radio frequency electric fields. J. Food Sci. 69:E134–E138

Ghuysen JM and Strominger JL (1963) Structure Of Cell Wall Of Staphylococcus Aureus, Strain Copenhagen.1. Preparation Of Fragments By Enzymatic Hydrolysis. Biochemistry 2:1110

Harrison STL (1991) Bacterial-Cell Disruption - A Key Unit Operation In The Recovery Of Intracellular Products. Biotechnol. Adv. 9:217–240

Heo J, Thomas KJ, Seong GH and Crooks RM (2003) A microfluidic bioreactor based on hydrogel-entrapped E. coli: Cell viability, lysis, and intracellular enzyme reactions. Anal. Chem. 75:22–26

Hofmann O, Murray K, Wilkinson AS, Cox T and Manz A (2005) Laser induced disruption of bacterial spores on a microchip. Lab Chip 5:374–377

Hong JW, Studer V, Hang G, Anderson WF and Quake SR (2004) A nanoliter-scale nucleic acid processor with parallel architecture. Nature Biotechnol. 20:435–439

Ivnitski D, Abdel-hamid I, Atanasov P and Wilkins E (1999). Biosensors for detection of pathogenic bacteria. Biosens. Bioelectron. 14:599–624

Kaspar CW and Tartera C (1990) Methods for detecting microbial pathogens in food and water. Methods Microbiol. 22:497–530

Kim J, Jang SH, Jia GY, Zoval JV, Da Silva NA and Madou MJ (2004) Cell lysis on a microfluidic CD (compact disc). Lab Chip 4:516–522

Kim KP, Jagadeesan B, Burkholder KM, Jaradat ZW, Wampler JL, Lathrop AA, Morgan MT and Bhunia AK (2006) Adhesion characteristics of Listeria adhesion protein (LAP)-expressing Escherichia coli to Caco-2 cells and of recombinant LAP to eukaryotic receptor Hsp60 as examined in a surface plasmon resonance sensor. FEMS Microbiol. Lett. 256:324–332

Lee JG, Cheong KH, Huh N, Kim S, Choi JW and Ko C (2006) Microchip-based one step DNA extraction and real-time PCR in one chamber for rapid pathogen identification. Lab Chip 6:886–895

Lee SW and Tai YC (1999) A micro cell lysis device. Sens. Actuators A. 73:74–79

Liu BF, Ozaki M, Hisamoto H, Luo QM, Utsumi Y, Hattori T and Terabe S (2005) Microfluidic chip toward cellular ATP and ATP-conjugated metabolic analysis with bioluminescence detection. Anal. Chem. 77:573–578

Liu RH, Yang JN, Lenigk R, Bonanno J and Grodzinski P (2004) Self-contained, fully integrated biochip for sample preparation, polymerase chain reaction amplification, and DNA microarray detection. Anal. Chem. 76:1824–1831

Malamy MH and Horecker BL (1964) Purification And Crystallization Of Alkaline Phosphatase Of Escherichia Coli. Biochemistry 3:1893

Manz A, Harrison DJ, Verpoorte EMJ, Fettinger JC, Paulus A, Ludi H and Widmer HM (1992) Planar Chips Technology For Miniaturization And Integration Of Separation Techniques Into Monitoring Systems - Capillary Electrophoresis On A Chip. J. Chromatogr. 593:253–258

Marentis TC, Kusler B, Yaralioglu GG, Liu SJ, Haeggstrom EO and Khuri-Yakub BT (2005) Microfluidic sonicator for real-time disruption of eukaryotic cells and bacterial spores for DNA analysis. Ultrasound Med. Biol. 31: 1265–1277

Miller HK, Valanju N and Balis ME (1965) Magnesium Requirement For Incorporation Of Cytidylic Acid Into Deoxyribonucleic Acid In Chick Embryo Extracts. Biochemistry 4:1295

Penman S, Darnell JE, Scherrer K and Becker Y (1963). Polyribosomes In Normal And Poliovirus-Infected Hela Cells And Their Relationship To Messenger-Rna. Proceedings of the National Academy of Sciences, USA 49:654

Peterson KA, Patel KD, Ho CK, Rohde SB, Nordquist CD, Walker CA, Wroblewski BD and Okandan M (2005) Novel microsystem applications with new techniques in low-temperature co-fired ceramics. Int. J. Appl. Ceram. Technol. 2:345–363

Prinz C, Tegenfeldt JO, Austin RH, Cox EC and Sturm JC (2002) Bacterial chromosome extraction and isolation. Lab Chip 2:207–212

Sanders GHW and Manz A (2000) Chip-based microsystems for genomic and proteomic analysis. Trac-Trend Anal. Chem. 19:364–378

Scandell CJ and Kornberg A (1971) Membrane-Bound Phospholipase-A1 Purified From Escherichia-Coli. Biochemistry 10:4447–4456

Schilling EA, Kamholz AE and Yager P (2002) Cell lysis and protein extraction in a microfluidic device with detection by a fluorogenic enzyme assay. Anal. Chem. 74:1798–1804

Sherman MR, Tuazon FB, Diaz SC and Miller LK (1976) Multiple Forms Of Oviduct Progesterone Receptors Analyzed By Ion-Exchange Filtration And Gel-Electrophoresis. Biochemistry 15:980–989

Swaminathan B and Feng P (1994) Rapid detection of food-borne pathogenic bacteria. Annu. Rev. Microbiol. 48: 401–426

Taylor MT, Belgrader P, Furman BJ, Pourahmadi F, Kovacs GTA and Northrup MA (2001) Lysing bacterial spores by sonication through a flexible interface in a microfluidic system. Anal. Chem. 73:492–496

Tietjen M and Fung DYC (1995) Salmonellae And Food Safety. Crit. Rev. Microbiol. 21:53–83

Tonshoff HK and Raschke HD (1977) Investigations Into Practical Reduction Of Noise-Levels In Manual Grinding Operations. WT-Z Ind. Fertigung 67:671–676

Verpoorte E (2002) Microfluidic chips for clinical and forensic analysis. Electrophoresis 23:677–712

Wang HY, Bhunia AK and Lu C (2006) A microfluidic flow-through device for high throughput electrical lysis of bacterial cells based on continuous dc voltage. Biosens. Bioelectron. 22:582–588

Waters LC, Jacobson SC, Kroutchinina N, Khandurina J, Foote RS and Ramsey JM (1998) Microchip device for cell lysis, multiplex PCR amplification, and electrophoretic sizing. Anal. Chem. 70:158–162

Zhou F, Rouse BT and Huang L (1991) An Improved Method Of Loading pH-Sensitive Liposomes With Soluble-Proteins For Class-I Restricted Antigen Presentation. J. Immunol. Methods 145:143–152

Zimmermann U, Scheurich P, Pilwat G and Benz R (1981) Cells With Manipulated Functions - New Perspectives For Cell Biology, Medicine, And Technology. Angew. Chem. Int. Ed. Engl. 20:325–344

31

Detection of Pathogens by On-Chip PCR

Pierre-Alain Auroux

Abstract

The purpose of this chapter is threefold: introducing microfluidics to the general audience, describing in detail the polymerase chain reaction (a technique used for DNA amplification), and reviewing the state-of-the-art methods regarding the detection of pathogens by on-chip PCR. The first section gives a brief introduction to the field of microfluidics. Although the microfluidic technologies have been developed substantially since 1990, their existence and applications are still unknown from the general public. The history and the applications of miniaturized total analysis systems (μTAS) are therefore summarized in the first section (Microfluidics). Secondly, the polymerase chain reaction (PCR) is described in detail. The second section (DNA amplification) therefore covers a brief history of DNA and the applications, requirements, and processes of PCR. As a conclusion of this section, the different techniques available to perform PCR (namely conventional PCR, real-time PCR and on-chip PCR) are compared. Lastly a mini-review presents the state-of-the-art in terms of detection of pathogens by on-chip PCR. The polymerase chain reaction is becoming recognized by official administrations as an acceptable method for the detection of pathogens. It is therefore no surprise that the microfluidic community is also developing devices to support this transition. The last section (Minireview) provides a snapshot of the most exquisite techniques available for the on-chip detection and analysis of pathogens.

1. Introduction

The toxicity of the common pathogens has been exemplified in previous chapters. The importance of their detection should therefore be obvious. However, as often is the case in science, there is not a unique approach to the issue. In this book a range of selected methods demonstrates such diversity. The object of the present chapter is to investigate the analysis of pathogens by amplifying their specific DNA using microfluidic chips. These microfluidic devices have critical dimensions in the micrometer range ($1\mu m = 10^{-6}m$), leading to volumes of solution in the order of the nanoliter ($1nL = 10^{-9}L$). The remainder of this chapter will outline the benefits of microfluidics (Auroux 2005), will provide an overview of DNA amplification by polymerase chain reaction (PCR) (Auroux 2005) and will briefly review on-chip pathogen detection by PCR.

Pierre-Alain Auroux, PhD • National Institute for Standards and Technology, EEEL, Semiconductor Electronics Division, Gaithersburg, MD, USA.

M. Zourob et al. (eds.), *Principles of Bacterial Detection: Biosensors, Recognition Receptors and Microsystems*,
© Springer Science+Business Media, LLC 2008

2. Microfluidics

2.1. History of Miniaturized Total Analysis System (μTAS)

The first miniaturized chemical analysis system was introduced in 1979 by Terry and co-workers (Terry et al. 1979). This gas chromatographic analyzer included a 1.5 m long separation column on a silicon chip. Despite its minute size and its remarkable capability to separate a simple mixture of chemicals in a matter of seconds, this revolutionary device received virtually no interest from the scientific community. It was only in 1990 that the relevance of microfluidic systems was recognized. It was then that Manz et al. presented a miniaturized open-tubular liquid chromatograph on a 5×5 (mm × mm) silicon chip (Manz et al. 1990a). The device comprised an open-tubular column coupled to a conductometric detector and was connected to an on-chip pressure system. In parallel, the same group introduced the concept of *miniaturized Total Analysis Systems* (μTAS) (Manz et al. 1990b). This concept was described as silicon analyzers combining several steps necessary for sample analysis (for example, sample handling, sample pre-treatment, separation, detection, etc.) onto one device. It was recognized that such an approach would eventually improve significantly the analytical performances of chemical sensors.

Because early considerations showed that electro-osmotic pumping had fundamental advantages when compared to conventional pumps (Manz et al. 1990a), such as the absence of high pressures needed to transport a fluid in small channels thereby inducing back pressure problems, further efforts were implemented to optimize injection and separation procedures by switching voltages (Harrison et al. 1991; Manz et al. 1991). Subsequently, the integration of an electrophoresis device into a silicon and glass substrate was demonstrated in 1992 (Harrison et al. 1992; Manz et al. 1992). This success excited the interest of the scientific community, and in 1993-1994 the number of papers related to μTAS increased drastically. Some research teams developed new methods or new protocols to improve existing devices, while others broadened the field of applications. A few examples include the reduction of separation time when using electrophoresis devices (from 30 seconds in 1993 (Harrison et al. 1993) down to hundreds of milliseconds in 1994 (Jacobson et al. 1994)), investigation of mixing conditions (Branebjerg et al. 1994), cell manipulation using an electric field (Fuhr and Wagner 1994), and implementation of a micro-system mass spectrometer (Feustel et al. 1994). Nowadays the application field of μTAS is extremely diverse and comprises, but is not limited to, micro-mixers, micro-reactors, electrophoretic devices using various separation methods, immunoassays and reaction chambers for on-chip polymerase chain reaction (PCR) (Vilkner et al. 2004).

2.2. Advantages of Miniaturized Analysis Systems

Most of the advantages of using miniaturized systems are directly linked to their inherently small critical dimensions, which are on the order of micrometers or lower. These benefits can be seen, for example, in the case of time-dependent physical and chemical processes, such as diffusion (Manz et al. 1990a). Indeed, when only diffusion is involved, the distance (d) covered by a molecule only depends on its diffusion coefficient (D) and on the period of time (t) given to the molecule to diffuse. This process is depicted by the Einstein-Smoluchowski equation:

$$d^2 = 2Dt.$$

It can be deduced from this equation that for small molecules (usually with diffusion coefficients in the order of $0.5 \times 10^{-5}\ cm^2/s$) the diffusion time over a $10\,\mu m$ distance is 0.1 s, whereas it will take the same molecule 17 min to cover 1 mm. A direct consequence is a great reduction in the time needed to reach completion for diffusion-controlled (bio)chemical reactions, producing faster systems. Reactions with immobilized molecules such as assays will also benefit from microdevices, as the surface-to-volume ratio is increased. Indeed, for the same number of molecules in solution, there are comparatively more molecules immobilized and thus more reactive sites at a micro-system surface than at a macro-scale level. A direct consequence is an enhanced detection limit, leading to a system that is both more accurate and faster (due to shorter diffusion time). Other time-limited processes such as longitudinal diffusion, transport, and analysis time also scale with the square of the size of the channel and thus participate to reduce the over-all reaction-time. Another process, thermal diffusion, also benefits from miniaturization. As the thermal properties of a system scale similarly to diffusion-controlled processes, thermal transport is accomplished in much shorter periods of time due to the large surface-to-area ratio of a microchannel.

Although microdevices are very attractive due to their potential to deliver faster and more accurate responses, there are also additional benefits when working with microstructures. For example, micro-size devices do not occupy as much footprint as their macro-size counterparts, which leads to a significant gain in laboratory space. Furthermore, the trend is to render the analysis systems smaller and more mobile. In particular, scientists want to be able to bring the apparatus to the location where it needs to be used (Opekun et al. 2002; Wahr et al. 1996). This concept is known as the point-of-care analysis, as well as the distributed diagnostic. In the case of disease diagnostics or therapy, implanting microchips directly into the human body has even been considered (Nagakura et al. 2003). Safety issues can also be solved when using μTAS because the amount of compounds used is minute. Indeed the consequences of a mishandled reaction are less dramatic than when using conventional equipment. For example, Wootton et al. (2001) describe the on-chip generation and reaction of highly reactive and potentially dangerous intermediates. In addition, Wallenborg and co-workers separated and detected a mixture of 14 different explosives, which would have been a delicate task at the macro-scale level (Wallenborg and Bailey 2000). From a more practical point of view the use of microdevices can also mean a reduction in expenses at different levels. For example, as the technology to make microdevices becomes suitable for mass-production, procedures have the potential to become more cost-efficient, while cheaper fabrication processes may curtail the price of microchips. In addition, sample consumption is also reduced, which in turn lowers the cost of each reaction. This is of particular importance for drug synthesis control. Additional examples of the benefits of microfluidics are available in the literature (de Mello and Wootton 2002).

To conclude, microdevices present numerous attractive features, such as the potential for low fabrication cost, faster response time and lower detection limits. Although these advantages are applicable to any field, it might be of interest to a potential user to carefully consider alternative options before applying miniaturization to a specific area (de Mello 2001).

3. DNA Amplification

3.1. A Brief History of DNA

Since early times people wondered how traits were inherited from one generation to another. It was well known from breeding cattle that, by controlling mating, some traits had higher chances to be passed on or were even enhanced. In 1865, a monk, Gregor Mendel, studied systematically the evolution of peas as they were being cultured over several years. Based on this research he found that individual traits are organized in "packages" nowadays called genes. The gene's location and storage conditions were still very unclear at that time. A small step toward the right direction was taken in 1869, when Friedrich Miescher discovered that a large molecule existed in the cell nucleus. He named it *nuclein* and even suggested after further study that it might be the support of genetic material. However, this belief was not shared at that time by the scientific community, who thought that proteins were more likely to be the genetic carriers. McCarty in 1944 proved this belief wrong while working in Oswald Avery's team. He showed that one benign pneumonia bacterium could be made infective when some DNA (deoxyribonucleic acid), and only DNA, was transferred into the cell. As infectiveness was known to be hereditary, DNA had to be the unit of inheritance. DNA was finally linked to Mendel's "trait packages" after nearly a century of research. This discovery triggered numerous works in molecular biology. Rosalind Franklin started using X-rays to study DNA, while Erwin Charga determined the "pairing rules" of DNA: Adenine (A) and Thymine (T) are always present in the same proportions, so are cytosine (C) and guanine (G). The work of both scientists considerably inspired James Watson and Francis Crick, who proposed in 1953 that the stable configuration for two strands of DNA is a double helix (Watson and Crick 1953). Maurice Wilkins helped to verify their model by using X-ray crystallography and shared with them a Nobel Prize in 1962.

From this moment on, more and more efforts were put into the study of DNA. In 1955, Seymour Benzer linked genes in bacteria to long stretches of DNA letters, while one year later Arthur Kornberg discovered the enzyme to copy DNA. In 1957 Crick discovered how DNA was used by the human body when he determined how the genetic material was translated into proteins. Frederick Sanger pioneered DNA sequencing when he radioactively tagged DNA bases and was able to determine the complete list of base pairs to produce the protein insulin. A major breakthrough was achieved in 1983 when Kary Mullis invented the polymerase chain reaction process. This method of replicating DNA was, and is still, cheaper and quicker than any previous DNA amplification methods. Thanks to this method a tremendous international effort to identify and study all human genes, the Human Genome Project, could be undertaken successfully. Currently, various studies involve DNA, such as mapping other species' genomes and the more controversial application of cloning.

3.2. PCR Characteristics and Applications

The polymerase chain reaction is a method by which a targeted region of DNA can be amplified in large quantities (Alberts et al. 1998). There are several advantages when performing PCR. First this reaction can be highly specific, the targeted region of DNA being determined by the primers (molecules indicating the beginning and the end of the area to be replicated). Choosing long primers will make the reaction highly selective. Second, only a very small amount of DNA sample is needed to perform the amplification, making it highly sensitive. Also, due to the exponential aspect of the amplification, very large amounts of DNA can be generated in a cost-effective manner. Finally, the versatility of the method renders it a universal technique for DNA amplification. The main drawback of this procedure is actually its high sensitivity: only a tiny amount of undesired DNA can lead to problematic contamination. However, the numerous advantages of this technique overcome its drawback, and a wide variety of applications benefits from PCR (Powledge 2004).

The most obvious applications are in medical research and clinical medicine. Indeed, PCR has proved to be a valuable tool for detection of infectious disease organisms. For example, targeting the AIDS virus' DNA in a PCR assay enables doctors and researchers to detect it at an earlier stage than by using a standard ELISA test (Thompson and Loeffelholz 1995). Furthermore, as PCR can distinguish between small variations of a DNA sequence, it is often used in genetics. Hereditary disorders can, of course, be diagnosed (Zaletaev et al. 2004), but PCR is also used by researchers to try to determine predictive tests: by studying mutations in genes we can attempt identify the deteriorating mechanism that leads to cancers in adulthood (Jacobson et al. 1996). The polymerase chain reaction can also be used in domains such as law and archeology. Indeed, in forensic science, DNA fingerprinting is a valuable tool (Connors et al. 1996). By applying PCR to a DNA sample collected at a crime scene, more DNA can be generated, and consequently, more tests can be performed. Although a matching DNA is not enough to inculpate a criminal, a mismatching one has proven valuable to demonstrate innocence. In archeology, PCR can be used to study ancient DNA. As a consequence, connections between different human groups can be assessed and migrating populations tracked. It can even be used to determine some human cultural practices. For example, analysis of the pigment used on some 4000-year-old paintings found bison blood to be one of its constituents. Considering that this animal did not live in the studied area, the painter had probably gone to some effort to have access to this ingredient, demonstrating that the painting was not a form of recreation but that it most probably had a religious or magical significance. It is also possible to apply PCR in ecological or animal behavior studies. Indeed, individuals can be identified, and population sizing, migrating or seed dispersal patterns can be determined.

Even though the present technology to perform PCR has reached maturity in terms of chemical reagent optimization, the new trend is to miniaturize the hardware instruments and to develop chip-based devices. It may not be long before diagnosing an infection, a genetic disorder or even a predisposition to cancer will be performed in a local doctor's office.

3.3. Components to Perform PCR

PCR is a basic and simple reaction in terms of understanding the requirement for the necessary reactants, namely the DNA, the enzyme, the primers, the deoxyriboNucleotide TriPhosphates (dNTPs) and a suitable buffer.

The first step toward collecting the DNA template is to separate the cells containing the DNA from the other components of the sample. The DNA is then extracted from the cells. Common procedures include phenol/chloroform precipitation, as well as salting out. The DNA template consists of two chains of nucleic acids that are complementary of each other. The template may come from various sources, such as an animal, a plant, a bacterium, or a virus. In a nucleic acid chain, each nucleotide is linked to the next one by a phosphodiester bond between the $3'$-hydroxyl group of one sugar and the $5'$-phosphate group of the other. Consequently, a strand of DNA has an overall chemical polarity as the $3'$ end carries an unlinked -OH group, whereas the $5'$ end carries a phosphate. This distinction between the two ends is essential as the DNA polymerase synthesizes from the $5'$ end to the $3'$ end direction, rendering the polymerase chain reaction a uni-directional process. An additional distinction can be made regarding the two strands of a DNA molecule, as they are not biologically equivalent (Forsdyle 1995). Indeed, a DNA segment encoding a protein has a sense strand and a complementary anti-sense strand that acts as a template for the RNA polymerase. Conventionally, the sense strand is also called the coding strand as it shares identical sequences with the mRNA molecule.

The primer set consists of two short chains of nucleic acids. One primer is complementary to one of the template strands, while the other primer is complementary to the other template strand, leaving a small region between the $3'$ ends of the primers. The primers can be considered as the "green lights" for the amplification: once the enzyme encounters an annealed primer (a primer hybridized to the template), it starts the replication. In addition, it is due to the primers that the amplification is specific. They are indeed designed by the user to specifically target the sequences of interest. Although longer primers will be more specific than short ones, they will also be more expensive, they will be more prone to degradation, and the annealing/extension steps will be slower. A compromise between these parameters needs to be found, and generally primers are 17 to 30 base-pairs long. Designing primers is not a trivial task as one should aim at three main characteristics: specificity, stability and compatibility (Liu 2003).

The enzyme, a DNA-dependent DNA-polymerase, catalyzes the synthesis of DNA strands. The most frequently used enzyme originates from a bacterium, *Thermus aquaticus* (Taq DNA polymerase), thriving in hot springs (Chien et al. 1976). It is thus a thermostable enzyme that does not deteriorate or lose its efficiency when the PCR is performed. A non-thermostable enzyme could also be used for the reaction; however, it would need to be replenished during thermocycling. This would have several consequences. First, the cost of the reaction would increase, as more enzyme would be needed. Second, the PCR would greatly lose in practicality as it would be difficult to replenish the enzyme for 96 samples in an adequate time scale. Last, but most importantly, the sample tubes would need to be opened during the reaction, considerably increasing risks of contamination. As can be seen, a thermostable enzyme offers many advantages, and, since the discovery of Taq polymerase, alternative thermostable enzymes have also been studied (Mattila et al. 1991; Lundberg et al. 1991).

In order to replicate the template, the enzyme needs building blocks: the deoxyriboNucleotide TriPhosphates (dNTPs). They are the subunits of DNA and consist of four different molecules: dATP, dTTP, dCTP and dGTP. Originally, the dNTPs are formed by incorporating three phosphate molecules to a nucleoside (A: Adenosine, T: Thymidine, C: Cytidine and G: Guanosine). During the template replication, dNTPs are added to the extending strand according to the pairing rule. Indeed, a dNTP, previously free in solution, is immobilized due to the formation of hydrogen bonds between itself and its complementary base situated on the template strand (p192, [50]). A-T pairs form 2-hydrogen bonds, whereas C-G pairs form 3-hydrogen bonds. Once the immobilization has occurred, the enzyme catalyses the incorporation of the sugar of the dNTP to the backbone of the DNA strand by breaking the diester bond between two phosphor molecules, namely $P\alpha$ and $P\beta$. The energy that is released by the formation of the pyrophosphate is then used to generate the phosphodiester bond that links the two consecutive nucleotides.

Finally the buffer is mainly used to maintain optimum conditions for the Taq DNA polymerase, and additional molecules might be present. This is the case, for example, of magnesium chloride. Magnesium ions are crucial for the polymerase chain reaction as they are enzymatic co-factors: their concentration will have an important influence on the shape, on the distribution charge, and on the binding properties of the Taq DNA polymerase. In addition, they also form soluble complexes with the dNTPs and promote their hybridization. Although the effect of magnesium ions is difficult to assess, an optimal concentration has to be determined for any PCR system. On one hand, a low concentration leads to poor reaction yields. On the other hand, a high concentration has also negative effects as it indirectly promotes the formation of non-specific products. Indeed, due to the increase in magnesium ion concentration, the dNTP's hybridization is facilitated, and they form unspecific substrates that are then amplified by the enzyme.

3.4. PCR Process

The polymerase chain reaction is a three-step procedure (see Fig. 31.1). Firstly the double stranded helix is separated into two single strands. Temperatures between 94°C and 96°C are commonly required to break the hydrogen bonds stabilizing the double helix structure. The temperature is then lowered to allow the primers to bind to the template: this is the annealing step. The annealing temperature is primer-dependent but varies often between 50°C and 65°C. Finally the DNA is replicated by the polymerase enzyme during the extension phase. The amplification yield is maximized by choosing an extension temperature close to the optimum working temperature of the enzyme used. A temperature of 72°C is usually used as a default value. The denaturation-annealing-extension cycle is repeated, usually between 25 and 40 times. Although these three steps are usually performed at three different temperatures, it is not uncommon to combine the annealing and the extension into one operation.

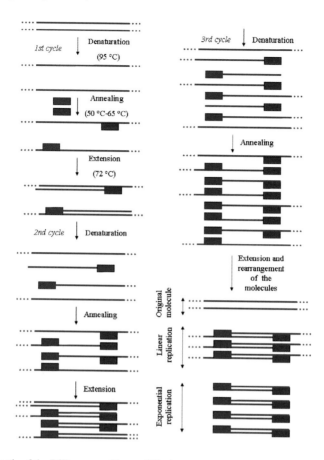

Figure 31.1. Schematic of the PCR process. The amplification process consists of three steps: denaturation (94°C – 96°C), annealing (50°C – 65°C), and extension (close to 72°C). (Reprinted from Auroux et al. (2004) with permission of the Royal society of Chemistry).

To avoid amplification of non-specific products, the PCR reaction needs to be optimized. Such a procedure should first involve the investigation of the annealing temperature, the magnesium chloride concentration and the duration of the extension step. The use of enhancers might also be considered if necessary. In case non-specificity is still an issue after parameter optimization, one can modify the PCR protocol and perform hot start PCR (the reaction is impeded until a high temperature is achieved (SuperArray–Bioscience Corporation 2004)), nested PCR (two sets of primers are used, the second set amplifying a target sequence within the sequence amplified by the first set of primers (Llop et al. 2000)), or touchdown PCR (the annealing step is performed at a temperature higher than the optimum annealing temperature, leading to enhanced specificity to the detriment of the yield (Don et al. 1991)). The reader is directed to the article by Auroux and co-workers for a more detailed review of DNA amplification by PCR (Auroux et al. 2004). Other replicating processes such as Loop-mediated isothermal AMPlification (LAMP), Rolling-Circle Amplification (RCA) and Solid Phase DNA Amplification (SPA) are also covered in this review.

3.5. Conventional PCR

Polymerase chain reaction is conventionally performed on thermocyclers. In simple terms, thermocyclers are machines that allow samples to be heated and cooled in a very accurate and controlled manner. The first thermocycler was originally manufactured by Perkin-Elmer, but numerous brands are available nowadays, including Applied Biosystems, Techne, Hybaid and MJ Research[1]. The basic requirements for an efficient thermocycler are fast heating and cooling rates (to minimize transition times from one temperature to another), accurate temperature control (a crucial requirement to be able to perform any PCR) and programmability (for the cycling conditions to be apparatus-controlled).

The samples are usually dispensed in individual tubes or in plates. The original format for the sample plate was of 96 samples. However, in the search for higher throughput, the trend has been to make thermocyclers compatible with 384 sample plates, and even 1536 sample plates. The material used for the tubes and the plates is usually polypropylene, as it is biologically inert and has acceptable thermal properties.

After amplification, the sample is analyzed. The most common analysis method is agarose gel electrophoresis coupled to UV detection. In gel electrophoresis, the separation is based on the particle sizes: the gel pores allow small solutes to elute easily, whereas bigger molecules will be slowed down. Incorporated into the gel mixture is an intercalator dye, such as ethidium bromide (EtBr). After having run the separation, the gel is illuminated with a UV lamp, and the DNA products are revealed due to EtBr. Their sizes are compared to a ladder, a mixture of DNA molecules of known and specific lengths. Although extremely popular in biology, agarose gel electrophoresis is a technique that lacks sensitivity and accuracy. Another drawback is that it requires large volumes of sample. Agilent Technologies developed an instrument, the Bio-analyzer 2100, to overcome these problems. The chip-based analysis procedure still uses gel electrophoresis, but only 1 μL of sample is required and much lower detection limits are achieved. However, this exquisite technique suffers from one drawback: it is primarily a research tool and is not equipped for high throughput. Interestingly, even if gel electrophoresis is the most applied technique for DNA separation ((Xu et al. 2002), or (Khandurina et al. 2000) for an integrated PCR-analysis device), new approaches only possible on chips are also being investigated, such as the entropic trap array proposed by Han and Craighead (2000).

3.6. Real-Time PCR: Apparatus and Detection Techniques

Polymerase chain reaction is, in theory, a readily applicable method to implement. However, the apparent simplicity is only theoretical, and PCR becomes rapidly challenging for the newcomer, as well as for the expert. For better monitoring of the reaction and a far greater accuracy in the detection, real-time PCR (RQ-PCR) is required. By including a substance (usually a fluorophore) in the PCR mixture, the DNA amplification is observed as it is being performed. Some commercially available thermocyclers, as well as different detection strategies, are described in this section. For further details on real-time PCR, Bustin (2002) and Auroux et al. (2004) provide excellent reviews.

Four of the widely available, commercially manufactured, real-time thermocyclers are ABI 7700 (Applied Biosystems), LightCycler (Roche Molecular Diagnostics), Smart Cycler (Cepheid) and Mx3000P (Stratagene). Although these machines are similar in their principle (to detect the amount of DNA produced by PCR in real-time), their characteristics differ significantly in terms of excitation source, detection method, throughput capability, and sizes (see Table 31.1).

The first detection technique to monitor real-time PCR was based on an intercalator dye (Higuchi et al. 1992, 1993), a fluorescent molecule that binds to double-stranded DNA (dsDNA). There are two requirements for an intercalator: its fluorescence must increase when

---

[1]Certain commercial equipment, instruments, or materials are identified in this chapter to foster understanding. Such identification does not imply recommendation or endorsement by the National Institute of Standards and Technology, nor does it imply that the materials or equipment identified are necessarily the best available for the purpose.

Table 31.1. Real-time PCR thermocyclers. This table compares the four main real-time thermocyclers in terms of excitation source, detection system, sample format, thermocycling system, and dimensions (including weight)

Characteristics of a Few Commercially Available Real-time Thermocyclers				
Product	ABI 7700	LightCycler	SmartCycler	M×3000P
Manufacturer	Applied Biosystems	Roche Molecular Diagnostics	Cepheid	Stratagene
Excitation Source	Argon laser (488 nm)	Blue LED (470 nm)	450 nm-495 nm 500 nm-550 nm 565 nm-590 nm 630 nm-650 nm	white light 350 nm-750 nm
Detection System	Spectrograph (500 nm-660 nm)	530 nm 645 nm 710 nm	510 nm-527 nm 565 nm-590 nm 606 nm-650 nm 670 nm-750 nm	350 nm-700 nm
Sample Format	≤ 96 wells	32 glass capillaries	16 wells	96 wells
Heating System	Peltier elements	Heating coil	I-CORE modules	Peltier elements
Cooling System	Peltier elements	Forced air	I-CORE modules	Forced air
Dimensions (W×H×L, cm)	84 × 64 × 72	30 × 40 × 45	31 × 31 × 26	46 × 43 × 33
Weight (kg)	82	19.2	10	20

bound to dsDNA, and it must not inhibit PCR. The two most popular intercalators are ethidium bromide and SYBR[R] Green. The selectivity of real-time PCR was greatly improved with the introduction of fluorogenic probes (Lee et al. 1993). A probe is an oligonucleotide whose sequence is complementary to part of the target molecule. At one end of the probe there is a fluorescent reporter (R); at the other is a quencher (Q). As long as the probe is intact, the reporter's emission upon excitation is quenched by Q. The quenching phenomenon is known as Fluorescence Resonance Energy Transfer (FRET) and corresponds to the transfer of energy from R to Q, leading to a light emission by Q at a higher wavelength. Several types of probes are based on the FRET principle, such as the TaqMan[R] probes (Lee et al. 1993), the molecular beacons (Tyagi and Kramer 1996; Bonnet et al. 1999; Marras et al. 2002), and the scorpion primers (Whitcombe et al. 1999).

3.7. On-Chip PCR

Although displaying reliable and efficient performances, conventional thermocyclers present some drawbacks. For example, even if efforts to render them more compact have been made, they are still quite bulky and heavy and are not suitable for field applications and point-of-care measurements. In addition the initial investment involves large sums of money, and such a decision needs to be taken with special care. Finally conventional thermocyclers also suffer from intrinsic pitfalls, such as the requirement of a considerable amount of sample and a lack of flexibility in terms of sample handling. In order to solve these problems, the trend has been to use micrometer-scale devices. Table 31.2 presents a comparison between conventional PCR and microfluidic-PCR. Regarding DNA amplification using miniaturised devices, two approaches have been favored: one is based on capillaries, the other on chips. The existing alternatives to conventional PCR, along with their advantages and limitations, are briefly presented in this section. For a more detailed overview, the reader is directed to the review by Auroux et al. (2004) covering miniaturized nucleic acid analysis.

Table 31.2. Comparison between conventional PCR and microfluidic PCR. The table above presents the main advantages and drawbacks of conventional PCR and DNA amplification based on miniaturized devices. Although conventional PCR is a well-established method, there are several drawbacks, such as contamination, requirements for important volumes of reagents and sample, bulkiness, and slow reaction time. Alternatively, micro-chip based PCR would answer these drawbacks. However, this technology is still at a development stage, and no devices are commercially available

Comparison Between Conventional and Microfluidic PCR		
	Conventional PCR	Microfluidic PCR

Drawbacks	Bulky
	Heavy
	Expensive
	Slow
	Prone to contamination
	High consumption Reagents
	                      Sample
		Research stage
		Post-PCR analysis difficult to
		   integrate
		Not yet automated

Advantages	Well established
	Automation
	Real-time PCR
		Already demonstrated:
		  Fast
		  Single-cell manipulation
		  Single-cell analysis
		  Improved LOD
		  Improved accuracy
		  Integration
		Potential:
		  Point-of-Care
		  Future automation
		  Reduced contamination
		    Mass-production
		Cost
		    Lower volumes

3.7.1. Capillary-Based Thermocyclers

Capillary-based systems for PCR were first introduced in 1994 by Nakano and co-workers (Nakano et al. 1994). Their device consisted of a capillary going through three different oil baths in a loop (see Fig. 31.2). The length of capillary in each bath determined the duration of the annealing, extension, and denaturation steps. The sample was continuously pumped through the capillary unidirectionally, and DNA amplification occurred as the sample was going through each temperature zone. An efficiency of 50 % and a reaction time of 10 % (when compared to conventional PCR) were reported.

Friedman et al. proposed a static sample approach in 1998 (Friedman and Meldrum 1998). They enclosed the PCR sample inside a capillary, the ends of which were blocked by rubber pads. Each capillary was coated with a transparent Indium Tin Oxide (ITO) layer, ensuring that thermo-cycling conditions could be independently controlled. Efficiencies comparable to conventional PCR were achieved.

Finally, Chiou and co-workers presented a bi-directional sample pumping system in 2001 (Chiou et al. 2001). In this device, the capillary was rested onto three heating blocks, and the sample was pumped back and forth to provide thermal cycling. An optical system stopped the pumping mechanism when the sample was detected to have reached the desired temperature zone. An efficiency of 78 % was reported. Although the capillary approach solves the issue

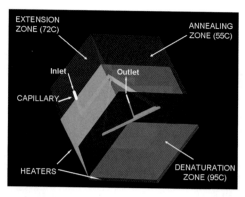

Figure 31.2. Capillary-based PCR system. A capillary is coiled in a loop going through three temperature-controlled oil baths. The black arrows indicate the direction of the flow. The length of the portion of the capillary immersed in a given bath determined the duration of the corresponding step.

of the sample volume requirement, it still presents several major drawbacks. Firstly, it lacks flexibility in terms of design. Indeed, capillaries have a fixed cylindrical cross-section. Secondly, the incorporation of additional integrated system features is complex, thus limiting the potential applications. Thirdly, implementing the parallel use of several structures (a process called parallelization) is not easily implemented and would require additional equipment. These two pitfalls are easily answered by lithographically manufactured microdevices; therefore capillary-based systems were not studied more extensively.

3.7.2. Microdevice-Based Thermocyclers

Thermocyclers based on microdevices can be categorized depending on their fluid handling properties. The static-sample approach consists of loading the sample in a well and then thermocycling the entire chip, mimicking conventional thermocyclers but on a micro-meter scale. In the dynamic-sample approach, the sample temperature changes are achieved by moving it over consecutive heating zones. Both types of systems are described in this section.

3.7.3. Static-Sample Systems

In 1993 Northrup et al. (1993, 1995) presented the first PCR-on-chip device, a silicon chip with a micro-well in which the sample was loaded (see Fig. 31.3). The entire substrate was heated and cooled to provide the adequate thermocycling conditions. Major improvements provided to this microchip have included integration of capillary electrophoresis (Woolley et al. 1996), as well as optimization of the reaction conditions (Chaudhari et al. 1998; Yoon et al. 2002) and providing faster PCR (under 7 minutes) (Belgrader et al. 1999). Additional applications have also been coupled to this chip, including drop-metering (Burns et al. 1998), single-molecule detection (Lagally et al. 2001), real-time applications (Taylor et al. 1997), multiplexing (Zou et al. 2002), Ligase Chain Reaction (Cheng et al. 1996a, 1996b), and DNA extraction (Waters et al. 1998; Caldarelli-Stefano et al. 1999). One major advantage of the static-sample approach is a simple layout that enables the fabrication of micro-arrays, therefore leading to higher throughputs (Zou et al. 2002; Poser et al. 1997; Cheng et al. 1996b).

Several static-sample-based devices differed significantly from the original design. One, introduced by Oda et al. (1998), circumvented the invasive heater elements by using infrared wavelengths. Alternatively, Pal and co-workers performed PCR using induction heating, another

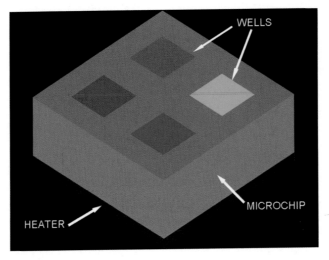

Figure 31.3. Well-based PCR. The sample is positioned in individual wells. The thermocycling is provided by heating or cooling to the desired temperature.

non-contact heating method (Pal and Venkataraman 2002). Induction heating was provided by mounting the chips onto a ferrous ring. By alternating the current going through the inductor, a copper wire wound around a ferrite core, magnetic fields were produced. Eddy currents were subsequently induced in the ferrous ring, therefore generating localized heat without any physical contact. Krishnan et al. (2003) relied mainly on convection to provide thermocycling. Two heaters, maintained at constant temperatures (namely 97°C at the bottom and 61°C at the top), sandwiched the Rayleigh-Bénard cell. A convection flow, based on the difference in temperatures, developed in the cell, with an upward (hot) flux in the center and downward (cold) fluxes on the sides. PCR was performed on the molecules following these fluxes.

Although these devices are more flexible than capillaries, in particular with regards to the ease of fabrication for customized use, they still do not take full advantage of lithographic techniques as their designs are very basic. In addition, they generally exhibit an unnecessary inertia as the entire chip needs to be heated/cooled to provide thermocycling conditions to the sample.

3.7.4. Dynamic-Sample Systems

A dynamic-sample system addresses some of the previously described pitfalls. Indeed, in the case of continuous-flow chips for example, the thermal inertia is minimal as the temperature transition of the sample only depends on the pumping rate and the time it needs to reach temperature equilibrium. Consequently thermal ramping is much faster and depends almost entirely on the adequacy of the sample to reach a temperature equilibrium. To date, the dynamic-sample systems are based on continuous-flow chips (Kopp et al. 1998) or on oscillatory devices (Auroux et al. 2005; Bu et al. 2003). The first continuous-flow PCR system was presented in 1998 by Kopp and co-workers. Their device consisted of a serpentine channel that passed over three heating zones (see Fig. 31.4). The remarkable features of the device enabled the detection of specifically amplified products in less than 1.5 minutes. This device has inspired many groups, leading to devices optimized for injection series—for example, Schneegaß et al. (2001). Additionally high throughput and single molecule amplification (Baker et al. 2003), infectious biological agent monitoring (Belgrader et al. 2003), and RNA Reverse-Transcription and PCR

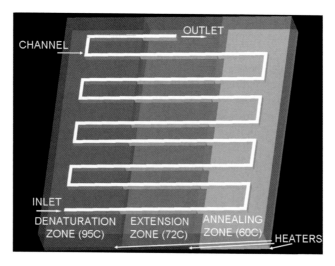

Figure 31.4. Continuous-flow PCR. The sample is pumped unidirectionally over the heating zones. The number of meanders determines the maximum number of cycles.

(Obeid et al. 2003) have also been performed on continuous-flow devices. It is to be noted that a completely different design has been proposed by Liu et al. whereby DNA amplification was performed in a rotary device in which the sample was passing in over the heating zones (Liu et al. 2002). Although continuous-flow devices provide fast heating and cooling rates—and thus improve product specificity—the design is restrictive in terms of number of cycles: the number of meanders (or turns) is decided during the design procedure and cannot be modified once the chip is manufactured. Furthermore, parallelization is not easily applicable and would significantly complicate the design.

The oscillatory approach was driven by the desire to combine the advantages of both static and dynamic systems. This system is based on shunting the sample back and forth over the heating zones in a straight channel (see Fig. 31.5). The number of cycles can easily

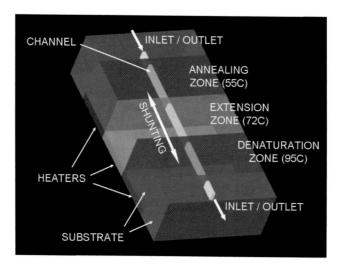

Figure 31.5. Sample-shunting PCR. The sample is pumped bidirectionally over the heating zones. The number of cycles is easily modified by increasing/decreasing the number of pumping strokes.

be modified by altering the pumping program, providing high flexibility. Transition times from one temperature zone to another can also be optimized by adapting pump velocities. In addition, parallelization is easily obtained due to the flexibility of lithographic methods. Finally, it is possible to incorporate post-PCR applications (such as assays) at the end of each channel, providing unique capabilities to apply simultaneously different treatments to the same original sample. This system has been previously presented at several conferences and was enthusiastically received, both for its simplicity and its originality (Auroux et al. 2002, 2003a, 2003b). A theoretical evaluation of a system based on this concept was provided by Bu and co-workers (Bu et al. 2003). A sample-shunting device using ferrofluidics for actuators has also been introduced by Hardt et al. (2004).

4. Minireview

The identification of pathogens is an issue rendered complex not only by the wide variety of existing pathogens, but also by the diversity of the possible environments. The different techniques to detect and analyze these micro-organisms reflect the multifaceted aspect of the problem. For example, the detection of seedborne pathogens includes macroscale techniques such as visual examination of the seeds or intentional growth of selected pathogens on specific media (Walcott 2003). Evidently the latter method is highly dependent on the response of the targeted organisms to the growth media, while the former method is unreliable and not reproducible. Additionally these techniques are time-consuming, as a standard growth phase requires several hours (4 to 7). Alternatively the interest for PCR-based methods has grown since improvements of speed (Wittwer and Garling 1991), selectivity (Toma et al. 2003), and sensitivity (Li et al. 1988) have been demonstrated. In particular DuPont Qualicon has developed the BAX® detection system (DuPont Qualicon 2007). This kit enables the detection by real-time PCR of several pathogens commonly found in food and environmental samples, such as *Salmonella, Listeria, E. Coli* O157:H7 or *Enterobacter sakazakii*. After an enrichment phase following standard procedures, the samples are then mixed with a lysing solution and heated. The lysed samples are subsequently transferred onto conventional plates containing dry PCR tablets made of the reagents required for the amplification. The final stage consists of putting the plates in a conventional thermocycler and performing real-time PCR. The company claims that concentrations as low as 10^4 cfu/ml (colony forming unit/ml) can be detected by this method, with sensitivity/specificity rates of at least 98 % in some cases. Several official administrations, such as the U.S. Food and Drug Administration or the federal agency Health Canada, have already approved the use of the BAX® system for specific applications. For example Health Canada has extended in December 2006 its application from ground beef and beef trim to fruits, vegetables, dairy, meat, animal feed, environmental samples, and dry products. Despite the success of the polymerase chain reaction, recurrent drawbacks associated with this method include lengthy sample preparation procedures and false positives due to the amplification of dead cells (Walcott 2003; Deisingh and Thompson 2004; Gilbert and Clay 1973). PCR is sensitive to contamination. In particular many molecules released during the lysing process are PCR inhibitors. It is therefore paramount to incorporate purification steps in the protocols. The techniques usually involved in DNA extraction and purification are time consuming. However this task was rendered less labor-intensive by the coupling of PCR with immunomagnetic separation (Molday and MacKenzie 1982). Magnetic beads coated with antibodies are used to target specific organisms. They are then captured using a magnetic field, and DNA is extracted for subsequent amplification. This method has been profusely applied (Molday and MacKenzie 1982; Safarikova and Safarik 1995; Jareo et al. 1997; Liakopoulos et al. 1997; Chalmers et al. 1998; Mavrou et al. 1998; Thiel et al. 1998; Caldarelli-Stefano et al. 1999; Deng et al. 2001;

Chronis et al. 2001; Walcott 2003; Prodelalova et al. 2004; Krizova et al. 2005; Olsvik et al. 1994). During pathogen detection false positives might be obtained from PCR analysis when a colony that contained a specific pathogen was present but not viable. To circumvent this issue, a Reverse-Transcription PCR (RT-PCR) can be performed: a strand of mRNA (only produced in viable cells and short-lived) is converted to a DNA strand that is subsequently amplified. Alternatively, the inclusion of an internal amplification control (IAC) can be used to assess the efficiency of PCR. An IAC is a DNA fragment that is co-amplified with the target sequence. Rodriguez-Lazaro et al. (2005) developed an IAC for real-time PCR that is not only an indication of PRC performance but is also highly quantifiable. The authors reported the use of this IAC, a chimeric double stranded DNA fragment of rapeseed with specific target sequences, in the quantitative real time PCR assay for *Listeria monocytogenes*. The assays were highly specific, identifying and differentiating between 49 *L. monocytogenes* isolates and 96 strains of non-target bacteria. Derzelle and Dilasser (2006) reported differentiation between 35 *Enterobacter sakazakii*, a foodborne pathogen and 180 non *E. sakazakii* strains using an IAC.

Based on these considerations PCR is a promising method for the detection of pathogens as it is more rapid, more sensitive, and more specific than conventional procedures. In addition, easily-implemented solutions exist to circumvent the potential drawbacks of this amplification process. It is therefore no surprise that scientists have been developing and optimizing protocols to perform DNA amplification, as well as sample purification and detection. The ultimate goal is to combine the above-mentioned advantages of PCR with the benefits of microfluidics described in Section 3, namely integration, reduction of reaction time, and reduction of reactant consumption. This association would lead to a miniaturized total analysis system integrating all modules necessary for the real-time analysis of untreated samples.

Several groups have proposed microfluidic chips to partially solve this challenge. As early as 1996 Woolley et al. (1996) developed a microfluidic-based assay integrating DNA amplification and capillary electrophoresis capable of detecting *Salmonella* in 45 minutes. In 1998, Professor Manz's group introduced the first continuous-flow PCR chip (Kopp et al. 1998). This device was used to investigate the amplification of *Neisseria gonorrhoeae* under different flow rate conditions. A 20-cycle amplification was successfully reported in 90s. In parallel, Belgrader and co-workers developed well-based chambers for the detection of several pathogens including *Erwinia herbicola, Bacillus subtili,* and *B. anthracis* (Belgrader et al. 1998). More recently Lee and co-workers demonstrated the use of a laser-irradiated magnetic bead system (LIMBS) for DNA extraction and real-time PCR for rapid pathogen analysis (Lee et al. 2006). A bead density of $0.9\,\mu g/\mu L$ was proved optimum, enabling the detection of twenty *E. Coli* cells at a concentration of 20 cells/μL in a 10μL sample. Highly sensitive detection of the equivalent of a mere 10 *E. coli* cells was accomplished by Yang et al. (2002) using a serpentine PCR microreactor in a plastic microchip. The microchip was sandwiched between two Peltier thermoelectric devices, which provided heat, and a cooling fan was attached to provide convective cooling. *E. coli* cells were thermally lysed initially, and 30 cycles of amplification were performed. Detection of the pathogen was performed off the chip using an Agilent Bioanalyzer. Further lysis and DNA amplification of four different pathogens with subsequent detection and differentiation were performed as well. Alternatively, Chang and co-workers used the electro-wetting-on-dielectric (EWOD) effect to perform procedures such as transportation, mixing, and PCR in order to detect the Dengue II virus (Chang et al. 2006). The EWOD effect consists of manipulating droplets by controlling the surface tension of the substrate (Pollack et al. 2000). By applying a voltage through an array of individually addressable electrodes, the meniscus of the droplet is deformed, generating a pressure gradient that leads to a bulk flow. The analysis of the Dengue II virus by Chang et al. led to a successful amplification with a 50 % reduction of the reaction time and a 70 % diminution of the sample consumption. In 2006 Huang et al. also worked on the detection of the Dengue II virus (Huang et al. 2006). They developed an integrated microfluidic chip for amplification, separation, and detection of DNA (*Streptococcus pneumoniae*) and RNA (Dengue-2 virus). PCR amplification is performed on this device in a separate chamber fabricated to reside below microheaters and temperature sensors. After amplification the sample is pumped to an electrophoresis segment of the device, and the amplified sample is then electrophoretically separated. Detection is accomplished using an embedded optical fiber with an index-matching oil in a side chamber to enhance signal intensity. The benefit to this system lies in the amplification chamber as separate from the electrophoretic chamber. Electrophoresis buffers can be dried through the thermocycling process and can sometimes decrease PCR efficiency (Kourkine et al. 2002). Microvalving allows the two chambers to be separated and for the movement of the sample from the isolated PCR chamber to the electrophoresis component of the chip. Tsai and co-workers have been developing a reverse-transcription/PCR chip based on a SU-8 substrate (Tsai and Sue 2006). Their particular interest was to investigate and optimize the heat transfer of a continuous-flow chip. Alternatively, Miyashita and co-workers developed a multiplex PCR assay for the detection of *Chlamydia pneumoniae, Mycoplasma pneumonia,* and *Legionella pneumophila* (Miyashita et al. 2004). Although the amplification was performed on a conventional thermocycler, the sample analysis was performed with the SV1210 Micro-Chip Electrophoresis Analysis System developed by Hitachi. Startis-Cullum et al. (2003) used an ELISA assay followed by a chip-based detection for the analysis of *Bacillus globigii* spores. Although harmless, these spores served as a model for *B. anthracis*, a warfare agent. A concentration of spores corresponding to 17 spores per liter of air could be detected. Liu and co-workers developed a micro-total analysis system comprising an extensive microfluidic system, heaters, PCR wells, and DNA sensors (Liu et al. 2004). They applied this microchip to the detection of pathogenic bacteria extracted from whole blood samples. A chip developed by Yeung et al. (2006) introduced a multiplex sample preparation, DNA amplification, and electrochemical detection in a single microchamber rather than chambers for each individual step in the process. In the microchamber a cell, in this case *E. coli,* was thermally lysed, and the DNA was captured on magnetic particles. Next, DNA was amplified using PCR, and the amplicons were then hybridized to complementary oligonucleotide capture probes, which were immobilized onto electrodes then labeled with gold nanoparticles. Electrochemical detection was accomplished by electrocatalytic deposition of silver onto the gold nanoparticles from a silver nitrate solution. The amount of deposited silver was determined by measurement of the oxidative silver dissolution response. In 2005 Kamei, et al. advanced the detection of pathogen genetic amplicons on microfluidic devices with the development of an integrated fluorescence detector (Kamei et al. 2005). They describe a multilayer optical interference detector, coupled with a half-ball lens and plasma-deposited annular a-Si:H photodiode. The detection platform was placed 4 mm from the microfluidic channel. Using this on-chip detection system, the researchers were successful in identifying *S. aureus* and *E. coli* pathogens.

5. Conclusions

The demand for real-time testing at point-of-care locations is increasingly pressing. The analysis of pathogens is of particular interest for scientists involved in a wide variety of fields, such as bioterrorism, food industry or environmental monitoring. As can be seen from this chapter, the association of microfluidics with the polymerase chain reaction is a powerful combination for the detection and identification of pathogens. In addition micro total analysis systems have the inherent potential to regroup different modules to perform multiple tasks. This capability offers for in-the-field analysis an enhanced flexibility, which is of considerable interest. However the delivery of sturdy and easy-to-handle devices that are capable of handling real-life samples has not been achieved yet. Nevertheless the scientific community involved in developing microfluidic devices is fast approaching the critical mass that will enable the successful transition from university-based research to industry-driven applications. It might, therefore, not be long before consumers are capable of state-of-the-art pathogen analysis in their own dining room.

Acknowledgements

The author would like to thank most particularly Brian Polk and Michael Halter for their insightful suggestions during the preparation of this chapter.

References

Alberts B, Bray D, Johnson A, Lewis JA, Raff M, Roberts K, Walter P (1998) Essential cell biology - An introduction to the molecular biology of the cell. Garland Publishing, Inc., New York and London

Auroux P-A (2005) Microfluidic devices used for shunting polymerase chain reactions. PhD thesis. Imperial College, London, UK

Auroux P-A, Day P, Manz A (2003) Sample-shunting based PCR microfluidic device. Gordon Research Conference—Physics and Chemistry of Microfluidics. Big Sky Resort, Montana, USA

Auroux P-A, Day PJ, Manz A (2005) Quantitative study of the adsorption of PCR reagents during on-chip bi-directional shunting PCR. International Conference on Miniaturized Systems for Chemistry and Life Sciences (MicroTAS) Boston, USA, 283–285

Auroux P-A, Day PJR, Niggli F, Manz A (2002) Microfluidic device for detection of low copy number nucleic acids. Nanotech 2002. Montreux, Switzerland

Auroux P-A, Day PJR, Niggli F, Manz A (2003) PCR micro-volume device for detection of nucleic acids. The nanotechnology conference and trade show. San Francisco, CA, USA, p. 55.

Auroux P-A, Koc Y, deMello AJ, Manz A, Day PJR (2004) Miniaturised nucleic acid analysis. Lab Chip 4:534–546

Baker J, Strachan M, Swartz K, Yurkovetsky Y, Rulison A, Brooks C, Kopf-Sill A (2003) Single molecule amplification in a continuous flow labchip device. International Conference on Miniaturized Systems for Chemistry and Life Sciences (MicroTAS) Squaw Valley, California, 1335

Belgrader P, Benett W, Hadley D, Long G, Mariella RJ, Milanovich F, Nasarabadi S, Nelson W, Richards J, Stratton P (1998) Rapid pathogen detection using a microchip PCR array instrument. Clin Chem 44:2191–2194

Belgrader P, Benett W, Hadley D, Long G, Mariella RJ, Milanovich F, Nasarabadi S, Nelson W, Richards J, Stratton P (1999) Infection disease: PCR detection of bacteria in seven minutes. Science 284:449

Belgrader P, Elkin CJ, Brown SB, Nasarabadi SN, Langlois RG, Milanovich FP, Colston BWJ, Marshall GD (2003) A reusable flow-through polymerase chain reaction instrument for continuous monitoring of infectious biological agents. Anal Chem 75:3114–3118

Bonnet G, Tyagi S, Libchaber A, Kramer FR (1999) Thermodynamics basis of the enhanced specificity of structured DNA probes. Proceedings of the National Academy of Science USA 96:6171–6176

Branebjerg J, Fabius B, Gravesen P (1994) Application of miniature analyzers - from microfluidic components to micro-TAS. International Conference on Miniaturized Systems for Chemistry and Life Sciences (MicroTAS) 141–151

Bu MQ, Melvin T, Ensell G, Wilkinson JS, Evans AGR (2003) Design and theoretical evaluation of a novel microfluidic device to be used for PCR. J Micromech Microeng 13:S125–S130

Burns MA, Johnson BN, Brahmasandra SN, Handique K, Webster JR, Krishnan M, Sammarco TS, Man PM, Jones D, Heldsinger D, Mastrangelo CH, Burke DT (1998) An integrated nanoliter DNA analysis device. Science 282:484–487

Bustin SA (2002) Quantification of mRNA using real-time reverse transcription PCR (RT-PCR): trends and problems. J Mol Endocrinol 29:23–39

Caldarelli-Stefano R, Vago L, Bonetto S, Nebuloni M, Costanzi G (1999) Use of magnetic beads for tissue DNA extraction and IS6110 Mycobacterium tuberculosis PCR. Journal of Clinical Pathology: Molecular Pathology 52:158–163

Chalmers JJ, Zborowski M, Sun LP, Moore L (1998) Flow through, immunomagnetic cell separation. Biotechnol Prog 14:141–148

Chang YH, Lee GB, Huang FC, Chen YY, Lin JL (2006) Integrated polymerase chain reaction chips utilizing digital microfluidics. Biomedical Microdevices 8:215–225

Chaudhari AM, Woudenberg TM, Albin M, Goodson KE (1998) Transient liquid crystal thermometry of microfabricated PCR vessel arrays. J Microelectromech Syst 7:345–355

Cheng J, Shoffner MA, Hvichia GE, Kricka LJ, Wilding P (1996a) Chip PCR II. Investigation of different PCR amplification systems in microfabricated silicon-glass chips. Nucleic Acids Res 24:380–385

Cheng J, Shoffner MA, Mitchelson KR, Kricka LJ, Wilding P (1996b) Analysis of ligase chain reaction products amplified in a silicon-glass chip using capillary electrophoresis. J Chromatogr A 732:151–158

Chien A, Edgar DB, Trela JM (1976) Deoxyribonucleic acid polymerase from the extreme thermophile Thermus Aquaticus. Journal of Bacteriology 127:1550–1557

Chiou J, Matsudaira P, Sonin A, Ehrlich D (2001) A Closed-Cycle Capillary Polymerase Chain Reaction Machine. Anal Chem 2018–2021

Chronis N, Lam W, Lee L (2001) A microfabricated bio-magnetic separator based on continuous hydrodynamic parallel flow. International Conference on Miniaturized Systems for Chemistry and Life Sciences (MicroTAS) 497–498

Connors E, Lundregan T, Miller N, McEwen T (1996) Convicted by juries, exonerated by science: case studies in the use of DNA evidence to establish innocence after trial. National Institute of Justice

de Mello AJ (2001) DNA amplification: does 'small' really mean 'efficient'? Lab on a Chip 1:24N

de Mello AJ, Wootton RCR (2002) But what is it good for? Applications of microreactor technology for the fine chemical industry. Lab on a Chip 1:7N

Deisingh AK, Thompson M (2004) Strategies for the detection of Escherichia coli O157 : H7 in foods. Journal of Applied Microbiology 96:419–429

Deng T, Whitesides GM, Radhakrishnan M, Zabow G, Prentiss M (2001) Manipulation of magnetic microbeads in suspension using micromagnetic systems fabricated with soft lithography. Appl Phys Lett 78:1775–1777

Derzelle S, Dilasser F (2006) A robotic DNA purification protocol and real-time PCR for the detection of Enterobacter sakazakii in powdered infant formulae. Bmc Microbiology 6:1–12

Don RH, Cox PT, Wainwright BJ, Baker K, Mattick JS (1991) 'Touchdown' PCR to circumvent spurious priming during gene amplification. Nucleic Acids Res 19:4008

DuPont Qualicon (2007) BAX (R) System Q7 - The power to do more. DuPontQualicon brochure

Feustel A, Muller J, Relling V (1994) A Microsystem Mass Spectrometer. International Conference on Miniaturized Systems for Chemistry and Life Sciences (MicroTAS) 299–304

Forsdyle DR (1995) Sense in anti-sense? Journal of Molecular Evolution 41:582–586

Friedman NA, Meldrum DR (1998) Capillary tube resistive thermal cycling. Anal Chem 70:2997–3002

Fuhr G, Wagner B (1994) Electric Field Mediated Cell Manipulation, Characterisation and Cultivation in Highly Conductive Media. International Conference on Miniaturized Systems for Chemistry and Life Sciences (MicroTAS) 209–214

Gilbert TR, Clay AM (1973) Determination of Ammonia in Aquaria and Sea Water using the ammonia electrode. Anal Chem 45:1757–1759

Han J, Craighead G (2000) Separation of long DNA molecules in a microfabricated entropic trap array. Science 288:1026–1029

Hardt S, Dadic D, Doffing F, Drese KS, Münchov G, Sörensen O (2004) Development of a slug-flow PCR chip with minimum heating cycle times. The nanotechnology conference and trade show. Boston, MA, USA, p. 55.

Harrison DJ, Glavina PG, Manz A (1993) Towards Miniaturized Electrophoresis and Chemical-Analysis Systems on Silicon - an Alternative to Chemical Sensors. Sens Actuators B 10:107–116

Harrison DJ, Manz A, Fan ZH, Ludi H, Widmer HM (1992) Capillary Electrophoresis and Sample Injection Systems Integrated on a Planar Glass Chip. Anal Chem 64:1926–1932

Harrison DJ, Manz A, Glavina PG (1991) Transducers '91 792–795

Higuchi R, Dollinger G, Walsh PS, Griffith R (1992) Simultaneous Amplification and Detection of Specific DNA-Sequences. Bio-Technology 10:413–417

Higuchi R, Fockler C, Dollinger G, Watson R (1993) Kinetic PCR Analysis - Real-Time Monitoring of DNA Amplification Reactions. Bio-Technology 11:1026–1030

Huang FC, Liao CS, Lee GB (2006) An integrated microfluidic chip for DNA/RNA amplification, electrophoresis separation and on-line optical detection. Electrophoresis 27:3297–3305

Jacobson DR, Xu JJ, Smith IC (1996) Lung cancer screening and diagnosis via k-ras mutation detection. International symposium on the impact of cancer biotechnology on diagnostics and prognostics indicators. Nice, France

Jacobson SC, Hergenröder R, Koutny LB, Ramsey JM (1994) High-Speed Separations on a Microchip. Anal Chem 66:1114–1118

Jareo PW, Preheim LC, Snitily MU, Gentry MJ (1997) Use of magnetic cell sorting to isolate blood neutrophils from rats. Lab Anim Sci 47:414–418

Kamei T, Toriello NM, Lagally ET, Blazej RG, Scherer JR, Street RA, Mathies RA (2005) Microfluidic genetic analysis with an integrated a-Si : H detector. Biomedical Microdevices 7:147–152

Khandurina J, McKnight TE, Jacobson SC, Waters LC, Foote RS, Ramsey JM (2000) Integrated system for rapid PCR-based DNA analysis in microfluidic devices. Anal Chem 72:2995–3000

Kopp MU, de Mello AJ, Manz A (1998) Chemical amplification: Continuous-flow PCR on a chip. Science 280: 1046–1048

Kourkine IV, Hestekin CN, Magnusdottir SO, Barron AE (2002) Optimized sample preparation methods for tandem capillary electrophoresis single -strand conformation polymorphism/heteroduplex analysis (CE-SSCP/HA), Biotechniques, 33:318–325.

Krishnan M, Ugaz VM, Burns MA (2003) PCR in a Rayleigh-Benard convection cell. Science 298:793

Krizova J, Spanova A, Rittich B, Horak D (2005) Magnetic hydrophilic methacrylate-based polymer microspheres for genomic DNA isolation. Journal of Chromatography A 1064:247–253

Lagally ET, Medintz I, Mathies RA (2001) Single-molecule DNA amplification and analysis in an integrated microfluidic device. Anal Chem 73:565–570

Lee JG, Cheong KH, Huh N, Kim S, Choi JW, Ko C (2006) Microchip-based one step DNA extraction and real-time PCR in one chamber for rapid pathogen identification. Lab on a Chip 6:886–895

Lee LG, Connell CR, Bloch W (1993) Allelic Discrimination by Nick-Translation PCR with Fluorogenic Probes. Nucleic Acids Res 21:3761–3766

Li HH, Gyllensten UB, Cui XF, Saiki RK, Erlich HA, Arnheim N (1988) Amplification and Analysis of Dna-Sequences in Single Human-Sperm and Diploid-Cells. Nature 335:414–417

Liakopoulos TM, Choi JW, Ahn CH (1997) A bio-magnetic bead separator on glass chips using semi-encapulated spiral electromagnets. Transducers '97 1:485–488

Liu J, Enzelberger M, Quake S (2002) A nanoliter device for polymerase chain reaction. Electrophoresis 23:1531–1536

Liu L (2003) Bioinformatics III: Primer Design. ICBR Molecular Biology International – Bioinformatic Workshop in Nicaragual.

Liu RH, Yang J, Lenigk R, Bonanno J, Grodzinski P (2004) Self-contained, fully integrated biochip for sample preparation, polymerase chain reaction amplification, and DNA microarray detection. Anal Chem 76:1824–1831

Llop P, Bonaterra A, Penalver J, Lopez MM (2000) Development of a highly sensitive nested-PCR procedure using a single closed tube for detection of Erwinia amylovora in asymptomatic plant material. Applied and Environmental Microbiology 66:2071–2078

Lundberg KS, Short JM, Sorge JA, Mathur EJ (1991) A New Thermostable Polymerase with High Fidelity. Faseb J 5:A1549

Manz A, Graber N, Widmer HM (1990b) Miniaturized Total Chemical-Analysis Systems - a Novel Concept for Chemical Sensing. Sens Actuators B 1:244–248

Manz A, Harrison DJ, Fettinger JC, Verpoorte E, Ludi H, Widmer HM (1991) Transducers 91:939–941

Manz A, Harrison DJ, Verpoorte EMJ, Fettinger JC, Paulus A, Ludi H, Widmer HM (1992) Planar Chips Technology for Miniaturization and Integration of Separation Techniques into Monitoring Systems - Capillary Electrophoresis on a Chip. J Chromatogr 593:253–258

Manz A, Miyahara Y, Miura J, Watanabe Y, Miyagi H, Sato K (1990a) Design of an Open-tubular Column Liquid Chromatograph Using Silicon Chip Technology. Sensors and Actuators B1:249–255

Marras SAE, Kramer FR, Tyagi S (2002) Efficiencies of fluorescence resonance energy transfer and contact-mediated quenching in oligonucleotide probes. Nucleic Acids Res 30:e122

Mattila P, Korpela J, Tenkanen T, Pitkanen K (1991) Fidelity of DNA-Synthesis by the Thermococcus-Litoralis DNA-Polymerase - an Extremely Heat-Stable Enzyme with Proofreading Activity. Nucleic Acids Res 19:4967–4973

Mavrou A, Colialexi A, Tsangaris GT, Antsaklis A, Panagiotopoulou P, Tsenghi C, Metaxotoy C (1998) Fetal cells in maternal blood isolation by magnetic cell sorting and confirmation by immunophenotyping and FISH. In Vivo 12:195–200

Miyashita N, Saito A, Kohno S, Yamaguchi K, Watanabe A, Oda H, Kazuyama Y, Matsushima T (2004) Multiplex PCR for the simultaneous detection of Chlamydia pneumoniae, Mycoplasma pneumoniae and Legionella pneumophila in community-acquired pneumonia. Respiratory Medicine 98:542–550

Molday RS, MacKenzie D (1982) Immunospecific ferromagnetic iron dextran reagents for the labeling and magnetic separation of cells. J Immunol Methods 52:353–367

Nagakura T, Maruo S, Ikuta K (2003) The study of micro blood sugar control device without energy supply for diabetes therapy. Transducers '03 2:1209–1212

Nakano H, Matsuda K, Yohda M, Nagamune T, Endo I, Yamane T (1994) High-Speed Polymerase Chain-Reaction in Constant Flow. Biosci Biotechnol Biochem 58:349–352

Northrup MA, Ching MT, White RM, Watson RT (1993) DNA amplification with a microfabricated reaction chamber. Transducers '03 924–926

Northrup MA, Gonzelez C, Hadley D, Hills RF, Landre P, Lehew S, Saiki R, Sinski JJ, Watson R, Whatson J (1995) A MEMS-based miniature DNA analysis system. Transducers 95 1:746–767

Obeid PJ, Christopoulos TK, Crabtree HJ, Backhouse CJ (2003) Microfabricated device for DNA and RNA amplification by continuous-flow polymerase chain reaction and reverse transcription-polymerase chain reaction with cycle number selection. Anal Chem 75:288–295

Oda RP, Strausbauch MA, Huhmer AFR, Borson N, Jurrens SR, Craighead J, Wettstein PJ, Eckloff B, Kline B, Landers JP (1998) Infrared-Mediated Thermocycling for Ultrafast Polymerase Chain Reaction Amplification of DNA. Anal Chem 70:4361

Olsvik O, Popovic T, Skjerve E, Cudjoe KS, Hornes E, Ugelstad J, Uhlen M (1994) Magnetic separation techniques in diagnostic microbiology. Clinical microbiology reviews 7:43–54

Opekun AR, Abdalla N, Sutton FM, Hammond F, Kuo GM, Torres E, Steinbauer J, Graham DY (2002) Urea breath testing and analysis in the primary care office. J Fam Pract 51:1030–1032

Pal D, Venkataraman V (2002) A portable battery-operated chip thermocycler based on induction heating. Sensors and Actuators A 102:151–156

Pollack MG, Fair RB, Shenderov AD (2000) Electro-wetting based actuation of liquid droplets for microfluidic applications. Appl Phys Lett 77:1725–1726

Poser S, Schulz T, Dillner U, Baier V, Kohler JM, Schimkat D, Mayer G, Siebert A (1997) Chip elements for fast thermocycling. Sens Actuator A-Phys 62:672–675

Powledge TM (2004) The polymerase chain reaction. Advances in Physiology Education, 28:44–50

Prodelalova J, Rittich B, Spanova A, Petrova K, Benes MJ (2004) Isolation of genomic DNA using magnetic cobalt ferrite and silica particles. Journal of Chromatoghraphy A 1056:43–48

Rodriguez-Lazaro D, Pla M, Scortti M, Monzo HJ, Vazquez-Boland JA (2005) A novel real-time PCR for Listeria monocytogenes that monitors analytical performance via an internal amplification control. Applied and Environmental Microbiology 71:9008–9012

Safarikova M, Safarik I (1995) Magnetic Separations in Biosciences and Biotechnologies. Chem Listy 89:280–287

Schneegaß I, Bräutigam R, Köhler JM (2001) Miniaturized flow-through PCR with different templates types in a silicon chip thermocycler. Lab Chip 1:42–49

Stratis-Cullum DN, Giffrin GD, Mobley J, Vass AA, Vo-Dinh T (2003) A miniature biochip for detection of aerosolized Bacillus globigii spores. Anal Chem 75:275–280

SuperArray–Bioscience Corporation (2004) The advantages of Hot-Start PCR technology, Newsletter – Pathway, 1(4):3.

Taylor TB, Winn-Deen ES, Picozza E, Woudenberg TM, Albin M (1997) Optimization of the performance of the polymerase chain reaction in silicon-based microstructures. Nucleic Acids Res 25:3164–3168

Terry SC, Jerman JH, Angell JB (1979) A gas chromatographic air analyzer fabricated on a silicon wafer, I.E.E.E.Transactions on Electron Devices, ED-26:1880–86.

Thiel A, Scheffold A, Radbruch A (1998) Immunomagnetic cell sorting - pushing the limits. Immunotechnology 4:89–96

Thompson C, Loeffelholz M (1995) Detection of HIV-1 infection by polymerase chain reaction (PCR). Hotlines - University of Iowa 34:1–2

Toma C, Lu Y, Higa N, Nakasone N, Chinen I, Baschkier A, Rivas M, Iwanaga M (2003) Multiplex PCR assay for identification of Human Diarrheagenic Escherichia coli. J Clin Microbiol 41:2669–2671

Tsai NC, Sue CY (2006) SU-8 based continuous-flow RT-PCR bio-chips under high-precision temperature control. Biosensors & Bioelectronics 22:313–317

Tyagi S, Kramer FR (1996) Molecular beacons: probes that fluoresce upon hybridization. Nat Biotechnol 14:303–308

Vilkner T, Janasek D, Manz A (2004) Micro total analysis systems. Recent developments. Anal Chem 76:3373–3386

Wahr JA, Lau W, Tremper KK, Hallock L, Smith K (1996) Accuracy and precision of a new, portable, handheld blood gas analyzer, the IRMA(R). J Clin Monit 12:317–324

Walcott RR (2003) Detection of seedborne pathogens. Horttechnology 13:40–47

Wallenborg SR, Bailey CG (2000) Separation and detection of explosives on a microchip using micellar electrokinetic chromatography and indirect laser- induced fluorescence. Anal Chem 72:1872–1878

Waters LC, Jacobson SC, Kroutchinina N, Khandurina J, Foote RS, Ramsey JM (1998) Microchip device for cell lysis, multiplex PCR amplification, and electrophoretic sizing. Anal Chem 70:158–162

Watson JD, Crick FHC (1953) A structure for deoxyribose nucleic acid. Nature 171:737–738

Whitcombe D, Theaker J, Guy SP, Brown T, Little S (1999) Detection of PCR products using self-probing amplicons and fluorescence. Nat Biotechnol 17:804–807

Wittwer CT, Garling DJ (1991) Rapid cycle DNA amplification: time and temperature optimization. BioTechniques 10:76–83

Woolley AT, Hadley D, Landre P, de Mello AJ, Mathies RA, Northrup MA (1996) Functional Integration of PCR Amplicfication and Capillary Electrophoresis in Microfabricated DNA Analysis device. Anal Chem 68:4081

Wootton RCR, Fortt R, de Mello AJ (2001) On-chip generation and reaction of unstable intermediates - monolithic nanoreactors for diazonium chemistry: Azo dyes. Lab on a Chip 2:5–7

Xu F, Jabasini M, Baba Y (2002) DNA separation by microchip electrophoresis using low-viscosity hydroxypropylmethylcellulose-50 solutions enhanced by polyhydroxy compounds. Electrophoresis 23:3608–3614

Yang JN, Liu YJ, Rauch CB, Stevens RL, Liu RH, Lenigk R, Grodzinski P (2002) High sensitivity PCR assay in plastic micro reactors. Lab on a Chip 2:179–187

Yeung SW, Lee TMH, Cai H, Hsing IM (2006) A DNA biochip for on-the-spot multiplexed pathogen identification. Nucleic Acids Research 34

Yoon DS, Lee YS, Lee YK, Cho HJ, Sung SW, Oh KW, Cha J, Lim G (2002) Precise temperature control and rapid thermal cycling in a micromachined DNA polymerase chain reaction. Journal of Micromechanics & Microengineering 12:813–823

Zaletaev DV, Nemtsova MV, Strelnikov VV, Babenko OV, Vasil'ev EV, Zemlyakova VV, Zhevlova AI, Drozd OV (2004) Diagnostics of epigenetics alterations in hereditary and oncological disorders. Molecular Biology 38:174–182

Zou Q, Miao Y, Chen Y, Sridhar U, Chong CS, Chai T, Tie Y, Teh CHL, Lim JS, Heng CK (2002) Micro-assembled multi-chamber thermal cycler for low-cost reaction chip thermal multiplexing. Sensors and Actuators A 102: 114–121

32

Micro- and Nanopatterning for Bacteria- and Virus-Based Biosensing Applications

David Morrison, Kahp Y. Suh and Ali Khademhosseini

Abstract

Current technologies capable of rapidly and accurately detecting the presence of infectious diseases and toxic compounds in the human body and the environment are inadequate and new, novel techniques are required to ensure the safety of the general population. To develop these technologies, researchers must broaden their scope of interest and investigate scientific areas that have yet to be fully explored. Lithography is a common name given to technologies designed to print materials onto smooth surfaces. More specifically, micropatterning encompasses the selective binding of materials to surfaces in organized microscale arrays. The selective micropatterning of bacteria and viruses is currently an exciting area of research in the field of biomedical engineering and can potentially offer attractive qualities to biosensing applications in terms of increased sensing accuracy and reliability. This chapter focuses on briefly introducing the reader to the fundamentals of bacterial and viral surface interactions and describing several different micropatterning techniques and their advantages and disadvantages in the field of biosensing. The application of these techniques in healthcare and environmental settings is also discussed.

1. Introduction

There is a great need for the development of techniques that can rapidly screen for the presence of infectious diseases or toxic compounds in the human body and the environment (Suh et al. 2004). Viral outbreaks, such as severe acute respiratory syndrome (SARS) and avian flu, have created a rising awareness about toxic and infectious agents in our surroundings. Antimicrobial agents have been developed to help treat patients suffering from exposure to harmful bacteria and viruses; however, bacteria, fungi, and viruses have all begun to develop resistance to the antimicrobial agents directed against them (Moellering 1995). Microbial resourcefulness and resilience have never been more evident than in their ability to develop resistance to chemotherapeutic agents (Wood and Moellering 2003). Bacteria accomplish this through the modification of their DNA by chromosomal mutation and then by acquiring resistance genes via conjugation, transformation, and even transduction (Wood and Moellering 2003). An example of this is the demonstration of transferable fluoroquinolone resistance genes in *Klebsiella*

David Morrison • Harvard–MIT Division of Health Sciences and Technology, Massachusetts Institute of Technology, Cambridge, MA.   **Kahp Y. Suh** • School of Mechanical and Aerospace Engineering, Seoul National University, Seoul, Korea.   **Ali Khademhosseini** • Center for Biomedical Engineering, Department of Medicine, Brigham and Women's Hospital, Harvard Medical School, Harvard–MIT Division of Health Sciences and Technology, Massachusetts Institute of Technology, Cambridge, MA.

M. Zourob et al. (eds.), *Principles of Bacterial Detection: Biosensors, Recognition Receptors and Microsystems*,
© Springer Science+Business Media, LLC 2008

pneumoniae shown by Tran (2002). Additionally, the emergence of past viruses that were previously eradicated through the delivery of vaccines reaffirms that faster and more reliable techniques to identify viruses and bacteria must be found. This can be seen in the devastating re-emergence of wild poliovirus in Namibia (Roberts 2006). This recent outbreak has resulted in many deaths and has the potential to become a very serious threat. This example also illustrates the increasing need for bacterial and viral detection on a global basis. With the recent advances in global transportation the ability of a lethal microbial infection to spread across the world is greater now than ever before.

Traditional laboratory diagnoses of viral and bacterial infections typically involve the use of direct fluorescent antibody assays, viral load testing and cell culturing (Whiley and Sloots 2005). Techniques such as Enzyme-Linked ImmunoSorbent Assay (ELISA) use two antibodies, one linked to an enzyme and the other specifically bound to an antigen, to identify virus presence (Ivanov and Dragunsky 2005). The enzyme acts as an amplifier, converting a substrate to a chemogenic or fluorescent signal that is easily detected by a spectrophotometer (Ivanov and Dragunsky 2005). This technique may be highly sensitive, however, it requires highly trained technicians and high maintenance equipment, which make it unfeasible in many underdeveloped regions of the world (Respess, Rayfield and Dondero 2001). Another method of cell culturing bacteria involves placing a sample in a plated environment and waiting for colonies of bacteria to form and grow. This technique is simple but too time consuming to be considered valuable.

Other viruses, such as HIV, can be detected through the use of viral load testing (Rich et al. 1999). Viral load testing detects cell-free plasma viral RNA through amplification techniques, such as polymerase chain reaction (PCR). Diagnostic techniques such as these provide quick results; however, they are generally not sensitive enough to provide reliable and consistent results (van et al. 2001).

Viruses and bacteria can also be detected through cell culturing methods. This involves obtaining a sample of interest and filtering it with a membrane capable of removing the microbes of interest. The microbes are subsequently detached from the membrane and transferred into a highly concentrated solution. Following this, a host cell is entrenched in the solution, and if there are microbes present, they will infect the host cell and initiate conformational changes that will be easily visible by light microscopy. These methods provide robust and sensitive results but are labor-intensive, time consuming and depend on optimal sample transport for virus isolation (Doller et al. 1992).

To minimize these potential difficulties, miniaturization of cell-based bioassays is currently under active study for the detection of toxic compounds in the environment (Dunn and Feygin 2000). Specifically, bacteria cells are useful in sensing applications because their analytic specificity is easily modified by genetic engineering and their dynamic structure as compared to mammalian cells (Rainina et al. 1996). Recent miniaturization of such assays has included the patterned collection of cells in microsystems, where cellular adhesion can be easily controlled through physical and chemical characteristics such as hydrophobicity, hydrophilicity and surface charge (Koh et al. 2003).

The application of micro and nanopatterning to the problems outlined above has the potential to eliminate the difficulties mentioned and generate newer, easier, cheaper and more reliable bacterial and viral detection methods. Micropatterning can be described as the selective binding of materials to surfaces in organized arrays at the micro and nanoscale. Currently used to pattern cells and proteins, micropatterning incorporates microfabrication techniques with materials science and surface chemistry to explore the interactions between cells that require surface attachment for proper function (Falconnet et al. 2006).

Micropatterning bacteria and viruses have the potential to offer three significant advantages not present in current detection technologies. First, screening sample volumes can be greatly minimized with the application of micro and nanotechnology. This technology has the potential to significantly reduce to required sample volume for detection testing, therefore making sample collection easier. Secondly, micro and nanopatterning may also be used to increase screening throughput. This is possible because of the potential to test for many different bacteria in a very small volume of sample. Many different bacteria and viruses could be patterned on a single surface, thereby increasing the number of simultaneous detections. Finally, micro and nanopatterning may also offer higher sensitivity than current technologies. This may be attributed to the ability to sense and track bacteria at much smaller scales.

The goal of this review is to briefly describe the fundamentals of bacterial and viral surface interactions, to discuss the emergence of micro and nanotechnologies used to pattern bacteria and viruses as well as give examples of specific applications. Furthermore, we will discuss challenges that exist in the development of this technology.

2. Fundamentals of Bacterial and Viral Surface Interactions

To understand the fabrication of micropatterned surfaces for bacteria, one must understand the mechanism of bacterial adhesion. When a bacterium initially comes in contact with a substrate surface it binds reversibly through non-specific interactions (Bonin, Rontani and Bordenave 2001). This leads to eventual irreversible binding through the formation of a protein-ligand interaction (Razatos et al. 1998). Therefore, the adhesion of bacterium, and most other cell types, is dependent on the formation of a protein layer on the surface of the substrate. As such, prevention of this protein layer can be applied to create biocompatible materials that resist bacterial adhesion. Such materials include polyethylene glycol (PEG) polymers, self assembling monolayers (SAMs) terminated by a PEG group, polysaccharides and phospholipids (Kingshott 1999). Lipids can also be involved in the initiation of bacterial adhesion, such as on the surface of hydrogel contact lenses (Franklin 1993); however, inhibiting protein formation is considered the primary objective when designing biocompatible materials (Kingshott 1999).

Viral infection of cells is initiated by the attachment of the virus to cell receptors present on the cell membrane (English and Hammer 2005). Therefore, by inhibiting receptor-mediated attachment, viral infections can potentially be reduced (English and Hammer 2005). The attachment is mediated by viral attachment proteins (VAPs) which are unique to each individual type of virus (Wickham et al. 1990). Several virus types have their VAPs and respective receptors identified. For example, Human Immunodeficiency VAP is a 120-kD glycoprotein that binds to a CD4 receptor present on the surface of T-cells (Wickham et al. 1990). Although many virus types bind through receptor-mediated binding, the nature of this binding varies quite significantly. This can be in terms of number of VAPs per virus and receptor types as well as binding affinity (Wickham et al. 1990). Although receptor-mediated linkages are considered the most common form of viral attachment to cells, there are obviously different methods with which to place viruses onto surfaces. Cheung et al. (2003) report that the fabrication of nanoscale virus arrays can be completed through the placement of a chemoselective linker to the surface of the virus, which enables the virus to attach to a patterned template that is manufactured using scanning probe nanolithography. In another study, Suh et al. (2006) tested the binding affinities of M13 viruses to surfaces using P3 and P9 antibodies. Fig. 32.1 shows scanning electron microscope (SEM) images of the M13 virus on a silicon wafer. Studies such as these may potentially be used to create virus arrays useful for the creation of miniaturized electronic devices or biosensors (Mao 2003).

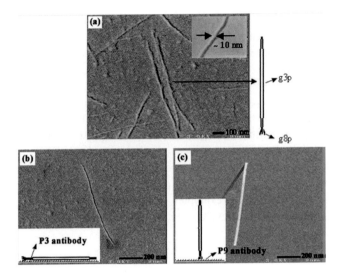

Figure 32.1. (a) SEM image showing the M13 viruses that were cast from a solution onto a silicon wafer. The width of the virus was increased from ~6 nm to ~10 nm because of the gold coating (inset). (b-c) SEM images showing the transition of virus morphology on the P3 and P9 antibody treated silicon wafer, respectively (Reproduced with permission from Nano Letters 2006, 6, 1198–1201. Copyright 2006 American Chemical Society).

3. Technologies for Patterning

3.1. Overview

There have been multiple techniques utilized in the development of bacterial and viral micropatterning. These techniques have been successfully adapted from other fields of research, ranging from microelectronic fabrication to miniaturization of patterning of mammalian cells (Falconnet et al. 2006). Photolithography, soft lithography and scanning probe lithography are described below and examples of relevant research are also included.

3.2. Photolithography

Historically, photolithography has been the most commonly used technique in micropatterning (Xia and Whitesides 1998). This technique was first implemented in the integrated circuit fabrication industry and has been adopted into the biomedical field (Xia and Whitesides 1998). Photolithography involves the placement of the precursor solution onto the substrate of choice. Although there are several variations to photolithography, the main characteristics are the use of a photoresist mask with the pattern of interest and illumination of the precursor solution under UV light.

Koh et al. (2003) report that substrate surface can be modified with the use of a polymer. They suggest the first step in photolithography is to make a precursor solution consisting of the polymer to be photopolymerized and a UV light photo-initiator. The precursor solution is then spin coated onto the surface of the substrate in order to ensure that the substrate is covered with a thin, uniform coat. Correspondingly, faster spin coat speeds result in thinner wells, which reduce the overall volume of the well (Koh et al. 2003). Photopolymerization is then performed under UV light, where only the exposed regions of the precursor solution experience photopolymerization. The areas covered by the mask can then be washed away, leaving the mask's projection patterned on the surface.

Disadvantages to photolithography include high costs associated with the equipment necessary to perform this technique, the required access to a clean room and the method's ineffectiveness at patterning non-planar surfaces (Xia and Whitesides 1998).

3.3. Micromolding (Soft Lithography)

Soft lithography is a technique that uses elastomeric molds to transfer patterns to the surface of the substrate (Whitesides et al. 2001). Soft lithography incorporates many different techniques, including replica molding (Ng et al. 2002), micro contact printing (μCP) (Takayama et al. 1999), microtransfer molding (Zhao, Xia and Whitesides 1996) and capillary force lithography (Suh et al. 2004). In most of the aforementioned techniques a stamp is created by molding an elastomer around a replica of the desired pattern, thus creating a negative replica. Most of the research occurring in soft lithographic techniques focuses on the use of PDMS as the elastomer (Ismagilov et al. 2001; Whitesides et al. 2001; McDonald and Whitesides 2002; Ng et al. 2002). PDMS has many of the desired characteristics of the elastomer, including being biocompatible and permeable to gases. In addition, it can be used for cell culture (Xia and Whitesides 1998). Softer materials such as PDMS offer exciting alternatives to photolithography due to their decreased costs, chemical versatility and potential biodegradability (Ismagilov et al. 2001; Whitesides et al. 2001).

3.3.1. Replica Molding

Replica molding (illustrated in Fig. 32.2a) is a very useful technique for replicating a patterned three-dimensional (3-D) solid surface into a reusable, elastomeric form. This process begins with the formation of a structure referred to as a *master* (Whitesides et al. 2001). The master is the original mold that was created through photolithography (as previously described). This master is initially cast in an elastomer, such as PDMS, and allowed to cure. This is accomplished by simply pouring the elastomer over the master mold. Once cured, the elastomer is peeled off of the master and the resulting "replica" can then be used in microcontact printing as a stamp. Elastomers are used as the replica material due to their soft structure (Whitesides et al. 2001). The nature of the master is typically rigid, and therefore the soft nature of the elastomer helps facilitate separation of the two. The master may also be exposed to $CF_3(CF_2)_6(CH_2)_2SiCl_3$ overnight to help ease the separation of the master and the elastomer replica (Whitesides et al. 2001).

3.3.2. Microcontact Printing

Microcontact printing (illustrated in Fig. 32.2b) is used in conjuction with replica molding to create micropatterned wells that can be used for bacteria and virus patterning (Whitesides et al. 2001). The stamp produced in replica molding is coated with an alkanthiol and brought into contact with a thin coat of gold that sits on top of a silicon substrate (Whitesides et al. 2001). When the stamp is removed the alkanethiol leaves a self-assembled monolayer on the surface of the gold. The monolayer self assembles due to the tendency of sulfur atoms in the thiol chain to coordinate the gold surface and expose the terminal groups of the alkanethiol. Sulfur will react similarly with silver; however, gold is only used in bacterial and viral applications because silver is cytotoxic (Whitesides et al. 2001). The stamp is then removed and the SAM is left on the surface of the silicon in the shape that is printed on the surface of the stamp. The concentration of the alkanthiol used can affect the quality of the SAM and the resolution of the printed features. Microcontact printing is advantageous in situations where it is desirable to transfer large patterns or series of patterns onto a substrate in one step (Whitesides et al. 2001).

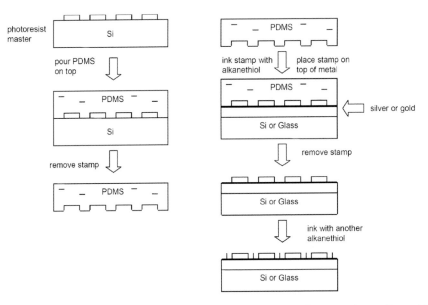

Figure 32.2. A schematic outline of micropatterning by preparation of PDMS stamp using replica molding (a), followed by pattern transfer by microcontact printing (b).

St. John et al. (1998) used microcontact printing to stamp an antibody grating pattern on a silicon surface in a way as to permit the binding of antigen to the antibody, thus creating a biosensor. The performance of the sensor was tested by encapsulating *E.coli* cells on the antibody stamped lines and measuring the intensity of a first-order diffraction beam resulting from the attachment. They found that the diffraction intensity increased with cell density on the surface of the antibody.

Through microcontact printing Morhard et al. were able to use a similar approach in their report of the attachment of covalently coupled antibodies to the surface of a gold substrate. These antibodies were able to selectively bind *E.coli* at well-defined adsorption sites. An incident laser beam was then used to measure light intensity changes in the diffraction pattern of the surface of the substrate, thus verifying the presence of the bacteria.

3.3.3. Microtransfer Molding

In microtransfer molding a patterned PDMS mold is used as a replica of the pattern of interest (Xia and Whitesides 1998). A drop of polymer solution is placed onto the PDMS mold and covered with a substrate. After heating or irradiating, the mold is peeled off of the liquid precursor, leaving a patterned microstructure present on the surface of the substrate. The most advantageous use of microtransfer molding is when molding on a non-planar surface (Xia and Whitesides 1998). This is particularly useful when constructing 3-D microstructures using a layer-by-layer methodology (Xia and Whitesides 1998). Another advantage of microtransfer molding is its ability to fabricate microstructures over large surface areas in a relatively short period of time (Xia and Whitesides 1998).

3.3.4. Capillary Force Lithography

Capillary force lithography is another soft lithographic technique that involves the placement of a PDMS stamp onto the surface of a spin-coated polymer film (Suh, Kim and Lee 2001). Depending on the capillary forces and the surface interactions of the polymer with the stamp, a variety of different polymeric shapes and structures could be formed. The placement of the stamp is followed by a short period of time in which the stamp and the substrate must remain undisturbed. During this time the polymer forms a negative replica of the stamp by spontaneously moving into the stamp's void space via capillary action. If the thickness of the film is small compared to the height of the stamp then the polymer moves to the edges of the stamp rather than fill the void space, where it localizes (Suh, Kim and Lee 2001).

Suh et al. (2004) demonstrated that bacterial cells could be micropatterned using host-parasite and virus-antibody interactions. In this procedure, virus-antibody interactions were introduced to enhance selectivity. It was determined that the adhesion of bacteria was significantly reduced on the surface of PEG when compared to that of silicon or glass (Suh et al. 2004). However, it was also determined that non-specific adhesion became evident when the concentration of bacteria was increased (Suh et al. 2004). This led to a reduction in specificity, which is very undesirable. Fig. 32.3 illustrates the procedure used for creating bacteria microarrays using capillary lithography.

3.4. Scanning Probe Lithography

Scanning probe lithography (SPL) encompasses a wide range of techniques that all share some similar qualities. Each involves the nanoscale surface modification of a substrate using

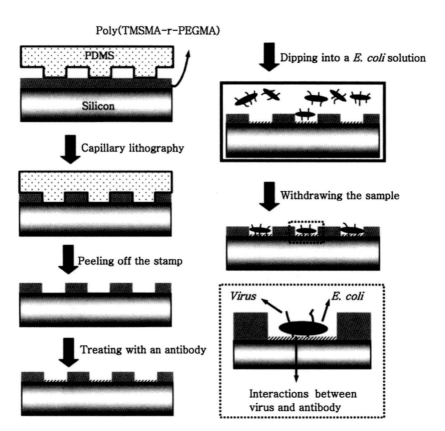

Figure 32.3. Experimental procedure of bacterial array fabrication using capillary lithography (Reprinted from Suh et al. 2004 with kind permission of Springer Science and Business Media).

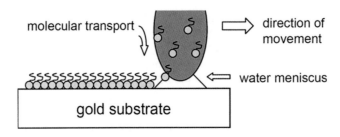

Figure 32.4. Schematic illustration of Dip-Pen nanolithography.

an atomic force microscope (AFM). The four most commonly used SPL techniques are dip-pen lithography, nanoshaving, nanowearing and nanografting.

As shown in Fig. 32.4, dip-pen nanolithography uses an AFM, operated in contact mode, to deposit thiolated molecules onto the surface of a gold substrate. The tip of the microscope is extremely thin and coated with a thiol, such as 1-octadecanethiol (ODT). When the tip of the microscope is covered with ODT and brought into contact with the surface of the gold, a water meniscus forms between the two. The size of the meniscus is controllable by the relative humidity present on the surface of the gold. The size of the meniscus affects the contact area between the tip and the gold, and correspondingly the resolution of the image. From the development of this novel technique, other forms of nanolithography have begun to emerge. These include nanoshaving, nanowearing and nanografting. These techniques permit scientists to create patterns of thiol groups in the gold substrates.

Nanoshaving uses an AFM to exert very high pressure onto the surface of molecules that have adsorbed to the surface of a substrate. This high pressure creates a high shear force on the attached molecules, which results in their immediate displacement. It is important to note that the load applied by the microscope must be higher than the displacement threshold of the attached molecules. Nanoshaving is commonly used because of its ability to fabricate holes and trenches in a single scan. Nanowearing is quite similar to nanoshaving, the key difference being that a lower force is applied to the molecules over an extended period of time. This allows groups to be gradually removed, a major advantage when working near the edge of a substrate and around sites that have been damaged or defected. Nanografting is another nanoscale technique that uses AFM technology. AFM tips remove molecules from the substrate surface by lightly brushing against their terminal chains. In the case of thiol groups attached to a gold substrate, if the surface of the AFM is immersed in a different thiol, it can be attached to the gold in the place of the removed thiol group.

All of the aforementioned techniques can be applied to the micro and nanopatterning of bacteria, viruses and proteins onto surfaces. These applications offer exciting new advances in the biomedical and biosensing fields, and shall be discussed in the following section.

4. Biosensing Applications and Examples

4.1. Overview

Biosensors can be adapted to many different applications, from sensing the presence of a toxin in the atmosphere to sensing an infectious disease in the body. The ability to selectively place bacteria and viruses on patterned surfaces presents tremendous opportunities in many different biomedical and biosensing applications. In conjunction with this, the ability to micro and nanopattern proteins capable of binding bacteria and viruses is equally important. Bacterial adhesion to substrate surfaces is typically accomplished by culturing solutions in LB broth and then applying them directly to the micropatterned substrate. The substrates must then be incubated in order to allow the bacteria time to adhere to the surface. Subsequent to the incubation period, the substrates are rinsed with fresh LB media and stained for imaging of the cells (Koh et al. 2003). Fig. 32.5 shows several different images of the process of bacterial patterning on a silicon substrate. Viral attachment is completed in much the same manner; however, recent work has been done using nanolithographic techniques and with chemoselective linkers to fabricate similar viral assemblies (Cheung et al. 2003). These methods can then be used to investigate how the interaction of intervirions affects such things as assembly morphology and kinetics.

Biosensors typically consist of a biologically specific recognition system, a signal emitted from this system when the target binds and a physical or electrochemical transducer to selectively and quantitatively convert this binding reaction into a machine-readable output signal (Hall 2002). The recent development of biosensor technology has reduced the interface between the biological recognition system and the transducer to the nanometer scale (Hall 2002). An example of this is the generation of a pH change at the surface of an electrode in response to

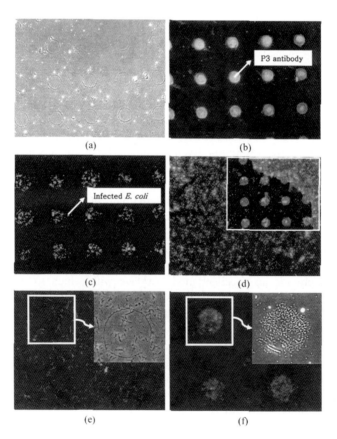

Figure 32.5. (a) An optical micrograph for the initial microstructure of 50μm circles; (b) a fluorescent micrograph for (a) stained with FITC-labeled secondary antibody; (c) an optical micrograph for aggregated arrays of E.coli onto the same pattern; (d) a fluorescent image of E. coli adsorption onto a bare silicon substrate without a predefined pattern (control) the inset shows the boundary between the patterned and non-patterned regions; (e) a fluorecent image of bacteria patterning without the aid of virus (inset: the corresponding optical image for the box region); and (f) a fluorescent image for the pattern stained with a different dye (inset: the corresponding optical image for the box region) (Reprinted from Suh et al. 2004 with kind permission of Springer Science and Business Media).

the release of CO_2 by urease (Cullum and Vo-Dinh 2000). This change can then be transduced by any range of highly sensitive electrochemical or fluorescent methods (Hall 2002). Dill et al. (1999) applied this principle for the detection of *Salmonella* in poultry using biotin–fluorescein-conjugated anti-*Salmonella* antibodies with an anti-fluorescein urease conjugate attached to the immunocomplex.

4.2. Healthcare Applications

Biosensors have a wide range of applications in healthcare. Miniaturized point-of-care tests that can be operated by consumers are projected to emerge in coming years over centralized laboratories with large-scale tests. An example of this is the development of glucose biosensors for patients with diabetes. Glucose biosensors have been generally found to exhibit high affinity, high sensitivity and fast response for the detection of glucose in the blood stream.

The adhesion of bacteria to mammalian cells is a common cause of infection within the human body. The ability to understand and inhibit this adhesion has the potential to be a valuable tool in the development of anti-bacterial therapeutics. There are several methods to evaluate the activity of potential inhibitors of bacterial adhesion, including hemagglutination inhibition assays and solid-phase binding (ELISA-type) assays (Qian et al. 2002). Hemagglutination inhibition assays are based on the conglomeration of erythrocytes in the presence of lectins and other proteins that are capable of the recognition of ligands on the surface of the cell (Goldhar 1995). The degree of complexity of the carbohydrates presented on the surface of erythrocytes often hinders agglutination; therefore, results from hemagglutination inhibition assays are not always reproducible (Qian et al. 2002). Solid-phase binding assays generally involve ligand or receptor non-covalent absorption to a surface. Ligand or receptor density is difficult to control, however, and bovine serum albumin (BSA) is used to block the uncoated surface sites and therefore reduce nonspecific adsorption. Even when the ligands are attached covalently, ligand density can vary, and blocking may be necessary.

In response to this, the use of SAMs on gold substrates is structurally well defined and offers great control over ligand density and environment (Ulman 1996). Previous results demonstrate that SAMs prepared from alkanethiols terminated with oligo(ethylene glycol) groups effectively resist the nonspecific adsorption of proteins and the nonspecific adhesion of mammalian cells (Prime and Whitesides 1991; Qian et al. 2002). Therefore, SAMs with alkanethiols terminated in oligo(ethylene glycol) groups and alkanethiols terminated with the ligand of interest provide an excellent model surface with which to study bio-interfacial problems (Qian et al. 2002). This was demonstrated by Qian et al. through the development of arrays of SAMs for the study of bacterial adhesion (2002). Specifically, the authors studied the adhesion of uropathogenic *E.coli*, which when bound to mammalian cells in the bladder and/or urinary tract, is a leading cause of urinary tract infections (Barnett and Stephens 1997). They suggest that SAMs consisting of alkanethiols terminated in oligo(ethylene glycol) groups provide biospecific surfaces that promote the adhesion of uropathogenic *E.coli* through inter-actions between multiple pili present on the surface of the bacteria and multiple copies of mannosides present on the SAM surface (Qian et al. 2002). This work helps to provide methods with which to study the effects of bacterial adhesion, which can then be applied to the devel-opment of new and novel therapeutics to prevent bacterial attachment and help eliminate bacterial based infections.

Bacterial infections may also be caused by the formation of a biofilm within the human body. Biofilms form on the surfaces of materials when bacteria first adsorb to a material's surface and then excrete a slimy, glue-like substance (commonly called an exopolysaccharide matrix) that anchors them in place (Costerton, Stewart and Greenberg 1999). Due to their inherent resistance to antimicrobial agents, the development of engineered biofilms could potentially be a valuable tool for understanding their mechanisms of formation and the development of techniques to treat the infections they cause (Stewart and Costerton 2001). Bos et al. (2000) began research in this area when they used photolithography and the formation of biofilms on teeth and voice prosthesis to illustrate that biofilm preferential growth did not exist between hydrophilic and hydrophobic surfaces. Cowan et al. (2001) continued this research with the development of poly-L-lysine (PLL) surfaces to enhance bacterial adhesion to surfaces of interest. Previously, the attachment of OmpR (oroitidine-5'-monophosphate) has been used to promote bacterial attachment to a surface (Vidal et al. 1998); however, laboratory strains of *E.coli* lack the required *ompr* gene, and therefore are non-adherent. Cowan et al.'s (2001) technique demonstrated that bacteria could be patterned onto surfaces, and artificial biofilms could be engineered. This technology could be potentially applied to fields where biofilm formation is preferred, such as in the reduction of biochemical oxygen demand in the treatment of wastewater (Nicolella, van Loosdrecht and Heijnen 2000).

4.3. Detection of Toxins in the Environment

As mentioned previously, there is an increasing need for the ability to detect toxic substances in the environment. The increased specificity offered by techniques such as SPL may make it possible to detect very small amounts of toxins in the air. The sources of these toxins could range from bioterrorists to extended growth of mold in a home. In either case, the fast, reliable detection of the toxin is required to maintain safety.

SAMs on gold substrates offer some unique characteristics that make them excellent for the application of biosensing. Firstly, SAM can be miniaturized on gold through the use of microcontact printing (Whitesides et al. 2001). Secondly, the high density and ordered nature of long-chain alkanethiols closely resembles the structure present in lipid bilayer membranes (Chaki 2001). This feature has the potential to provide novel substrates for biological systems (Chaki 2001). Furthermore, the compatibility of SAMs with metal substrates make electrochemical measurements simple and enable easy applications for biosensors requiring measurements of current or potential (Chaki 2001). Finally, the ability of the SAM to remain stable after interface with immobilizing molecules for biological sensing integrated with an electrochemical transducer make SAMs tremendous options for biosensor fabrication and development (Chaki 2001).

Conversely, the ability of bacteria to selectively attach to micropatterned substrates has also been studied for micro and nanosensing applications. Bacterial cells have been found to be relatively robust when compared to their mammalian counterparts, a characteristic that is important in biosensing applications. The increased density of selective attachment of bacterial cells could result in an increased amplification of signal transduction. This property is of significant interest in the field of biosensor development.

There has also been research into the development of methods in which spores can be micropatterned onto the surface of a glass substrate (Park et al. 2004). The increased stability of spores, in comparison with other conventional cells used in micropatterning, as well as their ease of manipulation could be of great benefit in the development of cell-based sensors and detectors (Park et al. 2004). Specifically, microcontact printing has been used to micropattern wells of Bacillus thuringiensis (BT) spores that displayed enhanced green fluorescent protein (EGFP) (Park et al. 2004). In the presence of nutrients such as sugars and amino acids the spores germinated into vegetative cells, thus suggesting that any microbial cell could be micropatterned by spatially addressing its spores onto a substrate and then allowing them to germinate. Such cells would include bacteria, yeasts and filamentous fungi, which are all spore-forming microbial cells.

4.4. Real Devices and Challenges

The examples given above show there has been a significant increase in the amount of research being done in the field of micropatterning of bacteria and viruses. The methods currently employed in the detection of viruses are considered unreliable, inconsistent, expensive and unfeasible in regions of the world where simplicity and cost effectiveness are mandatory. Research currently being completed is a direct result of the inefficiencies present in other approaches used, and initially the technologies previously described show great promise in bacterial and viral detection.

Traditional photolithographic techniques continue to be unavailable to regions of the world that do not have access to clean-room laboratories and large, expensive facilities. However, there are significant steps being taken toward approaches that focus on the use of photoresist masks and UV illumination without the use of conventional substrates and precursor solutions. These advents eliminate the need for clean rooms and suggest that further work may be completed to ease the use of photolithography in areas where it has previously been inaccessible.

Soft lithography has characteristics that are unique in microfabrication, and that enable the techniques it encompasses to be carried out conveniently, rapidly and relatively inexpensively (Whitesides et al. 2001). Procedures involving larger features (on the scale of $>1\mu m$) can also be conducted in an unprotected laboratory environment, thereby enhancing their usefulness in locations where normal microfabrication settings are available or where equipment costs must be minimized (Xia and Whitesides 1998). Generally, the aforementioned techniques are also based on concepts that do not require specialized technicians to perform them—another advantage over current technologies. Due to these advantages, soft lithographic approaches offer significant potential in the micro and nanopatterning applications.

Dip-pen lithography currently boasts a significant increase in resolution, when compared to other lithographic techniques; however, more development is necessary to make it feasible in areas of the world where skilled technicians and expensive laboratory equipment are not readily available. The techniques that offer this higher resolution at the nanoscale can be expected to be concentrated on in future research for the continued miniaturization of techniques and methods of patterning bacteria and viruses.

5. Conclusions and Future Perspectives

The development of the micropatterning techniques outlined above continues to be a major topic of interest in the field of biomedical engineering. The development of micropatterning methods has tremendous implications in the growth of biosensors and biomedical applications. These applications include the development of new and novel antibiotics as well as biomaterials capable of inhibiting bacterial adhesion, which makes them useful for human implantation. The understanding of biofilm formation may also make it possible to inhibit bacterial adhesion, therefore enhancing the capabilities of such biomaterials. The continued miniaturization of lithographic patterning techniques has the potential to enhance patterning resolution, thus creating more bacterial and viral specificity and effectively enhancing the capabilities of prospective biosensors. Correspondingly, continued miniaturization will also help the fight against bacteria that have recently been found to be resistant to traditionally prescribed antibiotics. These developments will be coupled with the goal of reducing the capital costs associated with the equipment necessary to perform these methods and the requirement for skilled technicians to operate this equipment. These aspects could potentially allow these techniques to be used on a worldwide platform, ranging from inside North American laboratories to the far-reaching portions of under-developed societies. This is a crucial aspect of the detection of bacteria and viruses because the spread of these potentially dangerous microbes is far easier today than even ten years ago. The probability of a foreign toxic agent reaching a developed country has become much greater with the advancement of transportation, and this technology is a vital component in the fight against such an attack. If history has taught us anything, it should be that bacterial and viral outbreaks can occur at any place and at any time, and it is in everyone's best interest to be ready for them.

References

Barnett BJ and Stephens DS (1997) Urinary tract infection: an overview. Am J Med Sci 314: 245–9

Bonin P, Rontani JF, et al. (2001) Metabolic differences between attached and free-living marine bacteria: inadequacy of liquid cultures for describing in situ bacterial activity. FEMS Microbiol Lett 194: 111–9

Bos R, van der Mei HC, et al. (2000) Retention of bacteria on a substratum surface with micro-patterned hydrophobicity. FEMS Microbiol Lett 189: 311–5

Chaki NK, Vijayamohanan K (2001) Self assembled monolayers as a tunable platform for biosensor applications. Biosensors and Bioelectronics 17: 1–12

Cheung CL, Camarero JA, et al. (2003) Fabrication of assembled virus nanostructures on templates of chemoselective linkers formed by scanning probe nanolithography. J Am Chem Soc 125:6848–9

Costerton JW, Stewart PS, et al. (1999) Bacterial biofilms: a common cause of persistent infections. Science 284: 1318–22

Cowan SE, Liepmann D and Keasling JD (2001) Development of engineered biofilms on poly-L-lysine patterned surfaces. Biotechnology Letters 23: 1235–1241.

Cullum BM and Vo-Dinh T (2000) The development of optical nanosensors for biological measurements. Trends Biotechnol 18:388–93

Dill K, Stanker LH, Young CR (1999) Detection of Salmonella in poultry using silicon chip-based biosensor. J. Biochem. Biophys. Methods 41:61–67

Doller G, Schuy W, et al. (1992) Direct detection of influenza virus antigen in nasopharyngeal specimens by direct enzyme immunoassay in comparison with quantitating virus shedding. J Clin Microbiol 30:866–9

Dunn DA and Feygin I (2000) Challenges and solutions to ultra-high-throughput screening assay miniaturization: submicroliter fluid handling. Drug Discov Today 5: 84–91

English TJ and Hammer DA (2005) The effect of cellular receptor diffusion on receptor-mediated viral binding using Brownian adhesive dynamics (BRAD) simulations. Biophys J 88: 1666–75

Falconnet D, Csucs G et al. (2006) Surface engineering approaches to micropattern surfaces for cell-based assays. Biomaterials 27:3044–63

Franklin VJ, Bright AM, Tighe B (1993) Hydrogel Polymers and Ocular Spoilation Processes. Trends in Polymer Science 1:9–16

Goldhar J (1995) Erythrocytes as target cells for testing bacterial adhesins. Methods Enzymol 253:43–50

Hall RH (2002) Biosensor technologies for detecting microbiological foodborne hazards. Microbes Infect 4:425–32

Ismagilov RF, Ng JM, et al. (2001) Microfluidic arrays of fluid-fluid diffusional contacts as detection elements and combinatorial tools. Anal Chem 73: 5207–13

Ivanov AP and Dragunsky EM (2005) ELISA as a possible alternative to the neutralization test for evaluating the immune response to poliovirus vaccines. Expert Rev Vaccines 4:167–72

Kingshott P, et al. (1999) Surfaces that resist bacterial adhesion. Current Opinion in Solid State and Materials Science 4: 403–412

Koh WG, Revzin A et al. (2003) Control of mammalian cell and bacteria adhesion on substrates micropatterned with poly(ethylene glycol) hydrogels. Biomedical Microdevices 5: 11–19

Mao CB, Qi JF, Belcher AM (2003) Building Quantum Dots into Solids with Well-Defined Shapes. Advanced Functional Materials 13:648–656

McDonald JC and Whitesides GM (2002) Poly(dimethylsiloxane) as a material for fabricating microfluidic devices. Acc Chem Res 35: 491–9

Moellering Jr. RC (1995) Past, present, and future of antimicrobial agents. Am J Med 99: 11S–18S

Ng JM, Gitlin I et al. (2002) Components for integrated poly(dimethylsiloxane) microfluidic systems. Electrophoresis 23: 3461–73

Nicolella C, van Loosdrecht MC et al. (2000) Wastewater treatment with particulate biofilm reactors. J Biotechnol 80:1–33

Park TJ, Lee KB et al. (2004) Micropatterns of spores displaying heterologous proteins. J Am Chem Soc 126:10512–3

Prime KL and Whitesides GM (1991) Self-assembled organic monolayers: model systems for studying adsorption of proteins at surfaces. Science 252:1164–7

Qian X, Metallo SJ et al. (2002) Arrays of self-assembled monolayers for studying inhibition of bacterial adhesion. Anal Chem 74:1805–10

Rainina EI, Efremenco EN et al. (1996) The development of a new biosensor based on recombinant E. coli for the direct detection of organophosphorus neurotoxins. Biosens Bioelectron 11:991–1000

Razatos A, Ong YL et al. (1998) Molecular determinants of bacterial adhesion monitored by atomic force microscopy. Proc Natl Acad Sci, USA 95:11059–64

Respess RA, Rayfield MA et al. (2001) Laboratory testing and rapid HIV assays: applications for HIV surveillance in hard-to-reach populations. Aids 15 Suppl 3:S49–59

Rich JD, Merriman NA et al. (1999) Misdiagnosis of HIV infection by HIV-1 plasma viral load testing: a case series. Ann Intern Med 130: 37–9

Roberts L (2006) Infectious disease. Polio experts strive to understand a puzzling outbreak. Science 312:1581

St John PM, Davis R et al. (1998) Diffraction-based cell detection using a microcontact printed antibody grating. Anal Chem 70:1108–11

Stewart PS and Costerton JW (2001) Antibiotic resistance of bacteria in biofilms. Lancet 358: 135–8

Suh KY, Khademhosseini A et al. (2006) Direct Confinement of Individual Viruses within Polyethylene Glycol (PEG) Nanowells. Nano Lett 6:1196–1201

Suh KY, Khademhosseini A et al. (2004) Patterning and separating infected bacteria using host-parasite and virus-antibody interactions. Biomed Microdevices 6:223–9

Suh KY, Kim YS et al. (2001) Capillary force lithography. Adv Mater 13:1386–1389

Takayama S, McDonald JC et al. (1999) Patterning cells and their environments using multiple laminar fluid flows in capillary networks. Proc Natl Acad Sci, USA 96: 5545–5548

Tran JH and Jacoby GA (2002) Mechanism of plasmid-mediated quinolone resistance. Proc Natl Acad Sci, USA 99:5638–42

Ulman A (1996) Formation and Structure of Self-Assembled Monolayers. Chem Rev 96:1533–1554

van Elden LJ, Nijhuis M et al. (2001) Simultaneous detection of influenza viruses A and B using real-time quantitative PCR. J Clin Microbiol 39:196–200

Vidal O, Longin R et al. (1998) Isolation of an Escherichia coli K-12 mutant strain able to form biofilms on inert surfaces: Involvement of a new ompR allele that increases curli expression. J Bacteriol 180:2442–9

Whiley DM and Sloots TP (2005) A 5′-nuclease real-time reverse transcriptase-polymerase chain reaction assay for the detection of a broad range of influenza A subtypes, including H5N1. Diagn Microbiol Infect Dis 53:335–7

Whitesides GM, Ostuni E et al. (2001) Soft lithography in biology and biochemistry. Annu Rev Biomed Eng 3: 335–73

Wickham TJ, Granados RR et al. (1990) General analysis of receptor-mediated viral attachment to cell surfaces. Biophys J 58:1501–16

Wood MJ and Moellering Jr. RC (2003) Microbial resistance: bacteria and more. Clin Infect Dis 36:S2–3

Xia YN and Whitesides GM (1998) Soft lithography. Angewandte Chemie-International Edition 37:551–575

Zhao XM, Xia XN et al. (1996) Fabrication of three-dimensional micro-structures: Microtransfer molding. Advanced Materials 8:837–840

33

Microfabricated Flow Cytometers for Bacterial Detection

Sung-Yi Yang and Gwo-Bin Lee

Abstract

Microfabricated flow cytometry has been extensively investigated recently. Miniaturization of flow cytometers by adopting microfabrication techniques to fabricate microchannels and micro-nozzles in silicon, glass/quartz, and even plastic substrates has been demonstrated. When compared with their large-scale counterparts, these micro flow cytometers are more compact in size, portable, cost-effective, user-friendly, and most importantly, could have almost comparable performance. In this chapter, microfabrication techniques for these micro flow cytometers were first reviewed. The operating principles for cell transportation, focusing, detection, and sorting inside these micro flow cytometers were briefly discussed. Finally, several promising applications including environmental monitoring, rapid assessment of bacterial viability, rapid analysis of bacteria levels in food, antibiotic susceptibility testing, and diagnosis of bacterial in blood and urine were reviewed. It can be envisioned that a portable flow cytometer system can be available for point-of-care applications if these issues can be addressed properly in the near future.

1. Introduction

Flow cytometry is a popular method for high throughput analysis of suspended cells, bacteria and other microorganisms. It has become a mature diagnostic tool for clinical and environmental applications. Typically, fluorescence-labeled cells or bacteria are hydrodynamically focused by surrounding sheath flows into a narrow stream to pass through a region where fluorescence emission or scattered light is collected by several sophisticated optical detection instruments. This technique is currently used in a wide variety of biomedical and environmental applications such as clinical hematology diagnosis (Brown and Wittwer 2000; Riley 2002), bacteria analysis (Gunasekera, Attfield and Veal 2000), gene diagnosis (Hrusak et al. 2004), transfusion medicine (Greve et al. 2004) and many others (Dallas and Evans 1990; Bartsch et al. 2004).

Profiling, sorting and measurement of various physical properties of cells and bacteria for biomedical and environmental applications have been widely performed with a flow cytometric approach. Recently, thanks to the rapid development of monoclonal antibody and fluorescent probes in the field of biotechnology, use of these technologies for cell-based analysis has made flow cytometry more and more popular in a wide range of applications in various fields. Flow cytometry has been demonstrated for measurement and analysis of various cell properties, including surface antigen, intracellular antigen, transgenic expression, immunoassay,

---

Sung-Yi Yang and Gwo-Bin Lee • Department of Engineering Science, National Cheng Kung University, Tainan, Taiwan 701.

M. Zourob et al. (eds.), *Principles of Bacterial Detection: Biosensors, Recognition Receptors and Microsystems*,
© Springer Science+Business Media, LLC 2008

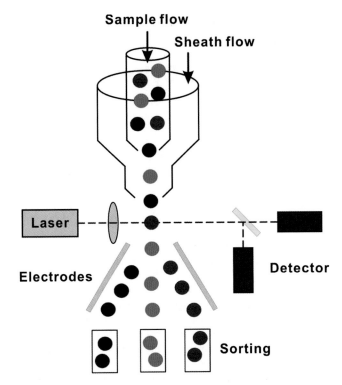

Figure 33.1. Schematic illustration of the operation principle of a large-scale flow cytometer for cell counting and sorting.

cell function, DNA analysis, apoptosis, cell cycle, cell activities, measurement of free radicals and many others. In addition, one can use the cell sorting function to separate and to collect any specific target cell for subsequent applications.

As shown in Fig. 33.1, large-scale flow cytometry normally uses an optical detection method to complete the measurement and analysis of multiple physical properties of cells. First, the sample flow containing microparticles is focused by a hydrodynamic flow focusing method by using the surrounding sheath flows. Then the sample flow passes through an optical detection region formed by the intersection of the flow with laser beams, and detectors use the scattered light from the cell or bacteria samples to analyze the types and sizes of the cells/bacteria. The fluorescence labeling method can also be used to collect induced fluorescent signals from the fluorescence-labeled cells/bacteria. The fluorescence emitted from the cells/bacteria through several optical components is then collected by photomultiplier tubes (PMTs) and then recorded by a computer for signal processing. Finally the electrodes are used to sort different cell samples using electrodes (Mandy, Bergeron and Minkus 1995; Shapiro 2003).

However, delicate optical components including focused laser beams, various optical detecting/filtering devices and complicated control circuits make the system relatively expensive. Not only is the price expensive but also the calibration process is relatively complicated, which usually requires an experienced, qualified user to operate. Therefore, at present, this equipment is mostly only used in general hospitals and research centers. In order to increase the mass-market appeal of cytometers, microfabricated flow cytometers using micro-electro-mechanical systems (MEMS) technology have been reported that reduce the size and price of these traditional large-scale flow cytometer systems (Fu et al. 1999; Sobek et al. 1993; Hodder, Blankenstein and Ruzicka 1997; Miyake et al. 1997; Koch, Evans and Brunnschweiler 1999; Schrum et al. 1999; Miyake et al. 2000; Cui, Zhang and Morgan 2002; Huh et al. 2002;

Kruger et al. 2002; Lin and Lee 2003; Tung et al. 2004). A microfabricated flow cytometer is more compact in size, more user-friendly, and relatively inexpensive when compared with its large-scale counterparts.

The complicated optical detection procedures make the large-scale flow cytometers ponderous and relatively expensive. MEMS techniques allow for the integration of aligned optical waveguides and fibers with the fluid channels. Besides, reduced sample/reagent consumption and the feasibility for integration into a portable device make them promising for on-site measurement. Several miniaturized flow cytometers have been reported using a micro-fabrication process by incorporating different materials including metal (Miyake et al. 1997; Miyake et al. 2000), silicon (Sobek et al. 1993; Hodder, Blankenstein and Ruzicka 1997; Koch, Evans and Brunnschweiler 1999), glass (Schrum et al. 1999; Lin and Lee 2003) and plastic materials (Fu et al. 1999; Cui, Zhang and Morgan 2002; Huh et al. 2002; Kruger et al. 2002; Tung et al. 2004) Successful counting and sorting of cells/bacteria have been demonstrated with these various designs.

1.1. Bio-MEMS

Rapid advancements in MEMS technology and microfabrication techniques have made substantial impacts on biomedical applications and attracted considerable interest for minia-turization of biomedical and chemical analytical instruments. Not only does this technology enable a miniaturized instrument, but it also provides access to information at a molecular level. Recently, many innovative microdevices or microsystems, especially in the fields of biomedical and chemical analysis, have been demonstrated using these technologies. Recently, a new technology called bio-micro-electro-mechanical-systems (Bio-MEMS), combining knowledge from biology and MEMS, has attracted considerable interest and has been used to fabricate biochips with a dimension of several centimeters. Since MEMS techniques can mass-produce microdevices with low-cost materials such as glass and polymers, the unit cost of these biochips can be drastically reduced and disposable chips to prevent cross-contamination become feasible. Bio-MEMS technology can be used to fabricate microbiochips with an excellent analytical capability at a lower unit cost. When scaling down the size of a bio-analytical instrument to a single chip, it provides several advantages including compactness of size, low sample/reagent consumption, fast reaction time, high precision, high sensitivity, portability, low power consumption, low unit cost, and the potential for automation and integration.

A micro-total-analysis-system (μ-TAS), which integrates sample pretreatment, trans-portation, reaction, mixing, separation, and detection functions on a small chip, can now be realized by combining functional microfluidic components manufactured by specific Bio-MEMS techniques (Sanders and Manz 2000). In 1990, the concept of a μ-TAS was first raised and demonstrated (Manz, Graber and Widmer 1990). Since then, microdevices and microsystems to realize the commercialization of μ-TAS have been extensively investigated. Briefly, μ-TAS combines various microfluidic devices to form a functional microsystem for measurement and analysis of biomedical samples. For example, a microsystem for fast detection of infectious pathogens can be realized on a single microchip by using this technology (Liao et al. 2005). Traditionally, biological laboratory processes, medical examination or biochem-istry analysis usually involve complicated, time-consuming and labor-intensive procedures. Besides, all these expensive, dedicated pieces of equipment take up valuable laboratory space. Microfabrication techniques can be used to miniaturize the instruments required for these tradi-tional analysis processes. Micropumps, microvalves, microfilters, micromixers, microchannels, microsensors and microreactors can be integrated on a single chip to automate these processes. For example, a sample pretreatment process can be performed by micromixing the cell samples with fluorescent dyes. Then these fluorescence-labeled cells can be transported in a microchannel and then detected by the optical detection system in a microfabricated flow cytometer. These cells can also be sorted and transported to the subsequent microfluidic devices for further post-processing. The μ-TAS can fit the requirement of a miniaturized point-of-care (POC) instrument, and may competitively enter into the existing biotechnology equipment market if several crucial issues can be properly addressed. For instance, reliability and reproducibility should be improved before they can be a viable substitute in the market. Potentially, a biochip made of Bio-MEMS technology promises to replace traditional, complicated detection systems, which may save a large amount of time and labor.

1.2. Review of Microfabrication Techniques

The feature dimensions of microfabricated devices are usually between 0.1 micrometer (μm) and 1 millimeter (mm). By integrating many crucial technologies in various fields, the applications of these microdevices have been widely expanded. Briefly, these fabrication techniques for MEMS can be divided into three major categories: bulk micromachining, surface micromachining and *lithographie galvanoformung abformtechik* (LIGA) processes. In addition to silicon, other materials such as glass, quartz and polymer materials are now commonly used for Bio-MEMS applications due to their biocompatibility and low cost.

1.2.1. Bulk Micromachining Technique

The bulk micromachining technique uses a silicon wafer as a base substrate material, and processes etch the wafer to form microstructures. Sculpturing of a bulk silicon substrate by using wet or dry etching methods, in combination with etch masks and etch stop techniques, can be used to form isotropic or anisotropic microstructures relative to the silicon crystal planes. Typically, wet chemical etching is the most commonly used bulk etching method. The dry etching process involves a deep reactive ion etching method, and is also used to form microstructures with a high aspect ratio (Madou 2002). The most commonly used masking materials for the wet etching process are thin-film silicon-dioxide and silicon-nitride layers. Alternatively, a thick photo-resist layer can be used as a masking material for dry etching. The bulk micromachining technique is popularly used to form a microchannel with an isotropic or a V-groove cross-sectional shape for micro flow cytometers.

1.2.2. Surface Micromachining Technique

Micromachined structures can be also formed by using a so-called surface micromachining technique, which deposits various thin-film layers on top of a silicon substrate and then selectively etches away these thin-film sacrificial layers, which are underneath the thin-film structural layers (Madou 2002). Furthermore, this method adopts thin-film deposition and etching processes from the integrated-circuit (IC) semiconductor industry. Therefore, integration of on-chip circuitry and electrical-based microstructures (either microsensors or microactuators) is feasible. Suspended or even movable microstructures can be successfully formed using this method. The surface micromachining technique is not popular for the formation of microfabricated flow cytometers since the depth of the microchannel is usually a few microns, and is limited by the thickness of the deposited thin films.

1.2.3. LIGA

LIGA is the German acronym for *lithographie galvanoformung abformtechnik*, which is a fabrication technique involving X-ray lithography, electroforming, and a micromolding process

(Madou 2002). Microstructures with a high aspect ratio can be formed by using this process. Metal or plastic microstructures with an almost straight sidewall and a high aspect ratio (easily over 100:1) can be made. The main fabrication processes include the following procedures: (1) An x-ray lithography technique is used to generate a patterned photo-resist structure with a high aspect ratio on a substrate (typically a silicon wafer); (2) the substrate is placed in an electroforming apparatus to electroplate metal or alloy microstructures. The region without the protection of the photo-resist pattern will be covered with the electroforming metal or alloy; (3) a metal or alloy mold with a high aspect ratio is formed after removing the photo-resist. Then the metal or alloy mold can be further used to replicate polymeric microstructures with an inverse image by using a micromolding process. This method requires a synchrotron X-ray source, which hinders its widespread practical use. Alternatively, ultraviolet (UV) lithography and thick UV-sensitive photo-resist can be used to form the electroplating mold. After the formation of the thick patterned photo-resist layer, the rest of the fabrication process is the same of the LIGA process. This modified process is popular for the formation of microflow cytometers with high-aspect-ratio channels.

1.2.4. Polymer-Based Micromachining Techniques for Microfluidic Devices

Microfluidic devices, including micro flow cytometers, have been fabricated from glass or fused silica substrates (Lin et al. 2001). These devices possess a high mechanical strength and well-known chemical surface properties for chemical analysis applications. Besides, the optical transparency of these materials is advantageous if an optical detection scheme is adopted. Alternatively, polymeric rather than glass (or fused silica) substrates is also a promising method for clinical applications as plastic chips are less expensive than silica-based substrates. Moreover, the inherent neutral hydrophilic nature of a polymer substrate allows direct use of the microfluidic chips for clinical analysis of bio-molecules without the need for surface modifications to reduce the wall adsorption. Additionally, polymers are promising materials for microfluidic chips since they are compatible with mature mass-production techniques such as injection molding (Cormick et al. 1997), hot embossing (Lee et al. 2001a), and casting methods (Unger et al. 2000). A template of the microfluidic structures can first be fabricated either using a bulk micromachining technique or the LIGA process. Then cheap and reproducible replicates can be mass fabricated using these aforementioned methods with polymeric materials. Their normally low cost and the development of reliable polymer fabrication as well as mass replication techniques have attracted considerable interest, especially in applications for disposable microfluidic devices.

For example, a microflow cytometer can be fabricated using the following procedures involving patterning photo-resist (SU-8, MicroChem, NANO™, USA) and polydimethyslsiloxane (PDMS) casting process (Wang and Lee 2005). First, the SU-8 is spin-coated onto a silicon wafer to form a photo-sensitive layer with a thickness from tens to hundreds of micrometers. Then it is soft-baked at 65°C for 10 minutes and at 95°C for 30 minutes. Afterwards, a standard UV lithography process is performed using a single-sided aligner, followed by a post-exposure-bake process at 65°C for 3 minutes and at 95°C for 10 minutes. Finally, a development process is performed to produce the finished SU-8 template with microfluidic structures. The PDMS casting process is then used to replicate inverse structures. A silicone elastomer and elastomer curing agent (Sil-More Industrial Ltd., USA Sylgard 184A and Sylgard 184B) of PDMS are mixed in a specific ratio and then poured onto the SU-8 microstructure mold. In order to prevent the formation of air pockets in the replicas, a vacuum process is usually required to remove the bubbles formed during the mixing process. The PDMS on the SU-8 template is then cured at 70°C in an oven overnight. The PDMS inverse structures are finally mechanically peeled off the SU-8 template. A microflow cytometer has been successfully fabricated using this approach (Yang et al. 2006).

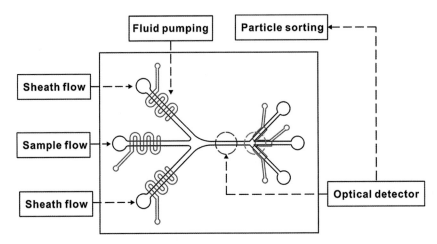

Figure 33.2. Schematic illustration of a microfabricated flow cytometer composed of three modules for cell transportation and focusing, cell detection, and cell sorting.

2. Operation Principles

Recently, miniaturized flow cytometers have been demonstrated and used for cell/particle counting and sorting. For example, a five-layer stainless steel and glass laminated sheath flow chamber (Miyake et al. 1991). A scattered light detector was used to detect the passage of the flow-focused particles. Similarly, the ability to construct a microfabricated flow cytometer on a silicon wafer was reported (Sobek et al. 1993). They also reported the integration of an optical waveguide on the microflow cytometer for detection of human blood cells. Sheath flows were employed to carry out hydrodynamic focusing in both cases. Alternatively, a micro flow cytometer which successfully adopted electrokinetic focusing of the particles/cells rather than hydrodynamic focusing (Schrum et al. 1999). With this approach, microelectrodes can be integrated with microfluidic structures for flow cytometry. These ingenious pioneering works have demonstrated that microfabricated flow cytometers are promising for cell/particle counting and sorting applications.

In this section, the operation principles of the microfabricated flow cytometers using Bio-MEMS technology will be briefly reviewed. Fig. 33.2 shows schematically a typical microfabricated flow cytometer. Three major modules for cell transportation and focusing, cell detection, and cell sorting, respectively, are commonly used to construct a microflow cytometer. A fluid pumping device, either a large-scale pump or a micropump, is commonly used for cell transportation and focusing. Sheath flows and the sample flow are driven through the microchannels and then pass through a cell detection area. When passing through the cell detection area, the fluorescence signal induced by an excited laser or the scattered signal deflected by the cell is detected by an optical detector. Then the optical signals are transformed into electrical signals, which will be used to trigger a particle-sorting device for cell sorting. With this approach, a specific group of cells can be counted and separated from the others and ready for subsequent cell-based applications. Table 33.1 compares a microflow cytometer with a conventional flow cytometry system. The microflow cytometer has comparable error rates for counting and sorting when compared with its large-scale counterpart. Commercially available large-scale flow cytometers can easily count and sort cells at a high throughput up to 10,000 cells/minute without affecting cell viability. However, the throughput of the microflow cytometer is much lower than the commercial flow cytometer because the flow rate

Table 33.1. Comparison of a microfabricated flow cytometer with a conventional flow cytometry system

System Performance	Micro flow cytometer (Yang et al. 2006)	Conventional Systems
Minimum sample size	> 1 μm	> 1 μm
Sorting throughput	About 350 cells/minute	About 1000–10000 cells/minute
Error rate of counting	< 1.5 %	< 1 %
Error rate of sorting	< 2 %	< 1 %
Flow rate	10~35 μl/min	Three operation modes Low : 12 μl/min Medium : 35 μl/min High : 60 μl/min

inside the microchannel is limited. Nonetheless, the microflow cytometer is still promising in other aspects due to its obvious advantages, including compactness, low-cost and portability. Table 33.2 lists a summary of microflow cytometers in development. Generally, in addition to the microfabricated flow cytometer chip, usually a microscope is incorporated for optical detection.

2.1. Cell Transportation and Focusing

Three major methods have been demonstrated in the literature for cell transportation and focusing, including hydrodynamic, pneumatic and electrokinetic approaches. The following sections will introduce these methods.

2.1.1. Hydrodynamic Approach

Control of focused streams inside the microchannel is crucial in various microfluidic applications. For example, the size of the focused cell stream can easily be reduced to the order of the cell size, which is suitable for cell counting and sorting for a microflow cytometer. Fig. 33.3 schematically represents the flow focusing in a microflow cytometer composed of three microchannels. The cell samples are injected from the center channel and focused hydrodynamically into a single cell stream constrained by flows from two sheath channels. The flow inside the microflow cytometer is considered to be laminar due to its low-Reynolds-number nature. The diffusion and mixing between the focused stream and the sheath flows are assumed negligible. The hydrodynamic focusing phenomenon can be then simply modeled by employing potential flow theory, resulting in a simple formula to predict the width of the focused stream inside the microflow cytometer (Lee et al. 2001b).

$$d = \frac{\rho_a D_a}{1.5 \left(\rho_1 \dfrac{\bar{v}_1}{\bar{v}_2} \dfrac{D_1}{D_2} + \rho_2 + \rho_3 \dfrac{\bar{v}_3}{\bar{v}_2} \dfrac{D_3}{D_2} \right)}, \tag{33.1}$$

where D_1, D_2, D_3, and D_a are width of the inlet channels and the outlet channel, respectively; $\bar{v}_1$, $\bar{v}_2$, and $\bar{v}_3$ are velocities in the inlet channels 1, 2, and 3; ρ_1, ρ_2, ρ_3, ρ_a are the densities of the fluids in the inlet channels and the outlet channel, respectively.

Table 33.2. Summary of the existing micro flow cytometer designs

Flow Pumping Method	Material	Type of Optical Detection	Dimensions of the Channels (DXW)	Counting Rate	Flow Rate	Sorting Method	Sorting Throughput	Reference
Syringe pump	PDMS	Microscope	10μm × 100μm	N.R.	$0.95 \times 10^4 \mu m/s$	N.R.	N.R.	Lee, Izuo and Inatomi 2004
Electrokinetics	PDMS + glass	Microscope	3μm × 4μm or 100μm	N.R.	N.R.	Electric switching	N.R.	Fu et al. 1999
Syringe pump	Glass	Microscope	25μm × 90μm	N.R.	N.R.	Electric switching	N.R.	Yao et al. 2004
Syringe pump	PDMS	Microscope	7.6μm × 25μm	N.R.	0.55–1.65 mm/s	Electric switching	0.3 ~ 1 bead/s	Dittrich and Schwille 2003
Vacuum pump	Glass	Microscope	25μm × 100μm	1KHz	3–4 mm/s	N.R.	N.R.	Chan et al. 2003
Syringe pump	SU-8 + glass	Microscope	50μm × 150μm	N.R.	N.R.	N.R.	200 bead/min	Kruger et al. 2002
Syringe pump	Glass	Microscope	20μm × 40μm	1000 bead/min	10 mm/s	Electrode	N.R.	Cheung, Gawad and Renaud 2005
Syringe pump	Silicon + glass	Microscope	50μm × 200μm	1400 beads/s	N.R.	Valve	12000 cell/s	Wolff et al. 2003
Syringe pump	PMMA+PDMS + SU-8	Waveguide + fiber microlens	90μm × 600μm	4~25 beads/s	N.R.	N.R.	N.R.	Wang et al. 2004
Syringe pump	PDMS + glass	Microscope	N.R.	N.R.	N.R.	Valve	N.R.	Bang et al. 2006
Pneumatics	Glass	Microscope	50μm × 150μm	N.R.	25 mm/s	Optical switching	106 cell/s	Wang et al. 2005
Electrokinetics	PDMS + glass	Fiber	80μm × 80μm	N.R.	5.5 mm/s	N/A	N/A	Xiang et al. 2005
Syringe pump	Glass	Microscope	40μm × 150μm	N.R.	1~10 mm/s	Electric switching	300 bead/s	Holmes et al. 2005
Syringe pump	PDMS	Fiber	100μm × 300μm	N.R.	15 ml/h	N.R.	N.R.	Tung et al. 2004
Syringe pump	Glass	Microscope	100μm × 160μm	N.R.	1~50 μl/min	N.R.	N.R.	Nieuwenhuis et al. 2003
Syringe pump	PDMS	Microscope	100μm × 300μm	55~113 beads/s	6 ml/h	N.R.	N.R.	Huh et al. 2002

Table 33.2. (continued)

Flow Pumping Method	Material	Type of Optical Detection	Dimensions of the Channels (DXW)	Counting Rate	Flow Rate	Sorting Method	Sorting Throughput	Reference
Pneumatics	PDMS	Microscope (image)	N.R.	N.R.	N.R.	Electrode	N.R.	Takahashil et al. 2004
Syringe pump	Glass	Fiber	40 μm × 80 μm	N.R.	0.5mm/s	Electrokinetic switching	N.R.	Fu et al. 2004
Syringe pump	Glass	Fiber	55 μm × 100 μm	N.R.	0.2 μl/min	N/A	N.R.	Lin and Lee 2003
Syringe pump	Glass	High-speed CCD	36 μm × 240 μm	1875 cells/s	1.5 mm/ s	Electrokinetic switching	N.R.	Lee, Lin and Chang 2005
Syringe pump	Glass	High-speed CCD	140 μm	N.R.	N.R.	Electrokinetic switching	N.R.	Yang et al. 2005
Syringe pump	Glass	Waveguide/fiber	25 μm × 140 μm	N.R.	0.05 μl/min	N.R.	N.R.	Lee, Lin and Chang 2003
Pneumatic micro-pump	PDMS	Microscope	50 μm × 100 μm	120 cell/min	35 μl/min	Pneumatic micro-valve	120 cell/min	Yang et al. 2006
Pneumatic micro-pump	PDMS	Fiber	50 μm × 100 μm	N.R.	10~35 μl/min	Pneumatic micro-valve	350 cell/min	Chang, Hsiung and Lee 2006

Note: N.R. = Not reported

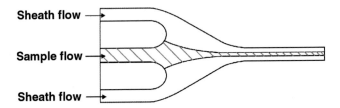

Figure 33.3. Schematic illustration of the flow focusing inside microchannels. The sample flow can be focused and squeezed into a narrow stream by the surrounding sheath flows.

This simple formula indicates that the width of the focused stream is inversely proportional to the sheath and sample flow velocity ratio and proportional to the volumetric flow rate of the sample flow. Usually, syringe pumps or vacuum pumps are used to drive the flows inside the microchannels. For example, syringe pumps are used to transport the fluid flows and to focus the central sample flow hydrodynamically (Lee et al. 2001b). With this approach, the width of the sample flow can be reduced to as small as 3 μm. The cell samples in the sample channel are then squeezed into a narrow line to pass through the detection area to achieve cell counting.

Typically, a single-channel flow-focusing scheme is prevalent in most microflow cytometry approaches. In order to provide a higher throughput for cell/particle counting, a multi-channel approach is required. The multiple sample streams allow for parallel processing of cell counting and can result in higher throughput since the cell-counting rate is directly proportional to the number of the sample streams. With four sheath flows and two sample flows, the sample flows are squeezed and focused successfully such that the counting rates of the microflow cytometer can be increased (Lee, Hwei and Huang 2001). A theoretical model for multi-channel flow focusing was also successfully derived using a similar approach. A demonstration of a multi-channel flow cytometer capable of multiple cell counting was reported (Lee, Lin and Chang 2005).

In order to miniaturize the dimensions of the microflow cytometer system, micropumps were used to provide the required hydrodynamic forces. Instead of using large-scale syringe or vacuum pumps, micropumps can be integrated on the flow cytometer chip to automate the entire cell transportation and flow focusing process. For example, a microflow cytometer integrated with pneumatic micropumps was reported (Chang, Hsiung and Lee 2006). The micropumps composed of double-layer PDMS structures include a serpentine-shaped (S-shaped) pneumatic microchannel and a straight sample flow channel located underneath. PDMS membrane structures with a thickness of 25 μm are located at each intersection area and are deformed when compressed air flows through the S-shape pneumatic microchannel. The time-phased deformation of successive membranes creates a continuous peristaltic effect that drives the fluid to flow along the sample flow channel. After inputting different frequencies to change the flow rate of each micropump, the width of the hydrodynamically focused sample flow can be precisely adjusted. By using pneumatic S-shaped micropumps to drive sample and sheath flows with different flow velocity ratios, hydrodynamic focusing to force cells to pass through detection regions in sequence has been demonstrated.

2.1.2. Pneumatic Approach

Pneumatic forces can be used for flow focusing inside a microflow cytometer. For example, a new method has been developed for achieving flow focusing inside an air-liquid two-phase flow cytometer (Huh et al. 2002). Compressed air is injected into the sheath flow channels such that liquid in the central sample channel can be focused pneumatically. A hydrophobic PDMS material is used for formation of this microflow cytometer. A stable liquid stream with a minimum focused width of 15 μm has been successfully achieved.

2.1.3. Electrokinetic Approach

In addition to hydrodynamic or pneumatic forces, electrokinetic forces can also be used for achieving flow focusing inside a microflow cytometer—since electrokinetic flow focusing only requires driving electrodes, which can be easily fabricated on the microfluidic devices, thus resulting in a compact system and an easier integration process. Typically, two different types of electrokinetic forces are commonly used for achieving flow focusing, including electro-osmotic and dielectrophoretic (DEP) forces. When using electro-osmotic forces, the sample flow can be electrokinetically confined into a narrow stream using two focusing sheath flows under electric fields (Fu et al. 2004). Both the sample and sheath flows are driven by generating electro-osmotic flows (EOF). Experimental data shows that the width of the focusing stream is inversely proportional to the relative sheath/sample focusing potential ratio. This also indicates that the width of the focused cell stream can be properly controlled using the electrokinetic forces. The width of the cell stream can be properly decreased to allow only one cell to pass at a time for cell counting. Likewise, the width of the focused stream is critical to the operation of a microflow cytometer driven by electro-osmotic forces. The electrokinetic focusing phenomenon can be theoretically modeled by using similar potential flow theory, which can be used to predict the width of the focused stream (Yang et al. 2005).

Another electrokinetic force, called the dielectrophoretic (DEP) force, can also be used to focus particles inside the sample flow. The DEP technique was reported and has been applied for manipulation of particles since the 1970s (Pohl 1978). Particles/cells may be moved, trapped, collected, aligned, or even rotated using induced dipole moments generated by microelectrodes. When particles are suspended in a spatially non-uniform electric field, the applied electric field induces a dipole moment in the particles. The interaction of the electric field with the induced charges on either side of the dipole generates a net force. Due to the inhomogeneous nature of the electric field, if the particle is more polarizable than the medium surrounding it, the dipole aligns with the electric field and there is a net motion towards the region with a higher electric field. This effect is called "positive DEP." Conversely, if the particle is less polarizable than the surrounding medium, the dipole aligns against the electric field and the particle is repelled from the higher electric field region, resulting in a so-called "negative DEP." DEP forces can also be used for manipulation of cells. The forces acting on the cells/particles can be used to focus them horizontally, aligning them along the center axis of the sample stream. Instead of moving fluids inside the microchannel using EOF, DEP forces move cells and focus them along the center of the channels. For example, using thick photo-resist (SU-8) and polyimide to fabricate the microchannels and microelectrodes for DEP focusing of particles (Cui and Morgan 2000), successful focusing of cells/particles has been demonstrated.

The DEP force can also be used for vertical focusing of cells/particles. Two sheath flows are used to focus the sample flow horizontally by means of hydrodynamic forces, and then two embedded planar electrodes can apply negative DEP forces to focus the particles/cells vertically. Focusing of particles/cells in the vertical direction inside a microflow cytometer is a critical issue while using an embedded optical detection system (such as buried optical waveguides and optical fibers) horizontally aligned with the microchannels. Even if the particles/cells have been focused centrally in the horizontal direction using co-planar sheath flows, appreciable errors may still arise if they are randomly distributed in the vertical direction. For example, scattered light signals can be dramatically changed if the cells move vertically away from the incident excitation light. Therefore, a pair of parallel microelectrodes can be deposited on the upper and bottom surface of the microfluidic channel to drive particles/cells into the vertical center of the sample flow. The three-dimensional focusing device using the DEP forces generated by the microelectrodes can improve the uniformity of the detected optical signals (Lin et al. 2004).

2.2. Cell Detection

Usually, measurement and detection of cells in a flow cytometer is achieved by using an optical detection system. There are several advantages to using an optical detection method to analyze biological samples, including high spatial resolution, rapid detection, and more importantly, no substantial damage to the biological samples caused by photo damage. This is particularly important for detection of viable cells since it does not interfere with their biological activities. The measurement and detection of cells can be achieved by measuring induced fluorescence or scattered light deflected from an individual cell. Fluorescence detection is one of the most popular detection schemes used for microflow cytometry. Cells have to be fluorescence-labeled prior to usage. When they pass through the optical detection area, fluorescence signals are induced by the incident laser light. The wavelengths of the induced fluorescence are usually longer than the excited laser light. For example, an argon ion laser with a wavelength of 488 nm is used to excite fluorescein isothiocyanate (FITC) and phycoerythrin (PE) with an induced peak emission wavelength of 530 nm and 570 nm, respectively. These peak emission wavelengths are usually used to count cells with different fluorescence labeling. The induced fluorescence signals are collected by photomultiplier tubes (PMTs) or avalanche photo-detectors (APDs) via mirrors, beam splitters and optical filters. Optical filters, including band-pass, long-pass, and short-pass filters, are commonly used to allow light with a specific bandwidth to enter the optical detector.

Alternatively, light scattering from cells occurs when cells pass the optical detection area and deflect the incident laser light. With this approach, intrinsic properties of cells can be acquired without a sophisticated dye-labeling process when scattered light is measured. When cells in the microchannel pass through the optical detection area, scattered light is dependent on the physical properties of the cells, including size, shape and surface topography. For flow cytometry, forward-scattered detection and side-scattered detection are two commonly used schemes. Usually, a forward-scattered detection scheme is used to detect cell surface topography or size. Alternatively, a side-scattering signal is related to cell granularity or internal complexity (Shapiro 2003). Fig. 33.4 shows a typical scattered signal collected by an APD when focused cells pass through the optical detection area. Each peak corresponds to a cell passing through the optical detection area. The number of the cells can be calculated by summing up the number of peaks (Yang et al. 2006).

Similar optical detection systems have been adapted for microflow cytometry. The coupling of the excitation and detection lights into and out of the cytometer chip is a crucial

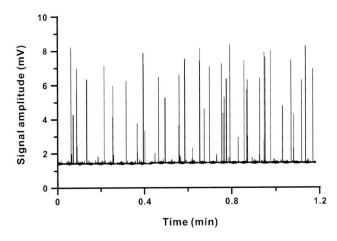

Figure 33.4. A typical signal collected by an APD when focused cells pass through the optical detection area.

issue to ensure the quality of the optical signals. Three major schemes, including optical waveguides, buried optical fibers, and large-scale optical systems, are commonly used. The first two approaches can integrate optical waveguides and optical fibers with microchannels such that delicate optical alignment procedures and apparatii are not necessary. The last approach still requires large-scale optical systems; however, it is a relatively mature technology and ready for commercialization.

2.2.1. Optical Waveguide Approach

Integration of optical waveguides with microfluidic devices can provide a promising approach for online optical detection. A large difference in the refractive index between cladding and core materials deposited on a substrate can be used as a planar waveguide. These optical waveguide structures can be integrated with microchannels for flow cytometry applications. The detection light can propagate in the waveguide structure by total internal reflection. Scattered or fluorescence light can be collected when cells pass through the integrated waveguide structures transmitting an excitation laser light. Therefore, cells can be measured and detected without any microscope or delicate optical alignment.

Several integrated waveguide structures fabricated by surface micromachining techniques were reported that used nitride/oxide or GeO_2/SiO_2 structures for planar waveguides (Leistiko and Jensen 1998; Kutter et al. 2000). However, the size of the core region was limited by the thin-film deposition process and may not meet the requirement for microflow cytometers, which usually require microchannels with a deeper depth for cell passage. As an alternative, polymeric channel waveguides are very attractive because it is possible to trim their core materials in order to achieve the desired optical properties (Ihlein et al. 1995).

In addition to planar waveguides, buried waveguide structures were reported to align the excitation and detection optical paths with the microchannel. Usually bulk micromachining techniques were adapted to fabricate a channel for formation of buried optical waveguides. For example, a SU-8 negative photo-resist was used as the core material of a buried optical waveguide with an organic-based spin-on-glass (SOG layer) as cladding, resulting in better light guiding efficiency (Lee, Lin and Chang 2003). Experimental results show that the optical loss is less than 15 dB for a 40 mm-long waveguide. With the integrated optical waveguides, a microflow cytometer capable of particle counting has been realized (Fig. 33.5). Data show that microparticles can be hydrodynamically focused and counted successfully without fluorescent labeling using the miniaturized flow cytometer with the integrated optical waveguides.

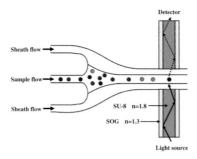

Figure 33.5. Schematic of the working principle of a microflow cytometer with optical waveguides. **a)** Micro-particles are hydrodynamically focused and pass through the optical waveguide structures. **b)** Detected light propagates in the waveguide structures, and light intensity change can be detected at the other waveguide. Light guiding efficiency is improved by SU-8/SOG double layers.

2.2.2. Buried Optical Fiber Approach

Optical fibers can be integrated with microchannels to form a micro flow cytometer to carry out on-line measurement and detection of particles and cells. Single-mode or multi-mode optic fibers are first etched such that they can be directly inserted into pre-etched fiber channels embedded within glass or polymer substrates. The advantage of this approach is that it is more straightforward to couple the light, which passes in and out of the microflow cytometer without fabricating optical waveguides, which usually required an extra fabrication process. The location of the fibers can be defined precisely using a standard lithography process, resulting in an enhanced optical performance and a superior detection resolution. The scattered light or induced fluorescence signals from cells are then detected by the buried optic fibers. This approach has the advantage that particles/cells can be measured and detected with or without the need for fluorescent labeling. The change of intensity or scatted light can be used directly for cell counting. Besides, delicate optical alignment procedures are not necessary since head-on optical fibers are buried inside the flow cytometer chip with high precision. For example, in a PDMS-based microflow cytometer integrated with optical fibers (Tung et al. 2004), two excitation laser lights combined with a multi-angle, optical-fiber layout can detect light signals induced by cells passing through the optical fibers. With this approach, successful cell detection with a high signal-to-noise ratio can be achieved with higher counting precision. Similarly, a microflow cytometer integrated with embedded etched optical fibers for the online detection of particles and cells was reported (Lin and Lee 2003). A simple yet effective fabrication process has been used to fabricate the microfluidic structures on soda-lime glass substrates. Etched optical fibers were inserted into the chip using a fiber insertion guide. The microflow cytometer has been used to count polystyrene beads and human blood cells successfully.

2.2.3. Large-Scale Optical System Approach

A large-scale optical system is widely used for flow cytometry. Most microflow cytometers are incorporated with a fluorescence microscope for cell measurement and detection. For example, in a microflow cytometer using a large-scale optical detection system (Fu et al. 1999), PMTs (or photo-detectors), mirrors, beam splitters and optical filters are commonly used to form an optical detection system for measurement and detection of the cells. Even

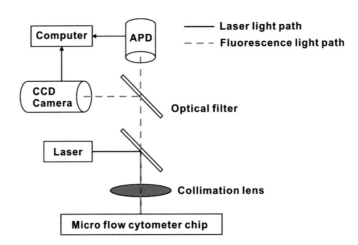

Figure 33.6. A schematic illustration of an optical detection system used for a hand-held microflow cytometer.

though the large-scale optical detection system is bulky and relatively expensive, it is relatively mature and can provide high quality optical signals. More compact designs for the optical detection system have been reported. For example, a hand-held microflow cytometer system has been demonstrated (Yang et al. 2006). As shown in Fig. 33.6, the optical detection system consists of several optical components such as a solid-state laser diode, collimation lenses, filters, an APD and a charge-coupled device (CCD) camera. A laser induced fluorescence (LIF) system was employed for the optical detection. The laser light source, with a 543 nm wavelength, was guided into the optical detection region and used to excite the fluorescent-dye-labeled cell samples while the cell samples flew through the optical detection area. Multiple fluorescent dyes can be used to label different cells so they can be differentiated. The fluorescence signals are then collected by an APD sensor, which transforms the fluorescence signal into an electric signal. After the optical detection process and signal processing, specific dye-labeled cell samples can be collected and sorted into sample collection chambers. In addition to detection of fluorescent signals, the CCD camera can also be used to verify the counting process.

2.3. Cell Sorting

Similar to their large-scale counterparts, a micro flow cytometer is equipped with a cell-sorting module for separation and collection of the cells. Typically, four major approaches have been demonstrated in the literature, including hydrodynamic, pneumatic, electrokinetic and magnetic sorting methods.

2.3.1. Hydrodynamic Sorting

The hydrodynamic flow switching of sample streams can be used for cell sorting. That differential pressures between inlet ports could be used to guide sample flow into any desired outlet-port (Blankenstein and Larsen 1998). Multiported fluidic switches were fabricated using micromachining techniques including etching of silicon wafers and bonding of Si to glass. A sample flow stream could be successfully guided to a desired outlet-port using the hydrodynamic forces. Sample injection into any one of the five possible outlet-ports has been demonstrated successfully in their study. With a similar approach, a microflow cytometer integrating two important microfluidic phenomena, namely hydrodynamic focusing and valveless flow switching inside multi-ported microchannels was reported (Lee et al. 2001c). The cell samples were injected into a center channel and hydrodynamically focused into a narrow stream constrained by two sheath flows. The "pre-focusing" process prior to the flow switching was used to inject the sample flow precisely into a desired outlet port without smearing into the other ports. Subsequently, the focused sample stream is then injected into a desired outlet port based on sheath and sample flow velocity ratios. A simple theoretical model based on the "flow-rate-ratio" method was proposed to predict the switching performance of the device.

In another ingenious approach using hydrodynamic forces for cell sorting (Takagi et al. 2005), continuous particle separation in a microchannel having asymmetrically arranged multiple branches was successfully achieved. By forming a drain channel to receive a large portion of the liquid flow, high separation performance could be achieved since the liquid flow near one sidewall could be effectively distributed. A mixture of 1-5 μm particles was successfully separated and sorted in a microchannel. Apparently, this method can also be used for cell sorting in a micro flow cytometer.

Likewise, the single-channel flow-sorting scheme is common in most micro flow cytometry approaches. In order to provide a higher throughput for cell/particle sorting, multiple sample switching was reported (Lee, Hwei and Huang 2001). Likewise, a pre-focused flow switch with multiple samples was designed and fabricated for multiple cell sorting in a micro flow cytometer (Lee, Lin and Chang 2005). A simple theoretical formula, also based on the "flow-rate-ratio" method and conservation of mass, was provided to predict the switching of the prefocusing multiple samples. Experimental data indicated that the multi-sample flows could be hydrodynamically prefocused and then guided into desired outlet ports precisely based on sheath and sample flow velocity ratios.

2.3.2. Pneumatic Sorting

A pneumatic flow-switching mechanism utilizing three pneumatic microvalves downstream of the sample flow channel can be used to achieve cell sorting (Chang, Hsiung and Lee 2006). Fig. 33.7 illustrates the operation principle of the cell sorting mechanism using pneumatic microvalve devices. Briefly, the pneumatic microvalve devices are composed of two PDMS layers and a membrane substrate to form the air control chamber and the liquid flow

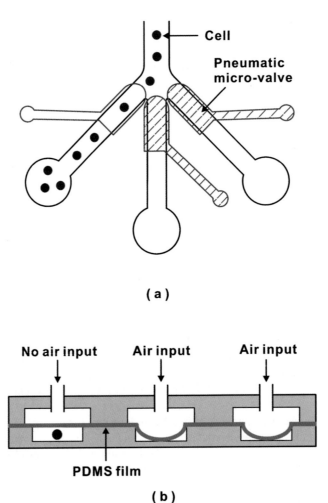

(a)

(b)

Figure 33.7. Schematic representation of the cell sorter, consisting of three micro-valves and three outlet ports. Pneumatic micro-valves are used for flow switching. The cells flow into one of the outlet ports when the other two micro-valves are closed.

channel underneath. When compressed air is injected into the upper air control chamber, a deformation of the PDMS membrane is generated to shut off the liquid flow channel. The air inlet of each pneumatic microvalve is controlled by an electromagnetic valve (EMV) device. The fluorescence signals detected by an APD module are used to trigger the EMV devices. Therefore, the specific fluorescence signal can trigger the corresponding EMV and directs the sample flow into the appropriate channel. Hence, by turning on two of the three micropneumatic valves and leaving only one outlet open, the sample flow can be directed into the specific outlet channel. Experimental results indicated the developed microvalve devices could switch the cells to the specific channel successfully.

2.3.3. Electrokinetic Sorting

Electrokinetic forces, including EOF and DEP, can be also used for cell sorting. EOF has been used for directing flows inside the microchannels. It is straightforward to use EOF for cell sorting applications. For example, EOF has been used for cell sorting and collecting (Dittrich and Schwille 2003). By activating perpendicular side channels electrokinetically, fluorescent cells and particles can be switched to any desired outlet channel, depending on the polarity of the electrodes. Similarly, in a microflow cytometer using EOF for cell sorting (Fu et al. 2004), the cell sample stream was driven electrokinetically through the inlet channel and then focused into a narrow stream by two focusing channel flows. Finally they were guided into the appropriate collective channels using a suitable control voltage.

Alternatively, DEP forces can be applied for cell sorting, for example, a continuous cell separation chip using a hydrodynamic DEP process (Doh, Seo and Cho 2005). Cells can be continuously directed to different outlet channels by applying positive or negative DEP forces with a three-electrode layout. Notably, DEP forces only act on the cells or particles to achieve sample sorting.

2.3.4. Magnetic Sorting

Magnetic forces can be used to sort cells attached to magnetic beads. A micro-immunomagnetic cell sorter (μ-IMCS) consisting of a micromixer and a separation reservoir equipped with an external permanent magnet was demonstrated (Rong, Choi and Ahn 2003; Miwa et al. 2005). Magnetic beads are first coated with an antibody, which attaches to specific surface antigens on the target cells. The cell-bead complexes are then separated in the buffer fluid by an external magnetic field, either generated by a large-scale permanent magnet or micromachined microcoils. With this approach, the cells attached to the magnetic beads can be sorted. Even though this method has been successfully used for cell separation and sorting, a microdevice capable of measurement and detection of the cells has not yet been integrated with the μ-IMCS and is still under extensive investigation.

3. Applications

Currently, bacterial detection involves the following methods, namely cell culture, antibody detection, molecular diagnosis, and flow cytometry. Cell culture methods are relatively cheap and reliable. However they are labor-intensive and time-consuming. Antibody detection methods are fast with a high specificity. Nevertheless, antibody detection is most effective at the later stages of bacterial infection and may require expensive equipment for detection. Molecular diagnosis methods using nucleic acids to identify specific marker genes associated with bacteria are both sensitive and specific. Nevertheless, they are relatively expensive when compared with other bacterial detection methods. Alternatively, flow cytometry can be used for bacterial detection with a high throughput. It is a popular technique for the rapid measurement and detection of biological cells and particles in a liquid. Fluorescence-labeled bacteria can be measured, analyzed and even sorted. Meanwhile, physical properties of the bacteria can be extracted from the scattered light signal. Bacterial detection using flow cytometry is especially suitable for counting slower growing bacterial cells such as mycobacterium and molds. The rapid development in molecular markers and fluorochromes has provided a powerful tool for bacterial detection using flow cytometry. They can be readily used for various applications, including environmental and clinical monitoring of bacteria. For instance, the clinical applications of flow cytometry for bacterial detection provide information about the interaction between pathogens and their corresponding antibiotic susceptibility by using fluorescent molecular and antibody markers to sensitively and specifically detect a small trace of bacteria in clinical samples.

Even though flow cytometry has been popularly used for bacterial detection, its complicated optical alignment process and relatively high cost hinders its practical applications. Recently, microfabricated flow cytometers have been fabricated and proven to be a promising approach for manufacturing a portable and cost-effective apparatus. Several pioneering works have been reported on using microflow cytometers for bacterial detection. Table 33.3 lists a summary of these microflow cytometers for bacterial detection. Important factors, including the flow pumping methods, materials, types of optical detection, feature sizes, flow rate, sorting methods, throughput and bacterial types, have been listed. Two major flow pumping methods were adapted, namely the hydrodynamic and electrokinetic approaches. Syringe pumps or vacuum pumps were used to provide the required hydrodynamic forces to transport and focus the sample flow. Alternatively, EOF was used to drive the sample and sheath flows electrokinetically. Silicon, glass, and polymeric materials were commonly used to fabricate the microflow cytometers. Wet chemical etching, injection molding, hot embossing and molding methods were used to fabricate microchannels. CCD, fluorescence and scattering detection were the most commonly used optical detection schemes. *Escherichia coli* (*E. coli*) cells are commonly used for demonstration of the capability of the developed microflow cytometers. In general, microflow cytometers can be used for the following bacterial detection applications.

(a) Environmental monitoring
(b) Rapid assessment of bacterial viability
(c) Rapid analysis of bacteria levels in food
(d) Antibiotic susceptibility testing
(e) Diagnosis of bacterial in blood and urine

3.1. Environmental Monitoring

Determining the total number of bacterial cells in water is important for many applications such as public health monitoring and water treatment. Flow cytometry has been used for measurement of bacterial cells by using the fluorescent-labeling method (Jepras et al. 1995; Lee, Izuo and Inatomi 2004). With a very similar approach, a microflow cytometer can be used as an environmental monitoring system such as for real time bacteria monitoring in a river or drainage system. There exists a great need to provide compact and portable flow cytometers for such on-site measurement and detection of bacteria. It will be even more useful if the detected information and telemetry can be transmitted immediately to a central lab wirelessly. Microflow cytometers use almost the same protocol as the large system for pretreatment of bacterial cells. After the fluorescence labeling of specific bacteria using several commercially available dyes, these suspended bacteria can be measured and detected. Useful information can be extracted for long-term environmental monitoring at a reasonable cost.

Table 33.3. Summary of micro flow cytometers for bacterial detection

Flow Pumping Method	Material	Optical Detection	Dimensions of Channels (WXD)	Flow Rate	Sorting Method	Sorting Throughput	Bacterial Type	References
Syringe pump	PDMS	APD	25 μm × 7.6 μm	0.55–1.65 mm/s	Electric switching	0.3 ~ 1 event/s	*E. coli* (BMH 71-18)	Dittrich and Schwille 2003
Syringe pump	PDMS + glass	N.R.	50 μm × 200 μm	0.3 mm/s	Magnetic	100 cell/s	*E. coli* (HB101 K-12)	Xia et al. 2006
EOF	PDMA	PMT	50 μm × 20 μm	36 Hz	N.R.	N.R.	*E. coli*	McClain et al. 2001
Syringe pump	PDMS	CCD camera	100 μm × 10 μm	50–100 nl/min	N.R.	N.R.	*E. coli*	Lee, Izuo and Inatomi 2004
Electrokinetics	PDMS + glass	PMT	3 μm × 4 μm	N.R.	Electric switching	20 cell/s	*E. coli* (HB101)	Fu et al. 1999
Vacuum pump	Glass	APD	75 μm × 25 μm	3–4mm/s	N.R.	N.R.	*Pseudomonas putida* (ATTC 12633) and *E. coli*	Sakamoto, Yamaguchi and Nasu 2005

Note: N.R. = Not reported

Many attempts have been made to measure bacteria concentration in water. For example, using a microflow cytometer and its application for bacterial detection in water, fluorescence-labeled *E. coli* cells can be successfully counted and sorted. The developed microflow cytometer is promising for rapid detection of bacteria in drinking water (Lee, Izuo and Inatomi 2004; Inatomi, Izuo and Lee 2006). The detection of bacterial cells in both a culture medium and natural river water samples were also reported (Sakamoto, Yamaguchi and Nasu 2005). Analysis of natural bacterial communities from a marine environment by using a microflow cytometer was reported (Gerdts and Luedke 2006). It is envisioned that these microflow cytometers may provide a fast analysis of bacteria concentration in water as an environmental monitoring system.

3.2. Rapid Assessment of Bacterial Viability

Flow cytometry has been used for rapid assessment of bacterial viability (Diaper, Tither and Edwards 1992). Rhodamine 123 (Rh123) stain was used to clearly differentiate viable from non-viable bacteria cells. Bacteria culture is still a routine laboratory procedure for bacterial detection. Traditional methods to detect and enumerate viable bacterial cells grown on laboratory media are time-consuming. Therefore, flow cytometry was reported to be a promising tool for rapid assessment of bacterial viability after culture. With an approach similar to traditional methods, the micro flow cytometer can be used for verification of bacteria culture and to provide a fast approach for quantification of culture. For example, a microfluidic flow cytometer has been used for detection of *E. coli* (McClain et al. 2001). Bacterial cells were electrokinetically transported and focused into a narrow stream along the center for single-cell scattering and fluorescence detection. Detection of scattering and fluorescence of *E. coli* labeled with a membrane-impermeable nucleic acid stain (Syto15) or a fluorescein-labeled antibody were successfully demonstrated. A viability testing of bacterial cells was performed with a counting rate of 30–85 cells/s.

For example, a microflow cytometer using hydrodynamic forces for flow focusing and electrokinetic forces for cell sorting has been developed (Dittrich and Schwille 2003). A specific recognition of surface proteins in the outer membranes of bacterial cells by a fluorescent ligand was demonstrated. This microflow cytometer is promising in its ability to count the growth rate of bacteria. The dispersed *E. coli* cells can be sorted into two different output channels. The sorting process is realized by a perpendicular deflection stream that can be switched electrokinetically. The microflow cytometer can be used to monitor the sorting of specific bacteria either bounded with a ligand or labeled with a specific fluorescein after cell culture. Thus quantification of the culture process can be realized by using a microflow cytometer.

3.3. Rapid Analysis of Bacteria Levels in Food

Rapid counting of yeast, molds and bacterial cells in the food industry is crucial since conventional time-consuming tests can lead to substantial delays for introduction of products to the market. Flow cytometry has been successfully reported to count low numbers of microbial contaminants in food (Laplace-Builhe et al. 1993). It can be a rapid method for process quality control of the food products. Likewise, the commercially-available Agilent 2100 Bioanalyzer (Agilent Technology, USA) has been used for rapid analysis of bacteria in food. Glass-based microfluidic devices allow cells to be moved by pressure-driven flow and analyzed individually by fluorescence detection. For example, a microflow cytometer system has been used to detect small amounts of milk-spoiling bacteria. *Pseudomonas* cells in milk were in-situ hybridized with a Cy5-labelled probe. The numbers of *Pseudomonas* cells in the stationary phase and in the starved state can be successfully determined by on-chip flow cytometry (Yamaguchi, Ohba and Nasu 2006).

3.4. Antibiotic Susceptibility Testing

A flow cytometer can be used for antibiotic susceptibility testing (Pore 1994). A flow cytometric approach using bacteria stained simultaneously with a membrane potential dye and a permeability indicator for investigating bacterial antibiotic susceptibility was reported (Shapiro 2001). With a very similar approach, it can also be used for drug resistance testing of cancer cells (Lacombe and Belloc 1996). Bacterial cells were tested under different antibiotic doses. Then viable and non-viable bacterial cells were fluorescent-labeled and counted by flow cytometry. The microbial viability can be measured and evaluated to determine minimum inhibitory and minimum bacterial concentrations for a given dose response. A commercial flow cytometer has been used to successfully detect the effects of antibiotics on *E. coli* (Gant et al. 1993). The usage of the microflow cytometer for antibiotic susceptibility testing is promising especially for the pharmaceutical industry.

3.5. Bacterial Detection in Blood and Urine

Bacterial screening in blood is important to reduce the risk of transfusion-transmitted sepsis. The flow cytometry can be used for detection of four transfusion relevant bacteria (Schmidt et al. 2006). Even though the sensitivity of the flow cytometry is not as good as the molecular diagnosis method, rapid screening of bacteria with a reasonable sensitivity was reported. Obviously, these applications can be realized using microfabricated flow cytometers.

Recently, another miniaturized, integrated microfluidic device that can attach living cells onto magnetic particles by applying a local magnetic field gradient, and subsequently selectively remove them from flowing biological fluids without any washing steps was developed (Xia et al. 2006). Using this microdevice, living *E. coli* bacteria adhered onto magnetic nano-particles were efficiently removed from solutions containing densities of red blood cells similar to that found in human blood. The on-chip microfluidic-micromagnetic cell separator has demonstrated its effectiveness for continuous cleansing of contaminant bacteria or particulates from biological fluids. This micro device can be useful if connected with a microflow cytometer since the separated bacteria can be counted afterwards. It is envisioned that these on-chip magnetic cell separators can be used for clinical applications similar to their large-scale counterparts.

4. Conclusions and Future Perspectives

Flow cytometry is a powerful and popular technique used in a wide variety of applications including hematology, immunology, genetics, food science, pharmacology, microbiology, oncology and many others. However, the conventional flow cytometer system tends to be bulky and relatively expensive. Besides, it also requires experienced personnel to run this delicate equipment. Therefore, there is a great need to develop a more compact, low-cost, cell counting/sorting device. Over the past decade, miniaturization of flow cytometers by adopting microfabrication techniques to fabricate microchannels and micronozzles in silicon, glass/quartz and even plastic substrates has been demonstrated. When compared with their large-scale counterparts, these microflow cytometers are more compact, portable, cost-effective and user-friendly. Most importantly, they could have almost comparable performance.

A microflow cytometer typically is composed of three major modules, namely flow transportation/focusing, cell detection and cell sorting devices. Hydrodynamic, pneumatic and electrokinetic forces are the three most commonly used driving methods. Successful flow transportation and focusing have been demonstrated in a microflow cytometer. Fluorescence or scattering signals were used for optical detection of the cells. Optical waveguides, optical fibers, or even a large-scale optical system, have been demonstrated to couple and to detect excitation signals in and out of the flow cytometer chip. After cells have been successfully counted, several methods have been reported to sort and collect cells, including hydrodynamic, pneumatic, electrokinetic, and magnetic forces.

These microflow cytometers are useful for bacterial detection. *E. coli* cells are commonly used to demonstrate the capability of these microflow cytometers. Microflow cytometers show promise in the following applications: environmental monitoring, rapid assessment of bacterial viability, rapid analysis of bacteria levels in food, antibiotic susceptibility testing and diagnosis of bacterial content in blood and urine.

The microflow cytometer is still in its development phase and not ready for general commercial applications. Most of them are still only used in laboratories and have not yet been proven in field or clinical applications. The system should be more user-friendly so that it can be extensively used for testing of daily activities. Besides, the complicated process for pre-treatment of bacteria samples, including the concentration/purification and fluorescent-labeling process is not alleviated by the miniaturization of the flow cytometry system and still hinders its practical applications. The integrated optical excitation and fluorescence/scattering detection systems should be more compact and cost-effective. The fouling of the microchannel should be properly solved. Multiple-wavelength excitation and detection schemes should be provided without significantly increasing the device's cost and complexity. It can be envisioned that a portable flow cytometer system may be available for point-of-care applications if these issues can be addressed properly in the near future.

Acknowledgements

The authors gratefully acknowledge the financial support provided for this study by the National Science Council of Taiwan and the MOE Program for Promoting Academic Excellence of Universities.

References

Bang H, Chung C, Kim JK, Kim SH, Chung S, Park J, Lee WG, Yun H, Lee J, Cho K, Han DC and Chang JK (2006) Microfabricated fluorescence-activated cell sorter through hydrodynamic flow manipulation. Microsyst Technol 12:746–753

Bartsch JW, Tran HD, Waller A, Mammoli AA, Buranda T, Sklar LA and Edwards BS (2004) An investigation of liquid carryover and sample residual for a high throughput flow cytometer sample delivery system. Anal. Anal. Chem. 76:3810–3817

Blankenstein G and Larsen UD (1998) Modular concept of a laboratory on a chip for chemical and biochemical analysis. Biosensors and Bioelectronics 13:427–438

Brown M and Wittwer C (2000) Flow cytometry: principles and clinical applications in hematology. Clin. Chem. 46:1221–1229

Chan DH, Luedkel G, Valer M, Buhlmann C and Precke T (2003) Cytometric analysis of protein expression and apoptosis in human primary cells with a novel microfluidic chip-based system. Cytometry Part A 55:119–125

Chang CM, Hsiung SK and Lee GB (2006) A micromachine-based flow cytometer chip integrated with micropumps/valves for multi-wavelength detection applications. Journal of Material Science Forum 505:637–642

Cheung K, Gawad S and Renaud P (2005) Impedance spectroscopy flow cytometry: on-chip label-free cell differentiation. Cytometry Part A 65:124–132

Cormick RM, Nelson RJ, Alonso-Amigo MG, Benvegnu DJ and Hooper HH (1997) Microchannel electrophoretic separations of DNA in injection-molded plastic substrates. Anal. Chem. 69:2626–2630

Cui L and Morgan H (2000) Design and fabrication of traveling wave dielectrophoresis structures. J. Micromech. Microeng. 10:72–79

Cui L, Zhang T and Morgan H (2002) Optical particle detection integrated in a dielectrophoretic lab-on-a-chip. J. Micromech. Microeng. 12:7–12

Dallas CE and Evans DL (1990) Flow cytometry in toxiciy analysis. Nature Publishing Group 345:557–558

Diaper JP, Tither K and Edwards C (1992) Rapid assessment of bacterial viability by flow cytometry. Appl Microbiol Biotechnology 38:268–272

Dittrich PS and Schwille P (2003) An integrated microfluidic system for reaction, high-sensitivity detection, and sorting of fluorescent cells and particles. J. Anal. Chem. 75:5767–5774

Doh I, Seo KS and Cho YH (2005) A continuous cell separation chip using hydrodynamic dielectrophoresis process. Sensors and Actuators A 121:59–65

Fu AY, Spence C, Scherer A, Arnold FH and Quake SR (1999) A microfabricated fluorescence-activated cell sorter. Nat. Biotechnol. 17:1109–1111

Fu LM, Lin CH, Yang RJ, Lee GB and Pan YJ (2004) Electrokinetically driven micro flow cytometers with integrated fiber optics for on-line cell/particle detection. Anal. Chim. Acta 507:163–169

Gant VA, Warnes G, Phillips I and Savidge GF (1993) The application of flow cytometry to the study of bacterial responses to antibiotics. Journal of Medical Microbiology 39:147–154

Gerdts G and Luedke G. (2006) FISH and chips: marine bacterial communities analyzed by flow cytometry based on microfluidics. Journal of Microbiological Methods 64:232–240

Greve B, Valet G, Humpe A, Tonn T and Cassens U (2004) Flow cytometry in transfusion medicine: development, strategies and applications. Transfusion Medicine and Hemotherapy 31:152–161

Gunasekera TS, Attfield PV and Veal DA (2000) A flow cytometry method for rapid detection and enumeration of total bacteria in milk. Appl. Environ. Microb. 66:1228–1232

Hodder PS, Blankenstein G. and Ruzicka J (1997) Microfabricated flow chamber for fluorescence-based chemistries and stopped-flow injection cytometry. Analyst 122:883–887

Holmes D, Sandison ME, Green NG and Morgan H (2005) On-chip high-speed sorting of micron-sized particles for high-throughput analysis. IEEE Proc. Nanobiotechnol. 152:129–135

Hrusak O, Vaskova M, Mejstrikova E, Kalina T, Trka J and Stary J (2004) Multiparametric flow cytometry is an ideal tool for diagnostic assessment of the top genes in acute leukemia. Tissue Antigens 64:339–340

Huh D, Tung YC, Wei HH, Grotberg JB, Skerlos SJ, Kurabayashi K and Takayama S (2002) Use of air-liquid two-phase flow in hydrophobic microfluidic channels for disposable flow cytometers. Biomed Microdevices 4:141–149

Ihlein G, Menges B, Mittler-Neher S, Osaheni JA and Jenekhe SA (1995) Channel waveguides of insoluble conjugated polymers. Optical Materials 4:685–689

Inatomi KI, Izuo SI and Lee SS (2006) Application of a microfluidic device for counting of bacteria. Letters in Applied Microbiology 43:296–300

Jepras RI, Carter J, Pearson SC, Paul EF and Wilkinson MJ (1995) Development of a robust flow cytometric assay for determining numbers of viable bacteria. Appl Environ Microbiol 61:2696–2701

Koch M, Evans AGR and Brunnschweiler A (1999) Design and fabrication of a micromachined coulter counter. J. Micromech. Microeng. 9:159–161

Kruger J, Singh K, O'Neill A, Jackson C, Morrison A and O'Brien P (2002) Development of a microfluidic device for fluorescence activated cell sorting. J. Micromech. Microeng. 12:486–94

Kutter JP, Mogensen KB, Friis P, Jorgensen AM, Pertersen NJ, Telleman P and Huebner J (2000) Integration of waveguides for optical detection in microfabricated analytical devices. Proc. SPIE 4177:98–105

Lacombe F and Belloc F (1996) Flow cytometry study of cell cycle, apoptosis and drug resistance in acute leukemia. Hematology and Cell Therapy 38:495–504

Laplace-Builhe C, Hahne K, Hunger W, Tirilly Y and Drocourt JL (1993) Application of flow cytometry to rapid microbial analysis in food and drinks industries. Biol Cell 78:123–128

Lee GB, Chen SH, Huang GR, Sung WC and Lin YH (2001a) Microfabricated plastic chips by hot embossing methods and their applications for DNA separation and detection. Sensors and Actuators B 75:142–148

Lee GB, Hung CI, Ke BJ, Huang GR and Hwei BH (2001c) Micromachined pre-focused 1×N flow switches for continuous sample injection. J. Micromech. Microeng. 11:567–573

Lee GB, Hung CI, Ke BJ, Huang GR, Hwei BH and Lai HF (2001b) Hydrodynamic focusing for a micromachined flow cytometer. ASME Journal of Fluids Engineering 123:672–679

Lee GB, Hwei BH and Huang GR (2001) Micromachined pre-focused MXN flow switches for continuous multi-sample injection. J. Micromech. Microeng. 11:654–661

Lee GB, Lin CH and Chang GL (2003) Micro flow cytometers with buried SU-8/SOG optical waveguides. Sensors and Actuators A 103:165–170

Lee GB, Lin CH and Chang SC (2005) Micromachine-based multi-channel flow cytometers for cell and particle counting and sorting. J. Micromech. Microeng. 15:447–454

Lee SS, Izuo SI and Inatomi KI (2004) A CAD study on micro flow cytometer and its application to bacteria detection. IEEE Sensor 1:308–311

Leistiko O and Jensen PF (1998) Integrated bio/chemical microsystems employing optical detection: a cytometer. Proc. Micro. Total Analysis Systems 291–294

Liao CS, Lee GB, Liu HS, Hsieh TM and Luo CH (2005) Miniature RT-PCR system for diagnosis of RNA-based viruses. Nucleic Acids Research 33:1–7

Lin CH and Lee GB (2003) Micromachined flow cytometers with embedded etched optic fibers for optical detection. J. Micromech. Microeng. 13:447–453

Lin CH, Lee GB, Fu LM and Hwey BH (2004) Vertical focusing device utilizing dielectrophoretic force and its application on micro flow cytometer. Journal of Microelectro Mechanical Systems 13:923–932

Lin CH, Lee GB, Lin YH and Chang GL (2001) A fast prototyping process for fabrication of microfluidic systems on soda-lime glass. J. Micromech. Microeng. 11:726–732

Madou MJ (2002) Fundamentals of microfabrication: the science of miniaturization. Boca Raton:CRC Press

Mandy FF, Bergeron M and Minkus T (1995) Principles of flow cytometry. Transfus. Sci. 16:303–314

Manz A, Graber N and Widmer HM (1990) Miniaturized total chemical analysis systems: a novel concept for chemical sensing. Sensors and Actuators B 1:244–248

McClain MA, Culbertson CT, Jacobson SC and Ramsey JM (2001) Flow cytometry of Escherichia coli on microfluidic devices. Anal. Chem. 73:5334–5338

Miwa J, Tan WH, Suzuki Y, Kasagi N, Shikazono N, Furukawa K and Ushida T (2005) Development of micro immunoreaction-based cell sorter for regenerative medicine. Int. Conf. Bio-Nano-Information Fusion 1:1–4

Miyake R, Ohki H, Yamazaki I and Takagi T (1997) Investigation of sheath flow chambers for flow cytometers. JSME Int. Journal B 40:106–113

Miyake R, Ohki H, Yamazaki I and Takagi T (2000) Flow cytometric analysis by using micro-machined flow chamber. JSME Int. Journal B 43:219–224

Miyake R, Ohki H, Yamazaki I and Yabe R (1991) A development of micro sheath flow chamber. IEEE MEMS 265–270 Nieuwenhuis JH, Bastemeijer J, Sarroc PM and Vellekoop MJ (2003) Integrated flow-cells for novel adjustable sheath flows. Lab on a Chip 3:56–61

Pohl HA (1978) Dielectrophoresis. Cambridge University Press, UK

Pore RS (1994) Antibiotic susceptibility testing by flow cytometry. Journal of Antimicrobial Chemotherapy 34:613–627

Riley RS (2002) Preface: flow cytometry and its applications in hematology and oncology. Hematol. Oncol. Clin. North Am. 16:xi–xii

Rong R, Choi JW and Ahn CH (2003) A functional magnetic bead/biocell sorter using fully integrated magnetic micro/nano tips. IEEE MEMS 530–533

Sakamoto C, Yamaguchi N and NasuM(2005) Rapid and simple quantification of bacterial cells by using a microfluidic device. Appl. Environ. Microbiol. 71:1117–1121

Sanders GHW and Manz A (2000) Chip-based microsystems for genomic and Proteomic analysis. Trends in Analytical Chemistry 19:364–378

Schmidt M, Hourfar MK, Nicol SB, Wahl A, Heck J, Weis C, Tonn T, Spengler HP, Montag T, Seifried E and Roth WK (2006) A comparison of three rapid bacterial detection methods under simulated real-life conditions. Transfusion 46:1367–1373

Schrum DP, Culbertson CT, Jacobson SC and Ramsey JM (1999) Microchip flow cytometry using electrokinetic focusing. Anal. Chem. 71:4173–4177

Shapiro HM (2001) Multiparameter flow cytometry of bacteria: implications for diagnostics and therapeutics. Cytometry 43:223–226

Shapiro HM (2003) Practical flow cytometry. Wiley & Liss, New Jersey

Sobek D, Young AM, Gray ML and Senturia SD (1993) A microfabricated flow chamber for optical measurements in fluids. IEEE MEMS 219–224

Takagi J, Yamada M, Yasuda M and Seki M (2005) Continuous particle separation in a microchannel having asymmetrically arranged multiple branches. Lab on a Chip 5:778–784

Takahashi K, Hattori A, Suzuki I, Ichiki T and Yasuda K (2004) Non-destructive on-chip cell sorting system with real-time microscopic image processing. Journal of Nanobiotechnology 2:1–8

Tung YC, Zhang M, Lin CT, Kurabayashi K and Skerlos SJ (2004) PDMS-based opto-fluidic micro flow cytometer with two-color multi-angle fluorescence detection capability using PIN photodiodes. Sensors and Actuators B 98:356–367

Unger MA, Chou HP, Thorsen T, Scherer A and Quake SR (2000) Monolithic microfabricated valves and pumps by multilayer soft lithography. Science 288:113–116

Wang CH and Lee GB (2005) Automatic bio-sampling chips integrated with micro-pumps and micro-valves for disease detection. Biosensors and Bioelectronics 21:419–425

Wang MM, Tu E, Raymond DE, Yang JM, Zhang H, Hagen N, Dees B, Mercer EM, Forster AH, Kariv I, Marchand PJ and Butler WF (2005) Microfluidic sorting of mammalian cells by optical force switching. Nature Biotechnology 23:83–87

Wang Z, El-Ali J, Engelund M, Gotsad T, Perch-Nielsen IR, Mogensen KB, Snakenborg D, Kutter JP and Wolff A (2004) Measurements of scattered light on a microchip flow cytometer with integrated polymer based optical elements. Lab on a Chip 4:372–377

Wolff A, Perch-Nielsen IR, Larsen UD, Friis P, Goranovic G, Poulsen CR, Kuttera JP, and Tellemana P (2003) Integrating advanced functionality in a microfabricated high-throughput fluorescent-activated cell sorter. Lab on a Chip 3:22–27

Xia N, Hunt TP, Mayers BT, Alsberg E, Whitesides GM, Westervelt RM and Ingber DE (2006) Combined microflu-idicmicromagnetic separation of living cells in continuous flow. Biomed Microdevices 8:299–308

Xiang Q, Xuan X, Xu B and Li D (2005) Multi - functional particle detection with embedded optical fibers in a poly(dimethylsiloxane) chip. Instrumentation Science and Technology 33:597–607

Yamaguchi N, Ohba H and Nasu M (2006) Simple detection of small amounts of pseudomonas cells in milk by using a microfluidic device. Letters in Applied Microbiology 43:631–636

Yang RJ, Chang CC, Huang SB and Lee GB (2005) Electrokinetically driven micro flow cytometers with integrated fiber optics for on-line cell/particle detection. J. Micromech. Microeng. 15:2141–2148

Yang SY, Hsiung SK, Hung YC, Chang CM, Liao TL and Lee GB (2006) A cell counting and sorting system incorporated with a microfabricated flow cytometer chip. Measurement Science and Technology 17:2001–2009

Yao B, Luo GA, Feng X, Wang W, Chena LX and Wanga YM (2004) A microfluidic device based on gravity and electric force driving for flow cytometry and fluorescence activated cell sorting. Lab on a Chip 4:603–607

34

Bacterial Concentration, Separation and Analysis by Dielectrophoresis

Michael Pycraft Hughes and Kai Friedrich Hoettges

Abstract

It has been known for millennia that electrostatic forces can be used to manipulate particles; recently developed techniques give sufficient sensitivity, selectivity and precision for the selective trapping and manipulation of bacteria. These forces are all defined from the attraction of charge in an electric field, but exploit it in different ways to achieve different ends. Such phenomena include electrophoresis, the electrostatic attraction of particles; dielectrophoresis, the force generated by the interaction of a particle with a non-uniform, time-variant electric field; and electro-osmosis, where micro-flows are induced in fluid by non-uniform field effects. These phenomena have demonstrated a number of benefits for bacterial study, including particle filtration, preconcentration and identification. These phenomena have much to offer for the field of bacterial detection and analysis. By treating the bacterial cell as an electronic object and hence considering its electrical properties, it is possible to gain important insights into the electrophysiology of the cell. Furthermore, differences in electrophysiology can be exploited to allow the separation or concentration of bacteria prior to analysis. As these technologies all rely on similar devices, they can be integrated into a single lab on a chip device, with many potential benefits including portability and disposability in addition to the ability to detect particles at lower concentrations than ever before.

1. Introduction

Since Thales of Miletus first described the phenomenon over two and a half millennia ago, it has been known that electrostatic forces can be used to manipulate particles, for example by using rubbed amber to trap dust motes. More recently developed techniques give far more sensitivity, selectivity and precision, and have been applied to the selective trapping and manipulation of bacteria. One such phenomenon is *dielectrophoresis* (Pohl 1978), the force generated by the interaction of a particle with a non-uniform, time-variant electric field. This has been demonstrated to have a number of benefits for bacterial analysis, including particle filtration, preconcentration and analysis.

The rapid detection of low concentrations of bioparticles such as bacteria, viruses or proteins is critical for a range of monitoring applications, from environmental study and water quality monitoring to the use of biological weapons, be it in the battlefield or incidences of bioterrorism. Since even low levels of pathogens can be fatal, detection systems need to be extremely sensitive to be of use. In order to measure particles at concentrations below the

Michael Pycraft Hughes and Kai Friedrich Hoettges • Centre for Biomedical Engineering, University of Surrey, Guildford, Surrey, UK.

M. Zourob et al. (eds.), *Principles of Bacterial Detection: Biosensors, Recognition Receptors and Microsystems*, sensor detection threshold, detection must be improved by processes where pathogenic particles can be filtered from other material that can foul the sensor, concentrated and manipulated onto a sensor surface in order to improve detection sensitivity. Many detection techniques (light scattering, quartz crystal microbalances, surface plasmon resonance) are based on a combination of surface-bound antibodies for identification combined with surface interaction for detection. These methods suffer from the problem of insufficient bioparticles adhering to the detection surface due to the particles being too small to move from suspension and attach to the functionalized sensor surface.

Recent technological advances in the field of microengineering have brought about a generation of devices known as "lab on a chip" (also known as micrometer–scale, total analytical systems or μTAS). These devices usually consist of integrated fluid handling, sorting and identification components on a single glass or plastic slide, with dimensions somewhere between those of a postage stamp and those of a credit card. The advantage of such devices is that the use of miniaturized sensing systems allows the investigation of small fluid samples far more quickly than large-scale laboratory analysis. This in turn enables the development of portable machines capable of rapidly detecting pathogens, be it in environmental samples to detect the use of biological weapons or in a blood sample for diagnosing conditions at the patient's bedside (so-called *point-of-care diagnostics*), for example allowing a physician to make appropriate clinical decisions rapidly. Beyond this, the technology also has benefits in veterinary medicine (for the diagnosis of animal disease) and for the detection of water-borne viruses in the environment, particularly in the third world, where laboratory facilities are scarce. All of these are cases where portable devices for rapid viral detection and identification are required. Furthermore, the system can be constructed so that the integrated fluid/sensing element (the "lab on a chip" itself) is disposable (allowing the maintenance of sterile conditions and avoiding contamination by eliminating re-use). Such a portable system would eliminate the need for transfer of the sample to a central laboratory, which inevitably leads to diagnostic delays. While in medical applications this can have implications for the prognosis of the individual, more seriously, in the detection of biological weapons it could influence the course and effect of an outbreak of disease.

The rapid detection of low concentrations of bioparticles such as bacteria, viruses or proteins is important for monitoring the use of biological weapons, be it in the battlefield or by terrorists. Efficient detection mechanisms also have applications in monitoring contamination of water supplies with pathogens. Since even low levels of pathogens can have a fatal effect, detectors have to be extremely sensitive to be of use. To measure particles at low concentrations, not only the detector itself is important, but also the whole process of analyzing the sample has to be reliable, sensitive and fast. In most cases, the measurement process consists of a number of steps; first a sample has to be taken be from the environment (e.g., air, water or soil), then prepared for the measurement by processes such as filtration, extraction, dissolution, adding of reagents, etc, and finally the concentration of one or more analytes is measured by a detector.

This chapter describes the application of two electrokinetic ("force from electricity") forces for the separation, concentration, detection and analysis of bacteria. *Dielectrophoresis* is the motion of particles caused by induced polarization effects in inhomogeneous electric fields. Depending on the electrical properties of the medium and the particle it can be attractive or repulsive, which we term negative and positive dielectrophoresis. Whether a particle experiences positive or negative dielectrophoresis depends mainly on the electrical properties of the particle, the surrounding medium and the applied field frequency. Since cells consist of a cell wall that is in most cases electrically insulating and an interior that is conductive, the impedance of the cell as a whole depends on capacitive coupling between the cell wall and interior. Therefore the conductivity of the cell is frequency dependent. When subjected to an inhomogeneous electric AC field, a cell can experience positive or negative dielectrophoresis depending on the frequency of the electric field. This allows particles to be trapped, for example to concentrate them on a surface, or separated.

Another significant phenomenon in electrically-induced fluid flow is *electro-osmosis*, which causes the motion of a liquid in contact with the surface of electrodes generating non-uniform electric fields. The combination of dielectrophoresis and AC-electro-osmotic flow causes a bulk fluid flow in the medium, which causes particles to be trapped on top of electrode surfaces. This effect appears at relatively low frequencies (<100KHz) and is perpendicular to the electrode edge, driving fluid onto the electrodes. In practice, this forms a vortex of liquid over the electrode edge that pulls particles out of the bulk liquid in a downdraft above the electrode edge and pushes the particles inwards along the electrode surface; since the updraft further inward is weaker, the particles collect on top of the electrode but away from the edge. However, the particles also experience positive dielectrophoresis when close to the electrode edge; the field pulls them closer to the electrode, thereby pulling them out of the vortex and preventing them from entering the stronger updraft above the electrode, and therefore increasing collection efficiency.

Furthermore, dielectrophoresis can itself be used for bacterial analysis. By varying the frequency of the applied electric field and monitoring the change in applied force, it is possible to characterize the electrical properties of the cells under study. This has a number of advantages for bacterial analysis, the most significant of which is that this technique allows rapid discrimination between live and dead bacteria. This has important applications in biotechnology, drug discovery and clinical diagnostics, where the assessment of the action of a drug on a bacterium can be used to identify a successful antibiotic, or a bacterium that has developed resistance to existing treatments.

2. Theory

The work described here involves the application of two electrokinetic ("force from electricity") forces that play a significant role in the induced motion of particles in liquid suspension. *Dielectrophoresis* is the motion of particles caused by induced polarization effects in inhomogeneous electric fields. Depending on the electrical properties of the medium and the particle it can be attractive or repulsive, which we term negative and positive dielectrophoresis. In the case of positive dielectrophoresis the particle moves toward the greater field inhomogeneity; in negative dielectrophoresis it moves away from the field inhomogeneity. Since electrodes induce the electric field, the field inhomogeneity is normally greatest at the edges of the electrodes; therefore, the particles move either toward or away from the electrodes (Fig. 34.1). Whether a particle experiences positive or negative dielectrophoresis depends mainly on the electrical properties of the particle, the surrounding medium and the applied field frequency. Since cells consist of a cell wall that is in most cases electrically insulating and an interior that is conductive, the impedance of the cell as a whole depends on capacitive coupling between the cell wall and interior. Therefore the polarizability of the cell is frequency dependent. When subjected to an inhomogeneous electric AC field, a cell can experience positive or negative dielectrophoresis depending on the frequency of the electric field.

The dielectrophoretic force, $\mathbf{F}_{\mathbf{DEP}}$, acting on a spherical, homogeneous body suspended in a local electric field gradient is given by the expression (Hughes et al. 2002):

$$\mathbf{F}_{\mathbf{DEP}} = 2\pi r^3 \varepsilon_m Re\left[K(\omega)\right] \nabla \mathbf{E}^2, \qquad (34.1)$$

where r is the particle radius, ε_m is the permittivity of the suspending medium, ∇ is the Del vector (gradient) operator, $\mathbf{E}$ is the *rms* electric field and $Re\left[K(\omega)\right]$ the real part of the

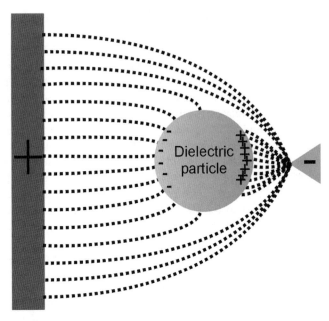

Figure 34.1. In an inhomogeneous electric field, a dielectric particle will be polarized. This induces a net force on the particle, either repelling from or attracting it toward the field gradient.

Clausius-Mossotti factor, given by:

$$K(\omega) = \frac{\varepsilon_p^* - \varepsilon_m^*}{\varepsilon_p^* + 2\varepsilon_m^*},\tag{34.2}$$

where ε_m^* and ε_p^* are the complex permittivities of the medium and particle respectively, and $\varepsilon^* = \varepsilon - \frac{j\sigma}{\omega}$ with σ the conductivity, ε the permittivity and ω the angular frequency of the applied electric field. The limiting (DC) case of equation 2 is

$$K(\omega = 0) = \frac{\sigma_p - \sigma_m}{\sigma_p + 2\sigma_m}.\tag{34.3}$$

The frequency-dependence of $Re[K(\omega)]$ indicates that the force acting on the particle varies with the frequency. The magnitude of $Re[K(\omega)]$ varies depending on whether the particle is more or less polarizable than the medium. If $Re[K(\omega)]$ is positive, then particles move to regions of highest field strength (positive dielectrophoresis); the converse is negative dielectrophoresis where particles are repelled from these regions.

It is possible to extend the model of dielectric behavior to account for more complex particle structure. Two cases of this will be considered here; the first is that of non-homogeneous (shelled) particles, the second case is that of non-spherical ellipsoids. In the first case, we can extend the models described above by replacing the Clausius-Mossotti factor for a homogeneous sphere by a more complex term representing the many dielectric materials in a layered or "shelled" object (such as a cell, where the inner cytoplasm is enclosed by an enveloping membrane). The model most widely used now was developed by Irimajiri (1979) and works by considering each layer as a homogeneous particle suspended in a medium, where that medium is in fact the layer surrounding it. So, starting from the core we can determine the dispersion at the interface between the core and the layer surrounding it, which we will call *shell 1*. This combined dielectric response is then treated as a particle suspended in shell 2, and a second dispersion due to that interface is determined; then a third dispersion due to the interface between shells 2 and 3 is determined, and so on. In this way, the dielectric properties of all the shells combine to give the total dielectric response for the entire particle. In order to examine this mathematically, let us consider a spherical particle with N shells surrounding a central core. To each layer i we assign an outer radius a_i, with a_1 being the radius of the core and a_{N+1} being the radius of the outer shell (and therefore the radius of the entire particle). Similarly each layer has its own complex permittivity given by:

$$\varepsilon_i^* = \varepsilon_i - j\frac{\sigma_i}{\omega}, \tag{34.4}$$

where i has values from 1 to N+1. In order to determine the effective properties of the whole particle, we first replace the core and the first shell surrounding it with a single, homogeneous "core." This new core has a radius a_2 and a complex permittivity given by:

$$\varepsilon_{1eff}^* = \varepsilon_2^* \frac{\left(\dfrac{a_2}{a_1}\right)^3 + 2\dfrac{\varepsilon_1^* - \varepsilon_2^*}{\varepsilon_1^* + 2\varepsilon_2^*}}{\left(\dfrac{a_2}{a_1}\right)^3 - \dfrac{\varepsilon_1^* - \varepsilon_2^*}{\varepsilon_1^* + 2\varepsilon_2^*}}. \tag{34.5}$$

This value provides an expression for the combined complex permittivity of the particle at any given frequency ω. It can also be combined with the complex permittivity of the medium to calculate the Clausius-Mossotti factor. Typical models of bacteria use either one or two shells (representing the membrane and cell wall, or an amalgam of the two) surrounding the cytoplasmic core.

The above expression specifically refers to the case of spherical particles. However a significant number of bacteria have a rod-like shape, varying between a short pill-like structure with the longer axis being marginally longer than the short axis, to long thin rods. No analytical expression exists for such shapes, but for the purposes of modeling, a rodlike bacterium can be approximated to a prolate ellipsoid. Prolate (football or rugby ball shaped) ellipsoids in electric fields have a dispersion along their long axes of different relaxation frequency to the dispersion across their short (but equal) axes, due to different polarization characteristics.

In addition to the dielectrophoretic force experienced by the particle, it will also experience a torque acting so as to align the longest non-dispersed axis with the field. This phenomenon, often observed in practical dielectrophoresis, is *electro-orientation*. When a non-spherical object is suspended in an electric field (for example, but not solely, when experiencing dielectrophoresis) it rotates such that the dipole along the longest non-dispersed axis aligns with the field. Since each axis has a different dispersion, the particle orientation will vary according to the applied frequency. For example, at lower frequencies, a rod-shaped particle experiencing positive dielectrophoresis will align with its longest axis along the direction of the electric field; the distribution of charges along this axis has the greatest moment and therefore exerts greatest torque on the particle to force it into alignment with the applied field. As the frequency is increased, the dipole along this axis reaches dispersion, but the dipole formed *across* the rod does not and the particle will rotate 90° and align perpendicular to the field. This smaller axis has a shorter distance for charges to travel between cycles and so the dispersion frequency will be higher; however, the shorter distance means the dipole moment is smaller. This will result in the force experienced by the particle being smaller in this mode of behavior.

When aligned with one axis parallel to the applied field, a prolate ellipsoid experiences a force given by the equation:

$$\mathbf{F}_{\mathbf{DEP}} = \frac{2\pi abc}{3} \varepsilon_m Re\left[X(\omega)\right] \nabla \mathbf{E}^2, \tag{34.6}$$

where

$$X(\omega) = \frac{\varepsilon_p^* - \varepsilon_m^*}{\left(\varepsilon_p^* - \varepsilon_m^*\right) A_\alpha + \varepsilon_m^*}, \tag{34.7}$$

where α represents either the x, y, or z axis and A is the *depolarization factor*, which represents the different degrees of polarization along each axis and takes a value between 0 and 1.

A related phenomenon to that of dielectrophoresis is that of *electro-osmosis*, sometimes referred to as the electro-hydro-dynamic (EHD) effect. This is caused by the motion of ions on the electrode surface, which acts to move the suspending medium away from interelectrode gaps. This causes more fluid to be drawn from above the electrodes, which is then in turn pushed away across the electrode surfaces by EHD. This ultimately causes the generation of vortices over interelectrode gaps that extend hundreds of microns over the electrodes.

One interesting aspect of this phenomenon is that when combined with dielectrophoresis, EHD can cause particles to be transported from the bulk medium and deposited on electrode surfaces. This was first observed by Price et al. (1988), who noted in early experiments using planar electrodes that, at low frequencies, particles trapped by positive dielectrophoresis moved to form diamond-shaped aggregations on the upper surface of the electrode arrays. The effect was subsequently explained by Green and Morgan (1998) as being due to the balance of dielectrophoretic force and the action of fluid flow. The source of this fluid flow was revealed by Ramos et al. (1999) who described how the electric field generated by planar electrodes is such that field lines pass tangentially through the electrical double layer surrounding the electrodes; this can be considered to consist of a component orthogonal to the electrode surface, plus a second component, parallel to the surface, which acts to move the charge accumulated in the double layer, creating an electro-osmotic flow. When particles collect on planar electrode arrays by dielectrophoresis alone, they do so at the points of highest electric field strength; that is, at the edges of the electrodes. However, as frequency is decreased, fluid flow due to electro-osmosis becomes increasingly prominent; as described above, the location where this is strongest is where the electric field intercepts the double layer at the sharpest angle, which is across the electrode surfaces where the electrodes are closest together. Therefore, those particles which have collected by positive dielectrophoresis are those that experience the greatest fluid motion, causing them to be "swept back" on to the electrode surface (Fig. 34.2). As they move further from the electrode edge, the angle of the electric field becomes more orthogonal and the fluid flow diminishes; eventually a "neutral point" is reached where the two processes are in equilibrium and the particles remain at rest; this is at the center of the array and is responsible for

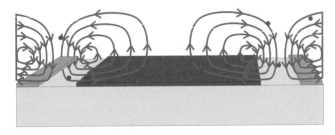

Figure 34.2. Electro-hydrodynamic forces form a bulk flow of liquid at the electrode edges. The liquid flows inward across the electrode surface and perpendicular to the electrode edges; if there is a narrow gap between two electrodes, the liquid flowing away from the electrode edge causes a downdraft above the inter-electrode gap. The liquid flows upward further away from the electrode edge and forms a vortex above the electrode edge, which pulls particles out of the bulk liquid and pushes them inward along the electrode surface. Since the updraft over the electrode surface is much weaker than the downdraft over the edge, most particles collect in a band parallel to the electrode edge.

anomalous collection behavior. The fluid-flow effect is frequency-dependent, being strongest at low frequencies where the double layer has time to form, diminish and reform with opposing polarity for every cycle of the electric field, but becoming limited at high frequencies where the electrode polarity changes too fast for a sufficiently charged double layer to form. Ramos et al. (2000) amd Green et al. (2001) demonstrated that the velocity profile v of the fluid, and hence the particles, follows a bell-shaped frequency dependence governed by the expression

$$v = \frac{1}{8} \frac{\varepsilon V_o^2 \Omega^2}{\eta x \left(1 + \Omega^2\right)^2}, \tag{34.8}$$

where ε represents the permittivity of the medium, V_o is the potential applied to the electrodes, η is the viscosity of the medium, x is the distance from the center of the inter-electrode gap, and Ω is a parameter given by the expression

$$\Omega = \frac{\omega x \kappa \varepsilon \pi}{2\sigma}, \tag{34.9}$$

where ω is the electric field frequency, σ represents the conductivity of the medium, and κ is the reciprocal double layer thickness.

3. Applications of Electrokinetics to Bacteria

The first report of the dielectrophoretic manipulation of bacteria regarded the analysis of the electrical properties of a species of micrococcus (Inoue et al. 1988). Since then, much effort has been expended in the pursuit of more advanced tools for the analysis of bacteria, for example as a method of distinguishing between different species, determining whether a drug has affected a cell, or forming the basis of separation techniques.

The early work on the dielectrophoresis of bacteria was performed at the lab of Professor Ronald Pethig at the University of Wales. It was in his group that the first characterization of the dielectric properties of bacteria was performed using dielectrophoresis, building on the work of others in the 1960s using other methods such as impedance spectroscopy. This was followed in the early 1990s with a more comprehensive study of the electrical properties of a larger number of bacterial species (Markx et al. 1994).

With this knowledge, the group was able to optimize conditions for dielectrophoretic separation; this is achieved by identifying the conditions where two populations experience dielectrophoresis of different signs, one being trapped by positive dielectrophoresis while the other is repelled. If a solution containing two such subpopulations flows through a flow cell such as the one shown in Fig. 34.3, then one will remain trapped at the electrode edges while the other flows to the outlet. Once the entire solution has been processed, a second fluid stream can be inserted and the field turned off, releasing the trapped population and allowing complete particle recovery while at the same time allowing filter regeneration. The first such separations to feature bacteria as one of the subpopulations were performed using *M. lysodeikticus* and yeast (Wang et al. 1993), whereas later work (Markx and Pethig 1995; Markx et al. 1996) involved optimizing conditions to allow the separation of, for example, Gram-positive and Gram-negative bacteria.

In 1998, a significant step forward was achieved by the U.S. company Nanogen, whose primary focus was the development of microarray-type technology (Cheng et al. 1998; Huang et al. 2001). Using dielectrophoresis, an integrated, lab-on-a-chip design was described for the first time wherein a bacterial sample could be concentrated onto a sensor and lysed, and its DNA analyzed in order to provide exact information identifying the bacterial species. Although this prototype system was effectively a work in progress, it has formed the basis of much

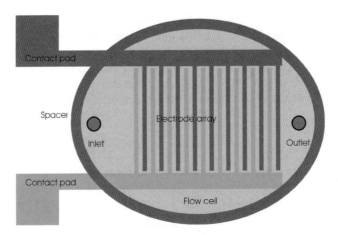

Figure 34.3. In order to use dielectrophoresis for particle separation, interdigitated electrode patterns such as this are used within a flow cell. The flow is orthogonal to the electrodes. Cells experiencing positive dielectrophoresis will collect in the inter-electrode gaps while those experiencing negative dielectrophoresis will pass through the chamber and be collected at the end. Removal of the field then allows recovery of the trapped cells.

technology development since, particularly in the use of dielectrophoresis to bring bacteria to a surface for lysis and subsequent analysis (e.g., Prinz et al. 2002; Yang et al. 2003; Lagally et al. 2005).

In the meantime, work has continued on the development of accurate dielectric models of bacteria in order to understand their electrophysiology. In 1999, Ralph Hölzel used electric fields up to 1GHz in frequency to analyze the electrical properties of the cytoplasm, membrane and cell wall of *E. Coli* with unprecedented accuracy. If the electrophysiology of the cell is known, then it can be used as a baseline to study the effects of agents that might affect it, such as antibiotics (Johari et al. 2003) or variations due to mutation (Castellarnau et al. 2006).

Following the anthrax attacks in New York, Washington and Boca Raton in the fall of 2001, a major driver of dielectrophoresis has been the requirement for improved detection of bacterial pathogens. There are three principal barriers to effective detection in which dielectrophoresis can be of benefit. First, current biosensor technology such as quartz crystal microbalances (QCM), evanescent light scattering, Raman spectroscopy and so forth all require a bacterium be brought to a sensor surface for identification and possible recognition using immunological componentry, which can be enhanced using dielectrophoresis. Second, the lower detection limit of concentration for many sensors is higher than the infectious concentration; that is, the sensors are unable to detect bacteria in concentrations up to two orders of magnitude above the infectious limit. Thirdly, a typical environmental sample is contaminated with non-bacterial material such as dust and diesel particles that can non-specifically bind to a sensor, fouling it and giving false positive results.

A novel method of particle concentration by dielectrophoresis is to use insulating objects between large electrodes to deform an otherwise-uniform electric field, rather than using complex electrode geometries. This was first demonstrated by Junya Suehiro and co-workers using glass beads to trap bacteria (Suehiro et al. 1993a), while a more advanced version that incorporated electrodeless dielectrophoresis to trap, concentrate, filter and transport bacteria was developed at Lawrence Livermore National Laboratory the following year (Lapizco-Encinas et al. 2004a,b).

To address the issue of attracting particles to the surface, a novel method was demonstrated by Hoettges et al. (2003a,b) in which dielectrophoresis and electrohydrodynamic forces were combined to pump particles onto an electrode, where they then collected at the electrode

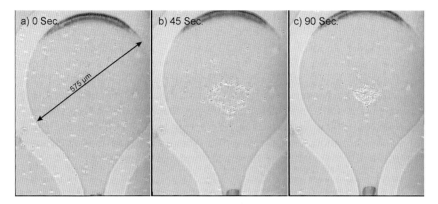

Figure 34.4. The number of spores collected on the inner portion of a $500\,\mu m$ zipper electrode pad, as a function of both particle concentration and time. As can be seen, an observable signal of 3-4 particles is still observable on the pad at concentrations of 5×10^3 particles/ml.

center. By fabricating electrodes with an approximately circular shape, such as the ones shown in Fig. 34.4, they were able to concentrate particles by up to one hundred times. The electrode is designed to ensure the collected bacteria are positioned at the center of the electrode, a position that is ideal for surface-based detection methods, and proved highly effective, forming the basis for other workers to develop the principle with other geometries and pushing to smaller particle sizes (Wong et al. 2004; Wu et al. 2005, Gangon and Chan 2005). Using more conventional dielectrophoretic means, Gomez-Sjoberg and colleagues (2005) concentrated particles for impedance spectroscopy analysis while Castellarneau et al. (2007) used dielectrophoresis to locate bacteria onto an ion-sensitive transistor for glucose monitoring.

For sample preparation, dielectrophoresis allows the selective removal of particles in the inlet stream that are not wanted downstream, such as contaminant particles. This can be achieved in different ways; Fatoyinbo et al. (2007) used conventional dielectrophoretic separation to remove diesel particles from a mixture containing *B. Globigii* spores, whereas James et al. (2006) used advanced microengineering techniques to develop dielectrophoretic "gates" that trap bacteria on electrodes by dielectrophoresis while repelling latex beads by negative dielectrophoresis.

More recently, the most significant contribution to the field of dielectrophoretic detection of bacteria has been Professor Junya Suehiro, whose DEPIM (dielectrophoretic-impedance analysis) has been demonstrated for a wide range of applications. The technique is fundamentally simple; an electric field is applied between two electrodes with a series of narrow inter-electrode gaps (commonly, interdigitated, castellated electrodes are used for this). Dielectrophoretic force is then used to attract bacteria from a sample into this electrode gap, and the change in impedance across the gap gives an indication of the presence and concentration of the bacteria (Suehiro et al. 2003b). Bacterial specificity was achieved using antibodies (Suehiro et al. 2003c; Suehiro et al. 2006).

In addition to the detection and analysis of samples, there are two other significant applications of electrokinetic phenomena to the study and manipulation of bacteria. The first of these is in the analysis of bacterial flagella, and more importantly, in the study of the flagellar motor. Flagellate bacteria use protein "tails" to achieve motion through their environment. These usually consist of protein helices tens of microns long, which are moved in a whip-like manner by a protein rotary motor—the only known rotary motor in nature. Although the proteins have been identified and their structures analyzed, the actual mechanism of motor action is still poorly understood. In order to better evaluate its characteristics, dielectrophoresis (and electrorotation, a related phenomenon of induced rotary motion) has been used to provide a barrier to the motor in order to evaluate stall characteristics (Berry et al. 1995; Berry and Berg 1996; Sugiyama et al. 2004), or force generated in the absence of drag (Washizu et al. 1993; Hughes and Morgan 1999).

The second application of note is the construction of bacterial colonies using dielectrophoresis. Pioneered by Gerard Markx, this approach uses dielectrophoresis to construct complex colonies of co-dependent bacteria by selectively placing them into inter-electrode gaps using dielectrophoretic force, and then immobilizing them on the surface (Alp et al. 2002; Verduzco-Luque et al. 2003; Andrews et al. 2006). This approach offers many benefits for the analysis of colony formation, for the creation of colonies for industrial applications such as drug synthesis, and for potential tissue constructs for tissue engineering.

4. Toward an Integrated Detection System

In the fall of 2001, five people were killed and several more suffered serious infection as a result of receiving letters containing powdered *Bacillus anthracis* spores, more commonly known as anthrax after the disease it causes. The inability to detect the origin of these letters, and the inability to detect which letters contained the spores once the alert had been raised, has highlighted a major weakness in defense programs; whereas chemicals and bombs can be identified by sensors and X-ray detectors, bioweapons are largely undetectable prior to release. With this in mind, there is an evident, urgent need for systems that can rapidly detect harmful bio-agents either before, or immediately after, deployment. Weaponized bacterial spores are commonly released into the air and can remain airborne for very long distances. Since most specific detection methods are based on chemical interactions, such as immunorecognition of some kind, there are difficulties associated with the direct detection of particles in air—hence samples collected can be re-suspended in liquid.

As highlighted in the previous sections, there are a number of problems faced in the detection of pathogens: the issue of selectively removing unwanted particles collected in the sample, a concentration stage to increase the sensitivity of the system to low concentrations of pathogens, and a detector stage where the bacteria can be identified. Dielectrophoresis can be used in all of these stages, and as there is increasing pressure to develop single-device technology for portable bioweapon detection, it can serve many purposes within the same device.

A complete lab-on-a-chip based analysis system resembles the image shown in Fig. 34.5. Here, a single device contains both filtering/preconcentration and detection stages. As described previously, filtering can be achieved using dielectrophoresis to attract contaminant particles to the electrode edges, while bacteria continue on their path through the chamber. Preconcentration can then take place; in one common approach to this, it is the bacteria that are this time attracted to the electrodes by dielectrophoresis. Using a common concentration technique called "trap and purge," particles are held for a period of time, after which they are all released simultaneously. This creates a signal "pulse" that travels along the microfluidic path to the next chamber where the articles are then detected. In terms of efficacy, it has been reported (Fatoyinbo et al. 2007) that contaminant matter can be removed with over 99 % efficiency in controlled laboratory experiments, while trap-and-purge concentration has been reported to increase local concentrations by more than two orders of magnitude.

In the second chamber, the final concentration and detection stage shown here is performed using a *zipper electrode* array; this series of interconnected teardrop shapes is energized at a relatively low frequency (1kHz) to use electrohydrodynamic flow to concentrate particles onto the surface. The zipper array has two advantages; the first is that it sweeps all particles in the chamber above it onto the surface, increasing local concentration by up to another two orders of magnitude. Secondly, it allows surface-based detection methods such as evanescent light scattering or quartz crystal microbalancing to be used. By using many interlocking electrodes in

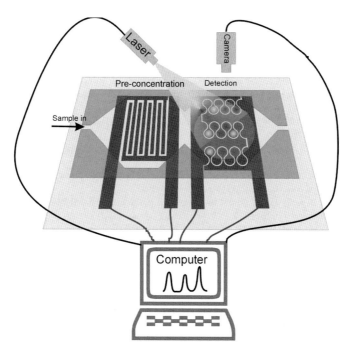

Figure 34.5. Proposed integrated detection system for bioparticles after a liquid sample is introduced into the chip. First a "trap and purge" pre-concentrator traps a range of bioparticles form the sample. After a sampling period the particles are released into a small volume that is passed on to the detector array. Here a number of zipper electrodes are patterned with different antibodies to detect a range of pathogens. A laser excites evanescent light scattering in the electrodes so electrodes that trap particles light up. If a particular pathogen is present a specific pad on the detector lights up and can be detected by an image processing system.

the array, it is possible to use many different antibody-based recognition systems; by applying and releasing the field, bacteria are able to interact with many different electrode "pads."

By using a preconcentration system to enhance the particle concentration delivered to the zipper structures, it may be possible to trap and hold observable numbers of particles with input concentrations of the order of 100 particles/ml. A number of questions remain; no complete system has yet been constructed that adequately meets these needs at a level deemed appropriate by the requisite authorities. However, given the advancement of technology outlined here, the delivery of such a system can only be seen to be imminent.

5. Conclusions and Future Perspectives

Dielectrophoresis—the motion of particles interacting with a non-uniform electric field—and electrohydrodynamics—the interaction of an ionic medium with a non-uniform electric field—both have much to offer the field of bacterial detection and analysis. By treating the bacterial cell as an electronic object and hence considering its electrical properties, it is possible to gain important insights into the electrophysiology of the cell. Furthermore, differences in electrophysiology can be exploited to allow the separation or concentration of bacteria prior to analysis. Finally, devices such as zipper electrodes allow the integration of concentration and focusing directly into the sensor surface. As these technologies all rely on similar devices. They can be integrated into a single lab-on-a-chip device, with many potential benefits including portability and disposability in addition to the ability to detect particles at lower concentrations than ever before.

References

Alp B, Stephens GM, Markx GH (2002) Formation of artificial, structured microbial consortia (ASMC) by dielectrophoresis. Enzyme and Microbial Technology 31:35–43

Andrews JS, Mason VP, Thompson IP et al. (2006) Construction of artificially structured microbial consortia (ASMC) using dielectrophoresis: Examining bacterial interactions via metabolic intermediates within environmental biofilms. Journal of Microbiological Methods 64:96–106

Berry RM, Turner L, Berg HC (1995) Mechanical limits of bacterial flagellar motors probed by electrorotation. Biophysical Journal 69:280–286

Berry RM, Berg HC (1996) Torque generated by the bacterial flagellar motor close to stall. Biophysical Journal 71:3501–3510

Castellarnau M, Errachid A, Madrid C, et al. (2006) Dielectrophoresis as a tool to characterize and differentiate isogenic mutants of Escherichia coli. Biophysical Journal 91:3937–3945

Castellarnau M, Zine N, Bausells J et al. (2007) Integrated cell positioning and cell-based ISFET biosensors. Sensors and Actuators B-Chemical 120:615–620

Cheng J, Sheldon EL, Wu L et al. (1998) Preparation and hybridization analysis of DNA/RNA from E-coli on microfabricated bioelectronic chips. Nature Biotechnology 16:541–546

Fatoyinbo HO, Hughes MP, Martin SP et al. (2007) Dielectrophoretic separation of Bacillus subtilis spores from environmental diesel particles. Journal of Environmental Monitoring 9:87–90

Gagnon Z, Chang HC (2005) Aligning fast alternating current electroosmotic flow fields and characteristic frequencies with dielectrophoretic traps to achieve rapid bacteria detection. Electrophoresis 26:3725–3737

Gomez-Sjoberg R, Morisette DT, Bashir R (2005) Impedance microbiology-on-a-chip: Microfluidic bioprocessor for rapid detection of bacterial metabolism. Journal of Microelectromechanical Systems 14:829–838

Green NG, Morgan H (1998) Separation of submicrometre particles using a combination of dielectrophoretic and electrohydrodynamic forces. Journal Of Physics D-Applied Physics 31:L25–L30

Green NG, Ramos A, Gonzalez A, Morgan H, Castellanes A (2000) Fluid flow induced by nonuniform ac electric fields in electrolytes on microelectrodes. I. Experimental measurements. Physical Review E 61:4011–4018 Part B

Hoettges KF, McDonnell MB, Hughes MP (2003a) Use of combined dielectrophoretic/electrohydrodynamic forces for biosensor enhancement. Journal of Physics D-Applied Physics 36:L101–L104

Hoettges KF, Hughes MP, Cotton A, et al. (2003b) Optimizing particle collection for enhanced surface-based biosensors. IEEE Engineering in Medicine and Biology Magazine 22:68–74

Holzel R (1999) Non-invasive determination of bacterial single cell properties by electrorotation. Biochimica et Biophysica Acta-Molecular Cell Research 1450:53–60

Huang Y, Ewalt KL, Tirado M et al. (2001) Electric manipulation of bioparticles and macromolecules on microfabricated electrodes. Analytical Chemistry 73:1549–1559

Hughes MP, Morgan H (1999) Measurement of bacterial flagellar thrust by negative dielectrophoresis. Biotechnology Progress 15:245–249

Hughes MP (2002) Nanoelectromechanics in Engineering and Biology. CRC Press, Boca Raton

Inoue T, Pethig R, Alameen TAK, Burt JPH, Price JAR (1988) Dielectrophoretic behavior of micrococcus-lysodeikticus and its protoplast. Journal of Electrostatics 21:215–223

Irimajiri A (1979) Dielectric theory of multi-stratified shell-model with its application to a lymphoma cell. Journal of Theoretical Biology 78:251

James CD, Okandan M, Galambos P et al. (2006) Surface micromachined dielectrophoretic gates for the front-end device of a biodetection system. Journal of Fluids Engineering-Transactions of the ASME 128:14–19

Johari J, Hubner Y, Hull JC et al. (2003) Dielectrophoretic assay of bacterial resistance to antibiotics. Physics in Medicine and Biology 48:N193–N198

Lagally ET, Lee SH, Soh HT (2005) Integrated microsystem for dielectrophoretic cell concentration and genetic detection. Lab on a Chip 5:1053–1058

Lapizco-Encinas BH, Simmons BA, Cummings EB et al. (2004a) Dielectrophoretic concentration and separation of live and dead bacteria in an array of insulators. Analytical Chemistry 76:1571–1579

Lapizco-Encinas BH, Simmons BA, Cummings EB et al. (2004b) Insulator-based dielectrophoresis for the selective concentration and separation of live bacteria in water. Electrophoresis 25:1695–1704

Markx GH, Huang Y, Zhou XF et al. (1994) Dielectrophoretic characterization and separation of microorganisms. Microbiology-UK 140:585–591

Markx GH, Pethig R (1995) Dielectrophoretic separation of cells—continuous separation. Biotechnology and Bioengineering 45:337–343

Markx GH, Dyda PA, Pethig R (1996) Dielectrophoretic separation of bacteria using a conductivity gradient. Journal of Biotechnology 51:175–180

Pohl HA (1978) Dielectrophoresis. Cambridge University Press, Cambridge

Price JAR, Burt JPH, Pethig R (1988) Applications of a new optical technique for measuring the dielectrophoretic behavior of microorganisms. Biochimica et Biophysica Acta 964:221–230

Prinz C, Tegenfeldt JO, Austin RH et al. (2002) Bacterial chromosome extraction and isolation. Lab on a Chip 2:207–212

Ramos A, Morgan H, Green NG et al. (1999) AC electric-field-induced fluid flow in microelectrodes. Journal Of Colloid And Interface Science 217:420–422

Suehiro J, Zhou GB, Imamura M et al. (2003a) Dielectrophoretic filter for separation and recovery of biological cells in water. IEEE Transactions on Industry Applications 39:1514–1521

Suehiro J, Hamada R, Noutomi D et al. (2003b) Selective detection of viable bacteria using dielectrophoretic impedance measurement method. Journal of Electrostatics 57:157–168

Suehiro J, Noutomi D, Shutou M et al. (2003c) Selective detection of specific bacteria using dielectrophoretic impedance measurement method combined with an antigen-antibody reaction. Journal of Electrostatics 58:229–246

Suehiro J, Ohtsubo A, Hatano T et al. (2006) Selective detection of bacteria by a dielectrophoretic impedance measurement method using an antibody-immobilized electrode chip. Sensors and Actuators B-Chemical 119: 319–326

Sugiyama S, Magariyama Y, Kudo S (2004) Forced rotation of Na+-driven flagellar motor in a coupling ion-free environment. Biochimica et Biophysica Acta-Bioenergetics 1656:32–36

Verduzco-Luque CE, Alp B, Stephens GM et al. (2003) Construction of biofilms with defined internal architecture using dielectrophoresis and flocculation. Biotechnology and Bioengineering 83:39–44

Wang XB, Huang Y, Burt JPH et al. (1993) Selective dielectrophoretic confinement of bioparticles in potential-energy wells. Journal of Physics D-Applied Physics 26:1278–1285

Washizu M, Kurahashi Y, Iochi H et al. (1993) Dielectrophoretic measurement of bacterial motor characteristics. IEEE Transactions on Industry Applications 29:286–294

Wong PK, Chen CY, Wang TH et al. (2004) Electrokinetic bioprocessor for concentrating cells and molecules. Analytical Chemistry 76:6908–6914

Wu J, Ben YX, Battigelli D et al. (2005) Long-range AC electroosmotic trapping and detection of bioparticles. Industrial & Engineering Chemistry Research 44:2815–2822

Yang JM, Bell J, Huang Y et al. (2002) An integrated, stacked microlaboratory for biological agent detection with DNA and immunoassays. Biosensors & Bioelectronics 17:605–618

35

Ultrasonic Microsystems for Bacterial Cell Manipulation

Martyn Hill and Nicholas R. Harris

Abstract

This chapter introduces the concept of using ultrasound for the manipulation of small particles in fluids for in vitro systems, and in particular how this can be applied to bacterial cells in suspension. The physical phenomena that lead to this effect are discussed, including radiation forces, cavitation, and streaming, thus allowing an appreciation of the limitations and applicability of the technique. Methods for generating ultrasound are described, together with practical examples of how to construct manipulation systems, and detailed examples are given of the current practical techniques of particle manipulation. These include filtration of particles for both batch and continuous systems, concentration of particles, cell washing from one fluid into another, fractionation of cellular populations, and trapping of material against flow. Concluding remarks discuss potential future applications of ultrasonic technology in microfluidic bacterial analysis and predict that it will be a significant tool in cell sample processing, with significant integration potential for Lab-On-Chip technologies.

1. Introduction

The use of ultrasound for medical imaging is well known, but its range of application in health, bioscience, and industrial applications is far wider than that (Leighton 2007). In the field of biosensing, ultrasonics has a number of potential uses (Kuznetsova and Coakley 2007), and this chapter will concentrate on the use of ultrasonics to manipulate micron scale particles, particularly bacterial cells. The majority of the techniques described use the radiation forces generated within an ultrasonic standing wave (USW) as shown diagrammatically in Fig. 35.1. If cells within a chamber are subject to a USW, the cells will first be driven by direct radiation forces to a nodal plane of the standing wave (generally a pressure node) as shown in Fig. 35.1b. Once cells have moved to the nodal plane they will typically be driven more slowly to specific points in that nodal plane (Fig. 35.1c). This agglomeration process is driven by lateral variations in the acoustic field and by secondary, inter-particle effects.

Martyn Hill • School of Engineering Sciences, The University of Southampton, Southampton, S017 1BJ, UK.
Nicholas R. Harris • School of Electronics and Computer Science, The University of Southampton, Southampton, UK.

M. Zourob et al. (eds.), *Principles of Bacterial Detection: Biosensors, Recognition Receptors and Microsystems*,
© Springer Science+Business Media, LLC 2008

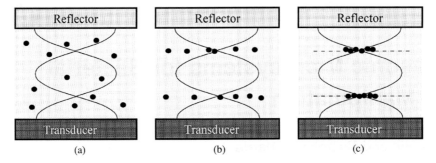

Figure 35.1. Cells within an ultrasonic standing wave **a** will typically be driven to pressure nodal planes within that standing wave **b**, and then tend to agglomerate at particular points within those nodal planes **c**.

1.1. Ultrasound and Bacterial Cells

1.1.1. Cell Viability

High intensity ultrasound can provide an effective means of destroying bacteria (Mason 2007), so an immediate consideration before manipulating bacteria within a USW is whether the viability of the cells will be maintained. There is a substantial body of evidence that suggests that the acoustic powers required to manipulate cells are not sufficient to reduce cell viability significantly.

Doblhoffdier et al. (1994) reported no significant loss in the viability of mammalian cells suspended in a large resonator at different power levels. This finding was reinforced by a more detailed study by Bohm et al. (2000) using a similar chamber and a review of USW cell retention systems from Shirgaonkar et al. (2004). Wang et al. (2004) held mammalian cells in a mesh using USW forces and demonstrated that they retained viability, and Gherardini et al. (2005) established that the viability of yeast cells was maintained for several days following treatment. Yasuda (2000) reported no significant damage to erythrocytes at field strengths high enough to induce mixing, and Hultström et al. (2006) trapped cells as 2D aggregates in a flowing culture medium within a microfluidic chip and showed that their ability to proliferate thereafter was maintained.

1.2. Ultrasound and Microfluidics

Ultrasonic standing waves have been used to manipulate cells and particles on a variety of scales. Frequencies of 1 MHz and above are commonly used to avoid cavitation, and this generates nodal planes (see Fig. 35.1b) that are separated by less than 1mm. This makes USW manipulation particularly attractive for use in microfluidic systems, particularly as particle and cell handling are important operations in many lab-on-a-chip type devices (Vilkner et al. 2004).

2. Relevant Ultrasonic Phenomena

2.1. Axial Radiation Forces

Discontinuities, including particles and cells, within acoustic fields typically experience a small, steady-state radiation force, in addition to the oscillatory acoustic forces. This phenomenon was studied by Rayleigh (1902), among others, and is a property of both progressive acoustic waves and acoustic standing waves (Gröschl 1998a). In general, the forces experienced by small particles in a standing wave are significantly higher than those experienced in a progressive wave, and systems for bacterial manipulation are typically based on an ultrasonic standing wave (USW). Throughout this chapter a "small" particle is defined as one that is significantly smaller than the acoustic wavelength.

King (1934) presented the first full derivation of acoustic radiation forces for small particles within standing waves. The theory was based on rigid spheres in an inviscid fluid, and it predicts that some particles will move toward the nodes of the standing wave while others will be driven to the antinodes. The direction of the force depends on what King calls "the relative density factor," which is a function of the ratio of the fluid and particle densities. For incompressible particles having a density of less than 0.4 of the value of the fluid density, the acoustic force will move particles toward the pressure antinode. For particle densities of above 0.4 of the fluid density (which will generally be true for real near-rigid particles), the acoustic radiation force will act toward the pressure node of the standing wave. King's predictions have proved to be in good agreement with experimental measurements of radiation forces on solid spheres within gases (Leung et al. 1981), but the predictions are significantly in error if the spheres have a compressibility that is not negligible in comparison with that of the fluid.

Yosioka and Kawasima (1955) extended King's theory to allow for bubbles and other compressible spheres. They demonstrated that the time-averaged radiation force on a sphere of radius a, at position x within a one dimensional acoustic standing wave could be expressed as:

$$F(x) = 4\pi k \varepsilon a^3 \Phi(\beta, \rho) \sin(2kx), \tag{35.1}$$

where ε is the acoustic energy density. The *acoustic contrast factor*, $\Phi(\beta, \rho)$, is:

$$\Phi(\beta, \rho) = \frac{\rho_p + \frac{2}{3}(\rho_p - \rho_f)}{2\rho_p + \rho_f} - \frac{\beta_p}{3\beta_f}, \tag{35.2}$$

where β and ρ are respectively the compressibility and the mass density of the fluid and the particle, indicated by subscripts f and p respectively. The wave number, k, is $2\pi/\lambda$ where λ is the wavelength and the compressibility may be expressed in terms of the speed of sound (c) in the medium by $\beta = 1/\rho c^2$.

The graph of $\Phi(\beta, \rho) = 0$ in Fig. 35.2 shows the direction in which a particle will tend to move in a particular fluid. The solution for a rigid sphere (with a compressibility of zero) corresponds with King's theory, but for compressible particles the acoustic force acts toward

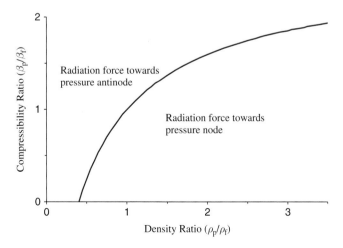

Figure 35.2. Solutions to $\Phi(\beta, \rho) = 0$ showing the boundary between particle combinations that move to the node and those that move to the antinode of a standing wave.

the pressure antinode at higher density ratios. For applications of primary interest here, such as bacterial cells in aqueous solution, the acoustic forces act toward the pressure node.

Gor'kov (1962) formulated the equation for radiation force in terms of the energy stored within a standing wave. He demonstrated that the force on a particle is dependent upon the spatial gradients of the time-averaged kinetic $(E_{kin}(x))$ and potential $(E_{pot}(x))$ energies, as shown below for the one-dimensional case:

$$F(x) = \frac{\partial}{\partial x} \left(\frac{4\pi a^3}{3} \left(\frac{3(\rho_p - \rho_f)}{(2\rho_p + \rho_f)} E_{kin}(x) - \left(1 - \frac{\beta_p}{\beta_f} \right) E_{pot}(x) \right) \right), \qquad (35.3)$$

The full three-dimensional version quoted by Gor'kov may be applied to resonant acoustic fields with a wide variety of geometries.

In predicting the position of particles within a field, it will typically be necessary to equilibrate the acoustic radiation force with forces due to buoyancy and fluid flow. Hence, in the case of horizontal flow through a horizontal resonator, such as that shown in Fig. 35.1, bacterial cells would typically agglomerate just below the acoustic node, due to gravitational effects. Higashitani et al. (1981) also considered diffusive forces on particles within a standing wave due to Brownian motion. This was expressed by using a probability distribution to describe particle concentration gradients around acoustic nodal planes.

The theories of King, Yosioka and Kawasima, and Gor'kov all assume an inviscid fluid. A number of authors, including Westervelt (1951) and Danilov and Mironov (2000) have extended this to allow for viscous fluids. A series of investigations by Doinikov (1997a, b, c) derives expressions for acoustic radiation forces for a number of geometric cases in fluids that are viscous and thermally conducting. It is shown that for small spheres in plane standing waves, corrections to Eq. (1) for viscosity and thermal effects are minor.

The models described above assume isolated small spheres within a standing wave. The influence of the particle, or cell, on the standing wave is local, and the overall energy distribution in the standing wave is assumed to be unchanged by the presence of the cell. However, even single particles may alter the resonant frequency of the standing wave (Leung et al. 1982), and as a uniform suspension of particles agglomerates at a pressure node, the effect on the standing wave characteristics increases and ultimately the standing wave itself may become sufficiently disrupted that the particles will no longer be held at the pressure node.

Another phenomenon that will change the standing wave characteristics is that of temperature change. Large resonators may require forced cooling (Doblhoffdier et al. 1994), but temperature rises reported for small devices (Bazou et al. 2005b) may be less than 0.5 K, although this is highly dependent on the design of the resonator. A change in temperature will lead to a change in acoustic characteristics, and hence the operating frequency, of an acoustic resonator. For small temperature changes in resonators with a broad operating peak (low Q-factor) the operational consequences may be negligible, but other cases may require frequency control to lock on to the required resonance frequency (Gaida et al. 1996; Gröschl 1998b). Such frequency control can also be used if high concentrations of particles cause a significant change in a resonator's operating frequency.

2.2. Lateral and Secondary Radiation Forces

The discussion of Sect. 2.1 assumes an idealized standing wave in which the field does not vary in the lateral plane of a plane axial field. However real acoustic fields will typically have lateral variations in the standing wave field due to phenomena such as a reduction in energy toward the edges of axial fields (Morgan et al. 2004; Bazou et al. 2005a), near field effects (Lilliehorn et al. 2005b), two or three dimensional enclosure modes (Townsend et al. 2006), excitation due to structural modes from the container boundaries (Haake et al. 2005b), or inhomogeneities in the transducer used to excite the standing wave (Gröschl 1998a).

These lateral variations will tend to move particles to preferred points within the nodal planes to which they initially migrate as shown in diagram (c) of Fig. 35.1. In some cases these lateral variations may degrade the performance of a device that is based upon standing wave manipulation (Townsend et al. 2006) while in other devices, the lateral forces may be essential for correct functioning of the device (Doblhoffdier et al. 1994; Bazou et al. 2005a). Particles will typically move rapidly to pressure nodes under the direct axial radiation force and then gather more slowly at certain points within the nodal plane under the influence of the lateral forces. The characteristics of lateral forces within resonators have been measured and compared with forces predicted using Eq. 3 and measurements of the lateral variations of the field (Whitworth and Coakley 1992; Woodside et al. 1997).

In addition to the primary radiations forces (axial and lateral) exerted by the acoustic field on each particle, particles will also experience secondary radiation forces due to acoustic interaction between particles within the field, also known as Bjerknes forces (Leighton 1995). These secondary forces increase the tendency of particles to agglomerate within a standing wave but are negligible unless the particles are very close together.

2.3. Acoustic Streaming

A simple linear concept of an ultrasonic field suggests that the acoustic energy causes elements of a fluid to oscillate around a mean position but does not induce any time-averaged momentum. However in reality, losses within the fluid or slip at fluid boundaries will induce steady state movement of that fluid, known as acoustic streaming (Lighthill 1978; Riley 2001). Zarembo (1971) identifies three types of streaming:

- Bulk fluid flow due to losses within the fluid itself, known as "Eckart streaming";
- Vortices generated on a wavelength scale within standing waves, known as "Rayleigh streaming";
- Microstreaming vortices within viscous boundary layers, known as "Schlichting streaming."

Any of the above may be observed in ultrasonic standing waves within microsystems, although it is Rayleigh and Schlichting streaming that are likely to predominate in microsystems with critical dimensions of less than a wavelength. Kuznetsova and Coakley (2004) studied the movement of particles under radiation forces and streaming currents in a sub-wavelength resonator. They observed that particles within ultrasonic standing waves moved rapidly to the pressure nodal planes, and that large particles (24 and 80 µm diameter) moved rapidly to the pressure node and were then concentrated at particular points within that field by lateral radiation forces. However, having moved to the nodal plane, the behavior of particles of 1 µm diameter was dominated by streaming convection. Much of the behavior can be attributed to Rayleigh streaming behavior, but the authors also note the existence of vortices rotating in a plane parallel to the transducer (i.e., perpendicular to the axis of the acoustic field) similar to the vortex shown in Fig. 35.3. This suggests that the minimum size of particles that can be manipulated by radiation forces may be limited by the susceptibility of smaller particles to streaming-induced drag forces. This has clear implications for the transport of small bacteria, with their successful manipulation becoming increasing difficult for cells smaller than 1 µm. Kuznetsova et al. (2005) demonstrate that the streaming patterns can be used to deposit sub-micron particles on an immuno-coated surface, enabling the capture of virus-scale 200 nm diameter particles to be significantly increased, and hence overcoming the normal diffusion-limited capture rate.

The ability of streaming forces to generate fluid movement can also be of benefit in systems that are otherwise highly laminar. Streaming can be used to overcome mass transport

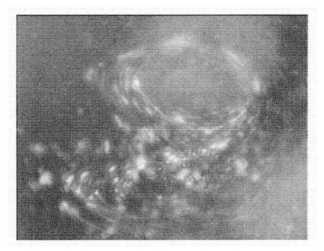

Figure 35.3. Streaming vortex observed in a sub-wavelength resonator, rotating anti-clockwise in a plane perpendicular to the axis of the acoustic field, similar to those described by Kuznetsova and Coakley (2004).

limitations and hence increase the reaction rate of biosensors. Wang et al. (2004) observed that the agitation created by a flexural plate wave gravimetric sensor reduced the time required to perform immunoassay tests for breast cancer antigens due to localized mixing. Streaming can also be used to increase the mixing rate of fluids within a microfluidic system (Bengtsson and Laurell 2004; Sritharan et al. 2006).

2.4. Cavitation

When a high amplitude ultrasonic field is excited within a liquid, a process known as cavitation may result (Leighton 1994). Cavitation is the generation of voids or bubbles within a liquid, during the low pressure period of the ultrasonic cycle. Typically these voids or bubbles will grow from pre-existing micro-bubbles and then collapse, and in so doing, release substantial quantities of localized (intense) energy. Cavitation is the dominant mechanism in ultrasonic cleaning, in which objects to be cleaned are placed in a solvent in an ultrasonic bath. The cavitation energy helps to loosen and remove impurities on the object. In general, inducing cavitation is undesirable when manipulating cells as it will disrupt the ordering of cells due to radiation forces and will tend to damage the cells. The onset of cavitation requires higher acoustic energy densities as the driving frequency increases, and this is an advantage of working in the MHz rather than the high kHz frequency domain when manipulating cells.

3. Applications of Ultrasonic Particle Manipulation

3.1. Practical Considerations

3.1.1. Transduction

The means of generating the ultrasonic field is an important consideration in the design of a resonator for cell manipulation. While ultrasound is defined as sound at frequencies above the threshold of hearing (usually taken to be 20 kHz), this frequency translates to a wavelength in water of 75mm and so is of a scale too large to be used in microfluidic devices. Typically frequencies of 1MHz or greater are used, helping to avoid cavitation and giving wavelengths of 1.5mm or less in water.

For the systems described in this chapter, the most common means of ultrasonic transduction utilizes a piezoelectric material. Piezoelectric materials are those which exhibit a strain with applied voltage, and conversely, an applied stress causes a charge to be released. Therefore if an alternating voltage is applied to a piezoelectric material, it will expand and contract in sympathy with this signal. This will cause the surrounding medium to move as well, and this is the basis of sound generation.

The most common material used as a transducer is PZT (Lead Zirconate Titanate), originally developed for use in sonar systems. It is a ceramic material with ferroelectric properties. Ferroelectric materials have dipoles that can be switched, allowing versatility in the shapes that can be made. The most common form is as plates or cylinders made from PZT powder that has been sintered, but other techniques such as screen printing are available (Lilliehorn and Johansson 2004; Harris et al. 2006). Once the material has been sintered, it requires polarization to align the domains in the material, and this is achieved by applying heat and a high electric field strength. This alignment of the dipoles gives the material an overall piezoelectric effect, and is similar to the process used to create magnets.

It is worth noting at this point that having used heat to impart the overall piezoelectric effect, heat can also be used to remove it. If a PZT material is heated too far, it depoles, and loses its effectiveness. This effectiveness is lost permanently—it does not return on cooling. The temperature at which this effect occurs is called the Curie temperature, and its value depends on the grade of PZT used.

There are many different grades of PZT, but these can be broken down into "hard" and "soft" grades (Rosen et al. 1992). The hard grades tend to be optimized for high power operation, particularly for sonar applications, and the soft grades are better suited to act as receivers, being optimized for their ability to respond accurately to incident acoustic energy (hydrophones). It must be said that there is no simple rule governing the choice of grade in many applications, and at low powers or for devices that combine a transmit and receive function, a soft grade may be better suited.

3.1.2. Mechanical Effects

Once movement is imparted to a PZT element, it will respond as a mechanical structure, and so may exhibit resonant characteristics depending on its physical dimensions compared with the wavelength of the operating frequency in the PZT itself. For example, the first thickness mode resonance of a disc or plate occurs when the plate is half a wavelength thick. Resonance can be used to magnify the movement of the transducer but can be a double-edged sword. If the transducer is being used to drive another system at this resonance, a coupling can occur between the two systems that can reduce the transmitted energy.

The standard model for a transducer operating near resonance consists of modelling the transducer as a series RLC (resistor, inductor, capacitor) circuit in parallel with a static capacitance (Zelenka 1986). More complex models can be used. For example 3-port models with either distributed or lumped parameters can model transient effects more accurately (Mason 1942; Krimholz et al. 1970; Sherrit 1999), but the standard RLC model is usually sufficient for narrow band continuous wave operation. It should be noted that bonding the transducer to the device needing to be actuated will affect these resonance conditions, so care is needed when making predictions of the efficiency of such structures.

3.1.3. Construction

It is common practice to bond PZT transducers to the device requiring actuation. Although this is acceptable, it does introduce some variability from device to device as an extra layer (glue) is introduced into the system, and this has been shown to have an effect on performance (Hill et al. 2002). It is also a relatively inefficient process, as bonding tends to be done on an individual basis, but for prototypes it represents the most convenient way of producing a device. An alternative production method is to use a thick-film deposition method. The thick-film process involves depositing pastes onto a carrier layer and then firing the pastes to create a hardened film. Usually the pastes take the form of conductors, resistors or dielectrics, but various special purpose pastes have been developed. One of these is a piezoelectric paste based on powdered PZT. This allows the actuator to be printed directly onto the fluidic device without the need to use a glue layer. The thick film process imposes some limitations on the choice of materials that can be used as a carrier layer, as traditional thick-film processing is a high temperature process (firing occurs at 800 degrees centigrade or above), but materials that have been successfully used include alumina ceramic, silicon, and certain stainless steels.

Figure 35.4 shows a thick-film PZT actuator printed on silicon. In this case the actuator is a 2-layer device, which gives improved performance over a single layer of equivalent thickness (Harris et al. 2004). The thick-film technique has been used to screen print 2-layer PZT actuators onto silicon for use in an ultrasonic clarifier (Harris et al. 2006).

An interesting extension to this is to produce free-standing thick-film transducers, where the substrate is removed after firing (Stecher, 1987), and a novel way of creating arrays of free standing transducers has been achieved by Lilliehorn et al. (2004). These transducers allow the creation of arrays as each can have dimensions of less than $500 \times 500\mu m$ and thickness resonances of between 10 and 15 MHz. The multilayer nature of these devices also results in low drive voltages being necessary. Such transducers have been used for ultrasonic particle trapping systems (Lilliehorn et al. 2005b).

The transducers described in the preceding paragraphs are used for directly generating acoustic energy in the fluidic device. There is another class of transducer that generates sound in fluids indirectly by using structural waves (such as surface waves or flexural waves) within a solid boundary that couple into the fluid. Although the field generated is evanescent, this does not preclude operation when the fluid chamber is small, and this type of excitation is well suited to microfabricated devices (Black et al. 2002). Haake and Dual (2004) excite bending waves in a glass plate that couples directly into a thin fluid layer. Under these conditions an acoustic wave is emitted into the fluid and is then reflected by the bottom of the channel, setting up a complex interaction between the modal behavior in the fluid and the plate. By modeling

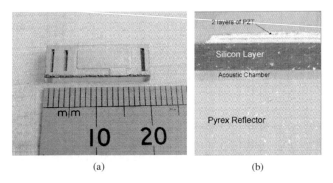

(a)   (b)

Figure 35.4. A two-layer PZT transducer on a silicon/Pyrex separator. **a)** View of whole microfluidic device. **b)** Cross section showing pzt layers.

dispersion relationships between the plate and fluid modes they establish conditions, in terms of the thickness of the plate and the operating frequency, that are required to establish the required acoustic pattern in the fluid. In a later paper the authors discuss different methods of controlling the sound field (Haake and Dual 2005) allowing particles to be levitated and then moved to specific lateral positions. The group has also demonstrated the movement of biological cells into lines and points without signs of significant cell damage (Haake et al. 2005a; Haake et al. 2005b), have modeled the field resulting from the fluid-structure interaction (Neild et al. 2007) and have developed a microfabricated device which allows the manipulation of particles using this technique prior to harvesting, using a microgripper (Neild et al. 2006).

3.2. Filtration and Fractionation of Cells

The ability to move particles under the influence of acoustic fields allows several effects to be achieved. These include

- Filtration and concentration (sedimentation, laminar flow filtration, ultrasonically enhanced filtration);
- Cell Washing;
- Fractionation;
- Cell Trapping.

The majority of the techniques described here discuss the manipulation of cells within a fluid, but USWs can also be used to concentrate, agglomerate and filter aerosols (Kaduchak et al. 2002; Gonzalez et al. 2003).

3.2.1. Filtration and Concentration

The potential of ultrasonic radiation forces to filter particles has been investigated by a number of authors. This can take two main forms:

- Removing unwanted particles from a fluid medium;
- Concentrating particles into a smaller volume.

Both of these processes have been reported in batch mode and in continuous mode. Although ultrasound has been used to enhance membrane filtration (see, for example, Caton and White (2001), in which ultrasound has been used to clean a filter element), this section will describe filtration techniques that rely primarily on ultrasonic radiation forces.

Filtration and concentration have very obvious applications in microfluidic systems for the detection and concentration of bacterial cells. For example, sample pre-treatment is commonly required for real samples contaminated by unwanted particulates. When a barrier filtration method is used, with the associated fouling and replacement of the barrier, the ability of microfluidic analysis systems to operate without intervention is by necessity limited. In addition its operating parameters may also change over time, as the pressure drop needed to overcome the barrier will rise as the barrier becomes clogged. USW filtration offers the opportunity to remove unwanted particles in a continuous, flow-through system, without the need for barriers. In addition, the potential to concentrate samples offers, for near neutrally buoyant samples in particular, a microfluidic equivalent of a centrifuge. Flow-through systems that manipulate particles within the flow can be used either as concentrators or as filters, and the techniques described below can be used as either.

3.2.1.1. Sedimentation

Enhanced sedimention was the most commonly reported ultrasonic filtration technique up to the late 1990s. This technique relies on particles agglomerating under ultrasonic standing wave forces as shown in Fig. 35.5. As agglomerates grow, the relative importance of gravitational forces over viscous drag forces grows until the clumps of particles sediment under gravity. Alternatively, once particles have been trapped, the field can be removed and the agglomerate of particles falls.

Although as described, this technique has obvious application in batch processing, it can be modified to work as a flow-through device. Examples of batch processing are given by Schram (1991) who describes a number of applications of this technique, including the use of orthogonal crossed standing waves to enhance agglomeration. Cousins et al used a similar technique for preparation of blood plasma (Cousins et al. 2000a; Cousins et al. 2000b) and for the concentration of bacteria (Cousins et al. 2000c).

A flow-through system requires careful fluidic design. Flow can be introduced from the left hand side at the bottom of the chamber shown in Fig. 35.5, with an exit at the top, so flow is in the vertical direction. As the flow progresses, the particles are trapped in the field and a clarified flow exits from the top. Once the agglomerates have achieved a certain critical size, they fall under gravity, past the fluid inlet. Such a system has been developed commercially, again on a macro-scale, for the retention of cells within a bioreactor (Gaida et al. 1996).

This technique reduces in efficiency as the size of particles of interest falls. Hawkes and Coakley (1996) show high separation efficiencies (up to 99 %) in a smaller device designed to have a small dead volume, and this was very successful with yeast cells. However, sedimentation of smaller bacterial cells was less successful, primarily as the sedimentation relies on particle agglomeration under USW forces overcoming the Stokes drag forces due to the fluid flow, which tends to dominate for smaller particles (Hawkes et al. 1997). However, this selectivity may make such devices attractive for pre-filtering applications in microfluidic systems for bacterial handling.

3.2.1.2. Laminar Flow Filtration

Another approach to filtration uses radiation forces to gather particles to a plane or planes within a channel's cross section and then to draw those particles off from a concentrated outlet and the clarified fluid from another outlet. This can be achieved on a large scale by holding particles within multiple nodal planes and allowing the fluid to flow diagonally through the resonator (Frank et al. 1993; Hill and Wood 2000). Then the particles are moved along the nodal planes towards one outlet and the now clarified flow crosses the nodes towards another. However working on a microfluidic scale lends itself to working with channels that are less than a wavelength and therefore contain a single nodal plane, so the cross flow technique has less application (Hawkes et al. 1998a; Hawkes et al. 1998b).

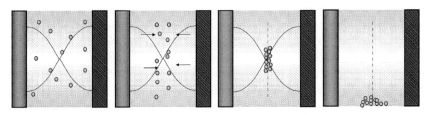

Figure 35.5. Enhanced sedimentation. Particles within an ultrasonic field (left) are agglomerated by the field (center) and then tend to sediment under gravity (right).

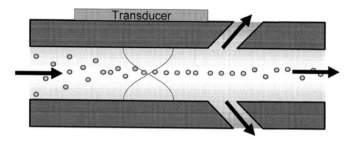

Figure 35.6. Half wavelength laminar flow filter.

A single node system is shown in Fig. 35.6. Particles are carried by a fluid into a channel containing a half-wavelength standing wave. Particles are forced to the nodal plane at the center of the chamber and, once out of the acoustic field, stay in the same position in the channel cross section due to the laminar nature of microfluidic systems. Thus, concentrated particles may be drawn off from the center of the flow and clarified fluid may be drawn from either the upper or lower outlet, or from both as shown here. Such a device was demonstrated by Hawkes and Coakley (2001) using a channel of 250 μm deep, machined from stainless steel and driven at about 3 MHz. The filter was tested with latex particles of between 1.5 and 25 μm and with yeast cells. Up to 1000-fold clarification was achieved with yeast.

This principle has also been demonstrated in a microfabricated flow through filter (Harris et al. 2003). Channels were wet etched into a Pyrex and silicon structure, so flow connections were only required on the lower face. Variations on the same basic design used both bulk PZT (Harris et al. 2005) and thick film PZT excitation (Harris et al. 2006). This microfabricated device achieved a 50-fold clarification of yeast cells, but only a 5-fold clarification of 1 μm latex beads, a limitation in terms of applicability to bacterial handling. The design of this filter also involved an investigation into different micro-channel geometries and their effect on the system's operation (Townsend et al. 2005).

Petersson et al. (2004, 2005a) also used the principle of half-wave filtration in a device designed to filter lipids from whole blood. This filter was based upon the different forces on red blood cells and lipids. Due to the different acoustic contrast factors, the red cells move to pressure nodes, while lipids move to the pressure antinodes. Hence the lipids are driven to the upper and lower faces of a channel such as that shown in Fig. 35.6 while the red cells are concentrated in the center. Over 70 % of the red cells were collected in one third of the original fluid volume with more than 80 % of the lipid particles being removed. The 350 μm wide channels were anisotropically etched into silicon, and the transducer was mounted in a plane perpendicular to the direction of the standing wave, enabling several channels to be excited from a single transducer. The authors envisage that the system could be scaled up to between 100 and 200 channels to achieve a surgically useful throughput.

3.2.1.3. Filtration Using Ultrasound Within a Porous Mesh

Radiation forces within a USW field may be used to modify the filtration characteristics of a porous mesh (Gupta and Feke 1998; Wang et al. 2004; Grossner et al. 2005). Particles or cells within a fluid flow through a polyester mesh that has a pore size of up to two orders of magnitude larger than the particles. Hence they pass through the mesh without being trapped. However when the fluid passing through the mesh is excited ultrasonically, acoustic radiation forces hold particles on the mesh elements, creating a filter. The mesh may then be cleaned by turning off the ultrasonic power and releasing the particles.

3.2.1.4. Cell Washing

An interesting particle washing technique is made possible by the laminar flow characteristics inherent in microfluidic devices. Two fluids may be brought into contact and can flow
alongside each other with no appreciable mixing (see Fig. 35.7). If the acoustic system is then
designed so that a node is positioned within fluid A, cells in fluid B can be moved by radiation
forces into fluid A while the two fluids remain largely unaffected. The fluids can then be
separated at the outlets and with the cells moving from fluid B to Fluid A on a continuous basis.

Hawkes et al. (2004a) have demonstrated a washing rate of 80 % using yeast particles.
In the device described, care needs to be taken when using higher powers as this can lead to
mixing fluid mixing due to streaming effects. This feature, in itself, may prove to be useful, as
mixing at a microfluidic scale has traditionally been a difficult task.

Blood washing using a similar approach has been investigated by Petersson et al.
(Petersson et al. 2005b). Initial tests were carried out with polyamide spheres in water, which
demonstrated a 95 % exchange efficiency, and when bovine blood was used, 98 % of the red
cells were "washed" into clean plasma. For the chamber used in this experiment, the flow rate
was of the order of 0.27 ml min^{-1}.

3.2.2. Fractionation of Cells

Fractionation allows a mix of different cells to be sorted according to a physical property
such as size or density. Ultrasound has the potential to fractionate on a continuous basis by
exploiting the feature that different size cells move at different speeds in a given acoustic field
(Masudo and Okada, 2006). One process that has been used with some success is field flow
fractionation, where the residence time of a population of particles is proportional to their size
(Giddings, 1993).

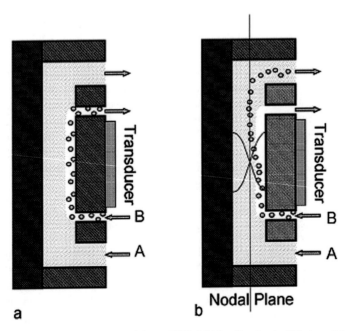

Figure 35.7. The principle of cell washing. In (**a**), the USW field is off and only diffusion mixing applies. In (**b**),
radiation forces drive cells across the fluid boundary (based on Hawkes et al. 2004a).

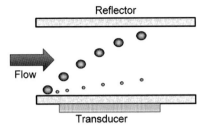

Figure 35.8. The principle of fractionation. Large particles experience larger forces than small particles.

For cells with similar acoustic properties, different sized cells will experience different forces. However, once a cell starts to move, it will also be affected by drag forces, which are also dependent on its size. Smaller cells have lower terminal velocities than larger ones, so a differential velocity is created, leading to separation. Aboobaker et al. (2003) make some predictions and measurements of particle velocity against size. Hence for a mixture of different sized particles, it is possible to differentiate particles by their velocity in a standing wave system, as indicated in Fig. 35.8.

Kumar et al. (2005) describe a continuous fractionator in which an ultrasonic field is used to drive a concentration of particles near one wall across the width of the chamber, similar to Fig. 35.8. For this to work, it is important that the particles all start off from the same point, so some mechanism for introducing the particles at the bottom of the chamber is necessary. At the outlet of the chamber, the particles are evenly spread across the height in size order if the residence time is correct. This requires a balance between residence time in the chamber and lateral particle velocity. If the residence time is too long, then all the particles will arrive at the same wall, albeit with the larger ones having arrived first. A variation of this technique, which has similarities to the particle washing concept above, is described by Kapishnikov et al. (2006). Larger particles are forced into a second fluid by a USW and are hence separated from smaller particles. Two transducers are used, and the phasing between them is altered to allow the node to be arbitrarily positioned. This work separated red blood cells from smaller cells in a plasma.

Other approaches to fractionation have been described in which the acoustic force is balanced against a force other than gravity. Electrostatic forces were used in addition to acoustic forces by Yasuda et al. (1996), and $10\,\mu$m and $20\,\mu$m polystyrene particles were separated from each other by balancing electrostatic forces against radiation forces. Wiklund et al. (2003) combined ultrasound with capillary electrophoresis to achieve separation for improving the concentration limit of detection in small samples.

3.2.3. Trapping of Cells

An ultrasonic field can be used to trap and hold a cell or cell agglomerate at a particular point within a fluid, providing an alternative to an optical or laser trap (Johann 2006). Hertz (1995) developed an ultrasonic trap based on a focused ultrasonic standing wave maintained with a pair of opposing concave transducers. The focusing generates a high lateral gradient in the acoustic field, causing particles or cells to be held at the transducer focus within a pressure node. This approach has been used to trap cells for fluorescence analysis within capillaries (Wiklund et al. 2001) and 96 well plates (Wiklund et al. 2004).

An alternative to the use of a focused field is described by Wiklund et al. (Wiklund 2006). A PMMA wedge refracts ultrasound from a PZT plate into a microfluidic channel formed between two plates of Pyrex. The field is such that the standing wave is set up across the width (rather than the depth) of the channel, with nodes running parallel to the direction of fluid flow. These nodes can be used to restrain particles laterally while further trapping and manipulation is then brought about using dielectrophoretic electrodes that cross the ultrasonic nodes. The combination of the ultrasonic and dielectrophoretic forces allows the authors to perform a variety of trapping and manipulation operations (trapping, sorting, selective switching between nodes, and fusion of different particle groups) on cells and particles of bacterial scale.

Also on a microfluidic scale, Lilliehorn et al. (2005b) used a thick-film technique to create sub-mm, multilayer, ultrasonic transducers, resonant at about 10 MHz. These were placed at selected points within microfluidic channels, and when excited, lateral energy variations in the near-field of the transducers trapped 5 μm particles against flows of up to 1 mm s^{-1}. A microfluidic array, based on this principle, has the potential to be expanded into a highly parallel protein analysis system (Lilliehorn et al. 2005a).

The use of ultrasonic standing wave fields to force particles into close proximity (and simultaneously increase mass-transfer between them by streaming) provides significant potential for enhancing bead-based immunoassays (Wiklund and Hertz 2006). This has been used to increase the detection rate of the Hepatitis B virus (Grundy et al. 1989) and Legionella pneumophila (Jepras et al. 1989). The technique has been shown to increase the speed of an agglutination test for meningitis based on antigen coated latex beads (Jenkins et al. 1997; Barnes et al. 1998; Gray et al. 1999; Sobanski et al. 2002). Thomas et al. (1999) investigated the detection rate and the analyte concentration limit of such particle agglutination immunoassays within USWs and demonstrated that coated silica beads performed significantly better than coated latex beads of similar size. A subsequent study (Sobanski et al. 2000) further investigated the influence of bead size, resonator geometry and acoustic streaming on USW assisted agglutination tests.

Sub-wavelength USW traps have also been used in cell–cell interaction studies (Bazou et al. 2004; Bazou et al. 2005a). Cell monolayers can be formed and suspended within a field so their interaction is not modified by adjacent boundaries.

Another reason for trapping cells may be to enhance their interaction with the surrounding fluid, rather than to enhance their interaction with each other. Once a cell, or cell agglomerate, is trapped, it can be held while fluid flows over it. Interaction is further enhanced by the presence of acoustic streaming. Morgan et al. (2004) showed that testing the toxic effects of a flowing toxicant on a multi-cellular spheroid can be accelerated within a trap.

A trap can also be used to increase the uptake of nanoparticles, such as retroviruses, by cells. The retroviruses are not held by the radiation forces that constrain the target cells, allowing an acceleration of genetic transfection (Lee and Peng 2005). If the amplitude of the ultrasound is increased, and particularly if a contrast agent is introduced, the onset of cavitation can further enhance transfection within a USW trap in the process of sonoporation (Khanna et al. 2003; Khanna et al. 2006).

3.3. Biosensor Enhancement by Forcing Cells to a Surface

Typically a fluid layer with plane, parallel, rigid boundaries would be expected to resonate at a frequency corresponding to a half wavelength and integer multiples thereof. At these frequencies the boundaries coincide with pressure maxima and velocity minima. However real boundaries are of finite extent and show their own resonant characteristics. Over certain frequency ranges, a finite thickness boundary will not offer a near zero-velocity boundary condition to the adjacent fluid. In such cases, the boundary will not be at a point of maximum pressure in a standing wave and could coincide with a pressure node (Hill 2003). When this occurs, typically when the boundary wall is at a half wavelength resonance, cells will be forced

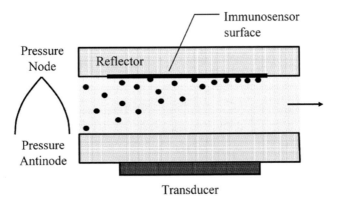

Figure 35.9. The use of a "quarter wave resonator" to force cells onto a biosensor surface.

toward the boundary, providing the potential to transport particles to a sensor surface and enhance the sensor system's sensitivity. Hawkes et al. (2004b) used an arrangement similar in principle to that shown in Fig. 35.9. A particle suspension flows into the chamber shown in the figure from the right. The chamber had a fluid layer of about 200 μm, a steel carrier layer and a glass reflector. A quarter wavelength resonance was excited in the chamber with a pressure node near the boundary with the reflector. The chamber was tested with both 2.8 μm diameter latex beads and *bacillus subtilis* var. *niger* (BG) spores. The surface of the reflector was coated with a BG antibody and the spores forced onto the surface by the USW. The authors report that the capture efficiency of the spores with the USW field activated was about 200 times that of the capture in the absence of ultrasound. Zourob et al. (2005) extended this principle by fabricating an optical metal-clad leaky waveguide (Zourob et al. 2003) on the reflector, which was in turn coated with the required antibody. This system was also tested using BG spores, and once again the efficiency of the system was significantly improved by the USW. It was shown that the detection limit for 3-minute exposure to BG spores was 10^3 spores/mL.

Martin et al. (2005) investigated the acoustics of a similar system in more detail. In this work, BG spores were forced onto an antibody coated surface using a 3 MHz USW. The system was tested in batch and flow-through modes, and it was found that the efficacy of capture was critically dependent on the thickness of the reflector layer. When a 980 μm thick reflector was used, there was almost no capture of the BG spores. Capture increased with a 1000 μm reflector, peaked with a thickness of about 1100 μm and had fallen away significantly with a reflector thickness of 1300 μm. The reason for this was revealed when the field was simulated. Fig. 35.10a shows 1D simulations of the acoustic pressure for different reflector thicknesses. With a thickness of 980 μm, the pressure node is in the fluid, away from the reflector boundary, so particles would be moved away from the antigen surface. With a 1000 μm reflector, the node is just in the reflector, so particles will be forced to the surface. A 1200 μm reflector places the node well into the reflector, but also brings a pressure antinode into the fluid, causing many particles to be force to the opposite boundary. Hence there is an optimum positioning of the node that is dependent on the reflector thickness and a significant decrease in capture efficiency on each side of this thickness. The authors added flow and particle tracking representation to the acoustic model (Townsend et al. 2004) and were able to predict the nature of the dependence of particle capture on reflector depth, as shown in Fig. 35.10b.

A similar approach (Kuznetsova et al. 2005) uses streaming patterns to increase the capture of sub-micron particles that are too small to be manipulated by radiation forces (see Sect. 2.3 above).

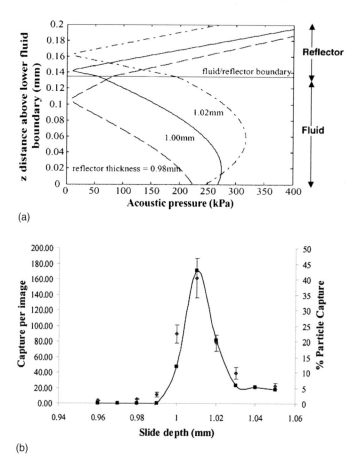

Figure 35.10. Acoustic simulation of the pressure profile across the chamber used by Martin et al. (2005) for different reflector thicknesses (**a**), and simulations of particle capture (line) compared with experimental data (**b**). (Reprinted from Martin et al. (2005), with permission of Elsevier).

4. Conclusions and Future Perspectives

The use of ultrasonic standing waves within microfluidic biosensor systems has significant potential. The magnitude and the scale of action of the forces generated make the approach appropriate for the trapping and manipulation of cells and cell agglomerates. The ultrasonic amplitudes required for manipulation do not appear to be detrimental to cell viability, the power requirements are low, and with appropriate design heating effects are small. Furthermore, the technology is relatively easy to integrate within microsystems and is complementary to techniques such as optical trapping and dielectrophoresis. The ability to increase mass transfer, cell-cell interaction, and cell-marker interaction by the inducing of acoustic streaming is a significant additional advantage, and may well prove to be one of the most important factors in the use of ultrasound in microfluidic biosensors in the future.

References

Aboobaker N, Meegoda JN, Blackmore D (2003) Fractionation and segregation of suspended particles using acoustic and flow fields. J. Environ. Eng.-ASCE 129:427–434

Barnes RA, Jenkins P, Coakley WT (1998) Preliminary clinical evaluation of meningococcal disease and bacterial meningitis by ultrasonic enhancement. Arch. Dis. Child. 78:58–60

Bazou D, Coakley WT, Meek KM, Yang M, Pham DT (2004) Characterisation of the morphology of 2-D particle aggregates in different electrolyte concentrations in an ultrasound trap. Colloid Surf. A-Physicochem. Eng. Asp. 243:97–104

Bazou D, Foster GA, Ralphs JR, Coakley WT (2005a) Molecular adhesion development in a neural cell monolayer forming in an ultrasound trap. Molecular Membrane Biology 22:229–240

Bazou D, Kuznetsova LA, Coakley WT (2005b) Physical environment of 2-D animal cell aggregates formed in a short pathlength ultrasound standing wave trap. Ultrasound In Medicine And Biology 31:423–430

Bengtsson M, Laurell T (2004) Ultrasonic agitation in microchannels. Analytical And Bioanalytical Chemistry 378:1716–1721

Black JP, White RM, Grate JW (2002) Microsphere capture and perfusion in microchannels using flexural plate wave structures. IEEE, Munich, Germany, p 475

Bohm H, Anthony P, Davey MR, Briarty LG, Power JB, Lowe KC, Benes E, Groschl M (2000) Viability of plant cell suspensions exposed to homogeneous ultrasonic fields of different energy density and wave type. Ultrasonics 38:629–632

Caton PF, White RM (2001) MEMS microfilter with acoustic cleaning. 14th IEEE International Conference on Micro Electro Mechanical Systems (Cat. No.01CH37090). IEEE, Interlaken, Switzerland, pp 479–482

Cousins CM, Holownia P, Hawkes JJ, Limaye MS, Price CP, Keay PJ, Coakley WT (2000a) Plasma preparation from whole blood using ultrasound. Ultrasound in Medicine and Biology 26:881–888

Cousins CM, Holownia P, Hawkes JS, Price CP, Keay P, Coakley WT (2000b) Clarification of plasma from whole human blood using ultrasound. Ultrasonics 38:654–656

Cousins CM, Melin JR, Venables WA, Coakley WT (2000c) Investigation of enhancement of two processes, sedimentation and conjugation, when bacteria are concentrated in ultrasonic standing waves. Bioseparation 9:343–349

Danilov, S.D., Mironov, M.A., 2000. Mean force on a small sphere in a sound field in a viscous fluid. Journal of the Acoustical Society of America 107, 143–153.

Doblhoffdier O, Gaida T, Katinger H, Burger W, Groschl M, Benes E (1994) A Novel Ultrasonic Resonance Field Device for the Retention of Animal-Cells. Biotechnology Progress 10:428–432

Doinikov AA (1997a) Acoustic radiation force on a spherical particle in a viscous heat-conducting fluid .1. General formula. Journal of the Acoustical Society of America 101:713–721

Doinikov AA (1997b) Acoustic radiation force on a spherical particle in a viscous heat-conducting fluid .2. Force on a rigid sphere. Journal of the Acoustical Society of America 101:722–730

Doinikov AA (1997c) Acoustic radiation force on a spherical particle in a viscous heat-conducting fluid .3. Force on a liquid drop. Journal of the Acoustical Society of America 101:731–740

Frank A, Bolek W, Groschl M, Burger W, Benes E (1993) Separation of suspended particles by use of the inclined resonator concept. Ultrasonics International 93, Conference Proceedings, pp 519–522

Gaida T, DoblhoffDier O, Strutzenberger K, Katinger H, Burger W, Groschl M, Handl B, Benes E (1996) Selective retention of viable cells in ultrasonic resonance field devices. Biotechnology Progress 12:73–76

Gherardini L, Cousins CM, Hawkes JJ, Spengler J, Radel S, Lawler H, Devcic-Kuhar B, Groschl M (2005) A new immobilisation method to arrange particles in a gel matrix by ultrasound standing waves. Ultrasound In Medicine And Biology 31:261–272

Giddings JC (1993) Field-flow Fractionation: Analysis of Macromolecular, Colloidal, and Particulate Materials Science 260:1456–1465

Gonzalez I, Gallego-Juarez JA, Riera E (2003) The influence of entrainment on acoustically induced interactions between aerosol particles - an experimental study. Journal of Aerosol Science 34:1611–1631

Gor'kov LP (1962) On the forces acting on a small particle in an acoustical field in an ideal fluid. Sov. Phys. Dokl. 6:773–775

Gray SJ, Sobanski MA, Kaczmarski EB, Guiver M, Marsh WJ, Borrow R, Barnes RA, Coakley WT (1999) Ultrasound-enhanced latex immunoagglutination and PCR as complementary methods for non-culture-based confirmation of meningococcal disease. J. Clin. Microbiol. 37:1797–1801

Gröschl M (1998a) Ultrasonic separation of suspended particles - Part I: Fundamentals. Acustica 84:432–447

Gröschl M (1998b) Ultrasonic separation of suspended particles - Part II: Design and operation of separation devices. Acustica 84:632–642

Grossner MT, Belovich JM, Feke DL (2005) Transport analysis and model for the performance of an ultrasonically enhanced filtration process. Chemical Engineering Science 60:3233–3238

Grundy MA, Coakley WT, Clarke DJ (1989) Rapid detection of hepatitis-B virus using a hemagglutination assay in an ultrasonic standing wave field. Journal of Clinical & Laboratory Immunology 30:93–96

Gupta S, Feke DL (1998) Filtration of particulate suspensions in acoustically driven porous media. Aiche Journal 44:1005–1014

Haake A, Dual J (2004) Positioning of small particles by an ultrasound field excited by surface waves. Ultrasonics 42:75–80

Haake A, Dual J (2005) Contactless micromanipulation of small particles by an ultrasound field excited by a vibrating body. Journal of the Acoustical Society of America 117:2752–2760

Haake A, Neild A, Kim DH, Ihm JE, Sun Y, Dual J, Ju BK (2005a) Manipulation of cells using an ultrasonic pressure field. Ultrasound In Medicine And Biology 31:857–864

Haake A, Neild A, Radziwill G, Dual J (2005b) Positioning, displacement, and localization of cells using ultrasonic forces. Biotechnology And Bioengineering 92:8–14

Harris NR, Hill M, Beeby SP, Shen Y, White NM, Hawkes JJ, Coakley WT (2003) A Silicon Microfluidic Ultrasonic Separator. Sens. Actuator B-Chem 95:425–434

Harris NR, Hill M, Torah RN, Townsend RJ, Beeby SP, White NM, Ding J (2006) A multilayer thick-film PZT actuator for MEMs applications. Sensors and Actuators A: Physical 132:311–316

Harris NR, Hill M, Townsend RJ, White NM, Beeby SP (2005) Performance of a micro-engineered ultrasonic particle manipulator. Sensors and Actuators B: Chemical 111:481–486

Harris NR, Hill M, White NM, Beeby SP (2004) Acoustic power output measurements for thick-film PZT transducers. Electronics Letters 40:636–637

Hawkes JJ, Barber RW, Emerson DR, Coakley WT (2004a) Continuous cell washing and mixing driven by an ultrasound standing wave within a microfluidic channel. Lab On A Chip 4:446–452

Hawkes JJ, Barrow D, Cefai J, Coakley WT (1998a) A laminar flow expansion chamber facilitating downstream manipulation of particles concentrated using an ultrasonic standing wave. Ultrasonics 36:901–903

Hawkes JJ, Barrow D, Coakley WT (1998b) Microparticle manipulation in millimetre scale ultrasonic standing wave chambers. Ultrasonics 36:925–931

Hawkes JJ, Coakley WT (1996) A continuous flow ultrasonic cell-filtering method. Enzyme And Microbial Technology 19:57–62

Hawkes JJ, Coakley WT (2001) Force field particle filter, combining ultrasound standing waves and laminar flow. Sensors and Actuators B-Chemical 75:213–222

Hawkes JJ, Limaye MS, Coakley WT (1997) Filtration of bacteria and yeast by ultrasound-enhanced sedimentation. J. Appl. Microbiol. 82:39–47

Hawkes JJ, Long MJ, Coakley WT, McDonnell MB (2004b) Ultrasonic deposition of cells on a surface. Biosensors & Bioelectronics 19:1021–1028

Hertz HM (1995) Standing-Wave Acoustic Trap For Nonintrusive Positioning Of Microparticles. Journal of Applied Physics 78:4845–4849

Higashitani K, Fukushima M, Matsuno Y (1981) Migration of suspended particles in plane stationary ultrasonic field. Chemical Eng Science 36:1877–1882

Hill M (2003) The selection of layer thicknesses to control acoustic radiation force profiles in layered resonators. JASA 114:2654–2661

Hill M, Shen Y, Hawkes JJ (2002) Modelling of layered resonators for ultrasonic separation. Ultrasonics 40:385–392

Hill M, Wood RJK (2000) Modelling in the design of a flow-through ultrasonic separator. Ultrasonics 38:662–665

Hultström J, Manneberg O, Dopf K, Hertz HM, Brismar H, Wiklund M (2006) Proliferation and viability of adherent cells manipulated by standing-wave ultrasound in a microfluidic chip. Ultrasound Med. Biol. 33:175–181

Jenkins P, Barnes RA, Coakley WT (1997) Detection of meningitis antigens in buffer and body fluids by ultrasound-enhanced particle agglutination. Journal of Immunological Methods 205:191–200

Jepras RI, Clarke DJ, Coakley WT (1989) Agglutination of legionella-pneumophila by antiserum is accelerated in an ultrasonic standing wave. Journal of Immunological Methods 120:201–205

Johann RM (2006) Cell trapping in microfluidic chips. Analytical And Bioanalytical Chemistry 385:408–412

Kaduchak G, Sinha DN, Lizon DC (2002) Novel cylindrical, air-coupled acoustic levitation/concentration devices. Review Of Scientific Instruments 73:1332–1336

Kapishnikov S, Kantsler V, Steinberg V (2006) Continuous particle size separation and size sorting using ultrasound in a microchannel. Journal Of Statistical Mechanics-Theory And Experiment Need publication data after journal title

Khanna S, Amso NN, Paynter SJ, Coakley WT (2003). Contrast agent bubble and erythrocyte behavior in a 1.5-MHz standing ultrasound wave. Ultrasound In Medicine And Biology 29:1463–1470

Khanna S, Hudson B, Pepper CJ, Amso NN, Coakley WT (2006) Fluorescein isothiocynate-dextran uptake by chinese hamster ovary cells in a 1.5 MHz ultrasonic standing wave in the presence of contrast agent. Ultrasound in Medicine And Biology 32:289–295

King LV (1934) On the acoustic radiation pressure on spheres. Proc R. Soc. London A147:212–240

Krimholz R, Leedom DA, Matthaei GL (1970) New Equivalent Circuit for Elementary piezoelectric transducers. Electronics Letters 6:398–399

Kumar M, Feke DL, Belovich JM (2005) Fractionation of cell mixtures using acoustic and laminar flow fields. Biotechnology And Bioengineering 89:129–137

Kuznetsova LA, Coakley WT (2004) Microparticle concentration in short path length ultrasonic resonators: Roles of radiation pressure and acoustic streaming. J. Acoust. Soc. Am. 116:1956–1966

Kuznetsova LA, Coakley WT (2007) Applications of ultrasound streaming and radiation force in biosensors. Biosensors & Bioelectronics 22:1567–1577

Kuznetsova LA, Martin SP, Coakley WT (2005) Sub-micron particle behaviour and capture at an immuno-sensor surface in an ultrasonic standing wave. Biosensors & Bioelectronics 21:940–948

Lee YH, Peng CA (2005) Enhanced retroviral gene delivery in ultrasonic standing wave fields. Gene Therapy 12: 625–633

Leighton TG (1994) The Acoustic Bubble. Academic Press, San Diego

Leighton TG (1995) Bubble population phenomena in acoustic cavitation. Ultrasonics Sonochemistry 2:S123

Leighton TG (2007) What is ultrasound? Progress in Biophysics & Molecular Biology 93:3–83

Leung E, Jacobi N, Wang T (1981) Acoustic radiation force on a rigid sphere in a resonance chamber. Journal of the Acoustical Society of America 70:1762–1767

Leung E, Lee CP, Jacobi N, Wang TG (1982) Resonance Frequency-Shift Of An Acoustic Chamber Containing A Rigid Sphere. Journal of the Acoustical Society of America 72:615–620

Lighthill J (1978) Acoustic Streaming. Journal of Sound and Vibration 61:391–418

Lilliehorn T, Johansson S (2004) Fabrication of multilayer 2D ultrasonic transducer microarrays by green machining. Journal of Micromechanics And Microengineering 14:702–709

Lilliehorn T, Nilsson M, Simu U, Johansson S, Almqvist M, Nilsson J, Laurell T (2005a) Dynamic arraying of microbeads for bioassays in microfluidic channels. Sensors And Actuators B-Chemical 106:851–858

Lilliehorn T, Simu U, Nilsson M, Almqvist M, Stepinski T, Laurell T, Nilsson J, Johansson S (2005b) Trapping of microparticles in the near field of an ultrasonic transducer. Ultrasonics 43:293–303

Martin SP, Townsend RJ, Kuznetsova LA, Borthwick KAJ, Hill M, McDonnell MB, Coakley WT (2005) Spore and micro-particle capture on an immunosensor surface in an ultrasound standing wave system. Biosensors and Bioelectronics 21:758–767

Mason TJ (2007) Developments in ultrasound—Non-medical Progress in Biophysics & Molecular Biology 93:166–175

Mason WP (1942) Electromechanical Transducers and Wave Filters. Van Nostrand, New York

Masudo T, Okada T (2006) Particle separation with ultrasound radiation force. Current Analytical Chemistry 2:213–227

Morgan J, Spengler JF, Kuznetsova L, Coakley WT, Xu J, Purcell WM (2004) Manipulation of in vitro toxicant sensors in an ultrasonic standing wave. Toxicol. Vitro 18:115–120

Neild A, Oberti S, Beyeler F, Dual J, Nelson BJ (2006) A micro-particle positioning technique combining an ultrasonic manipulator and a microgripper. Journal of Micromechanics and Microengineering 16:1562–1570

Neild A, Oberti S, Dual J (2007) Design, modeling and characterization of microfluidic devices for ultrasonic manipulation Sensors and Actuators B: Chemical 121:452–461

Petersson F, Nilsson A, Holm C, Jonsson H, Laurell T (2004) Separation of lipids from blood utilizing ultrasonic standing waves in microfluidic channels. Analyst 129:938–943

Petersson F, Nilsson A, Holm C, Jonsson H, Laurell T (2005a) Continuous separation of lipid particles from erythrocytes by means of laminar flow and acoustic standing wave forces. Lab On A Chip 5:20–22.

Petersson F, Nilsson A, Jonsson H, Laurell T (2005b) Carrier medium exchange through ultrasonic particle switching in microfluidic channels. Analytical Chemistry 77:1216–1221

Rayleigh JW (1902) On the pressure of vibrations. Philosophical Magazine 3:338–346

Riley N (2001) Steady streaming. Annu. Rev. Fluid Mech. 33:43–65

Rosen CZ, Hiremath BV, Newnham RE (1992) Piezoelectricity (Key Papers in Physics). AIP Press, New York

Schram CJ (1991) Manipulation of Particles in an Acoustic Field. In: Mason TJ (ed) Advances in Sonochemistry. Elsevier, Amsterdam

Sherrit S, Leary S, Dolgin B, Bar-Cohen Y (1999) Comparison of the Mason and KLM Equivalent Circuits for Piezoelectric Resonators in the Thickness Mode. 1999 IEEE Ultrasonics Symposium, Lake Tahoel, Nevada, pp 921–926

Shirgaonkar IZ, Lanthier S, Kamen A (2004) Acoustic cell filter: a proven cell retention technology for perfusion of animal cell cultures. Biotechnology Advances 22:433–444

Sobanski MA, Tucker CR, Thomas NE, Coakley WT (2000) Sub-micron particle manipulation in an ultrasonic standing wave: Applications in detection of clinically important biomolecules. Bioseparation 9:351–357

Sobanski MA, Vince R, Biagini GA, Cousins C, Guiver M, Gray SJ, Kaczmarski EB, Coakley WT (2002) Ultrasound enhanced detection of individual meningococcal serogroups by latex immunoassay. J. Clin. Pathol. 55:37–40

Sritharan K, Strobl CJ, Schneider MF, Wixforth A, Guttenberg Z (2006) Acoustic mixing at low Reynold's numbers. Applied Physics Letters 88

Stecher G (1987) Free Supporting Structures in Thick-film Technology:a substrate integrated pressure sensor. 6th European Micrelectronics Conference, Bournemouth, pp 421–427

Thomas NE, Sobanski MA, Coakley WT (1999) Ultrasonic enhancement of coated particle agglutination immunoassays: Influence of particle density and compressibility. Ultrasound in Medicine and Biology 25:443–450

Townsend RJ, Hill M, Harris NR, White NM (2004) Modelling of particle paths passing through an ultrasonic standing wave. Ultrasonics 42:319–324

Townsend RJ, Hill M, Harris NR, White NM (2006) Investigation of two-dimensional acoustic resonant modes in a particle separator Ultrasonics 44:e467–e471

Townsend RJ, Hill M, Harris NR, White NM, Beeby SP, Wood RJK (2005) Fluid modelling of microfluidic separator channels. Sensors and Actuators B: Chemical 111:455–462

Vilkner T, Janasek D, Manz A (2004) Micro total analysis systems. Recent developments. Analytical Chemistry 76:3373–3385

Wang ZW, Grabenstetter P, Feke DL, Belovich JM (2004) Retention and viability characteristics of mammalian cells in an acoustically driven polymer mesh. Biotechnology Progress 20:384–387

Westervelt PJ (1951) The theory of steady forces caused by sound waves. Journal of the Acoustical Society of America 23,:312–315

Whitworth G, Coakley WT (1992) Particle Column Formation In A Stationary Ultrasonic-Field. Journal of the Acoustical Society of America 91:79–85

Wiklund M, Günther C, Lemor R, Jäger M, Fuhr G, Hertz HM (2006) Ultrasonic standing wave manipulation technology integrated into a dielectrophoretic chip. Lab on a Chip 6:1537–1544

Wiklund M, Hertz HM (2006) Ultrasonic enhancement of bead-based bioaffinity assays. Lab on a Chip 6:1279–1292

Wiklund M, Nilsson S, Hertz HM (2001) Ultrasonic trapping in capillaries for trace-amount biomedical analysis. Journal of Applied Physics 90:421–426

Wiklund M, Spegel P, Nilsson S, Hertz HM (2003) Ultrasonic-trap-enhanced selectivity in capillary electrophoresis. Ultrasonics 41, 329–333.

Wiklund M, Toivonen J, Tirri M, Hanninen P, Hertz HM (2004) Ultrasonic enrichment of microspheres for ultrasensitive biomedical analysis in confocal laser-scanning fluorescence detection. Journal of Applied Physics 96:1242–1248

Woodside SM, Bowen BD, Piret JM (1997) Measurement of ultrasonic forces for particle-liquid separations. Aiche Journal 43:1727–1736

Yasuda K (2000) Non-destructive, non-contact handling method for biomaterials in micro-chamber by ultrasound. Sensors and Actuators B-Chemical 64:128–135

Yasuda K, Umemura S, Takeda K (1996) Particle separation using acoustic radiation force and electrostatic force. Journal of the Acoustical Society of America 99:1965–1970

Yosioka K, Kawasima Y (1955) Acoustic radiation pressure on a compressible sphere. Acoustica 5:167–173

Zarembo LK (1971) Acoustic Streaming. In: Rozenberg LD (ed) High Intensity Ultrasonic Fields. Plenum Press, New York

Zelenka J (1986) Piezoelectric Resonators and their Applications. Elsevier, Amsterdam

Zourob M, Hawkes JJ, Coakley WT, Brown BJT, Fielden PR, McDonnell MB, Goddard NJ (2005) Optical leaky waveguide sensor for detection of bacteria with ultrasound attractor force. Analytical Chemistry 77:6163–6168

Zourob M, Mohr S, Treves Brown BJ, Fielden PR, McDonnell M, Goddard NJ (2003) The development of a metal clad leaky waveguide sensor for the detection of particles Sensors and Actuators A: Physical 90:296–307

36

Recent Advances in Real-Time Mass Spectrometry Detection of Bacteria

Arjan L. van Wuijckhuijse and Ben L.M. van Baar

Abstract

   The analysis of bio-aerosols poses a technology challenge, particularly when sampling and analysis are done in situ. Mass spectrometry laboratory technology has been modified to achieve quick bacteria typing of aerosols in the field. Initially, aerosol material was collected and subjected off-line to minimum sample treatment and mass spectrometry analysis. More recently, sampling and analysis were combined in a single process for the real-time analysis of bio-aerosols in the field. This chapter discusses the development of technology for the mass spectrometry of bio-aerosols, with a focus on bacteria aerosols. Merits and drawbacks of the various technologies and their typing signatures are discussed. The chapter concludes with a brief view of future developments in bio-aerosol mass spectrometry.

1. Introduction

1.1. General

Real-time detection of biological material with absolute identity determination is the stuff of science fiction. In this case, science fiction actually represents the "market pull" that challenges a "technology push." As concerns the real-time analysis of biological aerosols, the challenge was taken up in the mid 1980s, when some of the technology seemed sufficiently mature for integration in a universally applicable real-time bio-aerosol mass spectrometer. Two decades later, direct bio-aerosol mass spectrometry is still in the early stages of technology development, with research going on in a few select institutes. In the same two decades, the "market pull" of science fiction was fortified by an increasing awareness of biological threat, propelled by possibly emerging pandemics (e.g., Perdue and Swayne 2005; Glass and Becker 2006), and by the fear of bioterrorist attacks (e.g., Hamburg 2002; Tegnell et al. 2006). By now, mass spectrometry (MS) has proven capable of producing complicated spectral signatures from biological materials within seconds. This positions MS as a core technology for meeting the challenge of bio-aerosol detection with adequate differentiation of detected agents. This chapter gives an overview of several lines of MS technology development and of the current state-of-the-art in real-time bio-aerosol MS. After the previous two decades of early technology development, the next five years will see the evolution of mature MS-based bio-aerosol detection.

Arjan L. van Wuijckhuijse and Ben L.M. van Baar • TNO Defence, Security and Safety, Rijswijk, the Netherlands.

M. Zourob et al. (eds.), *Principles of Bacterial Detection: Biosensors, Recognition Receptors and Microsystems*, © Springer Science+Business Media, LLC 2008

1.2. Scope

This chapter covers the MS of aerosols, in a broad sense, with a focus on the mass spectrometry of bio-aerosols, particularly of bacteria. Viruses and other agents are not considered, mainly because there are hardly any published reports on bio-aerosol MS of such agents. This section of the introduction deals with two lines of approach: MS of bacteria and MS of aerosols, in respective subsections. The next section discusses the current state of technology of bio-aerosol MS, after a brief introduction that ties MS of bacteria to MS of aerosols. The chapter concludes with a section on the outlook, focusing on the development of fieldable bio-aerosol mass spectrometers.

1.3. MS in the Whole Cell Analysis of Bacteria

Nowadays, several forms of MS are widely employed in the typing of bacteria. The information gained by MS can range from the profiling of intact cells to the detail of proteomics or nucleotide sequencing. However, chemical analysis by MS is an emerging method with respect to established microbiological typing methods, like serotyping, microscopy, and phage typing. Where identity of organisms is discussed between analytical chemists and microbiologists, misunderstanding often arises from a difference in perception of "identity" between the two disciplines. This issue is briefly addressed in the first subsection, to clarify the point that the many approaches to the identification of bacteria correspond to various levels of certainty. The second subsection gives a brief, non-exhaustive overview of applications of MS to the analysis of whole cell material from bacteria, with emphasis on aspects of potential screening and detection technology.

1.3.1. The Definition of 'Identity' of Bacteria

The understanding of "identity" in microbiology differs considerably from that of chemistry. In chemistry, identity is assigned on the basis of comparison of molecular structure, for example by spectroscopic methods. Although we like to think that live material is also subject to the rigidity of molecular structure, the adaptive and evolutionary qualities of life imply a certain fuzziness in chemical composition. Historically, classical microbiology and bacteriology set out to assign microorganism identity by descriptive comparison of a selected set of observable characteristics. These characteristics would be morphological, for example colony shape in culturing or cell shape in microscopy. Later methods of typing employed indirectly observable characteristics, for example particular enzymatic activity, chemical resistance or affinity, or immunological activity. With the advent of chemical methods of analysis, typing would also address parts of the chemical composition of microorganisms, like fatty acid distribution or excreted metabolites. The molecular biology revolution, which started around 1980, has provided typing methods based on hereditary properties as these are addressed through the analysis of DNA. It is envisaged that molecular biology may provide a Grand Unification Theory for typing of microorganisms. That will also resolve the practical problem of the evolution of "standards" because the abstract information can be documented on paper rather than in a collection of organisms. However, where the link between DNA sequence and biological function is still largely elusive, molecular biology cannot yet accomplish unification by the incorporation of existing typing schemes.

Without that Grand Unification scheme accomplished or even partly available, every bacterial typing method occupies its own dimension (see, for example, Spiegelman, Whissell, and Greer 2005). Studies employing multiple typing methods provide a link between the separate dimensions, often on an empirical basis. For example, if a bacteriophage susceptibility profile is obtained from a specifically serotyped *Salmonella* isolate, a single mapping term

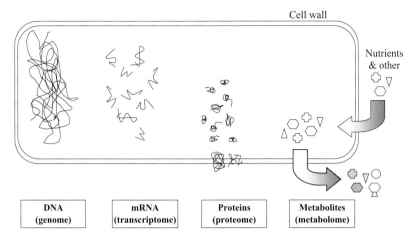

Figure 36.1. Representation of the distance of information from bacterial characteristics relative to the information in the genome.

is established between bacteriophage typing and serotyping. If mapping from a given typing method to other typing methods is not available, that typing method remains subjective. For practical purposes, for example in epidemiological comparison of bacterial strains from an outbreak, that subjectivity may be sufficient. When the identification problem is limited to a "threat list," subjectivity will be insufficient. Even when the typing profiles of all organisms on the "threat list" are acquired, it remains to be proven that no organisms on the list give the profile of one of the threat agents. For typing in any environment, a link to objective information, generally a DNA sequence, is eventually required to overcome the typing method dimension problem.

As MS methods for the typing of bacteria are concerned, the distance to the objective information of genomic DNA can be qualitatively evaluated. Fig. 36.1 gives the relative information distance of several characteristics of bacteria as they are accessible by chemical analysis.

1.3.2. Mass Spectrometry of Bacteria

Mass spectrometry is currently widely employed in the analysis of bacteria or their constituent compounds (as reviewed, see Van Baar 2000; Fenselau and Demirev 2001; Lay 2001; Fox 2006). However, many of the laboratory methods are not suitable for the detection of aerosolized bacteria, because sample treatment is too elaborate. Therefore, analysis of whole cells with little or no sample treatment is the only option, with the process applied to bulk material or to single cells in single particles. This subsection gives an overview of whole cell MS of bacteria, with emphasis on the signature obtained.

1.3.2.1. Pyrolysis MS

Early studies typically employed the mixture analysis capabilities of the then-newly emerging GC-MS technology, in pyGC-MS (Simmonds, Shulman, and Stembridge 1969; Simmonds 1970). A transition to direct inlet pyMS was quickly made for detection purposes because GC separation is relatively time consuming (Meuzelaar and Kistemaker 1973). Also, the use of field ionization (FI) (Schulten et al. 1973), chemical ionization (CI; Van der Greef, Tas, and Ten Noever de Brauw 1988), and metastable atom bomdardment (MAB; Wilkes et al. 2005) with pyMS was explored as an alternative to common electron ionization (EI).

Other studies addressed spectral comparison, for example through multivariate methods (Shute et al. 1984), artificial neural networks (Freeman, Sisson, and Ward 1995), or genetic algorithms (Taylor et al. 1998). PyMS was successfully applied in the differentiation of pathogens within specific families of bacteria, for example *Mycobacteria* (Wieten et al. 198) and *Salmonellae* (Van der Greef, Tas, and Ten Noever de Brauw 1988; Wilkes et al. 2005). Even though specific compounds were identified as pattern markers (e.g., DeLuca, Sarver, and Voorhees 1992), identification has always remained confined to a "fingerprinting" approach requiring the availability of a library or of a basis set of qualified spectra. Therefore, pyMS of bacteria is still typically used for quick screening, for example of clinical samples (Kyne et al. 1998; McCracken et al. 2000) or mail (Wilkes et al. 2006).

For many purposes, comparative matching of spectra or profiles is unsatisfactory because it does not allow a translation to other and independent knowledge available on the biological material at hand. For bacteria, this problem was approached by the application of MS/MS or by a minimum sample pretreatment. An example characteristic accessible by MS/MS is dipicolinic acid (DPA). This compound makes up 5-15 % of a *Bacillus* spore, and as such it is representative for the possible presence of *Bacillus anthracis*. Specific pyMS based methods for detecting *Bacillus* through DPA were developed (Beverly et al. 1996; Goodacre et al. 2000; Tripathi, Maswadeh, and Snyder 2001). Tandem mass spectrometry was employed to show that a more general characteristic of bacteria, their fatty acid profile, is accessible by pyMS/MS without prior chromatographic separation (DeLuca et al. 1990). A fast *in situ* chemical methylation reaction, to form fatty acid methyl esters (FAMEs), is included to allow better volatilization (Basile et al. 1998; Barshick, Wolf, and Vass 1999; Tripathi, Maswadeh, and Snyder 2001; Poerschmann et al. 2005). Although FAME profiling does not provide a complete typing scheme for bacteria, correlation with laboratory profiles can be employed in detection applications. Thus, the targeting of specific components by pyMS through MS/MS or fast *in situ* sample treatment obviates the need for elaborate full spectrum matching in detection.

1.3.2.2. LDI MS

LDI or LAMMA have not found much application in the typing of bacteria. The elemental analysis capability of LAMMA was employed to measure the Na^+/K^+ ratio in individual cells in order to establish the live or death status of *Mycobacteria* in response to external factors (Seydel et al. 1982; Seydel and Lindner 1988). Böhm et al. (1985) demonstrated laser ionisation of single bacterial cells from three *Bacillus* species. They concluded that high laser power densities ($>1000\,Wcm^{-2}$) were required to obtain positive or negative ions in pyMS-like spectra (see Fig. 36.2).

A discriminant analysis showed that typing by classification was poor, 70 to 75 %, even with this small *Bacillus* training set (Böhm et al. 1985). One later study, which compared several desorption/ionization methods in the analysis of cells, showed that the type of laser ionization employed in LDI and LAMMA gave signals of intact polar lipids from lysed *E. coli* (Heller et al. 1987). From these pioneering studies it is clear that LDI and LAMMA do not provide a sound basis for bacteria typing, because typing information is at best highly convoluted and of poorer quality than the information obtained from pyMS. In addition, the first reports on application of MALDI MS to bacteria, in the mid 1990s, left LDI and LAMMA obsolete as far as bacteria typing was concerned.

1.3.2.3. MALDI MS

Three pioneering studies on MALDI MS of vegetative whole bacteria were published independently in 1996 (Holland et al. 1996; Claydon et al. 1996; Krishnamurthy and Ross

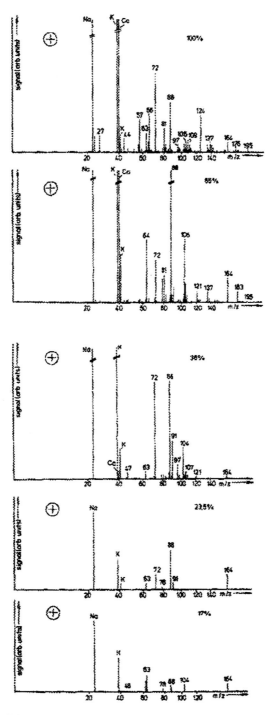

Figure 36.2. Typical LAMMA mass spectra of *Bacillus cereus* from pioloform foil with a copper grid, at a laser power density of 85 % (top) and 35 % (bottom) of the full Nd:YAG laser power (reprinted from Böhm et al. (1985), with permission of Elsevier).

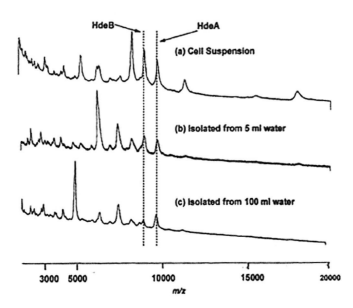

Figure 36.3. Typical MALDI mass spectra of *Shigella flexneri* with indication of the HdeA and HdeB protein signals (Reproduced from Holland et al. 2000, ©John wiley & Sons Limited).

1996). Three different sets of bacteria were employed in the three studies to arrive at a similar conclusion: Specific biomarker signals were repeatedly observed in the mass spectral profiles. Moreover, these signals allowed typing of the bacteria, at genus, species and strain level, by comparison to earlier recorded spectra or to newly acquired spectra from freshly cultured strains (see Fig. 36.3). Although these three studies used different mass ranges and different methods for sample application to the MALDI target, the work established a starting point for later studies.

The robustness of whole cell MALDI MS of bacteria was tested in many studies, which all show that sample preparation and details of the ionization process are important parameters for experimental control (e.g., Wang et al. 1998; Evason, Claydon, and Gordon 2001; Saenz et al. 2001; Williams et al. 2003). Bacterial growth conditions are beyond experimental control, at least in a real detection environment, but their effect on the variability of spectra has triggered further study. For example, it was observed that number and intensity of peaks varied considerably among spectra when *E. coli* K12 bacteria were sampled at selected stages between 6 to 84 h after inoculation of a culture (Arnold et al. 1999). However, the general trend is that protein fingerprinting in whole cell MALDI provides adequate identification despite the innate variability (e.g., Valentine et al. 2005; Wunschel et al. 2005a). Several studies with a wide scope have been reported, for example an interlaboratory study (Wunschel et al. 2005b) and some large scale typing studies for food pathogens (Mandrell et al. 2005), clinical isolates (Nilsson 1999; Marvin, Roberts, and Fay 2003; Keys et al. 2004), and in biotechnology (Jones et al. 2005).

Data analysis is an important and yet unsettled issue in whole cell MALDI MS. Most of the data analysis research focuses on the processing of spectra as "fingerprints." Because similar fingerprint type processing is also used in pyMS, procedures can be adapted for the purpose. However, plain spectrum fingerprint matching discards the more hidden information from the bacterial genome. The observed high-mass signals, over 1000 Da, are generally taken to represent peptides and proteins of which the mass is determined by the amino acid sequence. Therefore, these signals are considered "biomarkers" of the investigated bacteria. The biomarker link to genomes provides a more objective kind of information (Demirev et al. 1999), from a sequence database that is essentially independent of any analytical method. The notion of biomarkers implies that there is a potential for the kind of spectrum prediction that obviates the need for library accumulation. At present, fingerprinting is the established way for bacteria typing by MALDI MS, and the link to genome information is still under investigation. Both approaches are briefly discussed below.

Fingerprints may be accumulated as an internally consistent set, while the actual spectra are machine-dependent and sample dependent. Raw spectra are rarely used as fingerprints, and several condensation methods have been reported, for example cross-correlation (Arnold and Reilly 1998) and extraction of numerical fingerprints (Jarman et al. 1999; Jarman et al. 2000). Actual spectrum comparison is then achieved by methods also employed in matching of pyMS spectra, for example in the neural network approach demonstrated with SELDI (Schmid et al. 2005). The accumulation of a well-defined set of library spectra, possibly in a condensed form, is an elaborate job. However, it is generally assumed that translation of fingerprint data to other instruments and conditions is possible. This assumption was studied in an objective way by consideration of the protein distribution against the significance of an identification, for a limited set of organisms (Pineda et al. 2000). It was concluded that the cluttered and incomplete nature of the spectral data, as compared to bacterial proteomes, complicates or even compromises truly robust identification. At a practical level, the assumption of cross-instrument translation is supported by a single published interlaboratory study, which showed that tight control of sample treatment and of instrument conditions allows translation of MALDI MS fingerprints between instruments (Wunschel et al. 2005b). The assumption is further corroborated by the development of two commercially available MALDI MS platforms with libraries for bacteria typing: from Waters (employed, e.g., by Keys et al. 2004) and from Bruker (employed, e.g., by Maier et al. 2006). Although theoretical considerations and practical experience have not yet fully resolved the issue of method robustness, bacterial fingerprinting has become fairly well established as a method for bacteria typing in a laboratory setting and with a certain level of prior knowledge about target bacteria, for example in the screening for dairy pathogens.

Although the word *biomarker* appears in the pioneering papers on whole cell MALDI MS of bacteria, the link of signals to actually identified bacterial proteins was fairly speculative or even absent. Later studies provided more substantial evidence, for example for the identity of certain proteins from *Bacillus* spores (Hathout et al. 1999) or independent proof of protein identity for specific MALDI MS signals (e.g., Hathout et al. 2003; Dickinson et al. 2004). On the basis of this sporadic approach it was proposed by the group of Fenselau that it would be most likely that basic proteins, like many of the ribosomal proteins, are responsible for biomarker signals (Pineda et al. 2003). The proposal was corroborated by experiments that showed it is tenable as a working hypothesis, provided that post-translational modification of the proteins is accounted for (Demirev et al. 2001). Although this working hypothesis does not have any absolute predictive value, the basic protein approach is the most generalized proposal for biomarker attribution to date.

1.3.2.4. Comparison of MS Methods for Whole Cell Bacteria Typing

PyMS, LDI MS and MALDI MS have proven useful in rapid typing of bacteria in the laboratory, for which purpose they are operated with bulk samples and in a batchwise fashion. No direct experimental comparison of these methods is available. In addition, differences in ionization, in the bacterial compounds addressed, and in possible sample treatment hamper a theoretical comparison, for example of the sensitivity. For a qualitative consideration, a comparison can be made of the capability for generic bacteria typing to the attainable degree of resolution in typing (to the increasingly more detailed levels of bacteria/non-bacteria, bacteria *genus*, *species*, strain, and isolate).

PyMS, as it is applied in rapid analysis, will give information about the presence of a *Bacillus* genus marker (DPA), or it will produce a FAME profile. As a *Bacillus* detector, pyMS is not generic, and it has no further degree of typing resolution. From laboratory studies of FAME profiling without pyMS but by plain GC or GC-MS (e.g., Abel, DeSchmertzing, and

Peterson 1963; Moss 1990), it is clear that typing resolution may be highly specific within certain families of bacteria and fairly unspecific in other families. In addition, FAME profiles are known to change with changing environmental conditions, such as bacterial nutrition (see, e.g., Stoakes et al. 1991). As a FAME profiler, pyMS is a generic method, because such a profile can be generated from any bacteria. Overall, pyMS does cover some of the needs for rapid typing in the laboratory, while the method seems to have reached its full potential.

LDI MS and LAMMA, as they can be applied in rapid analysis, will not give very distinctive information. The near-atomization conditions of the direct laser ionization destroy the structure of biological compounds to an extent that highly specific molecular information is not observed. No recent reports have appeared on LDI MS or LAMMA investigations of bacteria in the laboratory. For laboratory typing of bacteria, LDI MS and LAMMA have been surpassed by MALDI MS at the present state of technology.

MALDI MS, as it is applied in rapid whole-cell analysis, will produce a biomarker profile in a single mass spectrum. Because this procedure can be done with any bacteria, the method is generic. Opinions on the specificity and usefulness of such a profile have come to some consensus with an increasing number of studies becoming available. The general conviction (see, e.g., Lay 2000) is that certain true biomarkers will always be present, whereas other biomarker signals will show major variations with environmental conditions such as culturing or growth stage. The principal advantage that biological molecules stay intact in MALDI, to retain the information content, is hardly employed, because most studies use plain profile matching. Overall, MALDI MS is well suitable to rapid typing of bacteria in the laboratory, while the method requires further exploration to assess its full potential.

1.4. Aerosol MS

MS analysis of aerosols can be accomplished in many ways. Some of the typical laboratory MS methods have been adapted to allow analysis of deposited aerosols, with sampling and chemical analysis offline. In addition, dedicated aerosol MS analysis technology was developed, to allow direct analysis of single particles from atmospheric aerosols without sample deposition. Although the deposition and direct analysis type methods share some of the technology, they are discussed separately.

1.4.1. MS of Deposited Aerosols

Bio-aerosols are often made amenable to mass spectrometry by particle collection. Although particle collection will allow further preparation for any desired MS analysis, subsequent sample treatment is generally minimized to allow quick analysis. Currently, one of three modes of MS analysis is typically employed: pyrolysis with EI or CI, laser desorption/ionization (LDI), and matrix-assisted LDI (MALDI). Because these modes of analysis produce distinct information from biological material, they are briefly discussed.

1.4.1.1. Pyrolysis MS

Pyrolysis generates volatile compounds from biological material, both by evaporation of any volatile compounds present and by the formation of new volatile compounds in elimination reactions under the influence of heat. The pyrolysis process can be conducted in a classical MS source to allow subsequent ionization. Many mass spectra can be acquired during a pyrolysis cycle to give a time-resolved profile. The spectra, the profile, or both can be employed as a characteristic for that particular biological material.

Pyrolysis MS (pyMS) initially evolved as a method for the chemical analysis of bacteria after it had been shown in the 1960s that pyrolysis gas chromatography (pyGC) provided "fingerprinting" of bacteria (see, e.g., Oyama 1963; Reiner 1965; Stern, Kotula, and Pierson 1979). Therefore, pyMS of bio-aerosols is discussed in more detail in the next section.

1.4.1.2. LDI MS

Laser desorption/ionisation (LDI) was developed in the late 1970s and early 1980s (e.g., Hillenkamp et al. 1975; Stoll and Röllgen 1979; Cotter 1980), when suitable laser technology became available. LDI was typically used for samples applied to a surface, and to direct surface analysis. In LDI, the ionizing laser energy is dissipated by the sample material and the corresponding energy density is generally so high that covalent chemical bonds easily dissociate. This laser ablation process burns away minute amounts of sample material. Therefore, a laser desorption spectrum will generally show signals of native ions such as Na^+ and NO_3^-, and of atomic cluster ions such as C_3^+ and C_3H^-, formed from the organic or biological compounds present. A typical example of such a spectrum is shown in Fig. 36.4. The high energy density in LDI makes the method less suitable for the ionization of relatively thermolabile organic and biochemical compounds. Nevertheless, occasional reports of successful LDI analysis of such compounds have appeared (see, for example, Balasanmugam et al. 1986; Posthumus et al. 1978). In many cases, the ionization of such thermolabile compounds involved preformed cationized and anionized species (Cotter 1981; Balasanmugam et al. 1981; Zakett et al. 1981).

The utility of LDI was greatly enhanced by combining the ionization with microscopic accuracy control of the ablation position to within 1 μm resolution (Hillenkamp et al. 1975; Wechsung et al. 1978). That form of LDI, generally known as Laser Ablation Microprobe Mass Analysis (LAMMA), was typically used for the analysis of samples on a surface. Initially, the capability for analysis of deposited aerosol particles was oddly reported among biomedical applications (Kaufman, Hillenkamp, and Wechsung 1979), but LAMMA was then quickly adopted by aerosol investigators (Wieser, Wurster, and Seiler 1988). The initial studies concerned technology development, for example for measuring the elemental composition of micron size particles (e.g., Bruynseels and Van Grieken 1984). This technology was then applied in atmospheric aerosol research (e.g., Bruynseels et al. 1988; Dierck et al. 1992; Hara et al. 1996). These aerosol studies generally encompassed inorganic analysis, where the particle

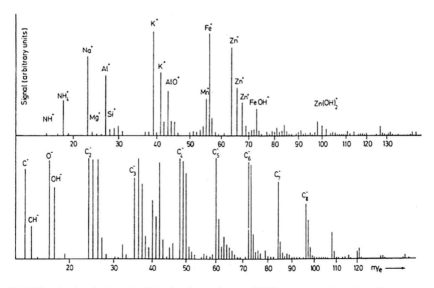

Figure 36.4. Negative ion (top) and positive ion (bottom) aerosol LDI spectra obtained from giant aerosol particles (reprinted from Wieser et al. (1980), with permission of Elsevier).

size range investigated was set by the aerosol fraction trapped by the offline sampler. In these studies, LAMMA MS is typically used for the analysis of the low Z elements, such as C, N, and S. Other methods, like electron probe X-ray microanalysis (EPXMA), turned out to be more suitable for the higher Z-elements, such as heavy metals (e.g., Van Malderen, Hoornaert, and Van Grieken 1996). Although only a single LAMMA study was reported in recent years, the available aerosol sampling technology and offline LAMMA analysis in a remote laboratory provided many new insights in environmental aerosol chemistry.

1.4.1.3. MALDI MS

MALDI MS evolved from LDI in the late 1980s (Tanaka et al. 1987; Karas and Hillenkamp 1988), when it turned out that the application of a specific "matrix" compound with a sample produced high mass ions upon UV laser irradiation. The matrix is thought to dissipate most of the incident UV energy. The matrix is also supposed to be a proton donor in positive ionization, and a proton acceptor in negative ionization. Although several studies were devoted to elucidation of the details of the MALDI ionization process (see, e.g., Knochenmuss and Zenobi 2003; Karas and Krüger 2003), there is no consensus theory or model. As a consequence, much of the chemistry involved in MALDI is still empirical.

The persistence of the empirical experimental component has not kept MALDI from gaining enormous acclaim as a method for MS of biological compounds. Empirical studies in MALDI MS covered issues like matrix crystallization (e.g., Westman et al. 1995; Dai, Whittal, and Li 1996; Luxembourg et al. 2003) and sample desalination (e.g., Kussman et al. 1997), to set the standard for a plethora of applied studies of large molecules in polymer science (e.g., Nielen 1999; Macha and Limbach 2002), microbiology (viruses, e.g., Lewis et al. 1998; Fenselau and Demirev, 2001; for bacteria, see above), and direct tissue analysis and imaging (e.g., Stoeckli, Farmer, and Caprioli 1999; Schwartz, Reyzer, and Caprioli 2003; Altelaar et al. 2005). MALDI MS also found application in aerosol analysis, as will be discussed later on.

For the purpose of this overview, surface enhanced laser desorption/ionization (SELDI) should also be included. SELDI was developed as a combination of MALDI with affinity capture chip technology on the target (Hutchens and Yip 1993; Caputo, Moharram, and Martin 2003). SELDI is the sophisticated, often commercialized variant of earlier academic experiments with on-target sample preparation (e.g., Brockman and Orlando 1995; Liang et al. 1998; Bundy and Fenselau 1999) for MALDI. The SELDI chip that serves as the laser target is coated with specific affinity molecules to allow on-target cleanup before matrix application and actual MS. Mass spectra are then obtained from any material retained by the affinity surface, for a rapid characterization of target material captured from the original sample. The chip technology makes SELDI particularly useful for quick screening of large amounts of samples, for example in clinical diagnosis of cancer by biomarkers (recently reviewed in Engwegen et al. 2006) and for bacteria typing (e.g. Schmid et al. 2005; Al Dahouk et al. 2006).

1.4.2. Direct MS of aerosols

In the 1980s LDI MS and LAMMA were proven to be of use in atmospheric (bio-)aerosol research. However, the required particle collection provided a cross-section picture of a certain particle size class, but any particle concentration information got lost in the process. In a separate development, Sinha et al. constructed an aerosol mass spectrometer for the analysis of individual particles. A beam of particles was continuously fed into the source of a quadrupole (Q) mass spectrometer, with in-source pyrolysis by a heated filament and electron ionization (Sinha et al. 1985). A key problem is that Q type mass analyzers require a constant ion input for a time span in the order of one second during a full mass range scan, whereas single particle

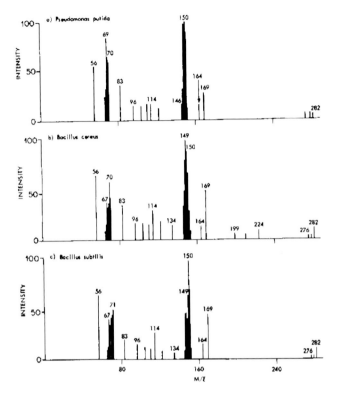

Figure 36.5. PyMS mass spectra of bacteria particles, *Pseudomonas putida* (top), *B. cereus* (middle), and *B. subtilis*, obtained with a particle beam quadrupole mass spectrometer (from Sinha et al. 1985; reproduced by permission of the American Society for Microbiology).

ionization produces only a short burst of ions during a period in the order of a millisecond. This problem was resolved by using the Q mass spectrometer as a band pass filter at a width of a few mass units. By stepping through multiple bands the 40–300 Da mass range was covered. This yielded mass spectra from a bacteria aerosol (Fig. 36.5), where a single spectrum would have required ionization of over 10^5 bacteria containing particles. This kind of sensitivity and mass range are not practically useful, but the work provided a most important proof of principle.

The concepts of the use of laser ionization (Sinha 1984) and of the use of time-of-flight (TOF) instead of Q mass spectrometers (Marijnissen, Scarlett, and Verheijen 1988) for single particle analysis were clear by the mid 1980s. However, the actual hardware development took until the early 1990s, when suitable lasers and mass spectrometers had finally become available. The technology was discussed in several reviews (Suess and Prather 1999; Noble and Prather 2000; Murphy 2006) and a summary description is given in the next section before the current state of the technology is addressed.

2. Current State of the Technology

2.1. Considerations on Aerosol MS of Bacteria

From the technology for aerosol MS and the application of MS to non-aerosolized bacteria it follows that the routine use of aerosol MS for the detection of bacteria is within reach of technology. It is also clear that the development and performance of such bio-aerosol MS technology is determined by several issues:

- aerosol deposition versus direct aerosol analysis;
- the possibility of applying any degree of sample treatment in the analysis process, either to single particles in the aerosol phase or after aerosol deposition;
- access to genome or proteome linked information *versus* accumulation of fingerprint type information.

At present, the field of bio-aerosol MS encompasses a variety of technology solutions, each of which deal with the above issues in their own way. This section discusses the available technology, as it was developed by the various research groups, with a clustering of similar technology.

2.2. Deposition and PyMS Based Technology

Pyrolysis MS based technology for the detection of bio-aerosols is currently best developed. PyMS analysis implies deposition of aerosol particles in a bulk, which process makes the particulate matter amenable to any form of elaborate sample treatment prior to the actual MS analysis (e.g., Szponar and Larsson 2001). However, because time is of the essence in bio-aerosol detection, two rapid methods with minimum sample treatment have found their way to field application: DPA confirmation and *in situ* derivatization with FAME profiling.

The proved presence of DPA in spores of the *Bacillus* family (see above) opened the way to a detection method for *Bacillus anthracis*, a main threat agent. Pyrolytic liberation of the acid and subsequent MS/MS in an ion-trap type instrument formed the key "biological" capability in the first version of the "Chemical and Biological Mass Spectrometer" detector (CBMS). The *in situ* methylation used for FAME profiling also methylates DPA. Thus, detection of dimethyl-DPA is still a key feature of the second version of the CBMS detector, the CBMS Block II (Hart et al. 2000; Lammert et al. 2002; see Fig. 36.6). Although the scope of DPA detection is limited to *Bacillus* species, it presently supports the only operational detector with a capability to differentiate to the level of an anthrax or closely related threat.

The development of a rapid process for *in situ* methylation, by heating with tetramethyl-hydroxide (TMH), formed the basis for FAME profiling in a field setting. This required some

Size		
Height	:	91 cm
Width	:	51 cm
Depth	:	36 cm
Weight	:	77 kg

Power		
Peak	:	1000 W
Continuous	:	500 W

Operating conditions		
Temperature	:	−32 − +49 °C
Humidity	:	5 − 95 %

Figure 36.6. Example of fieldable pyMS based equipment: the CBMS Block II instrument.

instrument and process optimization to allow sequential TMH treatment and pyrolysis in a single reaction vessel and to allow batchwise sample introduction. Several potentially transportable MS based detector systems have been under investigation for FAME profiling (e.g., Hart et al. 1999; Gardner et al. 2005), and the method has become the key mode of "biological detection" in the second version of the CBMS detector (Hart et al. 2000; Lammert et al. 2002). Some validation studies of the CBMS Block II system have appeared, to demonstrate much of its capabilities (e.g., Luo et al. 1999; Griest et al. 2001). For bacteria and fungi, the CBMS is presently the single operational biological detection system with a differentiation capability to the *genus* level and, sometimes, to the species level.

2.3. Deposition and MALDI MS Based Technology

The sample preparation process for on-target MALDI does not lend itself to easy automation in a batchwise process. To our knowledge, there is only a single line of development based on off-line aerosol entrapment and subsequent MALDI MS. In the mid 1990s it was demonstrated that a TOF mass spectrometer with a flight tube of a few centimeters long can have the performance of an instrument with a flight tube of over 1 meter in length (Cornish and Cotter 1992; Bryden et al. 1995). This small TOF mass spectrometer was the basis for interfacing with aerosol sampling and MALDI, as it was developed at the Johns Hopkins University Applied Physics Laboratory (JHU-APL) and described in a patent (Anderson and Carlson 1999; Anderson et al. 2003). Typically, aerosol is sampled onto a tape by an aerosol sample collector, where the sampling determines the overall duty cycle for detection. After completion of a sampling/deposition cycle, a matrix is applied and the tape is interfaced with the mass spectrometer by a sealable opening. With the sample spot on the tape in the correct position inside the mass spectrometer source, a common MALDI MS experiment can be conducted. Investigations on the performance of this system were not published in peer-reviewed journals, but aspects of system development were covered in JHU publications (e.g., Antoine et al. 2004). Although it is hardly possible to come to an evaluation of sensitivity and performance of this aerosol MALDI TOF MS system on the basis of the scant literature data, it is obvious that the technology has the full potential of MALDI MS for the identification of bacteria.

2.4. Single Particle LDI MS Technology

Following the pioneering work of Sinha (1984; Sinha et al. 1985) and a first concept of a single particle aerosol mass spectrometer (Marijnissen, Scarlett, and Verheijen 1988), first reports of actually built instruments appeared in the early 1990s (McKeown, Johnston, and Murphy 1991; Kievit et al. 1992). A typical design for a single particle MS instrument is given in Fig. 36.7 (Van Wuijckhuijse et al. 1998). Environmental aerosol is sampled into the mass spectrometer vacuum through a differentially pumped beam generator. The emerging particle beam is made to pass a low power continuous wave laser beam, where every single particle gives light scattering. The scattered light is detected and employed to fire a high power pulsed excimer UV laser, in order to ionize the particulate material. As the ionization events are located in the source of a TOF mass spectrometer, the ionization laser shot is also used to mark the start of the TOF MS process. Because this experimental set-up only allowed analysis of particles with a certain pre-set size, the demonstrator instruments served as a starting point for further exploration, pursued by several research groups.

The integration of aerodynamic sizing with triggering of the ionization laser, in a three-beam laser arrangement, made particle analysis more versatile (Weiss et al. 1993; Prather, Nordmeyer, and Salt 1994; Carson et al. 1995; Kievit et al. 1996). A double low power

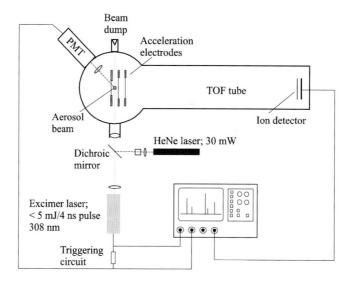

Figure 36.7. Typical design layout of a single particle aerosol MS instrument.

laser beam passage made ionization laser triggering a size independent process, and particle beam generation now became the main determinant for the particle size range. In addition, synchronous positive and negative ion TOF MS was implemented to improve detection capability (Hinz, Kaufmann, and Spengler 1994). Redesign to a transportable instrument was investigated (Gard et al. 1997). Also, the use of Ion Trap MS, instead of TOF MS, was explored (Dale et al. 1994; Yang et al. 1995). Finally, the TOF based technology was developed into a commercial instrument for the analysis of chemical aerosols (TSI Inc., USA; www.tsi.com).

Recently, it was shown that the same technology can be applied in the analysis of single particles of a bio-aerosol composed of *Bacillus* spores or *Mycobacteria* (Steele et al. 2003; Fergenson et al. 2004; see Fig. 36.8). Ionization was accomplished with the help of a 266 nm pulsed UV laser, for which it was considered that 266 nm coincides with the UV absorption maximum of the dipicolinic acid that is abundant in such spores. In a later study it was shown that the fluence threshold and the beam profile of the ionization laser are of prime importance to the spectrum quality and abundance (Steele et al. 2005). Data analysis in the *Bacillus* study showed that m/z 74^+ and m/z 173^- ions provided the main discriminant signals, where the positive ions were proposed to originate from trimethylglycine (Srivastava et al. 2005) and the negative ions from free arginine (Fergenson et al. 2004). However the robustness of these marker signals turned out to depend on the micro-organism growth conditions: Different growth conditions lowered the probability of detection of *B. atropheus* from 93 to 73 %, whereas the distinction from *B. thuringiensis* was lost (Fergenson et al. 2004). Specifically for *B. atropheus*, the impact of growth on the particle mass spectrum was investigated in more depth (Tobias et al. 2006). It was also shown that *Mycobacterium tuberculosis* can be distinguished from *Bacillus cereus*, *B. atropheus*, and *Mycobacterium smegmatis*, with a single marker signal as the determinant (Tobias et al. 2005). The observed negative ion marker, m/z 421^-, was tentatively attributed to a sulfolipid precursor earlier identified by independent methods as a component of the *M. tuberculosis*. Although there are just a few of these micro-organism studies, it is obvious that the focus in research on single particle LDI MS has shifted from technology exploration and development to the originally envisaged applications.

These few studies with biological material show that single particle LDI MS still holds some promise for bio-aerosol analysis. The sensitivity seems to be sufficient to give marker

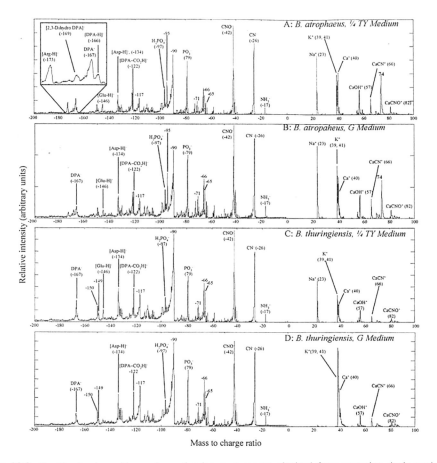

Figure 36.8. Combined axis positive and negative ion mass spectra obtained from averaging single particle LDI spectra of *B. thuringiensis* and *B. atropheus* after culturing on different growth media (Reprinted with permission from Fergenson et al. 2004. Copyright 2004 American Chemical Society).

signals from single particles. As the spectrum quality is concerned, there seems to be a strong analogy with the original pyrolysis mass spectra of bacteria, obtained in the 1970s and 1980s. Spectra typically display complicated patterns, with variable reproducibility and one or few biomarker signals. Many more microorganisms will have to be investigated to get to a more definite picture of single particle LDI capabilities.

2.5. Single Particle MALDI MS Technology

The step from single particle LDI MS to single particle MALDI MS seems logical by the historic analogy of the development of LDI and MALDI for bio-analysis. For single particle MS, this step is relatively easy when a matrix compound is mixed with the sample, prior to aerosolization. This was demonstrated with aerosols of bradykinin, gramicidin S, and myoglobin and α-cyano-4-hydroxycinnamic acid matrix (Murray et al. 1996), of leucine and of peptides (enkephalin and gramicidin S) mixed with several matrix compounds (Mansoori, Johnston, and Wexler, 1996) and with aerosols of gramicidin S with 3-nitrobenzyl alcohol as the matrix (Weiss 1997). At that time, the mass range and some other equipment limitations of typical single particle LDI mass spectrometers did not allow a full exploration of the premixed matrix approach. In addition, premixing of the matrix had little relevance in environmental aerosol research and in bio-aerosol detection. However, in the development of bio-analytical MS for laboratory applications there is a need for online coupling of MALDI MS with liquid flows from continuous separation methods (Foret and Preisler 2002; Gelpí 2002), for example with liquid chromatography (Murray et al. 1996; Preisler, Foret, and Karger 1998; Miliotis et al. 2000) or capillary electrophoresis (Zhang and Caprioli 1996). Therefore, the development of single particle MALDI MS follows two lines of sample treatment, either with matrix premixing prior to aerosolization or with online, in-flight coating of aerosol particles.

With the focus on bioaerosol mass spectrometry, interfacing of liquid based separations with single particle aerosol MALDI MS provides some insight into the performance of potential bio-aerosol mass spectrometer equipment. With various matrix compounds and several UV wavelengths employed, it is reported that the required matrix-to-analyte ratio lies between 10:1 and 1000:1 (Beeson, Murray, and Russell 1995; Mansoori, Johnston, and Wexler, 1996; He and Murray 1999). Although the quantity in a single particle of pure analyte lies between 100 amol and 10 fmol, the analyte loss by interfacing through aerosolization implies that the overall analysis method is relatively insensitive. However, this analyte loss depends strongly on the design of the interface, and actual numbers for the analyte loss are not available. Because MALDI MS of deposited material does not ionize all of the deposited analyte, aerosol MALDI MS is still able to compete favorably. At present, chromatography interfacing to mass spectrometry is still in the research stage of development.

Analysis of environmental bio-aerosol particles by single particle MALDI MS requires in-flight coating. Particle coating has found application in existing technology (see, e.g., Agarwal and Sem 1980). In principle, matrix coating by condensation can be accomplished by subsequent particle passage through a warm saturated vapor and a cold zone (Fig. 36.9). Upon cooling, the oversaturated vapor condenses on the aerosol particles, which are the only available condensation nuclei. These coated particles are then analyzed by a single particle TOF mass spectrometer (Stowers et al. 2000; Jackson, Mishra, and Murray 2004). In the coating process, the final particle size depends on the initial size, the temperature difference between hot and cold zone, the aerosol number concentration, and the sampling rate. These parameters were employed to tune the matrix-to-analyte ratio to produce MALDI mass spectra of acceptable quality. In a proof of principle, spectra were obtained from Gramicidine S and from *Bacillus subtilis* var niger (known as BG), and BG signals were tentatively assigned to peptidoglycan typical for BG cells (Stowers et al. 2000). This proof of principle showed that single particle MALDI TOF MS is possible.

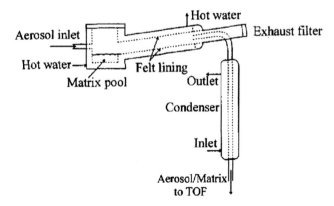

Figure 36.9. Schematic of an in-flight matrix coating apparatus for aerosol particles (from Van Wuijckhuijse et al. 2005b; reproduced with kind permission of Springer Science and Business Media).

The initial equipment was augmented with new ion optics to extend the mass range, and with a triggering mechanism for selective ionization of biological particles (Stowers et al. 2002; Van Wuijckhuijse et al. 2005a and 2005b). The earlier observations for BG were reproduced and additional signals were found in the 6000-7500 Da mass range. The improvements allowed regeistration of mass spectra in the 1 to 30 kDa range found relevant to whole cell bacteria typing in common MALDI (see above).

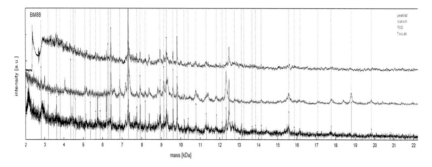

Figure 36.10. Spectra from "crushed crystallized" *E. coli* obtained on a common MALDI MS instrument (top trace; Biflex III MALDI mass spectrometer 337 nm ionization laser, average of 100 shots; Bruker Daltonics; Bremen, Germany) on two different direct MALDI ATOFMS instruments (middle: Delft University of Technology instrument with 308 nm ionization; bottom: instrument at TNO, with 337 nm ionization; both average of ~1000 single particle spectra); vertical lines represent the mass spectrum peak list.

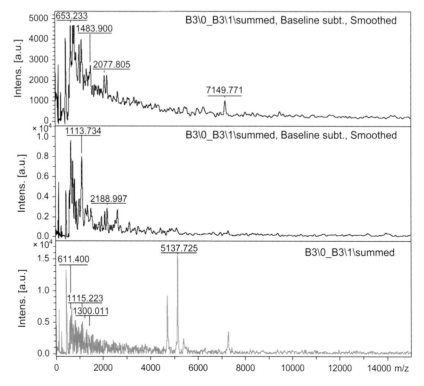

Figure 36.11. Direct Aerosol MALDI TOF mass spectra from *B. globigii* (top), *B. thuringiensis* (middle), and *B. cereus* (bottom) spores, obtained from the aerosolized spores and with spectrum averaging of ~500 particles.

Over the last year, work has been going on to improve matrix coating, to become reproducible, and to build a prototype transportable system (Lok, 2007; Van Wuijckhuijse et al. in preparation). This will show whether single particle MALDI MS gives signatures that compare to those from whole cell MALDI MS in the laboratory. In a first step, *Escherichia coli* was prepared for a laboratory MALDI experiment by the common addition of matrix. Part of the crystallized material was collected and crushed to allow subsequent aerosolization. As it is shown in Fig. 36.10, analysis of this material with two different aerosol TOF mass spectrometers and with a common laboratory instrument yielded similar spectra. Although this single experiment requires repetition with a few more organisms, the spectrum resemblance demonstrates that spectrum matching of spectra from a single particle MALDI MS instrument with those from common whole-cell MALDI MS will support bacterial typing. The improved performance of the present system (Van Wuijckhuijse in preparation) is also demonstrated by single particle MALDI MS averaged spectra from several *Bacillus* species: *B. thuringiensis*, BG, and *B. cereus* (Fig. 36.11). These 500 particle average spectra, evaluated in their entirety and not yet by biomarker signals, clearly discriminate between the three closely related *Bacillus* species, and they also discriminate between spores and vegetative cells.

3. Conclusions and Future Perspectives

Bio-aerosol MS is still in the early stages of development, with 10 to 20 research groups world wide investigating the various technologies. In general, technology development tends toward increasing transportability, ruggedness, and ease of operation. That tendency is driven by the potential applications, which all involve *in situ* measurement for quick and reliable detection: medical, environmental, as well as civil and military safety applications. Where potential military and civil safety bio-aerosol MS is pretty much limited to the scope of threat perception, medical and environmental bio-aerosol MS may open up a completely new view of the world. We will become able to "see" microorganisms in environments where we have been blind until now. Of course, this sense of science fiction translates into some practical requirements for science and technology development. In this outlook we do not pretend to be exhaustive, but we pick up on a few issues for improvement of bio-aerosol MS in the next few years.

For bio-aerosol analysis, pyMS is currently best developed among the aerosol MS technologies. Given the fact that, so far, pyGC-ion mobility spectrometry (IMS) was developed for field bio-aerosol detection (Snyder et al. 2004), the technology and chemistry of FAME pyMS can be further developed for bio-aerosol detection (Krebs et al. 2006; Prasad et al. 2006). Changing from MS to IMS technology does resolve design issues with regard to size and transportability of the overall system. Although pyMS has the advantage that it does not include relatively vulnerable lasers, the use of laser ionization may greatly aid the selectivity in ionization of specific pyrolysis products. For example, photoionization was demonstrated for the analysis of polymers by pyMS (see, e.g., Zoller et al. 1999), for the field analysis of chemical weapons agents (Syage, Hanning-Lee, and Hanold 2000), and for the direct analysis of biological material (Evans, Hanold, and Syage 2000; Nies, Evans, and Syage 2003). Alternatively, surface ionization might give useful results when applied to bio-aerosols. Surface ionization is already employed in aerosol MS (e.g., Jimenez et al. 2003), but the published applications were limited to organic and inorganic aerosols. Simplification of the instrument by going from MS to IMS and application of other ionization modes, such as photoionization or surface ionization, may provide new signatures for the analysis of bio-aerosol components.

For all bio-aerosol MS technologies, downsizing is an important issue on the way to increased transportability. Downsizing applies to all of the component technologies integrated in a bio-aerosol mass spectrometer. Downsizing of the mass spectrometer has the additional benefit that smaller vacuum volumes require less pumping capacity and, hence, smaller pumps. The application of Ion Trap MS instead of TOF MS will sometimes be an option in downsizing, and we see Ion Trap MS (e.g., Harris, Reilly, and Whitten 2006) as a viable line of development in bio-aerosol MS. However, both types of mass spectrometer have proved amenable to downsizing, as has been reviewed by Badman and Cooks (2000). Downsizing often involves application of different or refined physical principles, with a concomitant redesign of the mass analyzer. The application of higher order ion focusing and a redesign of the instrument has for example produced small "end cap" reflectron TOF instruments with capabilities comparable to those of common laboratory TOF instruments (e.g., Cornish and Cotter 1997; Fancher, Woods, and Cotter 2000). A completely different way was pursued with the redesign of the reflectron from the common multistage to a single stage element (Uphoff, Muskat, and Grotemeyer 2004). Also, developments in the manufacturing process have led to smaller TOF instruments, even down to suitcase size (Cornish, Ecelberger, and Brinckerhoff 2000; Ecelberger et al. 2004). Downsizing of Ion Traps to a hand-portable size was recently accomplished (Song et al. 2006; Gao et al. 2006), although it should be noted that the mass range of such small Ion Traps is not yet compatible with high mass range applications. For either TOF or Ion Trap instruments, downsizing of the mass spectrometer will require tailoring to the specific application. Downsizing of laser systems will also drastically reduce the system size of LDI or MALDI bio-aerosol MS systems, with the additional benefit that smaller lasers require less cooling. The development of lasers is mainly driven by other applications, but more specific requirements for (aerosol) mass spectrometry have recently been pointed out in great detail (Holle et al. 2006; Murphy 2006). In particular, the development of solid-state lasers with UV capability will help in downsizing bio-aerosol MS systems. Downsizing of laser equipment and mass analyzers will determine the overall system downsizing, because aerosol sampling and capture equipment will not be amenable to much downsizing given the required performance.

For all bio-aerosol technologies, data handling is of the essence. Spectrum comparison with library spectra is a routine application, but actual identification of bacteria from spectra requires more research. For any of the bio-aerosol analysis methods applied to bacteria, robustness of the signature towards environmental factors and phenotype requires more research. The signatures obtained from pyMS and LDI approaches will not have sufficient distinctive capability for adequate identification, because the ionized material is essentially a far derivative of biological material. Nevertheless, in the analysis of bacterial bio-aerosols, pyMS and LDI signatures may be useful as indicators rather than identifiers. As for the MALDI MS approach to bio-aerosol analysis, the obtained signatures relate directly to bacterial biomolecules present. However, more research is required to establish the link between aerosol MALDI MS signature and bacterial biomolecules. On the basis of the current state of the art in laboratory MALDI MS of bacteria, direct aerosol MALDI MS has the potential to become a rapid identification method.

In conclusion, bio-aerosol MS still poses a challenge to science and technology, with interesting promises of environmental, medical and safety applications to be fulfilled in the next few years.

References

Abel K, DeSchmertzing H and Peterson JI (1963) Classification of Microorganisms by Analysis of Chemical Composition .1. Feasibility of Utilizing GC. J. Bacteriol. 85:1039–1044

Agarwal JK and Sem GJ (1980) Continuous Flow, Single-particle-counting Condensation Nucleus Counter. J. Aerosol Sci. 11:343–357

Al Dahouk S, Nöckler K, Scholz HC, Tomaso H, Bogumil R and Neubauer H (2006) Immunoproteomic Characterization of Brucella abortus 1119–3 Preparations used for the Serodiagnosis of Brucella Infections. J. Immunol. Meth. 309:34–47

Altelaar AFM, Van Minnen J, Jimenez CR, Heeren RMA and Piersma SR (2005) Direct Molecular Imaging of Lymnaea stagnalis Nervous Tissue at Subcellular Spatial Resolution by Mass Spectrometry. Anal. Chem. 77:735–741

Anderson CW and Carlson MA (1999) A Time-of-Flight Mini-Mass Spectrometer: Aerosol Collection, Capture, and Load-Lock System. Johns Hopkins APL Tech. Dig. 20:352–362

Anderson CW, Scholl PF, Chappell RG, Bryden WA, Ko HW and Ecelberger SA (2003) Sample Collection Preparation for Time-of-flight Miniature Mass Spectrometer. US Patent 5,806,465

Antoine MD, Carlson MA, Drummond WR, Doss III OW, Hayek CS, Saksena A and Lin JS (2004) Mass Spectral Analysis of Biological Agents Using the BioTOF Mass Spectrometer. Johns Hopkins APL Tech. Dig. 25:20–26

Arnold RJ and Reilly JP (1998) Fingerprint matching of e. coli strains with matrix-assisted laser desorption ionization time-of-flight mass spectrometry of whole cells using a modified correlation approach. Rapid. Commun. Mass Spectrom. 12:630–636

Arnold RJ, Karty JA, Ellington AD and Reilly JP (1999) Monitoring the Growth of a Bacteria Culture by MALDI MS of Whole Cells. Anal. Chem. 71:1990–1996

Badman ER and Cooks RG (2000) Miniature Mass Analyzers. J. Mass Spectrom. 35:659–671

Balasanmugam K, Dang TA, Day RJ, and Hercules DM (1981) Some Cation and Anion Attachment Reactions in Laser Desorption Mass Spectrometry. Anal. Chem. 53:2296–2298

Balasanmugam K, Viswanadham SK and Hercules D (1986) Characterization of Polycyclic Aromatic Hydrocarbons by Laser Mass Spectrometry. Anal. Chem. 58:1102–1108

Barshick SA, Wolf DA and Vass AA (1999) Differentiation of Microorganisms Based on Pyrolysis – Ion Trap Mass Spectrometry Using Chemical Ionization. Anal. Chem. 71:633–641

Basile F, Beverly MB, Abbas-Hawks C, Mowry CD, Voorhees KJ and Hadfield TL (1998) Direct Mass Spectrometric Analysis of In Situ Thermally Hydrolyzed and Methylated Lipids from Whole Bacterial Cells. Anal. Chem. 70:1555–1562

Beeson MD, Murray KK and Russell DH (1995) Aerosol Matrix-Assisted Laser Desorption Ionization: Effects of Analyte Concentration and Matrix-to-Analyte Ratio. Anal. Chem. 67:1981–1986

Beverly MB, Basile F, Voorhees KJ and Hadfield TL (1996) A Rapid Approach for the Detection of Dipicolinic Acid in Bacterial Spores Using Pyrolysis/Mass Spectrometry. Rapid. Commun. Mass Spectrom. 10:455–458

Böhm R, Kapr T, Schmitt HU, Albrecht J and Wieser P (1985) Application of the Laser Microprobe Mass Analyser (LAMMA) to the Differentiation of Single Bacterial Cells. J. Anal. Appl. Pyrol. 8:449–461

Brockman AH and Orlando R (1995) Probe-immobilized Affinity Chromatography/Mass Spectrometry. Anal. Chem. 67:4581–4585

Bruynseels F and Van Grieken RE (1984) Laser Microprobe Mass Spectrometric Identification of Sulfur Species in Single Micrometer-size Particles. Anal. Chem. 56:871–873

Bruynseels F, Storms H, Van Grieken R and Van der Auwera L (1988) Characterization of North Sea Aerosols by Individual Particle Analysis. Atmosph. Environm. 22:2593–2602

Bryden WA, Benson RC, Ecelberger SA, Phillips TE, Cotter RJ and Fenselau C (1995) The Tiny-TOF Mass-Spectrometer for Chemical and Biological Sensing. Johns Hopkins APL Tech. Dig. 16:296–310

Bundy J and Fenselau C (1999) Lectin-based Affinity Capture for MALDI-MS Analysis of Bacteria., Anal. Chem. 71:1460–1463

Caputo E, Moharram R and Martin BM (2003) Methods for On-chip Protein Analysis. Anal. Biochem. 321:116–124

Carson PG, Neubauer KR, Johnston MV and Wexler AS (1995) On-line Chemical Analysis of Single Aerosol Particles by Rapid Single-particle Mass Spectrometry. J. Aeros. Sci. 26:535–545

Claydon MA, Davey SN, Edwards-Jones V and Gordon DB (1996) The Rapid Identification of Intact Microorganisms Using Mass Spectrometry. Nat. Biotechnol. 14:1584–1586

Cornish TJ and Cotter RJ (1992) A Compact TOF-MS for the Structural Analysis of Biological Molecules Using Laser Desorption. Rapid Comm. Mass Spectrom. 6:242–248

Cornish TJ and Cotter RJ (1997) High-order Kinetic Energy Focusing in an End Cap Reflectron Time-of-flight Mass Spectrometer. Anal. Chem. 69:4615–4618

Cornish TJ, Ecelberger S and Brinckerhoff W (2000) Miniature Time-of-flight Mass Spectrometer Using a Flexible Circuitboard Reflector. Rapid Commun. Mass Spectrom. 14:2408–2411

Cotter RJ (1980) Laser Desorption Chemical Ionization Mass Spectrometry. Anal. Chem. 52:1767–1770

Cotter RJ (1981) Cationized Species in Laser Desorption Mass Spectrometry. Anal. Chem. 53:719–720

Dai Y, Whittal RM and Li L (1996) Confocal Fluorescence Microscopic Imaging for Investigating the Analyte Distribution in MALDI Matrices. Anal. Chem. 68:2494–2500

Dale JM, Yang M, Whitten WB and Ramsey JM (1994) Chemical Characterization of Single Particles by Laser Ablation/Desorption in a Quadrupole Ion-Trap Mass-Spectrometer. Anal. Chem. 66:3431–3435

DeLuca S, Sarver EW, Harrington PD and Voorhees KJ (1990) Direct Analysis of Bacterial Fatty Acids by Curie-point Pyrolysis Tandem Mass Spectrometry. Anal. Chem. 62:1465–1472

DeLuca SJ, Sarver EW and Voorhees KJ (1992) Direct Analysis of Bacterial Glycerides by Curie-point Pyrolysis - Mass Spectrometry. J. Anal. Appl. Pyrolysis 23:1–14

Demirev PA, Ho YP, Ryzhov V and Fenselau C (1999) Micro-organism Identification by Mass Spectrometry and Protein Database Searches., Anal. Chem. 71:2732–2738

Demirev PA, Lin JS, Pineda FJ and Fenselau C (2001) Bioinformatics and Mass Spectrometry for Micro-organism Identification: Proteome-wide Post-translational Modifications and Database Search Algorithms for Characterization of Intact H. pylori. Anal. Chem. 73:4566–4573

Dickinson DN, La Duc MT, Haskins WE, Gornushkin I, Winefordner JD, Powell DH and Venkateswaran K (2004) Species Differentiation of a Diverse Suite of Bacillus Spores by Mass Spectrometry-Based Protein Profiling. Appl. Envir. Microbiol. 70:475–482

Dierck I, Michaud D, Wouters L and Van Grieken R (1992) Laser Microprobe Mass Analysis of Individual North Sea Aerosol Particles. Environ. Sci. Technol. 26:802–808

Ecelberger SA, Cornish TJ, Collins BF, Lewis DL and Bryden WA (2004) Suitcase TOF: A Man-Portable Time-of-Flight Mass Spectrometer. Johns Hopkins APL Tech. Dig. 25:14–19

Engwegen JYMN, Gast M-CW, Schellens JHM and Beijnen JH (2006) Clinical Proteomics: Searching for Better Tumour Markers with SELDI-TOF Mass Spectrometry. Tr. Pharmacol. Sci. 27:251–259

Evans MD, Hanold KA and Syage JA (2000) Rapid Response Chem/Bio Detection System Based on Photoionization Mass Spectrometry. Proc. 1st Joint Conf. for CB Defense, Oct 23–27, Williamsburg, Virginia

Evason DJ, Claydon MA and Gordon DB (2001) Exploring the Limits of Bacterial Identification by Intact Cell – Mass Spectrometry. J. Am. Soc. Mass Spectrom. 12:49–54

Fancher CA, Woods AS and Cotter RJ (2000) Improving the Sensitivity of the End-cap Reflectron Time-of-flight Mass Spectrometer. J. Mass Spectrom. 35:157–162

Fenselau C and Demirev PA (2001) Characterization of Intact Microorganisms by MALDI Mass Spectrometry. Mass Spectrom. Rev. 20:157–171

Fergenson DP, Pitesky ME, Tobias HJ, Steele PT, Czerwieniec GA, Russell SC, Lebrilla C, Horn J, Coffee K, Srivastava A, Pillai SP, Shih M-TP, Hall HL, Ramponi AJ, Chang JT, Langlois RG, Estacio PL, Hadley RT, Frank M and Gard E (2004) Reagentless Detection and Classification of Individual Bioaerosol Particles in Seconds. Anal. Chem. 76:373–378

Foret F and Preisler J (2002) Liquid Phase Interfacing and Miniaturization in Matrix-Assisted Laser Desorption / Ionization Mass Spectrometry. Proteomics 2:360–372

Fox A (2006) Mass Spectrometry for Species or Strain Identification After Culture or Without Culture: Past, Present, and Future. J. Clin. Microbiol. 44:2677–2680

Freeman R, Sisson PR and Ward AC (1995) Resolution of Batch Variations in Pyrolysis Mass Spectrometry of Bacteria by the Use of Artificial Neural Network Analysis. Antonie Van Leeuwenhoek 68:253–260

Gao L, Song Q, Patterson GE, Cooks RG and Ouyang Z (2006) Handheld Rectilinear Ion Trap Mass Spectrometer. Anal. Chem. 78:5994–6002

Gard E, Mayer JE, Morrical BD, Dienes T, Fergenson DP and Prather KA (1997) Real-time Analysis of Individual Atmospheric Aerosol Particles: Design and Performance of a Portable ATOFMS. Anal. Chem. 69:4083–4091

Gardner B. D., Donaldson, W., Chun, R., Lee, W. T., Tissandier, M. 2005. An Ion Trap Mass Spectrometer System for Continuous Monitoring of Biological and Chemical Backgrounds. Proc. 5th Harsh-Environment Mass Spectrometry Workshop, September 20–23, Sarasota, Florida

Gelpí E (2002) Interfaces for Coupled Liquid-Phase Separation / Mass Spectrometry Techniques. An Update on Recent Developments. J. Mass Spectrom. 37:241–253

Glass K and Becker NG (2006) Evaluation of Measures to Reduce International Spread of SARS. Epidemiol. Infect. 134:1092–1101

Goodacre R, Shann B, Gilbert RJ, Timmins EM, McGovern AC, Alsberg BK, Kell DB and Logan NA (2000) Detection of the Dipicolinic Acid Biomarker in Bacillus Spores Using Curie-point Pyrolysis Mass Spectrometry and Fourier Transform Infrared Spectroscopy. Anal. Chem. 72, 119–127

Griest WH, Wise MB, Hart KJ, Lammert SA, Thompson CV and Vass AA (2001) Biological Agent Detection and Identification by the Block II Chemical Biological Mass Spectrometer. Field Analyt. Chem. Technol. 5:177–184

Hamburg MA (2002) Bioterrorism: Responding to an Emerging Threat. Trends Biotechnol. 20:296–298

Hara K, Kikuchi T, Furuya K, Hayashi M and Fujii Y (1996) Characterization of Antarctic Aerosol Particles Using Laser Microprobe Mass Spectrometry. Environ. Sci. Technol. 30:385–391

Harris WA, Reilly PTA and Whitten WB (2006) Aerosol MALDI of Peptides and Proteins in an Ion Trap Mass Spectrometer: Trapping, Resolution and Signal-to-noise. Int. J. Mass Spectrom. 258:113–119

Hart KJ, Harmon SH, Wolf DA, Vass AA and Wise MB (1999) Detection of Chemical/Biological Agents and Simulants Using Quadrupole Ion Trap Mass Spectrometry. Proceedings of the 47th ASMS Conf. Mass Spectrom. All. Topics, June 13–17, Dallas, TX

Hart KJ, Wise MB, Griest WH and Lammert SA (2000) Design, Development and Performance of a Fieldable Chemical and Biological Agent Detector. Field Anal. Chem. Technol. 4:93–110

Hathout Y, Demirev PA, Ho Y-P, Bundy JL, Ryzhov V, Sapp L, Stutler J, Jackman J and Fenselau C (1999) Identification of Bacillus Spores by Matrix-assisted Laser Desorption Ionization – Mass Spectrometry. Appl. Environ. Microbiol. 65:4313–4319

Hathout Y, Setlow B, Cabrera-Martinez R-M, Fenselau C and Setlow P (2003) Small, Acid-Soluble Proteins as Biomarkers in Mass Spectrometry Analysis of Bacillus Spores. Appl. Envir. Microbiol. 69:1100–1107

He L and Murray KK (1999) 337 nm Matrix-Assisted Laser Desorption / Ionization of Single Aerosol Particles. J. Mass Spectrom. 34:909–914

Heller DN, Fenselau C, Cotter RJ, Demirev P, Olthoff JK, Honovich J, Uy M, Tanaka T and Kishimoto Y (1987) Mass Spectral Analysis of Complex Lipids Desorbed Directly from Lyophilized Membranes and Cells. Biochem. Biophys. Res. Commun. 142:194–199

Hillenkamp F, Unsöld E, Kaufmann R and Nitsche R (1975) A High Sensitivity Laser Microprobe Mass Analyzer. Appl. Phys. 8:341–348.

Hinz KP, Kaufmann R and Spengler B (1994) Laser-Induced Mass Analysis of Single Particles in the Airborne State. Anal. Chem. 66:2071–2076

Holland RD, Wilkes JG, Rafii F, Sutherland JB, Persons CC, Voorhees KJ and Lay JO (1996) Rapid Identification of Intact Whole Bacteria Based on Spectral Patterns Using Matrix Assisted Laser Desorption / Ionization with Time-of-flight Mass Spectrometry. Rapid. Commun. Mass. Spectrom. 10:1227–1232

Holland RD, Rafii F, Heinze TM, Sutherland JB, Voorhees KJ and Lay Jr. JO (2000) Matrix-Assisted Laser Desorption/Ionization Time-of-Flight Mass Spectrometric Detection of Bacterial Biomarker Proteins Isolated from Contaminated Water, Lettuce and Cotton Cloth. Rapid. Commun. Mass. Spectrom. 14:911–917

Holle A, Haase A, Kayser M and Höhndorf J (2006) Optimizing UV Laser Focus Profiles for Improved MALDI Performance. J. Mass Spectrom. 41:705–716

Hutchens TW and Yip TT (1993) New Desorption Strategies for the Mass Spectrometric Analysis of Macromolecules. Rapid Commun. Mass Spectrom. 7:576–580

Jackson SN, Mishra S and Murray KK (2004) On-line Laser Desorption/Ionization Mass Spectrometry of Matrix-coated Aerosols. Rapid Comm. Mass Spectrom. 18:2041–2045

Jarman KH, Daly DS, Petersen CE, Saenz AJ, Valentine NB and Wahl KL (1999) Extracting and Visualizing Matrix-assisted Laser Desorption/Ionization Time-of-flight Mass Spectral Fingerprints. Rapid. Commun. Mass. Spectrom. 13:1586–1594

Jarman KH, Cebula ST, Saenz AJ, Petersen CE, Valentine NB, Kingsley MT and Wahl KL (2000) An Algorithm for Automated Bacterial Identification Using Matrix-assisted Laser Desorption/Ionization Mass Spectrometry. Anal. Chem. 72:1217–1223

Jimenez JL, Jayne JT, Shi Q, Kolb CE, Worsnop DR, Yourshaw I, Seinfeld JH, Flagan RC, Zhang X, Smith KA, Morris JW and Davidovits P (2003) Ambient Aerosol Sampling Using the Aerodyne Aerosol Mass Spectrometer. J. Geophys. Res. 108:8425

Jones JJ, Wilkins CL, Cai Y, Beitle RR, Liyanage, R., and Lay Jr., J. O. 2005 Real-time Monitoring of Recombinant Bacterial Proteins by Mass Spectrometry. Biotechnol. Prog. 21:1754–1758.

Karas M and Hillenkamp F (1988) Laser Desorption Ionization of Proteins with Molecular Masses Exceeding 10,000 Daltons. Anal Chem. 60:2299–2301

Karas M and Krüger R (2003) Ion Formation in MALDI: The Cluster Ionization Mechanism., Chem. Rev. 103:427–439

Kaufmann R, Hillenkamp F and Wechsung R (1979) The Laser Microprobe Mass Analyzer (LAMMA): A New Instrument for Biomedical Microprobe Analysis. Med. Prog. Technol. 6:109–121

Keys CJ, Dare DJ, Sutton H, Wells G, Lunt M, McKenna T, McDowall M and Shah HN (2004) Compilation of a MALDI-TOF Mass Spectral Database for the Rapid Screening and Characterisation of Bacteria Implicated in Human Infectious Diseases. Infect. Genet. Evol. 4:221–242

Kievit O, Marijnissen JCM, Verheijen PJT and Scarlett B (1992) On-line Measurement of Particle Size and Composition. J. Aerosol Sci. 23, Suppl. 1:301–304

Kievit O, Weiss M, Verheijen PJT, Marijnissen JCM and Scarlett B (1996) The On-line Chemical Analysis of Single Particles Using Aerosol Beams and Time of Flight Mass Spectrometry. Chem. Eng. Commun. 151:79–100

Knochenmuss R and Zenobi R (2003) MALDI Ionization: The Role of In-plume Processes., Chem. Rev. 103:441–452

Krebs MD, Mansfield B, Yip P, Cohen SJ, Sonenshein AL, Hitt BA and Davis CE (2006) Novel Technology for Rapid Species-specific Detection of Bacillus Spores. Biomol. Eng. 23:119–127

Krishnamurthy T and Ross PL (1996) Rapid Identification of Bacteria by Direct Matrix-assisted Laser Desorption/Ionization Mass Spectrometric Analysis of Whole Cells. Rapid Commun. Mass Spectrom. 10:1992–1996

Kussman M, Nordhof E, Rahbek-Nielsen H, Haebel S, Rossel-Larsen M, Jakobsen L, Gobom J, Mirgorodskaya E, Kroll-Kristensen A, Palm L and Roepstorff P (1997) Matrix-assisted Laser Desorption / Ionization Mass Spectrometry Sample Preparation Techniques Designed for Various Peptide and Protein Analytes. J. Mass Spectrom. 32:593–601

Kyne L, Merry C, O'Connell B, Harrington P, Keane C and O'Neill D (1998) Simultaneous Outbreaks of Two Strains of Toxigenic Clostridium difficile in a General Hospital. J. Hosp. Infect. 38:101–112

Lammert SA, Griest WH, Wise MB, Hart KJ, Vass AA, Wolf DA, Burnett MN, Merriweather R and Smith RR (2002) A Mass Spectrometer-based System for Integrated Chemical and Biological Agent Detection – The Block II CBMS. Proceedings of the 50th ASMS Conf. Mass Spectrom. All. Topics, June 2–6, Orlando, Florida

Lay Jr. JO (2000) MALDI-TOF Mass Spectrometry and Bacterial Taxonomy. Tr. Anal. Chem. 19:507–516

Lay Jr. JO (2001) MALDI-TOF Mass Spectrometry of Bacteria. Mass Spectrom. Rev. 20: 172–194

Liang XL, Lubman DM, Rossi DT, Nordblom GD and Barksdale CM (1998) On-probe Immunoaffinity Extraction by Matrix-assisted Laser Desorption/Ionization Mass Spectrometry. Anal. Chem. 70:498–503

Lok JJ (2007) Dutch Detector Promises Swift BW Analysis. Jane's Int. Def. Rev. 40:4

Luo S, Mohr J, Sickenberger D and Hryncewich A (1999) Study of Purified Bacteria and Viruses by Pyrolysis Mass Spectrometry. Field Analyt. Chem. Technol. 3:357–374

Luxembourg SL, McDonnell LA, Duursma M, Guo X and Heeren RMA (2003) Effect of Local Matrix Crystal Variations in Matrix-Assisted Ionization Techniques for Mass Spectrometry. Anal. Chem. 75:2333–2341

Macha SF and Limbach PA (2002) Matrix-assisted Laser Desorption/Ionization (MALDI) Mass Spectrometry of Polymers. Curr. Opin. Sol. State Mat. Sci. 6:213–220

Maier T, Große-Herrenthey A, Krueger M, Kelly J and Kostrzewa M (2006) Automated Microorganism Identification Using a Database Software System and a High Quality MALDI-TOF Spectra Library. Proceedings of the 17th Int. Mass Spectrom. Conf., 27 Aug.–1 Sept., Prague, Czech Republic

Mandrell RE, Harden LA, Bates A, Miller WG, Haddon WF and Fagerquist CK (2005) Speciation of Campylobacter coli, C. jejuni, C. helveticus, C. lari, C. sputorum, and C. upsaliensis by Matrix-assisted Laser Desorption Ionization - Time of Flight Mass Spectrometry. Appl. Environ. Microbiol. 71:6292–6307

Mansoori BA, Johnston MV and Wexler AS (1996) Matrix-Assisted Laser Desorption/Ionization of Size- and Composition Selected Aerosol Particles. Anal. Chem. 68:3595–3601

Marijnissen J, Scarlett B and Verheijen P (1988) Proposed On-line Aerosol Analysis Combining Size Determination, Laser-induced Fragmentation and Time-of-flight Mass Spectroscopy. J. Aerosol Sci. 19:1307–1310

Marvin LF, Roberts MA and Fay LB (2003) Matrix-assisted Laser Desorption/Ionization Time-of-flight Mass Spectrometry in Clinical Chemistry. Clin. Chim. Acta 337:11–21

McCracken D, Flanagan P, Hill D and Hosein I (2000) Cluster of Cases of Mycobacterium chelonae Bacteraemia. Eur. J. Clin. Microbiol. Infect. Dis. 19:43–46

McKeown PJ, Johnston MV and Murphy DM (1991) Online Single-particle Analysis by Laser Desorption Mass-Spectrometry. Anal. Chem. 63:2069–2073

Meuzelaar HLC and Kistemaker PG (1973) A Technique for Fast and Reproducible Fingerprinting of Bacteria by Pyrolysis Mass Spectrometry. Anal. Chem. 45:587–590

Miliotis T, Kjellstrom S, Nilsson J, Laurell T, Edholm LE and Marko-Varga G (2000) Capillary Liquid Chromatography Interfaced to Matrix-Assisted Laser Desorption-Ionization Time-of-Flight Mass Spectrometry Using an On-Line Coupled Piezoelectric Flow-through Microdispenser. J. Mass Spectrom. 35:369–377

Moss CW (1990) Use of Cellular Fatty Acids for Identification of Microorganisms. In: Fox A, Morgan LS, Larsson L and Odham G (eds) Analytical Microbiology Methods: Chromatography and Mass Spectrometry. Plenum Press, New York, pp 59–69

Murphy DM (2006) The Design of Single Particle Laser Mass Spectrometers., Mass Spectrom. Rev. 26:150–165

Murray KK, Lewis TM, Beeson MD and Russell DH (1996) Aerosol Matrix-Assisted Laser-Desorption Ionization for Liquid-Chromatography Time-of-Flight Mass-Spectrometry. Anal. Chem. 66:1601–1609

Nielen MWF (1999) MALDI Time-of-flight Mass Spectrometry of Synthetic Polymers. Mass Spectrom. Rev. 18:309–344

Nies BJ, Evans MD and Syage JA (2003) Rapid Biological Weapons Monitoring by Pyrolysis/GC, Photoionization MS., PITTCON, March 9–14, Orlando, Florida

Nilsson CL (1999) Fingerprinting of Helicobacter pylori Strains by Matrix-assisted Laser Desorption/Ionization Mass Spectrometric Analysis. Rapid Commun. Mass Spectrom. 13:1067–1071

Noble CA and Prather KA (2000) Real-time Single Particle Mass Spectrometry: A Historical Review of a Quarter Century of the Chemical Analysis of Aerosols. Mass Spectrom. Rev. 19:248–274

Oyama VI (1963) Mars Biological Analysis by Gas Chromatography. Lunar Planetary Expl. Coll. Proc. 3:29–36

Perdue ML and Swayne DE (2005) Public Health Risk from Avian Influenza Viruses. Avian Dis. 49:317–327

Pineda FJ, Lin JS, Fenselau C and Demirev PA (2000) Testing the Significance of Micro-organism Identification by Mass Spectrometry and Proteome Database Search. Anal. Chem. 72:3739–3744

Pineda FJ, Antoine MD, Demirev PA, Feldman AB, Jackman J, Longenecker M and Lin JS (2003) Micro-organism Identification by Matrix-Assisted Laser/Desorption Ionization Mass Spectrometry and Model-Derived Ribosomal Protein Biomarkers. Anal. Chem. 75:3817–3822

Poerschmann J, Parsi Z, Gorecki T and Augustin J (2005) Characterization of Non-discriminating Tetramethylammonium Hydroxide – Induced Thermochemolysis – Capillary Gas Chromatography – Mass Spectrometry as a Method for Profiling Fatty Acids in Bacterial Biomasses. J. Chromatogr. A. 1071:99–109

Posthumus MA, Kistemaker PG, Meuzelaar HLC and Ten Noever de Brauw MC (1978) Laser Desorption-Mass Spectrometry of Polar Nonvolatile Bio-organic Molecules. Anal. Chem. 50:985–991

Prasad S, Schmidt H, Lampen P, Wang M, Güth R, Rao JV, Smith GB and Eiceman GA (2006) Analysis of Bacterial Strains with Pyrolysis - Gas Chromatography / Differential Mobility Spectrometry. Analyst 131:1216–1225

Prather KA, Nordmeyer T and Salt K (1994) Real-time Characterization of Individual Aerosol-Particles Using Time-of-Flight Mass-Spectrometry. Anal. Chem. 66:1403–1407

Preisler J, Foret F and Karger BL (1998) On-Line MALDI-TOF MS Using a Continuous Vacuum Deposition Interface. Anal. Chem. 70:5278–5287

Reiner E (1965) Identification of Bacterial Strains by Pyrolysis Gas-liquid Chromatography. Nature (London) 206:1272–1274

Saenz AJ, Petersen CE, Valentine NB, Gantt SL, Jarman KH, Kingsley M and Wahl KL (2001) Reproducibility of Matrix Assisted Laser Desorption/Ionization Time-of-flight Mass Spectrometry for Replicate Bacterial Culture Analysis. Rapid Commun. Mass Spectrom. 13:1580–1585

Schmid O, Ball G, Lancashire L, Culak R and Shah H (2005) New Approaches to Identification of Bacterial Pathogens by Surface Enhanced Laser Desorption/Ionization Time of Flight Mass Spectrometry in Concert with Artificial Neural Networks, with Special Reference to Neisseria gonorrhoeae. J. Med. Microbiol. 54:1205–1211

Schulten HR, Beckey HD, Meuzelaar HLC and Boerboom AJH (1973) High Resolution Field Ionization of Bacterial Pyrolysis Products Anal. Chem. 45:191–195

Schwartz SA, Reyzer ML and Caprioli RM (2003) Direct Tissue Analysis Using Matrix-assisted Laser Desorption/Ionization Mass Spectrometry: Practical Aspects of Sample Preparation. J. Mass Spectrom. 38:699–708

Seydel U, Lindner B, Seydel JK and Brandenburt K (1982) Detection of Externally Induced Impairments in Single Bacterial Cells by Laser Microbe Mass Analysis. Int. J. Lepr. Other Mycobact. Dis. 50:90–95

Seydel U and Lindner B (1988) Monitoring of Bacterial Drug Response by Mass Spectrometry of Single Cells. Biomed. Environ. Mass Spectrom. 16:457–459

Shute LA, Gutteridge CS, Norris JR and Berkeley RC (1984) Curie-point Pyrolysis Mass Spectrometry Applied to Characterization and Identification of Selected Bacillus Species. J. Gen. Microbiol. 130:343–355

Simmonds PG, Shulman GP and Stembridge CH (1969) Organic Analysis by Pyrolysis – Gas Chromatography Mass Spectrometry: A Candidate Experiment for the Biological Exploration of Mars J. Chromatogr. Sci. 7:36–41

Simmonds PG (1970) Whole Micro-organisms Studied by Pyrolysis Gas Chromatography – Mass Spectrometry: Significance for Extraterrestial Life Detection Experiments Appl. Microbiol. 20:567–572

Sinha MP (1984) Laser-Induced Volatilization and Ionization of Microparticles. Rev. Sci. Instrum. 55:886–891

Sinha MP, Platz RM, Friedlander SK and Vilker VL (1985) Characterization of Bacteria by Particle Beam Mass Spectrometry. Appl. Environ. Microbiol. 49:1366–1373

Snyder AP, Dworzanski JP, Tripathi A, Maswadeh WM and Wick CH (2004) Correlation of Mass Spectrometry Identified Bacterial Biomarkers from a Fielded Pyrolysis – Gas Chromatography – Ion Mobility Spectrometry Biodetector with the Microbiological Gram Stain Classification Scheme. Anal. Chem. 76:6492–6499

Song Y, Wu G, Song Q, Cooks RG, Ouyang Z and Plass WR (2006) Novel Linear Ion Trap Mass Analyzer Composed of Four Planar Electrodes. J. Am. Soc. Mass Spectrom. 17:631–639

Spiegelman D, Whissell G and Greer CW (2005) A Survey of the Methods for the Characterization of Microbial Consortia and Communities. Can. J. Microbiol. 51:355–386

Srivastava A, Pitesky ME, Steele PT, Tobias HJ, Fergenson DP, Horn JM, Russell SC, Czerwieniec GA, Lebrilla CB, Gard EE and Frank M (2005) Comprehensive Assignment of Mass Spectral Signatures from Individual Bacillus atrophaeus Spores in Matrix-Free Laser Desorption/Ionization Bioaerosol Mass Spectrometry. Anal. Chem. 77:3315–3323

Steele PT, Tobias HJ, Fergenson DP, Pitesky ME, Horn JM, Czerwieniec GA, Russell SC, Lebrilla CB, Gard EE and Frank M (2003) Laser Power Dependence of Mass Spectral Signatures from Individual Bacterial Spores in Bioaerosol Mass Spectrometry. Anal. Chem. 75:5480–5487

Steele PT, Srivastava A, Pitesky ME, Fergenson DP, Tobias HJ, Gard EE and Frank M (2005) Desorption/Ionization Fluence Thresholds and Improved Mass Spectral Consistency Measured Using a Flattop Laser Profile in the Bioaerosol Mass Spectrometry of Single Bacillus Endospores. Anal. Chem. 77:7448–7454

Stern NJ, Kotula AW and Pierson D (1979) Differentiation of Selected Enterobacteriaceae by Pyrolysis – Gas-liquid Chromatography. Appl. Environm. Microbiol. 38:1098–1102

Stoakes L, Kelly T, Schieven B, Harley D, Ramos M, Lannigan R, Groves D and Hussain Z (1991) Gas-Liquid Chromatography Analysis of Cellular Fatty Acids for Identification of Gram-Negative Anaerobic Bacilli. J. Clin. Microbiol. 29:2636–2638

Stoeckli M, Farmer TB and Caprioli RM (1999) Automated Mass Spectrometry Imaging with a Matrix-Assisted Laser Desorption Ionization Time-of-Flight Instrument. J. Am. Soc. Mass Spectrom. 10:67–71

Stoll R and Röllgen FW (1979) Laser Desorption Mass Spectrometry of Thermally Labile Compounds Using a Continuous Wave CO2 Laser. Org. Mass Spectrom. 14:642–645

Stowers MA, Van Wuijckhuijse AL, Marijnissen JCM, Scarlett B, Van Baar BLM and Kientz CE (2000) Application of Matrix-Assisted Laser Desorption/Ionization to On-line Aerosol Time-of-flight Mass Spectrometry. Rapid Comm. Mass Spectrom. 14:829–833

Stowers MA, Van Wuijckhuijse AL, Marijnissen JCM and Kientz CE (2002) Method and Device for Detecting and Identifying Bio-aerosol Particles in the Air. Patent WO/2002/052246

Suess DT and Prather KA (1999) Mass Spectrometry of Aerosols. Chem. Rev. 99:3007–3035

Syage JA, Hanning-Lee MA and Hanold KA (2000) A Man-portable Photoionization Mass Spectrometer. Field Anal. Chem. Technol. 4:204–215

Szponar B and Larsson L (2001) Use of Mass Spectrometry for Characterising Microbial Communities in Bioaerosols. Ann. Agric. Environ. Med. 8:111–117

Tanaka K, Ido Y, Akita S, Yoshida Y and Yoshida T (1987) Development of Laser Ionization Time of Flight Mass Spectrometer IV - Generation of Quasi-Molecular Ions from High Mass Organic Compound. Proc. Second Japan-China Joint Symposium on Mass Spectrometry, Sept. 15–18, Osaka, Japan

Taylor J, Goodacre R, Wade WG, Rowland JJ and Kell DB (1998) The Deconvolution of Pyrolysis Mass Spectra Using Genetic Programming: Application to the Identification of Some Eubacterium Species. FEMS Microbiol. Lett. 160:237–246

Tegnell A, Van Loock F, Baka A, Wallyn S, Hendriks J, Werner A and Gouvras G (2006) Development of a Matrix to Evaluate the Threat of Biological Agents Used for Bioterrorism. Cell Mol. Life Sci. 63:2223–2228

Tobias HJ, Schafer MP, Pitesky M, Fergenson DP, Horn J, Frank M and Gard EE (2005) Bioaerosol Mass Spectrometry for Rapid Detection of Individual Airborne Mycobacterium tuberculosis H37Ra Particles. Appl. Environ. Microbiol. 71:6086–6095

Tobias HJ, Pitesky ME, Fergenson DP, Steele PT, Horn J, Frank M and Gard EE (2006) Following the Biochemical and Morphological Changes of Bacillus atrophaeus Cells During the Sporulation Process using Bioaerosol Mass Spectrometry. J. Microbiol. Meth. 67:56–63

Tripathi A, Maswadeh WM and Snyder AP (2001) Optimization of Quartz Tube Pyrolysis Atmospheric Pressure Ionization Mass Spectrometry for the Generation of Bacterial Biomarkers. Rapid Commun. Mass Spectrom. 15:1672–1680

Uphoff A, Muskat T, Grotemeyer J (2004) Design, Setup and First Results of a Miniaturized Time-of-flight Mass Spectrometer with a Simple Reflector of a New Design. Eur. J. Mass Spectrom. 10:163–171

Valentine N, Wunschel S, Wunschel D, Petersen C and Wahl K (2005) Effect of Culture Conditions on Microorganism Identification by Matrix-assisted Laser Desorption Ionization Mass Spectrometry. Appl. Environ. Microbiol. 71:58–64

Van Baar BLM (2000) Characterisation of Bacteria by Matrix-assisted Laser Desorption/Ionisation and Electrospray Mass Spectrometry. FEMS Microbiol. Rev. 24:193–219

Van der Greef J, Tas AC, Ten Noever de Brauw MC (1988) Direct Chemical Ionization – Pattern Recognition: Characterization of Bacteria and Body Fluid Profiling. Biomed. Environm. Mass Spectrom. 16:45–50

Van Malderen H, Hoornaert S and Van Grieken R (1996) Identification of Individual Aerosol Particles Containing Cr, Pb, and Zn above the Noth Sea. Environ. Sci. Technol. 30:489–498

Van Wuijckhuijse AL, Grootveld CJ, Weiss M, Marijnissen JCM and Scarlett B (1998) Improvements to the TOF Aerosol Mass Spectrometer. J. Aerosol Sci. 29, Suppl. 1:443–444

Van Wuijckhuijse AL, Stowers MA, Kleefsman WA, Van Baar BL M, Kientz CE and Marijnissen JCM (2005a) Matrix-Assisted Laser Desorption/Ionisation Aerosol Time-of-flight Mass Spectrometry for the Analysis of Bioaerosols: Development of a Fast Detector for Airborne Biological Pathogens. J. Aerosol Sci. 36:677–687

Van Wuijckhuijse A, Kientz C, Van Baar B, Kievit O, Busker R, Stowers M, Kleefsman W and Marijnissen J (2005b) Development of Bioaerosol Alarming Detector. Proc. NATO Adv. Res. Workshop on Defense Against Bioterror: Detection Technologies, Implementation Strategies and Commercial Opportunities, 8–11 April 2004, Madrid, Spain, In: Morrison D, Milanovich F, Ivnitski D, Austin TR (eds) NATO Security through Science Series B: Physics and Biophysics, vol 1. Springer, Berlin, pp 119–128

Van Wuijckhuijse AL, Kientz CE, Van Baar BLM et al., paper in preparation

Wang ZP, Russon L, Li L, Roser DC and Long SR (1998) Investigation of Spectral Reproducibility in Direct Analysis of Bacteria Proteins by Matrix-assisted Laser Desorption/Ionization Time-of-flight Mass Spectrometry. Rapid Commun. Mass Spectrom. 12:456–464

Wechsung R, Hillenkamp F, Kaufmann R, Nitsche R, Unsöld E and Vogt H (1978) LAMMA – A New Laser-Microprobe-Mass-Analyzer. Microscopica Acta (Suppl) 2:281–296

Weiss M, Marijnissen JCM, Verheijen PJT and Scarlett B (1993) On-line Measurement of Particle Size and Composition. J. Aerosol Sci. 24 (Suppl) 1:201–202

Weiss M (1997) An On-line Mass Spectrometer for Aerosols. Masters Thesis, Delft University of Technology, pp 122–131

Westman A, Huth-Fehre T, Demirev P and Sundqvist BUR (1995) Sample Morphology Effects in Matrix-assisted Laser Desorption/Ionization Mass Spectrometry of Proteins. J. Mass Spectrom. 30:206–211

Wieser P, Wurster R and Seiler H (1980) Identification of Airborne Particles by Laser Induced Mass Spectroscopy. Atmospheric Environm. 14:485–494

Wieten G, Haverkamp J, Meuzelaar HLC, Bondwijn HW and Berwald LG (1981) Pyrolysis Mass Spectrometry: A New Method to Differentiate Between the Mycobacteria of the "Tuberculosis Complex" and Other Mycobacteria. J. Gen. Microbiol. 122:109–118

Wilkes JG, Rushing LG, Gagnon J-F, McCarthy SA, Rafii F, Khan AA, Kaysner CA, Heinze TM and Sutherland JB (2005a) Rapid Phenotypic Characterization of Vibrio Isolates by Pyrolysis Metastable Atom Bombardment Mass Spectrometry. Antonie van Leeuwenhoek 88:151–161

Wilkes JG, Rushing L, Nayak R, Buzatu DA and Sutherland JB (2005b) Rapid Phenotypic Characterization of Salmonella enterica Strains by Pyrolysis Metastable Atom Bombardment Mass Spectrometry with Multivariate Statistical and Artificial Neural Network Pattern Recognition. J. Microbiol. Methods 61:321–334

Wilkes JG, Rafii F, Sutherland JB, Rushing LG and Buzatu DA (2006) Pyrolysis Mass Spectrometry for Distinguishing Potential Hoax materials from Bioterror Agents. Rapid Commun. Mass Spectrom. 20:2383–2386

Williams TL, Andrzejewski D, Lay Jr. JO and Musser SM (2003) Experimental Factors Affecting the Quality and Reproducibility of MALDI TOF Mass Spectra Obtained from Whole Bacteria Cells. J. Am. Soc. Mass Spectrom. 14:342–351

Wunschel DS, Hill EA, McLean JS, Jarman K, Gorby YA, Valentine N and Wahl K (2005a) Effects of Varied pH, Growth Rate and Temperature Using Controlled Fermentation and Batch Culture on Matrix Assisted Laser Desorption/Ionization Whole Cell Protein Fingerprints. J. Microbiol. Meth. 62:259–271

Wunschel SC, Jarman KH, Petersen CE, Valentine NB, Wahl KL, Schauki D, Jackman J, Nelson CP and White 5th E (2005b) Bacterial Analysis by MALDI-TOF Mass Spectrometry: An Inter-laboratory Comparison. J. Am. Soc. Mass Spectrom. 16:456–462

Yang M, Dale JM, Whitten WB and Ramsey JM (1995) Laser Desorption Tandem Mass Spectrometry of Individual Microparticles in an Ion Trap Mass Spectrometer. Anal. Chem. 67:4330–4334

Zakett D, Schoen AE, Cooks RG and Hemberger PH (1981) Laser-Desorption Mass Spectrometry / Mass Spectrometry and the Mechanism of Desorption Ionization. J. Am. Chem. Soc. 103:1295–1297

Zhang H and Caprioli RM (1996) Capillary Electrophoresis Combined with Matrix-Assisted Laser Desorption/Ionization Mass Spectrometry, Continuous Sample Deposition on a Matrix-precoated Membrane Target. J. Mass Spectrom. 31:1039–1046

Zoller DL, Sum ST, Johnston MV, Hatfield GR and Qian K (1999) Determination of Polymer Type and Comonomer Content in Polyethylenes by Pyrolysis - Photoinization Mass Spectrometry. Anal. Chem. 71:866–872

Index